S82.232

THE ECOLOGY OF CYANOBACTERIA

AF735233

51959

The Ecology of Cyanobacteria

Their Diversity in Time and Space

Edited by

BRIAN A. WHITTON
University of Durham, Durham, England

and

MALCOLM POTTS
Virginia Tech., Blackburg, U.S.A.

KLUWER ACADEMIC PUBLISHERS

DORDRECHT / LONDON / BOSTON

A C.I.P. Catalogue record for this book is available from the Library of Congress.

ISBN 0-7923-4735-8 (HB)
ISBN 0-7923-4755-2 (PB)

Published by Kluwer Academic Publishers,
P.O. Box 17, 3300 AA Dordrecht, The Netherlands.

Sold and distributed in North, Central and South America
by Kluwer Academic Publishers,
101 Philip Drive, Norwell, MA 02061, U.S.A.

In all other countries, sold and distributed
by Kluwer Academic Publishers,
P.O. Box 322, 3300 AH Dordrecht, The Netherlands.

Printed on acid-free paper

All Rights Reserved
© 2000 Kluwer Academic Publishers
No part of the material protected by this copyright notice may be reproduced or
utilized in any form or by any means, electronic or mechanical,
including photocopying, recording or by any information storage and
retrieval system, without written permission from the copyright owner.

Printed in the Netherlands

List of Authors

DAVID G. ADAMS Department of Microbiology, University of Leeds LS2 9JT, UK
d.g.adams@leeds.ac.uk
REDHA H. AL-HASAN Department of Biological Sciences, Faculty of Science, University of Kuwait, PO Box 5969, Safat 13060, Kuwait
redha@kuc01.kuniv.edu.kw
DEVAKI BHAYA Carnegie Institute of Washington, Department of Plant Biology, 260 Panama St, Stanford, CA 94305, USA
devaki@andrew.stanford.edu
CRISTIANA CALLIERI Istituto Italiano di Idrobiologia, Largo Tonolli 50, 28922 Verbania Pallanza, Italy
c.callieri@iii.to.cnr.it
RICHARD W. CASTENHOLZ Department of Biology, College of Arts & Sciences, University of Oregon, Eugene OR 974103-1210, USA
rcasten@darkwing.uoregon.edu
JENNIFER S. CAVET Department of Biochemistry and Genetics, The Medical School, University of Newcastle, Newcastle-upon-Tyne NE2 4HH, UK
j.s.cavet@newcastle.ac.uk
CRAWFORD S. DOW Department of Biological Sciences, University of Warwick, Coventry CV4 7AL, UK
cd@dna.bio.warwick.ac.uk
GERTRUD CRONBERG Department of Ecology, Institute of Limnology and Marine Ecology, University of Lund, BOX 65, Lund, Sweden S-22100
gertrud.cronberg@limnol.lu.se
GEORGE G. GANF Department of Environmental Biology, University of Adelaide, SA 5005, Australia
gganf@botany.adelaide.edu.au
FERRAN GARCIA-PICHEL Department of Microbiology, Arizona State University, Tempe AZ 85287-2701, USA
ferran@asu.edu
ARTHUR R. GROSSMAN Carnegie Institute of Washington, Department of Plant Biology, 260 Panama St, Stanford, CA 94305, USA
arthur@andrew2.stanford.edu
NICHOLAS H. MANN Department of Biological Sciences, University of Warwick, Coventry CV4 7AL, UK
nm@dna.bio.warwick.ac.uk
RODERICK L. OLIVER Murray Darling Freshwater Research Centre, PO Box 921, Albury 2640, NSW 2640, Australia
oliverr@watson.canberra.edu.au
AHARON OREN Division of Microbial Molecular Ecology, Alexander Silverman Institute of Life Sciences, and the Moshe Shilo Minerva Center for Marine Biogeochemistry, The Hebrew University, 91904 Jerusalem, Israel
orena@cc.huji.ac.il

HANS W. PAERL Institute of Marine Sciences, University of North Carolina at Chapel Hill, 3431 Arendell Street, Morehead City, North Carolina 28557, USA
hans_paerl@unc.edu

ALLAN PENTECOST Division of Life Sciences, King's College London, Franklin-Wilkins Building, 150 Stamford Street, London SE1 8WA, UK
allan.pentecost@kcl.ac.uk

MATHEW R. POCOCK The Sanger Centre, Wellcome Trust Genome Campus, Hinxton, Cambridge CB10 1SA, UK

MALCOLM POTTS Virginia Tech Center for Genomics (VIGEN), Department of Biochemistry, Virginia Polytechnic Institute and State University, Blacksburg, VA 24061, USA
geordie@vt.edu

SAMIR RADWAN Department of Biological Sciences, Faculty of Science, University of Kuwait, PO Box 5969, Safat 13060, Kuwait

NIGEL J. ROBINSON Department of Biochemistry and Genetics, The Medical School, University of Newcastle, Newcastle-upon-Tyne NE2 4HH, UK
n.j.robinson@newcastle.ac.uk

JULIAN C. RUTHERFORD Department of Biological Chemistry, University of Utah, Salt Lake City, Utah 84112-5330, USA
julian.rutherford@hsc.utah.edu

J. WILLIAM SCHOPF IGPP Center for the Study of Evolution and the Origin of Life, Department of Earth and Space Sciences, and Molecular Biology Institute, University of California, Los Angeles, California 90095-1567 USA
schopf@ess.ucla.edu

RAKEFET SCHWARZ Carnegie Institute of Washington, Department of Plant Biology, 260 Panama St, Stanford, CA 94305, USA
schwarz@andrew.stanford.edu

LUCAS J. STAL Netherlands Institute of Ecology, Centre for Estuarine and Coastal Ecology (NIOO-CEMO), PO Box 140, NL-4401 AC Yerseke, The Netherlands
stal@cemo.nioo.knaw.nl

JOHN G. STOCKNER 2614 Mathers Avenue, West Vancouver, British Columbia, Canada V7V 2J4
stockner@axionet

CURTIS A. SUTTLE Department of Earth & Ocean Sciences, Botany, and Microbiology & Immunology,University of British Columbia, 6270 University Blvd, Vancouver V6T 1Z4, Canada
suttle@eos.ubc.ca

UTHAYA K. SWOBODA Department of Biological Sciences, University of Warwick, Coventry CV4 7AL, UK
uk@warwick.bio.warwick.ac.uk

LUISA TOMASELLI CSMA - CNR, p. le delle Cascine 27, I-50144 Firenze, Italy
tomaselli@csma.fi.cnr.it

WARWICK F. VINCENT Département de Biologie & Centre d'Etudes Nordiques, Université de Laval, Sainte-Foy, Québec G1K 7P4, Canada
warwick.vincent@bio.ulaval.ca

AVIGAD VONSHAK Microalgal Biotechnology, The Jacob Blaustein Institute for Desert Research, Ben-Gurion University of the Negev, Sede Boker Campus, Israel 84990
avigad@bgumail.bgu.ac.il

DAVID M. WARD Department of Land Resources and Environmental Science, Montana State University, Bozeman 59717, USA
umbdw@montana.edu
BRIAN A. WHITTON Department of Biological Sciences, University of Durham, Durham, DH1 3LE, UK
b.a.whitton@durham.ac.uk
DAVID D. WYNN-WILLIAMS British Antarctic Survey, High Cross, Madingley Road, Cambridge CB3 0ET, UK
ddww@pcmail.nerc-bas.ac.uk

Contents

Preface

The cyanobacteria have figured prominently in discussions between the editors over the years, in different countries and diverse settings, some more exotic than others. We have learned that Nature gives up her secrets with considerable reluctance. Many details and aspects of cyanobacterial ecology remain elusive. Our intention in assembling this book was to help all those trying to unravel their secrets and to reveal what is known to a wide audience.

The publication in 1994 of *The Molecular Biology of Cyanobacteria* edited by Donald Bryant summarized advances in not just molecular biology, but also in taxonomy, biochemistry, physiology and cellular differentiation. Although research in some topics progresses so rapidly that reviews become dated equally rapidly, Bryant's book is still the key starting point for anyone wanting to learn about these aspects of cyanobacteria. As soon as the present editors read his book, we realized that the literature on cyanobacterial ecology was even more fragmented than that on biochemistry and molecular biology had been before 1994. It was evident that there ought to be a sister volume on cyanobacterial ecology. The present book is the result.

It was Prof. G. E. Fogg's superb review (1956, Bacteriological Reviews 20: 148-165) on the *The comparative physiology and biochemistry of the blue-green algae* which first caught the interest, when an undergraduate student, of the more elderly of the present editors. Recognition of the prokaryote - eukaryote dichotomy and an explanation for the role of the heterocyst were still discoveries for the future. Nevertheless, even at this time, the review showed clearly the extent to which an understanding of the ecology of cyanobacteria could benefit from knowledge of their physiology and biochemistry.

The variety and complexity of populations of cyanobacteria on Aldabra Atoll, Indian Ocean, has left a permanent impression on both editors. One of them first saw the atoll as a PhD student in the mid 1970s. At that time molecular ecology was in its infancy, but its importance for understanding the interactions between cyanobacteria and their environment has become increasingly evident. Several chapters include molecular topics and many more aspects of cyanobacterial ecology are likely to benefit from molecular insight within the next few years. Thermal spring, desert and antarctic research are all areas where molecular ecology is only just starting, but where there is great potential for molecular studies. One-third of the references in the present molecular chapters has been published since Bryant's volume, showing how rapidly this subject is changing.

Unlike molecular biology, however, there is a great deal of older literature on the ecology of cyanobacteria which is still of value. Some is mentioned here and more can be found in earlier edited volumes (Carr and Whitton, 1973 and 1982, Blackwell; Fay and Van Baalen, 1987, Elsevier; Carr and Mann, 1992, Plenum). Hopefully this will be enough to persuade readers to check the older literature, rather than relying on a database starting in 1981. All too often new studies on physiological ecology are published which say little more than previous studies published thirty years ago. No doubt in most cases this is because modern researchers are unaware of the older studies. It would be excellent if someone took the initiative to produce a CD-ROM with the abstracts from all the older literature.

A number of chapters mention applied or commercial aspects. Practical problems related to cyanobacteria have often been tackled in the past with little thought about the ecology of the organisms. We hope that the book will encourage those dealing with such problems to make use of the information here. Topics which should benefit include the management of dense cyanobacterial populations in nature, such as fish-farms and toxic blooms, improvements in soil quality, and diverse commercial developments. We will regard the present book as a real success if it stimulates sufficient ecologically sound applied studies to justify within a few years a third book in this series, on *Practical Uses and Problems of Cyanobacteria*.

At the time of writing this the sequencing of one cyanobacterial genome is completed, and three others are in progress. One, that of *Nostoc*, is the largest microbial genome studied to date. We can ponder over the likely advances and developments these activities will bring to the appreciation of cyanobacteria.

Many people have helped us. It was enjoyable working with the various authors, especially when we had the chance to meet and argue about cyanobacterial science. We are also grateful to many other people. Several helped to review particular chapters, including George Bullerjahn, Ed Carpenter, Jean Houmard, Rocco Mancinelli, Jack Meeks and Tony Walsby. However, any errors which may have crept past the reviewing process are of course our fault. We thank all those who loaned colour slides for the plates, especially those who were not involved in writing chapters: S. Babic, Peter Baker, David Bellamy, B. Bergman, Michael Burch, John Davies, Earthrise Farms staff, Imre Friedmann, David Livingstone, Dieter Mollenhauer. A number of postdocs and research students also offered helpful comments and we particularly thank Martin Mühling. Much of the preparation of camera-ready copy and some redrawing of figures was done by John Daniell. Judith John prepared the Organism Index. Staff from Kluwer have been most enthusiastic about the book. René Mijs got it started, while Alison Bradshaw dealt with the later stages and André Tournois was in charge of production.

BRIAN WHITTON
MALCOLM POTTS

Colour Plates

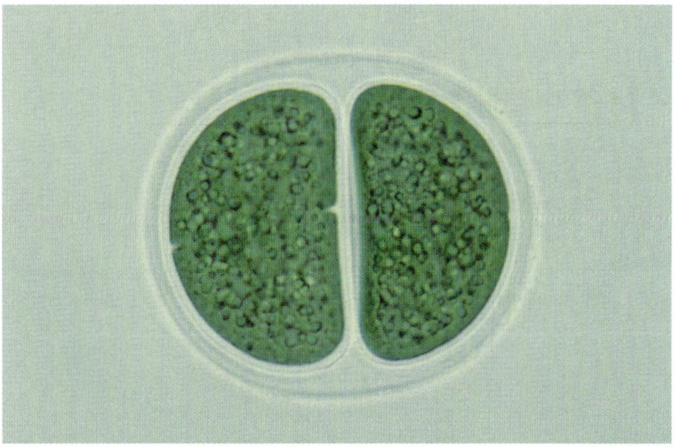

a

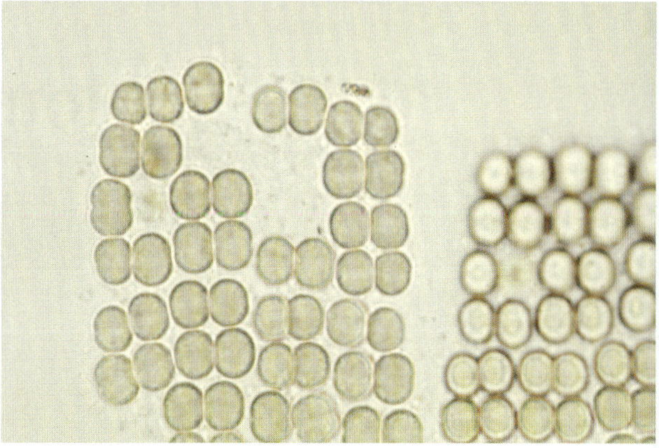

b

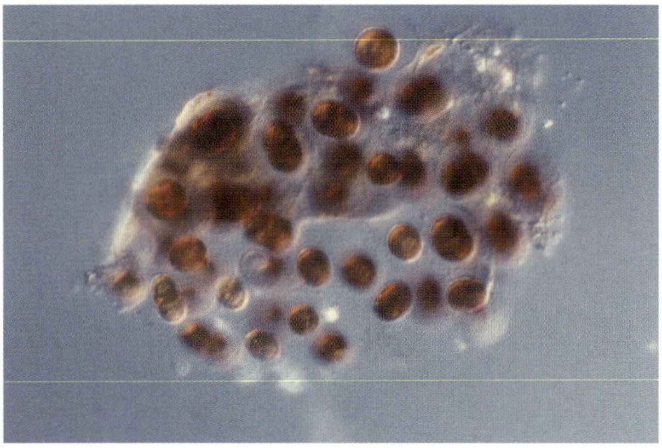

c

d

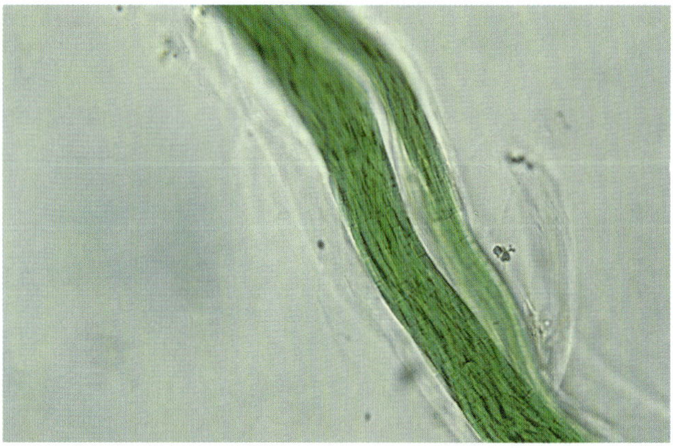

e

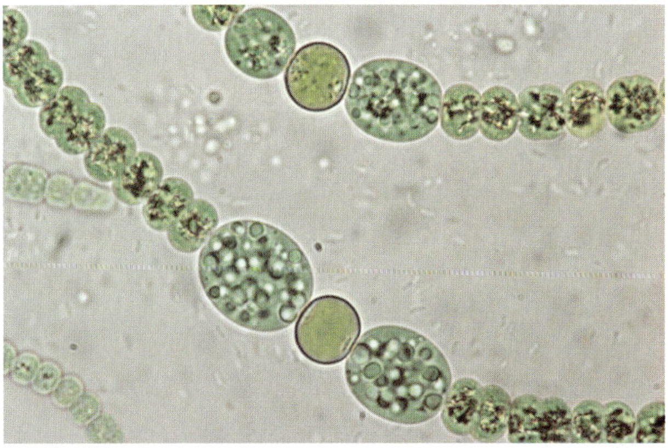

f

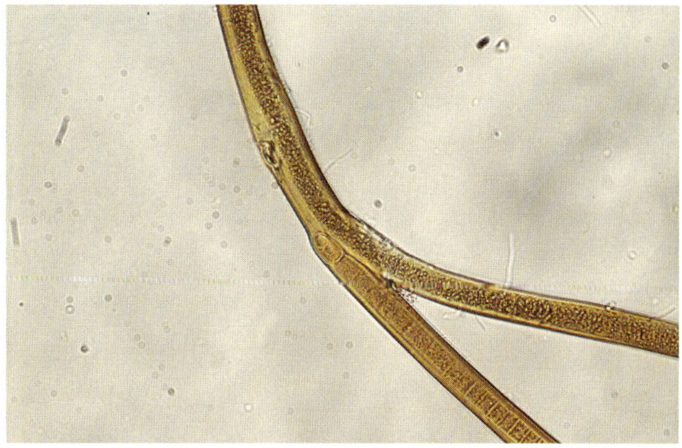

g

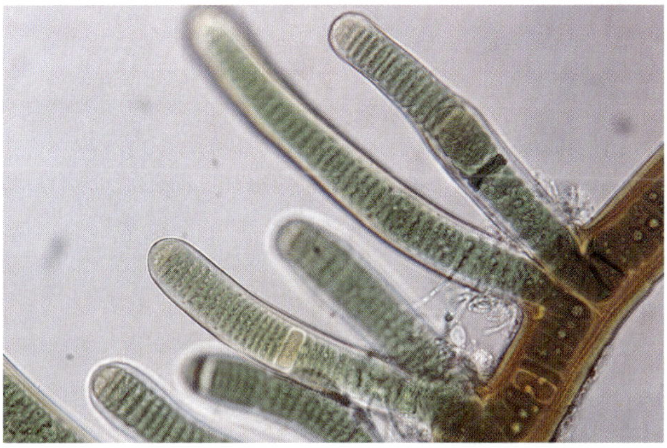

h

i

Plate 1

1 Diversity of form**.** **a** *Chroococcus.* **b** *Merismopedia*. **c** *Gloeocapsa* cf *sanguinea*. **d** *Microcystis aeruginosa* being consumed by amoeba. **e** *Schizothrix*. **f** *Anabaena*: planktonic form showing gas-vacuoles in vegetative cells, heterocysts and (in this species) akinetes developing either side of the heterocyst. **g** *Tolypothrix tenuis*, showing single 'false' branch;. **h** *Scytonema myochrous*, showing pairs of 'false' branches originating from main axis**.** **i** *Stigonema mamillosum*, with true branching.

(Photos **a**,**h** J. Davies & B.A. Whitton; **b**,**c** M. Potts; **d** H. W. Paerl; **e**,**f**,**g**,**i** P.N.G. Boulton & B. A. Whitton)

a

b

c

d

e

f

g

h

i

Plate 2

2 Diversity of habitat. **a**, **b**, **c**, **d**, **e** Aldabra Atoll, Indian Ocean (9° S): **a** Green silt dominated by *Aphanocapsa* and *Pleurocapsa* in intertidal region of lagoon. **b** Mollusc tracks in green silt. **c** Drier parts of terrestrial limestone are sometimes almost black due to *Tolypothrix byssoidea* with its dark brown sheaths. **d** Hemispherical colonies of *Phormidium hendersonii* on floor of shallow part of lagoon. **e** Section of *P. hendersonii* showing laminated structure due to daily differences in trichome orientation and consequent differences in particle trapping; colonies are eventually eaten by sea-slugs. **f** Stromatolites, Shark Bay, W. Australia. **g** *Chamaesiphon fuscus* colonies on rock from fast-flowing stretch of River Swale, England. **h** *Phormidium* in fast-flowing stream (Asir Mts, Saudi Arabia), showing pattern of many motile stream cyanobacterial populations under very high illumination: trichomes aggregate by day and disperse by night. **i.** *Nostoc* soup: *N. flagelliforme* harvested from the regions adjacent to the Gobi Desert is used widely in China for soups.

(Photos **a**,**b**,**c** B.A.Whitton & M. Potts; **d** B.A. Whitton & A. Donaldson; **e-i** B.A. Whitton; **f** D.J. Bellamy)

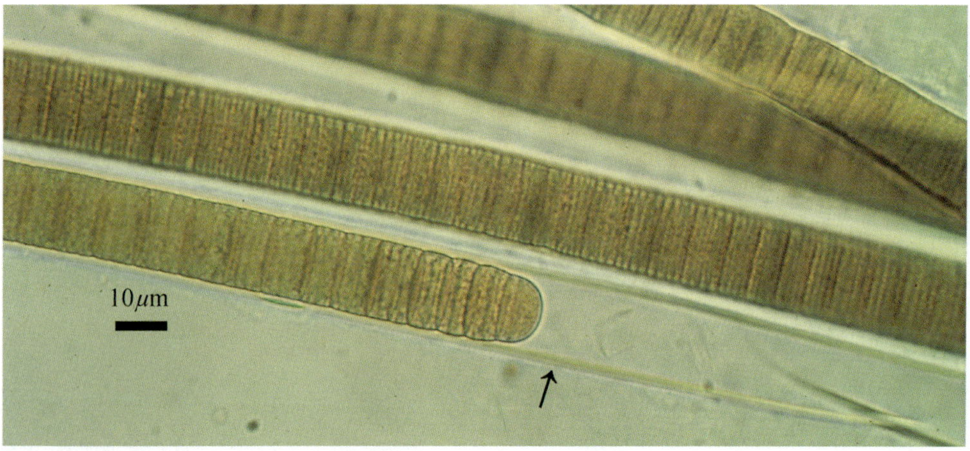

a

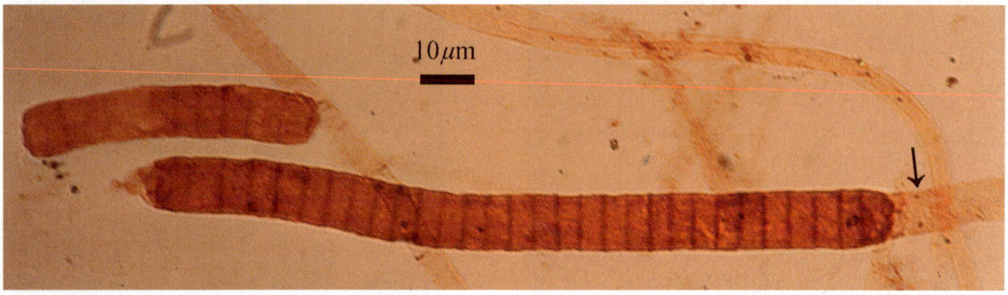

b

Plate 3

3 Comparison of modern and fossil cyanobacteria. **A** Modern *Lyngbya* sp. (Oscillatoriaceae), encompassed by a cylindrical mucilaginous sheath (arrow), from a mat-building stromatolitic community of northern Baja, Mexico (Schopf, 1994a). **B** Fossil *Palaeolyngbya helva*, similarly ensheathed (arrows), from siltstone of the ~ 950 Ma-old Lakhanda Formation of eastern Siberia, Russia (Hermann, 1981). (For references, see Chapter 2)

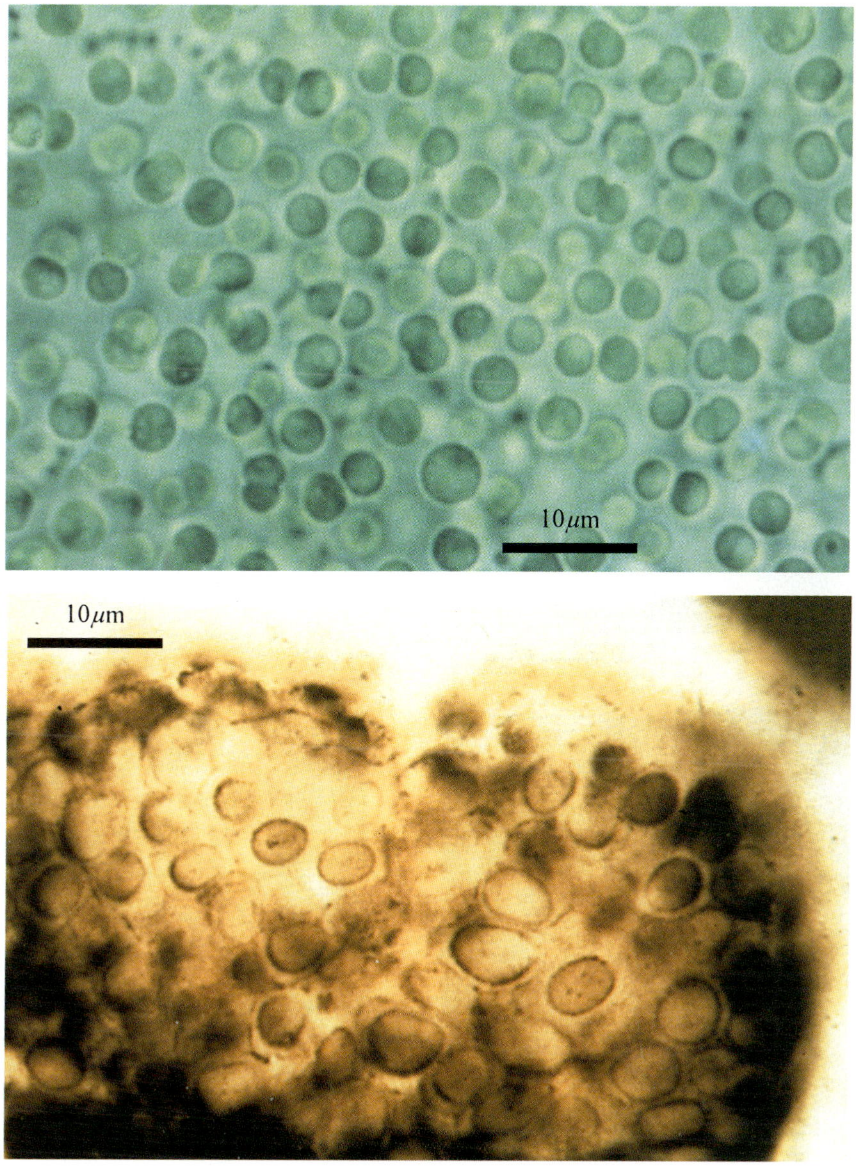

a

b

Plate 4

4 Comparison of modern and fossil cyanobacteria. **A** Modern *Entophysalis* sp. (Entophysalidaceae) from a mat-building stromatolitic community of northern Baja, Mexico (Schopf, 1994a). **B** *Eoentophysalis belcherensis*, an *Entophysalis*-like colonial cyanobacterium, from stromatolitic chert of the ~2150 Ma-old Kasegalik Formation of the Belcher Islands in southeastern Hudson Bay, Northwest Territories, Canada (Hofmann, 1976). (For references, see Chapter 2)

a

b

c

d

e

f

Plate 5

5 a, b, c Hunter's Hot Springs, Oregon: **a** One of >90°C sources showing the "V" of cooling to ~ 74° C, with green *Synechococcus* sp(p). followed at ~ 54° C by reddish-brown *Oscillatoria* cf. *terebriformis* mat; in some areas this mat has retracted, exposing undermat of *Chloroflexus*; **b** Edge of pool, with thick mat of *Oscillatoria* where it replaces green *Synechococcus* mat at ~ 54°C; ostracods (*Potamocypris* sp.: white objects, true size ~ 0.5 mm long) have grazed *Oscillatoria* below ~ 48°C; **c** Ostracods grazing on grazer-resistant, grazer-dependent *Pleurocapsa/Calothrix* community; glistening spot in each animal is the eye. **d, e, f** Yellowstone National Park. **d** Adult ephydrid flies grazing cyanobacterial mat in Serendipity Meadows (~40°C); eggs are laid within the mat, which may be almost destroyed when larvae are numerous. **e** Yellow-green *Synechococcus* biofilm in Octopus Spring on siliceous sinter, beginning at ~ 74°C, as indicated by "V" shape in main channel; source pool is 90-92° C. **f** Laminated *Synechococcus* mat in Octopus Spring at 55 to ~ 50°C.
(Photos D.M. Ward & R.W.Castenholz)

a

b

c

Plate 6

6 **a, b, c** Octopus Spring, Yellowstone National Park: **a** Overview, with letters indicating approximate locations of cyanobacterial features in Plates 5e, 5f, 6b, 6c; **b** streamers in high flow; **c** conical structures in quiescent pools at approximately 50°C, containing *Synechococcus* and *Phormidium*.
(Photos by D.M. Ward & R.W.Castenholz: **b**, **c** from Ward et al., 1992b)

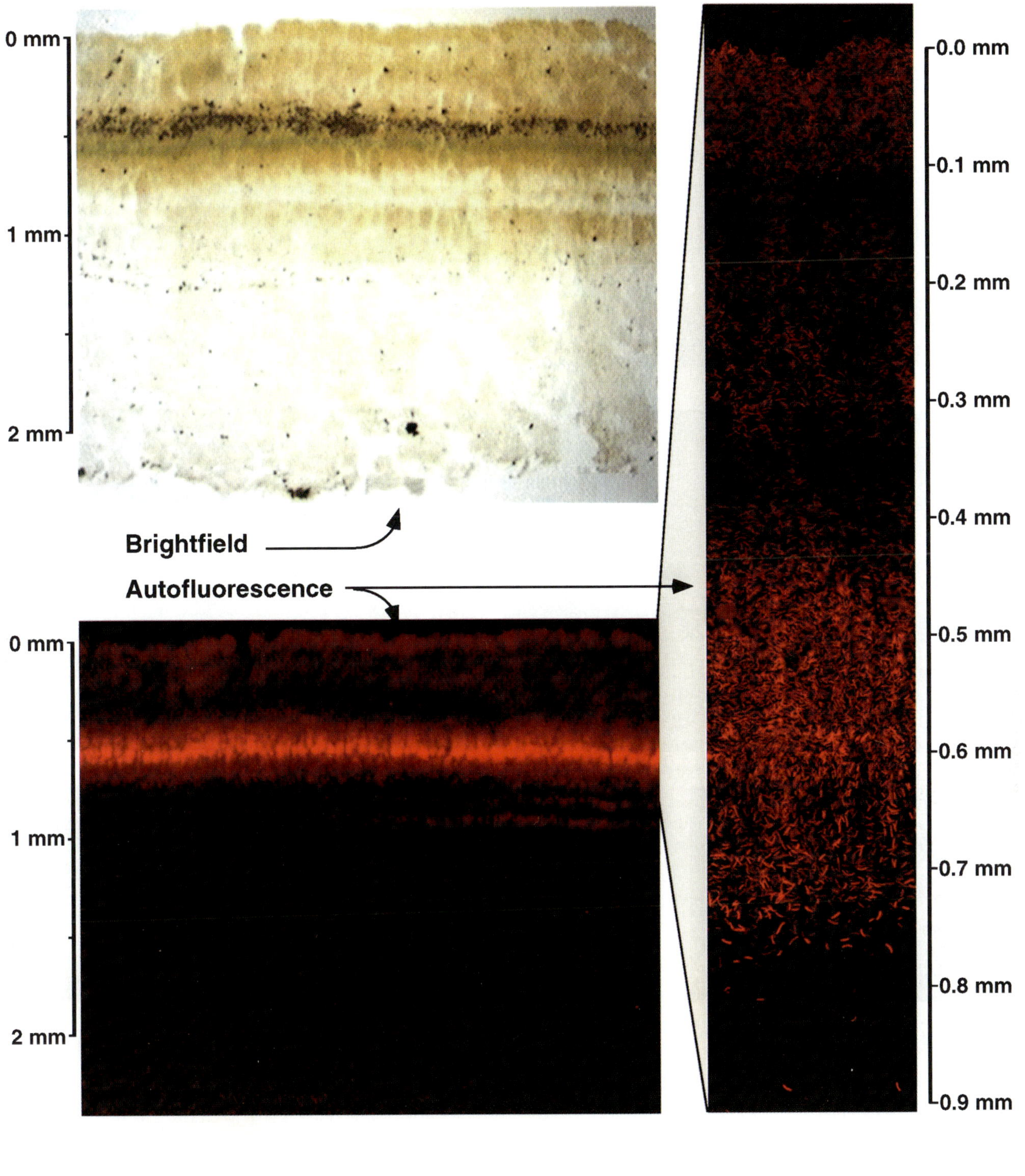

Plate 7

7 Vertical structure of a ~60°C laminated *Synechococus* mat as revealed by brightfield and autofluorescence microscopic images of cross-section of the top 0.9 to 2 mm of the Mushroom Spring 61°C cyanobacterial mat. (Photo by Niels Ramsing)

a

b

c

d

e

f

g

Plate 8

8 Microbial mats on intertidal flats of the Pacific coast of Guerrero Negro, Baja California Mexico. **a** Two types of mat develop close to each other, with smooth mat (shown as dark areas) in the lower intertidal and pustular mat in the higher intertidal. **b** Smooth mat, showing cracks caused by desiccation at low tide. **c** Pustular mat. **d** Differences between the mats shown in cross-section: smooth mat has laminated structure typical of microbial mats - with a thin and dense layer of cyanobacteria on top, next a layer of anoxygenic purple sulfur bacteria, then a black layer of FeS, indicating that the mat is permanently anoxic below the layer of cyanobacteria. **e** Pustular mat showing a much looser mat of cyanobacteria on top, while layers of purple sulfur bacteria and FeS are absent, indicating that the sediment below the cyanobacteria is predominantly oxic. **f** Smooth mat is composed of the non-heterocystous (but N_2-fixing) *Lyngbya aestuarii*, the trichomes of which are surrounded by a thick polysaccharide sheath and the organisms are embedded in a dense matrix of mucilage. **g** Pustular mat composed of *Calothrix*, which forms less mucilage; the terminal heterocysts are clearly visible.
(All photos L. Stal, in Stal, 1995)

a

b

c

d

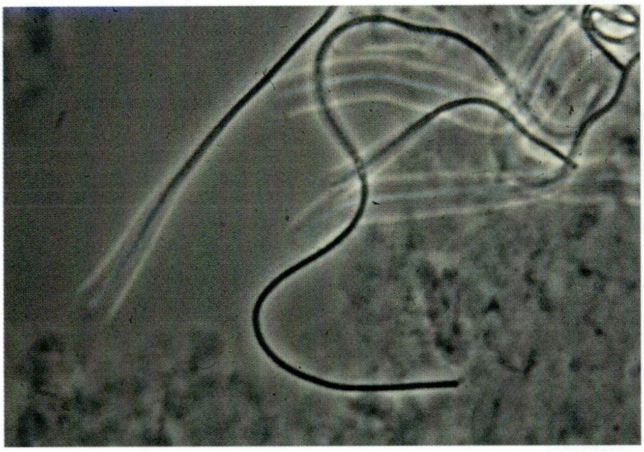

e

Plate 9

9 **a** Stromatolites formed by cyanobacteria in the intertidal of Exuma Cays, Bahamas: these structures are considered as modern examples of known fossil stromatolites. **b** A closer look at these lithified sedimentary structures, which consist of trapped carbonate sediment cemented by micritic (microcrystalline) carbonate. **c** Surface of stromatolites is covered by a cyanobacterial-algal mat, which are thought to be involved in formation of the micritic horizons. **d** Cross-section of the top part of the stromatolite shows a distinct green layer of cyanobacteria. **e** *Schizothrix* is the dominant cyanobacterium in these modern stromatolites. (Photos by L. Stal)

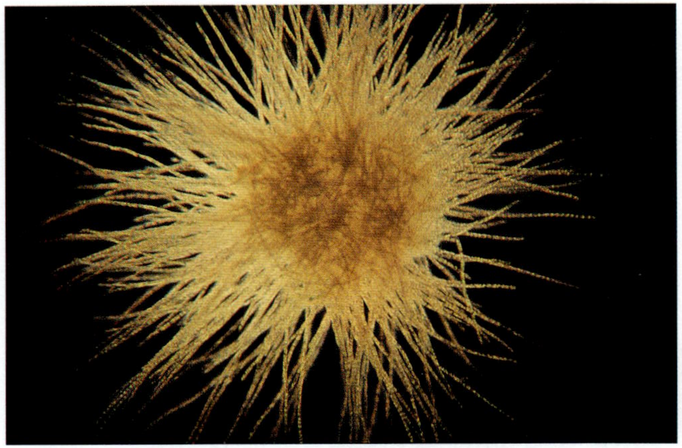

a

b

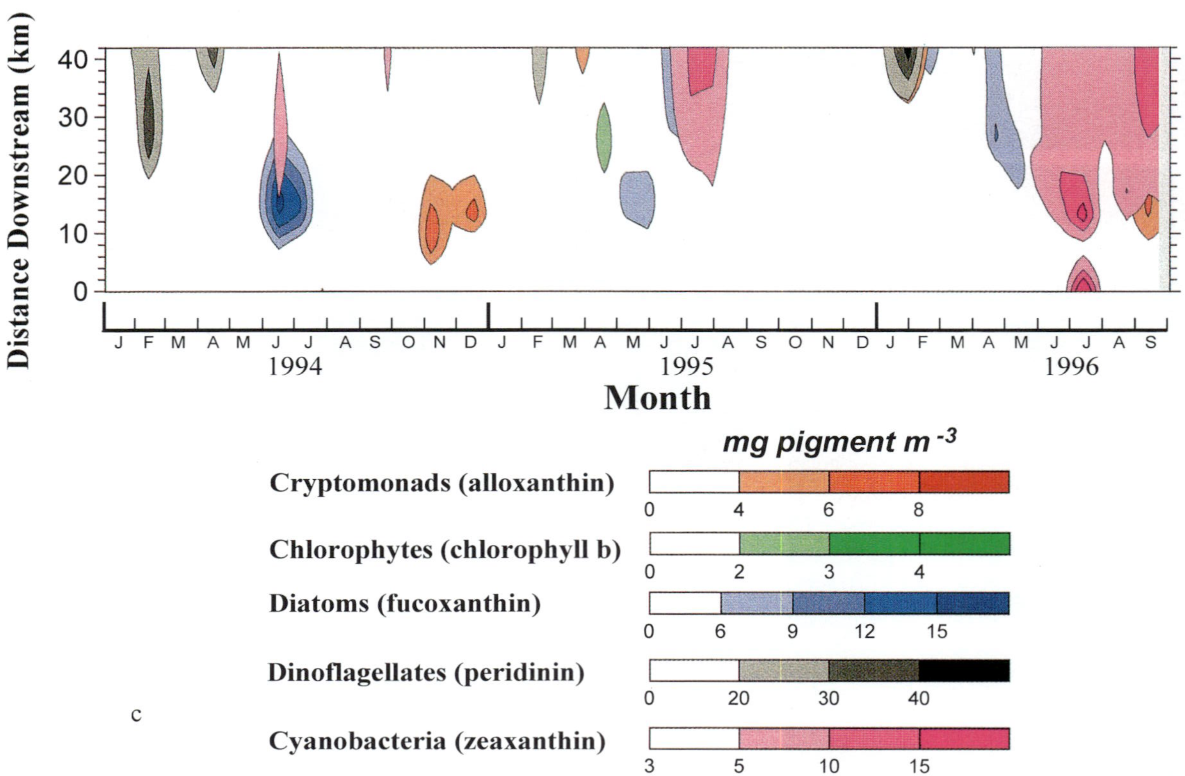
Distance Downstream (km)
40
30
20
10
0
J F M A M J J A S O N D J F M A M J J A S O N D J F M A M J J A S
1994
1995
1996
Month
mg pigment m -3
Cryptomonads (alloxanthin)
0 4 6 8
Chlorophytes (chlorophyll b)
0 2 3 4
Diatoms (fucoxanthin)
0 6 9 12 15
Dinoflagellates (peridinin)
0 20 30 40
Cyanobacteria (zeaxanthin)
3 5 10 15

c

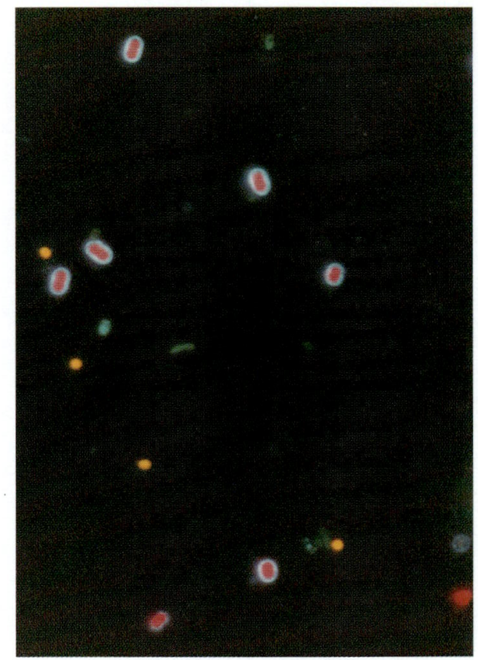

d

Plate10

10 Cyanobacteria in the sea. **a, b** *Trichodesmium thiebautii* from blooms in the western Atlantic Ocean: **a** radial 'puff' aggregate; **b** fusiform tuft aggregates from site 26 km off coast of North Carolina. **c** Contour plot of spatio-temporal distribution of plankton blooms along freshwater (0 km) to estuarine (41 km) salinity gradient in Neuse River Estuary, N. Carolina. Data based on biweekly sampling at 8 stations; carotenoid pigments quantified by HPLC.

d Marine *Synechococcus* tagged with fluorescently labelled viruses. *Synechococcus* strain BBC1 has been added to a natural marine microbial community and tagged with cyanophages labelled with the blue fluorescent strain POPO-1. They appear pink, surrounded by a blue halo. Untagged *Synechococcus* spp. fluoresce red or yellow, while a heterotriophic bacterium has been tagged with a virus that hs been labelled with a green fluorescent stain (YOYO-1).

(Photos **a,b,c** H.W. Paerl; **d,** C.A. Suttle)

Plate 11

11 Freshwater blooms in Australia. **a** Bloom of *Anabaena*. **b** Sampling a bloom of *Anabaena*. **c** *Microcystis aeruginosa.* **d** *Planktothrix (= Oscillatoria) perornata* (and *Anabaena*) **e** *Nodularia spumigena* from Lake Alexandrine, South Australia. **f** *Anabaena circinalis*.
(Photos **a**,**b**,**d** P.D. Baker; **c**,**e**,**f** P.D. Baker & M.D. Burch, Australian Water Quality Centre, South Australia)

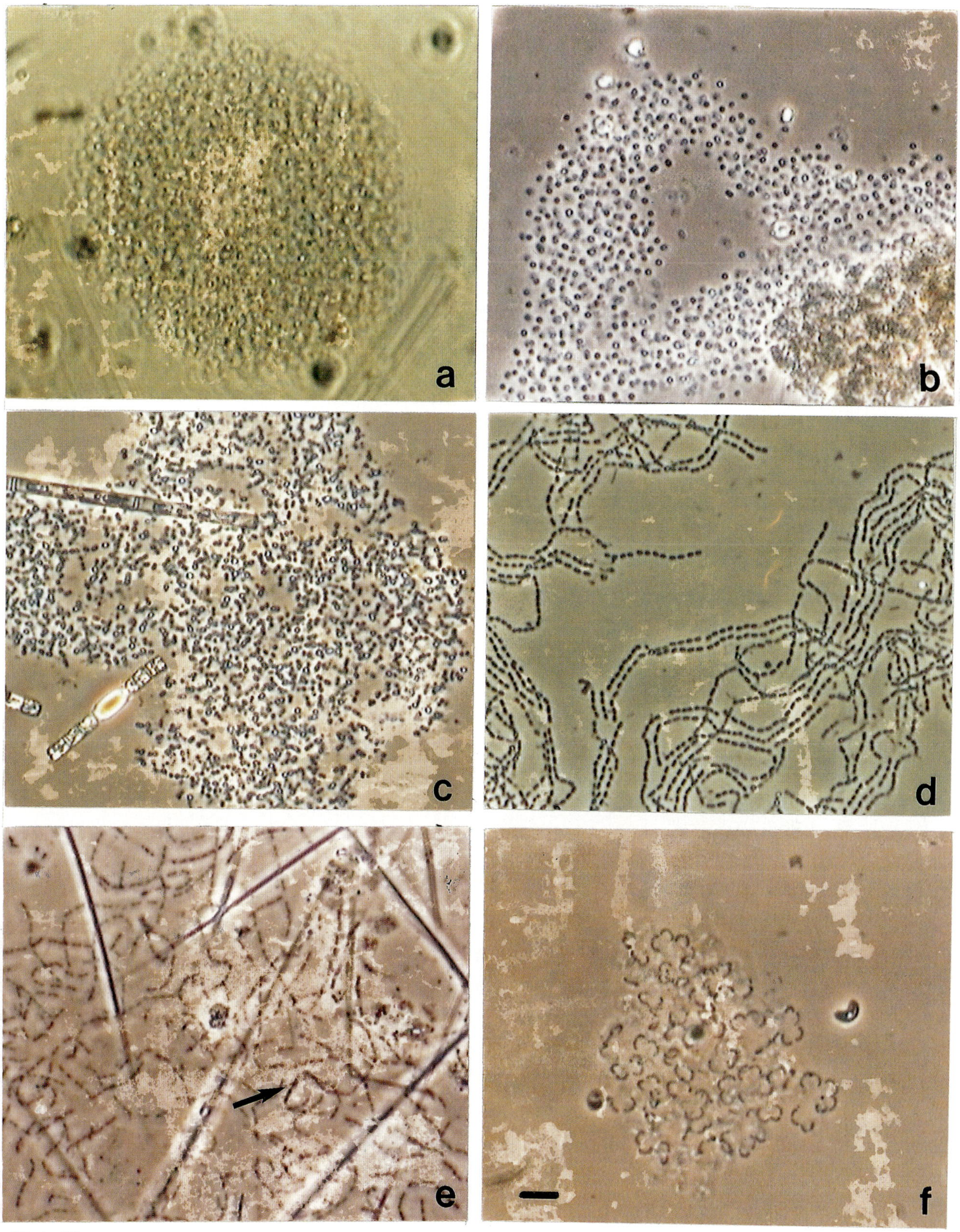

Plate 12

12 Examples of small-celled plankton forms. **a** *Aphanocapsa delicatissima*. **b** *Aphanocapsa holsatica*. **c** *Cyanodictyon imperfectum*. **d** *Cyanodictyon planktonicum*. **e** *Cyanodictyon filiforme*. **f** *Tetrarcus ilsteri* (All micrographs to same scale: bar = 10 μm) (Photos G. Cronberg)

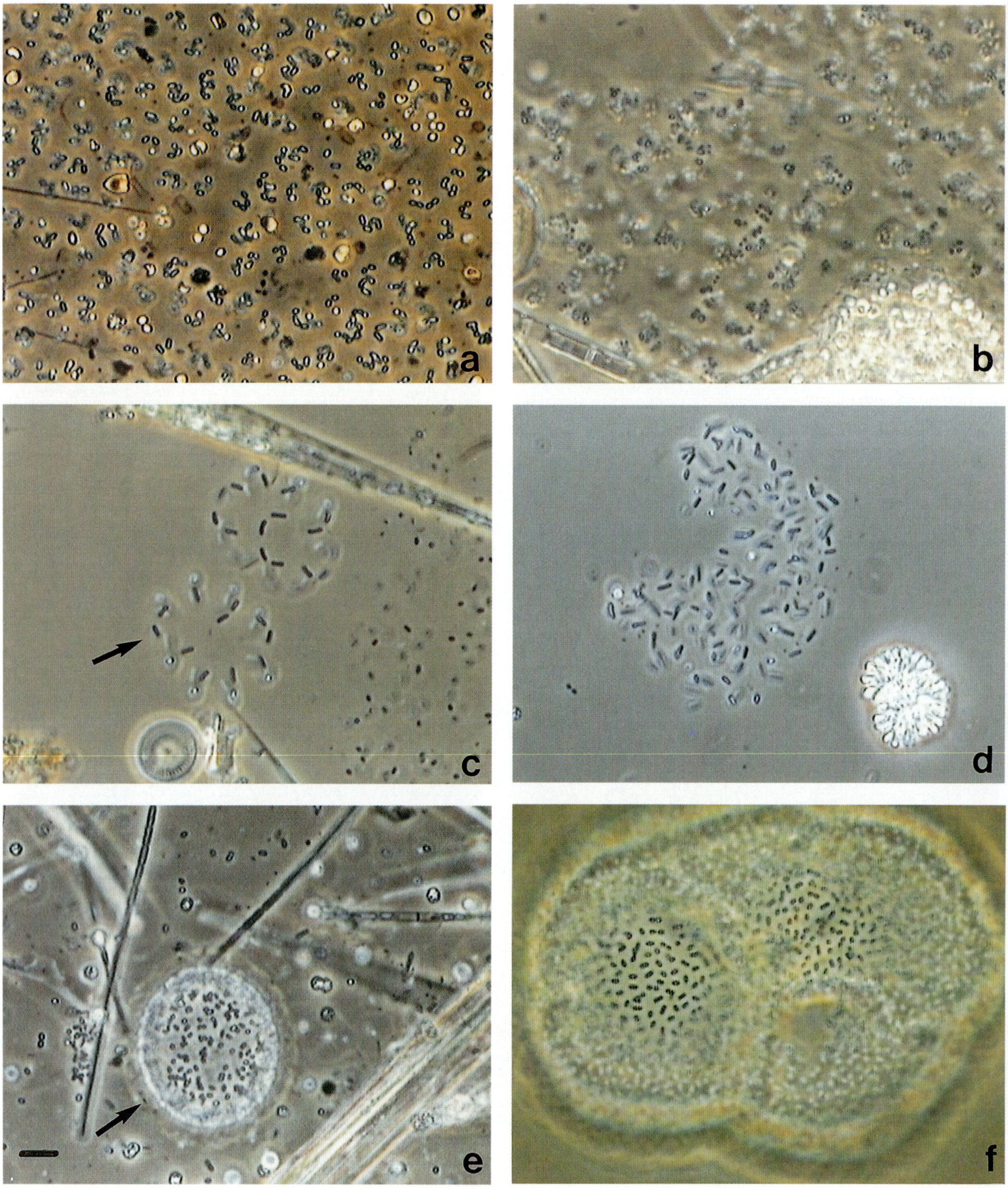

Plate 13

13 Examples of small-celled plankton forms (con.). **a** *Synechococcus* sp. **b** *Chroococcus microscopicus*. **c** *Cyanonephron styloides*. **d** *Aphanothece clathrata*. **e** *Lemmermanniella pallida*. **f** *Aphanothece bachmannii* (All micrographs to same scale: bar = 10 μm) (Photos G. Cronberg)

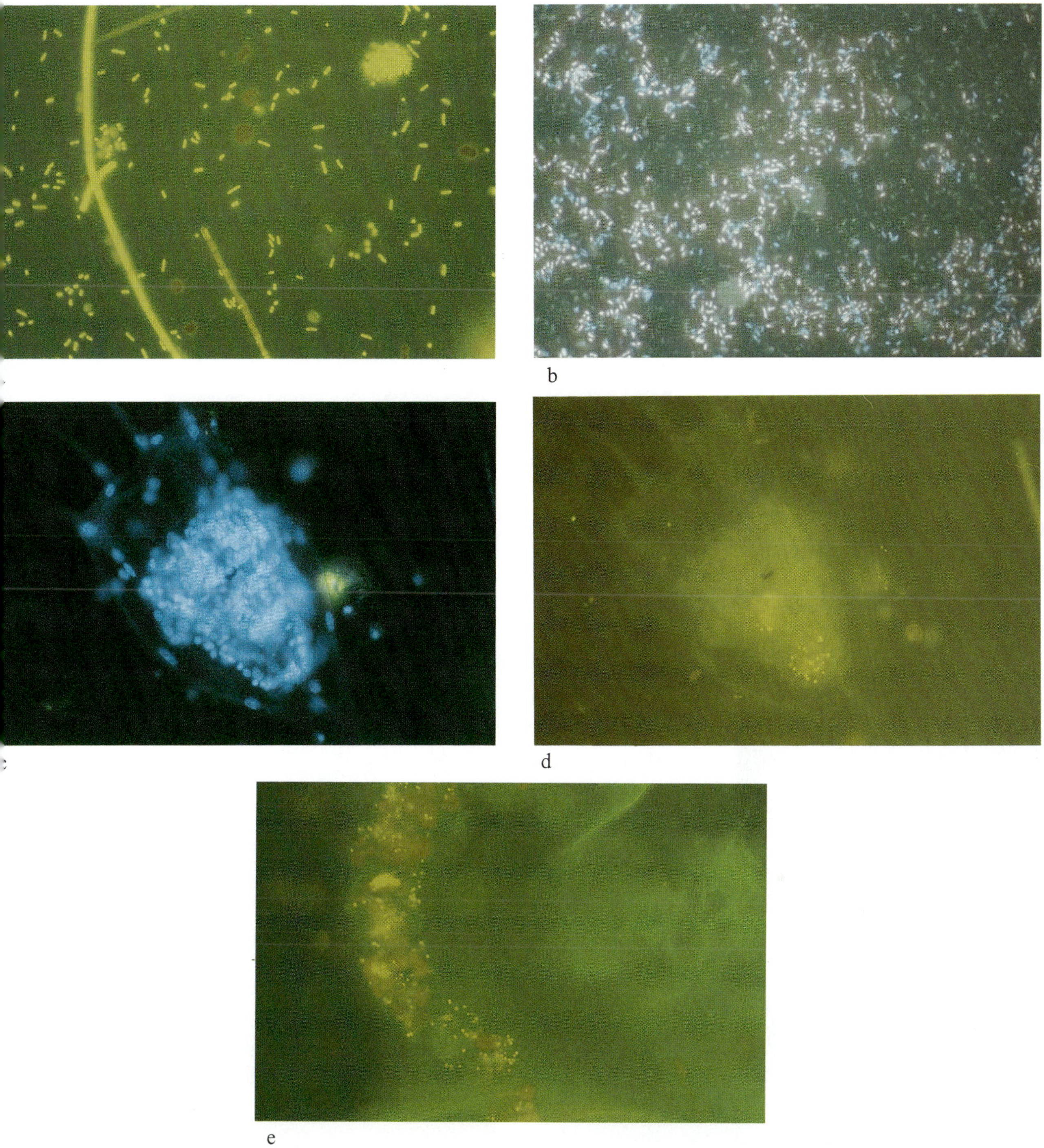

Plate 14

14 Use of fluorescence to study picoplankton. **a** Typical view of natural picoplankton, together with small aggregates, colonial forms and eukaryotic algae (780 x). **b** Association of bacteria (blue) and *Synechococcus* (pink) in a culture stained with DAPI (1250 x). **c** The ciliate *Vorticella* coloured with DAPI and visualized under blue excitation. **d** As previous, but visualized under UV excitation. **e** *Bosmina* gut full of autotrophic picoplankton, small colonial forms and eukaryotic cells. (Scale bars for **c**,**d**,**e** = 10 μm) (Photos C. Callieri)

a

b

c

d

e

f

Plate 15

15 Rice-fields. **a** Mixed cyanobacterial community on soil of transplanted rice-field. **b** Dense *Lyngbya* growth on rice-field soil (Manikganj, Bangladesh). **c**, **d** *Gloeotrichia* colonies on deepwater rice plants taken from edge of field at Sonargaon, Bangladesh. **e** *Azolla pinnata* grown with transplanted rice (San Fernando, Philippines). **f** *Gloeotrichia natans* floating on surface of rice-field drainage ditch two weeks after treatment with weed-killer (International Rice Research Institute, Philippines). (Photos **a**,**b** D. Livingstone; **c** - **f** B.A. Whitton)

b

d

e

Plate 16

16 Cyanobacteria and limestones. **a** Tintenstriche on hills by the Li Jiang, Guilin, China. **b** *Hyella* in beach-rock, Aldabra Atoll. **c**, **d**, **e** Travertine-depositing sites near Malham, England: **c** *Rivularia* colonies in small spring; **d** *Homoeothrix crustacea*, showing localization of colonies in more highly illuminated area; **e** *Schizothrix* (indicated by reddish colouration of sheaths) on moist travertine surface and adjacent to *Pinguicula vulgaris* and the moss *Gymnostomum recurvirostrum*, species with which it is often associated in north temperate regions.
(Photos **a**, **c**,**d**,**e** A. Pentecost; **b** M. Potts)

a

b

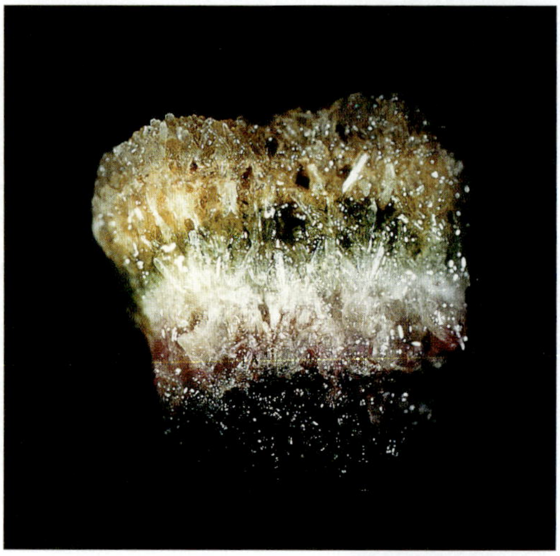

c

Plate 17

17 a b *Oscillatoria* mats in a hypersaline pool (39° C, pH 5.2, 170 g L^{-1} solids, 2.5 mM sulfide): Hamei Mazor, by the Dead Sea shore. **a** General view - white floating material is elemental sulfur. **b** Blue is *Oscillatoria*, pinkish areas are due to *Thiocapsa* and white patches are elemental sulfur. **c** Endoevaporite communities in a 4-cm gypsum crust covering bottom of a saltern pond in Eilat, Israel. Vertical zonation shows brownish-orange upper layer of unicellular cyanobacteria (*Aphanothece halophytica*, *Synechococcus* sp.) rich in carotenoids; the green layer below contains unicellular and filamentous (Oscillatoriaceae) cyanobacteria. The white, red-purple and black layers are characterized by colorless sulfur bacteria, purple sulfur bacteria (*Chromatium*, *Ectothiorhodospira*) and sulfate-reducing bacteria, respectively. (Photos by A. Oren, **c** together with M. Kühl)

a

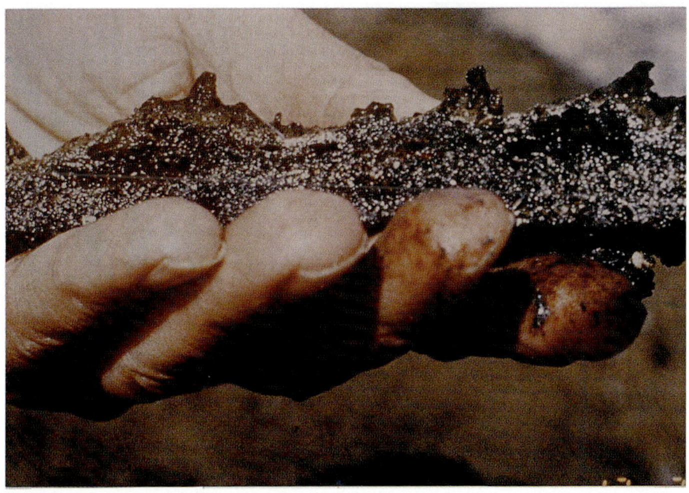

b

c

Plate 18

18 Cyanobacterial communities in oil-contaminated environment. **a** General view of cyanobacterial mats on top of oil polluting the intertidal zone of the Arabian Gulf. **b** Close-up of mat covering heavily contaminated sand. **c** Close-up of mat on top of contaminated muddy zone. (Photos S.S. Radwan & R.H. Al-Hasan)

a

b

c

Plate 19

19 Cyanobacterial communities in oil-contaminated environments (con.). **a** General view of non-dissected mats in contamined intertidal. **b** Close-up of polygonal mats strongly adhering to contaminated sediments. **c** Surface view of mat at early stage of development on contaminated sand in lower intertidal.
(Photos S.S. Radwan & R.H. Al-Hasan)

a

b

Plate 20

20 Cyanobacteria at southern Victoria Land in the Antarctic. **a** Cold desert landscape at Beacon Heights, upper Taylor Valley. **b** Sampling cyanobacteria-dominated cryptoendolithic microbial communities in beacon sandstone at Battleship Promontory. (Photos D.D. Wynn-Williams)

a

b

Plate 21

21 Cyanobacteria of Beacon Sandstone in the Antarctic. **a** Vertical fracture profile of cryptoendolithic community showing a distinct cyanobacterial zone ~ 6 mm from the surface of the rock; black zone (nearer the surface) contains pigmented lichen, while the white zone comprises mainly hyaline fungi with siome microalgae. **b** Artificially exposed cyanobacteria (green-brown colouration), hyaline fungi (white), black-pigmented lichen (nearest the surface) of an exfoliating cryptoendolithic community at Timber Peak, Priestley Glacier, central Victoria Land: white label is 6 cm wide. (Photos D.D. Wynn-Williams)

a

b

c

d

e

f

Plate 22

22 Cyanobacteria in hot deserts. **a** Rocky desert, Negev. **b** Desert pavement, Negev. **c** Desert pavement, close-up. **d** Desert pavement, with one stone overturned to reveal hypolithic cyanobacteria of stone-soil interface. **e** Surface of Nubian sandstone, with parts chipped off to reveal cryptoendolithic cyanobacterial colonization under the surface, Negev. **f** Chasmoendolithic colonization in granite shown in thin section of rock, Sonoran Desert, Mexico. (Photos E. I. Friedmann)

a

b

c

d

e

f

Plate 23

23 Old lead - zinc mining areas highly contaminated with Zn Cd and Pb: **a-e** Missouri, USA; **f** Einesberger Zentrale, Harz, Germany. **a** Elvins Tailings, lagoons (5 mg L^{-1} Zn) with planktonic *Synechococcus*. **b** Drainage stream (16 mg L^{-1} Zn) at same site showing brown colour due to *Plectonema* sheaths. **c** Close-up, with part of mat sectioned: community is similar to that at many other contaminated sites - mixture of narrow *Plectonema* and protonemal stage of the moss *Dicranella*. **d** *Schizothrix* community with reddish sheath typical of contaminated alternately wet - dry areas. **e** Floating mats of mixed populations of narrow sheathed Oscillatoriaceae typical of areas with standing water. **f** Part of old spoil heap only partially colonized by higher plants and showing dark cover of mixed *Plectonema* and moss protonema, which is apparently the typical terrestrial pioneer community of such contaminated areas worldwide. (Photos B.A. Whitton)

a

b

c

d

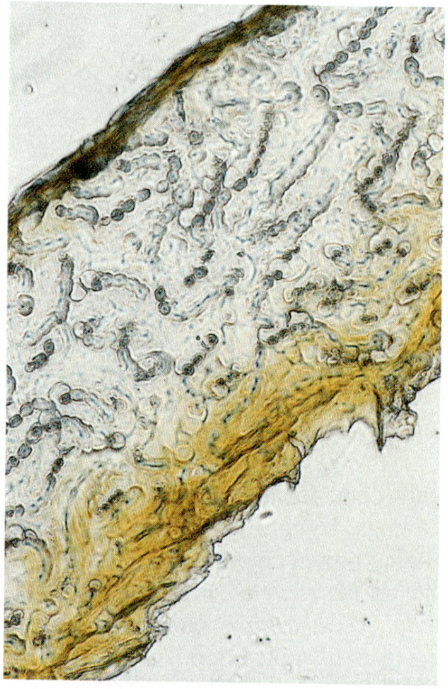
e

Plate 24

24 *Nostoc.* **a** *N. pruniforme.* **b** Solution hollow on limestone with moist *Nostoc commune* colonies and associated organic debris permitting growth of higher plant (*Portulaca*) (Aldabra Atoll): similar solution hollows with *N. commune* have been reported from many countries. **c** *Nostoc* colonies persisting at bottom of empty pool during the dry season on Aldabra Atoll: this 'crumbly' form takes up and loses water more slowly than the typical form. **d** Spherical colonies ("pearls") of *N. commune* growing on agar supplemented with calcium and magnesium carbonates. **e** Nomarski interference microscopy of thick section of *N. commune* thallus, showing brown and yellow layers due to scytonemin; individual filaments visible throughout the extracellular glycan.
(Photos **a** B.A. Whitton & R.W. Castenholz; **b,c** B.A. Whitton; **d,e** M.Potts)

a b c d e

Plate 25

25 *Scytonema* **a** Colony on upper part of mangrove pneumatophore (near Eilat, Israel). **b**, c Mats in upper intertidal inland regions of Andros Is., Bahamas: **b** polygons consist of laminated structure trapping particles; **c** mats developing on surface in sheltered area.. **d** Mats in young transplanted rice-field (Manikganj, Bangladesh). **e** Hemispherical colonies of *S. myochrous*, a characteristic species of moist travertine-depositing regions reported from many countries. (see also Plate 2h) (Photos **a** M. Potts; **b,c** B.A. Whitton; **d,e** D. Livingstone)

a

b

c

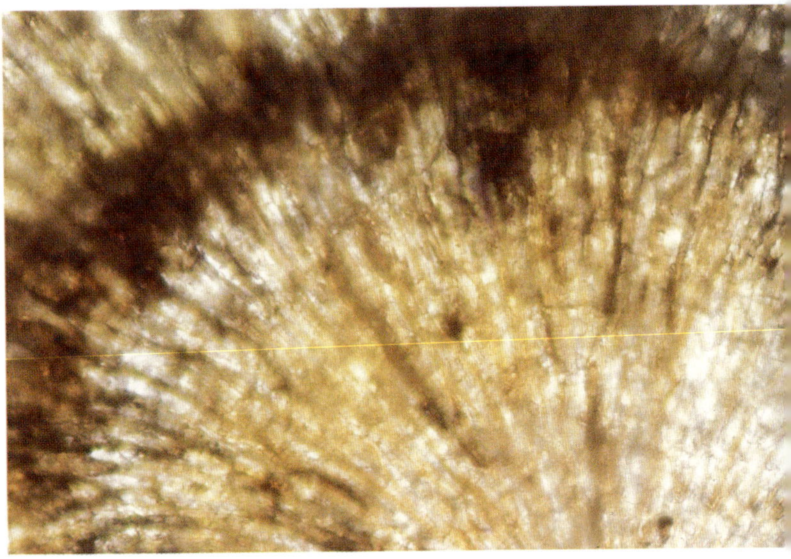

d

e

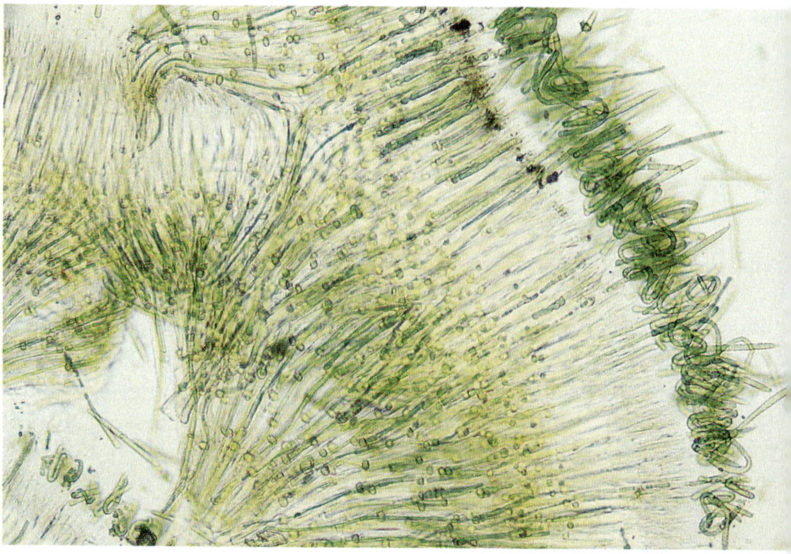

f

Plate 26

26 Rivulariaceae are characteristic of sites showing marked temporal or spatial variation in ambient phosphate. **a** Stream (Upper Teesdale, England: winter) draining region with limestone and peat and with extensive *Rivularia* cover throughout the year. **b** *Gloeotrichia natans* colonies (Bangladesh). **c** Oncoids (typically 1 - 1.5 cm) formed by small *Rivularia* colonies in calcareous flush (Sunbiggin, England). **d** Section of oncoid from this site. **e, f** Marine intertidal *Rivularia* (*R. atra*): **e**, section of typical colony, showing outer scytonemin-rich layer and intracellular location of heterocysts; **f** hormogonia formation following exposure to pulse of elevated phosphate. (Photos **a**,**b**,**c**,**e**,**f** B.A. Whitton; **d** G.A.L. Johnson)

a

b

c

d

e

Plate 27

27 "Spirulina" (*Arthrospira*). **a, b** Outdoor ponds at Earthrise Farms, California. **c** Processing "Spirulina" at Earthrise Farms. **d** Natural population of *Arthrospira* from African natron water: bar marker = 100 µm. **e** Trichome, showing gas vacuoles. (Photos **a,b,c** Earthrise Farms; **d** C. Sili; **e** M. Mühling.

a

b

c

d

e

f

Plate 28

28 Symbiotic associations - macroscopic views. **a** *Gunnera chilensis*, including area at base of leaves containing stem glands infected with *Nostoc*. **b** Cycads as an understorey in a newly burnt *Eucalyptus* forest in Australia. **c** Sporophyte of the fern *Azolla pinnata* viewed from above; the *Nostoc* symbiont is contained within swollen dorsal lobes of the leaves raised above the water surface. **d** The liverwort *Blasia pusilla* showing dark *Nostoc* colonies scattered around the prominent midrib. **e** The tripartite lichen *Peltigera aphthosa* composed of a fungus (seen as white areas at the curled edges), a green alga (colour of main thallus) and *Nostoc* contained in small dark spots (cephalodia) scattered over a thallus typically 5-8 cm in diameter. **f** The fungus *Geosiphon pyriforme* is coloured by its endosymbiotic *Nostoc*.
(Photos **a,c,d** D.G. Adams; **b,e** B. Bergman; **f** D. Mollenhauer)

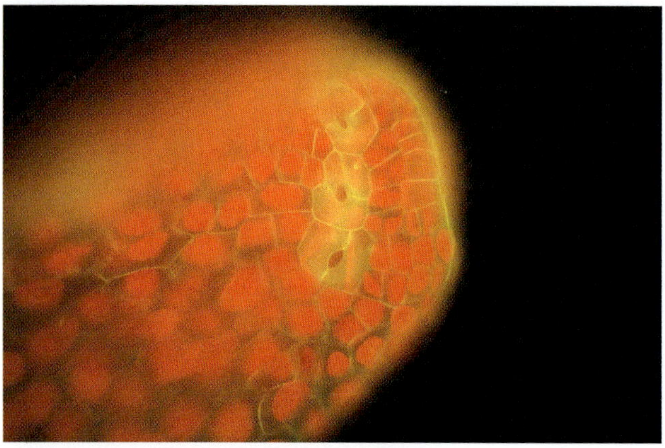

a

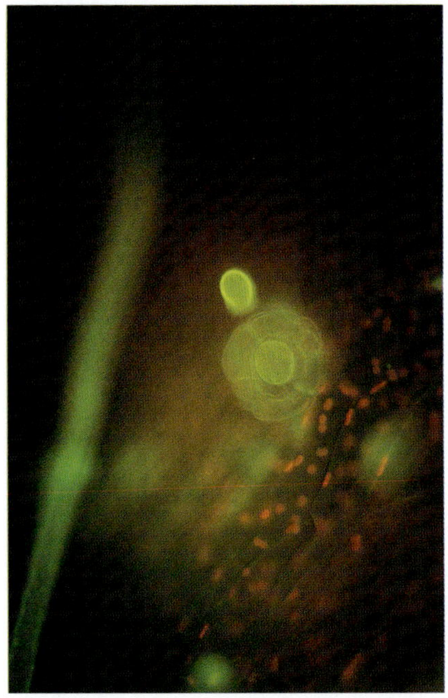

b

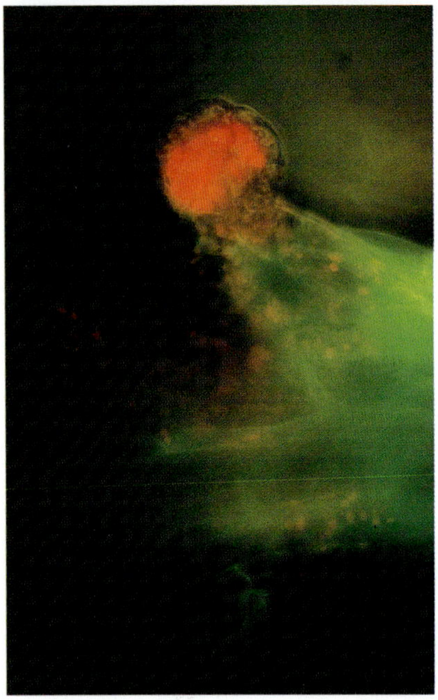

c

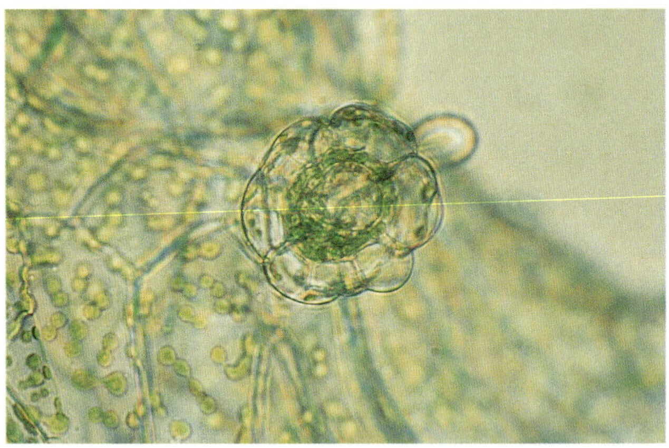

d

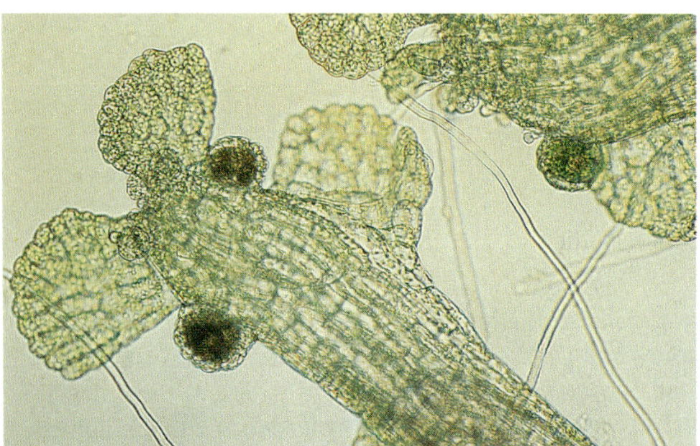

e

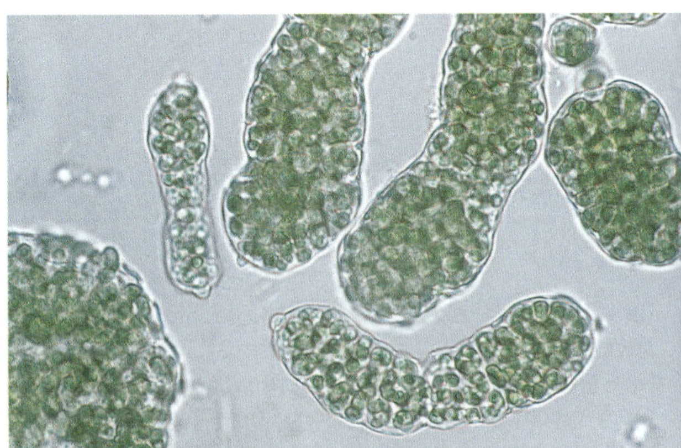

f

Plate 29

29 Symbiotic associations - microscopy. **a - c** Fluorescence studies with calcofluor white staining: **a** Ventral surface of *Phaeoceros*, showing three stomata-like entrances through which infecting hormogonia gain entry to slime cavities within the thallus; **b** Ventral surface of *Blasia*, showing a single auricle as an almost spherical structure on surface of thallus; a slime papilla almost fills the auricle except for a narrow space that appears as a darker, unstained region - this space becomes infected by cyanobacteria and the auricle enlarges as the symbiotic structure develops; **c** *Blasia*, showing the red autofluorescence of *Nostoc* within an enlarged auricle. **d** *Blasia* auricle, showing newly developed *Nostoc* colony surrounding the central slime papilla. **e** *Blasia* infected in laboratory with two different *Nostoc* strains, one blue-green due to chlorophyll and phycocyanin, and the other brown due to the additional phycoerythrin; the colours of the symbiotic colonies can be seen through the auricle walls. **f** *Nostoc* isolated from *Peltigera aphthosa*; heterocystous trichomes occur in packages surrounded by a pronounced mucilaginous sheath. (Photos **a,b,c,d, f** S. Babic; **e** B. Bergman)

a

b

c

Plate 30

30 a Scytonemin-rich mat of *Calothrix* in a tepid hot spring stream (Potts Basin, West Thumb of Yellowstone Lake, Yellowstone National Park). **b** Cliff face heavily streaked with sheath-pigmented cyanobacteria i.e. Tintenstriche (Numedal, Norway). **c** Scytonemin-containing *Nostoc* in melt stream originating from Canada Glacier, Taylor Valley, Antarctica.
(Photos R.W. Castenholz)

a

b

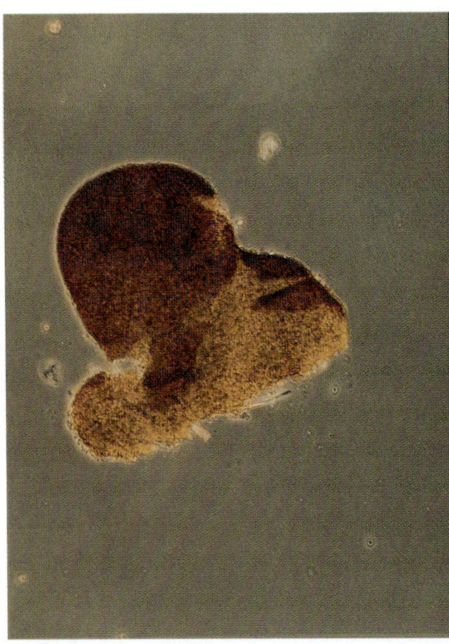

c

d

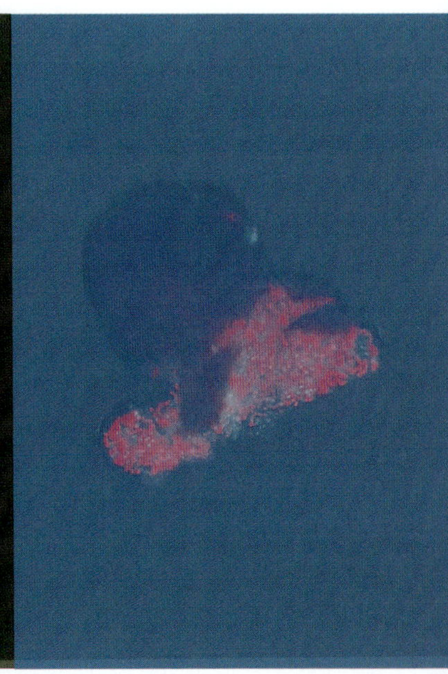

e

Plate 31

31 a Scytonemin-rich mat, mainly *Lyngbya* cf. *aestuarii*, in hypersaline intertidal flats, Laguna Guerrero Negro, Baja California S., Mexico. **b** Desert crust in Moab, Utah, showing experimental area with dark coloration due to the high amounts of scytonemin in *Nostoc commune*, *Scytonema* spp. and the cyanolichen *Collema cristata*; note bare soils in quadrats where crusts have been removed and in the walking paths. **c**, **d**, **e** The cyanolichen *Collema* cf. *coccophorum* (Arches National Park, Utah) contains high amounts of scytonemin synthesized by *Nostoc* in its upper lobe, so this region does not fluoresce due to the UVA shielding by scytonemin; lichen lobe is about 1 mm in diameter: **c** Lobe shown with natural coloration; **d** Autofluoresecence when excited with green light; **e** Autofluorescence when excited with UVA radiation. (Photos **a** R.W.Castenholz; **b** - **d** F. Garcia-Pichel)

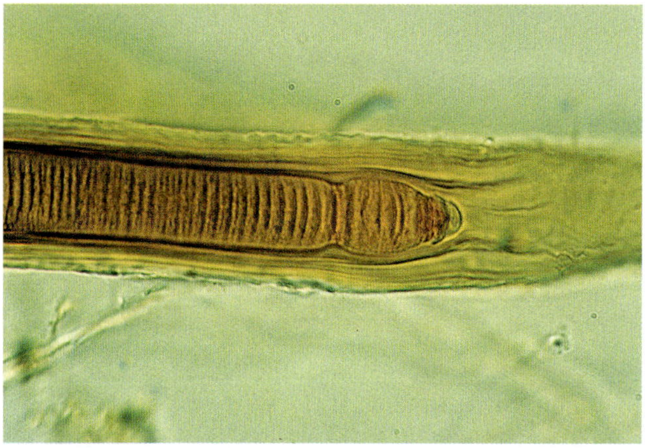

a

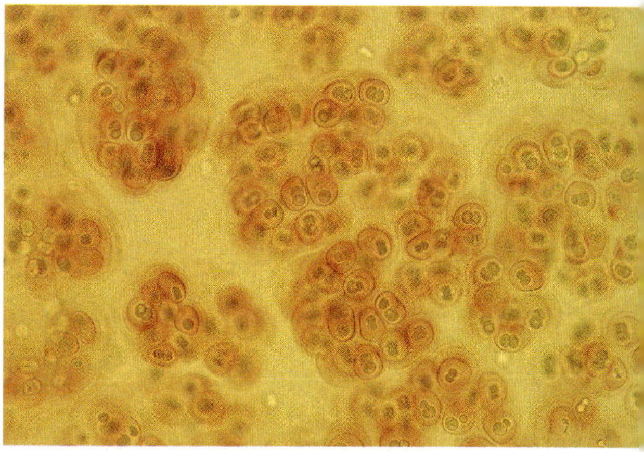

b

c

d

Plate 32

32 a *Lyngbya* with scytonemin-rich laminated sheath (trichome ~12 µm diam.). **b** *Gloeocapsa* cf. *sanguinea*.with reddish "gloeocapsin" in sheaths (cells ~5 µm diam.). **c** *Scytonema* sp. (Culture B-77-Scy.j) showing old filament grown under high irradiance (rich in sheath scytonemin) and new branches grown under low irradiance without scytonemin (axial filament ~14 µm diam.). **d** Laminated microbial mat from hypersaline pond at Guerrero Negro, Baja California, S. Mexico: upper section had remained under sun throughout the morning, while lower section had been covered with < 10 % transmittance filter for 1.5 h; dark coloration is caused by accumulation of "oscillatorian" cyanobacteria that have migrated to the surface. (Photos: **a**,**c**,**d** R.W. Castenholz; **b** F. Garcia-Pichel)

Chapter 1

Introduction to the Cyanobacteria

Brian A. Whitton[1] and Malcolm Potts[2]
[1]*Department of Biological Sciences, University of Durham, Durham DH1 3LE.*
[2]*Department of Biochemistry, Virginia Polytechnic Institute and State University, Blacksburg, VA 24061, USA*

Summary

This chapter is written largely for those who are not already specialists in the cyanobacteria. Features of these organisms are introduced by highlighting some of the topics described in the various chapters, together with other important subjects for which there was no space for a dedicated chapter. The reason is explained why cyanobacteria were formerly known as blue-green algae, and frequently still are known under this name by those involved in water management.

I. What are Cyanobacteria?

Cyanobacteria are photosynthetic prokaryotes possessing the ability to synthesize chlorophyll *a*. Typically water is the electron donor during photosynthesis, leading to the evolution of oxygen. Cyanobacteria have until recently also been characterized by their ability to form the phycobilin pigment, phycocyanin. It is the high concentration of this pigment occurring under some conditions which leads to the bluish colour of the organisms (Plate 2a, 2e) and hence both of the names by which the organisms are commonly known, cyanobacteria or blue-green algae. However, it has increasingly become clear that at least some oxygen-evolving prokaryotes such as *Prochlorothrix*, which do not possess phycobilins, but do form chlorophyll *b* in addition to chlorophyll *a* (Post, 1999; Tomitani et al., 1999), are quite closely related to organisms with phycocyanin. As the molecular evidence does not support the grouping together of all these chlorophyll-*b*-containing organisms, they are treated here (Chapter 5) as members of the cyanobacteria in the broadest sense (but see p. 532).

II. Ecological Diversity in the Past and Present

Geologists and geochemists all agree that cyanobacteria have had a long evolutionary history and most - though not all - agree that this extends to

B. A. Whitton and M. Potts (eds), The Ecology of Cyanobacteria, 1-11.
© 2000 *Kluwer Academic Publishers. Printed in the Netherlands.*

at least 3500 Ma ago (Chapter 2: Schopf), although speculation about the possible occurrence of cyanobacteria on Mars (Chapter 13: Wynn-Williams) raises even wider questions about their origin. Schopf and Walter (1982) have termed the Proterozoic Era (2500 - 570 Ma) the "Age of Cyanobacteria", because this is when they are most abundant in the fossil record. Derivatives of 2-methylbacteriohopanepolyols, which occur in many modern cyanobacteria, have been found in organic-rich sediments as old as 2500 Ma (Summons et al., 1999). Sometimes the filamentous structures found in Proterozoic rocks are strikingly like modern filaments (Plate 3). Laminated structures known as stromatolites, which were deposited widely then, appear to be quite similar to structures developing in a few places today, among which those at Shark Bay in western Australia (Plate 1f) are the best known. Grazing by invertebrates is responsible for the destruction of many modern laminated structures formed by cyanobacteria, such as *Phormidium hendersonii* (Plate 1e), and the lack of such activity in the Proterozoic Era is probably one of the reasons why so many structures from this era were able to persist as stromatolites (Schopf and Walter, 1982). Cyanobacteria were also important in the formation of many of the strata containing petroleum-rich deposits (Chapter 11: Radwan & Al-Hasan).

Some of the reasons for the success of cyanobacteria in modern habitats can be related to their long evolutionary history. Chapter 16 (Robinson, Rutherford, Pocock and Cavet) considers how the changing metal composition of early environments during the period when O_2 was first being released into the atmosphere may have influenced evolution of new metal resistance determinants and metal-utilizing proteins. Tolerance of low oxygen conditions is still widespread in cyanobacteria (Chapter 4: Stal) and free sulphide is tolerated by some strains at levels much higher than those tolerated by most eukaryotic algae (Padan and Cohen, 1982). Some can utilize H_2S as a hydrogen donor in addition to H_2O (Cohen et al., 1975), a feature which is apparently absent even in those eukaryotes, such as some diatoms, which can tolerate relatively high H_2S concentrations. Another feature of some strains is their high tolerance to ultraviolet-B and -C radiation (Chapter 21: Castenholz & Garcia-Pichel), a feature likely to have been especially important in the early evolution of cyanobacteria. The maximum temperature at which a cyanobacterium has been recorded is about 73°C (Chapter 3: Ward and Castenholz), a temperature reached by *Synechococcus* in many thermal springs in western North America. However, it is doubtful if this a relic from an earlier stage in the earth's history, because cyanobacteria elsewhere in the world never reach quite this temperature (Castenholz, 1978). It is not clear why they have failed to fill a photosynthetic niche at temperatures higher than 73°C; if they did so once, representatives apparently do not occur now.

Adaptation to low light and effective nutrient uptake kinetics at low ambient concentrations are important factors leading to modern-day picocyanobacteria dominating both biomass and production in the clear, nutrient-deficient waters which characterize vast segments of the world's oligotrophic oceans (Chapter 5: Paerl) and some large lakes (Chapter 7: Stockner, Callieri and Cronberg). However, there is still much to explain about these organisms, which were unknown before 1979 (Johnson and Sieburth; Waterbury et al.). For instance the density of picocyanobacteria reaches some of the highest values known for these organisms (up to 8×10^6 cells mL^{-1} in saline coastal lakes of Antarctica, whereas they are conspicuously absent or rare in the adjacent polar oceans (Chapter 12: Vincent).

Among the behavioural responses of aquatic cyanobacteria, the importance of circadian rhythms is becoming increasing clear (Chapter 14: Mann; Lepp and Schmidt, 1998; Kondo and Ishiura, 1999). Such rhythyms are a fundamental adaptation of living cells to the earth's daily fluctuation in light and temperature, and cyanobacteria are perhaps the most likely organisms for which the questions of genetic control of this important trait can be answered in the near future (Kondo et al., 1994).

Cyanobacteria are also important in many terrestrial environments and here their tolerance of desiccation and water stress is a key factor (Chapter 9: Pentecost and Whitton). As a consequence, cyanobacteria often play a key role in maintaining the stability of the surface crusts of semi-deserts and the fertility of soils used for farming in arid regions (Chapter 8: Whitton). In true deserts cyanobacteria, especially *Chroococcidiopsis*, sometimes form part of a microbial community several millimetres below the surface (Chapter 13: Wynn-Williams). These organisms are clearly growing near the limits for life in the dry deserts of Antarctica and it has been suggested that their doubling time may be as slow as ten thousand years (Nienow and Friedmann, 1993). This differs by a factor of 4×10^7 from the fastest known growth rate for a cyanobacterium, a doubling time of 2.1 h for "*Anacystis nidulans*"

(*Synechococcus* PCC 6301) (Kratz and Myers, 1955), an organism isolated from a freshwater site in Texas.

An important feature of many cyanobacteria is their ability to fix atmospheric nitrogen. In most well-oxygenated environments this takes place inside the heterocyst (Wolk et al., 1994), a thick-walled cell often with a nodule of cyanophycin (Plate 2c), a polymer of two amino-acids, at one or both ends of the cell. A few cyanobacteria have particular physiological strategies which permit them to fix nitrogen under well-oxygenated conditions even without a heterocyst, but this ability becomes more widespread under micro-oxic conditions (Chapter 4: Stal). The occurrence of microfossils resembling modern heterocystous genera is taken as evidence that nitrogen fixation occurred at the time (Schopf and Walter, 1982), though the process might well have been important much earlier, if nitrogen fixation first developed in non-heterocystous cyanobacteria.

Heterocysts usually develop in laboratory cultures only after the concentration of combined nitrogen in the medium has been lowered (Fogg, 1944; Castenholz and Waterbury, 1989). This fits with many field observations that nitrogen-fixing cyanobacteria become more important as the availability of combined nitrogen in the environment becomes a limiting factor for growth. Heterocyst differentiation has recently been shown to be controlled by a small diffusible peptide (Yoon and Golden, 1998) in a manner similar to physiological control of other processes in eukaryotic systems. However, no eukaryotic alga is known to fix molecular nitrogen, so nitrogen-fixing cyanobacteria have a major advantage at times when sources of combined inorganic nitrogen have been depleted from the water (Chapter 6: Oliver and Ganf). Hyenstrand et al. (1998) have argued that. in comparison with eukaryotic algae, cyanobacteria also a have high competitive ability to compete for ammonium under nitrogen-limiting conditions, but a low competitive ability for nitrate.

Numerous records for cyanobacteria in freshwaters and soils indicate that their diversity and abundance are greatest at higher pH values, though the explanation for their particular success under these conditions is still unclear. There are, however, a number of records at lower pH values, with planktonic picocyanobacteria (< 2 μm diameter) sometimes down to pH 4.5 (Steinberg et al., 1998) and several filamentous forms with true branching at about pH 4.0 (Chapter 8: Whitton). This has been considered to be the lower pH limit for cyanobacteria (Whitton, 1992). Many of the cyanobacteria found at lower pH values are heterocystous, so perhaps they are only able to compete effectively in an environment where the ability to fix nitrogen is important.

III. Morphological Diversity

Cyanobacteria show considerable morphological diversity. They may be unicells (e.g. *Chroococcus*: Plates 2a, 13b) or filaments (e.g. *Anabaena*: Plate 2f) and these may occur singly or grouped in colonies (Plate 13; *Schizothrix*: Plate 2e). Unicells may divide in one, two (*Merismopedia*: Plate 2b) or three planes. Some unicells dividing in one plane show asymmetric division; in *Chamaesiphon* the smaller cells glide down to the base of the larger cell and repeats of this process eventually give rise to a colony of many cells. The resulting colonies can often be seen as brownish spots on submerged rocks (e.g. *C. fuscus*: Plate 1g). Branching of filaments may be false or true. False branching is when division in a filament with a sheath leads to the formation of two separate sheathed filaments and one of these grows separately from the other (*Tolypothrix*: Plate 2g; *Scytonema*: Plate 2h). True branching occurs when cells in a filament have the potential for dividing in more one plane (e.g. *Hapalosiphon*). Maintenance of the colonial structure in many of these forms is aided by the presence of exopolysaccharides (Chapter 17: Potts), such as mucilage and/or a firm sheath. The presence or absence of a heterocyst is an important feature separating genera. However, gas-vacuoles (Chapter 6: Oliver and Ganf), structures which aid buoyancy, are found in species of many different genera (e.g. *Anabaena*: Plate 2f)

Several other terms are useful for interpreting the morphology of cyanobacteria. The trichome refers to the cellular part of a filament with a sheath (e.g. *Lyngbya*: Plate 3a), even though a similar morphological form occurring without a sheath is often termed a filament, as in *Oscillatoria*. In some false-branched filamentous forms, the filament grows apically and dies at the base. Where the sheath is persistent, this can lead to the development of a mass of old sheaths under the living material, as in some species of *Scytonema*. Another widespread morphological form is where cell division is localized nearer the base of a sheathed trichome and the apical cells become narrowed and develop into a multicellular hair; typically the cells lose their chlorophyll and the thylakoids split apart to form liquid-filled vacuoles (e.g. *Calothrix*).

Species that form true filaments all form hormogonia, structures which were defined by Desikachary (1959) as short, motile chains of rather uniform cells. However, not all hormogonia fit exactly within this definition. Essentially they are modified filaments associated with reproduction and dispersal. They are probably always rich in stores of nitrogen, phosphorus and perhaps other nutrients and, especially in heterocystous species, they are morphologically simpler than the other filaments (Rippka et al., 1979). Hormogonia play crucial roles in a number of important physiological processes in cyanobacteria (Tandeau de Marsac, 1994). Their dispersal is aided in some cases by their rapid motility on surfaces and in others by the formation of gas-vacuoles in species which otherwise do not possess these structures. The colonies of some genera (e.g. *Rivularia*) are typically formed by the aggregation of a number of hormogonia, whereas others (e.g. *Nostoc*) usually originate from a single hormogonium. Nothing is known of the signals involved in aggregation to form colonies in cyanobacteria, though studies have recently started on the attraction of hormogonia to potential symbiotic partners (Chapter 20: Adams).

The release of a hormogonium from the end of a heterocystous trichome with a basal - apical organization and the division of some non-heterocystous trichomes in two daughter trichomes takes place by means of a sacrificial cell, the necridium. Although such cells have been recognized for a long time, little is known of how a particular cell is determined or why this occurs in trichome division in some species, but not others.

Akinetes (Plate 2f) are cells formed under adverse conditions, though the particular conditions responsible for their formation differ among species, and perhaps even among strains of a single species (Whitton, 1992). They are usually much larger than a typical vegetative cell. Akinetes are largely restricted to heterocystous species, but occur also in a few non-filamentous species. It is not known whether the latter are nitrogen-fixers. The preface by Castenholz and Waterbury to the account of cyanobacteria in Bergey's Manual of Systematic Bacteriology (1989) provides an excellent guide to other morphological features and terms.

Vegetative cells of unicellular cyanobacteria range in diameter from about 0.4 µm to over 40 µm (*Chroococcus turgidus* UTEX 123) and in volume by a factor of at least 3 x 10^5. Filamentous forms belonging to the Oscillatoriaceae are occasionally reported with diameters in excess of 100 µm, but the cells are very short in relation to diameter and the maximum cell volume is not as great as in some unicells. This may perhaps be related to the typically short-lived mRNA occurring in prokaryotes (Demoulin and Janssen, 1981).

It seems likely that most, if not all, cyanobacteria possess multiple copies of the genome. Even the small-celled and morphologically simple forms, *Synechococcus* and *Synechocystis*, may contain six identical "chromosomes" (Mann and Carr, 1974; Binder and Chisholm, 1990; Castenholz, 1992). It is unclear how much the larger genome size characteristic of morphologically more complex cyanobacteria (Herdman et al., 1979; Simon, 1980) reflects a greater amount of information or is simply due to a greater degree of polyploidy. However, Herdman and Rippka (1988) concluded that the cells of hormogonia contain a much lower number of copies of the genome than typical vegetative cells. More needs to be known about this in order to understand how mutation and selection influence populations, especially in filamentous forms. It is well to keep in mind that a single filament of an organism like *Oscillatoria*, with perhaps more than a hundred cells, might contain many hundreds of copies of the genome. It should therefore be no surprize when clonal isolates of a filament sometimes give rise to phenotypically diverse filaments within a single culture flask.

The general acceptance of the cyanobacterial origin of chloroplasts played an important role in helping to establish the wider theory of the role of endosymbiosis in the evolution of the eukaryotic cell (Carr, 1999). Information about those modern organisms such as *Cyanophora paradoxa*, which have intracellular cyanelles with a slight resemblance to free-living cyanobacteria, is therefore of particular interest (Chapter 19: Adams). It is usually assumed that cyanelles represent an evolutionary dead-end and are not half-way stages to the development of a typical chloroplast. Nevertheless hardly anything is known of the ecological reasons for the success of the organisms which contain cyanelles.

IV. Taxonomy

It is clearly important that ecological accounts use names which can be interpreted easily and reliably by other researchers. Unfortunately this is often far from the case and there is no one book - or even series of books - which resolves all the problems. It is essential for anyone wanting to interpret and write their own accounts of cyanobacterial ecology to know something about the reasons for this. More detailed background is given by Castenholz and Waterbury (1989) and Whitton (1992).

The taxonomic system which developed during the 19th century was based almost entirely on morphology and the International Code of Botanical Nomenclature takes various papers published during the period 1886 - 1892 as the starting point for the valid publication of names of filamentous forms. Rather surprizingly, the starting point for the morphologically less complex forms is much earlier (eg. Linnaeus, 1753), although there was of course little understanding of their diversity at that time. A number of "floras" summarizing the known species of blue-green algae (the botanical name) in particular regions have been published during the 20th century. Several of these provide a lot of information about species occurring elsewhere in the world. In any case it is clear that many species have a very wide distribution. The best known flora is that of Geitler (1932), which focussed on central Europe, but is still an essential source of information for those wanting to name the organisms in any part of the world. Frémy's accounts are of special value because of his careful observations and extensive first-hand experience from equatorial (1929) and marine environments (1929-1933). Desikachary's (1959) flora for India could have reached a very wide audience because it is written in English, but unfortunately has long been out of print.

There have been two very different attempts to reform the classical botanical system. In both cases the authors provided detailed guides to the earlier literature, but differed totally in their approach to nomenclature. An account by Drouet and Daily (1956) followed by a series of monographs by F. Drouet (see Castenholz and Waterbury, 1989) reversed for "blue-greens" what had been done since the 1890s by the majority of microbiologists dealing with other groups (Whitton, 1992). Drouet reverted to the use of only a few simple morphological features: the more than 2000 species (of blue-green algae) described in the literature were eventually reduced to 62 species. The nomenclatural simplicity of the system tempted many ecological researchers in the USA during the 1960s and 1970s to use his names. No extensive critique of the system has ever been published, but it has quite rightly fallen into disuse, apart from the fact that two generic names, *Anacystis* and *Agmenellum*, were used for strains which have since become widely used in research.

Accounts of new species and a series of reviews by J. Komárek with either E. Kann or K. Anagnostidis have sought to bring the classical botanical system up to date in more conventional ways. Several of the earlier reviews provide comprehensive accounts of important genera, such as *Chamaesiphon* (Kann, 1972) and *Phormidium* (Kann and Komárek, 1970), but were published in journals with only a limited distribution and written in German and hence have often been overlooked. More recent accounts (e.g. Komárek and Anagnostidis, 1988, 1989) have been used more widely, although they introduce major nomenclatural changes for some genera, especially non-heterocystous filamentous forms like *Oscillatoria*. Perhaps half the descriptive accounts of plankton in European lakes made during the 1990s have adopted the nomenclature recommended by these authors. However, its use is hindered by the lack of detailed explanation of the principles underlying the changes and the problem of dealing with species for which the authors have not provided guidelines.

The recent account (Komárek and Anagnostidis, 1999) of the Chroococcales within a single volume overcomes the above problems for this group. A feature of the system is the logical arrangement of the organisms according to the various possible combinations of pattern of cell division, cell shape and organization of cells within bounding sheaths and/or mucilage; generic status is alloted to most of the combinations, so the number of species within genera tends to be low. The account is so detailed that it is likely to remain the definitive text on botanical species in this group for a long time. The amount of information needed for recognition of not only the Chroococcales, but the botanical species (of blue-green algae) as a whole is so great, that we suggest the only way for it to be accessed effectively by a large number of people is from a computer data-base. This has been done for the 320 species recorded from the British Isles and the information has been incorporated into an interactive identification system on CD-ROM (Whitton et al., 2000) using the LucID software developed at the University of Queensland, Australia. Most of the morphological and ecological information known for these species can be expressed in 113 characters or character states. About 150 characters would probably be sufficient to code the information for all described species. A somewhat similar approach has

Table 1. List of features in the orders of cyanobacteria recognized by Castenholz and Waterbury (1989). (A few additional details are included here.)

NON-FILAMENTOUS

Order Chroococcales

Unicellular or non-filamentous aggregates of cells held together by outer wall or gel-like matrix; binary division in one, two or three planes, symmetric or asymmetric; or by budding. Rarely form akinetes.

Order Pleurocapsales

Unicellular or non-filamentous aggregates of cells held together by outer wall or gel-like matrix; reproduction by internal multiple fissions with production of daughter cells smaller than parent; or by a mixture of multiple fission and binary fission. Rarely form akinetes.

FILAMENTOUS

Order Oscillatoriales

Binary division in one plane giving rise to 1-seriate trichomes, though sometimes with "false" branches; trichomes do not form heterocysts. Akinetes apparently not recorded.

Order Nostocales

Binary division in one plane giving rise to 1-seriate trichomes, though sometimes with "false" branches; one or more cells per trichome differentiate into a heterocyst, at least when concentration of combined nitrogen is low. Some also produce akinetes.

Order Stigonematales

Binary division periodically or commonly in more than one plane, giving rise to multiseriate trichomes, or trichomes with true branches or both; apparently always possess ability to form heterocysts; some also form akinetes.

been used to establish a similarity matrix for strains of *Anabaena* (Hiroki et al., 1998).

The taxonomic system introduced by R. Y. Stanier in the 1970s aimed to treat the organisms in the same way as other prokaryotes. A early step was to convince everyone of the need for a shift in name from blue-green algae to cyanobacteria. In practice the former term has been retained almost everywhere for water management and other practical purposes, whereas the latter is in general use for research studies.

The system is based on use of axenic, clonal cultures and takes information from a range of disciplines rather than just morphology. A detailed study of a large number of strains (Rippka et al., 1979) led to revised descriptions of some of the most important genera. The proposal (Stanier et al., 1978) that cyanobacteria be placed under the International Code of Nomenclature of Bacteria was intended to remove the organisms from the jurisdiction of the International Code of Botanical Nomenclature. However, taxa described under either of the codes are currently considered in accounts prepared according to the other code. For instance, the Chroococcales flora of Komárek and Anagnostidis (1999) includes genera which were first described according to the International Code of Nomenclature of Bacteria. The accounts (largely by R.W. Castenholz and J.B. Waterbury) in Bergey's Manual of Determinative Bacteriology (Staley et al., 1989) provide an overview using the bacteriological approach; although this is somewhat dated, it will soon be replaced in a new edition of the Manual. The features of the five orders recognized by Castenholz and Waterbury (Table 1 here) are quite similar to those recognized in older botanical texts.

Most ecologists still adopt the botanical approach for naming the organisms in field samples. This is partly because of the amount of time required to isolate and characterize representative strains from mixed samples, let alone strains chosen at random. A further problem is that it is not always easy to relate observations based on morphology of field samples to what is known from laboratory studies. Many - perhaps most - of the features used to identify classical blue-green algal taxa are based on the morphology of the organism growing under nutrient-limited conditions (Whitton, 1992) and may not be expressed in the typical laboratory media used to prepare biomass for biochemical and molecular studies. Although the requirement that combined nitrogen in the medium should be low or absent for heterocyst formation is well-known and has been considered in most studies (e.g. Rippka et al., 1979), the fact that P-limitation is required for the expression of other morphological features was not only overlooked in the original bacteriological studies, but often still is. For instance, features such as degree of tapering, formation of multicellular hairs and pattern of false branching are unlikely to show their characteristic field appearance when grown in batch culture using media with levels of phosphate much higher than likely to be encountered in the field.

Some of the molecular methods used for taxonomic studies of bacteria have also been applied to cyanobacteria (see Wilmotte, 1994), sometimes with modifications applicable to phototrophs. Recent approaches include: DNA-DNA hybridization (Wilmotte et al., 1997), fingerprinting based upon PCR with primers from short and long tandemly repeated elements (Rasmussen et al., 1998), classification of clone cultures based upon 16S rRNA sequences from the variable regions V6, V7 and V8 (Rudi et al., 1997) and amplified ribosomal DNA restriction analysis of the internally transcribed spacer (Scheldeman et al., 1999). One question which remains particularly controversial is the role of gene exchange (lateral transfer) and recombination in cyanobacterial evolution (Castenholz, 1992; Rudi et al., 1998). Other important questions are how far molecular studies will confirm the taxa recognized at present and how soon it will be possible to use molecular approaches routinely as a rapid means of characterizing cells from natural populations. Unfortunately the various "phylogenetic" trees illustrating similarities between strains based on similarities between particular parts of the genome often use results from strains whose generic and specific names are doubtful. This restricts the value of such trees for comparing possible evolutionary relationships with the taxonomic relationships suggested by classical names.

With regard to specific genetic markers that have been explored in taxonomy, different types of group I introns may have different evolutionary histories in cyanobacteria (Paquin et al., 1997), and some of these sequences may have been distributed through lateral transfer (Biniszkiewicz et al., 1994). Nevertheless, in the case of some lichen symbionts, group I intron analysis offers some promise as a useful technique (Paulsrud and Lindblad, 1998). Where species are delimited by only one or two morphological characters, molecular studies to date do not agree closely with the classical species.

A feature of some cyanobacterial populations is that they are often dominated by what appears to be a single morphological form, such as *Trichodesmium* (Chapter 5: Paerl), *Aphanothece halophytica* (Chapter 10: Oren) and *Nostoc commune* (Chapter 17: Potts). Because such forms may be readily identified, but occur in diverse localities, the concept of a "form species" is helpful in their study. Garcia-Pichel et al. (1996) concluded that *Microcoleus chthonoplastes* may in fact be considered a single cosmpolitan species, but the extent to which other such species form close genetic groupings remains to be evaluated critically. In the case of hot spring communities, recent studies indicate that their diversity is much more complex than earlier thought (Chapter 3: Ward and Castenholz).

V. Molecular Ecology

A review by Castenholz (1992) and the sister volume (Bryant, 1994) to the present one make clear how important is an understanding of cyanobacterial molecular biology not just as an aid to taxonomy, but for interpreting ecological phenomena in general. The widespread repetition in cyanobacteria of short sequences of DNA and its possible evolutionary significance are assessed in Chapter 16 (Robinson, Rutherford, Pocock and Cavet) How the cyanobacteria sense and respond to changes in their environment is described in Chapters 14 (Mann) and 15 (Bhaya, Schwarz and Grossman). Suboptimal light and nutrient conditions result in a number of responses that strongly influence the physiology of the cell. The responses can be striking or subtle and subsequent changes take place rapidly or very slowly. The key role of light means that the signal transduction pathways that permit an appropriate adaptive response to be set in train differ somewhat from most other prokaryotes. Cyanobacteria tend to show resistance to multiple environmental stresses, and it is probable that the reponse pathways to the different stimuli overlap (Chapter 15, Bhaya et al.). Caution is therefore needed in assuming that, because a particular stimulus brings about a characteristic morphogenetic response in the laboratory, it is the same stimulus which does so in nature. The ability of some cyanobacteria to withstand extremes of UV radiation (Chapter 21: Castenholz and Garcia-Pichel) and desiccation (Chapter 17: Potts) is aided by their capacity for efficient DNA repair. It will be of interest to understand if, and how, such repair plays a role in other responses to environmental stimuli such as nutrient limitation and cyanophage infection (Chapter 20: Suttle).

In 1995, the Kazuza DNA Research Institute of Japan, commenced sequence analysis of the entire genome of *Synechocystis* PCC 6803 (Kaneko et al., 1996). The project was completed within 16 months and the contiguous 3,573,470 base pairs of sequence, together with annotations of all 3168 genes were deposited in the CyanoBase database (http: //www.kazusa.or.jp:8080/cyano/). This was the first entire sequence for any photo-autotroph. CyanoBase provides an easy way of accessing the sequence data together with the annotation data by means of image maps, keyword searches and the gene category list. The rationale to use this strain included its comparatively small genome and its capacity for transformation. Cyanobase rapidly became a major resource not only for those interested in the cyanobacteria, but also for bacteriologists, plant physiologists and evolutionists. Currently, other projects underway include sequencing the genome of *Prochlorococcus* (in the USA) and *Anabaena* (in Japan). These may be compared (August 1999) with what has been done for other bacteria: 20 complete sequences plus 43 in progress. Developments of increasingly refined methods to sequence genomes and analyze sequence data will permit the genetic analysis, and manipulation, of morphologically more complex cyanobacteria, including global "form" species, such as *Trichodesmium*, which play such prominent roles in the earth's ecology.

VI. Interactions with Other Organisms

Symbiotic interactions between cyanobacteria and other organisms are surprisingly diverse, including examples from many of the major phyla of plants and animals. The associations with a wide range of lower and higher plants have especial potential as experimental systems (Chapter 19: Adams; Meeks, 1998). Many cyanobacterial symbionts share characteristics in common, including the formation of hormogonia, which often serve as the infective agent, and the presence of heterocysts with their ability to fix nitrogen. Some plants are known to produce chemical signals that both induce the cyanobacterium to form hormogonia and serve as chemoattractants to guide these hormogonia into the plant tissue. The application of molecular methods to such topics is revealing the exciting complexity of these dynamic systems (Campbell et al., 1998).

Among the infective agents potentially leading to cell lysis, cyanophage in the oceans have become a particular source of interest. Cyanophage that infect phycoerythrin-rich *Synechococcus* can be extremely abundant in coastal marine environments, where titres may exceed 10^6 mL^{-1} (Chapter 20: Suttle). The high density of both cyanophage and *Synechococcus* here result in high encounter frequencies and selection for *Synechococcus* populations that are largely resistant to infection, whereas the lower density of *Synechococcus* in oceanic waters and hence much lower encounter frequencies is probably responsible for the lower resistance of the *Synechococcus* to infection.

VII. Blooms and Toxins

Reports of dense populations of freshwater cyanobacteria forming water-blooms have increased in many countries and some of the factors leading to the formation of such blooms are understood quite well (Chapter 6: Oliver and Ganf). The fact that these blooms are often toxic (Chapter 22: Dow and Swoboda) has led to increasing awareness and concern over the environmental and health problems associated with them. In spite of the huge literature on toxins (Sivonen and Jones, 1999), it is not always easy to predict the toxicity of particular populations, let alone possible ways of minimizing the the level of toxicity present. Molecular approaches to understanding toxins have been slow to get underway, but the recent isolation of the gene cluster involved in the biosynthetic (non-ribosomal) pathway to the toxin microcystin provides a promising insight, which should eventually lead to the development of methods to minimize toxin levels.

VIII. Cyanobacteria as Health Food

The cyanobacterium often known as Spirulina, but taxonomically *Arthrospira*, is widely marketed as a health food (Belay et al., 1993). (The nomenclatural problem results from Geitler, 1932, who submerged what is a very different organism within the genus *Spirulina*.) Spirulina is grown under a wide range of systems, from crude ponds, to large-scale outdoor raceways (Plate 27a,b) and sophisticated bioreactors. As a result a lot is known about the physiological ecology of the organism in the laboratory and large-scale commercial culture (Chapter 18: Vonshak and Tomaselli), though less about how these features relate to growth of the organism in its natural environment. In contrast, little is known about the physiological ecology of *Aphanizomenon*, another cyanobacterium marketed on a large scale as a health food, and which is apparently at present obtained for this purpose solely from natural blooms in one lake, Klamath Lake in Oregon, USA. In view of its economic importance, it is surprizing how little has been published about this, though Drapeau and Gutermuth (1998) have provided a start.

IX. Use of Cultures and Culture Media

A strange feature of many studies on cyanobacteria is the care taken with certain aspects of research, but the sometimes casual approach to other aspects, such as the choice of name or culture medium. This can make it hard to assess the relevance of such studies to understanding ecological processes. A particularly serious problem is the choice of the strain to be used for a project. Early researchers on the physiology and biochemistry of cyanobacteria had a limited choice of strains and usually took ones which provided adequate biomass rapidly. Often the medium used for cultures contained much higher concentrations of many elements, and sometimes different forms of these elements, than likely to be found in natural environments; this was especially true for potassium phosphate (Whitton and Carr, 1982), which was for long included as part of the buffer system in the medium. Almost inevitably, the initial isolation and subsequent subculturing will have led to selection of strains which combine two properties - ability to grow exponentially and ability to make effective use of nutrients at very high ambient concentrations. "Spirulina" (*Arthrospira*) represents an extreme case, where the medium used for almost all laboratory studies (Zarrouk, 1966) has nutrient levels far in excess of those likely to be encountered by most natural populations. Carr (1999) has argued convincingly about the bias introduced to research by use of strains selected for their ability to grow exponentially. There is abundant evidence that morphological changes such as loss of colony structure and akinete formation, and changes in the properties of exopolysaccharides, have taken place during prolonged subculture and almost as much evidence for physiological changes.

Effective progress in cyanobacterial ecology requires shifts in attitudes to research, including a more critical approach to choice of strains, how these strains are maintained, the suitability of culture media and caution in interpreting the results of molecular studies. There are already many strains available which should be fairly representative of the environment from which they were taken, so there is no need to use old laboratory "weeds" if the work is directed to answering questions about the environment. Further, the problems of isolating strains have sometimes been exaggerated, at least in the case of filamentous forms. It is of course difficult to simulate high UV-B radiation or diel cycles of Eh and O_2 in the laboratory. Nevertheless in many cases all that is needed to isolate a culture of a filamentous species is an understanding of the biology of the organism, such as the factors leading to the formation of hormogonia or akinetes. The great majority of strains without gas-vacuoles survive well in liquid nitrogen for periods up to at least twenty years without any special precautions other than the presence of dimethylsulphoxide (authors, unpublished data). Only those organisms like many strains of "Spirulina", which have gas-vacuoles under all environmental conditions tested (M. Mühling, pers. comm.), pose a real problem for long-term maintenance.

Subcultures should normally be made from old batch cultures, where growth of the organism has been limited by the same factor that usually leads to growth limitation in the field. Because of the widespread ability of strains which are likely to encounter nitrogen limitation in the field to be able to fix nitrogen, it seems probable that growth of more species in nature will be limited by the availability of phosphate than by any other factor. It is therefore especially important to reduce the phosphate concentration of many widely used media.

X. The Future

A first look at this book may overwhelm the reader by what is already known, but it will soon become apparent how many important questions there are to answer. Cyanobacteria not only play key roles in the global cycles of several elements, but they are becoming increasingly important economically in both positive and harmful ways. A feature of research on cyanobacteria over the past forty years has been how often important phenomena have been overlooked or misinterpreted, when careful observation or simple experiments might have revealed this much earlier. It will be interesting to see which of the orthodoxies presented in this book is the first to be challenged by new thinking and research.

References

Belay A, Ota Y, Miyakawa K and Shimamatsu H (1993) Current knowledge on potential health benefits of *Spirulina*. J Appl Phycol 5: 235-241.

Binder BJ and Chisholm SW (1990) Relationship between DNA cycle and growth rate in *Synechococcus* sp. strain 6301. J Bacteriol 172: 2313-2319.

Biniszkiewicz D, Cesnaviciene E and Shub DA (1994) Self-splicing group I intron in cyanobacterial initiator methionine tRNA: evidence for lateral transfer of introns in bacteria. EMBO J 13: 4629-4635.

Bryant DA (ed.) (1994) The Molecular Biology of Cyanoacteria. 881 pp. Kluwer Academic Publishers, Dordrecht, The Netherlands.

Campbell EL, Brahamsha B, Meeks JC (1998) Mutation of an alternative sigma factor in the cyanobacterium *Nostoc punctiforme* results in increased infection of its symbiotic plant partner, Anthoceros punctatus. J Bacteriol 180: 4938-4941.

Carr NG (1999) Freiburg to Vienna - looking at cyanobacteria over the twenty-five years. In: Peschek GA, Löffelhardt W and Schmetterer G (eds) The Phototrophic Prokaryotes. Proceedings of the Ninth International Symposium held in Vienna, Austria, September 6–12, 1997. Kluwer Academic Publishers, Dordrecht, The Netherlands.

Castenholz RW (1978) The biogeogrpahy of hot spring algae through enrichment cultures. Mitt Int Verh Limnol 21: 296-315.

Castenholz RW (1992) Species usage, concept, and evolution in the cyanobacteria (blue-green algae). J Phycol 28: 737-745.

Castenholz RW and Waterbury JB (1989) Group 1. Cyanobacteria. Preface. In: Staley JT, Bryant MP, Pfennig N and Holt JG (eds) Bergey's Manual of Systematic Bacteriology. Volume 3: 1710-1727.

Cohen YJ, Padan E and Shilo M (1975) Facultative anoxygenic photosynthesis in the cyanobacterium *Oscillatoria limnetica*. J Bacteriol 123: 855-863.

Demoulin V and Janssen MP (1981) Relationship between diameter of the filament and cell shape in blue-green algae. Br Phycol J 16: 55-58.

Desikachary TV (1959) Cyanophyta. Indian Council of Agricultural Research, New Delhi.

Drapeau C and Gutermuth FB (1998) Turning nuisance algae into health food. 4th International Conference on Toxic Cyanobacteria, 27 September - 1 October 1998, Beaufort, North Carolina. Compilation of Abstracts, p. 89.

Drouet F and Daily WA (1956) Revision of the coccoid Myxophyceae. Butler Univ Stud 10: 1-218.

Fogg GE (1944) Growth and heterocyst production in *Anabaena cylindrica* Lemm. New Phytologist 43: 164-175.

Frémy P (1929-1933) Cyanophycées des Côtes d'Europe. Mém Soc Natn Sci Nat Math, Cherbourg 41: 1-236.

Frémy P (1929) Les Myxophycées de l'Afrique Equatoriale Française. Archives de Botanique III, Mémoire No. 2, 508 pp. (reprinted privately, Caen, 1930)

Garcia-Pichel F, Prufert-Bebout and Muyzer G (1996) Phenotypic and phylogenetic analyses show *Microcoleus chthonoplastes* to be a cosmopolitan cyanobacterium. Appl Environ Microbiol 62: 3284-3291.

Geitler L (1932) Cyanophyceae. In: Rabenhorst's Kryptogamen-Flora, Volume 14. Akademische Verlagsgesellschaft, Leipzig.

Herdman M, Janvier M, Rippka R and Stanier RY (1979) Genome size of cyanobacteria. J Gen Microbiol 111: 73-85.

Herdman M and Rippka R (1988) Hormogonia and baeocytes. Methods in Enzymology 167: 232-242.

Hiroki M, Shimizu A, Li R, Watanabe M and Watanabe MM (1998) Phycological Research 46, Suppl: 85-93.

Hyenstrand P, Blomqvist P and Pettersson A (1998) Factors determining cyanobacterial success in aquatic systems – a literature review. Arch Hydrobiol Spec Issues Advanc Limnol 51: 41-62.

Johnson PW and Sieburth JMcN (1979) Chroococcoid cyanobacteria in the sea: a ubiquitous and diverse phototrophic biomass. Limnol Oceanogr 24: 928-935.

Kaneko T, Sato S, Kotani H, Tanaka A, Asamizu, E, Nakamura Y, Miyajima N, Hirosawa M, Sugiura M, Sasamoto S, Kimura T, Hosouchi T, Matsuno A, Muraki A, Nakazaki N, Naruo K, Okumura S, Shimpo S, Takeuchi C, Wada T, Watanabe A, Yamada M, Yasuda M and Tabata S (1996) Sequence analysis of the genome of the unicellular cyanobacterium *Synechocystis* sp. strain PCC6803. II. Sequence determination of the entire genome and assignment of potential protein-coding regions. DNA Res 3: 109-136.

Kann E (1972) Zur Systematik und Ökologie der Gattung *Chamaesiphon* (Cyanophyceae) 1. Systematik; 2. Ökologie. Arch Hydrobiol (Suppl) 41: 117-11; 243-282.

Kann E and Komárek J (1970) Systematisch-Ökologische Bemerkungen zu den Arten Formenkreis *Phormidium autumnale*. Schweiz Z Hydrol 32: 495-518.

Komárek J and Anagnostidis K (1988) Modern approach to the classification system of cyanophytes. Arch Hydrobiol 73: 157-226.

Komárek J and Anagnostidis K (1989) Modern approach to the classification system of cyanophytes 4 - Nostocales. Arch Hydrobiol 82: 247-345.

Komárek J and Anagnostidis K (1999) Cyanoprokaryota. 1. Teil Chroococcales. Süßwasserflora von Mitteleuropa. Gustav Fischer: 548 pp.

Kondo T and Ishiura M (1999) The circadian clocks of plants and cyanobacteria. Trends in Plant Science 4(5): 171-176

Kondo T, Tsinoremas NF, Golden SS, Hirschie Johnson C, Kutsuna S and Ishiura M (1994) Circadian clock mutants of cyanobacteria. Science 266: 1233-1236.

Kratz WA and Myers J (1955) Nutrition and growth of several blue-green algae. Am J Bot 42: 275-280.

Lepp PW and Schmidt TM (1998) Nucleic acid content of *Synechococcus* spp. during growth in continuous light and light/dark cycles. Arch Microbiol 170: 201-207

Linnaeus C (1753) Species Plantarum, Exhibentes Plantas Rite Cognitas, et Genera Relatas, cum Differentus Specificis,. Nominibus Trivialibus, Synonymis Selectis, Locis Natalibus, Secundum Systema Sexuale Digestas II. Stockholm.

Mann NH and Carr NG (1974) Control of macromolecular composition and cell division in the blue-green alga, *Anacystis* nidulans. J Gen Microbiol 83: 399-405.

Meeks JC (1998) Symbiosis between nitrogen-fixing cyanobacteria and plants - The establishment of symbiosis causes dramatic morphological and physiological changes in the cyanobacterium. Bioscience 48: 266-276.

Neilan BA, Dittmann E, Rouhiaien L, Amanda Bass R, Schaub V, Sivonen K and Börner T (1999) Nonribosomal peptide synthesis and toxigenicity of cyanobacteria. J Bacteriol 181: 4089-4097.

Nienow JA and Friedmann EI (1993) Terrestrial lithophytic (rock) communities. In: Friedmann EI (ed.) Antarctic Microbiology, pp 353-412. Wiley-Liss, New York.

Padan E and Cohen Y (1982) Anoxygenic photosynthesis. In: Carr NG and Whitton BA (eds) The Biology of Cyanobacteria. Blackwell, Oxford, and University of California Press, Berkeley: 215-235.

Paquin B., S.D. Kathe, S.A. Nierzwicki-Bauer, and Shub DA (1997) Origin and evolution of group I introns in cyanobacterial tRNA genes. J Bacteriol 179: 6798-6806.

Paulsrud P and Lindblad P (1998) Sequence variation of the $tRNA^{LEU}$ intron as a marker for genetic diversity and specificity of symbiotic cyanobacteria in some lichens. Appl Environ. Microbiol 64: 310-315.

Post AF (1999) The prochlorophytes - an algal enigma. Biology of chlorophyll a/b containing photosynthetic prokaryotes. In: Seckbach J (ed.) Enigmatic Microorganisms and Life in Extreme Environments, pp 115-125. Kluwer Academic Publishers, Dordrecht, The Netherlands.

Rasmussen U and Svenning MV (1998) Fingerprinting of cyanobacteria based on PCR with primers derived from short and long tandemly repeated repetitive sequences. Appl Environ Microbiol. 64: 265-272.

Rippka R, Deruelles JB, Waterbury JB, Herdman M and Stanier RY (1979) Generic assignments, strain histories and properties of pure cultures of cyanobacteria. J Gen Microbiol 111: 1-61

Rudi K, Skulberg OM and Jakobsen KS (1998) Evolution of cyanobacteria by exchange of genetic material among phyletically related strains. J Bacteriol 180: 3453-3461.

Rudi K, Skulberg OM, Larsen F and Jakobsen KS (1997) Strain characterization and classification of oxyphotobacteria in clone cultures on the basis of 16S rRNA sequences from the variable regions V6, V7, and V8. Appl Environ Microbiol 63: 2593-2599.

Scheldeman P, Baurain D, Bouhy R, Scott M, Mühling M, Whitton BA, Belay A and Wilmotte A (1999) *Arthrospira* ('*Spirulina*') strains from four continents are resolved into only two clusters, based on amplified ribosomal DNA restriction analysis of the internally transcribed space. FEMS Microbiol Lett 172: 213-222.

Schopf JW and Walter MR (1982) Origin and early evolution of cyanobacteria: The geological evidence. In: Carr NG and Whitton BA (eds) The Biology of Cyanobacteria, pp 543-564, Blackwell, Oxford, and University of California Press, Berkeley.

Simon RD (1980) DNA content of heterocysts and spores of the filamentous cyanobacterium *Anabaena variabilis*. FEMS Microbiol. Lett. 8: 241-245.

Sivonen K and Jones G (1999) Cyanobacterial toxins. In: Chorus I and Bartram J (eds) Toxic Cyanobacteria in Water. A Guide to their Public Health Consequences, Monitoring and Management. 41-111. E & FN Spon, London & New York.

Staley JT, Bryant MP, Pfennig N and Holt JG (eds) Bergey's Manual of Determinative Bacteriology, Volume 3. Williams and Wilkins, Baltimore: 2298 + 18 pp.

Stanier RY, Sistrom WR, Hansen TA, Whitton BA, Castenholz RW, Pfennig N, Gorlenko VN, Kondratieva EMN, Eimhjellen KE, Whittenbury R, Gherna RL and Trüper HG (1978) Proposal to place the nomenclature of the cyanobacteria (blue-green algae) under the rules of the International Code of Nomenclature of Bacteria. Int J Syst Bacteriol 28: 335-336.

Steinberg CEW, Schafer H, Beisker W and Bruggemann R (1998) Deriving restoration goals for acidified lakes from ataxonomic phytoplankton studies. Restoration Ecology 6: 327-335.

Summons RE, Jahnke LL, Hope JM and Logan GA (1999) 2-Methylhopanoids as biomarkers for cyanobacterial oxygenic photosynthesis. Nature 400: 554-557.

Tandeau de Marsac N (1994) Differentiation of hormogonia and relationships with other biological processes. In: Bryant DA (ed.) The Molecular Biology of Cyanobacteria, pp 825-842. Kluwer Academic Publishers, Dordrecht, The Netherlands.

Tillett D, Dittmann E, Börner T and Neilan BA (1998) Structural analysis of the *Microcystis* genome associated with microcystin biosynthesis. In: Carmichael WW and Paerl HW (eds) (1998) Abstracts, 4th International Conference on Toxic Cyanobacteria, Beaufort, North Carolina: 49.

Tomitani A, Okada K, Miyashita H, Matthijs HCP, Ohno T and Tanaka A (1999) Chlorophyll *b* and phycobilins in the common ancestor of cyanobacteria and chloroplasts. Nature 400: 159-162.

Waterbury JB, Watson SW, Guillard RRL and Brand LE (1979) Widespread occurrence of a unicellular marine planktonic cyanobacterium. Nature 277: 293-294.

Whitton BA (1992) Diversity, ecology, and taxonomy of the cyanobacteria. In: Mann NH and Carr NG (eds) Photosynthetic Prokaryotes, pp 1-51. Plenum, New York.

Whitton BA and Carr NG (1982) Cyanobacteria: current perspectives. In: Carr NG and Whitton BA (eds) The Biology of Cyanobacteria, pp 1-8. Blackwell, Oxford, and University of California Press, Berkeley.

Whitton BA, Robinson PJ and Gemmell JJ (2000) Key to Blue-green Algae of the British Isles. Environment Agency (England and Wales) and University of Durham. Department of Biological Sciences, University of Durham, Durham, UK

Wilmotte A (1994) Molecular evolution and taxonomy of the cyanobacteria. In: Bryant DA (ed.) The Molecular Biology of Cyanobacteria, pp 1-25. Kluwer Academic Publishers, Dordrecht, The Netherlands.

Wilmotte A, Stam W and Demoulin V (1997) Taxonomic study of marine oscillatoriacean strains (Cyanophyceae, Cyanobacteria) with narrow trichomes. III. DNA-DNA hybridization studies and taxonomic conclusions. Algol Stud 87: 11-28.

Wolk CP, Ernst A and Elhai J (1994) Heterocyst metabolism and development. In: Bryant DA (ed.) The Molecular Biology of Cyanobacteria, pp 769-823. Kluwer Academic Publishers, Dordrecht, The Netherlands.

Yoon HS and Golden JW (1998) Heterocyst pattern formation controlled by a diffusible peptide. Science 282: 935-938.

Zarrouk C (1966) Contribution à l'étude d'une cyanophycée. Influence de divers facteurs physiques et chimiques sur la croissance et la photosynthèse de *Spirulina maxima* (Setch. et Gardner) Geitl. Ph.D. Thesis, Paris.

Chapter 2

The Fossil Record: Tracing the Roots of the Cyanobacterial Lineage

J. William Schopf
IGPP Center for the Study of Evolution and the Origin of Life,
Department of Earth and Space Sciences, and Molecular Biology Institute,
University of California, Los Angeles, California 90095-1567 USA

Summary

Since the mid-1960s, following a century of unrewarded search, impressive progress has been made toward deciphering the Precambrian fossil record, evidence of life extant during the earliest seven-eighths of geologic time. Hundreds of fossiliferous units have been discovered containing thousands of microbial fossils — dominantly but not exclusively cyanobacterial — and the documented antiquity of life has been extended to an age roughly three-quarters that of the Earth. Mutually reinforcing lines of evidence, paleontological, geological, and isotopic geochemical, indicate that stromatolitic microbial ecosystems, evidently including cyanobacteria and other members of the bacterial domain, were extant ~3500 Ma ago; methanogenic archaeans by ~2800 Ma ago; and Gram-negative sulfate-reducing bacteria at least as early as ~2700 Ma ago. The discrepancy between these dates and those suggested for emergence of these groups by a recently proposed amino acid-based "molecular clock" is too great and too consistent to be ignored. The challenge is to unify the molecular data with the increasingly well-established paleobiologic record.

B. A. Whitton and M. Potts (eds), The Ecology of Cyanobacteria, 13-35.
© 2000 *Kluwer Academic Publishers. Printed in the Netherlands.*

I. Tracing the Roots of Cyanobacteria: Progress and Problems

A. Darwin's Dilemma: The "Missing" Precambrian Fossil Record

In 1859, when Charles Darwin unveiled his monumental monograph, *The Origin of Species*, major chapters in the history of life had already been deciphered. The familiar progressions from seaweeds to land plants, from marine invertebrates to higher mammals provided Darwin a fossil-based foundation for his masterfully unifying thesis. This evolutionary sequence from water to land is the history of Phanerozoic life dating from 545 million years (Ma) ago and the appearance of calcareous algae and shelled invertebrates that marks the beginning of the Cambrian Period of geologic time. What were the forerunners of these early algae and primitve invertebrates? How did evolution proceed during the Precambrian Eon, the pre-Phanerozoic seven-eighths of Earth history?

To these questions Darwin had no answers — indeed, until a scant three decades ago, Precambrian biologic history was *terra incognita*. But Darwin did know that if answers were not forthcoming his theory was in jeopardy:

"If the theory [of evolution] be true it is indisputable that before the lowest Cambrian stratum was deposited ... the world swarmed with living creatures. [Yet] to the question why we do not find rich fossiliferous deposits belonging to these earliest periods ... I can give no satisfactory answer. The case at present must remain inexplicable; and may be truly urged as a valid argument against the views here entertained" (Darwin, 1859, Chapter X).

Though Darwin posed the problem, it was not until a century later — in the mid-1960s with the birth of a new field of science, Precambrian paleobiology — that this earliest "missing" fossil record began to be uncovered (Cloud, 1983; Schopf, 1992a). Progress since the 1960s has been impressive. The documented record of life has been extended steadily and now stretches to nearly 3.5 billion years (Ga) ago, a date in the geologic past approaching the age (4.5 Ga) of the Earth itself.

Abbreviations: BIF – Banded Iron Formation; Ga – billion (10^9) years; Ma – million (10^6) years

B. The Universal Tree of Life

As the early fossil record came increasingly into focus during the 1970s and 1980s, benchmark advances were underway in studies of the molecular biology of living systems. By the late 1980s, these had coalesced into powerful means to determine the evolutionary, phylogenetic relations among *all* organisms alive today (Woese, 1987).

The bases of these determinations, and for constructing from them a Universal Tree of Life, are straightforward:

(1) All organisms biosynthesize proteins on ribosomes which are composed largely of ribonucleic acids.

(2) The placement of each nitrogenous base in a molecule of ribosomal RNA is determined by information encoded in chromosomal DNA; mutations in this DNA are reflected by substitution of bases in the synthesized rRNA.

(3) Because closely related organisms have closely similar DNAs, their rRNAs are also closely similar, and the more distantly related are any two organisms the more dissimilar their rRNAs.

(4) Thus, comparison of the base sequences of rRNA from diverse organisms provides a simple, logical, universally applicable method to determine how closely or distantly related are all organisms of the living world.

The Universal Tree of Life (based on 16S rRNA) resulting from this type of analysis is well illustrated in Olsen and Woese (1993). In such trees, the length of each branch corresponds quantitatively to the number of base sequence changes that have occurred in that lineage since its divergence from its nearest neighbor.

In rRNA phylogenic trees, four important features stand out. First, despite the seeming dominance in the living world of plants and animals, they are minor components of the total biota, only two of many principal evolutionary branches. Second, the Tree of Life is overwhelmingly composed of microscopic organisms; of all major lineages, only three (plants, fungi, and animals) contain megascopic members, and each of these includes microscopic forms as well. Third, all organisms alive today are members of but three major groups (formally, "domains"; Woese et al., 1990) — the *Eucarya* (organisms having cells in which chromosomes are encapsulated in a distinct membrane-bounded nucleus); the *Archaea* (non-nucleated microorganisms including methanogenic procaryotes and many extremophiles, microbes that thrive in exceptionally acidic high temperature settings); and the *Bacteria*, the domain that includes the cyanobacteria and bacteria of traditional classifications. Fourth, the Eucarya is more closely related to the Archaea than the Bacteria, and the last common ancestor of all organisms living today — the root stock of the Universal Tree of Life — lies between the Archaea and the Bacteria (Iwabe et al., 1989).

The Universal Tree of Life shows the order of branching of the major biotic lineages. When in the geologic past did these various branches emerge? Unfortunately, the answer cannot be provided by rRNA trees. Trees of this type are based wholly on organisms that live today, so if they were precise clocks of evolutionary history the branch tips of all lineages would extend to more or less the same height (a level corresponding to the present) and the top of the tree would be essentially flat rather than markedly undulating as it actually is. As Woese (1987) points out, different lineages have evolved at different rates: Those with relatively long branches have evolved faster (for example, within the Bacteria, flavobacteria and green non-sulfur bacteria; and within the Eucarya, microsporidia and diplomonads) whereas the rRNA in lineages with relatively short branches (several in the Archaea and eucaryotic ciliates, plants, and fungi) has changed more slowly.

C. Limitations of the Early Fossil Record

Accurate dating of the times of origin of the major biologic lineages has been a long-standing goal in Precambrian paleobiology (Schopf, 1970; Schopf et al., 1983; Schopf, 1992b). This is a young field, however, and the early fossil record is too incompletely known to provide precise answers. Moreover, even under the best of circumstances fossil evidence can record only the first *detected* occurrence of a biologic lineage, not its first *actual* occurrence. Together, fossils and associated geological and geochemical indicators of biologic activity (that is, "paleobiologic" evidence) can establish a minimum age for a lineage, but they cannot reveal how much earlier the lineage actually existed.

D. Timing Evolution by the Amino Acid "Clock"

Another way to solve this problem, based on the amino acid composition of proteins, has recently been proposed by Doolittle et al. (1996). Proteins evolve over time, for much the same reason as do molecules of rRNA: Information encoded in chromosomal DNA determines which of 20 commonly occurring amino acids is emplaced at each position in a protein polymer, and changes (mutations) in this code can result in replacement of amino acids in the protein product. There are many different families of proteins (cytochromes, hemoglobins, and the like) and members of the same family occur in diverse organisms, some in lineages of all three biotic domains. Because each biologic lineage has a unique evolutionary history of mutation-driven amino acid replacement, the proteins of any given family vary somewhat from one lineage to another.

On the basis of these considerations, Doolittle et al. (1996) reasoned that a molecular "clock" by which to date times of lineage emergence might be established by comparing the sequences of amino acids in proteins of the same families in diverse groups of organisms. Their study was thorough: it took into account fast and slowly evolving lineages and compared 531 amino acid sequences in 57 families of enzymatic proteins from 15 major groups of organisms. Their analysis yielded an internally consistent amino acid-based tree in which the branches are all more or less the same length and have a branching order in good agreement with the rRNA Universal Tree of Life.

Strangely, however, this amino acid clock meshes not at all well with the Precambrian fossil record. As noted above, paleobiologic evidence can yield only a minimum date for the emergence of a biologic lineage, so the molecular clock would be expected to have yielded *older* ages than those documented by fossils. Yet the Doolittle et al. (1996) study concluded that all early evolving lineages originated at *younger*, not older, ages than those indicated by paleobiology, and the differences are substantial. For example, as shown in Fig. 1, the amino acid data suggest that cyanobacteria (and closely allied Gram-positive bacteria) originated about 1500 Ma ago, two billion years later than inferred from fossil evidence. Similarly, the origins of Gram-negative bacteria and methane-producing archaeans are placed a billlion or more years later than indicated by paleobiology, and the last common ancestor — the root of all present-day life — is dated at only ~2000 Ma ago whereas cellular fossils of diverse morphologies have been found in rocks nearly 3500 Ma old (Schopf, 1992c, 1993).

This lack of consonance is unsettling, Something is amiss. Perhaps the ancient fossils have been

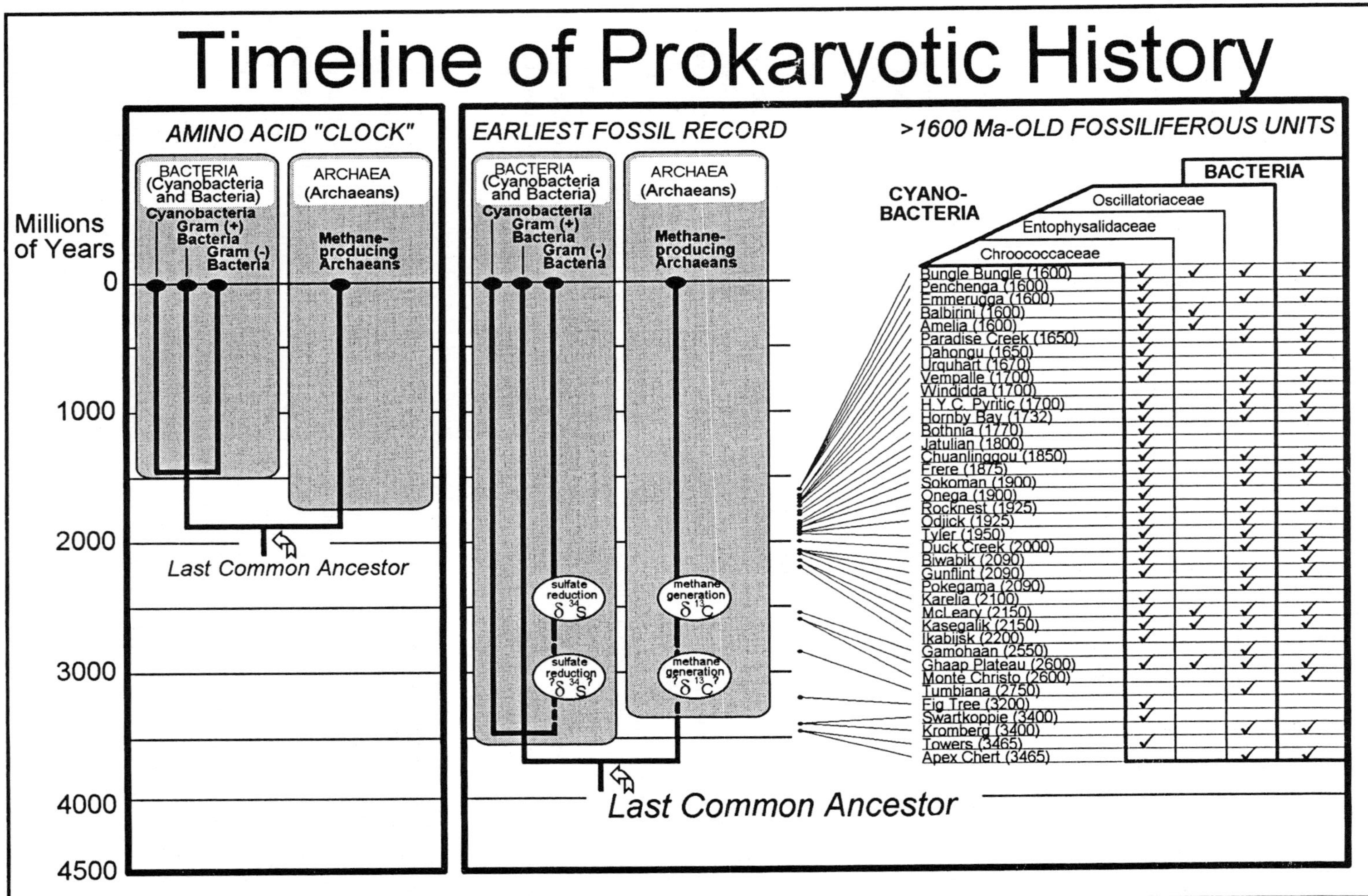

Fig 1. Timeline of life's history inferred from amino acid sequences in proteins (left; data from Doolittle et al., 1996) compared with that based on the known Precambrian paleobiologic record (right; data from Schopf and Walker, 1983; Hofmann and Schopf, 1983; Schopf et al., 1983; Schopf, 1992b).

misinterpreted and are unrelated to present-day lineages. Perhaps life originated, became extinct, then originated a second time about 2000 Ma ago. Or, perhaps, like the rRNA trees, the amino acid data yield reliable evidence only of the branching order of evolution, not the timing of that branching.
How firmly established *is* the early fossil evidence? What amount of data would have to be overturned for the amino acid clock to prove correct?

II. Ancient Cyanobacteria

A. Fossil and Modern "Look-alikes"

The best documented early branch of the Tree of Life is the cyanobacterial lineage, represented in the Precambrian fossil record by ensheathed solitary and colonial unicells (referred chiefly to the living cyanobacterial family Chroococcaceae; Figs. 2-4) and cellular microscopic filaments (for the most part regarded as belonging to the Oscillatoriaceae, taxonomically the most diverse extant cyanobacterial family; Figs. 5-7). Well preserved examples (Plates 3, 4) are essentially indistinguishable in morphology from modern cyanobacteria, an identity first shown nearly three decades ago in detailed studies of the microbial community of the ~850 Ma-old Bitter Springs Formation of central Australia (Fig. 6, J-S; Schopf, 1968; Schopf and Blacic, 1971). Since that time, such similarities have been encountered so frequently that it is now common practice among Precambrian paleobiologists to name fossil taxa after their modern morphological counterparts by adding appropriate prefixes (palaeo-, eo-) or suffixes (-opsis, -ites) to the names of present-day cyanobacterial genera (Mendelson and Schopf, 1992; Schopf, 1994a). More than 40 such namesakes (for example, *Palaeolyngbya*, *Eomicrocoleus*, *Aphanocapsaopsis*, and *Oscillatorites*) are in usage worldwide.

The striking similarity of Precambrian microfossils to modern cyanobacteria is characteristic of a wide range of morphologic types, among which are fossils that are hundreds of millions of years older than the date of ~1500 Ma suggested by the amino acid clock for the origination of the group. For example, *Eoentophysalis belcherensis* (Plate 4b) — compared with species of modern *Entophysalsis* (Plate 4a) and on this basis of referred to the living cyanobacterial family Entophysalidaceae (Golubic and Hofmann, 1976) — is preserved in microbially laminated stromatolites of Northwest Territories, Canada, that are ~2150 Ma old (Hofmann, 1976). Though the fossil taxon is decidedly "too old" to be cyanobacterial according to the amino acid clock, the fossil and modern species are morphologically indistinguishable (in cell shape, and in form and arrangement of originally mucilaginous cell-encompassing envelopes); exhibit similar frequency distributions of dividing cells and essentially identical patterns of cellular development (resulting from cell division in three perpendicular planes); form microtexturally similar stromatolitic structures in comparable intertidal to shallow marine environmental settings; undergo essentially identical postmortem degradation sequences; and occur in microbial communities that are comparable in both species composition and biological diversity (Golubic and Hofmann, 1976).

Many other cyanobacterial look-alikes, virtually all from intertidal to shallow marine stromatolitic settings like those inhabited by their living morphological counterparts, are "too old" by the amino acid data. Among these is *Oscillatoriopsis majuscula* (Knoll et al., 1988), a ~2000 Ma-old filamentous taxon more than 70 mm in diameter and thus very much broader than almost all noncyanobacterial procaryotic filaments (95% of which are < 5 mm in diameter; Schopf, 1996a). Numerous other such examples can be cited. Indeed, as summarized in Fig. 1, fossils interpreted as cyanobacteria and referred to the Chroococcaceae, Entophysalidaceae, and Oscillatoriaceae occur in many geologic units older than permitted by the amino acid clock — if *any* of these thousands of fossils and the more than 120 cyanobacterial species to which they have been referred has been identified correctly, the protein-based amino acid clock is in error.

B. Are Fossil "Cyanobacteria," Cyanobacteria?

What about the idea that life originated, became extinct, then originated a second time, leading to the advent of cyanobacteria ~1500 Ma ago? Perhaps Precambrian paleobiologists have so accepted the presumed cyanobacterial affinity of the fossils found they have neglected to consider other possibilities, especially for very ancient forms. Is there a break in the fossil record of "cyanobacteria" ~1500 Ma ago? One way to address this problem objectively — free of taxonomic predilections — is through detailed morphometric analyses of the fossil record. Such studies have been carried out on 55 communities of

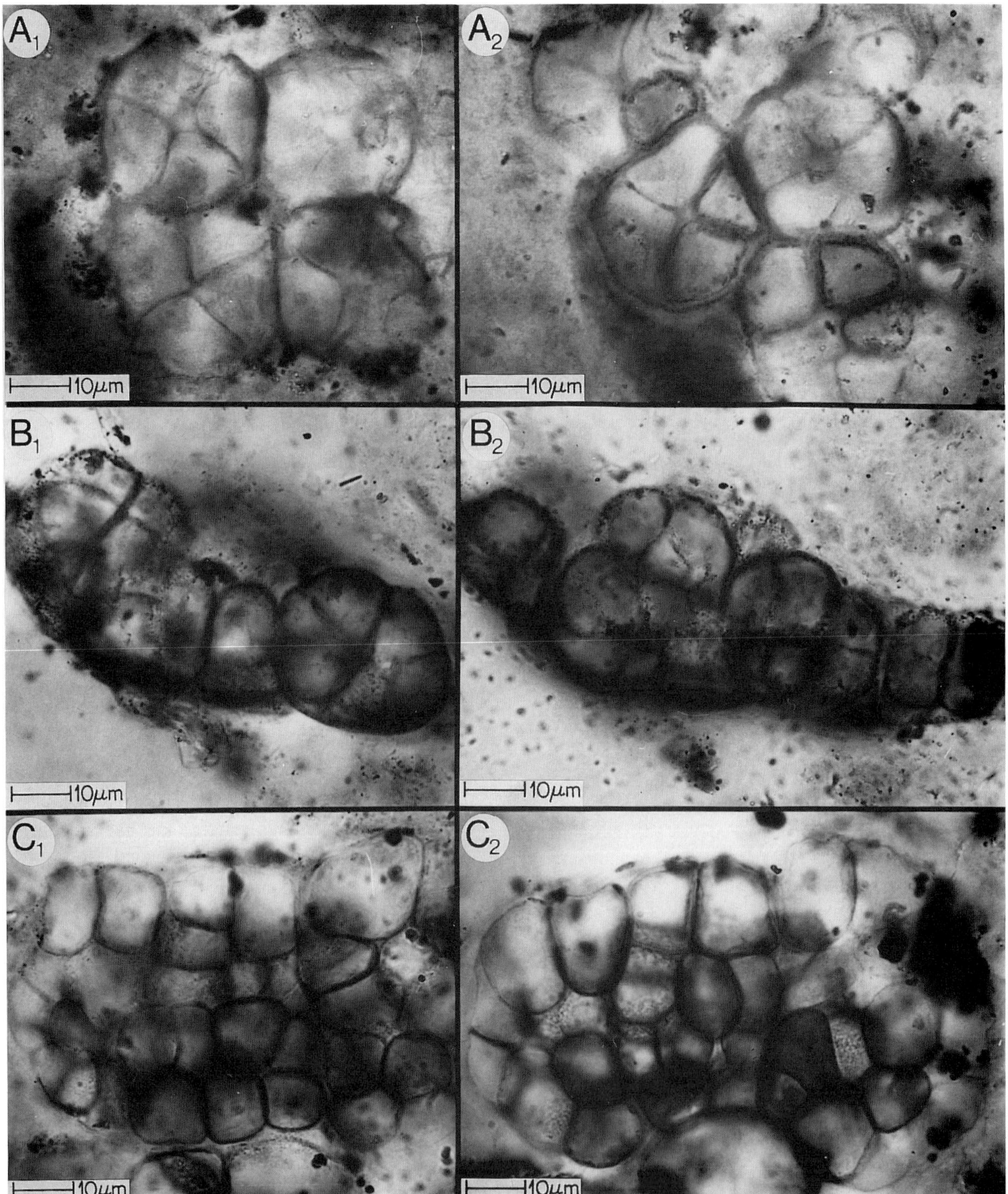

Fig. 2. Ensheathed colonies of coccoid fossil cyanobacteria from stromatolitic chert of the ~650 Ma-old Chichkan Formation of the Zhanatas region, southern Kazakhstan. (***A***) *Tetraphycus bistratosus*, Chroococcaceae (Ogurtsova and Sergeev, 1987). (***B***) *Palaeopleurocapsa fusiforma*, Pleurocapsaceae (Ogurtsova and Sergeev, 1987). (***C***) *Palaeopleurocapsa reniforma*, Pleurocapsaceae (Ogurtsova and Sergeev, 1987).

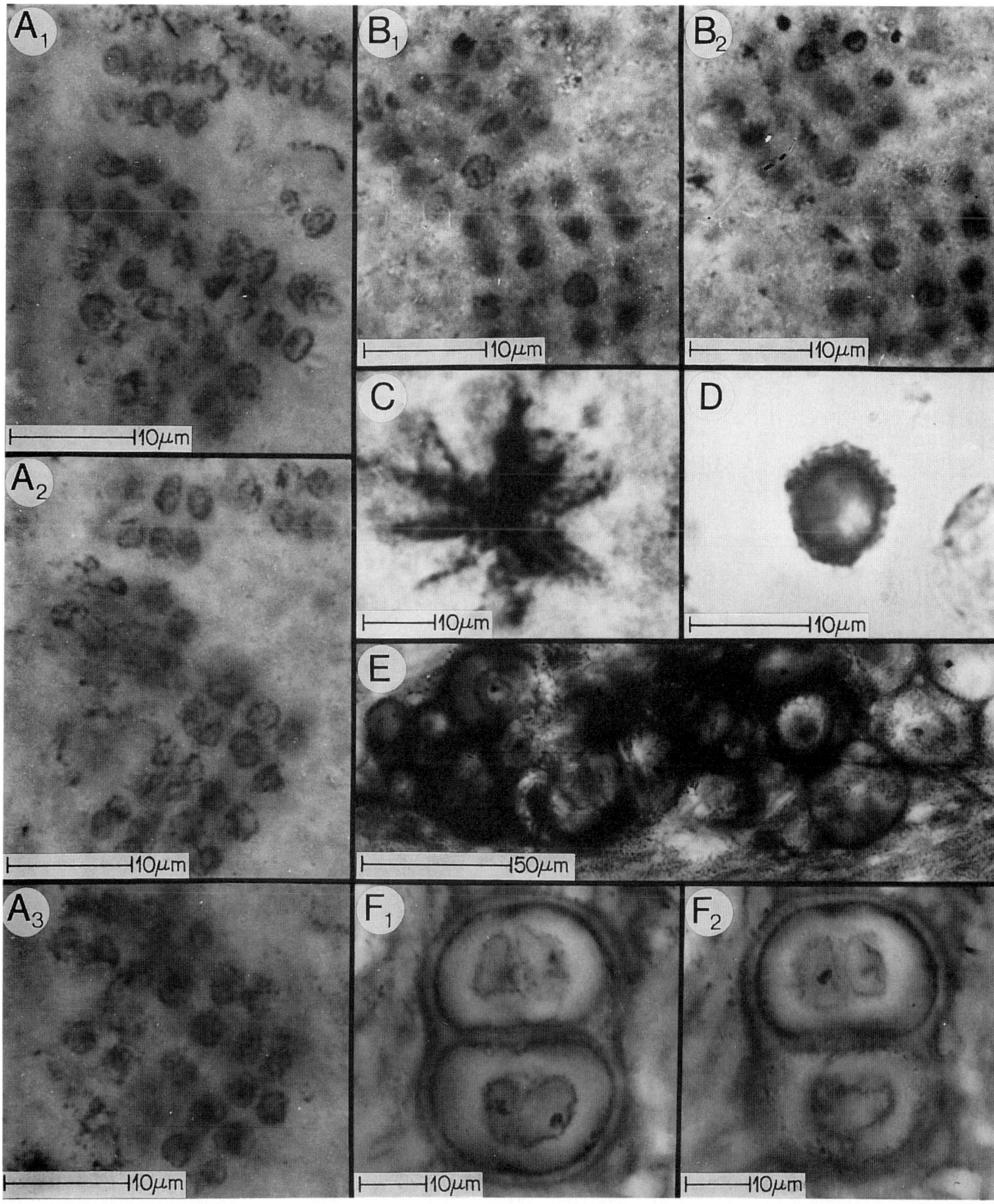

Fig. 3. Colonial (*A*, *B*, *E* and *F*) and unicellular (*D*) fossil cyanobacteria, and a colonial fossil bacterium (*C*), from siltsone (*D*) and stromatolitic cherts (*A-C*, *E* and *F*) of the ~1650 Ma-old Paradise Creek Formation of east-central Queensland, Australia (*A-C*) and the ~1550 Ma-old Satka Formation of southern Bashkiria, Russia (*D-F*). (***A*** and ***B***) Cuboidal *Eucapsis*-like colony, Chroococcaceae (Licari et al., 1969). (***C***) *Metallogenium* sp. = *Eoastrion* sp., a mycoplasma-like iron- and manganese-oxidizing bacterium (Cloud, 1976). (***D***) *Eomarginata striata*, Chroococcaceae (Jankauskas 1980). (***E***) *Palaeopleurocapsa kelleri*, Pleurocapsaceae (Krylov and Sergeev, 1987). (***F***) *Gloeodiniopsis uralicus*, Chroococcaceae (Krylov and Sergeev, 1987).

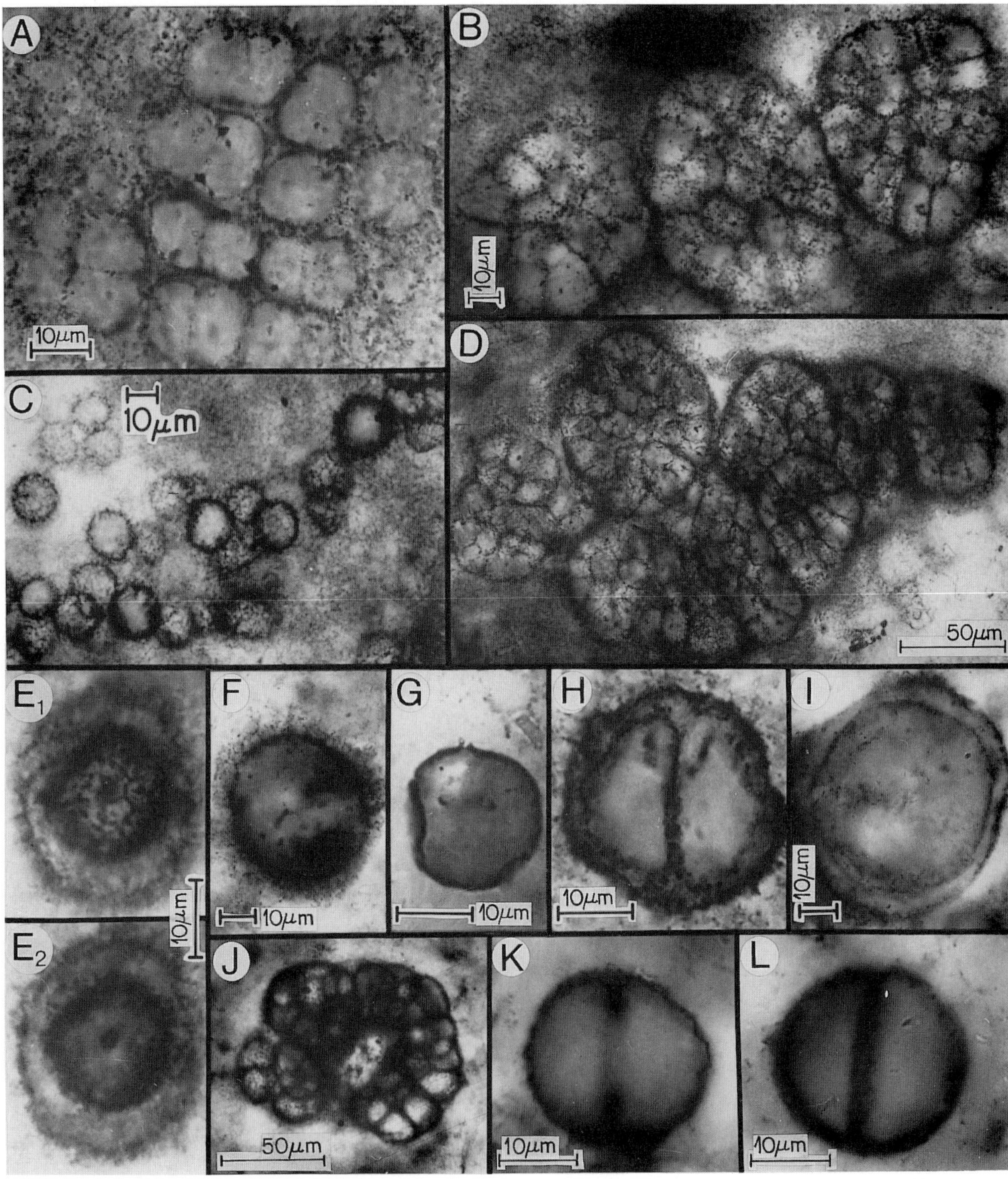

Fig. 4. Fossil cyanobacterial colonies (*A-D* and *J*), unicells (*E-G* and *I*), and ensheathed paired unicells (*H, K* and *L*) from stromatolitic chert of the ~850 Ma-old Skillogalee Dolomite of central South Australia, Australia. (*A*) Tabular *Merismopedia*-like colony, Chroococcaceae (Schopf and Fairchild, 1973). (***B, D*** and ***J***) Rosette-like colonies similar to *Myxosarcina*, Pleurocapsaceae (Fairchild, 1975). (***C, E, F-I, K*** and ***L***) Chroococcacean microfossils (Schopf and Fairchild, 1973; Fairchild, 1975; Schopf, 1977).

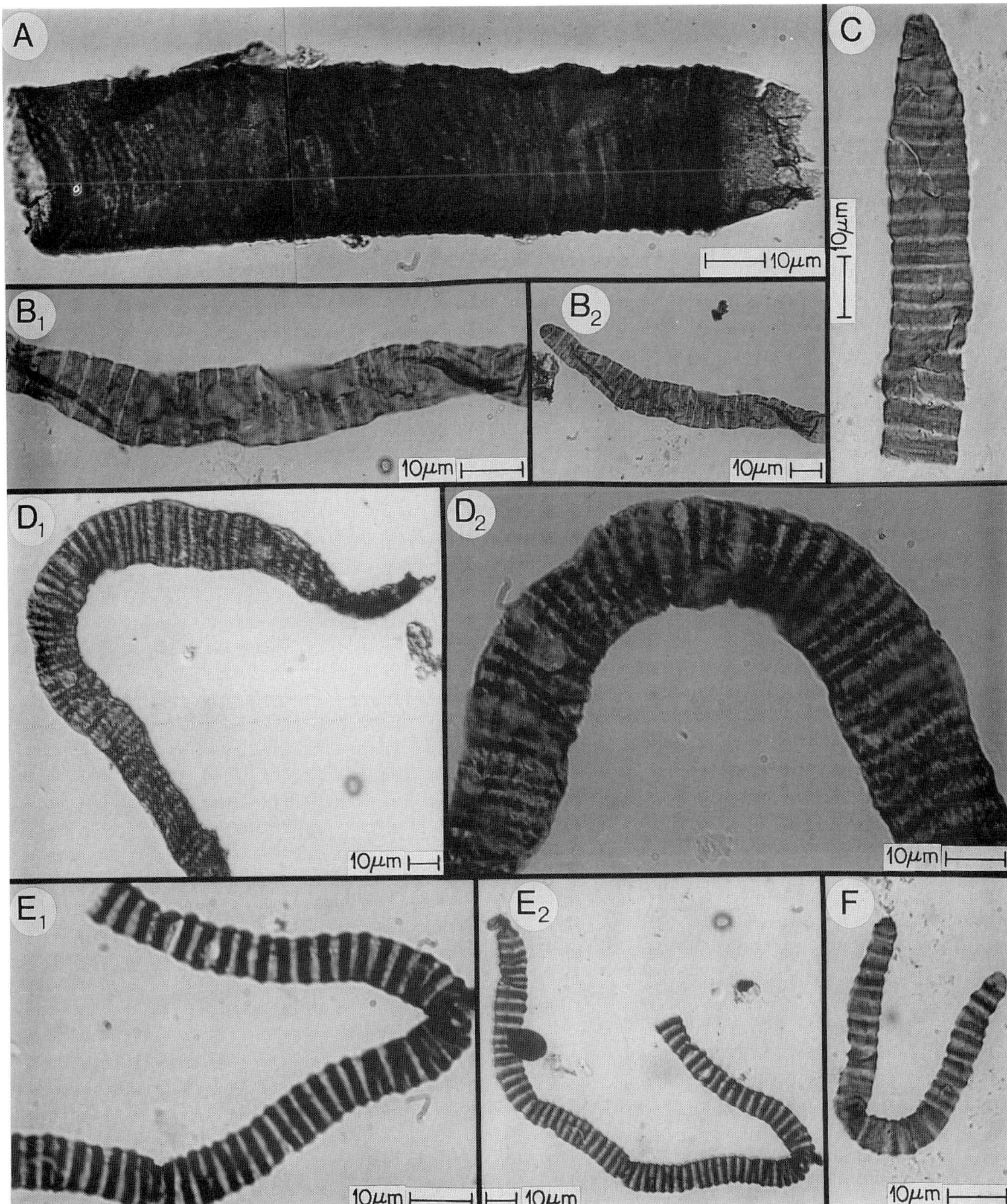

Fig. 5. Fossil cyanobacterial tubular sheath (*A*) and cellular trichomes (*B-F*) from siltstones of the ~900 Ma-old Shtandin Formation of southern Bashkiria, Russia. (***A***) *Siphonophycus costatus*, Oscillatoriaceae (Jankauskas, 1980). (***B*** and ***D***) *Tortunema* sp., Oscillatoriaceae (Jankauskas, 1980). (***C***) *Calyptothrix geminata*, Oscillatoriaceae (Jankausakas, 1980). (***E*** and ***F***) *Calyptothrix alternata*, Oscillatoriaceae (Jankauskas, 1980).

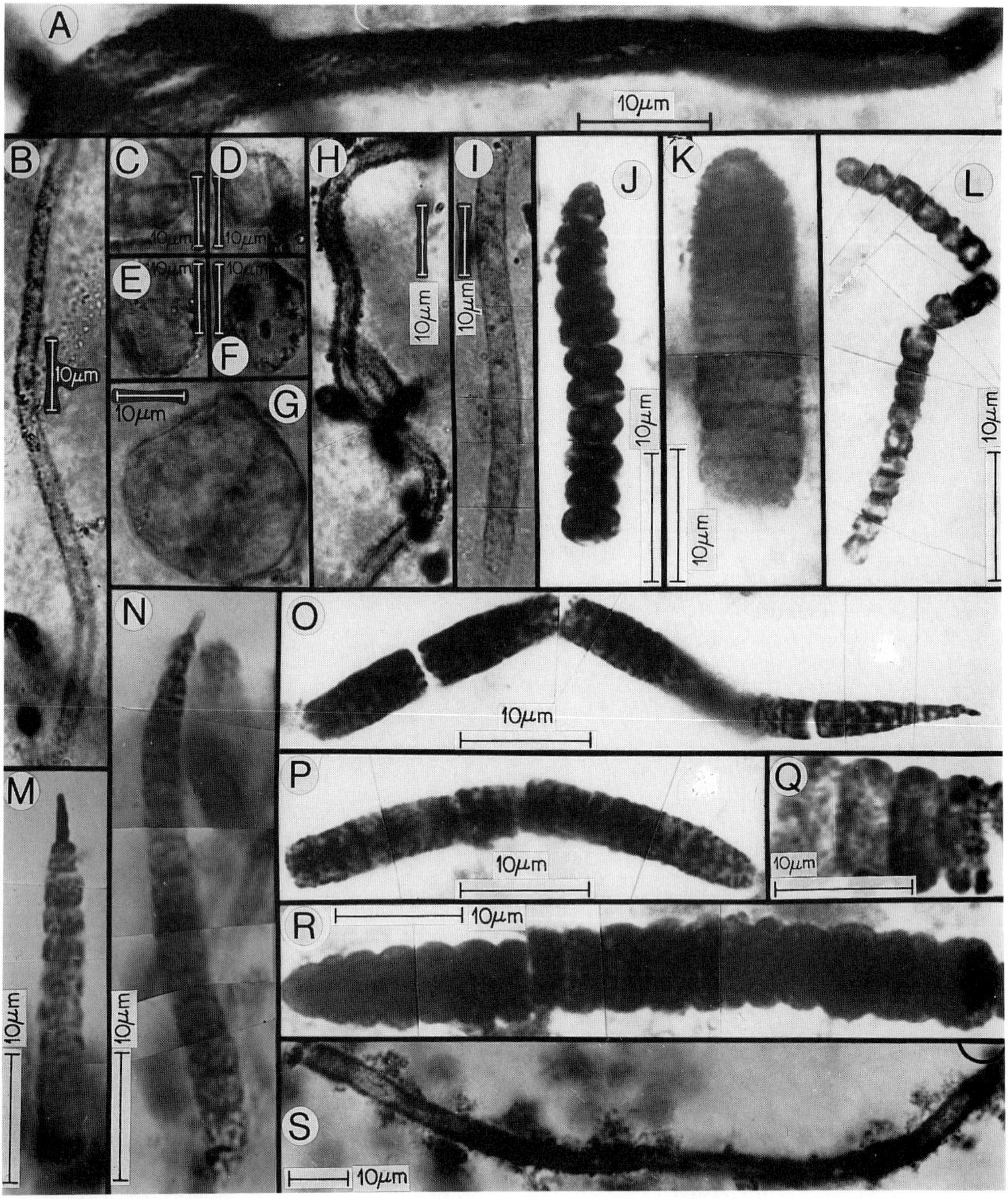

Fig. 6. Filamentous (*A, B, H-S*) and coccoidal (*C-G*) cyanobacterial fossils from stromatolitic cherts of the ~850 Ma-old Kwagunt Formation of nothern Arizona, USA (*A-I*) and the ~850 Ma-old Bitter Springs Formation of southern Northern Territory, Australia (*J-S*). (***A, B, H, I*** and ***S***) *Eomycetopsis* spp., tubular sheaths of oscillatoriacean filaments (Schopf, 1968; Schopf et al., 1973). (***C-G***) Chroococcacean unicells (Schopf et al., 1973). (***J***) *Filiconstrictosus diminutus*, Oscillatoriaceae (Schopf and Blacic, 1971). (***K***) *Palaeolyngbya minor* (Schopf and Blacic, 1971). (***L***) *Veteronostocale amoenum* (Schopf and Blacic, 1971). (***M*** and ***N***) *Caudiculophycus rivularioides*, ?Rivulariaceae (Schopf 1968). (***O***) *Caudiculophycus acuminatus*, ?Rivulariaceae (Schopf and Blacic, 1971). (***P***) *Partitiofilum gongyloides*, Oscillatoriaceae (Schopf and Blacic, 1971). (***Q***) *Palaeolyngbya barghoorniana*, Oscillatoriaceae (Schopf, 1968). (***R***) *Filiconstrictosus majusculus*, Oscillatoriaceae (Schopf and Blacic, 1971).

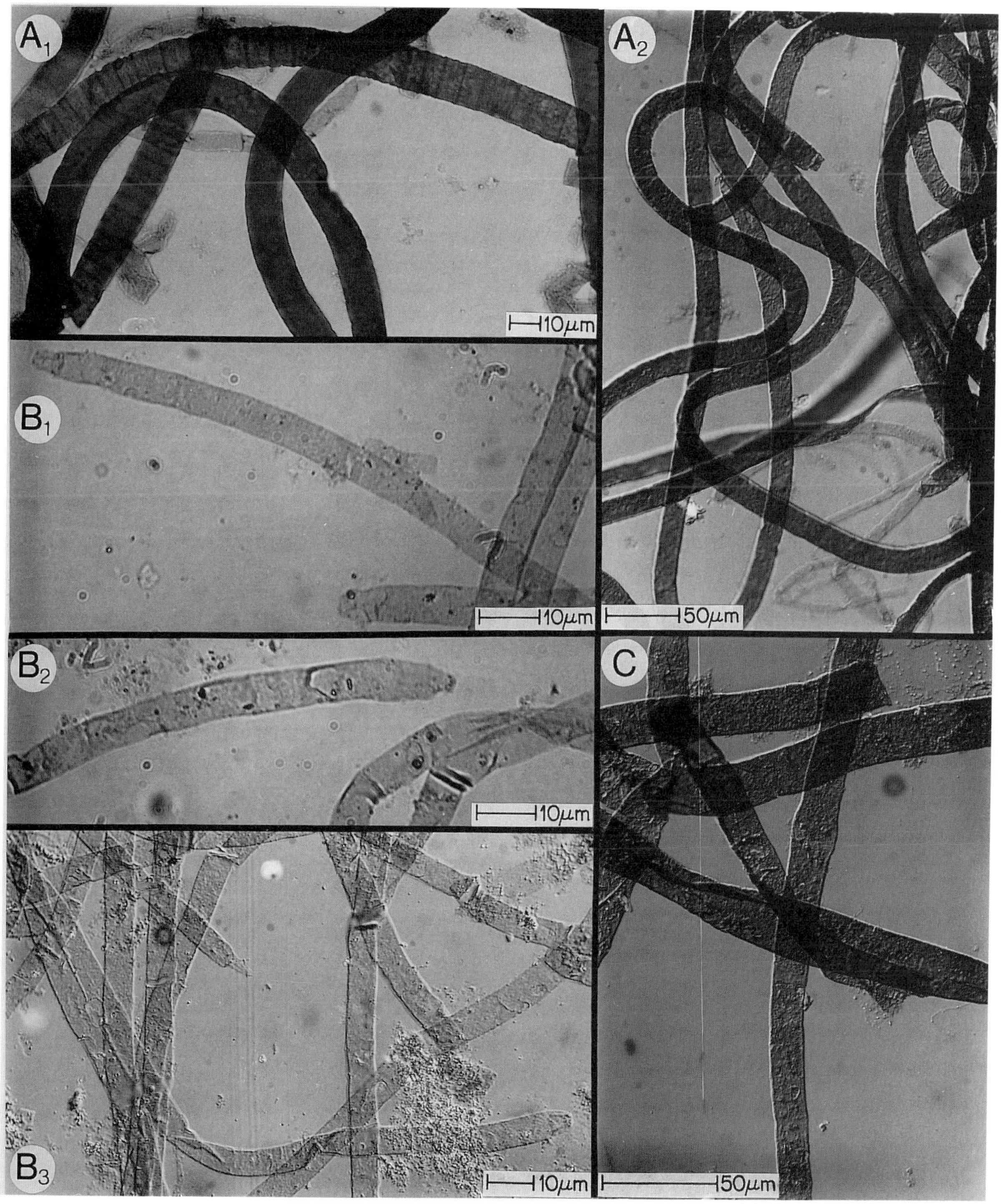

Fig. 7. Fossil cellular cyanobacterial trichomes (*A*) and originally tubular filamentous sheaths (*B, C*) from siltstones of the ~950 Ma-old Lakhanda Formation of eastern Siberia, Russia. (***A***) *Palaeolyngbya helva*, Oscillatoriaceae (Hermann, 1981). (***B***) *Leiothrichoides tenuitunicatus*, Oscillatoriaceae (Hermann, 1981). (***C***) *Palaeovaucheria clavata*, ?Oscillatoriaceae (Hermann, 1981).

coccoidal or filamentous microfossils ~2300 to 1500 Ma in age and compared with data for 73 such communities preserved in immediately younger strata (1500 to 900 Ma in age); numerous morphometric traits (cell size, shape, and range of variability; colony form; sheath thickness and structure) were measured for 690 occurrences of coccoidal species ≤60 mm in diameter (if cyanobacteria, members of the Chroococcaceae) and 270 occurrences of cellular filamentous oscillatoriacean-like taxa (Schopf, 1992d). No discontinuity in community structure, biotic composition, or environments inhabited was detected at 1500 Ma ago or any other time-horizon of the ~1.4 Ga-long period analyzed.

Any marked biotic discontinuity occurring during this period — like that predicted by the amino acid data — should have been apparent. As a group, cyanobacteria differ distinctly in morphology both from the Archaea and noncyanobacterial Bacteria, and the diagnostic characters required to draw such distinctions are commonly preserved. Many archaeans are morphologically unlike most other procaryotes (occurring as lobed cocci, rods in clusters, discs, discs with fibers, and so forth) and virtually all of the 43 species now recognized (Stetter, 1996) are readily distinguishable from cyanobacteria. Most other noncyanobacterial procaryotes can also be distinguished from cyanobacteria. The median diameter of coccoidal non-cyanobacterial Bacteria is <1 μm, less than one-fourth that of cyanobacterial analogues, and for cellular filamentous taxa is ~1.6 μm, less than one-third that of similar cyanobacteria (Schopf, 1996a). If cyanobacteria originated ~1500 Ma ago as suggested by the amino acid clock, this event should have left its mark in the fossil record by an easily detectable increase in median cell size, but this demonstrably did not occur (Schopf, 1992d).

Two families of cyanobacteria dominate modern intertidal to shallow marine environments (Golubic, 1976a, 1976b), the Chroococcaceae and Oscillatoriaceae. Fossils referable to both of these families are abundant in these same settings in Precambrian rock units both older (Fig. 1) and younger than 1500 Ma in age. In fact, the great majority of Precambrian cyanobacterium-like fossils both older and younger than 1500 Ma are referable to *living genera* of chroococcaceans or oscillatoriaceans. Moreover, 37% of coccoidal taxa (26 of 70) older than 1,500 Ma and 28% (31 of 109) of those younger are indistinguishable in morphology from *living species* of chroococcaceans; 35% of filamentous taxa (18 of 51) older than 1500 Ma and 37% (56 of 152) of those younger have *living species-level* morphological counterparts among the Oscillatoriaceae; and virtually all colony forms exhibited by living chroococcaceans occur also among chroococcacean-like fossils both older and younger than 1500 Ma (Figs. 2-4; Schopf, 1992d, 1992e, 1992f).

III. How Old is the Cyanobacterial Lineage?

The evidence from cellularly preserved fossils is unambiguous: Cyanobacteria originated well before 1500 Ma ago and were present earlier than 2000 Ma ago (the clock date for the divergence of present-day biotic lineages from their last common ancestor). How much farther into the geologic past does this evolutionary continuum extend?

As shown in Fig. 1, the documented fossil record is abundant, diverse, and continuous back to about 2200 Ma ago. However, only a few fossiliferous deposits have been discovered in older geologic terrains, and the microbial assemblages of these tend to be meager and poorly preserved (see, for example, Altermann and Schopf, 1995). Though much of the older rock record is paleontologically unexplored, these deficiencies are largely a result of normal geologic processes. Because of erosion and geological recycling, few rock units from these earlier times have survived to the present and most of these are severely metamorphosed. Significantly, however, the oldest microbial community now known — nearly 3500 Ma in age — is also the most diverse and among the best preserved of these very ancient assemblages. This community, from the Apex chert of northwestern Western Australia, holds the key to understanding early stages in the history of life and, very likely, of the cyanobacterial lineage.

A. Evidence From the ~3.5 Ga-old Apex Chert

Though there is reason to question some reports of exceedingly ancient "microfossils" (Schopf and Walter, 1983), such uncertainty does not apply to the microbial assemblage recently described from the ~3500 Ma-old conglomeratic Apex chert (Schopf, 1992c, 1993):

(1) The geographic and stratigraphic source of the fossiliferous horizon is known with certainty (fossil-bearing samples having been collected from outcrop

at the locality by numerous workers on multiple occasions).

(2) The ~3465 Ma age (specifically, >3458 $\pm$1.9 Ma; <3471 $\pm$5 Ma) of the chert is well established, based on U-Pb zircon ages of immediately overlying and stratigraphically underlying units of the rock sequence (Blake and McNaughton, 1984; Thorpe et al., 1992).

(3) The fossils are demonstrably indigenous to the chert, as shown by their occurrence in petrographic thin sections (Figs. 8-10).

(4) The fossils are preserved in transported and redeposited rounded pebbles that are assuredly syngenetic with deposition of the fossiliferous silicified sedimentary conglomerate (and the fossils themselves therefore predate deposition of this bedded chert unit).

(5) Carbon isotopic data (Schopf, 1993, 1994b) and the morphological complexity and carbonaceous composition of the fossils establish that they are unquestionably biogenic — 11 filamentous species have been identified, ranging from 0.5 to 19.5 µm in diameter and exhibiting rounded (Figs. 8, 10) to conical terminal cells (Fig. 9); quadrate, disc-shaped, or barrel-shaped medial cells; taxon-specific degrees of filament attenuation; and evidence of cell division like that occurring in living filamentous procaryotes.

The Apex assemblage stands out not only because of its great age but also because of its diversity. Indeed, the next oldest known comparably diverse microbiotas (Barghoorn and Tyler, 1965; Hofmann, 1976; Altermann and Schopf, 1995) are a billion or more years younger. Moreover, the Apex fossils provide unequalled insight into the evolutionary status of the early biota. In particular, of the 11 species identified in the assemblage, seven (comprising ~63% of measured specimens; Schopf, 1993) are notably similar in cellular organization to oscillatoriacean cyanobacteria and in size range, median dimensions, and pattern of size distribution are much more like oscillatoriaceans than other filamentous procaryotes. Several of these taxa are essentially indistinguishable at the species level from common well-known oscillatoriaceans, both fossil (*Oscillatoriopsis* spp.) and modern (*Oscillatoria* spp.).

The presence of oscillatoriacean cyanobacteria in the Apex assemblage is consistent also with four other lines of evidence. First, models of the early global ecosystem and trace element (cerium and europium) concentrations in Apex-age sediments show that aerobic respiration and O_2-producing photosynthesis — processes characteristic of cyanobacteria, the latter not exhibited by any other procaryotes — had probably evolved by this early stage in Earth history (Towe, 1990, 1991). Second, the isotopic compositions of organic and carbonate carbon in the Apex chert and associated sedimentary units evidence the occurrence of photosynthetic CO_2-fixation like that occurring in extant cyanobacterial populations grown in CO_2-rich environments (Fig. 9; Schopf, 1994b). Third, the occurrence within the Apex filaments of bifurcated cells and cell-pairs very likely reflects the original presence of partial septations and, thus, cell division like that occurring in extant oscillatoriaceans (Schopf, 1992c, 1993). Fourth, as shown by analyses of rRNA, the Oscillatoriaceae is among the earliest evolved of extant cyanobacterial families (Giavannoni et al., 1988; Wilmotte, 1994).

Taken together these various lines of evidence seem persuasive: There is little doubt that if the Apex filaments had been discovered in younger Precambrian sediments — where fossil oscillatoriaceans are well known and widespread — or if they had been detected in a modern microbial community and morphology were the only criterion by which to infer biologic relations, most would be interpreted as oscillatoriacean cyanobacteria.

IV. Paleobiology: Fossils, Geology and Geochemistry

Traditionally, paleontology has focused almost entirely on fossils — their morphology, anatomy, communities, affinities, evolution, paleoecology, paleogeography, biostratigraphy. Other paleobiologic issues have drawn less attention, among them biotic interactions with long-term environmental change and the physiology and biochemistry of ancient life. For example, though the Phanerozoic is punctuated by repeated ice ages and fluctuations in sea level, aridity, volcanism, and the like, all these are episodic relatively short-term events. Truly major changes in the global environment — in the composition of the Earth's atmosphere, for example — are too long-term to fall within the scope of Phanerozoic studies. Similarly, because virtually all Phanerozoic fossils are readily relatable to living biologic groups, questions regarding their basic metabolic characteristics do not arise. Uncertainty may exist whether one or another group of dinosaurs was more or less warm-blooded, but it is safe to assume all were oxygen-requiring obligate aerobes, like their reptilian

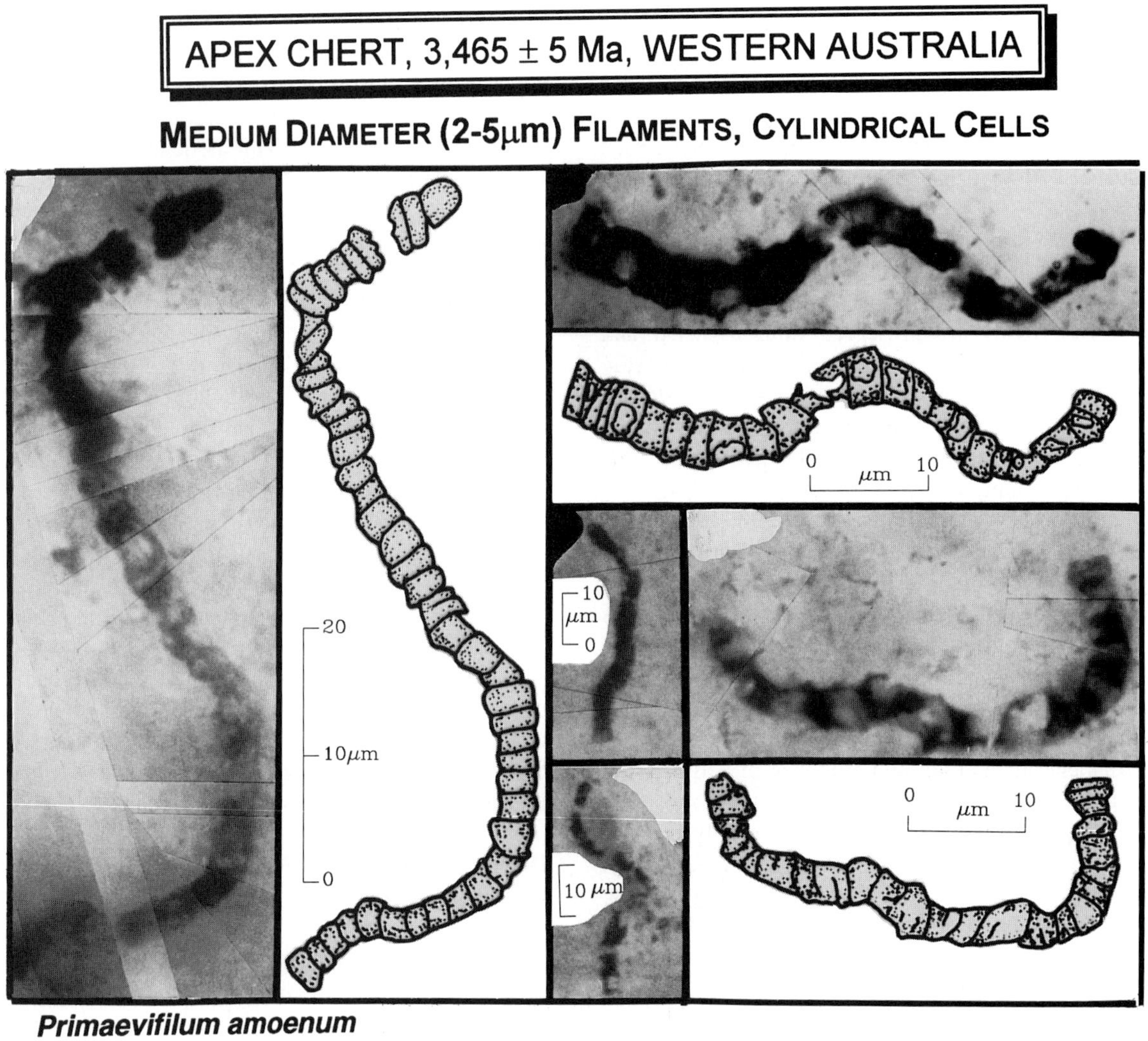

Fig. 8. Cellular filamentous cyanobacterium-like fossils (*Primaevifilum amoenum*; ?Oscillatoriaceae) from the ~3465 Ma-old Apex chert of northwestern Western Australia, Australia (Schopf 1992c, 1993).

and avian descendants; and early-evolving land plants assuredly were oxygen-producing photosynthesizers, even though this lineage, like the dinosaurs, is long extinct.

The Precambrian Eon, however, is nearly an order of magnitude longer than the Phanerozoic and events that are minor or even imperceptible over Phanerozoic time can have cumulative impact if they span appreciable portions of earlier Earth history. For example, the dramatic change from an early anoxic global environment to the current oxygen-rich (~21%) atmosphere, if spanning the time from the Apex fossils to the present would have required sustained addition of only 0.006% O_2 per million years. Yet were this same rate of increase to date from the beginning of the Phanerozoic, the total oxygen rise would be little more than 3%, roughly half that needed to support aerobic higher animals. Moreover, most major events in early evolution centered around the development of metabolic processes (Schopf, 1996b), but because prokaryotes are highly diverse physiologically — and those of similar morphology can differ markedly in metabolism — the sorting out of such capabilities requires more than traditional fossil-focused

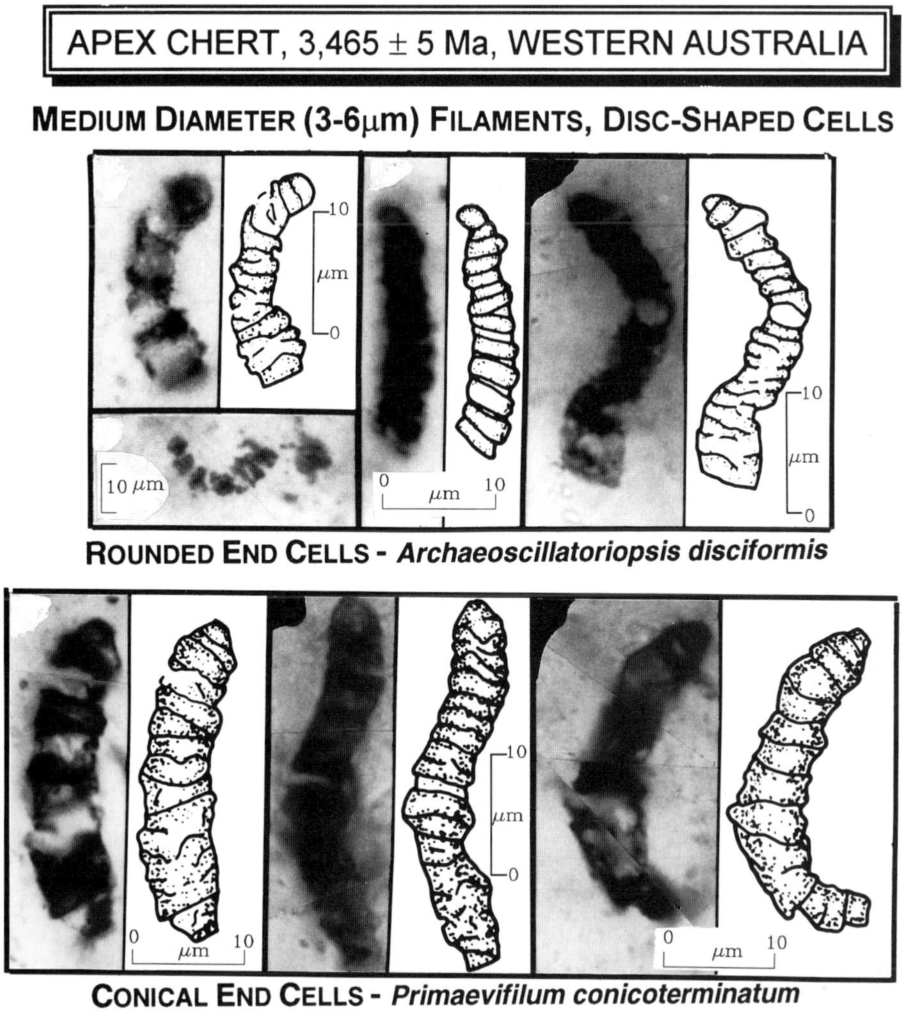

Fig. 9 **Cellular filamentous cyanobacterium-like fossils** (*Archaeocillatoriopsis disciformis*, above; ***Primaevifilum conicoterminatum***, below), ?Oscillatoriaceae, from the ~3465 Ma-old Apex chert of northwestern Western Australia, Australia (Schopf 1992c, 1993).

paleontology. Data from geology and geochemistry are also crucial to unraveling the early history of life.

A. Geologic Evidence of Cyanobacterial Photosynthesis

The capability to carry out oxygen-producing photosynthesis is a universal characteristic of cyanobacteria that distinguishes them from all other procaryotes. If the inference is correct that cyanobacteria are represented among the Apex fossils, the Apex sequence and coeval geologic units should contain evidence of the reactants required for oxygenic photosynthesis (H_2O and CO_2) as well as the products produced (reduced organic carbon and O_2).

Both of the reactants are well-evidenced in the early rock record. The fossiliferous Apex chert is a bedded, water-laid, conglomeratic deposit, part of an interbedded sequence of volcanic and sedimentary rock that contains abundant evidence of subaqueous deposition (Groves et al., 1981). The scene was dominated by extensive shallow seas, with scattered volcanic islands fringed by sedimentary debris (river gravels and sands), mud flats, and interspersed evaporitic lagoons (Barley et al., 1979). Further

APEX CHERT, 3,465 ± 5 Ma, WESTERN AUSTRALIA

BROAD (6-9μm) FILAMENTS, CYLINDRICAL CELLS

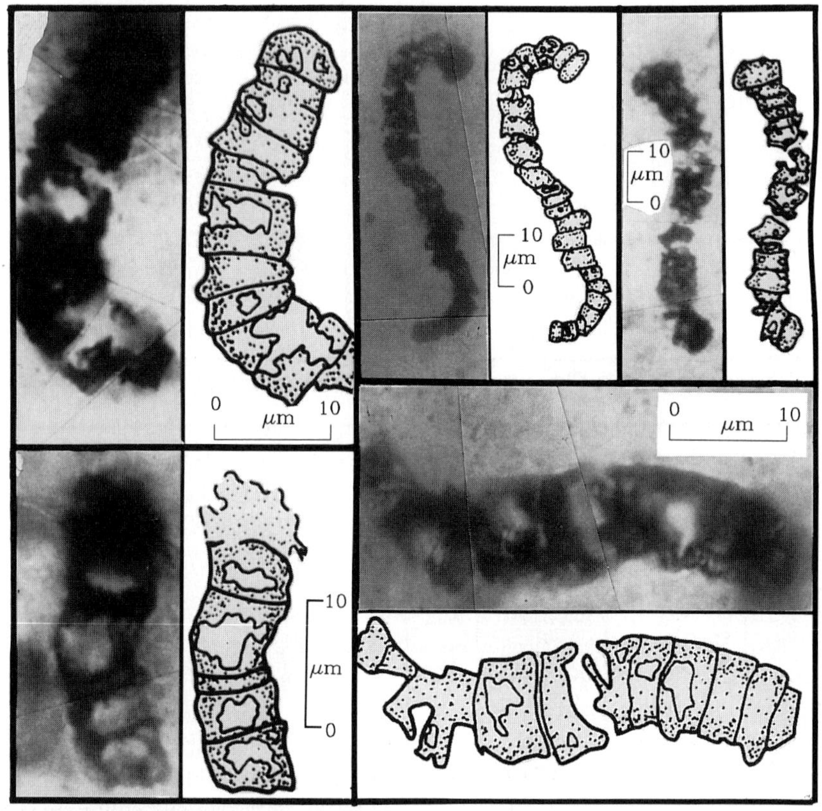

Primaevifilum laticellulosum

Fig. 10. Cellular filamentous cyanobacterium-like fossils (*Primaevifilum laticellulosum*; ?Oscillatoriaceae) from the ~3465 Ma-old Apex chert of northwestern Western Australia, Australia (Schopf 1993).

evidence of liquid water is provided by ripple marks and pillow lavas, well represented in the Apex sequence. Similarly, there is ample evidence of the other reactant of photosynthesis, carbon dioxide, in the form of $CaCO_3$-rich limestones deposited as a result of aqueous reaction between Ca^{2+}, derived from weathering of the land surface, and bicarbonate (HCO_3^-), produced by dissolution of atmospheric CO_2.

The two products of cyanobacterial photosynthesis, organic matter and O_2, are also evidenced in the early rock record. Not only is particulate degraded organic carbon, kerogen, ubiquitous and abundant (ranging from ~0.5 to ~0.8% by weight; Strauss and Moore, 1992) in sediments of the Apex sequence but it also makes up the carbonaceous cell walls of the petrified microscopic fossils (Figs. 8-10). And at least small amounts of free oxygen were present in the environment, reflected by the occurrence of iron-oxide rich sedimentary units known as banded iron formations (BIFs), the world's major source of iron ore. Units of this type are widespread in geologic terrains older but not younger than about 2,000 Ma and, together with uranium-rich pyritic conglomerates, provide evidence of the early history of the atmosphere.

The banding in BIFs is produced by an alternation of iron-rich and iron-poor layers, and because the ferruginous beds are composed of fine rustlike particles of hematite (Fe_2O_3) and, in some deposits, magnetite (Fe_3O_4), BIFs have a characteristic dull to bright red color. The iron evidently owes its origin to the hydrothermal circulation of seawater through oceanic crust, primarily at deep submarine ridge systems. In a dissolved (ferrous) state, it then circulated upward into shallower reaches of the water column — often seasonally, giving rise to the distinctive millimetric banding — where it was oxidized to ferric iron, chiefly by reaction with dissolved molecular oxygen, and precipitated as a fine rusty rain of minute insoluble iron oxide particles.

The Earth's atmosphere is chiefly an accumulated product of volcanic outgassing of the planetary interior over geologic time, but unlike the other principal components of the atmosphere (N_2, H_2O, CO_2), molecular oxygen (O_2) is not released from rocks when they are heated. Other inorganic sources of molecular oxygen (for example, thermal disssociation or UV-induced photodissociation of water vapor) similarly are wanting, leaving oxygenic photosynthesis as the only quantitatively plausible source for the enormous amount of oxygen sequestered in Precambrian BIFs. The presence of these deposits, however, does not indicate the environment was oxygen-rich. On the contrary, BIF-containing basins are large, typically several hundred kilometers in length and breadth, and dissolved ferrous iron could not have been distributed over such broad areas unless at least the lower portions of the water column were anoxic. Thus, two cardinal inferences can be drawn from the continuous presence of abundant widespread BIFs >3500 to ~2000 Ma ago (Klein and Beukes, 1992):

(1) Copious amounts of oxygen were being pumped into the environment by oxygenic (cyanobacterial) photosynthesis; but

(2) the oceanic water column was not fully oxygenated (nor was the atmosphere) because the molecular oxygen was removed from the system by its reaction with iron to form sedimented iron oxide minerals.

The existence of a low-O_2 environment up to ~2,000 Ma ago is supported also by other geologic evidence. Resistance of some minerals to weathering is strongly affected by the presence of O_2. Uraninite (UO_2) and pyrite (FeS_2), both minerals oxidized and then dissolved in the presence of molecular oxygen, are good examples. Extensive detrital accumulations of these minerals do not occur in sediments today because the minerals are easily destroyed during weathering in the present oxygen-rich atmosphere. Nevertheless, large sedimentary ore deposits that contain detrital pebbles of uraninite and pyrite occur in geologic units older but not younger than ~2300 Ma (Holland, 1994). The persistence of pebbles of uraninite and pyrite during weathering, fluvial transport, and deposition in conglomeratic deposits more than 2300 Ma ago is consistent with a low-O_2 atmosphere up to this time as are other geologic data, most notably the mineralogy of paleosols (ancient soil horizons). Taken together, these and related lines of evidence support the conclusion that the O_2 content of the atmosphere increased dramatically between 2200 and 1900 Ma ago, evidently from $\leq$1% to $\geq$15% of the present atmospheric level (Holland, 1994).

Geologic and paleontologic evidence thus agree with the inferred existence of oxygen-producing cyanobacteria as early as ~3500 Ma ago, about two billion years earlier than suggested by the amino acid clock. Data from isotopic geochemistry are also consistent with this interpretation and, like the record of cellularly preserved noncyanobacterial procaryotes (Fig. 1), evidence the presence of other early-evolved microbial lineages as well.

B. Geochemical Evidence of Early-Evolving Lineages

Studies of the stable isotopic geochemistry of several biologically important elements (C, H, O, N, S) have proven indispensable in paleobiology. Among these, analyses of carbon and sulfur stand out as providing powerful evidence of the metabolic capabilities, and thus the phylogenetic relations, of the early evolving biota.

Consider carbon, for example. The carbon isotopic composition of the organic matter in any organism depends on the composition of the source(s) utilized and isotope effects associated with carbon assimilation. Primary among these effects in autotrophs are enzyme-mediated reactions that result in the biosynthesized organic matter being enriched in the lighter stable isotope, ^{12}C, relative to the $^{13}C/^{12}C$ composition of the source carbon. As a result, various types of autotrophic procaryotes typically exhibit ranges of carbon isotopic compositions that are phylogenetically characteristic but somewhat variable due to the varying influences of pH, temperature, CO_2 and metal ion availability, and other factors (O'Leary, 1981). Hence, relative to the carbon isotope PDB standard, *cyanobacteria*, oxygenic photosynthesizers, tend to be ^{12}C-enriched by 16 ±8‰; anoxygenic *photosynthetic bacteria*, by 24 ±6‰; and chemoautotrophic *methanogenic archaeans*, by 28 ±10‰ (Fig. 11). Isotopic signatures of such autotrophs are preservable in the geologic record where in relatively young sediments they can be linked to specific metabolic sources. A prime example is the Eocene (47 Ma-old) Messel Shale of Germany in which biomarker indicators of oxygenic photosynthesis (chlorophyll derivatives), anoxygenic bacterial photoautotrophy (degradation products of bacteriochlorophyll *d*), and methanogenic chemoautotrophy (derivatives of phytyl and biphytyl ethers) are ^{12}C-enriched relative to PDB by 22, 24, and 30‰, respectively (Hayes et al., 1987).

Unfortunately, biomarkers like those extracted from the Messel Shale degrade over time and chemically combine with biologic polymers to form insoluble particulate carbonaceous matter (the "geopolymer" known as kerogen), so isotopic studies of extractable metabolically distinctive molecules cannot be carried out on most Precambrian materials (Summons and Hayes, 1992). Moreover, though carbon isotopic analyses of the kerogen making up the cell walls of *individual* fossil procaryotes might provide means for determining their metabolic characteristics, conventional methods of mass spectrometry lack the sensitivity to analyze carbon isotopes at the requisite microscopic scale. Almost all relevant data are therefore derived from analyses of bulk samples, typically of acid-resistant residues of kg-sized carbonaceous rocks (Strauss et al., 1992). The carbon isotopic compositions of Precambrian kerogens relative to PDB measured by such analyses (expressed as $\delta^{13}C_{PDB}$ values) provide strong evidence for the presence of both photoautotrophic and chemoautotrophic procaryotes decidedly earlier than permitted by the amino acid data.

1. Carbon Isotopic Evidence of Prokaryotic Photoautotrophs

The carbon isotopic compositions of oxygenic (cyanobacterial) and anoxygenic (photosynthetic bacterial) procaryotic photoautotrophs are rather well characterized. Data are available for a large number of modern cyanobacterium-dominated stromatolitic communites (Des Marais et al., 1992) as well as for individual species of chroococcaceans (*Agmenellum, Anacystis, Coccochloris, Synechococcus*) and oscillatoriaceans (*Microcoleus, Oscillatoria, Schizothrix*) cultured under diverse conditions (Fig. 11). With regard to photosynthetic bacteria, carbon isotopic fractionation occurring during photoautotrophy has been characterized in cultures for representatives of the Chlorobiaceae (*Chlorobium*), Chloroflexaceae (*Chloroflexus*), Rhodospirillaceae (*Rhodospirillum*), and Chromatiaceae (*Chromatium*), as well as in naturally occurring photosynthetic bacterium-dominated biocoenoses (Fig. 11).

As a result of isotopic discrimination during CO_2-fixation, photosynthetically produced organic matter sequestered in sedimentary rocks as kerogen is ^{12}C-enriched relative to the carbon isotopic ratio of concurrently deposited carbonate minerals (which reflect the isotopic composition of atmospheric and dissolved CO_2), a difference typically of about 25 ±10‰ (Schidlowski et al., 1983; Strauss et al., 1992). The occurrence of such ^{12}C-enriched kerogen is therefore consistent with the existence of photosynthesis (whether cyanobacterial and oxygenic, or bacterial and anoxygenic) and, if coupled with the evidence of oxidation provided by BIFs, the presence both of oxygenic photosynthesis and aerobic respiration (Towe, 1990, 1991; Schopf, 1996b). Although the degree of isotopic discrimination can be affected markedly by CO_2 concentrations (Fig. 11) — minimum fractionation occuring at low CO_2 levels (associated commonly with high population densities) and maximum fractionation at high CO_2 concentrations (1.5 to 3% = 50 to 100 times the present atmospheric level), like those postulated for the early and mid-Precambrian environment (Kasting, 1992) — analyses of kerogenous organic matter in more than a thousand relatively unmetamorphosed Precambrian samples establish that the carbon isotopic signature of photosynthesis extends to at least ~3500 Ma ago (Schidlowski et al., 1983; Strauss et al., 1992; Hayes, 1994; Schopf, 1994b). These data, together with the geologic and paleontologic evidence outlined above, support the conclusion that the cyanobacterial, photosynthetic bacterial, and Gram-positive bacterial lineages have been extant since at least ~3,500 Ma ago, about two billion years earlier than suggested by the amino acid clock (Fig. 1).

2. Sulfur Isotopic Evidence of Sulfate-reducing Bacteria

Geochemical evidence also records the presence of Gram-negative bacteria at a time much "too old" by the amino acid clock. This lineage includes

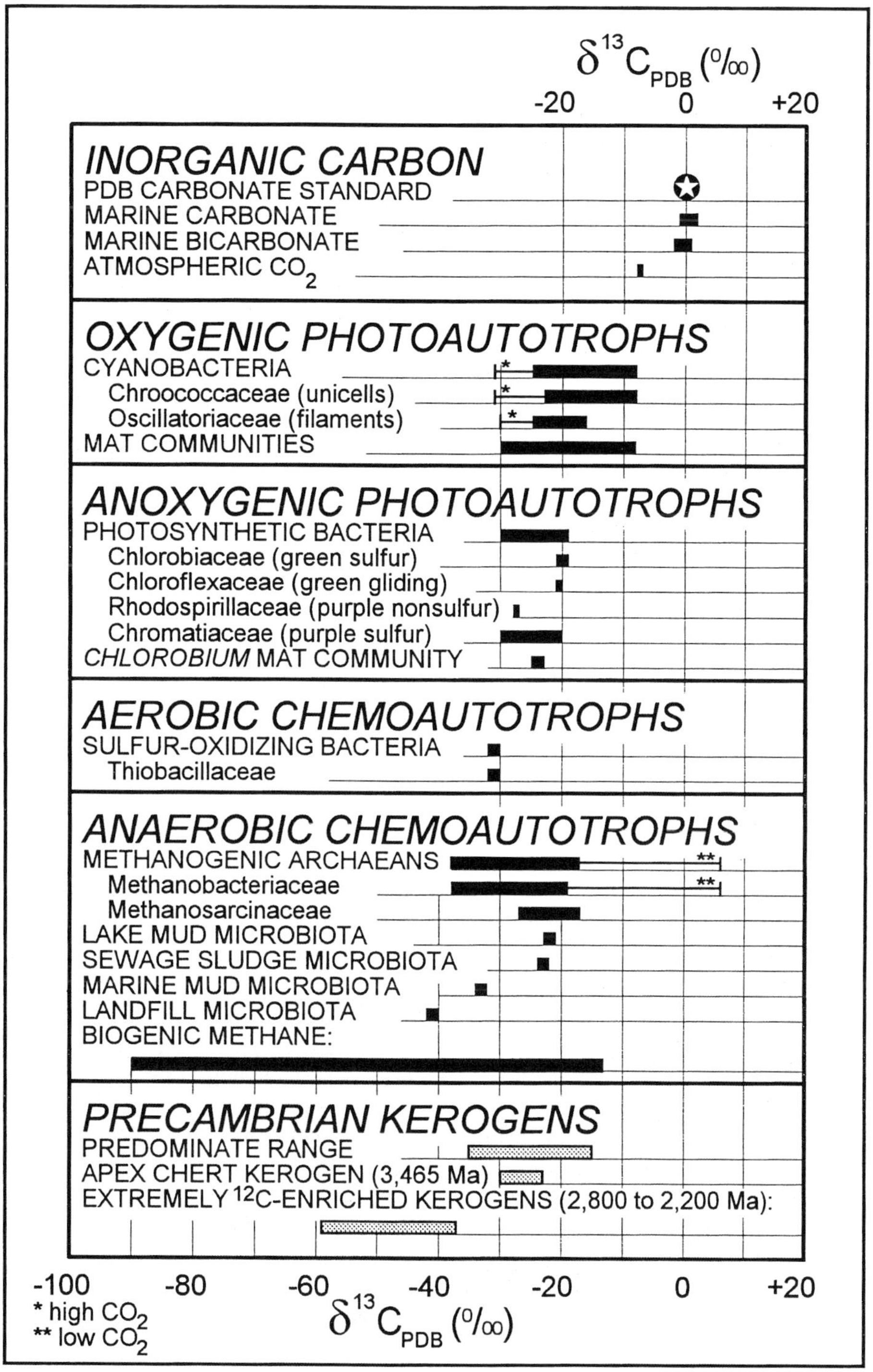

Fig. 11. Carbon isotopic compositions of inorganic carbon reservoirs, procaryotic autotrophs, and Precambrian kerogens. Fractionation ranges for cultures and anaerobic microbiotas are recalculated for a CO_2 source with $\delta^{13}C_{PDB}$ = -7‰. Data from: Abelson and Hoering (1961); Calder and Parker (1973); Wong et al. (1975); Games and Hayes (1976); Pardue et al. (1976); Fuex (1977); Barghoorn et al. (1977); Sirevåg et al. (1977); Fuchs et al. (1979); Mizutani and Wada (1982); Belyaev et al. (1983); Hayes (1983); Hayes et al. (1983); Schidlowski et al. (1983); Holo and Sirevag (1986); Rubey et al. (1987); Des Marais et al. (1992); Strauss and Moore (1992); and Strauss et al. (1992).

dissimilatory sulfate-reducing bacteria (such as *Desulfovibrio*, *Desulfotomaculum*, and *Desulfomonas*) that derive energy from hydrogenation of oceanic sulfate, SO_4^{2-}, to sulfide, H_2S (that is, reduction of S^{6+} to S^{2-}), a form of anaerobic respiration. As in photosynthesis, this process involves an enzymatic fractionation of two stable isotopes but of sulfur rather than carbon such that the sulfide generated is enriched in the lighter isotope, ^{32}S, relative to the $^{32}S/^{34}S$ ratio of the SO_4^{2-} sulfur source (Schidlowski et al., 1983). The bacterially generated H_2S reacts with iron-bearing minerals in sediments to form pyrite, FeS_2, which consequently is also enriched in the lighter sulfur isotope. Biogenic pyrite grains in modern sediments typically have ratios of sulfur isotopes that are highly variable, encompassing a ~60‰ range of $\delta^{34}S$ values, and are thus distinguishable from the sulfide minerals in igneous rocks which fall generally within a narrow range of about 5‰. The isotopic signature of microbial sulfate reduction, and thus of Gram-negative bacteria, can be traced back to at least 2700 Ma ago (Schidlowski et al., 1983) and the lineage may well have been extant as early as 3400 Ma ago (Ohmoto et al., 1993; Kakegawa et al., 1994), nearly two billion years "too early" by the amino acid clock (Fig. 1).

3. Carbon Isotopic Evidence of Methanogenic Archaeans

In addition to bacterial lineages, geochemical evidence records the presence of the Archaea much earlier than permitted by the amino acid data. The cell walls and cytoplasm of anaerobic chemoautotrophic methanogens, like those of aerobic chemoautotrophs (Rubey et al., 1987), are typically enriched in ^{12}C relative to PDB by about 30‰, having $\delta^{13}C_{PDB}$ values ranging from -17 to -38‰ (Fig. 11). Although these values overlap with those of photoautotrophic procaryotes, and therefore cannot be used to distinguish between the groups, the methane produced by methanogenic archaeans leaves a telltale signature in the geologic record.

Metabolic studies demonstrate that the various carbon sources utilized in methane production include carbon monoxide, carbon dioxide, formic acid, methanol, and acetate (Wolfe, 1971), with hydrogenation of the first two being energetically most productive:

$$4CO + 2H_2O \rightarrow CH_4 + 3CO_2 + 50\text{ kcal}$$
$$CO_2 + 4H_2 \rightarrow CH_4 + 2H_2O + 32\text{ kcal}$$
$$4HCO_2H \rightarrow CH_4 + 3CO_2 + 2H_2O$$
$$4CH_3OH \rightarrow 3CH_4 + CO_2 + 2H_2O$$
$${}^*CH_3CO_2H \rightarrow {}^*CH_4 + CO_2$$

The methane produced has the largest variation of the $^{13}C/^{12}C$ ratio of any known biogenic material, with $\delta^{13}C_{PDB}$ values ranging from -13 to -90‰ (Fuex, 1977). Maximum isotopic fractionation attainable during photosynthesis, both observed (Wong et al., 1975; Pardue et al., 1976; Sirevåg et al., 1977) and calculated (Summons and Hayes, 1992; Hayes, 1993), is $\leq$36‰. Thus, highly ^{12}C-enriched biogenic methane can be distinguished readily from products of photosynthesis and its incorporation into sedimented organic matter leaves an unmistakable signature in the preserved kerogen (Kaplan and Nissenbaum, 1966). The presence of kerogens with $\delta^{13}C_{PDB}$ values ranging from -38 to -59‰ in at least 16 geologic units ~2800 to ~2200 Ma in age (Strauss and Moore, 1992) has therefore been interpreted as strong evidence for the presence of biogenic methane, and thus of methanogenic archaeans, at least 2800 Ma ago (Hayes, 1983, 1994).

V. Paleobiology: Final Arbiter of Competing Theories

Taken together, paleontology, geology, and geochemistry provide convincing evidence of the antiquity and evolutionary continuity of the bacterial and archaean domains. The lines of paleobiologic evidence are distinct yet mutually reinforcing, the data voluminous. Stromatolitic microbial ecosystems, evidently including cyanobacteria, photosynthetic bacteria, and other members of the bacterial domain, were extant ~3500 Ma ago; methanogenic archaeans by ~2800 Ma ago; and Gram-negative sulfate-reducing bacteria at least as early as ~2700 Ma ago. The discrepancies between these dates and the times suggested by the amino acid clock (Fig. 1) are too great and too consistent to be ignored.

The amino acid clock is evidently in error, perhaps for one or more of the following reasons. Doolittle et al. (1996) were forced to assume that amino acids are replaced in different parts of a given protein at the same constant rate, but site to site variation and different degrees of multiple replacement at the same site no doubt occur, both of which lead to underestimation of lineage divergence times.

Similarly, if detectable amino acid replacement was slower in the Precambrian and more rapid in the Phanerozoic (due, for example, to procaryotic-eucaryotic differences in the efficacy of protein repair) the clock dates would be too young. And, of course, the amino acid clock might be dating episodes of horizontal gene transfer and the endosymbiotic origin of eucaryotes, rather than times of lineage divergence and the age of the last common ancestor.

As Darwin clearly recognized, the sole source of direct evidence of the nature, products, and timing of the evolutionary process is the rock record. Fossils, geology, and geochemistry, together — the science of paleobiology — is the court of last resort, the final arbiter of competing theories about the history of life.

Acknowledgments

This paper is a revised and expanded version of a contribution prepared earlier for a different audience (Schopf, In Press). In preparing it, I have had the benefit of thoughtful correspondence from Russell Doolittle, which I greatly appreciate. Preparation of this manuscript has been supported by NASA Grant NAGW-2147.

References

Abelson PH and Hoering T (1961) Carbon isotope fractionation in formation of amino acids by photosynthetic organisms. Proc Nat Acad Sci USA 47: 623—632

Altermann W and Schopf JW (1995) Microfossils from the Neoarchean Campbell Group, Griqualand West Sequence of the Transvaal Supergroup, and their paleonvironmental and evolutionary implications. Precambrian Res 75: 65—90

Barghoorn ES, Knoll AH, Dembricki H and Meinschein WG (1977) Variation in stable carbon isotopes in organic matter from the Gunflint Iron Formation. Geochim Cosmochim Acta 41: 425—430

Barghoorn ES and Tyler SA (1965) Microorganisms of the Gunflint chert. Science 147: 563—577

Barley ME, Dunlop JSR, Glover JE and Groves DI (1979) Sedimentary evidence for an Archean shallow-water volcanic-sedimentary facies, eastern Pilbara Block, Western Australia. Earth Planet Sci Lett 43: 74—84

Belyaev SS, Wolkin R, Kenealy WR, DeNiro MJ, Epstein S and Zeikus J G (1983) Methanogenic bacteria from the Bondyuzhskoe Oil Field: General characterization and analysis of stable-carbon isotopic fractionation. App Env Microbiol 45: 691—697

Blake TS and McNaughton NJ (1984) A geochronological framework for the Pilbara region. In: Muhling DK, Groves DK and Blake TS (eds) Archean & Proterozoic Basins of the Pilbara, Western Australia: Solution and Mineralization Potential (Publication 9), pp. 1—22. University of Western Australia Geology Department and University Extension, Perth, Australia

Calder JA. and Parker PL (1973) Geochemical implications of induced changes in C^{13} fractionation by blue-green algae. Geochim Cosmochim Acta 37: 133—140

Cloud P (1983) Early biogeologic history: The emergence of a paradigm. In: Schopf JW (ed) Earth's Earliest Biosphere, Its Origin and Evolution, pp. 14—31. Princeton University Press, Princeton, NJ

Cloud P (1976) Beginnings of biospheric evolution and their biogeochemical consequences. Paleobiol 2: 351—387

Darwin CR (1859) The Origin of Species by Means of Natural Selection. John Murray, London

Des Marais DJ, Bauld J, Palllmisano AC, Summons RE and Ward DM (1992) The biogeochemistry of carbon in modern microbial mats. In: Schopf JW and Klein C (eds) The Proterozoic Biosphere, A Multidisciplinary Study, pp. 299—308. Cambridge University Press, New York

Doolittle RF, Feng D-F, Tsang S, Cho G and Little E (1996) Determining divergence times of the major kingdoms of living organisms with a protein clock. Science 271: 470—477

Fairchild TR (1975) The Geologic Setting and Paleobiology of a Late Precambrian Stromatolitic Microflora from South Australia. Ph.D. Thesis, Department of Geology, University of California, Los Angeles

Fuchs G, Thauer R, Ziegler H and Stichler W (1979) Carbon isotopic fractionation by *Methanobacterium thermo-autotrophicum*. Arch Microbiol 120: 135—139

Fuex AN (1977) The use of stable carbon isotopes in hydrocarbon exploration. J Geochem Explor 7: 155—158

Games LM and Hayes JM (1976) On the mechanisms of CO_2 and CH_4 production in natural anaerobic environments. In: Nriagu JO (ed) Environmental Biogeochemistry Volume 1, Carbon, Nitrogen, Phosphorus, Sulfur and Selenium Cycles, pp 51—73. Ann Arbor Science, Ann Arbor, MI

Giavannoni SJ, Turner S, Olsen GJ, Barns S, Lane DJ and Pace NR (1988) Evolutionary relationships among cyanobacteria and green chloroplasts. J Bacteriol 170: 3584—3592

Golubic S (1976a) Organisms that build stromatolites. In: Walter MR (ed) Stromatolites, Developments in Sedimentology 20, pp. 113—126. Elsevier, Amsterdam

Golubic S (1976b) Taxonomy of extant stromatolite-building cyanophytes. In: Walter MR (ed) Stromatolites, Developments in Sedimentology 20, pp. 127—140. Elsevier, Amsterdam

Golubic S and Hofmann HJ (1976) Comparison of Holocene and mid-Precambrian Entophysalidaceae (Cyanophyta) in stromatolitic mats: Cell division and degradation. J Paleontol 50: 1074—1082

Groves DI, Dunlop JSR and Buick R (1981) An early habitat of life. Scient Am 245: 64—73

Hayes JM (1983) Geochemical evidence bearing on the origin of aerobiosis: A speculative hypothesis. In: Schopf JW (ed) Earth's Earliest Biosphere, Its Origin and Evolution, pp. 291—301. Princeton University Press, Princeton, NJ

Hayes JM (1993) Factors controlling ^{13}C contents of sedimentary organic compounds: Principles and evidence. Marine Geol. 113: 111—125

Hayes JM (1994) Global methanotrophy at the Archean-Proterozoic transition. In: Bengtson S (ed) Early Life on Earth, pp. 220—236. Columbia University Press, New York

Hayes JM, Kaplan IR and Wedeking KM (1983) Precambrian organic geochemistry, preservation of the record. In: Schopf JW (ed) Earth's Earliest Biosphere, Its Origin and Evolution, pp. 93—134. Princeton University Press, Princeton, NJ

Hayes, JM, Takigiku R, Ocampo R, Callot HJ and Albrecht P (1987) Isotopic compositions and probable origins of organic molecules in the Eocene Messel Shale. Nature 329: 48—51

Hermann TN (1981) Nitchatye mikroorganizmy lakhandinskoj svity reki Mai [Filamentous microorganisms from the Lakhanda Formation on the Maya River]. Paleontol Zhur 1981(2): 126—131 (in Russian)

Hofmann HJ (1976) Precambrian microflora, Belcher Islands, Canada: Significance and systematics. J Paleontol 50: 1040—1073

Hofmann HJ and Schopf JW (1983) Early Proterozoic microfossils. In: Schopf JW (ed) Earth's Earliest Biosphere, Its Origin and Evolution, pp. 321—360. Princeton University Press, Princeton NJ

Holland, HD (1994) Early Proterozoic atmospheric change. In: Bengstson S (ed) Early Life on Earth, pp. 237—244. Columbia University Press, New York

Holo H and Sirevåg R (1986) Autotrophic growth and CO_2 fixation of *Chloroflexus aurantiacus*. Archiv Microbiol 145: 173—180

Iwabe N, Kuma K, Hasegawa M, Osawa S and Miyata T (1989) Evolutionary relationships of archaebacteria, eubacteria and eukaryotes inferred from phylogenetic trees of duplicated genes. Proc Nat Acad Sci USA 86: 9355—9359

Jankauskas TV (1980) Novye vodorosli iz verkhnego rifeya yuzhnogo Urala i priural'ya [New algae from the upper Riphean of the southern Urals and adjacent areas]. Paleontol Zhur 1980(4): 107—113 (in Russian)

Kakegawa T, Kawai H and Ohmoto H (1994) Biological activities and hydrothermal activity recorded in the ~2.5 Mount McRae Shale, Hamersley District, Western Australia. II. Sulfur isotopic composition of pyrite. Resource Geol 44: 284—285

Kaplan IR and Nissenbaum A (1966) Anomalous carbon isotope ratios in nonvolatile organic material. Science 153: 744—745

Kasting, JF (1992) Proterozoic climates: The effect of changing atmospheric carbon dioxide concentrations. In: Schopf JW and Klein C (eds) The Proterozoic Biosphere, A Multidisciplinary Study, pp. 165—168. Cambridge University Press, New York

Klein C and Buekes NJ (1992) Time distribution, stratigraphy, sedimentologic setting, and geochemistry of Precambrian iron-formations. In: Schopf JW and Klein C (eds) The Proterozoic Biosphere, A Multidisciplinary Study, pp. 139—146. Cambridge University Press, New York

Knoll AH, Strother PK and Rossi S (1988) Distribution and diagenesis of microfossils from the Lower Proterozoic Duck Creek Dolomite, Western Australia. Precambrian Res 38: 257—279

Krylov IN and Sergeev VN (1987) Rifeiskie mikrofossilii uzhnogo Urala v raione goroda Kusa [Riphean microfossils of the southern Urals near the town of Kusa]. Paleontol Zhur 1987(2): 107—116 (in Russian)

Licari GR, Cloud, PE, Jr and Smith, WD (1969) A new chroococcacean alga from the Proterozoic of Queensland. Proc Nat Acad Sci USA 62: 56—62

Mendelson CV and Schopf JW (1992) Proterozoic and selected Early Cambrianmicrofossils and microfossil-like objects. In: Schopf JW and Klein C (eds) The Proterozoic Biosphere, A Multidisciplinary Study, pp. 865—951 Cambridge University Press, New York

Mizutani H and Wada E (1982) Effect of high atmospheric CO_2 on $\delta^{13}C$ of algae. Origins Life 12: 377—390

Ogurtsova RN and Sergeev VN (1987) Mikrobiota chichkanskoj svity verkhnego dokembriya Malago Karatau (Yuzhnyj Kazakhstan) [Microbiota of the Chichkan Formation, Little Karatau Range (South Kazakhstan)]. Paleontol Zhur 1987(2): 107—116 (in Russian)

Ohmoto H, Kakegawa T and Lowe DR (1993) 3.4-billion-year-old biogenic pyrites from Barberton, South Africa: Sulfur isotope evidence. Nature 262: 555--557

O'Leary MH (1981) Carbon isotopic fractionation in plants. Phytochem 20: 553—567

Olsen GJ and Woese CR (1993) Ribosomal RNA: A key to phylogeny. FASEB J 7: 113—123

Pardue JW, Scalan RS, Van Baalen C and Parker PL (1976) Maximum carbon isotope fractionation in photosynthesis by blue-green algae and a green alga. Geochim Cosmochim Acta 40: 309—312

Rubey EG, Jannasch HW and Deuser WG (1987) Fractionation of stable carbon isotopes during chemoautotrophic growth of sulfur-oxiding bacteria. App Env Microbiol 53: 1940—1943

Schidlowski M, Hayes JM and Kaplan IR (1983) Isotopic inferences of ancient biochemistries: Carbon, sulfur, hydrogen, and nitrogen. In: Schopf JW (ed) Earth's Earliest Biosphere, Its Origin and Evolution, pp. 149—186. Princeton University Press, Princeton, NJ

Schopf JW (1968) Microflora of the Bitter Springs Formation, Late Precambrian, central Australia. J Paleontol 42: 651—688

Schopf JW (1970) Precambrian micro-organisms and evolutionary events prior to the origin of vascular plants. Biol Rev Cambridge Phil Soc 45: 651—688

Schopf JW (1977) Biostratigraphic usefulness of stromatolitic Precambrian microbiotas: A preliminary analysis. Precambrian Res 5: 143—173

Schopf JW (1992a) Historical development of Proterozoic micropaleontology. In: Schopf JW and Klein C (eds) The Proterozoic Biosphere, A Multidisciplinary Study, pp. 179—183. Cambridge University Press, New York

Schopf JW (1992b) Times of origin and earliest evidence of major biologic groups. In: Schopf JW and Klein C (eds) The Proterozoic Biosphere, A Multidisciplinary Study, pp. 587—593. Cambridge University Press, New York

Schopf JW (1992c) Paleobiology of the Archean. In: Schopf JW and Klein C (eds) The Proterozoic Biosphere, A Multidisciplinary Study, pp. 25—39. Cambridge University Press, New York

Schopf JW (1992d) Proterozoic prokaryotes: Affinities, geologic distribution, and evolutionary trends. In: Schopf JW and Klein C (eds) The Proterozoic Biosphere, A Multidisciplinary Study, pp. 195—218. Cambridge University Press, New York

Schopf JW (1992e) Informal revised classification of Proterozoic microfossils. In: Schopf JW and Klein C (eds) The Proterozoic Biosphere, A Multidisciplinary Study, pp. 1119—1168. Cambridge University Press, New York

Schopf JW (1992f) Atlas of representative Proterozoic microfossils. In: Schopf JW and Klein C (eds) The Proterozoic Biosphere, A Multidisciplinary Study, pp. 1055—1117. Cambridge University Press, New York

Schopf JW (1993) Microfossils of the Early Archean Apex chert: New evidence of the antiquity of life. Science 260: 640—646

Schopf JW (1994a) Disparate rates, differing fates: Tempo and mode of evolution changed from the Precambrian to the Phanerozoic. Proc Nat Acad Sci USA 91: 6735—6742

Schopf JW (1994b) The oldest known records of life: Stromatolites, microfossils, and organic matter from the Early Archean of South Africa and Western Australia. In: Bengtson S (ed) Early Life on Earth, pp. 193—206. Columbia University Press, New York

Schopf JW (1996a) Cyanobacteria: Pioneers of the early Earth. Nova Hedwigia 112: 13—32

Schopf JW (1996b) Metabolic memories of Earth's earliest biosphere. In: Marshall CR and Schopf JW (eds) Evolution and the Molecular Revolution, pp. 73—107. Jones & Bartlett, Boston, MA

Schopf JW (In Press) Tracing the roots of the universal tree of life. In: Brack A (ed) The Molecular Origins of Life: Assembling the Pieces. Cambridge University Press, New York

Schopf JW and Blacic JM (1971) New microorganisms from the Bitter Springs Formation (Late Precambrian) of the north-central Amadeus Basin, Australia. J Paleontol 45: 925—961

Schopf JW and Fairchild TR (1975) Late Precambrian microfossils: A new stromatolitic biota from Boorthanna, South Australia. Nature 242: 537—538.

Schopf JW, Ford TD and Breed, WJ (1973) Microorganisms from the Late Precambrian of the Grand Canyon, Arizona. Science 179: 1319—1321

Schopf JW, Hayes JM and Walter MR (1983) Evolution of Earth's earliest ecosystems: Recent progress and unsolved problems. In: Schopf JW (ed) Earth's Earliest Biosphere, Its Origin and Evolution, pp. 361—384. Princeton University Press, Princeton, NJ

Schopf JW and Walter MR (1983) Archean microfossils: New evidence of ancient microbes. In: Schopf JW (ed) Earth's Earliest Biosphere, Its Origin and Evolution, pp. 214—239. Princeton University Press, Princeton, NJ

Sirevåg R, Buchanan BB, Berry JA and Troughton JH (1977) Mechanisms of CO_2 fixation in bacterial photosynthesis studied by the carbon isotope fractionation technique. Archiv Microbiol 112: 35—38

Stetter KO (1996) Hyperthermophilic prokaryotes. FEMS Microbiol Rev 18: 149—158

Strauss H, Des Marais DJ, Hayes JM and Summons R E (1992) The carbon-isotopic record. In: Schopf JW and Klein C (eds) The Proterozoic Biosphere, A Multidisciplinary Study, pp. 117—127. Cambridge University Press, New York

Strauss H and Moore TB (1992) Abundances and isotopic compositions of carbon and sulfur species in whole rock and kerogen samples. In: Schopf JW and Klein C (eds) The Proterozoic Biosphere, A Multidisciplinary Study, pp. 709—798. Cambridge University Press, New York

Summons RE and Hayes JM (1992) Principles of molecular and isotopic biogeochemistry. In: Schopf JW and Klein C (eds) The Proterozoic Biosphere, A Multidisciplinary Study, pp. 83—93. Cambridge University Press, New York

Thorpe RI, Hickman AH, Davis DW, Mortensen JK and Trendall AF (1992) U-Pb zircon geochronology of Archaean felsic units in the Marble Bar region, Pilbara Craton, Western Australia. Precambrian Res 56: 169—189

Towe KM (1990) Aerobic respiration in the Archaean? Nature 348: 54—56

Towe KM (1991) Aerobic carbon cycling and cerium oxidation: Significance forArchean oxygen levels and banded iron-formation deposition. Palaeogeog Palaeoclimatol Palaeoecol 97: 113—123

Wilmotte A (1994) Molecular evolution and taxonomy of the Cyanobacteria. In: Bryant DA (ed) The Molecular Biology of Cyanobacteria, pp. 1—25. Kluwer Academic Publishers, Dordrecht, The Netherlands

Woese CR (1987) Bacterial evolution. Microbiol Rev 51: 221—271

Woese CR, Kandler O and Whellis ML (1990) Towards a natural system of organisms: Proposal for the domains Archaea, Bacteria, and Eucarya. Proc Nat Acad Sci USA 87: 4576—4579

Wolfe RS (1971) Microbial formation of methane. Adv Microbial Physiol 6: 107—146

Wong W, Sackett WM and Benedict CRB (1975) Isotope fractionation in photosynthetic bacteria during carbon dioxide assimilation. Plant Physiol 55: 475—479

Chapter 3

Cyanobacteria in Geothermal Habitats

David M. Ward[1] & Richard W. Castenholz[2]
[1]*Department of Microbiology, Montana State University, Bozeman, MT, 59717, USA.*
[2]*Department of Biology, University of Oregon, Eugene, OR, 97403, USA*

Summary

Recent advances in molecular biology enabled a much more detailed view of cyanobacteria that inhabit well-studied hot spring habitats. What seemed on the basis of microscopy and culture methods to be a relatively simple story became one which is more complex, yet more ordered. For example, 16S rRNA studies of one well-studied Yellowstone hot spring cyanobacterial mat, often thought to be constructed by a single cosmopolitan *Synechococcus* species, revealed many cyanobacterial populations, most probably *Synechococcus* spp., whose genetic diversity is considerable. Some closely related populations exhibited orderly distributions along thermal and vertical gradients, and provided evidence that "speciation" of thermophilic cyanobacteria may have resulted,

B. A. Whitton and M. Potts (eds), The Ecology of Cyanobacteria, 37-59.
© 2000 *Kluwer Academic Publishers. Printed in the Netherlands.*

in part, from adaptive radiation of specialized ecotypes. These results correspond with previous descriptions of *Synechococcus* temperature "strains" cultivated from a well-studied Oregon hot spring mat, raising interesting questions which may be answered through molecular analysis. For example, are the same cyanobacteria found in geographically isolated springs, or does limited dispersal also cause divergence of cyanobacteria with subsequent adaptive radiations in independent lineages ? Advances in microelectrode technology provided detailed views of the vertical distribution and dynamics of light, oxygen and sulfide within the photic zone in such mats, which may range from ~0.5 mm to > 1 cm in thickness. These methods increased our knowledge of diel chemical changes within the mats, including the basis for dynamic migrations of motile cyanobacteria. Microelectrode studies of one Yellowstone mat provided additional evidence for specialization of *Synechococcus* populations, and led to a new view of oxygenic photosynthesis in such habitats. Intensive localized photosynthesis strongly influences microenvironmental chemistry, which in turn enhances photorespiration. While very active, these *Synechococcus* populations partition the majority of photosynthate into polyglucose, as opposed to macromolecules needed for growth. Photoexcretion of glycolate and dark fermentation of polyglucose result in a diel cross feeding of most of the fixed carbon to heterotrophs. Thorough investigation of a few hot spring systems, using both contemporary and traditional methods, is providing a more sophisticated view of the biodiversity, ecology and evolution of thermophilic cyanobacteria.

I. Introduction

Previous reviews of thermophilic cyanobacteria in geothermal habitats described in detail those taxa that are defined by phenotypic, mainly morphological, characters (Brock, 1978; Castenholz, 1969, 1973a, 1978, 1981, 1984, 1996; Ward et al., 1987, 1989a, 1992b). A similar overview is first provided here, but the focus is to place taxa within the local context of two microbial mat systems under long-term investigation. We compare this view with one provided by the application of molecular approaches to the detection of cyanobacterial populations that inhabit natural thermal mats. These two considerations, and a review of experiments which were conducted mainly on the two model systems, provide the current understanding of the physiological and behavioral ecology of thermophilic cyanobacteria.

II. Distribution of thermophilic cyanobacteria based on morphology and enrichment culture

Obvious differences in cyanobacterial populations between springs can be easily discerned by simple macro- or micro-scopic examination, based on distinctive morphological (Fig. 1), or physiological (Table 1) characteristics. Differences observed among springs a few meters or kilometers from each other are likely due to differences in chemical composition, temperature, or exposure to solar irradiance. Those differences observed among widely separated springs may be the consequence of geographic isolation and limitations of dissemination.

Abbreviations: cDNA complementary deoxyribonucleic acid; DGGE denaturing gradient gel electrophoresis; HPLC high-performance liquid chromatography; HTF high temperature form, PCR polymerase chain reaction, UVR ultraviolet radiation

A. Geographic distribution

Since geothermal springs are scattered like islands or archipelagos on all continents (except Antarctica; where fumaroles are present) and many island groups, often with great distances separating them, it is not unreasonable to believe that there should be endemic species of thermophiles, restricted to certain hot spring clusters as a result of geographic isolation and possibly evolutionary divergence. The concept should not be strange, since dispersal of obligate thermophiles from rare and distant point sources is certainly limited for some taxa, and the time between successful long-distance disseminations may be enough to allow speciation to take place in some hot spring clusters. For instance, some species of *Synechococcus* (Fig. 1a and b) are clearly restricted in geographical distribution. All forms of thermophilic *Synechococcus* are absent from Icelandic hot springs (although numerous springs exist that appear chemically suitable). Morphologically different *Synechococcus* exist in New Zealand and European springs, but only forms which grow up to temperature limits of 62° and 58° C, respectively (Castenholz, 1969, 1976, 1978, 1996; Table 1). Forms of *Synechococcus* that grow in nature up to 73-74°C in the western contiguous

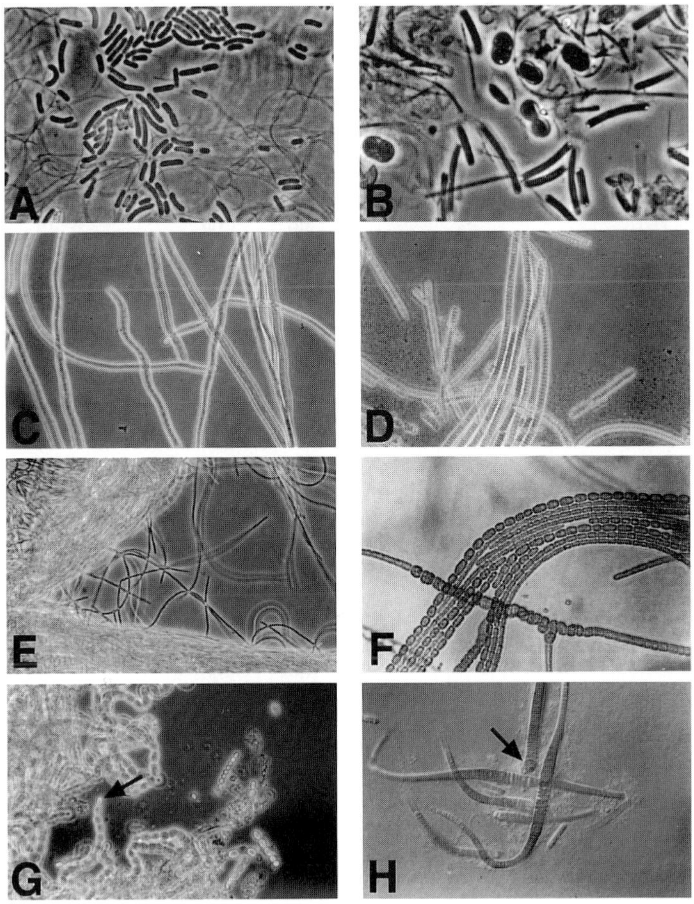

Fig. 1. Representative cyanobacteria of hot springs. **A.** Co-culture of *Synechococcus* cf. *lividus* (thermotype I, culture H-53-s from Hunter's Hot Springs, OR) plus filamentous green nonsulfur bacterium *Chloroflexus aurantiacus* (culture J-10-fl). Width of *Synechococcus* is ~1.5 µm. **B.** Collected cells from 54°C at Kah-nee-ta Hot Springs, OR. The narrow rods (~1.5 µm wide) are *Synechococcus* cf. *lividus*. The wider, squat cells are *Cyanothece* (*Synechococcus*) cf. *minervae*. **C.** *Oscillatoria* cf. *terebriformis* collected from Hunter's Hot Springs, OR. Trichome width is ~4 µm. **D.** Collection of *Spirulina* cf. *labyrinthiformis* from ~45°C at Mammoth Hot Springs, Yellowstone National Park. Width of tight coil ~2.5 µm. **E.** Culture of *Phormidium* cf. *laminosum* (culture OH-1-p) from Hunter's Hot Springs, OR. Trichome width ~1.3 µm. **F.** Clonal culture of *Fischerella* (*Mastigocladus*) cf. *laminosus* (culture NZ-69-m from Ohinemutu Hot Springs, New Zealand) with one primary trichome showing a true branch. Other trichomes are hormogonia in various stages of steelement and cell enlargement. Trichome on lower right is ~4.5 µm wide. **G.** Clonal culture of high temperature form, "HTF *Chlorogloeopsis*" (culture I-15-HTF) from Geysir, Iceland) grown in medium free of combined nitrogen. Diameter of heterocyst (end of arrow) is ~3 µm. **H.** Clonal culture of *Calothrix* sp. from Potts Basin, Yellowstone National Park (culture Y-WT-94-cal 2). Width of heterocyst (end of arrow) is ~8 µm.

United States (so far not found in Alaska), and probably south into Central and South America, are absent in the geographic regions mentioned above, although they apparently extend into eastern Asia and possibly central Africa (Castenholz, 1996). In contrast, some thermophilic cyanobacteria, such as *Mastigocladus* cf. *laminosus* (Fig. 1f) and the high temperature form (HTF) of "*Chlorogloeopsis*" (Fig. 1g) appear to be cosmopolitan in distribution (Table 1; Castenholz, 1996).

B. Distribution Determined by Chemistry

In hot springs worldwide, cyanobacteria are not observed below pH 4.0, and their diversity seems quite limited below pH 6 (Brock, 1973). In Yellowstone (Clearwater Springs and Norris Basin) *Synechococcus* spp. occur in hot springs with daytime pH levels as low as ~5.2, and "HTF *Chlorogloeopsis*" populations occur at levels as low as ~ pH 4.5. The *Synechococcus* cf. *lividus* (clone Y-7C-s) isolated from a pH 5.5 spring in the Clearwater

Table 1. Representative cyanobacteria of hot springs based on morphology and physiology

Genus/species	Upper temp. limit, °C	pH	N_2-fix.	Sulfide tolerance[1]	Known locations
Synechococcus cf. *lividus* (HTF)[2]	74	>6	no	low (<50 μM)	W. of N. and S. America except Alaska; China (E. Asia ? Africa ?)
Synechococcus cf. *lividus*[3]	~58-66	>5	no	low (varies)	W. Americas, New Zealand, Asia, Europe, Africa
Cyanothece (*Synechococcus*) cf. *minervae*	62	>6 ?	no	low (?)	W. of N. America + ?
"*Chlorogloeopsis* HTF"[4]	64	>4.5	yes	low (~0.15 mM)	W. Americas, New Zealand, Europe, Iceland, vocanic islands
Phormidium spp. (cf. *P. laminosum*)	~62	>6.5	no	mod. (?)	Hot springs world-wide ?
Mastigocladus (*Fischerella*) cf. *laminosus*[5]	58	>5	yes	mod. (~0.25 mM)	Hot springs world-wide
Oscillatoria cf. *amphigranulata*[6]	~56	>6.5	no	high (~3 mM)	New Zealand
Oscillatoria cf. *terebriformis*	55	>6	no	high (~1 mM)	W. Americas, E. Asia, Saudi Arabia
Spirulina cf. *labyrinthiformis*[7]	51	>6	no	mod. (~0.1 mM)	Hot springs world-wide ?
Calothrix spp.[8]	~50	?	yes	low (?)	Hot springs world-wide ?

[1]Field observations, not from studies of sulfide effects on growth of pure cultures.
[2]The "high temperature form" of *Synechococcus* cf. *lividus* may consist of more than one species. Studies were carried out in Yellowstone and E. Oregon springs.
[3]Many "look-alike" species of lower temperature forms apparently exist, although some show some morphological differences (Fig. 1-4 in Castenholz, 1996).
[4]Formerly referred to as "*Mastigocladus* HTF". Although this widespread cyanobacterium has a morphological life cycle similar to *Chlorogloeopsis fritschii*, 16S rRNA sequence data suggest that a separate generic designation will be required (A. Wilmote, pers. comm.).
[5]According to 16S rRNA sequence data, more than one species may be involved (R. Rippka, pers. comm.).
[6]This is not a close phylogenetic relative of other representatives of the large "form-genus" *Oscillatoria* (Wilmotte, 1994). Also it is apparently not related to the planktonic *Oscillatoria* (*Limnothrix*) *amphigranulata*, its morphological namesake.
[7]See Castenholz, 1977.
[8]Several apparent species of *Calothrix* occur in or around the edges of hot springs. The species in Hunter's Hot Springs grows up to a temperature of ~ 50°C (Wickstrom and Castenholz, 1978).

group grew at maximal rates only at pH levels above pH 7 and thus appeared acidotolerant not acidophilic (Kallas and Castenholz, 1982 a, b). The actual pH within cyanobacterial mats may vary above and below the pH of overlying water due to photosynthetic CO_2 uptake and respiratory CO_2 production (see below).

Temperature, in combination with availability of combined nitrogen and/or concentration of free sulfide (i.e. H_2S, HS^-, S^{2-}), also determines cyanobacterial species composition. When the outflows of neutral to alkaline, non-sulfidic hot springs that contain combined nitrogen have cooled to 74°C there is the likelihood (at least in some geographic regions) that a high temperature form (HTF) of *Synechococcus* will be present as a biofilm or mat which may, in turn, influence the chemistry downstream where other species of cyanobacteria enter the thermal gradient. For example, the combined nitrogen (usually as NH_4^+) in the spring source may be largely removed by the non-nitrogen fixing *Synechococcus* which may then be succeeded downstream below ~58°C by N_2-fixing cyanobacteria, most commonly heterocystous *Mastigocladus (Fischerella)* cf. *laminosus* (but possibly "HTF *Chlorogloeopsis*" below ~64°C or *Calothrix* spp. (Fig. 1h) below ~50°C). Nitrogen fixation has been measured below ~60°C in hot spring cyanobacterial mats, and especially at lower temperatures where nitrogen-fixing *Calothrix* occurs (Stewart, 1970; Wickstrom, 1980). In contrast, a spring may be rich enough in combined nitrogen to exclude *Mastigocladus* in favor of *Synechococcus* and other non-N_2 fixing cyanobacteria (e.g., *Phormidium*: Fig. 1e) which may dominate at lower temperatures. In other cases, combined nitrogen may be very low at the source, even in high-temperature springs, and heterocystous cyanobacteria may be the upper-temperature species at the specific upper temperature limit for growth (e.g. "HTF *Chlorogloeopsis*" at 64°C; *Mastigocladus* at 58°C).

Many neutral to alkaline geothermal springs contain primary soluble sulfide in the source water. Sulfide is an effective inhibitor of photosynthesis and possibly other physiological processes in the majority of cyanobacteria, but may be used as a photosynthetic electron donor in sulfide-tolerant species (Cohen et al. 1986, Castenholz & Utkilen 1984; See Chapter 10). Present evidence indicates that no thermophilic cyanobacteria with the capacity to grow above 56°C are capable of growing in waters with more than ~10 μM sulfide (Castenholz, 1976, 1977; Garcia-Pichel and Castenholz, 1990). In sulfide-rich springs of New Zealand the upper-temperature, sulfide-tolerant and sulfide-utilizing cyanobacterium is an *Oscillatoria* cf. *amphigranulata* morphotype (Castenholz, 1976). In some hot springs of Mammoth Terraces, Yellowstone Park, where source temperatures are ~52°C or below, *Spirulina* cf. *labyrinthiformis*, with an upper temperature of 51-52°C, predominates near the source (Castenholz, 1977; Fig. 1d). In higher temperature springs, with similar sulfide concentrations at the source, waters usually lose all detectable sulfide by the point where the outflow reaches 52°C and are dominated by less sulfide-tolerant species at that temperature and below. Non-photosynthetic sulfide-oxidizing bacteria (e.g. *Thermothrix* sp.) inhabit zones above 56°C in several sulfidic hot springs (e.g. upper terraces of Mammoth Hot Springs, Yellowstone National Park) and are likely to be partly responsible for sulfide removal in the upper zone. Icelandic and Yellowstone Park springs with primary sulfide often have essentially pure mats of photoautotrophic *Chloroflexus* at temperatures of ~66°C down to a temperature where surface water sulfide disappears (<~60°C; Castenholz, 1973b; Giovannoni et al., 1987). In a few interesting Icelandic mats, sulfide-oxidizing photosynthetic bacteria on the mat surface, which resemble *Chloroflexus aurantiacus*, remove sulfide and permit sulfide-sensitive cyanobacteria to grow beneath them in the lower part of the photic zone (Jørgensen and Nelson, 1988.).

C. Well-Studied Mat Systems

1. Hunter's Hot Spring, Oregon

The privately owned Hunter's Hot Spring, 3.5 km north of Lakeview, Oregon (elev. 1470 m), were studied periodically by Castenholz and students since the early 1960s. They consist of a series of springs initiating from an active deep fault system, many issuing at temperatures above 90°C at a pH of 8.0-8.4. The major ions are sodium (~170-210 mg L^{-1}), chloride (~180 mg L^{-1}), silicate (~140 mg L^{-1}), bicarbonate (~75 mg L^{-1}) and sulfate (260-285 mg L^{-1}) (specific conductance of ~1.1 mS). The chemistry and microbiota of Hunter's Hot Springs is characteristic of many hot springs of the Great Basin (Wingard and Castenholz, unpublished data; Mariner et al., 1974)

Initially, besides identifying the conspicuous cyanobacteria, the objective of early studies was to explain the abrupt upper and lower temperature boundaries of distinctive phototrophic populations incontinuous and linear temperature gradient. These distributions are illustrated in Fig. 2 and Plate 5a, and are described in Castenholz (1969, 1973a) and Wickstrom and Castenholz (1978, 1985). Briefly, the upper temperature limit for cyanobacteria, and for global photosynthesis, almost certainly is 73-74°C (Table 1) and this boundary is easily seen when the temperature remains relatively constant at a particular point in the drainway (Fig. 2, Plate 5a). A green cover, composed of a cyanobacterial form-species identifiable morphologically as *Synechococcus lividus* Copeland (Fig. 1 a), occurs as the top cover of the mat in most of the Hunter's drainways to about 54-55°C where it is abruptly replaced by a dark red-brown cover of *Oscillatoria* cf. *terebriformis* (Fig. 1c), a distinctive form identified by morphology, physiology, and ecology (Castenholz, 1978, 1996; Fig. 2 and Plate 5a). The uniform green cover of *Synechococcus*, however, was found to consist of at least 4 stable temperature-defined ecotypes which

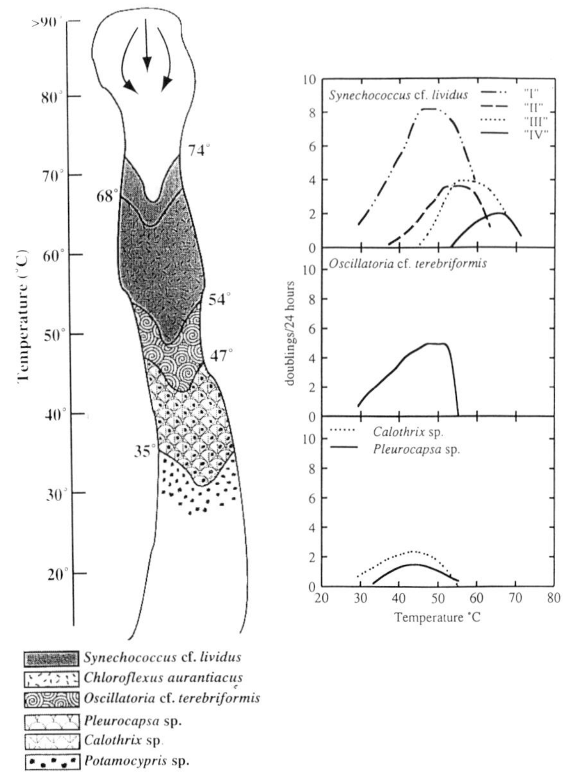

Fig. 2. On the left: idealized horizontal distribution of mat surface organisms in outflows of Hunter's Hot Springs, Oregon, with symbols depicting the predominant species. On the right: the approximate growth rates in culture of the principal cyanobacterial species relative to temperature. The roman numerals of *S.* cf. *lividus* refer to the designations in Peary and Castenholz (1964). *Potamocypris* sp. is the as yet unnamed thermophilic ostracod in Wickstrom and Castenholz (1973, 1985).

appeared morphologically identical (Peary and Castenholz, 1964; Fig. 2). In culture, the most thermotolerant clone grew at a maximum rate at 63-68° with temperatures over 68° to 72° being supraoptimal (Meeks and Castenholz, 1971; Fig. 2). In culture, with continuous illumination, the maximum growth rate of this strain was about 2 doublings per 24 h, a rate considerably slower than the lower temperature strains (Fig. 2). *Oscillatoria* cf. *terebriformis* has an upper growth temperature limit of 54.5°C in culture and grows at a maximal rate to 0.5°C of this limit (Castenholz, 1968, 1973a; Fig. 2). Since it is highly motile (by gliding), the upper edge of the mat adjusts its position to its upper limit (Plate 5a and 5b). Since the mat is generally thick enough to absorb over 95% of visible radiation, its growth abruptly limits the visible lower edge of the *Synechococcus* cover (which is only slightly motile), presumably because *O.* cf. *terebriformis* outcompetes other cyanobacteria for light. All clones of *O.* cf. *terebriformis* had respectable growth rates well below 48°C, although in most of the springs of E. Oregon and N. Nevada, the *Oscillatoria* mat ended abruptly at about 47-48°C (Fig. 2, Plate 5a). This is a result of voracious populations of the thermophilic and "herbivorous" ostracod *Potamocypris* sp. that have an almost ubiquitous distribution in the same geographic region (Castenholz, 1973a, Wickstrom and Castenholz, 1985; Fig. 2; Plates 5b, 5c).

Potamocypris can survive and reproduce at temperatures as high as 48-49°C (Wickstrom and Castenholz, 1973), and will eliminate the delicate trichomes of *Oscillatoria* at a rapid rate. As a consequence the cyanobacterial populations below this temperature are composed primarily of the more grazer-resistant species, *Pleurocapsa* sp. and *Calothrix* sp. (Fig. 1h). The tapered filaments of *Calothrix* are embedded within the seemingly amorphous mass of *Pleurocapsa* cells (Wickstrom and Castenholz, 1978). The ostracods appear to graze primarily on the exposed lawn of *Calothrix* filament tips which continue to grow through the *Pleurocapsa* as well as on trichome masses of *O.* cf. *terebriformis* which are frequently washed downstream (Wickstrom and Castenholz, 1985). Combined nitrogen is probably limiting in the wash water over this portion of the hot spring mat, and both the *Calothrix* and *Pleurocapsa* are capable of nitrogen fixation, with no significant numbers of photosynthetic nitrogen-fixers above this temperature. By comparing the drainways of similar springs with and without ostracods it became obvious that the *Pleurocapsa/Calothrix* community was not only quite grazer-resistant, but was also grazer-dependent. In spring outflows without the thermophilic ostracod, the *Oscillatoria* cf. *terebriformis* cover extended downstream to about 35°C.

2. Octopus Spring, Yellowstone National Park

Octopus Spring is located in the White Creek drainage of the Lower Geyser Basin, Yellowstone National Park (elevation 2,237 m), and was studied extensively by T.D. Brock, Castenholz, Ward and their students since the late 1960s. It is an alkaline spring (pH ~8.3) of volcanic origin. Like Hunter's Hot Springs, the major ions are sodium (~320 mg L^{-1}), chloride (~250 mg L^{-1}) and silicate (~250 mg L^{-1}). Unlike Hunter's Hot Springs, Octopus Spring has a very low sulfate content (~20 mg L^{-1}; Brock, 1978). The chemistry and microbiota of Octopus Spring are typical of many alkaline springs in Yellowstone National Park.

A landscape view of Octopus Spring and features of its cyanobacterial populations is presented as Plate 6a. As the 90-92°C source water flows into the two effluent channels and cools, a thin film of yellow to green *Synechococcus* sp. resembling *S.* cf. *lividus* (Fig. 1a) develops on the sinter in a fluctuating environment of 75-66°C (Plates 6a, site A; and 5e). Octopus Spring surges slightly, which accounts for the regular variation in temperature in the effluents. For example, in summer the biofilm at a mean temperature of about 70°C experiences 75 to 65°C over about a 2min period (Miller et al., unpublished data). At slightly lower temperature (below about 65°C), a thicker laminated *Synechococcus* mat (Plate 6a, site B; and 5f) develops with a thin green layer on top of lower orange mat layers. The latter are comprised largely of filamentous bacteria that may correspond to green nonsulfur bacteria (e.g., *Chloroflexus aurantiacus* or *Chloroflexus* sp.), as judged by their growth characteristics and the high bacteriochlorophyll *c* content of the mat (see below). Mat thickness increases from < 1-2 mm at higher temperatures to a maximum of several cm as temperature decreases along the effluent channel, although mat formation and decomposition is probably maximized in the upper few millimeters (Brock, 1978; Ward et al., 1987). Below ~57°C *Phormidium* sp. (Fig. 1e) may occur together with *Synechococcus,* where it is especially obvious in streamers and as raised columns or pinnacles (Plates 6a, sites C and D; 6b and 6c, respectively) which appear predominantly in strong flow or quiescent pools, respectively. *Phormidium*-dominated mats are usually orange in summer. The laminated mats and conical structures were studied extensively as analogs or homologs of their fossil equivalents; planar stromatolites and the pinnacle Conophyton forms (see Walter et al., 1972; Brock, 1978; Awramik and Vanyo, 1986; Vanyo et al., 1986). Below approximately 43°C the larvae and adults of ephydrid flies graze upon the cyanobacterial mat (Plate 5d). The thermophilic ostracod, *Potamocypris* sp. which occurs mainly in hot springs of the Great Basin region (e.g. Hunter's Hot Springs), was not found in any of Yellowstone's springs. In the cooler waters farther downstream (~40°C-ambient), N_2-fixing *Calothrix* sp. (Fig. 1h) forms dark brown scytonemin-containing mats, which adhere to the siliceous substratum both in the flow and on the moist edges of the effluent.

III. Distribution of Cyanobacteria Based on Molecular Analysis

An intensive effort was undertaken, using molecular approaches, to analyze the native populations of bacteria that inhabit alkaline siliceous hot spring *Synechococcus* mats, especially those in Octopus Spring and nearby Mushroom Spring. The detection of populations was based primarily on retrieval of 16S rRNA sequences from mat RNA or DNA, by cloning and sequencing either cDNAs made from 16S rRNA templates (Weller and Ward, 1989; Weller et al., 1991) or products of PCR amplification from 16S rRNA genes (Kopcyznski et al., 1994). More recently, populations were identified through the use of denaturing gradient gel electrophoresis (DGGE) to provide a profile of community composition and to purify 16S rRNA gene fragments obtained from PCR amplification before sequencing (Ferris et al., 1996a). The technique utilizes a gradient of denaturants within an electrophoretic gel to separate 16S rRNA gene fragments of different sequences based on differences in their melting behavior (Muyzer et al., 1993). To a more limited extent lipid biomarkers of mat populations were also examined. Since cyanobacteria are predominant members of these mat communities it was not surprising that their 16S rRNA sequences and lipid biomarkers were readily detected in molecular analyses. Identification of 16S rRNA sequences as being of cyanobacterial origin was facilitated by the monophyletic evolutionary descent of the oxygenic phototrophic prokaryotes (Giovannoni et al., 1988; Wilmotte, 1994).

A. Detection of 16S rRNA-Defined Cyanobacterial Populations

At the outset of molecular studies it was thought that a single widespread *Synechococcus* species, *S. lividus*, may have constructed the *Synechococcus*-type hot spring mats which are common in western North America (Brock, 1978; Castenholz, 1973a, 1978). Genetically-stable temperature-specialized "strains" were cultivated from Hunter's Springs (see Section II.C.1) but the genetic differences among the "strains" were not measured (Peary and Castenholz, 1964). Similarity, examination of the mol% G+C contents of thermophilic *Synechococcus* isolates led to the inference that thermophilic *Synechococcus* was not a genetically diverse group (Waterbury and Rippka, 1989). In fact, the 16S rRNA sequence of *S.* cf. *lividus* was expected to serve as an internal standard in analyses of the Octopus Spring mat, and its the detection was expected to verify molecular procedures. To date, the 16S rRNA sequence of this organism was not detected in analyses of Octopus Spring mat nucleic acids, but isolates which contain this sequence were recovered through enrichment culture methods (Table 2). Several lines of evidence provided a high degree of confidence that the detected sequences were not artifacts of molecular methods, but constituted real cyanobacterial populations. First, within the limitations of the evaluation, none of the sequences showed evidence of being chimeric (Robinson-Cox et al., 1995). Second, sequences were often detected in replicate and/or by different molecular methods (Ward et al., in press). Third, pure *Synechococcus* cultures were isolated which had sequences identical to some of those detected in the mat (Table 2). Finally, different sequences showed different environmental distributions (see below). The latter two observations also ruled out the possibility that different sequences represented different 16S rRNA genes within the same species.

B. Genetic Diversity Among 16S rRNA-Defined Cyanobacterial Populations

The genetic diversity among 16S rRNA-defined Octopus Spring cyanobacterial populations is impressive; the variation among their 16S rRNA sequences is similar to that observed for all cultured cyanobacterial species (Fig. 3). Six populations (types A, A', A", A"', B and B') are closely related, exhibiting less than 3% difference in their 16S rRNA sequences, yet are ecologically distinct (see below). These occur as two distinct phylogenetic clusters referred to as A-like (type A, A', A" and A"') and B-like (types B and B') cyanobacterial sequences. Given that 16S rRNA sequences are highly conserved (Woese, 1987), it is possible that other thermophilic cyanobacterial ecotypes may exist which can only be discovered using comparisons of more rapidly evolving genetic elements.

The previous idea that thermophilic *Synechococcus* exhibited limited genetic diversity may have resulted from the strong selection for a single genotype in liquid enrichment culture (e.g. strain C1 in Table 2). As shown by Ferris et al. (1996b) all culture collection strains of thermophilic *Synechococcus* cf. *lividus* had very similar, if not identical, 16S rRNA sequences. Successful enrichment and cultivation of a strain with the same genotype (C1) verified its presence in the Octopus Spring mat, apparently at a density too low to be detected by direct molecular analysis (Ferris et al., 1996b). Dilution of inoculum to near extinction before enrichment prevented competitive exclusion of other *Synechococcus* spp. by the C1 population, and led to the recovery of more numerically relevant and novel *Synechococcus* genotypes (Ferris et al., 1996b; Ward et al., 1997). Isolates from highly diluted inocula (containing approximately 8 to 250 *Synechococcus* cells) included three strains. Sequences from these three strains were detected in the mat (B, B', P), verifying that multiple *Synechococcus* populations exist in the mat. One *Synechococcus* isolate (C9) was cultivated from an enrichment inoculated with approximately 250 *Synechococcus* cells) and may be a subdominant population; its 16S rRNA sequence has yet to be detected in the mat through molecular analysis. The cultivation of strains C9 and C1 raised to eleven the number of genotypically unique cyanobacterial populations detected in the Octopus Spring mat; at least five of which are known to exhibit a *S.* cf. *lividus*-like morphology (Table 2). The four

Table 2. Unique cyanobacterial 16S rRNA sequence types found in the Octopus Spring laminated cyanobacterial mat or in cyanobacteria isolates from the mat.

Sequence type	16S rNA in Detected Mat[1]	GenBank Accession No.	Cultivated	ATTC Accession No.	Known or possible phenotype based on cultivation and distribution	Ref.[2]
A"	X	U88069			very high-temperature adapted ?	1
A'	X	U42374			high-temperature adapted ?	1,2
A	X	X52544			moderate- to high-temperature adapted ? adapted to 300-800 μm mat depth (low light, UV ?) ?	1,2,3, 4
B'	X	U42375	X[3]		*S.* cf. *lividus* morphology; moderate- to low-temperature adapted ? adapted to upper 500 μm of mat (high light, UV ?) ?	1,2,4, 6
B	X	X52545 M62776	X	700245 700246	*S.* cf. *lividus* morphology; low-temperature adapted ?	1,2,3, 6
A'''	X	U88530			colonist after disturbance	5
I	X	L04709			unknown	7
J	X	L04710			low-temperature adapted ?	7,8
P	X	L35331	X[3]		*S.* cf. *lividus* morphology	7,9
C9		L35481-3	X	700244	*S.* cf. *lividus* morphology; phototaxis positive at 250 lux but negative at 2000 lux	9,10
C1		L35345 L35479-80	X[4]	700243[4]	*S.* cf. *lividus* morphology; phototaxis positive at 2000 lux	9,10

[1]Sequence detected by cloning or DGGE
[2]References: 1. Ferris et al. (1996a) 2. Ferris and Ward (1997) 3. Ward et al. (1990) 4. Ramsing et al. (in preparation) 5. Ferris et al. (1997) 6. Ward et al. (1997) 7. Weller et al. (1992) 8. Ruff-Roberts et al. (1994) 9. Ferris et al. (1996a) 10. Ramsing et al. (1997)
[3]strain lost, but limited DNA may be available in the lab of DMW
[4]partial 16S rRNA sequence identical to those of all thermophilic *Synechococcus* culture collection strains (see Ferris et al., 1996b for ATTC Nos.)

uncultivated A-like populations may exhibit this morphology because only *S.* cf. *lividus*-like cells were observed by microscopy in the mat at the high temperatures where these populations were detected. Strains C1 and C9 were genetically dissimilar from populations detected by molecular means; Synechococcus strain C9 is currently the most deeply divergent cyanobacterium known, though the statistical support for this inference is not strong (Ferris et al., 1996b).

C. Distribution of 16S rRNA-Defined Cyanobacterial Populations

The distributions of some of these cyanobacterial populations were studied using oligonucleotide hybridization probes and DGGE.

1. Temperature Distribution

Analyses using DGGE suggest that five closely related cyanobacterial populations (A", A', A, B' and B) occur along the thermal gradient from high to low temperature, respectively (Fig. 4). Hybridization probe studies suggested that the type J population was also found at the lowest temperatures (Ruff-Roberts et al., 1994). Hybridization probe experiments also demonstrated that during a one-week *in situ* incubation, in which samples were shifted from low to high temperature, A-like 16S rRNA increased, while B-like (and J) 16S rRNA decreased, suggesting that these derive from populations with different

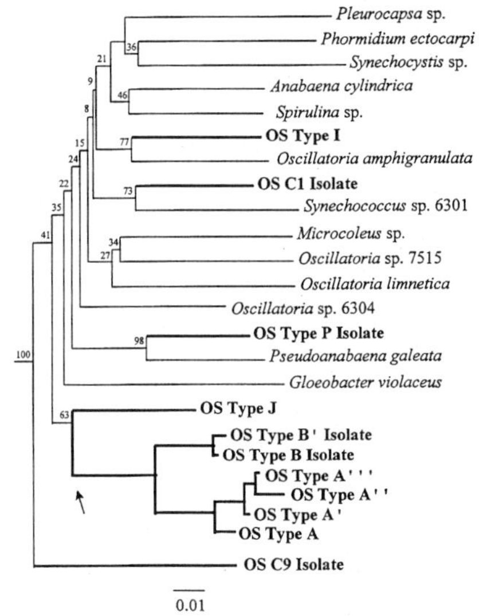

Fig.3 Distance matrix phylogenetic tree of cyanobacterial 16S rRNA sequences detected in the Octopus Spring cyanobacterial mat or cyanobacterial isolates from the mat (thick lines) compared to representative cyanobacterial lineages (thin lines). A tree with all lines except Octopus Spring types A, A', A", A"' and B' was first constructed using nucleotides that align with *E. coli* postions 332 to 452, 480 to 507, 712 to 892 and 1140 to 1364; the tree was rooted with the 16S rRNA sequences of *Thermotoga maritima, Chlorobium vibrioforme* and *Escherichia coli*. The consensus values at the forks indicate the number of times the group consisting of the sequences to the right of the fork occurred among 100 trees inferred from the bootstrapped data set of 606 nucleotides. A tree for the remaining sequences, which also included their closest relative, the type B sequence, was constructed using *E. coli* positions 1071 to 1367. The two trees were connected at the point where the type B sequence diverges from other sequences in the first tree (arrow). Bar indicates 0.01 fixed point mutations per sequence position.

temperature optima (Ruff-Roberts et al., 1994). These data reinforced the concept of temperature-adapted "strains" introduced by Peary and Castenholz (1964; Fig. 2). Whether these should be considered strains of one species, or different species, depends on one's concept of "species" (Castenholz, 1992). Temperature distribution data suggest that these populations represent different ecotypes, and ecologically distinct asexual populations are sometimes considered to be separate species (Mayr, 1982). The one surviving temperature "strain" (OH-53-s = PCC6716 and ATCC 27179) cultivated from Hunter's Springs by Peary and Castenholz (1964) had a *Synechococcus* C1-like genotype (Ferris et al., 1996b). One possible inference is that temperature-adapted ecotypes of the *Synechococcus* C1 genotype exist in Hunter's Hot Springs mats. This would be significant as it would indicate that temperature-adapted populations evolved in two widely-separated cyanobacterial lineages (Fig. 3). Alternatively, as in the Octopus Spring mat, *Synechococcus* genotype C1 may be a minor population in Hunter's Hot Springs mats and its recovery may be favored by liquid enrichment culture from low-dilution inocula. While adaptive radiation to temperature may at least partially explain the evolution of the A- and B-like genotypes, the evolutionary pressures which caused divergences among the more deeply branching lines of thermophilic cyanobacterial descent (e.g., A/B-like, C1, C9, I, J and P) remain obscure without further distribution analysis or phenotypic characterization.

2. Vertical Distribution

Several types of observation suggest that one possible evolutionary stimulus for adaptation is irradiance (e.g. Pierson and Castenholz, 1990) including both intensity and quality of visible and/or ultraviolet radiation. *Synechococcus* rich in chlorophyll *a* occur underneath the uppermost *Synechococcus* layer (Brock, 1978), which is more yellow. Microscopy studies of vertical thin sections of this layer in the 60°C mat of Mushroom Spring revealed vertical stratification of *Synechococcus* populations (Plate 7). Steep light gradients were measured in such mats (Fig. 5). Cyanobacteria (as measured by chlorophyll *a* and phycocyanin absorption troughs and intervening peaks) appeared to occur mainly in the upper 1 mm. The presence of bacterichlorophyll *c*-containing

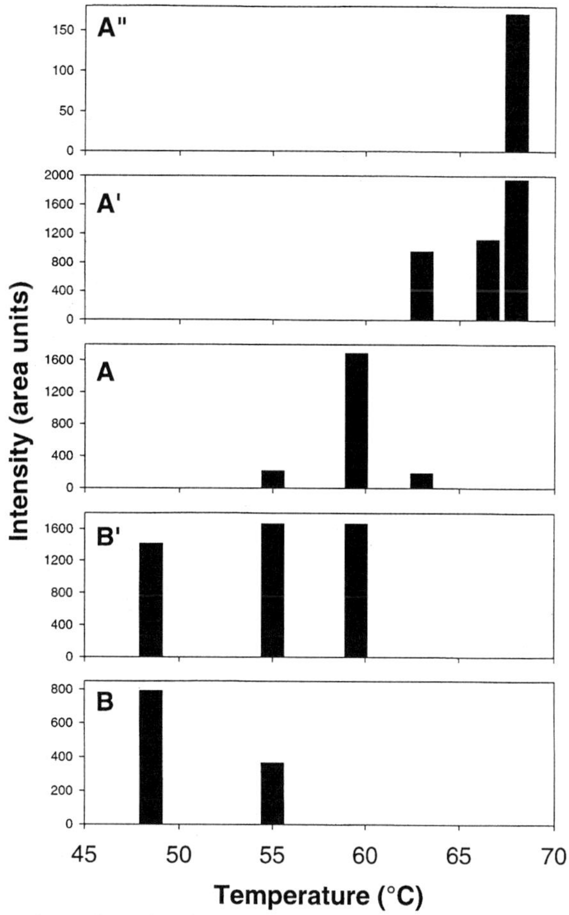

Fig. 4 Temperature distribution of cyanobacterial 16S rRNA sequence types detected in the Octopus Spring cyanobacterial mat by DGGE analysis. Intensities of DGGE bands were determined by integrating scanned gel images, and adding the intensity of homoduplex and one-half of the intensity of heteroduplex bands containing each sequence type [see Ferris and Ward (1997].

organisms (e.g., *Chloroflexus* spp. or green sulfur bacteria) and bacterichlorophyll *a*-containing organisms (purple bacteria and *Heliothrix*-like bacteria) was also quite obvious, especially at 0.5 mm and below. With this very steep light gradient, it is reasonable to predict that either *Synechococcus* populations at various depths are physiologically acclimatized to different light intensities or, that genetically distinct *Synechococcus* populations, adapted to different light regimes, predominate at different depths. Analysis of 100 µm-thick sections of the upper portion of the mat using DGGE suggested the presence of cyanobacterial populations A and B'. This finding was consistent with temperature distribution data for Octopus Spring where population B' appeared to reside above population A (Fig. 6). A population, phylogenetically related to green sulfur bacteria (type E''), was detected in the layers which contained the type A cyanobacterial populations. Another population (type C), phylogenetically related to green nonsulfur bacteria (e.g., *Chloroflexus*), was detected in increasing amounts below ~400 mm (Ramsing et al., unpublished data).

Other kinds of evidence are consistent with the possibility that genetically distinct *Synechococcus* populations are adapted to different light intensities. For example, hybridization probe studies revealed differential responses of mat populations to reduced light intensity (Ruff-Roberts et al., 1994). Also, genetically distinct *Synechococcus* cf. *lividus* isolates exhibited different phototactic movements in response to light of different intensity (Ramsing et al., 1997). Furthermore, light-adapted ecotypes of the moderately thermophilic cyanobacterium *Plectonema* sp. were described (Sheridan, 1976). It is hypothesized that *Synechococcus* populations which are adapted to different light intensities and/or ultraviolet radiation exist along the thermal gradient,

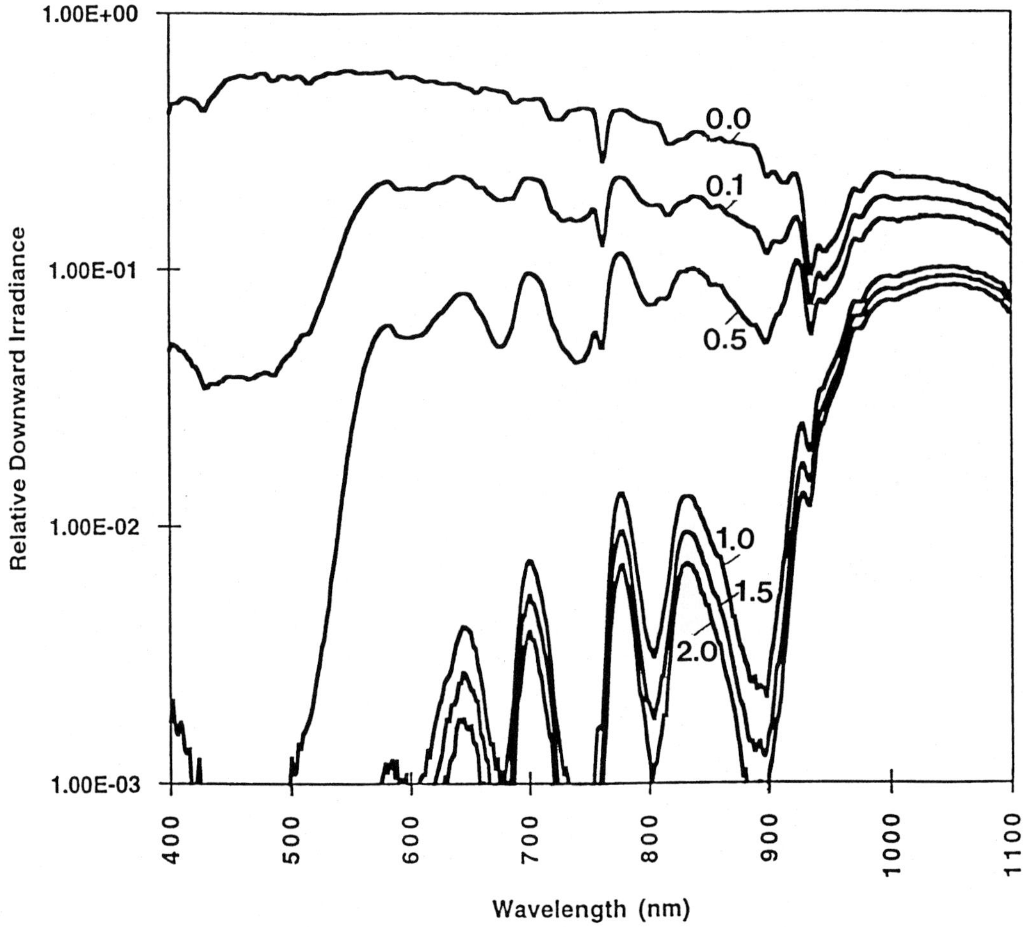

Fig. 5 . Logarithmic plot of downward irradiance in 57°C Octopus Spring laminated *Synechococcus* mat measured with a LI-Cor spectroradiometer and a 0.5 mm diameter fiber optic tip (see Pierson et al., 1990). Depth of each trace marked in mm below surface. Depressions in spectrum at : ~610-620nm, phycocyanin; ~670-680 nm, chlorophyll a; ~740 nm, bacteriochlorophyll *c*; ~800 and 890 nm, bacterichlorophyll *a*. [kindly provided by B.K. Pierson]

so that adaptations to temperature and irradiance may explain much of the genotypic diversity found in *Synechococcus* mats. It is also possible that other parameters which vary vertically in the mat were responsible for the specialization of mat cyanobacterial populations.

3. Temporal Distribution

It was suggested that thermophilic cyanobacteria adapted to different light intensities may shift dominance in mats in different seasons (Sheridan, 1979). In contrast, the pattern of temperature distribution within the Octopus Spring mat community (Fig. 4) was similar in all seasons (Ferris and Ward, 1997), which implies that there may be seasonal stability of those cyanobacterial populations. The shift of the same communities occurs upstream as the thermal gradient steepens in winter however, and is a commonly observed phenomenon in hot springs such as Hunter's (e.g. Wickstrom and Castenholz, 1985). Cyanobacterial populations in the 55° to 62°C mat of Octopus Spring changed over a 20- to 40-day period following physical disturbance of the upper green mat layer (Ferris et al., 1997). Populations initially present in the mat were displaced by others which were initially below detection. In particular, the A' population, usually found at higher temperatures, and a new population (A'''), were effective colonists. The basis for specialization as a colonist remains unknown.

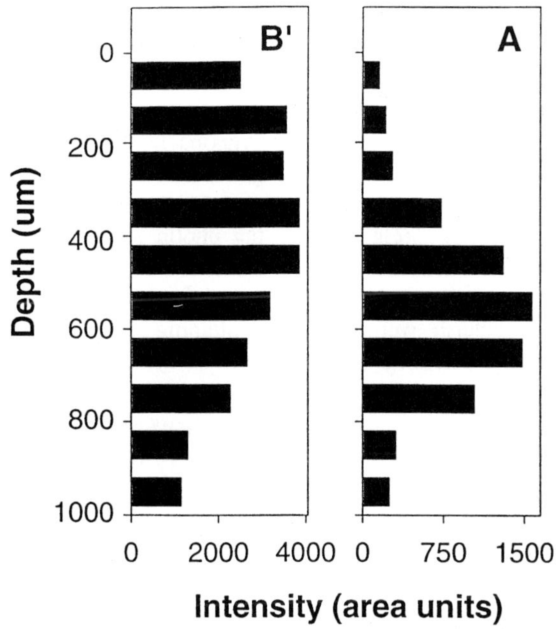

Fig. 6. Vertical distribution of cyanobacterial 16S rRNA sequence types detected by DGGE analysis of 100 μm thick cryotome sections of the same mat shown in Fig. 10. Intensities of DGGE bands were determined by integrating scanned gel images, and adding the intensity of homoduplex and one-half of the intensity of heteroduplex bands containing each sequence type. [from Ramsing et al.]

4. Distribution Among Springs

Another possible evolutionary driving force for diversity among thermophilic cyanobacteria is pH. Since pH may vary over a diel period within a single mat over a range of ~3 to 4 units (see below), adaptations to pH may be limited to extremes of the pH range found in different springs inhabited by thermophilic cyanobacteria (see Table 1). The 16S rRNA of a *S.* cf. *lividus* isolate (C1 genotype) was readily detected in the pH 6.1 Clearwater Springs pool from which it was originally cultivated; and where B-like populations were also detected (Ruff-Roberts et al., 1994). Springs at lower pH (e.g. down to pH 5.0) that contained *Synechococcus*-shaped cells did not contain detectable 16S rRNA of the C1 or A-like, B-like or J cyanobacterial populations, which may suggest the possible existence of low-pH adapted *Synechococcus* populations.

Relatively little is know about the geographical distribution of cyanobacterial populations, defined through molecular analyses, in neutral to alkaline springs which may or may not differ in other chemical properties. As mentioned above, DGGE analysis revealed the similarity of cyanobacterial populations found in ~60°C mats of Octopus Spring and Mushroom Spring (both pH 8.3), which are separated by only a few hundred meters. DNA hybridization analysis revealed that neutral to alkaline springs, separated by up to tens of kilometers, contained A-like (pH 6.7-8.3) and B-like (pH 6.2-9.1) cyanobacterial populations at temperatures comparable to those at which they were found in the Octopus Spring mat; type-J cyanobacterial populations were observed only in 48 to 62°C mats in Octopus and Twin Butte Vista Springs (pH 8.4-9.1), which are separated by approximately 100 meters (Ruff-Roberts et al., 1994; Ward et al., 1994a). So far, no molecular analyses were performed on hot spring cyanobacterial populations of greater geographical separation; such genotypic information may prove to be essential for the evaluation of the importance of allopatric speciation in thermophilic cyanobacteria.

D. Lipid Biomarker Studies

Lipid biomarkers appear to be much less specific than 16S rRNA sequences. However, because of the greater persistence of lipids over geological time, they may be useful for recognizing cyanobacteria as a group (as opposed to specific cyanobacterial ecotypes) in deposits of geologically significant age. For instance, the mid-chain-branched mono- and dimethylalkanes found in Precambrian oils appear to be typical of cyanobacteria found in thermal mats

(Shiea et al., 1990). Free lipids (Dobson et al., 1988; Shiea et al., 1990, 1991), polar lipid fatty acids (Zeng et al. 1992a and b; Ward et al., 1989b) and complex polar lipids (Ward et al., 1994b), typical of those in *S.* cf. *lividus,* dominate lipid extracts from the Octopus Spring mat. These lipids were difficult to distinguish from lipids of green nonsulfur bacteria (e.g., *C. aurantiacus*), but the presence of sulfoquinovosyl diglycerides and the predominance of C_{16} over C_{18} fatty acids suggested that cyanobacteria were the main source. Fatty acid chain length and degree of saturation varied with mat temperature. It is uncertain whether this reflected population variation along the thermal gradient (Fig. 4), acclimation to temperature, or both.

E. Cyanobacterial Mat Community Structure

The greater abundance of lipids characteristic of cyanobacteria and *Chloroflexus,* compared to lipids typical of nonphototrophic mat inhabitants, is consistent with a trophic structure model in which the most predominant organisms are cyanobacteria and green nonsulfur bacteria, with successively smaller populations of aerobic and anaerobic chemo-organotrophic bacteria, and terminal anaerobic respiratory bacteria (sulfate reducers and methanogens; Ward et al., 1989b; Des Marais et al., 1992). The vertical positioning of these lipid biomarkers was also consistent with the maximization of cyanobacteria within the upper green layer, the maximization of green nonsulfur bacteria in the subsurface few millimeters, and the maximization of other anaerobes beneath this zone (Zeng et al., 1992b).

Analysis of 16S rRNA sequences and cultivation studies revealed more detail within the oxygenic photosynthetic "guild" (i.e., cyanobacteria). One must, however, be cautious in interpreting molecular data quantitatively, because primer-dependent reactions like cDNA synthesis and PCR may be selective (Reysenbach et al., 1992; Rainey et al., 1994) and there is no guarantee that PCR is quantitative (Suzuki and Giovannoni, 1996; Ferris and Ward, 1997). Nevertheless, the observations that A- or B-like cyanobacterial 16S rRNA sequences predominate cDNA- and PCR-based clone libraries (Ward et al., 1992a) and DGGE profiles, combined with the enrichment and isolation of B and B' *Synechococcus* cultures from samples diluted to near extinction (inocula contained as few as 8 *Synechococcus* cells), suggest that these are likely to be the predominant Octopus Spring mat cyanobacteria. Changes in specific A- and B-like cyanobacterial populations along thermal and vertical gradients apparently reflect specialization to temperature and light (or other parameters that vary in the photic zone). Other cyanobacterial populations exist in the mat; some of which (e.g., C9 and P) may be subdominant (based on recovery from enrichments containing 25 to 250 *Synechococcus* cells); some may even be zymogenous (e.g., C1).

IV. Physiological and Behavioral Ecology of Cyanobacteria of Geothermal Habitats

Earlier work on the activities and behavior of thermophilic cyanobacteria was reviewed by Castenholz (1969) and Brock (1978). Here, we emphasize more recent work which resulted from microprobe and detailed radiolabelling analyses. Experiments were conducted on natural material or on cultivated cyanobacteria and the data must be interpreted with caution, given our increasing understanding of the population biology of mat cyanobacteria.

A. Photosynthesis

Rates of oxygenic photosynthesis in Yellowstone *Synechococcus* mats, as measured by oxygen microelectrodes and light-dependent incorporation of $^{14}CO_2$, are among the highest reported for any aquatic habitat (Revsbech and Ward, 1984a). Use of microautoradiography confirmed that most of the incorporated $^{14}CO_2$ was associated with *Synechococcus* cells (Bateson and Ward, 1988). Photosynthetic rates were maximal in the 55 to 60°C temperature range of *Synechococcus* mats (Revsbech and Ward, 1984a), and the rates were optimal at the temperature from which samples were collected along the entire thermal gradient (Brock, 1967). However, multiple experiments indicated that the highest temperature populations of *Synechococcus* from Octopus Spring, and nearby non-fluctuating Pine Spring, were at several degrees lower (~65°C) than their optimum in the (~68-74°C; Miller et al., 1998). This was also shown in cultivated high temperature *Synechococcus* from Hunter's Hot Springs (Meeks and Castenholz, 1971). The pattern of distribution of photosynthesis along most of the thermal gradient in hot springs probably reflects the coordinate activities of a mixture of co-existing *Synechococcus* populations adapted to different temperatures. Further evidence for this was provided by microelectrode measurements of oxygenic photosynthesis in the surface of a mat core taken from a fluctuating 50° to 70°C region of Octopus Spring after incubation at different temperatures (Fig. 7). Two, possibly three, peaks for integrated gross photosynthesis were observed at different incubation temperatures. These results are consistent with molecular evidence for two, and sometimes three, 16S rRNA-defined populations at a given site (Figs 4 and 6).

Analyses of vertical profiles through mats showed that oxygenic photosynthesis occurred within the upper 1-mm green layer with maximum rates a few hundred micrometers below the mat surface (Fig.8a). Here, an O_2 maximum was observed (Fig. 8b) and intensive consumption of CO_2 was thought to cause a rise in pH of nearly 2 units greater than that of the water overflowing the mat (Fig. 8c). At night pH decreased to nearly 2 units lower than that of the overflowing water (Revsbach and Ward, 1984a). The intensive consumption of CO_2 during photosynthesis may have limited stable carbon isotope discrimination during autotrophic synthesis of mat organic matter, causing the stable isotopic signature ($^{13}C/^{12}C$) of photosynthetically fixed carbon to be heavier than in other aquatic settings (DesMarais et al., 1992). Gradients of CO_2 or other nutrients (e.g., combined nitrogen, phosphate) might also influence vertical positioning of *Synechococcus* populations in these mats. Such a trend is typical of oxygenic photosynthesis in other thermal mats such as that in Hunter's Hot Springs (see below), but it was altered in Octopus Spring mats by removal of the upper *Synechococcus* layer and subsequent recolonization

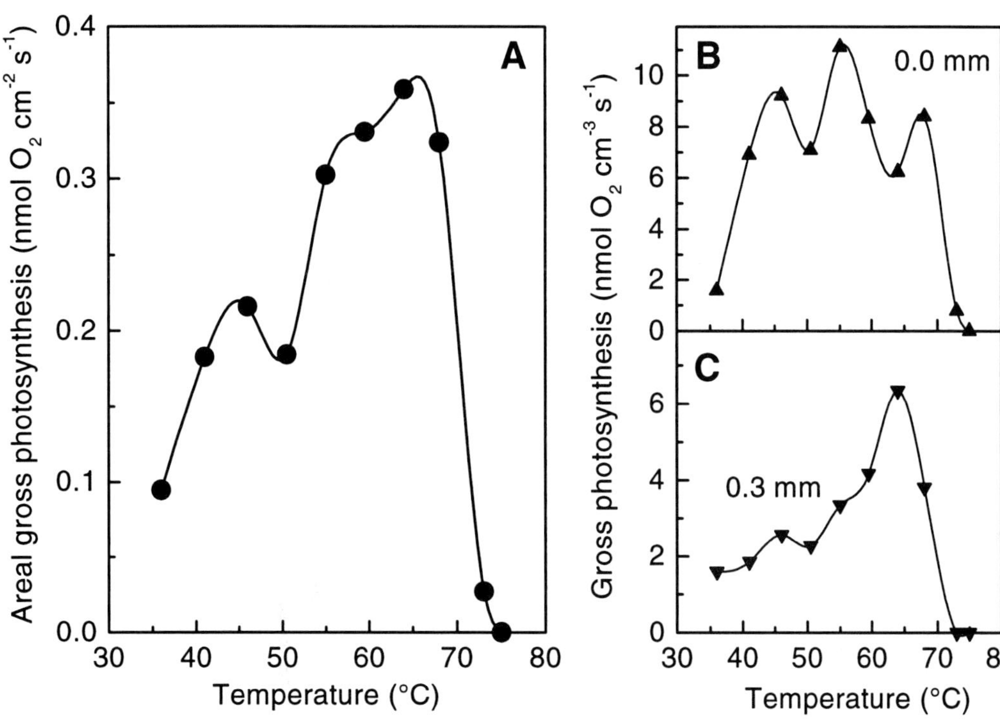

Fig. 7. Microelectrode measurements of oxygenic photosynthesis in a laminated *Synechococcus* mat from the shoulder region of Octopus Spring, incubated at different temperatures. (A) Depth-integrated rates, (B) rates in the surface 0-0.1 mm zone, (C) rates at 0.3 mm depth. [kindly provided by M. Kuhl and F. Garcia-Pichel]

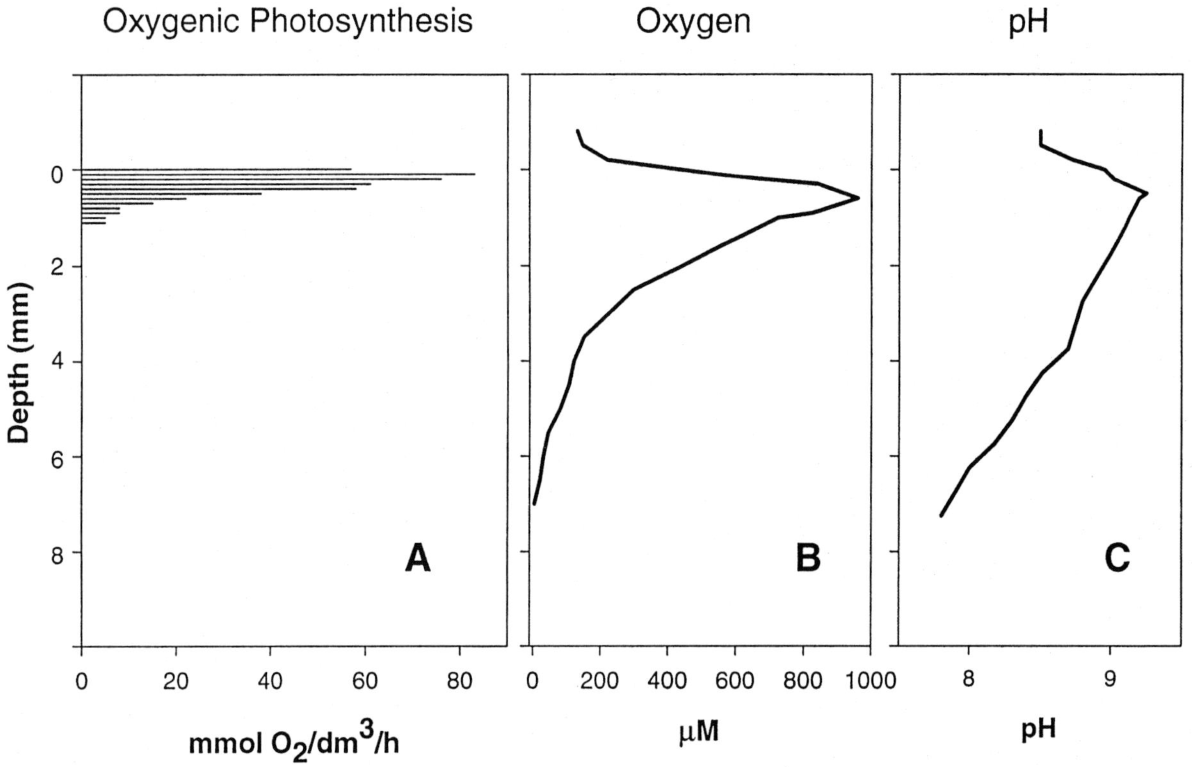

Fig. 8. Microelectrode measurements illustrating vertical distributions of (A) oxygenic photosynthesis, (B) oxygen and (C) pH for laminated Octopus Spring *Synechococcus* mat in full mid-day sunlight [from Revsbech and Ward (1984a)]

(Ferris et al., 1997). After such a disturbance the photic zone became thicker (approximately 2 mm), causing greater light penetration, and a bimodal vertical profile of oxygenic photosynthesis was observed. These results may suggest possible light adaptation by different cyanobacterial populations. Also, results from light reduction experiments are consistent with either acclimation (Madigan and Brock, 1977) and/or adaptation (Ruff Roberts et al., 1994) of *Synechococcus* populations to reduced light intensity. Further evidence of depth stratification of cyanobacterial populations derives from observations of differences in temperature optima for oxygenic photosynthesis near the surface or at 300 mm depth in the Octopus Spring mat (Fig. 7b and c) and from molecular analysis (Fig. 6).

It is interesting to speculate that coexisting cyanobacterial populations adapted to different temperature or light regimes may help stabilize photosynthesis during fluctuating conditions of light and temperature. The different specialized cyanobacterial populations can be said to belong to the same "guild" (i.e. oxygenic photosynthesis), so that intra-guild diversity might stabilize guild function, which in turn may benefit other community members which are dependent on photosynthesis.

B. Effect of Ultraviolet Radiation on Photosynthesis

UVB (280-320 nm) and UVA (320-400 nm) are detrimental to photosynthesis either by direct lesions of molecular targets or indirectly by production of reactive oxygen metabolites. Partial protection is afforded by UVR-screening compounds such as scytonemin and probably mycosporine-like amino acids, reactive oxygen metabolite "quenchers" such as carotenoids, superoxide dismutase, catalase or peroxidases, and by various synthetic and DNA repair processes such as excision repair and photoreactivation (Chapter 21). In hot springs only *Calothrix* spp. and *Pleurocapsa* sp., which occur at temperatures below ~50° C, are known to possess scytonemin. Cyanobacteria that grow at higher temperatures, such as *Synechococcus* spp. and *Mastigocladus* cf. *laminosus*, appear to lack screening compounds and artificial removal of UVR by filters greatly enhanced photosynthesis (Miller et al., unpublished data). Quenching and possibly dark repair processes are probably necessary for survival of the high-temperature forms under normal conditions.

C. Partitioning of Photosynthate in Synechococcus *Cells*

Most of the photoautrophically fixed carbon retained by Yellowstone *Synechococcus* populations is stored as polysaccharide (specifically polyglucose; Konopka, 1992; Nold and Ward, 1996; Fig. 9). Incorporation into rRNA was undetectable and incorporation into proteins and lipids was far slower than one would have expected for populations growing rapidly and exponentially. Previous measurements of *in situ Synechococcus* growth rates (Brock, 1978; Brock and Brock, 1968) and mat accretion rates (Brock 1978; Doemel and Brock, 1977) suggested that *Synechococcus* spp. grew much more slowly in natural mats than in culture (doubling times of 4 to 13 days *in situ*, as opposed to 2 to 10 hours in pure culture, as reported by Peary and Castenholz, 1964). However, the field experiments might not have provided accurate measures of *in situ* growth rates. Growth rates estimated from rates of decrease of *Synechococcus* cells after darkening the mats (Brock and Brock, 1968) required the *assumption* that populations were growing at a rate in balance with washout and decomposition. Even non-growing cells would have decreased in numbers due to washout or decomposition after exhaustion of their energy reserves. Attempts to measure upward accretion of mat above silicon carbide layers (Doemel and Brock, 1977) may have caused a physical disturbance leading to migration (of motile cells, see below) or more rapid growth following recolonization (Brock, 1978). After disturbance, fixed carbon was distributed more into growth-related macromolecules (protein and low molecular weight metabolites) and less into polysaccharide (Fig. 9; Ferris et al., 1997). At high cell densities, *Synechococcus* spp. may enter into an active, but non-growing or very slow growing phase similar to that found in other high density bacterial populations (e.g., bioluminescent bacteria in light organs (Ruby and Asato, 1993) and N_2-fixing bacteria in root nodules (Werner, 1992). In these cases control of physiological state appears to be linked to autoregulatory substances (Fuqua et al., 1994). Perhaps the same is true of cyanobacterial mats.

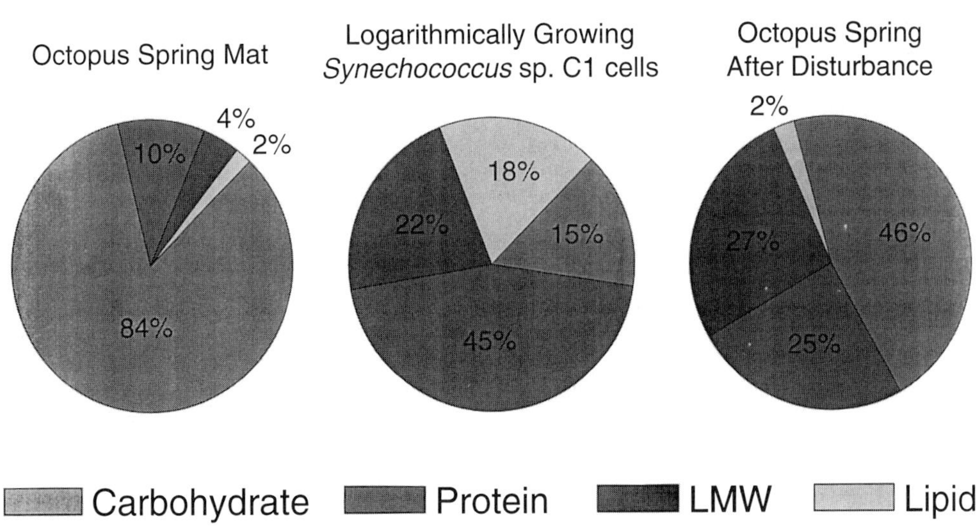

Fig. 9. Partitioning of photoautotrophically fixed CO_2 into macromolecular and low molecular weight metabolites in laminated Octopus Spring *Synechococcus* mat, exponentially growing *Synechococcus* sp. C1 cells, and mat 5 days following removal of the cyanobacterial surface layer. [from data of Nold and Ward (1996) and Ferris et al. (1997)]

D. Photoexcretion, Fermentation and the Cross-Feeding of Heterotrophs

Photoautotrophs must play a major role in cross feeding of fixed carbon to heterotrophic mat populations (Chapter 5). In pure culture, *S.* cf. *lividus* isolates from hot springs in Oregon and Yellowstone Park excreted a large variety of compounds which were identified by HPLC. The major products excreted under different light intensities were fructose, adonitol, oxalate, fumarate, succinate, acetate, formate, and β-hydroxybutyrate; a small amount of glycolate was released under high light intensity (~100 W m^{-2}; Teiser, 1993; Teiser and Castenholz, unpublished data). Under the high-light, extremely oxic and presumably low-CO_2 conditions in which maximum photosynthesis occurs *in situ* in the Octopus Spring mat, photorespiration and glycolate excretion were found to account for at least 60% of photoexcreted carbon. The latter can represent at least 11-12% of photoautotrophically fixed carbon (Bateson and Ward, 1988; Bauld and Brock, 1974). In Hunter's Hot Springs the excretion of total fixed carbon by suspensions of field populations of *S.* cf. *lividus* ranged from 2 to 10%; the higher percentages occurring in populations pre-exposed to a low light intensity (Teiser, 1993; Teiser and Castenholz, unpublished data). Analysis with microautoradiography demonstrated that glycolate was photoheterotrophically incorporated into *Chloroflexus*-like filaments that inhabited the mat (Bateson and Ward, 1988; Ward et al., 1987).

The fermentative capacity of pure cultures of thermophilic cyanobacteria was demonstrated by Richardson and Castenholz (1987a). Nold and Ward (1996) demonstrated that this was an important process in carbon and energy cycling within several Yellowstone *Synechococcus* mats. Most of the ^{14}C-polyglucose synthesized from $^{14}CO_2$ during photoautotrophic metabolism was fermented during a subsequent twelve-hour dark period, with the production of ^{14}C-acetate, ^{14}C-propionate and $^{14}CO_2$. In previous studies of mat fermentation it was presumed that obligately heterotrophic fermentative bacteria were responsible for production of fermentation products (Anderson et al., 1987). From the limited decrease in specific activity of ^{14}C-acetate produced from ^{14}C-polyglucose during dark anaerobic incubation it appears that *Synechococcus* spp. were a significant source of this intermediate, although other sources of acetate may have existed (perhaps from anaerobic fermenters or acetogenic bacteria; see Bateson and Ward, 1988). Volatile fatty acid products of fermentation accumulate during darkness and are later photoassimilated by filamentous photoheterotrophs (Anderson et al., 1987). Thus, through a diel cycle, *Synechococcus* spp. in mats appear to function like a major carbon and energy pump, cross-feeding most of the fixed carbon and energy to photoheterotrophs. It is not known if H_2 is also produced in *Synechococcus* fermentations, although this was shown to be the major electron donor for methanogenesis and sulfate reduction - the dominant terminal anaerobic processes - in low-sulfate and high-sulfate mats, respectively (Sandbeck and Ward, 1981; Ward and Olsen, 1980; Ward et al., 1984). Sulfide dynamics in a high-sulfate mat system relative to light, oxygen and pH are represented in Fig. 10. Sulfide variation over a diel cycle may influence the spatial positioning of motile populations and/or the activity of oxygen-sensitive populations.

E. Movement

1. Filamentous, Motile Cyanobacteria in Oregon Hot Springs

Many cyanobacteria are capable of motility and photomovements and the best-studied example involving thermophiles is that of *Oscillatoria* cf. *terebriformis* in Hunter's Hot Springs. The coherent, but non-cohesive, mat of this species covers the substratum up to a temperature of about 55°C, which is the maximum constant temperature tolerated (Plate 5a). If the temperature of the water increases at that boundary the entire mat edge will normally retreat (thermophobic response; Castenholz, 1968). Since the rate of movement of this cyanobacterium is commonly about 5-6 μm s^{-1}, this easily allows those movements required to avoid slow changes in stream temperature. In addition to this type of movement, large masses of these continuously motile trichomes contract into thicker masses during periods of high light intensity (Castenholz, 1968; Richardson and Castenholz, 1987b; See Color Plate 5b). This behavior may be interpreted as an escape from continuous exposure to high irradiance, since trichomes are continuously moving in and out with respect to the surface of the clumps. Thinner masses of these trichomes migrate downward into soft undermats during periods of high radiation intensity (e.g. late morning to late afternoon in summer) and reappear on the surface during late afternoon. This is a common phenomenon among many species of

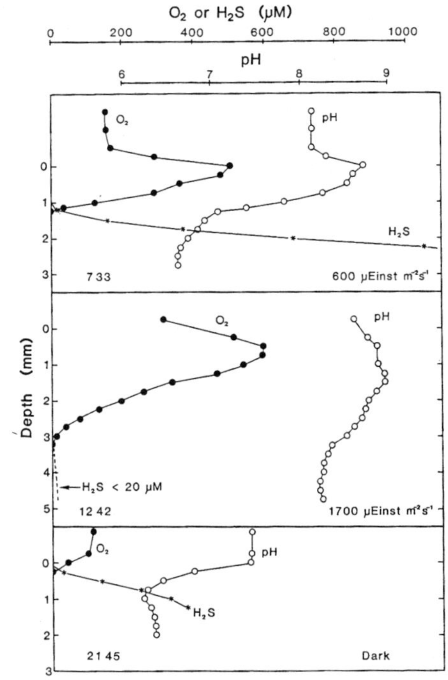

Fig. 10. Microelectrode measurements of vertical distributions of oxygen, sulfide and pH in a 39 to 45°C cyanobacterial mat in Hunter's Hot Springs at different times of day (and light intensities), as indicated. [from Revsbech and Ward, 1984b]

"oscillatorian" cyanobacteria (e.g. Castenholz et al., 1991; Garcia-Pichel et. al., 1994). Most of these cyanobacteria will return to the surface if the mat is completely darkened. However, in *Oscillatoria* cf. *terebriformis,* there is a second retreat downward into sulfide-rich undermats after dusk with a return to the surface in early morning (Richardson and Castenholz, 1987b; Jørgensen et al., 1992). Neither the cues nor the reasons for this unusual nighttime behavior are fully understood. Other studies showed that the daytime avoidance of high solar irradiance, especially the UVR component, prevented the photoinhibition or photodestruction of these migrating species (Garcia-Pichel et al., 1994, Garcia-Pichel and Castenholz, 1994; Kruschel and Castenholz, unpublished data). In one study, involving hypersaline mat species of *Oscillatoria* and *Spirulina*, there was evidence that not only photoinhibitory intensities were avoided, but that exposure to optimal intensities of photosynthetically active light was achieved during most of the day by downward movement (Garcia-Pichel et al., 1994). Vertically migrating, filamentous cyanobacteria also retain a high content of light-harvesting pigments, which under high light can also act as lethal photosensitizing pigments. However, this high pigment content also allows motile cyanobacteria to take advantage of low irradiance under overcast conditions or during early morning and late afternoon.

The cue for daytime downward and upward movements of motile cyanobacteria is intensity of solar irradiance, and when UVR is isolated as a factor, this appears to be the principal controlling region of the spectrum (Garcia-Pichel and Castenholz, 1994; Bebout and Garcia-Pichel, 1995; Kruschel and Castenholz, unpublished data). Although UVR was shown to be quite inhibitory to movement and photoorientation in gliding cyanobacteria (Häder, 1984), this damage occurs after many minutes or hours and may not affect the rapid migratory response. In some cases of vertical migration in mats, step-up photophobic responses appear to maintain populations below the mat surface, but true phototactic, chemotactic and aerotactic responses are probably involved in some of the movements, such as upward or downward movements during darkness.

2. *Phormidium in Vertical Structures*

Both conical ("Conophyton"-like) and columnar vertical structures composed largely of *Phormidium* sp. (or spp.) occur in slow flowing thermal streams or in hot spring pools (Plate 6c; Walter et al., 1972). Many are silicified or calcified, and it was suggested that they form as a result of upward photo-oriented movements of the principal filamentous species of *Phormidium* present. The upward orientation usually begins on small nodules or blisters in the mat which are often formed by O_2 bubbles during daytime. It was also suggested that the conical and columnar structures were heliotropic, pointing somewhat southward in the northern hemisphere (Awramik and Vanyo, 1986). The formation of columnar and conical cyanobacterial structures, as well as laminations in hot spring mats were reviewed by Brock (1978).

3. *Synechococcus*

Some thermophilic *Synechococcus* strains appear to exhibit a twitching, jerky or pivoting gliding motion (Castenholz, 1973a; Brock 1978). Unicells of cyanobacteria glide and show phototactic responses (e.g. *Synechococcus* spp., Stanier et al., 1971; baeocytes of some Pleurocapsales: Waterbury and Stanier, 1978). Recently, it was demonstrated quantitatively, by using time-lapse digital image analysis, that some Octopus Spring *Synechococcus* isolates were capable of slow movement (ca. 0.1 to 0.3 $\mu m\ s^{1}$) and showed positive phototaxis (Ramsing et al.,1997). The mechanism of movement in these non-flagellated unicells is unknown, though it requires contact between cells and a solid surface and/or other cells. Different *Synechococcus* strains exhibited different responses to light of different intensity. For instance, strain C1 was positively phototactic at 2000 lux, whereas strain C9 was positively phototactic at 250 lux, but negatively phototactic at 2000 lux. Native mat *Synechococcus* populations also exhibited light-directed motility. The speed of movement was sufficient to permit migration through the photic zone within a few hours. Shading experiments led to the observation that *Synechococcus* mats became dark green within hours of a reduction in light intensity (Madigan and Brock, 1976; Brock, 1978). This may reflect upward vertical migration of chlorophyll-rich, low-light adapted populations which normally reside deeper in mats.

V. Conclusion

The work described here shows the value of focusing investigations on a limited group of organisms that occur within a limited set of habitats. We also hope that it illustrates a point made by the organic geochemist Geoffrey Eglinton (1981) in his acceptance speech upon receipt of the Alfred Treibs Award, who found the relationship between the work of past and present students of the same systems;

akin to that of the masons who built the great cathedrals of Europe...[which] *took, in many instances, several hundred years to build. The successive master masons gave their whole lives to them, building on, adding to, modifying, and restoring the work of their predecessors.*

The reductionism provided by molecular and microelectrode analyses is beginning to give an increasingly detailed view of the biodiversity, ecology and evolution of thermophilic cyanobacteria. However, it is comforting that the results from such analyses are consistent with those of previous mentors, colleagues and their students who studied these organisms in these habitats with more traditional approaches. Future work on such topics, such as the role of adaptive radiation and geographic isolation as drivers of biodiversity in this group of microorganisms, will hopefully find that a similar solid base exists upon which embellishments can be added. The study of hot spring microbial communities serves as a model for microbial community ecology in general, and illustrates the value of combining traditional and contemporary approaches, rather than promoting new methods over old.

Acknowledgments

We thank Niels Ramsing, Michael Kuhl, Ferran Garcia-Pichel and Beverly Pierson for kindly sharing their unpublished figures, and Mary Bateson for help with figure construction. We thank the U.S. National Park Service for their support of, and assistance with, our work. The preparation of this chapter and many of its contents was funded by grants from the National Science Foundation (BSR-9209677 to DMW and IBN-9219273 and IBN-9630674 to RWC) and the National Aeronautics and Space Administration (NAGW-2764 to DMW).

References

Anderson KL, Tayne TA and Ward DM (1987) Formation and fate of fermentation products in hot spring cyanobacterial mats. Appl Environ Microbiol 53: 2343-2352

Awramik SM and Vanyo JP (1986) Heliotropism in modern stromatolites. Science 231: 1279-1281

Bateson MM and Ward DM (1988) Photoexcretion and consumption of glycolate in a hot spring cyanobacterial mat. Appl Environ Microbiol 54: 1738-1743.

Bauld J and Brock TD (1974) Algal excretion and bacterial assimilation in hot spring algal mats. J Phycol 10: 101-106

Bebout BM and Garcia-Pichel F (1995) UVB-induced vertical migration of cyanobacteria in a microbial mat. Appl Environ Microbiol 61: 4215-4222

Brock TD (1967) Microorganisms adapted to high temperatures. Nature 214: 882-885

Brock TD (1973) Lower pH limit for the existence of blue-green algae: evolutionary and ecological implications. Science 179: 480-483

Brock TD (1978) Thermophilic microorganisms and life at high temperatures. Springer-Verlag, Berlin.

Brock TD and Brock ML (1968) Measurement of steady-state growth rates of a thermophilic alga directly in nature. J Bacteriol 95: 811-815

Castenholz RW (1968) The behavior of *Oscillatoria terebriformis* in hot springs. J Phycol 4: 132-139

Castenholz RW (1969) Thermophilic blue-green algae and the thermal environment. Bacteriol Rev 33: 476-504

Castenholz RW (1973a) Ecology of blue-green algae in hot springs. In: Carr NG and Whitton BA (eds) The Biology of Blue-green Algae, pp 379-414. Blackwell Scientific Publications, Oxford

Castenholz RW (1973b) The possible use of sulfide by the filamentous phototrophic bacteria of hot springs. Limnol Oceanogr 18: 863-867

Castenholz RW (1976) The effect of sulfide on the bluegreen algae of hot springs. I. New Zealand and Iceland. J Phycol 12: 54-68

Castenholz RW (1977) The effect of sulfide on the blue-green algae of hot springs. II. Yellowstone National Park. Microbial Ecol 3: 79-105

Castenholz RW (1978) The biogeography of hot spring algae through enrichment cultures. Mitt Internat Verein Limnol 21: 296-315

Castenholz RW (1981) Isolation and cultivation of thermophilic cyanobacteria. In: Starr MP, Stolp H, Truper HG, Balows A and Schlegel HG (eds) The Prokaryotes: a handbook on habitats, isolation and identification of bacteria, pp 236-246. Springer-Verlag, Berlin

Castenholz RW (1984) Composition of hot spring microbial mats: a summary. In: Cohen Y, Castenholz RW and Halvorson H (eds) Microbial Mats: Stromatolites, pp 101-119. Alan R. Liss, NY

Castenholz, RW (1992) Species usage, concept, and evolution in the cyanobacteria (blue-green algae). J Phycol 28: 737-745

Castenholz, RW (1996) Endemism and biodiversity of thermophilic cyanobacteria. Nova Hedwigia Beih, 112: 33-47

Castenholz RW and Utkilen HC (1984) Physiology of sulfide tolerance in a thermophilic cyanobacterium. Arch Microbiol 138:306-309

Castenholz RW, Jørgensen BB, D'Amelio E and Bauld J (1991) Photosynthetic and behavioral versatility of the cyanobacterium *Oscillatoria boryana* in a sulfide-rich microbial mat. FEMS Microbial Ecol 86: 43-58

Cohen Y, Jørgensen BB, Revsbeck NP, and Poplawski R (1986) Adaptation to hydrogen sulfide of oxygenic and anoxygenic photosynthesis among cyanobacteria. Appl Environ Microbiol 51: 398-407

DesMarais DJ, Bauld J, Palmisano AC Summons RE and Ward DM (1992) The biogeochemistry of carbon in modern microbial mats. In: Schopf, JW and Klein C (eds) The Proterozoic Biosphere: A Multidisciplinary Study, pp 299-308. Cambridge University Press, Cambridge

Dobson G, Ward DM, Robinson NR and Eglinton G (1988) Biogeochemistry of hot spring environments: free lipids of a cyanobacterial mat. Chem Geol 68: 155-179

Doemel WN and Brock TD (1977) Structure, growth, and decomposition of laminated algal-bacterial mats in alkaline hot springs. Appl Environ Microbiol 34:433-452

Eglinton G (1981) Acceptance speech for the Alfred E. Treibs Award. Geochim Cosmochim Acta 46: 1139-1140

Ferris MJ, Muyzer G and Ward DM (1996a) Denaturing gradient gel electrophoresis profiles of 16S rRNA-defined populations inhabiting a hot spring microbial mat community. Appl Environ Microbiol 62: 340-346

Ferris MJ, Nold SC, Revsbech NP and Ward DM (1997) Population structure and physiological changes within a hot spring microbial mat community following disturbance. Appl Environ Microbiol 63: 1367-1374

Ferris MJ, Ruff-Roberts AL, Kopczynski ED, Bateson MM and Ward DM (1996b) Enrichment culture and microscopy conceal diverse thermophilic *Synechococcus* populations in a single hot spring microbial mat habitat. Appl Environ Microbiol 62: 1045-1050

Ferris MJ and Ward DM (1997) Seasonal distributions of 16S rRNA-defined populations in a hot spring microbial mat examined by denaturing gradient gel electrophoresis. Appl Environ Microbiol 63: 1375-1381

Fuqua C, Winans SC and Greenberg EP (1994) Quorum sensing in bacteria: the LuxR-LuxI family of cell density-responsive transcriptional regulators. J Bacteriol 176: 269-275

Garcia-Pichel F and Castenholz RW (1990) Comparative anoxygenic photosynthetic capacity in 7 strains of a thermophilic cyanobacterium. Arch Microbiol 153: 344-351

Garcia-Pichel F and Castenholz RW (1994) On the significance of solar ultraviolet radiation for the ecology of microbial mats. In: Stal L and Caumette P (eds) Microbial Mats, pp 77-84. NATO ASI Series, Vol G 35, Springer-Verlag, Berlin Heidelberg.

Garcia-Pichel F, Mechling M and Castenholz RW (1994) Diel migrations of microorganisms within a benthic, hypersaline mat community. Appl Environ Microbiol 60:1 500-1511

Giovannoni SJ, Revsbech NP, Ward DM, and Castenholz RW (1987) Obligately phototrophic *Chloroflexus*: primary production in anaerobic hot spring microbial mats. Arch Microbiol 147:80-89

Giovannoni SJ, Turner S, Olsen GJ, Lane DJ and Pace, NR (1988) Evolutionary relationships among cyanobacteria and green chloroplasts. J Bacteriol 170: 3584-3592.

Häder D-P (1984) Effects of UV-B on motility and photoorientation in the cyanobacterium, *Phormidium uncinatum*. Arch Microbiol 140:34-39

Jørgensen BB and Nelson DC (1988) Bacterial zonation, photosynthesis, and spectral light distribution in hot spring microbial mats of Iceland. Microb Ecol 16: 133-147

Jørgensen BB, Castenholz RW and Pierson BK (1992) The microenvironment within modern microbial mats. In: Schopf JW and Klein C (eds) The Proterozoic Biosphere: A Multidisciplinary Study, pp 271-278. Cambridge University Press, Cambridge

Kallas T and Castenholz RW (1982a) Internal pH and ATP-ADP pools in the cyanobacterium *Synechococcus* sp. during exposure to growth-inhibiting low pH. J Bacteriol 149: 229-236

Kallas T and Castenholz RW (1982b) Rapid transient growth at low pH in the cyanobacterium *Synechococcus* sp. J Bacteriol 149: 237-246

Konopka A (1992) Accumulation and utilization of polysaccharide by hot-spring phototrophs during a light-dark transition. FEMS Microb Ecol 102: 27-32.

Kopczynski ED, Bateson MM and Ward DM (1994) Recognition of chimeric small-subunit ribosomal RNA sequences composed of genes from uncultivated microorganisms. Appl Environ Microbiol 60: 746-748

Madigan MT and Brock TD (1976) Quantitative estimation of the bacteriochlorophyll *c* in the presence of chlorophyll *a* in aquatic environments. Limnol Oceanogr 21: 462-467

Madigan MT and Brock TD (1977) Adaptation of hot spring phototrophs to reduced light intensity. Arch Microbiol 113: 111-120

Mariner RH, Rapp JB, Willey LM, and Presser TM (1974) The chemical composition and estimated minimum thermal reservoir temperatures of selected hot springs in Oregon, 27 pp, U.S. Geol. Survey, Open File Report, March 1994, Menlo Park, California.

Mayr E. (1982) The growth of biological thought: diversity, evolution, and inheritance. Harvard University Press, Cambridge, Mass.

Meeks JC and Castenholz RW (1971) Growth and photosynthesis in an extreme thermophile, *Synechococcus lividus* (Cyanophyta). Arch Mikrobiol 78: 25-41

Miller SR, Wingard CE and Castenholz RW (1998) Effects of visible light and UV radiation on photosynthesis in a population of a hot spring cyanobacterium, a *Synechococcus* sp., subjected to high temperature stress. Appl Environ Microbiol 64: 3893-3899

Muyzer G, De Waal EC and Uitterlinden AG (1993) Profiling of complex microbial populations by denaturing gradient gel electrophoresis analysis of polymerase chain reaction-amplified genes coding for 16S rRNA. Appl Environ Microbiol 59: 695-700.

Nold SC and Ward DM (1996) Photosynthate partitioning and fermentation in hot spring microbial mat communities. Appl Environ Microbiol 62: 4598-4607

Peary JA and Castenholz RW (1964) Temperature strains of a thermophilic blue-green alga. Nature (London) 202: 720-721

Pierson BK, Sands VM and Frederick JL (1990) Spectral irradiance and distribution of pigments in a highly layered marine microbial mat. Appl Environ Microbiol 56: 2327-2340

Rainey FA, Ward N, Sly LI and Stackebrandt E (1994) Dependence of the taxonomic composition of clone libraries for PCR-amplified naturally occurring 16S rDNA on the primer pair and the cloning system used. Experientia 50: 789-801

Ramsing NB, Ferris MJ and Ward DM (1997) Light-induced motility of thermophilic *Synechococcus* isolates from Octopus Spring, Yellowstone National Park Appl Environ Microbiol 63:2347-2354

Revsbech NP and Ward DM (1984a) Microelectrode studies of interstitial water chemistry and photosynthetic activity in a hot spring microbial mat. Appl Environ Microbiol 48: 270-275

Revsbech NP and Ward DM (l984b) Microprofiles of dissolved substances and photosynthesis in microbial mats measured with microelectrodes. In: Cohen Y, Castenholz RW and Halvorson HO (eds) Microbial Mats: Stromatolites, pp 174—188. A.R. Liss, Inc, NY

Reysenbach AL, Giver LJ, Wiskham GS and Pace NR (1992) Differential amplification of rRNA genes by polymerase chain reaction. Appl Environ Microbiol 58: 3417-3418

Richardson LL and Castenholz RW (1987a) Enhanced survival of the cyanobacterium *Oscillatoria terebriformis* in darkness under anaerobic conditions. Appl Environ Microbiol 53:2 151-2158

Richardson L and Castenholz RW (1987b) Diel vertical movements of the cyanobacterium *Oscillatoria terebriformis* in a sulfide-rich hot spring microbial mat. Appl Environ Microbiol 53: 2151-2158

Robison-Cox J, Bateson MM and Ward DM (1995) Evaluation of nearest-neighbor methods of detection of chimeric small-subunit rRNA sequences. Appl Environ Microbiol 61: 1240-1245

Ruby EG and Asato LM (1993) Growth and flagellation of *Vibrio fischeri* during initiation of the spiolid squid light organ symbiosis. Arch Microbiol 159: 160-167

Ruff-Roberts AL, Kuenen JG and Ward DM (1994) Distribution of cultivated and uncultivated cyanobacteria and *Chloroflexus*-like bacteria in hot spring microbial mats. Appl Environ Microbiol 60: 697-704

Sandbeck KA and Ward DM (1981) Fate of immediate methane precursors in low sulfate hot spring algal-bacterial mats. Appl Environ Microbiol 41: 775-782

Sheridan RP (1976) Sun-shade ecotypes of bluegreen algae in a hot spring. J Phycol 12:279-285

Sheridan RP (1979) Seasonal variation in sun-shade ecotypes of *Plectonema notatum* (Cyanophyta). J Phycol 15:223-226

Shiea J, Brassell S and Ward DM (1990) Mid-chain branched mono- and dimethylalkanes in hot spring cyanobacterial mats: a direct biogenic source for branched alkanes in ancient sediments. Org Geochem 15: 223-231

Shiea J, Brassell S and Ward DM (1991) Comparative analysis of free lipids in hot spring cyanobacterial and anoxygenic photosynthetic bacterial mats. Org Geochem 17: 309-319

Stanier RY, Kunisawa R, Mandel M and Cohen-Bazire G (1971) Purification and properties of unicellular blue-green algae (Order Chroococcales). Bacteriol Rev 35: 171-205

Stewart WDP (1970) Nitrogen fixation by blue-green algae in Yellowstone thermal areas. Phycologia 9: 261-268

Suzuki MT and Giovannoni SJ (1996) Bias caused by template annealing in the amplification of mixtures of 16S rRNA genes by PCR. Appl Environ Microbiol 62: 625-630

Teiser MLO (1993) Extracellular low molecular weight organic compounds produced by *Synechococcus* sp. and their roles in the food web of alkaline hot spring microbial mat communities. PhD thesis, University of Oregon, Eugene, 197 pp

Vanyo JP, Hutchinson RA and Awramik SM (1986) Helotropism in microbial stromatolitic growths at Yellowstone National Park: geophysical inferences. EOS 67: 153-156

Walter MR, Bauld J and Brock TD (1972) Siliceous algal and bacterial stromatolites in hot spring and geyser effluents of Yellowstone National Park. Science 178: 402-405

Ward DM, Beck E, Revsbech NP, Sandbeck KA and Winfrey MR (l984) Decomposition of hot spring microbial mats. In: Cohen Y, Castenholz RW and Halvorson, HO (eds) Microbial mats: Stromatolites, pp 191—214. A.R. Liss, Inc, NY

Ward DM and Olson JG (1980) Terminal processes in the anaerobic degradation of an algal-bacterial mat in a high sulfate hot spring. Appl Environ Microbiol 40: 67-74

Ward DM, Shiea J, Zeng YB, Dobson G, Brassell S and Eglinton, G. (1989b) Lipid biochemical markers and the composition of microbial mats. In: Cohen Y and Rosenburg E (eds.) Microbial Mats: Physiological Ecology of Benthic Microbial Communities, pp 439—454. Am Soc Microbiol, Wash DC

Ward DM, Tayne TA, Anderson KL and Bateson MM (1987) Community structure, and interactions among community members in hot spring cyanobacterial mats. Symp Soc Gen Microbiol 41: 179-210

Ward DM, Weller R, Shiea J, Castenholz RW and Cohen Y (1989a) Hot spring microbial mats: anoxygenic and oxygenic mats of possible evolutionary significance. In: Cohen Y and Rosenburg E (eds) Microbial Mats: Physiological Ecology of Benthic Microbial Communities, pp 3-15. Am Soc Microbiol, Washington DC

Ward DM, Bateson MM, Weller R and Ruff-Roberts AL (1992a) Ribosomal RNA analysis of microorganisms as they occur in nature. Adv Microbial Ecol 12: 219-286

Ward DM, Bauld J, Castenholz RW and Pierson BK (1992b) Modern phototrophic microbial mats: anoxygenic, intermittently oxygenic/anoxygenic, thermal, eucaryotic and terrestrial. In: Schopf JW and Klein C (eds) The Proterozoic Biosphere: A Multidisciplinary Study, pp 309-324. Cambridge University Press, Cambridge

Ward DW, Ferris MJ and Bateson MM (in press) Organization of native populations within hot spring microbial mat communities: need for a more ecological approach. In: ISME Symposium Proceedings, Santos, Brazil

Ward DM, Ferris MJ, Nold SC, Bateson MM, Kopczynski ED and Ruff-Roberts AL (1994a) Species diversity in hot spring microbial mats as revealed by both molecular and enrichment culture approaches-relationship between biodiversity and community structure. In: Stal LJ and Caumette P (eds) Microbial mats: Structure, Development and Environmental Significance, Series G: Ecological Sciences, vol 35, NATO/ASI Series, pp 33—44. Springer-Verlag, Heidelburg

Ward DM, Panke S, Kloeppel KD, Christ R and Fredrickson H (1994b) Complex polar lipids of a hot spring cyanobacterial mat and its cultivated inhabitants. Appl Environ Microbiol 60: 3358-3367

Ward DM, Santegoeds CM, Nold SC, Ramsing NB, Ferris MJ and Bateson MM (1997) Biodiversity within hot spring microbial mat communities: molecular monitoring of enrichment cultures. Antonie van Leeuwenhoek 71: 143-150

Ward DM, Weller R and Bateson MM (1990) 16S rRNA sequences reveal numerous uncultured microorganisms in a natural community. Nature 345: 63-65

Waterbury JB and Ripkka R (1989) Subsection I. Order Chroococcales Wettstein 1924, Emend. Rippka et al., 1979. In: Staley JT, Bryant MP, Pfennig N, and Holt JG (eds) Bergey's Manual of Systematic Bacteriology, Vol 3. pp 1728-1746. Williams and Wilkins, Baltimore

Waterbury JB and Stanier RY (1978) Patterns of growth and development in Pleurocapsalean cyanobacteria. Microbiol Rev 42: 2-44

Weller R, Bateson MM, Heimbuch BK, Kopczynski ED and Ward DM (1992) Uncultivated cyanobacteria, *Chloroflexus*-like inhabitants, and spirochete-like inhabitants of a hot spring microbial mat Appl Environ Microbiol 58: 3964-3969

Weller R and Ward DM (1989) Selective recovery of 16S ribosomal RNA sequences from natural microbial communities in the form of cDNA. Appl Environ Microbiol 55: 1818-1822

Weller R, Weller FW and Ward DM (1991) 16S rRNA sequences retrieved as randomly primed cDNA from a hot spring cyanobacterial mat community. Appl Environ Microbiol 57: 1146-1151

Werner D (1992) Symbiosis of Plants and Microbes. Chapman and Hall, New York

Wickstrom CE (1980) Distribution and physiological determinants of blue-green algal nitrogen fixation along a thermogradient. J Phycol 16: 436-443

Wickstrom CE and Castenholz RW (1973) Thermophilic ostracod: aquatic metazoan with highest known temperature tolerance. Science 181: 1063-1064

Wickstrom CE and Castenholz RW (1978) Association of *Pleurocapsa* and *Calothrix* (Cyanophyta) in a thermal stream. J Phycol 14:84-88

Wickstrom CE and Castenholz RW (1985) Dynamics of cyanobacterial and ostracod interactions in an Oregon hot spring. Ecology 66:1024-1041

Wilmotte A (1994) Molecular evolution and taxonomy of the cyanobacteria. In: Bryant DA (ed) The Molecular Biology of Cyanobacteria, pp 1—25. Kluwer Academic Publishers, The Netherlands

Woese CR (1987) Bacterial evolution. Microbiol Rev 51: 211—271

Zeng YB, Ward DM, Brassell S and Eglinton G (1992a) Biogeochemistry of hot spring environments. 2. Lipid compositions of Yellowstone (Wyoming, U.S.A.) cyanobacterial and *Chloroflexus* mats. Chem Geol 95:327-345

Zeng YB, Ward DM, Brassell S and Eglinton G (1992b) Biogeochemistry of hot spring environments. 3. Apolar and polar lipids in the biologically active layers of a cyanobacterial mat. Chem Geol 95: 347-360

Chapter 4

Cyanobacterial Mats and Stromatolites

Lucas J. Stal
Netherlands Institute of Ecology, Centre of Estuarine and Coastal Ecology, PO Box 140, NL-4400 AC Yerseke, The Netherlands

B. A. Whitton and M. Potts (eds), The Ecology of Cyanobacteria, 61-120.
© 2000 *Kluwer Academic Publishers. Printed in the Netherlands.*

Summary

Cyanobacteria are often the key organisms in microbial mats. They form dense micrometer-scale communities in which the full plethora of microbial metabolism can be present. Such mats are therefore excellent model systems and because of their analogy with Precambrian stromatolites they are also attractive for evolutionary studies. Growth and metabolism of the phototrophic cyanobacteria enrich the sediment with organic matter. However, in mature mats net growth of cyanobaceria appears to be of minor importance. Most of the organic matter produced from photosynthetic CO_2 fixation is liberated in the sediment by one of the following: fermentation, photorespiration, pouring out of solutes or secretion of mucus. This organic matter is degraded by chemotrophic microorganisms, among which sulfate-reducing bacteria are particularly important. The combined activities of the cyanobacteria and sulfate-reducing bacteria result in steep and fluctuating gradients of sulfide and oxygen. Cyanobacteria may thus have to cope with high concentrations of sulfide and oxygen but also with anoxic conditions. The physicochemical gradients force different functional groups of microorganisms to particular vertical stratified positions in the mat. This, and the fact that accretion of sediment fluctuates, gives rise to one of the most conspicuous properties of microbial mats, namely their laminated structure. Modern microbial mats have this laminated structure in common with Precambrian stromatolites. However, the great majority of modern mats do not lithify. Only a few examples of recent calcifying stromatolithic microbial mats are known. A hypothesis has been developed which conceives a role for extracellular polysaccharides in calcification. These molecules may serve as inhibitors of calcification. Extracellular polysaccharides in cyanobacterial mats are probably produced as a result of unbalanced growth caused by nitrogen limitation. Although nitrogen fixation is an important process in many marine microbial mats, in only a few cases the specialized nitrogen-fixing heterocystous cyanobacteria are present. Non-heterocystous diazotrophic cyanobacteria are less efficient in nitrogen fixation. The possible reasons for the exclusion of heterocystous species in many mats are discussed.

I. Introduction

Microbial mats are generally formed by filamentous, entangled organisms that produce a macroscopically 'mat-like' structure. In some cases such mats can indeed be peeled off from the sediment as a large, coherent piece. However, benthic microbial communities of unicellular organisms, that usually do not form such coherent structures, are also termed microbial mats. Microbial mats come in a great variety and they may include mats of diatoms and other biofilms of immobilized microorganisms, including dental plaque (Rosenberg, 1989).

More generally the term microbial mat is used for multilayered microbial communities growing on sediments in a variety of different environments such as tidal sand flats, hypersaline ponds, hot springs and others. Although this review focuses on mats formed by cyanobacteria, a range of other types of organisms may form mats, including diatoms and green algae. Diatoms may also often be an important component of cyanobacterial mats (Bauld, 1984). Purple and sometimes green sulfur bacteria are normal components of most cyanobacterial mats (Nicholson et al., 1987; Pierson et al., 1987).

The reason why cyanobacteria are the most successful mat-building organisms may be found in the combination of a number of important characteristics of this unique group. Cyanobacteria are the only oxygenic phototrophic prokaryotes. Eukaryotic organisms are few or excluded altogether from many environments with extreme conditions. One reason for this is the wide spectrum of metabolic capabilities of prokaryotes and the great capacity they display to adapt to changes and fluctuations in environmental conditions. As their predominant metabolism is oxygenic photosynthesis, cyanobacteria use light as an energy source, water as an electron donor and CO_2 as a carbon source. These are the major requirements for growth and they are abundant in the environments where most microbial mats are found. Another important property of many cyanobacteria, which is not shared by eukaryotic algae, is their capacity for using dinitrogen as a nitrogen source. Thus, such organisms grow independently of a source of combined nitrogen. Photosynthesis in cyanobacteria saturates at low light intensity; they have high affinity for light and maintenance requirements are extremely low (Van Liere and Mur, 1979). These properties allow photosynthesis even under unfavorable light conditions. Moreover, in addition to oxygenic photosynthesis, several species are capable of sulfide-dependent anoxygenic photosynthesis (Garlick et al., 1977). Mat-forming cyanobacteria are well adapted for life under anoxic conditions. In addition to the normal aerobic dark respiration, virtually all species of cyanobacteria in microbial mats are capable of fermentation (Stal and Moezelaar, 1997). These properties of cyanobacteria are essential for life in microbial mats in which environmental conditions strongly fluctuate.

A typical property of microbial mats is their laminated structure in which different functional groups of microorganisms occur in vertically stratified layers (Stal et al., 1985). In addition to this biological stratification a biomineralogical stratification can be distinguished (Monty, 1976). This type of lamination can be attributed to different growth periods, seasonal events, periodical events (e.g. tides) or episodic or erratic events (e.g. storms). Often this laminated pattern is restricted, since most of the organic matter of the mat is degraded. Microbial mats, when conditions allow, precipitate minerals, mainly calcite (Golubic, 1973; Monty, 1976; Krumbein, 1979a). The precipitation of these minerals is strongly associated with microbial metabolism. Therefore this precipitation of minerals also may give rise to the formations of distinct laminae, which eventually may give rise to consolidated rock. Laminated rocks dating from the Precambrian and later eras are known as stromatolites (Krumbein, 1983). Present day microbial mats built by cyanobacteria show remarkable similarities with fossil stromatolites. Stromatolites are known dating back to 3.5 billion years (Mason and Von Brunn, 1977; Lowe, 1980; Walter et al, 1980; Orpen and Wilson, 1981; Chapter 2). In some of these fossil formations well-preserved microfossils occur with, in some cases, a remarkable resemblance to present-day cyanobacteria (Schopf and Walter, 1982; Awramik, 1984; Chapter 2). It is attractive to consider modern microbial mats as analogues of Precambrian stromatolites; however, structural differences do not always seem to justify the comparison. A major problem is the fact that the great majority of present day microbial mats does not form consolidated rock.

Studies on microbial mats have been published in a number of books (Cohen et al., 1984; Cohen and Rosenberg, 1989; Stal and Caumette, 1994). Detailed information on stromatolites can be found in books edited by Walter (1976), Schopf (1983) and Schopf and Klein (1992). A general overview of benthic cyanobacteria in marine littoral environments can be found in the book chapter by Whitton and Potts (1982). This review will discuss the metabolic activities of cyanobacteria that allow them to form microbial mats and stromatolites.

II. Microbial Mats, Stromatolites and their Environments

A. What are Microbial Mats and Stromatolites ? Some Definitions

Krumbein (1983), referring to the work of Kalkowsky (1908), proposed the following definition of a stromatolite: *"Stromatolites are laminated rocks, the origin of which can clearly be related to the activity of microbial communities, which by their morphology, physiology, and arrangement in space and time interact with the physical and chemical environment to produce a laminated pattern which is retained in the final rock structure"*. This definition includes fossil as well as recent formations. Modern stromatolites that fit the definition of Krumbein are rare. Awramik and Margulis (in Walter, 1976) defined stromatolites as: *"Organosedimentary structures produced by sediment trapping, binding and/or precipitation as a result of the growth and metabolic activity of microorganisms, principally cyanophytes"*. This definition includes fossil and recent consolidated stromatolites as well as unconsolidated microbial mats. Both definitions, however, emphasize the role of microbial mats and the activity of the organisms therein for the formation of stromatolites. Walter (1976) formulated the following conditions necessary to form a microbial mat:

(i) the environmental conditions must allow growth of the mat-building microorganisms

(ii) the growth rate of the mat-building organisms must be faster than consumption by grazing organisms

(iii) sedimentation rates should not be exceedingly high to allow stabilized colonization of the surface by the mat-building organisms

(iv) destructive forces such as burrowing organisms and mechanical and chemical erosion must be absent or at least not prevent accretion of organisms.

In order to produce a stromatolite, preservation of the structure must occur. In many environments today unconsolidated microbial mats are formed, i.e. systems that do not have the potential to preserve its structure, defined by Krumbein (1983) as: *"Unconsolidated laminated systems, clearly related to the activity of microbial communities, often called recent stromatolites or living stromatolites are defined as potential stromatolites"*. Indeed, stromatolites *sensu* Krumbein are still formed today. Excellent examples of consolidated, well-laminated stromatolites that still grow through the growth and metabolic activity of a microbial mat are found in the Exuma Cays, Bahamas (Reid and Browne, 1991; Pinckney et al., 1995) (Plate 9).

Since their discovery in 1961 in Shark Bay, Western Australia, poorly laminated, consolidated calcareous stromatolites have been presented as strikingly similar to Precambrian stromatolites (Logan, 1961). Recent calcareous stromatolites are also formed in Polynesian atolls (Kopara) (Défarge et al., 1994a, b) and other lacustrine and perimarine settings (Kempe et al., 1991; Kempe and Kazmierczak, 1993). Such stromatolites can be called 'recent stromatolites' to distinguish them from fossil formations. It is not necessary to name unconsolidated microbial mats 'potential stromatolites' since they are not stromatolites (*sensu* Krumbein) and most of these microbial mats will never become one. The term 'potential' suggests that these unconsolidated microbial mats keep in themselves the capacity for consolidation. For the majority of microbial mats this has neither been proven nor is it likely. Calcification is the general process for consolidation and preservation of microbial mats. In many microbial mats calcification does not occur and the reasons for this are discussed later on in this chapter. Many consolidated rocks without a clear lamination are formed by microbial communities. It may be that laminations were lost during the process of diagenesis or that neither vertical, stratified communities of microorganisms nor clear, seasonal variations were involved in growth and metabolic activity. Another mechanism may be that carbonate sand accretes through trapping and binding in the microbial mat without *in situ* calcification. Such clotted and poorly laminated stromatolites are also known as thrombolites (Kennard and James, 1986). The stromatolites of the Exuma Cays, Bahamas, in contrast with other recent formations possess a fine laminated structure. In addition to trapping and binding of carbonate sand, *in situ* precipitation of calcium carbonate produces distinct layers of cement (Reid and Browne, 1991).

B. Microbial Mats and Stromatolites: the Geological Evidence (see also Chapter 2)

During the Hadean era from the origin of the Earth 4.5×10^9 years before present to 3.9×10^9 there is no rock record. The oldest rock known from the early Archean is probably not of biogenic origin. The oldest stromatolites are known from the Archean about 3.5×10^9 years ago. Only a few examples are known from this era, but they were undoubtedly biologically produced. Microfossils have been found in these stromatolites, but it is premature to identify them as cyanobacteria. From measurements of carbon isotope compositions in these rocks it has been deduced that autotrophic microorganisms must have been active at that time. However, it could have been chemoautotrophs that fixed the CO_2, rather than photoautotrophs. The morphology of the microfossils from these oldest stromatolites also does not give an unequivocal clue about the identity of the organisms. Cyanobacteria are a group of oxygenic phototrophic organisms and it is well established that the Archean atmosphere did not contain oxygen.

During the Proterozoic, which started about 2.5×10^9 years ago, stromatolites became abundant (Fig. 1) and occur in a wide variety of facies. They occupied every major ecological niche, marine and lacustrine, shallow and deep water. Most of limestones, dolomites and magnesites as well as many phosphorites and iron formations of the Proterozoic contain stromatolites. Like modern microbial mats it seems certain that the Proterozoic stromatolites were produced by growth and metabolic activity of cyanobacteria. The Proterozoic stromatolites contain a wealth of very well preserved microfossils that show an astonishingly good similarity to present day cyanobacteria. Over 1100 microfossils have been described from 190 stromatolite formations (Walter et al., 1992). Many of these fossils are preserved in the cherty parts of stromatolites. The best preservation occurred following early silicification of the stromatolites. Because of the absence of diatoms during this era it is thought that seawater might have contained much higher concentrations of silica.

Oxygen was present in the atmosphere at 2.3×10^9 years before present. It is now well accepted that the oxygenation of the atmosphere was the result of oxygenic photosynthesis. It might have taken

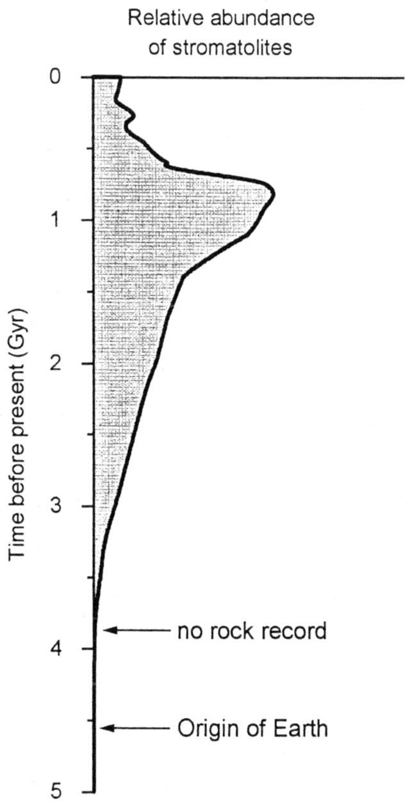

Fig. 1. The relative abundance of stromatolites plotted against time (after Awramik, 1984).

considerable time after the origin of oxygenic photosynthesis until the atmosphere became oxygenated, since a large amount of reduced compounds had to be oxidized. Banded iron formations (BIFs) are known from 2.5×10^9 years before present. These are huge formations consisting of oxidized iron and they have been taken as evidence for the presence of oxygenic photosynthesis. However, iron oxidation could also have taken place by the activity of anoxygenic phototrophic bacteria under anaerobic conditions (Widdel et al., 1993; Ehrenreich and Widdel, 1994) or perhaps even through iron-dependent anoxygenic photosynthesis by cyanobacteria (Cohen, 1989). It seems evident that oxygenic photosynthesis must have evolved in the beginning of the Proterozoic. However, many cyanobacteria in present day microbial mats are capable of anoxygenic photosynthesis and it seems likely that cyanobacteria were anoxygenic phototrophs before they evolved oxygenic photosynthesis (Pierson and Olson, 1989).

The morphology of stromatolites of the 3.1×10^9 old Insuzi group of South Africa hint at the involvement of filamentous organisms that might have had a phototactic response (Mason and Van Brunn, 1977). However, a chemotactic response could also explain the structure of this formation (Schopf and Walter, 1982).

Of the many different morphological forms of microfossils, many can be traced back from present day cyanobacteria such as *Oscillatoria* and *Lyngbya* (Schopf and Walter, 1982). These organisms are common in modern microbial mats, particularly in systems that fix nitrogen. It is of course not possible to determine whether these ancient mats were diazotrophic. However, microfossils resembling cyanobacteria of the heterocystous genera *Nostoc* and *Scytonema* were abundant as well (Schopf and Walter, 1982), and this could be taken as evidence that nitrogen fixation might indeed have been important. However, it is remakarble that heterocystous cyanobacteria are not often found in modern microbial mats.

Proterozoic stromatolites reached a maximum in number and diversity towards the end of this era after which they showed a rapid decline (Walter and Heys, 1985) (Fig. 1). It has been postulated that metazoa that then appeared on Earth were responsible for this decline (Walter and Heys, 1985). The grazing activity of these animals would prevent the accumulation of the benthic mat-forming organisms and destroy the fabric of microbial mats by bioturbation. After the appearance of metazoa microbial mats would be much more limited in their distribution and develop in environments in which metazoa are largely excluded (so-called extreme environments). However, other explanations have been offered such as the appearance of eukaryotic algae that in many environments would compete successfully for light and nutrients. Benthic eukaryotic microalgae such as diatoms or green algae often produce less coherent mats.

Gebelein (1976) proposed that sea level changes, caused by changes of climate in addition to tectonic processes, may have been involved in the sudden decrease in stromatolite abundance. But another possibility is that seawater in the Proterozoic was greatly oversaturated with calcium carbonate (alkaline soda ocean) (Kempe and Kazmierczak, 1990a).

Proterozoic stromatolites probably formed through one or more of the following (Walter et al, 1992):

i) *in situ* precipitation as cement;

ii) *in situ* precipitation as micrite either accreted passively from suspension or through trapping and binding of the grains by the mat microorganisms;
iii) precipitation of micrite imported from adjacent environments.

The fine and distinct lamination of Proterozoic stromatolites hints at *in situ* precipitation. Most Phanerozoic stromatolites show poor or no lamination. These formations are probably produced by trapping and binding of carbonate and sand grains (Cloud and Semikhatov, 1969).

C. Stratification and Structure of Microbial Mats and Stromatolites

Microbial mats are characterized by the vertical stratification of different functional groups of microorganisms. This structure is the result of the physicochemical gradients that are present in mats and in fact produced by the metabolic activity of the mat organisms themselves (Jørgensen et al., 1983). The typical structure of a microbial mat build by cyanobacteria is schematically depicted in Fig. 2.

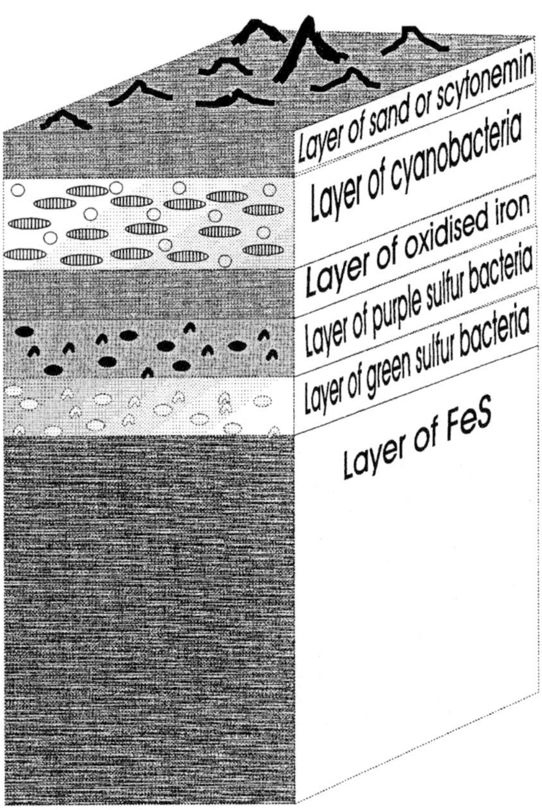

Fig. 2. Schematic drawing of a building of a typical microbial mat formed by cyanobacteria. The layer of green sulfur bacteria has been observed on only a few occasions.

Cyanobacteria evidently form the top layer of microbial mats. They need to harvest light for photosynthesis and are essentially aerobic organisms. However, the layer of cyanobacteria is sometimes overlain by a film of diatoms. The cyanobacteria may further be covered by a layer of sand or sediment of varying thickness or be covered by a organic-rich mucilaginous layer which may contain pigments such as scytonemin. Scytonemin is produced by cyanobacteria in mats that are exposed to high light intensities. It occurs predominantly in the extracellular polysaccharide sheaths. Scytonemin is highly recalcitrant remaining in the empty sheaths that are left behind by the cyanobacteria. Scytonemin protects the underlying community from damage caused by UV irradiation (sunglass effect) (Garcia-Pichel and Castenholz, 1991). The organic matter introduced in the sediment through the photosynthetic activity of the cyanobacteria is decomposed by a variety of chemotrophic microorganisms. The degradation of organic matter and the accompanying demand for oxygen result in permanent anoxic conditions below the layer of cyanobacteria (Fig. 3). Obligately anaerobic sulfate-reducing bacteria play a major role in the decomposition of organic material in marine cyanobacterial mats and other sulfur dominated environments. These bacteria produce sulfide which is used by anoxygenic phototrophic bacteria.

Purple sulfur bacteria are very common in microbial mats and are often seen as a pink layer below the cyanobacteria. Purple sulfur bacteria are essentially anaerobic bacteria, but species that occur in microbial mats are usually metabolically versatile (van Gemerden, 1993). Anoxygenic photosynthesis in purple sulfur bacteria saturates at even lower light intensities than photosynthesis in cyanobacteria. In addition, these organisms use a different part of the electromagnetic spectrum, one not used by cyanobacteria (Pierson et al., 1987). This far red light is also least attenuated by the sediment (Fig. 4) (Stal et al., 1985; Jørgensen and Des Marais, 1988). This biological stratification is thus the result of gradients of light, oxygen and sulfide and is found in virtually all cyanobacterial mats (Fig. 3). In some rare cases a layer of green anoxygenic phototrophic bacteria can be found underneath the purple bacteria (Nicholson et al., 1987).

A distinct layer of oxidized iron may be present between the cyanobacteria and the purple sulfur bacteria (Stal, 1994). It is not clear how this layer is formed. It may be formed by chemical oxidation by

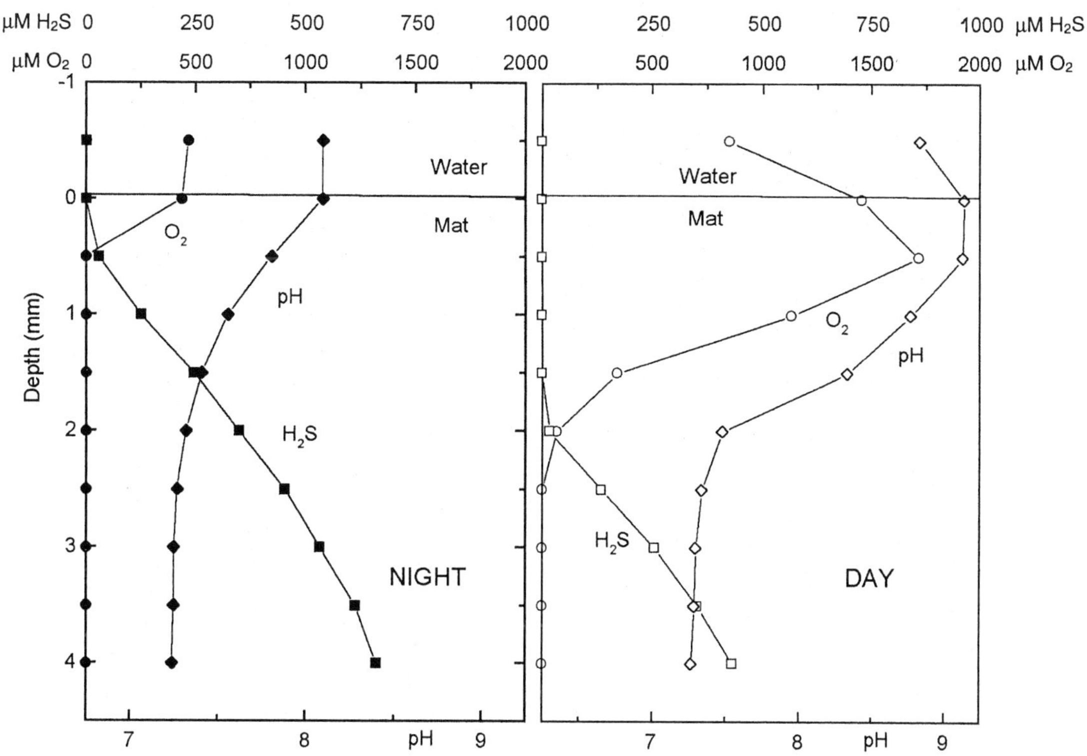

Fig. 3. Vertical profiles of oxygen, sulfide and pH at night (left panel) and during the day (right panel) in a mat of *Microcoleus chthonoplastes* from Solar Lake, Sinai. (Redrawn from Revsbech et al., 1983).

the oxygen produced during photosynthesis. An alternative explanation is the anaerobic oxidation of iron by anoxygenic photosynthesis in a newly discovered group of purple bacteria (Widdel et al., 1993; Ehrenreich and Widdel, 1994). Aerobic oxidation of iron by chemotrophic bacteria seems unlikely at the alkaline pH levels that are usually encountered in microbial mats (Fig. 3) although it should also not be excluded as a possibility as shown by Emerson and Revsbech (1994a, b).

The deeper layer of the mat is often black or gray, indicating the presence of iron sulfide (FeS) or pyrite (FeS_2). This layer has often been referred to as the layer of the sulfate-reducing bacteria. However, it has become clear that these bacteria in fact do not form a distinct layer and occur throughout the sediment (Visscher et al., 1992; Stal, 1993). They are both abundant and highly active in the top layers of microbial mats. At first sight this distribution of sulfate-reducing bacteria is unexpected and odd. However, their substrates are mainly produced by the cyanobacteria and it is certainly beneficial to the

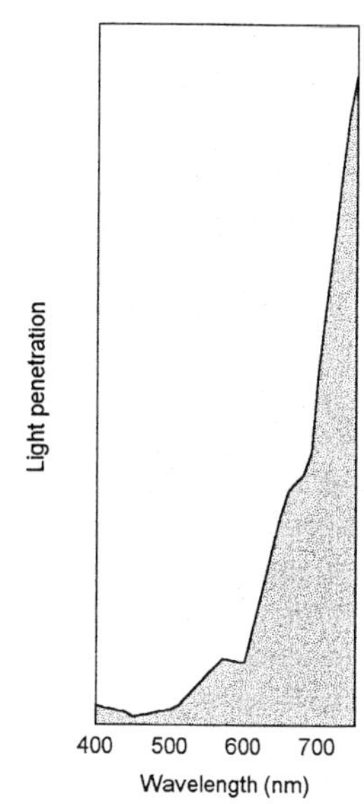

Fig. 4. Spectral light penetration through a 1.5 mm mat of *Microcoleus chthonoplastes* of the North Sea island of Mellum (Germany). Far-red light is least attenuated by the mat. The effects of light absorption by chlorophyll *a* and phycobiliproteins at respectively 680 and 600 nm are clearly present.

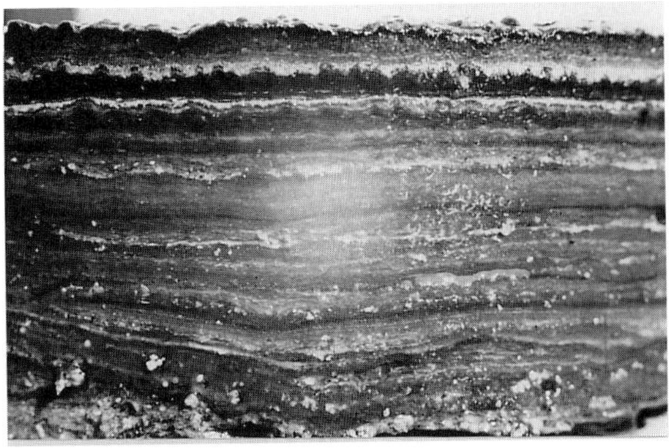

Fig. 5. The mat of *Microcoleus chthonoplastes* of hypersaline pond 5 of the saltern of Exportadora de Sal, S.A., Guerrero Negro, Baja California Sur, Mexico, showing the typical multilaminated structure.

bacteria to be close to the site of production. At night when photosynthesis ceases the cyanobacterial mat turns anoxic (Fig. 3), providing the appropriate environment for sulfate-reducing bacteria. Sulfate reducing bacteria appear to be quite tolerant to oxygen and some are even capable of aerobic respiration (Dilling and Cypionka, 1990; Marschall et al., 1993). It has also been shown that sulfate reduction in a microbial mat can occur in the presence of oxygen (Canfield and Des Marais, 1991; Fründ and Cohen, 1992).

Chemotrophic bacteria can also oxidize sulfide in the mat. This group of organisms has been shown to be important in microbial mats. Numbers of colorless sulfur bacteria as high as 2×10^9 cm^{-3} sediment have been detected in the top layer of microbial mats (Visscher et al., 1992). These bacteria may therefore be quantitatively most important in microbial mats. Colorless sulfur bacteria are essentially aerobic and gain energy from the aerobic oxidation of sulfide. They are autotrophic organisms and fix CO_2 through the reductive pentose phosphate pathway (Calvin cycle). Colorless sulfur bacteria have high affinities for their substrates and their presence cause highly dynamic oxygen and sulfide gradients, thereby overruling the chemical oxidation of sulfide. Since the sulfide-oxygen interface is highly dynamic and not fixed at a certain depth in the sediment (Fig. 3), these bacteria also do not form a distinct layer, but they are clearly most abundant in the top layer (Visscher et al., 1992).

The biological stratification in microbial mats may however be far more complex than described above. Cyanobacteria may be sandwiched between layers of anoxygenic phototrophic bacteria and even perform oxygenic photosynthesis there. Microbial mats may be also just 'inverted', with cyanobacteria occurring underneath the layer of anoxygenic phototrophic bacteria (Van Gemerden et al., 1989). This type of microbial mat can develop when much organic matter is deposited on the sediment and its degradation results in very high rates of sulfide production.

The joint metabolic activity of microorganisms in microbial mats results in steep physicochemical gradients, e.g. of light, oxygen, sulfide, carbon dioxide and pH; these gradients shift markedly during a 24-h cycle (Fig. 3) (Jørgensen et al., 1979; Revsbech et al., 1983). However, these gradients also respond immediately to fluctuations of incident light. All microorganisms in microbial mats must therefore be highly versatile and flexible to cope with these extreme changes in environmental conditions.

The organic matter produced by photosynthesis seems to be recycled immediately in many microbial mats. In these mats net growth is often virtually absent (Nold and Ward, 1996). However, when growth occurs seasonally, a new mat may grow on top of the old one. This could result in a 'historical' lamination (Monty, 1976). In many coastal environments the organic matter is fully degraded and a historical lamination is absent. However, in other mats, particularly those growing in hypersaline environments, degradation may be incomplete. Examples of such mats are the well-investigated hypersaline mats of 'pond 5' in Guerrero Negro, Baja California, Mexico and those of Solar Lake, Sinai, Egypt (D'Antoni D'Amelio et al., 1989). The Guerrero Negro pond 5 mats are about 5-6 cm thick and are well-laminated (Fig. 5). This mat grows at a rate of about 1 cm per year. However, net accretion of the mat is virtually zero so that it its thickness remains about the same. This means that decomposition of the mat also must proceed at a rate of approximately 1 cm per year. This mat therefore seems to be in steady state (Canfield and Des Marais, 1994). The mineralization of the mat of Solar Lake is incomplete, although up to 99% of the primary production is immediately recycled in the mat, leaving only 1% for net accretion (Jørgensen and Cohen, 1977; Krumbein et al., 1977). The Solar Lake microbial mat is about 1m thick and the historical lamination goes back almost 2000 years.

D. Environments in which Cyanobacterial Mats occur

1. Coastal Microbial Mats

Microbial mats, including those produced by cyanobacteria, usually form in environments in which grazing activities by metazoa are largely excluded. It is often the case that the mat-forming cyanobacteria in these environments are not thriving optimally. They may also grow well under more moderate conditions, but in such cases grazing will prevent the accumulation of sufficient biomass and the accretion of sediment.

Coastal tidal sand flats often are excellent environments for microbial mats to develop, particularly when the flats extend over a large area and when the slope of the flat is low (Fig. 6a). Large areas will be covered by water for only a short period during the tide and often the sediment is not inundated for several days during neap tides. Such sediments often experience large fluctuations in water content, salinity and temperature, resulting in extreme conditions that limit the range of organisms able to inhabit this environment. The near absence of grazing organisms allows mat-building organisms such as cyanobacteria to accumulate. Coastal sand flats are usually nutrient-poor, but the phototrophic cyanobacteria have low nutrient demands. Moreover, most cyanobacteria resist long periods of drought, tolerating large fluctuations of salinity and temperature. Often these coastal microbial mats are composed of filamentous cyanobacteria that form a dense entangled mass which traps and binds sediment particles. Such mats are clearly visible to the naked eye as massive structures that to a large extent resist erosion (Fig. 6b). Their sediment stabilizing effect is of great importance for coastal morphogenesis. Typical examples are found in tidal sand flats of islands of the southern North Sea (e.g. Mellum, Germany), along the east coast of North America (e.g. Great Sippewissett Marsh, Cape Cod, Massachusetts; Bird Shoal, North Carolina), Pacific Coast (e.g. Guerrero Negro, Mexico) (Plate 8) and Shark Bay and Spencer Gulf in, respectively, Western and South Australia. A more complete list is given by Pierson (1992). Most such coastal microbial mats are not stromatolites, but examples of stromatolites in coastal sediments can be found in El Hamira Bay, Sinai, where stromatolitic beachrock is formed (Krumbein, 1979a) and along the Exuma Cays of the Bahamas intertidal and subtidal stromatolites are formed (Reid and Browne, 1991) (Plate 9). *Schizothrix* sp. can settle there in spite of the high wave energy to which the Atlantic coast of the Bahamas is exposed. Calcification of the Exuma Cays microbial mats renders stability to the system, which is necessary to cope with the extreme wave energy.

Other coastal mats are present in protected lagoons and are semi-permanently inundated. Examples of such coastal lagoons in which benthic microbial mats develop are found in many countries. Mats develop in the shallow parts of coastal lagoons, where large fluctuations of temperature and salinity may occur (Stal et al., 1996; Villbrandt et al., 1996).

2. Hypersaline Microbial Mats

Hypersaline environments can be found in shallow and sheltered coastal lagoons and tidal channels with high rates of evaporation and low precipitation. In the Mediterranean, hypersaline lagoons may form when they have narrow connections to the open sea and exchange of water is limited because a tide is virtually absent. Alternatively they can be totally disconnected from the sea and are fed by sea water through a sand barrier as is the case in Solar Lake on the Red Sea coast of Sinai, Egypt. When virtually all water in such coastal lagoon environments evaporates, a natural salt pond develops, forming structures known in the Sinai desert as Sabkhas. In those geographical regions where the combination of sun and wind result in sufficient evaporation, artificial salt ponds have been constructed. Other hypersaline environments are inland seas which can be found at many different locations on the globe (Oren, 1988) and in shallow lagoons of many of these lakes cyanobacterial mats develop (e.g. Zavarzin et al., 1993). The best studied hypersaline microbial mats are from the salt ponds in Guerrero Negro, Baja California Sur, Mexico and Solar Lake, Sinai, Egypt (D'Antoni D'Amelio et al., 1989; Des Marais et al., 1992). Salts and brines are discussed more fully by Oren in Chapter 10.

3. Hot Spring Mats of Cyanobacteria

Thermal hot springs are environments in which the combination of high temperature in combination with H_2S or acidic conditions decreases biodiversity enormously. Cyanobacterial mats are most common in hot springs at near neutral or alkaline conditions and are described more fully in Chapter 21. Thermal springs that contain sulfide may limit the formation mats of since thermophilic cyanobacteria do not

a

b

Fig. 6. *a*: The extended tidal sand flat of the island of Mellum (Southern North Sea, Germany) at low tide, covered with microbial mats. *b*: Mature mats of *Microcoleus chthonoplastes* of the island of Mellum have accreted and fixed so much sediment that they are often not submerged at high tide. This decreases the grazing pressure.

tolerate the combination of high temperature and high levels of sulfide (Castenholz, 1976; 1977). At moderate concentrations of sulfide, mats of *Oscillatoria* spp. have been shown to lower the sulfide concentration through anoxygenic photosynthesis by the cyanobacterium (Cohen et al., 1986; Ward et al., 1989). Another strategy is found in the so-called inverted mats (Castenholz, 1976). Here mats of the anoxygenic phototroph *Chloroflexus* overlay the cyanobacterial mat (Castenholz, 1976). Anoxygenic photosynthesis scavenges the sulfide and protects the underlying mat of the oxygenic heterocystous cyanobacterium *Chlorogloeopsis* sp. (Jørgensen and Nelson, 1988).

4. Terrestrial Cyanobacterial Mats

Terrestrial cyanobacterial mats can be found in a variety of different environments. De Winder et al. (1989a, b) described a cyanobacterial-algal crust in coastal dunes. Sand dunes have a poor capacity for retaining water and are therefore extremely dry environments that are characterized by a low biodiversity. Under certain conditions a mat of *Crinalium epipsammum*, a unique band-shaped filamentous cyanobacterium, develops (Fig. 7); it has an unusual cell envelope, exceptionally well-adapted to desiccation (De Winder et al., 1990). This species was found to be important in the Netherlands in stabilizing and fixing the dune sand and protecting it from wind erosion. Once this organism has established a matrix the community is taken over by the green alga *Klebsormidium flaccidum*.

Desiccation and life under low water potential are also the controlling factors for the development of cyanobacterial mats and stromatolites in the hot desert. *Microcoleus chthonoplastes*, which occurs in some desert crusts (Brock, 1975), has a polysaccharide sheath which plays an important role in protection from desiccation. After re-wetting the sheath absorbs water and the cyanobacterium resumes activity immediately (Campbell, 1979). The sheaths of the unicellular desert *Chroococcus* sp. and *Chroococcidiopsis* sp. also have this function and these species are found hypolithically on rocks in the Negev desert (Potts and Friedmann, 1981; Potts et al., 1983; Caiola et al., 1993, 1996). Cyanobacterial mats are particularly well investigated in the Negev desert in Israel (Friedmann et al., 1967; Berner and Jensen, 1982). Krumbein and Giele (1979) demonstrated calcification in a mat of a unicellular cyanobacterium producing stromatolitic structures in the desert.

Fig. 7. A scanning electron micrograph of *Crinalium epipsammum*, a new species of a cyanobacterium that forms phototrophic microbial crusts on coastal dunes. This organism is unusual because of the flat trichomes. *C. epipsammum* possesses an unusual cell wall, containing cellulose. The scale bar is 3 μm.

Cyanobacterial mats also seem to be involved in the formation of rock varnish in the desert. Desert rock varnish is composed of iron and manganese oxides that are precipitated by the metabolic activity of mat microorganisms. The cyanobacterial mat is usually present underneath this hard brownish layer where they are protected from direct sunlight and are capable of retention of some water (Krumbein and Jens, 1981).

The unicellular nitrogen-fixing *Gloeothece* (*Gloeocapsa*) forms mats on the walls of carbonate caves where daylight is low (Cox et al., 1981; Griffiths et al., 1987). Another example of such a low-light terrestrial environment is the mats of *Leptolyngbya* sp. described by Albertano and Kovacik (1996) on the walls of Casa Aureum in Rome. *Nostoc* occurs at the entrance of the carbonate caves where *Gloeothece* occurs (Griffiths et al., 1987). Terrestrial mats of *Nostoc* have been reported from a variety of desert environrments (Chapter 17), including the cold desert in Antarctica (Davey, 1983; Davey and Marchant, 1983).

III. The Organisms: Cyanobacteria that build Microbial Mats

Cyanobacteria that build microbial mats belong to a range of species, filamentous as well as unicellular. The filamentous non-heterocystous *Microcoleus chthonoplastes* is the dominant in marine intertidal microbial mats all over the world (Stal et al., 1985),

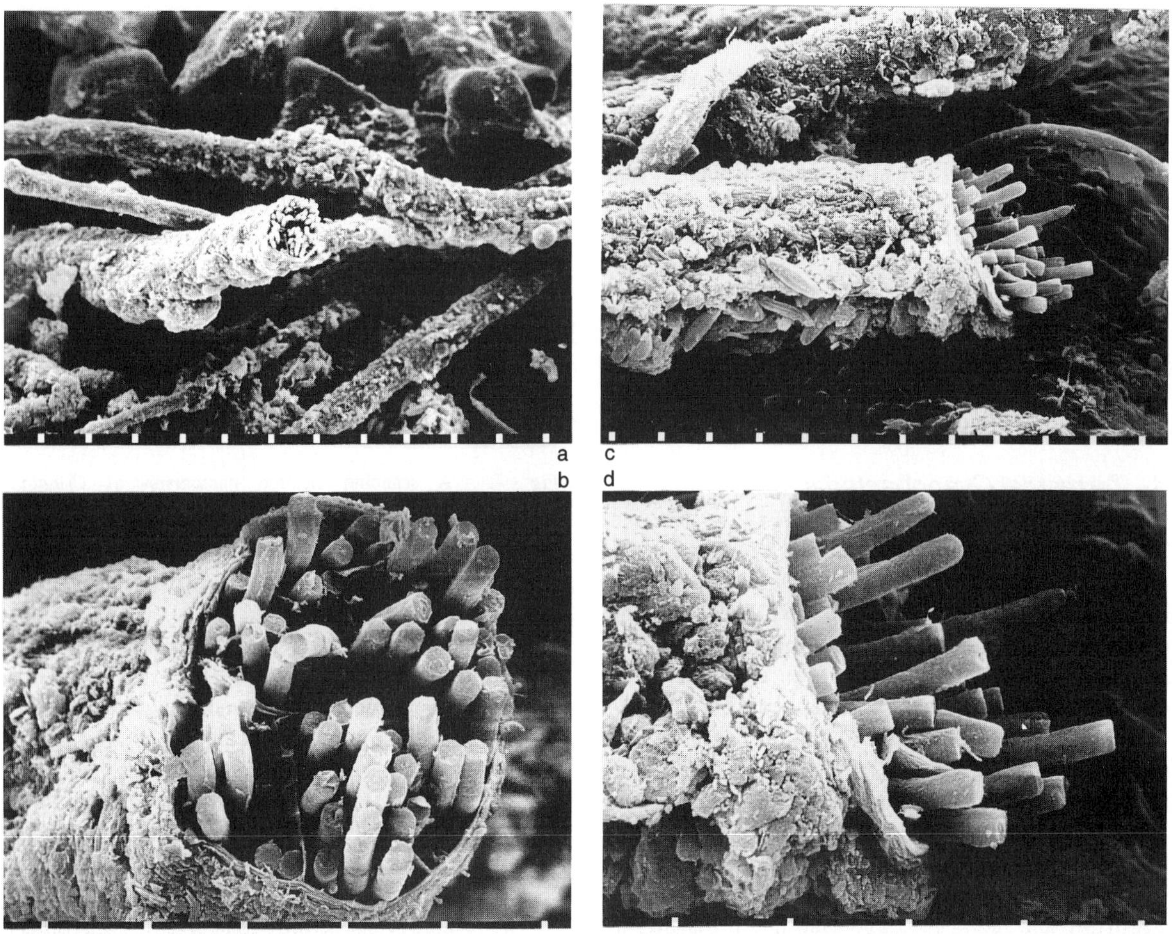

Fig. 8. Scanning Electron Microscopy photographs of a mat of *Microcoleus chthonoplastes* of the island of Mellum, Germany.
a: Overview of the mat showing the large polysaccharide ensheathed bundles of trichomes. Scale bar 30 μm. *b*: Detail showing one end of a bundle of *M. chthonoplastes*. The inner room of the bundle is composed of different compartments separated by polysaccharide walls. This bundle contains a large number of trichomes. Scale bar 10 μm. *c*: A side view of a bundle of *M. chthonoplastes*. The outside of the bundle is colonized with other microorganisms, among which are diatoms. Scale bar 10 μm. *d*: A detail of the opening of the bundle with the individual trichomes sticking out. The trichomes can move freely in and out of the bundle. Scale bar 10 μm.

hypersaline environments (Garcia-Pichel et al., 1996) and in the hot desert (Campbell, 1979). A considerable amount of work has been concentrated on this organism or on mats where it is a constituent. A notable characteristic is its occurrence in bundles containing many trichomes, often twisted like a rope. The bundles are enclosed in a common polysaccharide sheath (Fig. 8). These sheath bundles when thick may be partitioned in different compartments by polysaccharide walls (Fig. 8b). In some cases several dozen trichomes may be present in one sheath. The cells are quadratic or elongated with constrictions between the cells. The width of the trichomes is 2.5 - 6 μm and the apical cells are conical (Geitler, 1932). In the field this species is readily recognized by its morphological characteristics. Garcia-Pichel et al. (1996) recently investigated and compared several cultures isolated from a variety of mats from geographically distant locations, both marine and hypersaline. Based on a number of morphological and genetic characteristics, the authors concluded that all these isolates were closely related and belonged at least to the same genus and probably the same species. Two other strains that have been assigned in the literature as *M. chthonoplastes*, SAG 3192 (Collection of Algae, University of Göttingen, Germany), also known as strain 11 and isolated from a tidal mat of the North

Sea island Mellum (Stal and Krumbein, 1985) and 10mfx, originally isolated from a mat in Australia, did not belong to this group. The most conspicuous characteristic of *M. chthonoplastes*, the formation of a bundle enclosed by a polysaccharide sheath, is often lost in culture. The cultures described by Garcia-Pichel et al. (1996) possessed this sheath in culture, but some strains lost this property after maintaining the culture in the laboratory for 2 years. Although strain SAG 3192 does not normally form a bundle enclosed by a sheath, it could be induced by cultivating the strain on sand in a laboratory model of an tidal sediment (Krumbein et al., 1991). The strain isolated by Pearson et al. (1979) and assigned by them to *M. chthonoplastes*, formed bundles, but these were not surrounded by a common sheath. This strain was re-assigned as *Symploca* sp. (R. Rippka, pers. comm.), a genus that shows great similarity with *Schizothrix*, common in microbial mats. An interesting feature of *M. chthonoplastes* (*Symploca*) strain Pearson is its capacity for nitrogen fixation (Pearson et al., 1979). Strain SAG 3192 is capable of inducing nitrogenase under anoxic conditions (Stal and Krumbein, 1985). It has been noted that mats of *M. chthonoplastes* fix nitrogen (Potts et al., 1978) and in early work on microbial mats it has been suggested that the common sheath of this non-heterocystous cyanobacterium would help to provide an anoxic micro-environment for the oxygen-sensitive nitrogenase. Pearson et al. (1979) showed that bundle formation in *M. chthonoplastes* was not involved in nitrogen fixation. Moreover, strain SAG 3192 was isolated from a diazotrophic microbial mat, but only capable of nitrogen fixation under anoxic conditions. None of the strains investigated by Garcia-Pichel et al. (1996) and assigned by them to the tightly delimited group of *M. chthonoplastes* was capable of synthesizing nitrogenase under aerobic or anaerobic conditions (Villbrandt and Stal, unpublished). Strain PCC 7420 belongs to the *M. chthonoplastes* group *sensu* Garcia-Pichel et al. (1996) and was originally isolated from a salt marsh in Woods Hole (USA), it is also incapable of inducing nitrogenase (Rippka and Waterbury, 1977). This is remarkable since 64% of the strains of the LPP-group (*Lyngbya, Plectonema, Phormidium*), to which *M. chthonoplastes* belongs, have the capacity of synthesizing nitrogenase. This is another property which separates strain SAG 3192 from the core group of *Microcoleus chthonoplastes*. Future investigations may reveal that SAG 3192 should be assigned to *Symploca*. *Symploca* and *Schizothrix* are both cosmopolitan mat-forming cyanobacteria that share some morphological characteristics with *M. chthonoplastes*. A 'true' *M. chthonoplastes* was isolated from the same mats as SAG 3192 (Garcia-Pichel et al., 1996). It is not clear which of the two species is the more important in these mats. In fact, more than 20 different morphotypes of cyanobacteria were isolated from these mats of the intertidal sediments of the North Sea island Mellum (Stal and Krumbein, 1985). These also included heterocystous species that were never observed in the field in this specific type of mat. These mats showed a high population dynamics. Sometimes the mats consisted virtually exclusively of *Spirulina* or *Merismopedia* (Palinska et al., 1996). Often the mats were composed of mixtures of different species (Fig 9). An important species that

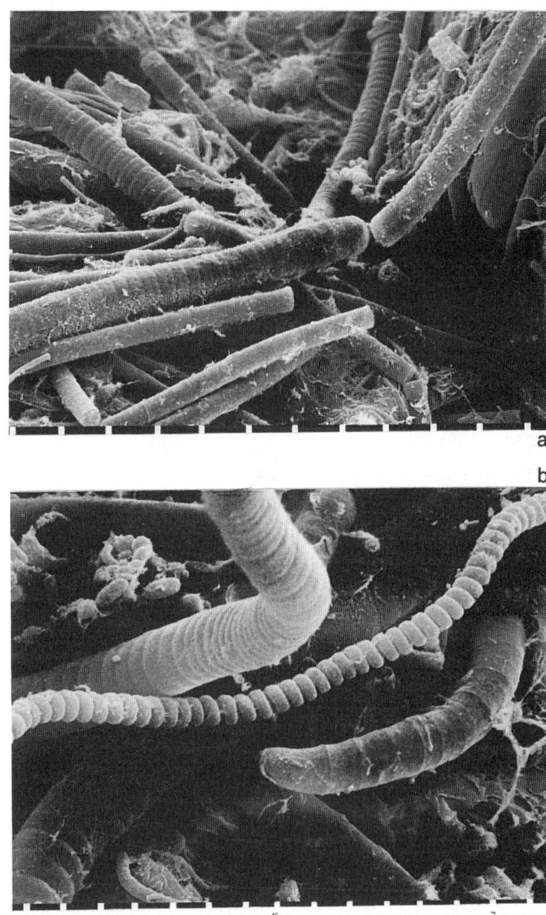

Fig. 9. a: Scanning electron microscopic (SEM) photograph of a nitrogen-fixing microbial mat of *Oscillatoria* spp. and other cyanobacteria. *b*: SEM photograph of a detail of a mat of *Oscillatoria limosa* with the typical coiled filament of *Spirulina subsalsa* which is a typical component of these mats and a single trichome of *Microcoleus chthonoplastes*.

was always present and was frequently dominant, was *Oscillatoria limosa* (Stal and Krumbein, 1985), which was shown to be the diazotrophic component of these mats. This strain was capable of aerobic nitrogen fixation in culture (Stal and Krumbein, 1981). The taxonomic position of *O. limosa* (strain 23) isolated from Mellum is also uncertain. The morphotype *O. limosa* is observed frequently in microbial mats all over the world and as far as known all of these mats are diazotrophic. However, often this cyanobacterium is identified as *Lyngbya aestuarii*. *L. aestuarii* is a filamentous cyanobacterium with disk-shaped cells and relatively wide trichomes (8 - 24 μm). The trichomes have a thick polysaccharide sheath which is often pigmented (Plate 8). The organism is not motile. Rippka et al. (1979) assign this organism to the group LPP A, which differs from *Oscillatoria* mainly by the absence of a sheath and motility in the latter. Strain 23 possesses a thin sheath and is slightly motile. Mats of *Lyngbya/Oscillatoria* can be found in geographically distant locations and are characterized by high rates of nitrogen fixation (Fig. 10).

Although many microbial mats are evidently growing diazotrophically, heterocystous forms are rarely the dominant component. However, the fact they can sometimes be isolated proves their presence, but apparently factors are present that prevent their proliferation. A few exceptions are known. In parts of the tidal flat in Guerrero Negro (Baja California Sur, Mexico) extensive mats of the heterocystous *Calothrix* sp. are present (Stal, 1995) (Plate 8). In a coastal lagoon near Bordeaux, France, mats of *Anabaena* sometimes develop (Villbrandt and Stal, 1996). Mangroves often support extensive mats of the heterocystous *Scytonema* sp. (Potts, 1979).

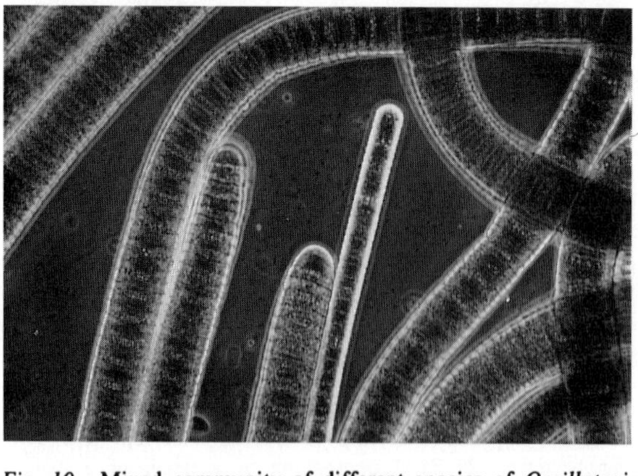

Fig. 10. **Mixed community of different species of *Oscillatoria* from a nitrogen-fixing microbial mat in a coastal lagoon on the Mediterranean coast of France.**

Calothrix sp. is also known to form mats on rocks in the spray zone at the seashore (Whitton and Potts, 1992). The development of these heterocystous diazotrophic mats is discussed further in Section VI.

Hot spring microbial mats such as Octopus Spring in the Lower Geyser Basin of Yellowstone National Park in the USA and similar microbial mats are dominated by the rod-shaped unicellular cyanobacterium *Synechococcus lividus* (Brock, 1978). Until recently strain *Synechococcus* sp. Y-7c-s was the only cultivated species. This strain was isolated from the 50 - 55°C Octopus Spring mat and considered to be representative for all thermophilic *Synechococcus* species since they possessed DNA with almost identical G+C ratios (Waterbury and Rippka, 1989). In fact, at least seven different strains, all with identical morphology, are present (Ward et al., 1994). Y-7c-s was only detected in Clearwater Spring, which is slightly acidic (Ruff-Roberts et al., 1994; Ferris et al., 1996) and from which this strain may have been originally isolated. Hybridization probes have shown that *Synechoccoccus* Y-7c-s was present in the Octopus Spring mats in extremely low frequency. Ferris et al. (1996) demonstrated that enrichment cultures select for this strain. However, by diluting inocula prior to enrichment these authors were able to isolate two new strains of *Synechoccus* in axenic cultures of which one was shown to be identical as the one occurring in the natural community. The different strains of hot-spring *Synechococcus* sp. have growth optima at different temperatures (Ward et al., 1994). Other adaptations may include pH and light. Although morphological indistinguishable, the populations of *Synechococcus* sp. in Octopus hot spring microbial mats belong to a phylogenetically diverse group. The work of Ward and co-workers who used oligonucleotide probe analysis demonstrated that isolation of cyanobacteria from microbial mats may selectively enrich a species which is of minor importance in the mat.

In the hypersaline mats of Pond 5 of the Guerrero Negro saltern and the shallow flat mat of Solar Lake *Microcoleus chthonoplastes* is the dominant species and forms gelatinous organic mats (Fig. 5). Other cyanobacteria that may occur in these hypersaline mats are *Oscillatoria* sp. and *Spirulina* sp. Unicellular cyanobacteria may also occur. The Pond 5 mat of Guerrero Negro grows at salinities from 60 - 95‰. The salinity in the shallow flat mat of Solar Lake ranges from 45 - 180‰. At salinities that are permanently above 100‰ *M. chthonoplastes* does not proliferate well. *Spirulina subsalsa* may be found at salinities up to 150‰. At higher salinity the unicellular *Aphanothece halophytica* usually dominates the mats. This species occurs up to a salinity of 200‰ (Dor and Paz, 1989). The salinity tolerances for cyanobacteria seem to be higher in mats of the Sabkha, where *A. halophytica* occurs at a salinity of 250-330‰, which is close to saturation, while *S. subsalsa* is present at a salinity of 205‰ (Dor and Paz, 1989). The reason for this difference is unclear. Salinity tolerance may be influenced by temperature. Although the sediment surface of the Sabkha may become hot from the solar radiation, the submerged mats in Solar ponds may also be exposed to high temperatures.

Lithified microbial mats found in the Exuma Cays, Bahamas, are built by the filamentous cyanobacterium *Schizothrix* sp. (Pinckney et al., 1995) (Plate 9). This cyanobacterium forms thin trichomes of about 1 μm wide, with cells 2-5 times as long as wide. The trichomes may be enclosed by a thick sheath. Communities of *Schizothrix* sp. may form dense and tough mats that are often associated with calcium carbonate precipitation. The lithified microbial mats of Exuma Cays, Bahamas result in the formation of stromatolites, a process which is continuous today (Reid and Browne, 1991; Pinckney et al., 1995).

IV Motility, Chemo- and Phototaxis of Cyanobacteria in Microbial Mats

Motility is an extremely important property for most mat-forming cyanobacteria and occurs by gliding, which can be defined as a self-propulsion along a surface. This surface can also be the interior of the polysaccharide sheath. Trichomes may move forwards and backwards in their sheaths and may move out of it, leaving an empty sheath behind. Trichomes may also glide along each other. The hypotheses to explain gliding motility that have received most attention are:

(i) the secretion of mucilage

(ii) contractile structures that cause surface undulations.

Some motile cyanobacteria possess pores in the cell wall from which it was assumed that mucilage could be secreted (Pankratz and Bowen, 1963; Pringsheim, 1968). According to this hypothesis the secretion of mucilage through the pores would be the propulsive force that moves the trichome. However, theoretical considerations indicate that the amount of polysaccharide needed for gliding would greatly exceed the amount that could possibly be provided by the organism (Holton and Freeman, 1965). Moreover, the highly motile *Phormidium uncinatum* does not possess pores, although it secretes polysaccharide (Häder, 1987b). The secretion of mucilage as a model for the propulsion of cyanobacteria also does not explain the rotation along the long axis in some Oscillatoriaceae. Halfen and Castenholz (1971) and Castenholz (1973) suggested that contractile microfibrils that they found in the external layers of the cyanobacterium *Oscillatoria princeps* could produce a gliding movement.

Microbial mats are characterized by steep and fluctuating physicochemical gradients. In order to experience optimum conditions at all times, cyanobacteria must position themselves continuously in the mat. Microbial mats also often occur in environments with high sedimentation rates. This demands a light-oriented motility, in order to prevent permanent burial.

Cyanobacteria possess essentially three different ways in which they respond to light: phototaxis, photokinesis and photophobic response (Häder, 1987a, b). Phototaxis is a movement, which orients itself to the direction of the light. Phototaxis can be either positive or negative. Positive phototaxis is towards the direction of light whereas negative phototaxis is the movement away from the light. Both positive and negative phototaxis are important for cyanobacteria in microbial mats. Most cyanobacteria are adapted to growth at low light intensities. Excess light may result in photooxidative stress and can cause damage. The combination of positive and negative phototaxis will allow the organism to obtain an optimum position in the mat. Most of the research on light responses of cyanobacteria has been carried out on *Phormidium* and *Anabaena*. Little work has been carried out on light responses in cultures of mat-forming cyanobacteria.

Photokinesis is the term used for the phenomenon where speed of movement increases with light intensity. This is because of the greater supply of energy. Only positive photokinesis is known (negative photokinesis would be the decrease of speed at higher light intensities).

The photophobic response is the reversal of the direction of movement as a result of a sudden change in light intensity. This response is very important for cyanobacteria. Both step-down and step-up responses are known (Häder, 1987a). A step-down response causes the accumulation of the organisms in the light. At very high light intensities a step-up response may result in the accumulation of the organisms in a shaded area. Photophobic responses are clearly related to photosynthesis as can be concluded from action spectra. (Häder, 1988).

It is not certain how important these three different responses to light are in microbial mats. Ramsing and Prufert-Bebout (1994) concluded from light measurements in mats made by fiber-optic micro light sensors that light fields in microbial mats are uniform. This is caused by scattering of light, and it means that there is in fact no direction of light. Moreover, light intensity will not be subject to sudden changes in microbial mats. These authors therefore conceived that phototactic and photophobic responses would not be especially beneficial for mat-forming cyanobacteria. Studies with *M. chthonoplastes* indicated that the strategy of this organism is to minimize movement when conditions are favorable. Instead of varying the speed of movement (photokinesis) it moves less frequently. *M. chthonoplastes* also reverses its movement frequently. This is not a photophobic response because this would imply a step-down or step-up change in light intensity which is not the case. Ramsing and Prufert-Bebout (1994) further observed that *M. chthonoplastes* bends more frequently at optimum light intensity. In the long-term this could lead to curling of trichomes into bundles. Motility in such bundles is likely to be restricted. Such cyanobacteria are likely to be confined to a fixed position in the mat (Fig. 11).

Garcia-Pichel et al. (1994) demonstrated that mat-forming cyanobacteria *Oscillatoria* sp. and *Spirulina subsalsa* migrated up and down in the mat in a daily manner (Fig. 11). At sunset these cyanobacteria moved towards the mat surface and stayed there throughout the night. At sunrise they migrated downwards. The depth to which they migrated appeared to be related to the light intensity, reaching the maximum depth in the mat at midday when the light intensity was highest. Interesting was also that *Oscillatoria* sp. and *S. subsalsa* contained unusually high amounts of chlorophyll *a* (3.9% d. wt). A unicellular cyanobacterium in this mat was non-motile. This species contained only 0.3% chlorophyll *a* (Garcia-Pichel et al., 1994). It was calculated that if *Oscillatoria* and *S. subsalsa* did not migrate they would be photoinhibited for most of the time, whereas the daily movement guaranteed optimum photosynthesis throughout the light period. Many cyanobacteria move deeper into the sediment at high light intensities (Pentecost, 1984; Whale and Walsby, 1984; Richardson and Castenholz, 1987a) (Fig. 11).

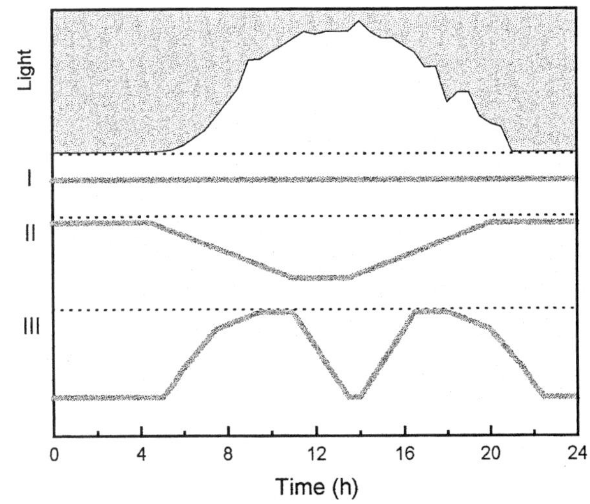

Fig. 11. Movements of the layer of cyanobacteria in microbial mats during a 24-h period. In the upper panel an example of the daily sinusoidal light curve is shown. Example I is a mat which does not displace itself during a day-night cycle. This is either the case with unicellular cyanobacteria that are not motile and grow at optimal light intensity or by species that minimize movement when conditions are favorable, which may be the case in some populations of *Microcoleus chthonoplastes*. Example II is a mat that moves toward the surface at sunset and moves down into the sediment during the day. Upwards movement may be controlled by chemical factors such as oxygen or sulfide. Downwards movement is in most cases attributed to negative phototaxis. Mats of *Oscillatoria* spp. often show this type of displacement during a 24-h cycle. Example III is exhibited by the hot-spring *Oscillatoria terebriformis*. In the dark the organism moves randomly but motility is inhibited by sulfide, which eventually results in the accumulation of the population in the sulfide-rich layer deep in the sediment. In the light it shows a positive phototaxis at low light and negative phototaxis at high light. This forces the organism to move deeper into the sediment during midday.

Other researchers have also noticed that cyanobacteria in a mat migrate to the surface during the night or when the mat is shaded. Migration occurs also during the dark and Whale and Walsby (1984) therefore concluded that this upward movement was not controlled by light. Since these authors could not find any evidence for geotactic or magnetotactic responses, they assumed that chemotaxis occurred and that chemical gradients controlled the direction of movement of cyanobacteria in the dark. On the other hand, not all cyanobacteria are capable of moving in the dark. Malin and Walsby (1985) observed that motility of *Oscillatoria* sp. was strictly dependent on light and gliding stopped in the dark after a short period, presumably because energy reserves were exhausted. These authors demonstrated responses of *Oscillatoria* sp. to oxygen (aerotaxis) and CO_2 and bicarbonate. A light-dependent response to CO_2 would be advantageous. Photosynthetic activity in microbial mats causes depletion of CO_2 and the high pH usually encountered in these environments as a result of photosynthetic activity and CO_2 fixation will shift carbonate equilibria resulting in even lower concentrations of CO_2. A light-dependent positive response to oxygen seems to be less advantageous. High concentrations of oxygen in the light may cause photo-oxidative reactions (Eloff et al., 1976) and photorespiration with therefore less efficient CO_2 fixation (Lorimer, 1981; Reinhold et al., 1991).

Aerotaxis would be a useful strategy for dark migration. This would allow aerobic degradation of endogenous storage carbohydrate. The migration of *Microcoleus chthonoplastes* (Whale and Walsby, 1994), *Oscillatoria* sp. and *S. subsalsa* (Garcia-Pichel et al., 1994) to the mat surface during the dark can be explained by a positive aerotaxis (Fig. 11). Alternatively, this behaviour can be explained by a negative response to sulfide. In the dark the concentration of sulfide will increase because anoxygenic photosynthesis is absent and no oxygen is available for biological or chemical oxidation (Fig. 3). Sulfide is very toxic and a negative response would certainly be advantageous. Castenholz (1982) therefore suggested a chemophobic response to sulfide in cyanobacteria that move to the mat surface during the dark.

A totally different behavior is encountered in the cyanobacterium *Oscillatoria terebriformis*. This cyanobacterium occurs in hot spring microbial mats and has a light-oriented motility. In the dark the organism continues to move, but randomly. It may thus happen that it moves down into the sediment reaching the sulfide layer. Sulfide seems to inhibit motility in *O. terebriformis* and therefore the organism is trapped in this layer (Richardson and Castenholz, 1987a) (Fig. 11). Under laboratory conditions 0.7 mM sulfide completely inhibited its gliding motility. Sulfide inhibited motility only in the dark or in the light when photosystem II was blocked by (3,-(3,4-dichlorophenyl)-1,1-dimethylurea (DC MU). This inhibition was reversible and was abolished in the light. Since every individual organism has a high probability to become trapped in the sulfide layer, virtually the whole population will end up there. During the day the sulfide horizon will move down into the sediment relieving the inhibition of motility and at the same time the organism will move towards the light to the mat surface. At midday, when light intensity is high, *O. terebriformis* shows negative phototaxis and moves deeper into the sediment in order to prevent photo-oxidative damage (Fig. 11). The majority of motile mat-forming cyanobacteria will prefer low light intensities and move deeper into the sediment during the day. The trapping of *O. terebriformis* in the sulfide layer during the dark is unusual, but essential for this organism to survive the dark period. In the presence of oxygen dark respiration will cause a rapid depletion of the endogenous storage carbohydrate which will result in the death of the organism in a matter of hours. The sulfide layer is of course anoxic. *O. terebriformis* is capable of fermentation and this process is slow, allowing for an extended period of energy generation. Indeed, the amount of energy that can be generated is small, but it is apparently sufficient to cover its maintenance requirements (Stal and Moezelaar, 1997). Most cyanobacteria, including mat-forming species, have low rates of dark respiration, allowing them to overcome long periods in the dark in the presence of oxygen. In the dark many microbial mats are virtually anoxic up to the surface, and fermentation is probably the only metabolism possible for the majority of cyanobacteria in the mat, with the exception of those that are exposed to the air. When the mat is submersed, oxygen decreases to zero within the diffusive boundary layer and no oxygen will be available to the mat (Fig. 3).

Several cyanobacteria are capable of assimilating organic compounds such as glucose and fructose in the light (photoheterotrophy) and some even display a fully chemoorganotrophic metabolism (Smith, 1982). *O. terebriformis* is capable of fermenting extracellular compounds such as fructose and glucose (Richardson and Castenholz, 1987b). Chemotactic responses of cyanobacteria to organic compounds are largely unknown. Fechner (1915) reported a negative chemotactic response to organic acids and Richardson and Castenholz (1989) observed inhibition of gliding of *O. terebriformis* by fructose. This effect was similar to that observed for sulfide. Glucose, the other substrate for this organism, did not have an effect, nor lactate which is one of the fermentation products produced by *O. terebriformis*.

In contrast to what had been assumed, it was demonstrated by Garcia-Pichel and Bebout (1996) that ultraviolet radiation penetrates well in microbial mats. The amount of penetration varies with the type of sediment on which microbial mats developed. Silty mud absorbed UV light most and quartz sand the least. Mats that are mainly organic take an intermediate position. UV light was absorbed in these mats more or less exponentially, in a similar way to visible light. There were two important aspects of the penetration of UV light in microbial mats, regardless of their sedimentological characteristics. In some mats the intensity of UV-B just below the surface is considerably higher than the incident intensity; this is caused by scattering. Secondly, the total amount of UV-B in the euphotic zone of the mat ranged from 15 - 33% of incident irradiance which is high, particularly when compared with aquatic systems, where this number varies from 3 - 9%. These measurements carried out by Garcia-Pichel and Bebout (1996) were the first to demonstrate unequivocally that cyanobacterial mats develop under UV stress. Garcia-Pichel and Castenholz (1994) and Bebout and Garcia-Pichel (1995) provided also strong evidence that vertical migrations are partly under control of UV light. Garcia-Pichel and Castenholz (1994) reported that only 1.3 W m^{-2} of UV-A (315 – 400 nm) was sufficient to keep the cyanobacteria *Oscillatoria* sp. and *Spirulina subsalsa* deep in the sediment. This intensity is only 3 - 4% of the level that these organisms would experience at midday. These cyanobacteria responded by negative phototaxis. In another study of microbial mats in Solar Lake (Sinai) it was shown that *M. chthonoplastes* responds clearly to UV-B light (310 nm). Exposure of the mat to 0.35 - 0.79 W m^{-2} was sufficient to cause a downwards migration of the cyanobacteria. The effect of UV-B was about two orders of magnitude stronger than normal visible light. Also UV-A had this effect but was about 5 times less efficient than UV-B (Bebout and Garcia-Pichel, 1995). It was concluded from these experiments that *M. chthonoplastes* is capable of sensing UV light, particularly UV-B. There is no doubt that UV light causes serious damage to oxygenic phototrophic organisms (Cullen and Neale, 1994) and has therefore negative effects on primary productivity (Smith et al, 1992). A mat-forming cyanobacterium will therefore benefit from the capability of sensing low levels of UV radiation and combining it with a negative taxis. This will nevertheless result in a negative effect on total gross photosynthesis and productivity during exposure to UV light, but the effect is largely reversible (Bebout and Garcia-Pichel, 1995). Due to the downwards migration of the cyanobacterium, the biomass at the surface, and thus gross photosynthesis, decreases. In addition, surface photosynthesis may be partly inhibited by UV irradiation. Because in the deeper layers more biomass accumulates gross photosynthesis is even higher but due to the low level of photosynthetic active radiation (PAR), biomass specific photosynthesis is low. Not all cyanobacteria exhibit negative phototaxis with respect to UV light. Donker and Häder (1991) and Donkor et al. (1993) showed that motility in the cyanobacteria they investigated was rather impaired by UV-B (280 – 315 nm). This may also have been the case in a mat of *M. chthonoplastes* of the temperate southern North Sea, where photosynthesis was strongly inhibited by UV-B radiation and did not recover during the subsequent 3 h when UV was excluded (Garcia-Pichel and Castenholz, 1994). The responses of cyanobacteria to ultra-violet irradiation are discussed in more detail in Chapter 21.

V. Carbon Metabolism

A. Introduction

Cyanobacteria are the principal primary producers in the majority of microbial mats. Oxygenic photosynthesis and sometimes anoxygenic photosynthesis drives CO_2 fixation. Cyanobacteria enrich the microbial mat with organic matter. CO_2 fixation primarily results in the formation of the structural biomass of the cyanobacteria. This organic matter may become available to other organisms in the mat by the death and subsequent lysis of the cyanobacteria. However, it appears that, in spite of the high rates of photosynthesis usually observed in microbial mats, net growth of the cyanobacteria is often low in mature mats (Nold and Ward, 1996). Hence, other processes must be involved in order to divert photosynthate to the mat community. As will be discussed below, these processes may include the excretion of glycolate during photorespiration, the excretion of compatible solutes after an osmotic down shock, the excretion of fermentation products during dark anoxic conditions and the secretion of extracellular polymeric substances (EPS).

B. Oxygenic Photosynthesis

Oxygenic photosynthesis requires the presence of two photosystems (PS I and II). Cyanobacteria contain chlorophyll *a* in the reaction centers of both PS I and II, but the former contains about 2-3 times as many molecules of chlorophyll *a*. This chlorophyll may also contribute to the light harvesting, but the phycobiliproteins are far more important pigments as light-harvesting antennae. Jørgensen et al. (1987) demonstrated by recording photosynthetic action spectra in cyanobacterial mats that chlorophyll *a* hardly contributed to the action spectra even when additional 600 nm light was given to excite PS II.

Light is strongly attenuated in microbial mats, both by the sediment and by absorption by the dense phototrophic community. Sediments are transparent to light of long wavelengths (Stal et al., 1985). Dry sediments consisting of fine sandy quartz attenuate light much stronger than the same sediment saturated with water (Stal et al., 1985). Through the upper 1 mm of the latter more than 10% of surface irradiance penetrated, while this was only 2.5% of the dry sediment. The photic depth of the uncolonized wet fine sandy sediment was about 4 mm. Through a 1.5 mm mat of cyanobacteria (0.5 g chlorophyll *a* m^{-2}), 0.45% of photosynthetically active radiation (PAR) penetrated. However, due to the specific absorption of the mat, wavelengths that would support oxygenic photosynthesis are specifically attenuated and oxygenic photosynthesis would not be possible below the cyanobacterial mat (Stal et al., 1985; Jørgensen et al., 1987; Pierson et al., 1987; Jørgensen and Des Marais, 1988; Pierson et al., 1990) (Fig.4). Sediment and cyanobacterial mats are relatively transparent to light of wavelengths above 700 nm, which explains the occurrence of communities of bacteriochlorophyll *a*-containing anoxygenic phototrophic purple-sulfur bacteria (Pierson et al., 1987; 1990). These findings were confirmed by using newly developed fiber-optic microprobes connected to a diode array detector (Kühl and Jørgensen, 1992) and by a scalar irradiance microsensor which for the first time allowed spectral light measurements in sediments at the scale of the phototrophic microorganisms (Lassen et al., 1992a). More than 50% of the incident irradiance of 800 nm light penetrated a 1 mm thick microbial mat (Ploug et al., 1993). Lassen et al. (1992b) used this technique to study photosynthesis and photosynthetic efficiency in a microbial mat in Limfjorden, Denmark. This mat consisted of a top layer of diatoms and a cyanobacterial layer (*Oscillatoria* spp.) underneath. Using an oxygen micro-sensor, two peaks of oxygenic photosynthesis were found, corresponding to the diatom biofilm and the second deeper maximum corresponded with the layer of cyanobacteria. This latter maximum at 1 mm depth occurred at a light intensity of only 12 μmol photon m^{-2} s^{-1} i.e. 1.5% of incident light intensity. However, photosynthetic efficiency (rate of photosynthesis at a specific depth divided by the scalar irradiance at that depth) appeared to be 10-fold higher in the cyanobacterial mat compared to the diatom film. This increased photosynthetic efficiency at low light intensity is the result of both the concentration of cyanobacteria at the depth of the second maximum of photosynthesis, as well as a likely increased efficiency with which the available light is absorbed by the organisms (Lassen et al., 1992b).

The dense biomass of cyanobacteria in the upper photic zone of microbial mats results in high rates of photosynthesis, and on a surface basis it compares to the productivity of rain forests, which are usually considered as the most productive ecosystems on Earth (Guerrero and Mas, 1989) (Table 1). Revsbech et al. (1983) measured a total daily photosynthesis in a cyanobacterial mat in Solar Lake (Sinai) of 156 mmol O_2 m^{-2} and similar rates were found by Villbrandt et al. (1990) for a cyanobacterial mat in temperate region (North Sea). The daily rates of photosynthesis measured by Revsbech et al (1983) in Solar Lake microbial mats followed the light intensity during the day. The rates of photosynthesis were the same when measured in the morning or in the afternoon. The rate of photosynthesis increased with increasing light intensity. The efficiency of photosynthesis was highest at low light intensity (up to 120 μmol photon m^{-2} s^{-1}). Photosynthesis was not inhibited at high light intensity. The same was found

Table 1. Comparison of primary productivity in microbial mats with other ecosystems.

Ecosystem	Primary productivity (mg C m^{-2} d^{-1})
Microbial Mat	
Mellum (North Sea)	6200
Solar Lake (Sinai)	5000
Sea and Ocean	
Mediterranean	60-500
Coastal Upwellings	1000-4000
Ocean	<100
Lakes	
Oligotrophic lakes	40-80
Eutrophic lakes	300-3000
Hypertrophic lakes	2000-5000
Mangrove forests	5600
Rain forest	6000

after Stal (1993)

by Villbrandt (1992) when diurnal photosynthesis data on several days were plotted against light intensity (Fig. 12a). Photosynthesis increased in a linear way with light intensity. This photosynthesis versus light intensity curve of a microbial mat is completely different from isolated cultures of cyanobacteria (Fig. 12b). At low light intensity photosynthesis the curve is steep, depending on photosynthetic efficiency of the organism. At a certain, usually moderate, light intensity photosynthesis saturates (P_{max}) and decreases again at higher intensity as a result of photoinhibition. The different P versus I curve of a mat is explained by the fact that all light is absorbed and used for

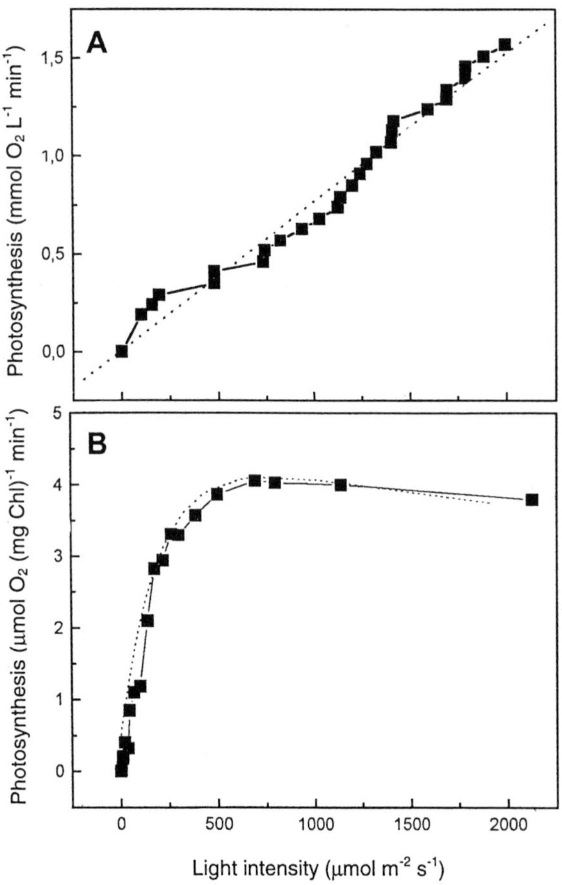

Fig. 12. A: Photosynthesis versus light curve in a microbial mat of the North Sea island of Mellum (Germany) (data from Villbrandt, 1992). A large number of depth integrated measurements of photosynthesis recorded at different days and during different times of the day at the same location in the mat were used to plot this curve. The curve was fitted by linear regression: P (mmol O_2 L^{-1} min^{-1}) = 0.01062 + (7.58 10^{-4})I (R=0.99 ± 0.08; N=27 P<0.0001). *B*: Typical photosynthesis versus light intensity curve of a culture of a cyanobacterium.

photosynthesis. At higher light intensity cyanobacteria in the deeper parts of the mat will exhibit higher rates of photosynthesis. The diurnal variation of photosynthesis on a bright, cloudless day in July in a mature mat of *M. chthonoplastes* on the island of Mellum, North Sea (chlorophyll *a* content typically 0.3 g m^{-2}: Stal et al., 1985), showed a different pattern as the one measured by Revsbech et al. (1983). The rates of photosynthesis were highest during the morning hours and showed a sharp drop after midday (Villbrandt et al., 1990) (Fig. 13b). This midday drop in photosynthesis has been observed by other workers both in microphytobenthos as well as in phytoplankton (Paerl et al., 1989; Storch et al., 1990). A conclusive explanation for this observation is not yet available, but it may be that in the morning hours the concentration of DIC in the pore water of the mat is high as a result of decomposition of organic matter during the preceding night. After some hours of photosynthetic CO_2 fixation the pore water becomes depleted of DIC, which could explain the much lower rates of photosynthesis measured in the afternoon. A similar pattern as the one measured by Revsbech et al (1983), was measured at another site of tidal flat on the island of Mellum. This site was characterized by freshly colonized sediment with *Oscillatoria limosa* as the dominant species and with only 1/10 of the biomass compared to the mature mat (chlorophyll content typically 0.03 g m^{-2}, Stal et al., 1985) (Villbrandt, 1992) (Fig 13a). Photosynthesis on a surface basis was much lower which is of course due to the much lower biomass and therefore the supply of DIC may have been sufficient throughout the day. Photosynthesis on a biomass basis was about twice as high as in the mature mat with comparable surface incident light intensity. This is attributed to the much smaller attenuation of light when biomass is low, i.e. the individual cyanobacteria receive much more light in the freshly colonized sediment compared to the mature mat.

The high rates of photosynthesis often observed in microbial mats may cause such high levels of supersaturation that oxygen bubbles are formed. The dense organic matrix in which the mat cyanobacteria are embedded represent a diffusion barrier that limits gas exchange to a considerable extend. Oxygen gas bubbles that eventually develop may also be trapped in this matrix. When such mats are then inundated they may become buoyant and lift off from the sediment. Erosion of microbial mats as a result of this phenomenon can be regularly observed. Pieces of mat may be carried to vegetated areas and become

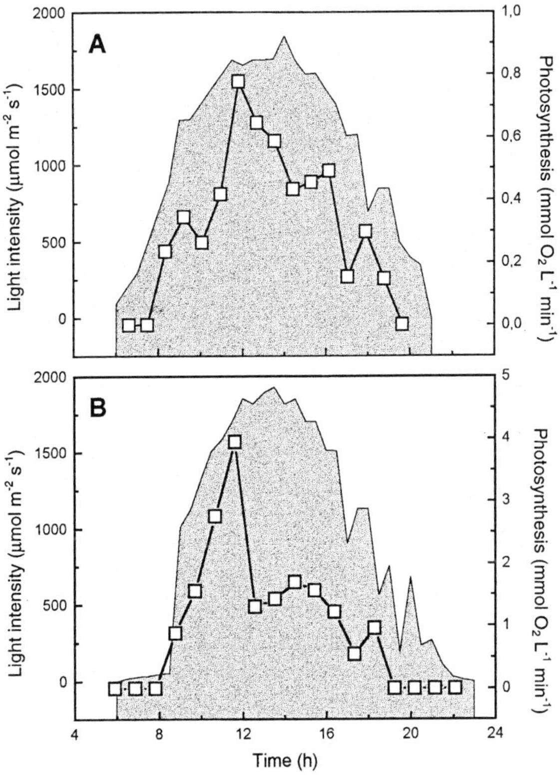

Fig. 13. The daily light curve (shaded area) and depth integrated photosynthesis of (*A*) a freshly colonized sediment with *Oscillatoria limosa* as the dominant cyanobacterium and (*B*) a mature mat of *Microcoleus chthonoplastes*. Both mats were located on tidal sand flats on the North Sea island of Mellum, Germany. Data from Villbrandt et al. (1990).

desiccated when the tide has gone. Such desiccated pieces may be transported by wind over long distances. The first time this phenomenon was reported was in 1686 and became known as "Meteorpapier" because it was believed that it came from space. In his publication "Über das im Jahre 1686 in Curland vom Himmel gefallene Meteorpapier und über dessen Zusammensetzung aus Conferven und Infusorien" Ehrenberg (1838) identified this meteor paper as desiccated microbial mats.

C. Anoxygenic Photosynthesis

Anoxygenic photosynthesis in microbial mats is not the exclusive trait of purple- or green sulfur bacteria. Some species of cyanobacteria are capable of a typical bacteria-like anoxygenic photosynthesis in which only photosystem I is involved. As with phototrophic sulfur bacteria, anoxygenic photosynthesis in cyanobacteria depends on sulfide as the electron donor. Roughly two categories of cyanobacteria can be distinguished with respect to the capacity of anoxygenic photosynthesis. In one group oxygenic photosynthesis is inhibited at low concentrations of sulfide and anoxygenic photosynthesis is induced. Inhibition of oxygenic photosynthesis by sulfide is probably at the level of the manganese-containing, water-splitting enzyme (Oren et al., 1977). Both types of photosynthesis are mutually exclusive in these organisms. In the other group anoxygenic and oxygenic photosynthesis occur concurrently. At low sulfide concentrations oxygenic photosynthesis is more important and with increasing sulfide concentrations anoxygenic photosynthesis gradually takes over.

In both types of cyanobacteria anoxygenic photosynthesis must be induced, a process which depends on a certain threshold of sulfide concentration and on light. Induction of anoxygenic photosynthesis in some organisms may take several hours. Therefore, cyanobacteria that possess the capability of carrying out oxygenic and anoxygenic photosynthesis concurrently have an ecological advantage in environments in which sulfide concentrations fluctuate, as is the case for instance in many marine and hypersaline microbial mats. Cyanobacteria that can carry out only one type of photosynthesis at a time are typical of environments with a constant supply of sulfide, as in certain hot spring microbial mats with an indigenous supply of sulfide. Moreover, these cyanobacteria tolerate higher concentrations of sulfide.

Anoxygenic photosynthesis is defined as photosystem I mediated fixation of CO_2 with sulfide as an electron donor. This means that this process can also take place when photosystem II is inhibited by 3-(3,4-dichlorophenyl)-1,1-dimethylurea (DCMU) or by far red light (> 700 nm) which is not used by photosystem II. Cyanobacteria that perform anoxygenic photosynthesis oxidize sulfide to elemental sulfur according to the following stoichiometric equation:

$$12\,H_2S + 6\,CO_2 \longrightarrow C_6H_{12}O_6 + 12\,S^o + 6\,H_2O$$

The elemental sulfur that is produced is deposited outside the cell but is often found attached to the outer sheath of the cyanobacterium as finely dispersed particles. The elemental sulfur is not further oxidized, as is the case in for instance purple sulfur bacteria. In the mat-forming *Microcoleus chthonoplastes* thiosulfate was found to be the product of sulfide oxidation (De Wit and Van Gemerden, 1987):

$$6\,H_2S + 3\,H_2O + 6\,CO_2 \longrightarrow C_6H_{12}O_6 + 3\,H_2S_2O_3$$

While the oxidation of sulfide to sulfur yields only two electrons, the oxidation to thiosulfate yields four electrons per sulfide oxidized. Thus, the oxidation of sulfide to thiosulfate in *M. chthonoplastes* seems to be twice as efficient as in other cyanobacteria in which zerovalent sulfur is the product. Rabenstein et al. (1995) reported that sulfite was the intermediate in those cyanobacteria that oxidized sulfide to thiosulfate. It is possible that thiosulfate is formed in a chemical reaction of sulfite with sulfide.

It is not clear why most anoxygenic cyanobacteria oxidize sulfide only to elemental sulfur. Anoxygenic phototrophic sulfur bacteria usually oxidize sulfide to sulfate which yields eight electrons. However, in these bacteria the oxidation to elemental sulfur is usually much faster than the further oxidation to sulfate. Depending on the species this elemental sulfur is accumulated intra- or extracellularly. It can be hypothesized that the rapid oxidation of sulfide to elemental sulfur would be advantageous, since it assures a rapid removal of the toxic sulfide. Moreover, in anoxygenic phototrophic bacteria, no matter whether they store the sulfur produced intra- or extracellularly, the sulfur is not available for other organisms (Van Gemerden, 1987). In that way they continue anoxygenic photosynthesis when sulfide is depleted without competition with other organisms. In cyanobacteria the rapid consumption of sulfide most likely serves for detoxification since they can immediately switch to oxygenic photosynthesis when sulfide has been completely oxidized. In addition, cyanobacteria can also use elemental sulfur as an electron acceptor in anaerobic dark metabolism. *O. amphigranulata* is capable of using elemental sulfur for assimilatory purposes (Castenholz and Utkilen, 1984). In the unicellular *Anacystis nidulans* low rates of CO_2 fixation are supported by the oxidation of thiosulfate (Utkilen, 1976; Peschek, 1978).

Microcoleus chthonoplastes belongs to the group that perform oxygenic and anoxygenic photosynthesis concurrently. In this organism the growth rate decreased exponentially with increasing concentrations of sulfide. At a concentration of 1 mM sulfide (pH8.0) growth was completely inhibited. Oxygenic photosynthesis was gradually inhibited with increasing concentrations of sulfide. The relative contribution of anoxygenic photosynthesis to total photosynthesis was >95% at concentrations of sulfide exceeding 0.35 mM (pH8.0). The inhibition of growth at 1 mM of sulfide is probably caused by the compete inhibition of oxygenic photosynthesis, rather than anoxygenic photosynthesis. It was shown that *M. chthonoplastes* requires oxygen for growth. When oxygenic photosynthesis was inhibited by DCMU, *M. chthonoplastes* was not capable of sulfide-dependent anoxygenic phototrophic growth unless some oxygen was present. Therefore, a small contribution of oxygenic photosynthesis may be necessary in order to provide the essential oxygen. Oxygen may be required for the oxidation of fatty acids. *Anacystis halophytica,* for instance, possesses an oxygen-dependent desaturation mechanism (Padan and Cohen, 1982). Several cyanobacteria contain polyunsaturated fatty acids (Kenyon et al., 1972) and Padan and Cohen (1982) suggested that such cyanobacteria may be incapable of anaerobic growth. *O. limnetica* does not contain polyunsaturated fatty acids which may explain its capacity for anaerobic growth (Padan and Cohen, 1982). *M. chthonoplastes* SAG 3192 contains considerable amounts of linoleate (18:2), linolenate (18:3) and tetradecadienate (14:2), regardless of whether the culture was grown in the presence or absence of sulfide (De Wit et al., 1988). The affinity of *M. chthonoplastes* for sulfide is extremely low. The K_m for sulfide has been calculated as 974 µM, approximately the concentration at which growth of *M. chthonoplastes*, ceases (De Wit and Van Gemerden, 1987). These authors analyzed the data of Jørgensen et al. (1986), who investigated the transition of anoxygenic photosynthesis to oxygenic photosynthesis in a mat of *M. chthonoplastes*. A K_m of 710 µM was calculated for sulfide oxidation in this mat. This affinity is close the one calculated in culture by De Wit and van Gemerden (1987). These affinities for sulfide are extremely low when compared to the value of 5 µM for an anoxygenic phototrophic purple sulfur bacterium as *Thiocapsa roseopersicina*, that is frequently present in microbial mats (De Wit and Van Gemerden, 1988). The similarity of the K_m of sulfide oxidation estimated in a culture to that estimated in a natural microbial mat of *Microcoleus chthonoplastes* indicates that this organism was probably responsible for the sulfide oxidation in the microbial mat. This was also suggested by Jørgensen et al. (1986) although they concluded this from the fact that purple sulfur bacteria constituted only a minor fraction in that mat. The results on anoxygenic photosynthesis in a culture of *M. chthonoplastes* obtained by De Wit and van Gemerden confirmed those for a natural mat by Jørgensen et al. (1986). Jørgensen et al. (1986) found that oxygenic and anoxygenic photosynthesis were carried out concurrently. However, at higher concentrations of sulfide oxygenic photosynthesis was insignificant. Oxygenic photosynthesis in this mat started when the sulfide concentration decreased to about 0.3 mM, which is in agreement with the results obtained by De Wit and van Gemerden. The experiments of Jørgensen and co-workers showed that oxygenic photosynthesis could occur even when the microbial mat was exposed to 5-6 mM of sulfide in the overlying water. An oxygen peak was sandwiched between layers of sulfide. This sandwiching of cyanobacteria in microbial mats is often observed. Apparently, *M. chthonoplastes* is capable of resisting high concentrations of sulfide and recovers oxygenic photosynthesis. It is also capable of oxidizing this high concentration of sulfide in the light, but this does not mean that the organism is indeed growing or even fixing CO_2. Taking together the extremely low affinity of *M. chthonoplastes* for sulfide, the low growth rate with anoxygenic photosynthesis and the fact that this organism cannot grow in the absence of oxygen, the major function seems to be the detoxification of sulfide.

D. CO_2 Fixation

Carbon dioxide is the most important source of carbon for cyanobacteria and it is therefore crucial for the functioning of microbial mats. Cyanobacteria use the energy and low potential reductant (NADPH) produced during photosynthesis to fix CO_2. Cyanobacteria fix carbon dioxide through the reductive pentose phosphate pathway (Calvin cycle). The same pathway in the opposite direction, the oxidative pentose pathway, is used for the oxidation of storage carbohydrate during the dark in combination with aerobic respiration (Smith, 1982; Schmetterer, 1994).

High rates of photosynthesis will deplete the sediment of CO_2 and raise the pH. The pH may even reach values of over 9.5 (Revsbech et al., 1983) (Fig. 3) and any dissolved inorganic carbon will be present as bicarbonate or carbonate. Cyanobacteria are capable of adapting to growth at extremely low concentrations of dissolved inorganic carbon. Both CO_2 and bicarbonate are taken up by cyanobacteria. However, CO_2 is the substrate for RubisCO, the key enzyme of the Calvin cycle and responsible for the fixation of CO_2. As in other autotrophic organisms in cyanobacteria this enzyme has a relative low affinity for CO_2, which means that both a high concentration of CO_2 and of RubisCO are prerequisites for efficient fixation of carbon dioxide. Cyanobacteria possess an inorganic carbon-concentrating mechanism (CCM) which may result in up to 1000-fold accumulation of inorganic carbon in the cell. A tentative model of this CCM in cyanobacteria proposes that either bicarbonate or CO_2 is taken up but that the former is the predominant species of inorganic carbon in the cytoplasm (Kaplan et al., 1994). Bicarbonate enters the carboxysome, a cell inclusion in autotrophic prokaryotes also known as polyhedral bodies. Carboxysomes contain virtually all RubisCO in organisms that possess these inclusions. The importance of carboxysomes for the CCM is also shown by the observations of Turpin et al. (1984) and McKay et al. (1992) that the number of of these inclusions increases during adaptation of cyanobacteria to low CO_2 concentrations.

The fixation of CO_2 in microbial mats can be investigated by measuring the $^{12}C/^{13}C$ carbon isotope ratio in the organic matter. RubisCO discriminates between carbon isotopes with a slight preference for the lighter isotope ^{12}C. This fractionation factor α equals 1.029 (Roeske and O'Leary, 1984), which means that organic matter may become 29‰ depleted in the heavy isotope ^{13}C when its origin is from RubisCO mediated CO_2 fixation. This isotopic discrimination is only achieved when the CO_2 concentration is sufficiently high. This is generally not the case. Moreover, the measured value may differ from the expected one because other organisms responsible for RubisCO independent CO_2 fixation may have been present in the system. Also cyanobacteria may fix significant amounts of CO_2 via alternative pathways such as PEP carboxylase or carbamylphosphate. Furthermore the DIC produced from the decomposition of organic matter may be recycled and give rise to a different net isotope discrimination. At the low concentrations of CO_2 that usually occur in cyanobacterial mats active transport of HCO_3^- becomes important (Badger and Andrews, 1982) which results in a much smaller isotope discrimination than in the case of CO_2 uptake (Des Marais and Canfield, 1994). Microbial mats are usually not much depleted in ^{13}C ($\delta^{13}C_{mat}$ is not very negative) because the pool of DIC is small compared to the rate of CO_2 fixation. This minimizes the isotope discrimination. The most negative values of $\delta^{13}C_p$ (photosynthate) are expected when CO_2 does not become depleted from the medium and when exchange between the medium and the site of fixation is rapid. Des Marais and Canfield (1994) investigated the carbon isotope discrimination in two microbial mats in Guerrero Negro, Baja California, Mexico. The $\delta^{13}C_{mat}$ in these mats was not very negative (-7‰). In the mat of *Lyngbya aestuarii* this value corresponded with the fractionation factor 1.007. This low value was evidently attributed to the closed reservoir behaviour of the system. The DIC that was produced by the mat during the night possessed to same negative value and therefore no changes in the $\delta^{13}C_{mat}$ were expected in the *L. aestuarii* mat. In the mat of *M. chthonoplastes* photosynthesis did not discriminate between the lighter and heavier isotopes. At present a conclusive explanation for the negative value of $\delta^{13}C_{mat}$ in the *M. chthonoplastes* mat is not available (Des Marais and Canfield, 1994). Processes such as excretion, fermentation and respiration do not change isotopic discrimination. Diagenesis of organic matter does not alter its isotopic composition (Des Marais et al., 1992) and the $\delta^{13}C_{DIC}$ is similar as $\delta^{13}C_{mat}$ (Bauer et al., 1991). It is known that the 'pond 5' mats of *M. chthonoplastes* are more or less in 'steady state', i.e. most of the organic matter that is produced by photosynthesis is mineralized in the mat. It is likely that photosynthesis scavenges very efficiently the DIC that is produced in the mat, thereby limiting net isotope fractionation. In fossil Proterozoic stromatolites $\delta^{13}C$ is much more negative than in present day microbial mats and stromatolites. This may reflect the higher levels of DIC in the Precambrian compared to today's concentrations (Kempe and Kazmierczak, 1990a) or a more negative $\delta^{13}C_{DIC}$.

E. Photorespiration and Glycolate Excretion

In the light the dense phototrophic biomass in the cyanobacterial mat depletes CO_2 and accumulates oxygen, which sometimes may reach high supersaturation. Such conditions will support photorespiration. RubisCO also possesses oxygenase activity and can oxidize ribulose-1,5-bisphosphate to one molecule of each 2-P-glycolate and 3-P-glycerate (Lorimer et al., 1973; Lorimer, 1981; Miziorko and Lorimer, 1983). In fact, Rubisco has a much better affinity for oxygen as substrate than for carbon dioxide (Pierce, 1988) and it has been suggested that the original function of the enzyme was an oxygenase rather than a carboxylase (Tabita, 1988). Warburg (1920) discovered that O_2 inhibited CO_2 fixation in algae. Schau et al. (1950) showed that glycolate was a product of CO_2 fixation and Warburg and Krippahl (1960) demonstrated that its synthesis could be stimulated by oxygen. Glycolate is produced from 2-P-glycolate by phosphoglycolate phosphatase and is metabolized via the glycine-serine pathway (Renström-Kellner and Bergman, 1990), resulting in the production of CO_2 and NH_3. This light-dependent oxygen uptake and CO_2 evolution is called photorespiration. Photorespiration may represent a serious loss of fixed carbon which may be as high as 15-50% of net photosynthesis (Artus et al., 1986; Gerbaud and Andre, 1987). It may be asked which function of photorespiration would be so important to justify this loss of productivity. To date no clear answer can be given to this question. In microbial mats it may help to prevent photooxidative damage, particularly under conditions of high oxygen and low carbon dioxide concentrations in the light. Other hypotheses include the synthesis of metabolic intermediates, the lowering of O_2 and the production of CO_2 (Husic et al., 1987).

Many microbial mats are characterized by oxygen supersaturation in the light and CO_2 depletion and by very high pH (sometimes above 10) They also subject to high light intensities and are chronically nitrogen depleted. All these factors will force the cyanobacteria to maximum rates of photorespiration. Glycolate metabolism, and therefore photorespiration, is closely associated with nitrogen metabolism. Renström-Kellner and Bergman (1989) demonstrated that the excretion of glycolate by *Anabaena cylindrica* decreased drastically in the presence of a source of nitrogen such as NH_4Cl or glutamate. As was shown by these authors, nitrogen-fixing cyanobacteria could lose up to 60% of photosynthetic fixed CO_2 as glycolate. Since heterocystous cyanobacteria are capable of providing themself with sufficient nitrogen through the fixation of N_2, most mat-forming cyanobacteria are non-heterocystous and probably grow under severe nitrogen limitation. This suggests that these organisms may even lose the greater part of net photosynthesis. Bateson and Ward (1988) showed the importance of glycolate in microbial mats. Glycolate may serve as a substrate for sulfate-reducing bacteria in microbial mats, even in the presence of oxygen (Fründ and Cohen, 1992). More recently glycolate-oxidizing sulfate-bacteria were indeed isolated from marine sediments (Friedrich and Schink, 1993; 1995), but these organisms were strictly anaerobic.

F. Organic Compatible Solutes

In marine and hypersaline environments microorganisms accumulate solutes in order to obtain a sufficient turgor pressure necessary to allow cell division and growth (Taiz, 1984). The cytoplasmic membrane is permeable to water and an organism that is exposed to an elevated salt concentration in the surrounding medium would tend to lose water. In order to retain water inside itself the cell can either take up ions until an osmotic equilibrium with the environment is obtained or accumulate low molecular weight organic solutes. High concentrations of inorganic ions are not compatible with the metabolism of cyanobacteria and cause inhibition of enzyme activity (Warr et al., 1984).

Cyanobacteria can be subdivided into three groups with respect to the type of organic solute they accumulate in response to osmotic stress (Reed et al., 1986a). Halotolerant freshwater cyanobacteria accumulate disaccharides (either sucrose or trehalose). Marine cyanobacteria accumulate the heteroside glucosylglycerol (2–O–α–D–glucopyranosylglycerol) and the very halotolerant hypersaline cyanobacteria accumulate quaternary ammonium compounds (glycine betaine and in one case glutamate betaine) (Mackay et al., 1984). The type of osmolyte accumulated by cyanobacteria seems to be predominantly related to the degree of halotolerance of the organism. Halotolerance increases in the order disaccharide < heteroside < quaternary ammonium compounds (Reed et al., 1986a). There is no clear link between the type of solute and the taxonomic group of cyanobacteria, although all strains of *Anabaena* that were screened accumulated sucrose in response to osmotic stress. A habitat relationship is suggested among species of the unicellular *Synechococcus.* Of the 33 strains investigated all originating from freshwater environments accumulated sucrose, those isolated from marine systems accumulated glucosylglycerol and those from hypersaline habitats without exception betaine (Reed et al., 1986a). Stal and Reed (1987) screened 25 strains of cyanobacteria isolated from a microbial mat in the North Sea and found glycosylglycerol as well as trehalose and sucrose as osmolytes, suggesting no habitat relationship with the type of solute. Glucosylglycerol was nevertheless typically the dominant osmolyte in this marine ecosystem. The two dominant cyanobacteria in this mat, *Microcoleus chthonoplastes* and *Oscillatoria limosa* accumulated glucosylglycerol and trehalose, respectively. This property has been used to estimate the respective biomass of both species in these microbial mats (Stal and Reed, 1987). Betaine seems to be limited to cyanobacteria from hypersaline environments.

Although cyanobacteria normally accumulate a single, low-molecular weight organic compound in response to osmotic stress, many species may produce a secondary compound. The synthesis of disaccharide is much faster than glucosylglycerol. Within 8 h of an osmotic upshock the disaccharide pool has reached 90% of its maximum, while with glucosylglycerol this is only the case after 24 – 48 h (Reed and Stewart, 1988). Therefore the synthesis of disaccharide as secondary osmolyte may help for a quicker response to salt stress. Thus cyanobacteria that thrive under relative constant salinities may prefer glucosylglycerol while those that are exposed to fluctuating salinities may be better off with trehalose for example. This difference could explain why the pioneer in microbial mats, *Oscillatoria* sp., contains trehalose while the typical organism in established microbial mats, *Microcoleus chthonoplastes,* contains glucosylglycerol.

For the same reason hypersaline species contain sucrose in addition to betaine. Since betaine is a nitrogen-containing compound, nitrogen deficiency may also lead to the accumulation of non-nitrogenous sucrose as secondary osmolytes (Trüper and Galinski, 1989).

Osmotic down shock exerted on betaine-containing *Aphanothece halophytica* results in the release of this osmolyte into the environment (Reed and Stewart, 1988). This may have important consequences for an ecosystem such as a microbial mat because it may allow chemotrophic bacteria that can not synthesize betaine to take it up from the environment (Reed and Stewart, 1988). Moreover, betaine may serve as substrate for sulfate-reducing bacteria and the product of its metabolism, trimethylamine (TMA) is known as a so-called non-competitive substrate for methanogenic bacteria (Heijthuisen and Hansen, 1989).

Osmotic downshock in *Rivularia atra* resulted in a corresponding decrease of the osmoticum trehalose but only 10% was recovered from the medium and the rest was apparently metabolized or converted to glycogen (Reed and Stewart, 1983). The glucosyl glycerol accumulating *Synechocystis* PCC6714 and the sucrose-containing *Synechococcus* PCC6311 released 50% of their carbohydrates and over 70% of their amino acids after hypo-osmotic shock (Reed et al., 1986b). However, in some other cyanobacteria there is no evidence for the release of low molecular weight compounds upon hypo-osmotic shock (Reed and Stewart, 1988). The release by cyanobacteria of low molecular weight compounds into a microbial mat would have a great impact on the ecosystem. The cellular concentration of these osmolytes is considerable and at full seawater salinity it may amount to as much 270 mM. These carbohydrates are easy accessible substrates for chemotrophic bacteria in the mat. Except in the case of a hypo-osmotic shock, which may occur in exposed microbial mats after a rain shower for instance, osmotica will also be liberated after death and lysis of the organism.

Microbial mats have often been found to evolve dimethylsulfide (DMS), a sulfur-containing organic volatile compound. It is known that DMS can be produced from dimethylsulfoniopropionate (DMSP) by microbial activity or by chemical decomposition at high pH (Kiene and Visscher, 1987). DMSP occurs in a number of eukaryotic algae where its most likely function is that it serves as an osmoprotectant ((Turner et al., 1988). Visscher and Van Gemerden (1991) suggested that the cyanobacterium *Microcoleus chthonoplastes* may produce DMSP as a secondary osmolyte and could be the source of DMS in microbial mats. However, Van Bergeijk and Stal (1996) found a correlation between the number of diatoms in these mats and the amount of DMS that evolved from it. Some benthic diatoms accumulate large amounts of DMSP as osmoticum.

G. Fermentation

When in microbial mats photosynthesis ceases, they may rapidly, sometimes even within minutes, turn anoxic. Cyanobacteria are essentially aerobic organisms that during the dark normally have a respiratory metabolism in which the endogenous storage carbohydrate glycogen is degraded (Smith, 1982). When oxygen is absent, aerobic respiration is evidently not an option. Many cyanobacteria die and lysis occurs within 2-3 h after transfer to dark anoxic conditions. However, mat-forming cyanobacteria survive dark anoxic conditions for much longer time, often for several days. A number of these cyanobacteria were investigated in more detail and it was discovered that they were capable of fermenting glycogen. This is an important aspect in microbial mats because the excretion of fermentation products supplies other microorganisms, notably the sulfate-reducing bacteria, of substrate.

Stal and Moezelaar (1997) reviewed fermentation in cyanobacteria. Table 2 lists cyanobacteria capable of fermentation. The phenomenon was first discovered in *Oscillatoria limnetica* (Oren and Shilo, 1979), which occurs in the sulfide-rich hypolimnion of Solar Lake, Sinai, and is typically adapted to anaerobic growth. In the dark this organism ferments glycogen to lactate. Since no other fermentation products were found, it was assumed that the homolactic acid pathway was used in this organism. In the non-heterocystous diazotrophic mat-building *Oscillatoria limosa*, heterolactic acid fermentation was found (Heyer et al., 1989). This organism produced equimolar amounts of lactate and ethanol from glycogen. In addition, it is capable of homoacetic fermentation, for which its osmoprotectant trehalose was used as substrate. Trehalose was degraded to approximately 5-6 acetate and some hydrogen and CO_2. The occurrence of homoacetate fermentation in cyanobacteria is remarkable since it further only occurs in a group of specialized anaerobic bacteria, the acetogenic bacteria. Homoacetate fermentation is energetically efficient. More recently, the occurrence homoacetate fermentation has been proposed in a number of other cyanobacteria (De Philippis et al., 1996). The degradation of the osmoprotectant in *O. limosa* was another unexpected phenomenon. Of course, trehalose represents a large amount of energy, which may be important for the organism to use under a situation of severe starvation. The question of how the organism compensates for the loss of compatible solute has not been answered, but it has been suggested that this may be through a temporary accumulation of inorganic ions such as K^+ (Stal and Moezelaar, 1997). Also the mat building cyanobacterium *Microcoleus chthonoplastes* has been shown to ferment part of its osmoprotectant (Moezelaar et al., 1996). *M. chthonoplastes* accumulates the heteroside glucosyl glycerol which is only degraded in cultures that contain low amounts of glycogen. Unlike in *O. limosa*, *M. chthonoplastes* possesses only one fermentation pathway. Glycogen and the glucose part of glucosyl glycerol are fermented via a mixed acid fermentation, resulting in the formation of formate, acetate, ethanol, lactate, H_2 and some CO_2.

There is a great diversity of fermentation pathways in cyanobacteria. In some cases the pathways have been elucidated by the demonstration of the enzymes involved. In *O. limosa* the key enzymes of the homoacetate pathway have been demonstrated as well (Heyer et al., 1989). In the majority of pathways the

Table 2. Cyanobacteria capable of fermentation

ORGANISM	STRAIN, ORIGIN	FERMENTATION PATHWAY	PRODUCTS*
Anabaena azollae AaL	Symbiont from *Azolla caroliniana*	Homoacetate	acetate (lactate, CO_2, H_2)
Anabaena azollae AaN	Symbiont from *Azolla caroliniana*	Homoacetate	acetate (lactate, CO_2, H_2)
Anabaena azollae AaS	Symbiont from *Azolla filiculoides*	Homoacetate	acetate (lactate, CO_2, H_2)
Anabaena siamensis As1	Paddy field	Homoacetate	acetate (CO_2, H_2)
Cyanothece	PCC 7822 (Inst. Pasteur)	Mixed acid	H_2, ethanol, lactate, formate, acetate
Microcoleus chthonoplastes	Microbial mat	Mixed acid	H_2, ethanol, lactate, formate, acetate
Microcystis aeruginosa	PCC 7806 (Inst. Pasteur)	Mixed acid	H_2, ethanol, acetate
Nostoc sp. Cc	Symbiont from *Cycas circinalis*	Homoacetate	acetate (lactate, CO_2, H_2)
Nostoc sp. Al2	Symbiont from *Anthoceros laevis*	Homoacetate	acetate (lactate, CO_2, H_2)
Nostoc sp. Ef1	Symbiont from *Encephalartos ferox*	Homoacetate	acetate (lactate, CO_2, H_2)
Nostoc sp. MAC	Symbiont from *Macrozamia lucida*	Homoacetate	acetate (lactate, CO_2, H_2)
Nostoc sp. Mm1	Symbiont from *Macrozamia moorei*	Homoacetate	acetate (lactate, CO_2, H_2)
Nostoc sp. M1	Symbiont from *Macrozamia* sp.	Homoacetate	acetate (CO_2, H_2)
Nostoc sp. Gm	Symbiont from *Gunnera manicata*	Homoacetate	acetate (lactate)
Nostoc sp. T1	Paddy field	Homoacetate	acetate (formate, CO_2, H_2)
Nostoc sp. Bali	Paddy field	Homoacetate	acetate (CO_2, H_2)
Oscillatoria limnetica	Hypolimnion Solar Lake	Homolactate	lactate
Oscillatoria limosa	Microbial mat	Heterolactate Homoacetate	lactate, ethanol, acetate
Oscillatoria sp.	Microbial mat	not known	lactate, ethanol, acetate, formate
Oscillatoria terebriformis	Hot spring microbial mat	Homolactate?	?
Spirulina platensis	not known	Mixed acid	H_2, ethanol, acetate, formate, lactate
Spirulina minosa	not known	not known	lactate, acetate

*compounds in brackets are produced in minor quantities. From Stal and Moezelaar (1997)

Embden-Meyerhof-Parnas pathway (glycolysis) was involved in fermentation. Only the heterolactate fermentation makes use of parts of the oxidative pentose phosphate pathway. The oxidative pentose phosphate pathway is used by cyanobacteria during aerobic dark respiration and it is essentially the reverse of the reductive pentose pathway, which serves CO_2 fixation in the light (Smith, 1982). In all cyanobacteria capable of fermentation the capacity for fermentation appears to be constitutive (Stal and Moezelaar, 1997). Fermentation as a constitutive property would have also a number of advantages. In the first place it would greatly increase the reactivity of the organism. Microbial mats are generally environments in which steep gradients of light and oxygen occur and these factors fluctuate strongly. If oxygen disappears rapidly, fermentation can immediately provide energy for maintenance, allowing the organism to survive.

Hydrogen is often a product of fermentation in cyanobacteria. Hydrogenases in cyanobacteria have been extensively reviewed by Houchins (1984). Cyanobacteria possess different hydrogenases. Nitrogen-fixing cyanobacteria produce hydrogen as a by-product of nitrogenase. Because nitrogenase obligatory produces hydrogen during nitrogen fixation, aerobic nitrogen-fixing cyanobacteria usually possess an uptake hydrogenase. This enzyme carries out an oxy-hydrogen reaction. The third type of hydrogenase in cyanobacteria is reversible hydrogenase. This enzyme is frequently found in obligately anaerobic bacteria. Depending on the conditions it catalyses either the uptake or the production of hydrogen at approximately equal rates. Although its function in cyanobacteria has been debated for some time, reversible hydrogenase plays certainly an important role in fermentation (Stal and Moezelaar, 1997).

Elemental sulfur may serve as an electron acceptor in cyanobacteria. Many cyanobacteria have been shown to be able to reduce elemental sulfur to sulfide. In *Oscillatoria limnetica* evidence has been provided that sulfur reduction might represent a type of anaerobic respiration (Oren and Shilo, 1979). However, in other cyanobacteria the advantage of the reduction of sulfur is probably that it serves as an electron sink, allowing the formation of a more oxidized product (acetate) which results in a higher amount of substrate phosphorylation.

Stal and Moezelaar (1997) have discussed the energetics of fermentation in a number of different cyanobacteria on which sufficient information is available. Although evidently the amount of energy that is generated during fermentation is low, calculations showed that it usually exceeded the minimum amount required for maintenance. This remaining energy could even drive metabolic processes. For instance, *O. limosa* is capable of maintaining a considerable rate of nitrogen fixation under anaerobic conditons in the dark (Stal and Heyer, 1987).

H. Extracellular Polymeric Substances (EPS)

Extracellular polysaccharides are important components in microbial mats. They are involved in the attachment of cyanobacteria to the substrate and also produce a matrix in which the orgamisms are embedded. This polysaccharide matrix fulfills a number of other important functions in microbial mats, which will discussed below.

Cyanobacteria produce polysaccharides which can be roughly categorized in three groups:

(i) endogenous polysaccharides that serve as storage compounds
(ii) cell envelope polysaccharides
(iii) extracellular polysaccharides.

The endogenous polysaccharides in cyanobacteria can be found in the so-called α-granules, which are composed of a branched glycogen-like polymer. This polymer consists of α(1-4) and α(1-6) linked glucose molecules. The cell envelope consists of the cell wall polysaccharides and the external layers (glycocalyx). The glycocalyx can be subdivided into:

(i) the well-structured polysaccharide sheath
(ii) a polysaccharide capsule which extends outside the sheath but is clearly associated with the organism and is less structured
(iii) mucilage polysaccharide.

The latter is not or very loosely associated with the organism. In fact, the polysaccharides that form the glycocalyx should all be considered as extracellular polysaccharides or exopolysaccharides (Fig. 14). The different fractions are often poorly defined and mostly based on the different extraction procedures. Relatively little is known about cyanobacterial exopolysaccharides (Bertocchi et al., 1990; Drews and Weckesser, 1982).

Microbial exopolymers, including those produced by cyanobacteria are high molecular weight mucous secretions that often have a complex structure. The molecular weight is often more than 100000 dalton.

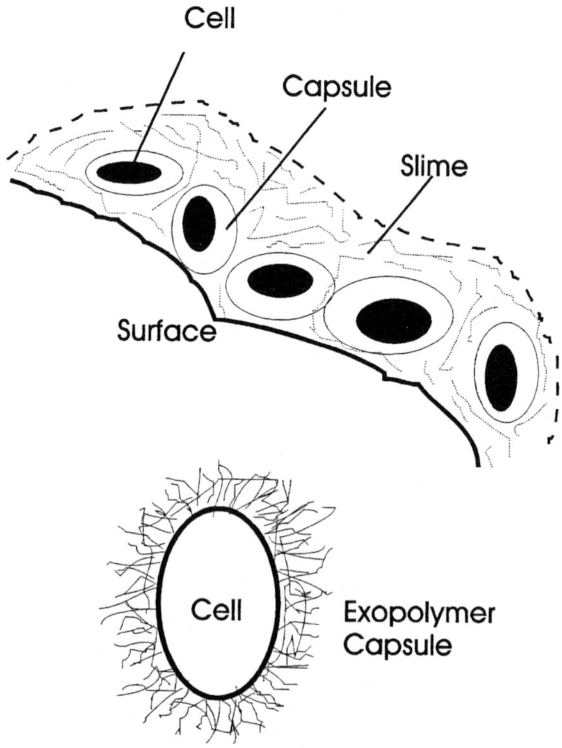

Fig. 14. Capsular and slime extrapolymeric substances (EPS) and the formation of a microbial mat. After Decho (1990).

Although polysaccharides are quantitatively the most important part of these exopolymers, other components are present as well. Particularly proteins make up a significant part of the exopolymers (Decho, 1990). These polymers are also known as extracellular polymeric substances (EPS). The composition and structure of EPS vary widely among different microorganisms (Tago and Aida, 1977; Bertocchi et al., 1990; Decho, 1994; Stal, 1994) and even one single strain may produce more than one type EPS simultaneously or at different stages of growth (Christensen et al., 1985). Most of the polysaccharides in EPS are heteropolysaccharides that are composed of variety of different monosaccharides, arranged in repeating units. EPS often contain uronic acids such as D-glucuronic acid, D-galacturonic acid and D-mannuronic acid. These are important functional groups because they contain carboxyl groups that are responsible for interactions with other EPS molecules or the binding of metals. However, other types of EPS are composed of neutral sugars. Extracellular polymeric substances may be hydrophilic or hydrophobic. Many are hydrophilic and may contain over 95% water by weight (Decho, 1994). Depending on the chemical composition and the functional groups present the tertiary structure of EPS is determined. The tertiary structure of EPS determines whether it is a cohesive gel or in a colloidal form. An intermediate form could be described as a looser mucilage (Decho, 1994). The tertiary structure of EPS not only depends on the chemical composition but also strongly on temperature. Microbial mats and intertidal mudflats during emersion are subject to large variations in temperature and this will thus effect the cohesiveness and rheological properties of the sediment.

A large number of functions have been ascribed to EPS (Decho, 1990). These include adhesion and immobilisation of the organism, protection against desiccation, protection from grazing, protection from toxic substances, scavenging of trace metals, (anti) calcification and other. Some of these functions will be discussed below as far as they are relevant to microbial mats.

Organisms in microbial mats are often subject to desiccation. EPS may retain large amounts of water, and organisms that produce it may thus overcome long periods of drought (Potts and Friedman, 1981; Potts et al. 1983; Shephard, 1987; Caiola et al., 1996). Particularly when EPS contain uronic acids or hydrophobic proteins, they may be important to micro-organisms, including cyanobacteria and diatoms, enabling these to attach to surfaces (Robins et al., 1986). For benthic organisms it is important to stay on surfaces when conditions are optimal for growth. Some cyanobacteria are capable of modifying EPS from hydrophobic to hydrophilic and they may thus detach from a surface when conditions become inappropriate (Bar-Or et al., 1985). Benthic communities of diatoms may attach to the surface of intertidal mudflats by the production of hydrated and hydrophilic exopolymers during periods of emersion. During immersion these polymers go into solution releasing the diatoms into the water column (Talbot et al., 1990). Benthic cyanobacteria may secrete flocculants, exopolymers that produce flocs with detritus and other material in the overlying water. These flocs eventually sediment, thereby clearing the overlying water and hence improving the conditions for these benthic phototrophs (Bar-Or and Shilo, 1987; 1988).

Mat-forming cyanobacteria, that excrete EPS, produce a matrix that stabilizes the sediment (Fig. 14). This is also the case with benthic films of diatoms that grow on intertidal mudflats (Paterson, 1989; Stal, 1994; Yallop et al., 1994). Uronic acids are important components of EPS because these charged groups would interact with sediment particles. Thus EPS with a large content of uronic acids are more efficient in the stabilization of sediments (Martin, 1971; Stal, 1994). EPS may also contain sulfated sugars. As the uronic acids, sulfate groups are also important for the tertiary structure of the polysaccharide and influence the stability of the microbial mat matrix (Decho, 1990). Uronic acids as well as sulfate groups interact with a variety of metals. This may either result in the immobilization of toxic metals or scavenge trace metals that form important nutrients for the organisms. The uronic acid groups of polysaccharides may be involved in the regulation of calcification. Sulfated polysaccharides are often encountered in eukaryotic algae but rarely in prokaryotes, including cyanobacteria (Bertocchi et al., 1990). Nevertheless, sulfated polysaccharides have been found in cyanobacteria (Tease et al., 1991; Ortega-Calvo and Stal, 1994) and a more thorough investigation of mat-forming cyanobacteria may reveal that such polysaccharides are more common in this group of organisms than previously thought.

The polysaccharide produced by the mat-forming cyanobacteria fulfils an important function as a matrix for exoenzymes, plasmids and DNA (Decho, 1990). Extracellular DNA is protected against DNases in the sediment (Romanowski et al., 1991) and may give rise to natural transformation in these ecosystems (Lorenz and Wackernagel, 1990; 1994). Hence, in microbial mats considerable gene exchange may take place.

A considerable number of functions can be attributed to EPS but it is not clear what controls the formation of this polysaccharide and how this relates to one or more of the possible functions. It may very well be that the production of mucilage by cyanobacteria is the result of unbalanced growth caused by nutrient deficits (Lange, 1976). Particularly a shortage or deficiency of nitrogen and sulfur results in the stagnation of protein synthesis while the full photosynthetic capacity remains. Under such conditions cyanobacteria accumulate large amounts of glycogen (Allen and Smith, 1969; Lehmann and Wöber, 1976). The capacity of the cell to store glycogen is limited and any additional polysaccharide may be excreted as mucilage. Old starved cultures often become viscous as a result of excess mucilage production.

Little is known about the fate of EPS in mats. Some polysaccharides seem to be recalcitrant to microbiological degradation, whereas others are not.

VI. Calcification in Mats and Stromatolites

Calcification is responsible for the lithification of microbial mats and is the basis of the formation of stromatolites. In most cases calcification seems under a stringent biological control, but the mechanisms by which living organisms influence the precipitation of calcium carbonate are poorly understood. Whereas the function of calcium carbonate precipitation in many organisms is obvious (e.g. shell formation) this is not the case in microorganisms, including algae and cyanobacteria. Calcification in bloom-forming algae such as the coccolithophore *Emiliania huxleyi* would predominantly serve the production of CO_2:

$$2\,HCO_3^- + Ca^{2+} \longleftrightarrow CaCO_3 + CO_2 + H_2O$$

The so-called 'whitings', clouds of aragonite needles, that often occur in tropical lagoons and that have been considered as inorganic precipitates may be produced by the photosynthetic activity of dense communities of picoplankton (Robbins and Blackwelder, 1992). Calcification in microbial mats may serve as a mechanism of producing CO_2. Due to the dense phototrophic biomass and high rates of photosynthesis and the alkaline conditions it is likely that mats become depleted in CO_2. Another function that has been proposed is the detoxification of intracellular calcium. But whatever the function, calcification can be generally inferred from the changes in the concentration of inorganic carbon and from the low solubility product of calcium carbonate.

Calcium carbonate is rather insoluble. Aragonite, which is often thought to be a product of biologically produced calcium carbonate, has a solubility product of $10^{-8.22}$, and the more stable form calcite $10^{-8.42}$ (Ehrlich, 1996). Dolomite has a solubility product of even $10^{-16.7}$. This means that when the product of the ion activities (molar concentrations multiplied by their activity coefficients) of Ca^{2+} and CO_3^{2-} in a solution exceeds $10^{-8.22}$ calcium carbonate may precipitate.

$$Ca^{2+} + CO_3^{2-} \longrightarrow CaCO_3$$

In normal seawater the concentration of calcium-ion is about 10^{-2} M (Ehrlich, 1996). When the concentration of calcium ion is assumed to be constant under certain conditions, then the carbonate concentration determines calcification. CO_2 reacts with water to form bicarbonate and this dissociates according to the following reversible reactions:

$$CO_2 + H_2O \longleftrightarrow H^+ + HCO_3^-$$
$$HCO_3^- \longleftrightarrow H^+ + CO_3^{2-}$$

In microbial mats a number of biological processes influence the equilibria of carboxy species and hence may control calcification (Krumbein, 1979b; Ehrlich, 1996). These processes include particularly metabolisms in which CO_2 is consumed, such as photosynthesis, chemosynthesis and, less importantly, heterotrophic CO_2 fixation. Metabolisms in which CO_2 is produced such as respiration and fermentation may cause an acidification of the medium and eventually result in a dissolution of calcium carbonate rather than a precipitation. Nevertheless, Krumbein (1974) demonstrated the formation of aragonite on the surface of marine bacteria as the result of their metabolism of substrates such as glucose, sodium acetate and sodium lactate. However, this also strongly depends on the environmental conditions that apply (Canfield and Raiswell, 1991). A variety of metabolic processes influence the equilibrium of inorganic carbon by the production of acids and bases:

1. Photosynthesis is an important process in the vast majority of microbial mats. Because this process

usually involves CO_2 fixation it has often been considered important for calcification (Golubic, 1973; Krumbein and Giele, 1979; Pentecost, 1988). The fixation of CO_2 from a bicarbonate solution will result in an increase in carbonate ion:

$$2\,HCO_3^- \longleftrightarrow CO_2 + H_2O + CO_3^{2-}$$

Other metabolic processes that are important in microbial mats and that remove CO_2 are chemosynthesis and heterotrophic CO_2 fixation. Organisms that carry out chemosynthesis include colorless sulfur oxidizing bacteria, nitrifying bacteria, autotrophic sulfate-reducing bacteria as well as methanogenic- and acetogenic bacteria. Heterotrophic CO_2 fixation occurs in virtually all organisms but is limited and negligible compared to the CO_2 produced during the oxidation of organic compounds. This process is therefore not important for calcification. Many reports mention calcification as a result of photosynthesis by cyanobacteria (e.g. Golubic, 1973; Krumbein and Giele, 1979; Pentecost, 1988; Pentecost and Bauld, 1988). Calcification associated with anoxygenic photosynthesis or anoxygenic phototrophic bacteria has not been reported. Precipitation of calcium carbonate may be indirectly associated with photosynthesis and due to an increase of pH and/or a shift in the equilibrium of inorganic carbon (Golubic, 1973; Krumbein and Cohen, 1977). Anoxygenic phototrophic bacteria, which use sulfide as an electron donor, do not exert such effects on their environment during photosynthesis. The same holds for most chemosynthetic metabolisms and for heterotrophic CO_2 fixation. On the other hand, Chafetz and Buczynski (1992) demonstrated that calcification in stromatolithic microbial mats was associated with heterotrophic bacteria rather than with the cyanobacteria.

2. Aerobic or anaerobic oxidation of organic compounds results in the production of CO_2 and/or HCO_3^- and affect pH and consequently causes a shift in the equilibrium of inorganic carbon. Organic matter possessing the "Redfield" stoichiometry of C:N:P of 106:16:1 is oxidized by O_2, NO_3^- and SO_4^{2-} according to the following reactions (Boudreau and Canfield, 1993):

$$(CH_2O)_{106}(NH_3)_{16}H_3PO_4 + 138O_2 \rightarrow 122\,H_2O + 106\,CO_2 + 16\,NO_3^- + 16\,H^+ + H_3PO_4$$

$$(CH_2O)_{106}(NH_3)_{16}H_3PO_4 + 94.4NO_3^- \rightarrow 84.8\,H_2O + 13.6\,CO_2 + 92.4\,HCO_3^- + 55.2\,N_2 + H_3PO_4$$

$$(CH_2O)_{106}(NH_3)_{16}H_3PO_4 + 53SO_4^{2-} \rightarrow 106\,HCO_3^- + 53\,H_2S + 16\,NH_3 + H_3PO_4$$

The formation of CO_2 and the acidification of the medium could result in the dissolution of calcium carbonate rather than cause its precipitation. Anaerobic respiration results in the formation of bicarbonate and could give rise to supersaturation of calcium carbonate. The precipitation of sulfide as iron sulfide acts as a pH buffer. The effects of the sequential oxidation of organic matter by the three electron acceptors oxygen, nitrate and sulfate on porewater pH and calcium carbonate saturation are complex and depend on the prevailing conditions (Boudreau and Canfield, 1993).

In the microbial mats of Solar Lake (Sinai) it has been shown that sulfate reduction and $CaCO_3$ formation were stoichiometrically related and organic carbon was transformed into a number of different carbonate minerals (Jørgensen and Cohen, 1977; Krumbein and Cohen, 1977; Krumbein et al, 1977). However, whether sulfate reduction in microbial mats in reality results in calcium carbonate precipitation depends largely on a variety of conditions that prevail in these microbial mats. Most important are the development of alkaline conditions, the removal of excess CO_2 or the presence of a suitable buffer (Ehrlich, 1996). In the absence of iron, sulfate reduction produces equal amounts of H^+ and HCO_3^-, which will thus cause a decrease of carbonate saturation. In many microbial mats high rates of sulfate reduction occur, but despite this calcification is totally absent.

The biological control over calcium carbonate precipitation in the ocean leads to overproduction. It is estimated that 5 Gt calcium carbonate is annually produced in the ocean of which 3 Gt is removed from the system by incorporation and accumulation in sediments, while the other 2 Gt is dissolved (Milliman, 1993). The weathering of rock on the continents causes a continuous runoff of calcium and carbonate into the sea. Therefore the oceans tend to be supersaturated with calcium carbonate. In order to maintain a steady state, the amount of calcium carbonate removed from the oceans must be the same as that entering. However, it is estimated that twice as much calcium is removed from the ocean by calcium carbonate precipitation than is brought in

(Milliman, 1993). This means that the ocean is not in equilibrium or that sources and sinks are respectively under- or overestimated. The equilibrium of calcium carbonate in the oceans could be maintained by the dissolution of the excess calcium carbonate. Part of this dissolution is biologically controlled because it acts as a pH buffer for respiratory and fermentative processes. Another part dissolves in the deep sea, which is under saturated with calcium carbonate. Some calcium carbonate leaves the system by sinking as fecal pellets to the ocean floor or by fast burial. Although the surface waters of the ocean are supersaturated with calcium carbonate spontaneous precipitation does not normally occur.

Recent microbial mats are often considered as the structural analogues of stromatolites. By definition stromatolites are lithified laminated formations that are known from the Archaean, 3.6 Gyr b.p. to the present day (Margulis et al., 1980) and were particularly abundant in the late Proterozoic (Awramik, 1984). Precambrian stromatolites were formed in shallow marine areas. Lithification of present day microbial mats is extremely rare and it is still an enigma why this should be so. Kempe and Kazmierczak (1990a, b) and Kazmierczak et al. (1996) investigated stromatolites in the sea-linked Satonda Crater Lake in Indonesia and alkaline Lake Van in Turkey, both formed under extreme alkaline conditions. They hypothesized that the greater abundance of stromatolites during the Precambrian should be attributed to the much greater alkalinity of the marine environment during that era (Kempe and Kazmierczak, 1990a). The hypothesis of a Precambrian soda ocean may certainly offer an explanation for the greater abundance of stromatolites and the discovery of modern calcifying stromatolites in alkaline seas supports this. Nevertheless, calcification in these stromatolites is still under biological control rather than being a spontaneous occurence. Moreover, other recent stromatolites are formed under less alkaline or normal marine conditions such as those found in the French Polynesian atolls (Défarge et al., 1994a, b) or in the Bahamas (Reid and Browne, 1991). Even if the early oceans were more alkaline, the marine environment today is still supersaturated with calcium carbonate. Furthermore, in microbial mats several biological processes predominate which presumably increase the concentration of carbonate ion, which theoretically should lead to calcium carbonate precipitation. As a result of active photosynthesis and CO_2 fixation in the top layer of cyanobacterial mats the pH in these mats may reach values as high as 9.5 (Fig. 3) (Revsbech et al., 1983). Notwithstanding these factors that would normally promote calcification, most marine microbial mats do not lithify.

Hence spontaneous calcification does not seem to be important in stromatolites. Biological control of calcification may not only exist in the change of the carbonate ion concentration and equilibrium but also in a mechanism that inhibits calcification (anti-calcification) (Westbroek et al., 1994).

Very little is known about anti-calcification mechanisms. Biologically controlled calcification must distinguish between supersaturation of calcium carbonate in a solution (which is the thermodynamic force) and those factors that influence the kinetics of the process. The latter may be either inhibitory or stimulatory factors. Supersaturation of calcium carbonate in the ocean is the primary driving force of calcification that can be dramatically increased in the immediate vicinity of phototrophic organisms. Because uncontrolled calcification in or around organisms may cause damage it will be clear that some mechanism must exist that inhibit the process. Crystal poisons such as Mg^{2+} and PO_4^{3-} are not sufficient and additional mechanisms must be postulated. It is known that some small acidic molecules may inhibit crystallization. An example is the binding of Ca^{2+} to oxalate (Verrecchia et al, 1990). Acidic polysaccharides are also very effective in binding Ca^{2+} or interact with it (Decho, 1994). These will doubtless influence calcification. Fig. 15 depicts the way in which such polysaccharides could influence crystallization or crystal growth (Westbroek et al, 1994). The association of a polyanion with Ca^{2+} ions will inhibit nucleation with CO_3^{2-} ions and prevent subsequent crystallization. Likewise, polyanions may associate with a growing calcium carbonate crystal and prevent its further growth. A layer of charged polymers may bind calcium carbonate crystal and arrest its growth. The structure of such polymers may determine crystal shape. Evidence has been obtained that this latter mechanism is involved in the formation and morphology of coccoliths in the coccolithophore *Emiliania huxleyi* (Borman et al, 1982, 1987).

In microbial mats it is hypothesized that extracellular polymeric substances (EPS) which are mainly composed of polysaccharides serve as agents that inhibit calcification. EPS produced by cyanobacteria are often rich in uronic acids and contain other acidic groups (Bertocchi et al, 1990, Decho, 1990, 1994). Many microbial mats are

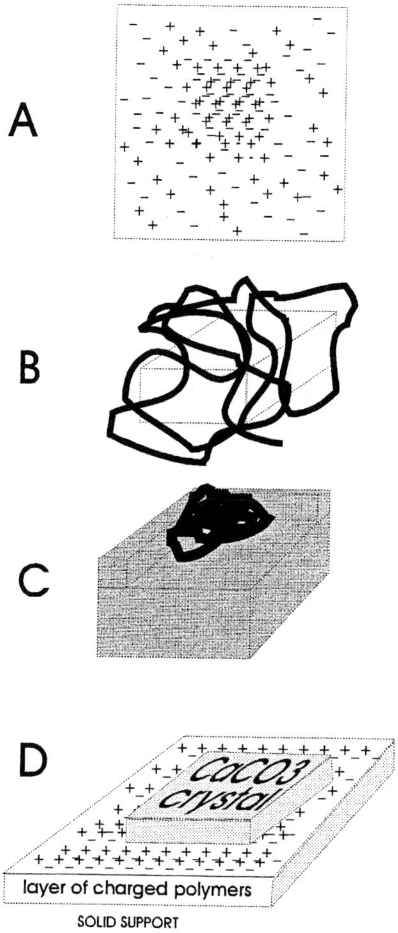

Fig. 15. Simplified model of the possible interactions of charged extracellular polymeric substances with calcium carbonate. *A*: nucleation of calcium (+) and carbonate (-) ions. *B*: Inhibition of nucleation of polyanion. *C*: Inhibition of crystal growth by association of a crystallization nucleus with a polyanion. *D*: Calcium carbonate crystal bound to a layer of charged polymers. The growth of the crystal may be arrested. The charged polymer may determine the eventual shape of the calcium carbonate crystal. After Westbroek et al. (1994).

composed of vast amounts of EPS in which the cyanobacteria and other organisms are embedded. It is possible that this EPS acts as an anti-calcification agent. When heterotrophic bacteria decompose this EPS, high concentrations of calcium carbonate may exist locally, leading to precipitation. In some non-lithifying microbial mats such as in Solar Lake (Sinai) aragonite needles are formed in the deeper layers of the mat, where the organic matter is subject to degradation. Several others have observed the association of calcification with bacterial activity (Chafetz and Buczynski, 1992; Krumbein, 1979a; Krumbein and Giele, 1979).

Mucilage EPS is often produced by cyanobacteria as an overflow metabolism when they grow under nutrient limitation. This is particularly the case with nitrogen as the limiting factor, a situation that is normally occurring in the marine environment. In microbial mats, where extreme dense communities of cyanobacteria are present there is a high demand for nitrogen, while there is often a shortage of this important nutrient. Therefore, many microbial mats are diazotrophic, i.e. the cyanobacteria that build these mats fix atmospheric nitrogen. However, most diazotrophic mats consist of non-heterocystous cyanobacteria. As argued in the section on nitrogen fixation these cyanobacteria are not efficient nitrogen fixers because the process is seriously hindered by oxygen in these organisms. It is likely that these cyanobacteria are in fact still nitrogen limited. Cyanobacteria that grow under nitrogen limited conditions tend to produce a lot of mucilage (Ortega-Calvo and Stal, 1994). Perhaps Precambrian microbial mats as well as modern calcifying mats do not grow nitrogen-limited and therefore produce less mucilage. Some of these mats may for other reasons receive sufficient combined nitrogen for growth. Other microbial mats are composed of heterocystous cyanobacteria which are extremely well adapted to grow diazotrophically and are seldom likely to experience nitrogen limitation under normal conditions. One example from freshwater environments is the Rivulariaceae. This group of heterocystous cyanobacteria produces extent calcium carbonate formations (Whitton, 1987). The marine *Calothrix* belongs to this taxonomic group and in some exceptional cases also forms microbial mats.

A model for calcification and the development of stromatolites in the Exuma Cays (Bahamas) is presented in Fig. 16. Subtidal and intertidal stromatolites that can be found in the Exuma Cays (Bahamas) are characterized by mats of the cyanobacterium *Schizothrix* sp. The model for calcification in these mats is based on a number of observations and assumptions. Two types of mats of *Schizothrix* can be distinguished. Lithifying microbial mats of *Schizothrix* are usually characterized by low ratios of photosynthesis to respiration while the opposite is true in non-lithifying mats (Pinckney et al., 1995). Moreover, calcium carbonate was not associated with the cyanobacteria but with heterotrophic bacteria (Chafetz and Buczynski, 1992; Chafetz, 1994). It was further assumed that extracellular polymeric substances (EPS) may interfere by binding calcium and magnesium ions,

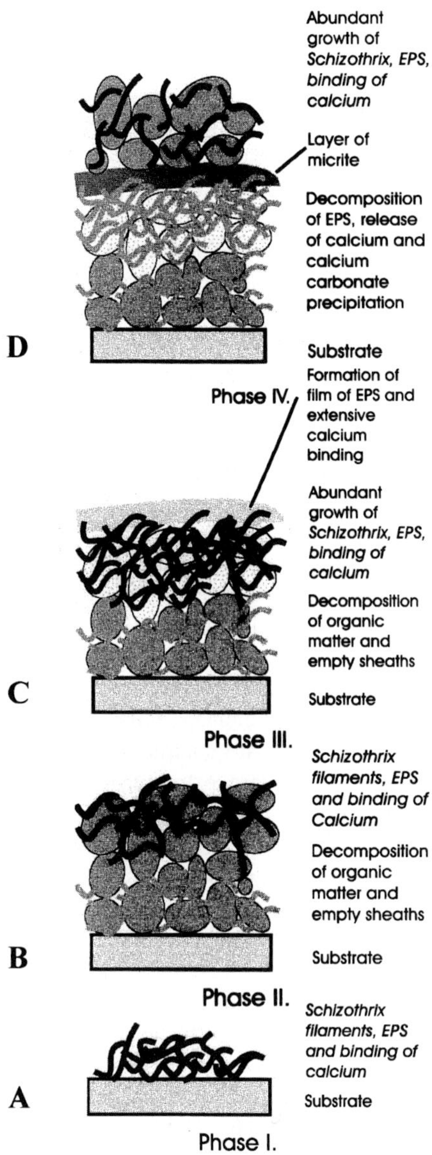

Fig. 16. Simplified scheme of the development of lithified micritic layers in Bahamas stromatolites. See text for explanation. This hypothesis was developed by P. Reid, J. Pinckney, H.W. Paerl, A.W. Decho and L.J. Stal during a research cruise in the Bahamas in 1994.

thus locally inhibiting carbonate precipitation (Borman et al., 1982; 1987; Westbroek et al., 1994).

Exuma Cays stromatolites are formed at high energy sites. Intertidal stromatolites can be found on the Atlantic Ocean coast and are exposed to high wave energy. Subtidal stromatolites are almost exclusively encountered in channels with high currents. The cyanobacterium *Schizothrix* sp. is a filamentous organism composed of thin (often less than 1 µm wide) trichomes which are enveloped by a thin rigid polysaccharide sheath. This organism is capable of colonizing a solid substrate. Because grazing pressure will be low in these high-energy areas, a community of *Schizothrix* may develop. Under conditions of low sedimentation rate a mat of *Schizothrix* and associated microorganisms will develop (Fig. 16a). These mats are rigid and tightly associated with the underlying substrate. The cyanobacteria will grow and produce sheath material and possibly some mucus. It is assumed that this EPS will bind Ca^{2+} or that uronic acids prevent further growth of crystallization nuclei (Borman et al., 1982). During a period of sedimentation, *Schizothrix* will move rapidly upwards by phototaxis and continue growth in the top layers where optimum light conditions prevail. The trichomes form a dense network in which carbonate sand grains are trapped and agglutinated by EPS, while Ca^{2+} is further bound. Empty sheaths and other organic matter that has been abandoned deeper in the sediment will be decomposed (Fig. 16b). During a subsequent period of low rates of sedimentation a dense mat of *Schizothrix* sp. will develop in the top layer of the sediment. This layer is characterized by active growth of the cyanobacteria and may be associated with abundant production of EPS. The matrix of EPS in which the mat is embedded may bind Ca^{2+} efficiently and condenses EPS to the gel state (Rees, 1969; Decho and Moriarty, 1990; Decho, 1994). It is conceived that this will lower the activity of this ion so that calcium carbonate will not precipitate. Depending on the physicochemical gradients that typically develop in microbial mats due to phototrophic and heterotrophic activities some dissolution and re-precipitation of $CaCO_3$ and re-crystallization of the carbonate sediment grains may occur (Fig. 16c). During the next stage of development, *Schizothrix* sp. moves upward after another period of high rate of sedimentation. While growth of the cyanobacterium and the production of EPS in the new top layer trap and agglutinate the carbonate sand, the large amount of organic matter left behind is decomposed by heterotrophic bacteria. Because EPS is also decomposed, Ca^{2+} that was bound will be released. This will locally cause supersaturation of calcium carbonate resulting in the formation of a microcrystalline crust of precipitated carbonate (Fig. 16d). A similar type of bacterial calcification occurs during the degradation of calcium oxalate. Oxalate is a product of metabolism of fungi and other organisms and is capable of immobilizing calcium (Verrecchia et al., 1990).

This model explains that *in situ* calcium carbonate precipitation and lithification of the mat is controlled by biology. It is indirectly associated with the cyanobacteria but degradation of organic matter, notably EPS, by heterotrophic bacteria is required for this process. Alternating periods with high and low rates of sedimentation are responsible for the formation of the laminated structure of the lithified microbial mats, which could therefore be termed stromatolites.

VII. Nitrogen Metabolism and Nitrogen Fixation

A. Introduction

In cyanobacteria nitrogen content may amount up to about 10% of dry weight and is quantitatively the third most important element. Any shortage of it will immediately affect the amount of phycobiliproteins and, consequently, the efficiency of light harvesting for photosynthesis (Allen and Smith, 1969). Cyanobacteria may produce a unique nitrogenous compound known as cyanophycin or multi-L-arginyl-poly(L-aspartic acid). Its high nitrogen content means that it can serve as a nitrogen reservoir (Mackerras et al, 1990a, b).

Cyanobacteria use a variety of nitrogen sources (Flores and Herrero, 1994). Ammonia can be taken up by passive diffusion or the protonated form ammonium (NH_4^+) by a specific uptake system (Fig. 17). The amino acids arginine, asparagine and glutamine have been reported to serve as nitrogen sources in cyanobacteria (Flores and Herrrero, 1994). Nitrate and nitrite are important sources of nitrogen for cyanobacteria. This involves the uptake of nitrate or nitrite and its subsequent reduction to ammonia. This process involves ferredoxin as an electron donor and is therefore intimately associated with photosynthesis. Many cyanobacteria are capable of using dinitrogen (N_2) as the source of nitrogen.

B. The Nitrogen Cycle in Microbial Mats

Nitrogen occurs in different chemical oxidation states, varying from its most reduced form ammonia (NH_3) (-3), to hydroxylamine (NH_2OH) (-1), dinitrogen (N_2) (0), nitrous oxide (N_2O) (+1), nitric oxide (NO) (+2), nitrite (NO_2^-) (+3) to its most oxidized form nitrate (NO_3^-) (+5). All of these oxidation states are biologically significant and microorganisms may carry out reduction and oxidation reactions transforming one form into another. The element nitrogen therefore is subject to microbiological cycling in nature. In microbial mats all steps of the nitrogen cycle may be present and cyanobacteria play a particular important role (Fig. 17).

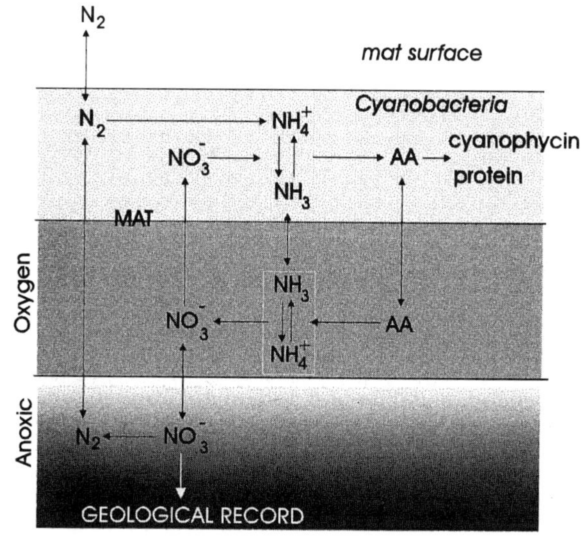

Fig. 17. The nitrogen cycle in a cyanobacterial mat. Cyanobacteria take up and assimilate ammonium into amino acids (AA) which are used for protein synthesis or can be stored as cyanophycin. Amino acids and ammonia may leak out the cells and be oxidized to nitrate by nitrifying bacteria. In the anoxic part of the mat nitrate is converted to dinitrogen by dinitrifying bacteria. Nitrate can be taken up by the cyanobacteria and assimilated. Nitrogen-fixing cyanobacteria reduce N_2 to ammonium.

Ammonia is assimilated into amino acids that are used for the synthesis of proteins. Luxury uptake of nitrogen may occur and be stored as cyanophycin. Ammonia and amino acids may leak out of the cell. When oxygen is present, ammonium may be oxidized via nitrite to nitrate by nitrifying bacteria. Nitrate may be taken up by the cyanobacteria and assimilated or under anoxic conditions converted to gaseous nitrogen by denitrifying bacteria. Hence, this process represents a loss of combined nitrogen, which is counteracted by the capacity of some cyanobacteria to fix nitrogen (Joye and Paerl, 1994).

In microbial mats the decomposition of organic matter may be incomplete which could mean that part of the nitrogen is not recycled and so enters the geological record (Fig. 17). Hence this nitrogen is withdrawn from the microbial nitrogen cycle. It is not clear how important this process is since even in mats in which a net accretion of organic matter occurs, up to 99% of the produced organic matter may be recycled within the mat (Krumbein et al., 1977). In exceptional cases organic nitrogen produced by nitrogen-fixing cyanobacteria can be transformed to nitrate deposits known as guano or caliche nitrates (Ehrlich, 1996).

Most of the nitrogen in the biosphere is present in the atmosphere in the form of dinitrogen (N_2), which amounts to 3.9×10^{18} kg N. In the oceans and on land the amount of combined nitrogen (organic and inorganic) amounts each about 10^{15} kg N. The amount of nitrogen in living biomass on earth amounts only 1.3×10^{13} kg (Ehrlich, 1996). It is generally assumed that primary production in the marine environment is limited by nitrogen. In particular marine microbial mats with their dense and compressed biomass often experience a shortage of nitrogen. The majority of organisms can not use the most abundant form of nitrogen, N_2. Only nitrogen-fixing organisms or diazotrophs are capable of using N_2. All of these organisms possess nitrogenase. Cyanobacteria are among the most important nitrogen-fixing organisms and in the majority of marine microbial mats that have been investigated to date, high rates of nitrogen fixation have been observed.

C. Nitrogenase

In all nitrogen-fixing organisms the enzyme complex nitrogenase is present. This enzyme is very similar in all organisms that contain it. The complex is composed of two enzymes: dinitrogenase reductase which is a dimer of identical subunits and also termed the iron-protein, and dinitrogenase, a tetramer composed of two different subunits ($\alpha_2\ \beta_2$). Dinitrogenase is also known as the molybdenum-iron protein (Howard and Rees, 1994).

Nitrogenase catalyzes the following reaction:

$$N_2 + 8\,e^- + 8\,H^+ + 16\,MgATP \longrightarrow$$
$$2\,NH_3 + H_2 + 16\,MgADP + 16\,P_i$$

Reduced ferredoxin is the electron donor of nitrogenase. The equation shown above makes clear that the fixation of nitrogen is at the expense of considerable amount of energy and low potential electrons. This high energy demand of nitrogenase presents often a problem for diazotrophic organisms except for cyanobacteria who produce reduced ferredoxin and convert light energy into chemical energy during photosynthesis. However, all nitrogenases are extremely sensitive to oxygen and therefore diazotrophic organisms must provide an anaerobic environment in order to be able to fix nitrogen. Cyanobacteria as oxygenic, phototrophic and principally aerobic organisms need special adaptations.

D. Nitrogen-Fixing Cyanobacteria in Microbial Mats

Diazotrophic cyanobacteria are capable of using dinitrogen (N_2) as the sole source of nitrogen for growth. These organisms can be subdivided in three main groups (Table 3).

Group I consists of heterocystous cyanobacteria. These filamentous organisms differentiate special cells, heterocysts, which have lost the capacity of oxygenic photosynthesis and have evolved a modified thick cell envelope. Heterocysts are the site of nitrogen fixation in these cyanobacteria. The thick cell wall contains special lipopolysaccharides and forms a diffusion barrier for gases, limiting the entry of oxygen. Respiration scavenges the little oxygen that enters the heterocyst. Since photosystem II is absent from the heterocyst, no photosynthetic oxygen is evolved in these cells. Therefore the heterocyst is virtually anoxic and provides an excellent environment for the oxygen-sensitive nitrogenase. Photosystem I mediated conversion of light energy in the heterocyst provides nitrogenase with ATP. However, for reducing equivalents nitrogenase depends on the importation of carbohydrates from the neighboring vegetative cells. The strategy that heterocystous cyanobacteria have developed in order to be able to grow diazotrophically can be best described as the spatial separation of the two incompatible processes of nitrogen fixation and oxygenic photosynthesis. Among the cyanobacteria heterocystous species are the ultimate adapted organisms for nitrogen fixation. The vast majority of heterocystous cyanobacteria can be found in freshwater or terrestrial systems, both free-living and as symbionts. Heterocystous cyanobacteria occur in some brackish basins but are rare in the marine environment, including microbial mats. The heterocystous *Calothrix* has been found as the dominant organism in microbial mats in the tidal area of the Pacific coast in Baja California Sur, Mexico. *Calothrix* is also known from a variety of other marine and brackish habitats such as the spray zone

Table 3. Types and characteristics of nitrogen-fixing cyanobacteria

Type I Heterocystous Cyanobacteria
* exclusively filamentous species that differentiate special cells: heterocysts
* strategy: spatial separation of N_2 fixation and oxygenic photosynthesis and protection of nitrogenase in the heterocyst
* diazotrophic growth under fully aerobic conditions
* examples: *Anabaena, Nostoc, Aphanizomenon, Nodularia, Calothrix, Scytonema*
* occurrence: waterblooms (freshwater lakes and brackish seas), paddy fields, microbial mats, symbiotic with a variety of different organisms

Type II Anaerobic N_2-Fixing Non-Heterocystous Cyanobacteria
* filamentous and unicellular species
* strategy: avoidance (of oxygen)
* induction and maintenance of nitrogenase only under anoxia or low oxygen; sulfide may be necessary in order to inhibit oxygenic photosynthesis
* examples: *Plectonema boryanum, Oscillatoria limnetica, Synechococcus* sp., many other cyanobacteria
* occurrence: many different environments, particularly in microbial mats

Type III Aerobic N_2-Fixing Non-Heterocystous Cyanobacteria
* filamentous and unicellular species
* strategy not precisely known (possibly: temporal separation of N_2 fixation and oxygenic photosynthesis in concert with other oxygen protection mechanisms)
* diazotrophic growth possible under fully aerobic conditions
* examples: *Gloeothece, Oscillatoria, Trichodesmium, Lyngbya, Microcoleus*
* occurrence: tropical ocean (*Trichodesmium*), carbonate cave walls and paddy fields (*Gloeothece*), microbial mats (*Oscillatoria, Lyngbya, Microcoleus*)

after Stal (1995)

of rocky shores (Jones and Stewart, 1969; Whitton and Potts, 1982). Mats of the heterocystous cyanobacterium *Anabaena* have been found in a coastal lagoon in southwest France (Villbrandt and Stal, 1996). However, these are exceptions rather than a rule. The vast majority of microbial mats are built by non-heterocystous cyanobacteria, notwithstanding the fact that in many cases nitrogen fixation is a crucial process in these systems.

Group II consists of filamentous and unicellular cyanobacteria that do not show cell differentiation and are capable of nitrogen fixation only under virtually anoxic conditions with no oxygenic photosynthesis occuring. These organisms, although possessing the genetic capacity of synthesizing nitrogenase, have obviously not evolved a mechanism to protect effectively nitrogenase from oxygen inactivation. Consequently, their strategy can be characterized as avoidance of oxygenated environments. Such environments usually also prevent oxygenic photosynthesis. Among the non-heterocystous cyanobacteria up to about 50% may belong to this group of organisms but for virtually all of them it is uncertain whether they live diazotrophically in their natural environment. Non-heterocystous cyanobacteria that are capable of inducing nitrogenase activity under anaerobic conditions can be found in many environments, including microbial mats. However, most environments in which these cyanobacteria occur are permanently oxygenated and therefore diazotrophic growth is unlikely. In contrast, microbial mats are often characterized by steep and fluctuating gradients of oxygen and sulfide. Anoxia frequently occurs in microbial mats; this as a rule coincides with high levels of sulfide, a very potent inhibitor of oxygenic photosynthesis. Thus, it is not surprising that recently evidence was obtained that anaerobic nitrogen-fixing cyanobacteria were growing diazotrophically in microbial mats in which H_2S was present (Villbrandt and Stal, 1996).

Group III cyanobacteria also comprise non-heterocystous filamentous and unicellular cyanobacteria but they are remarkable as they possess the capacity of inducing nitrogenase and growing diazotrophically under fully aerobic conditions. To date our knowledge of how these organisms are protecting their undoubtedly oxygen-sensitive nitrogenase is incomplete. Although the species that possess this capability are still relatively rare, the numbers are increasing at steady pace. Examples can be found in terrestrial environments such as cave-walls and paddy fields and in the marine environment. Freshwater lakes apparently do not harbour aerobic nitrogen-fixing non-heterocystous cyanobacteria. In the ocean the planktonic colony-forming *Trichodesmium* spp. (*Oscillatoria* spp.) is known as an efficient diazotrophic growing, non-heterocystous cyanobacterium. In microbial mats aerobic nitrogen-fixing non-heterocystous cyanobacteria are reported to belong predominantly to the morphologically closely related genera *Oscillatoria* and *Lyngbya*.

The first report of a culture of a filamentous non-heterocystous aerobic nitrogen-fixing cyanobacterium was by Pearson et al. (1979). This organism was originally identified as *Microcoleus chthonoplastes* but later re-named as *Symploca* sp. (R. Rippka, pers. comm.). *Symploca* is also morphologically related to *Oscillatoria*, and was isolated from a tidal microbial mat (Pearson et al., 1979; Malin and Pearson, 1988). It has been proposed that the strategy of aerobic non-heterocystous cyanobacteria, in analogy with the heterocystous species, is temporal separation of the incompatible processes of photosynthesis and nitrogen fixation. The latter would than occur during the dark (Mullineaux et al., 1981; Stal and Krumbein, 1987). However, not all species in this group follow this strategy. *Trichodesmium* spp. fix nitrogen during the day (Capone et al., 1990). Moreover, all species that have been cultured are capable of growing diazotrophically under continuous light and, in the unicellular *Gloeothece* sp., culture conditions can be chosen under which nitrogen fixation occurs during the light period of a light dark cycle (Ortega-Calvo and Stal, 1991).

E. Daily Variation of Nitrogen Fixation in Microbial Mats

Nitrogen fixation is a process with a high demand of energy and low-potential reducing equivalents. For the oxygenic phototrophic cyanobacteria light is the source of ATP generation and electrons are derived from water and transferred to ferredoxin mediated by photosynthetic electron transport. Thus, in cyanobacterial mats nitrogen fixation ought to be directly linked to light. However, since oxygen exerts a negative effect on nitrogenase, daily variations of nitrogen fixation in microbial mats can be expected. The patterns of these daily variations will depend on the type of diazotrophic cyanobacterium and on the dynamics of light and oxygen in the mat. In Fig. 18 five daily patterns of nitrogenase activity, measured in different microbial mats, are depicted.

Nitrogen fixation in heterocystous cyanobacteria is intimately linked to light. The heterocyst is not capable of CO_2 fixation and therefore does not accumulate storage carbohydrate, as is the case in vegetative cells. Dark energy generation in heterocysts will be limited because at one hand the reducing equivalents must be imported from the vegetative cells while at the other hand the oxygen entry in the heterocyst is limited as a result of the diffusion barrier provided by the cell wall (Walsby, 1985). Therefore it is not surprising that daily variations of nitrogen fixation in communities of heterocystous cyanobacteria are strongly light dependent (Griffiths et al., 1987; Storch et al., 1990; Stal, 1995) (Fig. 18a). However, considerable dark nitrogenase activity may occur in such communities. The ratio of light over dark nitrogenase activity in different populations of heterocystous cyanobacteria varies considerably and is possibly dependent on the species, the light history or other conditions.

In microbial mat communities composed of non-heterocystous cyanobacteria the daily pattern of nitrogen fixation is less predictable (Paerl et al., 1989; 1996) (Fig. 18b-e). It depends largely on the type of organism and on the prevailing conditions in the mat. Moreover, due to the fact that these conditions may also vary from day to day (tidal movement, light and overcast, temperature and other factors), the daily pattern of nitrogen fixation may change considerably.

The daily pattern of nitrogen fixation in non-heterocystous cyanobacteria is the result of the combined effects of oxygen, light and, in some cases, sulfide. As in heterocystous cyanobacteria, non-heterocystous species must supply nitrogenase with sufficient energy and low-potential reducing equivalents. This condition is satisfied in the light but the serious drawback is the evolution of oxygen. In such mats photosynthesis obviously must occur at daytime and nitrogen fixation is confined to the night (Fig. 18b). For instance this is the case in mats of *Gloeothece* and *Oscillatoria*. Whereas during the day microbial mats often become extremely supersaturated with oxygen because the diffusion of this gas is limited, at night they may turn anoxic within minutes (Stal, 1995). The microbial community, including the cyanobacteria, consumes oxygen in the dark by respiration. Obviously, anoxic

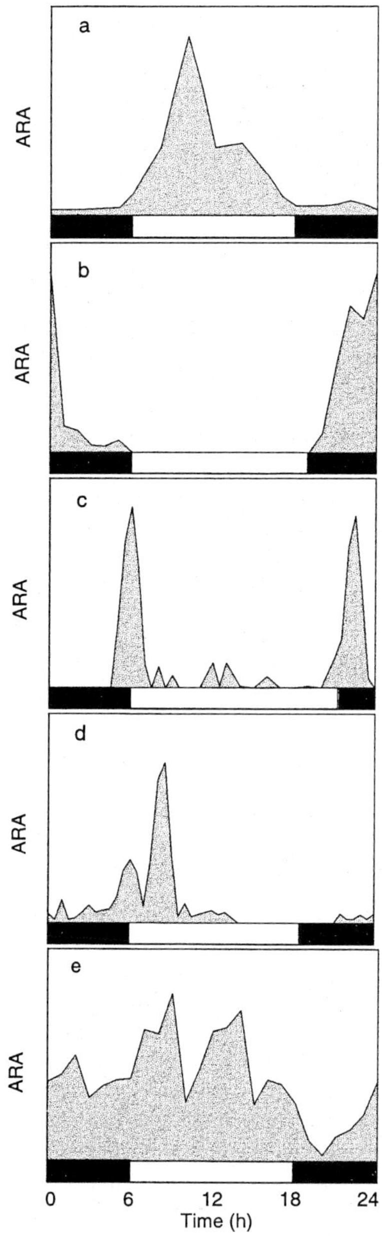

Fig. 18. Five typical patterns of daily variations of nitrogen fixation (acetylene reduction activity, ARA, relative units) in microbial mats. *A*: mat of the heterocystous cyanobacterium *Calothrix* sp. in Baja California, Mexico (data from Stal et al., 1994). *B*: mat of the unicellular cyanobacterium *Gloeothece* sp. on the wall of a carbonate cave (data from Griffiths et al., 1987). *C*: mixed mat of the non-heterocystous cyanobacteria *Microcoleus chthonoplastes* and *Oscillatoria limosa* from a tidal flat on the North Sea island of Mellum (data from Villbrandt et al., 1990). *D*: mat dominated by *Oscillatoria limosa* (location as *C*) (data from Villbrandt et al., 1990). *E*: mat of the non-heterocystous cyanobacterium *Lyngbya aestuarii* (location as *A*) (Stal, unpublished).

conditions are ideal for nitrogen fixation, but pose a problem with respect to the supply of energy and reducing equivalents. However, all cyanobacteria isolated from marine microbial mats and tested appeared to be capable of fermentation of endogenous storage carbohydrate (Stal and Moezelaar, 1997). Although the energy generation by fermentation is undoubtedly small, it has been shown that it exceeds many times the extremely low maintenance requirements of cyanobacteria (Stal and Moezelaar, 1997). It has also been shown that dark anoxic conditions supported considerable nitrogenase activity in the filamentous, non-heterocystous cyanobacterium *Oscillatoria limosa* (Stal and Heyer, 1987). In microbial mats in which this cyanobacterium occurred, daily patterns of nitrogen fixation were found in which this activity was low but totally confined to the dark (Villbrandt et al., 1990). However, other patterns were also observed at different times in the same mats with the same organism. For instance, it could often be seen that nitrogenase activity peaked around sunset and sunrise (Fig. 18c). This was confirmed by experiments with *O. limosa* grown in the laboratory under an alternating light dark cycle and with anoxic conditions established 1 h after the onset of the dark period and oxic conditions 1 h after the onset of the light period (Stal and Heyer, 1987). Highest nitrogenase activities in these cultures were obtained in the light in the absence of oxygen. Also in natural samples it has been observed that highest nitrogenase activities occurred at sunrise (Villbrandt et al., 1990) (Fig. 18d). This is because light is available while oxygen is still absent. After sunset oxygen may have been present for some time, allowing energy generation through aerobic respiration. Once anoxic conditions are established only low rates of nitrogenase activity can be supported by the lower rate of fermentative energy generation. Vertical profiles of oxygen measured during a 24 h period have shown that in the mat which possessed this type of fluctuating nitrogenase activity, oxygen was indeed present during the first hours after sunset and that it appeared again in the morning only hours after sunrise.

In freshly colonized sediments of North Sea tidal sand flats *O. limosa* is often the pioneer cyanobacterium (Stal et al., 1985). This is most likely because of its capacity to grow diazotrophically. In this pioneer state biomass is low and therefore so is the oxygen demand in the dark. Such sediments normally do not turn anoxic. However, during the light they may accumulate oxygen up to several fold saturation (Villbrandt et al., 1990). Nitrogen fixation in such systems is typically confined to the night, peaking at sunrise when light becomes available but at oxygen levels well below air saturation (Fig. 18d). In other systems such as in mats of the unicellular nitrogen-fixing *Gloeothece,* which grows on carbonate cave walls, a peak of nitrogenase activity is observed immediately after sunset and then decreasing gradually until hardly any activity is detectable at the end of the night (Fig. 18b). This organism depends on oxygen for respiratory energy generation and it is possible that in the course of the dark period this organism depletes its endogenous storage carbohydrate (Maryan et al., 1986).

Another type of daily pattern of nitrogen fixation in microbial mats of non-heterocystous cyanobacteria is more or less constant activity or fluctuations scattered throughout the day and night (Fig. 18e). This often the case when Group 2 diazotrophic cyanobacteria are involved. These cyanobacteria are only capable of fixing nitrogen under anoxic conditions or at least when oxygen concentrations are low and oxygenic photosynthesis is inhibited. Such a situation can be expected in microbial mats in which high concentrations of sulfide inhibit oxygenic photosynthesis. In the light, sulfide at the same time may serve as an electron donor for nitrogenase in these situations. In most cases oxygenic photosynthesis is continuing in the surface layers of the mat. Sulfide in inhibitory concentrations for photosynthesis is usually present in the deeper layers where light intensity will also be low. In the dark nitrogen fixation may be supported by fermentation of endogenous storage carbohydrate. Thus both in the light and in the dark relatively low nitrogenase activities can be expected.

F. Vertical Distribution of Nitrogen Fixation in Microbial Mats

Little is known about the vertical distribution of nitrogen fixation in microbial mats, but the vertical distribution and dynamics of factors that control it such as light, oxygen and sulfide, have been investigated in considerable detail. Light is attenuated strongly in microbial mats. The wavelengths that are absorbed by the cyanobacteria in the top layers are obviously attenuated most strongly. Far red light (< 700 nm), however, is absorbed by the cyanobacteria to only a small extent; also the attenuation of this light in (wet) sediment is small compared to shorter wavelengths. Far red light does not support oxygenic photosynthesis but anoxygenic photosynthesis depends on it. It can thus be assumed that the cyanobacteria in the lower part of the mat are not capable of oxygenic photosynthesis. This has been shown by microelectrode measurements of oxygen concentration and photosynthesis. Such measurements have also shown that in some microbial mats sulfide is present in these layers. In an attempt to measure potential nitrogenase activity in microbial mats it was shown that in a mat of 3 mm maximum surface related nitrogenase activity occurred in the depth horizon of 1 - 2 mm (Stal et al., 1984). However, when nitrogenase activity was expressed on the basis of chlorophyll *a* highest specific nitrogenase activity was present in the lowest layer of the cyanobacterial mat (2 – 3 mm) (Fig. 19). Cyanobacterial biomass was highest in the top layer, decreasing gradually until about 3 mm depth. Thus it is likely that a spatial separation of nitrogen fixation and oxygenic photosynthesis had occurred in this mat. The top layer carries out oxygenic photosynthesis and CO_2 fixation, while nitrogen is fixed in the lower layers.

G. Effects of Anoxia and Sulfide on Nitrogen Fixation in Microbial Mats

Among the non-heterocystous diazotrophic bacteria those that are capable of nitrogen fixation under fully aerobic conditions are rare (Bergman et al., 1997). Since they grow by oxygenic photosynthesis, these organisms do not normally perform nitrogen fixation. It has been questioned whether this capacity of nitrogen fixation is of any importance in the natural environment (Rippka and Waterbury, 1977). In their chapter on anoxygenic photosynthesis Padan and Cohen (1982) mention that the facultative anoxygenic photosynthetic cyanobacterium *Oscillatoria limnetica* is capable of nitrogen fixation when carrying out sulfide-dependent anoxygenic photosynthesis. Villbrandt and Stal (1996) investigated the effect of sulfide on nitrogen fixation in cyanobacterial mats and on cultures of cyanobacteria isolated from these mats. They compared a mat dominated by a heterocystous cyanobacterium (*Anabaena*) with another dominated by non-heterocystous filamentous organisms (*Oscillatoria* and *Phormidium*). Sulfide inhibited nitrogenase activity in the mat of *Anabaena* but greatly stimulated it in the mat of the non-heterocystous cyanobacteria (Fig. 20). Both light and dark nitrogenase activity was inhibited by sulfide in

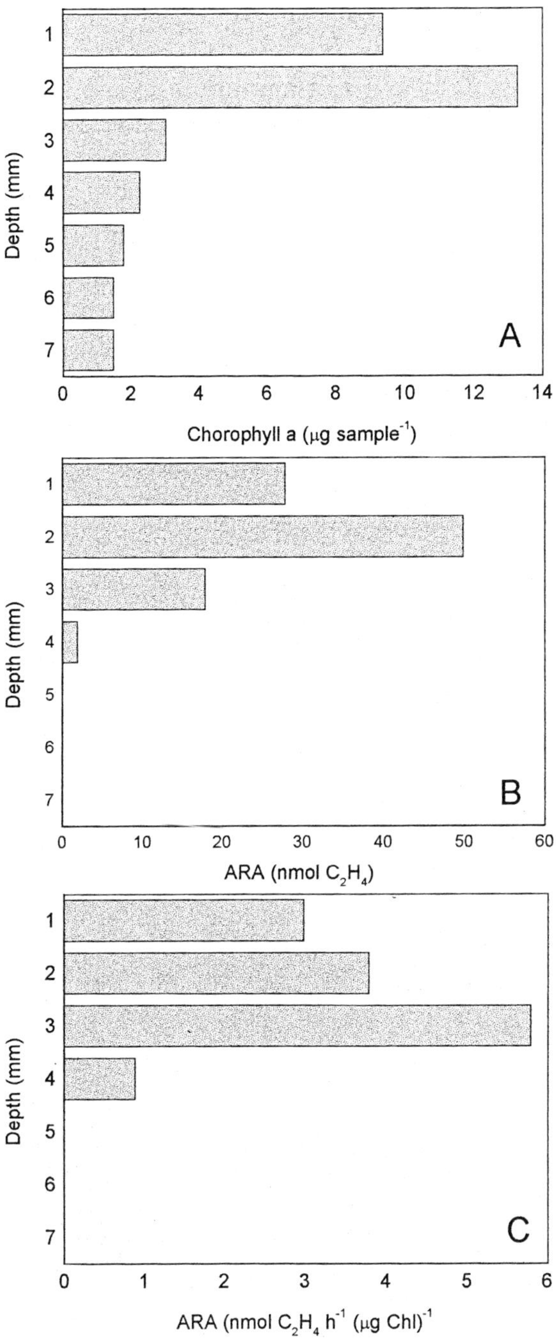

Fig. 19. Vertical distribution of chlorophyll *a* (*A*), potential nitrogenase activity (acetylene reduction, ARA) (*B*) and specific, chlorophyll *a*-based ARA (*C*) in a microbial mat (data from Stal et al., 1984).

the mat of *Anabaena* sp. but when DCMU was added in order to inhibit oxygenic photosynthesis virtually no effect of sulfide on nitrogen fixation was seen in this mat. Therefore the effect of sulfide was mainly through the inhibition of oxygenic photosynthesis and respiration. Only the addition of 10 mM sulfide resulted in the almost complete inhibition of dark nitrogenase activity which depends on respiratory energy generation. In the light this sulfide concentration resulted in a decrease of nitrogenase activity to the level obtained in the presence in DCMU, which in this case was about 40% of the control. This demonstrated that this amount of sulfide caused the complete inhibition of oxygenic photosynthesis. Due to the large amount of iron in this mat the actual concentration of free sulfide was probably much lower. Moreover, as was shown in a laboratory culture of *Anabaena* sp. isolated from this mat, the effect of sulfide strongly depended on the pH. At pH9.5 a total sulfide concentration of 5 mM had no effect on nitrogen fixation but at pH6.5 this concentration almost completely inhibited nitrogenase activity. This shows that the effect of sulfide is through the gaseous species H_2S. This gas will passively diffuse into the cell. Because the pH in these mats is usually high, very little H_2S will be present, even when the total concentration of sulfide is high.

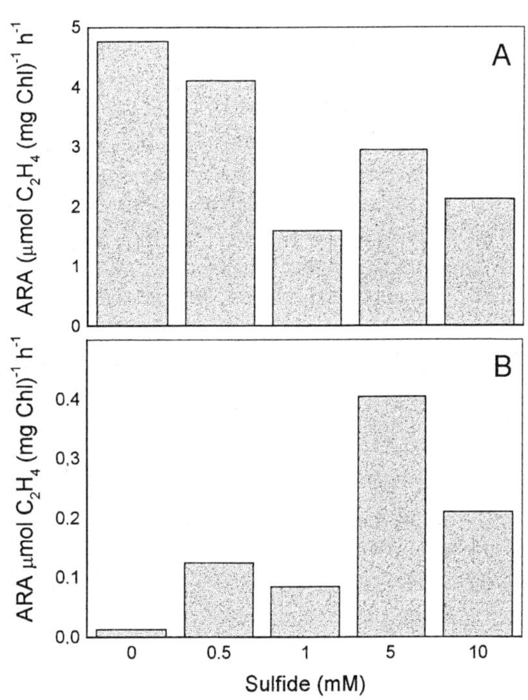

Fig. 20. Effect of sulfide on nitrogenase activity (acetylene reduction, ARA) in (*A*) a mat of the heterocystous cyanobacterium *Anabaena* sp. (lagoon, Atlantic coast of France) and (*B*) a mat of non-heterocystous cyanobacteria (*Oscillatoria* sp. and *Phormidium* sp.) (lagoon, Mediterranean coast of France) (data from Villbrandt and Stal, 1996).

In the non-heterocystous mat the situation is totally different. In the control, without sulfide, nitrogenase activity is low in the light. In the dark or when oxygenic photosynthesis is inhibited by DCMU, nitrogenase activity is greatly stimulated. This can obviously be explained by the sensitivity on nitrogen fixation in these organisms for photosynthetic and atmospheric oxygen. Sulfide stimulated nitrogenase activity in the light, in the dark and with DCMU. Stimulation was most marked in the light and reached a maximum at 5 mM (Fig. 20b). However, even at 10 mM sulfide nitrogenase activity was about 10-fold the control. The stimulation in the dark was small (it doubled) and reached already maximum at 0.5 mM. Also, with DCMU, stimulation was maximum at 0.5 mM, the same order of magnitude as the effect in the dark (Villbrandt and Stal, 1996). The effect of sulfide on the light activity of nitrogenase is best explained by its inhibition of oxygenic photosynthesis in concert with a lowering of environmental oxygen concentration.

None of the cyanobacteria isolated from this mat possessed the capacity of aerobic nitrogen fixation but all of the strains were capable of inducing nitrogenase under anaerobic conditions (Villbrandt and Stal, 1996). In experiments with *Phormidium* it was shown that sulfide (total concentration up to 8 mM) had no effect on nitrogenase activity when this was induced anaerobically with DCMU. Therefore it was concluded that sulfide did not act as an electron donor to nitrogenase and that the stimulatory effect observed in the mat was most probably due to the scavenging of environmental oxygen. Sulfide very efficiently induced nitrogenase in *Phormidium* and other non-heterocystous cyanobacteria with anaerobic nitrogenase. About 4 mM total sulfide was sufficient for full induction of nitrogenase. However, in contrast with what was seen with the heterocystous cyanobacterium and to what was expected, it appeared that induction of nitrogenase with sulfide was optimal at high pH, which means that the ions HS^- and/or S^{2-} were more efficient than the gas H_2S. This suggested that *Phormidium* could actively take up sulfide ion. Thus the uptake of sulfide ion may be essential to allow diazotrophic growth in these organisms.

H. Oxygen Protection of Nitrogenase in Microbial Mats

Because cyanobacterial mats often contain a high density of biomass and have low rates of molecular diffusion, they may become markedly supersaturated with oxygen. This poses the question of oxygen protection of nitrogenase in microbial mats. Although in heterocystous cyanobacteria nitrogenase is confined to the heterocysts, and protected from oxygen under normal atmospheric conditions, it has been shown that nitrogen fixation may be seriously impaired at oxygen pressure well above atmospheric levels. In the majority of cases non-heterocystous cyanobacteria are the dominant organisms in mats. Many of these species possess only the capacity of anaerobic nitrogenase activity because the lack of an adequate oxygen protection mechanism. They may be able to grow diazotrophically under anaerobic conditions and when sulfide inhibits oxygenic photosynthesis. This may under circumstances lead to a vertical spatial separation of oxygenic photosynthesis in the top layer of the mat and nitrogen fixation in the deeper parts. Paerl and Prufert (1987) and Paerl et al. (1995) emphasized the importance for nitrogen fixation of anoxic microzones in microbial mats and other systems. In a few cases microbial mats have been shown to be built by non-heterocystous cyanobacteria that are capable of nitrogen fixation under fully aerobic conditions (Pearson et al., 1979; Stal et al., 1984; Villbrandt et al., 1990; Gallon et al., 1991; Paerl et al., 1991). Since nitrogenase in these organisms is as sensitive to oxygen as in any other organism, these cyanobacteria obviously must possess a protection mechanism. Despite a large amount of research on this problem the precise mechanism by which these species protect nitrogenase from oxygen inactivation is still not known (Bergman et al., 1997). In fact, in all cases in which aerobic nitrogen-fixing cyanobacteria form microbial mats, nitrogen fixation is confined to the night. Thus, a temporal separation of nitrogen fixation and photosynthesis (respectively during the night and during the day) is maintained (Stal, 1995). Because these mats turn anoxic during the night, there is no need for oxygen protection. The problem of oxygen protection of nitrogenase in microbial mats is therefore hardly relevant.

Aerobic nitrogen-fixing non-heterocystous cyanobacteria isolated from microbial mats include strains of the filamentous *Oscillatoria*, *Lyngbya*, *Microcoleus*, *Gloeothece*, *Cyanothece* and *Synechococcus* (Bergman et al., 1997). Among the different mechanisms that have been proposed for oxygen protection of nitrogenase, the uptake and reduction of oxygen seems to be the most promising.

Such systems may act in concert with enzymes that remove oxygen radicals.

In Table 4 the effects of different treatments of a diazotrophic microbial mat composed of *Oscillatoria* on nitrogenase activity are shown. When these mats were incubated in the laboratory and exposed to elevated salinity, phosphate fertilization or to a tidal movement of the water, all these treatments resulted in a dramatic increase of nitrogenase activity. The application of a tidal movement (alternating immersion and emersion of the mat) resulted in a 2 orders of magnitude increase in nitrogenase activity. The vertical profiles of oxygen in these mats, measured at the same time, showed that the increase of nitrogen fixation was probably the result of markedly decreased concentrations of oxygen (Fig. 21). The reference showed oxygen super saturation, peaking at about 250 μm depth, typical for these mats. When the mats were subject to increased salinity, phosphate fertilization or to a tidal movement, oxygen profiles decreased dramatically. The decrease of oxygen concentration was most pronounced in the mats subject to both phosphate fertilization and tidal movement. These treatments also resulted in the strongest stimulation of nitrogenase activity. It is obvious that the decreased oxygen concentration is associated with the increased potential to fix nitrogen. However, although

Table 4. Effect of different treatments of a nitrogen-fixing microbial mat of *Oscillatoria* sp. from the island of Texel, The Netherlands. Sediment cores containing the mat were incubated in the laboratory in aquaria filled with seawater (Instant Ocean). The seawater was aerated. Illumination was by 75 W halogen lamps applied at a 16-8 h light-dark cycle. Heating of the mats was prevented by a heat filter and fans. The reference cores were incubated in such a way that the mat was just exposed while the water level was just underneath the mat surface. The mat surface was moist. This incubation mimics the natural situation most closely. In another aquarium the seawater was pumped in and out in 6 h intervals, mimicking a tidal movement. At each high water the mat was covered by 5 cm of water. The tidal range was about 15 cm. In the third incubation the salinity was increased to twice the normal value (3%). In the fourth treatment, phosphate concentration in the seawater was increased to 100 μM. The cores were incubated for 1 week; vertical oxygen profiles (Fig. 25) and nitrogenase activity (acetylene reduction) were measured.

Treatment	Nitrogenase activity (nmol C_2H_4 cm^{-2} h^{-1})
Reference	2 ± 1
High salinity	31 ± 5
Phosphate fertilization	164 ± 18
Tidal movement	234 ± 21

it is possible that other metabolic processes may have been responsible for the decrease of oxygen, this may also have been the result of the increased nitrogenase activity in the first place.

One possibility that must now be investigated is the capacity of nitrogenase in these cyanobacteria to reduce oxygen (autoprotection) (Bergman et al., 1997). This causes the reduction of O_2 to H_2O_2, which can be further reduced by peroxidases.

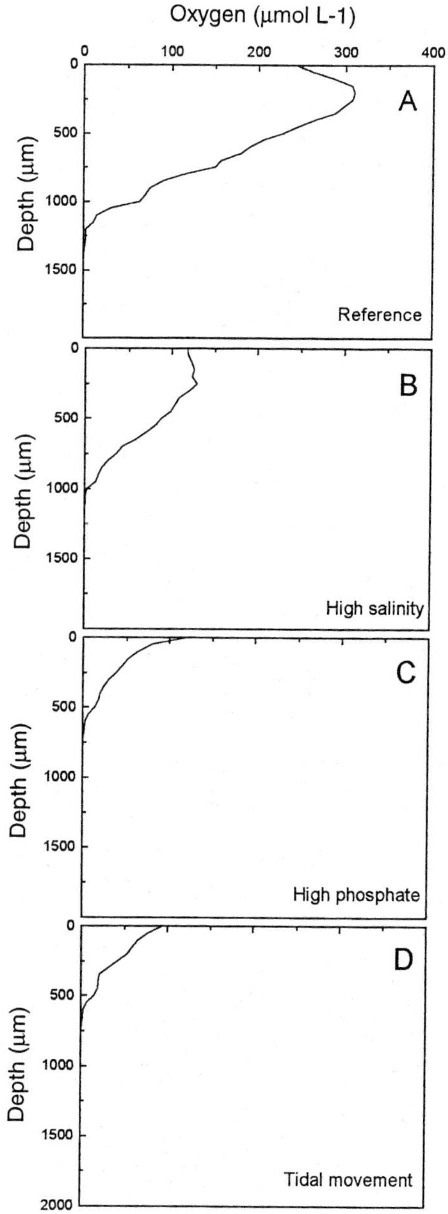

Fig. 21. Vertical profiles of oxygen in diazotrophic mats of *Oscillatoria* sp. subjected to different treatments (See Table 4).

I. Heterocystous Versus Non-Heterocystous Cyanobacteria in Microbial Mats

There is no doubt that heterocystous cyanobacteria are particularly well adapted for diazotrophic growth. They can fix nitrogen in the light while carrying out oxygenic photosynthesis. In this way they make optimal use of light energy to cover the large demands of nitrogenase. Oxygen protection of nitrogenase in these organisms is virtually perfect. Anoxic conditions usually result in scarcely higher nitrogenase activities and the inhibition of oxygenic photosynthesis by DCMU invariably results in lower activities, apparently because it cuts off the flow of reduction equivalents from the vegetative cells. Non-heterocystous cyanobacteria either cannot fix nitrogen at all in the presence of oxygen or those that can invariably can much better in the dark or when transferred to anoxic conditions (Stal, 1995). Often the inhibition of oxygenic photosynthesis by DCMU also stimulates nitrogenase activity considerably. Notwithstanding these facts, the vast majority of marine microbial mats are composed of non-heterocystous cyanobacteria. Thus the question is raised as to why heterocystous cyanobacteria are not more common in these mats.

On the tidal flats in Guerrero Negro, Baja California Sur, Mexico, two types of microbial mats can be found in close vicinity of each other (Stal et al., 1994; Stal, 1995) (Plate 8). The smooth mat is composed of the non-heterocystous cyanobacterium *Lyngbya aestuarii* and covers the lower areas of the tidal flat. On the upper tidal flat and on slightly elevated spots a pustular mat develops which is composed of the heterocystous cyanobacterium *Calothrix*. Both mats fix nitrogen but show distinct differences in their daily nitrogenase patterns. *Calothrix* fixes predominantly during the day while nitrogenase activity in the mats of *L. aestuarii* is confined to the night. Due to their locations on the tidal flat the mats of *L. aestuarii* are covered more often and during longer periods of time at high tide than the mats of *Calothrix*. During inundation of the mats of *L. aestuarii* diffusion is limited. This causes oxygen supersaturation during the period of photosynthesis and anoxic conditions at night. These anoxic conditions also allow the development of a community of sulfate-reducing bacteria. This mat has a very dense biomass and is characterized by steep gradients of oxygen and sulfide, typical for microbial mats. Due to this dense mat structure the gradients of oxygen and sulfide exist, regardless whether the mat is inundated or not. The situation in the *Calothrix* mat is totally different. This pustular mat has a porous structure. Due to this structure in this mat there is a free exchange of oxygen with the atmosphere and oxygen super saturation or anoxic conditions are not usually the case. In exceptional cases when the mat is inundated for a prolonged period of time anoxic conditions or oxygen super saturation may occur but normally the mat will be inundated for short periods or not at all. Stal et al. (1994) hypothesized that heterocystous cyanobacteria would not be able to maintain themselves in an environment in which either dark anoxic conditions or high concentrations of sulfide occur. Cyanobacteria incapable of fermentation will die within 2-3 hours of dark anoxic conditions (Stal and Moezelaar, 1997). However, there is no reason why heterocysts should be incapable of fermentation and this has been demonstrated in a number of symbiotic *Nostoc* spp. (Margheri and Allotta, 1993; De Philippis et al, 1996).

In order to investigate the possibility of sulfide as a selecting factor, Villbrandt and Stal (1996) compared a heterocystous and non-heterocystous nitrogen-fixing mat in two coastal lagoons in France. The mat of heterocystous cyanobacteria was found in a lagoon with exceptional high amounts of iron, while this was not the case in the other system (Stal et al., 1996). As a result of the high amount of iron the sediment on which this microbial mat was found did not contain any free sulfide because it precipitated as iron sulfide (Schaub and Van Gemerden, 1996). Villbrandt and Stal (1996) hypothesized that the absence of sulfide would allow the proliferation of heterocystous species. It was indeed demonstrated that nitrogen fixation in heterocystous cyanobacteria was sensitive to sulfide. On the other hand, unrealistically high concentrations of sulfide (10 mM) were required to obtain full inhibition. Sulfide inhibition of nitrogenase in heterocystous cyanobacteria depended on H_2S, which in microbial mats is present in very low concentrations as a result of the alkaline conditions. However in heterocystous cyanobacteria it is uncertain whether other metabolic processes than nitrogen fixation are more severely influenced by sulfide.

Another reason why heterocystous cyanobacteria are absent from the majority of microbial mats could lie in the fact that such organisms generally are not motile and that the link between the heterocyst and the vegetative cell is weak (Stal et al., 1994). Non-heterocystous cyanobacteria that form microbial mats are mostly motile by gliding movement. This is an important property since it facilitates optimal vertical positioning. It allows the cyanobacteria to compensate for the rapidly shifting physicochemical gradients in microbial mats. Gliding motility is also important because microbial mats often develop in environments that are characterized by high rates of sedimentation. In the rare cases that heterocystous cyanobacteria dominate microbial mats their filaments are orientated in a uniform manner at the mat surface. The tapered trichomes of *Calothrix* are orientated vertically with the terminal heterocysts situated away from the surface. These filaments do not glide freely. The same is the case for another mat-building heterocystous cyanobacterium, *Scytonema*, which likewise reveals a vertical orientation. In this organism intercalary heterocysts are formed located in the center of the aggregates. It is likely that shear forces produced during gliding in these highly compressed microbial mats would result in the breakage of the weak link between heterocysts and the neighboring vegetative cells. Hence, although gliding motility is an essential property for cyanobacteria in microbial mats, it has at the same time a serious disadvantage for heterocystous species. It is therefore expected that mats of such cyanobacteria can only develop in environments with low rates of sedimentation and relatively constant physicochemical gradients.

VIII. Cyanobacteria and the Sulfur Cycle in Microbial Mats

The sulfur cycle may have a large impact on microbial mats either when sulfate is present and the end-oxidation of organic matter is carried out by sulfate reducing bacteria, or when the ecosystem receives primary sulfide as is the case in sulfur springs. Seawater contains abundant sulfate (28 mM) and therefore sulfate reduction is usually important in coastal microbial mats. Sulfate-reducing bacteria are essentially anaerobic micro-organisms that oxidize simple organic compounds using sulfate as an electron acceptor, which results in the formation of sulfide. A variety of different micro-organisms are capable of reducing elemental sulfur to sulfide. Sulfide can be eventually oxidized back to sulfate. This can be done anaerobically by anoxygenic phototrophic bacteria such as purple and green sulfur bacteria. Elemental sulfur is produced as an intermediate in this process. Some cyanobacteria are capable of sulfide-dependent anoxygenic photosynthesis but they oxidize sulfide only to elemental sulfur or to thiosulfate (Fig. 22). Colorless sulfur bacteria oxidize sulfide aerobically to sulfate but a few species can carry out this oxidation anaerobically using nitrate as an electron acceptor (denitrification). Other processes in the sulfur cycle in microbial mats include the disproportionation of thiosulfate, sulfite and elemental sulfur. In these reactions one part of the molecule is oxidized while the other is reduced (Bak and Pfennig, 1987; Canfield and Thamdrup, 1996).

In microbial mats a large amount of organic matter is liberated into the environment (dissolved organic matter, DOM) by a variety of different mechanisms. Some of these processes such as fermentation and photorespiration by the cyanobacteria result in the excretion of compounds that can immediately serve as substrate for sulfate reducing bacteria. Photorespiration results in the formation and excretion of glycolate. Sulfate-reducing bacteria can use glycolate (Fründ and Cohen, 1992; Friedrich and Schink, 1995) (Fig. 22). Degradation of DOM, which is produced as a result of cell lysis or the exudation of extracellular polymeric substances (EPS), requires the action of several different microorganisms. Hence, the metabolic activity of the mat cyanobacteria can directly influence sulfate reduction. This, however, poses a problem since sulfate reduction is considered to be an anaerobic process.

It has been assumed that sulfate-reducing bacteria are present only below the euphotic depth in the microbial mat because only there, conditions are permanently anoxic. This layer is often recognized by its black color that indicates the presence of FeS. There are now several lines of evidence that this may be incorrect. Sulfate-reducing bacteria are found throughout the sediment as is sulfate reduction (Visscher et al., 1992; Stal, 1993). Thus it seems that sulfate-reducing bacteria can co-exist with cyanobacteria. It was already known for some time that many sulfate-reducing bacteria were much less oxygen-sensitive than assumed and recently, sulfate-reducing bacteria are known that are even capable of aerobic respiration (Cypionka et al., 1985; Dilling and Cypionka, 1990; Marschall et al., 1993). There have even been reports of sulfate reduction itself taking place under oxygenated conditions (Canfield and Des Marais, 1991); however, cultures with this property have not been isolated thus far. The majority of sulfate-reducing bacteria are obligate anaerobes. In fact, 16S rRNA analysis of microbial mats have

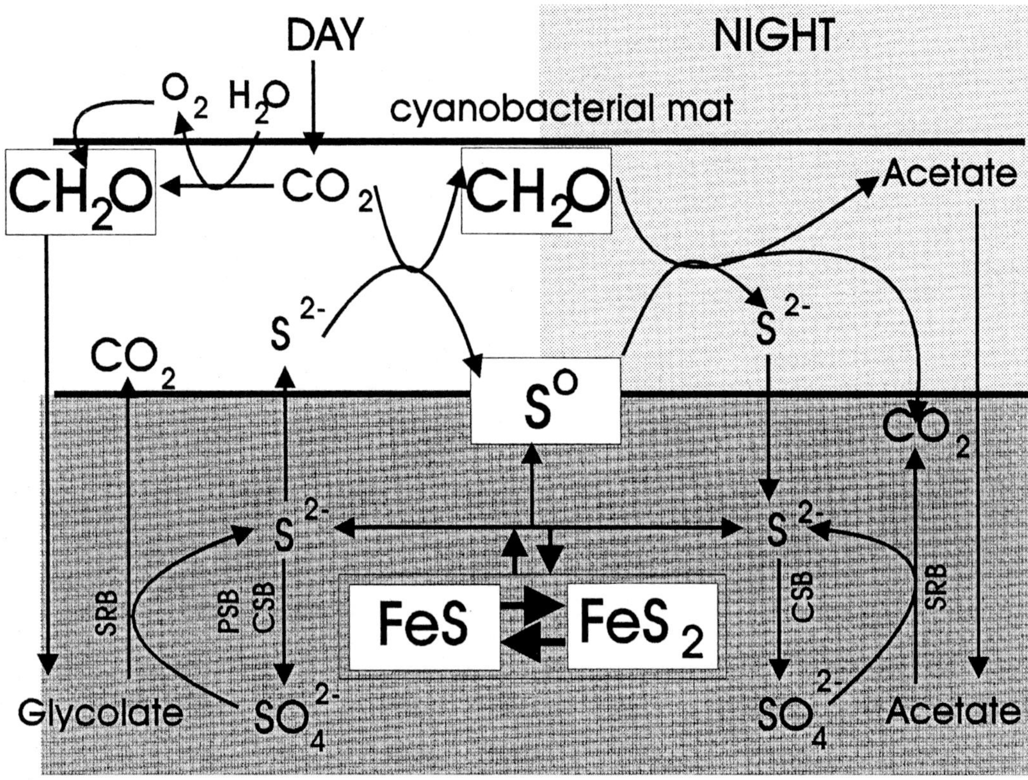

Fig. 22. A simplified scheme showing the role of cyanobacteria in the cycle of sulfur in a microbial mat. SRB, CSB and PSB are sulfate reducing bacteria, colorless sulfur bacteria and photosynthetic sulfur bacteria, respectively. See text for further explanation.

demonstrated that the oxygen tolerant sulfate-reducing bacteria are predominantly found in the top layers of the mat while the obligate anaerobic species are found in the deeper layers of the mat (Risatti et al., 1994). Thus a vertical stratification of different groups of sulfate-reducing bacteria is likely. More importantly for the cyanobacteria is the fact that sulfate reduction is likely to occur in the top photic layer of the mat, in their immediate vicinity. Sulfate reduction takes place in the cyanobacterial layer during the dark, when the mat is anoxic, but it could also take place during the day. Cyanobacteria ferment glycogen and excrete fermentation products under anaerobic conditions in the dark, but it is possible that fermentation is constitutive (Stal and Moezelaar, 1997). Probably even more important is the excretion of glycolate during the day. Hence, the substrates for sulfate reducing bacteria are produced and excreted by the cyanobacteria day and night. This could well explain the fact that 99% of the primary production is immediately recycled in the mat (Krumbein et al., 1977).

Sulfate-reducing bacteria are not the only organisms in the mat that produce sulfide. Mat-forming cyanobacteria are capable of reducing elemental sulfur to sulfide (Oren and Shilo, 1979; Stal, 1991; Moezelaar et al., 1996) (Fig. 22). In the presence of elemental sulfur part of the electrons produced in the course of the degradation of glycogen are transferred to elemental sulfur. This process has not been investigated in much detail. Although thermodynamically possible, the reduction of sulfur in cyanobacteria itself does not generate biochemical energy, but it will allow a greater proportion of acetate to be produced which is associated with the formation of ATP. An exception may be *Oscillatoria limnetica* that may be capable of true sulfur respiration (Oren and Shilo, 1979). However, for the majority of cyanobacteria sulfur serves as an electron sink during fermentation. Other microorganisms in microbial mats that contribute to the production of sulfide include fermentative and dissimilatory sulfur-reducing bacteria and sulfur, thiosulfate and sulfite disproportionating bacteria (Jørgensen, 1990;

Canfield and Thamdrup, 1996). Also anoxygenic phototrophic bacteria are known to reduce elemental sulfur to sulfide (Van Gemerden, 1993). All these bacteria are known to be present in microbial mats but their contribution to total sulfide production is uncertain.

Sulfide is an extremely toxic compound even to those organisms that depend on it such as anoxygenic phototrophic sulfur bacteria, colorless sulfur bacteria or organisms that produce it, such as the sulfate-reducing bacteria. Sulfide reacts with iron-containing compounds in the cell such as cytochromes and haemo-proteins. These compounds are particularly important with electron transport and therefore sulfide interferes with photosynthesis and respiration. In addition, it is known that sulfide is a potent inhibitor of photosystem II. This inhibition is irreversible, although in the mat-forming cyanobacterium *O. amphigranulata* oxygenic photosynthesis recovered after initial inhibition by moderate concentrations of sulfide. This recovery occurred at high light intensities and depended on *de novo* protein synthesis (Castenholz and Utkilen, 1984). Moreover, different degrees of recovery after initial inhibition by sulfide exist in different strains of cyanobacteria (Garcia-Pichel and Castenholz, 1990). Even sulfide-dependent anoxygenic photosynthesis is inhibited by sulfide (De Wit and Van Gemerden, 1988). In part, sulfide toxicity in cyanobacteria may also be indirect since growth and metabolism in some species is negatively effected by low redox potential. The toxicity of sulfide in cyanobacteria is probably the result of a combination of effects and it can be expected that the degree of inhibition differs greatly between different strains.

Cohen et al. (1986) distinguished four groups of cyanobacteria with respect to the degree of sulfide inhibition and the possibility to carry out sulfide-dependent anoxygenic photosynthesis (Table 5). Cyanobacteria belonging to Group 1 are extremely sulfide sensitive. Oxygenic photosynthesis is inhibited at low levels of sulfide (< 0.1 mM) and these species are not capable of anoxygenic photosynthesis. Cyanobacteria belonging to this group are evidently not important in marine microbial mats but are likely to be found in freshwater lakes or terrestrial systems in which sulfide is absent or present at insignificant concentrations. Examples of such cyanobacteria are *Anacystis nidulans* and *Plectonema boryanum* in which CO_2 fixation was inhibited at 60 and 75 μM (Cohen et al., 1986). In Group 2 cyanobacteria are represented that are incapable of anoxygenic

Table 5. Groups of cyanobacteria with different types of adaptation to sulfide.

Group 1.	Sulfide-sensitive oxygenic photosynthesis only
Group 2.	Sulfide-resistant oxygenic photosynthesis only
Group 3.	Sulfide-insensitive oxygenic photosynthesis concurrent with sulfide-dependent anoxygenic photosynthesis
Group 4.	Sulfide-sensitive oxygenic photosynthesis replaced by sulfide-dependent anoxygenic photosynthesis

After Cohen et al. (1986) and Stal (1995)

photosynthesis but that resist considerable levels of sulfide. Oxygenic photosynthesis in these organisms is often stimulated at moderate (< 1 mM) sulfide concentration. This type of adaptation is typical for marine microbial mats with fluctuating sulfide concentrations. The mat-forming and diazotrophic cyanobacterium *Oscillatoria limosa* is a typical example of this group (Stal, 1995). Also Group 3 cyanobacteria are typically found in marine microbial mats. These cyanobacteria are characterized by sulfide-insensitive oxygenic photosynthesis concurrent with sulfide-dependent anoxygenic photosynthesis. The cosmopolitan mat-forming cyanobacterium *Microcoleus chthonoplastes* belongs to this group (De Wit and Van Gemerden, 1988). Oxygenic photosynthesis in Group 4 cyanobacteria is as sensitive to sulfide as in those cyanobacteria belonging to Group 1. The difference is in their capacity for carrying out sulfide-dependent anoxygenic photosynthesis. The sulfide tolerance of this group of cyanobacteria varies considerably from less than 1 mM to 10 mM. *Oscillatoria limnetica* is the best-studied cyanobacterium belonging to that group. Photosystem II in this cyanobacterium is switched off when exposed to < 0.1 mM of sulfide. Anoxygenic photosynthesis is induced in a process requiring protein synthesis. *O. limnetica* tolerates up to 9.5 mM sulfide but anoxygenic photosynthesis is gradually inhibited at concentrations exceeding 4 mM.

In microbial mats most of the sulfide is present as 'acid-volatile sulfide' (AVS) which is mostly in the form of ferrous sulfide (FeS) (Fig. 22). In this form sulfide is virtually insoluble. Only free sulfide may be toxic. Free dissolved sulfide occurs as hydrogen sulfide or sulfide ions in a pH-dependent equilibrium.

$$H_2S \longleftrightarrow HS^- \longleftrightarrow S^{2-}$$

Below pH7, H_2S becomes gradually more important while above pH9 this is S^{2-}. Between pH7 and 9 virtually all sulfide is present as HS^-. H_2S is a gas that can enter the cell by passive diffusion. It has been shown that in heterocystous cyanobacteria sulfide is much more toxic at low pH (Howsley and Pearson, 1979; Villbrandt and Stal, 1996). This can only be explained when the ions HS^- and S^{2-} can not be taken up by these organisms. However, cyanobacteria capable of anoxygenic photosynthesis are apparently capable of uptake of the sulfide ion. Induction of nitrogenase by sulfide in non-heterocystous cyanobacteria is also a function of the concentration of total sulfide rather than H_2S. The uptake of sulfide ion may even be facilitated at high pH (Villbrandt and Stal, 1996).

Sulfide may also react with elemental sulfur to form polysulfides. It was thought that this process could occur only in microbial mats in which the amount of iron is not sufficient to keep the level of free sulfide low (Jørgensen and Cohen, 1977). However, Visscher (1992) measured very high concentrations of polysulfides in a cyanobacterial mat, indicating that this compound may be more common than previously assumed. While on the one hand polysulfides are an order of magnitude more toxic for most organisms than sulfide, on the other hand they may serve as the form of elemental sulfur that is transported in cells (Steudel et al., 1990).

IX. Interactions of Cyanobacteria with Iron

Iron is one of the most abundant elements on Earth and it has several important functions in microbial mats. Iron occurs in three oxidation states: elemental iron Fe^o, ferrous (reduced iron), Fe^{2+} and ferric (oxidized iron), Fe^{3+}. Ferric iron is virtually insoluble and in the presence of oxygen ferrous iron is readily oxidized, except under acidic conditions (pH< 2). Elemental iron is not stable in nature because it will be readily oxidised. Thus, in the presence of oxygen at physiological pH iron is hardly available for organisms and aerobic microorganisms often produce compounds that have a high affinity for iron. These siderophores bind iron and transport it into the cells. The redox behavior of iron gives rise to its biological importance. Iron is an essential micronutrient for virtually all organisms. It occurs in a number of enzymes that act as electron carriers, such as cytochromes, in respiratory electron transport chains and ferredoxins which serve as electron donors to a variety of processes (including nitrogen fixation, nitrate and sulfate reduction) in the cell, and indirectly, CO_2 fixation. Moreover, iron is an important co-factor in enzymes such as nitrogenase and nitrate reductase. In addition to this assimilatory metabolism of iron, the dissimilatory iron metabolism is of importance in microbial mats and other environments. Ferric iron may serve as an electron acceptor in anaerobic respiration (Lovley, 1991). Under acid conditions, ferrous iron can be oxidized aerobically by the chemolithotrophic, autotrophic bacterium *Thiobacillus ferrooxidans* (Leduc and Ferroni, 1994). Under neutral conditions ferrous iron is rapidly oxidised by oxygen; however, *Gallionella ferruginea* is capable of competing with the chemical reaction and oxidizes iron to support an autotrophic mode of metabolism (Hallbeck and Pedersen, 1991). The third biologically controlled iron transformation is the formation of magnetite in magnetotactic bacteria and in a variety of other organisms (Stolz, 1993).

In microbial mats high amounts of iron are often present. In coastal microbial mats and hypersaline microbial mats suspended iron oxides present in seawater precipitate in the sediment. In microbial mats iron may precipitate either as oxides and hydroxides, siderite ($FeCO_3$) or iron sulfide (FeS) and pyrite (FeS_2). Iron readily reacts with sulfide:

$$2\,Fe^{3+} + H_2S \longrightarrow 2\,Fe^{2+} + S^o + 2\,H^+$$

$$2\,Fe^{2+} + 2\,H_2S \longrightarrow 2\,FeS + 4\,H^+$$

$$2\,Fe^{3+} + 3\,H_2S \longrightarrow 2\,FeS + S^o + 6\,H^+$$

FeS is virtually insoluble. Thus both ferric and ferrous iron are important in immobilizing toxic sulfide. Ferrous iron including FeS will react with oxygen both chemically as well as biologically. In microbial mats oxygen super saturation may present a problem for cyanobacterial growth and the presence of ferrous iron may aid in keeping the partical pressure of oxygen low. Moreover, the oxidation of iron in siderite will result in the liberation of CO_2.

In many microbial mats a layer of oxidized iron is often observed between the layer of cyanobacteria and the anoxic layers below (Fig. 2). When purple sulfur bacteria are present, this layer of oxidized iron usually separates them from the cyanobacteria. The origin of this layer of oxidized iron is not clear. Since this layer is generally found in the anoxic part of the mat it is not likely that aerobic iron-oxidizing bacteria are responsible. Very recently it was discovered that anaerobic denitrifying bacteria could be responsible for anaerobic iron oxidation (Straub et al., 1996). However, in many microbial mats the amount of nitrate is probably too low to allow for this process. Cohen (1989) offered another explanation. He supposed that some mat-forming cyanobacteria were capable of iron-dependent anoxygenic photosynthesis. Although this process may be the origin of the observed layer of oxidized iron, it seems more likely that it is produced by specialized anoxygenic phototrophic purple bacteria that use iron as an electron donor. Such organisms have recently been discovered and isolated from freshwater and marine sediments, including intertidal mud, that are capable of iron-dependent anoxygenic photosynthesis (Widdel et al., 1993; Ehrenreich and Widdel, 1994). It is therefore likely that these bacteria are responsible for the layer of ferric iron found in microbial mats. This layer may present an efficient barrier between the aerobic and anaerobic parts of the system. Ferric iron would scavenge any sulfide and protect the cyanobacteria from its toxic effect. Moreover, ferrous iron will react with oxygen and keep the partial pressure of oxygen sufficient low to allow efficient photosynthesis by the cyanobacteria. Also it may protect the underlying community of purple sulfur bacteria from oxygen.

The formation of distinct layers of oxidized iron in microbial mats may well have resulted in the formation of so-called Banded Iron Formations (BIFs). Banded Iron Formations are finely layered sedimentary rocks composed mainly of silica and ironoxides (James and Trendall, 1982). BIFs were deposited over large areas and several thousands are known. Although the majority are only a few meters thick and cover a limited area, others are several hundreds of meters thick and extend over many thousands of square kilometers (James and Trendall, 1982). The iron content of BIFs is typically in the range of 24 - 35%, which is 5-7 times more than normally found in the crust. These iron formations are therefore of great economic importance. The silica content of BIF's is about 45%. Together iron oxides and silica may make up to 90% of the weight of BIF. Iron oxides and silica (chert) occur in alternating layers. The cherty banded iron formation of Hamersley Basin, Australia, is one of the largest in the world and is characterized by stratification at different scales. At the millimeter scale microbands of iron minerals are recognized, separated at the centimeter scale by mesobands of chert. Regular banding is seen at the meter scale (macrobands) (James and Trendall, 1982). BIFs were formed during the Archean and Proterozoic ages. Over 90% of the deposits are from the early Proterozoic age (2500-1900 My). Although it is tempting to assume a biological basis for the genesis of these cherty iron stromatolites, so far evidence of biogenesis has not emerged. Because of the fact that BIFs were overwhelmingly present during the early Proterozoic this has also been taken as evidence for the oxygenation of the earth's atmosphere which started 2300 My ago. One mechanism for BIF formation may be the chemical oxidation of ferrous iron with oxygen evolved by oxygenic photosynthesis, most likely by cyanobacteria. The huge amounts of ferrous iron in the earth's crust would act like a buffer and prevent the oxygenation of the atmosphere until most of the iron was oxygenated. Although less abundant, the fact that BIF's are also known from the mid Archean (3400-2900 My) might indicate other mechanisms. Geological evidence is available that indicates that until 2000 My ago the earth's atmosphere was free of oxygen. High energy solar UV irradiation (200 - 300 nm range) could freely reach the surface of the earth where it could be absorbed by ferrous iron, resulting in the formation of ferric iron and H_2 which would escape into the atmosphere (Cairns-Smith, 1978). This reaction has been experimentally proven to be a possible explanation for the precipitation of ferric iron.

It has been suggested that both ferrous and ferric iron play an important role in protection from UV irradiation because they provide a effective UV screen (Pierson and Olson, 1989). Ferrous and ferric iron strongly absorb in the region 220-270 nm (Pierson and Olson, 1989), which is particularly deleterious for organisms. It is assumed that the flux of UV irradiation that reached the earth surface during the early Precambrian was very high since the oxygen-free atmosphere would scarcely attenuate it. It is also known that microbial life developed on earth during this period, particularly in stromatolites. This life was apparently not arrested by the high UV flux. Although UV-C light does not reach the earth surface because it is completely absorbed by the earth's atmosphere, it is interesting that some mat-forming cyanobacteria such as *Microcoleus chthonoplastes* accumulate large amounts of iron at the outer polysaccharide sheath (Stal, 1994). Iron is bound to negatively charged polysaccharides, particularly through the presence of uronic acids (Bender et al.,

1994). The function is not precisely known but one of the possibilities could be that it represents an ancient UV screen, which was retained during evolution because it also has additional ecologically important effects. The accumulation of iron by mat-forming cyanobacteria has been shown to protect the organism from sulfide produced either by sulfate-reducing bacteria living in the immediate vicinity of the cyanobacteria or by themselves through the reduction of elemental sulfur. Another possibility is that ferrous iron will react with oxygen, keeping its concentration low and minimizing the effects of photorespiration. *M. chthonoplastes* is also capable of reducing ferric iron (Stal, 1994).

X. Phosphorus in Microbial Mats

Few studies have addressed the role of phosphorus in microbial mats. This is remarkable because phosphate is involved in a variety of geochemical reactions that are significant in mats and it is indispensible for growth and metabolic activity for all forms of life, particularly also for cyanobacteria. The almost complete ignorance of phosphorus in the study of microbial mats is also in strong contrast with the attention it receives in the study of phytoplankton. Generally, nitrogen or phosphorus limits growth of phytoplankton and most likely this applies also to cyanobacteria that form microbial mats.

Typically only about 0.6% of the dry mass of cells of cyanobacteria consist of phosphorus, but they sometimes can store phosphate as polyphosphate. Cyanobacteria take up orthophosphate (H_3PO_4) which is the most common form of inorganic phosphorus.

The solubility of orthophosphate is controlled by elements such as Ca^{2+}, Mg^{2+}, Fe^{2+}, Fe^{3+} and Al^{3+}. In seawater, the solubility of orthophosphate is predominantly controlled by Ca^{2+}, which at a suitable pH (7.4 – 8.1) produces the virtually insoluble hydroxyapatite ($Ca_{10}(PO_4)_6(OH)_2$; solubility product 1.53×10^{-112}) (Ehrlich, 1996). In addition, phosphate may also form an insoluble precipitate with ferric iron ($FePO_4 \cdot 2H_2O$, strengite, solubility product 1.35×10^{-18}). Phosphate may be liberated from these insoluble minerals by microbial activity. The mechanisms include:

(i) production of organic acids
(ii) production of chelators
(iii) dissimilatory reduction of ferric iron
(iv) production of sulfide (Ehrlich, 1996).

The latter can react with ferric ironphosphate according to:

$$2\,FePO_4 + 3\,H_2S \longrightarrow 2\,FeS + 2\,H_3PO_4 + S^o$$

All these processes are likely to occur in microbial mats, but the phosphate liberated will be taken up immediately by the microbial community. Hence, the occurrence of free orthophosphate ion in microbial mats is expected to be negligible. Any organic phosphates must be cleaved hydrolytically by phosphatases.

As argued in Section VI, most cyanobacterial mats in coastal environments are probably limited by nitrogen. However, microbial mats formed by heterocystous cyanobacteria are more likely to become phosphate-limited because nitrogen fixation by these organisms provides all the nitrogen needed for growth. Mats built by non-heterocystous cyanobacteria are probably still nitrogen-limited as a result of impairment of nitrogen fixation by oxygen. However, as is shown in Table 4 (Section VI.H), phosphate fertilization of such a mat resulted in a dramatic increase of nitrogen fixation. This indicated that the mat was probably also phosphate limited. It is known that nitrogen fixation requires a certain amount of phosphate for optimal performance (de Nobel et al., 1997). The effect of phosphate fertilization on nitrogen fixation may also have been indirect since it also caused a strong decrease in dissolved oxygen in the mat (Fig. 21). The latter explanation is supported by the fact that nitrogen fixation was also stimulated by other treatments that resulted in a decrease of oxygen (Table 4; Fig. 21). Phosphate fertilization may also have stimulated heterotrophic bacterial activity and consequently oxygen uptake. Phosphate fertilization of these coastal mats resulted in a considerable increase of chlorophyll *a* and a shift in cyanobacterial species composition from an *Oscillatoria*-dominated community to one with mainly *Phormidium*-type forms (Stal, unpublished).

Although phosphorus may occur in other oxidation states (from +5 to –3), it is not important in redox reactions, as is the case with nitrogen and sulfur. Bacteria readily oxidize any reduced phosphorus, both aerobically and anaerobically. The reduction of orthophosphate is thermodynamically not favorable and is therefore not important. Hence, the microbial phosphorus cycle consists predominantly of the uptake of inorganic phosphate and the liberation by excretion or autolysis of organic phosphate, which is subsequently mineralized by phosphatases.

Phosphate may be stored in mineral deposits such as phosphorite, apatite, strengite, and other forms.

Phosphorite deposits are usually found in coastal waters or shallow seas. They can be formed authigenically when soluble phosphate reacts with calcium to form calcium phosphate, or by diagenesis when phosphate replaces carbonate in calcareous concretions (Ehrlich, 1996). Both processes are probably biologically controlled. The model of Piper and Codispoti (1975) explains phosphorite formation in the marine environment from the mineralization of organic matter below the oxygen minimum layer, where it is coupled to denitrification. This results in excess inorganic phosphate compared to combined nitrogen. Upwelling transports phosphate to the sea surface, where it precipitates with calcium. This model could also apply to microbial mats where the same processes take place. Dahanayake and Krumbein (1985) reported phosphorite formed by a microbial mat, but concluded that fungi rather than cyanobacteria produced this particular fossil mat.

Clearly, more research is definitely required to complete our understanding of the formation of microbial mats and stromatolites.

XI. Conclusions

Laminated microbial mats are often considered to be recent analogues of fossil Precambrian stromatolites. Stromatolites are laminated lithified structures that have been formed by growth and metabolism of microorganisms. Studies of carbon isotope ratios provide evidence that photosynthesis was involved in the formation of stromatolites and the discovery of microfossils supports the idea that cyanobacteria have built these formations. However, modern microbial mats rarely lithify and doubts have been raised as to whether these systems really can be considered as analogues. Moreover, the sedimentary record is biased and contains only examples of lithified mats. It may well be that non-lithifying mats have existed throughout geological history. There are a limited number of examples of microbial mats that calcify and form more or less laminated lithified structures which have morphologies very similar to the Precambrian examples. The comparison of lithifying and non-lithifying microbial mats has given deeper understanding of the factors that determine the processes leading to lithification.

In the majority of examples of microbial mats, cyanobacteria play a key role in their formation. Cyanobacteria are oxygenic phototrophic prokaryotes and many species are capable of using dinitrogen (N_2) as their only source of nitrogen. Hence, these organisms have a minimum requirement to proliferate, which is important considering the harsh conditions in which microbial mats often develop. Only extreme conditions will limit the biodiversity and exclude higher grazing organisms so that cyanobacteria accumulate to the dense community that produces a mat. Cyanobacteria have a number of additional properties that make them very suitable for forming microbial mats. Many species are motile through gliding movement, which allows them to position themselves under optimal conditions. Light and possibly chemical factors serve as signals to directed the movement of the organisms. Many cyanobacteria are further characterized by a high affinity for light and reach maximum rate of photosynthesis at very low light intensities and have low compensation points. The requirement for energy for maintenance purposes is low. Cyanobacteria often have high affinities for nutrients and perhaps even more important, they possess storage possibilities for a variety of growth factors. In addition, their metabolic versatility and reactivity are important properties of cyanobacteria. For instance, cyanobacteria are not only photoautotrophs that perform oxygenic photosynthesis but many are also capable of anoxygenic photosynthesis. The majority of mat-forming cyanobacteria are even capable of performing oxygenic and anoxygenic photosynthesis in concert, allowing maximum flexibility and reactivity to quickly changing environmental conditions. Whereas aerobic respiration of endogenous glycogen seems to be the normal metabolism in the dark, this does not usually occur in microbial mats, which often are devoid of oxygen during the night. However, most mat forming cyanobacteria seem to be capable of fermentation.

Growth and metabolic activity of the cyanobacteria introduce organic matter in the microbial mat system and its degradation will drive the growth of other micro-organisms in microbial mats. Although some organic matter may become liberated into the environment by death and lysis of the cyanobacteria, this seems not to be most important. In mature microbial mats there is hardly any growth of cyanobacteria notwithstanding the fact of often extremely high rates of photosynthesis. Organic matter may become liberated as a result of photorespiration, fermentation, excretion of organic solutes and the secretion of extracellular polymeric substances (EPS), notably polysaccharides. Cyanobacteria may produce a well-defined polysaccharide sheath. This is often a structural component of the cell envelope of cyanobacteria. However, cyanobacteria may also produce vast amounts of mucilage which is not or only partly associated with the organism. Mucilage is often composed of recalcitrant polysaccharides. It produces a matrix in which the microbial mat is embedded. This material is sticky and it will glue sediment particles and organisms together, giving a high stability to the sediment surface. Since this matrix can not be mixed it presents a diffusive barrier. The polysaccharide matrix is therefore responsible for the accumulation and supersaturation with oxygen in the light and likewise for the anoxic conditions that occur during the night. Of particular importance is the role that EPS probably plays in calcification. It is likely that EPS inhibits this process and that it serves as an anti-calcification agent. This may be either by preventing growth of small calcite crystallization nuclei or by binding Ca^{2+}. Therefore cyanobacterial mats in which a high amount of mucus is produced will not calcify. The production of mucus in cyanobacteria may be related to unbalanced growth. Unbalanced growth occurs when one growth factor is in shortage. Nitrogen limitation is well known as a factor that stimulates the secretion of mucus. Marine microbial mats are often developing under conditions of nitrogen shortage. This is evidenced by the fact that these mats are built by nitrogen-fixing cyanobacteria. These mats usually consist of non-heterocystous diazotrophic cyanobacteria that are inefficient in fixing nitrogen because they lack an effective mechanism to protect nitrogenase from oxygen. This is particularly the case when during the day very high levels of oxygen occur in the mat. At night oxygen is absent, therefore nitrogen fixation in these mats occurs predominantly then. However, the limited amount of energy and low-potential electrons that can be generated under such conditions will not allow the fixation of ample nitrogen. Heterocystous cyanobacteria are optimally equipped for nitrogen fixation in the light and will not usually face nitrogen-limited growth. However, as stated above, such cyanobacteria are largely excluded from microbial mats. Possibly heterocystous cyanobacteria cannot tolerate high levels of sulfide or dark anoxic conditions. Also the absence of gliding motility in heterocystous cyanobacteria and the weak connection between the heterocyst and the vegetative cell may exclude these organisms from environments with high rates of sedimentation. It seems reasonable to assume that Precambrian microbial mats did not face nitrogen limitation and thus might have produced much less mucus, calcification therefore not being inhibited. Future research should investigate the nitrogen state of modern calcifying microbial mats and the role of EPS as anti-calcification agents in non-lithifying mats.

Acknowledgements

I would like to thank Dr P. Westbroek who draw my attention to the concept of 'anticalcification' and Drs A.W. Decho, H.W. Paerl, J. Pinckney and P. Reid for the discussions during a research cruise in the Bahamas in 1994 that led to the formulation of a model for calcification in the Exuma Cays stromatolites. I am also grateful to Dr J. Middelburg for his constructive comments,

References

Albertano P and Kovacik L (1996) Light and temperature responses of terrestrial sciaphilous strains of *Leptolyngbya* sp. in cross-gradient cultures. Algol Stud 83: 17-28

Allen MM and Smith AJ (1969) Nitrogen chlorosis in blue-green algae. Arch Mikrobiol 69: 114-120

Artus NN, Somerville SC and Somerville CR (1986) The biochemistry and cell biology of photorespiration. CRC Crit Rev Plant Sci 4: 121-147

Awramik SM (1984) Ancient stromatolites and microbial mats. In: Cohen Y, Castenholz RW and Halvorson HO (eds) Microbial Mats: Stromatolites pp. 1-22 Alan R. Liss Inc., New York

Badger MR and Andrews TJ (1982) Photosynthesis and inorganic carbon usage by the marine cyanobacterium, *Synechococcus* sp. Plant Physiol 70: 517-523

Bak F and Pfennig N (1987) Chemolithotrophic growth of *Desulfovibrio sulfodismutans* sp. nov. by disproportionation of inorganic sulfur compounds. Arch Microbiol 147: 184-189

Bar-Or Y, Kessel M and Shilo M (1985) Modulation of cell surface hydrophobicity in the benthic cyanobacterium *Phormidium* J-1. Arch Microbiol 142: 21-27

Bar-Or Y and Shilo M (1987) Characterization of macromolecular flocculants produced by *Phormidium* sp. strain J-1 and by *Anabaenopsis circularis* PCC 6720. Appl Environ Microbiol 53: 2226-2230

Bar-Or Y and Shilo M (1988) The role of cell-bound flocculants in coflocculation of benthic cyanobacteria with clay particles. FEMS Microbiol Ecol 53: 169-174

Bateson MM and Ward DM (1988) Photoexcretion and fate of glycolate in a hot spring cyanobacterial mat. Appl Environ Microbiol 54: 1738-1743

Bauer MR, Haddad RI and Des Marais DJ (1991) Method for determining stable isotope ratios of dissolved organic carbon in interstitial and other natural marine waters. Mar Chem 33: 335-351

Bauld J (1984) Microbial mats in marginal marine environments: Shark Bay, Western Australia, and Spencer Gulf, South Australia. In: Cohen Y, Castenholz RW and Halvorson HO (eds) Microbial Mats: Stromatolites pp. 39-58 Alan R. Liss Inc., New York

Bebout BM and Garcia-Pichel F (1995) UV B-induced vertical migrations of cyanobacteria in a microbial mat. Appl Environ Microbiol 61: 4215-4222

Bender J, Rodriguezeaton S, Ekanemesang UM and Phillips P (1994) Characterization of metal-binding bioflocculants produced by the cyanobacterial component of mixed microbial mats. Appl Environ Microbiol 60: 2311-2315

Bergman B, Gallon JR, Rai AN and Stal LJ (1997) N_2 fixation by non-heterocystous cyanobacteria. FEMS Microbiol Rev 19: 139-185

Berner R and Jensen T (1982) Ultrastructure of two hypolithic cyanobacteria from the Negev desert of Israel. Cytobios 35: 7-18

Bertocchi C, Navarini L and Cesaro A (1990) Polysaccharides from cyanobacteria. Carbohydrate Polymers 12: 127-153

Borman AH, De Jong EW, Huizinga M, Kok DJ, Westbroek P and Bosch L (1982) The role in $CaCO_3$ crystallization of an acid Ca^{2+}-binding polysaccharide associated with coccliths of *Emiliania huxleyi*. Eur J Biochem 129: 179-183

Borman AH, De Jong EW, Thierry R, Westbroek P and Bosch L (1987) Coccolith-associated polysaccharides from cells of *Emiliania huxleyi* (Haptophyceae). J Phycol 23: 118-123

Boudreau BP and Canfield DE (1993) A comparison of closed- and open-system models for porewater pH and calcite-saturation state. Geochim Cosmochim Acta 57: 317-334

Brock TD (1975) Effect of water potential on a *Microcoleus* (Cyanophyceae) from a desertcrust. J Phycol 11: 316-320

Brock TD (1978) Thermophilic Microorganisms and Life at High Temperatures. Springer Verlag, New York, 465 pp.

Caiola MG, Billi D and Friedmann EI (1996) Effect of desiccation on envelopes of the cyanobacterium *Chroococcidiopsis* sp. (Chroococcales). Eur J Phycol 31: 99-105

Caiola MG, Ocampo-Friedmann R and Friedmann EI (1993) Cytology of long-term desiccation in the desert cyano-bacterium *Chroococcidiopsis* (Chroococcales). Phycologia 32: 315-322

Cairns-Smith AG (1978) Precambrian solution photochemistry, inverse segregation, and banded iron formations. Nature 276: 807-808

Campbell SE (1979) Soil stabilization by a prokaryotic desert crust: implications for Precambrian land biota. Origins of Life 9: 335-348

Canfield DE and Des Marais DJ (1991) Aerobic sulfate reduction in microbial mats. Science 251: 1471-1473

Canfield DE and Des Marais DJ (1994) Cycling of carbon, sulfur, oxygen and nutrients in a microbial mat. In: Stal LJ and Caumette P (eds) Microbial Mats. Structure, Development and Environmental Significance pp. 255-263 Springer Verlag, Heidelberg

Canfield DE and Raiswell R (1991) Carbonate precipitation and dissolution. Its relevance to fossil preservation. In: Allison PA and Briggs DEG (eds) Taphonomy: Releasing the Data Locked in the Fossil Record. Topics in Geobiology Vol. 9 pp. 411-453 Plenum Press, New York

Canfield DE and Thamdrup B (1996) Fate of elemental sulfur in an intertidal sediment. FEMS Microbiol Ecol 19: 95-103

Capone DG, O'Neil JM, Zehr J and Carpenter EJ (1990) Basis for diel variation in nitrogenase activity in the marine planktonic cyanobacterium *Trichodesmium thiebautii*. Appl Environ Microbiol 56: 3532-3536

Castenholz RW (1973) Movements. In: Carr NG and Whitton BA (eds) The Biology of Blue-Green Algae pp. 320-339 Blackwell Scientific Publications, Oxford

Castenholz RW (1976) The effect of sulfide on the blue green algae of hot springs. I. New Zealand and Iceland. J Phycol 12: 54-68

Castenholz RW (1977) The effect of sulfide on the blue-green algae of hot springs. II. Yellowstone National Park. Microbial Ecol 3: 79-105

Castenholz RW (1982) Motility and taxes. In: Carr NG and Whitton BA (eds) The Biology of Cyanobacteria pp. 413-439 Blackwell Scientific Publishers, Oxford

Castenholz RW and Utkilen HC (1984) Physiology of sulfide tolerance in a thermophilic *Oscillatoria*. Arch Microbiol 138: 299-305

Chafetz HS (1994) Bacterially induced precipitates of calcium carbonate and lithification of microbial mats. In: Krumbein WE, Paterson DM and Stal LJ (eds) Biostabilization of Sediments pp. 149-163 BIS Verlag, Oldenburg

Chafetz HS and Buczynski C (1992) Bacterially induced lithification of microbial mats. Palaios 7: 277-293

Christensen BE, Kjosbakken J and Smidsrod O (1985) Partial chemical and physical characterization of two extracellular polysaccharides produced by marine periphytic *Pseudomonas* sp. strain NCMB 2021. Appl Environ Microbiol 50: 837-845

Cloud PE and Semikhatov MA (1969) Proterozoic stromatolite zonation. Am J Sci 267: 1071-1061

Cohen Y (1989) Photosynthesis in cyanobacterial mats and its relation to the sulfur cycle: a model for microbial sulfur interactions. In: Cohen Y and Rosenberg E (eds) Microbial Mats. Physiological Ecology of Benthic Microbial Communities pp. 22-36 ASM, Washington

Cohen Y, Castenholz RW and Halvorson HO (1984) Microbial Mats: Stromatolites. Alan R. Liss, Inc, New York

Cohen Y, Jørgensen BB, Revsbech NP and Poplawski R (1986) Adaptation to hydrogen sulfide of oxygenic and anoxygenic photosynthesis among cyanobacteria. Appl Environ Microbiol 51: 398-407

Cohen Y and Rosenberg R (1989) Microbial Mats. Physiological Ecology of Benthic Microbial Communities. 494 pp. ASM, Washington

Cox G, Benson D and Dwarte DM (1981) Ultrastructure of a cave-wall cyanophyte *Gloeocapsa* NS4. Arch Microbiol 130: 165-174

Cullen JJ and Neale PJ (1994) Ultraviolet radiation, ozone depletion, and marine photosynthesis. Photosynth Res 39: 303-320

Cypionka H, Widdel F and Pfennig N (1985) Survival of sulfate-reducing bacteria after oxygen stress and growth in sulfate-free oxygen-sulfide gradients. FEMS Microbiol Ecol 31: 39-45

Dahanayake K and Krumbein WE (1985) Ultrastructure of a microbial mat-generated phosphorite. Mineral Deposita 20: 260-265

D'Antoni D'Amelio E, Cohen Y and Des Marais DJ (1989) Comparative functional ultrastructure of two hypersaline submerged cyanobacterial mats: Guerrero Negro, Baja California Sur, Mexico, and Solar Lake, Sinai, Egypt. In: Cohen Y and Rosenberg E (eds) Microbial Mats. Physiological Ecology of Benthic Microbial Communities pp. 97-113 ASM, Washington

Davey A (1983) Effects of abiotic factors on nitrogen fixation by blue-green algae in Antarctica. Polar Biology 2: 95-100

Davey A and Marchant HJ (1983) Seasonal variation in nitrogen fixation by *Nostoc commune* Vaucher at the Vesthold Hills, Antarctica. Phycologia 22: 377-385

De Philippis R, Margheri MC and Vincenzini M (1996) Fermentation in symbiotic and free-living cyanobacteria. Arch. Hydrobiol. 83: 459-468

De Winder B, Matthijs HCP and Mur LR (1989a) The role of water retaining substrata on the photosynthetic response of three drought tolerant phototrophic micro-organisms isolated from a terrestrial habitat. Arch Microbiol 152: 458-462

De Winder B, Pluis J, De Reus L and Mur LR (1989b) Characterization of a cyanobacterial, algal crust in the coastal dunes of the Netherlands. In: Cohen Y and Rosenberg E (eds) Microbial Mats. Physiological Ecology of Benthic Microbial Communities pp. 77-83 ASM, Washington

De Winder B, Stal LJ and Mur LR (1990) *Crinalium epipsammum* sp. nov.: a filamentous cyanobacterium with trichomes composed of elliptical cells and containing poly-beta-(1,4) glucan (cellulose). J Gen Microbiol 136: 1645-1653

De Wit R, Van Boekel WHM. and Van Gemerden H (1988) Growth of the cyanobacterium *Microcoleus chtonoplastes* on sulfide. FEMS Microbiol Ecol 53: 203-209

De Wit R and Van Gemerden H (1987) Oxidation of sulfide to thiosulfate by *Microcoleus chthonoplastes*. FEMS Microbiol Ecol 45: 7-13

De Wit R and Van Gemerden H (1988) Interactions between phototrophic bacteria in sediment ecosystems. Hydrobiol Bull 22: 135-145

Decho AW (1990) Microbial exopolymer secretions in ocean environments: their role(s) in food webs and marine processes. Oceanogr Mar Biol Ann Rev 28: 73-153

Decho AW (1994) Molecular-scale events influencing the macroscale cohesiveness of exopolymers. In: Krumbein WE, Paterson DM and Stal LJ (eds) Biostabilization of Sediments pp. 135-148 BIS Verlag, Oldenburg

Decho AW and Moriarty DJW (1990) Bacterial exopolymer utilization by a harpacticoid copepod: A methodology and results. Limnol Oceanogr 35: 1039-1049

Défarge C, Trichet J and Coute A (1994a) On the appearance of cyanobacterial calcification in modern stromatolites. Sed Geol 94: 11-19

Défarge C, Trichet J, Maurin A and Hucher M (1994b) Kopara in Polynesian atolls: early stages of formation of calcareous stromatolites. Sed Geol 89: 9-23

Des Marais DJ and Canfield DE (1994) The carbon isotope biogeochemistry of microbial mats. In: Stal LJ and Caumette P (eds) Microbial Mats. Structure, Development and Environmental Significance. pp. 289-298 Springer Verlag, Heidelberg

Des Marais DJ, D'Amelio E, Farmer JD, Jørgensen BB, Palmisano AC and Pierson BK (1992) Case study of a modern microbial mat-building community: the submerged cyanobacterial mats of Guerrero Negro, Baja California Sur, Mexico. In: Schopf JW and Klein C (eds) The Proterozoic Biosphere. A Multidisciplinary Study pp. 325-333 Cambridge University Press, New York

De Nobel WT, Snoep JL, Westerhoff HV and Mur LR (1997) Interaction of nitrogen fixation and phosphorus limitation in *Aphanizomenon flos-aquae* (Cyanophyceae). J Phycol 33: 794-799

Dilling W and Cypionka H (1990) Aerobic respiration in sulfate-reducing bacteria. FEMS Microbiol Lett 71: 123-127

Donkor VA, Amewowor DHAK and Häder DP (1993) Effects of tropical solar radiation on the motility of filamentous cyanobacteria. FEMS Microbiol Ecol 12: 143-148

Donkor V and Häder DP (1991) Effects of solar and ultraviolet radiation on motility, photomovement and pigmentation in filamentous, gliding cyanobacteria. FEMS Microbiol Ecol 86: 159-168

Dor I. and Paz N (1989) Temporal and spatial distribution of mat microalgae in the experimental solar ponds, Dead Sea area, Israel. In: Cohen Y and Rosenberg E (eds) Microbial Mats. Physiological Ecology of Benthic Microbial Communities pp. 114-122 ASM, Washington

Drews G and Weckesser J (1982) Function, structure and composition of cell walls and external layers. In: Carr NG and Whitton BA (eds) The Biology of Cyanobacteria pp. 333-357 Blackwell Scientific Publications, Oxford

Ehrenberg CG (1838) Über das im Jahre 1686 in Curland vom Himmel gefallene Meteorpapier und über dessen Zusammensetzung aus Conferven und Infusorien. Abhandlungen der Königlichen Akademie der Wissenschaften Berlin, Physikalische Klasse: 44-60.

Ehrenreich A and Widdel F (1994) Anaerobic oxidation of ferrous iron by purple bacteria, a new type of phototrophic metabolism. Appl Environ Microbiol 60: 4517-4526

Ehrlich HL (1996) Geomicrobiology. 719 pp. Marcel Dekker Inc , Basel

Eloff JN, Steinitz Y and Shilo M (1976) Photooxidation of cyanobacteria in natural conditions. Appl Environ Microbiol 31: 119-126

Emerson D and Revsbech NP (1994a) Investigation of an iron-oxidizing microbial mat community located near Aarhus, Denmark: field studies. Appl Environ Microbiol 60: 4022-4031

Emerson D and Revsbech NP (1994b) Investigation of an iron-oxidizing microbial mat community located near Aarhus, Denmark: laboratory studies. Appl Environ Microbiol 60: 4032-4038

Fechner R (1915) Die Chemotaxis der Oscillarien und ihre Bewegungs-erscheinungen überhaupt. Zeitschr Bot 7: 289-364

Ferris MJ, Ruff Roberts AL, Kopczynski ED, Bateson MM and Ward DM (1996) Enrichment culture and microscopy conceal diverse thermophilic *Synechococcus* populations in a single hot spring microbial mat habitat. Appl Environ Microbiol 62: 1045-1050

Flores E and Herrero A (1994) Assimilatory nitrogen metabolism and its regulation. In: Bryant DA (ed.) The Molecular Biology of Cyanobacteria pp. 487-517 Kluwer Academic Publishers, Dordrecht

Friedmann EI, Lipkin Y and Ocampo-Paus R (1967) Desert algae of the Negev (Israel). Phycologia 6: 185-200

Friedrich M and Schink B (1993) Hydrogen formation from glycolate driven by reversed electron transport in membrane vesicles of a syntrophic glycolate-oxidizing bacterium. Eur J Biochem 217: 233-240

Friedrich M and Schink B (1995) Isolation and characterization of a desulforubidin-containing sulfate-reducing bacterium growing with glycolate. Arch Microbiol 164: 271-279

Fründ C and Cohen Y (1992) Diurnal cycles of sulfate reduction under oxic conditions in cyanobacterial mats. Appl Environ Microbiol 58: 70-77

Gallon JR, Hashem MA and Chaplin AE (1991) Nitrogen fixation by *Oscillatoria* spp under autotrophic and photoheterotrophic conditions. J Gen Microbiol 137: 31-39

Garcia-Pichel F and Bebout BM (1996) Penetration of ultraviolet radiation into shallow water sediments: High exposure for photosynthetic communities. Mar Ecol Prog Ser 131: 257-262

Garcia-Pichel F and Castenholz RW (1990) Comparative anoxygenic photosynthetic capacity in 7 strains of a thermophilic cyanobacterium. Arch Microbiol 153: 344-351

Garcia-Pichel F and Castenholz RW (1991) Characterization and biological implications od scytonemin, a cyanobacterial sheath pigment. J Phycol 27: 395-409

Garcia-Pichel F and Castenholz RW (1994) On the significance of solar ultraviolet radiation for the ecology of microbial mats. In: Stal LJ and Caumette P (eds) Microbial Mats. Structure, Development and Environmental Significance pp. 77-84 Springer Verlag, Heidelberg

Garcia-Pichel F, Mechling M and Castenholz RW (1994) Diel migrations of microorganisms within a benthic, hypersaline mat community. Appl Environ Microbiol 60: 1500-1511

Garcia-Pichel F, Prufert-Bebout L and Muyzer G (1996) Phenotypic and phylogenetic analyses show *Microcoleus chthonoplastes* to be a cosmopolitan cyanobacterium. Appl Envirn Microbiol 62: 3284-3291

Garlick S, Oren A and Padan E (1977) Occurence of facultative anoxygenic photosynthesis among filamentous and unicellular cyanobacteria. J Bacteriol 129: 623-629

Gebelein CD (1976) Open marine subtidal and intertidal stromatolites (Florida, The Bahamas and Bermuda). In: Walter MR (ed.) Stromatolites pp. 381-388 Elsevier, Amsterdam

Geitler L (1932) Cyanophyceae. Rabenhorst's Kryptogamen-Flora von Deutschland, Österreich und der Schweiz. 14 pp. 1196 Akad. Verlagsgesellschaft, Leipzig

Gerbaud A and Andre M (1987) An evaluation of the recycling in measurements of photorespiration. Plant Physiol 83: 933-937

Golubic S (1973) The relationship between blue-green algae and carbonate deposits. In: Carr NG and Whitton BA (eds) The Biology of Blue-Green Algae pp. 434-473. Blackwell, Oxford, and Univ Calif Press, Berkeley

Griffiths MSH, Gallon JR and Chaplin AE (1987) The diurnal pattern of dinitrogen fixation by cyanobacteria in situ. New Phytol 107: 649-657

Guerrero R and Mas J (1989) Multilayered microbial communities in aquatic ecosystems: growth and loss factors. In: Cohen Y and Rosenberg E (eds) Microbial Mats. Physiological Ecology of Benthic Microbial Communities pp. 37-51 ASM, Washington

Häder D-P (1987a) Photomovement. In: Fay P and Van Baalen C (eds) The Cyanobacteria pp. 325-345 Elsevier, Amsterdam

Häder D-P (1987b) Photosensory behavior in procaryotes. Microbiol Rev 51: 1-21

Häder D-P (1988) Signal perception and amplification in photoresponses of cyanobacteria. Biophys Chem 29: 155-159

Halfen LN and Castenholz RW (1971) Gliding motility in the blue-green alga *Oscillatoria princeps.* J Phycol 7: 133-145

Hallbeck L and Pedersen K (1991) Autotrophic and mixotrophic growth *of Gallionella ferruginea.* J Gen Microbiol 137: 2657-2661

Heijthuijsen JHFG and Hansen TA (1989) Betaine fermentation and oxidation by marine *Desulfuromonas* strains. Appl Environ Microbiol 55: 965 - 969

Heyer H, Stal LJ and Krumbein WE (1989) Simultaneous heterolactic and acetate fermentation in the marine cyanobacterium *Oscillatoria limosa* incubated anaerobically in the dark. Arch Microbiol 151: 558-564

Holton RW and Freeman AW (1965) Some theoretical and experimental considerations of the gliding movement of blue-green algae. Am J Bot 52: 640

Houchins JP (1984) The physiology and biochemistry of hydrogen metabolism in cyanobacteria. Biochim Biophys Acta 768: 227-255

Howard JB and Rees DC (1994) Nitrogenase: a nucleotide-dependent molecular switch. Ann Rev Biochem 63: 235-264

Howsley R and Pearson HW (1979) pH dependent sulphide toxicity to oxygenic photosynthesis in cyanobacteria. FEMS Microbiol Lett 6: 287-292

Husic DW, Husic HD and Tolbert NE (1987) The oxidative photosynthetic carbon cycle or C_2 cycle. CRC Crit Rev Plant Sci 5: 45-100

James HL and Trendall AF (1982) Banded Iron-Formation: Distribution in time and Paleoenvironmental significance. In: Holland HD and Schidlowski M (eds) Mineral Deposits and the Evolution of the Biosphere. pp. 199-217 Springer Verlag, Heidelberg

Jones K and Stewart WDP (1969) Nitrogen turn over in marine and brackish habitats. III. The production of extracellular nitrogen by *Calothrix scopulorum.* J Mar Biol Ass UK 49: 475-488

Jørgensen BB (1990) A thiosulfate shunt in the sulfur cycle of marine sediments. Science 249: 152-154

Jørgensen BB and Cohen Y (1977) Solar Lake (Sinai) .5. The sulfur cycle of the benthic cyanobacterial mats. Limnol Oceanogr 22: 657-666

Jørgensen BB, Cohen Y and Des Marais DJ (1987) Photosynthetic action spectra and adaptation to spectral light distribution in a benthic cyanobacterial mat. Appl Environ Microbiol 53: 879-886

Jørgensen BB, Cohen Y and Revsbech NP (1986) Transition from anoxygenic to oxygenic photosynthesis in *a Microcoleus chthonoplastes* cyanobacterial mat. Appl Environ Microbiol 51: 408-417

Jørgensen BB and Des Marais DJ (1988) Optical properties of benthic photosynthetic communities: Fiber-optic studies of cyanobacterial mats. Limnol Oceanogr 33: 99-113

Jørgensen BB and Nelson DC (1988) Bacterial zonation, photosynthesis, and spectral light distribution in hot spring microbial mats of Iceland. Microb Ecol 16: 133-147

Jørgensen BB, Revsbech NP, Blackburn TH and Cohen Y (1979) Diurnal cycle of oxygen and sulfide microgradients and microbial photosynthesis in a cyanobacterial mat sediment. Appl Environ Microbiol 38: 46-58

Jørgensen BB, Revsbech NP and Cohen Y (1983) Photosynthesis and structure of benthic microbial mats: Micro-electrode and SEM studies of four cyanobacterial communities. Limnol Oceanogr 28: 1075-1093

Joye SB and Paerl HW (1994) Nitrogen cycling in microbial mats: rates and patterns of denitrification and nitrogen fixation. Mar Biol 119: 285-295

Kalkowsky E (1908) Oolith und Stromatolith im Norddeutschen Buntsandstein. Zeitschr Deutsche Geol Gesell 60: 68-125

Kaplan A, Schwarz R, Lieman-Hurwitz J, Ronen-Tarazi M and Reinhold L (1994) Physiological and molecular studeis on the response of cyanobacteria to changes in the ambient inorganic carbon concentration. In: Bryant DA (ed.) The Molecular Biology of Cyanobacteria. pp. 469-485 Kluwer Academic Publishers, Dordrecht

Kazmierczak J, Coleman ML, Gruszczynski M and Kempe S (1996) Cyanobacterial key to the genesis of micritic and peloidal limestones in ancient seas. Acta Palaeont Pol 41: 319-338

Kempe S and Kazmierczak J (1990a) Chemistry and stromatolites of the sea-linked Satonda Crater Lake, Indonesia: A recent model for the Precambrian sea? Chem Geol 81: 299-310

Kempe S and Kazmierczak J (1990b) Calcium carbonate supersaturation and the formation of in situ calcified stromatolites. In: Ittekkot V, Kempe S, Michaelis W and Spitzy A (eds) Facets of Modern Biogeochemistry. pp. 255-278 Springer Verlag, Heidelberg

Kempe S and Kazmierczak J (1993) Satonda crater lake, Indonesia: Hydrogeochemistry and biocarbonates. Facies 28: 1-32

Kempe S, Kazmierczak J, Landmann G, Konuk T, Reimer A and Lipp A (1991) Largest known microbialites discovered in Lake Van, Turkey. Nature 349: 605-608

Kennard JM and James NP (1986) Thrombolites and stromatolites: two distinct types of microbial structures. Palaios 1: 492-503

Kenyon CN, Rippka R and Stanier RY (1972) Fatty acid composition and physiological properties of some filamentous blue-green algae. Arch Microbiol 83: 216-236

Kiene RP and Visscher PT (1987) Production and fate of methylated sulfur compounds from methionine and dimethylsulfoniopropionate in anoxic salt marsh sediments. Appl Environ Microbiol 53: 2426-2434

Krumbein WE (1974) On the precipitation of aragonite on the surface of marine bacteria. Naturwissenschaften 61: 167

Krumbein WE (1979a) Photolithotropic and chemoorganotrophic activity of bacteria and algae as related to beachrock formation and degradation (Gulf of Aqaba, Sinai). Geomicrobiol J 1:139-203

Krumbein WE (1979b) Calcification by bacteria and algae. In: Trudinger PE and Swaine DJ (eds) Biogeochemical Cycling of Mineral-Forming Elements. pp. 47-68 Elsevier Scientif Publ, Amsterdam, Oxford, New York

Krumbein WE (1983) Stromatolites. The challenge of a term in space and time. Precambrian Res 20: 493-531

Krumbein WE, Carius RT and Stal LJ (1991) On the interaction of the two mat forming cyanobacteria *Microcoleus chthonoplastes* and *Oscillatoria limosa* and laboratory experiments on bundle formation and *de novo* genesis of microbial mats. Kieler Meeresforsch Sonderheft 8: 146-151

Krumbein WE and Cohen Y (1977) Primary production, mat formation and lithification: contribution of oxygenic and facultative anoxygenic cyanobacteria. In: Flügel E (ed.) Fossil Algae pp. 38-56 Springer Verlag, Berlin

Krumbein WE, Cohen Y and Shilo M (1977) Solar Lake (Sinai).4. Stromatolitic cyanobacterial mats. Limnol Oceanogr 22: 635-656

Krumbein WE and Giele C (1979) Calcification in a coccoid cyanobacterium accociated with the formation of desert stromatolites. Sedimentology 26: 593-604

Krumbein WE and Jens K (1981) Biogenic rock varnishes of the Negev Desert (Israel) an ecological study of iron and manganese transformation by cyanobacteria and fungi. Oecologia 50: 25-38

Kühl M and Jørgensen BB (1992) Spectral light measurements in microbenthic phototrophic communities with a fiber-optic microprobe coupled to a sensitive diode array detector. Limnol Oceanogr 37: 1813-1823

Lange W (1976) Speculations on a possible essential function of the gelatinous sheath of blue-green algae. Can J Microbiol 22: 1181-1185

Lassen C, Ploug H and Jørgensen BB (1992a) Microalgal photosynthesis and spectral scalar irradiance in coastal marine sediments of Limfjorden, Denmark. Limnol Oceanogr 37: 760-772

Lassen C, Ploug H and Jorgensen BB (1992b) A fibre-optic scalar irradiance microsensor - Application for spectral light measurements in sediments. FEMS Microbiol Ecol 86: 247-254

Leduc LG and Ferroni GD (1994) The chemolithotrophic bacterium *Thiobacillus ferrooxidans*. FEMS Microbiol Rev 14: 103-119

Lehmann M and Wöber G (1976) Accumulation, mobilization and turn-over of glycogen in the blue-green bacterium *Anacystis nidulans*. Arch Microbiol 111: 93-97

Logan BW (1961) Cryptozoon and associate stromatolites from the recent of Shark Bay, Western Australia. J Geol 69: 517-533

Lorenz MG and Wackernagel W (1990) Natural genetic transformation of *Pseudomonas stutzeri* by sand-adsorbed DNA. Arch Microbiol 154: 380-385

Lorenz MG and Wackernagel W (1994) Bacterial gene transfer by natural genetic transformation in the environment. Microbiol Rev 58: 563-602

Lorimer GH (1981) The carboxylation and oxygenation of ribulose 1,5-biphosphate: the primary events in photosynthesis and photorespiration. Ann Rev Plant Physiol 32: 349-383

Lorimer GH, Andrews TJ and Tolbert NE (1973) Ribulose diphosphate oxygenase. II. Further proof of reaction products and mechanisms of action. Biochemistry 12: 18-23

Lovley DR (1991) Dissimilatory Fe(III) and Mn(IV) reduction. Microbiol Rev 55: 259-287

Lowe DR (1980) Stromatolites 3,400-Myr old from the Archean of Western Australia. Nature 284: 441-443

Mackay MA, Norton RS and Borowitzka LJ (1984) Organic osmoregulatory solutes in cyanobacteria. J Gen Microbiol 130: 2177-2191

Mackerras AH, De Chazal NM and Smith GD (1990b) Transient accumulations of cyanophycin in *Anabaena cylindrica* and *Synechocystis* 6308. J Gen Microbiol 136: 2057-2065

Mackerras AH, Youens BN, Weir RC and Smith GD (1990a) Is cyanophycin involved in the integration of nitrogen and carbon metabolism in the cyanobacteria *Anabaena cylindrica* and *Gloeothece* grown on light/dark cycles? J Gen Microbiol 136: 2049-2056

Malin G and Pearson HW (1988) Aerobic nitrogen fixation in aggregate-forming cultures of the nonheterocystous cyanobacterium *Microcoleus chthonoplastes*. J Gen Microbiol 134: 1755-1763

Malin G, Walsby AE (1985) Chemotaxis of a cyanobacterium on concentration gradients of carbondioxide, bicarbonate and oxygen. J Gen Microbiol 131: 2643-2652

Margheri MC and Allotta G (1993) Homoacetic fermentation in the cyanobacterium *Nostoc* sp. strain Cc from *Cycas circinalis*. FEMS Microbiol Lett 111: 213-217

Margulis L, Barghoorn ES, Ashendorf D, Banerjee S, Chase D, Francis S, Giovannoni S and Stolz J (1980) The microbial community in the layered sediments of laguna Figueroa, Baja California Mexico: Does it have Precambrian analogues? Precambr Res 11: 93-123

Marschall C, Frenzel P and Cypionka H (1993) Influence of oxygen on sulfate reduction and growth of sulfate-reducing bacteria. Arch Microbiol 159: 168-173

Martin JP (1971) Decomposition and binding action of polysaccharides in soil. Soil Biol Biochem 3: 33-41

Maryan PS, Eady RR, Chaplin AE and Gallon JR (1986) Nitrogen fixation by *Gloeothece* sp. PCC 6909: respiration and not photosynthesis supports nitrogenase activity in the light. J Gen Microbiol 132: 789-796

Mason TR and Von Brunn V (1977) 3-Gyr-old stromatolites from South Africa. Nature 266: 47-49

McKay RML, Gibbs SP and Espie GS (1992) Effect of dissolved inorganic carbon on the expression of carboxysomes, localization of RubisCO and the mode of inorganic carbon transport in cells of the cyanobacterium *Synechocystis* UTEX 625. Arch Microbiol 159: 21-29

Milliman, JD (1993) Production and accumulation of calcium carbonate in the ocean: budget of a nonsteady state. Global Biogeochem. Cycles 7: 927-957

Miziorko HM and Lorimer GH (1983) Ribulose-1,5-biphosphate carboxylase-oxygenase. Ann Rev Biochem 52: 507-535

Moezelaar R, Bijvank SM and Stal LJ (1996) Fermentation and sulfur reduction in the mat-building cyanobacterium *Microcoleus chthonoplastes*. Appl Environ Microbiol 62: 1752-1758

Monty CLV (1976) The origin and development of cryptalgal fabrics. In: Walter MR (ed.) Stromatolites pp.139-249 Elsevier, Amsterdam

Mullineaux PM, Gallon JR and Chaplin AE (1981) Acetylene reduction (nitrogen fixation) by cyanobacteria grown under alternating light-dark cycles. FEMS Microbiol Lett 10: 245-247

Nicholson JAM, Stolz JF and Pierson BK (1987) Structure of a microbial mat at Great Sippewissett Marsh, Cape Cod, Massachusetts. FEMS Microbiol Ecol 45: 343-364

Nold SC and Ward DM (1996) Photosynthate partitioning and fermentation in hot spring microbial mat communities. Appl Environ Microbiol 62: 4598-4607

Oren A (1988) The microbial ecology of the Dead Sea. Adv Microb Ecol 10: 193-229

Oren A, Padan E and Avron M (1977) Quantum yields for oxygenic and anoxygenic photosynthesis in the cyanobacterium *Oscillatoria limnetica*. Proc Natl Acad Sc USA 74: 2152-2156

Oren A and Shilo M (1979) Anaerobic heterotrophic dark metabolism in the cyanobacterium *Oscillatoria limnetica*: Sulfur respiration and lactate fermentation. Arch Microbiol 122: 77-84

Orpen JL and Wilson JF (1981) Stromatolites at ca. 3,500 Myr and a greenstone granite unconformity in the Zimbabwean Archaean. Nature 291: 218-220

Ortega-Calvo JJ and Stal LJ (1991) Diazotrophic growth of the unicellular cyanobacterium *Gloeothece* sp PCC 6909 in continuous culture. J Gen Microbiol 137: 1789-1797

Ortega-Calvo JJ and Stal LJ (1994) Sulphate-limited growth in the N_2-fixing unicellular cyanobacterium *Gloeothece* (Nageli) sp PCC 6909. New Phytol 128: 273-281

Padan E and Cohen Y (1982) Anoxygenic photosynthesis. In: Carr NG and Whitton BA (eds) The Biology of Cyanobacteria pp. 215-235 Blackwell Scientific Publications, Oxford

Paerl HW, Bebout BM and Prufert LE (1989) Naturally occurring patterns of oxygenic photosynthesis and N_2 fixation in a marine microbial mat: Physiological and ecological ramifications. In: Cohen Y and Rosenberg E (eds) Microbial Mats. Physiological Ecology of Benthic Microbial Communities pp. 326-341, ASM, Washington, DC.

Paerl HW, Fitzpatrick M and Bebout BM (1996) Seasonal nitrogen fixation dynamics in a marine microbial mat: Potential roles of cyanobacteria and microheterotrophs. Limnol Oceanogr 41: 419-427

Paerl HW, Pinckney JL and Kucera SA (1995) Clarification of the structural and functional roles of heterocysts and anoxic microzones in the control of pelagic nitrogen fixation. Limnol Oceanogr 40: 634-638

Paerl HW and Prufert LE (1987) Oxygen-poor microzones as potential sites of microbial N_2 fixation in nitrogen-depleted aerobic marine waters. Appl Environ Microbiol 53: 1078-1087

Paerl HW, Prufert LE and Ambrose WW (1991) Contemporaneous N_2 fixation and oxygenic photosynthesis in the nonheterocystous mat-forming cyanobacterium *Lyngbya aestuarii*. Appl Environ Microbiol 57: 3086-3092

Palinska KA, Liesack W, Rhiel E and Krumbein WE (1996) Phenotype variability of identical genotypes: the need for a combined approach in cyanobacterial taxonomy demonstrated on *Merismopedia*-like isolates. Arch. Microbiol. 166: 224-233

Pankratz HS and Bowen CC (1963) Cytology of blue-green algae. I. The cells of *Symploca muscorum*. Amer J Bot 50: 387-399

Paterson DM (1989) Short-term changes in the erodibility of intertidal cohesive sediments related to the migratory behaviour of epipelic diatoms. Limnol Oceanogr 34: 223-234

Paterson DM, Crawford RM and Little C (1990) Subaerial exposure and changes in the stability of intertidal estuarine sediments. Est Coast Shelf Sci 30: 541-556

Pearson HW, Howsley R, Kjeldsen CK and Walsby AE (1979) Aerobic nitrogenase activity associated with a non-heterocystous filamentous cyanobacterium. FEMS Microbiol Lett 5: 163-169

Pentecost A (1984) Effects of sedimentation and light intensity on mat-forming Oscillatoriaceae with particular reference to *Microcoleus lyngbyaceus* Gomont. J Gen Microbiol 130: 983-990

Pentecost A (1988) Growth and calcification of the cyanobacterium *Homoeothrix crustacea*. J Gen Microbiol 134: 2665-2671

Pentecost A and Bauld J (1988) Nucleation of calcite on the sheaths of cyanobacteria using a simple diffusion cell. Geomicrobiol J 6: 129-135

Peschek GA (1978) Reduced sulfur and nitrogen compounds and molecular hydrogen as electron donors for anaerobic CO_2 photoreduction in *Anacystis nidulans*. Arch Microbiol 119: 313-322

Pierce J (1988) Prospects for manipulating the substrate specificity of ribulose bisphosphate carboxylase/oxygenase. Physiol Plant 72: 690-698

Pierson BK (1992) Introduction. In: Schopf JW and Klein C (eds) The Proterozoic Biosphere. A Multidisciplinary Study pp. 247-251 Cambridge University Press, New York.

Pierson B, Oesterle A and Murphy GL (1987) Pigments, light penetration, and photosynthetic activity in the multi-layered microbial mats of Great Sippewisset salt marsh, Massachusetts. FEMS Microbiol Ecol 45: 365-376

Pierson BK and Olson JM (1989) Evolution of photosynthesis in anoxygenic photosynthetic procaryotes. In: Cohen Y and Rosenberg E (eds) Microbial Mats. Physiological Ecology of Benthic Microbial Communities. pp. 402-427 ASM, Washington

Pierson BK, Sands VM and Frederick JL (1990) Spectral irradiance and distribution of pigments in a highly layered marine microbial mat. Appl Environ Microbiol 56: 2327-2340

Pinckney J, Paerl HW, Reid RP and Bebout B (1995) Ecophysiology of stromatolitic microbial mats, Stocking Island, Exuma Cays, Bahamas. Microb Ecol 29: 19-37

Piper DZ and Codespoti LA (1975) marine phosphorite deposits and the nitrogen cycle. Science 179: 564-565

Ploug H, Lassen C and Jørgensen BB (1993) Action spectra of microalgal photosynthesis and depth distribution of spectral scalar irradiance in a coastal marine sediment of Limfjorden, Denmark. FEMS Microbiol Ecol 102: 261-270

Potts M (1979) Nitrogen fixation (acetylene reduction) associated with communities of heterocystous and non-heterocystous blue-green algae in mangrove forests of Sinai. Oecologia 39: 359-373

Potts M and Friedmann EI (1981) Effects of water stress on crypto-endolithic cyanobacteria from hot desert rocks. Arch Microbiol 130: 267-271

Potts M, Krumbein WE and Metzger J (1978) Nitrogen fixation rates in anaerobic sediments determined by acetylene reduction, a new ^{15}N field assay, and simultaneous total N ^{15}N determination. In: Krumbein WE (ed.) Environmental

Biogeochemistry and Geomicrobiology 3: Methods, Metals and Assessment pp. 753-769 Ann Arbor Science, Ann Arbor

Potts M, Ocampo-Friedmann R, Bowman MA and Tozun B (1983) *Chroococcus* S24 and *Chroococcus* N41 (cyanobacteria): morphological, biochemical and genetic characterization and effects of water stress on ultra structure. Arch Microbiol 135: 81-90

Pringsheim EG (1968) Cyanophyceen-Studien. Arch Mikrobiol 63: 331-355

Rabenstein A, Rethmeier J and Fischer U (1995) Sulphite as intermediate sulphur compound in anaerobic sulphide oxidation to thiosulphate by marine cyanobacteria. Zeitschr Naturforsch C 50: 769-774

Ramsing NB and Prufert-Bebout L (1994) Motility of *Microcoleus chthonoplastes* subjected to different light intensities quantified by digital image analysis. In: Stal LJ and Caumette P (eds) Microbial Mats. Structure, Development and Environmental Significance. pp. 183-191 Springer Verlag, Heidelberg

Reed RH, Borowitzka LJ, Mackay MA, Chudek JA, Foster R, Warr SCR, Moore DJ and Stewart WDP (1986a) Organic solute accumulation in osmotically stressed cyanobacteria. FEMS Microbiol Rev 39: 51-56

Reed RH and Stewart WDP (1983) Physiological responses of *Rivularia atra* to salinity: osmotic adjustment in hyposaline media. New Phytol 95: 595-603

Reed RH and Stewart WDP (1988) The responses of cyanobacteria to salt stress. In: Rogers LJ and Gallon JR (eds) Biochemistry of Algae and Cyanobacteria pp. 217-231 Clarendon Press, Oxford

Reed RH, Warr SRC, Kerby NW and Stewart WDP (1986b) Osmotic shock-induced release of low molecular weight metabolites from free-living and immobilized cyanobacteria. Enzyme Microbiol. Technol 8: 101-104

Rees DA (1969) Structure, conformation and mechanism in the formation of polysaccharide gels and network. Adv Carbohydr Chem Biochem 24: 267-332

Reid RP and Browne KM (1991) Intertidal stromatolites in a fringing Holocene reef complex, Bahamas. Geology 19: 15-18

Reinhold L, Kosloff R and Kaplan A (1991) A model for inorganic carbon fluxes and photosynthesis in cyanobacterial carboxysomes. Can J Bot 69: 984-988

Renstrom-Kellner E and Bergman B (1989) Glycolate metabolism in cyanobacteria. III. Nitrogen controls excretion and metabolism of glycolate in *Anabaena cylindrica*. Physiol Plant 77: 46-51

Renstrom-Kellner E and Bergman B (1990) Glycolate metabolism in cyanobacteria. IV. Uptake, growth and metabolic pathways. Physiol Plant 78: 285-292

Revsbech NP, Jørgensen BB, Blackburn, TH and Cohen Y (1983) Microelectrode studies of the photosynthesis and O_2, H_2S and pH profiles of a microbial mat. Limnol Oceanogr 28: 1062-1074

Richardson LL and Castenholz RW (1987a) Diel vertical movements of the cyanobacterium *Oscillatoria terebriformis* in a sulfide-rich hot spring microbial mat. Appl Environ Microbiol 53: 2142-2150

Richardson LL and Castenholz RW (1987b) Enhanced survival of the cyanobacterium *Oscillatoria terebriformis* in darkness under anaerobic conditions. Appl Environ Microbiol 53: 2151-2158

Richardson LL and Castenholz RW (1989) Chemokinetic motility responses of the cyanobacterium *Oscillatoria terebriformis*. Appl Environ Microbiol 55: 261-263

Rippka R, Deruelles J, Waterbury JB, Herdman M and Stanier RY (1979) Generic assignments strain histories and properties of pure cultures of cyanobacteria. J Gen Microbiol 111: 1-61

Rippka R and Waterbury JB (1977) The synthesis of nitrogenase by non-heterocystous cyanobacteria. FEMS Microbiol Lett 2: 83-86

Risatti JB, Capman WC and Stahl DA (1994) Community structure of a microbial mat: the phylogenetic dimension. Proc Nat Acad Sci USA 91: 10173-10177

Robbins LL and Blackwelder PL (1992) Biochemical and ultrastructural evidence for the origin of whitings: a biologically induced calcium carbonate precipitation mechanism. Geology 20: 464-468

Robins RJ, Hall DO, Shi DJ, Turner RJ and Rhodes MJC (1986) Mucilage acts to adhere cyanobacteria and cultured plant cells to biological and inert surfaces. FEMS Microbiol Lett 34: 155-160

Roeske CA and O'Leary M (1984) Carbon isotope effects on the enzyme-catalyzed carboxylation of ribulose bisphosphate. Biochemistry 23: 6275-6284

Romanowski G, Lorenz MG and Wackernagel W (1991) Adsorption of plasmid DNA to mineral surfaces and protection against DNAse-I. Appl Environ Microbiol 57: 1057-1061

Rosenberg M (1989) Microbial films in the mouth: Some ecologically relevant observations. In: Cohen Y and Rosenberg E (eds) Microbial Mats. Physiological Ecology of Benthic Microbial Communities. pp. 246-251 ASM, Washington, D.C.

Ruff-Roberts AL, Kuenen JG and Ward DM (1994) Distribution of cultivated and uncultivated cyanobacteria and *Chloroflexus*-like bacteria in hot spring microbial mats. App Environ Microbiol 60: 697-704

Schau L, Benson AA, Bassham JA and Calvin M (1950) The path of carbon in photosynthesis. XI. The role of glycolic acid. Physiol Plant 3: 487-495

Schaub BEM and Van Gemerden H (1996) Sulfur bacteria in sediments of two coastal ecosystems: the Bassin d'Arcachon and the Étang du Prevost, France. Hydrobiologia 329: 199-210

Schmetterer G (1994) Cyanobacterial respiration. In: Bryant DA (ed.) The Molecular Biology of Cyanobacteria pp. 409-435 Kluwer Academic Publishers, Dordrecht

Schopf JW (ed.) (1983) Earth's Earliest Biosphere. Its Origin and Evolution. 543 pp. Princeton University Press, Princeton, New Jersey

Schopf JW and Klein C (1992) The Proterozoic Biosphere. A Multidisciplinary Study. 1348 pp. Cambridge University Press, New York

Schopf JW and Walter MR (1982) Origin and early evolution of cyanobacteria: the geological evidence. In: Carr NG and Whitton BA (eds) The Biology of Cyanobacteria pp. 543-564 Blackwell Scientific Publications, Oxford

Shephard KL (1987) Evaporation of water from the mucilage of a gelatinous algal community. Br Phycol J 22: 181-185

Simon RD (1987) Inclusion bodies in the cyanobacteria: cyanophycin, polyphosphate, polyhedral bodies. In: Fay P and Van Baalen C (eds) The Cyanobacteria pp. 199-225 Elsevier Science Publishers, Amsterdam

Smith AJ (1982) Modes of cyanobacterial carbon metabolism. In: Carr NG and Whitton BA (eds) The Biology of Cyanobacteria. pp. 47-85 Blackwell, Oxford

Smith RC, Prezelin BB, Baker KS, Bidigare RR, Boucher NP, Coley T, Karentz D, Macintyre S, Matlick HA, Menzies D, Ondrusek M, Wan Z and Waters KJ (1992) Ozone depletion - Ultraviolet radiation and phytoplankton biology in Antarctic waters. Science 255: 952-959

Stal LJ (1991) The sulfur metabolism of mat-building cyanobacteria in anoxic marine sediments. Kieler Meeresforsch 8: 152-157

Stal LJ (1993) Mikrobielle Matten. In: Meyer-Reil L-A and Köster M (eds) Mikrobiologie des Meeresbodens pp. 196-220 Gustav Fischer, Jena

Stal LJ (1994) Microbial mats in coastal environments. In: Stal LJ and Caumette P (eds) Microbial Mats. Structure, Development and Environmental Significance pp. 21-32 Springer Verlag, Heidelberg

Stal LJ (1995) Physiological ecology of cyanobacteria in microbial mats and other communities. New Phytol 131: 1-32

Stal LJ, Behrens SB, Villbrandt M, Van Bergeijk S and Kruyning F (1996) The biogeochemistry of two eutrophic marine lagoons and its effect on microphytobenthic communities. Hydrobiol 329: 185-198

Stal LJ and Caumette P (eds) (1994) Microbial Mats. Structure, Development and Environmental Significance. 463 pp. Springer Verlag, Heidelberg

Stal LJ, Grossberger S and Krumbein WE (1984) Nitrogen fixation associated with the cyanobacterial mat of a marine laminated microbial ecosystem. Mar Biol 82: 217-224

Stal LJ and Heyer H (1987) Dark anaerobic nitrogen fixation (acetylene reduction) in the cyanobacterium *Oscillatoria* sp. FEMS Microbiol Ecol 45: 227-232

Stal LJ and Krumbein WE (1981) Aerobic nitrogen fixation in pure cultures of a benthic marine *Oscillatoria* (cyanobacteria). FEMS Microbiol Lett 11: 295-298

Stal LJ and Krumbein WE (1985) Isolation and characterization of cyanobacteria from a marine microbial mat. Bot Mar 28: 351-365

Stal LJ and Krumbein WE (1987) Temporal separation of nitrogen fixation and photosynthesis in the filamentous, non-heterocystous cyanobacterium *Oscillatoria* sp. Arch Microbiol 149: 76-80

Stal LJ and Moezelaar R (1997) Fermentation in cyanobacteria. FEMS Microbiol Rev 21: 179-211

Stal LJ, Paerl HW, Bebout B and Villbrandt M (1994) Heterocystous versus non-heterocystous cyanobacteria in microbial mats. In: Stal LJ and Caumette P (eds) Microbial Mats. Structure, Development and Environmental Significance pp. 403-414 Springer Verlag, Heidelberg

Stal LJ and Reed RH (1987) Low-molecular mass carbohydrate accumulation in cyanobacteria from a marine microbial mat in response to salt. FEMS Microbiol Ecol 45: 305-312

Stal LJ, Van Gemerden H and Krumbein WE (1985) Structure and development of a benthic marine microbial mat. FEMS Microbiol Ecol 31: 111-125

Steudel R, Holdt G, Visscher PT and Van Gemerden H (1990) Search for polythionates in cultures of *Chromatium vinosum* after sulfide incubation. Arch Microbiol 153: 432-437

Stolz JF (1993) Magnetosomes. J Gen Microbiol 139: 1663-1670

Storch TA, Saunders GW and Ostrofsky ML (1990) Diel nitrogen fixation by cyanobacterial surface blooms in Sanctuary Lake, Pennsylvania. Appl Environ Microbiol 56: 466-471

Straub KL, Benz M, Schink B and Widdel F (1996) Anaerobic, nitrate-dependent microbial oxidation of ferrous iron. Appl Environ Microbiol 62: 1458-1460

Tabita FR (1988) Molecular and cellular regulation of autotrophic carbon dioxide fixation in microorganisms. Microbiol Rev 52: 155-189

Tago Y and Aida K (1977) Exocellular mucopolysaccharide closely related to bacterial floc formation. Appl Environ Microbiol 34: 308-314

Taiz L (1984) Plant cell expansion: regulation of cell wall mechanical properties. Ann Rev Plant Physiol 35: 585-657

Talbot MMB, Bate GC and Campbell EE (1990) A review of the ecology of surf-zone diatoms, with special reference to *Anulus australis*. Oceanogr Mar Biol Annu Rev 28: 155-175

Tease B, Jürgens UJ, Golecki JR, Heinrich UR, Rippka R and Weckesser J (1991) Fine-Structural and Chemical Analyses on Inner and Outer Sheath of the Cyanobacterium *Gloeothece* sp PCC-6909. Antonie Van Leeuwenhoek Int J Gen Mol Microbiol 59: 27-34

Trüper HG and Galinski EA (1989) Compatible solutes in halophilic phototrophic procaryotes. In: Cohen Y and Rosenberg E (eds) Microbial Mats. Physiological Ecology of Benthic Microbial Communities pp. 342-348 ASM, Washington

Turner SM, Malin G, Liss PS, Harbour DS and Holligan PM (1988) The seasonal variation of dimethyl sulfide and dimethylsulfonio-propionate concentrations in nearshore waters. Limnol Oceanogr 33: 364-375

Turpin DH, Miller AG and Canvin DT (1984) Carboxysome content of *Synechococcus leopoliensis* (Cyanophyta) in response to inorganic carbon. J Phycol 20: 249-253

Utkilen HC (1976) Thiosulphate as electron donor in the blue-green alga *Anacystis nidulans*. J Gen Microbiol 95: 177-180

Van Bergeijk SA and Stal LJ (1996) The role of oxygenic phototrophic microorganisms in production and conversion of dimethylsulfonio-propionate and dimethylsulfide in microbial mats. In: Kiene RP, Visscher PT, Keller MD and Kirst GO (eds) Biological and Environmental Chemistry of DMSP and Related Sulfonium Compounds pp. 369-379 Plenum Press, New York

Van Gemerden H (1987) Competition between purple sulfur bacteria and green sulfur bacteria: role of sulfide, sulfur and polysulfides. Acta Academiae Aboensis 47: 13-27

Van Gemerden H (1993) Microbial mats: A joint venture. Mar Geol 113: 3-25

Van Gemerden H, Tughan CS, De Wit R and Herbert RA (1989) Laminated microbial ecosystems on sheltered beaches in Scapa Flow, Orkney Islands. FEMS Microbiol Ecol 62: 87-102

Van Liere L and Mur LR (1979) Growth kinetics of *Oscillatoria agardhii* Gomont in continuous culture, limited in its growth by the light energy supply. J Gen Microbiol 115: 153-160

Verrecchia EP, Dumant J-L and Collins KE (1990) Do fungi building limestone exist in semi-arid regions? Naturwissenschaften 77: 584-586

Villbrandt M (1992) Interactions of nitrogen fixation and photosynthesis in marine cyanobacterial mats (Mellum, Southern North Sea). PhD thesis, University of Oldenburg, Oldenburg, Germany

Villbrandt M and Stal LJ (1996) The effect of sulfide on nitrogen fixation in heterocystous and non-heterocystous cyanobacterial mat communities. Algol Stud 83: 549-563

Villbrandt M, Stal LJ and Krumbein WE (1990) Interactions between nitrogen fixation and oxygenic photosynthesis in a marine cyanobacterial mat. FEMS Microbiol Ecol 74: 59-72

Visscher PT (1992) Microbial sulfur cycling in laminated marine ecosystems. PhD Thesis University of Groningen, Groningen

Visscher PT, Prins RA and Van Gemerden H (1992) Rates of sulfate reduction and thiosulfate consumption in a marine microbial mat. FEMS Microbiol Ecol 86: 283-293

Visscher PT and Van Gemerden H (1991) Production and consumption of dimethylsulfoniopropionate in marine microbial mats. Appl Environ Microbiol 57: 3237-3242

Walsby AE (1985) The permeability of heterocysts to the gases nitrogen and oxygen. Proc Royal Soc Lond B 226: 345-366

Walter MR (ed.) (1976) Stromatolites. 790 pp. Elsevier, Amsterdam

Walter MR, Buick R and Dunlop JSR (1980) Stromatolites 3,400-3,500 Myr. old from the North Pole area, Western Australia. Nature 284: 443-445

Walter MR, Grotzinger JP and Schopf JW (1992) Proterozoic stromatolites. In: Schopf JW and Klein C (eds) The Proterozoic Biosphere. A Multidisciplinary Study pp. 253-260 Cambridge University Press, New York

Walter MR and Heys GR (1985) Links between the rise of metazoa and the decline of stromatolites. Precambrian Res 29: 149-174

Warburg O (1920) Über die Geschwindigkeit der photochemischen Kohlensäurezersetzung in lebenden Zellen. II. Biochem Zeitschr 103: 188-217

Warburg O and Krippahl G (1960) Glykolsäurebildung in *Chlorella*. Zeitschr Naturforsch B15: 197-199

Ward DM, Ferris MJ, Nold SC, Bateson MM, Kopcynski ED and Ruff-Roberts AL (1994) Species diversity in hot spring microbial mats as revealed by both molecular and enrichment culture approaches - relationship between biodiversity and community structure. In: Stal LJ and Caumette P (eds) Microbial Mats. Structure, Development and Environmental Significance pp. 33-44 Springer Verlag, Heidelberg

Ward DM, Weller R, Shiea J, Castenholz RW and Cohen Y (1989) Hot spring microbial mats: Anoxygenic and oxygenic mats of possible evolutionary significance. In: Cohen Y and Rosenberg E (eds) Microbial mats. Physiological Ecology of Benthic Microbial Communities pp. 3-15, ASM, Washington

Warr SCR, Reed RH and Stewart WDP (1984) Osmotic adjustment of cyanobacteria: the effects of NaCl, KCl, sucrose and glycine betaine on glutamine synthetase activity in a marine and a halotolerant strain. J Gen Microbiol 130: 2169-2175

Waterbury JB and Rippka R (1989) Subsection I. Order Chroococcales. In: Staley JT, Bryant MP, Pfennig N and Holt JG (eds) Bergey's Manual of Systematic Bacteriology Vol. 3 pp. 1728-1746 Williams and Wilkins, Baltimore, MD

Westbroek P, Buddemeier B, Coleman M, Kok DJ, Fautin D and Stal LJ (1994) Strategies for the study of climate forcing by calcification. In: Doumenge F (ed.) Past and Presence Biomineralization Processes pp. 37-60 Musee Oceanographique, Monaco

Whale GF and Walsby AE (1984) Motility of the cyanobacterium *Microcoleus chthonoplastes* in mud. Br Phycol J 19: 117-123

Whitton BA (1987) The biology of Rivulariaceae. In: Fay P and Van Baalen C (eds) The Cyanobacteria pp. 513-534 Elsevier Science Publishers, Amsterdam

Whitton BA and Potts M (1982) Marine littoral. In: Carr NG and Whitton BA (eds) The Biology of Cyanobacteria pp. 515-542 Blackwell Scientific Publications, Oxford

Widdel F, Schnell S, Heising S, Ehrenreich A, Assmus B and Schink B (1993) Ferrous iron oxidation by anoxygenic phototrophic bacteria. Nature 362: 834-836

Yallop ML, De Winder B, Paterson DM and Stal LJ (1994) Comparative structure, primary production and biogenic stabilization of cohesive and non-cohesive marine sediments inhabited by microphytobenthos. Est Coast Shelf Sci 39: 565-582

Zavarzin GA, Gerasimenko LM and Zhilina TN (1993) Cyanobacterial communities in hypersaline lagoons of Lake Sivash. Microbiology 62: 645-652

Chapter 5

Marine Plankton

Hans W. Paerl
Institute of Marine Sciences, The University of North Carolina at Chapel Hill
3431 Arendell Street, Morehead City, North Carolina 28557, USA

Summary

Cyanobacteria are a taxonomically diverse, productive and biogeochemically important component of oceanic, coastal and estuarine phytoplankton communities. Recent molecular and analytical characterization of these communities, together with a growing number of cruise-based and remote-sensing surveys of the world's ocean surface waters, indicate that cyanobacteria play a highly significant role in marine carbon and nitrogen cycling. Vast segments of the world's ocean surface waters are nitrogen deplete and planktonic N_2-fixing cyanobacteria can compete effectively in this large niche. Surface blooms of the filamentous diazotrophic genera *Trichodesmium* and *Richelia* (endosymbiont of the diatom *Rhizosolenia*) achieve magnitudes of growth in excess of hundreds of square kilometers in what are some of the most nutrient deplete, oligotrophic waters known. The N_2-fixing enzyme complex, nitrogenase, is oxygen-sensitive and requires protection from ambient oxygen levels as well as from intracellular O_2 evolved from photosynthetic activities of the cyanobacteria. Numerous molecular, physiological and structural adaptations ensure compatibility between an oxygenic phototrophic mode of growth and the need to fix N_2. These adaptations include: spatial and temporal separation of photosynthesis and nitrogen fixation, formation of aggregates which can support O_2 deplete microzones and respiratory and other enzymatic, O_2 consuming mechanisms. Coccoid cyanobacteria such as *Synechococcus, Synechocystis* and *Prochlorococcus* are common and often dominant in the phytoplankton community in subsurface waters where they may account for more than 50% of the biomass. The picoplanktonic (< 5 µm) and nanoplanktonic (5 – 20 µm) forms generally do not have the capacity to fix nitrogen; however, their small size, high surface to volume ratio, and ability to grow effectively in poorly illuminated, nutrient-rich deep waters ensure their access to nitrogen and other essential nutrients. Cyanobacteria that occupy and periodically dominate oligotrophic waters evolved various physiological, morphological and ecological strategies to optimize their capacities for nitrogen fixation, photosynthesis and nutrient (P, Fe) sequestration. Strategies include buoyancy regulation to ensure access to sunlight (energy), aggregation in colonies and consortia in which there are numerous interactions with an array of microheterotrophs and invertebrate grazers, close coupling of production and nutrient cycling

B. A. Whitton and M. Potts (eds), The Ecology of Cyanobacteria, 121-148.
© 2000 *Kluwer Academic Publishers. Printed in the Netherlands.*

within these communities, development of endosymbioses, intracellular nitrogen and phosphorus storage and the development of capacities for heterotrophy, photoheterotrophy, and pigment adaptation to counteract potential photooxidative conditions in surface waters. As a functional component of the phytoplankton, cyanobacteria exhibit remarkable ecophysiological flexibility and capacity for adaptation in response to anthropogenic alteration of the marine environment. In addition to their ability to thrive in oligotrophic waters, cyanobacteria can exploit nutrient-enriched estuaries and seas such as the Baltic, sometimes as persistent nuisance blooms which may be toxic and a source of hypoxic and anoxic conditions.

I. Introduction

Cyanobacterial dominance, in the form of dense surface blooms, was clearly and commonly recognized in nutrient-enriched, freshwater ecosystems (Reynolds, 1987; Paerl, 1988; Chapter 6). It became apparent only recently, however, that the diversity, densities and activities of cyanobacteria have an impact on microbial production, nutrient cycling and community structure in the world's estuarine, coastal and open ocean (pelagic) waters. The marine environment provides a cryptic, yet biogeochemically and trophically important range of habitats in which cyanobacteria are ubiquitous and periodically dominant as blooms. Our rather rudimentary understanding of the ecology, physiology and population dynamics of marine planktonic cyanobacteria stems from the fact that a large segment of the world's oceans is a nutrient deplete, unproductive or oligotrophic, "desert-like" environment. Here, there is a paucity of biomass at the primary producer level. Moreover, oceanic primary production is dominated by microscopic single-celled phytoplankton less than 5 mm in size (picoplankton: Johnson and Sieburth, 1979; Waterbury et al., 1979; Platt and Li, 1986). Before the advent and application of fluorescence microscopy and electron microscopy in the late 1970s, this size fraction was largely overlooked and hence was not included in biomass and activity assessments of planktonic primary producers. Since then, it became apparent that picoplanktonic, nanoplanktonic (5 - 20 mm) and microplanktonic (> 20 mm) cyanobacteria play key biogeochemical and trophic roles in the marine environment.

Cyanobacteria influence marine production and nutrient cycling in two principal ways. First, they are important, and at times dominant, phototrophic fixers of CO_2 and contribute significantly to primary production and C cycling. Cosmopolitan examples include the picoplanktonic genera *Synechococcus, Prochlorococcus* and *Synechocystis*. These forms are responsible for a large fraction (30 to > 50%) of the phytoplankton biomass and the primary production in waters ranging from the oligotrophic open ocean to more eutrophic coastal and estuarine ecosystems (Itturiaga and Mitchell, 1986; Booth, 1988; Chisholm et al., 1988). Second, some cyanobacterial genera can convert dinitrogen into the biologically-useful form ammonia (NH_3) via nitrogen fixation (Fogg, 1944, 1974; Stewart, 1973). This process is of particular relevance and importance in the vast stretches of the world's oceans that exhibit nitrogen (N) deficiencies (Dugdale, 1967); ammonia alleviates N limited growth and represents a source of new N to support the primary and secondary production of microbes, higher plants and animals. The capacity to fix nitrogen permitted certain marine planktonic cyanobacterial genera to proliferate as dense surface blooms in chronically N deficient oceanic, coastal and estuarine waters (e.g. *Trichodesmium;* Carpenter, 1983a, 1983b) (*Nodularia, Aphanizomenon*; Huber, 1986; Kononen et al., 1996). Diazotrophic cyanobacteria play key roles in mutualistic and symbiotic associations with microalgae (e.g. *Rhizosolenia-Richelia*: Villareal, 1992), macroalgae (e.g. with the siphonous green seaweed *Codium* and the planktonic phaeophyte *Sargassum*: Carpenter, 1973; Rosenberg and Paerl, 1980; Phlips and Zeman, 1990), as well as higher rooted plants (mangroves, seagrasses: Gallon et al., 1991; Capone, 1983) and animals (Stewart, 1973).

This chapter examines the biogeochemical, trophic, ecological and evolutionary importance of marine planktonic cyanobacteria as well as environmental controls which influence their abundance and activities. The focus is on ecophysiological, molecular and structural strategies which marine planktonic cyanobacteria developed to cope with environmental constraints on primary production and N_2 fixation. These strategies were developed in a range of marine ecosystems from nutrient-enriched estuaries and ultra-oligotrophic open ocean waters. A particularly enigmatic problem is the challenging, and thus far unresolved, paradox that N-limited marine waters represent a vast, yet apparently unfilled niche for this taxonomically, physiologically and ecologically diverse group of diazotrophs.

II. Key Functional Groups of Marine Planktonic Cyanobacteria

Marine planktonic cyanobacteria are classified into three morphological and functional groups: unicellular, non-heterocystous filamentous and heterocystous filamentous (Fig.1). Representative genera from each group are listed in Table 1 and are discussed in the context of their roles in marine production and nutrient cycling, as well as ecological, evolutionary and environmental dynamics. Within each group there are aggregated (colonial) and non-aggregated taxa. In addition, certain members of the unicellular and non-heterocystous groups are known to fix N_2, while all heterocystous genera have this capacity (Stewart 1973; Fogg 1982; Gallon, 1989).

A. Filamentous Forms

Members of the filamentous groups are generally considered as a relatively large collection of microplanktonic taxa and as such were identified through microscopy well before the turn of the 20th century (Ostenfeld and Schmidt, 1901; Karsten, 1905, 1907; Lemmermann, 1905). They were thoroughly classified according to morphology (Drouet, 1968; Sournia, 1970). Historically, filamentous cyanobacterial blooms were observed in trophically diverse waters; the biblical description of the "blood-red waters" of the Red Sea presumably refers to blooms of the non-heterocystous filamentous genus *Trichodesmium*, *"tricho"* being derived from the Greek for hair and *"desmos"* Greek for chain (Fogg, 1969; Plates 10a, b). In his account of the voyages of the Beagle (1831-1836), Darwin (1839) provided detailed descriptions of large and widely-distributed blooms of *Trichodesmium* in becalmed pelagic and near-shore southwestern Atlantic Ocean waters off South America. Similar reports of "sea sawdust," attributed to surface blooms of *Trichodesmium*, were

Table 1. Morphological groups and respresentative genera of marine planktonic cyanobacteria. N_2-fixing, ecological and trophic characteristics of genera are listed.

GROUPS	GENERA	N_2 FIXATION	CHARACTERISTICS
Unicellular			
(solitary)	*Synechococcus* *Prochlorococcus*	Most (-)	Solitary, common in oligotrophic oceans, form deep chlorophyll maxima, can form blooms in hypersaline waters
	Synechocystis	(-)	Solitary, bloom forming in hypersaline lagoons/bays
	Aphanothece	Some (+)	Solitary, picoplanktonic, oligohaline
(colonial)	*Merismopedia*	(-)?	Prefers nutrient enriched waters
Filamentous			
(non-heterocystous)	*Lyngbya*	Some (+)	Aggregates, often associated with protists, metazoans
	Oscillatoria	Some (+)	A few blooms reported, some associated with planktonic macroalgae and higher plants
	Phormidium	(-)	Solitary, and associated with other protists
	Spirulina	(-)	Solitary, hypersaline and nutrient enriched waters
	Trichodesmium	(+)	Aggregares form blooms in tropical and subtropical waters, associated with protists, invertebrate grazers
(heterocystous)	*Anabaena*	(+)	Solitary, in lagoons, bays
	Aphanizomenon	(+)	Aggregated bloom former in nutrient enriched waters
	Nodularia	(+)	Bloom former in coastal, nutrient enriched waters
	Richelia	(+)	Endosymbiont in the diatoms *Rhizosolenia* and *Hemiaulus* in tropical and subtropical waters

a

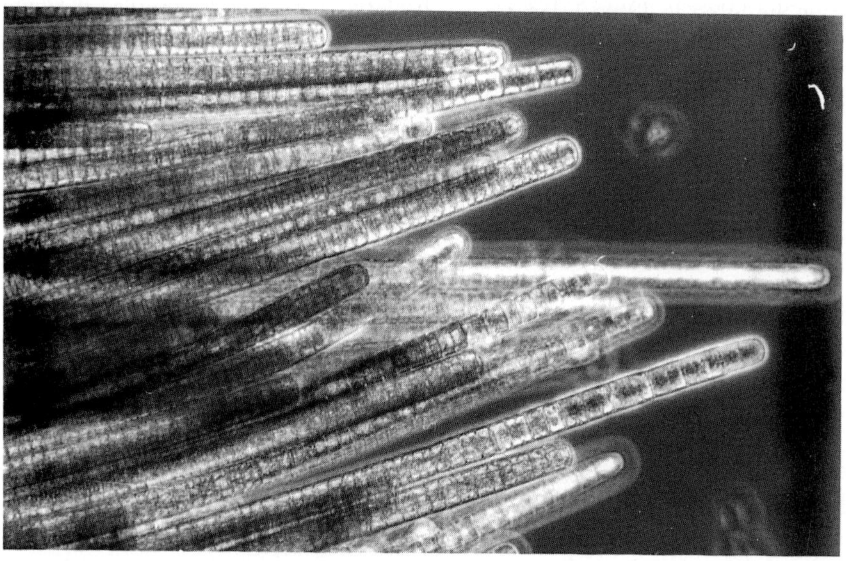

b

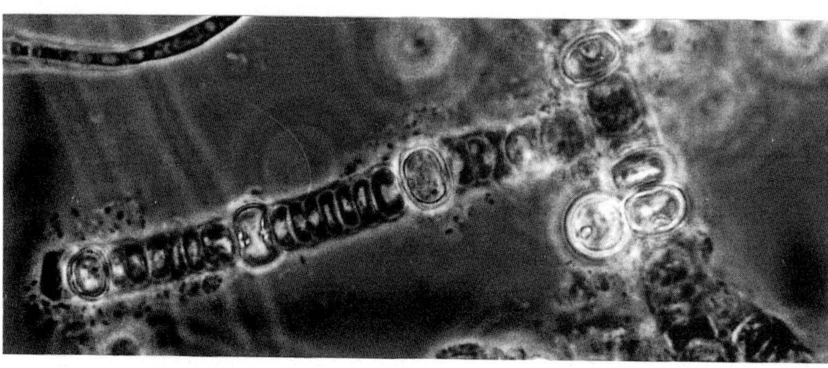

c

Fig. 1. Examples of planktonic cyanobacteria viewed with light microscopy (brightfield and phase contrast). Tuft-shape aggregates of the non-heterocystous, N_2-fixing, oceanic bloom former *Trichodesmium thiebautii* are shown in Plate 10b.

a *Synechocystis*, a common coccoid estuarine/coastal form found as solitary cells or aggregates (as shown). Sample from Bogue Sound, North Carolina, a full-salinity (32 ppt), oligotrophic coastal sound.

b Aggregated *Trichodesmium thiebautii* trichomes from bloom in the Gulf Stream (western North Atlantic Ocean), approximately 75 km southeast of Beaufort, North Carolina.

c Heterocystous coastal/estuarine bloom-former *Nodularia:* note thick-walled heterocysts and bacterial epiphytes.

logged by Captain J. Cook and his naturalist colleague J. Banks during cruises of H.M.S. Endeavour in the South Pacific during 1768-71 (Beaglehole, 1955). Many of these observations document occurrences in waters well-away from land masses and reflect the presence of cyanobacterial blooms in oligotrophic ocean regions. While not realized at the time, it is now well established that *Trichodesmium* and other frequently occurring heterocystous cyanobacteria in blooms such as *Richelia,* or as an endosymbiont in the pelagic diatoms *Rhizosolenia* and *Hemiaulus*, fix nitrogen and are significant contributors to oceanic N budgets (Carpenter, 1983b; Mague et al., 1977; Martinez et al., 1983; Carpenter and Romans, 1991). In coastal waters that undergo eutrophication, such as the Baltic, the heterocystous diazotrophic genera *Nodularia* and *Aphanizomenon* can contribute as much as 20 - 30% of the biologically-available "new" N (Sørensson and Sahlsten, 1987; Wulff 1988). In addition to their biogeochemical significance major bloom taxa, such as surface-dwelling *Nodularia* blooms in the Baltic Sea, can alter spectral qualities, increase light absorption, and thus contribute to the heating of surface waters (Kahru et al., 1993).

Thick-walled, poorly-pigmented heterocysts are a diagnostic feature of the filamentous heterocystous (Fig. 1) genera listed in Table 1. Heterocysts are differentiated, O_2 deplete cells, which harbor the O_2 labile N_2-fixing enzyme complex of nitrogenase (Donze et al. 1972; Wolk 1982). Heterocysts offer a clear advantage in the aerobic, N deplete surface waters which characterize much of the world's oceans. In addition, heterocysts often selectively attract heterotrophic O_2-respiring bacterial epiphytes which facilitate removal of O_2 in their vicinity (Paerl and Kellar, 1978).

One particularly striking aspect of marine planktonic heterocystous cyanobacteria is their confinement to specific geographic regions. Free-floating genera such as *Nodularia* and *Aphanizomenon* can be found, often as blooms, in certain estuarine and coastal environments, including periodically-stratified regions of the nutrient enriched Baltic Sea, and physically isolated estuarine and lagoon systems where there is limited exchange with the ocean (Kononen et al., 1996). The best documented example is the stratified, hypersaline Peel-Harvey estuary in Western Australia (Huber, 1986).

Some common features emerge when one considers the reasons for this rather exclusive confinement of heterocystous taxa: sustained (days to weeks) periods of either temperature and/or salinity stratification, adequate supplies of phosphorus and trace metals (particularly iron) for growth and relatively low N:P ratios (< 10) of total dissolved N and P known to select for N_2-fixing cyanobacterial taxa (Niemi, 1979; Smith, 1983, 1990). Potential environmental controls on heterocystous cyanobacteria are considered later in this chapter.

In sharp contrast to specific coastal and estuarine habitats, open ocean (pelagic) waters are virtually free of planktonic heterocystous cyanobacterial blooms despite the fact that the latter systems exhibit highly-favorable N:P ratios (< 5). Heterocystous forms can, however, be very abundant and widespread as endosymbionts. The best known of these is *Richelia* which inhabits the large pennate diatoms *Rhizosolenia* and *Hemiaulus* in tropical and subtropical, N deplete, oligotrophic waters (Taylor, 1983; Villareal, 1992). Relatively high rates of N_2 fixation are attributed to these associations (Mague et al., 1977; Martinez et al., 1983), especially in large surface mats growing in calm seas. During such times, the *Richelia-Rhizosolenia* symbiosis may coexist with the non-heterocystous N_2 fixer *Trichodesmium* (Villareal, 1992). Low levels of turbulence (small-scale shear and stress; large scale vertical mixing) favor surface accumulations of actively growing blooms of both assemblages (Carpenter, 1983; Paerl et al., 1995). The ways in which N_2 fixation is carried out by these morphologically-distinct diazotrophs may be quite different however, and are the subject of considerable research.

The fact that *Richelia* is found exclusively as an endosymbiont in pelagic waters suggests strongly that it requires a protective and possibly nutritionally enriched refuge. Open ocean waters are extremely nutrient deprived, are exposed to chronic wind mixing and shear, and thus represent a relatively hostile environment for filamentous cyanobacteria, which generally prefer calm, low wind and low shear-stress conditions (Carpenter, 1973; Bryceson and Fay, 1981; Fogg, 1982). It is likely that to escape from shear stress, and to ensure adequate light and nutrients, *Richelia* took refuge in the micro-environment provided by the large diatom *Rhizosolenia*. *Rhizosolenia* benefits from having access to biologically available N in what are some of the most N deficient waters on earth. In addition, the host diatom provides buoyancy and thereby maintains the endosymbiont in the illuminated euphotic zone where radiant energy can support its nitrogen-fixing activities.

Interestingly, the other inhabitant of these nutrient-deplete waters, *Trichodesmium*, appears to employ a very different, yet effective, strategy for meeting its N and C requirements. Rather than forming heterocysts, this undifferentiated diazotroph fixes CO_2 and N_2, and evolves O_2 simultaneously during daylight, a feat not shared with other non-heterocystous filamentous diazotrophic cyanobacteria such as *Lyngbya* and *Oscillatoria*. The latter strains fix most of their N_2 at night when O_2 evolution ceases (Stal and Krumbein, 1985; Gallon and Stal, 1992). Several mechanisms were proposed to account for this unique ability of *Trichodesmium*. Most mechanisms involve the need to exclude O_2 from nitrogenase on either intra- or inter-cellular levels. These mechanisms are discussed in Section III.C.

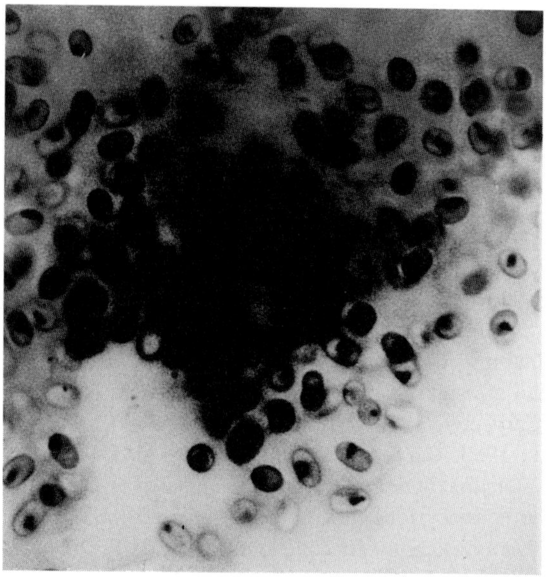

Fig. 2. Picoplanktonic (<2 μm) cyanobacteria concentrated by micro-filtration from the Sargasso Sea. The cells were incubated for 6 h with the tetrazolium salt iodonitrotetrazolium (INT); dark cellular inclusions are colored formazan crystals that form upon reduction. INT is commonly used as an indicator of metabolically-active (i.e. growing) cells.

B. Picoplankton and Nanoplankton

An additional component of the marine plankton is the unicellular photosynthetic picoplankton (Waterbury et al., 1979; Johnson and Sieburth, 1979), This < 5 mm fraction accounts for a significant fraction (30 to > 50%) of global ocean primary production (Platt et al. 1983) and plant biomass (Booth,1988; Itturiaga and Mitchell 1986). Picoplanktonic cyanobacterial unicells are relatively nondescript (coccoid to ovoid/rod shaped; Fig. 2) and owing to their prokaryotic nature are difficult to discriminate.

The use of fluorescence microscopy (Davis and Sieburth, 1982), flow cytometry coupled to fluorescence detection (Olson et al., 1985; Chisholm et al., 1988), photopigment analyses (Gieskes and Kraay, 1986; Veldhuis and Kraay, 1990; Goericke and Welshmeyer, 1993b) and immunological and molecular characterization (Giovannoni et al., 1990; Britschgi and Giovannoni, 1991; Zehr and McReynolds, 1989; Zehr et al., 1993) enabled investigators to obtain a more thorough and comprehensive picture of the ubiquity, the spatial and temporal distribution as well as the genetic diversity and physiological capabilities of the cyanobacterial pico- and nano-plankton. In addition, ^{14}C-based primary productivity studies, coupled to size-selective filtration (fractionation) and microautoradiography helped to assess the contribution of pico- and nano-plankton to total phytoplankton primary productivity (Watt, 1971; Paerl, 1978; 1996). Findings based on these techniques led to revised estimates of the diversity, geographic distributions and overall contributions of this cyanobacterial group to marine primary production and nutrient cycling dynamics.

Biochemical and molecular techniques were instrumental in delineating the taxonomy of cyanobacterial picoplankton and nanoplankton. A distinct biochemical feature of some picoplankton is the presence of phycoerythrin that can be detected fluorometrically (i.e. orange fluorescence at an excitation wavelength of 540 nm). The presence of phycoerythrin serves to identify members of the ubiquitous genus *Synechococcus* (Carr and Mann, 1994). Parallel RFLP analysis and 16S rRNA sequence analysis suggest further possible taxonomic divisions into phycoerythrin-positive and -negative sub-groups. With more extensive applications of these molecular techniques in geographically and trophically distinct waters, it is expected that additional strains may be identified.

Pico- and nano-plankton are major contributors to marine phytoplanktonic primary production and biomass across both geographic and trophic gradients (Itturiaga and Mitchell, 1985; Chisholm et al., 1988; Booth, 1988). These groups dominate primary production, nutrient assimilation and trophodynamics in the oligotrophic ocean and are the major biomass component of the "deep chlorophyll maxima" which characterizes the lower 75 - 125m of the euphotic zone. The latter is the upper region of the water column where photosynthetic CO_2 fixation exceeds respiratory CO_2 generation, resulting in net primary production. Owing to their small size (Olson et al., 1985; Chisholm et al., 1988), effective nutrient uptake kinetics at low ambient concentrations and adaptation to low light (Glover et al., 1987; Prezelin et al., 1989), the cyanobacterial picoplankton and nanoplanktonic are well-suited for dominating biomass and production in the clear, nutrient-deficient waters which characterize vast segments of the world's oligotrophic oceans. In addition, picoplanktonic cyanobacteria can assimilate organic compounds at trace concentrations which suggests they have the capacity for heterotrophy under natural conditions (Paerl, 1991). A variety of amino acids and sugars were readily incorporated by forms in waters ranging from mesohaline estuaries to the ultra-oligotrophic open ocean (Sargasso Sea). It was shown that organic matter uptake was higher under illuminated than dark conditions, and that the photosystem II inhibitor 3-(3,4-dichlorophenyl)-1,1-dimethyl urea (DCMU) was ineffective at blocking light-stimulated uptake. These data may suggest that some picoplankton are photoheterotrophic (Neilson and Lewin, 1974). Since N-containing amino acids are readily assimilated by this group, photoheterotrophy may be advantageous in oligotrophic waters, especially where there are chronic N deficiencies and where dissolved organic nitrogen (DON) concentrations may exceed dissolved inorganic nitrogen (DIN) by several orders of magnitude.

Cyanobacteria are of paramount importance in the shaping and the supporting of marine food webs because they are ubiquitous, they are a significant source of primary production, and they have the capacity for heterotrophy. The past two decades witnessed increased research to investigate the trophic fate of picoplanktonic production (Table 2). This research led to the recognition and characterization of microbial grazers (i.e. protozoan ciliates, phyto-flagellates) and heterotrophic bacteria as prime consumers and processors of picoplankton biomass (Azam et al., 1983; Itturiaga and Mitchell, 1986).

Table 2. A chronology of major discoveries, technical and conceptual contributions to the understanding of the roles cyanobacteria play in marine planktonic community structure and function.

DATE	CONTRIBUTION	REFERENCE
mid- to late 1800s	Filamentous, bloom-forming taxa described throughout temperate, tropical and subtropical oceans	Darwin, 1886; Karstens, 1905; Ostenfeld & Schmidt, 1901
early 1900s	Nanoplankton discovered in the sea	Lohman, 1911; Sournia, 1970
1960s	N_2 fixation associated with filamentous non-heterocystous and heterocystous bloom-forming taxa	Dugdale et al.,1961; Goering et al., 1966
late 1960s	Acetylene reduction and $^{15}N_2$ assimilation assays enable large-scale assessments of oceanic N_2 fixation: it is significant in places	Taylor et al., 1973; Mague et al., 1974; Carpenter, 1973; Maruyama, 1975
1970s	Application of fluorescence microscopy in biological oceanography	
late 1970s	Picoplanktonic taxa found to be ubiquitous and abundant	Johnson & Sieburth, 1979; Waterbury et al., 1979
early 1980s	Use of flow cytometry coupled to fluorescence detection allows spatial characterisation and quantitation of picoplanktonic bacteria	Olson et al., 1985; Chisholm et al., 1988
late 1980s – early 1990s	Molecular characterisation of taxa and functional genes facilitates community structural and functional analysis	Giovannoni et al., 1990; Wyman et al., 1994; Zehr et al., 1995

III. Planktonic Cyanobacterial Habitats: Physical, Chemical and Biotic Considerations

Planktonic cyanobacteria exhibit remarkable physiological, morphological and ecological adaptation and diversification in response to global evolutionary change (Fogg et al., 1973; Carr and Whitton, 1982; Gallon, 1992). As opportunistic phototrophs, planktonic cyanobacteria occur in waters from polar to tropical regions. They are present in extreme environments including geothermal and sulfide-rich, anoxic waters, which are toxic to most eukaryotes, as well as nutrient poor ultraoligotrophic open ocean waters, where deep-living populations occur near the compensation point (i.e. the depth below which no net photosynthesis takes place).

Cyanobacteria exploited recent anthropogenic alterations of aquatic environments, most notably nutrient enhanced eutrophication (Paerl, 1988; Paerl and Tucker, 1995). A striking example of cyanobacterial ecological opportunism is the development and persistence of dense blooms in a wide range of nutrified lakes, reservoirs and rivers (Fogg, 1969; Reynolds, 1987; Reynolds and Walsby, 1975). These blooms not only dramatically alter phytoplankton community structure and function, they can also influence C and N cycling significantly by contributing a substantial, and at times a dominant fraction of the input of C and N to aquatic ecosystems (Horne and Goldman, 1972; Horne, 1977).

The most visible and notorious cyanobacterial blooms involve nuisance genera, including freshwater and brackish *Anabaena, Aphanizomenon, Microcystis, Nodularia* and *Oscillatoria* (Fogg, 1969; Reynolds and Walsby, 1975; Paerl, 1988; Paerl and Tucker, 1995). Extensive surface blooms often pose serious problems for water quality management and fisheries including deoxygenation of underlying waters, foul odors e.g. H_2S, undesirable tastes and fish kills. Some nuisance strains are toxic to a variety of algal consumers and animals (including man) that use affected waters for drinking purposes (Carmichael, 1991; 1995). Food webs can also be altered markedly (often negatively) by blooms due to the poor edibility and food quality of bloom species (Porter and Orcutt, 1980; Fulton and Paerl, 1988). In this context a particular problem are those cyanobacterial forms that meet their N requirements through N_2 fixation. These cyanobacteria enjoy an obvious competitive advantage in N deficient waters, and there are numerous examples of opportunism among diazotrophic bloom formers, which exploit N deficient waters where there are sufficient amounts of phosphorus, Fe and essential trace elements. Using an extensive data set, collected from a range of freshwater lakes and reservoirs, Smith (1983) showed a strong correlation between total N:P ratios (by weight) and the prevalence of cyanobacterial bloom genera in freshwater environments. Ratios of N:P < 20 were conducive to the development and periodic persistence of N_2-fixing genera. Despite the general applicability and utility of this apparent correlation there is a conspicuous scarcity, and in most instances an absence, of planktonic N_2-fixing cyanobacteria in estuarine, coastal and pelagic ocean waters. It is important to note that such water bodies may exhibit N:P ratios far below 20 and they are often chronically N limited (Nixon, 1986; D'Elia et al., 1986).

Estuaries and coastal waters are generally N limited and experience unprecedented eutrophication in response to urbanization, agricultural and industrial growth in the coastal zone (Nixon 1995). Estuaries thus appear to provide ideal conditions for the development of blooms. There are notable exceptions however, including mesohaline regions of the nutrient enriched Baltic Sea and nutrient rich euhaline and hypersaline lagoon systems such as the Peel-Harvey Estuary, Australia (Niemi, 1979; Huber, 1986; Kononen et al., 1996). In waters that receive agricultural and urban nutrient inputs the frequencies and magnitudes of cyanobacterial blooms are increasing. Recent examples include occurrences in nutrified mangrove forests in Brazil (toxic *Microcystis*), coastal lagoons and embayments in Colombia (*Anabaena, Aphanizomenon*), and shallow coastal lagoons which drain arable lands in Spain, Portugal and New Zealand (*Anabaena, Microcystis, Nodularia*). Despite the growing number of cases of nuisance events in nutrified waters, many P sufficient, N limited estuarine waters remain devoid of N_2-fixing cyanobacterial blooms. From an ecological perspective, it appears that these waters represent an unoccupied and potential niche for diazotrophs.

Freshwater studies conducted during the past three decades on bloom-forming genera helped identify the physical, chemical and biotic controls of cyanobacterial dominance and bloom dynamics (Table 3). These studies point to key environmental controls that help clarify the notable absence of these genera in marine waters with stoichiometric nutrient ratios thought to favor bloom development; the "empty niche" paradox. Potential controls include:

Table 3. Physical/chemical factors potentially controlling cyanobacterial blooms.

FACTORS	IMPACTS
Physical	
Flushing/Altered water residence time	Potential removal mechanism for blooms, if flushing exceeds growth rates of bloom taxa
Large scale vertical mixing	Counteracts near-surface accumulations of buoyant bloom populations. May enhance competition for light and nutrients with non-bouyant eukaryotic taxa
Small scale turbulence (shear)	May disrupt filaments, colonies, aggregates and associations with other microflora and micro/macro-fauna
Shading (reduced surface irradiance)	Can alter phytoplankton community composition and can negatively affect cyanobacterial surface bloom taxa
Temperature	Generally, temperatures in excess of 20°C accompanied by stratification and high nutrient loading can promote certain blooms
Chemical	
Nutrient (N and P) inputs	Long-term (months/years) reductions in *both* N and P inputs in estuaries may reduce cyanobacterial bloom potentials. Low N to P loading ratios (< 20) acompanied by high P may enhance bloom potentials of diazotrophic taxa.
Salinity	Salinity in excess of a few ppt (as NaCl) can be an effective barrier to development and persistence of freshwater nuisance species. Other species tolerate salinity and hypersalinity.
Trace metals	Under high N and P loading conditions or when N_2-fixing taxa prevail, restricted availability of Fe may control growth. Some cyanobacteria are able to compete for low levels of Fe. No convincing evidence for other trace metal limitations.

elevated salinity (Thomas et al., 1988); nutrient deficiencies, specifically phosphorus (Doremus, 1985) and trace metals (Fe, Mo; Howarth et al. 1988; Paerl et al., 1987); low supply rates and concentrations of organic matter (Fogg 1969); relatively high and potentially-inhibitory (to nitrogenase) oxygen tensions in marine surface waters; inadequate genetic diversity (i.e. inocula) for potential colonization by diazotrophic cyanobacteria in estuaries (Zehr et al., unpublished data); grazing by zooplankton and/or fish as well as bacterial and viral controls on cyanobacteria in estuaries (Fulton and Paerl, 1988; Sellner, 1997); high levels of turbulence in the marine environment, especially in wind- and tide-exposed large estuaries, coastal and open ocean waters (Table 3).

A. Salinity and Genetic Constraints

Salinity of waters can influence N_2 fixation and is hence a potential barrier to the establishment and proliferation of diazotrophic cyanobacteria. Nitrogen fixation seems particularly susceptible to osmotic stress, and organisms unable to adjust by the production of compatible solutes show inhibition of activity at increasing salt concentrations (DuBois and Kapusta, 1981; LeRudulier et al., 1984). Cyanobacteria introduced into an estuarine environment from terrestrial or freshwater habitats may not be able to compensate for increasing salinities and osmotic stress (Paerl et al., 1983). Indigenous populations are often able to adjust to varying salinities by the production of compatible osmolytes (Reed and Stewart, 1985). Apart from the work of Dicker and Smith (1980) that described wide salt tolerance for an *Azotobacter* sp. isolated from a salt marsh, there has been relatively little consideration of salinity as a factor which regulates estuarine N_2 fixation. If allochthonous populations were responsible for "seeding" diazotrophs in estuaries, salinity might represent a considerable barrier to their establishment.

With respect to ionic strength, certain bloom-forming cyanobacteria may be highly sensitive to a slight change in salinity when they enter estuarine waters (Paerl et al., 1983). In addition, Howarth and Cole (1985) and Howarth et al. (1988) proposed that relatively high concentrations of sulfate (SO_4^{2-}) - a structural analogue of molybdate (MoO_4^{2-}) and present in seawater and saline lakes - might competitively inhibit N_2 fixation. While competitive inhibition of MoO_4^{2-} uptake by the SO_4^{2-} concentrations present in seawater is operative (Cole et al., 1993), the N_2 fixing potentials of marine diazotrophs appear unaffected by this competition (Paulsen et al., 1991). Most likely, the small cellular Mo requirements for N_2 fixation are met though reduced but sufficient uptake and storage.

Estuarine and oceanic waters are *a priori* not barriers to the establishment and proliferation of cyanobacterial diazotrophs. Nitrogen deficient estuaries and coastal lagoon environments worldwide support benthic non heterocystous and heterocystous taxa that show relatively high N_2 fixation rates. Oceanic coral reef, shelf and coastal mangrove habitats are often richly endowed with diverse non-heterocystous (e.g. *Oscillatoria, Lyngbya, Microcoleus*) and heterocystous (e.g. *Scytonema, Nostoc, Anabaena, Cylindrospermum, Calothrix*) genera (Frémy, 1933; Potts, 1980; Whitton and Potts, 1982; Potts and Whitton, 1979a; 1979b). In addition, numerous diazotrophic bacteria have been isolated from marine sediments, suspended aggregates, and suspended plants and animals (Paerl,1990). This makes it difficult to invoke either salinity, atypical ionic ratios, or paucity of genetic potential as barriers *per se* to the development and proliferation of diazotrophs. It is striking, however, that virtually all examples cited above are all benthic or otherwise attached populations.

A notable exception to the preponderance of benthic N_2-fixing cyanobacteria is the planktonic, aggregated filamentous genus *Trichodesmium*. This surface bloom-forming, non-heterocystous diazotroph is widely distributed in oligotrophic tropical and subtropical pelagic waters and is considered a major contributor to oceanic N and C budgets (Carpenter and Romans, 1991; Karl et al., 1992; Capone et al., 1997). Like some freshwater bloom formers such as *Anabaena* and *Aphanizomenon*, *Trichodesmium* cells contain gas vesicles which impart buoyancy and enable this genus to occupy a niche in N deficient oligotrophic surface waters.

Non-diazotrophic cyanobacterial genera are broadly distributed throughout estuarine, coastal and pelagic waters. Common planktonic genera include coccoid (e.g. *Synechococcus, Synechocystis, Merismopedia*), and a variety of non-heterocystous filamentous (*Oscillatoria, Phormidium*) forms. Because they are generally small, and do not form surface blooms, coccoid cyanobacteria are less conspicuous than their freshwater nuisance counterparts, and they were generally overlooked and underestimated as contributors to phytoplankton biomass and productivity in these waters. Detailed examinations using epifluorescence microscopy (Davis and Sieburth, 1982; Li, 1986; Itturiaga and Mitchell, 1986), flow cytometry coupled to fluorescence detection (Olson et al., 1985), high performance liquid chromatography (HPLC) and diagnostic photopigment (carotenoids) analyses (Pinckney et al., 1996, 1998) led to a greater appreciation of the prevalence and importance of these taxa.

Recent studies in the eutrophic Neuse River estuary, North Carolina, USA, exemplified the growing awareness of the importance of non-bloom taxa in estuarine and coastal waters (Pinckney et al., 1998). Prior to the late 1980s, coccoid and filamentous cyanobacteria were largely thought to be confined to the upper, freshwater components of this estuarine system. The introduction of HPLC as a qualitative and quantitative analytical tool in the late 1980s increased our conceptual and functional appreciation of these groups in the lower, more saline segments of the estuary. Cyanobacteria, like many important eukaryotic phytoplankton groups, possess unique accessory photopigments including the carotenoids zeaxanthin, myxoxanthophyll, echinenone, aphanxanthin and oscillaxanthin (Rowan, 1989; Millie et al., 1993). Routine monitoring and analyses of chlorophylls *a*, *b* and *c* and carotenoids indicated that cyanobacteria were a highly significant, and at times a dominant component of the phytoplankton community (Plate 10; Pinckney et al., 1998). Analytical techniques were developed further to include radiodetection and quantitation of photopigments that incorporated ^{14}C during measurements of primary productivity (Redalje, 1993; Georicke and Welshmeyer, 1993a; 1993b). Using this approach, the rate of synthesis of photopigments was used as an indicator of group-specific growth rates (Pinckney et al., 1996). These studies pointed to cyanobacteria as dynamic and quantitatively important contributors to community productivity and composition (Plates 10c, d, e). Nutrient driven eutrophication of this and other estuarine and coastal waters may be enhancing the expansion of cyanobacterial production and dominance (Paerl, 1996b).

The past four years of analyses in the Neuse River estuary indicate a trend of increasing biomass and dominance by cyanobacteria. On a seasonal basis, this trend is most noticeable during summer months, when elevated water temperatures enhance cyanobacterial growth. Microscopic observations conducted over the past decade appear to confirm this trend. The readily detectable filamentous, non-heterocystous genera *Oscillatoria* and *Phormidium*, which were seldom reported in phytoplankton surveys during the 1970s and 80s were more commonly observed throughout the estuary in the early 90s. These genera characteristically do not form surface blooms and hence are more susceptible to being overlooked in field surveys.

In conclusion, cyanobacteria are a quantitatively important component of estuarine, coastal and pelagic phytoplankton biomass and primary production. The relative contribution of the cyanobacterial planktonic fraction may be increasing in response to anthropogenic nutrient enrichment in estuarine and coastal waters (Paerl, 1997). This aspect of marine eutrophication and water quality alteration warrants further research and management.

B. Nutritional Constraints

With respect to nutrient supply rates, concentrations and ratios, numerous estuaries should be able to support cyanobacteria, including heterocystous N_2-fixing genera. The latter, however, are confined to periodically stratified waters (e.g. *Nodularia*), or may occur as endosymbionts in diatoms (i.e. *Rhizosolenia-Richelia*). Interestingly, endosymbiotic heterocystous cyanobacteria can be found even in well mixed ultra-oligotrophic surface waters such as the Sargasso Sea. Therefore, it seems unlikely that widespread inorganic nutrient (P, Fe, Mo, other trace metals) deficiencies readily explain the "unfilled niche" phenomenon in estuaries.

Specific nutrient deficiencies may, however, selectively restrict the activities and geographic extent of both diazotrophic and non-diazotrophic cyanobacteria. Of those inorganic nutrients, excluding nitrogen, which may control oceanic phytoplankton primary productivity and growth, geochemists often mentioned P (Doremus, 1985; Smith, 1984). This view was based on the assumption that biological N_2 fixation was capable of satisfying oceanic N requirements over long time scales (i.e. millions of years). For biologically relevant time scales (i.e. days to years) there is little experimental evidence to support this assumption and on the whole, the oceans remain N limited (Carpenter and Capone, 1982). Furthermore, oceanic N_2 fixation is controlled by a complex interplay of physical-chemical and biotic factors, going far beyond mere P supply alone (Paerl, 1990). Important factors include turbulence, O_2 tension, organic matter supply, trace metal (particularly Fe) availability, establishment of symbioses, etc. Phosphorus supply rates have, however, been shown to control production and N_2 fixation in some localized, highly productive benthic habitats, including reefs, seagrass beds, microbial mats, carbonate sediments and mangrove communities (Paerl et al., 1981; Capone, 1983). In addition, habitats which receive elevated levels of anthropogenically generated nutrients (e.g. N rich sewage outfalls, agricultural and urban runoff, atmospheric deposition) can at times, exhibit P limitation (Smith 1984).

In regard to nutrient requirements for growth, coastal and pelagic waters may at times have high N and P concentrations; other nutrients, such as silicon (Si), may thus control production (Glibert et al., 1995). While Si limitation may apply to eukaryotic phytoplankton (diatoms, silicoflagellates), cyano-bacteria exhibit very low requirements for this nutrient, and hence are not affected by its relative scarcity. In contrast, the availability of specific trace metals known to serve as cofactors for key assimilatory enzymes (RUBISCO, nitrogenase, nitrate and nitrite reductase) and for the synthesis of cellular constituents may be restricted in waters where N and P are in excess. Iron has received much consideration, in part because it is essential for growth, and its biological availability is restricted due to chemical precipitation, chelation and other natural binding/loss processes which occur in seawater (Bruland et al., 1991).

Much of the recent focus on Fe limitation in the oceans was on the "excess N and P" regions of the oligotrophic ocean, including the North Pacific gyre, mid-Atlantic, and Southern and Antarctic Oceans (Martin et al., 1990; 1994), waters where picoplanktonic cyanobacteria can account for a sizable fraction of phytoplankton production (Itturiaga and Mitchell, 1986; Chisholm et al., 1988). Here, there is a paradox with respect to major phytoplankton nutrient concentrations vs. primary productivity. It has been shown repeatedly that despite relatively high soluble N (especially nitrate) and P and Si concentrations, phytoplankton productivity and biomass are anomalously low. These data suggest that environmental constraints other than limitation by major nutrients are important. Martin et al. (1990) hypothesized that because Fe inputs were extremely low in these pelagic waters and Fe availability was severely restricted (for the above mentioned reasons), phytoplankton production was Fe limited. During the past 5 years, this hypothesis has been successfully tested by Martin and his colleagues (1994) and others (Takeda et al., 1995; Zhuang et al., 1995). These studies indicate that a mechanistic basis for Fe limitation is the inability of phytoplankton to meet the Fe requirements for the synthesis of nitrate and nitrite reductases, enzymes essential for the assimilation of these oxidized forms of combined N. Iron is an essential component of nitrogenase and occurs in the Fe-S centers of different components of the nitrogenase complex (Fogg, 1974; Yates, 1980; Dean and Jacobson, 1992). As such, diazotrophic microorganisms exhibit a relatively high demand for Fe. It has been suggested that constraints on Fe availability in N limited waters (i.e. those regions of the oceans not exhibiting "excess N and P") may restrict the geographic distributions and activities of pelagic diazotrophs, including *Trichodesmium* and the *Richelia-Rhizosolenia* symbiosis (Rueter et al., 1992). This possibility was evaluated in N deficient waters of the western Atlantic (Gulf Stream; Sargasso Sea) where these communities occur (Paerl et al., 1994). Using bioassays of naturally occurring and cultured (on Sargasso Sea water) *Trichodesmium* populations, it was shown that both chelated (EDTA) and non-chelated forms of Fe (as $FeCl_3$) stimulated N_2 fixation and growth (as chlorophyll *a* increase) relative to untreated controls. The strongest stimulation was observed in response to FeEDTA (Fig. 3). Interestingly, EDTA alone also led to stimulation. Control experiments indicated the response was not due to detoxification of potentially-toxic (and hence inhibitory) metals such as Cu (Paerl et al., 1994). Possibly, the added EDTA was capable of chelating existing forms of Fe in the water column, thereby minimizing their "loss" via precipitation and thus enhancing their availability.

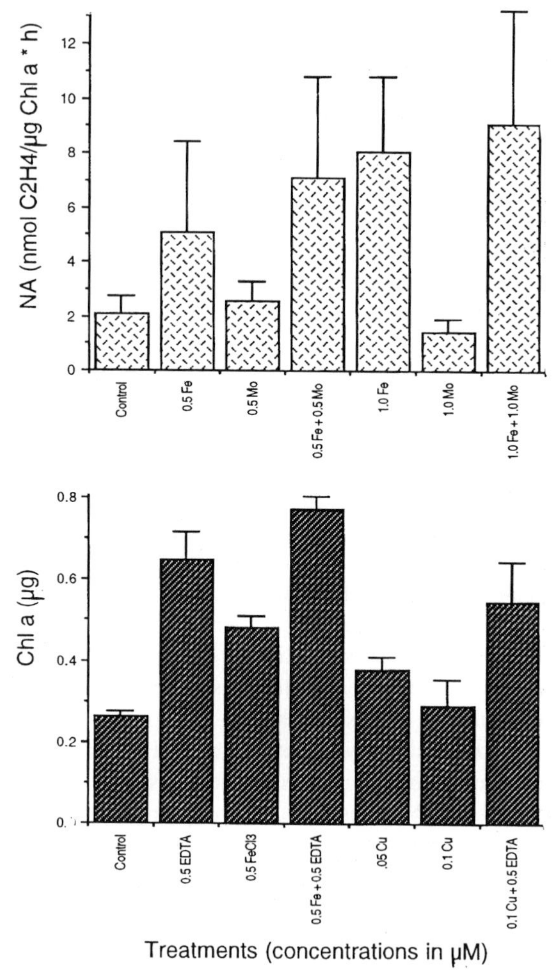

Fig.3. Effects of iron, either as $FeCl_3$ or EDTA-chelated $FeCl_3$ (equimolar) and molybdenum, as $NaMoO_4$, on nitrogenase activity (NA) and chlorophyll *a* of *Trichodesmium thiebautii*. A freshly-collected (North Carolina, W. Atlantic Gulf Stream) *T. thiebautii* (97% of total phytoplankton biomass) bloom was examined in a 5 day bioassay (Paerl et al., 1994). Error bars represent standard deviations of triplicate additions. Final mM concentrations of various forms of Fe and other nutrients are shown. Seawater used for bioassays initially contained 0.15 mM dissolved inorganic N (NO_3^-/NO_2^- + NH_4^+) and 0.2 mM PO_4^{3-}. Note that in the upper frame all Fe additions were in EDTA-chelated form (EDTA and Fe added in equimolar amounts) while $FeCl_3$ and EDTA were added individually and combined in the lower frame. Using ANOVA, it was shown that both nonchelated and chelated forms of $FeCl_3$ significantly ($p<0.01$) stimulated nitrogenase activity (upper frame) and growth (lower frame). EDTA by itself significantly stimulated growth (lower frame). This was not due to Fe contamination of EDTA, which proved free of residual Fe (K. Bruland personal communication; source of EDTA and its chemical analysis). The positive of EDTA was also not due to *in situ* copper (Cu) inhibition, since fairly large additions of Cu (0.1 mM) with and without EDTA did not negatively affect growth of *Trichodesmium*. It is therefore concluded that increasing Fe availability (either via enrichment or by supplying EDTA) stimulated N_2 fixation and growth of this cyanobacterial bloom-former.

In a related hypothesis, Rueter et al. (1992) suggested that by forming large buoyant aggregates near the water's surface, *Trichodesmium* was able to intercept and trap particulate Fe in the web-like matrix of trichomes. The Fe could originate from a range of sources including volcanic emissions, dust generated from desertification (e.g., the sub-Sahara), and air pollution from industrial and automotive emissions (Church et al., 1984; Duce and Tindale, 1991). The atmosphere is considered to be the main route by which Fe can be resupplied to the open ocean. Therefore, strategies that optimize capture and retention of Fe from the atmosphere would be highly advantageous in these waters.

Freshwater diazotrophic cyanobacteria such as *Anabaena* spp. produce potent chelators that are capable of sequestering Fe from ambient waters at exceedingly low levels, providing a competitive advantage over eukaryotic phytoplankton which require combined N (Murphy et al., 1976). This mechanism may play a similar role in marine environments where there is both chronic N limitation and exceedingly low external Fe inputs.

Dissolved organic matter (DOM) content was mentioned as a possible modulator of cyanobacterial growth and bloom potential. Early studies (Fogg, 1969) cite DOM as a factor potentially controlling cyanobacterial blooms. The hypothesized mechanism for DOM-stimulated cyanobacterial growth is that DOM conditions the water for cyanobacteria, possibly by inducing the synthesis of nutrient assimilatory enzymes and heterotrophy (Antia, 1991). Dissolved inorganic matter may also provide a source of energy and nutrition for closely-associated heterotrophic bacteria which are known to form synergistic interactions with cyanobacteria (Paerl and Pinckney, 1996). When examining freshwater lakes on a local scale, positive relationships between trophic state (i.e. oligotrophic systems containing low DOM, to eutrophic having relatively high DOM) and cyanobacterial dominance can at times be observed. Regional examinations, however, yield no consistent trend and other factors, such as pH, alkalinity and hardness hinder attempts to establish simple direct mechanistic relationships. In addition, it was pointed out by Fogg (1969) and others (Lange, 1967; Walsby, 1974; Paerl, 1990) that elevated DOM is more than likely a result rather than a cause of cyanobacterial blooms. In brackish estuarine and full salinity coastal waters, no clear relationship between DOM content and cyanobacterial biomass and/or bloom frequencies is evident.

C. The Oxygen Problem

As oxygenic phototrophs, cyanobacteria are faced with a formidable ecophysiological problem; molecular oxygen (O_2) evolved during photosynthesis is a potent inhibitor of nitrogenase, the enzyme complex which mediates N_2 fixation (Yates, 1980; Bothe, 1982; Gallon, 1992). Nitrogen limitation generally prevails in oxic near-surface waters, where bloom-forming diazotrophic cyanobacteria are often responsible for creating and maintaining O_2 saturated and supersaturated conditions. To solve this problem dominant freshwater filamentous planktonic bloom genera (e.g. *Anabaena, Aphanizomenon, Gloeotrichia, Nodularia*) and benthic diazotrophs (e.g. *Nostoc, Calothrix, Scytonema*) form O_2 deplete heterocysts which are capable of harboring O_2 sensitive nitrogenase (Wolk, 1982; Wolk et al., 1994).

Heterocysts have thicker cell walls and envelopes and are larger and less pigmented than the vegetative cells from which they differentiate. The differentiation process is greatly enhanced under N deficient conditions and the frequency of heterocysts is often closely related to N_2 fixation rates, clues noted in early studies suggesting that this process was heterocyst-associated (Fogg, 1944). Subsequent biochemical studies proved the heterocyst was the site of N_2 fixation (Donze et al., 1972; Wolk, 1982; Wolk et al., 1994). Typically, phosphorus enriched, nitrogen deplete waters promote the development and periodic proliferation (i.e. as blooms) of heterocystous cyanobacteria. Heterocystous species often exhibit the highest frequencies of heterocysts during periods of maximum N depletion (Horne and Goldman, 1972). Heterocysts are well-suited for supporting N_2 fixation, in large part because they do not evolve O_2, but retain the capacity for photophosphorylation (Donze et al., 1972; Wolk, 1982; Bothe, 1982; Wolk et al., 1994). In this manner, chemical energy (ATP) can be generated to support N_2 fixation. Because components of photosystem II are missing, heterocysts are unable to transfer reducing equivalents (from the photolysis of H_2O) to the N_2-fixing enzyme complex nitrogenase. Thus reducing power is derived from photosynthate imported from neighboring vegetative cells (Wolk et al., 1994).

The heterocyst represents an adaptation to ambient oxic conditions, representative of much of the world's N deplete, near surface ocean and estuarine waters. Nevertheless, not all diazotrophic cyanobacteria possess heterocysts. In this respect, it seems puzzling that this adaptation has not radiated across all diazotrophic cyanobacterial groups (Chapter 17). Instead, several parallel, perhaps coevolutionary, morphological and biochemical adaptive strategies exist to cope with ambient oxygen concentrations in marine planktonic (and benthic) populations of cyanobacteria.

Some cyanobacteria that lacked heterocysts but which otherwise had the capacity to fix nitrogen evolved structurally, biochemically and ecologically diverse strategies aimed at exploiting oxygenated, N limited waters. The main genera include filamentous forms such as *Oscillatoria*, *Lyngbya* and *Trichodesmium*, and coccoid forms, including *Synechococcus* and *Gloeothece*. These cyanobacteria exist in planktonic colonies, aggregated bundles, laminated mats and biofilms (Fogg, 1982, Paerl, 1990; Chapter 4). Such cyanobacteria are often dominant diazotrophs in submerged sediments, intertidal sand and mudflats, reefs and fouling layers (biofilms) where localized O_2 depletion can occur.

Non-heterocystous cyanobacteria fix N_2 by employing a variety of ecophysiological strategies. The most common strategies involve either spatial or temporal separation of photosynthesis (i.e. O_2 evolution) from N_2 fixation (Stewart, 1973; Stal and Krumbein, 1985; Paerl, 1990), examples of which are discussed below. Spatial separation of these processes was observed among the filamentous and colonial forms, while temporal separation occurs in single- and multiple-celled taxa.

Immunochemical studies on a variety of coccoid and filamentous non-heterocystous cyanobacteria suggest a broad capacity for N_2 fixation (Paerl et al., 1989; Bergman and Carpenter, 1991; Stal and Bergman, 1990). The ability to fix nitrogen is, therefore, most likely to be controlled by a variety of ecophysiological, genetic and environmental factors including the ability to create localized O_2-free conditions within cells or filaments, high intracellular respiration rates, regulation of transcription and translation and periodic hypoxia/anoxia (Paerl and Carlton, 1988; Paerl and Prufert, 1987; Paerl et al., 1988). In some coccoid and non-heterocystous filamentous species, photosynthesis (i.e. O_2 evolution) occurs during the day, while N_2 fixation can proceed at night when cellular pO_2 decreases below the level at which nitrogenase is inhibited. This temporal separation of N_2 and CO_2 fixation was identified under alternating light/dark cycles in cultures of the coccoid diazotrophs *Synechococcus*, *Synechocystis*, *Erythrospira* and *Gloeothece* (Leon et al., 1986; Mitsui et al., 1986; Waterbury et al., 1988; Gallon and Stal, 1992). More recent studies indicate the presence of an endogenous rhythm among some members of these coccoid genera, in which photosynthesis and N_2 fixation alternate under continuous (albeit low levels) of irradiance (Mitsui et al. 1986; Huang et al. 1990; Gallon and Stal 1992). The latter situation strongly suggests genetic regulation, coupled to cellular accumulations of specific stimulatory and inhibitory metabolites acting with positive and negative effect on nitrogenase activity. Temporal separation of these processes was observed in certain non-heterocystous filamentous species belonging to the genera *Oscillatoria* and *Lyngbya* (Stal and Krumbein, 1985; Stal and Bergman, 1990; Paerl et al., 1991; Currin et al., 1990). Marine cyanobacterial mat communities dominated by these forms often show pronounced diel patterns of photosynthesis during the day and N_2 fixation maxima at night (Bebout et al., 1987; Paerl et al., 1991; Villbrandt et al., 1990). When grown either under low turbulence conditions (like plankton) or as benthic mats, the cosmopolitan species *Lyngbya aestuarii* shows optimization of N_2 fixation at night, but also low levels of N_2 fixation during the day (Paerl et al. 1991). In this diazotroph, two distinct strategies appear to operate. The nocturnal optimization strategy is similar to that described for *Oscillatoria* spp. (Stal and Krumbein, 1985). In addition, in the day, low-level N_2 fixation appears possible through spatial partitioning of photosynthesis and N_2 fixation in single filaments. This is accomplished by confinement of CO_2 fixation to the interior, or proximal cells and confinement of N_2 fixation to the terminal, or distal cells; despite the fact that there is no obvious ultrastructural differentiation between both cell types (Paerl et al. 1991). This pattern is somewhat similar to that observed in heterocystous cyanobacteria, although a vast majority of the diel N_2 fixation occurs during the night. Microautoradiographs of $^{14}CO_2$ incorporation indicated that virtually all the photosynthetic activity was confined to proximal cells while some, but not all, distal cells were devoid of photosynthetic activity (Fig. 4). Distal cells that lacked photosynthetic activity were not senescent; they were capable of reducing the tetrazolium salt 2,3,5–triphenyl-3-tetrazolium chloride (TTC), which suggested the presence of active metabolism and highly reduced (O_2-devoid) conditions (Fig. 4).

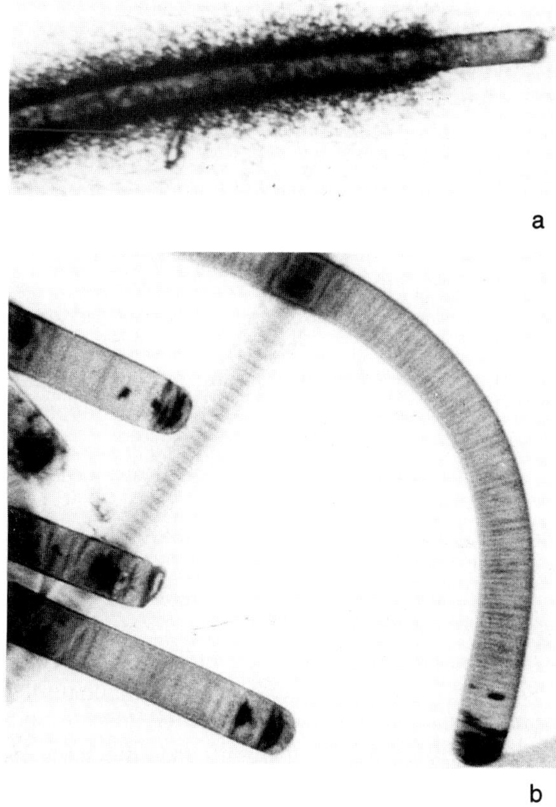

Fig. 4. Microscopic examinations of localized $^{14}CO_2$ fixation in *Lyngbya aestuarii*, using microautoradiography (Paerl 1991), and cellular reduction (as colored formazan crystals) of the tetrazolium salt 2,3,5-triphenyl-3-tetrazolium chloride (TTC) in this cosmopolitan N_2 fixing non-heterocystous cyanobacterium
a: Filaments incubated under 200 μmol photon m^{-2} sec^{-1} PAR with $^{14}CO_2$ (as $NaH^{14}CO_3$) for 30 min. Samples were then preserved in 2% buffered (pH 7.5) formalin and prepared for microautoradiography. The photosynthetic incorporation of ^{14}C is shown as exposed (black) silver grains superimposed on the filaments. Distal (terminal) of most (<70%) filaments fail to incorporate ^{14}C, indicating an absence of photosynthetic activity.
b: Patterns of TTC reduction, following a 3 h incubation (under illuminated conditions), followed by preservation with 2% formalin. Note distal cells show reduction of TTC, indicating relatively high metabolic activities and low O_2 concentrations (i.e. net O_2 consumption) associated with these cells. Immunological determinations for the presence of the N_2 fixing enzyme complex nitrogenase indicate its presence in photosynthetically-devoid distal cells (Paerl et al. 1991).

Immunogold labeling of *Lyngbya aestuarii* suggests that both distal and proximal cells contain the N_2-fixing enzyme complex nitrogenase and may imply that N_2 fixation can occur in both regions (Paerl et al., 1991). Presumably, with the absence of O_2-evolving photosynthesis in certain distal cells, *L. aestuarii* should be able to fix N_2 (albeit in limited amounts) during daylight. This spatial partitioning of N_2 and CO_2 fixation is not consistently observed in *L. aestuarii*; it appears dependent on growth phase and diel periodicity.

The bloom-forming *Trichodesmium* shows contemporaneous daytime maxima of N_2 fixation, CO_2 fixation and O_2 evolution (Carpenter, 1983a; 1983b; Capone et al., 1997). This property distinguishes *Trichodesmium* from all other marine non-heterocystous cyanobacterial diazotrophs. Capone et al. (1990) demonstrated that even when supplied with light well into the hours of night, N_2 fixation (but not CO_2 fixation) ceased. These observations suggest a strong daytime dependence and a diel rhythm in its diazotrophic behavior. Molecular studies indicate that the nitrogenase enzyme complex of *Trichodesmium* is highly conserved and similar to that found in other diazotrophs (Zehr et al., 1993; Capone et al., 1997). Furthermore, the nitrogenase of *Trichodesmium* is O_2 sensitive like the enzymes of other diazotrophs. How then, does *Trichodesmium* fix N_2 in the presence of oxygenic photosynthesis ?

In its oceanic habitat, *Trichodesmium* occurs as buoyant macroscopic (a few mm in length) radial (puff) or fusiform (tuft) aggregates (Plates 10c, d, e). When calm seas prevail, aggregates can frequently accumulate in surface waters, imparting brownish-yellow slicks which can extend kilometers or more. Field studies indicated that *Trichodesmium* fixed N_2 at maximum rates when it was present as surface-dwelling aggregates (Carpenter, 1983; Paerl and Bebout, 1988). Aggregates have compact, dense, dark-pigmented central cores, while their peripheral regions appear diffuse and hairlike (Fig. 5). The contrast between core and peripheral regions is evident when viewing suspended aggregates under natural conditions (i.e. in the upper region of the water column). The difference in compaction and resultant change in optical density between inner and outer regions of *Trichodesmium* colonies is suggestive of morphological and functional differentiation. Fogg (1974) hypothesized that CO_2 and N_2 fixation may be partitioned. He suggested that carbon dioxide fixation was confined to external regions subject to high irradiance and N_2 fixation was confined to dense, self-shaded internal regions. If it is assumed that photosynthate and recently synthesized organic N compounds are exchanged between inner and outer regions of aggregates, then

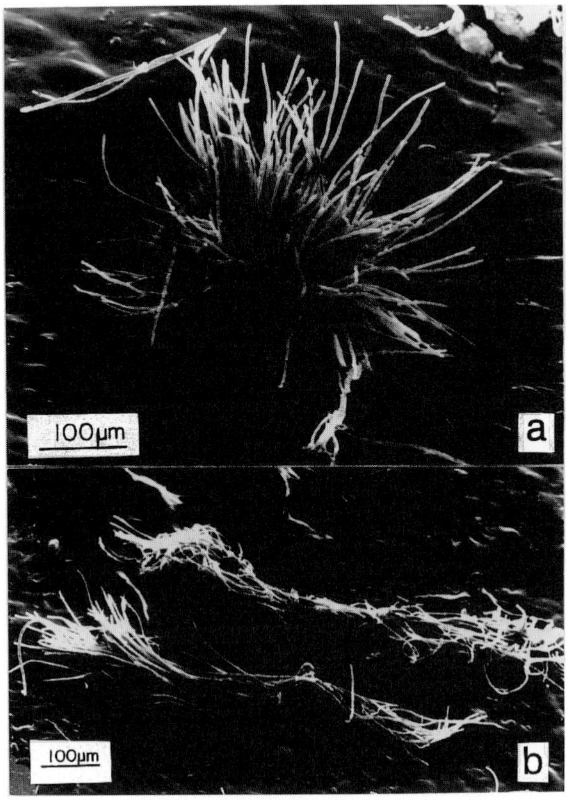

Fig. 5. Morphology of *Trichodesmium thiebautii* aggregates viewed with scanning electron microscopy, showing puff- and tuft- shaped colonies commonly encountered in oceanic near-surface bloom populations. Samples were obtained from the Western Atlantic Ocean, near San Salvador Island, The Bahamas.

this may suggest a mechanism by which CO_2 and N_2 could be fixed simultaneously (Fogg, 1974). Carpenter and Price (1976) generated evidence in support of Fogg's hypothesis by using microautoradiography to show that external regions of individual trichomes were photosynthetically much more active than internal regions. They also showed a strong positive correlation between calm sea state (facilitating aggregation) and relatively high rates of N_2 fixation in Caribbean Sea *Trichodesmium* populations. Bryceson and Fay (1981) also showed close relationships between aggregate shape, size and N_2 fixation potentials in *Trichodesmium* sampled off the east coast of Africa.

An examination of tetrazolium dye (TTC) reduction showed that highly reduced (O_2 deplete) zones were present in core regions of actively N_2-fixing western Atlantic (Gulf Stream) *Trichodesmium* aggregates (Paerl and Bland, 1982). Additional evidence for the presence of small (mm scale) O_2 deplete internal regions came from studies with microelectrodes (Pearl and Bebout, 1988). In further support of Fogg's hypothesis these studies documented how the oxygen tension of the external environment of *Trichodesmium* aggregates was high, yet simultaneously the photosynthetically-active cells maintained a low intracellular pO_2.

Subsequent studies revealed heterogeneity in O_2 distributions and photosynthetic CO_2 incorporation among naturally occurring aggregates (Paerl, 1994). Examination of photosynthetic $^{14}CO_2$ fixation in trichomes within aggregates showed uniform CO_2 fixation rates along the length of some trichomes; others exhibited much higher rates of $^{14}CO_2$ fixation in terminal rather than internal regions. Regions where there was no $^{14}CO_2$ fixation appeared metabolically active, since they were able to reduce TTC to its highly reduced (E'_o=-0.4V) red formazan. These observations suggest that only certain cells were engaged in photosynthesis to permit the creation of low pO_2 microzones.

Several laboratories have been able to culture and maintain naturally occurring strains of *Trichodesmium* spp. (Ohki and Fujita, 1988; Prufert et al., 1993). This facilitated well defined, controlled examinations of CO_2 and N_2 fixation in this cyanobacterium. While small puff and tuft aggregates can be cultured, single trichomes frequently predominate (Ohki and Fujita, 1988; Prufert et al., 1993). Single trichomes are capable of lightmediated N_2 fixation, albeit at low rates (per amount of chlorophyll *a*) relative to aggregates (Fig. 6) and this may suggest that N_2 fixation and CO_2 fixation occur contemporaneously within trichomes. This situation however, can only be maintained under relatively low levels of irradiance (< 100 µmol photon $m^{-2} s^{-1}$). Under higher irradiances, cultures of single trichomes cease to fix N_2 (Paerl, 1994; Fig. 6). If they survive increased levels of irradiance at all, they tend to form small aggregates, capable of N_2 fixation. It has not been possible to maintain cultured populations at irradiance levels > 300 µmol photon $m^{-2} s^{-1}$. Apparently, cultured populations are not capable of re-adapting to the high levels of irradiance (> 1500 µmol photon $m^{-2} s^{-1}$) commonly experienced in oceanic surface waters, where *Trichodesmium* is most often encountered as densely packed aggregates (Fig. 5).

The ability of single trichomes to fix both CO_2 and N_2 in culture may be related to the low levels of photosynthesis (O_2 production) experienced at reduced irradiance (<100 µmol photon m^{-2} s^{-1}).

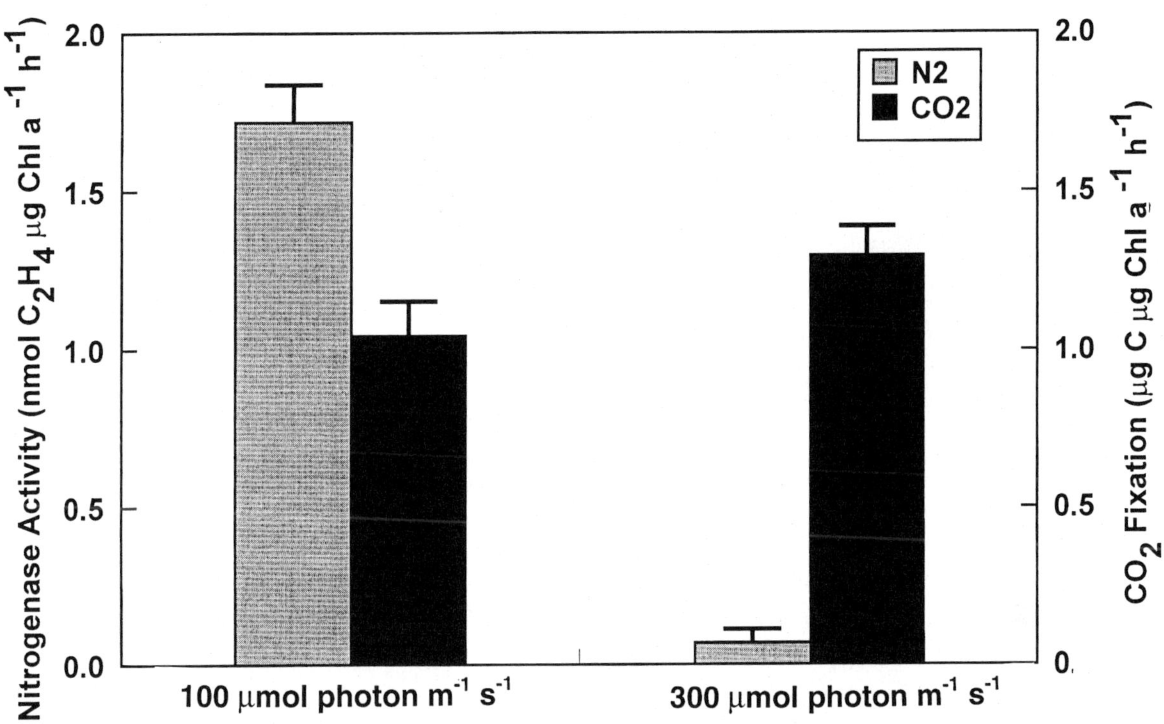

Fig. 6. Photosynthetic CO_2 (based on $^{14}CO_2$ incorporation) and N_2 fixation (nitrogenase activity) in single trichome cultures of a recent *Trichodesmium* isolate (IMS 101) derived from Atlantic Ocean waters (Prufert-Bebout et al., 1992). At <100 µmol photon m^{-2} s^{-1} PAR this isolate is capable of simultaneous CO_2 and N_2 fixation. At higher irradiances (>300 µmol photon m^{-2} s^{-1}), CO_2 fixation continues but N_2 fixation ceases, leading to N starvation and death in these N-deplete waters.

Apparently, N_2 fixation can accompany photosynthesis in this cyanobacterium, as long as cellular rates of O_2 production do not exceed a critical threshold which is inhibitory to nitrogenase. These conditions prevail in the lower third of the water column, where the potential for single trichome populations of *Trichodesmium* to exist may be greatest.

Immunogold labeling of cultured and natural *Trichodesmium* populations demonstrated the ability of cells to synthesize nitrogenase (Paerl et al. 1989a; Bergman and Carpenter, 1991). This indicates that the genetic potential for N_2 fixation may be broadly distributed in trichomes and aggregates. These findings, combined with the observations of single-trichome N_2 fixation in culture and aggregate-level N_2 fixation in nature, suggest *nif* gene expression may be controlled by environmental conditions such as irradiance, turbulence and mixing. Environmental stress, including high surface irradiance, may lead to morphological adaptations such as aggregation which facilitates contemporaneous photosynthesis and N_2 fixation on the trichome and/or aggregate level. Why this planktonic cyanobacterium does not carry out a bulk of its N_2 fixation at night, in order to minimize the "O_2 problem," remains unclear, since this appears to be a feature of other non-heterocystous forms (*Lyngbya*, *Oscillatoria*).

Close associations between cyanobacteria and bacterial or algal epiphytes are commonplace in marine cyanobacterial assemblages (Paerl et al., 1989b; Paerl, 1992; Fig. 7). In the case of the heterocystous genera *Nodularia, Aphanizomenon* and *Anabaena*, such bacterial associations may enhance N_2 fixing capacity (Paerl and Kellar, 1978; Kononen et al., 1996). Non-heterocystous genera, including *Trichodesmium*, *Oscillatoria* and *Lyngbya* exhibit close associations with diverse microflora. These associations may help to optimize the N_2 fixation potentials of cyanobacteria, perhaps by enhancing localized O_2 consumption and nutrient regeneration

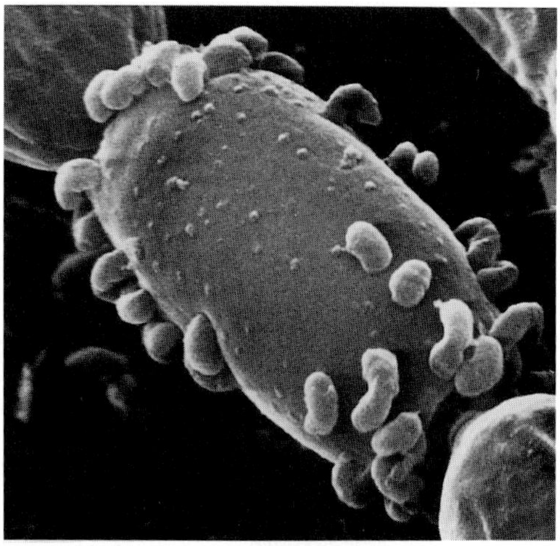

a

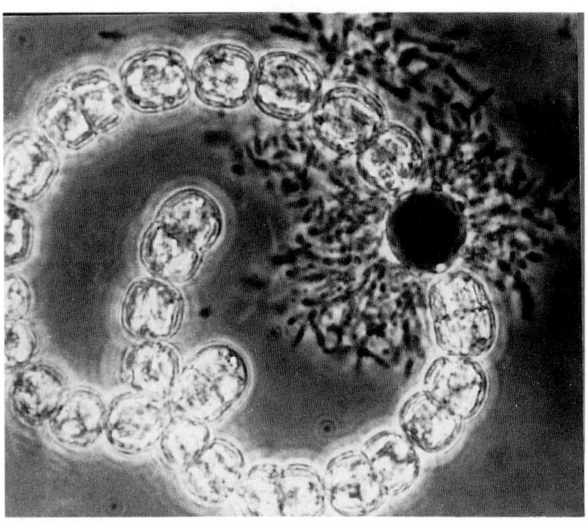

b

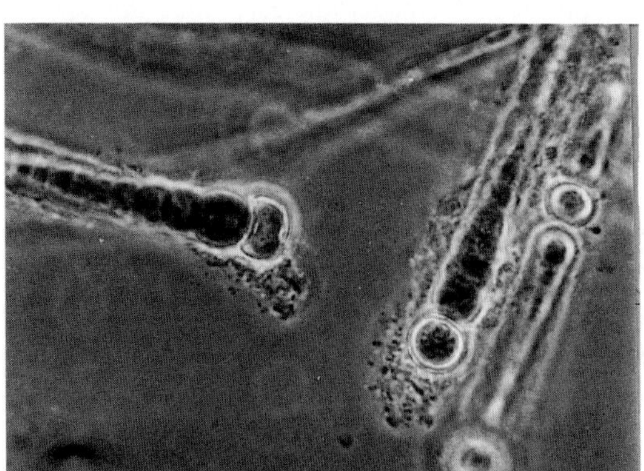

c

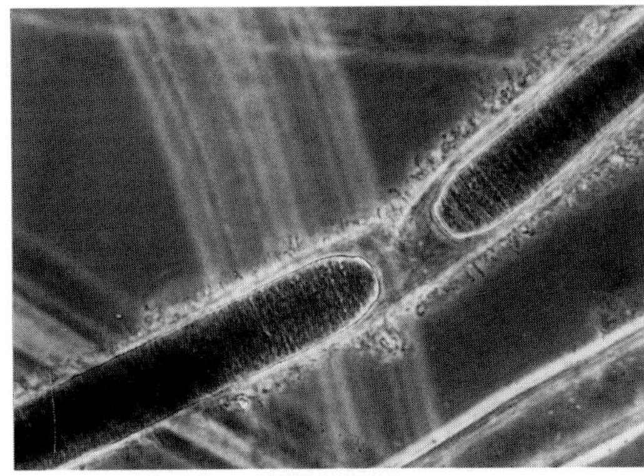

d

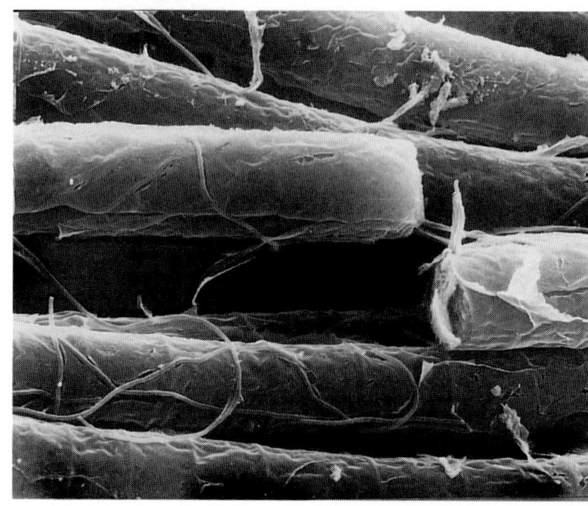

e

Fig. 7. Scanning electron and light microscopy observations of planktonic cyanobacterial-bacterial interactions. Heterocyst-specific bacterial associations in: a) *Aphanizomenon flos-aquae*, b) *Anabaena spiroides* and c) *Nodularia* sp. d) Bacterial associations with non-heterocystous cyanobacteria, including *Lyngbya aestuarii, e) Trichodesmium thiebautii.* Note close bacterial associations with mucilaginous sheaths and slime layers surrounding trichomes.

(Paerl and Kellar, 1978; Paerl and Pinckney, 1996). Many marine (and freshwater) cyanobacteria are not cultivable under bacteria-free (axenic) conditions (Fogg, 1982). It is well known that bacterial epiphytes commonly observed on aquatic cyanobacteria utilize cyanobacterial excretion products (Chróst and Brzeska, 1978; Paerl, 1984). Furthermore, bacteria associated with cyanobacteria were shown to be chemotactically attracted to a range of organic compounds known to be common cyanobacterial excretion products (Paerl and Gallucci, 1985). These associations may benefit the cyanobacteria because growth rates of the host are often enhanced and maximized in the presence of bacterial epiphytes. These associations appear to have evolved to maximize the physiological activities of the two organisms and to counter the problem caused by elevated oxygen levels.

D. Biotic Interactions

Marine planktonic cyanobacteria are associated with diverse microorganisms (Whitton, 1973; Paerl and Kellar, 1978; Caldwell and Caldwell, 1978; Paerl, 1992). Associated microorganisms can include heterotrophic bacteria, fungi, phytoflagellates, ciliated and amoeboid protozoans, and viruses. Cyanobacterial-microbial associations are commonly observed inside colonies, aggregates of filaments and within the fibrillar, mucilaginous sheaths, capsules and slimes located ouside cyanobacterial cell walls (Fig. 7). Collectively, these regions comprise what is referred to as the phycosphere.

The diversity of phycosphere associated microbial associations and specificities of microbial attachment sites are striking. Bacterial, flagellate or protozoan species may appear randomly attached to cyanobacterial filaments and colonies, or scattered throughout mucilaginous matrices which border cells. In contrast, there are other highly specific associations, where a single microbial species may exclusively associate with differentiated cyanobacterial host cells such as akinetes or heterocysts (Fig. 7).

Cyanobacterial-microbial associations occur during all stages of growth of the cyanobacteria. During a seasonal bloom cycle, both the intensity and specificity of microbial epiphytism can vary dramatically (Whitton, 1973; Paerl, 1982). Diel periodicity was observed with respect to the intensity and frequency of bacterial attachment (Paerl and Kellar, 1978). Bacterial isolates originally obtained from cyanobacteria were shown to be chemotactically attracted to the heterocysts, and to a variety of organic substrates, including some sugars and amino acids. It was shown that $^{15}N_2$ fixed by *Anabaena oscillarioides* was rapidly transferred to heterocyst-associated bacteria (Paerl, 1984). Axenic isolates of *A. oscillarioides* exhibited maximum growth and N_2 fixation rates when reinoculated with bacterial strains derived from the host (Paerl and Kellar, 1978).

Marine cyanobacterial bloom species are notoriously difficult to culture axenically (Fogg, 1982; Ohki and Fujita, 1988; Prufert et al., 1993). The literature on freshwater forms include many reports on differential growth obtained under axenic vs. bacterized conditions; with few exceptions, bacterized freshwater cyanobacterial genera exhibited higher growth rates and were easier to maintain in culture than axenic strains. This general finding was summarized by Gibson and Smith (1982): *"Experience at our laboratory exemplifies this problem (obtaining and maintaining cyanobacterial growth under axenic conditions); when both axenic* Oscillatoria redekei *and* O. agardhii *have been isolated, we have obtained consistent growth only with* O. redekei. *The* O. agardhii *always appears to grow better in the presence of contaminant heterotrophic bacteria".*

It was suspected for a long time that the reliance of cyanobacteria on bacterized conditions for optimal growth was related to both the exchange of mutually beneficial metabolites and growth factors as well as the detoxifying roles that phycosphere associated bacteria may play. Kuentzel's (1969) and Lange's (1971) early, albeit incomplete work suggested that CO_2 recycling by such bacteria proved essential for optimal growth of cyanobacteria under inorganic carbon limited conditions which can occur in dense blooms. Although cyanobacteria, as a group, appear to depend little upon exogenous vitamin supplies (Gibson and Smith, 1982), specific growth factors supplied by associated bacteria may be responsible for optimization of bloom development. Caldwell (1979) suggested that phycosphere associated bacteria detoxify cyanobacterial extracellular metal (Fe) chelates which may otherwise (under axenic conditions) prove autotoxic to host cyanobacteria. Lastly, the activities of phycosphere associated bacteria lead to decreased O_2 tensions near cyanobacterial cells and filaments engaged in activities prone to the adverse effects of high oxygen tensions.

Specific cyanobacterial-microbial interactions can characterize, if not typify, periods of maximum bloom development and proliferation. Both extensive bacterial colonization of mucilaginous layers surrounding estuarine *Microcystis aeruginosa* colonies, as well as the presence of amoeboid protozoan grazers within such colonies, are most profound during periods of optimal CO_2 fixation. It was shown that *M. aeruginosa* cells, which bordered regions being grazed, exhibited higher cell-specific rates of CO_2 fixation than cells more distant from grazers (Table 4). Thus, even though protozoan grazing represents consumption of cyanobacterial

Table 4. Microautoradiographic determination of cell-specific rates of photosynthetic CO_2 fixation in colonies of *Microcystis aeruginosa*. Original stocks were derived from a bloom on the Neuse River, NC. They were cultured on ASM + N medium and exposed to 250 μmol photon m^{-2} s^{-1} PAR at 25°C. Determinations were made on axenic (bacteria and protozoa-free), *Amoeba* spp.-associated and bacteria-associated (bacteria largely confined to mucilaginous sheaths) populations of *M. aeruginosa*. Mean of 3 independent experiments ± SE (data from Paerl, 1995).

Date	pgC $cell^{-1}$ h^{-1}		
	Axenic	Amoeba-Assoc.	Bacteria-Assoc.
10 Oct. 1985	0.11 ± 0.012	0.25 ± 0.015	0.23 ± 0.012
12 Nov. 1985	0.14 ± 0.010	0.31 ± 0.012	0.22 ± 0.014
25 Nov. 1985	0.12 ± 0.008	0.27 ± 0.017	0.31 ± 0.013

biomass, it also enhances growth potentials of ungrazed cells.

These observations may suggest that cyanobacterial-microbial interactions are not necessarily indicative of senescence in host cyanobacteria. Rather, they may lead to parallel optimization of growth and bloom potentials among microbial epiphytes and cyanobacterial hosts, and they may reflect mutually-beneficial ecological adaptations and physiological exchange processes enabling certain cyanobacterial species to periodically dominate planktonic communities as blooms. Such observations indicate that in addition to the well known physical-chemical factors that promote cyanobacterial nuisance blooms (nutrient (particularly phosphorus) enrichment, water column stability, high levels of irradiance and surface water heating, etc: Fogg 1969; Reynolds and Walsby, 1975), biotic phycosphere-scale interactions may play an additional regulatory role during the development, proliferation and maintenance of such blooms.

E. The Roles of Buoyancy and Turbulence.

While many factors were implicated in periodic cyanobacterial dominance (e.g. N_2 fixation, colony formation, symbiosis), the ability to utilize buoyancy for remaining in the radiant energy-rich euphotic zone imparts a distinct and critically important, ecophysiological advantage (Walsby, 1972; Reynolds and Walsby, 1975; Klemer and Konopka, 1980). Excessive buoyancy is thought to be responsible for surface scum formation and resultant nuisance conditions (Reynolds and Walsby, 1975; Reynolds, 1987). Not surprisingly, irradiance and nutrient supply are key environmental controls of buoyancy. Reduced irradiance was shown to be responsible for increasing buoyancy in a number of bloom-forming species (Klemer and Konopka, 1980). Reduced CO_2 availability (Paerl and Ustach, 1982) and high inorganic nitrogen concentrations (Klemer and Konopka, 1980) can also increase buoyancy. There is evidence that surface accumulation facilitates interception of atmospheric CO_2 (Paerl and Ustach, 1982), but this may only play a role in poorly-buffered brackish waters where dissolved inorganic carbon (DIC) levels can be low and pH can rise several units in response to high photosynthetic demand for CO_2. In full salinity waters where ambient DIC is consistently high (~2 mM), pH remains quite constant (~8.3) and inorganic N concentrations are chronically low (< 0.5 mM); irradiance is likely to be a dominant variable controlling buoyancy.

Irradiance regulates buoyancy by controlling photosynthesis. This is accomplished in two ways. High rates of photosynthesis (i.e. high irradiance) can increase intracellular turgor pressure, which leads to the collapse of gas vesicles; gas-filled cytoplasmic inclusions which provide buoyancy (Walsby 1972). Conversely, low levels of irradiance lead to suboptimal photosynthetic rates, which in turn reduce cell turgor pressure and promote gas vesicle formation and buoyancy. In this manner, cells are capable of regulating their position in the water column in relation to photosynthetic needs. Active photosynthesis can also lead to the accumulation of relatively dense polysaccharides which decrease buoyancy (Klemer and Konopka, 1980). Both mechanisms interact to control buoyancy in natural waters where irradiance, transparency, inorganic carbon (CO_2/HCO_3^{2-}) and nitrogen supplies can vary rapidly in time and space.

Buoyancy responses to environmental change vary substantially among planktonic cyanobacterial genera. The pelagic N_2 fixer *Trichodesmium* spp. is buoyant even when large changes in irradiance are encountered. Persistent buoyancy was attributed to extremely strong membrane-bound gas vesicles. Despite its ability to remain buoyant, and its capacity to remain in near surface waters under a variety of environmental conditions, *Trichodesmium* does exhibit some regulation of buoyancy due to synthesis of carbohydrates (Romans et al., 1994). This attribute may explain the periodic occurrence of populations below the euphotic zone at relatively deep levels.

Over daily to weekly time scales, persistent turbulent conditions which characterize wind-exposed, highly mixed, N limited estuarine, coastal and oceanic surface waters negatively affect buoyancy, N_2 fixation and to a lesser extent photosynthetic performance of cyanobacteria (Paerl et al., 1995). Both large-scale (cm – m) and small-scale (mm – cm) turbulence can affect cyanobacterial metabolism and growth. To some extent, buoyancy regulation can counteract large-scale vertical mixing. However, under strong, continuously mixed conditions, buoyancy compensation is overridden, giving way to vertical displacement in the water column. Wind or tide induced vertical mixing forces cyanobacteria to mix throughout the water column and leads to competition for light and nutrients with eukaryotic algae. Under highly mixed conditions eukaryotes frequently grow faster than cyanobacteria and thus may outcompete them for nutrients and light energy. Non-buoyant planktonic marine cyanobacteria, including the picoplanktonic genera *Synechococcus* and *Synechocystis*, often compete effectively with eukaryotes for nutrients and light under highly mixed conditions. They generally exhibit high affinities for nutrients at low concentrations, and as such can form a substantial fraction of the phytoplankton biomass in well-mixed estuaries, coastal and pelagic waters which experience chronic N and other nutrient limitations.

Small scale turbulence, or shear, was shown to be an impediment to optimal N_2 fixation and growth rates in stirred cultures of heterocystous cyanobacteria (Fogg, 1969; Kucera et al., unpublished data). The negative effects may be due to breakage of cyanobacterial filaments, specifically at the delicate heterocyst-vegetative cell junction (Lang and Fay, 1971), promotion of O_2 inactivation of nitrogenase in heterocysts (Fogg, 1969) or disruption of associations between bacteria and cyanobacteria which are known to promote the potential of cyanobacteria for N_2 fixation and growth (Paerl and Kellar, 1978; Paerl and Pinckney, 1996).

With few exceptions, systems devoid of heterocystous cyanobacteria share the common characteristic of relatively high turbulence in the surface mixed layer. This has fostered the hypothesis that turbulent conditions may adversely affect cyanobacteria; the idea is supported by research showing the effects of turbulence on phytoplankton productivity (Madden and Day, 1992). Studies of the effects of turbulence on microorganisms focused on small scale shear forces and this, in turn, lead to more accurate quantification of the scales most relevant to phytoplankton (Thomas and Gibson, 1990). Identification of the small-scale forces, coupled with improved techniques of evaluation, improved our understanding of the controlling effects of turbulence on microorganisms. The influence of turbulence as a mitigating condition for cyanobacterial dominance (i.e. effects on cyanobacterial N_2 fixation, CO_2 fixation and growth) were investigated through the use of Couette chambers which are capable of exerting well-defined small-scale shear forces on particles the size of cyanobacteria (Kucera, 1996; Kucera and Paerl, unpublished data).

Phytoplankton are generally much smaller than the smallest eddies in turbulent waters and are thus subjected to a uniform straining motion that increases with distance from the organism (Thomas and Gibson, 1990). Straining is defined as a change in size relative to the original size and is proportional to the stress at the object's surface. Structurally, phytoplankton are not affected by large-scale eddies. Only at very small length scales does the shear stress (force per unit area) of moving water become relevant. Because microalgae are influenced by straining, the studying of the effect of turbulence can be simplified by generating shear stresses without fully turbulent flow in a Couette chamber (Taylor, 1936).

Couette chambers were used recently to produce shear forces that represent the range of turbulence present in estuarine and coastal waters (Paerl, unpublished data). Published values in natural systems are reported as the energy dissipation rate, e, having units $cm^2\,s^{-3}$ or watts kg^{-1}. A precise measurement of shear rate (G) in the Couette chamber can be related to naturally occurring energy dissipation rate (e) by the expression:

$$G = (e/h)^{\frac{1}{2}}$$

(Camp and Stein, 1943)

Assuming the dynamic viscosity (h), has a value of 10^{-2} cm^2 s^{-1}, a low value of e $=10^{-6}$ cm^2 s^{-3} for open water environments and a high value of e = 1 cm^2 s^{-3} for tidal channels, then G is expected to range between 10^{-2} s^{-1} and 10 s^{-1}, respectively (Thomas and Gibson, 1990; Dam et al., 1994). Our Couette Chambers can generate shear rates from 0 to 50 s^{-1}.

For illustrative purposes, consider a comparison of N_2 fixation (nitrogenase activity) and CO_2 fixation (primary productivity) in representative heterocystous species, *Anabaena oscillarioides* and *A. circinalis*, exposed to small-scale shear representative of estuarine (Neuse River estuary) mixed layer conditions. High shear rates negatively affect N_2 and CO_2 fixation (Table 5). After 4 - 6 hours, the influence of high shear treatment on *A. oscillarioides* was significantly greater than the influence of the low shear treatment and the control. Carbon dioxide fixation and N_2 fixation of *A. circinalis* were significantly lower in the high shear treatment relative to the control (Table 5). Since N_2 fixation, which is reliant on optimal photosynthetic performance, is known to confer a selective advantage to cyanobacterial bloom genera in N deplete waters, the detrimental effects of high shear on these processes could potentially negate this advantage. Persistent high shear, representative of wind and tide exposed waters, may therefore act as a impediment to the establishment and proliferation of diazotrophic bloom genera.

Table 5. Impact of small (µm scale) shear stress, representative of wind shear on estuarine and coastal surface waters (mixed layer), on CO_2 and N_2 fixation (nitrogenase activity) in *Anabaena oscillarioides* and *A. circinalis* (Kucera and Paerl, unpublished data). Well-defined shear forces were imparted by Couette chambers. Control conditions represent no shear stress, low shear stress represents moderately windy conditions and high shear stress represents highly windy conditions. Figures are mean ± SE

Shear stress	*A. oscillarioides*		*A. circinalis*	
	N_2 Fixation (nmol C_2H_4 mg chl^{-1} h^{-1})	CO_2 Fixation (mg C mg chl^{-1} h^{-1})	N_2 Fixation (nmol C_2H_4 mg chl^{-1} h^{-1})	CO_2 Fixation (mg C mg chl^{-1} h^{-1})
control	4.98 ± 1.8	2.31 ± 0.5	8.52 ± 1.2	1.75 ± 0.6
low stress	4.26 ± 1.5	2.03 ± 0.4	7.27 ± 1.1	1.65 ± 0.4
high stress	3.29 ± 1.5	1.34 ± 0.5	4.92 ± 0.9	1.32 ± 0.3

IV. Synthesis: The Ecosystem Perspective

While mixing and shear play intimate roles in the regulation of cyanobacterial metabolism, growth and bloom potential, these physical controls are not independent of other chemical and biotic constraints. On the contrary, physical, chemical and biotic controls act in a coordinated, often synergistic fashion over a range of spatial and temporal scales, relevant to the habitat or system in question (Paerl, 1996b; Sellner, 1997). The contemporaneous interplay of physical (i.e. irradiance, temperature, water column circulation, large- and small-scale turbulence), chemical (nutrient, salinity, toxins) and biotic (grazing, mutualism, symbiosis) controls of marine planktonic cyanobacterial growth and bloom dynamics is illustrated in Fig. 8.

Both structurally and functionally, marine planktonic habitats are closely coupled to the atmosphere, nearby land masses and their hydrology (runoff, riverine and groundwater inputs), as well as underlying sediments. In order to fully understand and appreciate environmental controls on planktonic cyanobacterial ecophysiology, productivity and community dynamics, the rates of exchange of organic matter, inorganic nutrients (e.g. N, P, metals), gases and growth-altering substances (toxins, chelators, secondary metabolites) between these ecosystem components are of fundamental importance. Of prime consideration are the relative contributions of terrigenous, atmospheric and advective (from deep water) nutrient loading to overall nutrient budgets. Relevant information includes:

rate of sediment-regenerated (recycled) vs. externally-loaded (new) total nutrient input;

sediment-mediated nutrient loss and input mechanisms, including sedimentation and burial; denitrification; N_2 fixation; lithification;

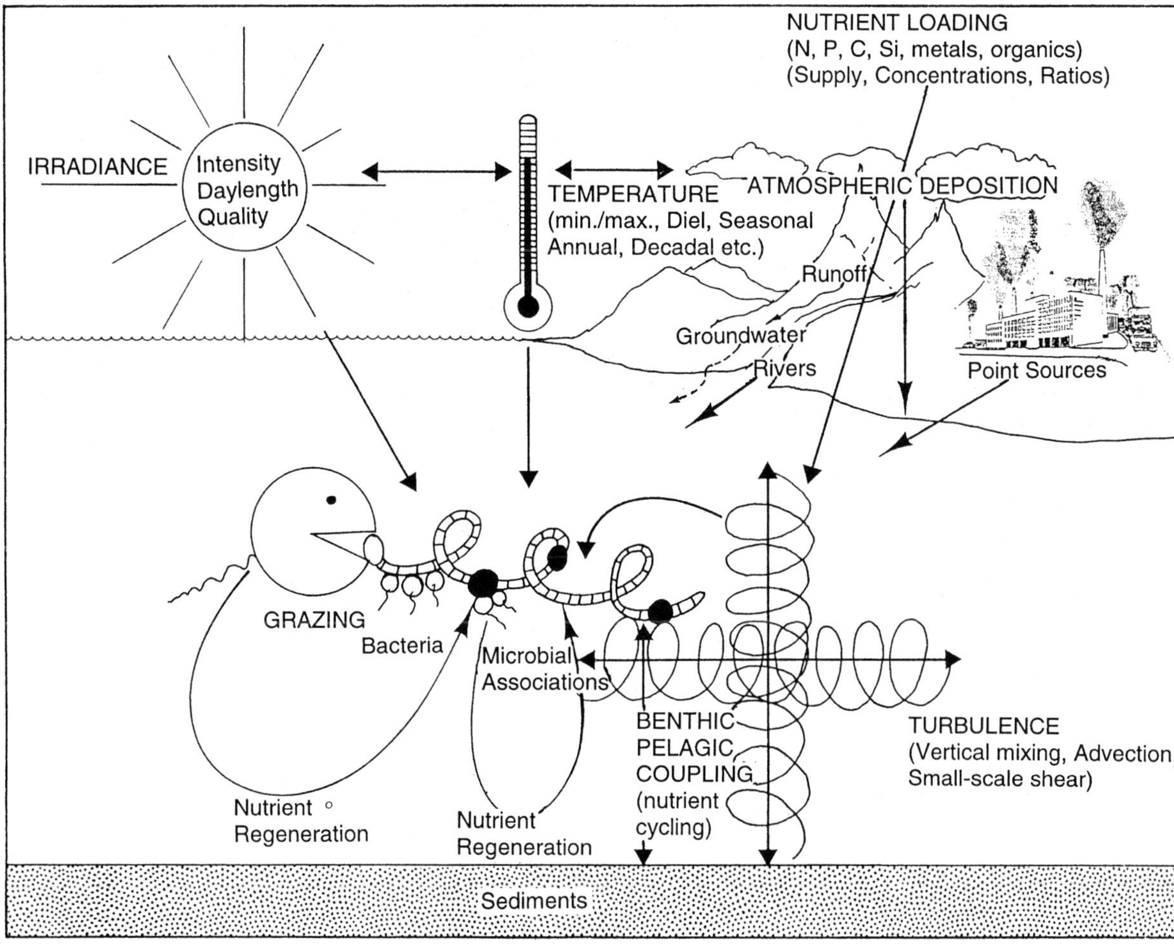

Fig. 8. Conceptual diagram of contemporaneous and contiguous interacting physical, chemical and biotic controls on planktonic cyanobacterial growth and bloom dynamics in the marine environment (after Paerl and Millie, 1996).

other forms of precipitation;

benthic-pelagic coupling of nutrient and biotic exchanges as mediated by vertical mixing and lateral transport

rates of gas exchange at air-water and sediment-water interfaces;

roles of benthic habitats as attachment substrates for cyanobacterial communities.

The relationships between the physical and optical depth of the water column, the mixed layer and the euphotic zone, and the resultant vertical physical-chemical stratifications and gradients as they occur in relation to benthic nutrient cycling and production dynamics, are key determinants of ecosystem trophic state; specifically, its potential for the support of cyanobacterial populations. It is well known that shallow, eutrophic freshwater environments that experience intense benthic-pelagic coupling are more likely to experience periodic cyanobacterial dominance and blooms than are deep oligotrophic lakes (Reynolds, 1987; Paerl, 1988, 1996b). Vertical stratification may be intermittent or prolonged. Stratification influences optical and physical depth, and plays a role in the transport and regeneration of nutrients from sediments as well as new nutrient inputs. Depending on the interactions between these factors positively-buoyant bloom species or neutrally buoyant picoplankton may exhibit more or less prevalence and/or dominance.

Biotic interactions known to play central roles in cyanobacterial growth and physiology include grazing, synergistic and antagonistic interactions with other microbes, and mutualistic and symbiotic associations. These interactions, together with prevailing physical-chemical conditions, may yield a wide range of conditions which may positively or adversely influence both the structure and function of cyanobacterial communities, and the capacity of those communities to compete with the eukaryotic component of the phytoplankton community. Such complex interactions must be considered carefully during future efforts that use modeling to enhance our understanding of how cyanobacteria come, or do not come to dominate these marine ecosystems. Despite their "primitive" structural and evolutionary properties, the cyanobacterial phytoplankton represents an ecophysiologically complex group which in many respects remains enigmatic and challenging as we strive to better understand structure and function of marine ecosystems.

Acknowledgments

I thank J. Pinckney, J. Zehr, E. Carpenter, D. Capone and D. Millie for technical assistance, discussions and critical reviews of this chapter. Research and documentation support was provided by the National Science Foundation, Projects DEB 92-20886, DEB 94-08471, IBN 96-15772, OCE 94-15985, The North Carolina Sea Grant Program RMER/30, and the US Department of Agriculture, Project 9600509.

References

Antia NJ, Harrison PJ and Oliveira L (1991) The role of dissolved organic nitrogen in phytoplankton nutrition, cell biology and ecology. Phycologia 30: 1-89

Azam F, Fenchel T, Field JG, Gray JS, Meyer-Reil LA and Thingstadt F (1983) The ecological role of water-column microbes in the sea. Mar Ecol Prog Ser 10: 257-263

Beaglehole JC (ed.) (1955) The Journals of Captain James Cook on his Voyages of Discovery. Vol. 1. The Voyage of the Endeavour 1768-1771. Cambridge University Press, Cambridge

Bebout BM, Paerl HW, Crocker KM and Prufert LE (1987) Diel interactions of oxygenic photosynthesis and N_2 fixation (acetylene reduction) in a marine microbial mat community. Appl Environ Microbiol 53: 2353-2362

Bergman B and Carpenter EJ (1991) Nitrogenase confined to randomly distributed trichomes in the marine cyanobacterium *Trichodesmium thiebautii*. J Phycol 27: 158-165

Bothe H (1982) Nitrogen fixation. In: Carr N G and Whitton B A (eds) The Biology of Cyanobacteria, pp 87-104. Blackwell Scientific Publications, Oxford

Booth BC (1988) Size classes and major taxonomic groups of phytoplankton at two locations in the subarctic Pacific Ocean in May and August 1984. Mar Biol 97: 275-286

Britschgi TB and Giovannoni SJ (1991) Phylogenetic analysis of a natural marine bacterioplankton population by rRNA gene cloning and sequencing. Appl Environ Microbiol 57: 1707-1713

Bruland KW, Donat JR and Hutchins DA (1991) Interactive influences of bioactive trace metals on biological production in oceanic waters. Limnol Oceanogr 36: 1555-1577

Bryceson I and Fay P (1981) Nitrogen fixation in *Oscillatoria (Trichodesmium) erythraea* in relation to bundle formation and trichome differentiation. Mar Biol 61: 159-166

Caldwell DE (1979) Associations between photosynthetic and heterotrophic prokaryotes in plankton. In: Nichols JM (ed.) Abstracts of Third International Symposium on Photosynthetic Prokaryotes

Calwell DE and Caldwell SJ (1978) A *Zoogloea* sp. associated with blooms of *Anabaena flos-aquae*. Can J Microbiol 24: 922-931

Camp TR and Stein PC (1943) Velocity gradients and internal work in fluid motion. J Boston Soc Civil Eng 30: 219-237

Capone DG (1983) Benthic nitrogen fixation. In: Carpenter EJ and Capone DG (eds) Nitrogen in the Marine Environment, pp 105-137. Academic Press, New York

Capone DG and Carpenter EJ (1982) Nitrogen fixation in the marine environment. Science 217: 1140--1142

Capone DG, O'Neil JM, Zehr J and Carpenter EJ (1990) Basis for diel variation in nitrogenase activity in the marine planktonic cyanobacterium *Trichodesmium thiebautii*. Appl Environ Microbiol 56: 3532-3536

Capone DG, Zehr JP, Paerl HW, Bergman B and Carpenter EJ (1997) *Trichodesmium*, a globally significant marine cyanobacterium. Science 276: 1221-1229.

Carmichael WW (1991) Toxic freshwater blue-green algae (cyanobacteria): An overlooked health threat. Health Environ Digest 5: 1-4

Carpenter EJ (1973) Nitrogen fixation by *Oscillatoria (Trichodesmium) thiebautii* in the southwestern Sargasso Sea. Deep Sea Res 20: 285-288

Carpenter EJ (1983a) Physiology and ecology of marine plankton *Oscillatoria (Trichodesmium)*. Mar Biol Lett 4: 69-85

Carpenter EJ (1983b) Nitrogen fixation by marine *Oscillatoria (Trichodesmium)* in the world's oceans. In: Carpenter EJ and Capone DG (eds) Nitrogen in the Marine Environment, pp 65-104. Academic Press, New York

Carpenter EJ and Price CC (1976) Marine *Oscillatoria (Trichodesmium)*: Explanation for aerobic nitrogen fixation without heterocysts. Science 191: 1278-1280

Carpenter EJ and Romans K (1991) Major role of the cyanobacterium *Trichodesmium* in nutrient cycling in the North Atlantic Ocean. Science 254: 1356-1358

Carr NG and Mann NH (1994) The oceanic cyanobacterial picoplankton. In: Bryant DA (ed.) The Molecular Biology of Cyanobacteria, pp 27-48. Kluwer Academic Publishers, Dordrecht

Carr NG and Whitton BA (eds) (1982) The Biology of Cyanobacteria. Blackwell Scientific Publications, Oxford

Chisholm SW, Olson RJ, Zettler ER, Goericke R and Waterbury JB (1988) A novel free-living prochlorophyte abundant in the oceanic euphotic zone. Nature 334: 340-343

Chróst RJ and Brzeska D (1978) Extracellular release of organic products and growth of bacteria in *Anabaena cylindrica* (blue-green alga) culture. Acta Microbiol Pol 27: 287-295

Church TM, Tramontano JM, Scudlark JR, Jickells TD, Tokos JJ and Knapp AM (1984) The wet deposition of trace metals to the western Atlantic Ocean at the mid-Atlantic coast and on Bermuda. Atmos Environ 18: 2657-2664

Cole JJ, Lane JM, Marino R and Howarth RW (1993) Molybdenum assimilation by cyanobacteria and phytoplankton in freshwater and salt water. Limnol Oceanogr 38: 25-35

Currin CA, Paerl HW, Suba G and Alberte RS (1990) Immunofluorescence detection, and characterization of N_2 fixing microorganisms from aquatic environments. Limnol Oceanogr 35: 59-71

D'Elia CF, Sanders JG and Boynton WR (1986) Nutrient enrichment studies in a coastal plain estuary: Phytoplankton growth in large scale, continuous cultures. Can J Fish Aquat Sci 43: 397-406

Dam HG, Drapeau DT and Grenier G (1994) An improved flocculation design for use in particle aggregation experiments. Limnol Oceanogr 39: 23-729

Darwin C (1839) Journal of the Researches into the Geology, and Natural History of Various Countries Visited by the H.M.S. Beagle. Hafner Publishing Co, New York (1952)

Davis PG and Sieburth J McN (1982) Differentiation of phototrophic and heterotrophic nanoplankton populations in marine waters by epifluorescence microscopy. Ann Inst of Oceanogr Paris 58 (S): 249-260

Dean DR and Jacobson MR (1992) Biochemical genetics of nitrogenase. In: Stacey G, Burris RH and Evans HJ (eds) Biological nitrogen fixation, pp 763-834. Chapman & Hall, New York

Dicker HJ and Smith DW (1980) Acetylene reduction (nitrogen fixation) in a Delaware, USA salt marsh. Mar Biol 57: 241-250

Donze W, Haveman J and Schiereck P (1972) Absence of photosystem 2 in heterocysts of the blue-green alga *Anabaena*. Biochim Biophys Acta 256: 157-161

Doremus C (1985) Geochemical control of dinitrogen fixation in the open ocean. Biol Oceanogr 1: 429-436

Drouet F (1968) Revision of classification of *Oscillatoriaceae*. Monogr Acad Nat Sci Phil 15: 1-334

Dubois JD and Kapustka LA (1981) Osmotic stress effects on the $N_2(C_2H_2)$ ASA activity of aquatic cyanobacteria. Aquat Bot 11: 11-20

Duce RA and Tindale NW (1991) Atmospheric transport of iron and its deposition in the ocean. Limnol Oceanogr 36: 1715-1726

Dugdale RC (1967) Nutrient limitation in the sea: Dynamics, identification and significance. Limnol Oceanogr 12: 685-695

Fogg, GE (1944) Growth and heterocyst production in *Anabaena cylindrica* Lemm. New Phytol 43: 164-175

Fogg GE (1969) The physiology of an algal nuisance. Proc Roy Soc Lond B 173: 175-189

Fogg GE (1974) Nitrogen fixation: In: Stewart WDP (ed.) Algal Physiology and Biochemistry, pp 560-582. Blackwell Scientific Publications, Oxford

Fogg GE (1982) Marine plankton. In: Carr NG and Whitton BA (eds) The Biology of Cyanobacteria, pp 491-514. Blackwell Scientific Publications, Oxford

Fogg GE, Stewart WDP, Fay P and Walsby AE (1973) The Blue-Green Algae. Academic Press, London

Frémy P (1933) Les cyanophycées des côtes d'Europe. Mémoirs Soc Nat Sci Nat Math Cherbourg 41: 1-235

Fulton RS and Paerl HW 1988. Effects of the blue-green alga *Microcystis aeruginosa* on zooplankton competitive relations. Oecologia 76: 383-389

Gallon JR (1989) The physiology and biochemistry of N_2 fixation by nonheterocystous cyanobacteria. Phykos 28:18-46

Gallon JR (1992) Reconciling the incompatible: N_2 fixation and O_2. New Phytol 122:571-609

Gallon JR, Hashem MA and Chaplin AE (1991) Nitrogen fixation by *Oscillatoria* spp. under autotrophic and photoheterotrophic conditions. J Gen Microbiol 137: 31-39

Gallon JR and Stal LJ (1992) N_2 fixation in non-heterocystous cyanobacteria: An overview. In: Carpenter EJ, Capone DG and Rueter JG (eds) Marine Pelagic Cyanobacteria: *Trichodesmium* and other Diazotrophs, pp 115--139. Kluwer Academic Publishers, Dordrecht, The Netherlands

Gibson CE and Smith RV (1982) Freshwater plankton. In: Carr NG and Whitton BA (eds) The Biology of Cyanobacteria, pp. 463-450. Blackwell Scientific Publications, Oxford

Gieskes W and Kraay G (1986) Floristic and physiological differences between the shallow and deep nanophytoplankton community in the euphotic zone of the tropical Atlantic revealed by HPLC analysis of pigments. Mar Biol 91: 567-576

Giovannoni SJ, Britschgi TB, Moyer CL and Field KG (1990) Genetic diversity in Sargasso Sea bacterioplankton. Nature 344: 60-63

Glibert P, Conley D, Fisher T, Harding L and Malone T (1995) Dynamics of the 1990 winter/spring bloom in Chesapeake Bay. Mar Ecol Progr Ser 122: 27-43

Glover HE, Keller MD and Spinrad RW (1987) The effects of light quality and intensity on photosynthesis and growth of marine planktonic prokaryotic phytoplankton clones. J Exp Mar Biol Ecol 105: 137-159

Goericke R and Welshmeyer NA (1993a) The marine prochlorophyte *Prochlorococcus* contributes significantly to phytoplankton biomass and primary production in the Sargasso Sea. Deep-Sea Res 40: 2283-2294

Goericke R and Welshmeyer NA (1993b) The chlorophyll-labeling method: measuring specific rates of chlorophyll a synthesis in cultures and the open ocean. Limnol Oceanogr 38: 80-95

Healy FP (1982) Phosphate. In: Carr NG and Whitton BA (eds) The Biology of Cyanobacteria, pp. 105-124. Blackwell Scientific Publications, Oxford

Horne AJ (1977) Nitrogen fixation - a review of this phenomenon as a polluting process. Progr Water Technol 8: 359-372

Horne AJ and Goldman CR (1972) Nitrogen fixation in Clear Lake, CA: 1. Seasonal variations and the role of heterocysts. Limnol Oceanogr 17:678-692

Howarth RW and Cole JJ (1985) Molybdenum availability, nitrogen limitation, and phytoplankton growth in natural waters. Science 229: 653-655

Howarth RW, Marino R, Lane J and Cole JJ (1988) Nitrogen fixation in freshwater, estuarine and marine ecosystems. 1. Rates and importance. Limnol Oceanogr 33: 619-687

Huber AL (1986) Nitrogen fixation by *Nodularia spumigena* Mertens (Cyanobacteria). I. Field studies on the contribution of blooms to the nitrogen budget of the Peel-Harvey Estuary, Western Australia. Hydrobiologia 131: 193-203

Huang T-C, Tu J, Chow T-J and Chen T-H (1990) Circadian rhythm of the prokaryote *Synechococcus* sp. RF-1. Plant Physiol 92: 531-533

Itturiaga R and Mitchell BG (1986) Chrococcoid cyanobacteria: A significant component in the food web dynamics of the open ocean. Mar Ecol Progr Ser 28: 291-297

Johnson PW and Sieburth JMcN (1979) Chrococcoid cyanobacteria in the sea: a ubiquitous and diverse phototrophic biomass. Limnol Oceanogr 24: 928-935.

Kahru M, Leppänen, J-M and Rud O (1993) Cyanobacterial blooms cause heating of the sea surface. Mar Ecol Progr Ser 101: 1-7

Karl DM, Letelier R, Hebel DV, Bird DF and Winn CD (1992) *Trichodesmium* blooms and new nitrogen in the North Pacific gyre. In:: Carpenter EJ, Capone DG and Rueter JG (eds.) Marine Pelagic Cyanobacteria: *Trichodesmium* and Other Diazotrophs, pp. 219-238. Kluwer, Dordrecht, The Netherlands

Karsten G (1905) Das phytoplankton des atlantischen oceans nach dem material der deutschen Tiefsee-Expedition 1898-1899. Wissenschaften Ergen. Deutschen Tiefsee-Expedition "Valdiva" 1898-1899 2: 137-219

Karsten G (1907) Das indische Phytoplankton nach dem Material der deutschen Tiefsee-Expedition 1898-1899. Wissenschaften

Ergen. Deutschen Tiefsee-Expedition "Valdiva" 1898-1899 2: 221-548

Klemer AR and Konopka AE (1980) Causes and consequences of blue-green algal (cyanobacterial) blooms. Lake Res Manag 5: 9-19

Kononen K, Kuparinen J, Mäkelä K, Laanemets J, Pavelson J and Nomann (1996) Initiation of cyanobacterial blooms in a frontal region at the entrance to the Gulf of Finland, Baltic Sea. Limnol Oceanogr 41: 98-112

Kucera SA (1996) Effects of small-scale chear forces on nitrogen fixing cyanobacteria. M.Sc Thesis, 129p. University of North Carolina, Chapel Hill

Kuentzel EL (1969) Bacteria, carbon dioxide, and algal blooms. J Water Poll Cont Fed 41: 1737-1747

Lang NJ and Fay P (1971) The heterocysts of blue-green algae II. Details of ultrastructure. Proc Roy Soc Lond B 178: 193-203

Lange W (1967) Effects of carbohydrates on the symbiotic growth of the planktonic blue-green algae with bacteria. Nature 215: 1277-1278

Lange W (1971) Enhancement of algal growth in cyanophyta-bacteria systems by carbonaceous compounds. Can J Microbiol 17: 303-314

Lemmemann E (1905) Die Algenflora der Sandwich-Inseln. Ergebnisse einer Reise nach dem Pacific, H. Schauinsland 1996/97. Englers Bot J 34: 607-663

León C, Kumazawa S and Mitsui A (1986) Cyclic appearance of aerobic nitrogenase activity during synchronous growth of unicellular cyanobacteria. Curr Microbiol 13: 149-153

LeRudilier DT, Bernard T, Goas G and Hamelin J (1984) Osmoregulation in *Klebsiella* pneumoniae: Enhancement of anaerobic growth and nitrogen fixation under stress by proline betaine, m-butyrobetaine, and other related compounds. Can J Microbiol 30: 299-305

Li WKW (1986) Experimental approaches to field measurements: methods and interpretations. In: Platt T and Li WW (eds.), Photosynthetic Picoplankton, pp 251-286. Can Bull Fish Aquat Sci Vol 214, Ottawa, Canada

Madden CJ and Day JW (1992) Induced turbulence in rotating bottles affects phytoplankton productivity measurements in turbid waters. J Plankt Res 14: 1171-1191

Mague TH, Mague FC and Holm-Hansen O (1977) Physiology and chemical composition of nitrogen-fixing phytoplankton in the central North Pacific Ocean. Mar Biol 41: 213-227

Mague TH, Weare NM and Holm-Hansen O (1974) Nitrogen fixation in the North Pacific Ocean. Mar Biol 24: 109-119

Martin JH, Fitzwater SE and Gordon RM (1990) Iron deficiency limits phytoplankton growth in Antarctic waters, Globa Biogeo Cycl 4: 5-12

Martin JH and 43 co-authors (1994) Testing the iron hypothesis in ecosystems of the equatorial Pacific Ocean. Nature 28: 159-182

Martinez L, Silver MW, King JM and Alldredge AL (1983) Nitrogen fixation by floating datm mats: A sources of new nitrogen to oligotrophic ocean waters. Science 221: 152-154

Millie DF, Paerl HW and Hurley J (1993) Microalgal pigment assessments using high performance liquid chromatography: a synopsis of organismal and ecological applications. Can J Fish Aquat Sci 50: 2513-2527

Mitsui A, Kumazawa S, Takahashi S., Ikemoto A, Cao S and Arai T (1986) Strategy by which nitrogen-fixing unicellular cyanobacteria grow phototrophically. Nature 323: 720-722

Murphy TO, Lean DRS and Nalewajko C (1976) Blue-green algae: Their excretion of selective chelators enables them to dominate other algae. Science 221: 152-154

Neilson AH and Lewin RA (1974) The uptake and utilization of organic carbon by algae: An essay in comparative biochemistry. Phycologia 13: 227-264

Niemi A (1979) Blue-green algal blooms and N:P ratio in the Baltic Sea. Acta Bot Fenn 110: 57-61

Nixon SW (1986) Nutrient dynamics and the productivity of marine coastal waters. In: Halwagy R, Clayton D and Behbehani M (eds), Marine Environment and Pollution, pp 97-115. The Alden Press, Oxford

Nixon SW (1995) Coastal marine eutrophication: A definition, social causes, and future concerns. Ophelia 41: 199-219

Ohki K and Fujita Y (1988) Aerobic nitrogenase activity measured as acetylene reduction in the marine non-heterocystous cyanobacterium *Trichodesmium*spp. grown under artificial conditions. Mar Biol 98: 111-114

Olson R, Vaulot R and Chisholm S (1985) Marine phytoplankton distributions measured using shipboard flow cytometry. Deep-Sea Res 32: 1273-1280

Ostenfeld CH and Schmidt J (1901) Plankton fra det Rrde hav og Adenbugten, Vidensk Meddel Naturh Forening i Kbhvn. 141: 141-182

Paerl HW (1978) Nannoplankton vs. netplankton photosynthetic and heterotrophic activities in Fijian waters of the South Pacific Ocean. Roy Soc New Zeal Bull 17: 211-221

Paerl HW (1984) Transfer of N_2 and CO_2 fixation products from *Anabaena oscillarioides* to associated bacteria during inorganic carbon sufficiency and deficiency. J. Phycol 20: 600-608

Paerl HW (1988) Nuisance phytoplankton blooms in coastal, estuarine, and inland waters. Limnol Oceanogr 33: 895-905

Paerl HW (1990) Physiological ecology and regulation of N_2 fixation in natural waters. Adv Microb Ecol 11: 305-344

Paerl HW (1991) Ecophysiological and trophic implications of light-stimulated amino acid utilization in marine picoplankton. Appl Environ Microbiol 57: 473-479

Paerl HW (1992) Cyanobacterial epi- and endobiotic interactions In: Reisser W (ed.) Algal Symbioses, pp 537-566. Biopress Ltd, Bristol, UK

Paerl HW (1994) Spatial segregation of CO_2 fixation in *Trichodesmium* sp.: Linkage to N_2 fixation potential. J Phycol 30: 790-799

Paerl HW (1996a) Microscale physiological and ecological studies of aquatic cyanobacteria: Macroscale implications. Microsc Res Tech 33: 47-72

Paerl HW (1996b) A comparison of cyanobacterial bloom dynamics in freshwater, estaurine and marine environments. Phycologia 35: 25-35

Paerl HW (1997) Coastal eutrophication and harmful algal blooms: Importance of atmospheric deposition and groundwater as ìnewî nitrogen and other nutrient sources. Limnol Oceanogr 42:1154-1165

Paerl HW and Bebout BM (1988) Direct measurement of O_2-depleted microzones in marine *Oscillatoria*: Relation to N_2 fixation. Science 242: 441-445

Paerl HW, Bebout BM and Prufert LE (1988) Naturally-occurring patterns of oxygenic photosynthesis and N_2 fixation in a marine microbial mat: Physiological and ecological ramifications. In: Cohen Y and Rosenberg E (eds), Microbial Mats, pp 326-341. ASM Publications, Washington DC

Paerl HW, Bebout BM and Prufert LE (1989b) Bacterial associations with marine *Oscillatoria* sp. (*Trichodesmium* sp.) populations: Ecophysiological implications. J Phycol 25: 773-784

Paerl HW and Bland PT (1982) Localized tetrazolium reduction in relation to N_2 fixation, CO_2 fixation and H_2 uptake in aquatic filamentous cyanobacteria. Appl Environ Microbiol 43: 218-226

Paerl HW, Bland PT, Tucker J and Blackwell J (1983) The effects of salinity on the potential of a blue-green algal (*Microcystis aeruginosa*) bloom in the Neuse River Estuary, NC. NC Sea Grant Rep 83-1, 84 p

Paerl HW and Carlton RG (1988) Control of nitrogen fixation by surface-associated microzones. Nature 332: 260-262

Paerl HW, Crocker KM and Prufert LE (1987) Limitation of N_2 fixation in coastal marine waters: Relative importance of molybdenum, iron, phosphorus and organic matter availability. Limno Oceanogr 32: 525-536

Paerl HW and Gallucci KK (1985) Role of chemotaxis in establishing a specific nitrogen-fixing cyanobacterial-bacterial association. Science 227: 647-649

Paerl HW and Kellar PE (1978) Significance of bacterial-*Anabaena* (Cyanophyceae) associations with respect to N_2 fixation in freshwater. J Phycol 14: 254-260

Paerl HW and Millie DW (1996) Physiological ecology of toxic cyanobacteria. Phycologia 35: 169-167

Paerl HW and Pinckney JL (1996) Microbial consortia: Their role in aquatic production and biogeochemical cycling. Microb Ecol 31: 225-247

Paerl HW, Pinckney JL and Kucera SA (1995) Clarification of the structural and functional roles of heterocysts and anoxic microzones in the control of pelagic nitrogen fixation. Limnol Oceanogr 40: 634-638

Paerl HW, Priscu JC and Brawner DL (1989a) Immunochemical localization of nitrogenase in marine *Trichodesmium* aggregates: Relationship to N_2 fixation potential. Appl Environ Microbiol 55: 2965-2975

Paerl HW and Prufert LE (1987) Oxygen-poor microzones as potential sites of microbial N_2 fixation in nitrogen-depleted aerobic marine waters. Appl Environ Microbiol 53: 1078-1087

Paerl HW, Prufert LE and Ambrose WW (1991) Contemporaneous N_2 fixation and oxygenic photosynthesis in the non-heterocystous mat-forming cyanobacterium *Lyngbya aestuarii*. Appl Environ Microbiol 57: 3086-3092

Paerl HW, Prufert-Bebout LE and Guo C (1994) Iron-stimulated N_2 fixation and growth in natural and cultured populations of the planktonic marine cyanobacterium *Trichodesmium*. Appl Environ Microbiol 60: 1044-1047

Paerl HW and Tucker C (1995) Ecology of blue-green algae in aquaculture ponds. J World Aqua Soc 26: 1-53

Paerl HW and Ustach JF (1982) Blue-green algal scums: an explanation for their occurrence during freshwater blooms. Limnol Oceanogr 28: 847-857

Paerl HW, Webb KL and Wiebe W (1981) Nitrogen fixation in waters. In: Broughton WJ (ed.), Nitrogen Fixation. Volume 1: Ecology, pp 193-240. Oxford Publications, Oxford

Paulsen, DM, Paerl HW and Bishop PE (1991) Evidence that molybdenum-dependent nitrogen fixation is not limited by high sulfate in marine environments. Limnol Oceanogr 36: 1325-1334

Phlips EJ and Zeman C (1990) Photosynthesis, growth and nitrogen fixation by epiphytic forms of filamentous cyanobacteria from pelagic *Sargassum*. Bull mar Sci 47: 613-621

Pinckney JL, Millie DF, Howe K, Paerl HW and Hurley J (1996) Flow scintillation counting of ^{14}C-labeled microalgal photosynthetic pigments. J Plankton Res18(10): 1867-1880

Pinckney JL, Paerl HW, Harrington MB, Howe KE (1998) Annual cycles of phytoplankton community structure and bloom dynamics in the Neuse River Estuary, NC (USA). Mar Biol 131:371-382

Platt T and Li WKW, Eds. (1986) Photosynthetic Picoplankton. Can Bull Fish Aqua Sci 214: 1-583, Ottawa, Canada

Platt T, Subba Rao DV and Irwin B (1983) Photosynthesis of picoplankton in the oligotrophic ocean. Nature 301: 702-704

Porter KG and Orcutt JD (1980) Nutritional adequacy, manageability, and toxicity as factors that determine the food quality of green and blue-green algae for *Daphnia*. Am Soc Limnol Oceanogr Spec Symp Vol3: 268-281

Potts M (1980) Blue-green algae (cyanophyta) in marine coastal environment of the Sinai Peninsula: distribution, zonation, stratification and taxonomic diversity. Phycologia 19: 60-73

Potts M and Whitton BA (1979a) pH and E_h on Aldabra Atoll. 1. Comparison of marine and freshwater environments. Hydrobiologia 67: 11-17

Potts M and Whitton BA (1979b) pH and E_h on Aldabra Atoll.2. Intertidal photosynthetic microbial communities showing zonation. Hydrobiologia 67: 99-105

Prezelin BB, Glover HE, Ver Hoven B, Steinberg D, Matlick HA, Schofield O, Nelson N, Wyman M and Campbell L (1989) Blue-green light effects on light-limited rates of photosynthesis: Relationship to pigmentation and productivity estimates for *Synechococcus* populations from the Sargasso Sea. Mar Ecol Progr Ser 54: 121-136

Prufert LE, Paerl HW and Lassen C (1993) Growth, nitrogen fixation and spectral attenuation in cultivated *Trichodesmium*. Appl Environ Microbiol 59: 1350-1359

Redalje D (1993) The labeled chlorophyll *a* technique for determining photoautotrophic carbon-specific growth rates and biomass. In: Kemp P (ed.) Handbook of Methods in Aquatic Microbial Ecology, pp 563-572. Lewis Publishers, Boca Raton, FL

Reed RH and Stewart WDP (1985) Osmotic adjustment and organic solute accumulation in unicellular cyanobacteria from planktonic and marine habitats. Mar Biol 88: 1-9

Reynolds CS (1987) Cyanobacterial water blooms. Adv Bot Res 13: 67-143

Reynolds CS and Walsby AE (1975) Water blooms. Biol Rev 50: 437-481

Romans KM, Carpenter EJ and Bergman B (1994) Buoyancy regulation in the colonial diazotrophic cyanobacterium *Trichodesmium tenue*: ultrastructure and storage of carbohydrate, polyphosphate and nitrogen. J Phycol 30: 935-942

Rosenberg G and Paerl HW (1980) Nitrogen fixation by blue-green algae associated with the siphonous green seaweed *Codium decorticatum*: Effects on ammonium uptake. Mar Biol 61: 151-158

Rowan K (1989) Photosynthetic Pigments of Algae. Cambridge University Press, New York, 343 p

Rueter JG, Hutchins DA, Smith RW and Unsworth NL (1992) Iron nutrition in *Trichodesmium*. In: Carpenter EJ, Capone DG and Rueter JG (eds) Marine Pelagic Cyanobacteria: *Trichodesmium* and Other Diazotrophs, pp 289-306. Kluwer Academic Publishers, Dordrecht, The Netherlands

Sellner K (1997) Physiology, ecology, and toxic properties of marine cyanobacterial blooms. Limnol Oceanogr 42: 1089-1104

Smith SV (1984) Phosphorus vs. nitrogen limitation in the marine environment. Limnol Oceanogr 29: 1149-1160

Smith VH (1983) Low nitrogen to phosphorus ratios favor dominance by blue-green algae in lake phytoplankton. Science 221: 669-671

Smith VH (1990) Nitrogen, phosphorus, and nitrogen fixation in lacustrine and estuarine ecosystems. Limnol Oceanogr 35: 1852-1859

Sørenssen F and Sahlsten E (1987) Nitrogen dynamics of a cyanobacterial bloom in the Baltic Sea: New versus regenerated production. Mar Ecol Progr Ser 37:277-284

Sournia A (1970) Les cyanophycées dans le plancton marin. Ann Biol 9: 63-66
Stal LJ and Bergman B (1990) Immunological characterization of nitrogenase in the filamentous non-heterocystous cyanobacterium *Oscillatoria limosa*. Panta 182: 287-291
Stal LJ and Krumbein WE (1985) Nitrogenase activity in the non-heterocystous cyanobacterium *Oscillatoria* sp. grown under alternating light-dark cycles. Arch Microbiol 143: 67-71
Stewart WDP (1973) Nitrogen fixation: In: Carr NG and Whitton BA (eds) The Biology of Blue-Green Algae, pp 260-278, Blackwell Scientific Publications, Oxford
Takeda S, Kamatani A and Kawanobe K (1995) Effects of nitrogen and iron enrichments on phytoplankton communities in the northwest Indian Ocean. Mar Chem 50: 229-241
Taylor BF, Lee CC and Bunt JS (1973) Nitrogen fixation associated with the marine blue-green alga *Trichodesmium* as measured by the acetylene reduction technique. Arch Microbiol 88: 205-212
Taylor DL (1983) Symbioses. In: Carpenter EJ and Capone DG (eds) Nitrogen in the Marine Environment, pp 679-697. Academic Press, New York
Taylor GI (1936) Fluid friction between rotating cylinders. Part I. Torque measurements. Proc Roy Soc Lond A. 157: 546-564
Thomas JS, Apte K and Reddy BR (1988) Sodium metabolism in cyanobacterial nitrogen fixation and salt tolerance. In: Bothe H, de Bruyn FJ and Newton WE (eds) Nitrogen Fixation: Hundred Years After, pp 195-201. Gustav Fischer, Stuttgart
Thomas WH and Gibson CH (1990) Effects of small-scale turbulence on microalgae. J Appl Phycol 2: 71-77
Veldhuis, MJW and Kraay GW (1990) Vertical distribution and pigment composition of a picoplanktonic prochlorophyte in the subtropical North Atlantic: A combined study of HPLC-analysis of pigments and flow cytometry. Mar Ecol Prog Ser 68: 121-127
Villareal TA (1992) Marine nitrogen fixing diatom-cyanobacterial symbioses. In: Carpenter EJ, Capone DG and Rueter JG (eds) Marine Pelagic Cyanobacteria: *Trichodesmium* and other Diazotrophs, pp 163-175. Kluwer Academic Publishers, Dordrecht, The Netherlands
Villbrandt M, Stal LJ and Krumbein WE (1990) Interactions between nitrogen fixation and oxygenic photosynthesis in a marine cyanobacterial mat. FEMS Microbial Ecol 74: 59-72
Walsby AE (1972) Structure and function of gas vacuoles. Bacteriol Rev 36: 1-32
Walsby AE (1974) The extracellular products of *Anabaena cylindrica* Lemm. I. Isolation of macromolecular pigment-peptide complex. Brit Phyc J 9: 371-381
Waterbury, JB, Watson S, Guillard RRL and Brand LE (1979) Widespread occurrence of a unicellular, marine, planktonic, cyanobacterium. Nature 277: 293-294
Waterbury JB, Watson SW and Valois FW (1988) Temporal separation of photosynthesis and dinitrogen fixation in the marine unicellular cyanobacterium *Erythrospira marina*. Eos 69: 1089
Watt WD (1971) Measuring the primary production rates of individual phytoplankton species in natural mixed populations. Deep-Sea Res 18: 329-339
Whitton BA (1973) Interactions with other organisms. In: Carr NG and Whitton BA (eds) The Biology of Blue-Green Algae, pp 415-432. Blackwell Scientific Publications, Oxford
Whitton BA and Potts MA (1982) Marine littoral. In: Carr NG and Whitton BA (eds) The Biology of Cyanobacteria, pp 515-542. Blackwell Scientific Publications, Oxford
Wolk CP (1982) Heterocysts. In: Carr NG and Whitton BA (eds) The Biology of Cyanobacteria, pp 359-386. Blackwell Scientific Publications, Oxford
Wolk CP, Ernst A and Elhai J (1994) Heterocyst metabolism and development. In : Bryant DA (ed.) The Molecular Biology of Cyanobacteria, pp 69-823. Kluwer Academic Publishers, Dordrecht
Wulff F (1988) Understanding the Baltic Sea: Systsems ecology in theory and practice. In: Drepper WS (ed.) Ecodynamics, pp. 113-126, Springer, New York
Wyman M, Zehr JP and Wyman DG (1966) Temporal variability in nitrogenase gene expression in natural populations of the marine cyanobacterium *Trichodesmium thiebautii*. Appl Environ Microbiol 62: 1073-1075
Yates MG (1980) The biochemistry of nitrogen fixation. In: Miflin BJ (ed.) The Biochemistry of Plants. Vol 5, Academic Press, NY
Zehr, JP and McReynolds LA (1989) Use of degenerate oligonucleotides for amplification of the *nifH* gene from the marine cyanobacterium *Trichodesmium thiebautii*. Appl Environ Microbiol 55: 522-2526
Zehr JP, Wyman M, Miller V, Duguay L and Capone DG (1993) Modification of the Fe protein of nitrogenase in natural populations of *Trichodesmium thiebautii*. Appl Environ Microbiol 59: 669-676
Zhuang G, Yi Z and Wallace (1995) Iron(II) in rainwater, snow, and surface seawater from a coastal environment. Mar Chem 50: 41-50

Chapter 6

Freshwater blooms

Roderick L Oliver[1] & George G Ganf[2]
[1]Co-operative Research Centre for Freshwater Ecology, Murray Darling Freshwater Research Centre, PO Box 921, Albury, NSW 2640, Australia.
[2]Department of Environmental Biology, The University of Adelaide, SA 5005, Australia

B. A. Whitton and M. Potts (eds), The Ecology of Cyanobacteria, 149-194.
© 2000 *Kluwer Academic Publishers. Printed in the Netherlands.*

I. Introduction

The successful net growth of a species depends on its ability to optimise resource capture, to utilise efficiently these resources and to minimise losses. It is unlikely that any one organism will have the flexibility to excel under all circumstances, but the appearance of a dominant suggests it has characteristics needed to maximise net growth under the prevailing environmental conditions. At times cyanobacteria may come to dominate the phytoplankton of lakes, reservoirs and rivers. The purpose of this chapter is to appraise the physiological and ecological characteristics of planktonic cyanobacteria which enable them to dominate the phytoplankton. When environmental conditions are appropriate for growth freshwater blooms occur, but which species dominates depends on the interaction between the organism and its habitat.

The term "bloom" is poorly defined, but generally it describes a phytoplankton biomass significantly higher than the lake's average. By this definition even oligotrophic waters may have blooms, although this stretches the concept beyond its general meaning. Blooms are usually comprised of only one or two species and identified by the dominant phytoplankton type eg. cyanobacterial bloom, diatom bloom, *Anabaena* bloom etc. In potable and recreational waters a bloom is frequently defined in terms of cell concentrations that cause a nuisance to humans and a lower limit may be set at ca. 10 mg m^{-3} of chlorophyll-a (ca. 20,000 cells mL^{-1}). The occurrence of a bloom is a function of the environmental conditions and the resource requirements of the organism and many phytoplankton can form blooms under suitable circumstances.

In contrast, surface blooms are restricted to those organisms that are buoyant or motile and on occasions accumulate at the water surface to form a scum. Usually surface blooms are comprised of cyanobacteria made buoyant by the presence of gas-filled cell inclusions called gas-vacuoles, and it is these (Plate 2f) that have historically been referred to as "water blooms" (Reynolds and Walsby, 1975). A few non-cyanobacterial species can also form surface blooms, notably the green alga *Botryococcus braunii* which becomes buoyant by producing and storing oils. Occasionally flagellates such as *Euglena* also form surface blooms.

Cyanobacterial blooms have been recorded from early history (Reynolds and Walsby, 1975) and for some decades those involved in water supply have been aware of their economic impacts due to impairment of water treatment processes including filter blockage, increased disinfection costs and taste and odour problems. Cyanobacterial blooms also degrade the recreational value of surface waters, particularly where thick surface scums reduce the use of amenities for contact sports or large concentrations cause deoxygenation of the water leading to fish kills. Concern about the detrimental effects of freshwater cyanobacteria on water quality was heightened during the 1980s and 1990s as information accumulated on the potency of their toxins (Gorham and Carmichael, 1988; Carmichael, 1994; Codd, 1994; Falconer, 1993). The presence of cyanobacterial toxins has long been known (Francis, 1878), but these were generally associated with the death of domestic animals. Recognition of the potentially harmful effects of these toxins on humans led to re-assessment of these organisms as a threat to water supplies. An apparent increase in the occurrence of cyanobacterial blooms over the last few decades, coupled with the heightened concern about toxins, has created the need for a better understanding of the environmental conditions supporting the growth of the gas-vacuolate cyanobacteria to provide a basis for improved control and management of their occurrence and abundance.

II. Bloom-Forming Cyanobacteria

The cyanobacteria principally responsible for forming blooms are gas-vacuolate species. They are distributed across a number of genera and vary in form and size from small filaments to large globular colonies (Table 1).

The filamentous forms occur as straight, spiral or twisted chains of cells, and sometimes take on a secondary morphology as a result of the aggregation or entanglement of many filaments (Reynolds and Walsby, 1975; Lewis, 1976). A consequence of this complex secondary structure is the size of the biomass unit. Filamentous forms differ significantly

Table 1. The major genera of gas-vacuolate, planktonic cyanobacteria

	N_2-fixer	Family	Order
Filamentous			
Anabaena	+	Nostocaceae	Nostocales
Anabaenopsis	+	Nostocaceae	
Aphanizomenon	+	Nostocaceae	
Nodularia	+	Nostocaceae	
Cylindrospermopsis	+	Nostocaceae	
Gloeotrichia	+	Rivulariaceae	
Oscillatoria	??	Oscillatoriaceae	Oscillatoriales
Spirulina	-	Oscillatoriaceae	
Non-filamentous			
Microcystis	--	Chroococcaceae	Chroocococcales
Gomphosphaeria	--	Chroococcaceae	
Coelosphaerium	--	Chroococcaceae	

in size, from small individual filaments such as *Anabaena minutissima* (4 μm wide and up to 104 μm long: Walsby et al., 1989) or *Oscillatoria agardhii var. isothrix* (3.5 μm wide and up to 125 μm long: Reynolds, 1984a) to spirally coiled filaments as in *Anabaena circinalis* that may be 220 μm long and can aggregate to form macroscopic colonies visible to the naked eye. *Aphanizomenon* filaments aggregate into rafts that are reminiscent of grass blades, while in *Gloeotrichia* the filaments are clustered around a central node to form large (1-2 mm) urchin-like balls.

The colonial cyanobacteria also vary in size and form. The globular colonies usually alter size as a result of growth and reduced colony disaggregation, although loose assemblages or aggregations of separate colonies can occur (Reynolds et al., 1981). *Microcystis aeruginosa* ranges in size from single cells of 5 - 6 μm diameter, the form most often found in culture, to large globular or semi-spherical colonies several millimetres in diameter containing tens of thousands of cells per colony (Reynolds et al., 1981). This represents a change in unit volume of more than three orders of magnitude and has important implications for a variety of processes including buoyancy regulation, nutrient uptake, gas exchange, light interception and susceptibility to grazing.

Surface blooms can appear rapidly, often within hours, and to the casual observer without prior warning of the presence of the organisms. Historically this imbued the blooms with a somewhat mystical character and a notion developed that the cyanobacteria could grow extremely rapidly. In fact the sudden appearance results from the upward migration of an existing dispersed population (Reynolds, 1971) and is not a consequence of rapid cell growth. Their sudden appearance is often associated with calm conditions and reduced turbulence that allows buoyant migration to the water surface. Consequently surface blooms occur only if there is an existing cyanobacterial population and its severity depends, in part, on the size of the pre-existing population. However, the notion that the pre-existing population must be of bloom proportions is mistaken because a dispersed population becomes greatly concentrated as it floats to the surface. Indeed the pre-existing population need not be particularly large at all. For example, if a population that is homogeneously dispersed through a 5 m water column with a cell concentration of 2×10^3 mL^{-1} were to float up into a 2 cm surface layer it would form a surface bloom with a concentration of 0.5×10^6 cells mL^{-1}.

III. Distribution

Cyanobacteria are a common feature of many aquatic systems including tropical and temperate lakes, rivers and estuaries but cell densities, species composition, vertical distribution, longevity and timing of the population maxima differ. To a large degree this may be explained by climatic and meteorological conditions which influence the degree of stratification and mixing as well as light and nutrient availability. It is this physical and chemical setting that provides the stage upon which competitive interactions between species are enacted.

In deep, monomictic, temperate lakes the strong seasonal climatic signal results in a progression of phytoplankton from diatoms in early spring as thermal stratification commences through populations of green algae to culminate during summer in populations of cyanobacteria and dinoflagellates. The two major environmental variables stimulating the species progressions are changes in the stability of stratification and declining nutrient availability (Reynolds, 1984b). Over the growing season the intensity of stratification increases to a maximum in summer when apparently the mixing intensity is insufficient to help maintain heavy phytoplankton, such as diatoms, in suspension. The separation of the water column into an upper epilimnion where light is available for growth and a dark hypolimnion leads to nutrient depauperate conditions developing in the surface layers as a result of phytoplankton growth and sedimentation. It is during calm weather in summer and autumn that surface blooms of cyanobacteria frequently develop, often associated with minimum nutrient concentrations in the surface layer. A secondary peak of diatoms can be associated with the onset of meromixis in autumn before phytoplankton populations are reduced to low levels in winter. This simplified general over-view of the responses observed in deep temperate lakes will be strongly modified by local conditions so that seasonal progressions are altered (Round, 1971). Reynolds 1980 described periodic progressions from one dominant assemblage to another in lakes and enclosure of the English Lake District that are broadly characteristic of the chemical and physical environments that they inhabit. In the more eutrophic lakes the sequence was from diatoms through Volvocales, Nostocales to dinoflagellates or *Microcystis* and in mesotrophic lakes from diatoms through chrysophytes and *Sphaerocystis* to dinoflagellates. Gas-vacuolate cyanobacteria are more typical of eutrophic lakes.

In shallow, well-mixed eutrophic lakes of the temperate northern hemisphere the cyanobacteria that dominate during summer are commonly species of *Oscillatoria*. These lakes are typically very turbid, and when winters are not too cold dominance can persist throughout the year (Scheffer et al. 1997; Sas 1989).

In some clear water temperate lakes, where light penetrates beyond the depth of the epilimnion, species such as *Aphanizomenon flos-aquae* and *Oscillatoria agardhii* form metalimnetic populations of single filaments. In Crooked Lake, Indiana peak concentrations of 25-50 μg chlorophyll-a L^{-1} occurred at a depth of ca. 3 - 4 m, where there were opposing gradients of irradiance and nutrient availability (Konopka 1989). If conditions become unsuitable these filaments may aggregate together and move towards the surface as was observed in Lake Gjersjøen, Norway (Walsby et al. 1983).

In the tropics cyanobacterial blooms and surface scums can form at almost any time of the year, as the annual solar input and air temperature are relatively constant. However, despite this relative constancy, there are major seasonal hydrographic and meteorological changes that alter the phytoplankton community structure. Marked similarities occur between large tropical lakes such as Lake Victoria in East Africa (Talling, 1987) and Lake Lanao in the Philippines (Lewis, 1978). Diatom peaks coincide with marked depressions of the thermocline brought about by meteorological events, while both the chlorophytes and the cyanobacteria decline during these periods of diatom growth. As the thermocline re-establishes the phytoplankton follow a similar progression to that observed in temperate lakes but over a shorter time scale. The chlorophytes dominate first when the nutrient stress is less severe followed by the cyanobacteria during periods of severe nutrient depletion. Lewis (1978) divided the habitat of Lake Lanao through time on the basis of light and nutrient availability as deduced from the growth pulses of individual species. He concluded there was a predictable successional pathway from diatoms through to cyanobacteria and dinoflagellates which reflected adaptations to nutrient and light availability, and adaptations to minimise sinking rate. The diatoms and cryptomonads plus the small single-celled cyanobacterium *Dactylococcopsis* showed growth pulses in habitats characterised by high nutrients and mixing and consequently relatively low light availability. Turbulence was reduced with the onset of thermal stratification and the light climate improved while nutrient availability decreased, the community became dominated by green algae as well as the cyanobacterium, *Aphanothece*, which is characterised by many small cells embedded in a thick gelatinous matrix. As nutrients became less available and the light climate remained favourable cyanobacteria showed pulses of growth alongside the chlorophytes *Sphaerocystis* and *Selenastrum*. *Gymnodinium*, a dinoflagellate, was found in association with the larger cyanobacteria and occupies the habitat where light availability is high, but nutrients and turbulence are minimal. This successional sequence may be interrupted at any time by storm events or other exceptional climatic conditions which either reset the successional clock or cause elements to be skipped.

At offshore stations in Lake Victoria during periods of thermal stratification the cyanobacteria *Anabaenopsis tanganyikae* and *Anabaena flos-aquae* dominate the community but give way to diatoms (eg *Melosira nyassensis* var. *victoriae*) as the lake becomes isothermal. The depth distribution during periods of thermal stratification also differs between species. During periods of relatively strong stratification *Lyngbya circumcreta* and *Aphanocapsa elachista* maxima are confined to the top 20 m (the depth of the euphotic zone in the offshore stations was ca. 13 - 18 m) and decline to very small numbers in the deeper water. In contrast diatom species are usually confined to or below the zone of thermal discontinuity. As thermal stratification breaks down both cyanobacteria become more evenly distributed with depth, although the larger colonial *Aphanocapsa elachista* is most abundant in the upper 20 m.

These temporal and spatial distribution patterns illustrate two important points. Firstly, the natural process of stratification and destratification influences the occurrence of cyanobacteria, but the response is not uniform between species. Secondly, the contrasting depth distribution of diatoms and cyanobacteria illustrate how the lower sinking rates of the cyanobacteria enhance their ability to remain in the epilimnion.

The behaviour and composition of the phytoplankton in Lake Victoria contrasts with that in nearby Lake George (Viner and Smith, 1973). Although Lake George is shallow (2.4 m), light does not penetrate to the sediments and the depth of the euphotic zone is usually less than 0.8 m. The lake stratifies diurnally: at dawn the water column is isothermal (25°C), but heats during the day so by late afternoon as much as 10°C may separate top and bottom waters. The surface water becomes supersaturated with oxygen (ca. 250%) due to the photosynthetic activity of the dense phytoplankton community (ca. 250 mg m^{-3} chlorophyll-a) by late afternoon and the pH rises to > 9. As nocturnal mixing occurs the thermal stratification breaks down, the pH returns to ca 7 and the oxygen concentration falls below 100%. The climatic constancy of Lake George superimposed on the over-riding diel pattern of stratification and destratification appears to be the ideal habitat for permanent, dense populations of the cyanobacteria *Microcystis aeruginosa* and *M. flos-aquae*, *Anabaenopsis* spp., *Aphanizomenon* sp. and *Lyngbya* spp. Surprisingly, the other major cyanobacterial component of the phytoplankton community, *Anabaena flos-aquae*, showed a pronounced but unexplained seasonal periodicity. *Microcystis* spp. appear well suited to the diel pattern of stratification and destratification, a thermal periodicity often found in the surface waters of deeper lakes overlying a summer thermocline. The rapid vertical migration of the large colonies of *Microcystis* provides them with the ability to exploit both the illuminated surface waters and the potentially nutrient rich shallow sediments.

Talling (1992) identified three cyclical environmental changes, each involving many variables, that structure the phytoplankton communities in these lakes. Systems like Lake George are primarily influenced by the diel cycle with relatively little annual change in water volume. The other two systems are influenced by annual cycles, but differ depending on whether or not there are major water volume changes. These comparisons emphasise the importance of both hydrological and hydrographic features as principal determinants of the species composition of phytoplankton communities, and in general diatoms and cyanobacteria occupy complementary positions related to nutrient availability, vertical mixing and water retention.

Many studies have addressed the question of why cyanobacteria should be so successful in such a wide range of environmental conditions. Explanations proposed for cyanobacterial dominance (Steinburg and Hartmann, 1988; Shapiro, 1990; Blomqvist, 1994) include traits to take advantage of warmer water temperatures, to capture reduced photosynthetic photon flux densities (PPFD), to utilise low TN:TP ratios or to access low dissolved carbon dioxide concentrations. Other characteristics suggested as being advantageous include buoyancy regulation, reduced zooplankton grazing and a capacity to store phosphorus. Occasionally one or other of these characteristics has been offered as the sole or principal reason for cyanobacterial dominance, but the distinct morphological, physiological and ecological characteristics of individual species suggests that the factors which promote one will not necessarily promote another. For example, although *Microcystis aeruginosa* may co-exist with *Anabaena flos-aquae*, it does not co-exist with *Oscillatoria rubescens*. Nevertheless, a unique characteristic of all bloom- forming species is the presence of gas vesicles, and there can be little doubt that the buoyancy provided by these structures is a significant attribute.

IV. Gas Vacuoles, Gas Vesicles, Buoyancy and its Regulation

The success of gas vacuolate cyanobacteria is often attributed to their buoyancy and to their ability to regulate buoyancy in response to changing environmental conditions (Reynolds and Walsby 1975; Ganf and Oliver 1982; van Rijn and Shilo 1985; Walsby 1987; Reynolds et al. 1987; Walsby 1994). Advantages associated with buoyancy and its regulation include a reduction in sedimentation losses (Reynolds 1984b), an improvement in the supply of light as buoyant cells move into the well illuminated surface layers (Humphries and Lyne 1988; Walsby et al. 1997) and access to improved nutrient supplies that are available at depth, particularly in waters that are thermally stratified with vertically separated sources of nutrients and light (Ganf and Oliver 1982).

A. Gas Vacuole Structure

Gas vacuoles were discovered by the German microbiologist Klebahn in 1895. Seventy years later Bowen and Jensen (1965) showed that the gas vacuoles were made up of numerous, cylindrical vesicles which were called gas vesicles. The molecular structure, morphology and physical properties of gas vesicles have been reviewed by Walsby (1994). They are hollow, but rigid, proteinacous, cylinders capped at either end by a cone and synthesised under the direction of specific gas vacuole genes which encode for the various proteins required (Walsby, 1994; Oliver, 1994). The gas vesicle wall allows the free passage of gases, but is impermeable to water due to the presence of amino acids with hydrophobic aliphatic side-chains exposed at the inner gas-facing surface of the protein wall.

Since gas vesicles are small, many are needed to provide buoyancy and estimates of 10000 per cell have been made. Gas vesicles are not randomly dispersed throughout the cytoplasm, but are ordered into gas vacuoles to occupy minimal space and provide maximum buoyancy. To achieve this the cylindrical gas vesicles are stacked in hexagonal arrays with the cones interdigitating. In *Anabaena* where the gas vesicle density is 120 kg m^{-3}, if the intervening space (15%) is filled with water at a density of 1000 kg m^{-3}, then the overall gas vacuole density will be 252 kg m^{-3}, one-fourth the density of water and an efficient mechanism to provide lift (Walsby 1994).

The common perception that gas vesicles are equivalent to balloons or bubbles in the cytoplasm is incorrect. Walsby (1994) has suggested a more appropriate analogy would be a pair of old fashioned earthenware flower pots, coated with oil on the inner surface to represent the hydrophobic inner layer of the gas vesicle, and glued together at the rims. Gas vesicles, as with the flower pots, are permeable to gas and therefore contain air at atmospheric pressure, but the air is not required to maintain the hollow space as the gas vesicle walls are rigid. If the gas vesicle "analogy" is submerged, water will seep into the wall, but will be prevented from entering the vesicle cavity by the hydrophobic layer. Air will diffuse back and forth between the cavity and the surrounding medium and the vesicle will be filled with air of composition similar to that in the surrounding liquid.

B. Pressures Acting on Gas Vesicles

The net pressure (p_n) experienced by a vesicle contained within a submerged cell will be the sum of the hydrostatic pressure (p_h), plus the turgor pressure (p_t) plus the pressure of the overlying atmosphere (p_f), minus the gas pressure (p_g) inside the vesicle which is determined by the concentration of gases dissolved in the surrounding liquid and is usually in balance with p_f (Fig. 1). It is important to recognise that the vesicle wall has a finite strength upon which hydrostatic and turgor pressure act, and if the combination of these exceed the strength of the wall the vesicle will collapse and the cell will lose buoyancy (Walsby, 1971).

Hydrostatic pressure increases with water depth at a rate of 0.1 MPa every 10m and for cells to remain buoyant their gas vesicles must be strong enough to withstand the pressures generated during episodes of deep mixing. Increases in turgor pressure result from enhanced rates of photosynthesis producing increased levels of soluble organic intermediates (Grant and Walsby, 1977; Konopka, 1984), coupled with an associated light-dependent uptake of potassium salts (Allison and Walsby, 1981). Turgor pressure can be measured as the difference in the applied pressure (p_c) required to collapse vesicles of cells suspended in 0.5 M sucrose which removes the turgor pressure, compared with the pressure (p_a) required for turgid cells suspended in filtered lake water or culture medium: $p_t = p_c - p_a$. Turgor pressures can vary from 0 to 0.5 MPa in different organisms (Walsby 1994) and increases can lead to gas vesicle collapse.

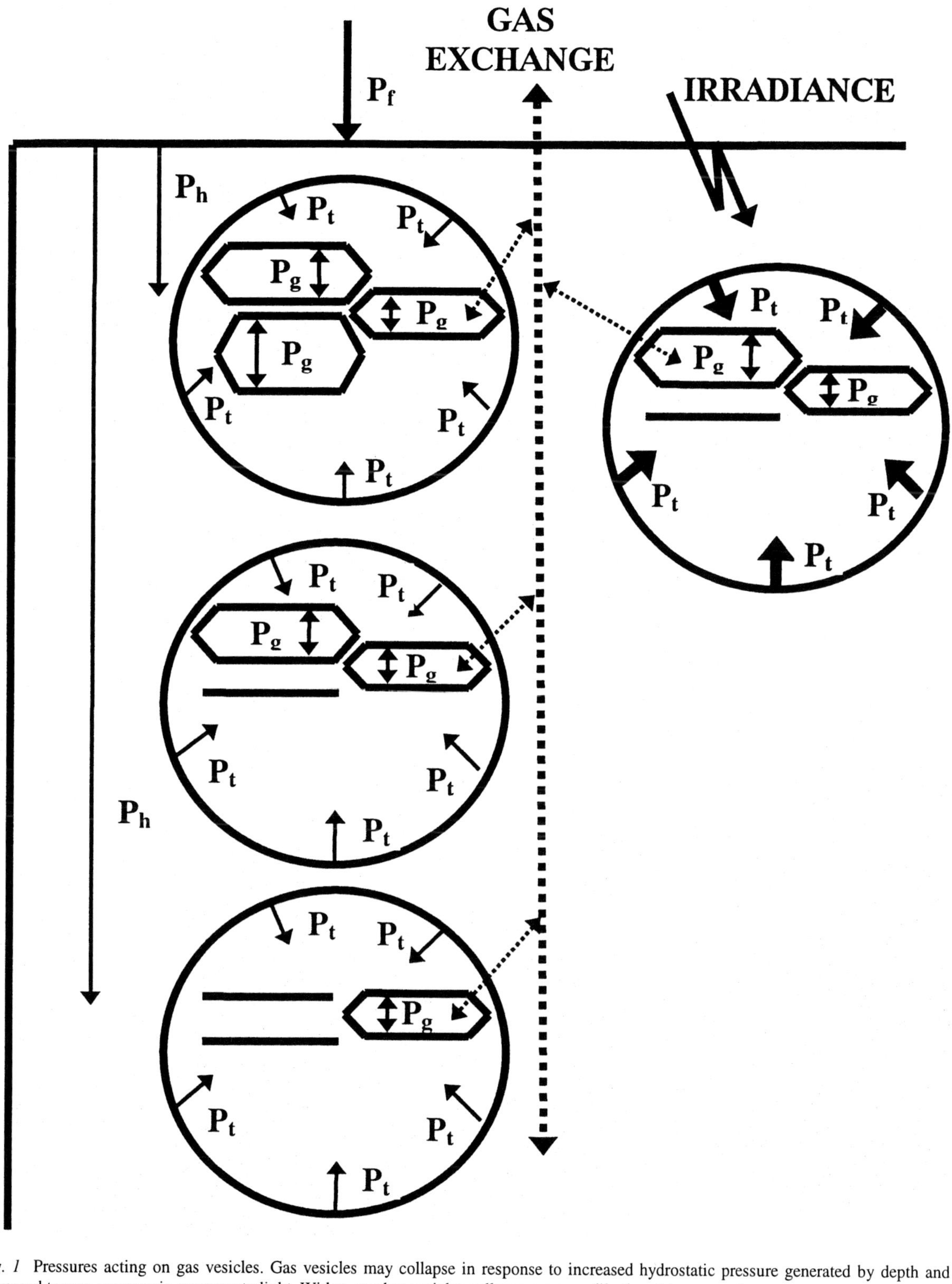

Fig. 1 Pressures acting on gas vesicles. Gas vesicles may collapse in response to increased hydrostatic pressure generated by depth and increased turgor pressure in response to light. Wider, weaker vesicles collapse more readily than thinner, weaker ones.
Key: p_h hydrostatic pressure, p_t turgor pressure, p_f pressure due to overlying atmosphere, p_g pressure of gas in vesicle.

C. Gas Vesicle Collapse

The critical pressure (p_c) at which 50 % of gas vesicles are collapsed varies widely between gas-vacuolate organisms within the range 0.1 to 3.5 MPa. In the bloom-forming freshwater cyanobacteria the range is less (ca. 0.35 to 0.95 MPa), but different species still have particular limits. In gas vesicles isolated from *Anabaena flos-aquae* the critical collapse pressures ranged from 0.45 to 0.85 MPa (Walsby 1980), while for natural populations of *Microcystis aeruginosa* fo. *aeruginosa* the range was 0.5 to 1.2 MPa (Brookes et al., 1994). The critical collapse pressure depends on gas vesicle strength, which is independent of the vesicle length but varies inversely with the cylinder radius (r in nanometres), as described approximately by the expression; $p_c = 275(r)^{-1.67}$ MPa (Walsby, 1994). The width of gas vesicles in various strains of cyanobacteria appears to be determined by the balance between efficient provision of buoyancy and the strength required to withstand hydrostatic pressures (Hayes and Walsby, 1986; Walsby and Bleything ,1988). While buoyancy is provided most efficiently by wider gas vesicles because they have a smaller surface area to volume ratio and provide more buoyancy per unit protein, the gas vesicle strength is reduced as radius increases and vesicles are more susceptible to hydrostatic pressure. *Trichodesmium thiebautii* inhabits the deep oceans, where hydrostatic pressure can be considerable, and has a gas vesicle width of only 45 nm (Gantt et al., 1984), while *Oscillatoria agardhii*, which inhabits deep lakes has a gas vesicle width of 62 nm, and *Microcystis aeruginosa*, *Anabaena flos-aquae* and *Aphanizomenon flos-aquae*, which are found principally in shallow lakes or in lakes where the mixing depth is reduced by thermal stratification have gas vesicle diameters of 67, 84 and 78 nm, respectively (Walsby and Bleything, 1988). *Dactylococcopsis salina* has the broadest gas vesicles (109 nm) and inhabits shallow saline pools where hydrostatic pressures are minimal.

Although gas vesicle strength is not affected by its length, few are longer than 1000 nm and the average length measured in eight species was 470 nm. To answer the question of why on average gas vesicles are not longer, Walsby used geometrical analysis to show that for a gas vesicle of given width the increase in the ratio of the gas vesicle volume to wall volume approaches a plateau at a length of ca. 5 times the width (see Fig 25 of Walsby 1994). He concluded that there would be little point in investing the protein required for a longer gas vesicle, since it would not much improve the provision of buoyancy.

D. Buoyancy Regulation

The buoyancy of gas-vacuolate organisms is dependent on the extent to which the lift provided by gas vesicles counteracts cellular density. Cyanobacteria have the ability to regulate their buoyancy in response to environmental conditions (Reynolds and Walsby, 1975; Walsby and Reynolds, 1980; Reynolds, 1987; Oliver, 1994; Walsby, 1994), either by modifying the degree of gas vacuolation or by altering the extent to which dense components such as carbohydrate (density ca. 1600 kg m^{-3}) and protein (density ca. 1300 kg m^{-3}) accumulate in the cell (Fig. 2). Three mechanisms have been described by which gas-vacuolate cyanobacteria regulate their buoyancy. One is where cell density (p') is changed through alterations in cellular composition. Of particular importance to this mechanism is the accumulation of carbohydrate reserves through photosynthesis and their reduction either through respiration or by conversion to less dense protein. Differences in the relative rates of accumulation and processing cause cell density to increase or decrease in response to environmental conditions (Oliver, 1994; Walsby, 1994). Changes in other cell components, including storage materials such as polyphosphate granules, can also alter cell density and affect buoyancy.

In the two other buoyancy regulating mechanisms the degree of gas vacuolation is altered by either the collapse of gas vesicles due to increased turgor pressure or a reduction in gas vesicle synthesis and their subsequent dilution by growth (Oliver 1994; Walsby 1994). Turgor pressure is generated within the organism as a result of photosynthesis, the magnitude of the increase depending upon the previous light history. The reliance on photosynthesis to generate turgor pressure forms a connection between this mechanism and buoyancy regulation brought about by alterations in cell composition. Whether the accumulation of polysaccharide exerts its influence on cell density through a change in cellular composition, or through gas vesicle collapse due to a rise in turgor pressure, depends on the fate of recently fixed photosynthate and this is a function of the physiological condition of the cells (Gibson, 1978; Kromkamp et al., 1986; Konopka et al., 1987; Klemer et al., 1988). For example, carbohydrate is more efficiently accumulated and stored in response

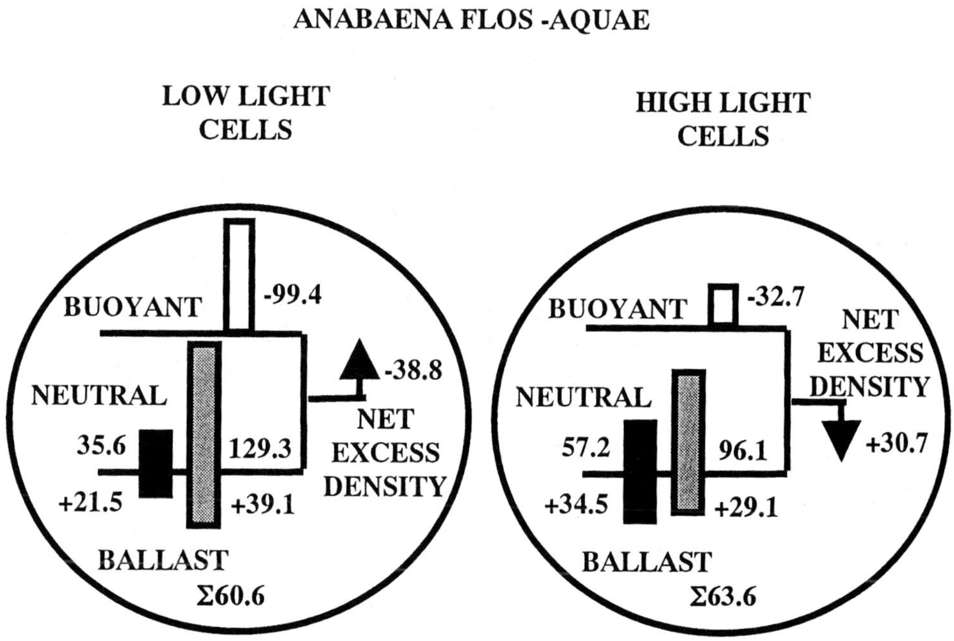

Fig. 2 Diagrammatic balance sheet demonstrating changes in cell density of *Anabaena flos-aquae* in response to increased illumination. Contributions to cell density made by carbohydrate (solid bar) and protein (hatched bar) are calculated from the total weight of each component per unit of cell volume (indicated by length of the bar and in $\mu g\ \mu L^{-1}$). The proportion that is offset by displacement of water is shown as having a neutral effect leaving the excess weight to contribute to cell ballast (21.5 + 39.1 $\mu g\ \mu L^{-1}$, and 34.5 + 29.1 $\mu g\ \mu L^{-1}$, low and high light cells, respectively). The buoyant lift provided by gas vesicles (open bar) is similarly calculated from the weight of water displaced by the total gas vesicle space per unit cell volume (indicated by the length of the bar and in $\mu g\ \mu L^{-1}$). The sum of the ballast and the buoyant lift gives the net excess density of the cell due to these components (-38.8 and +30.7 $\mu g\ \mu L^{-1}$). The excess density is the difference between the density of the cell and that of water. Increased illumination caused an increase in the ballast due to carbohydrate, but this was offset by a decrease in protein ballast and the cells changed from floating to sinking because of a decrease in gas vesicle volume. (Data from Oliver and Walsby, 1984).

to increases in irradiance in cells acclimatised to shorter light periods (Foy and Smith, 1980), whereas in cells grown under continuous light an increase in irradiance results in a more significant production of soluble intermediates and a greater turgor rise (Kromkamp et al., 1986).

The synthesis of gas vesicles is regulated at two levels, the molecular level through control of gene expression, and the physiological level by the availability of energy and structural components required for gas vesicle assembly. Evidence for molecular control has come from several sources. In light-limited cultures of *Aphanizomenon flos-aquae* the gas vesicle content per unit protein was regulated by the supply of energy so that cells were non-buoyant at all growth rates exceeding twenty percent of the maximum growth rate (Konopka et al., 1987; Kromkamp et al., 1988). Similarly, when *Oscillatoria agardhii* was shifted from low to higher light intensities the resulting buoyancy loss was explained by a dilution of gas vesicles as cell volume increased but without a matching increase in gas vesicle synthesis (Utkilen et al., 1985). In *Pseudanabaena* sp. the expression of a gene for one of the gas vesicle proteins was found to be regulated at the transcriptional level, with the abundance of mRNA inversely correlated to irradiance intensity (Damerval et al., 1991). In each of these cases photo-regulation of gas-vesicle genes may explain the altered rate of production of gas vesicles and the resulting change in buoyancy. However, much more information is needed on these molecular mechanisms before their relative importance can be ascertained, as not all cyanobacteria react in the same way. For example, cultures of *Microcystis aeruginosa* responded to an increase in the limiting energy supply by increasing gas vesicle protein synthesis in proportion to total protein synthesis (Thomas and Walsby, 1985; Kromkamp et al., 1988).

Physiologically induced alterations in gas vesicle synthesis include the effects of nitrogen limitation on the production of proteins for gas vesicle assembly. Sustained limitation results in reduced gas vacuolation and a loss of buoyancy (Klemer et al., 1982). Sustained carbon limitation can also result in reduced cell buoyancy due to a restriction in the energy available for synthesis of gas vesicles (Klemer, 1991).

The relative importance of these various responses differs not only between strains of cyanobacteria, but also within strains, depending on both their physiological status and the prevailing environmental conditions. Of key importance is the relative availability of irradiance and the major nutrients, carbon, nitrogen and phosphorus. Although this complexity makes it difficult to provide a functional description appropriate for all gas-vacuolate cyanobacteria under all conditions, the interplay of factors can be described to demonstrate the nuances of buoyancy regulation (Fig. 3). A useful starting point is the observation that the accumulation of carbohydrate as polysaccharide is the result of excess photosynthate production over its incorporation into other compounds (Gibson, 1978; Konopka, 1984). The balance between accumulation or incorporation is altered by light intensity and nutrient deficiency (Healey, 1978).

Short-term, periodic carbon limitation leads to increased buoyancy as cells utilise the carbohydrate reserves accumulated during prior periods of nutrient limitation or excess energy supply (Fig. 3). These cells either already have sufficient gas vesicles for buoyancy, or use the energy reserves to synthesise the extra gas vesicles required (Klemer, 1991). As noted earlier, sustained carbon limitation can result in reduced gas vesicle synthesis and reduced buoyancy through a reduction in gas vesicle synthesis (Fig. 3).

Under nutrient limited conditions the level of excess energy that is accumulated as photosynthate is a function of the irradiance captured relative to the energy requirements of the cells at their nutrient-restricted growth rate. The relative growth rate (RGR) provides a means of comparing the effects of subsaturating levels of two essential growth factors on cell physiology (Konopka, 1989). The specific growth rate μ is limited by one factor, while the potential maximum growth rate under the environmental conditions μ_{max} is set by the second factor. When light and nutrient supply rates are balanced, then the relative growth rate given by μ/μ_{max} is high, even if the specific growth rate is low.

If growth is restricted by a limiting nutrient then energy capture exceeds that utilised in the nutrient restricted growth. As light supply is greater than that required to maintain the RGR_{max} carbohydrate is stored (Fig. 3). Konopka and Schnur (1980) obtained carbohydrate to protein ratios four to seven times higher in cultures limited by nitrogen, phosphorus or sulphur, than in non-limited cultures or those limited by carbon. In general, buoyancy decreases when major nutrients such as phosphorus or nitrogen limit cell growth because carbohydrate accumulates, but associated turgor pressure increases can also collapse gas vesicles especially if there is an accompanying rise in hydrostatic pressure due to cell sedimentation (Reynolds and Walsby, 1975; Klemer, 1978, 1991; Konopka, 1984; Walsby, 1987). The degree of gas vacuolation may also be reduced through molecular controls on gas vesicle production but this is species specific (SS) as shown in Fig. 3. If nutrient limitation greatly depresses growth rate, then carbohydrate accumulation and buoyancy loss can occur even at low light intensities. Under severe and sustained nitrogen limitation reduced gas vesicle synthesis results in reduced buoyancy (Fig. 3).

When all nutrients, including carbon, are present in abundance, then the buoyancy of the organisms is largely a function of the irradiance intensity relative to the growth requirements of the cells (Fig. 3). If the irradiance captured is less than that required to achieve the maximum growth rate under the prevailing environmental conditions, then energy supply will be low relative to what could be utilised and carbohydrate reserves will be reduced and cell buoyancy increased. In organisms like *Aphanizomenon flos-aquae*, where the degree of gas vacuolation is a function of the limiting energy supply, molecular processes will increase gas vesicle synthesis and enhance the positive buoyancy response (Utkilen et al., 1985; Konopka et al., 1987; Kromkamp et al., 1988; Damerval et al., 1991). As discussed earlier for *Microcystis aeruginosa*, not all organisms necessarily show this molecular response to light changes and in Fig. 3 it is depicted as being species specific (SS).

If irradiances are sufficient to support growth rates close to maximum under nutrient sufficient conditions, then cellular composition will approach a mean elemental ratio similar to the Redfield ratio (106 C:16 N:1 P by atoms) indicative of cells growing without nutrient limitation (Hecky and

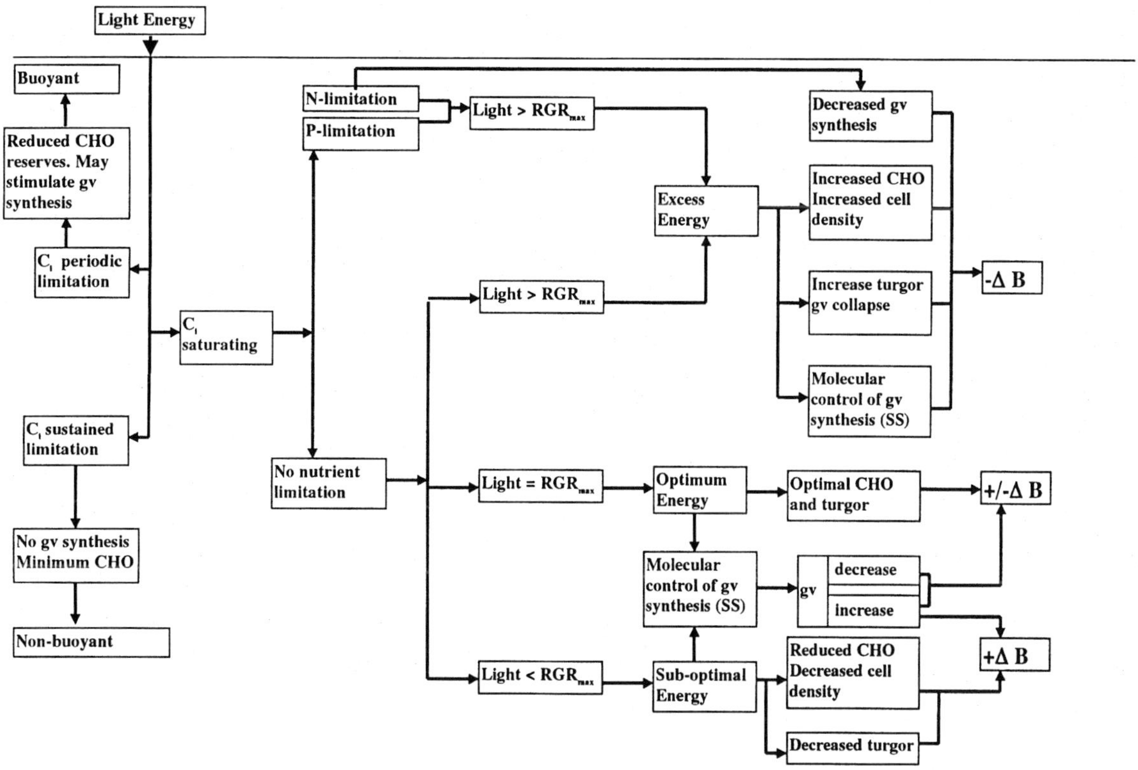

Fig. 3 Flow chart of the major factors influencing gas vesicle production and buoyancy regulation.
Key: C_i inorganic carbon, CHO carbohydrate, gv gas vesicle, N nitrogen, P phosphorus, *RGR_{max} maximum relative growth rate, SS species specific indicating the response is not universal, ΔB change in buoyancy which can be positive (+), negative (-) or either (+/-) depending on species.

Kilham, 1988; Hecky et al., 1993). The buoyancy status of cells under these balanced growth conditions is species-specific (Fig. 3), with *Aphanizomenon flos-aquae* being non-buoyant (Konopka et al., 1987; Kromkamp et al., 1988) and *Microcystis aeruginosa* buoyant (Kromkamp et al., 1988). Presumably the buoyancy status is set by molecular controls on cellular structure.

At irradiances above those saturating nutrient sufficient cell growth the responses are similar to those found under nutrient limitation. The carbohydrate store increases with irradiance up to a maximum when photosynthesis is saturated. This results in an enlarged carbohydrate store which increases cell density, while molecular controls in some organisms reduce the rate of gas vesicle synthesis (Fig. 3). If turgor pressure increases sufficiently then gas vesicle collapse can occur, although this may not be sufficient on its own to reduce buoyancy in cells acclimatised to normal light:dark cycles (Oliver and Walsby, 1984; Kromkamp et al., 1986).

The relationships between nutrients, energy, growth and buoyancy described in Fig. 3, show that gas vesicle regulation is not just related to cell growth, but also depends on the factors controlling growth (Konopka et al., 1987). At times the interaction between these factors is complicated, particularly under natural conditions where there can be large changes in nutrient and irradiance conditions over short periods that affect the utilisation and re-supply of cellular stores of nutrients and energy.

V. Mixing Regimes and Cyanobacteria

Buoyancy maintains cyanobacterial cells in suspension, while its regulation enables them to move vertically through the water in response to changing growth conditions. However the extent to which gas-vacuolate cyanobacteria can control their vertical distribution is also a function of the turbulent mixing regime. Steinberg and Hartmann (1988) analysed cyanobacterial distributions across a number of water bodies and came to the conclusion that turbulence in lakes and rivers should be regarded as a special quasi-resource that can be differentially exploited by various phytoplankton in a manner analogous to nutrients or light (Reynolds and Walsby, 1975; Harris, 1986). Over the last two decades research on mixing processes has led to significant advances in understanding of the effects of turbulent mixing on the growth and distribution of phytoplankton, and particularly in selecting for gas-vacuolate cyanobacteria.

A. Floating and Sinking under Quiescent and Turbulent Mixing Conditions

In quiescent waters the sinking or floating rate of phytoplankton can be calculated from the Stokes equation, modified if necessary by the inclusion of a "form factor" that adjusts for the non-spherical shape of some organisms:

$$v = 2gr^2(\rho' - \rho) / 9\eta\theta \qquad [1]$$

where:

the terminal velocity of the organism (v), is dependent on gravitational acceleration (g), the size of the organism represented as the radius (r) of a sphere of equal volume, the density of the organism (ρ') and the density (ρ) and the viscosity (η) of the medium. The form resistance factor ϕ is defined as v/v_S where v_S is the terminal velocity of a sphere of equal volume and density to that of the organism.

For some shapes the form factor is known from empirical relationships (McNown and Malaika, 1950; Davey and Walsby, 1985), but generally it has to be determined after measuring all other variables (Oliver et al., 1981). The Stokes equation is suitable for calculating sinking and floating velocities if the assumption of laminar flow is not violated, and this will be the case provided the particle-Reynolds number ($Re=2rv\rho/\eta$) does not exceed 0.5 (Walsby and Reynolds, 1980; Reynolds, 1987). As a consequence, the equation will not be reliable for large phytoplankton colonies where $r > 300\ \mu m$ (Reynolds, 1987).

The Stokes function explicitly identifies major features that influence the floating and sinking rate of phytoplankton. Velocity is greatly enhanced by a large size as it is related to the square of the particle radius. The direction of movement, as well as the velocity, is a function of the density difference between particle and surrounding medium, while the shape of the particle may enhance or retard its motion. The large size range of the cyanobacteria, coupled with their ability to alter density, is reflected in a wide range of floating and sinking velocities (Fig. 4). In contrast most freshwater eukaryotic micro-algae have a cell density greater than that of water and only the motile, flagellate species have any means of negating their propensity to sink. A notable exception to this is *Botryococcus* that can become buoyant after producing and accumulating oils.

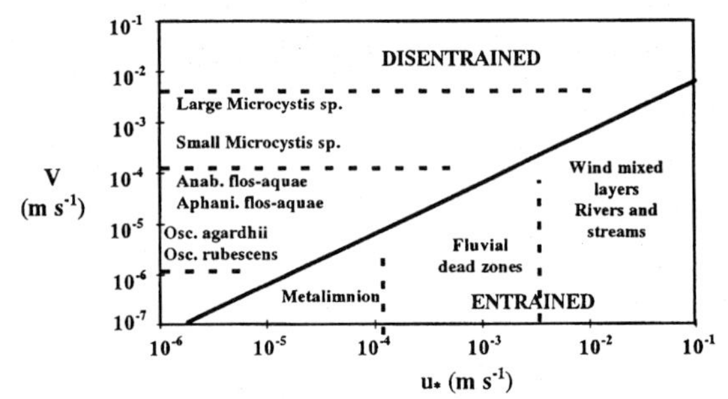

Fig. 4 A comparison of floating and sinking velocities (v) of selected cyanobacteria with the characteristic velocity of turbulence (u). The solid line separates regions of particle entrainment and disentrainment as defined by Eqn. 5. Values of u in different habitats provide perspective. (after Reynolds, 1994a).

B. Turbulence Intensity and the Mixed Layer Depth

Rarely will natural waters be so quiescent that the observed vertical rate of phytoplankton movement matches that expected from Equation 1. In open waters, where edge effects are minimal, vertical mixing is attributed to turbulence generated either by wind stress on the water surface, or to convective cooling due to fluctuations of temperature in the surface water (Spigel and Imberger 1987). Turbulent mixing in the water column acts to homogenise the vertical distribution of the phytoplankton by entraining slowly moving particles within the motion of the turbulence. If mixing is sufficiently intense it may negate advantages derived by populations using buoyancy regulation to make controlled vertical movements.

The characteristic velocity of turbulence can be envisaged either as the root mean square of the time-averaged vertical velocity fluctuations, $u = (\langle u'^2 \rangle)^{1/2}$ where angle brackets denote averaging, or as the rate of turbulent energy dissipation. Direct measurements of u' rely upon instruments of exceptionally high resolution such as laser or accoustic Doppler velocimeters and few direct measurements have been made in lakes. However, when wind speeds are moderate to low and the lake is not loosing heat the characteristic velocity (u) within the diurnal surface layer (Monismith et al. 1990) can be equated to the water friction velocity u_* estimated from the wind speed (Denman and Gargett 1983):

$$u_*^2 = \rho_a c U_{10}^2 / \rho \qquad [2]$$

where:

ρ_a is the air density (ca. 1.2 kg m^{-3}), ρ is the water density, c is the dimensionless drag coefficient (1.3 x 10^{-3}) and U_{10} is the wind speed at a height 10 m above the surface.

For typical densities, equation 2 simplifies to $u_* = 0.001U_{10}$. It must be emphasised that this estimate only holds when the lake is not losing heat and when wind speeds are low to moderate (<6 to 8 m s^{-1}). MacIntyre and her collaborators (pers. Com.), working with data from Lake Victoria (East Africa) and Lake Calado (Brazil), have shown that u will be 2 to 5 times higher than predicted by u, if a lake is losing heat or if windy conditions prevail.

The alternative approach is to take advantage of the relationship between the energy dissipation rate (ε) and u, $\varepsilon = u^3 / l$ where l is the scale of the overturning eddy (MacIntyre 1993). The rate of energy dissipation is routinely measured using temperature gradient or shear microstructure profilers, a technique more common in oceanography but gaining acceptance in limnology. The variability of turbulence in the upper mixed layer and thermocline over diurnal cycles is described in Brainerd and Gregg (1993), Imberger (1985), and MacIntyre (1993, 1996, 1998). Microstructure profiling allows discrimmination of the parts of the upper mixed layer that are mixing from those that are not and allows assessment of the variability of u as a function of depth.

The vertical penetration of mixing energy is resisted by the formation of a heated, buoyant surface layer (Spigel and Imberger, 1987; Reynolds et al., 1987; Reynolds 1989a, b, 1990). The depth of this layer is a function of the extent of solar energy penetration, heat capture and the degree of mixing. The depth of the mixed layer will tend to a point where the buoyant energy of the surface layer and the kinetic energy of the surface wind stress are balanced. The instantaneous balance between these two processes is described by the Wedderburn number (Imberger and Hamblin, 1982), which is the ratio of the energy per surface area from buoyancy to that generated by the shear forces;

$$w = gh^2 \Delta\rho_w / \rho u_*^2 L \qquad [3]$$

In this expression $\Delta\rho_W$ is the density difference between the water below the mixed layer and the water in the mixed layer, L is the length of the lake at the base of the mixed layer in the direction of the wind, and h is the depth of the diurnal thermocline (Spigel and Imberger, 1987).

When W>1 the structure is robust and resistant to further deepening unless there is a substantial change in $\Delta\rho_W$ or u_*, whereas if W<1 then the mixed layer deepens rapidly until h and $\Delta\rho_W$ are large enough to make W>1 (Spigel and Imberger, 1987; Reynolds, 1994a). Estimates of the depth of mixing (z_m) have been obtained from this function by setting W = 1 (Reynolds, 1989a); Ibelings et al., 1991a). More recently the suitability of equation 3 to predict the depth of the mixed layer has been questioned since it more correctly estimates whether or not the pycnocline will undergo full or partial upwelling and whether boundary mixing is likely. MacIntyre (pers. Com.) has suggested that the regions where the water column is mixing and the mixing intensity can be obtained more appropriately from modelling or from microstructure profiling.

The ratio between the depth to which 1% of light penetrates (euphotic depth, z_{eu}) and the depth of mixing is an important indicator of the light regime encountered by cells captured in the water motion. As the z_{eu}/z_m ratio declines below a value of one the proportion of time that the cells spend in the light decreases until eventually cells encounter insufficient light to grow.Consequently, in shallow turbid waters that are uniformly and continually mixed light may limit population growth as at any one time only a small proportion of the population can gain access to the euphotic zone. Conversely, in clear deep lakes the onset of thermal stratification increases the mean irradiance as the mixing depth decreases. If the z_{eu}/z_m ratio is greater than one in a thermally stratified lake then cells moving with the water are always in the

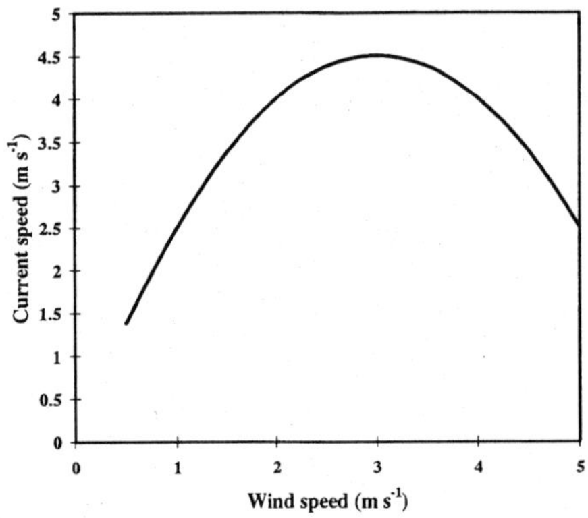

Fig. 5 Changes in the surface current speed (c_S) as a function of wind speed (U_a).

illuminated zone and light penetrates below the mixed layer into the thermocline.

C. Sinking Organisms

In considering the effect of turbulent mixing on the distribution of sinking organisms, Reynolds (1979) carried out field experiments measuring the settling of Lycopodium spores in large lake enclosures. The results demonstrated that the measured *effective* sinking rate of the spores under turbulent conditions was significantly less than the *intrinsic* sinking rate measured in quiescent water columns in the laboratory. Reynolds proposed a mixing model that envisaged a water column that was periodically fully mixed, but with a trapping zone at the bottom from which cells could not be resuspended. Smith (1982) developed a simple model to formalise these concepts and derived a general equation to calculate the number of particles remaining in suspension (N_t) after time t for conditions where organisms are fully entrained in the turbulence:

$$N_t = N_0 e^{-t/t_1} \qquad [4]$$

Here N_o is the original number of particles in suspension and t' is the column clearance time, which for particles with a sinking velocity V in a water column of height H is t' = H/V.

Under turbulent conditions an infinite time is required to completely clear the water column of particles and a small number of cells may remain in suspension to seed subsequent growth phases (Smith 1982). Replacing t' with H/V in Eqn. 4 provides a means for measuring the intrinsic sinking or floating velocity of particles in a fully turbulent system (Smith, 1982; Reynolds, 1984a). Hutchinson and Webster (1994) have described and tested an apparatus for making these measurements.

The influence of a reduction in sinking rate on the number of cells retained within a mixed column of 5 m depth is illustrated in Fig. 6 for sinking velocities of 0.0036, 0.072 and 0.167 m h^{-1}. After two days at the slowest sinking rate 96% of the cells still remain in the water column, while at the fastest sinking rate only 20% remain. This example illustrates that gas vacuoles can provide a substantial ecological advantage at the population level, even if they do not provide sufficient lift to make cells positively buoyant but simply reduce the sinking rate. This may provide an explanation for the presence of gas-vacuoles in *Oscillatoria redekei* that are restricted in occurrence to the ends of the cells and do not appear to confer positive buoyancy.

The loss of sedimenting particles due to the presence of trapping zones at the lower boundary of a mixed layer occurs at all intensities of mixing, and represents a continual loss to non-buoyant or non-motile phytoplankton populations. In natural waters, particles will be removed from the mixing zone if they sediment into the thermocline where turbulence

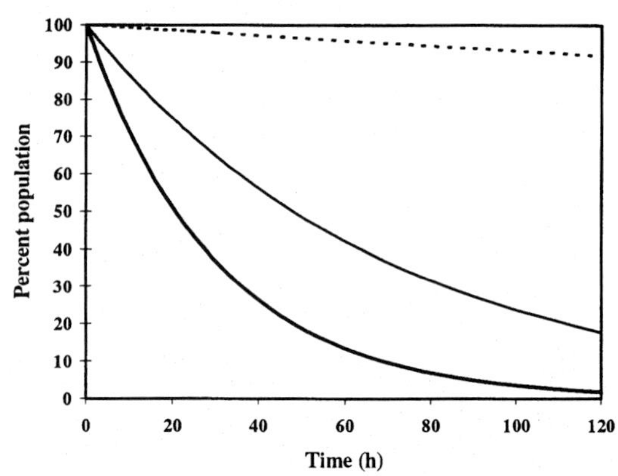

Fig. 6 Percent of the population retained within a 5 m, fully mixed water column. The three curves represent three sinking velocities; 0.0036 (upper), 0.072 (middle) and 0.167 m h^{-1} (bottom), over 120 h.

is small, or settle onto the substratum where water velocities approach zero. Buoyant cyanobacteria can avoid this loss entirely and so buoyancy alone, without the need for its precise regulation, can provide the cyanobacteria with a significant advantage over sedimenting phytoplankton. Furthermore, if a deep mixing episode moves part of the phytoplankton community to a depth below the level of the thermocline, then once stratification reforms buoyant cells can move back into the surface mixed layer while sedimenting cells cannot (Humphries and Lyne, 1988).

D. Entrainment and Vertical Distributions

Mixing that fully entrains a phytoplankton population will homogenise its vertical distribution, but the degree of entrainment varies depending on the intrinsic velocity of the phytoplankton relative to the velocity of the turbulent motion. A function describing this interaction, and used to distinguish the extent of entrainment (ψ), was proposed by Humphries and Imberger (1982);

$$\psi = 15|v| / u \qquad [5]$$

where:

$|v|$ is the modulus (ie. the directional sign has been removed) of the sinking or floating velocity of the phytoplankton unit and u is the velocity of the turbulent eddies that transport the cells through the mixed layer (Humphries and Lyne, 1988).

This formulation differs from some published versions where the equation has been incorrectly transcribed (Humphries and Imberger, 1982; Reynolds, 1994a).

The size of u can be estimated by the shear velocity of u_* (see previous section). When $\psi < 1$, then water velocities are sufficiently large relative to the intrinsic sinking or floating velocities of the phytoplankton to entrain the cells within the water motion so that turbulence homogenises the vertical distribution of the population as envisaged in Eqn. 4 (Humphries and Imberger, 1982; Humphries and Lyne, 1988). When $\psi > 1$, then the intrinsic velocity of the phytoplankton plays an increasing role in the population distribution. This begins to occur when the turbulent velocity falls below 15 times the phytoplankton velocity. In Fig. 5 the floating or sinking speed of selected organisms is compared with the mixing velocity using equation 5 to distinguish the degree of entrainment Estimated values of u_* in different habitats are indicated for comparison (Reynolds, 1994a). Single filaments of *Oscillatoria* with slow floating rates are entrained at all u values greater than those found in the metalimnion. In contrast large colonies of *Microcystis* can continue to utilise buoyancy to alter their vertical distribution within the mixing zone up to u values typical of reasonably well mixed layers. At u values in excess of 10^{-2} m s^{-1} all forms of cyanobacteria are entrained in the water motion and will have uniform vertical distributions.

Field data to test critically this relationship are scarce, but in Lake Vinkeveen in The Netherlands, Ibelings et al. (1991a) found 30% less floating colonies of the gas-vacuolate cyanobacterium *Microcystis* near the surface than at the bottom of the mixed layer, when the shear velocity u was only five times the maximum floating velocity of the *Microcystis*. Presumably the poor entrainment of colonies in the water motion had resulted in a separation between the top and bottom populations.

When $\psi > 1$ floating or sinking rate controls cell distribution, while if $\psi < 1$ diffusive transport becomes increasingly important in determining cell distribution (Reynolds et al., 1987). Buoyant colonies of the gas vacuolate cyanobacterium *Microcystis* sp. can have rapid floating rates (up to 250 m d^{-1}) compared to the generally slow sinking rates of many non-buoyant micro-algae (ca. 1 m d^{-1}). Simulation studies predicted the occurrence of top-heavy vertical profiles of buoyant *Microcystis* for which $\psi > 1$, while under the same conditions the diffusion dominated distribution of sinking cells ($\psi < 1$) resulted in an even distribution with depth (Humphries and Imberger 1982; Humphries and Lyne 1988). Assuming an equivalent growth response to changes in light intensity the floating populations had higher production rates than sinking populations since a larger proportion encountered increased irradiance in the upper layers. In addition the simulations showed that the depth integrated growth rate of floating species closely followed the changes in mixing depth, while sinking cells were dispersed by deep mixing episodes and suffered a significant population loss when the water column re-stratified. The model predicted that diurnal cycles of mixing and stratification would provide rapidly floating cells with a significant growth advantage. Walsby et al (1997) carried out a quantitative analysis of the benefit of flotation to the primary production of *Aphanizomenon flos-aquae* over a 9-day period in the Baltic Sea. The analysis demonstrated that, averaged over the alternating periods of calm and mixing, buoyancy provided by gas vacuoles increased the daily net areal photosynthesis of the cyanobacterial population by nearly two-fold.

Currents generated by wind mixing not only influence the vertical distribution of phytoplankton, but also advect particles from one location to another causing horizontal and vertical heterogeneity of algal populations on both a local and basin scale (George and Edwards, 1976; Reynolds, 1984a; Stauffer 1988). The influence of wind-induced currents on particle distributions is a function of particle buoyancy and water velocity. Neutrally buoyant particles will be randomly distributed, while positively buoyant particles tend to become concentrated in regions of downwelling water, and negatively buoyant particles tend to concentrate in upwelling regions (Stommel, 1949; George and Edwards, 1976; Webster, 1990). This horizontal heterogeneity has been described empirically by George and Edwards (1976) for Eglwys Nynydd, a shallow eutrophic lake in Wales, and modelled by Webster (1990) and Webster and Hutchinson (1994). The model describes the dependence of the horizontal and vertical distribution of particles on their floating or sinking rate, the wind speed, and the depth and fetch of the lake. This heterogeneity can be a major problem to monitoring programs aimed at describing seasonal population changes in phytoplankton and quantifying biomass dependent measurements such as phytoplankton production (Horne and Commins, 1989).

E. Surface Accumulations and Turbulence

The concepts describing particle sedimentation losses from mixed layers into low turbulence trapping-zones can also apply to floating particles provided there is a surface "trapping" layer of reduced vertical dissipation, such that buoyant cells are not re-entrained into underlying turbulent layers (Webster and Hutchinson, 1994). It has long been recognised that wind speeds of ca. 3 m s^{-1} mark a significant transition in the near-surface mixing regime that can be observed as a change in surface roughness (George and Edwards, 1976), the formation of waves (Reynolds, 1989a, b; Webster and Hutchinson, 1994), and the development of Langmuir circulations (Reynolds 1987).

George and Edwards (1976) observed in Eglwys Nynydd that at high wind velocities mixing was sufficient to suppress the development of vertical patchiness in cyanobacterial populations and they were homogeneously distributed throughout the water column. Local surface concentrations of cyanobacteria only appeared at wind speeds below 4 m s^{-1}, and increased in occurrence as wind velocity declined. Webster and Hutchinson (1994) modelled mathematically the distribution of phytoplankton under different wind stress conditions, and estimated that a wind speed of >2-3 m s^{-1} was required to mix floating phytoplankton cells away from the water surface. They calculated that the surface viscous boundary layer at a wind speed of 2 m s^{-1} was 4 mm, sufficient to contain large colonies of buoyant cyanobacteria.

Although wind speeds less than ca. 3 m s^{-1} can lead to trapping of buoyant cells within the surface layer, accumulations will not necessarily become obvious. Apart from the capacity of the cells to increase their density and sink out of the surface layer, they can also be dispersed horizontally. At wind speeds below ca. 3 m s^{-1} the surface is hydrodynamically smooth and the vertical turbulent transfer of momentum is weak, so a larger proportion of the wind stress is converted to horizontal surface water velocities. Above ca. 3 m s^{-1} the vertical transfer of turbulence into the water column increases and surface current speeds are a smaller fraction of the wind speed (George and Edwards, 1976). As a result a wind speed of 2 m s^{-1} generates a surface current of 6 cm s^{-1}, while a wind speed of 5 m s^{-1} generates a surface current of only 3 cm s^{-1} (Webster and Hutchinson, 1994). The ratio of the surface current speed to the wind speed increases approximately linearly as wind speeds decline from 5 m s^{-1} to 0.5 m s^{-1} and this relationship was described empirically by George and Edwards (1976). Recasting their equation gives:

$$c_s = 3.002U_a - 0.5U_a^2$$

For

$$0.5 < U_a < 5$$

where:

c_S is the surface current speed (cm s^{-1}) and U_a is the wind speed (m s^{-1}).

The current speeds given by this empirical equation are similar to those calculated by Webster and Hutchinson (1994). Within its range the equation indicates that a maximum surface current speed of 4.5 cm s^{-1} (162 m h^{-1}) occurs at a wind speed of 3 m s^{-1}, with surface current velocities decreasing either side of this peak (Fig. 4). However, even a wind speed of 1 m s^{-1} generates a surface current of 2.5 cm s^{-1} (90 m h^{-1}) so that surface accumulations are quickly transported horizontally. Depending on lake size and the time required for buoyancy regulation, organisms may be transported to shallow waters before they can sink out of the surface layer. Consequently surface blooms are frequently seen at the down wind edges of lakes, and only under very calm conditions will they be widely spread over the lake surface (Plate 11b). Accumulations at the lake edge (Plate 11a) create the highest risk for livestock poisoning and wind speeds of 0.5 to 3 m s^{-1} are critical to their formation.

F. Surface Blooms

The gas-vacuolate cyanobacteria that commonly produce surface blooms are those species that form large biomass units, either colonies as in *Microcystis*, or aggregates of filaments as in *Anabaena* and *Aphanizomenon*. These units are often more than 0.2 mm across and at times greater than 1 mm, so that their sinking or floating velocities are enhanced, and under calm conditions intense surface concentrations appear rapidly. Short-term, periodic surface blooms can occur as a result of responses to daily meteorological events or cyclical changes in cell density. For example, under calm conditions surface blooms frequently occur in the early morning as the respiratory demands during the hours of darkness consume the carbohydrate which acts as ballast against the upward lift provided by the gas vesicles. As the colonies encounter favourable light conditions, carbohydrate is accumulated and the units steadily gain weight until the gas vesicles can no longer provide the lift required for positive buoyancy and the colonies descend out of the surface scum. This explains why blooms sometimes tend to "disappear" in the afternoon and reappear in the morning as observed by Ganf (1974) in Lake George, Walsby and McAllister (1987) in Lake Okaro and Walsby et al (1983) in Lake Gjersjøen. Under certain conditions the continuous exchange of colonies at the water surface may give the mistaken impression of a persistent bloom, a scenario described in the computer model of Kromkamp and Walsby (1990). However, on occasions surface blooms do persist for several weeks and in some cases, such as the infamous hyperscums of Hartbeespoort Dam in South Africa, they can last for many months (Robarts and Zohary 1984; Zohary and Robarts 1989). In between these extremes surface blooms occur at various intervals depending on the interplay between buoyancy regulation and environmental conditions.

The occurrence of persistent surface blooms has been variously interpreted as a failure of buoyancy regulation due to physiological damage and senescence (Walsby 1994), a mechanism by which cyanobacteria can dominate surface waters (Paerl and Ustach, 1982; Paerl 1988a, b; Ganf et al. 1989), or the result of physical obstruction where the sinking colonies that have lost their buoyancy in the surface layers are impeded by the presence of underlying buoyant colonies (Ibelings and Mur 1992, Walsby 1994). This physical restraint on vertical movement has been demonstrated by Ibelings and Mur (1992) in very dense surface scums of *Microcystis* where surface colonies in the upper most millimetres lost buoyancy but buoyant colonies below acted as a barrier to their downward movement.

Surface blooms frequently occur in calm conditions following a period of deep mixing when the cells have become highly buoyant as a result of the low mean irradiance (Reynolds 1984a). Because of their excess buoyancy it takes longer to negate the lift provided by the gas vesicles and the cells may spend sufficient time at the surface to be physiologically damaged by the high light intensities. Such damage includes photoinhibition, photo-oxidation (Abeliovich and Shilo 1972) and dehydration (Zohary and Pais-Madeira 1990) and leads to cell senescence. This reduces population growth and photosynthesis (Ibelings and Mur 1992) and disables buoyancy regulation. The reduced ability to regulate buoyancy at the surface may also be related to inorganic carbon limitation in the thick scums. In this model surface blooms are due to an inability of organisms to adapt their buoyancy regulation to a sudden change in hydrological conditions.

The formation of surface blooms has also been attributed to inorganic carbon limitation. Klemer et al. (1996) using continuous cultures of *Microcystis aeruginosa* showed that CO_2 limitation can promote buoyancy in the short term by preventing both gas vesicle collapse and the accumulation of carbohydrate ballast. This supports the results of Walsby and Booker (1980), who found that carbon limitation caused *Anabaena flos-aquae* to float up and form a surface bloom in a stratified laboratory water column. An extension of this scenario is that cells in the centre of large colonies may be deprived of an adequate supply of inorganic carbon (Paerl, 1983) providing sufficient gas vesicle lift to overwhelm the ballast accumulated by cells on the periphery of the colony. Once carbon-limited cells are at the surface, physiological damage as a result of the high light intensities could then disable buoyancy regulation resulting in a persistent surface bloom.

Paerl and Ustach (1982) proposed that surface blooms were not just a consequence of poorly acclimated cells suffering from excess buoyancy, but were part of an ecological strategy aimed at making optimal use of photosynthetically active radiation and atmospheric carbon dioxide. From investigations on *Aphanizomenon flos-aquae* and *Anabaena oscillarioides* they concluded that CO_2 was the preferred inorganic carbon source for sustaining high rates of photosynthesis and that the formation of surface blooms would provide access to the preferred form at the air-water interface. In poorly buffered waters, where pH values are frequently in excess of 9 indicating diminishing supplies of CO_2 due to sustained periods of intense photosynthesis, this would provide an ecological advantage over non-buoyant species. Paerl et al. (1985) showed that natural populations of *Microcystis* had optimal photosynthetic rates and resistance to photoinhibition at surface irradiances. However, Ibelings and Mur (1992) working at a very fine scales (0 - 3000 μm) suggest that the case is not yet proven and provided evidence to show that photosynthesis in *Microcystis* scums can be inhibited at the surface.

Surface blooms markedly influence the depth to which light penetrates the water column, largely as a result of scattering by gas vesicles (Walsby 1994). Scattering from the vertical increases the path length of photons through horizontal water layers thereby enhancing the probability of absorption within the near surface layers. Those organisms which have taken up preferential residence within the euphotic zone, such as buoyant cyanobacteria, therefore have a greater probability of intercepting the light. For example, Ganf et al (1989) showed that 80% of the light scattered in Mt. Bold Reservoir was due to *Microcystis aeruginosa* colonies and concluded that this provided *Microcystis* with a distinct advantage when in competition with non-buoyant species.

G. Metalimnetic Populations

Some species of gas-vacuolate cyanobacteria (eg *Oscillatoria* spp.) form deep water concentrations often at or near the thermocline. The size of the biomass unit forming these metalimnetic populations is small so that even though they regulate their buoyancy the rate at which they move is slow and the amplitude of the daily vertical movement is often restricted to a few centimetres. A general requirement for the formation of metalimnetic populations is that the depth of the illuminated zone exceeds the depth to which the water column is mixed by wind or convective cooling ($z_{eu} > z_m$). It has been proposed that the cyanobacteria stratify at a depth where the opposing gradients of light and nutrient availability result in neutrally buoyant cells (Klemer, 1976, 1978; Konopka, 1984, 1989). Species that form metalimnetic populations are low-light adapted and sensitive to variations in light intensity. Konopka et al (1993) found that at intensities of > 15 μmol m^{-2} s^{-1} buoyant filaments of *O. agardhii* in Deming Lake, Minnesota lost buoyancy within a few hours. Buoyancy also responds to nutrient conditions. If nutrient limitation becomes more severe then the depth at which the cells stratify increases because the light level provides an excess of energy for the reduced growth rate (Fig. 3) resulting in an increased carbohydrate accumulation and loss of buoyancy (Klemer 1976, 1978; Konopka 1984). Conversely, in metalimnetic populations artificially enriched with either nitrogen or phosphorus, the increased availability of a limiting nutrient caused the population to stratify at a shallower depth (Klemer 1976, 1978; Konopka 1989).

Metalimnetic populations do sometimes form into colonies by aggregation of filaments and this increases their sinking or floating velocities enabling them to perform vertical migrations comparable with the colonial species. This was observed for *Oscillatoria agardhii* in Lake Gjersjøen (Walsby et al. 1983). The variable responsible for inducing the filament aggregation has not been identified.

H. Major Habitats Structured by Turbulent Mixing

Reynolds and Walsby (1975) suggested that different forms of gas-vacuolate cyanobacteria are adapted to occupy water masses with different mixing regimes. Reynolds et al (1987) reviewed the occurrence of these cyanobacteria and suggested that interactions between the size of the morphological unit, its shape and its density were selected for by habitats with the hydraulic conditions which optimised for these features. Subsequently, evidence supporting this notion has accumulated (Steinberg & Hartmann, 1988) permitting some further refinement of the concept (Walsby et al., 1989).

On the basis of the z_{eu}/z_m ratio water bodies can be divided into those where $z_{eu} \leq z_m$ and those where $z_{eu} > z_m$.

A. In waters where the euphotic depth is approximately equal to, or less than, the mixing depth but the z_{eu}/z_m ratio is not so small that light is insufficient to support growth, two principal patterns of mixing support different cyanobacterial populations:

1(a). Shallow, productive waters that are frequently fully mixed support populations of non-aggregated, filamentous cyanobacteria (Plate 11d) such as

Oscillatoria spp. (Gibson et al., 1988; Scheffer et al., 1997).

1(b). An analogous mixing regime occurs in deeper lakes that are thermally stratified and have an intense seasonal thermocline, but where the surface layer is consistently well mixed. These too are frequently inhabited by cyanobacteria consisting of small solitary filaments, a unique example being *Anabaena minutissima* in Lake Rotongaio, New Zealand (Walsby et al., 1989).

2(a). In shallow, productive waters, where thermal stratification of the water column occurs diurnally or occasionally persists for longer periods, the larger morphological forms of cyanobacteria occur, including colonial forms such as *Microcystis aeruginosa* (Ganf, 1974; Ibelings et al., 1991b), and aggregated or contorted filamentous forms such as *Aphanizomenon* and *Anabaena* (Plate 11f).

2(b). An analogous mixing regime occurs in those deeper lakes where intense thermal stratification of the upper mixed layer is common, often occurring on a diurnal or even longer time period, and where mixing episodes are contained within the depth of the seasonal thermocline. These environments also contain the larger morphological forms of cyanobacteria.

B. In less productive lakes where the euphotic depth exceeds the extent of the surface mixed layer ($z_{eu} > z_m$) and a strong seasonal thermocline develops within the euphotic depth, a stable, illuminated zone is formed below the mixed layer. Under these conditions metalimnetic populations of cyanobacteria can develop that are concentrated over a narrow depth range. Usually these are comprised of solitary filamentous forms eg. species of *Oscillatoria*, *Anabaena* or *Aphanizomenon* (Edmondson, 1970; Klemer, 1976; Walsby et al., 1983; Konopka et al., 1993), but metalimnetic populations of other forms including *Microcystis*, have been recorded (Ward and Wetzel, 1980).

These patterns suggest that the smaller forms of cyanobacteria occur in conditions where the importance of buoyancy regulation is reduced, either in situations where buoyancy regulation is overwhelmed by mixing or when the organisms are holding station below the depth of the mixed layer. In contrast, when the duration of thermal stratification is prolonged and the utility of buoyancy regulation is enhanced, then the larger forms of cyanobacteria become more prevalent. In essence morphology reflects the advantage provided by buoyancy regulation and this is dependent on the mixing regime.

The genera of gas-vacuolate cyanobacteria do not separate out strictly on the basis of these three habitat categories as variations in form and function occur within and between taxa. In particular some of the filamentous species, such as *Oscillatoria agardhii* (Walsby et al., 1983) and *Aphanizomenon flos-aquae* (Lynch, 1980; Ganf, 1983), can occur either as single filaments in metalimnetic populations or as large aggregates in epilimnetic populations (Walsby et al., 1983). Even within a morphological type there is a mix of species with quite different physiological characteristics. For example, some filamentous forms can fix molecular nitrogen, while others cannot. Differences between the three species *Oscillatoria rubescens*, *O. agardhii* and *O. redekei* appear to separate their occurrence between mixing regimes. *O. rubescens* occurs more commonly as metalimnetic populations, while *O. agardhii* can form metalimnetic populations (Edmondson 1970, Klemer 1976, Konopka et al. 1993), but also occurs in well mixed layers (Brook et al. 1971, Gibson et al. 1988). *O. redekei* appears to be restricted to shallow, unstratified lakes, probably because it has too few gas vesicles for positive buoyancy (Whitton and Peat 1969, Meffert and Krambeck 1977). *O. agardhii* and *O. redekei* may co-occur as in the shallow, completely mixed Lough Neagh (Gibson et al. 1988).

VI. Physical Control of Cyanobacteria

The interaction between mixing processes, buoyancy regulation and species composition, suggests that lake restoration via manipulation of the turbulent environment is a viable lake management strategy.

A. Collapsing Gas Vesicles in Lakes

If cyanobacterial cells are artificially or naturally circulated to a sufficient depth then hydrostatic pressure will cause gas vesicle collapse. A cell circulating to 40 m would experience a hydrostatic pressure of 0.4 MPa which could be sufficient to collapse a significant proportion of the gas vacuoles if cells were highly turgid. Cells with lower turgor would require a proportionately greater depth of mixing. If a lake is continuously mixed the cells without gas vacuoles will be entrained in the motion and the effective sinking rate will be reduced. The time to eliminate particles from a mixed column is ca. 4.6 times longer than predicted from still water settling (Reynolds 1984). If mixing is artificially induced then intermittent circulation maybe more appropriate than continuous circulation (Steinberg and Gruhl 1992; Visser et al. 1996). This would allow non-buoyant cells to sediment more rapidly below the zone of circulation and into darkness where the synthesis of vesicles is much slower being dependent upon stored energy reserves. Deacon and Walsby (1990) found that at an optimal irradiance of 30 μmol m^{-2} s^{-1} it took *Microcystis* three days before buoyancy recovered. Intermittent circulation would also provide an opportunity for buoyant cells to move towards the illuminated surface layers where turgor pressure is likely to increase prior to the next mixing event. Kinsman et al. (1991) found that turgor pressure increased from 0.3 to 0.54 in *Anabaena flos-aquae* after 16 h at a photon irradiance of 135 μmol m^{-2} s^{-1}.

An alternative strategy to lake mixing and adopted by the former Suffolk Water Board in England is to abstract water from the lake and circulate it through a 86 m deep pipe where the hydrostatic pressure at the bottom is sufficient to collapse all gas vacuoles in *Microcystis aeruginosa* (Clarke and Walsby 1988). Alternatively Walsby (1992) noted that Menday and Buck (1972) had successfully used explosions to create shockwaves to collapse gas vacuoles, but the side-effects of this treatment were not always acceptable, particularly due to their effect on other organisms.

B. Artificial Mixing and Cyanobacterial Growth

Steinberg and Hartmann (1988) concluded from an analysis of a number of water bodies that above a threshold total-phosphorus concentration of 10 mg m^{-3} the occurrence of cyanobacteria was dependent largely on physical factors, particularly the degree of water column stability. On the basis of their investigations they supported the widely held view that physical manipulation of the mixing regime is likely to be a quicker and more effective means of managing cyanobacterial populations than either nutrient reductions or biomanipulation. In particular manipulation of the turbulent environment in an appropriate manner should promote a species shift away from cyanobacteria towards other species.

A number of papers have reported a reduction in the growth of *Microcystis aeruginosa* in response to mixing (Toetz, 1981; Reynolds et al., 1984; Visser et al., 1996). Lake Nieuwe Meer is hypertrophic and dominated by *Microcystis* during the summer. To overcome this problem artificial mixing was installed (1992 & 1993) and the results compared with the previous two years (1990 and 1991) without artificial mixing (Visser et al., 1996). The concentration of chlorophyll fell from an average of 23.3 to 8.9 mg m^{-3}, but this was principally due to a dilution effect as the mixed depth increased. When the results were expressed as mg m^{-2} the artificially mixed years had an average chlorophyll of 208 mg m^{-2} compared with 110 mg m^{-2}. The increased overall biomass was accompanied by a major shift in species composition. In the two years without artificial mixing cyanobacteria dominated from July to September but with the introduction of mixing *Scenedesmus*, centric diatoms and flagellates dominated.

In 1993 artificial mixing had been continuous from March to September, whereas during 1994, to reduce energy costs, mixing was applied intermittently and controlled by the surface and bottom temperatures and oxygen concentrations. If greater than 2°C or 2 g m^{-3}, respectively, the compressor was switched on, and if the difference was <1°C or 1 g m^{-3}, it was switched off. This gave an annual saving of 27%. *Microcystis* failed to develop in either of the years when artificial mixing took place and *Anabaena* numbers per m^{-2} decreased but the numbers of *Aphanizomenon* and to a less extent *Aphanocapsa* were greater. This observation suggests that uncritical adoption of artificial destratification to manage cyanobacteria is unwise, although it can cause major shifts in species composition (Steinberg and Zimmermann 1988). In a review of the effectiveness of artificial destratification in 52 Australian reservoirs McAuliffe and Rosich (1989) found that it failed to satisfactorily control phytoplankton in over 60% of cases.

Steinberg and Gruhl (1992) reported on a series of experiments investigating physical measures to inhibit planktonic cyanobacteria in Fischkaltersee a small Bavarian lake (3.4 ha and 5.7 m mean depth). Permanent destratification initially resulted in positive responses with almost complete removal of the cyanobacteria but nearly a doubling in the biomass of chlorophytes and diatoms. However, *Limnothrix* reappeared in the third year and reached a peak biomass in the fourth year of the treatment that was higher than had occurred prior to destratification being used. It was concluded that although the continuous mixing produced relatively constant mixing patterns with a low mean irradiance the lake was not sufficiently deep to reduce the mean irradiance below that required by the Oscillatoriaceae. Intermittent destratification was instituted on the basis that it should elevate biomass losses via sedimentation and improve the light climate as a consequence of diminished phytoplankton biomass and this would curtail the low light adapted cyanobacteria. Under this regime *Limnothrix* peaked only for a few days and was outcompeted by the chlorophytes and diatoms as had occurred initially with continuous destratification. In contrast to permanent mixing the cyanobacteria did not reappear and the biomass of algae remained low.

C. Mixing in Regulated Rivers

Manipulation of mixing regimes is not confined to still water bodies but can also be used in regulated rivers. Burch et al (1994) and Webster et al (1996) have shown that the occurrence of *Anabaena circinalis* is a function of discharge rates in regulated rivers. At Maude Weir on the Murrumbidgee River, Australia, flows below 500 to 1000 M L d^{-1} caused a decline in the population density of the diatom *Aulacoseira* (*Melosira*) *granulata*, while *Anabaena circinalis* increased (Sherman et al.,1998). This shift in species composition was explained by changes in thermal stratification and mixing. Discharges < 1000 M L d^{-1} permitted the development of persistent stratification (6°C over 5m), and the cooler river water at the upstream end of the pool plunged beneath the stratified epilimnion discharging via the underflow weir. These conditions resulted in epilimnetic water velocities close to zero providing a retention time sufficient to permit *A. circinalis* growing at a net rate of 0.37 d^{-1} to develop a significant biomass over a period of one to two weeks. With the persistent stratification the buoyant *Anabaena* accumulated in the top 1-2 m, while negatively buoyant *Aulacoseira* sank and accumulated in the bottom waters. As the river has a euphotic zone of only 1.5 m and a maximum depth of 6 m ($z_{eu}/z_m = 0.25$), this provided a distinct advantage to the cyanobacterium.

The emergence of *Anabaena* as the dominant genus was attributed to nitrogen limitation which had been demonstrated in the *Aulacoseira* population prior to its demise using a physiological assay based on chlorophyll-a fluorescence (Wood and Oliver 1995). It was suggested that nitrogen limitation provided conditions suitable for *Anabaena* to dominate rather than a non-nitrogen fixing cyanobacterium. During a similar low flow period the *A. circinalis* population reached a plateau as the total reactive and filtrable reactive phosphorus fell below <10 mg m^{-3} suggesting that phosphorus availability may have controlled the population maximum (Webster et al. 1996).

Burch et al (1994) in the South Australian section of the River Murray demonstrated that populations of *A. circinalis* occurred in the river when discharge was small enough to permit thermal stratification. More importantly the drop in river levels (due to irrigation demand at low flows) allowed water from the neighbouring wetlands (billabongs) to flow back into the main channel carrying with it significant numbers of *A. circinalis* (P. Baker, pers. comm.). In this system it appears as though the main inoculum of *A. circinalis* comes from the calm, adjacent wetlands.

VII. Cell Size, Growth Rate and Temperature

In searching for clues to explain the occurrence and abundance of gas-vacuolate cyanobacteria efforts have been made to identify adaptive advantages that they might have over competing eukaryotic micro-algae. This has led to a number of generalisations which more recent studies now suggest are inaccurate or inappropriately applied.

The generalisation that cyanobacteria are smaller than their eukaryotic counterparts is not supported by comparative measurements of planktonic units (cells, filaments or colonies) since they fall within the same size range (Foy 1980; Reynolds 1989a, 1993). This is an important similarity as the rates of many cellular functions are strongly governed by the surface area to volume ratio.

The maximum growth rates of cyanobacteria and micro-algae, standardised for their surface area:volume ratio and measured in cultures at 20°C are similar (Reynolds 1989a, 1993), ranging from ca. 0.4 to 2.0 d^{-1}. The changes in specific growth rate between 10 and 20°C for light and nutrient saturated cultures of cyanobacteria and micro-algae (q_{10} values) are also similar (Reynolds, 1989a), although for any given surface area to volume ratio the q_{10} for cyanobacteria was on average lower than that of green algae, but overlapped that of diatoms. A marked exception to this was *Microcystis* which had a large q_{10} value of ca. 9, and also a higher minimum temperature for growth (Thomas and Walsby, 1986; Robarts and Zohary, 1987).

The belief that cyanobacteria prefer higher temperatures is based mainly on field studies and seasonal correlations, but these may be spurious as high temperatures are also associated with thermal stratification and changes in turbulent mixing that provide an environment conducive to the gas-vacuolate cyanobacteria. Foy et al. (1976) concluded that the temperature optima of cultures of *Anabaena flos-aquae*, *Aphanizomenon flos-aquae* fo. *gracile*, *Oscillatoria agardhii* and *Oscillatoria redekei* were similar to those of other planktonic autotrophs.

It is not surprising that there are similarities between cyanobacteria and micro-algae, as both groups are adapted to planktonic growth. However, within each group, individual species have developed attributes suitable for particular environmental niches and the task is to distinguish the significant adaptations that differentiate between competitors. This necessarily requires an understanding of the environmental conditions pertinent to the growth of the organisms. The observation that cyanobacteria can at times dominate waters to the virtual exclusion of micro-algae suggests that the search for characteristics of the cyanobacteria that set them apart is still a valid pursuit.

VIII. Light Capture

In addition to its crucial role in controlling buoyancy regulation in cyanobacteria, direct competition for light energy will also influence their likely success. Cyanobacteria can be distinguished from all other eubacteria by their ability to carry out oxygenic photosynthesis in a manner closely resembling that of eukaryotic photoautotrophs. As in the eukaryotic micro-algae, cyanobacteria have two connected photosystems that provide energy and reductant largely for fixation of CO_2 by the Calvin cycle. Although the chlorophyll-a containing reaction centres of photosystem I (PSI) and photosystem II (PSII) are similar in cyanobacteria and micro-algae, the major antennae, or light harvesting complexes (LHC), that capture the incident PAR are quite different (Ormerod, 1992; Grossman et al., 1995). In the micro-algae the antenna is integral to the thylakoid membrane and comprised largely of accessory chlorophylls (b and c), whereas the major LHC in the cyanobacteria is the phycobilisome (PBS), a hemispherical structure attached to the periphery of the thylakoid membranes. The PBS has markedly different attributes compared to the antennae of micro-algae (Grossman et al., 1995).

The blue-green colour typical of many freshwater cyanobacteria is due to the presence of pigments called phycobilins. These pigments are associated with proteins and arranged in the phycobilisome in a distinctive order (Glazer, 1982; Ormerod, 1992; Grossman et al., 1995). The three major phycobilins, and their absorption maxima, are allophycocyanin (AP, A_{max}= 650 nm), phycocyanin (PC, A_{max}= 620 nm) and phycoerythrin (PE, A_{max}= 565 nm). All cyanobacterial phycobilisomes contain allophycocyanin and phycocyanin. In some cyanobacteria, especially the red coloured phycoerythrin containing species, the proportions of these pigments can be altered to increase the absorption of light of specific wavelengths (Tandeau de Marsac and Houmard, 1993).

The phycobiliproteins absorb PAR over a much wider range of wavelengths than the antennae of the micro-algae (Glazer et al., 1994), particularly in the region between the absorption bands of the accessory chlorophylls b and c, and the carotenoids (Fig. 7). This is a fundamental difference between cyanobacteria and the eukaryotic micro-algae and is likely to be of distinct advantage in situations where either the spectral quality of the underwater light is

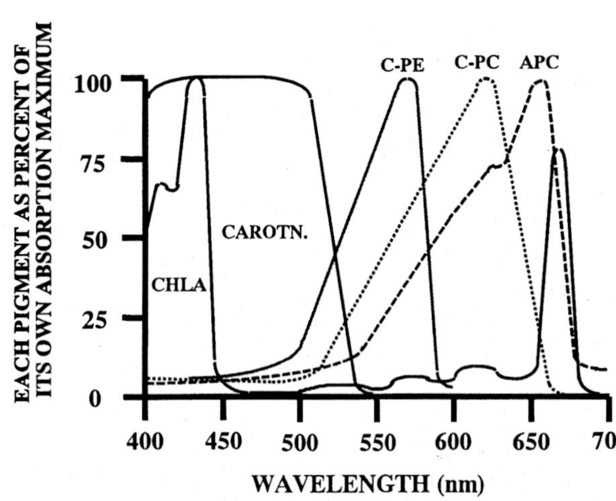

Fig. 7 Absorption spectra of chlorophyll-a (chla), carotenoids (carotn.), and the phycobiliproteins, c-phycoerythrin (C-PE), c-phycocyanin (C-PC) and allophycocyanin (APC) compiled from a number of sources. The absorption spectrum for each pigment is shown relative to its absorption peak.

concentrated in these wavebands, or when there are substantial fluctuations in light quality over time.

The underwater light climate changes both in quantity and in quality with depth. The intensity decreases exponentially as a result of absorption and scattering by particles and coloured compounds, and the selective removal of wavelengths causes shifts in the spectral distribution (Kirk, 1983; Oliver and Ganf, 1988; Oliver, 1990; Kirk and Oliver, 1995). Water absorbs strongly in the red so that in marine systems and very clear inland waters where the majority of light attenuation is due to water, the irradiance becomes dominated at depth by shorter wavelengths. In contrast, many inland and coastal waters contain dissolved organic compounds and suspended inorganic particles that absorb strongly in the blue, causing a shift towards longer wavelengths with depth (Fig. 8). Phytoplankton also modify the spectral distribution of light. For example, high concentrations of green algae leave an orange-green window of irradiance where their pigments absorb poorly (Kirk, 1983). These wavelengths are suitable for absorption by the phycobiliproteins and as a result microalgae may modify the spectral distribution of light at depth to the advantage of cyanobacteria. The converse argument is not so forceful. Cyanobacteria will not modify the spectral distribution of light to the distinct advantage of micro-algae because they absorb over a wide range of wavelengths including those utilised by the major chlorophyll pigments of the micro-algae.

As a result of these depth-dependent variations in irradiance, phytoplankton cells that move vertically, either as a result of turbulent mixing or through buoyancy regulation or motility, will encounter changes in the intensity and spectral distribution of photosynthetically active radiation (PAR). The vertical extent, and duration of the movement will also affect the periodicity of the PAR supply. If cells remain within the euphotic zone during mixing (ie. z_{eu} • z_m), then the periodicity of the irradiance supply is determined by day length, whereas if mixing is sufficient to move the cells below the euphotic depth ($z_{eu} < z_m$), then shorter light-dark cycles will be superimposed on a daily light cycle.

The problem of quantifying the light field experienced by individual calls is complicated by the depth dependent nature of both the light field and the characteristic velocity of turbulence (u) as well as the intrinsic sinking or floating velocities of cyanobacteria. Although not always appreciated, the upper mixed layer often does not circulate fully. On calm days, only the phytoplankton near the very surface will circulate due to slight wind mixing and will experience fluctuating irradiances (MacIntyre 1996, 1998). MacIntyre (1993) showed that in a shallow, turbid, productive lake in Western Australia the part of the water column that was mixing could be sub-divided into two zones each with different characteristics. An upper layer 0.3 to 1.5 m deep was activley mixing and a deeper layer where turbulence

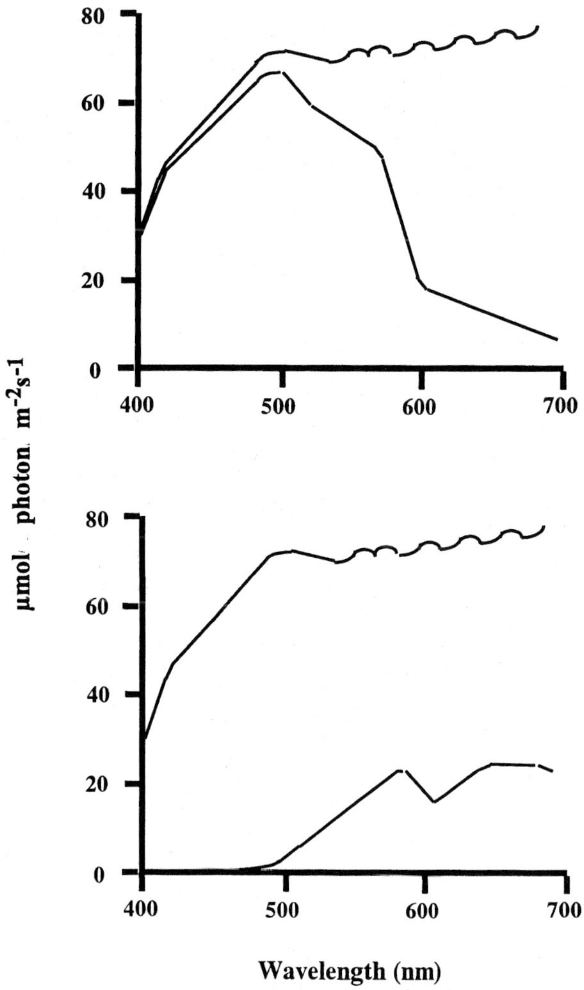

Fig. 8 Spectral distribution of quanta calculated for clear natural water at the surface and at 5m depth (top figure) and for organically coloured water from Mt Bold Reservoir, South Australia, at the surface and at 2m depth (bottom figure).

was constrained by buoyancy. In the upper layer *Microcystis* could experience a light gradient from 90 to 5% of surface irradiance in 3 to 4 min. Whether or not cyanobacteria adapt to rapidly fluctuating irradiances will depend upon the ratio of the time scale for photoadaptation to that of turbulent mixing (Lewis et al., 1984a,b).

To maintain their light-harvesting efficiency, phytoplankton have developed mechanisms for adjusting to alterations in the intensity, spectral distribution and periodicity of the PAR supply (Falkowski and LaRoche, 1991). Two major strategies are employed for adjusting to irradiance intensity. The first involves alterations in the size of the light harvesting antennae that serve the photosystems, and the second is a change in the total number of photosynthetic units (Falkowski and LaRoche, 1991). If photosynthesis becomes limited by the rate of delivery of light energy to the photosystems, as under low irradiance, then an increase in antenna size provides one means of increasing the photon supply. If the supply of photons from the antenna approaches the maximum turnover rate of the photosystem, then an increase in the number of photosynthetic units will increase the total supply of energy to the cell for photosynthesis and growth (Falkowski and LaRoche, 1991). This adaptation could be particularly important if cells were regularly exposed to saturating light intensities for short periods of time, for example, if mixed rapidly between the surface and aphotic zone, as the increase in the number of photosynthetic units would help maximise the capture of energy during the brief light period (Tilzer, 1987). When cells are exposed to high irradiance intensities for extended periods their is an increased risk of photodamage to the photosystems. This risk can be curtailed by a reduction in antenna size, and further, if the rate of energy capture exceeds the capacity of the cell machinery then the number of photosynthetic units can be reduced. Changes in antenna size and/or changes in the number of photosynthetic units are photo-acclimation strategies employed by both cyanobacteria and micro-algae under various conditions.

It might be expected from comparisons of antenna structure and spectral absorption characteristics that the cyanobacteria would respond quite differently to changes in PAR compared to the micro-algae. To date there is little experimental data to assess this expectation as the majority of laboratory studies have used white light and investigated responses to changing irradiance intensity. The different characteristics of the photosynthetic units suggests that increased effort should be given to describing the effects of spectral changes on photosynthesis and growth. Perhaps an indication of the importance of this is the common observation that cyanobacteria are frequently associated with organically-rich waters. Although often interpreted as indicating a role for (photo)heterotrophy, it might also reflect a more conducive spectral distribution of PAR for cyanobacteria.

A. Light Intensity

Cyanobacteria show an inverse correlation between pigment content and irradiance intensity, a response commonly found in phototrophic organisms (Konopka and Schnur, 1980; Foy and Gibson, 1982; Richardson et al., 1983, Wyman and Fay, 1986a, Tandeau de Marsac and Houmard, 1993). This response occurs in cyanobacteria whether they are growing under a short light:dark cycle (L:D) (Foy, 1993), a long L:D cycle (Post et al., 1985a), or continuous illumination (Van Liere and Mur, 1980; Post et al., 1986).

In eight strains of cyanobacteria (*Anabaena flos-aquae*, *Anabaena solitaria*, *Anabaena circinalis* (x 2), *Oscillatoria redekei*, *Oscillatoria agardhii*, *Microcystis aeruginosa*, *Gloeotrichia echinulata*) studied by Wyman and Fay (1986a) the ratio of chlorophyll-a to phycobiliproteins (chl:PBP) remained constant during light limited growth, despite the reduction in pigments as irradiance intensity increased. This indicated that the number of photosynthetic units per cell declined in response to increasing irradiance (Wyman and Fay, 1986a, Post et al., 1986; Tandeau de Marsac and Houmard, 1993). Once maximum growth rate was attained, the chl:PBP ratio increased with further increases of irradiance in all species except the phycoerythrin-rich strains (*Oscillatoria agardhii* and *Gloeotrichia echinulata*). This change in ratio indicated a decline in the size of the PBS (Wyman and Fay, 1986a), presumably reducing exciton delivery to PSII and helping postpone the onset of photoinhibition and damage to the reaction centre. Although the two phycoerythrin-rich strains did not increase their chl:PBP ratio at irradiances in excess of that required for maximum growth rate, the phycoerythrin content of the PBS declined, reducing its effective spectral size. These same adjustments were observed on shifting *Synechococcus* PCC 6301 from high to low light, with an initial increase in the PBS antenna size followed by an increase in PBS per area of thylakoid (Tandeau de Marsac and Houmard, 1993).

Although similar changes in pigment content occur in the eukaryotic algae in response to alterations in light intensity, there are preliminary indications that the acclimation to photoperiod is different (Kromkamp, 1987; Flameling and Kromkamp, 1997). In *Scenedesmus protuberans* the pigment content increased as expected when the growth intensity was decreased under a 16/8 L:D cycle (Gons and Mur, 1979; Post et al., 1985b). However, the response was modified when the incident irradiance on a culture was increased but the total light dose kept constant by introducing a light:dark cycle. The cellular chlorophyll content and the size of photosynthetic units declined as expected but the number of photosynthetic units increased (Flameling and Kromkamp, 1997). This increased the maximum photosynthetic capacity and enabled the organism to take better advantage of the periodic high light intensities. Ibelings (1992) also suggested that the ability of *Scenedesmus* to acclimatise more readily to high photon densities and to take advantage of short duration peaks in irradiance by increasing its maximum rate of photosynthesis was responsible for its competitive advantage over *Microcystis* in a highly turbulent environment. It was found that the time for adaptation to high light intensities was longer for *Microcystis* so that if the two species were rapidly circulating through a strong vertical light gradient *Scenedesmus* would out-compete *Microcystis.*

In the hypertrophic Bautzen Reservoir, *M. aeruginosa* contributes >70% of the total phytoplankton biomass from June to September (Kohler, 1992). The growth of *Microcystis* was restricted to periods when the water column was stratified. Under mixing conditions primary production fell, although the biomass remained high. Surface photosynthesis was inhibited at intensities of > 1100 μmol m^{-2} s^{-1} during and immediately after periods of mixing but inhibition was not observed after two or more calm days. Kohler (1992) suggests that *Microcystis* takes two days to adapt to the high irradiances characteristic of surface blooms which is in agreement with the observations of Ward and Wetzel (1980)

Photoinhibition occurs when phytoplankton are shifted to irradiances substantially above those to which they have been acclimatised. High light intensities can reduce the functionality of reaction centres and may be of particular significance to buoyant cyanobacteria, especially during surface blooms when extreme intensities may have to be endured for extended periods. Although photoinhibition of photosynthesis has been demonstrated in the field and the laboratory, its impact on natural populations is difficult to assess. Most field measurements involve enclosure of samples in bottles during extended incubations in surface layers and it is not apparent just how relevant these measurements are to unfettered cells constantly redistributed by turbulence, or moving of their own volition. Paerl et al. (1983; 1985) did not observe surface inhibition of photosynthesis in natural populations of *Microcystis aeruginosa* from the Neuse River when they were incubated in glass and quartz bottles. He related the lack of photoinhibition to the presence of carotenoids which protected the photosystems from photooxidation and served as accessory pigments. In contrast laboratory cultures of *Microcystis aeruginosa* with low levels of carotenoids displayed significant reductions in photosynthesis when exposed to sunlight in the same way. Although photoinhibition reduces the rate of photosynthesis its impact varies with light intensity and exposure period and may be quickly reversed on return to low light. Consequently the degree of photoinhibition will be influenced by water mixing and the vertical movement of organisms. Furthermore, as maximum rates of photosynthesis generally occur at irradiances exceeding that required to saturate growth, the impact of photoinhibition on growth rates is unlikely to be direct, although it is generally presumed that photoinhibition reduces growth rates.

Laboratory experiments using white light have generally indicated that at very low irradiance intensities cyanobacteria have lower light requirements for growth than green algae due to a lower maintenance energy (Mur, 1983; Richardson et al., 1983). It would be predicted from this that cyanobacteria would outcompete the micro-algae when light was in very limiting supply. However, because of spectral differences in energy capture and apparent changes in the maintenance energy requirement depending on light history, this might not always be the case (Gibson 1985) and further investigation is required.

Within the cyanobacteria there are differences between species in the light requirements for growth. Foy et al. (1976) compared the growth of *Anabaena flos-aquae*, *Aphanizomenon flos-aquae*, *Oscillatoria agardhii* and *Oscillatoria redekei* under different light intensities on a 6h:18h light:dark cycle. The results showed that *Oscillatoria redekei* is adapted to use low light intensities and would be expected to dominate the other species only at light intensities below ca.12-18 μmol photon m^{-2} s^{-1} (assuming 1 lux ≈ 0.018 μmol photon m^{-2} s^{-1}: van Liere and Walsby, 1982). At higher intensities the cell growth of *Oscillatoria redekei* was photoinhibited. These results are in accord with field observations from Lough Neagh where *Oscillatoria redekei* has its maximum population in early spring and declines during the summer months. In contrast, *Oscillatoria agardhii* reaches its population maximum in mid-summer (Foy et al., 1976). Across the species there was a decreasing efficiency of light-limited growth in the order *O. redekei*, *O. agardhii*, *Anabaena flos-*

aquae, *M. aeruginosa*, with *Oscillatoria* spp. being regarded as particularly low-light adapted organisms.

Scheffer et al. (1997) analysed 55 shallow turbid lakes < 3 m deep to assess the importance to the presence of Oscillatoriaceae of TN, TP, TN:TP ratio, lake depth, secchi depth and a shade factor defined as the product of the vertical attenuation coefficient and average lake depth. The shade factor is inversely proportional to the ratio of euphotic depth to mixed depth (z_{eu}/z_m). Their analysis showed no correlation between the relative abundance of Oscillatoriaceae and nutrient concentrations, nutrient ratios or lake depth, but there was a significant correlation with secchi depth, and an even more significant correlation between the relative abundance of *Oscillatoria* and the shade factor. This suggests that the Oscillatoriaceae respond more to the changing light climate than to changes in nutrient availability and is favoured by low levels of irradiance, supporting the findings of Foy et al. (1976). The detailed analysis of Mur et al. (1993 in Scheffer et al., 1997) further supports this view and demonstrates that the decline of *Oscillatoria* populations is more closely correlated with increases in the mean irradiance as Z_{eu}/Z_{mix} rises above a critical value of ca. 0.4 - 0.5 than with declining TP concentrations. As Sas (1989) has noted, a characteristic of *O. agardhii* populations is that they retreat to the metalimnion as the light climate improves in response to nutrient and turbidity reduction and may disappear altogether as water clarity improves further. Scheffer et al. (1997) used a simple graphical model based upon field observations to demonstrate that with increasing TP concentrations, turbidity increases due to both the presence of other algae and suspended particulate matter. At a critical attenuation coefficient (high shade) *Oscillatoria agardhii* is favoured and will suddenly dominate the community. If the nutrient load is then decreased *O. agardhii* will continue to dominate until the biomass is reduced to a point where the critical attenuation coefficient is reached, but because of self-shading this will be at a lower TP than that which was present when the bloom was initiated. These critical break points offer a possible explanation of why *Oscillatoria* lakes switch so rapidly from one state to another and why in some lakes with a high background turbidity *O. agardhii* may not disappear in response to nutrient reductions.

B. Spectral Changes

Wyman and Fay (1986b) grew eight strains of cyanobacteria under equivalent photon fluxes of red, green, blue and white light and found large differences in the cell concentrations of photosynthetic pigments. In red light there was a decline in chlorophyll and phycobiliprotein content, but all strains grew at a significantly faster growth rate than under an equivalent photon flux of white light. For example, *Anabaena solitaria* grew 2.9 times faster in red light than in white light. Under green light the pigment compositions were similar to those in white light, but only the two phycoerythrin-rich strains (*Oscillatoria agardhii* and *Gloeotrichia echinulata*) grew significantly faster than in white light, all other strains growing at 60 to 75% of their white light rate. In blue light the pigment compositions were again similar to those in white light although a majority of the phycocyanin-rich strains showed a reduction in chlorophyll content. The phycocyanin rich forms had growth rates<50% of their white light rate, while the phycoerythrin rich strains, *O. agardhii* and *G. echinulata*, were able to maintain growth rates of 65% and 100% of the white light growth rate respectively. Direct comparisons with micro-algae were not made, but, as they contain chlorophylls b and c, they are expected to capture and utilise blue light with greater efficiency (Richardson et al., 1983).

The substantial changes in cyanobacterial growth rates in response to the spectral distribution of incident light would suggest that spectral changes in the underwater light field could play a major role in structuring community composition. Field data to assess this hypothesis do not appear to be available.

IX. Nutrients

General responses of phytoplankton to nutrient limitation include, carbohydrate accumulation (discussed further for cyanobacteria under buoyancy regulation), a reduction in the cell-specific quantum yield of photosynthesis (Turpin, 1985, 1991), a reduction in the cellular content of the limiting nutrient (Droop, 1973; Riegman and Mur, 1984) and an increase in the nutrient specific uptake rate of the limiting nutrient (Gotham and Rhee, 1981; Riegman and Mur, 1984; Kromkamp, 1987). Nutrient limitation stimulates the storage of non-limiting nutrients as a result of their relative excess compared to the reduced requirements of the cell. Nutrient storage is a valuable attribute, enabling cells to utilise pools of nutrients that are spatially and temporally separated so that growth is maintained during periods of nutrient scarcity.

X. Phosphorus

Under phosphorus limiting conditions cellular phosphorus concentrations decline as phosphorus limited growth rate declines, while the phosphorus uptake potential increases. As a consequence, a pulse of phosphorus delivered to phosphorus limited cells results in substantial formation of polyphosphate reserves, (the polyphosphate "overplus" phenomenon), with cellular phosphorus levels able to exceed those occurring under steady state maximum growth rates (Allen, 1984; Riegmann and Mur, 1984). Most phytoplankton are able to store surplus phosphorus, usually in the form of polyphosphate (PP), and these reserves can be sufficient for several cell doublings. It has been suggested that phosphorus storage in cyanobacteria may be larger than in micro-algae providing them with a competitive advantage (Sommer 1985), but the storage capacity of some micro-algae seems equally large (Lund, 1965). There do not seem to be any consistent phylogenetic differences between micro-algae and cyanobacteria in the range of values for phosphorus uptake and the kinetics appear to be species specific (Healey, 1982; Tilman et al., 1982; Kromkamp, 1987; Reynolds, 1993). Riegman and Mur (1984) for example found the half saturation constant for growth under P-limitation to be similar for *Oscillatoria agardhii* and the two diatoms *Cyclotella meneghiniana* and *Asterionella formosa*.

Rather than a consistent phylogenetic difference in phosphorus uptake and accumulation characteristics, the gas-vacuolate cyanobacteria appear to be advantaged by behavioural features of their life-history that enable them to utilise specific phosphorus conditions more effectively. For example, *Microcystis* has a high V_{max} for phosphorus uptake, a low minimum P content and a large capacity to accumulate phosphorus (Kromkamp et al., 1989). These attributes suggest it is a storage specialist and this is in accord with its ability to regulate buoyancy to gain access to phosphorus in deeper water layers when epilimnetic concentrations are low (Ganf and Oliver, 1982). In contrast *Oscillatoria agardhii* had a similar phosphate uptake rate, but a larger minimum P content, a lower V_{max} and was less adept at accumulating phosphorus. In competition experiments with a pulsed supply of phosphorus at saturating concentrations *Microcystis* outcompeted *Oscillatoria* because of its larger V_{max} and more efficient use of phosphorus (Kromkamp et al., 1989).

A more extreme example of behavioural adaptation is afforded by *Gloeotrichia echinulata* where often a large percentage of planktonic populations are comprised of colonies newly recruited from the sediments (Barbiero and Welch, 1992; Istvánovics et al., 1993; Perakis et al., 1996). *G. echinulata* is a nitrogen-fixing, heterocystous, filamentous cyanobacterium which forms spherical colonies up to 1 or 2 mm diameter. Large numbers of colonies and akinetes are found in the phosphorus enriched sediments of some shallow lakes that are mildly eutrophic, but with relatively low soluble reactive phosphorus concentrations in solution. Examples include Green Lake with soluble reactive phosphorus concentrations in the range 1 – 4 $\mu g\ L^{-1}$ (Barbiero and Welch, 1992) and Lake Erken with soluble reactive phosphorus concentrations usually less than 5 $\mu g\ L^{-1}$ P (Pettersson et al., 1993). In Lake Erken 5×10^5 colonies m^{-2} of *G. echinulata* were found in the top 4 cm of sediments in areas of the lake with a depth < 10m (Pettersson et al., 1993).

During *G. echinulata* blooms the direct contribution to increases in the planktonic population by newly recruited colonies from the sediments can be over 100% (Barbiero and Welch, 1992; Perakis et al., 1996) suggesting that they require large, sustained inputs of colonies from the sediments to persist. The massive recruitment of benthic *G. echinulata* colonies to the plankton is associated with a large transfer of phosphorus into the water column. Istvánovics et al. (1993) investigated the phosphorus uptake characteristics of *G. echinulata* in Lake Erken and showed that it was unable to utilise the low epilimnetic phosphorus concentrations present during the bloom. It did not seem to regulate buoyancy to access phosphorus rich sub-surface water like *Microcystis* but instead remained continuously buoyant and they concluded that the colonies assimilate phosphorus in the sediments prior to their migration into the plankton and use this internal store to support planktonic growth. On the basis of these results *G. echinulata* was described as an extreme storage-adapted species, with its P-assimilation and growth phases completely separated in time and space (Istvánovics et al., 1993).

Recruitment of populations from the sediments occurs in several other bloom-forming cyanobacteria, including *Microcystis*, *Anabaena*, *Aphanizomenon* and *Coelosphaerium* and occasionally these are large enough to influence their population dynamics (Perakis et al. 1996). However, in general these genera increase from growth in the water column (Preston et al., 1980; Trimbee and Harris, 1984; Barbiero and Welch, 1992; Perakis et al., 1996) and are reliant on obtaining on-going supplies of phosphorus.

XI. Nitrogen

Nitrogen is of particular significance to the gas-vacuolate cyanobacteria, as it is an essential component in the synthesis of their gas-vesicles. Consequently nitrogen limitation may not only impact on cell growth, but also on cell buoyancy and the ability to regulate buoyancy. Nitrogen limitation will be particularly detrimental to the non-nitrogen fixing bloom-forming cyanobacteria and may be a critical factor in their replacement by other phytoplankton species.

It was previously thought that inorganic nitrogen metabolism differed between micro-algae and cyanobacteria (Gibson and Smith, 1982), but more recent studies have demonstrated that in many respects they are similar (Guerrero and Lara, 1987; Turpin, 1991; Garcia-Gonzalez et al., 1992; Coronil et al., 1993; Tapia et al., 1996). Nitrogen can be acquired as NO_3^-, NO_2^-, or NH_4^+, and also as N_2 in thosecapable of nitrogen fixation. The order of preference is $NH_4^+ > NO_3^- > N_2$ (Tandeau de Marsac and Houmard, 1993), and when NH_4^+ is available cyanobacteria and micro-algae do not assimilate alternative N sources (Turpin, 1991; Ochoa de Alda et al., 1996). Combined inorganic nitrogen is actively assimilated through a series of steps dependent on the N-source. Nitrate is reduced by nitrate reductase to nitrite and nitrite is reduced by nitrite reductase to ammonium (Guerrero and Lara, 1987). In both micro-algae and cyanobacteria the most important pathway for ammonium assimilation is via the glutamine synthetase-glutamine synthase (GS-GOGAT) enzyme systems.

Nitrogen metabolism is closely connected with carbon fixation, as both processes compete for energy and reductant generated by the light reactions of photosynthesis. The assimilation of carbon dioxide to carbohydrate requires four electrons while the formation of amino-N from nitrate requires 10 electrons (Guerrero and Lara, 1987; Turpin, 1991). At a cellular C/N ratio of ca. 5, up to 50% of the reductant generated through the light reactions will be used to assimilate nitrogen (Guerrero and Lara, 1987).

The interaction with carbon metabolism is further enhanced by the need for carbon skeletons to incorporate nitrogen into protein. In nutrient replete cells carbohydrate stores are small and assimilation of combined inorganic nitrogen is strongly dependent on recent CO_2 fixation (Guerrero and Lara, 1987; Turpin, 1991). Under these conditions reductions in photosynthesis, for example due to darkness or CO_2 deprivation, will reduce nitrogen assimilation. In contrast, N-limited cells accumulate carbohydrate reserves that can be utilised as a source of energy and carbon skeletons for nitrogen assimilation both in the dark and the light (Guerrero and Lara, 1987; Turpin, 1991; Garcia-Gonzalez et al., 1992; Tapia et al., 1996). However, cells growing under N-limited conditions increase their capacity for nitrogen uptake so that the re-supply of inorganic nitrogen causes a large demand for reductant and carbon skeletons. If this demand cannot be met by photosynthesis, then carbon skeletons are supplied through glycolysis of the existing carbohydrate reserves (Turpin 1991; Tapia et al., 1996). The assimilation of nitrate and nitrite into amino-N is limited by the availability of reductant to form ammonium and this reduces their requirement for carbon skeletons compared with ammonium assimilation. In comparison, the large demand for carbon skeletons generated by the assimilation of ammonium can significantly reduce carbohydrate reserves (Turpin, 1991; Garcia-Gonzalez et al., 1992; Tapia et al., 1996). As a result of these interactions nitrogen assimilation influences the rate of CO_2 fixation, the fate of newly fixed carbon, and the level of carbohydrate reserves (Guerrero and Lara, 1987; Turpin, 1991; Garcia-Gonzalez et al., 1992; Tapia et al., 1996) with major effects expected on cell growth, cell turgor pressure and cell density (see section on buoyancy regulation).

Metabolic processes that alter rapidly the size of carbohydrate reserves will be of major significance to buoyancy regulating cyanobacteria as these reserves provide ballast to offset the lift due to gas vesicles. The results of laboratory studies (Turpin, 1991; Garcia-Gonzalez et al., 1992; Tapia et al., 1996) suggest that where cyanobacteria move between the well illuminated, nutrient-poor surface layers and nutrient-rich aphotic zones, the source of available nitrogen at depth could have a significant effect on rates of buoyancy regulation. For example, the increased availability of ammonium common in the hypolimnion of stratified lakes may cause a reduction in the carbohydrate reserves of sedimenting cyanobacteria leading to a quicker reversal of cell buoyancy and a reduction in the extent of vertical migration. Detailed studies of this interaction are required.

Blomqvist et al. (1994) noted from measurements on the oligotrophic, low-alkaline, clear water Lake Njupfatet and the mesotrophic, alkaline Lake Erken, that the development of cyanobacterial populations did not commence until nitrate was almost depleted. Based on results from a series of enclosure experiments enriched with either ammonium or nitrate, they postulated that non-nitrogen fixing cyanobacteria are favoured by ammonium, eukaryotic algae by nitrate-nitrogen and nitrogen fixing cyanobacteria by nitrogen deficiency. The suggestion of nitrate and ammonium being favoured by different organisms has not been well substantiated although laboratory studies provide some support.

In cultures of micro-algae kept under light limiting conditions growth on nitrate is equivalent to, or better than, growth on ammonium (Syrett, 1981; Thompson et al., 1989; Levasseur et al., 1993). In the cyanobacterium *Anacystis nidulans* assimilation and growth on nitrate or ammonia is comparable if light is saturating (Guerrero and Lara, 1987), but at light intensities half saturating to photosynthesis nitrate assimilation is reduced while ammonium assimilation remains unchanged (Garcia-Gonzalez et al., 1992). Ward and Wetzel (1980) showed that the lowest light intensity at which growth of *Aphanizomenon flos-aquae*, *Microcystis aeruginosa* and *Anabaena flos-aquae* would occur was determined by the nitrogen source. Of the three tested (ammonia, nitrate and nitrogen gas) ammonia supported the highest growth rate under all light regimes. These comparisons indicate a preference for NH_4^+ by cyanobacteria particularly at low light intensities, but they do not support the notion that this enables them to dominate the micro-algae when ammonium is the major source of nitrogen. Similarly the results do not support the contention that the micro-algae have a particular preference for nitrate. However they do indicate that under suboptimal light conditions cyanobacterial growth is reduced when the nitrogen source is nitrate, whereas growth rates of micro-algae are not affected. So rather than non-nitrogen fixing cyanobacteria being favoured by ammonium and eukaryotic algae by nitrate nitrogen (Blomqvist et al., 1994), it would seem that the cyanobacteria may be disadvantaged by using nitrate-nitrogen under low light conditions.

A. Nitrogen Fixation

Some cyanobacteria are able to utilise N_2 when sources of combined inorganic-N are depleted. Numerous reviews have described the biochemistry, physiology and molecular biology of cyanobacterial nitrogen fixation (see Bothe, 1982; Van Baalen, 1987; Howarth et al., 1988; Tandeau de Marsac and Houmard, 1993; Flores and Herrero, 1994). No known micro-algae that can fix molecular nitrogen, so the nitrogen-fixing cyanobacteria have a major advantage at times when sources of combined inorganic nitrogen have been depleted from the water.

The common bloom-forming cyanobacteria that are nitrogen fixers are the heterocystous, filamentous members of the Nostocales, including *Anabaena*, *Aphanizomenon* and *Gloeotrichia*. When combined nitrogen is absent, these organisms differentiate thick walled cells called heterocysts that isolate the nitrogen-fixing enzyme system, nitrogenase, from inactivation by oxygen. The heterocyst provides this protection by enhanced respiration, and by the barrier of the heterocyst envelope (Wolk et al., 1994). Some non-heterocystous species are also able to fix N_2, including species of *Oscillatoria*, but the extent of this in natural systems has not been quantitatively estimated.

Howarth et al. (1988), who summarised field data on nitrogen fixation in both marine and freshwater environments, concluded that cyanobacteria are responsible for most planktonic nitrogen fixation in freshwaters and that rates are only high when these organisms are present in large numbers. A comparison of seven eutrophic lakes showed that nitrogen fixation accounted for 6 - 82% of the nitrogen load, differing markedly between lakes. Even when N_2-fixation contributes only a small percentage of the total load to a lake, the direct supply of nitrogen to the bloom-forming cyanobacteria still is of major importance to their success. In particular nitrogen fixation enables continued production when nitrogen supplies are depleted, and the observation that phosphorus is the nutrient frequently found to control the development of phytoplankton biomass even in lakes where nitrogen is initially limiting is a result of nitrogen deficits being compensated for through nitrogen fixation (Schindler, 1977; Howarth et al., 1988). However, nitrogen deficits that commonly occur in waters where nutrient loadings have low N : P ratios are not always counteracted by nitrogen fixation. In tropical Lake Valencia nitrogen fixation rates were high, but insufficient to restore the low N:P ratio to the Redfield ratio, suggesting that nitrogen limitation controlled the phytoplankton biomass (Levine and Lewis, 1987).

Frequently N:P ratios are used to assess the likelihood of nitrogen-fixing cyanobacteria occurring, but as Horne and Commins (1987) have stressed, nutrient ratios alone are not reliable indicators as the critical limiting level of total inorganic nitrogen required for the induction of nitrogen fixation must be reached regardless of the N:P ratio. They reviewed the literature on laboratory and field experiments and concluded that total inorganic nitrogen needs to fall below 50-100 mg m^{-3} to induce nitrogenase activity.

B. Nitrogen Storage

Unlike eukaryotic micro-algae, the cyanobacteria have a capacity to store significant amounts of nitrogen in excess of their immediate requirements. The two storage components are cyanophycin, a co-polymer of aspartate and arginine, and the phycobiliprotein, phycocyanin. Whereas the only function of cyanophycin is to store nitrogen and perhaps energy, phycocyanin is also a major pigment component of the light-harvesting antenna, but under conditions of nitrogen limitation it acts as a nitrogen reserve (Kromkamp, 1987).

Cyanophycin and phycocyanin are both at low concentrations in nitrogen-limited cells (Allen, 1984), but even in non-limited, rapidly growing cells, the amount of nitrogen stored as cyanophycin is comparatively low relative to when cell growth is limited by other requirements. Cyanophycin is accumulated when cells are starved of light, phosphorus or sulphur, and when grown at low temperatures (Allen, 1984). In a manner reminiscent of the phosphorus overplus phenomenon, cyanophycin accumulates on the addition of a utilisable nitrogen source to N-limited cells (Simon, 1987).

In response to nitrogen starvation the cyanophycin granules are first degraded, followed by cell bleaching due to degradation of components of the phycobilisome including phycocyanin (Tandeau de Marsac and Houmard, 1993). Nitrogen stores are also utilised when low light cells are shifted to high light, with cyanophycin and phycocyanin both decreasing.

XII. Responses of Cyanobacteria to N and P

A. Whole Lake Phosphorus Enrichment

The classical work of a number of authors in the 1960s (eg Sakomoto, 1966; Vollenweider, 1968) led to the recognition of the importance of increased phosphorus loadings in the process of eutrophication of lakes. These studies on phosphorus, and later on the interaction between nitrogen and phosphorus (eg Smith, 1983), led to ecological research focused on the manipulation of whole lakes or portions of them to explore the responses of phytoplankton abundance and community structure to nutrient conditions (Schindler, 1971; Lund and Reynolds, 1982).

An implication of phosphorus loading models is that discharging nutrient-rich waters into a water body will increase productivity, and if physical conditions permit, lead to the proliferation and eventual dominance of bloom-forming cyanobacteria. Pearsall (1932) suggested a relationship between increasing nutrients, dissolved organic matter, and the presence of bloom-forming cyanobacteria in English lakes. Pick and Lean (1987) recalculated data from Gorham et al. (1974) for lakes in northern England and came to the conclusion that there was a significant relationship between the mean annual phytoplankton biomass and that of the bloom-forming cyanobacteria, with the proportion of cyanobacteria increasing as the total phytoplankton biomass rose above 5 - 10 g m^{-3} fresh weight. Reynolds (1987), Steinberg and Hartmann (1988) and Steinberg and Gruhl (1992) suggest eutrophication, especially by phosphorus, often leads to significant shifts in phytoplankton species composition towards bloom-forming cyanobacteria.

Additional evidence to support this conclusion comes from Europe and North America (Harris, 1986; Cullen and Forsberg, 1988; Sas, 1989; Cooke et al., 1993) as well as from Australia and Africa. The Peel Harvey Estuary in Western Australia and the Hawkesbury River in New South Wales both illustrate that increases in nutrient loads from either agricultural run-off or sewage treatment plants results in extensive populations of bloom-forming cyanobacteria, in these examples *Nodularia spumigenea* (Plate 11e) and *Microcystis aeruginosa* / *Anabaena* sp., respectively (George & Bradby, 1993; Cullen, 1994; Humphries & Robinson, 1995).

The relationship between eutrophication and an increased biomass of gas-vacuolate cyanobacteria has been attributed to the requirement that sufficient nutrients be available, either in the water or from internal recycling, when physical conditions eventually become suitable to provide the cyanobacteria with a competitive advantage. In temperate systems if nutrients are depleted by phytoplankton growth during spring and early summer then the bloom-forming cyanobacteria are faced with depauperate nutrient conditions when the physical environment is most suitable for their growth. Similar arguments can be proffered for tropical waters but on cycles driven by both meteorological events as well as seasonal conditions (Lewis, 1978).

B. Biomass Response to P-Removal

Lake McIlwaine, near the city of Harare, Zimbabwe, experienced anthropogenic eutrophication as a result of sewage discharge into the lake (Thornton, 1982). As nitrogen and phosphorus concentrations rose 5 to 10 times their original levels, chlorophyll-a concentrations peaked at ca. 150 mg m^{-3} and did not fall below ca. 50 mg m^{-3} throughout 1968-69. During this period the lake was dominated almost exclusively by *Microcystis aeruginosa* and *Anabaena flos-aquae*. After diversion of the sewage to pasture irrigation between 1970 and 1975, the soluble reactive phosphate levels fell by an order of magnitude and chlorophyll-a concentrations fell to a mean of 15 mg m^{-3}. Although, the lake still supported populations of cyanobacteria, as well as *Melosira granulata*, there was a re-appearance of significant populations of other eukaryotic genera.

The Lake McIlwaine experience suggests that when nutrient loads are reduced phytoplankton biomass decreases, but the conclusion that cyanobacteria will also disappear is a convenient rather than a realistic one. Nevertheless, there are situations where this does occur. For example, as a result of the diversion of sewage from Lake Washington (Edmondson and Lehman, 1981; Seip et al., 1992) the percentage of total phosphorus from sewage was reduced from ca. 70% to zero and the total phosphorus content of the lake fell from 200 x 10^3 kg to 59 x 10^3 kg. This resulted in a decrease in maximum chlorophyll-a concentrations from ca. 45 mg m^{-3} (1962 - 1965) to less than 10 mg m^{-3} (1976 - 1978). Concurrently the proportion of the total phytoplankton biomass attributable to filamentous cyanobacteria (*Oscillatoria rubescens*) in surface samples fell from > 90% to < 20%.

In contrast, there are examples where a reduction in the phosphorus load has had no apparent influence on water quality. Talling and Heaney (1983) recommended the removal of sewage borne phosphorus which contributed 47 - 67% of the total phosphorus loading to Esthwaite Water. Heaney et al. (1992) reported the outcome of this remedial action and concluded that, although the impact of phosphorus removal had yet to be fully realised, it was questionable if water quality would ever improve with the current phosphorus loading. This was due in part to the internal loading from the P-rich sediments compensating for the reduction in the external load as the lake approached a new equilibrium state, and partly due to the continued input of phosphorus from an adjacent fish farm. One of the interesting aspects of Esthwaite Water is that the most significant changes in phytoplankton species composition occurred prior to the remedial action. In the early 1970s and 1980s the summer algal community was dominated by *Ceratium* spp., but due to intense parasitism by *Aphanomycopsis cryptica*, a biflagellate fungus, the *Ceratium* diminished from 1983 was were replaced by *Aphanizomenon flos-aquae* fo. *gracile*, *Anabaena flos-aquae* and *A. solitaria*. Subsequent changes in the timing and maxima of the cyanobacteria were attributable to the intense grazing by the protozoan ciliate *Nassula aurea* (Canter et al., 1990) and not to the reduction in P-load.

These examples, plus many others (Reynolds and Walsby, 1975; Reynolds, 1984b,1987, 1989a, b, 1992b), illustrate that relationships between nutrient concentrations and phytoplankton biomass and species composition are not simple. Indeed it may appear as though each lake should be considered as an individual water body and this reductionist view may be correct. However there is evidence to suggest general relationships between the physical and chemical characteristics of lakes and phytoplankton communities and in particular the conditions that promote the wax and wane of cyanobacteria (Round 1971).

To understand the complexity of the relationships Sas and his co-workers (Sas, 1989; Seip et al., 1992; Reynolds, 1992; Cooke et al., 1993) analysed data from 18 water bodies in Europe ranging in are (0.03 to 503 ha), depth (0.5 to 177m) and theoretical retention time (0.1 to 11 years). The total phosphorus (TP) concentration range was 6 to 1440 mg m^{-3} and inorganic nitrogen 2 to 2000 mg m^{-3}. All water bodies had undergone a reduction in the P input principally, but not exclusively, via P-precipitation at sewage treatment plants. To analyse the data, criteria were set to identify lakes where nutrient limitation (N or P) occurred. If filtrable reactive phosphate (FRP) concentration was consistently >10 mg m^{-3} then P-limitation to growth was not considered likely, whereas if the lake-FRP concentration was < 10 mg m^{-3} either (1) on average over the entire growing season, or (2) absolutely during at least half of the period of the growing season, then phytoplankton growth was assumed to be P-limited during the growing season. Similarly, the threshold value below which N-limitation occurred was assumed to be 100 mg m^{-3} inorganic nitrogen.

Despite phosphorus removal, seven of the 18 water bodies studied by Sas (1989) did not show a significant decrease in phytoplankton biomass as measured by either chlorophyll-a concentration or biovolume. Four of these were deep lakes (z_M = >18 m) which were incompletely mixed and three were shallow lakes (z = < 5 m) where mixing was a regular event. The four deep lakes were characterised by relatively low initial chlorophyll concentrations and it appears that the principal aim was to reduce the likelihood of future increases in cyanobacteria rather than to prevent the bloom-forming species appearing.

Of the shallow lakes only Lake Hylke experienced periods of P-limitation with FRP < 10 mg m^{-3} for two weeks during the growing season, while inorganic nitrogen always exceeded 100 mg m^{-3}. Lake Sobygard, dominated by *Scenedesmus*, never experienced P- limitation and, although N-limitation was reported for 2 to 4 weeks, the growth season average inorganic nitrogen concentrations were well in excess of 100 mg m^{-3}. On the basis of the critical value for phosphorus, the third shallow lake (Lough Neagh) with a relatively low, mean, growing season FRP concentration (ca 25 mg m^{-3}) experienced P limitation for an average of six weeks of the growing season (1976-1986). However inorganic nitrogen fell below the critical level for 2 - 5.5 months in eight of the eleven years monitored, suggesting that nitrogen was more likely to be limiting the phytoplankton. Apparently the lack of response of these three shallow lakes to phosphorus reduction indicated that phosphorus had not been reduced sufficiently to decrease the phytoplankton biomass.

The responses of the remaining 11 lakes to P-removal were manifested by both a reduction in the mean growth-season chlorophyll concentration and less frequently by a shift in species composition. The time lag for a reduction in chlorophyll to occur varied between lakes. In Lake Wahnbach where the chlorophyll declined from ca 30 mg m^{-3} in 1969 to < 5 mg m^{-3} in 1985, the response was immediate and followed that predicted by the Vollenweider/OECD model. In Lake Veluwe the chlorophyll concentration pre-restoration (1975) was > 200 mg m^{-3} and fell to ca 50 mg m^{-3} by 1982, but with a two year lag time. In Lake Schlachten the average summer chlorophyll concentration remained high (64-85 mg m^{-3}) for the first four years after restoration but as phosphorus was further reduced chlorophyll fell to 14 mg m^{-3}. In each case the reduction in chlorophyll with phosphorus availability had a slope close to 1.0 and the time lags reflected the magnitude of the sediment release of phosphorus. In addition to the influence of nutrients, grazing was considered important to the overall reduction in the algal biomass for five of the lakes and light was implicated in nine of the lakes.

C. Four-Stage Response to P-Removal

The significance of the study by Sas (1989) was that quantitative comparisons were made between lakes, across years and validation was an integral part of the analysis. The conclusion was that the phytoplankton biomass and the cyanobacterial component responded to remedial action in four stages (Fig. 9)

- Stage 1 no biomass reduction as phosphorus in excess to requirements
- Stage 2 declining amount of unused phosphorus, small reduction in biomass
- Stage 3 phytoplankton biomass falls, minimal unused phosphorus
- Stage 4 further decline in biomass and changes in composition of the phytoplankton.

The first stage occurred in lakes with a nutrient rich pre-restoration phase where nutrients were never growth limiting and there was an unused fraction of the total phosphorus (Fig. 9). In these cases there was no immediate effect on phytoplankton biomass or species composition. In general the influence of P-limitation did not occur until the FRP in the trophogenic layer had fallen on average during the growth season to < 10 mg m^{-3}. Evidence from a number of the responsive lakes showed that as the chlorophyll concentration fell the difference between total phosphorus and the particulate fraction became small which indicated that FRP was virtually exhausted (Reynolds, 1992) and all the P was incorporated into algal cells (Fig. 9). This interpretation would need to be modified in turbid

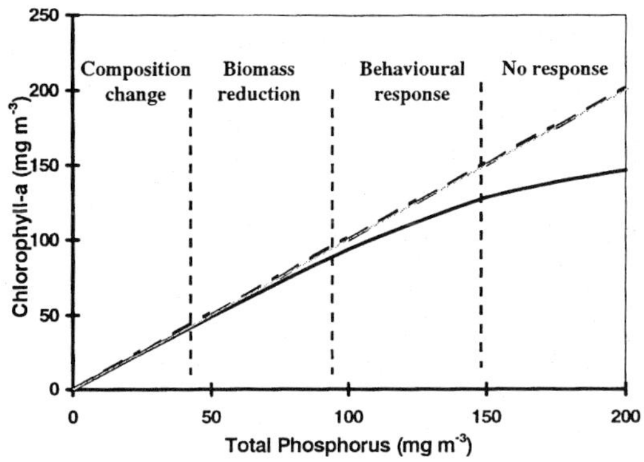

Fig. 9 Hypothetical example of the response of lake chlorophyll-a concentration to reductions in phosphorus concentration according to Sas (1989). The theoretical chlorophyll-phosphorus relationship is shown by the straight line (dashed) while the actual response follows the solid line. At high phosphorus concentrations the difference between the theoretical and actual chlorophyll concentrations is due to restriction of the algal biomass by some other factor and phosphorus is present in excess.

lakes where a large fraction of the bioavailable-P may be associated with the inorganic particulate fraction and not measured as FRP, and a proportion of the particulate total phosphorus may be unavailable for incorporation into phytoplankton (Oliver, 1993). The second stage of recovery depends upon the behaviour of the phytoplankton and the morphometry and mixing regime of the lake. In those lakes which have a persistent thermocline (usually the deep, calm lakes) the phytoplankton community is dominated by motile algae such as dinoflagellates and buoyant cyanobacteria.

As a consequence of the reduced nutrient load these phytoplankton disperse to greater depths as they seek additional nutrients.

This causes a marked increase in the water transparency even though the biomass per unit area may not fall. In contrast, shallow, vertically mixed lakes do not afford a deep refuge zone and consequently this behavioural aspect of phytoplankton is not realised. The recovery process is not buffered by access to previously unused phosphorus sources and it skips to the third phase.

In stage 3 the P-concentration continues to decline as a consequence of both the internal and external reduction of the P-loading (Fig. 9). The overall result was a significant decrease in the phytoplankton biomass as P-limitation began to take effect, described by the equation:

$$C_A/C_B = (P_A/P_B)^m$$

where:

C_B and C_A denote the growth-seasonal average chlorophyll-a concentrations pre- and post-restoration respectively, P_B and P_A the pre- and post-restoration values of the whole-lake annual mean total phosphorus concentration and *m* the empirical exponent derived from the correlation (Sas 1989).

The relationship between decreasing P-load and phytoplankton biomass reduction conformed closely with the Vollenweider/OECD model such that the ratio of the growth-season average chlorophyll post- to pre-restoration was equal to the ratio of the whole lake annual mean total phosphorus post- to pre-restoration raised to the exponent 1.0 compared with the Vollenweider/OECD model of 0.96 ± 0.12. This relationship was further subdivided into shallow lakes where the exponent was 1.4 ± 0.3 and deep lakes where the exponent was 0.7 ± 0.5. Although the small number of lakes involved in the study and the differences between the sampling procedures for phytoplankton biomass cause some inconsistencies it was nevertheless possible to discern a qualitative difference between deep and shallow lakes.

The fourth stage of recovery once the lake reaches its new equilibrium state entails a change in species composition. For the perennial species such as *Oscillatoria* spp the data suggest that the trend of increasing cell numbers with increasing nutrient supply is more or less reversible as the nutrient load decreases below threshold values. In shallow lakes this threshold appears to be 50 - 100 mg m^{-3} FRP, but is rather lower for deep lakes at 10-20 mg m^{-3} (Sas 1989).

D. Responses of Cyanobacteria to P-Removal

The slope of the linear log-log relationship between concentrations of chlorophyll-a and total phosphorus were used to assess whether or not cyanobacteria responded to changes in the TP concentration in a similar or different manner to micro-algae. The frequency of occurrence of particular slopes was used to compare the trajectory of the relationship between chlorophyll-a and TP for cyanobacteria and for micro-algae (Seip et al., 1992). Positive slopes of 45° indicated a 1:1 relationship between chlorophyll-a and TP. Slopes near 0° indicate small ratios that suggest changes in TP correspond with much smaller changes in chlorophyll-a, such as occurs when there are significant time lags between action and response. Angles > 78°, equivalent to ratios greater than 4.7, indicate large changes in chlorophyll with small changes in TP, while higher ratios with angles close to 90° indicate that factors other than TP influence the trajectory. Negative angles suggest that chlorophyll increases with decreasing TP or vice versa. For deep lakes cyanobacteria had very similar responses to other algae. For shallow lakes the occurrence of low ratios with angles near to 0° is more pronounced for micro-algae than for cyanobacteria indicating there are fewer instances where cyanobacterial chlorophyll does not respond to an increase or decrease in TP. There were a significant number of angles < 0° or close to 90° which highlights the probability of factors other than nutrients influencing species composition.

Shifts in species composition were most noticeable for *Oscillatoria* spp. which either decreased significantly, disappeared, or retreated to the metalimnion as water transparency increased in response to reductions in biomass. Less obvious was the response of cyanobacterial species such as *Aphanizomenon* spp, *Anabaena* spp and *Microcystis* spp. that can form surface blooms. In a significant number of cases where temperatures were favourable, these tended to increase as *Oscillatoria* spp decreased in abundance, and this was correlated with a decline in the ratio of TP: z_{eu}/z_{mix} (ie. the ratio of phosphorus to available light).

This species response to lake restoration is a prime avenue for future research. It is evident that species changes are not simply a function of the relative growth rates of different species although this may be contributory, rather it is the interaction between a number of biological features and the stability of the water column as well as the quality of the underwater light field.

E. Bioassays, Lake Enrichment and Nitrogen Limitation

Elser et al. (1990) reviewed 62 North American lakes from which data were available on the response of algal biomass (chlorophyll, ^{14}C-fixation, cell counts, in vivo chlorophyll fluorescence) to nutrient enrichment of N, P or N+P versus a control (no addition) in enrichment bioassays and whole-lake experiments. The average concentrations added were 46.3 μM N and 2.63 μM P. Simultaneous N and P enrichment in bioassays nearly always elicited a positive growth response (86% of cases), whilst single nutrient additions of either N or P produced positive results in 40% and 47%, respectively. This suggests that in the 62 lakes studied nitrogen was just as frequently in limiting supply as phosphorus. Results obtained for lakes and rivers in south-eastern Australia using both growth assays and physiological assays (Wood and Oliver, 1995; Fink and Oliver, submitted) support this view.

Of the studies reviewed by Elser et al. (1990) whole-lake fertilisations provided a total of 80 lake-years of data. Of these, only 2 of 31 lake-years showed a positive response to single nutrient additions (+ P: Lake Maryjo and ELA 261). In agreement with the bioassay experiments, 39 lake-years showed a positive response to the simultaneous addition of N and P, while 10 lake-years showed no significant response to any of these additions (Elser et al., 1990). Interpretation of whole-lake data was hampered by a lack of factorial design for N and P additions and the authors concluded that even for this extensive data set the effects of N and P had not been separated adequately and urged scientists in future to implement designs which evaluate the interactive roles of N and P. In addition they suggested that the time scales of experiments may be inadequate to assess the extent to which N_2-fixing cyanobacteria impact on the nitrogen budget of a lake. The frequent requirement for combine N and P enrichment consistently to produce a substantial growth response confirmed the results of the bioassays and indicated a more important role for nitrogen limitation in freshwaters than previously recognised.

F. Lake and Laboratory Studies on the Influence of TN:TP Ratios

In recognition of the influence of nitrogen availability on phytoplankton growth the interpretation of log-log plots of P-loading and annual mean chlorophyll-a concentrations have been modified by the inclusion of TN:TP ratios to account for situations of nitrogen limitation and provide an indication of the probability of nitrogen fixing cyanobacteria appearing in the plankton (Smith 1982, 1983). From a study of 17 lakes world-wide Smith concluded that bloom-forming cyanobacteria tended to dominate in lakes where the TN:TP mass ratio was less than 29. This has led to claims that increasing the mass ratio above 30 will reduce the proportion of cyanobacteria as a fraction of the total algal biomass. Many substantial reviews that have discussed the impact of TN:TP ratios on phytoplankton populations (Harris, 1986; Pick and Lean, 1987; Elser et al., 1990; Jensen et al., 1994; Scheffer et al., 1997) have found little evidence to support the contention that TN:TP ratios are an important determinant of cyanobacterial dominance. Pick and Lean (1987) suggest that neither laboratory nor whole-lake studies provide conclusive evidence that N:P ratios play a major role in cyanobacterial dominance. Others (Trimbee and Prepas, 1987; Scheffer et al., 1997) have suggested that even when a response is observed it maybe spurious and due to increasing P concentrations rather than a decrease in the N:P ratio.

The problem appears to be that the majority of experiments and lake manipulations are done without assessing whether major nutrients are limiting. Horne and Commins (1987) and Reynolds (1992) have pointed out that the ratio is immaterial if the nutrient concentrations are in excess of those limiting to growth. Any discussion on the influence of TN:TP ratios on the occurrence of cyanobacteria must start from the premise that the cells are, or will become limited by either nitrogen or phosphorus.

Two examples which support a central role for TN:TP in stimulating cyanobacterial growth are those by Schindler (1977) and Stockner and Shortreed (1988) both carried out in oligotrophic Canadian lakes. In the first experiment, Lake 227 was fertilised at a ratio of 14:1 (by weight). During the entire six year period the lake was dominated by *Scenedusmus* and other algae unable to fix atmospheric nitrogen. However, when the N:P ratio was cut back to 5:1, a bloom of the nitrogen-fixing *Aphanizomenon gracile* occurred, which fixed significant amounts of nitrogen. A similar response was noted in Lake 226 when fertilised with a TN:TP ratio of 5:1, althoughhere the dominant genus was *Anabaena*.

In the second example (Stockner and Shortreed, 1988), the Clayoquot Arm of oligotrophic Kennedy Lakee was enriched with solutions of NH_4NO_3 and $(NH_4)_2HPO_4$ (o to 7.6 mg P m^{-2} week $^{-1}$) to give N:P molar ratios that rose from 10 (ca. 5:1 by weight) in 1978, the first year of fertiliasation, to 15 in 1979-1981, and then further increased to 26 and 35 over the next three years. *Anabaeuna circinalis* concentrations were <20 cells ml^{-1} prior to fertilisation, but reached >30000 cells ml^{-1} in 1981 and subsequently fell to an average cell concentration of 25 ml^{-1} in 1983-85. Concurrently the growing season average secchi depths decreased frpm 8.1 down to 3.5 m in 1981, before deepening again to 9m in 1985. During this period the total phosphorus concentration increased to a maximum of only 6 mg m^{-3}. The authors concluded that the low initial N:P ratios in the presence of rising phosphorus concentrations promoted *A. circinalis* but, as the N:P ratio rose and nitrogen was no longer limiting, then other species, predominantly *Synechococcus* sp., became dominant.

G. Cyanobacteria, Water Column Stability, and TN:TP Ratios

Few attempts have been made to combine information on the degree of water mixing and nutrient conditions and to relate this with the occurrence of particular cyanobacteria. The correlative analysis of 435 US lakes by Harris (1986) attempted to distinguish between the distribution of species on a basis of TN:TP ratios and M, a measure of water column stability ($M = D_{th} /z$) where D_{th} is the mixed depth estimated from the location of the thermocline and z is the mean depth of the lake. M < 1 indicates stable conditions in summer as the mean depth of the lake exceeds the depth of the thermocline. M >> 1 indicates strong vertical mixing as the mixed depth greatly exceeds the mean depth of the lake. Table 2 gives the maximum percent probabilities of occurrence of four species of cyanobacteria under different mixing conditions and the prevailing TN:TP ratio within decadal intervals from <10 to >50 (Harris, 1986; US EPA data).

Table 2. The percent probabilities of occurrence of four cyanobacteria under different mixing conditions (M) and the prevailing TN:TP ratios (Harris 1986, US EPA data).

Species	M 0-1	TN:TP	M 1-2	TN:TP	M 2-3	TN:TP	M 3-4	TN:TP	M > 4	TN:TP
Anabaena	31	<20	24	<10	26	<10	35	<10	**38**	<10
Aphanizomenon	**67**	<10	49	<10	30	<10	40	<10	14	<10
Microcystis	45	>50	**50**	>50	21	<10	0		0	
Oscillatoria	15	<20	40	<20	50	>50	**100**	<40	36	<10

Anabaena's distribution spanned the complete range of M from stable to highly turbulent water columns but was favoured by TN:TP ratios of <10, suggesting that it competes best in habitats where the flux of N into the water maybe limiting. *Aphanizomenon* also occurred in habitats with low TN:TP ratios, consistent with the observation that when the loading was reduced from 15 to 5:1 in the Experimental Lakes Area a bloom of *Aphanizomenon* developed, whereas it had not been observed before (Schindler, 1977). *Aphanizomenon* had a clear preference for stable summer conditions, although it also occurred at $M > 2$. The growth of the two filamentous species over a wide range of mixing regimes may reflect their ability to occur both as single filaments and as large aggregated colonies and is crudely in accord with the classification discussed in the earlier section. Habitats structured by turbulent mixing. *Oscillatoria* appeared independent of TN:TP ratios, but preferred less stable water columns, suggesting that most samples were from well mixed lakes and that those with metalimnetic populations were not well represented in the data. In contrast, *Microcystis* occurred in stable habitats where there was little evidence for nitrogen limitation.

XIII. Inorganic Carbon

The common observation that cyanobacteria frequently dominate lake phytoplankton at times when the pH is high has led to the hypothesis that these organisms are able to outcompete the eukaryotic micro-algae in situations where carbon dioxide concentration is low and that this is fundamental to their dominance (King, 1970). Shapiro (1990) has reviewed the field and laboratory data supporting this notion and considers that it provides a more robust explanation than most other hypotheses accounting for cyanobacterial dominance. However, in a later publication he concludes low CO_2 concentrations or high pH do not intiate cyanobacterial blooms, but rather their abundance reduces CO_2 to levels that only they can utilise (Shapiro, 1997). The concentration and speciation of dissolved inorganic carbon is strongly linked to pH through equilibrium reactions between the species CO_2, H_2CO_3, HCO_3^- and CO_3^{2-}. The proportion of CO_2 declines from a maximum at pH 4 to only 0.003% of the total inorganic carbon concentration at pH 9, with the actual CO_2 concentration dependent on the total inorganic carbon content of the water.

Phytoplankton photosynthesis removes CO_2 from solution resulting in disruption to the equilibrium and a rise in pH, the magnitude of which depends on the buffering intensity of the water (Shapiro, 1990). It has been shown by manipulating CO_2 concentration and pH in enclosure and lake experiments that cyanobacteria are better adapted to take advantage of low CO_2 concentrations than some, but not all microalgae (Talling, 1976; Shapiro, 1990). Some cyanobacteria have been shown to utilise HCO_3^- (Talling, 1976), but as they generally contain carbonic anhydrase that catalyses the dehydration of bicarbonate it is likely that CO_2 is used within the cell. However these features are not unique to the cyanobacteria and some eukaryotic algae are also able to utilise very low CO_2 concentrations (Talling, 1976) and some can access HCO_3^- as they also contain carbonic anhydrase. The debate over the importance of competition for inorganic carbon has been further complicated by the discovery of the carbon concentrating mechanism in phytoplankton. It is still unclear whether this mechanism is more beneficial in cyanobacteria or microalgae and so the role of inorganic carbon limitation on competition is difficult to assess.

Field experiments where manipulations of carbon and pH have resulted in shifts in species dominance to cyanobacteria under conditions of inorganic carbon limitation (Shapiro, 1990) suggest that a low availability of CO_2 can advantage the gas-vacuolate cyanobacteria. This advantage would be re-enforced during periods of large cyanobacterial blooms, assisting these organisms to maintain dominance. However, the bloom-forming cyanobacteria are not restricted to eutrophic waters and can dominate the communities of less productive lakes, even though the total phytoplankton biomass is low. Conversely, cyanobacteria can also be found to dominate in some well-mixed shallow waters where inorganic carbon limitation is considered unlikely (Steinberg and Gruhl, 1992; Scheffer et al., 1997). These observations would argue against CO_2 limitation being essential for the formation of cyanobacterial blooms, but whatever the final outcome regarding direct competition for inorganic carbon, any limitation will still have a major influence through its impact on buoyancy regulation. Undoubtedly more information is required on the inorganic carbon requirements of the phytoplankton.

XIV. Grazing

Grazing can be a major loss factor modifying the biomass and community composition of the phytoplankton. For example, Gliwicz (1968) and Haney (1973) estimated that the zooplankton community ingested 48 - 162% of their biomass per day, and daily community grazing rates could process 10 to > 100% of the volume they occupy.

The four major groups of animals represented in the zooplankton community; the rhizopods, ciliophorans, rotifers and crustaceans, have diverse means of selecting, obtaining and ingesting food organisms (Reynolds, 1994b). Although the specialist protozoans and rotifers impact significantly on their particular food sources, it is the generalist feeders that often make the greatest impact on the phytoplankton and particularly on the cyanobacteria. Above certain threshold concentrations of prey the grazing impact moves towards the less selective filter feeding rotifers and cladocera (Reynolds, 1994b). Of these it is populations of *Daphnia* species that frequently have the greatest impact.

In Lake Mendota, Wisconsin, the phytoplankton community was regulated by both grazing and nutrients but the influence of each varied seasonally (Vanni and Temte, 1990). In general grazing impacted most on the phytoplankton community during spring while nutrient limitation was more severe in summer. This change was in part due to the replacement of the dominant edible phytoplankton species that occurred in spring by more resistant species in summer. Presumably the appearance of the more resistant species was a result of selection during the earlier grazing period. The composition of the zooplankton community also altered during this period. In early spring the community was dominated by cyclopoids while the clear water period in late spring was dominated by *Daphnia galeata mendotae* which continued into early summer before becoming sub-dominant to calanoids and cylcopoids. *Daphnia* had the largest impact on the spring phytoplankton and was considered mainly responsible for the clear water phase. It was less effective on the large, inedible phytoplankton species that were dominant in summer, particularly cyanobacteria and the dinoflagellate *Ceratium*.

In Lakes Hume and Dartmouth, both man-made lakes located in south-eastern Australia, manipulations of the large cladocerans (*Daphnia* and *Diaphanosoma*) and the large copepod (*Boeckella*) had negative effects on the phytoplankton biomass, while smaller copepods (*Mesocyclops* and *Calamoecia*) had little impact or occasionally a positive effect (Matveev and Matveeva, 1997). In both reservoirs the variation in *Daphnia carinata* alone could account for 50% of the variance in total phytoplankton biovolume. Colonies of *Microcystis* < 50 μm diameter were grazed in feeding trials by both *Daphnia* and *Boeckella* without any suggestion of selectivity. When *Microcystis* dominated in the lakes enclosure experiments showed a negative response of biomass to *Daphnia* and *Boeckella* concentrations.

These studies are in general accord with many others that have found correlations between the abundance of zooplankton grazers and the timing of cyanobacterial blooms (Haney, 1987). In general the reduced grazing on cyanobacteria is associated with large size, high density, allelopathic compounds and poor assimilability. The maximum size of food particles (y) taken in by *Daphnia* species is a function of the animal's length (L) as described by Burns (1968),

$$y = 22L + 4.87$$

which indicates a maximum particle size of 49 μm for a 2 mm animal (Reynolds, 1994b). In general the optimum food sources are small planktonic algae whereas larger colonial and filamentous cyanobacteria have a depressive effect on filter feeding due to mechanical interference with the feeding apparatus (Reynolds, 1994b).

Enclosure experiments have been used to investigate these relationships and Burns (1987) summarised the experiments describing interactions between zooplankton and cyanobacteria by making the following points:

- There was only circumstantial evidence from field experiments for decreased fecundity and growth in response to food limitation effects or toxicity.
- In long term enclosure experiments dominated by large herbivores the more edible cyanobacteria showed an inverse relationship with grazer density such that there appeared to be a threshold density of ca. 5×10^4 grazers m^{-2}, above which filamentous cyanobacteria could not withstand grazing but large inedible colonies could (Lynch and Shapiro 1981). A similar result was reported by Ganf (1983) comparing the ungrazed *Aphanizomenon flos-aquae* which occurs as a large flake and dominates in the presence of *Daphnia pulex*, with filamentous *Aphanizomenon elenkinii* which dominated only where *D. pulex*

was absent. Reductions in the edible species required a threshold concentration of grazers of ca. 12 L^{-1}.

- Zooplankton grazing could suppress cyanobacterial growth by altering light and nutrient conditions, in particular by increasing transparency, reducing primary productivity and pH, and increasing nutrient re-supply (Schoenberg and Carlson 1984).

It seems from this that zooplankton grazing can reduce the biomass of cyanobacterial populations provided they are present before the cyanobacteria attain a size larger than the animals can manage. If the phytoplankton species can reach a large size prior to substantial increases in the most effective grazers then the likelihood of control by grazing will be diminished. The grazing impact will therefore depend on the dynamics of the phytoplankton and zooplankton communities.

A number of indices have been suggested for assessing zooplankton grazing effects at the whole lake level (Reynolds, 1994b) including zooplankton number (Lynch and Shapiro, 1981), total zooplankton biomass, and the ratio of biomass of zooplankton to phytoplankton. Matveev and Matveeva (1997) showed using enclosure experiments that grazing was significant when the Cladocera/Phytoplankton biomass ratio was greater than 0.1, and found in both Lakes Hume and Dartmouth that clear water phases occurred when this ratio was exceeded, even when cyanobacteria were present.

XV. Concluding remarks

We have examined many of the attributes of gas-vacuolate cyanobacteria that have been proposed to account for their success. These have been compared between species of cyanobacteria, and between the cyanobacteria and the eukaryotic micro-algae in an attempt to assess their relative importance. The following comments summarise these considerations:

- Buoyancy and its regulation provide gas-vacuolate cyanobacteria with a significant advantage over the micro-algae. In deep waters where turbulence intensity is low buoyancy regulation is an asset, as it enables the cyanobacteria to maximise their growth conditions, largely by circumventing the vertical separation in resources that develops in thermally stratified waters. When the velocity of turbulent eddies is more than 15 times the floating velocity then the population will be evenly distributed and the advantage of buoyancy for accessing the illuminated surface layers is negated. However, even in well mixed surface layers of stratified lakes and in moderately turbulent shallow waters the attribute of buoyancy can reduce losses by sedimentation and provide the cyanobacteria with an advantage over non-buoyant micro-algae. The benefit of buoyancy is intimately linked with the nature of the turbulent mixing regime.
- Within the cyanobacteria there are major differences between species which influence their sinking/floating velocity, in particular the size of the biomass unit varies from small, single filaments in *Oscillatoria* to complex, large colonies in *Microcystis*. The small size and slow sinking/floating velocities of *Oscillatoria* permits it to position itself at its preferred low light environment where it often forms deep water maxima in stratified lakes. The large colonial forms of *Microcystis* can move quickly enabling it to take advantage of habitats which oscillate between stratified and mixed conditions.
- There are marked differences in the light harvesting complexes and pigment composition of micro-algae and cyanobacteria, however, it is still unclear whether cyanobacteria respond differently to micro-algae to changing light intensity, day length and spectral composition. Laboratory studies suggest that micro-algae do not respond to changing photoperiod but to light intensity. Field studies suggest that certain cyanobacteria (eg *Microcystis*) may take two days to adapt to high irradiances, but green algae (eg *Scenedesmus*) respond more quickly which provides an explanation of why *Scenedesmus* may dominate in turbid, continuously mixed, shallow water bodies. However, the energy required to synthesise gas vesicles in the competing cyanobacteria may be sufficient to push the energy balance in favour of green algae under low light conditions. The phycoerythrin rich species such as *Oscillatoria agardhii* and *Gloeotrichia echinulata* are low light adapted having a wide spectral activity and have an advantage over both micro-algae and phycoerythrin poor cyanobacteria, especially in water bodies where the most penetrating waveband is 560-570 nm.
- There is no evidence to support the hypothesis that bloom-forming cyanobacteria prefer higher temperatures than micro-algae. However, in *Microcystis* carbohydrate metabolism is depressed

at temperatures below 8-12°C, which influences buoyancy regulation.

- There does not appear to be any consistent phylogenetic differences in phosphorus kinetics between the cyanobacteria and the micro-algae. Some cyanobacteria are able to store significant quantities of phosphorus, but in the case of *Microcystis* and *Gloeotrichia* their success appears just as dependent on behavioural features that enable them to utilise this attribute.
- The general response to phosphorus reduction is a lowering of the cyanobacterial biomass. However, the extent and nature of the response is dependent upon the internal phosphorus load and the magnitude of the persistent external load. Where initial phosphorus concentrations are in excess of requirements, the effects of P-removal may be minimal. If internal loads are significant then responses may take many years to eventuate. If the FRP in the trophogenic layer during the growing season falls below 10 mg m^{-3}, a significant reduction of biomass occurs as the community becomes P-limited.
- The source of inorganic nitrogen available to cyanobacteria may influence their success. It has been proposed that non-nitrogen fixing cyanobacteria are favoured by ammonium, micro-algae by nitrate and the nitrogen-fixing cyanobacteria by nitrogen deficiency. The evidence for this is not conclusive particularly for organisms using ammonia, but laboratory studies do support the suggestion that the cyanobacteria are disadvantaged when using nitrate under low light conditions.
- Significant nitrogen fixation in the plankton is restricted to the heterocystous cyanobacteria and provides them with a distinct advantage when inorganic nitrogen concentrations fall below ca. 100 mg m^{-3}; nevertheless, these populations will often be limited ultimately by the available phosphorus.
- The overwhelming evidence is that TN:TP ratios do not *per se* influence the occurrence of planktonic cyanobacteria. However, nitrogen fixers (eg *Anabaena* spp.) may occur if the TN:TP ratio is low and correctly predicts that inorganic nitrogen is limiting. On the other hand *Gloeotrichia echinulata* may dominate when TN:TP ratios are high, but the P-content of the water column is insufficient to support growth.
- Very few data are available to test critically the proposal that cyanobacteria are better able to utilise low inorganic carbon concentrations than the micro-algae. Again this does not appear to be a consistent phylogenetic difference, but more extensive measurements are required.
- Grazing can reduce cyanobacterial populations if they are comprised of sufficiently small units (filaments or colonies), but if they reach a large size before substantial increases in the most effective grazers then the likelihood of control is diminished. In general the reduced grazing on cyanobacteria is due to their large size and high density and this can provide an advantage over small microalgae.

In conclusion, we suggest that the occurrence and abundance of various types of gas-vacuolate cyanobacteria is not reliant on any one particular environmental stimulus, but depends on a complex interplay of factors. A flow chart has been used in an attempt to portray these interactions and to highlight the role of environmental conditions in supporting the growth of particular species of gas-vacuolate cyanobacteria (Fig. 10). The flow chart is not definitive and is unlikely to be a dependable tool for predicting either the likelihood of cyanobacterial blooms or their identity; it is presented simply for illustrative purposes. It consists of a series of questions assessing the major environmental conditions believed to influence the success of gas-vacuolate cyanobacteria and in this way it summarises what we view as key interactions. The critical values that are proposed at each level to discriminate between the success or not of gas-vacuolate cyanobacteria must be viewed as hypotheses only and are unlikely to be robust. Most of these are covered in more detail within the text.

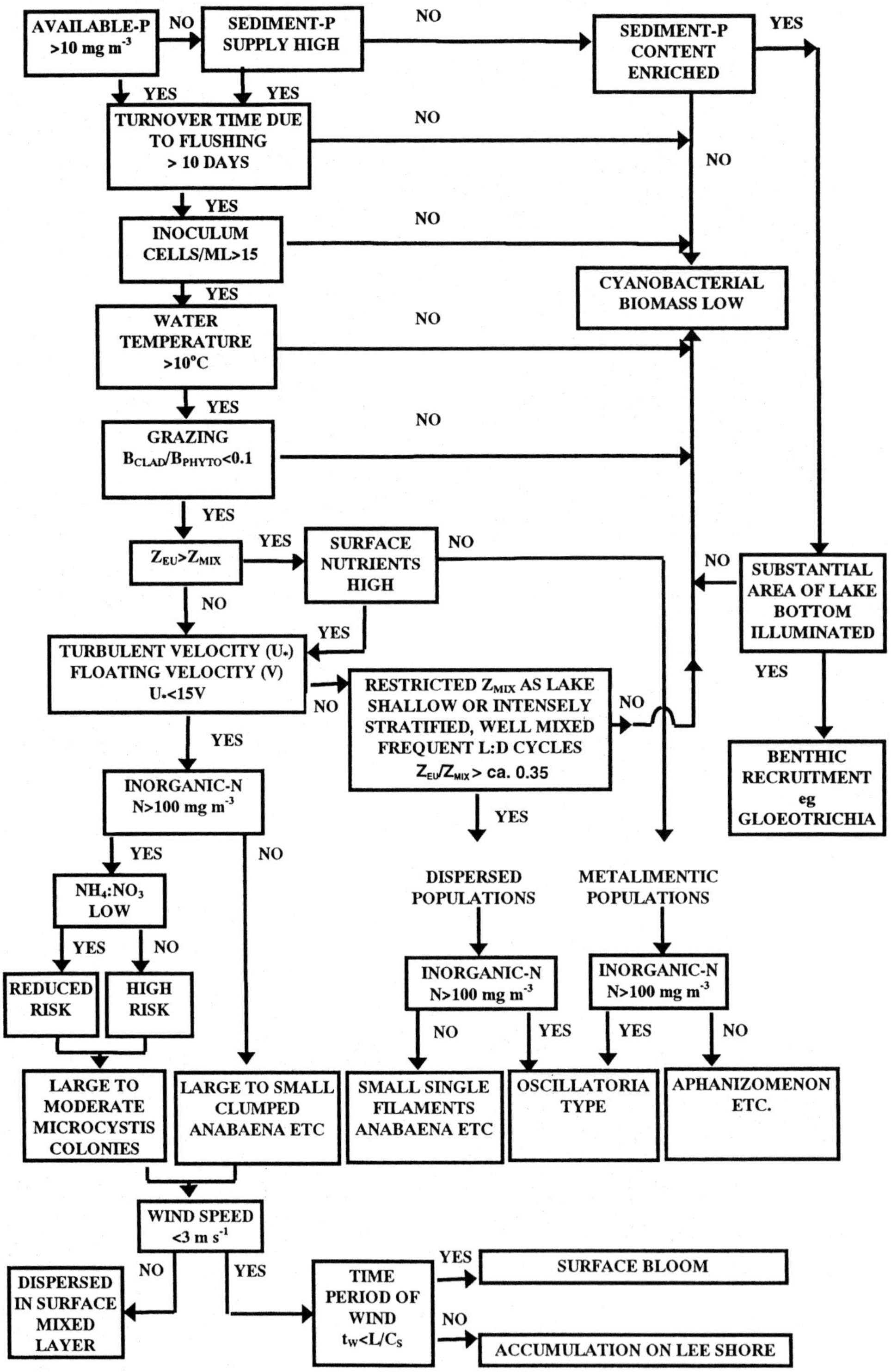

Fig. 10 Flow chart summarising prominent environmental characteristics supporting the development of cyanobacterial blooms and selecting for particular genera. The text provides further detail on some of these components. Key: B_{CLAD} biomass of cladocerans, B_{PHYTO} biomass of phytoplankton, z_{eu} euphotic depth, z_{mix} depth of mixing, u_* shear velocity, V floating or sinking velocity of cyanobacteria, t_w time that the wind blows, L lake fetch, c_s surface current speed.

References

Abeliovich A and Shilo M (1972) Photooxidative death in blue-green algae. J Bacteriol 111: 682-689

Allen MM (1984) Cyanobacterial cell inclusions. Annu Rev Microbiol 38: 1-25

Allison EM and Walsby AE (1981) The role of potassium in the control of turgor pressure in a gas-vacuolate blue-green alga. J Exp Bot 32: 241-249

Barbiero RP and Welch EB (1992) Contribution of benthic blue-green algal recruitment to lake populations and phosphorus translocation. Freshwater Biol 27: 249-260

Blomqvist P, Pettersson A and Hyenstrand P (1994) Ammonium-nitrogen: A key regulatory factor causing dominance of non-nitrogen-fixing cyanobacteria in aquatic systems. Arch Hydrobiol 132: 141-164

Bothe H (1982) Nitrogen fixation. In: Carr NG and Whitton BA (eds) The Biology of Cyanobacteria: pp 87-104. Blackwell Scientific Publications, Oxford

Bowen CC and Jensen TE (1965) Blue-green algae: fine structure of the gas vacuoles. Science 147: 1460-1462

Brainerd KE and Gregg MC (1993) Diurnal restratification and turbulence in the oceanic surface mixed layer. 1. Observations. J Geophys Res 98: 22645-22656

Brook AJ, Baker AL and Klemer AR (1971) The use of a turbidimetry in studies of the population dynamics of phytoplankton populations, with special reference to *Oscillatoria agardhii* var *isothrix*. Mitt Verh Int Verein Limnol 19: 244-252

Brookes JD, Ganf GG and Burch MD (1994) Separation of forms of *Microcystis* from *Anabaena* in mixed populations by the application of pressure. Aust J Mar Freshwater Res 45:863-868

Burch MD, Baker PD, Steffensen DA, Bursill DB, Bain DB, Ganf, G.G. and Brookes JD (1994). Critical flow and blooms of the cyanobacterium *Anaebaena circinalis* in the River Murray, South Australia. *Proceedings of the Environmental Flows Seminar.* Australian Water and Wastewater Association, Canberra: 17-21.

Burns CW (1968) The relationship between body size of filter feeding cladocera and the maximum size of particle ingested. Limnol Oceanogr 13:675-678

Burns CW (1987) Insights into zooplankton-cyanobacteria interactions derived from enclosure studies. N Z J Mar Freshwater Res 21:477-482

Canter HM, Heaney SI and Lund JWG (1990) The ecological significance of grazing on planktonic populations of cyanobacteria by the ciliate *Nassula*. New Phytol 17:247-263

Carmichael WW (1994) The toxins of cyanobacteria. Scientific American January:64-72

Clarke KB and Walsby AE (1988) Improvements in the treatment of water to remove gas vacuolate cyanobacteria. UK patent application 2 196 330 A

Codd GA (1994) Blue-green algal toxins: water-borne hazards to health. In: Golding AMB, Noah N and Stanwell-Smith R (eds) Water and Public Health, pp 271-278. Smith-Gordon, UK

Cooke GD, Welch EB, Peterson SA and Newroth PR (1993) Restoration and Management of Lakes and Reservoirs. Second edn. Lewis Publishers, Boca Raton, pp 548

Coronil T, Lara C and Guerrero MG (1993) Shift in carbon flow and stimulation of amino-acid turnover induced by nitrate and ammonium assimilation in *Anacystis nidulans*. Planta 189: 461-467

Cullen P (1994) Report from expert seminar on algal blooms in the Hawkesbury River. A report to the Hawkesbury-Nepean Catchment Management Trust, NSW, Australia 19 pp.

Cullen P and Forsberg C (1988) Experiences with reducing point sources of phosphorus to lakes. Hyrobiologia 170: 321-336

Damerval T, Castets A-M, Houmard J and Tandeau de Marsac N (1991) Gas vesicle synthesis in the cyanobacterium *Pseudanabaena* sp.: occurrence of a single photoregulated gene. Mol Microbiol 5: 657-664

Davey MC and Walsby AE (1985) The form resistance of sinking algal chains. Br Phycol J 20: 243-248

Deacon C and Walsby AE (1990) Gas vesicle formation in the dark, and in light of different irradiances, by the cyanobacterium *Microcystis* sp. Br Phycol J 25: 133-139

Denman KL and Gargett AE (1983) Time and space scales of vertical mixing and advection of phytoplankton in the upper ocean. Limnol Oceanogr 28: 801-815

Droop MR (1973) Some thoughts on nutrient limitation in algae. J Phycol 9: 264-272

Edmondson WT (1970) Phosphorus, nitrogen and algae in Lake Washington after diversion of sewage. Science 196: 690-691

Edmondson WT and Lehman JT (1981) The effect of changes in the nutrient income on the condition of Lake Washington. Limnol Oceanogr 26:1-29

Elser JJ, Marzolf ER and Goldman CR (1990) Phosphorus and nitrogen limitation of phytoplankton growth in the freshwaters of North America: a review and critque of experimental enrichments. Can J Fish Aquat Sci 47: 1468-1477

Falconer IR (1993) Algal toxins in seafood and drinking water. Academic Press, London

Falkowski PG and LaRoche J (1991) Acclimation to spectral irradiance in algae. J Phycol 27: 8-14

Flameling IA and Kromkamp J (1997) Photoacclimation of *Scenedesmus protuberans* (Chlorophyceae) to fluctuating irradiances simulating vertical mixing. J Plankton Res 19: 1011-1024

Flores E and Herrero A (1994) Assimilatory nitrogen metabolism and its regulation. In: Bryant DA (cd), The Molecular Biology of Cyanobacteria: pp 487-517. Kluwer Academic Publishers, Dordrecht, The Netherlands

Foy RH (1980) The influence of surface to volume ratio on the growth rates of planktonic blue-green algae. Br Phycol J 15: 279-289

Foy RH (1993) The phycocyanin to chlorophyll a ratio and other cell components as indicators of nutrient limitation in two planktonic cyanobacteria subjected to low-light exposures. J Plankton Res 15:1263-1276

Foy RH and Gibson CE (1982) Photosynthetic characteristics of planktonic blue-green algae: the response of twenty strains grown under high and low light. Br Phycol J 17: 169-182

Foy RH, Gibson CE and Smith RV (1976) The influence of daylength, light intensity and temperature on the growth rates of planktonic blue-green algae. Br Phycol J 11: 151-163

Foy RH and Smith RV (1980) The role of carbohydrate accumulation in the growth of planktonic *Oscillatoria* species. Br Phycol J 15: 139-150

Francis G (1878) Poisonous Australian Lake. Nature (London) 18: 11-12

Ganf GG (1974) Diurnal mixing and the vertical distribution of phytoplankton in a shallow equatorial lake (Lake George, Uganda). J Ecol 62: 611-629

Ganf GG (1983) An ecological relationship between *Aphanizomenon* and *Daphnia pulex*. Aust J Mar Freshwater Res 34: 755-773

Ganf GG and Oliver RL (1982) Vertical separation of light and available nutrients as a factor causing replacement of green algae by blue-green algae in the plankton of a stratified lake. J Ecol 70: 829-844

Ganf GG, Oliver RL and Walsby AE (1989) Optical properties of gas-vacuolate cells and colonies of *Microcystis* in relation to light attenuation in a turbid, stratified reservoir (Mount Bold Reservoir, South Australia). Aust J Mar Freshwater Res 40: 595-611

Gantt E, Ohki K and Fujita Y (1984) *Trichodesmium thiebautii*; structure of a nitrogen-fixing marine blue-green alga (Cyanophyta). Protoplasma 119:188-196

Garcia-Gonzalez M, Sivak MN, Guerrero MG, Preiss J and Lara C (1992) Depression of carbon flow to the glycogen pool induced by nitrogen assimilation in intact cells of *Anacystis nidulans*. Physiol Plant 86: 360-364

George DG and Edwards RW (1976) The effect of wind on the distribution of chlorophyll a and crustacean plankton in a shallow eutrophic reservoir. J Appl Ecol 13: 667-690

George PR and Bradby K (1993) The Peel-Harvey Catchment Management Program. Fertilizer Research 36: 185 -192

Gibson CE (1978) Carbohydrate content as an ecological tool in the study of planktonic blue-green algae. Verh Int Verein Limnol 20: 630-635

Gibson CE (1985) Growth rate, maintenance energy and pigmentation of planktonic cyanophyta during one-hour light:dark cycles. Br. Phycol. J 20: 155-161

Gibson CE and Smith RV (1982) Freshwater plankton. In: Carr NG and Whitton BA (eds) The Biology of Cyanobacteria. pp 463-489. Blackwell Scientific Publications, Oxford

Gibson CE, Smith RV and Stewart DA (1988) A long-term study of the phosphorus cycle in Lough Neagh, Northern Ireland. Int Rev Ges Hydrobiol 23:

Glazer AN (1982) Phycobilisomes: structure and dynamics. Annu Rev Microbiol 36: 173-198

Glazer AN, Gindt YM, Chan CF and Sauer K (1994) Selective disruption of energy flow from phycobilisomes to photosystem I. Photosynth Res 40: 167-173

Gliwicz ZM (1968) The use of anaesthetizing substrates in studies on the food habits of zooplankton communities. Ekol Polska 16: 279-295

Gons HJ and Mur LR (1979) Growth of *Scenedesmus protuberans* Fritsch in light limited continuous cultures with a light-dark cycle. Archiv Hydrobiol 85:41-56

Gorham PR and Carmichael WW (1988) Hazards of freshwater blue-green algae (cyanobacteria). In: Lembe, CA and Waaland, JR (eds) Algae and Human Affairs, pp 403-431. Cambridge University Press, Cambridge

Gorham E, Lund JWG, Sanger JE and Dean WE, Jr (1974) Some relationships between algal standing crop, water chemistry and sediment chemistry in the English Lakes. Limnol Oceanogr 19:601-617

Gotham IJ and Rhee G-Y (1981) Comparative kinetic studies of nitrate-limited growth and nitrate uptake in phytoplankton in continuous culture. J Phycol 17:309-314

Grant NG and Walsby AE (1977) The contribution of photosynthate to turgor pressure rise in the planktonic blue-green alga *Anabaena flos-aquae*. J Exp Bot 28:409-415

Grossman AR, Bhaya D, Apt KE and Kehoe DM (1995) Light-harvesting complexes in oxygenic photosynthesis:Diversity, control, and evolution. Annual Review of Genetics 29:231-288

Guerrero MG and Lara C (1987) Assimilation of inorganic nitrogen. In: Fay P and van Baalen C (eds) The Cyanobacteria, pp 163-186. Elsevier , Amsterdam

Haney JF (1973) An *in situ* examination of the grazing activities of natural zooplankton communities. Arch Hydrobiol 72: 87-132

Haney JF (1987) Field studies on zooplankton-cyanobacteria interactions. N Z J Mar Freshwater Res 21: 467-475

Harris GP (1986) Phytoplankton Ecology: Structure, Function and Fluctuation. Chapman and Hall, London, 384 pp.

Hayes PK and Walsby AE (1986) The inverse correlation between width and strength of gas vesicles in cyanobacteria. Br Phycol J 21:191-197

Healey FP (1978) Physiological indicators of nutrient deficiency in algae. 21: 34-41

Healey FP (1982) Phosphate. In: Carr, NG and Whitton, BA (eds) The Biology of Cyanobacteria. pp. 105-124. Blackwell Scientific Publications, Oxford

Heaney SI, Corry JE and Lishman JP (1992) Changes of water quality and sediment phosphorus of a small productive lake following decreased phosphorus loading. In:Sutcliffe DW and Jones JG (eds) Eutrophication: Research and Application to Water Supply, pp 119-131. Freshwater Biological Association, Ambleside, UK

Hecky RE, Campbell P and Hendzel LL (1993) The stoichiometry of carbon, nitrogen, and phosphorus in particulate matter of lakes and oceans. Limnol Oceanogr 38:709-724

Hecky RE and Kilham P (1988) Nutrient limitation of phytoplankton in freshwater and marine environments: a review of recent evidence on the effects of enrichment. Limnol Oceanogr 33:796-822

Horne AJ and Commins ML (1987) Macronutrient controls on nitrogen fixation in planktonic cyanobacterial populations. N Z J Mar Freshwater Res 21:413-423

Horne AJ and Commins ML (1989) Phytoplanktonic monotony 1 Low spatial heterogeneity for nitrogen fixation, algae, nutrients, and temperature, but not for zooplankton, in Lake Rotongaio, New Zealand. Arch Hydrobiol Beih Ergebn Limnol 32:83-97

Howarth RW, Marino R, Lane J and Cole JJ (1988) Nitrogen fixation in freshwater, estuarine, and marine ecosystems. 1. Rates and importance. Limnol Oceanogr 33:669-687

Humphries R and Robinson S (1995) Assessment of the success of the Peel-Harvey estuary system management strategy - a Western Australian attempt at integrated management. Wat Sci Tech 32:255-264

Humphries SE and Imberger J (1982) The influence of the internal structure and the dynamics of Burrinjuck Reservoir on phytoplankton blooms. Centre for Water Research University of Western Australia. Evironmental Dynamics Report No ED-82-023

Humphries SE and Lyne VD (1988) Cyanophyte blooms : the role of cell buoyancy. Limnol Oceanogr 33: 79-91

Hutchinson PA and Webster IT (1994) On the distribution of blue-green algae in lakes: Wind-tunnel tank experiments. Limnol Oceanogr 39:374-382

Ibelings BW (1992) Cyanobacterial waterblooms: the role of buoyancy in water columns of varying stability. Ph.D thesis Laboratory of Microbiology, University of Amsterdam pp 1-171

Ibelings BW and Mur LR (1992) Microprofiles of photosynthesis and oxygen concentration in *Microcystis* sp scums. FEMS Microb Ecol 86:195-203

Ibelings BW, Mur LR, Kinsman R and Walsby AE (1991a) Microcystis changes its buoyancy in response to the average irradiance in the surface mixed layer. Arch Hydrobiol 120:385-401

Ibelings BW, Mur LR and Walsby AE (1991b) Diurnal changes in the buoyancy regulation and vertical distribution in populations of Microcystis in two shallow lakes. J Plankton Res 13:419-436

Imberger J (1985) The diurnal mixed layer. Limnol Oceanogr 30: 737-770

Imberger J and Hamblin PF (1982) Dynamics of lakes, reservoirs, and cooling ponds. Annual Review of Fluid Mechanics14: 153-187

Istvánovics V, Pettersson K, Rodrigo MA, Pierson D, Padisák J and Colom W (1993) *Gloeotrichia echinulata*, a colonial cyanobacterium with a unique phosphorus uptake and life strategy. J Plankton Res 15: 531-552

Jensen JP, Jeppesen E, Olrik K and Kristensen P (1994) Impact of nutrients and physical factors on the shift from cyanobacterial to chlorophyte dominance in shallow Danish lakes. Can J Fish Aquat Sci 51: 1692-1699

King DL (1970) The role of carbon in eutrophication. J Water Poll Contr Fed 42: 2035—2051

Kinsman R, Ibelings BW and Walsby AE (1991) Gas vesicle collapse by turgor pressure and its role in buoyancy regulation by *Anabaena flos-aquae*. J Gen Microbiol 137:1171-1178

Kirk JTO (1983) Light and photosynthesis in aquatic ecosystems. Cambridge University Press, Cambridge

Kirk JTO and Oliver RL (1995) Optical closure in an ultraturbid lake. Journal of Geophysical Research 100:13221-13225

Klebahn H (1895) Gasvakuolen, ein Bestandtiel der Zellen der Wasserblutebildenden Phycochromaceen. Flora, Jena 80: 241-282

Klemer AR (1976) The vertical distribution of *Oscillatoria agardhii* var. *isothrix*. Arch Hydrobiol 78: 343-362

Klemer AR (1978) Nitrogen limitation of growth and gas vacuolation in *Oscillatoria rubescens*. Verh Int. Verein Limnol 20:2293-2297

Klemer AR (1991) Effects of nutritional status on cyanobacterial buoyancy, blooms, and dominance, with special reference to inorganic carbon. Can J Bot 69:1133-1138

Klemer AR, Cullen JJ, Mageau MT, Hanson KM and Sundell RA (1996) Cyanobacterial buoyancy regulation: The paradoxical roles of carbon. J Phycol 32: 47-53

Klemer AR, Detenbeck N, Grover J and Fung Basino T (1988) Macronutrient interactions involved in cyanobacterial bloom formation. Verh Int Verein Limnol 23: 1881-1885

Klemer AR, Feuillade J and Feuillade M (1982) Cyanobacterial blooms: carbon and nitrogen limitation have opposite effects of the buoyancy of *Oscillatoria*. Science 215: 1629-1631

Kohler J (1992) Influence of turbulent mixing on growth and primary production of *Microcystis aeruginosa* in the hypertrophic Buatzen Reservoir. Arch Hydrobiol 123: 413-429

Konopka A (1984) Effect of light-nutrient interactions on buoyancy regulation by planktonic cyanobacteria. In: Reddy CA and Klug MJ (eds) Current Perspectives in Microbial Ecology : Proceedings of the Third International Symposium on Microbial Ecology, Michigan State University, 1983 Aug 7-1983 Aug 12, pp 41-48. American Society of Microbiology, Washington

Konopka A (1989) Metalimnetic cyanobacteria in hard-water lakes: Buoyancy regulation and physiological state. Limnol Oceanogr 34: 1174—1184

Konopka A, Kromkamp J and Mur LR (1987) Regulation of gas vesicle content and buoyancy in light- or phosphorus-limited cultures of *Aphanizomenon flos-aquae* (Cyanophyta). J Phycol 23: 70-78

Konopka A and Schnur M (1980) Effect of light intensity on macromolecular synthesis in cyanobacteria. Microb Ecol 6: 291-301

Konopka AE, Klemer AR, Walsby AE and Ibelings BW (1993) Effects of macronutrients upon buoyancy regulation by metalimnetic *Oscillatoria agardhii* in Deming Lake, Minnesota. J Plankton Res 15:1019-1034

Kromkamp J (1987) Formation and functional significance of storage products in cyanobacteria. N Z J Mar Freshwater Res 21: 457-465

Kromkamp J, Konopka A and Mur LR (1986) Buoyancy regulation in a strain of *Aphanizomenon flos-aquae* (Cyanophyceae): the importance of carbohydrate accumulation and gas vesicle collapse. J Gen Microbiol 132:2113-2121

Kromkamp J, Konopka A and Mur LR (1988) Buoyancy regulation in light-limited continuous cultures of *Microcystis aeruginosa*. J Plankton Res 10:171-183

Kromkamp J, Van Den Heuvel A and Mur LR (1989) Phosphorus uptake and photosynthesis by phosphate-limited cultures of the cyanobacterium *Microcystis aeruginosa*. Br Phycol J 24: 347-355

Kromkamp J and Walsby AE (1990) A computer model of buoyancy and vertical migration in cyanobacteria. J Plankton Res 12: 161-183

Levasseur M, Thompson PA and Harrison PJ (1993) Physiological acclimation of marine phytoplankton to different nitrogen sources. J Phycol 29: 587-595

Levine SN and Lewis WM (1987) A numerical model of nitrogen fixation and its application to Lake Valencia, Venezuela. Freshwater Biol 17: 265-274.

Lewis WM (1976) Surface/volume ratio:implications for phytoplankton morphology. Science 192: 885-887

Lewis WM (1978) Dynamics and succession of the phytoplankton in a tropical lake: Lake Lanao, Philippines. J Ecol 66: 849—880

Lewis MR, Cullen JJ and Platt T (1984a) Relationships between vertical mixing and photoadaptation of phytoplankton: similarity criteria. Mar Ecol Progr Ser 15: 141-149

Lewis MR, Horne EPW, Cullen JJ, Oakey NS and Plat T (1984b) Turbulent motions may control phytoplankton photosynthesis in the upper ocean. Nature 311: 49-50.

Lund JWG (1965) The ecology of the freshwater phytoplankton. Biological Reviews 40: 231-293

Lund JWG and Reynolds CS (1982) The development and operation of large, limnetic enclosures in Blelham Tarn, English Lake District, and their contribution to phytoplankton ecology. Prog Phycol. Res. 1: 1-65

Lynch M (1980) Aphanizomenon blooms: alternate control and cultivation by *Daphnia pulex*. In: Kerfoot W C (ed.) Evolution and Ecology of Zooplankton Populations, pp 299-304. University Press, New England, USA

Lynch M and Shapiro J (1981) Predation, enrichment, and phytoplankton community structure. Limnol Oceanogr 26: 86-102

MacIntyre S (1993) Mixing in the euphotic zone of a shallow, turbid lake: Consequences for the phytoplankton. Limnol Oceanogr 38: 798-817

MacIntyre S (1996) Turbulent eddies and their implications for phytoplankton within the cuphotic zone of Lake Biwa, Japan. Japanese Journal of Limnology 57: 395-410

MacIntyre S (1998) Turbulent mixing and resource supply to phytoplankton. Coastal and Estuarine Studies, Special Issue on Physical Processes in Lakes and Oceans 54: 539-567

Matveev V and Matveeva L (1997) Grazer control and nutrient limitation of phytoplankton biomass in two Australian reservoirs. Freshwater Biology 38: 49-65

McAuliffe TF and Rosich RS (1989) Review of artificial destratification of water storages in Australia. Urban Water Research Assoc. of Australia, Research Report No 9, 235 pp.

McNown JS and Malaika J (1950) Effects of particle shape of settling velocity an low Reynolds numbers. Transactions American Geophysical Union 31: 74-82

Meffert M-E and Krambeck H-J (1977) Planktonic blue-green algae of the *Oscillatoria redekei* group. Arch Hyrobiol 79: 149-171

Menday DC and Buck AA (1972) A shock-wave technique to collapse the vacuoles of blue-green algae. Water Res 6: 279-284

Monismith SG, Imberger J and Morison ML (1990) Convective motions in the sidearm of a small reservoir. Limnol Oceanogr 35: 1676-1702

Mur LR (1983) Some aspects of the ecophysiology of cyanobacteria. Annales Microbiology 134B:61-72

Ochoa de Alda JAG, Tapia MI, Fabrice F, Llama MJ and Serra JL (1996) Changes in nitrogen source modify distribution of excitation energy in the cyanobacterium *Phormidium laminosum*. Physiol Plant 97:69-78

Oliver RL (1990) Optical properties of waters in the Murray-Darling Basin, south-eastern Australia. Aust J Mar Freshwater Res 41:581-601

Oliver RL (1993) Phosphorus dynamics and bioavailability in turbid waters. In: Proceedings of the National Workshop on Phosphorus in Australian Freshwaters, Wagga Wagga, 1993 Jun 10-1993 Jun 11, pp 93-103. Charles Sturt University, Environmental and Analytical Laboratories, Wagga Wagga, NSW

Oliver RL (1994) Floating and sinking in gas-vacuolate cyanobacteria. J Phycol 30:161-173

Oliver RL and Ganf GG (1988) The optical properties of a turbid reservoir and its phytoplankton in relation to photosynthesis and growth (Mount Bold Reservoir, South Australia). J Plankton Res 10:1155-1177

Oliver RL, Kinnear AJ and Ganf GG (1981) Measurements of cell density of three freshwater phytoplankters by density gradient centrifugation. Limnol Oceanogr 26:285-294

Oliver RL and Walsby AE (1984) Direct evidence for the role of light-mediated gas vesicle collapse in the buoyancy regulation of *Anabaena flos-aquae* (cyanobacteria). Limnol Oceanogr 29: 879-886

Ormerod JG (1992) Physiology of the photosynthetic prokaryotes. In: Mann NH and Carr NG (eds) Photosynthetic Prokaryotes, pp 93-120. Plenum Press, New York

Paerl HW (1983) Partitioning of CO2 fixation in the colonial cyanobacterium *Microcystis aeruginosa*: mechanism promoting formation of surface scums. Appl Environ Microbiol 46: 252-259

Paerl HW (1988) Growth and reproductive strategies of freshwater blue-green algae (cyanobacteria). In:Sandgren CD (ed.) Growth and Reproductive Strategies of Freshwater Phytoplankton, pp 261-315. Cambridge University Press, Cambridge UK

Paerl HW (1988) Nuisance phytoplankton blooms in coastal, estuarine, and inland waters. Limnol Oceanogr 33: 823-847

Paerl HW, Bland PT, Bowles ND and Haibach ME (1985) Adaptation to high-intensity, low-wavelength light among surface blooms of the cyanobacterium *Microcystis aeruginosa*. Appl Environ Microbiol 49:1046-1052

Paerl HW, Tucker J and Bland PT (1983) Carotenoid enhancement and its role in maintaining blue-green algal (*Microcystis aeruginosa*) surface blooms. Limnol Oceanogr 28:847-857

Paerl HW and Ustach JF (1982) Blue-green algal scums: an explanation for their occurrence during freshwater blooms. Limnol Oceanogr 27:212-217

Pearsall WH (1932) Phytoplankton in the English Lakes II. The composition of the phytoplankton in relation to dissolved substances. J Ecol 20:241-262

Perakis SS, Welch EB and Jacoby JM (1996) Sediment-to-water blue-green algal recruitment in response to alum and environmental factors. Hydrobiologia 318:165-177

Pettersson K, Herlitz E and Istvánovics V (1993) The role of *Gloeotrichia echinulata* in the transfer of phosphorus from sediments to water in Lake Erken. Hydrobiologia 253:123-129

Pick FR and Lean DRS (1987) The role of macronutrients (C, N, P) in controlling cyanobacterial dominance in temperate lakes. N Z J Mar Freshwater Res 21: 425-434

Post AF, De Wit R and Mur LR (1985a) Interactions between temperature and light intensity on growth and photosynthesis of the cyanobacterium *Oscillatoria agardhii*. J Plankton Res 7: 487-495

Post AF, Eijgenraam F and Mur LR (1985b) Influence of light period length on photosynthesis and synchronous growth of the green alga *Scenedesmus protuberans*. Br Phycol J 20: 391-397

Post AF, Loogman JG and Mur LR (1986) Photosynthesis, carbon flows and growth of *Oscillatoria agardhi* Gomont in environments with a periodic supply of light. J Gen Microbiol 132: 2129-2136

Preston I, Stewart WDP and Reynolds CS (1980) Bloom-forming cyanobacterium *Microcystis aeruginosa* overwinters on sediment surface. Nature 288:365-367

Reynolds CS (1971) The ecology of the planktonic blue-green algae in the North Shropshire meres. Field Studies 3: 409—432

Reynolds CS (1979) Seston sedimentation: experiments with *Lycopodium* spores in a closed system. Freshwater Biol 9: 55-76

Reynolds CS (1980) Phytoplankton assemblages and their periodicity in stratifying lake systems. Holarctic Ecology 3:141-159

Reynolds CS (1984a) The Ecology of Freshwater Phytoplankton. Cambridge University Press, Cambridge

Reynolds CS (1984b) Phytoplankton periodicity: the interactions of form, function and environmental variability. Freshwater Biol. 14:111-142

Reynolds CS (1987) Cyanobacterial water-blooms. In: Callow JA (ed.) Advances in Botanical Research, Vol 13, pp pp 67-143. Academic Press, London

Reynolds CS (1989a) Physical determinants of phytoplankton succession. Sommer U (ed.) Plankton Ecology, pp 9-56. Springer Verlag

Reynolds CS (1989b) Relationships among the biological properties, distribution and regulation of production by planktonic cyanobacteria. Toxicity Assessment 4:229-255

Reynolds CS (1990) Temporal scales of variability in pelagic environments and the response of phytoplankton. Freshwater Biol 23: 25-53

Reynolds CS (1992) Eutrophication and the management of planktonic algae: what Vollenweider couldn't tell us. In:Sutcliffe DW and Jones JG (eds) Eutrophication: Research and Application to Water Supply, pp 4-29. Freshwater Biological Association, Ambleside, UK

Reynolds CS (1993) Swings and roundabouts: Engineering the environment of algal growth. In: White KN and others (eds) Urban Waterside Regeneration: Problems and Prospects, 330-349. Ellis Horwood, New York

Reynolds, CS (1994a) The role of fluid motion in the dynamics of phytoplankton in lakes and rivers. In: Giller PS, Hildrew AG and Raffaelli DG (eds) Aquatic Ecology: Scale, Pattern and Process, pp 141-187. Blackwell Scientific Publications, Oxford

Reynolds CS (1994b) The ecological basis for the successful biomanipulation of aquatic communities. Arch Hydrobiol 130: 1-33

Reynolds CS, Jaworski GHM, Cmiech HA and Leedale GF (1981) On the annual cycle of the blue-green alga *Microcystis aeruginosa* Kütz Emend Elenkin. Phil Trans R Soc Lond B 293: 419-477

Reynolds CS, Oliver RL and Walsby AE (1987) Cyanobacterial dominance : the role of buoyancy regulation in dynamic lake environments. N Z J Mar Freshwater Res 21: 379-390

Reynolds CS and Walsby AE (1975) Water-blooms. Biological Reviews 50: 437-481

Reynolds CS, Wiseman SW and Clarke MJO (1984) Growth- and loss-rate responses of phytoplankton to intermittent artificial mixing and their potential application to the control of planktonic algal biomass. J Appl Ecol 21: 11-39

Richardson K, Beardall J and Raven JA (1983) Adaptation of unicellular algae to irradiance: an analysis of strategies. New Phytol 93: 157-191

Riegman R and Mur LC (1984) Regulation of phosphate uptake kinetics in *Oscillatoria agardhii.* Arch Microbiol 139:28-32

Riegman R, Rutgers M and Mur LR (1985) Effects of photoperiodicity and light irradiance on phosphate-limited *Oscillatoria agardhii* in chemostat cultures. Arch Microbiol 142:66-71

Robarts RD and Zohary T (1984) *Microcystis aeruginosa* and underwater light attenuation in a hypertrophic lake (Hartbeespoort Dam, South Africa). J Ecol 72:1001-1017

Robarts RD and Zohary T (1987) Temperature effects on photosynthetic capacity, respiration, and growth rates of bloom-forming cyanobacteria. N Z J Mar Freshwater Res 21:391-399

Round FE (1971) The growth and succession of algal populations in freshwaters. Mitt Internat Verein Limnol 19:70-99

Sakamoto M (1966) Primary production by phytoplankton community in some Japanese lakes and its dependence upon lkae depth. Arch Hydrobiol 62: 1-28

Sas H (1989) Lake restoration by reduction of nutrient loading: expectations, experiences, extrapolations. Academia Verlag Richarz, St Augustin pp 497

Scheffer M, Rinaldi S, Gragnani A, Mur LR and van Nes EH (1997) On the dominance of filamentous cyanobacteria in shallow, turbid lakes. Ecology 78: 72-282

Schindler DW (1971) Carbon, nitrogen and phosphorus and the eutrophication of freshwater lakes. J Phycol 7: 321-329

Schindler DW (1977) Evolution of phosphorus limitation in lakes. Science 195: 260-262

Schoenberg SA and Carlson RE (1984) Direct and indirect effects of zooplankton grazing on phytoplankton in a hypertrophic lake. Oikos 42:291-302

Seip KL, Sas H and Vermu S (1992) Nutrient-chlorophyll trajectories across trophic gradients. Aquatic Science 54: 58-76

Shapiro J (1990) Current beliefs regarding dominance by blue-greens: The case for the importance of CO_2 and pH. Verh Int Verein Limnol 24:38-54

Shapiro J (1997) The role of carbon dioxide in the initiation and maintenance of blue-green dominance in lakes. Freshwater Biology 37:307-323

Sherman BS, Webster, IT, Jones GJ, and Oliver RL (1998) Transitions between *Aulacoseira* and *Anabaena* dominance in a turbid river weir pool. Limnol Oceanogr 43: 1902-1915

Simon RD (1987) Inclusion bodies in the cyanobacteria: cyanophycin, polyphosphate, polyhedral bodies. In: Fay P and van Baalen C (eds) The Cyanobacteria, pp 199-225. Elsevier, Amsterdam

Smith IR (1982) A simple theory of algal deposition. Freshwater Biol 12: 445-449

Smith VH (1982) The nitrogen and phosphorus dependence of algal biomass in lakes: An empirical and theoretical analysis. Limnol Oceanogr 27: 1101-1112

Smith VH (1983) Low nitrogen to phosphorus ratios favour dominance by bluegreen algae in lake phytoplankton. Science 221:669-671

Sommer U (1985) Comparison between steady state and non-steady state competition: experiments with natural phytoplankton. Limnol Oceanogr 30:335-346

Spigel RH and Imberger J (1987) Mixing processes relevant to phytoplankton dynamics in lakes. N Z J Mar Freshwater Res 21:361-377

Stauffer RE (1988) Sampling strategies and associated errors in estimating epilimnetic chlorophyll in eutrophic lakes. Water Resources Research 24:1459-1469

Steinberg CEW and Gruhl E (1992) Physical measures to inhibit planktonic cyanobacteria. In: Sutcliffe DW and Jones JG (eds) Eutrophication: Research and application to water supply, pp 163-184. Freshwater Biological Association, Cumbria

Steinberg CEW and Hartmann HM (1988) Planktonic bloom-forming cyanobacteria and the eutrophication of lakes and rivers. Freshwater Biol 20: 279-287

Steinberg CEW and Zimmerman GM (1989) Intermittent destratification: a therapy against cyanobacteria in lakes. Environmental Technology Letters 9: 337-350

Stockner JG and Shortreed KS (1988) Response of *Anabaena* and *Synechococcus* to manipulation of nitrogen:phosphorus ratios in a lake fertilization experiment. Limnol Oceanogr 33: 1348-1361

Stommel H (1949) Horizontal diffusion due to oceanic turbulence. J Mar Res 8: 199-255

Syrett PJ (1981) Nitrogen metabolism of microalgae. Can Bull Fish Aquat Sci 210: 182-210

Talling JF (1976) The depletion of carbon dioxide from lake water by phytoplankton. J Ecol 64: 79-121

Talling JF (1987) The phytoplankton of Lake Victoria (East Africa). Arch Hydrobiol Beih Ergebn Limnol 25: 229-256

Talling JF and Heaney SI (1983) Evaluation of historical data on the nutrient status of Esthwaite Water, Cumbria. Commissioned Report from the Freshwater Biological Association to the Nature Conservancy Council, pp 44

Tandeau de Marsac N and Houmard J (1993) Adaptation of cyanobacteria to environmental stimuli: new steps towards molecular mechanisms. FEMS Microbiol Rev 104: 119-190

Tapia MI, Ochoa de Alda JAG, Llama MJ and Serra JL (1996) Changes in intracellular amino acids and organic acids induced by nitrogen starvation and nitrate or ammonium resupply in the cyanobacterium *Phormidium laminosum.* Planta 198:526-531

Thomas RH and Walsby AE (1985) Buoyancy regulation in a strain of *Microcystis.* J Gen Microbiol 131:799-809

Thomas RH and Walsby AE (1986) The effect of temperature on recovery of buoyancy by *Microcystis*. J Gen Microbiol 132: 1665-1672

Thompson PA, Levasseur ME and Harrison PJ (1989) Light-limited growth on ammonium vs nitrate: What is the advantage for marine phytoplankton? Limnol Oceanogr 34: 1014-1024

Thornton JA (1982) Lake McIlwaine; the eutophication and recovery of a tropical African man-made lake. dr W Junk Publishers The Hague, pp 251

Tilman D, Kilman SS and Kilham P (1982) Phytoplankton community ecology: the role of limiting nutrients. Annu Rev Ecol Syst 13:349-372

Tilzer MM (1987) Light-dependence of photosynthesis and growth in cyanobacteria: implications for their dominance in eutrophic lakes. N Z J Mar Freshwater Res 21: 401-412

Toetz DW (1981) Effects of whole lake mixing on water quality and phytoplankton. Water Res 15:1205-1210

Trimbee AM and Harris GP (1984) Phytoplankton population dynamics of a small reservoir : use of sedimentation traps to quantify the loss of diatoms and recruitment of summer bloom-forming blue-green algae. J Plankton Res 6: 897-918

Trimbee AM and Prepas EE (1987) Evaluation of total phosphorus as a predictor of the relative biomass of blue-green algae with emphasis on Alberta Lakes. Can J Fish Aquat Sc 44: 1337-1342

Turpin DH (1991) Effects of inorganic N availability on algal photosynthesis and carbon metabolism. J Phycol 27:14-20

Utkilen HC, Oliver RL and Walsby AE (1985) Buoyancy regulation in a red *Oscillatoria* unable to collapse gas vacuoles by turgor pressure. Arch Hydrobiol 102:319-329

van Baalen C (1987) Nitrogen fixation. In: Fay P and van Baalen C (eds) The Cyanobacteria, pp 187-198. Elsevier, Amsterdam

van Liere L and Mur LR (1980) Occurrence of *Oscillatoria agardhii* and some related species, a survey. In Baric J and Mur LR (eds) Hypertrophic Ecosystems, pp 67-77. Junk, The Hague

van Liere L and Walsby AE (1982) Interactions of cyanobacteria with light. In: Carr NG and Whitton BA (eds) The Biology of Cyanobacteria, pp 9-45. Blackwell Scientific Publications, Oxford

van Rijn J and Shilo M (1985) Carbohydrate fluctations, gas vacuolation, and vertical migration of scum-forming cyanobacteria in fishponds. Limnol Oceanogr 30: 1219-1228

Vanni MJ and Temte J (1990) Seasonal patterns of grazing and nutrient limitation of phytoplankton in a eutrophic lake. Limnol Oceanogr 35: 697-709

Viner AB and Smith IR (1973) Geographical, historical and physical aspects of Lake George. Proc R Soc Lond B184: 235-270

Visser PM, Ibelings BW, Van Der Veer B, Koedood J and Mur LR (1996) Artificial mixing prevents nuisance blooms of the cyanobacterium Microcystis in Lake Nieuwe Meer, the Netherlands. Freshwater Biol 36: 435-450

Vollenweider RA (1968) Scientific Fundamentals of the Eutrophication of Lakes and Flowing Waters, with Particular Reference to Nitrogen and Phosphorus as factors in Eutrophication. OECD, Paris

Walsby AE (1971) The pressure relationships of gas vacuoles. Proc Ro Soc Lond B 178: 301-326

Walsby AE (1980) The water relations of gas-vaculate prokaryotes. Proc Ro Soc Lond B 208: 73-102

Walsby AE (1987) Mechanisms of buoyancy regulation by planktonic cyanobacteria with gas vesicles. Fay P and Van Baalen C (eds) The Cyanobacteria: A Comprehensive Review, pp 376-392. Elsevier, Amsterdam

Walsby AE (1992) The control of gas-vacuolate cyanobacteria. In:Sutcliffe DW and Jones JG (eds) Eutrophication: Research and Application to Water Supply, pp 150-162. Freshwater Biological Association, Ambleside, UK

Walsby AE (1994) Gas vesicles. Microbiol Rev 58: 94-144

Walsby AE and Bleything A (1988) The dimensions of cyanobacterial gas vesicles in relation to their efficiency in providing buoyancy and withstanding pressure. J Gen Microbiol 134: 2635-2645

Walsby AE and Booker MJ (1980) Changes in buoyancy of a planktonic blue-green alga in response to light intensity. British Phycological Journal 15: 311-319

Walsby AE, Hayes PK, Boje R and Stal LJ (1997) The selective advantage of buoyancy provided by gas vesicles for planktonic cyanobacteria in the Baltic Sea. New Phytol. 136:407-417

Walsby AE and Reynolds CS (1980) Sinking and floating. Morris IG (ed.) The Physiological Ecology of Phytoplankton, pp 371-412. Blackwell Scientific, Oxford

Walsby AE, Reynolds CS, Oliver RL and Kromkamp J (1989) The role of gas vacuoles and carbohydrate content in the buoyancy and vertical distribution of *Anabaena minutissima* in Lake Rotongaio, New Zealand. Ergeb Limnol 32:1-25

Walsby AE, Utkilen HC and Johnsen IJ (1983) Buoyancy changes of a red coloured *Oscillatoria agardhii* in Lake Gjersjøen, Norway. Arch Hydrobiol 97:18-38

Ward AK and Wetzel RG (1980) Interactions of light and nitrogen source among planktonic blue-green algae. Arch Hydrobiol 90:1-25

Webster IT (1990) Effect of wind on the distribution of phytoplankton cells in lakes. Limnol Oceanogr 35:989-1001

Webster IT and Hutchinson PA (1994) Effect of wind on the distribution of phytoplankton cells in lakes revisited. Limnol Oceanogr 39:365-373

Webster IT, Jones GJ, Oliver RL, Bormans M and Sherman B (1996) Control strategies for cyanobacterial blooms in weir pools. CSIRO Centre for Environmental Mechanics Technical Report No 119

Whitton BA and Peat A (1969) On *Oscillatoria redekei* van Goor. Arch Microbiol 68:362-376

Wolk CP, Ernst A and Elhai J (1994) Heterocyst metabolism and development. In: Bryant DA (ed.), The Molecular Biology of Cyanobacteria, pp 769-823. Kluwer Academic Publishers, Dordrecht, The Netherlands

Wood MD and Oliver RL (1995) Fluorescence transients in response to nutrient enrichment of nitrogen- and phosphorus-limited *Microcystis aeruginosa* cultures and natural phytoplankton populations : a measure of nutrient limitation. Aust J Plant Physiol 22:331-340

Wyman M and Fay P (1986a) Underwater light climate and the growth and pigmentation of planktonic blue-green algae (Cyanobacteria) I. The influence of light quantity. Proc R Soc Lond B 277:367-380

Wyman M and Fay P (1986b) Underwater light climate and the growth and pigmentation of planktonic blue-green algae (Cyanobacteria) II. The influence of light quality. Proc R Soc Lond B 227:381-393

Zohary T and Pais Madeira AM (1990) Structural, physical and chemical characteristics of *Microcystis aeruginosa* hyperscums from a hypertrophic lake. Freshwater Biol 23:339-352

Zohary T and Robarts RD (1989) Diurnal mixed layers and the long-term dominance of *Microcystis aeruginosa.* J Plankton Res 11:25-48

Chapter 7

Picoplankton and Other Non-Bloom-Forming Cyanobacteria in Lakes

John Stockner[1], Cristiana Callieri[2] and Gertrud Cronberg[3]

[1]Eco-Logic Ltd, 2614 Mathers Avenue, West Vancouver, British Columbia, Canada V7V 2J4
[2]C.N.R. - Istituto Italiano di Idrobiologia, Largo V. Tonolli 50, 28048 - Pallanza, Italy
[3]Department of Ecology\Limnology, University of Lund, Sweden

Summary

Studies on the picocyanobacteria, the major non-bloom forming group in lakes, have increased markedly in the past two decades. These cyanobacteria fall into two major groups, those that predominantly occur as solitary cells (here termed Pcy) and those that occur primarily as colonies (CPcy). The single-celled picoplankton have received most ecological study, though the colonial forms are better known taxonomically. The two groups are distributed worldwide and are ubiquitous in all types of lakes of varying trophic condition. The single-celled Pcy populations tend to be predominant in large, deep oligo-mesotrophic lakes, while the colonial species (CPcy) find optimal conditions in warmer, shallower and more nutrient rich (meso-eutrophic) lakes during the summer. The two groups are euryphotic and appear capable of adapting to a wide variety of light conditions, with CPcy more surface oriented and Pcy often reaching sub-surface peaks at irradiance levels 20 - 50% of surface values. Growth of the Pcy ranges from 0.1 d^{-1} to about 3.0 d^{-1} with doubling times from 7 h to 7 d. Their natural

B. A. Whitton and M. Potts (eds), The Ecology of Cyanobacteria, 195-231.
© 2000 *Kluwer Academic Publishers. Printed in the Netherlands.*

population growth rates are highly variable and appear in most cases to be in balance with loss rates, primarily from grazing. Pcy contribute substantially to total primary production within the euphotic zone, most significantly in ultra-oligotrophic lakes, but also in some meso- and eutrophic lakes, showing optimal growth at high N:P ratios (>20 molar) and more limitation by nitrogen than phosphorus.

Pcy seem less resistant to UV-B radiation than larger-celled algae, and their photosynthesis may be more severely impacted because of their small size and high metabolic activity. Both Pcy and CPcy are generally most common in lakes with neutral to slightly alkaline pH conditions, and Pcy disappear from lakes below pH: < 6.0. The CPcy appear to be resistant to grazing and may operate as energy shunts or 'sinks' in lakes. Pcy on the other hand are an important food source for many protozoan (nanoflagellates and ciliates) and microzooplankton (rotifers and nauphli) grazers in microbial food webs. A plea is made to researchers to increase their interest in CPcy populations so we may in future better understand their role in food webs and energy flows in lakes. We conclude with a comment on the role of Pcy and CPcy pelagic food webs under a warmer, more nutrient deplete and strongly stratified surface layer, of the type that may be more prevalent in a warmer climate expected in the early decades of the next century.

I. Introduction

This chapter discusses a group of cyanobacteria, the "non-bloom formers", that are common in lakes throughout the world, and abundant across a wide spectrum of trophic conditions. Yet, despite their ubiquity (Hawley and Whitton, 1991a), they are not a well-known group and, compared with bloom formers, little is known about their ecology. Though their abundance in meso-eutrophic lakes is often high enough to reduce transparency and cause water discoloration, they seldom create the blooms that are commonly associated with the larger, colonial cyanobacteria. Although studies of the ecology of colonial non-bloom formers are few, there have been sufficient studies within the past 15 years of picocyanobacteria in lakes and their role in food webs to warrant synthesis and review (Stockner, 1991; Weisse, 1993).

The non-bloom formers exhibit two common morphologies: single cells, (coccoid, rods or loose aggregates of cells) and colonies with diverse colonial morphology. In this chapter we designate the two groups as picocyanobacteria (Pcy), which are single celled (0.2 - 2.0 μm) and a major component of the autotrophic picoplankton (APP) community, or colonial picocyanobacteria (CPcy), which include all species whose predominant morph is colonial and have single cell sizes ranging from 0.5 - 3.0 μm. The separation of groups is not perfect because some Pcy can form loose aggregates or small colonies; however, if the predominant morph is a single-celled type, then they are considered Pcy. Furthermore, some non-bloom forming colonies overlap in size with the larger bloom-formers colonies and lie within the microphytoplankton size range (20 - 200 μm). However, if the average cell size in the colony is between 0.5 - 3.0 μm, they are treated as non-bloom forming CPcy.

Until the past few decades most research on non-bloom formers focused on the CPcy, because their colonies were easily seen by conventional microscopy, and their ubiquity in meso-eutrophic lakes of Northern Europe attracted the attention of early descriptive taxonomists (Lemmermann, 1904; Naumann, 1924; Skuja, 1948). CPcy also seem to have received more thorough systematic descriptions, along with some comment on habitat preference and distribution (Komárek, 1958; 1996; Cronberg, 1991; Komárková-Legnerová and Cronberg, 1994). Unfortunately the contribution of non-bloom-forming CPcy to carbon flow and food webs in meso- and eutrophic lakes has not received sufficient study to permit even a brief synthesis. With the advent of epifluorescence microscopy by the late 1970s, a more precise enumeration of Pcy populations became possible (Waterbury et al., 1979; Johnson and Sieburth, 1979), and their ubiquity and major contributions to carbon production and pelagic ecosystem function in both oceans and lakes was heralded in several reviews (Platt and Li, 1986; Fogg, 1986; Stockner and Antia, 1986). These reviews present little evidence for there being any major divergence between the physiological ecology of freshwater and marine forms of Pcy.

The emergence of Pcy as an important research topic for limnologists and oceanographers provides an opportunity to discuss the extent to which this

Abbreviations: Pcy - picocyanobacteria, CPcy - colonial picocyanobacteria, APP - autotrophic picoplankton, PE - phycoerthyrin containing Pcy, PC - phycocyanin containing Pcy, FDC – frequency of dividing cells, Chl - chlorophyll *a*, HNF - heterotrophic nanoflagellates, MFW - microbial food webs.

large and diverse group of cyanobacteria share a common ecology. We conclude the review with a plea for limnologists to pay more attention to these organisms, because the extent of their abundance and distribution in lakes may provide an important message about the changes likely to occur in pelagic community structure in the warmer world expected in the 21st century (Mann, 1993; Stockner, 1998).

II. Sampling, Preservation and Enumeration

Informed comment on the ecology of non-bloom formers relies on the investigator's commonsense in the selection of sampling techniques, preservation of the samples and a reliable protocol for counting the cells. The first step in sampling is to choose representative sites within the chosen basin(s) of a study lake, and then to obtain replicate samples (Kirchman, 1993). Most non-bloom formers show considerable variation in vertical distribution within the euphotic zone, so the second step is to choose between integrated or discrete depth sampling to best depict their spatial distribution (Weisse and Kenter, 1991; Fahnenstiel and Carrick, 1992; Hall and Vincent, 1994).

Because of their small size and neutral density, most Pcy and CPcy are buoyant, and it is important to add preservatives to reduce buoyancy, whether for Utermöhl microscopy to observe colonies or for epifluorescence microscopy to count single cells concentrated on membrane filters (Smayda, 1974; Wetzel and Likens, 1990). Fixation of Pcy with a preservative does not always ensure that cells will sink within a chamber, so the method of choice for enumeration of Pcy and APP has become epifluorescence microscopy (MacIsaac and Stockner, 1993). Preservation of Pcy cells also aids in enumeration by preventing cell destruction, and in ensuring cells settle in the oil layer between membrane and cover slip. Different fixation protocols using both formaldehyde and glutaraldehyde have been used but there was no consensus as to the most appropriate preservatives to use (Hall, 1991; Kuuppo-Leinikki and Kuosa, 1989; Bloem et al., 1986; Hayat, 1981). Of the three filters commonly used for Pcy collection and enumeration, the *Anopore*® results in the best retention of small particles (Stockner et al., 1990; Bertoni, 1997), and the most homogeneous distribution of cells on the filter surface (R. Bertoni, unpublished). *Anopores*® do not need to be blackened or pre-stained for observation of autofluorescing cells (Bertoni and Callieri, 1989), so it is possible to count Pcy cells on the filter immediately after addition of Cargille FF immersion oil, or, if necessary, to freeze them immediately for long-term storage (MacIsaac and Stockner, 1993).

Phycoerythrin (PE) and phycocyanin (PC) pigments in Pcy can easily be observed by epifluorescence microscopy under blue (BP450-490, FT510, LP520) and green (510-560, FT580, LP590) excitation. The autofluorescence of the pigments (Chl and phycobilins) is the result of the concordance of excitation wavelengths with the optimum light absorption. It is therefore important to use the suitable filter combinations as recommended by MacIsaac and Stockner (1993), as otherwise there is a risk of missing important components of the APP community.

Like the single-celled Pcy, live CPcy cells are neutrally buoyant and will remain floating in the plankton chamber, so cannot be counted without preservation using iodine (Lugol's solution). The cells become heavier and after some hours settle to the bottom of the chamber. The settled CPcy can then be examined with an inverted microscope using phase-contrast microscopy. Most CPcy cells are embedded in a diffuse mucilage which can be difficult to see using standard microscopy, hence staining with Indian ink or methylene-blue is often necessary to observe the different, structured mucilage of the CPcy colony. It is also possible to disintegrate the CPcy colonies with ultrasound, then filter the solution onto membrane filters and count the individual cells as rods or spheres in different size classes, but once the colony is destroyed it becomes difficult to separate the different taxa from one another. Samples that are subjected to ultrasound and then counted on membrane filters should first be preserved with formalin to 1 - 2% final solution.

Pcy generally appear as single cells, but sometimes occur as clusters or loosely aggregated colonies. To count the Pcy aggregates it is recommended that Lugol-preserved samples are used and allowed to settle overnight in a 25 mL chamber. The sample is counted with an inverted microscope using the standard Utermöhl method. Different types of aggregates are counted and the number of cells per colony is estimated.

III. The Non-Bloom Formers: What are They?

A. Species

Some of the most important species of Pcy and CPcy are listed in Table 1, together with salient morphological and ecological information. The single-celled Pcy were missed by most early investigators because of their minute size, or, if they were seen, they were either poorly described or simply ignored (Stockner and Antia, 1986; Stockner, 1991). They have been called 'μ algae' (Lund, 1961; Overbeck, 1962), described as 'little round green things' (LRGT) (Pearl, 1977) or as small coccoid or *Chlorella*-like cells. Recently, Komárek (1996) has proposed a new classification for the solitary living pico-cyanoprokaryotes (Pcy) based on cyto-morphological, molecular and biochemical approaches. He separated three taxonomic groups at the generic level - *Cyanobium*, *Synechococcus* and *Cyanothece dianae/cedrorum* types. He indicates that the first cluster is the one comprising the majority of picoplankton or Pcy organisms that most authors for lack of a better name call *Synechococcus*. While most pelagic Pcy are yet be described (Stockner and Antia, 1986), some *Synechococcus* species have been well studied (Bailey-Watts and Komárek, 1991; Maeda et al., 1992).

Most of the small pico-cell-sized CPcy belong to the chroococcal blue-green algae (cyanobacteria). The cell size ranges between 0.5 - 2 or 3 μm in diameter and the cell form is generally round or oval. The cells occur in colonies of different morphology and these colonial morphologies are often species specific, and can be used in separating species. The cells inside the CPcy colony can be loosely or densely packed, or can form pseudo-filaments or other net-like structures. In some species the cells are attached to mucilaginous stalks, which are centred in the middle of the colony (e.g. *Cyanonephron, Snowella*). In those lakes where CPcy are common, there is usually a mixture of species each with distinctive colony structures, which are quite readily identifiable. CPcy are found throughout the entire spectrum of lake trophic conditions, however, most tend to occur in more productive meso-eutrophic lakes. Some of the most common CPcy in fresh waters are species belonging to the genera *Aphanocapsa, Aphanothece, Chroococcus, Coelosphaerium, Cyanodictyon, Merismopedia, Snowella* and *Tetrarcus* (Table 1).

B. Pcy Community

It is important to identify the composition of all components of APP communities to get a better understanding of species interactions, spatial and temporal succession and survival strategies of mixed APP populations (Stockner, 1991). Because Pcy are numerically the predominant species in APP communities and have received the most attention in community studies, there has been a tendency to overlook pico-eukaryotes. Weisse (1993) suggested that the factors controlling pico-eukaryote distribution differ markedly from those affecting Pcy both in space and time, largely because of different nutritional and light requirements and potential growth rates. Pico-eukaryotes are often about one order of magnitude less abundant than Pcy (Table 2), and in temperate regions they tend to show a single population peak in spring or early summer (Stockner, 1991; Fahnenstiel et al, 1991a; Callieri, 1994a); they are predominant in humic, acidic lakes (pH < 6.2) (Søndergaard, 1989; Stockner and Shortreed, 1991; Findley and Kasian, 1990). However, in large, deep oligotrophic and mesotrophic lakes Pcy populations are always dominant and represent the major species within the APP community (Table 2) (Weisse, 1988; Callieri and Pinolini, 1995; Padisák et al., 1997).

IV. A Common Ecology?

A. Pcy Seasonal Distribution.

The seasonal cycle of Pcy populations in temperate lakes has been studied in lakes of all trophic types, and a variety of successional patterns have been noted (Weisse, 1993). Sufficient information is now available to permit some generalization to emerge about Pcy succession in oligo-mesotrophic temperate lakes. Typically there is a spring and late summer peak of abundance separated by low densities in June and July. Temperature and the onset of stratification are triggering variables for the inception of the spring peak, while grazing is largely responsible for the early summer decline (Weisse, 1993). This recurrent bimodal seasonal cycle appears to be dominated by at least two different populations or 'strains' of Pcy, each responsible for one of the peaks. This new perspective, validated by recent findings on the genetic diversity of morphologically similar

Table 1. Common picocyanobacteria (Pcy) from different habitats.

Species	Structure of colonies	Single cell shape	Single cell size (μm)	Ecological niche	Country	Reference
Aphanocapsa delicatissima W. & G. S. West	Spherical to irregular colonies in diffuse mucilage, < 50 μm. Cells evenly spread in the colony	spherical	0.5-1	probably cosmopolitan, in eutrophic lakes	temperate zone, England, Norway, Sweden, North America	Komárková-Legnerová & Cronberg, 1994
A. incerta (Lemmermann) Cronberg & Komárek	Colonies spherical to irregular sometimes flattened. Cells densely, irregularly packed in the colony.	spherical	0.5-2 (2.7)	eutrophic lakes and ponds	cosmopolitan	Cronberg & Komárek, 1994
A. holsatica (Lemmermann) Cronberg & Komárek	Colonies irregularly shaped, clathrate with cells ± densely aggregated	spherical	about 1	eutrophic lakes and ponds	cosmopolitan, very common in Denmark, Sweden, Finland, Germany	Cronberg & Komárek, 1994
A.elachista W. & G.S. West	Colonies spherical to oval, few-celled. Cells solitary or in pairs, sparsely positioned.	spherical	1.5-1.8 (2)	eutrophic waters	tropical distribution, in temperate zone during summer	W. & G. S. West, 1894
A. nubilum Komárek & Kling	Colonies irregular with cells ±densely packed	spherical	1.2-1.5	mesotrophic lakes	Africa, cosmopolitan	Komárek & Kling 1991
A. planctonica (G. M. Smith) Komárek & Anagnostidis	Colonies irregular with cells sparsely distributed	spherical	2-3	oligo-to eutrophic lakes	N. America, Europe, temperate zone	Komárek & Kling 1991
Aphanothece minutissima (W. West) Komárková-Legnerová & Cronberg	Colonies of irregular form, cloudy clusters <120 μm in diam. Cells in diffluent, colourless mucilage	widely oval or shortly rod-like	0.8-1 (1.6) x 1-1.5-2	freshwater, from oligotrophic to eutrophic lakes, ponds and pools	temperate zones, Sweden, Finland, Germany	Komárková-Legnerová & Cronberg, 1994
A. bachmannii Komárková-Legnerová & Cronberg	Colonies usually flat, clathrate with elongate cells in ± parallel rows.	elongate, oval to cylindrical	0.5-1 x 0.8-2	freshwater and brackish water, meso- to eutrophic lakes and ponds	temperate zones, Denmark, Finland, Sweden, N. Germany, Baltic Sea	Komárková-Legnerová & Cronberg, 1994
A. clathrata W. & G.S. West	Colonies irregular, large, flat, clathrate with cells ± evenly distributed	rod-like, straight or slightly curved	0.5-0-7-1 x 2.5-3.5-4	freshwater and brackish water, oligo-eutrophic lakes and ponds	cosmopolitan, common in Finland, Sweden, Germany, Baltic Sea	Komárková-Legnerová & Cronberg, 1994

Table 1 (continued). Common picocyanobacteria (Pcy) from different habitats.

Species	Structure of colonies	Single cell shape	Single cell size (µm)	Ecological niche	Country	Reference
A. smithii Komárková-Legneerová & Cronberg	Colonies spherical to oval, sometimes elongated, of varying size. Cells densely packed	short cylindrical to oval	1-1.2-1.5 x 1..8-2.2-3.5	freshwater, oligo- to eutrophic lakes and ponds	Canada, Finland, Sweden, N. Germany	Komárková-Legnerová & Cronberg 1994
Chroococcus microscopicus Komárková-Legnerová & Cronberg	Colonies cloud-like with cells regularly arranged in groups of 4-6 cells, surrounded by sheath.	spherical or hemispherical after division	0.7-1	meso-to eutrophic lakes,	Known only from Sweden, but probably with wider distribution	Komárková-Legnerová & Cronberg 1994
C. aphanocapsoides Skuja	Colonies spherical, oval or irregular, size up to 100 µm in diam. with cells gathred in groups of 2-8 cells.	spherical to hemispherical after division, with individual sheath	1.8-2	oligo- to eutrophic freshwater lakes	Known only from Sweden, but probably with wider distribution	Skuja,1964;
C. minimus (Keissler) Lemmermann	Colonies with 2-8 cells. Cells regularly arranged in groups within coulourless mucilage.	spherical to hemispherical after division, ±individual sheath	1.7-3	oligo- to mesotrophic lakes	known from the temperate zone of northern hemisphere	Cronberg & Komárek 1994
Coelosphaerium punctiferum Komárek & Komárková-Legnerová	Colonies spherical to sub-spherical, up to 80µm in diam. with cells situated just beneath the colony surface	spherical	± 1	tropical or warmer regions, mesotrophic lakes	Canada, Africa	Komárek & Komárková-Legnerová 1992
C. minutissimum Lemmermann	Colonies planktic, spherical to oval, 20-30 (170) µm in diam. with cells just beneath the colony surface	spherical	0.8-1.2	oligo- slightly eutrophic waters, also brackish waters	northern Europe, also in the Baltic Sea	Komárek & Komárková-Legnerová1992
C. subarcticum Komárek & Komárková-Legnerová	Colonies spherical or oval, sometimes as two hemispherical colonies. Cells regularly arranged in ± one layer near the colony surface	spherical	1.2-1.6	in oligo- to eutrophic waters	temperate zone	Komárek & Komárková-Legnerová,199 2
Cyanocatena planctonica Hindák	Colonies irregular, spherical, oval with hyaline mucilage, < 20 µm diameter	cells cylindrical to oval with ring-like precipitates on the surface	0.5-1.2 x 1-2.5	planktic eutrophic ponds and lakes	Austria, Czech Republic, Germany Greece, Slovakia	Hindák 1975

Table 1 (continued). Common picocyanobacteria (Pcy) from different habitats.

Species	Structure of colonies	Single cell shape	Single cell size (µm)	Ecological niche	Country	Reference
Cyanodictyon imperfectum Cronberg & Weibull	net-like colonies made of small, loose pseudofilaments	spherical, just before division slightly elongate, precipitate ferric rings	0.4-0.8.-1	planktic in eutrophic lakes	Germany, Greece, Sweden, Canada, East African lakes, Laos, probably cosmopolitan	Cronberg & Weibull 1981
C. reticulataum (Lemmermann) Geitler	Colonies initially formed by rows of single cells, older colonies net-like, three-dimensional.	spherical	1-1.5	freshwater, meso-eutrophic lakes	Austria, Denmark, Germany, Greece, Russia, Canada	Komárková-Legnerová & Cronberg 1994
C. tubiforme Cronberg	Colonies with uniserial to multiserial rows of cells, sometimes clathrate, < 400 µm diameter	spherical, hemispherical after division, or widely oval to rod-shaped	1.9-2.2 x 2.2-3.8	freshwater, planktic in shallow eutrophic lakes	Sweden, N. Germany, the Baltic states	Cronberg 1988
C. planctonicum Meyer	Colonies three-dimensional and irregular elongated net-like, < 150 µm	oval to almost rod-like	0.8-1 x 1.5	freshwater, planktic in eutrophic lakes	Denmark, Sweden,N. Germany	Meyer 1994
C. filiforme Komárková-Legenerová & Cronberg	Filaments unbranched sometimes forming loose net-like bundles	rod-like cells in short filaments	0.2-0.5 x 1.5x2.5	freshwater, planktic, meso- and eutrophic lakes	Sweden, Canada	Komárková-Legnerová & Cronberg 1994
Cyanogranis ferruginea (Wawrik) Hindák	Colonies with 3-50-(100) cells, <12 µm diameter	spherical to oval with black ferric precipitate	0.4-1 x 0.6-1.5	planktic in freshwaters, fish ponds, slightly alkaline	Austria, Czech Republic, Germany Greece, Slovakia	Hindák 1982
Cyanonephron styloides Hickel	Spherical to ellipsoidal colonies with cells sitting on stalks	cells elongate, kidney-shaped	0.8-1.2 x 2.3-4.5	freshwater, brackish water, meso- to hypertrophic lakes	Finland, N. Germany, Sweden, Baltic Sea	Hickel 1985
Lemmermanniel -la parva Hindák	Spherical colonies with cells beneath the surface, 55-120 (180) µm in diameter	cells short cylindrical or oval	0.8-1 x 1-1.5 (1.8)	freshwater, brackish water, meso-eutrophic	Slovakia, the Baltic Sea	Hindák 1985
L. pallida (Lemmermann) Geitler	Spherical colonies, with cells beneath the surface, <85 µm diameter	rod-like, cylindrical	0.5-1.6 x (0.7) 1.1-3.7 (4,3)	freshwater, brackish water, planktic meso- to eutrophic lakes	Sweden, N. Germany, Denmark, Finland, Russia, Baltic Sea	Komárková & Cronberg 1985
Merismopedia warmingiana Lagerheim	Colonies regular, flat. Cells grouped in quadrates, slightly irregular, 4-16 (64) cells in colony	spherical to slightly elongate before division	0.5-1 (1.2)	eutrophic, polluted waters also in brackish and saline ponds, sometimes in masses	cosmopolitan, Europe	Lagerheim, 1883

Table 1 (continued). Common picocyanobacteria (Pcy) from different habitats.

Species	Structure of colonies	Single cell shape	Single cell size (μm)	Ecological niche	Country	Reference
M. tenuissima Lemmermann	Colonies flat, ± rectangular. Cells grouped in quadrates, regular, 16 - 100 cells per colony	spherical or oval, after division hemispherical	0.4-1.6-(2-2.5?)	eutrophic, stagnant freshwaters, fish ponds, in brackish waters,	cosmopolitan, common in Europe	Komárková-Legnerová & Cronberg 1994
Pannus spumosus Hickel	Colonies ± spherical, clathrate. Cells densely arranged with individual fine sheath	spherical	1-1.5	from brackish stagnant waters	N. Germany, Sweden, Baltic Sea	Hickel,1991
Romeria elegans (Woloszynska) Koczwara	Trichomes 20-45 μm long, easily fragmented into single cells, witout mucilage	cylidrical, slightly bent cells with rounded ends	1-2 x 3.5-7	common in eutrophic waters, also in brackish water	temperate zone, also in the Baltic Sea	Geitler 1932
R. leopoliense (Raciborski) Koczwara	Trichomes with 4-6 cells, twisted in half circles, no mucilage sheath	elongated cells, slightly bent with cut ends	0.8-1.2 x 3-5	common in eutrophic waters,	temperate zone	Geitler 1932
Snowella atomus Komárek & Hindák	Colonies planktic, ±spherical, up to 25 μm in diam. Cells sitting on the outer end of muci-laginous stalks joined at the centre of the colony	spherical, after division hemispherical	0.6-1.4	mesotrophic lakes	central Europe	Komárek & Komárková-Legnerová, 1992
Synechococcus bacillaris Butcher	Cells solitary or forming short chains	spherical, oval to cylindical	(1.5)-1,7-4.5	planktic in reservoirs, seawater	England	Butcher 1952
S. gaarderi Ålvik	Cells solitary or in pairs	spherical to ellipsoidal	1.2-1.5 x 1.5-2.8	planktic in brackish and marine waters close to the coast	Norway	Ålvik 1934
S. elongatus Nägeli	Cells solitary or in clusters, sometimes making small chains	cells oval to cylindrical, straight or slightly curved	1-2-3 x 2-9	mostly sub-aerophytic,	temperate zone	Komárek 1976
Synechococcus nidulans (Pringsheim) Komárek	Cells solitary, sometimes mass-devlopment	rod-like without mucilage	0,4-1.3 -2.2 x 1.5-8.5	freswater, pools and small ponds, eutrophic	temperate zone	Bourrelly, 1985
S. plancticus Drews et al.	Cells solitary	oval to rod-like	0.9-1.1 x 1.5-3	planktic , freshwater, waste water ponds	Czech Republic. Germany	Drews et al. 1961
Synechococcus sp. sensu Waterbury et al.	Cells solitary	spherical, oval to rod-like, red-coloured	0.8-1.7 x 1.8-2.2	planktic, marine	Pacific Ocean, Northern Gulf Stream, Sargasso Sea, Arabian Gulf	Waterbury et al. 1979

Table 1 (continued). Common picocyanobacteria (Pcy) from different habitats.

Species	Structure of colonies	Single cell shape	Single cell size (μm)	Ecological niche	Country	Reference
Tetrarcus ilsteri Skuja	Colonies with 2-4 cells in ring-like groups dispersed in colour-less mucilage, < 150 μm idiameter	Cells half-moon formed with yellow-green to blue-viol&t colour	1-1.5 x 2,5-4	humic, oligotrophic lakes and ponds	known from Lithuania, Finland, Sweden, Switzerland	Skuja 1932

Table 2. Average (or maximum when specified) number of picoplanktonic cells divied into prokaryotic (Pcy), eukaryotic and colonial (CPcy) in lakes of different trophic conditions. The percentage contribution of APP to total algal biomass is reported as a mean or as a range.

Lakes	Picoprokaryotes		Picoeukaryotes		Colonial forms of Syn- type cells		APP % on algal biomass	References
	cells mL^{-1}	%	cells mL^{-1}	%	cells mL^{-1}	%		
Oligotrophic lakes								
Lake Baikal	max 2000 x 10^3	> 80	max 400 x 10^3	< 20	irreg. clumps		9 (POC)	Nagata et al. 1994
Lake Maggiore, 1992	48.7 x 10^3	89	6.1 x 10^3	11	small clumps		0.3 - 27	Callieri & Pinolini 1995
Lake Maggiore, 1995	70 x 10^3	89	4.9 x 10^3	7	3.6 x 10^3	5		Callieri (unpublished)
Stechlinsee, Germany	199 x 10^3	95	6 x 10^3	3	4 x 10^3	2	4 - 38	Padisák, 1997
B.C.Clear south coastal	29 x 10^3	83	0.9 x 10^3	3	4.8 x 10^3	14		Stockner & Shortreed, 1991
Dystrophic north coastal	0.1 x 10^3	1	8 x 10^3	98	0.1 x 10^3	1		
B.C. interior plateau	23.6 x 10^3	86	1 x 10^3	4	2.7 x 10^3	10		
Chilko lake	58.9 x 10^3	88	0.6 x 10^3	1	7.2 x 10^3	11	29	Stockner & Shortreed, 1994
Lake Huron	25.4 x 10^3	60	9 x 10^3	21	7.4 x 10^3	17	0.5 - 50	Fahnenstiel & Carrick, 1992
Lake Michigan	32.1 x 10^3	54	14.5 x 10^3	24	12.6 x 10^3	21	0.5 - 50	
Mesotrophic lakes								
Lake Constance	93.5 x 10^3		?	< 10	?		10	Weisse & Kenter, 1991
Lake Biwa	max 703 x 10^3	95	max 37 x 10^3	5	?		?	Nagata et al., 1996
Eutrophic lakes								
Lake Kinneret	max 800 x 10^3	99	max 5.7 x 10^3	1	?		7	Malinsky-Rushansky et al.1995
Lake Aydat, France	594 x 10^3	63	329 x 10^3	35	17 x 10^3	2	?	Sime-Ngando, 1995
Hypereutrophic lakes								
Lake Apopka, Florida	13600 x 10^3		?		?		38	Carrick & Schelske, 1993

Synechococcus in Lake Constance (Ernst et al., 1995; Postius et al., 1996), suggests a possible adaptive response by Pcy strains to the seasonally variable physico-chemical factors in the euphotic zone.

In a 4 year study of APP communities in Lake Maggiore the abundance of Pcy gradually increased as the lake's nutrient loads declined, and the bimodal abundance pattern with a large spring peak is a prominent feature of the seasonal cycle (Fig. 1; Callieri and Stockner, in press). Large spring peaks are also common in eutrophic and hypereutrophic lakes (Sime-Ngando, 1995; Vörös et al., 1991), but not in warm-monomictic lakes (Stockner and Shortreed, 1991; Malinsky-Rushansky et al., 1995). The seasonal patterns found in seven Danish lakes (Søndergaard, 1991), several Canadian lakes (Pick and Agbeti, 1991), Lake Biwa, Japan (Maeda et al., 1992), and in the northern English lakes (Hawley and Whitton, 1991b) all lack the spring peak, there being only a summer or autumn maximum. Such findings, obtained from abundance estimates at only one selected depth, are not really representative of the euphotic zone and provide, at best, only a 'snapshot' description of seasonal variation in surface waters. As well demonstrated by a 4 year time series from Lake Maggiore, only by using integrated samples in longer-term studies can a clear picture of Pcy seasonal and depth distribution patterns emerge (Fig. 2). This kind of sampling should be accompanied by a study of the vertical distribution with frequent samples down the water column.

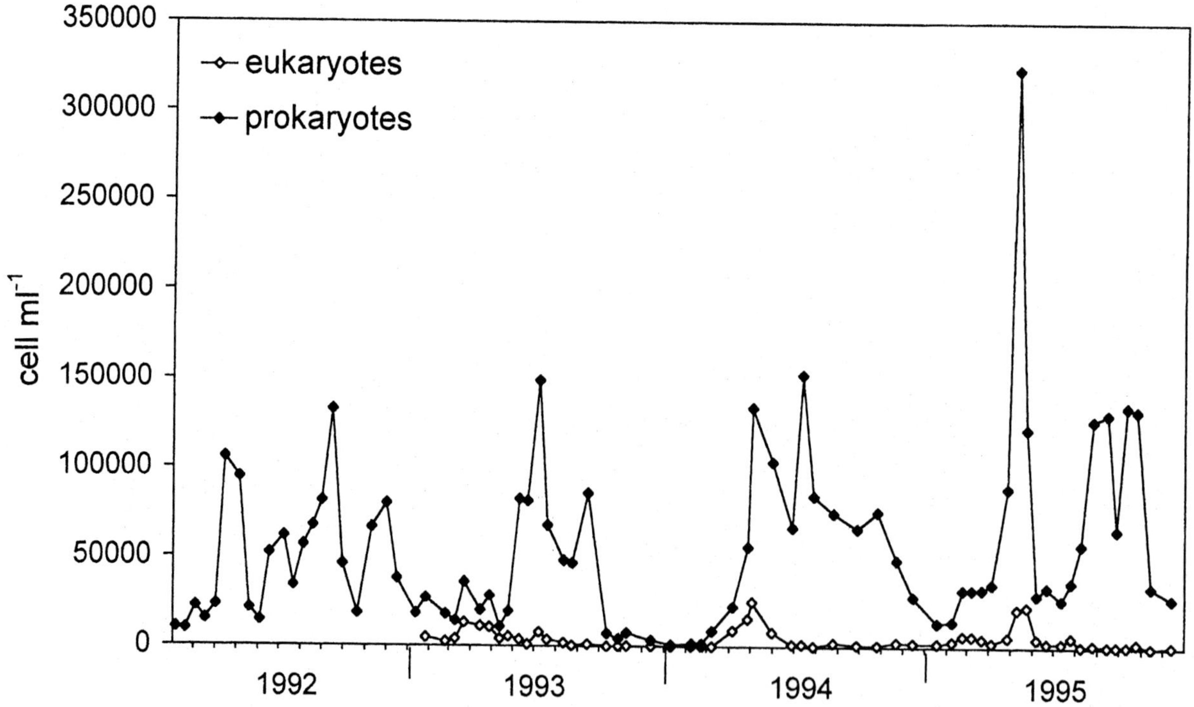

Fig. 1. Lake Maggiore APP abundance from 1992 to 1995. Eukaryotes and prokaryotes (Pcy) were counted separately (Callieri and Stocker, in press).

B. Pcy Vertical Distribution.

Studies of the vertical distribution of populations of Pcy have provided important information about their response to changing physical and biological variables within the euphotic zone. Though Pcy cells are small and their settling rate negligible, their abundance and distribution within the water column can change rapidly with different thermal and light regimes, or with the presence or absence of predators (Weisse and Kenter, 1991; Pernthaler et al., 1996b). Increases of turbulence in the mixed layer generally affect only phytoplankton whose settling velocity is higher than ~1 m d^{-1} (Ruiz et al., 1996), but because of the low sinking velocity of Pcy, their distribution within the water column is not likely to be affected by the lake's thermal structure. During thermal stratification, peak Pcy abundance has been noted in the lower metalimnion or upper hypolimnion in Lake Huron and Michigan (Fahnenstiel and Carrick, 1992), in the metalimnion below the thermocline in Lakes Constance, Maggiore and Stechlin (Weisse & Schweizer, 1991; Callieri and Pinolini, 1995; Padisák et al., 1997), in the metalimnion in Lake Baikal (Nagata et al., 1994), in the epilimnion in Lake Kinneret (Malinsky-Rushansky et al., 1995), and in the surface waters of Lake Biwa (Maeda et al., 1992) With the possible exception of Lake Biwa, where Pcy attain great abundance and their abundance is well correlated with the temperature curve, most studies report little or no relation between temperature profiles and Pcy abundance (Stockner and Shortreed, 1991).

Theoretically, underwater irradiance should affect the vertical Pcy distribution, as there is good experimental evidence relating the optimum growth rate of *Synechococcus* cultures to specific light irradiances (Waterbury et al., 1986). However, Pcy have been classified as euryphotic, and are

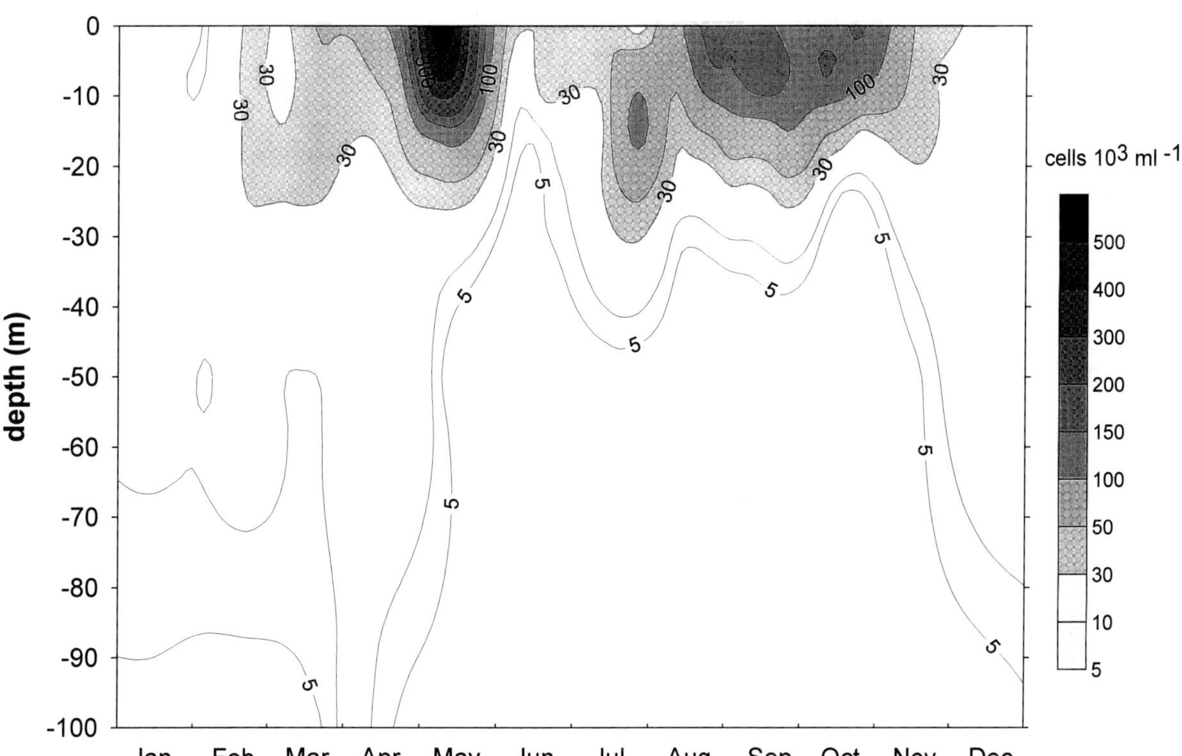

Fig. 2. Isoplethes of Pcy abundance in Lake Maggiore during 1995 (Passoni et al.,1997).

capable of growth in a wide range of light intensities (Kana and Glibert, 1987a). As a result of photo-adaptation, Pcy peaks have been found at different underwater light irradiances and depths in various types of lakes around the world. In Lake Maggiore at the spring and autumn peaks the maximum in cell abundance and photosynthetic activity was found at 25 - 50% of surface irradiance, corresponding to 200 µmol photon m^{-2} s^{-1} (Passoni et al. 1997; C. Callieri, unpublished). The suggestion by Pick and Agbeti (1991) that light penetration may determine the vertical distribution of Pcy is in contrast with Joint's (1990) conclusions that Pcy are extremely adaptable organisms and can grow in a variety of light conditions. Light, like temperature, is an important deterministic factor for Pcy survival, but its influence on vertical distribution is difficult to isolate from those related to biological interactions, i.e. grazing.

In Lake Maggiore the ciliate and flagellate vertical distributions during summer stratification are inversely related with Pcy abundance (Fig. 3), which suggests, that 'top down' factors (grazing) exert more control of Pcy populations and vertical distribution patterns than physical factors in large oligo- mesotrophic lakes (Weisse and Schweizer, 1991; Stockner and Shortreed, 1994), and the same seems also to be true in small mesotrophic lakes and reservoirs (Simek et al., 1996). However, there is compelling evidence from whole-lake fertilisation experiments that in ultra-oligotrophic British Columbia lakes nutrient limitation or 'bottom up' factors may be more important than light or grazing in establishing Pcy vertical distribution patterns within the euphotic zone (Suttle et al., 1991; Stockner and Shortreed, 1991).

C. CPcy Distribution

Many of the CPcy show a cosmopolitan distribution, while others are apparently more restrictive and found only in the temperate latitudes of the northern hemisphere e.g. *Aphanothece minutissima, A bachmannii, Lemmermanniella pallida*. Most colonial CPcy have been recorded from meso- to eutrophic lakes or ponds with very few species found in oligotrophic lakes (Table 1). Only the taxon *Tetrarcus ilsteri* has been found exclusively in oligotrophic and poly-humic lakes in Finland, Sweden and the Baltic States (Skuja, 1932; G. Cronberg, unpub.), but in such localities, when present, it can be very abundant. Many CPcy co-

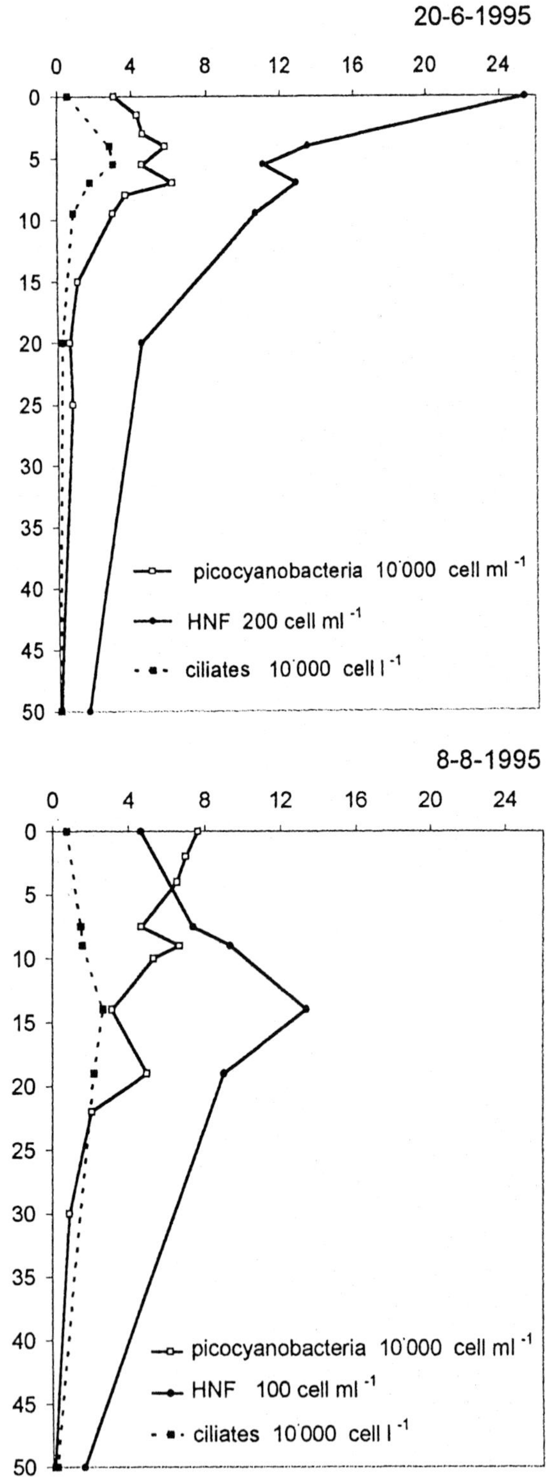

Fig. 3. Vertical distribution of Pcy cell numbers and of important grazers during summer stratification in oligotrophic Lake Maggiore. In June ciliates graze on pico and HNF and in August HNF are the most important Pcy grazer.

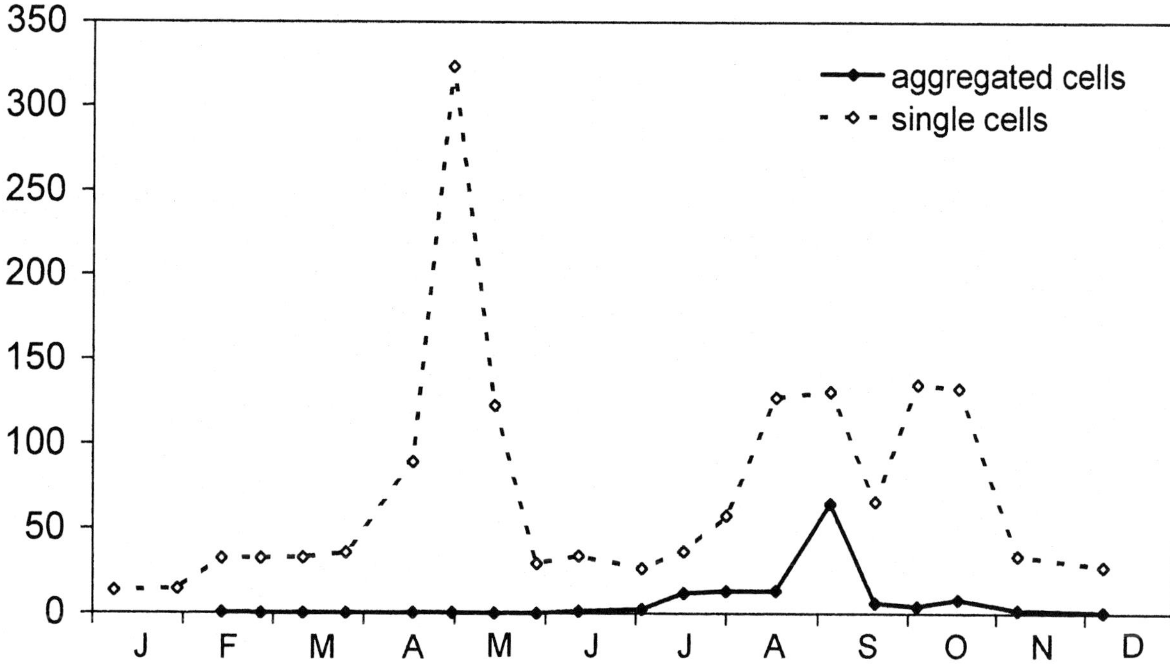

Fig. 4. Pcy abundance in a single and aggregated colony form in Lake Maggiore, 1995.

occur with larger blue-greens e.g. *Anabaena, Aphanizomenon, Microcystis, Planktothrix* and *Woronichinia* in water-blooms, but blooms dominated exclusively by CPcy are rare, though they can occasionally occur in successive years (G. Cronberg, unpublished). Most CPcy show optimal growth and maximum abundance during the warm, summer season in temperate latitudes, usually at temperatures >10 °C. However, again there are exceptions, the most notable where Laamanen (1996) found several CPcy species under ice in the Baltic sea at temperatures around 0 - 1°C. A few species seem to have a preference for much warmer waters, >25 °C, and are found in the tropics, e.g. *Aphanocapsa nubilum, A. elachista* and *Coelosphaerium punctiferum.*

Aphanocapsa, *Aphanothece* and *Chroococcus* are common in eutrophic lakes in Scandinavia, in Finland, the Baltic states and northern Germany (Komárková-Legnerová and Cronberg, 1994). In fact, most of the common CPcy species have been described from this part of the world (Table 1). Only a few of these taxa have been cultured in the laboratory under controlled conditions, thus the variability within or among these species is not well known. The Swedish Scanian lakes, where members of most of the well-known genera have been found, are eutrophic to hypertrophic, with TP values ranging from 50 - 170 μg L^{-1}, TN from 1.1 - 4.2 mg L^{-1}, pH ranging from 8 -9 and conductivity from 27 – 35 mS m^{-1}. These common CPcy genera quite often co-occur in lakes but the species can be separated from each other if typical colonies are found. However, the identification can be extremely difficult if transition forms appear, i.e. occurring neither as true colonies or as single celledmorphs. Based on current knowledge of their distribution, many CPcy species seem to have a narrow distribution pattern, but it is likely that this is not necessarily due to a narrower niche specificity, but rather to a paucity of sufficient studies of CPcy in other parts of the world.

D. Pcy Colony Formation

Small Pcy colonies are common in most lakes and usually reach highest abundance during periods of severe nutrient limitation in summer and autumn (Fig. 4; Plate 13) (Stockner, 1991; Klut and Stockner, 1991).However, there is a paucity of literature on the subject and without a better comprehension of the process, studies of Pcy seasonal succession will remain equivocal (Stockner, 1991; Weisse, 1993).

The presence of loose or tightly clustered colonies appears to have adaptive significance, perhaps as a strategy for more efficient nutrient recycling during periods of scarcity (Klut and Stockner, 1991; Pedrós-Alió and Brock, 1983), or possibly as an anti-predator mechanism (Stockner, 1991). Little is known as to whether colonies are more readily grazed than single cells, but it is thought that the large external polysaccharide layer is avoided by micro- and macrozooplankton (Klut and Stockner, 1991).

Studies on marine *Synechococcus* strains have shown a novel form of 'swimming' motility without flagella whose ecological significance may be related to aggregate formation, and their ability to exhibit chemotaxis and move to higher nutrient zones, e.g. marine snow, nitricline boundaries (Waterbury et al., 1985; Willey and Waterbury , 1989). It is quite likely that similar 'motility' occurs in freshwater *Synechococcus* strains, which could explain why colonies are more prevalent during periods of nutrient depletion in ultra oligotrophic systems (Klut and Stockner, 1991). The ability of Pcy to move and congregate around aggregates would be a plausible mechanism of colony formation. The production of sticky exopolymeric substances or mucus when the cell number increases could facilitate the clump formation (Kiørboe and Hansen, 1993). Finally, formation of aggregates or colonies should alter specific density and increase sinking velocity, thereby allowing Pcy to be transported downwards to the base of the euphotic zone and zones of higher nutrient concentration.

Recent studies permit some generalisation about colony distribution:

1) colonies are generally present throughout the euphotic zone, albeit in low abundance in all oligotrophic lakes (Passoni and Callieri, in press)
2) peak abundance of Pcy colonies appears in summer or autumn (Fahnenstiel and Carrick, 1992; Klut and Stockner, 1991), and in Lake Maggiore can represent 25% of the total population (Passoni and Callieri, in press)
3) in many lakes abundance of colonies is higher at the base of the euphotic zone than in surface layers (Stockner and Shortreed, 1991; Sime-Ngando, 1995).

E. Growth

Cells of *Synechococcus* have been widely used for laboratory experiments on growth rate and cell cycles, and results from these studies have been applied to natural populations (Waterbury et al., 1986; Chisholm et al., 1986; Campbell and Carpenter, 1986a; Sweeney and Borgese, 1989; Armbrust et al., 1989; Fahnenstiel et al., 1991b; Binder and Chisholm, 1995; Callieri et al., 1996). The processes of cell growth and division are as tightly coupled as photosynthesis and growth rate and are light dependent (Kana and Glibert, 1987a, b; Chisholm et al., 1986). There is little difference between marine and fresh water strains of *Synechococcus* in both cell division and growth, with cell division reaching a maximum in the afternoon, triggering an increase in cell number that proceeds in the dark cycle (Fig. 5) (Chisholm et al., 1986).

These light/dark cycles produce rhythmic cell divisions and related growth rate and photosynthetic activity that are all driven by prevailing light conditions. Experimental laboratory evidence of the influence of light on growth rate is well documented by Fahnenstiel et al. (1991b) for freshwater *Synechococcus* strains and by Campbell and Carpenter (1986a) for marine strains. These investigators have measured growth rates at light intensities up to 75 and 120 μmol photon m^{-2} s^{-1}, respectively, simulating natural irradiance levels. Kana and Glibert (1987b) have extended the light intensity limit up to 2000 μmol m^{-2} s^{-1}, demonstrating that growth rate becomes light saturated at 200 μmol m^{-2} s^{-1}, but *Synechococcus* has a mechanism of photo-adaptation which permits cell growth and photosynthetic activity to continue also at very high irradiances. Other environmental conditions such as nutrient concentration and temperature can also affect cell specific growth rates. Growth rates of APP in lakes along a trophic gradient ranged from a low of 0.10 to a high of 2.14 d^{-1} (Weisse 1993). Recent estimates of Pcy growth rates from Lakes Biwa and Baikal, 0.65 and 0.3 - 0.4 d^{-1}, respectively, (Nagata et al., 1994, 1996) and from Lake Kinneret, 0.29 - 0.60 d^{-1}, (Malinsky-Rushansky et al., 1995) fall within published ranges. The maximum net growth rates of unicellular cyanobacteria in oligotrophic Lake Stechlin was 0.23 d^{-1} (Padisák et al., 1997), while in the Lake Balaton it was 2.27 d^{-1} (Mastala et al., 1996). Pcy growth in Lake Maggiore lies between 0.28 - 1.14 d^{-1} as a net growth rate and 0.91 - 2.36 d^{-1} as a potential growth using the FDC method (Callieri et al., 1996). Loss rates, estimated as a difference between potential and net growth rate, show the disappearance of > 50% of the new APP production. This result is obtained by taking into account potential growth rate and therefore the calculated losses are those due to biological processes such grazing, lysis, parasitism, aggregation and sedimentation. Similar results from *in situ* experiments have also demonstrated a good balance between growth and grazing rates for Pcy populations (Weisse, 1993). A high abundance of virus-like particles were noted in an ultra-oligotrophic British Columbia lake, and were also found within Pcy cells in various stages of infection and lysis (Klut and Stockner, 1991). Suttle (1994) has implicated cyanophages as potentially responsible for 5 - 7% of *Synechococcus* lysis in marine systems, and this proportion can increase markedly, depending on host abundance, temperature and phosphate status of the cell (Wilson et al., 1996).

The issue of Pcy cell division regulation by a true circadian cycle has previously been noted (Pick and Berube, 1992; Weisse, 1993). Though Sweeney and Borgese (1989) concluded that there are endogenous circadian rhythms in cell division, their results could just as well be interpreted as a demonstration of the dependence of cell division on photosynthetic activity and growth. In other words, to keep cultures under continuous illumination does not stop cell division because the cells can grow and photosynthesise, and consequently must divide when the critical size is reached. Nevertheless, recent studies present further evidence that circadian rhythms do operate in *Synechococcus* (Huang et al., 1990; Kondo et al., 1993).

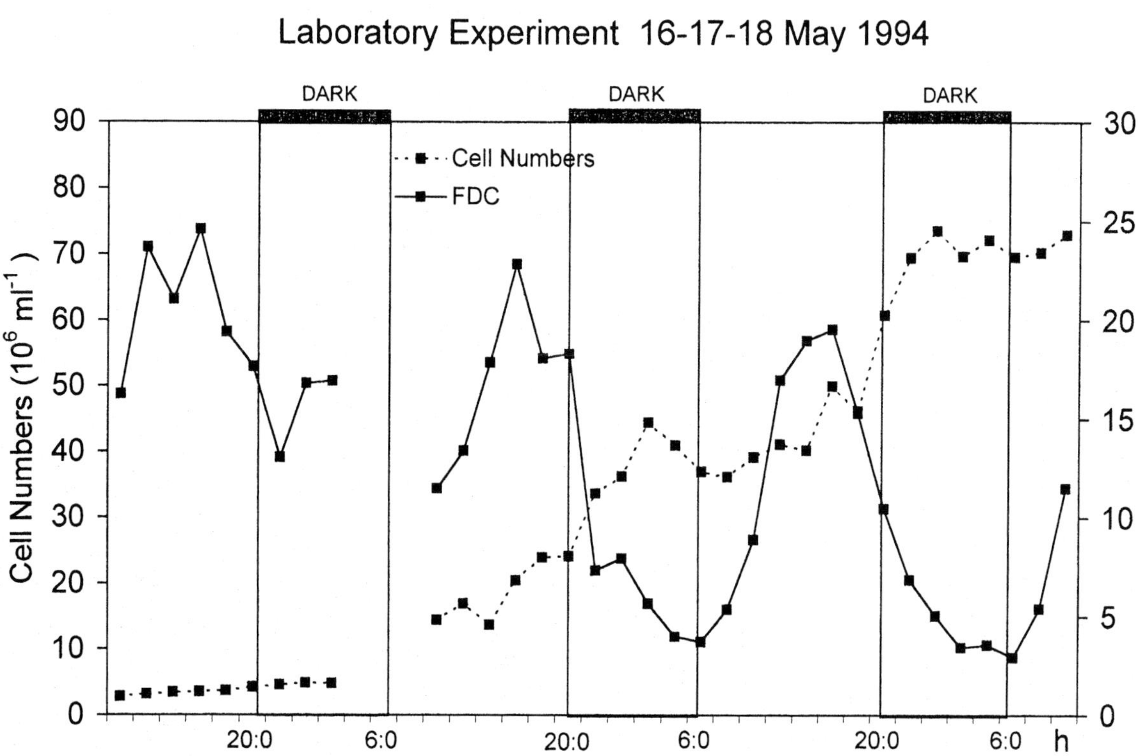

Fig. 5. Diel pattern of the cell numbers and of the percentage of dividing cells (FDC) of a *Synechococcus* culture isolated from Lake Maggiore. (Callieri et al., 1996; reproduced with permission).

The strong correlation between the frequency of dividing cells (FDC) and both net growth rate in laboratory cultures (Fahnenstiel et al., 1991b) and potential growth rate μ_{FDC} d^{-1} in Lake Maggiore (Fig. 6) (Callieri et al., 1996), demonstrates the good predictive value of the FDC method in estimating *in situ* growth rates. Some consider that the presence of two gaps during the cell cycle to be light-dependent, blocking the population in a doublet stage during the night at a constant FDC (Armbrust et al., 1989), but Binder and Chisholm (1995) have suggested the possible use of the FDC technique for field studies, despite some uncertainties regarding the cell cycle. Data on *Synechococcus* growth rate from FDC when regressed against maximum FDC% from lakes Maggiore, Huron and Michigan shows a significant correlation (Fig. 6) which allows one to obtain an estimate of Pcy growth rate from calculations of FDC% from natural populations.

Based on present knowledge of Pcy growth rates in lakes using a variety of available methods we can conclude:

- doubling times of *Synechococcus* natural populations range from 7 h to 7 d
- light influences growth which in turn is tightly coupled with the diel cycle
- the diel cycle could be triggered by an endogenous clock
- growth rates are discontinuous and largely controlled by predation, cell lysis and other loss processes.

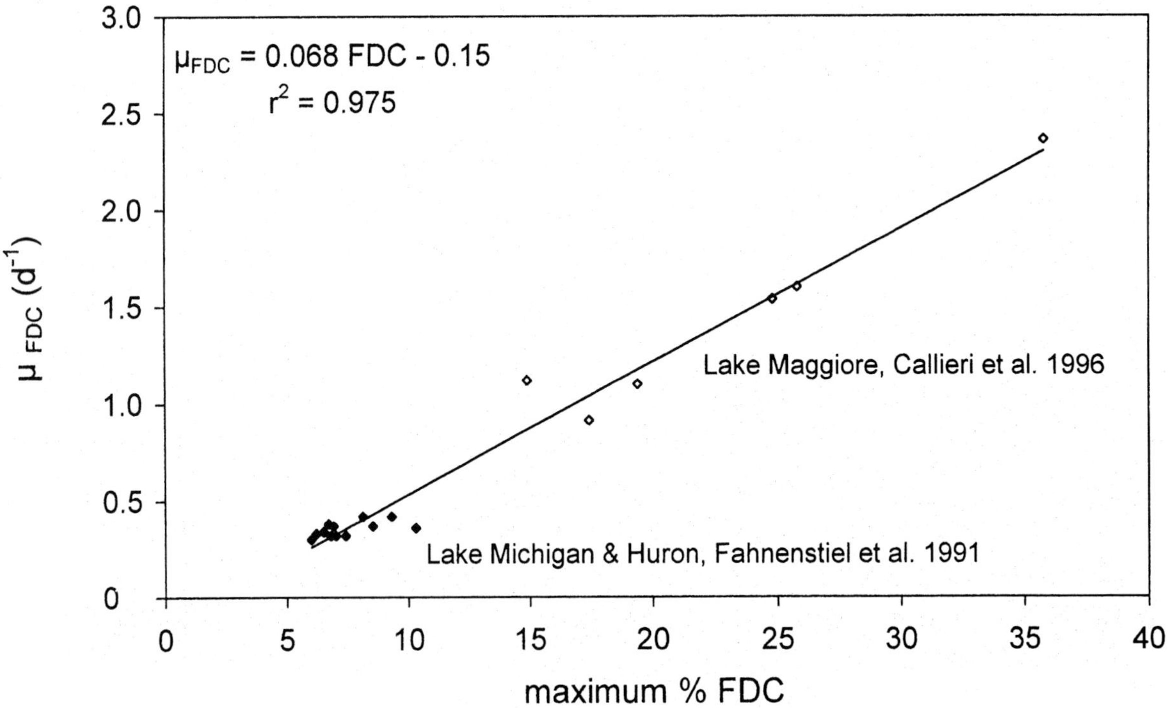

Fig. 6. Maximum percent of dividing cells of natural population of *Synechococcus* from Lakes Maggiore, Michigan and Huron and their *in situ* growth rates - μ_{FDC} (d^{-1}). (C. Callieri, unpublished data).

F. Pcy Contributions to Carbon Production and Biomass

Pcy contributions to total phytoplankton production and biomass along a trophic gradient of lakes has been the ultimate challenge to recent research efforts aimed at providing a better understanding of the ecological role of APP in freshwater (Stockner, 1991). Photosynthetic carbon uptake is affected by availability of nutrients, the light and temperature regime, and phytoplankton community structure, which can vary widely among lakes. Unicellular Pcy and pico-eukaryotes can out-compete the larger phytoplankton in the ultra-oligotrophic extreme of the trophic gradient (Suttle et al., 1987, 1988). One of the advantages of small cell size in low nutrient environments is to be less limited by molecular diffusion of nutrients because of the increase of the surface-to-volume ratio (Raven, 1986; Chisholm, 1992). The prokaryotic structure of the Pcy cell gives them the lowest costs for maintenance metabolism, and this factor has been cited as the primary reason for their success in oligotrophic conditions (Weisse, 1993).

Although considerations based on physiological studies tend to over generalise the importance and relative contribution of Pcy in most oligotrophic lakes, the experimental *in situ* measurements from different lakes are often in disagreement. The model outlined by Stockner (1991) of an increase in the contribution of APP production and biomass with the decrease of phosphorus concentration in lakes has been widely accepted, but in the light of recent data needs further specifications and refinement. The origin of the model applied to lake trophic gradient lies in the observations that marine *Synechococcus* are relatively more abundant in oligotrophic oceans that are nutrient depleted and largely regenerative than along the more productive coastal zones (Stockner and Antia, 1986).

The model of APP contribution to production in freshwater is largely based on results from a study of eight New Zealand lakes (Petersen, 1991). In a trophic gradient, expressed as increasing chlorophyll concentration from 0.57 - 103 μg Chl L^{-1}, Petersen found an inverse relationship between picoplankton contribution to total carbon fixation and lake trophic state. But there is considerable variability in his relationship. For example in some large oligo-mesotrophic lakes (Lakes Huron and Michigan), the average contribution of APP to phytoplankton production was unusually low at 17% (Fahnenstiel and Carrick, 1992), while in the ultra-oligotrophic Lake Baikal it was very high at 80% (Nagata et al., 1994). In eleven oligotrophic lakes of western Canada the relative contribution of picoplankton to total photosynthesis ranged from 29 to 53% (Stockner and Shortreed, 1991; 1994). A wider range has been found in Lake Constance were the ^{14}C incorporation varied between 0.1 and 7.5 mg C m^{-3} h^{-1} with a relative contribution of 5 - 65% of total phytoplankton production. Recently, researchers have shown that Pcy can also contribute significantly to total production in eutrophic and even to hypertrophic lakes but in a variable way (Vörös et al., 1991).

Some of these results demonstrate that the Stockner model (1991) cannot explain the wide variability of APP or Pcy contributions to total production in all lakes and leaves questions unanswered that require further research. Two interpretations of the model are possible, either:

1) that APP communities can adapt so quickly to changing physico-chemical regimes (e.g. light, temperature, nutrient concentrations, turbulence) that it is impossible to predict the relationship on a longer-term, more generalised basis or,

2) the methodologies and data elaboration used by various researchers are so different that data cannot be easily compared.

Furthermore, few large data sets exist that clearly document the seasonal variations of freshwater Pcy or APP primary production (Fahnenstiel and Carrick, 1992; Weisse and Schweizer, 1991), and their relative contributions to total phytoplankton production is not often expressed as the yearly mean calculated from the euphotic zone. As previously discussed, if values are not expressed on an annual basis the percentage contributions can vary widely, depending on the dynamics of short-term variations.

A good example of how the choice of data expression or presentation can influence the interpretation is provided by Lake Maggiore APP production measured in two successive years (Fig. 7). If expressed as a yearly mean calculated using the Platt model (1971), the values are very similar in 1994 (15 g C m^{-2} y^{-1}, 10% of total PP) and in 1995 (14 g C m^{-2} y^{-1}, 11% of total PP). Similar results are obtained if we compare the seasonal variation of the photosynthetic rate calculated on a daily basis as mg C m^{-2} d^{-1} or expressed on an hourly basis as mg C m^{-3} h^{-1} (Fig. 7). However, by examining the vertical profile of carbon the APP percentage contribution to total production ranges from 0 to 40% in 1994 and

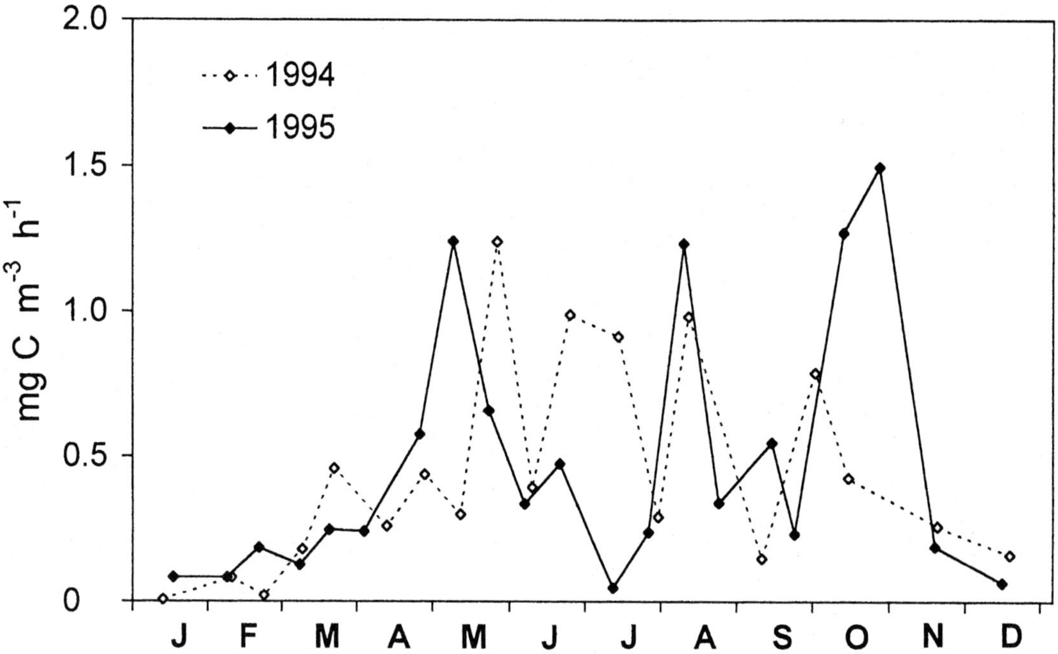

Fig. 7. Hourly rate of carbon fixation by APP community in Lake Maggiore during 1994-1995.

from 0 to 80% in 1995. This example illustrates two points:

1) short-term events can go unrecognised and contribute for many years before resulting in evident transformations of the system; therefore information obtained from short-term data sets cannot explain the evolution of the system

2) general models of the contribution to carbon production across trophic gradients should be obtained using data from more than a single year data set.

It appears that Pcy populations are increasing their importance in this lake as annual nutrient loads decline (Bertoni and Callieri, 1997; Callieri and Stockner, in press).

With differential filtration and the traditional ^{14}C uptake methods it is not possible to estimate the actual Pcy production because of the interference of the pico-eukaryotes. For this reason frequent microscopic examination for the presence of pico-eukaryotes would be helpful to critically evaluate the results. In most lakes the majority of APP are *Synechococcus* Pcy cells (Stockner, 1991), but in some cases red fluorescing pico-eukaryotes can represent a large fraction of picoplankton (Fahnenstiel et al., 1991a). In addition, questions concerning the accuracy of post-incubation differential filtration to measure APP primary production (Fahnenstiel et al., 1994), give rise to the need for a critical re-evaluation of the estimates obtained. The FDC method has been proposed as an alternative to radiochemical techniques because of its selectivity (Pcy are recognised under the microscope), and because sample incubation is not required (Carpenter and Campbell, 1988; Affronti and Marshall, 1994). The *in situ* photosynthetic rate of Pcy estimated by FDC and by the ^{14}C incorporation method have recently been compared in fresh water (Callieri et al., 1996), and the results are very similar, despite the presence of marked fluctuations. The use of the FDC method for *in situ* measurement of Pcy production should be considered, balancing the advantages of the method with its problems (Callieri et al., 1996). Therefore, in any study of the ecology of Pcy it is important that investigators use methods that selectively count only Pcy, and, if possible, measure only their photosynthetic activity, growth rate and grazing losses; otherwise the contributions of pico-eukaryotes will continue to confound interpretations. The seasonal dynamics and relative contribution of Pcy to total carbon flows when evaluated in this way should bring new interpretations to statements about

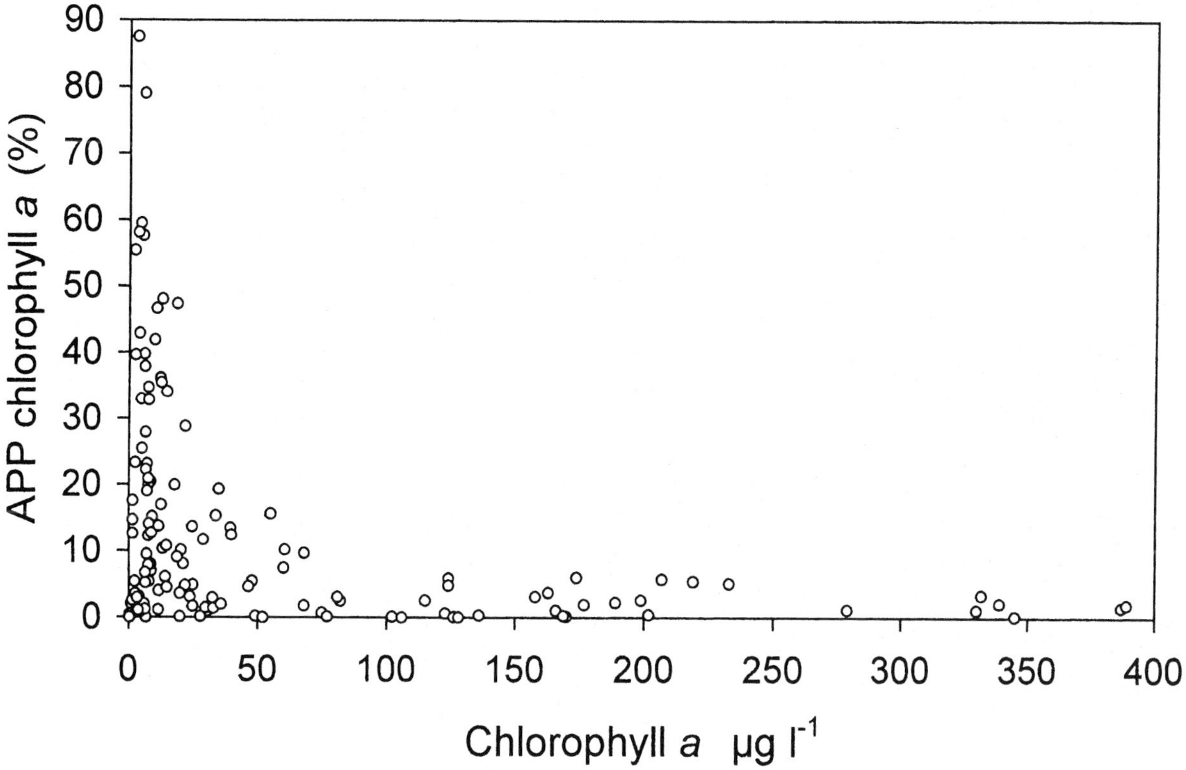

Fig. 8. The relation between APP CHL contribution (%) and total CHL a (from Vörös et al., 1997).

roles of APP populations in lacustrine plankton communities. This has recently been shown to be the case in marine ecosystems (Lindell and Post, 1995; Charpy and Blanchot, 1996).

Stockner's (1991) model applies quite well to the observed APP contribution to total phytoplankton biomass along a trophic gradient. Søndergaard (1991) found that although the actual values of APP biomass were rather similar among lakes, the relative importance was highest in more oligotrophic lakes and lower in eutrophic lakes. The seven Danish lakes he studied (one acidic, the others shallow hypertrophic and deep meso-eutrophic) exhibit Chl concentrations ranging from 7 - 200 µg L^{-1}.

It is interesting to compare Søndergaard's data with similar data obtained in a study of 32 lakes that included high altitude clear lakes, deep large subalpine lakes, large shallow lakes, small reservoirs and fish ponds (Fig. 8) (Vörös et al., 1998). Across such a trophic gradient Chl concentrations ranged between 0.2 and 390 µg L^{-1}. The relative contribution of Pcy can be more than 70% of the total phytoplankton biomass in lakes with Chl below 10 µg L^{-1}, whereas above the threshold of 100 µg Chl L^{-1} the contribution of the Pcy fraction does not exceed 10%. Vörös et al. (1998) have noted the absence of a linear variation of the contribution of APP or Pcy to total biomass along this lake trophic gradient, with a logarithmic inverse curve providing a better fit (Fig. 8).

The pattern of Pcy cell abundance across trophic gradients remains poorly understood. We conclude that the few studies done in freshwater (Burns and Stockner, 1991; Hawley and Whitton, 1991b) rely on gradients not sufficiently wide to permit a generalisation at this time. Considering a Chl range from 0.2 to 390 µg L^{-1} it is clear that the relation between Pcy abundance and the trophic state of lakes cannot be fully explained with a simple two-variable model (Fig.9) (Vörös et al., 1998).

Using the more extensive data base of Vörös et al. (1998) we have attempted to indicate a general trend in a schematic concept, illustrating the areas of possible occurrence of Pcy numbers and their

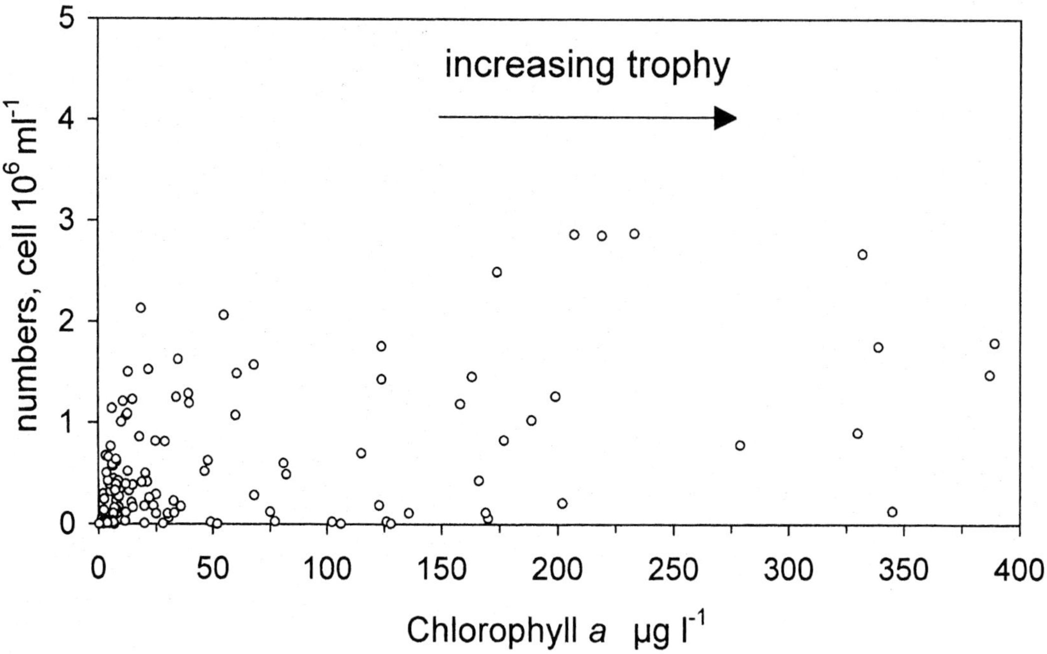

Fig. 9. Pcy abundance and euphotic zone CHL a concentrations in lakes of increasing trophic state (from Vörös et al., 1997).

contribution to total phytoplankton biomass (Chl) within a broad trophic gradient (Fig. 10). We suggest that there is not a linear relationship driving the presence and importance of Pcy in lakes of differing trophic status, and that the success of Pcy in oligotrophic lakes is not a certainty, only a potentiality. In some oligotrophic lakes, if conditions like light climate, grazing impact and competition are favourable Pcy can grow rapidly, out-compete their competitors and become very abundant, but there is a host of factors that can influence the outcome of this competition, and ultimately influence Pcy success in lakes of all trophic types.

G. Factors Affecting Pcy Growth, Production and Distribution

1. Light

Underwater light conditions are extremely important for the growth and production of pelagic photoautotrophs, and the change in light quantity and quality modulates the algal growth response. The great abundance of Pcy cells in most lakes demonstrates the efficiency of their acclimation mechanisms to changing light conditions, and necessitates a brief review of their complex photo-adaptation response.

The response of Pcy to different light intensity has been studied both in laboratory experiments and *in situ*, and it has been shown that the optimum growth rate of *Synechococcus* occurs under low light conditions (Waterbury et al., 1986), notably at a quantum flux of 45 µmol photon $m^{-2}s^{-1}$, where highest growth has been observed (Morris and Glover, 1981). These findings agree with field observations where the maximum peak abundance has been found deep in the Atlantic mixed layer (Glover et. al., 1985), and in the DCM (deep chlorophyll maximum) of Lake Stechlin below the euphotic zone (Gervais et al., 1997). The response by *Synechococcus* to low light results in an increase in the ratio phycobiliprotein/Chl, but with high variability in the observed response among different strains tested (Kana and Glibert, 1987a). However, in lakes, Pcy and other pico-eukaryotes have been found at a variety of depths and light irradiance (Fahnenstiel and Carrick, 1992; Nagata, 1994; Callieri and Pinolini, 1995), confirming the classification of *Synechococcus* (marine strain) as an euryphotic organism (Kana and Glibert, 1987b). In laboratory experiments Kana and Glibert (1987b) obtained growth of *Synechococcus* cells with irradiances up to 2000 µmol photon $m^{-2}s^{-1}$

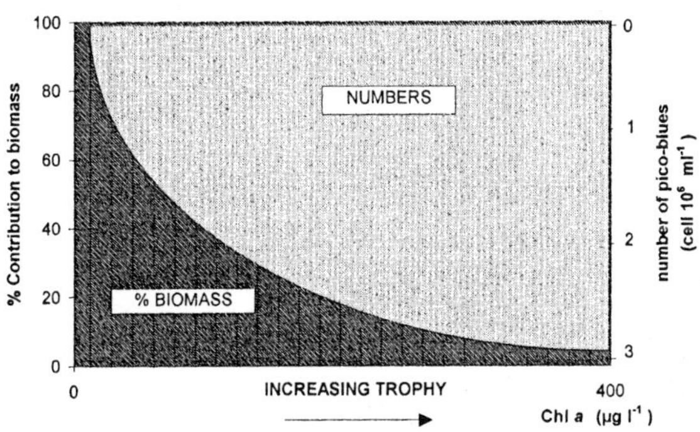

Fig.10. Schematic draft of the areas of occurrence of Pcy abundance and their relative contribution to total phytoplankton biomass.

with a growth saturation at 200 µmol photon $m^{-2}s^{-1}$ (Fig. 11). This latter value corresponds to the light irradiance where in natural systems the maximum cell abundance and photosynthetic activity was found (C. Callieri, unpublished). The Kana and Glibert (1987a, b) results evidently contradict those of many of the previous studies and the prevalent belief that Pcy are shade tolerant organisms. However, it is important to note that Kana and Glibert used *Synechococcus* cultures already adapted to high irradiance in their experiments, giving them the necessary time to acclimate, in this case longer than a generation time.

The success of Pcy under low light conditions is tightly coupled with competition for limiting nutrients, and low light seems to give *Synechococcus* the advantage, but only under low-P conditions (Wehr 1993). Another typical response to low light is an increase in accessory pigments relative to Chl *a* (Falkowski and LaRoche, 1991) and nitrogen limitation may be responsible for a low PE/Chl *a* ratio (McMurter and Pick, 1994).Even though many genetically distinct *Synechococcus* strains have been found (Ernst et al., 1995), it is still helpful to broadly classify Pcy into the two cell-types: the first with yellow autofluorescing phycoerythrin (PE cells), and the second with red autofluorescing phycocyanin (PC cells) as the major light-harvesting pigment with maximal emissions at 570 nm and 630 nm excitation respectively (Wood et al., 1985; Callieri et al., 1996) A type 1 phycoerythrin, containing both phycoerythrobilin (PEB) and phycourobilin (PUB) chromophores, and type 2 phycoerythrin containing only PEB chromophores have been described (Wood et al., 1985).

The fluorescence characteristics of Pcy based on phycobiliprotein spectra have been an easy and convenient way to distinguish between the 2 groups of Pcy (McMurter and Pick, 1994). For example the difference between PE and PC containing *Synechococcus* is evident from fluorescence emission spectra, since the PE show an emission maximum at 578 nm when excited at 520 nm, while the PC emit maximally at 648 nm when excited at 600 nm (Ernst, 1991; Callieri et al., 1996).

Freshwater strains of PE and PC type Pcy have been used to compare the effect of increasing light on strains with different pigment, and PC cells have shown a higher growth rate than PE cells when tested at all irradiances (Fig. 12) (Callieri et al., 1996).

The influence on light intensity seems more related to the structural change in Photosystem II (PSII) than on the different pigment synthesis (Clark et al., 1995). Recent studies have discovered that during a shift in light intensity in a *Synechococcus* culture, there is a rapid interchange from a light sensitive form of the D1 protein of PSII centres (Form I) to a form less susceptible to photo-inhibition (Form II) (Clark et al., 1995). It has also been shown that during adaptation to high light intensity there is gene regulation to encode the Form II of D1 protein (Kulkarni and Golden, 1995). Therefore the photo-inhibition can be efficiently counteracted and eventually overcome (Clark et al., 1995). It is important to keep in mind that these laboratory experiments were run using a single *Synechococcus* strain (PCC 7942), and similar work must be extended to other strains before generalising these results to natural systems.

It appears that in Pcy the fluctuation of the relative pigment content may depend on light climate (Wyman and Fay, 1986; Hauschild et al., 1991), and that only cyanobacteria capable of PE synthesis can undergo complementary chromatic adaptation (Tandeau de Marsac, 1977). Recent studies by Hauschild et al. (1991) distinguished three physiological groups of PE cells: Group I with no

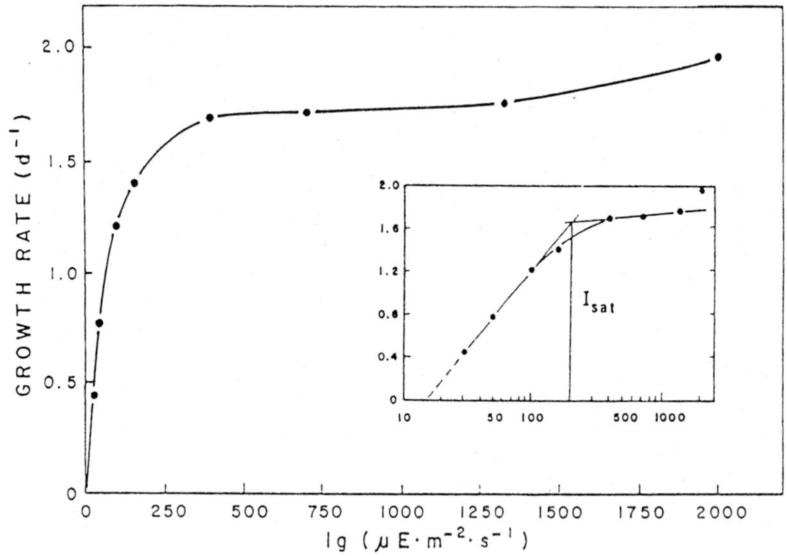

Fig. 11. The effect of irradiance on growth rate of *Synechococcus* WH7803. (from Kana & Glibert, 1987a).

adaptation, Group II with regulation of PE, and Group III with regulation of PE and PC pigment. Marine *Synechococcus* strains exhibited chromatic adaptation of the Group II type with an increase in PE concentration and accelerated growth. The comparison of the growth response of two freshwater strains of *Synechococcus*, one of PE type the other of PC type, demonstrates the selective value of red light in stimulating PC and in suppressing PE (Fig. 13) (Callieri et al., 1996; Callieri, 1998). The importance of red light for PC and biomass production has been

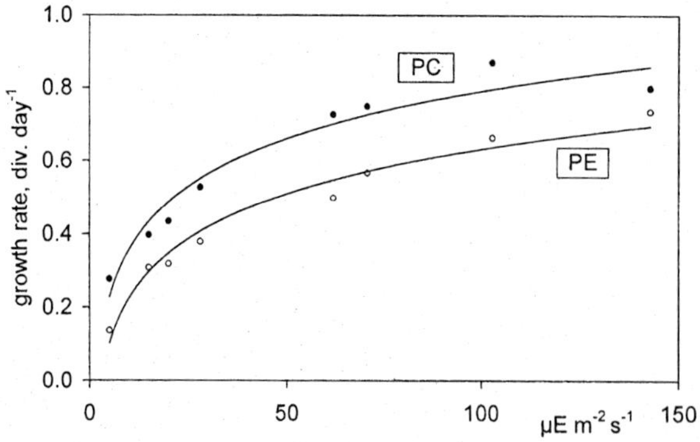

Fig. 12. Exponential growth rate (K: div d^{-1}) as a function of light intensities for PE and for PCPcy cultures. (from Callieri et al., 1996).

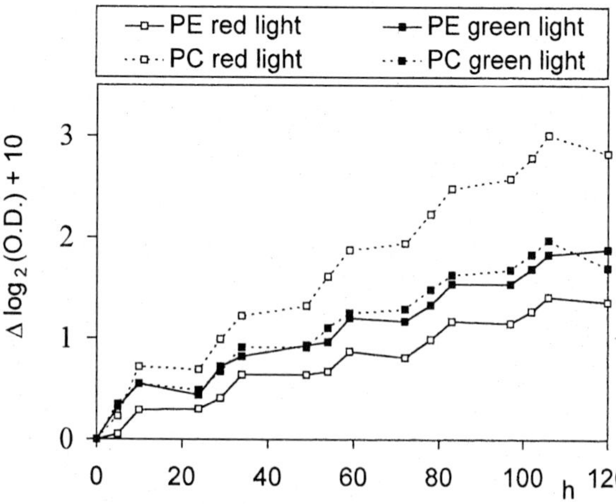

Fig.13. Growth rates of two strains of *Synechococcus* with PE and PC pigments exposed to green and red light of an equal quantum flux (20 μmol photon $m^{-2}s^{-1}$) (from Callieri et al., 1996).

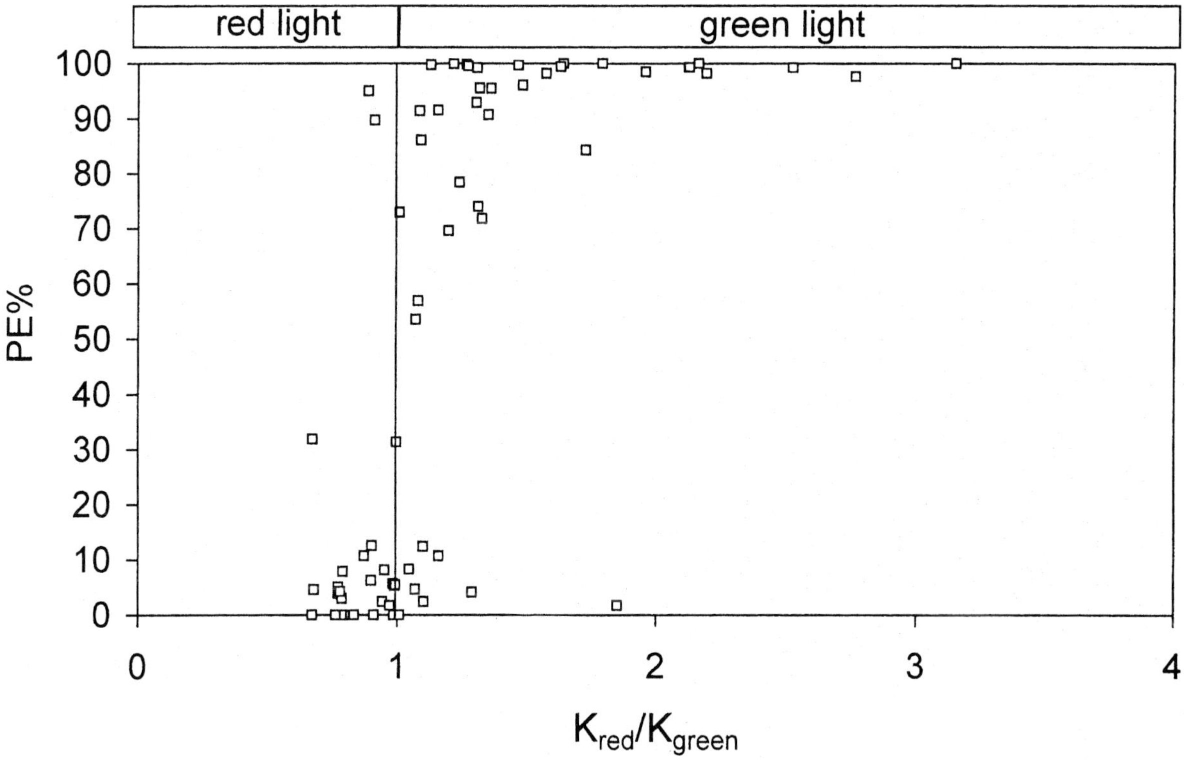

Fig. 14. Relative contribution of PE cells to total Pcy abundance in lakes with underwater light climate expressed as K_{RED} / K_{GREEN} (from Vörös et al., 1997).

ascertained in a *Synechococcus* strain of PC type in a recent laboratory experiment (Takano et al., 1995). Blue and green are more efficiently used than red light of a similar intensity by *Synechococcus* of PE type (Glover et al., 1986).

In highly coloured lakes, non PE Pcy cells were dominant while in oligotrophic, hard-water lakes PE Pcy cells were the most abundant (Pick, 1991). The influence of underwater light quality on the selection of Pcy types with different pigments has been studied in many lakes along a broad trophic gradient (Callieri et al., 1996; Callieri, 1994b; Vörös et al., 1998). When the K_{RED} /K_{GREEN} ratio is >1 the extinction of red light is high and the dominant underwater light is green (and blue). Very low values of K_{RED}/K_{GREEN} ratio indicate a red dominant underwater radiation. In oligotrophic lakes with low Chl concentrations green light dominates, while red light prevails in more eutrophic lakes. Vörös et al. (1998) found that the percentage of PE cells in the total Pcy community increased with increasing values of the K_{RED}/K_{GREEN} ratio, while concurrently the total Chl decreases and the waters become more transparent and less productive (Fig. 14).

2. *UV-B*

The effect of ultraviolet radiation on aquatic organisms has been widely reviewed (Holm-Hansen et al., 1993; Häder, 1993a, b), but only recently has the detrimental effect of UV-B radiation on the rate of phytoplankton productivity and on food webs in aquatic ecosystems been emphasised (Williamson, 1995). Many laboratory experiments have been done to quantify the effects of UV radiation on different kinds of aquatic organisms from marine and freshwaters (Smith and Baker, 1989; Hessen et al., 1995; Cullen et al., 1992). The action spectra for the UV-B inhibition of photosynthesis have demonstrated that there is a variety of responses among species and groups (Häder, 1993b; Vincent and Roy, 1993). Many aquatic organisms react promptly to UV-B stress by producing protective substances such as mycosporine-like amino acid compounds (MAAs), which have absorption maxima ranging from 310 nm to 359 nm (Carreto et al., 1990; Karentz et al., 1991). Cyanobacteria also produce an extracellular yellow-brown pigment – scytonemin – that absorbs most strongly in the UV-A spectral region (Garcia-Pichel and Castenholz, 1991; Proteau et al., 1993). The sunscreen capacities of MAAs and scytonemin are higher if they are present simultaneously, and their production is considered an adaptive strategy for photo-protection against UV irradiance (Garcia-Pichel and Castenholz, 1993.

Häder (1993b) expressed concern about the lack of information on UV-B effects on the small-sized pico- and nano-planktonic algae which are important links in microbial food webs, but Wangberg et al. (1996) have shown that it is not possible to find any clear relation between cell size and sensitivity to UV-B irradiation. The few data available on UV-B and Pcy cells show that, as for other algal groups, UV-B inhibits the PSII reaction centre activity, altering the structure of D1/D2 polypeptides and the light-harvesting complex (Rajagopal and Murthy, 1996).

Additional information from Garcia-Pichel and Castenholz (1991) on UV-B effects on cultured Pcy strains constitutes the current data base upon which to interpret Pcy UV-B results.

Until recently, efforts to quantify the effects of increases in UV-B radiation on aquatic micro-organisms have been focused on the photo-damage to the total phytoplankton and/or bacterioplankton community (Müller-Niklas et al. 1995; Häder, 1996; Wanberg et al., 1996; Vinebrooke and Leavitt, 1996; Jeffrey et al., 1996; Bertoni et al., 1997). The carbon uptake of natural marine phytoplankton was inhibited by UV-B at 50 cm depth if irradiated with 1.7 W m^{-2} in the band 280 - 320 nm (Nielsen and Ekelund, 1995). UV-B affected carbon uptake of epilimnetic algae from four lakes in Pennsylvania reducing the uptake by 82 - 95% (Moeller, 1994), but in this study the UV-B effects were difficult to separate from those of UV-A. As far as we know, no data have been published specifically on Pcy carbon uptake inhibition in lakes. But recent experimental work on the effect of UV-B irradiation on natural APP populations is useful and relevant to the present discussion (Bertoni and Callieri, submitted). When compared with larger phytoplankton, the carbon uptake by the pico size fraction (dominated by Pcy in Lake Maggiore) were more UV-B photo-inhibited (Fig. 15), and this is even more evident when one considers that Pcy cells are more photosynthetically active. Others have also noted that very active cells in the exponential growth phase are the most sensitive to UV-B irradiation (Wang and Chai, 1994). The sensitivity of natural populations of Pcy to UV-B radiation is limited to the first few meters of the water column (Fig. 16). At 0.5m, with 50% of surface UV-B irradiance, Pcy produced significantly less carbon in UV-B unscreened than in UV-B screened tubes. At this depth Pcy showed more sensitivity than larger algae to photo-inhibition as shown by the decrease of their contribution to the total carbon uptake - 6% compared to 23% in the protected tubes. At a depth of 2m, where the UV-B decreased to 1% of the surface irradiance, there was no apparent effect on Pcy populations. These results could also indicate an absence of mixing and the consequent stratification of the detrimental UV-B effects on the population more exposed to UV-B radiation.

At this early stage of UV-B research many results are still preliminary, nonetheless it is now possible to conclude that there is definitely an effect of UV-B radiation on Pcy, probably causing structural changes in the PS II and in other functional parts of the cell resulting in reduced Pcy cell abundance and photosynthetic activity. Nevertheless this effect is evident only when natural UV-B irradiance, measured underwater at mid-latitudes, is artificially increased to reach $9 - 10\ \mu W\ cm^{-2}\ nm^{-1}$ (Bertoni and Callieri, submitted). As has already been shown, algal species react differently to UV irradiation, some producing sunscreen substances, while others do not, so attempts to compare the different components of the microbial community, e.g. bacteria, Pcy, CPcy, microflagellaes, ciliates, etc., are going to be difficult.

However, preliminary results suggest that the smaller organisms are more sensitive to UV-B photo-inhibition, or are slower in adaptation or in the activation of the reaction mechanisms necessary for survival (Fig. 17), so the microbial community may be more seriously impacted by UV than larger organisms.

3. pH

Synechococcus strains are most often grown in a medium at a neutral pH (Stanier et al., 1971). Their preference for neutral or slightly alkaline conditions is also evident in their abundance and distribution patterns in fresh water ecosystems (Stockner, 1991). In western Canadian oligotrophic lakes, Pcy populations have been found to rapidly decline with decreasing pH, and below pH6 they are generally absent from the APP community (Fig. 18) (Stockner and Shortreed, 1991).

The trend towards Pcy disappearance with decreasing pH and their replacement in the APP community by pico-eukaryotes has been noted in three dystrophic Canadian lakes (Stockner and

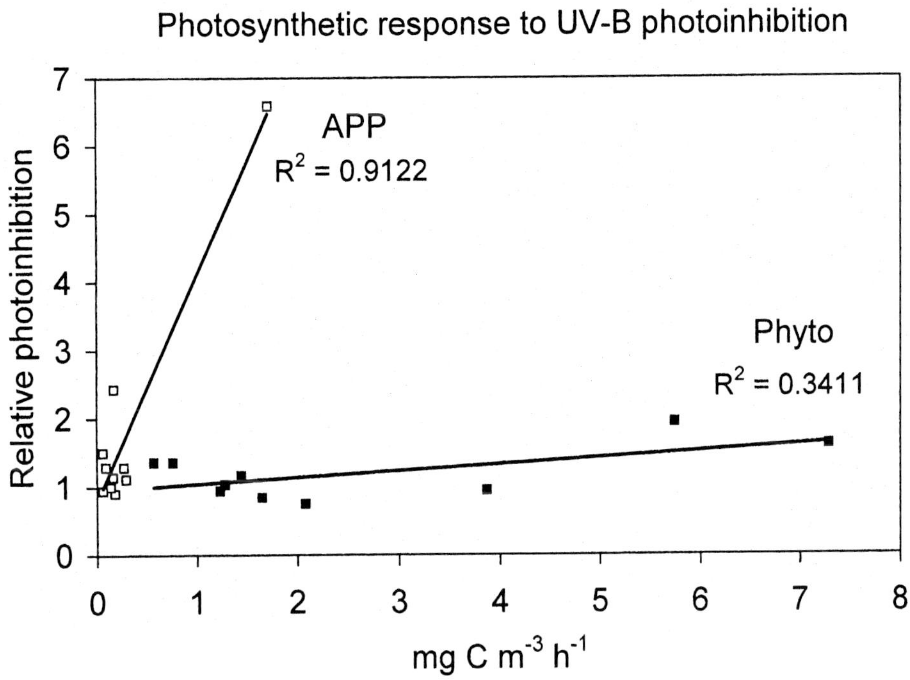

Fig.15. Photosynthetic response to UV-B photo-inhibition measured in two cell size fractions (<2μm and >2μm) in Lake Maggiore, Northern Italy. The relative photo-inhibition is calculated by the ratio UV-B exposed : UV-B protected carbon uptake (Bertoni and Callieri, submitted).

Shortreed, 1991) and in several low pH, humic Danish lakes (Søndergaard, 1991). The effect of lake acidification on the microbial community can be indirect through alteration of community structure and hence carbon flows to higher trophic levels, or direct by induction of physiological stress. In recent studies on a Swedish acidified lake before and after liming, a non-edible CPcy, *Merismopedia tenuissima* was the dominant species in the late summer phytoplankton community in the naturally acidic lake. However, this population was removed by the liming (Bell and Tranvik 1993; Blomqvist, 1996). Unfortunately the authors present no data on Pcy abundance in this lake, however, they suggest a probable allelopathic mechanism to explain the population dynamics of *Merismopedia tenuissima* (Bloomqvist, 1996; Vrede 1996). In an Italian acidic alpine lake Pcy populations are very low, and their contribution to the microbial food web appears to be negligible (C Callieri, unpublished). The presence of a shift from Pcy to net plankton has been described in mesocosm experiments (Havens and Heath, 1991), and it has been noted that as pH declines the proportion of larger species tends to increase and become dominant (Schindler, 1990). Nevertheless, no experimental studies on the influence of pH on Pcy strains have been done, so at this stage it is difficult to discuss ranges of pH tolerance by Pcy or physiological mechanisms of adaptation to low pH in lakes. The only possible generalisation at this stage is that Pcy and many CPcy are not common in lakes with a pH < 6.0, and are seldom mentioned or included in studies on acidic lakes because they are probably in low abundance or absent.

4. Nutrients

Data from both field and laboratory have demonstrated that Pcy are seldom P-limited, but do show increased growth and production responses when N is added (Suttle et al., 1991; Wehr, 1989, 1991). Pcy and some CPcy prefer ammonium N for growth (Mastala et al., 1996), and have shown a lower competitiveness for nitrate N; nitrate reductase

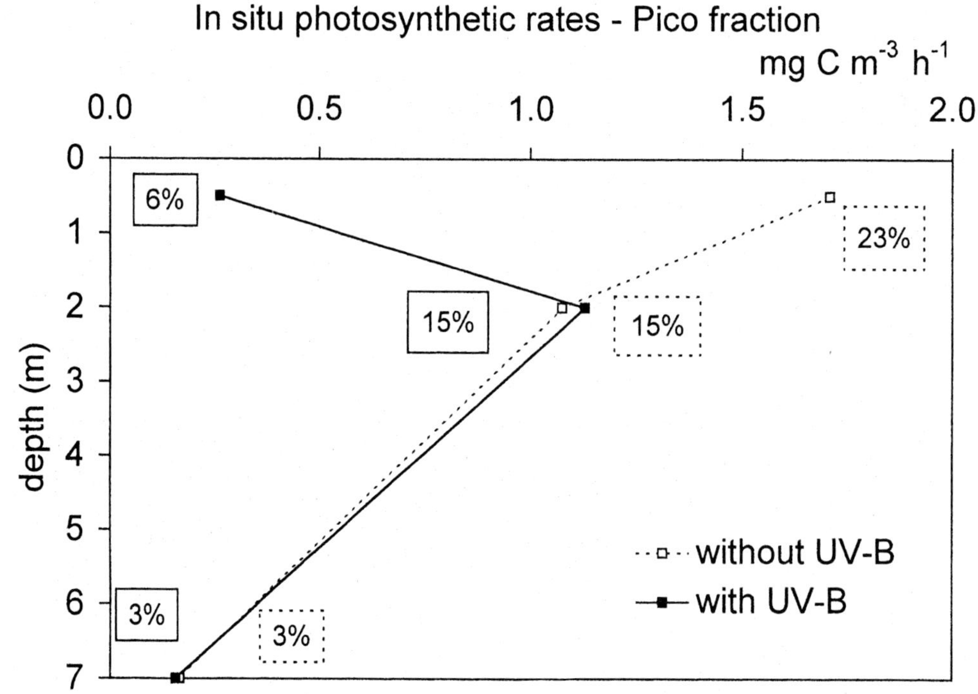

Fig.16. In situ photosynthetic rates of Pcy measured with C^{14} technique and incubated for 4 h in quartz and mylar tubes in Lake Maggiore, Northern Italy (C. Callieri, unpublished data).

activity is induced more rapidly in eukaryotic than in prokaryotic cells (Blomqvist et al., 1994). The response of Pcy to nutrient additions will always be mediated by light conditions and grazer abundance. Wehr (1991) suggested that the flux of P can be regulated by Pcy only under P-limited conditions and without grazing pressure. The difficulty with studies examining the impact of limiting resource interactions is that sometimes the effect of grazing on the depletion or uptake of nutrients is indirect and yields equivocal results difficult to interpret. For example, recent studies have shown that the alleviation of one limiting resource may induce competition for a second limiting resource (Wehr, 1993). Thus the pulsed addition of P can have an interactive effect, because it increases the prevalence of larger algae that can alter the light climate, thereby increasing light limitation which will enhance the growth of Pcy.

In studies on the effect of P addition on the *Synechococcus* cell cycle in marine systems, Vaulot et al. (1996) obtained results suggesting the importance of P as a limiting nutrient only during certain times of the year, which are contrary to Wehr's (1991; 1993) results, and to the general consensus for Pcy in freshwater (Stockner, 1991). However, caution must be exercised in interpreting the results of Vaulot et al. (1996), because when P is added, the cell cycle variables respond very quickly so the applicability of their findings to *in situ* situations in the epilimnion of lakes may be tenuous at best.

There is good evidence to suggest that the N:P ratio may determine the dominance of Pcy over nano- or microphytoplankton (Stockner and Shortreed, 1988; 1994; Shortreed and Stockner, 1990). Other accounts (Wehr, 1991; Takamura and Nojiri, 1994; Mastala, et al., 1996) have also stressed the importance of high N:P ratios (> 25:1 molar) and not necessarily the trophic state in controlling the abundance of Pcy in lakes.

The important question on the effects of limiting nutrients and their ecological role in influencing phytoplankton community size structure has not yet been reconciled and will require further research efforts to be solved. Major problems remaining are:

1) selection of realistic parameters to monitor, which respond rapidly in nutrient bioassays
2) the proper evaluation of interactions among different limiting resources and biotic factors, e.g. grazing.

5. Grazing

Studies of picoplanktivory have markedly increased in the last few years prompted by new methodologies or modifications of older ones. Since the 1980s, the rates of Pcy removal by grazers have been measured using five basic techniques:
1) metabolic inhibitors (Campbell and Carpenter, 1986b)
2) diffusion chambers and the dilution technique (Landry et al., 1984)
3) fluorescent labelled particles (Sherr and Sherr, 1987)
4) direct cell counts (Waterbury et al., 1986)
5) radioisotope-labelled prey (Iturriaga and Mitchell, 1986).

Many of these methods have been improved over the last decade. In the past, various growth inhibitors were tested, including the eukaryote inhibitors colchicine and cycloheximide, and ampicillin, which

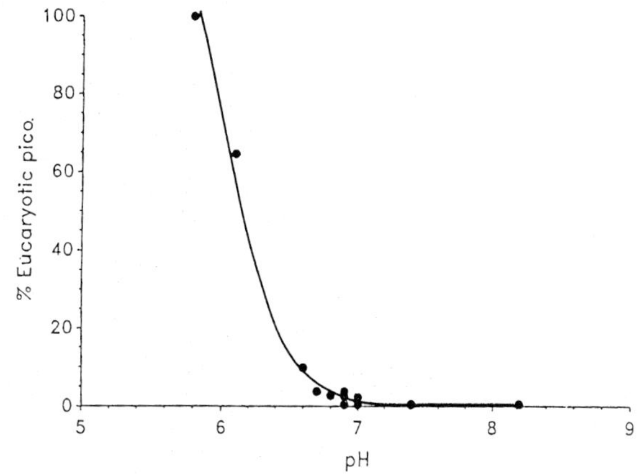

Fig.18. Progressive disappearance of Pcy in the APP community in temperate lakes with a decreasing pH (Stockner and Shortreed, 1991).

has been used to stop protozoan grazing activity on Pcy (Campbell and Carpenter 1986b; Caron et al., 1991). Recently, Liu et al. (1995) chose kanamicin as an effective growth inhibitor of *Synechococcus* and *Prochlorococcus*, to estimate growth and grazing rates. Using this new approach Pcy grazing mortality has been estimated to range from 43 - 87% of growth rate in marine systems (Liu et al., 1995). Landry et al. (1995) have proposed a refined dilution technique to overcome the assumption that grazing impact varies in direct proportion to the dilution of grazer population density. In the new protocol they introduce fluorescent labelled cells and flow cytometry to measure the relative grazing activity in each dilution treatment. Ayukai (1996) has suggested the use of FDC to control and monitor the growth rate in dilution measurements, a technique that has not yet been applied in fresh water systems. The method of fluorescently labelled particles in grazing experiments gained new impetus when inert, and less effective microspheres were substituted for fluorescently marked bacteria and Pcy (FLB & FLC) (Sherr et al., 1991; Sherr and Sherr, 1993).

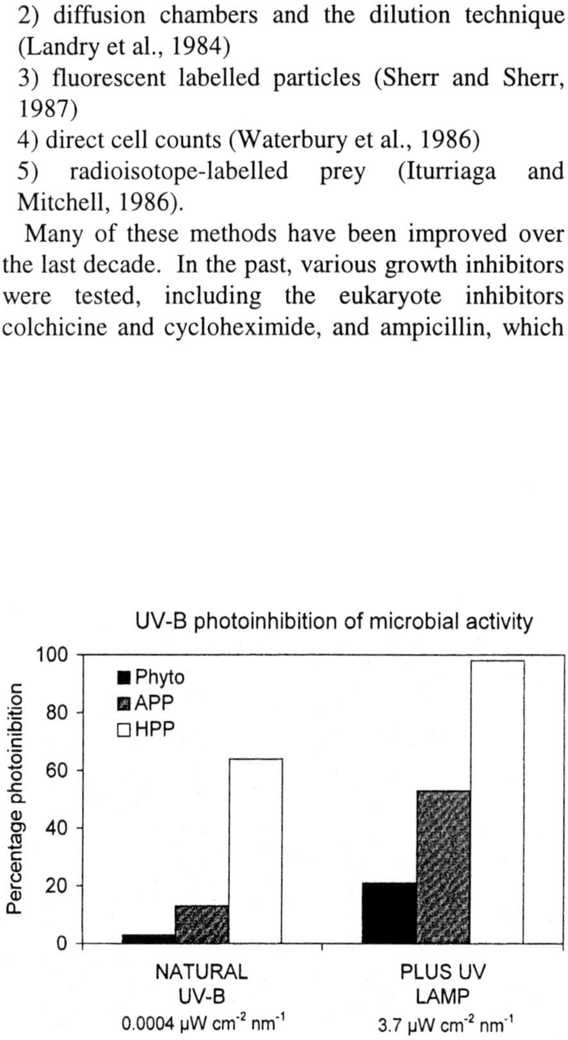

Fig.17. Percentage UV-B photo-inhibition of natural autotrophic and heterotrophic micro-organisms exposed to natural and artificially increased UV-B radiation (C. Callieri, unpublished data).

It is not surprising that the existence and continuing development of so many methodological approaches to measuring grazing has produced such different and often contradictory results in the literature. For example Sherr et al., (1991) estimated that in Lake Kinneret the ciliate carbon requirement cannot be obtained only by the Pcy energy source and suggested that pico-eukaryotic cells must be added to complete the requirements. But recently, Šimek et al. (1996) have shown that some of the most common freshwater ciliate species can survive solely on a diet of Pcy. These discrepancies and contrasting results have enhanced the discussion of grazing impact on Pcy, to such an extent that it is now possible to focus more clearly on salient points that improve our understanding of the impact of grazing on Pcy.

It is now generally recognised that protozoa are the most important Pcy consumers in lakes (Stockner and Antia, 1986; Stockner and Porter, 1988). Protozoa in the pelagic zone of lakes are represented by nano-flagellates (2 – 20 μm) and ciliates (15 – 200 μm) (Fenchel, 1987). Among the picoplanktivore flagellates there are the obligate heterotrophic forms and the mixotrophic forms (Stockner and Porter, 1988; Sanders and Porter, 1988). Heterotrophic and mixotrophic nanoflagellates and small ciliates are the most important Pcy grazers (Sanders et al., 1989; Bird and Kalff, 1987; Christoffersen, 1994; Šimek et al., 1995). Recent studies have demonstrated that heterotrophic flagellates grazing could account for 90% of the protozoan picoplankton mortality, while ciliates accounted for only 10% (Pernthaler et al., 1996a). Their results show that flagellates and some ciliates seem to select Pcy as prey over bacteria; however, others have noted feeding selectivity for Pcy in only one protozoan species (Caron et al., 1991). Similar results to those of Pernthaler et al. (1996a) have been obtained in marine systems, where a feeding preference of protozoa (flagellates plus ciliates) for Pcy was observed with removal rates ranging from 6 - 32% of Pcy cells d^{-1} (James et al., 1996). Pernthaler et al. (1996a) emphasise the influence of community composition and taxa-specific clearance rates on the grazing impact on bacteria and Pcy, e.g. the prevalence of highly specialised bacteria feeders, such as choanoflagellates, among the heterotrophic flagellate community, results in a higher consumption of bacteria instead of APP.

Among ciliates, oligotrich species and some scuticociliates (< 30 μm) can be important picoplanktivores in lakes (Simek et al., 1995; Callieri, in press). Šimek et al., (1996) have summarised three ecological categories of freshwater ciliates with different feeding strategies and a decreasing importance of pico-size prey in their diet. Among the most efficient suspension feeders there are some very active Pcy grazers, e.g. *Vorticella aquadulcis*, *Halteria grandinella*, *Cyclidium* and *Strobilidium hexachinetum*. These protozoa are able to graze 560, 210, 80, 76 cells h^{-1}, respectively, with clearance rates reaching values of 3150 nL cells h^{-1} (Šimek et al., 1996). Some raptorial feeders such as *Urotricha* and *Balanion planktonicum* have grazing rates of 2 and 0.2 APP cells h^{-1}, respectively, and clearance rates of about 10 nL cells h^{-1} (Šimek et al., 1996).

There are many link-organisms, grazing on Pcy and bacteria that provide important connections between the classic pelagic food web and microbial food webs (Plate 14). Current models of the impacts of pelagic microbial food webs on material and energy flows are based on the presence or absence of *Daphnia*, or other large 'keystone' grazers that can not only interfere with other Pcy predators (flagellates, ciliates, microzooplankton), but also have the capability to use them as prey (Stockner & Porter 1988, Weisse and Stockner, 1993; Jurgens, 1994; Callieri et al., in press). Together with *Daphnia* several other cladoceran genera such as *Bosmina*, *Eubosmina* and *Ceriodaphnia* are able to ingest Pcy (Gophen and Geller, 1984; Weisse, 1993). Whether or not copepods can efficiently ingest and then assimilate Pcy has remained problematic since this was initially proposed (Stockner and Antia, 1986). Nonetheless, it seems that amounts of Pcy cells transferred through flagellates to copepods is very small, about 4% of the initial or total carbon production in Lake Biwa (Nagata et al., 1996).

Rotifers can either act directly on Pcy populations by grazing or indirectly by preying on nanoflagellates and small ciliates (Stockner and Shortreed, 1989; Arndt, 1993; Pernthaler et al., 1996b). It would be interesting to apply the recently developed method of fluorescently labelled HNF to verify the extent of rotifer predation on HNF in lakes (Cleven, 1996). As Stockner and Antia (1986) asserted, Pcy are within the size range suitable for grazing by nauplii and early copepodite stages of copepods. This possibility has been partially confirmed by the direct estimate of the grazing rate on Pcy and bacteria by a copepod naupliar stage in a marine system (Roff et al., 1995). Thus, another intermediary link between the classic and microbial pelagic food webs can now be recognised.

At present there are few studies, and these are all in marine systems on the role of bivalve and gastropod larvae as possible picoplanktivores. In one of these, Bell (1991) has demonstrated that *Crepidula aculeata* and *Littoraria scabra*, two gastropod larvae, can thrive on a Pcy and bacteria diet. Furthermore, the larvae of a bivalve, *Mercenaria mercenaria*, have recently been shown to ingest and grow on a *Synechococcus* strain (Gallager et al., 1994). Thus, there now appears to be some coupling between microbial and littoral benthic food webs in aquatic ecosystems.

Grazing rates are usually balanced by similar growth rates, and can have a pronounced diel pattern related to prey size. If and to what extent the grazing activity is growth rate dependent, and which are the more important regulating mechanisms involved, still remains a matter of conjecture. Nonetheless, this is an important issue in the study of picoplanktivory and some authors have devoted their efforts primarily to measuring directly *in situ* grazing and growth rates to ascertain whether these are in balance, and how their relative importance varies with changing environmental conditions (Fahnenstiel et al.,1991a; Weisse and Schweizer, 1991; Nagata et al., 1994; Nagata et al., 1996). Some workers make the generalisation that grazing loss rates are similar to or somewhat lower than growth rates (Weisse, 1993), while others have estimated that grazing mortality of Pcy is in balance with growth rate (Nagata et al.,1994). In Lakes Huron and Michigan Pcy grazing losses ranged from 0.1 - 0.7 d^{-1} (Fahnenstiel et al., 1991b), and in Lake Constance values ranged from 0.05 - 1.27 d^{-1} (Weisse, 1988). So it seems that it is now possible to conclude that Pcy grazing losses seem to show large seasonal variability and are tightly coupled and sometimes in balance with Pcy growth rates.

To better comprehend the significance of grazing in controlling Pcy populations in lakes it is necessary to understand the influence of the Pcy diel cell cycle on grazing. The mechanisms which connect grazing to growth can be explained in the light of modifications to the cells during their diel cycle. The most evident change in Pcy cells is that their volume increases before division. Many have noted the influence of predator selection on prey size and various feeding mechanisms have been discussed (Vanderploeg, 1990; Burns and Schallenberg, 1996). Recent work has shown a marked effect of protistan grazing on size structure of natural bacterioplankton (Pernthaler et al., 1996b), but few studies have been done that specifically target the size range of Pcy. Grazing studies by Monger and Landry (1992) report that clearance rates increase approximately linearly with prey diameter (in the prey range 0.7 - 1.4 μm), and Pernthaler et al. (1996a) have shown, by a selectivity index, that Pcy are ingested preferentially over bacteria by protozoan predators, and they have interpreted this result as a size effect more than a food quality effect. However, a possible error in this approach could be the difference in palatability of Pcy and bacteria. Other authors circumvented the problem by estimating grazing rate on Pcy during the day and during the night (Ning and Vaulot, 1992; Liu, 1990). They have obtained higher grazing rates during the day when the cell size is larger, similar to the findings of Pernthaler et al. (1996b). It would be interesting to verify whether predation can influence the FDC of Pcy as has been done for bacteria (Sherr et al., 1992). Nevertheless, if the hypothesis of a size control of grazing processes is attractive, we cannot lose sight of the possibility of a chemosensory behaviour or allelopathic response to potential predators as well (Christoffersen, 1994; Vrede 1996).

There is a surprising lack of published accounts of grazing on CPcy by herbivorous zooplankton in lakes. Whether this is because they are poor quality food for macrozooplankton or simply due to the fact that nobody has bothered to study the predator-prey interaction in CPcy is unclear. Zaret and Suffern (1976), who studied the grazing effect of zooplankton on phytoplankton in Gatun Lake, Panama, found that chroococcal CPcy were not grazed, and they suggested that this was probably due to the fact that the cells were embedded in a gelatinous matrix. Recently, Bloomqvist (1996) and Vrede (1996) have reported *Merismopedia tenuissima* to be resistant to grazing by the dominant grazer *Eudiaptomus* in a clear water oligotrophic lake in central Sweden. Thus, from what little is known at this time, we can conclude that it is likely that the mucilaginous colonial morphology as found in larger colonial cyanobacteria, is an effective anti-grazing adaptation that, perhaps when coupled with the possibility of allelopathic strategies in CPcy colonies, create the perfect, coupled anti-predation device for these CPcy species. This may very well be the reason for their ubiquity and abundance in lakes.

H. Microbial Food Webs

A glimpse at the APP and Pcy literature from the early 1980s shows a marked escalation of interest by aquatic ecologists in picoplankton and their functional role within, and contribution to, pelagic MFW (Sieburth et al., 1978; Williams, 1981; Azam et al., 1983). Our intention here is neither to review the large literature on MFW in lakes that has accrued over the last 2 decades (Stockner and Porter, 1988; Weisse and Stockner, 1993; Reimann and Christoffersen, 1993), nor to list the functional roles of major players as this has already been reviewed in Section 5 above. Rather we wish to briefly focus attention on Pcy as important templates at the base of MFW. In many MFW studies growth and production of heterotrophic bacteria are the focal point of template food production with little attention being given to the co-occurrence and contribution of the Pcy community (Ducklow 1991; Bell and Tranvik, 1993). Though less abundant that free living bacteria, the biomass and production of the APP community, particularly the Pcy, is a more important carbon source than bacteria in MFW (Weisse, 1993). It has recently been shown in whole lake fertilisation experiments that increased Pcy carbon production flows through MFW to positively affect production at all trophic levels, including fish (Stockner and MacIsaac, 1996). These results are contrary to the arguments of Ducklow et al. (1986) who maintained that bacterioplankton can create large carbon pools or 'sinks' with large maintenance costs for coastal marine pelagic ecosystems. Several studies have suggested that material and energy flux within MFW in lakes can be enhanced by having a intermediate 'link' or 'keystone' grazer like *Daphnia* present in the system to move carbon from MFW to fish (Stockner, 1987; Porter et al., 1988; Pace et al., 1990). However, as Stockner and MacIsaac (1996) have shown, even in the absence of the keystone grazer, MFW can funnel carbon, albeit inefficiently, to all trophic levels. Thus, we urge limnologists not to ignore the contributions of Pcy and also CPcy to energy flows in planktonic MFW of lakes of all trophic types.

V. Conclusions

New perspectives on Pcy and CPcy ecology will ultimately rely on the improvement and incorporation of new sampling and enumeration protocols, as well as the application of new molecular methods to study genetic diversity in Pcy and CPcy communities. These advances should improve our comprehension and understanding of Pcy spatial and temporal variation within the framework of a new clonal succession theory. Recent findings suggest that the seasonal changes of Pcy assemblages in lakes are the result of the development of a succession of different strains or clones, interacting with multiple biotic and abiotic variables (Ernst, 1991; Ernst et al., 1995). Looking at the concept of phytoplankton species succession in lakes, Pcy communities may now assume heightened significance for the comprehension of "species associations" (or clones in the case of Pcy) (Reynolds 1988). It can now be argued that Pcy seasonal succession in lakes is really a manifestation of clonal succession.

Recent results, albeit preliminary, on effects of UV-B radiation on Pcy populations in fresh water suggests severe impairment of photosynthetic function and resultant loss of an important carbon source for microbial food webs in surface waters of lakes. New ecological perspectives on Pcy communities and their key role at the base of microbial food webs in pelagic ecosystems suggests that they may be good sentinels or "canaries in the coal mine", signalling both impending changes in the way nutrients are recycled, and energy flows through pelagic food webs as phytoplankton communities move from reliance on exogenous nutrient supplies to support new production to recycled nutrients (Weisse and Stockner, 1993). Will the gradual decline of diatom based food chains or 'classic' food webs, and the increased importance of Pcy based 'microbial' food webs portend a growing oligotrophication of our once productive coastal ocean margins, upwelling zones and large deep lakes ? With Pcy dominance come regenerative communities and less efficient carbon flows resulting in declining fish production in lakes and oceans (Henderson et al., 1992; Mann, 1993). Finally, we argue that we desperately need to improve our knowledge of non-bloom forming Pcy, especially the CPcy, whose ecology is so poorly understood. If we fail to do so, we will predict less reliably their population responses to an uncertain but probably much warmer climatic future in the next century. How will Pcy and CPcy adapt to this global warming and ozone layer thinning with the possibility of higher UV-A and B radiation ? With the growing likelihood of a warmer and more strongly stratified surface layer in lakes and coastal oceans, coupled with more severe nutrient depletion in the euphotic zone, how can current levels of pelagic carbon production be sustained ? How will global fisheries be affected ? Will the pelagic ecosystem of the temperate oceans and inland fresh waters of the next century return to a Pcy or CPcy dominance ? These are fair questions because Pcy and CPcy are our most ancient extant survivors of primordial seas, capable of adapting to extreme environments. These are crucial questions that need answers if we are to preserve the quality and integrity of fresh water and global ocean fisheries, let alone the viability of the human species on earth.

Acknowledgements

The authors wish to thank several people for their help. Firstly, the Department of Fisheries and Oceans for their support of limnological food chain research in BC lakes over the past two decades, and for the collaborative efforts and unwaivering support of colleagues in the Plankton Ecology Laboratory, West Vancouver, especially the extra efforts of Messrs Ken Shortreed, Erland MacIsaac and Bruce Nidle. We also thank Dr Roberto Bertoni for his comments on the manuscript throughout its evolution and for providing laboratory facilities in support of plankton research at the Istituto Italiano di Idrobiologia, Verbania-Pallanza, and Mr Mario Contesini for technical assistance and field work on Lago Maggiore.

References

Affronti LF and Marshall HG (1994) Using frequency of dividing cells in estimating autotrophic picoplankton growth and productivity in the Chesapeake Bay. Hydrobiologia 284: 193-203

Ålvik G (1934) Plankton-Algen norwegischer Austernpollen I. Systematik und Vorkommen der Arten. Bergens Mus Årb 1934 (6): 47 pp

Armbrust EV, Bowden JD, Olson RJ and Chisholm SW (1989) Effect of light on the cell cycle of a marine *Synechococcus* strain. Appl Environ Microbiol 55: 425-432

Arndt H (1993) Rotifers as predators on components of the microbial web (bacteria, heterrotrophic flagellates, ciliates) – aa review. Hydrobiologia 225/226: 231-246

Ayukai T (1996) Possible limitation of the dilution technique for estimating growth and grazing mortality rates of picoplanktonic cyanobacteria in oligotrophic tropical waters. J Exp Mar Biol Ecol 198: 101—111

Azam F, Fenchel T, Field JG, Gray JS, Meyer-Reil LA and Thingstad F (1983) The ecological role of water-column microbes in the sea. Mar Ecol Progr Ser 10: 257-263

Bailey-Watts AE and Komárek J (1991) Towards a formal description of a new species of *Synechococcus* (Cyanobacteria/Cyanophyceae) from freshwater picoplankton. Archiv Hydrobiol, Suppl 88, Algological Studies 61: 5-19

Bell JL (1991) Patches and picoplankton. Effects on larval life spans on gastropod larvae. Am Zool 31: 6A

Bell RT and Tranvik L (1993) Impact of acidification and liming on the microbial ecology of lakes. Ambio 22(5): 325-330

Bertoni R (1997) The routine use of Anodisc™ filters with automatic CHN analysers. Mem Ist Ital Idrobiol 56: 157-161

Bertoni R and Callieri C (1989) Organic matter and decomposers in Lago Maggiore: a pluriannual study. Mem Ist Ital Idrobiol 46: 145-172

Bertoni R and Callieri C (1997) Microbial loop and organic carbon in large southern Alpine lakes. In: Mosello R and Giussani G (eds) Documenta Ist. ital. Idrobiol. 61: 201-224

Bertoni R and Callieri C (submitted) The effects of UV-B radiation on freshwater autotrophic and heterotrophic picoplankton in a subalpine lake.

Bertoni R, Pugnetti A and Barbarewicz C (1997) La radiazione UV-B nei laghi ed i suoi effetti sulle comunità microbiche. Proc 12th AIOL Congress. Vol 1 Piccazzo M. (ed.): 245-254

Binder BJ and Chisholm SW (1995) Cell cycle regulation in marine *Synechococcus* sp. strains. Appl Environ Microbiol 61: 708-717

Bird DF and Kalff J (1986) Bacterial grazing by planktonic lake algae. Science 231: 493-495

Bird DJ and Kalff J (1987) Algal phagotrophy: regulating factors and importance relative to photosynthesis in Dinobryon (Chrysophyceae). Limnol Oceanogr 32: 277--284

Bloem J, Bar-Gilissen MJB and Cappenberg TE (1986) Fixation, counting and manipulation of heterotrophic nanoflagelates. Appl Environ Microbiol 52:1266-1272

Blomqvist P (1996) Late summer phytoplankton responses to experimental manipulations of nutrients and grazing in unlimed and limed Lake Njupfatet, central Sweden. Arch Hydrobiol 137: 425--455

Blomqvist P, Petterson A and Hyenstrand P (1994) Ammonium-nitrogen: a key regulatory factor causing dominance of non-nitrogen-fixing cyanobacteria in aquatic systems. Arch Hydrobiol 132:141-164

Bourrelly P (1985) Les Algues d'Eaux Douce. N. Boubée, Paris

Burns CW and Stockner JG (1991) Picoplankton in six New Zealand lakes: abundance in relation to season and trophic state. Int Revue ges Hydrobiol 76: 523-536

Burns CW and Schallenberg M (1996) Relative impacts of copepods, cladocerans and nutrients of the microbial food web of a mesotrophic lake. J Plankton Res 18(5): 683-714

Butcher RW (1952) Contributions to our knowledge of smaller marine alga. J marine Biol Assoc 31: 610-652

Callieri C (1994a) Primi approcci allo studio del picoplancton del Lago Maggiore. Atti dell'11th Congresso dell'Associazione Italiana di Oceanologia e Limnologia. G Albertelli, De Maio A and Piccazzo M (eds): 247-257.

Callieri C (1994b) Extinction coefficient of red, green and blue light and its influence on picocyanobacterial types in lakes at different trophic levels. Mem Ist ital Idrobiol 54: 135-142

Callieri C (1998) Pico-blues and their interactions with PAR and UV-B. Atti dell'12th Congresso dell'Associazione Italiana di Oceanologia e Limnologia. Vol 2 Piccazzo M (ed.): 149-158

Callieri C (in press) Carbon flux from picocyanobacteria to ciliates in a subalpine lake. Ver Internat Verein Limnol 27

Callieri C, Bertoni R, Amicucci E, Pinolini ML and Jasser I (1996) Growth rates of freshwater picocyanobacteria measured by FDC: problems and potentials for the estimation of picoplankton organic carbon synthesis. Arch Hydrobiol Spec Issues Advances Limnol 48: 93-103

Callieri C and Pinolini ML (1995) Picoplankton in Lake Maggiore, Italy. Int Revue ges Hydrobiol 80: 491-501.

Callieri C and Bertoni R (in press) Organic carbon and microbial asssemblages in a oligotrophic alpine lake.. Mem Ist Ital Idrobiol

Callieri C, Pugnetti A and Maanca M (in press) Carbon partitioning in the food web of a high mountain lake: from bacteria to zooplankton. Mem Ist Ital Idrobiol

Callieri C and Stockner JG (in press) Pico-blues success across freshwater trophic gradient. Lakes & Reservoirs.

Campbell L and Carpenter EJ (1986a) Diel pattern of cell division in marine *Synechococcus* spp. (Cyanobacteria): use of frequency of dividing cell technique to measure growth rate. Mar Ecol Progr Ser 32: 139-148

Campbell L and Carpenter EJ (1986b) Estimating the grazing pressure of heterotrophic nanoplankton on *Synechococcus* spp. using the seawater dilution and selective inhibitor techniques. Mar Ecol Progr Ser 33: 121-129

Caron DA, Lim EL, Miceli G, Waterbury JB and Valois FW (1991) Grazing and utilization of chroococcoid cyanobacteria and heterotrophic bacteria by protozoa in laboratory cultures and coastal plankton community. Mar Ecol Progr Ser 76: 205-217

Carpenter EJ and Campbell L (1988) Diel pattern of cell division and growth rate of *Synechococcus* spp. in Long Island Sound. Mar Ecol Progr Ser 47: 179-183

Carreto JI, Carignan MO, Daleo G and De Marco SG (1990) Occurrence of mycosporine-like amino acids in the red tide dinoflagellate *Alexandrium escavatum*: UV-photoprotective compounds? J Plankton Res 12: 909-921

Charpy L and Blanchot J (1996) *Prochlorococcus* contribution to phytoplankton biomass and production of Takapoto Atoll (Tuamotu Archipelago). Comptes Rendus de l'Academie des Sciences 319 (2): 131-137

Chisholm S,. Armbrust EV and Olson RJ (1986) The individual cell in phytoplankton ecology: cell cycle and applications of flow cytometry. In: Platt T and Li WK (eds) Photosynthetic Picoplankton. pp 343-369. Can Bull Fish Aquat Sci 214

Chisholm SW (1992) Phytoplankton size. In: Falkowski PG and Woodhead AD (eds) p 213-237 Primary Productivity and Biogeochemical Cycles in the Sea

Clarke AK, Campbell D, Gustafsson P and Oquist G (1995) Dynamic responses of photosystem II and phycobilisomes to changing light in the cyanobacterium *Synechococcus* pcc 7942. Planta 197 (3): 553-562

Cleven EJ (1996) Indirectly fluorescently labelled flagellates (IFLF): a tool to estimate the predation on free-living heterotrophic flagellates. J Plankton Res 18 (3): 429-443

Craig SR (1987) The distribution and contribution of picoplankton to deep photosynthetic layers in some meromictic lakes. Acta Acad Abo 47: 55-81

Cristoffersen K (1994) Variation of feeding activities of heterotrophic nanoflagellates on picoplankton In: Arndt H (ed.) Protozooplankton Ecology Methods and Seasonal Successions. pp 111-123 Vol 8 (1-2)

Cronberg G (1988) *Cyanodictyon tubiforme* a new chroococcal blue-green alga from Lake Börringesjön, Scania, Sweden. Arch Hydrobiol Suppl 80(1-4*),* Algological Studies 50-53: 191-194

Cronberg G (1991) *Cyanothamnos planctícus* gen. et sp. nov., a new colonial cyanophyte from an eutrophic Scanian lake, Sweden. Arch Hydrobiol Suppl 92(1-4*),* Algological Studies 64: 61-70

Cronberg G and Komárek J (1994) Planktic Cyanoprokaryotes found in South Swedish lakes during the XIIth International Symposium on Cyanophyte Research, 1992. Algological Studies 75: 323-352

Cronberg G and Weibull C (1981) *Cyanodictyon imperfectum*, a new chroococcal alga from Lake Trummen, Sweden. Arch Hydrobiol/Suppl 60, Algological Studies 27: 101-110

Cullen JJ, Neale PJ and Lesser PM (1992) Biological weighting function for the inhibition of phytoplankton photosynthesis by ultraviolet radiation. Science 258: 646—650

Drews G, Prauser H and Uhlmann D (1961) Massenvorkommen von *Synechococcus planctícus* nov. spec., einer solitären planktischen Cyanophyceae, in einem Abwasserteich Betrag zur Kenntnis der sogenannten "µ-Algen". Arch Mikrobiol 39: 101-115

Ducklow HW (1991) The passage of carbon through microbial food webs: results from network flow analysis. Mar Microb Food Webs 5:129-144

Ducklow HW, Purdie DA, LeB-Williams PJ and Davies JM (1986) Bacterioplankton: a sink for carbon in a coastal marine plankton community. Science 232: 865-867

Ernst A (1991) Cyanobacterial picoplankton from Lake Constance. I. Isolation by fluorescence characteristics. J Plankton Res 13: 1307-1312

Ernst A, Marschall P and Postius C (1995) Genetic diversity among *Synechococcus* spp. (cyanobacteria) isolated from the pelagial of Lake Constance. FEMS-Microbiol Ecol 17(3): 197-204

Fahnenstiel GL and Carrick HJ (1992) Phototrophic picoplankton in lakes Huron and Michigan: abundance, distribution, composition and contribution to biomass and production. Can J Fish Aquat Sci 49 (2): 379-388

Fahnenstiel GL, Carrick HJ, Rogers CE and Sicko-Goad L (1991a) Red fluorescing phototrophic picoplankton in the Laurentian Great Lakes: what are they and what are they doing? Int Revue ges Hydrobiol 76: 603-616

Fahnenstiel GL, Patton TR, Carrick HJ and McCormick MJ (1991b) Diel division cycle and growth rates of *Synechococcus* in lakes Huron and Michigan. Int Revue ges Hydrobiol 76: 657-664

Fahnenstiel GL, Redalje D and Lohrenz S (1994) Has the importance of photoautotrophic picoplankton been overstimated? Limnol Oceanogr 39(2): 432-438

Falkowski P and LaRoche J (1991) Acclimation to spectral growth irradiance in algae. J Phycol 25: 8-14

Fenchel T (1987) The Ecology of Protozoa. Springer Verlag, New York, NY

Findley DL and Kasian SEM (1990) Phytoplankton communities of lakes experimentally acidified with sulphuric and nitric acids. Can J Fish Aquat Sci 47: 1378-1386

Fogg GE (1986) Picoplankton. Proc R Soc Lond B 228: 1-30

Gallager SM, Waterbury JB and Stoecker DK. (1994) Efficient grazing and utilization of the marine cyanobacterium *Synechococcus* by larvae of the bivalve *Mercenaria mercenaria.* Mar Biol 119 (2): 251-259

Garcia-Pichel F and Castenholz RW (1991) Characterization and biological implications of scytonemin, a cyanobacteria sheath pigment. J Phycol 27: 395-409

Garcia-Pichel F and Castenholz RW (1993) Occurrence of UV-absorbing mycosporine-like compounds among cyanobacterial isolates and an estimate of their screening capacity. Appl Environ Microbiol 59 (1): 163-169

Geitler L (1932) Cyanophyceae In Rabenhorst´s Kryptogamenflora Vol 14 pp 1069. Leipzig

Gervais F, Padisak J and Koschel R (1997) Do light quality and low nutrient concentration favour picocyanobacteria below the thermocline of the oligotrophic Lake Stechlin? J Plankton Res 19 (6): 771-781.

Glover HE, Keller MD and Guillard RRL (1986) Light quality and oceanic ultraphytoplankters. Nature 319: 142-143

Glover HE, Phinney DA and Yentsch CS (1985) Photosynthetic characteristics of picoplankton compared with those of larger phytoplankton populations in various water masses in the Gulf of Maine. Biol Oceanogr 3: 223-248

Gophen M and Geller W (1984) Filter mesh size and food particle uptake by *Daphnia*. Oecologia 64: 408-412

Häder DP (1993a) Effects of enhanced solar ultraviolet radiation on aquatic ecosystems. In: Tevini M (ed.) UV-B Radiation and Ozone Depletion, Effects on Humans, Animals, Plants, Microorganisms and Materials. pp 155-191

Häder DP (1993b) Risks of enhanced solar ultraviolet radiation for aquatic ecosystems. Pro Phycol Res 9: 1-45

Häder DP (1996) Effect of enhanced solar UV-B radiation on phytoplankton. Scentia Marina 60 (Suppl 1): 59-63

Hall JA (1991) Long-term preservation of picophytoplankton for counting by fluorescence microscopy. Br Phycol J 26: 169-176

Hall JA and Vincent WF (1994) Vertical and horizontal structure of the picophytoplankton community in a stratified coastal system off New Zealand. NZ J Mar Freshwater Res 28 (3): 299-308

Hauschild CA, McMurter HJG and Pick FR (1991) Effect of spectral quality on growth and pigmentation of picocyanobacteria. J Phycol 27: 698-702

Havens KE and Heath RT (1991) Increased tranparency due to changes in the algal size spectrum during experimental acidification in mesocosms. J Plankton Res 13(3): 673-679

Hawley GRW and Whitton BA (1991a) Survey of algal picoplankton from lakes in five continents. Verh Internat Verein Limnol 24: 1220-1222.

Hawley GRW, Whitton BA (1991b) Seasonal changes in chlorophyll-containing picoplankton populations of ten lakes in northern England. Int. Rev. ges. Hydrobiol. 76: 545-554.

Hayat MA (1981) Fixation for Electron Microscopy. Academic Press NY

Henderson M, Stockner JG and Levy DA (1992) Probable consequences of climate change on freshwater aspects of the production of Adams River sockeye salmon (*Oncorhynchus nerka*). Geo Journal 28: 51-59

Hessen DO, VanDonk E and Andersen T (1995) Growth responses, P-uptake and loss of flagellae in *Chlamydomonas reinhardtii* exposed to UV-B. J Plankton Res 17(1): 17-27

Hickel B (1985) *Cyanonephron styloides* gen et sp nov a new chroococcal blue-green algae (Cyanophyta) from a brackish lake. Arch Hydrobiol/Suppl 71, Algological Studies 38/39: 99-104

Hickel B (1991) Two new chroococcal cyanophytes from a brackish environment, (Schlei-Fjord) Germany Algological Studies 64: 97-104

Hindák F (1975) Einige neue und interessante Planktonblaualgen aus der Westslowake. Arch Hydrobiol/Suppl 46 Algological Studies 13: 330-335

Hindák F (1982) On some planktonic coccoid blue-green algae characteristic of Fe-precipitates. Arch Hydrobiol Suppl 63(3) Algological Studies 32: 241-258

Hindák F (1985) The cyanophycean genus *Lemmermanniella* Geitler 1942. Arch Hydrobiol Suppl 71(3) Algological Studies 40: 393-401

Holm-Hansen O, Lubin D and Helbling EW (1993) Ultraviolet radiation and its effects on organisms in aquatic environments. In Young AR (ed.) Environmental UV Photobiology. pp 379-425 Plenum Press New York

Huang TC, Tu J Chow JT and Chen TH (1990) Circadian rhythm of the prokaryote *Synechococcus* sp. RF-1. Plant Physiol 92: 531-533

Iturriaga R and Mitchell BG (1986) Chroococcoid cyanobacteria: a significant component in the food web dynamics of the open ocean. Mar Ecol Progr Ser 28: 291-297

James MR, Hall JA and Barrett DP (1996) Grazing by protozoa in marine coastal and oceanic ecosystems off New Zealand. N Z J Mar Freshwater Res 30: 313-324

Jeffrey WH, Pledger RJ, Aas P, Hager S, Coffin RB, Vonhaven R and Mitchell DL (1996) Diel and depth profiles of DNA photodamage in bacterioplankton exposed to ambient solar ultraviolet radiation. Mar Ecol Progr Ser 137 (1-3): 238-291

Johnson PW and Sieburth J McN (1979) Chrococcoid cyanobacteria in the sea: a ubiquitous and diverse phototrophic biomass. Limnol Oceanogr 24: 928-935

Joint IR (1990) The response of picophytoplankton to light. In: Herring PJ, Campbell AK Whitfield M and Maddock L (eds) Light and Life in the Sea, pp 105-115. Cambridge University Press, UK

Jurgens K, Gasol JM Massana R and Pedrós Alió C (1994) Control of heterotrophic bacteria and protozoans by *Daphnia pulex* in the epilimnion of Lake Cisò. Arch Hydrobiol 131: 55-78

Kana TM and Glibert PM (1987a) Effect of irradiances up to 2000 μE $m^{-2}s^{-1}$ on marine *Synechococcus* WH7803 - I. Growth, pigmentation, and cell composition. Deep Sea Res 34: 479-495

Kana TM and Glibert PM (1987b) Effect of irradiances up to 2000 μE m^{-2} s^{-1} on marine *Synechococcus* WH7803-II. Photosynthetic responses and mechanisms. Deep Sea Res 34 (4): 497-516

Karentz D, McEuen FS, Land MC and Dunlap CW (1991) Survey of mycosporine-like amino acid compounds in Antarctic marine organisms: potential protection from ultraviolet exposure. Mar Biol 108: 157-166

Kiørboe T and Hansen J (1993) Phytoplankton aggregate formation: observations of patterns and mechanisms of cell sticking and the significance of exopolymeric material. J. Plankton Res. 15: 993-1018

Kirchman DL (1993) Statistical analysis of direct counts of microbial abundance.In: Kemp PF, Sherr BF, Sherr EB and Cole JJ (eds) Handbook of Methods in Aquatic Microbial Ecology. pp 117-121 Lewis Publ,

Klut, EM and Stockner JG (1991) Picoplankton associations in an ultra-oligotrophic lake on Vancouver Island, British Columbia. Can J Fish Aquat Sci 48: 1092-1099

Komárek J (1958) Die taxonomische Revision der planktischen Blaualgen der tschechoslovakei Komárek J Ettl H Algological Studies pp 10-106

Komárek J (1976) Taxonomic review of the genera *Synechocystis* SAUV. 1892, *Synechococcus* Näg. 1849 and *Cyanothece* gen nov (Cyanophyceae) Arch Protistenk 118: 119-179

Komárek J (1996) Towards a combined approach for the taxonomic and species delimitation of picoplanktic cyanoprokaryotes. Algological Studies 83: 377-401

Komárek J and Kling H (1991) Variation in six planktonic cyanophyte genera in Lake Victoria (East Africa) Algological Studies 61: 21-45

Komárek J and Komárková-Legnerová J (1992) Variability of some planktic gomphosphaeriod cyanoprocaryotes in northern lakes. Nord J Bot. 12: 513-524

Komárková J and Cronberg G (1985) *Lemmermanniella pallida* (LEMM) GEITL from South Swedish lakes. Arch Hydrobiol Suppl 71(3) Algological Studies 40: 403-413

Komárková-Legnerová J and Cronberg G (1994) Planktic blue-green algae from lakes in South Scania Sweden Part I: Chroococcales. Algological Studies 72: 13-51

Kondo T, Strayer CA, Kulkarni RD, Taylor W, Ishiur M, Golden SS and Johnson CH (1993) Circadian rhythms in prokaryotes: luciferase as a reporter of circadian gene expression in cyanobacteria. Proc Natl Acad Sci US 90: 5672-5676

Kulkarni RD and Golden SS (1995) Form II of D1 is important during transition from standard to high light intensity in *Synechococcus* sp strain PCC7942. Photosynthesis Res 46: 435-443

Kuuppo-Leinikki P and Kuosa H (1989) Preservation of picoplanktonic cyanobacteria and heterotrophic nanoflagellates for epifluorescence microscopy. Arch Hydrobiol 114: 631-636

Laamanen M (1996) Cyanoprokaryotes in the Baltic Sea ice and winter plankton Algological Studies 83: 423-433

Landry MR, Haas LW and Fagerness VL (1984) Dynamics of microbial plankton community: experiments in Kaneohe Bay Hawaii. Mar Ecol Progr Ser 16:127-133

Landry MR Kirshtein J and Constantinou J (1995) A refined dilution technique for measuring the community grazing impact of microzooplankton with experimental tests in the Central Equatorial Pacific. Mar Ecol Progr Ser 120 (1-3): 53-63

Lagerheim G (1883) Bidrag till Sveriges algflora Öfv Kgl Vetensk-Akad Förhandl 40 (2)

Lemmermann E (1904) Das Plankton schwedischer Gewasser. Ark Bot 2(2):1-209

Lindel D and Post AF (1995) Ultraphytoplankton succession is triggered by deep winter mixing in the Gulf of Aqaba (Eilat) Red Sea. Limnol Oceanogr 40: 1130-1141

Liu Z (1990) Estimating the grazing pressure of heterotrophic nanoplankton on Synechococcus spp. in the surface coastal water using selective inhibitor techniques. Dongay-Mar Sci 8 (3): 49-54

Liu H, Campbell L and Landry MR (1995) Growth and mortality rates of *Prochlorococcus* and *Synechococcus* measured with a selective inhibitor technique. Mar Ecol Progr Ser 116: 277-287

Lund JWG (1961) The periodicity of μ-algae in three English lakes. Verh Intern Verh Limnol 14: 147-154

MacIsaac ER and Stockner JG (1993) Enumeration of phototrophic picoplankton by autofluorescence microscopy. In: Sherr B and Sherr E (eds) The Handbook of Methods in Aquatic Microbial Ecology. pp 187-197 CRC Press, Boca Raton Fl

Maeda H, Kawai A and Tilzer MM (1992) The water bloom of cyanobacterial picoplankton in Lake Biwa Japan. Hydrobiologia 248: 93-103

Malinsky-Rushansky N, Berman T and Dubinsky Z (1995) Seasonal dynamics of picophytoplankton in Lake Kinneret Israel. Freshwater Biol 34: 241-254

Mann KH (1993) Physical oceanography, food chains, and fish stocks: a review. ICES J Mar Sci 50: 105-119

Mastala Z, Herodek S, V-Balogh K, Borbély G, Shafik HM and Vörös L (1996) Nutrient requirement and growth of a *Synechococcus* species isolated from Lake Balaton. Int Revue ges Hydrobiol 81: 503-512

McMurther HJC and Pick F (1994) Fluorescence characteristics of a natural assemblage of freshwater picocyanobacteria. J Plankton Res 16: 911-925

Meyer B (1994) A new species of *Cyanodictyon* (Cyanophyceae, Chroococcales) planktic in eutrophic lakes. Algological Studies 75: 183-188

Moeller RE (1994) Contribution of ultraviolet radiation (UV-A UV-B) to photoinhibition of epilimnetic phytoplankton in lakes of differing UV transparency. Arch Hydrobiol Beih 43: 157-170

Monger BC and Landry MR (1992) Size selectivity grazing by heterotrophic nanoflagellates: an anlysis using live-stained bacteria and dual-beam flow cytometry. In: Bjoersen PK and Riemann B (eds) Microbial Ecology of Pelagic Environments. 37 pp 173-185

Morris I and Glover HE (1981) Physiology of photosynthesis by marine coccoid cyanobacteria-some ecological implications. Limnol Oceanogr 26: 957-961

Müller-Niklas G, Heissenberger A, Puskaric S and Herndl GJ (1995) Ultraviolet-B radiation and bacterial metabolism in coastal waters. Aquat Microb Ecol 9: 111-116.

Nagata T, Takai K, Kawanobe K, Kim D, Nakazato R, Guselnikova N, Bondarenko N, Mologawaya O, Kostrnova T, Drucker V, Satoh Y and Watanabe Y (1994) Autotrophic picoplankton in southern Lake Baikal: abundance growth and grazing mortality during summer. J Plankton Res 16 (8): 945-959

Nagata T, Takai K, Kawabata K, Nakanishi M and Urabe J (1996) The trophic transfer via a picoplankton-flagellate-copepod food chain during a picocyanobacterial bloom in Lake Biwa. Arch Hydrobiol 137: 145-160

Naumann E (1924) Sotvattnets plankton Stockholm, Sweden pp 267

Nielsen T and Ekelund NGA (1995) Influence of solar ultraviolet radiation on photosynthesis and motility of marine phytoplankton. FEMS Microbiol ecology. 18(4): 281-288

Ning X and Vaulot D (1992) Estimating *Synechococcus* spp growth rates and grazing pressure by heterotrophic nanoplankton in the English channel and the Celtic Sea. Acta-Oceanol Sin 11 (2): 255-273

Overbeck J (1962) Das Nannoplankton (μ-Algen) der Rugenschen Brackwasser als Hauptproduzent in Abhangigkeit vom Salzgehalt. Kieler Meeresforschungen 18: 157-171

Pace ML, McManus GB and Findlay SEG (1990) Plankton community structure determines the fate of bacterial production in a temperate lake. Limnol Oceanogr 35:795-808

Padisák J, Krienitz L, Koschel R and Nedoma J (1997) Deep-layer autotrophic picoplankton maximum in the oligotrophic Lake Stechlin, Germany: origin, activity, development and erosion. Eur J Phycol 32: 403-416.

Padisák J, Krienitz L, Scheffler W, Koschel R, Kristiansen J and Grigorszky I (1998) Phytoplankton succession in the oligotrophic Lake Stechlin (Germany) in 1994 and 1995. Hydrobiologia 369/370: 179-197

Paerl HW (1977) Ultraphytoplankton biomass and production in some New Zealand Lakes. N Z J Mar Freshwater Res 11: 297-305

Passoni S, Callieri C and Heinimaa S (1997) Dinamiche di distribuzione del picoplancton autotrofo nel Lago Maggiore. Proc.12th AIOL Congress. Vol 1 Piccazzo M. (ed.): 109-118

Passoni S and Callieri C (in press) Picocyanobacteria single forms, aggregates and micro-colonies: survival strategy or species succession? Ver Internat Verein Limnol 27

Pedrós-Alió C and Brock TD (1983) The importance of attachment to particles for planktonic bacteria. Arch Hydrobiol 98: 354-379

Pernthaler J, Šimek K, Sattler B, Schwarzenbacher A, Bobkova J and Psenner R (1996a) Short-term changes of protozoan control on autotrophic picoplankton in an oligo-mesotrophic lake. J Plankton Res 18: 443-462

Pernthaler J, Sattler B, Šimek K, Schwarzenbacher B and Psenner R (1996b) Top-down effects on the size-biomass distribution of a freshwater bacterioplankton community. Aquat Microb Ecol 10: 255-263.

Petersen R (1991) Carbon-14 uptake by picoplankton and total phytoplankton in eight New Zealand lakes. Int Revue ges Hydrobiol 76: 631-641

Pick FR (1991) The abundance and composition of freshwater picocyanobacteria in relation to light penetration. Limnol Oceanogr 36: 1457-1462

Pick FR and Agbeti DM (1991) The seasonal dynamic and composition of photosynthetic picoplankton communities in temperate lakes in Ontario, Canada. Int Revue ges Hydrobiol 76: 565-580

Pick FR and Bérubé C (1992) Diel cycles in the frequency of dividing cells of freshwater picocyanobacteria. J Plankton Res 14: 1193-1198

Platt T (1971) The annual production by phytoplankton in St. Margaret's Bay Nova Scotia. J Cons int Explor Mer 33(3): 324-333

Platt T and Li WKW (1986) Photosynthetic picoplankton. Can Bull Fish Aquat Sci 214

Porter KG, Pearl H, Hodson R, Pace M, Priscu J, Riemann B, Scavia D and Stockner J (1988) Microbial interactions in lake food webs. In: Carpenter SR (ed.) Complex Interactions in Lake Communities. pp 209-228. Springer Verlag, New York

Postius C, Ernst A, Kenter U and Böger P (1996) Persistence and genetic diversity among strains of phycoerythrin-rich cyanobacteria from the picoplankton of Lake Constance. J Plankton Res 18: 1159-1166

Proteau PJ, Gerwick WH, Garcia-Pichel F and Castenholz RW (1993) The structure of scytonemin an ultraviolet sunscreen pigment from the sheaths of cyanobacteria. Experientia 49: 825-829

Rajagopal S and Murthy SDS (1996) Ultraviolet-B radiation induced alterations in photosynthetic electron transport activities of the cyanobacterium *Synechococcus* 6301. Photosynthetica 32 (2): 281-284

Raven JA (1986) Physiological consequences of extremely small size for autotrophic organisms in the sea. In: Platt T and Li WKW (eds) Photosynthetic Picoplankton. Can Bull Fish Aquat Sci 214

Reimann B and Christoffersen K (1993) Microbial trophodynamics in temperate lakes. Marine Microbial Food Webs. 7: 69-100

Reynolds CS (1988) The concept of ecological succession applied to seasonal periodicity of freshwater phytoplankton. Verh internat Verein Limnol 23: 683-691

Roff JC, Turner JT, Webber MK and Hopcroft RR (1995) Bacterivory by tropical copepod nauplii extent and possible significance. Aquatic Microbial Ecology 9 (2): 165-175

Ruiz J, Garcia CM and Rodriguez J (1996) Sedimentation loss of phytoplankton cells from the mixed layer: effects of turbulence levels. J Plankton Res 18 (9): 1727-1734

Sanders RW and Porter KG (1988) Phagotrophic phytoflagellates. Adv Micr Ecol 10: 167-192

Sanders RW, Porter KG, Bennet SJ and DeBiase AE (1989) Seasonal patterns of bacterivory by flagellates, ciliates, rotifers and cladocerans in a freshwater planktonic community. Limnol Oceanogr 34:673-687

Schindler DW (1990) Experimental perturbations of the whole lakes as tests of hypotheses concerning ecosystem structure and function. Oikos 57: 25-41

Sherr EB and Sherr BF (1987) High rates of consumption of bacteria by pelagic ciliates. Nature 325: 710-711

Sherr EB and Sherr BF (1993) Protistan grazing rates via uptake of fluorescently labeled prey. In: Kemp P, Sherr B, Sherr E and Cole J (eds) Handbook of Methods in Aquatic Microbial Ecology. pp 695-701. Lewis Pub Boca Raton, FL

Sherr EB, Sherr BF, Berman T and Hadas O (1991) High abundance of picoplankton-ingesting ciliates during late fall in Lake Kinneret Israel. J Plankton Res 13: 789-799

Sherr BF, Sherr EB and McDaniel J (1992) Effect of protistan grazing on the frequency of dividing cells in bacterioplankton assemblages. Appl Environ Microbiol 58 (8): 2381-2385

Shortreed KS and Stockner JG (1990) Effect of nutrient additions on lower trophic levels of an oligotrophic lake with a seasonal hypolimnetic chlorophyll maximum. Can J Fish Aquat Sci 47: 262-273

Sieburth JMcN, Smetacek V and Lenz J (1978) Pelagic ecosystem structure: heterotrophic compartments of the plankton and their relationship to plankton size fractions. Limnol.Oceanogr. 23: 1256-1263

Sime-Ngando T (1995) Population dynamics of autotrophic picoplankton in relation to environmental factors in a productive lake. Aquat Sci 57: 91-105

Šimek K, Bobkova J, Macek M, Nedoma J and Psenner R (1995) Ciliate grazing on picoplankton in a eutrophic reservoir during the summer phytoplankton maximum: a study at the species and community level. Limnol Oceanogr 40(6): 1077-1090

Šimek K, Macek M, Pernthaler J, Straskrabová V and Psenner R (1996) Can freshwater planktonic ciliates survive on a diet of picoplankton? J Plankton Res18: 597-613

Skuja H (1932) Vorarbeiten zu einer Algenflora von Lettland. Bibliotheca Phycologica 26: 1-302

Skuja H (1948) Taxonomie des Phytoplanktons einiger Seen in Uppland, Schweden. Symb bot upsal 9(3): 1-399

Skuja H (1964) Grundzüge der Algenflora und Algenvegetation der Fjeldgegenden um Abisko in schwedishch-Lappland. Nova Acta R Soc Sci Upsal Ser. 4 18(3):1-465

Smayda TJ (1974) Some experiments on the sinking characteristics of two freshwater diatoms. Limnol Oceanogr 19: 628-635

Smith RC and Baker KS (1989) Stratospheric ozone middle ultraviolet radiation and phytoplankton productivity. Oceanography 2: 4-11

Sommaruga R and Robarts (1997) The significance of autotrophic and heterotrophic picoplankton in hyperrtrophic ecosystems. FEMS Microbiol Ecol 24: 187-200

Søndergaard M (1991) Phototrophic picoplankton in temperate lakes: seasonal abundance and importance along a trophic gradient. Int Revue ges Hydrobiol 76: 505-522

Stanier RY, Kuniswawa R, Mandel R and Cohen-Bazire G (1971) Purification and properties of unicellular blue-green algae (Order Chroococcales). Bacteriol Rev 35:171-205

Stockner JG (1987) Lake fertilization: the enrichment cycle and lake sockeye salmon (*Oncorhynchus nerka*) production. In: Smith HD Margolis L and Wood CC (eds) Sockeye Salmon (*Oncorhynchus nerka*) Population Biology and Future Management, pp 198-215. Can Spec Publ Fish Aquat Sci 96

Stockner JG (1991) Autotrophic picoplankton in freshwater ecosystems: the view from the summit. Int Revue ges Hydrobiol 76: 483-492

Stockner JG (1998) Global warming, picocyanobaceria and fisheries decline: is there a connection? Proc.12°th AIOL Congress. Vol 2 Piccazzo M. (ed.): 29-37

Stockner JG and Antia NJ (1986) Algal picoplankton from marine and freshwater ecosystems: a multidisciplinary perspective. Can J Fish Aquat Sci 43: 2472-2503

Stockner JG, Klut ME and Cochlan WP (1990) Leaky filters: a warning to aquatic ecologists. Can J Fish Aquat Sci 47: 16-23

Stockner JG and MacIsaac EM (1996) British Columbia lake enrichment programme: two decades of habitat enhancement for sockeye salmon. Reg Rivers 12: 547-561

Stockner JG and Porter KG (1988) Microbial food webs in freshwater planktonic ecosystems. Carpenter SR (ed.) Complex Interactions in Lake Communities. pp 69-83 Springer-Verlag New York

Stockner JG and Shortreed KS (1988) Response of *Anabaena* and *Synechococcus* to manipulation of nitrogen:phosphorus ratios in a lake fertilization experiment. Limnol Oceanogr 33(6): 1348-1361

Stockner JG and Shortreed KS (1989) Algal picoplankton production and contribution to food webs in oligotrophic British Columbia lakes. Hydrobiologia 173:151-166

Stockner JG and Shortreed KS (1991) Phototrophic picoplankton: community composition abundance and distribution across a gradient of oligotrophic British Columbia and Yukon Territory lakes. Int Revue ges Hydrobiol 76: 581-601

Stockner JG and Shortreed KS (1994) Autotrophic picoplankton community dynamics in a prealpine lake in British Columbia, Canada. Hydrobiologia 274:133-142

Suttle CA (1994) Dynamics and distribution of cyanophages and their effect on marine *Synechococcus* sp. Appl Env Microbiol 60: 3167-74

Suttle CA, Cochlan WP and Stockner JG (1991) Size-dependent ammonium and phosphate uptake and N:P supply ratios in an oligotrophic lake. Can J Fish Aquat Sci 48: 1226-1234

Suttle CA, Stockner JG and Harrison PJ (1987) Effects of nutrient pulses on community structure and cell size of a freshwater phytoplankton assemblage in culture. Can J Fish Aquat Sci 44: 1768-1774

Suttle CA, Stockner JG, Shortreed KS and Harrison PJ (1988) Time-course studies of size-fractionated phosphate uptake: are larger cells better competitors for pulses of phosphate than smaller cells? Oecologia 74: 571-576

Sweeney BM and Borgese MB (1989) A circadian rhythm in cell division in a prokaryote, the cyanobacterium *Synechococcus* WH7803. J Phycol 25: 183-186

Takamura N and Nojiri Y (1994) Picophytoplankton biomass in relation to lake trophic state and the TN:TP ratio of lake water in Japan. J Phycol 30: 439-444

Takano H, Arai T, Hirano M and Matsunaga T (1995) Effects of intensity and quality of light on phycocyanin production by a marine cyanobacterium *Synechococcus* sp NKBG 042902. Appl Microbiol Biotechnol 43(6): 1014-1018

Tandeau de Marsac N (1977) Occurrence and chromatic adaptation in cyanobacteria. J Bacteriol 130: 82-91

Vanderploeg HA (1990) Feeding mechanisms and particles selection in suspension-feeding zooplankton. In: Wotton RS (ed.) The biology of particles in aquatic systems. pp 183-212 CRC Press Boston

Vaulot D, LeBot N, Marie D and Fukai E (1996) Effect of phosphorus on the *Synechococcus* cell cycle in surface Mediterranean waters during summer. Appl Environ Microbiol 62 (7): 2527-2533

Vincent WF and Roy S (1993) Solar ultraviolet-B radiation and aquatic primary production: damage, protection and recovery. Environ Rev 1: 1-12

Vinebrooke RD and Leavitt PR (1996) Effects of ultraviolet radiation on periphyton in an alpine lake. Limnol Oceanogr 41(5): 1035-1040

Vörös L, Callieri C Balogh KV and Bertoni R (1998) Freshwater picocyanobacteria along trophic gradient and light quality range. Hydrobiologia 369/370: 117-125

Vörös L, Gulyas P and Nemeth J (1991) Occurrence, dynamics and production of picoplankton in Hungarian shallow lakes. Int Revue ges Hydrobiol 76: 617-629

Vrede K (1996) Regulation of bacterioplankton production and biomass in an oligotrophic clearwater lake - the importance of the phytoplankton community. J Plankton Res 18: 1009-1032

Wanberg SA, Selmer JS and Gustavson K (1996) Effects of UV-B radiation on biomass and composition in marine phytoplankton communities. Scentia Marina 60: 81-88

Wang KS and Chai TJ (1994) Reduction in omega-3 fatty acids by UV-B irradiation in microalgae. J Appl Phycol 6: 415-421

Waterbury JB, Watson SW, Guillard RR and Brand LE (1979) Widespread occurrence of a unicellular, marine planktonic cyanobacterium. Nature 277: 293-294

Waterbury JB, Watson SW, Valois FW and Franks DG (1986) Biological and ecological characterisation of the marine unicellular cyanobacterium *Synechococcus*. Can Bull Fish Aquat Sci 214: 17-120

Waterbury JB, Willey JM, Franks DG, Valois FW and Watson SW (1985) A cyanobacterium capable of swimming motility. Science 230: 74-76

Wehr JD (1989) Experimental tests of nutrient limitation in freshwater picoplankton. Appl Environ Microbiol 55(6): 1605-1611

Wehr JD (1991) Nutrient and grazer-mediated effects on picoplankton and size structure in phytoplankton communities. Int Revue ges Hydrobiol 76: 643-656

Wehr JD (1993) Effects of experimental manipulations of light phosphorus supply on competition among picoplankton and nanoplankton in an oligotrophic lake. Can J Fish Aquat Sci 50: 936-945

Weisse T (1988) Dynamics of autotrophic picoplankton in Lake Constance. J Plankton Res 10: 1179-1188

Weisse T (1993) Dynamics of autotrophic picoplankton in marine and freshwater ecosystems. In: Jones JG (ed.) Advances in Microbial Ecology. Vol 13, pp 327-370 Plenum Press, New York

Weisse T and Kenter U (1991) Ecological characteristics of autotrophic picoplankton in a prealpine lake. Int Revue ges Hydrobiol 76: 493-504

Weisse T and Schweizer A (1991) Seasonal and interannual variation of autotrophic picoplankton in a large prealpine lake (Lake Constance). Verh Internat Verein Limnol 24: 821-825

Weisse T and Stockner JG (1993) Eutropication: the role a microbial food webs. Mem Ist ital Idrobiol 52: 133-150

West W and West GS (1894) On some algae from the West Indies. J Linn Soc 30(208): 264-280

Wetzel RG and Likens GE (1990) Limnological analyses. Springer-Verlag

Willey JM and Waterbury JB (1989) Chemotaxis toward nitrogenous compounds by swimming strains of marine *Synechococcus* spp. Appl Environ Microbiol. 55: 1888-1894

Williams PJ LeB (1981) Incorppration of microheterotrophic processes into the classical paradigm of the planktonic food web. Kiel Meeresforsch 5: 1-28

Williamson CE (1995) What role does UV-B radiation play in freshwater ecosystems? Limnol Oceanogr 40 (2): 386-392

Wilson WH, Carr NG and Mann NH (1996) The effect of phosphate status on the kinetics of cyanophage infection in the oceanic cyanobacterium *Synechococcus* sp. WH7803. J Phycol 32: 506-516

Wood AM, Horan PK, Muirhead K, Phinney DA, Yentsch CM and Waterbury JB (1985) Discrimination between types of pigments in marine *Synechococcus* spp. by scanning

spectroscopy, epifluorescence microscopy and flow cytometry. Limnol Oceanogr 30: 1303-1315
Wyman M and Fay P (1986) Underwater light climate and the growth and pigmentation of planktonic blue-green algae (Cyanobacteria) I. the influence of light quality. Proc R Soc Lond 227: 381-393
Zaret M and Suffem KL (1976) Vertical migration in zooplankton as a predator avoidance mechanism. Limnol Oceanogr 21: 804-813

Chapter 8

Soils and Rice-Fields

Brian A. Whitton
Department of Biological Sciences, University of Durham, Durham DH1 3LE, UK

Summary

Cyanobacteria are an important component of many soils, including the surface crusts that sometimes cover extensive areas in semiarid regions and mine spoil wastes. They are also abundant in many areas which are wet or submerged for part of the year, especially rice-fields. Most soils forms have sheaths or mucilage and this polysaccharide has important effects on the soil, mostly beneficial, such as improved soil structure, but sometimes adverse where a dense surface layer impedes drainage. Nitrogen-fixing species often constitute half or more of the species present in soils not enriched with nitrogenous fertilizer and these can contribute combined nitrogen in several ways to adjacent vascular plants.

Attempts to enhance crop yield by adding cyanobacteria to soils have mostly focussed on paddy rice. Although many studies have reported positive effects of such 'algalization', the number of locations where it has been adopted as routine practice appear to be few, in contrast to the relatively widespread use of *Azolla* with rice culture. Algalization is most successful where local species are used to prepare the inoculum, but there is considerable scope for other improvements. It is important to obtain a much more detailed understanding of cyanobacterial population dynamics over the whole annual cycle in agricultural systems where rice is grown for only part of the year.

B. A. Whitton and M. Potts (eds), The Ecology of Cyanobacteria, 233-255.
© 2000 *Kluwer Academic Publishers. Printed in the Netherlands.*

I. Introduction

Cyanobacteria have been reported from a wide range of soils, both on and below the surface. They are also often a characteristic feature of other types of subaerial environment and many intermittently wet ones such as rice-fields. This chapter gives a brief overview of their features. Other relevant information is included in the chapters on limestones, salts and deserts and some of the features of cyanobacteria discussed there contribute to their ability to compete in soils.

According to Granhall (1975), the major factors (in addition to light) influencing the occurrence of cyanobacteria in soil are moisture, pH, mineral nutrients and combined nitrogen. Moisture was especially important in favouring cyanobacterial high cyanobacterial density in the loamy soil studied by Zimmerman et al. (1980), where wet depressions contained much higher populations than surrounding dry areas. Tolerance of desiccation and water stress is, however, widespread and cyanobacteria are among the most successful organisms in highly saline environments (Chapter 10). Cyanobacteria have been reported to be infrequent in most soils below pH 6.0 (Lund, 1947; Granhall and Henriksson, 1969), though they are more frequent at low pH values in tropical soils (Moore, 1963). Their frequent occurrence on the surface of tropical soils may also be favoured by the fact that the temperature optimum for cyanobacteria is often higher by at least several degrees than for most eukaryotic algae (Castenholz and Waterbury, 1989). The ability to fix dinitrogen is widespread in soil cyanobacteria (e.g. Roger and Kulasooriya, 1980), species with this ability having a competitive advantage where combined nitrogen levels are low (Roger and Kulasooriya, 1980; Howarth et al., 1988). Many cyanobacteria tolerate high levels of ultra-violet irradiation (Chapter 21) permitting them to survive at the soil surface, whereas others photosynthesize efficiently at low photon flux densities (van Liere and Walsby, 1982), permitting effective photosynthesis just below the surface.

Studies on the role of cyanobacteria in soils and rice-fields have been hindered by a number of problems. Most of these are considered in more detail later in the chapter, but it is useful to be aware of some of them straight away. The difficulty in naming species has been a particular problem, because many soil forms are morphologically simple; what is probably the same organism may have been given a variety of names, while the same name may have been used for obviously different organisms. In contrast, flooded rice-fields often have a diverse flora of morphologically distinctive forms and this has led to an extensive descriptive literature; a great deal of useful information is held in such accounts, although they tend to be ignored in the mainstream literature. A lack of suitable methods for quantifying the occurrence of cyanobacteria in soils and rice-fields was at one time a problem, though the situation is much improved (Whitton, 1996). Finally, because of the importance of nitrogen-fixing cyanobacteria in these environments, there have been many applied studies aimed at enhancing their contribution to nitrogen fixation. These have provided important and interesting results, but, perhaps because of the funding sources for such work, sometimes they have led to uncritical papers and reports.

II Soils

A. Taxonomic Composition

A detailed review of soil algae by Metting (1981) gives many references to cyanobacteria, including a list of those 37 genera with records for soil species. Both unicellular and filamentous forms are represented, but the biomass usually consists largely of filamentous forms. Examples of studies that contain descriptions and figures of soil species are given in Table 1. Anand's (1990) taxonomic account of rice-field species in south India provides a particularly useful guide to species likely to be found in rice-fields everywhere.

B. Environmental Factors

1. Light

Cyanobacteria growing on the soil surface often show dark colorations - blue-black, brown, red-brown or red. The darker colours usually result from the presence of brown sheaths surrounding the typical photosynthetic trichome and this colour is much more pronounced in populations in open than in shaded positions. The pigment reponsible for the brown colouration, scytonemin, absorbs strongly in the near ultra-violet region of the spectrum (c 370 nm *in vivo*: Garcia-Pichel and Castenholz, 1991) and the evidence strongly suggests that scytonemin production is an adaptive strategy for photoprotection against short-wavelength solar irradiation (Chapter

Table 1. Some examples of studies with detailed taxonomic and/or floristic information on soil cyanobacteria. (See also Chapter 13, Table 1).

Region	Subject	Reference
Argentina	Paddy fields	de Halperin et al. (1992)
Australia	Cryptobiotic crusts	Eldridge & Greene (1994)
Bangladesh	Rice-field soils	Khan et al. (1994)
Czech Republic, Russia	Taxonomic account with illustrations	Desertová (1974)
Greece	Morphology of soil species	Economou et al. (1984)
India	Usar (alkaline) land floristics	Prasad & Srivastava (1968)
	Rice fields	Gupta (1966)
	Non-heterocystous species in rice-fields	Tiwari (1975)
	Nostocaceae in rice-fields	Tiwari & Pandey (1976)
	Floristic account for rice-fields near Pusa	Jha et al. (1986)
	Floristic account for Kerala rice-fields	Anand & Hopper (1987)
	Detailed taxonomic account of rice-field species	Anand (1990)
	Floristic account for Arunachal Pradesh rice-fields	Singh et al. (1997a)
	Floristic account for Nagaland rice-fields	Singh et al. (1997b)
Iraq	Floristic account of rice-fields	Al-Kaisi (1976)
Romania	Survey of wide range of soil types	Gruia (1964)
Russia and other former USSR	Guide to the older literature	Forest (1965)
UK	Illustrations and floristic account	Bristol (1920)
USA, Utah	Soil flora of semi-desert	Anderson & Rushforth (1976)
		Ashley et al. (1985)
		Johansen (1993)

21). The darker colour may also enhance warming of the substratum at times which are ecologically important for some populations, as discussed by Belnap and Harper (1995) (see below).

A number of studies have been made of the vertical distribution of cyanobacteria (and eukaryotic algae) in soil profiles. Most of the biomass usually occurs at the surface, with some cells or filaments penetrating several millimetres into the soil (Schwabe, 1963), though the peak biomass sometimes occurs just below the surface. Light penetration through the substratum may be sufficient for growth of filaments inside quartz-rich soils or endolithic forms in deserts (Bell et al., 1986; Palmer and Friedmann, 1990); the latter are described in some detail in Chapter 13. The most detailed study on light penetration is that of Garcia-Pichel and Belnap (1996) on two desert soil crusts in S-E. Utah formed by *Microcoleus vaginatus*, *Nostoc* and *Scytonema*, where strong attenuation meant that the euphotic zone of recently wetted crusts was only about 1 mm below the surface of the crust. In frost-sorted polygons on Signy Island, South Orkney Islands, a large proportion of the microflora occurred in the zone 0 -1 mm below the surface and few algae occurred on the soil surface (Davey and Clarke, 1991). The authors suggested that such subsurface colonization may be a desiccation - avoidance strategy. Chlorophyll degradation products occurred to depths of up to 8 cm.

There are many reports of cyanobacteria occurring at depths some way below the surface in agricultural and other soils (Metting, 1981), but it is not clear to what extent natural populations are able to persist or even increase in the absence of light. Some soil cyanobacteria can grow photoheterotrophically using simple sugars, though apparently not fatty acids; for instance, 8 out of 14 isolates from the waterlogged bank of a Senegal lake showed photoheterotrophic abilities (Reynaud and Franche, 1986). Some strains can also grow heterotrophically (Khoja and Whitton, 1971; Rippka et al., 1979), but *Calothrix marchica* was the only one of four Spanish rice-field soil isolates capable of doing so (Prosperi et al., 1992).

The filamentous forms are probably all capable of movement under some conditions and phototaxis has been demonstrated in many cyanobacteria , including strains isolated from soil (Castenholz, 1982). Species differ, however, in the extent to which filaments exhibit motility in nature. Actively growing filaments of *Phormidium autumnale* and some other Oscillatoriaceae are apparently motile for much of the time, whereas motility in many typically heterocystous forms (e.g. *Nostoc* and *Calothrix*) is restricted to hormogonia, which lack heterocysts and are formed under specific conditions, such as addition of phosphate to P-limited filaments (Whitton, 1992). The ability of some soil cyanobacteria to move permits them to be viewed in a semi-natural form by placing cover-slips in close contact with the surface of moistened soil and then incubating in the light (e.g. Broady, 1979; Davey, 1991). Motile organisms frequently attach themselves to the cover-slip within a few hours, while new colonies of these and non-motile forms may develop within a week.

Most species occurring below the soil surface are probably attached to soil particles. Davey et al. (1991) investigated the influence of morphology, mucilage production and soil texture as factors influencing attachment of cyanobacteria and eukaryotic algae isolated from Antarctic fellfield soils. *Phormidium autumnale* showed the highest attachment of any taxon under simulated flow conditions; in contrast to the eukaryotic alga *Zygnema* sp., *Pseudanabaena catenata* showed greater retention to the substratum with the least rugosity.

2. Water and Desiccation

Air-dried terrestrial and soil cyanobacteria can survive prolonged dry periods, as has been shown in many studies. Some of these have been based on material stored deliberately, such as Trainor's (1985) use of 25-yr old material, but most have been made on samples kept for other purposes. Widely quoted samples are an 87-yr record for *Nostoc commune* (Lipman, 1941) and the 7 genera of cyanobacteria which grew from 16 samples of partially air-dried soils from England kept in sealed bottles for 23 - 70 yr (Bristol, 1919). Parker et al. (1969) reviewed old literature and carried out further studies on materials from the Missouri Botanical Garden. Some of the older studies were not rigorous enough to rule out the possibility of aerial contaminants, but it is clear that some akinete- and non-akinete-forming cyanobacteria can survive for very long periods. The most frequently mentioned genera are *Anabaena*, *Aulosira*, *Cylindrospermum*, *Fischerella*, *Lyngbya*, *Nostoc*, *Plectonema* and *Stigonema*. A number of green algae and a few diatoms are also mentioned, so this property is not unique to cyanobacteria.

This ability can make it relatively easy to store field samples, enrichment cultures or even pure cultures in a dry state. Roger and Ardales (1991) reported on studies conducted at the International Rice Research Institute in the Philippines. Strains of cyanobacteria, which had been inoculated onto sterile soil and then allowed to dry, showed high survival in viability tests: of 70 strains, 67 regrew after 20 months. Some dried and powdered samples of mass cultures regrew after 8 yr. Cultures distributed from the International Rice Research Institute were sent as dried material on paper strips, because these remain viable for several months and are easy to mail but this method is less efficient than the other two methods. Viability under these conditions was relatively short. Of 136 N_2-fixing strains, 30 were lost after 16 months storage and 129 after 30 months. J. W. Simon and the author (unpublished data) tested the ability of 20 axenic isolates from Bangladesh and Thailand rice-fields to survive after the samples had been allowed to dry on 2-cm lengths of washed cotton: all the filamentous forms, but not a *Synechococcus* strain, survived for at least one year.

Most soil cyanobacteria survive periods of desiccation with little obvious morphological change other than that resulting from loss of water. Those genera which form akinetes (e.g. *Anabaena*) do so largely as a response to nutrient or other limitation rather than desiccation (Whitton, 1987). Physiological features of desiccation-tolerant cyanobacteria have been reported in a number of studies. For instance, two drought-resistant strains accumulated sugars to high concentrations when matric water stress was applied, whereas two strains not showing drought resistance did not do so (Hershkovitz et al., 1991). The most detailed studies have been done on *Nostoc* (Chapter 17).

Although many soil and desert cyanobacteria are highly tolerant of desiccation, they require to be rewetted thoroughly for full metabolic processes to resume. A study (Palmer and Friedmann, 1990) of cryptoendolithic communities in deserts showed that cyanobacteria photosynthesize only at high matric water potentials (> - 6.9 MPa, 90% relative humidity at 20° C) relative to green algal-containing lichens, which began to photosynthesize at a much lower potential (- 46.4 MPa, 70% relative humidity at 8° C). The rewetting of dried *Nostoc commune* leads to physiological responses in the sequence, respiration, photosynthesis and, lastly, nitrogen fixation (Scherer et al., 1984). A non-colonial culture of *N. commune*, after storage at -99.5 MPa, showed increased nitrogenase activity and size of intracellular ATP pool, when rewetted; the upshift in nitrogenase activity was preceded by a lag (Potts and Bowman, 1985).

3. *Salinity*

There are many accounts of saline (Singh, 1961; Ali and Sandhu, 1972) and semiarid (e.g. Smith et al., 1990) soils containing a rich cyanobacterial flora. Where the soil becomes thoroughly wetted at intervals, crusts of photosynthetic microorganisms frequently occur at or near the surface and cyanobacteria are usually an important component of such crusts (Section III). Singh (1950) proposed that enhancement of cyanobacterial growths on saline (usar) soils in India could provide a means of improving soil quality and eventually reversing the trend to increased salinity. His proposals were later expanded into a book (Singh, 1961). *Microcoleus* appears to be the most widespread cyanobacterium on saline surfaces throughout the world (Whitton, 1990), where it can play an important part in stabilization of the underlying soil (Anderson and Rushforth, 1976). Buttars et al. (1998) have developed a method for inoculating sterilized soil crusts, which involves the escape of *Microcoleus* from alginate beads and its subsequent growth on the soil. If this method can be applied on a large scale, it might be possible to accelerate the rate of recovery of highly disturbed soils and add to the more conventional methods suggested by Singh (1961).

4. *pH*

As mentioned above, cyanobacteria are infrequent below pH 6.0 in most temperate soils: they were, for instance, entirely absent below pH 5.4 in several samples from Ireland (Dooley and Houghton, 1973). However, there are situations where they occur at substantially lower pH values. These include tropical soils (Moore, 1963) and the edges of small pools (Section II.C). Values down to pH 4.2 or even slightly lower have been recorded in both cases (author's observations). It seems probable that the soils at pH values just above 4.0 which include cyanobacteria are all ones with weak buffering capacitiy. There are also scattered records of cyanobacteria in highly acidic environments, but in those cases where the sites have been revisited the organisms have proved to be *Cyanidium* (or perhaps one of the other small phycobilin-containing eukaryotes). Nevertheless, the possible occurrence of cyanobacteria at highly acidic sites still requires thorough re-evaluation.

Cyanobacteria may also influence the pH due to their metabolic activity. In addition to the marked changes which can take place in paddy fields and temporary pools overlying soil, they can also occur within surface crusts. Localized values of 10, 2-3 units above the soil pH, have been found in desert crusts as a result of photosynthetic activity (Garcia-Pichel and Belnap, 1996).

5. *Herbicides and Pesticides*

The influence of herbicides and pesticides on cyanobacteria has been investigated in many studies (Padhy, 1985; Roger, 1996b; Vaishampayan et al., 1998), though most have been restricted to laboratory cultures. Cyanobacteria are in general quite sensitive to herbicides, because they share many of the physiological features of higher plants (Leganés and Fernandez-Valiente, 1992), which form the site of herbicide action. The pre-emergence herbicide, benthiocarb, was highly toxic to strains of *Nostoc* and *Anabaena* (Zaitseva, 1979). However, N_2-fixing cyanobacteria are mostly relatively tolerant to 2,4D (2,4-dichlorophenoxy acetic acid), at least under field conditions (Stratton, 1987; Leganés and Fernandez-Valiente, 1992). Differences have been found between the tolerance to herbicides of cyanobacteria and that of other organisms. *Anabaena*, *Nostoc* and *Oscillatoria* formed visually obvious growths in Italian rice-fields treated with Fentin derivatives and sodium dithiocarbamate to control green algae (Bisiach, 1970). A study (Peterson et al., 1997) carried out in liquid culture showed that hexazinone was less toxic to cyanobacteria than to green algae, diatoms and a duckweed, whereas the cyanobacteria were more sensitive to diquat than the green algae. A number of herbicide-resistant mutants of cyanobacteria have been isolated from rice-fields or obtained in laboratory studies (Singh et al., 1986; Modi et al., 1991; Tiwari et al., 1991). However, there appear to be no data on the rate at which resistance can build up in the whole population of a species at a particular site.

Fungicides may also be quite toxic (Tarar and Shewale, 1984), though resistant mutants are known to occur (Vaishampayan and Prasad, 1981). There are also a few reports of insecticide-tolerant strains, but the extensive literature on pesticide tolerance of cyanobacterial strains largely fails to consider whether the particular strains used for experimental studies have developed genetic tolerance to the pesticide in question. However, it seems clear that the influence of the pesticides used to combat insect and nematode problems is usually much less than that of herbicides and probably also fungicides, though it differs according to the agent applied and the species or strain tested. For instance, a comparison of the effect of the carbamate pesticide, carbofuran, on nine heterocystous cyanobacteria showed that *Anabaena fertilissima* and *Nostoc commune* were the most sensitive (Rath and Adhikary, 1995). Three other pesticides (based on carbaryl, dimethoate or endosulfan) tested on heterocystous genera also showed wide differences between species (Das and Adhikary, 1996). The application rates recommended for field use probably do not significantly reduce the growth of cyanobacteria, but double the levels of Sevin (carbaryl) and Hildan (endosulfan) might well be harmful. Several field studies (e.g. Grant et al., 1983; Simpson et al., 1994) have indicated that the application of pesticides to paddy rice can enhance cyanobacterial growths, presumably due to the reduced activity of grazers.

C. Nitrogen Fixation

Cyanobacteria growing at the soil surface usually form distinct colonies or aggregates. It is therefore easy to collect material for assays of N_2 fixation using $^{15}N_2$ or of nitrogenase activity using the acetylene reduction methodology (see Section VB). Many soil species have been reported to be N_2-fixers (e.g. Kabli et al., 1997) and some of the lichens making up the crust communities of arid regions also contain nitrogen-fixing cyanobacteria (Belnap, 1996; Lange et al., 1998).

The literature includes quantitative data on N_2 fixation by soil communities from diverse sites, though the values have to be treated with caution because of the near impossibility of making sufficient measurements to consider all aspects of temporal change and spatial heterogeneity. Another problem is that it is seldom clear how much of the measured activity is due to the cyanobacteria and how much to closely associated heterotrophic bacteria. Crusts or mats dominated by *Microcoleus* have been reported to fix nitrogen in desert (Belnap and Harper, 1995), estuarine and marine populations (Pearson et al., 1979; Stal, 1988).

Although Pearson et al. (1979) included studies with an axenic strain, Steppe et al. (1996) failed to find evidence for a *nifH* gene sequence in four isolates from a terrestrial crust and marine mats and the authors concluded that *Microcoleus* spp. appear to be incapable of fixing N_2 and that it is epiphytic bacteria which are responsible for N_2 fixed in soil crusts and marine mats dominated by *Microcoleus*. One possibility is that the strain used by Pearson et al. should be referred to another genus (Chapter 3), but more isolates need to be studied before it is concluded that there are no N_2-fixing strains of *Microcoleus*. Some of the illustrations of the organisms isolated by Smith et al. (1990) from arid soils in Central Australia) and which were shown to fix nitrogen, do look quite like *Microcoleus*. Soils inoculated with *Microcoleus* from pellets eventually showed significant nitrogenase activity (Buttars et al., 1998). It is clear that N_2 fixation is important in some *Microcoleus*-dominated crusts and mats, even if it is due to heterotrophic bacteria.

The majority of nitrogen-fixers reported from soils are heterocystous forms. The factors influencing nitrogenase activity in *Nostoc commune* have been investigated at many sites. A study (Solheim et al., 1996) at Kongsflorden, Spitsbergen (79 °N), showed that heterocystous cyanobacteria were the most important sources of biologically fixed N in the area. This was due either to *Nostoc* colonies on moist sparsely vegetated ground or to cyanobacteria epiphytic on mosses. In another study on Spitsbergen, the activity of sheets of *Nostoc commune* assayed *in situ* was shown to have a linear relationship with their moisture content and the assay temperature (Liengen and Olsen, 1997). Among nitrogen-fixing cyanobacteria lacking heterocysts (Gallon et al., 1991), most soil forms appear to be associated with conditions such as waterlogging, which lead to reduced ambient oxygen concentration (Rother et al., 1988; Rother and Whitton, 1989).

Among the wetland area communities which would repay detailed study is the association of *Hapalosiphon* with submerged growths of *Sphagnum*. This was widely reported in the older literature (e.g. West and Fritsch, 1927), yet appears to have received no critical study. Growths of *Hapalosiphon* are frequently associated with *Sphagnum cuspidatum* at the edge of small pools in the 'flow country' of N-E. Scotland, where pH values of of pool waters mostly lie in the range 4.0 - 4.3 (V. Standen and the author, unpublished data). There are apparently no records of *Hapalosiphon* among *Sphagnum* in northern England. In view of the relative absence of

Fig. 1. Colonies of cyanobacteria (*Anabaena, Cylindrospermum, Lyngbya, Porphyrosiphon notarisii, Scytonema mirabile* and *Tolypothrix byssoidea*) developing on moist soil in a deepwater rice field at Sonargaon, Bangladesh, several weeks before the arrival of the flood water. Visually obvious communities will cover about half the surface of the field by the time the flood arrives. (Width of figure is 30 cm.)

atmospheric pollution in the flow country and the high levels of deposition of atmospheric N in northern England, it would be of interest to establish whether atmospheric N deposits can lead to the loss of the nitrogen-fixing cyanobacteria associated with some species of *Sphagnum*.

Nostoc commune in and on a loamy soil in Russia was found to fix 3.3 kg N ha^{-1} yr^{-1} (Pankratova and Vakrushev, 1971). Apart from *Nostoc*-dominated sites, most studies have been on the whole community, including cyanobacteria and heterotrophic bacteria. The importance of the cyanobacteria in a moist Bangladesh rice-field prior to the arrival of the floods (Fig. 1) was evident from the fact that nitrogenase activity at mid-day was one to three orders higher where there was a cyanobacterial cover than where the soil was bare (Rother and Whitton, 1989). *Nostoc* always showed higher activity than *Tolypothrix*, whether expressed per unit area or biomass. Most values for paddy fields summarized by Roger and Kulasooriya (1980) and Kaushik (1998) fall in the range 10 - 30 kg N ha^{-1}, though rates up to 80 kg N ha^{-1} were reported from Mali (Traore et al., 1978). Values for Spanish rice fields ranged from 0.23 to 75.5 kg N ha^{-1} yr^{-1} (Quesada et al., 1997), though dense cyanobacterial growths were excluded, so the rates for the whole community were probably sometimes even higher. Studies on rice fields generally do not consider N_2 fixation during the period of the year when the land is not cultivated for rice, so the annual total may be higher than that quoted. According to Belnap (1996), the reported values for the crust communities (Section III) of S-W. USA range from 0.02 - 365 kg N ha^{-1} yr^{-1}. (Belnap's summary is based on the following: Mayland et al., 1966; MacGregor and Johnson, 1971; Rychert and Skujins, 1974; Eskew and Ting, 1978; Jeffries et al. 1992.)

D. Influence on Soil Properties and Roots

Roger and Kulasooriya (1980) described the properties of rice-field soils which may be enhanced by the development of cyanobacteria, though they seem equally likely to apply to other soils. The properties include improved organic content, improved water-holding capacity, addition of combined nitrogen, release of vitamins or plant-stimulating hormones, formation of extracellular polysaccharides leading to improved soil aggregation and solubilization of phosphates. Some of these effects have been studied in considerable detail, though mainly for rice-field soils (Roger and Kulasooriya, 1980), whereas others (e.g. vitamins and hormones) require further study before being accepted as proven.

Changes in the properties of the upper 0.7 cm of experimental columns of a brown earth silt loam incubated in the light were attributed to an increase in the cyanobacterial population by Rao and Burns (1990). The cyanobacterial counts eventually reached about 10^6 g^{-1} dry soil, whether or not the columns were inoculated deliberately. Significant improvements were measured in soil aggregation properties, and increases in dehydrogenase, urease and phosphatase activities were also detected. All three enzyme activities were however low compared with activities in arable soils, perhaps reflecting the adverse effect of waterlogging.

The production of extracellular polysaccharides by cyanobacteria has also been shown to improve soil properties such as aggregation in a number of other studies (Bailey et al., 1973; Roychoudhury et al., 1980; de Winder et al,. 1989; de Caire et al., 1997). A study (Falchini et al., 1996) of the effects of *Nostoc* strains on two clay soils from Tuscany showed not only the beginning of aggregation, but also protection of soil porosity due to reduction of the damaging effect of the addition of water. However, there was no significant improvement in the stability of the water - soil structure. Extracellular polysaccharides can also have an adverse effect, if they are associated with thick surface mats which reduce the penetration of water into the soil. A study (Katznelson, 1989) of effluent penetration into the ground at the Dan Wastewater Reclamation Project, Israel, showed that *Phormidium autumnale* had a very marked clogging capacity. This was ascribed to the release of copious mucus; two other cyanobacteria, which formed less polysaccharide, had a much lower clogging capacity.

An ability to mobilize insoluble forms of inorganic phosphate is apparently widespread. All but one of 18 strains tested by Bose et al. (1971) solubilized tricalcium phosphate; other materials utilized as P sources by cyanobacteria include Mussorie rock phosphate (Roychoudhury and Kaushik, 1989) and hydroxyapatite (Cameron and Julian, 1988). Several authors (e.g. Natesan and Shanmugasundaram, 1989) have implicated extracellular phosphatases in solubilization of fixed soil phosphates, but more critical studies are needed on this topic. However, most cyanobacteria can mobilize organic phosphates in their environment by means of cell-bound ('surface') phosphatases and often also extracellular phosphatase; this is usually an inducible activity (Healey, 1982). All 50 strains, including some soil isolates, assayed for their ability to grow with organic phosphates as their sole P source used phosphate monoesters, almost all used diesters, but only some used phytic acid (Whitton et al., 1991). Most strains released part of their phosphomonoesterase extracellularly, but never the phosphodiesterase. There is some evidence that cyanobacteria may be particularly effective at increasing phosphorus availability in saline soils (Singh, 1961; Kaushik and Subhashini, 1985), but it is unclear which cyanobacterial features are responsible for this property.

Cyanobacterial growths in flooded rice soils have been reported (Das et al., 1991) to influence the forms in which Mn, Fe and perhaps also Zn occur in the soil. Their presence led to a decrease in ammonium acetate - extractable forms of Mn and Fe and increases in other forms of these elements. These changes were considered to be due to release of oxygen, addition of organic matter and especially extracellular material. The decomposition of the cyanobacterial biomass led to further changes in the various fractions, which were ascribed to the development of reducing conditions and the formation of organic acids. A decreased content of readily available Fe might help to minimize Zn deficiency in rice.

Examples have been reported of higher plants which stimulate or inhibit the growth of soil cyanobacteria (e.g. Parks and Rice, 1969). Several laboratory studies have shown that cyanobacteria may not only be associated closely with the surface of some roots, but may also occur intracellularly in rice (Kozyrovskaya, 1990), wheat and other species. The most detailed studies have done on wheat. In addition to loose associations of *Anabaena* with root hairs of wheat, Gantar et al. (1991a, b) found that there were much tighter associations with some *Nostoc* isolates; the latter penetrated both the root epidermis and cortex. Experiments on joint cultivation of cyanobacteria and crop plants were extended by Svircev et al. (1997) to corn (maize), bean, sugar beet and two rice cultivars. Where effects were found, they tended to be greater in liquid than sand cultures.

Cyanobacteria are sometimes conspicuous on the surface of aquatic roots of deepwater rice (Whitton et al., 1989) and there is evidence from $^{15}N_2$ studies (Kulasooriya et al., 1980) that part of this nitrogen reaches the rice plant either as the result of release of extracellular combined nitrogen or indirectly following grazing or parasitism. These various observations suggest that careful studies should be made to determine whether or not intracellular cyanobacteria are of quantitative importance in the roots of plants at field sites where cyanobacteria are conspicuous at the soil surface.

There are numerous other reports of free-living cyanobacteria influencing the growth of adjacent vascular plants. Many of the positive effects have been ascribed to enhanced nitrogen availability, either due to release extracellularly or following decay of the cyanobacteria. The majority of studies have been with agricultural crops, especially rice (Section IV), but there have also been studies on natural ecosystems. In the case of slow-growing communities, convincing evidence depends on the use of $^{15}N_2$, as used to demonstrate transfer of fixed nitrogen from a semiarid crust (Mayland et al., 1966) and from a sand-dune crust (Stewart, 1967). Kleiner and Harper (1972, 1977) found more extractable P in soils with a cyanobacterial cover than in nearby soils without such a cover.

The gelatinous sheath material of half the cyanobacterial species studied by Lange (1976) was able to chelate elements essential for their growth (Fe, Cu, Mo, Zn, Co, Mn) and Belnap and Harper (1995) considered the possibility that the sheaths may also influence the availability of elements to other organisms. Four of the five genera listed by Lange as having this ability are represented by common species in the soil crusts of western North American desert soils. Belnap and Harper (1995) showed that the effects of the soil crust on semiarid soil in S-E. Utah (Section III) on vascular plant species depended on both the element and the plant. The elements N, P, K, Mg, Ca and Fe were present in significantly greater concentrations in *Festuca octoflora* growing on soils heavily encrusted with cyanobacteria and cyanolichens than in plants on the same soil where the crust had been destroyed. In the case of *Mentzelia multiflora*, N, Mg and Fe were present at significantly higher concentrations, but P was present at significantly lower concentrations. The authors suggested a variety of possible reasons for these differences. As described above, cyanobacterial N_2 fixation is likely to increase the availability of N to adjacent plants. The reduced uptake of P by *Mentzelia* was probably due to direct competition of its near surface root layer with the surface crust. Cyanobacterial sheaths reduce particle erosion (Section III) and may adsorb charged nutrient cations Chelating compounds present in sheaths may be responsible for the enhanced uptake of Fe. Well-developed crusts are much darker than areas without them, so soil surface temperatures may be warmer and hence metabolic activity higher at the time of year when usable moisture is most likely to be available.

Many of the earlier studies which showed an influence of cyanobacteria on seed germination or growth of young rice (see bibliography in Kulasooriya and Roger, 1980) were ascribed to growth-regulating substances such as hormones and vitamins. However, there has been no study in which a cyanobacterial regulator has been isolated and characterized. It also seems likely that only positive results have been published, following the screening study by Pedurand and Reynaud (1987) on 135 non-axenic cyanobacterial isolates on the germination and growth of rice. 30% strains had no effect on germination and 30% caused inhibition. Among 8 strains of *Anabaena* stimulating growth, only one remained effective after it had been made axenic.

III Subaerial Habitats

1. Development of Subaerial Communities

Subaerial communities of cyanobacteria and algae were first described in detail from Sri Lanka (Ceylon) by Fritsch (1907a,c) and his general account (1922) of sites in a number of countries is still worth reading. He concluded that the first colonizers of rock surfaces are colonial Chroococcales (*Gloeocapsa*, *Gloeothece* and *Aphanocapsa*), often accompanied by *Nostoc*. The stratum formed by these genera is sooner or later colonized by filamentous forms (*Lyngbya*, *Scytonema*, *Stigonema* and *Hapalosiphon*), usually possessing sheaths which are firmer and less gelatinous. These filamentous forms initially produce a dense tangle on the rock, but often eventually give rise to a tufted mode of growth, the surface of which in extreme cases acquires a velvety appearance.

It is unclear how much the succession described by Fritsch really does represent the sequence of changes taking place on rock surfaces in the wet tropics or merely reflects a comparison of different types of community with increasingly favourable conditions for growth. An extensive cover of tufted forms, for instance, appears to be restricted to sites which are very humid throughout the year. However, in the case of lava and ash deposits from recent volcanic activity, it is possible to observe true succession. The widely quoted observation of Treub (1888) that species of *Tolypothrix*, *Anabaena*, *Symploca* and *Lyngbya* formed a gelatinous layer over the surface of lava and cinders on Krakatau within three years of eruption has led some ecological texts to regard this as a general phenomenon. While cyanobacteria certainly are important in the early stages of colonization of volcanic deposits and other bare surfaces such as mine spoils, they are not always the dominants. During the colonization of Surtsey (off the coast of Iceland), cyanobacteria were important in some areas such as depressions, but overall were less important than on Krakatau (Brock, 1973). They were absent altogether during the pioneer stage on the volcanic ash of Katmai, Alaska (Griggs, 1933). Brock attributed these differences in the relative importance of cyanobacteria to temperature.

An enrichment study of surface materials from different zones around a volcano in Hawaii did not indicate that cyanobacteria were especially abundant in early stages here (Carson and Brown, 1978), in spite of the tropical location. The diversity and quantity of soil cyanobacteria increased with nutrient levels and organic status at the sites, with cyanobacteria mainly being found in climax forest. However, three genera were found nearer the volcano, including *Tolypothrix* which was observed directly among volcanic ash. This study would be well worth repeating, this time including controls to estimate the contribution from the air spora and using a much less nutrient-rich medium.

2. Dispersal in the Air

Since cyanobacteria are frequent components of the air spora (Schlichting, 1961, 1964; Tiberg et al., 1983), they are likely often to be recovered from soil enrichment cultures. Both the species of cyanobacteria (*Phormidium ambiguum*, *P. laminosum*) found on styrofoam blocks in a British Columbia tree nursery also occurred in the local air flora (Ross and Puritch, 1981). Sixty-three cyanobacterial species were found during the monsoon season in the air flora at Poona, India (Balakrishnan and Gunale, 1980). Some aerial organisms probably originate from soil surfaces (Brown et al. 1964), but, especially in the tropics, many are likely to come from aerial habitats such as the surfaces of buildings and trees (John, 1988; Garty, 1990; Mrozinska, 1990). Novelo-Maldonado and Gonzalez-Gonzalez (1981) attempted to distinguish between the soil and airborne algal (including cyanobacteria) floras at a site in Tehuacan, Mexico, by sampling simultaneously the air, superficial and deep soil floras.

3. Crust Communities in Semiarid Regions

The most detailed accounts of subaerial communities with abundant cyanobacteria come not from the wet tropics, but from arid regions, such as the volcanic soils in S-W. USA, where *Microcoleus* and *Porphyrosiphon* are important (Shields, 1957). Crusts consisting of cyanobacteria, eukaryotic algae, lichens, mosses and heterotrophs are conspicuous components of the surfaces of arid and semiarid soils worldwide (West, 1990). In cold deserts the crusts cover interspaces between and under vascular plants, and often constitute 70% of more of the living ground cover (Belnap, 1996). (Belnap terms the deserts of some parts of S-W. USA 'cold', whereas the sames ones are termed 'hot' in Chapter 13 to provide a contrast with the deserts of Antarctica.) Such communities (often termed 'cryptobiotic') are initiated by the growth of cyanobacteria during the episodic events supplying moisture, leading to a network of filaments and a matrix of extracellular slime (Belnap and Gardner, 1993). If the largely cyanobacterial communities remain undisturbed, they may eventually give rise to the more complex cryptobiotic communities. The presence of these communities has marked effects on the properties of the soil. Three of these have already been discussed, the soil temperature at times critical for plant growth (Section II.B.1), nitrogen fixation (Section II.C) and ion chelation (Section II.D).

Belnap (1995) has shown how important the cryptobiotic communities are for maintaining soil stability and nutrient cycles and how harmful are activities which disrupt the communities, such as cattle and human trampling and off-road vehicle use. Soil compaction and disturbance of the cryptobiotic communities can result in decreased water availability to vascular plants through decreased water infiltration and increased albedo with possible decreased erosion and decreased diversity and abundance of the soil biota. The marked effects of disturbance on the erodibility of these communities were made clear by comparing the threshold values of wind velocity leading to disruption of disturbed and undisturbed sites in S-E. Utah (Belnap and Gillette, 1997) and New Mexico (Marticorena et al., 1997; Belnap and Gillette, 1998). In the last study the threshold values for well-developed communities were shown to be well above those of wind forces occurring at the sites, whereas the values for disturbed sites were below them and hence these sites can be eroded by wind action. The use of a wind tunnel to compare the ability of various phototrophic communities growing on sand to resist erosion showed that only cyanobacteria, especially *Nostoc commune*, provided considerable protection (Neuman et al., 1996). The need to manage semiarid regions carefully in order to minimize damage to the soil crust community also became evident in a study of rangeland in S-W. Queensland (Hodgins and Rogers, 1997).

The frequency of disturbance by a different environmental factor, burning, has a similar effect on soil shear strength in coppice woodland in Zimbabwe (Belnap et al., 1996). The effects of the following treatments were compared: no-burn; burn every 2 years; burn and mattock every 2 years; burn every 4 years; burn and mattock every 4 years. (Mattocking involves chopping woody vegetation into small pieces and distributing across the plot.) The 4 year burn-mattock treatment resulted in the greatest cyanobacterial cover, the greatest cyanobacterial biomass and the greatest soil shear strength.

Nutrient cycles are also altered when cryptobiotic crusts are degraded. Changes include the reduction of C and N inputs, slowing of the breakdown of organic matter and a reduction in the nitrogen-fixing capacity of the soil (Belnap, 1995). The last may take at least 50 years to recover. In a study (Belnap, 1996) at sites near Moab, Utah, the nitrogenase activity of crusts dominated by *Microcoleus vaginatus* on sandy soils was more susceptible than crusts with the lichen *Collema tenax*. Although disturbance may actually lead to an increase in the abundance of free-living cyanobacteria, as shown for sites on the Colorado Plateau (Evans and Belnap, 1999), it leads to a decrease in the abundance of lichens, some of which are also N_2-fixers (Section II.C), and thus an overall decrease in nitrogenase activity. Nitrogenase activity was 250% greater at pristine sites on the Colorado Plateau than at ones intermittently disturbed 30 years ago. The decrease in N input from fixation and an enhanced loss of gaseous N has caused a 25 to 75% decrease in soil N content.

Although it has been widely considered that crust communities dominated by cyanobacteria are also the major source of fixed nitrogen in hotter deserts, Zaady et al. (1998) showed the importance of heterotrophic N_2 fixation associated with litter breakdown in the vicinity of vascular plants. For instance, where macrophyte patches covered 25% of the soil surface, they contributed 40% of the total N_2 fixed in this part of the desert.

4. Heavy Metal and Other Contamination

Cyanobacteria are sometimes abundant on and near the surface of soils contaminated by various pollutants or naturally enriched by heavy metals. Most records for heavy metals are for wastes associated with lead-zinc mining (Whitton, 1980; Whitton et al., 1981), but there are also records for soils rich in copper and other metals (Ernst, 1974; Rana et al., 1971). The dominants are usually narrow forms of Oscillatoriaceae such as *Plectonema* and *Schizothrix* and tend to occur mainly at the surface (Plate 23d). Older communities in Zn-rich areas develop a mixture of *Plectonema* and protonema of species of the moss *Dicranella* and this in turn gives rise to intermingled *Plectonema* and leafy *Dicranella* (Plate 23f), and sometimes also lichens, especially *Collema* spp. The younger stages of such communities have many similarities with the cryptobiotic crusts of semiarid regions (see above), although they can develop under conditions of high rainfall. Old sheaths of *Plectonema* and decaying parts of the moss appear to play the main role in developing a true soil on many spoil heaps, though species of *Collema* and *Peltigera* with cyanobacterial symbionts sometimes occur at these sites and contribute fixed nitrogen to the ecosystem (author's unpublished data).

There are many records of cyanobacteria tolerating contamination by oil or perhaps even being favoured by its presence, though these deal mostly with shallow waters rather than soil. Several studies have noted the occurrence of the nitrogen-fixer *Nostoc commune* in or on soils naturally contaminated by oil (tar sands of Alberta: M. Hickman and the author, unpublished data) or polluted by petroleum products (e.g. Atlas et al., 1976). The extensive list of cyanobacteria growing on coal tar reported by Barhate and Tarar (1982) does not include any heterocystous forms, but no studies were done to establish whether any of the species listed were non-heterocystous nitrogen-fixers.

5. Sports Turf

Soil cyanobacteria in golf courses and other amenity grasslands may reach densities sufficient to have favourable effects on soil properties, but surface growths can also be extensive enough to cause management problems (Baldwin and Whitton, 1992). Two different types of situation have been recognized. One is where the development of *Nostoc commune* colonies is large enough to cause a risk of people slipping. (Eukaryotic algae are, however, in general a bigger problem in this respect.) The other is associated with the phenomenon of 'black layer', which sometimes results from unsatisfactory drainage when coarse sand is used to cover a poorly draining soil. An anoxic black subsurface layer develops with a characteristic bacterial community (Lindenback and Cullimore, 1989).

In some situations a dark layer including filamentous non-heterocystous cyanobacteria also develops at the surface in the region where there is an underlying anoxic layer (Hodges, 1987a, b). The arguments concerning the role of cyanobacteria in this phenomenon are not fully resolved. It has been suggested that the initial development of surface cyanobacterial growths as a layer occurs subsequent to the formation of an underlying anoxic zone. The presence of cyanobacteria may subsequently enforce the stability of this underlying zone (Baldwin and Whitton, 1992), presumably in a similar manner to that described above by Katznelson (1989). However, experimental studies by Hodges (1992) with two *Oscillatoria* spp., *Nostoc* sp. and *Desulfovibrio* spp. in sand columns showed that blackening occurred only when cyanobacteria and *Desulfovibrio* were both present. Blackening was most intense in response to added sulphur, chelated iron or a mixture of the two.

IV Rice-fields

Cyanobacteria were first reported to be abundant in rice-fields by Fritsch (1907a, c) and their importance in helping to maintain rice-field fertility due to N_2 fixation was suggested by De (1939). Cyanobacteria are especially evident in wetland rice-fields, which supply 86% of the world requirement for rice (Ladha and Reddy, 1995). Many rice-field soils not only contain a high density of cyanobacteria, but possess visually obvious growths of cyanobacteria at (or floating above) the surface, at least during some seasons. Typically, about half the cyanobacterial genera present are heterocystous (*Anabaena, Aulosira, Calothrix, Cylindrospermum, Fischerella, Gloeotrichia, Nostoc, Scytonema, Tolypothrix, Wollea*). However, there is a lack of critical studies on changes during growth of the rice crop and Kulasooriya (1998) has stressed how this makes it difficult to compare results from different sites. However, general comments can be made when the data comes from a wide enough range of sites. For instance, a survey of 102 soils from four countries showed that the abundance of heterocystous forms was correlated positively with pH and the available P content of the soils (Roger et al., 1987), though nitrogen-fixing cyanobacteria can still be important in acid rice soils (Kulasooriya, 1991). Mandal et al. (1993) also showed that the abundance of heterocystous species in Bangladesh rice-field soils was significantly correlated with available P in the soil.

The influence of light on cyanobacterial communities in rice-fields near Valencia, Spain, was investigated by Quesada et al. (1998). Light had a significant effect on the relative abundance of the dominant genera. The abundance of non-heterocystous forms was three times more at high than low (about 52% versus 7%) incident irradiation. The three main heterocystous genera (*Anabaena, Calothrix, Nostoc*) all responded differently to levels of irradiation. Higher abundance of *Nostoc* coincided with lower abundance of the other two genera. The nitrogenase activity of the communities became adapted to the light regime to which they had been exposed for the previous month, though the community exposed to, and subsequently assayed at, the highest irradiation showed the highest activity.

In view of the many floristic studies on rice-fields, there have been surprisingly few studies on the ecology of particular species, though this information would be useful for assessing the nutrient status of fields and also how to manage them. A comparison of the cyanobacterial flora of rain-moistened fields in Bangladesh (Fig. 1) with that present after the flood waters had arrived showed that most species were different (Rother and Whitton, 1989; Whitton et al., 1989). However, mats of *Scytonema mirabile* were sometimes abundant both on the moist soil and floating in the same field when flooded. Many of the rice-field species form akinetes, so it would be useful to know what factors lead to akinete formation in particular species. A strain of *Fischerella muscicola* isolated from a field at the Cuttack Rice Research Institute formed akinetes under P limitation (Mishra, 1985) as do about half the strains of cyanobacteria isolated from diverse habitats (Whitton, 1992). However, not all rice-field species form akinetes in response to P limitation (Reddy, 1983). A study (Whitton, 1991) of surface phosphomonoesterase activity of macroscopic colonies in paddy fields in the Philippines showed that *Aphanothece stagnina* tended to have low activity, whereas *Gloeotrichia* always showed high activity. Among *Nostoc*, *N. commune* had high activity, whereas other species tended to have less activity. As phosphomonoesterase activity is enhanced in all these species under conditions of P limitation, it was concluded that *Aphanothece* typically grew under P-rich conditions, whereas *Gloeotrichia* grew under conditions where growth was P-limited.

Surface application of fertilizers to rice-field soils without standing water generally leads to conspicuous growths of algae (Roger and Kulasooriya, 1980), with cyanobacteria usually forming an important component unless nitrogenous fertilizer is in excess. Replacement of the cyanobacteria as dominants by masses of filamentous green algae is an indication of high quantities of nitrogenous fertilizer available at the surface (Whitton and Roger, 1989). According to Roger et al., (1980) deep placement of nitrogen fertilizer in the form of urea supergranules not only prevents the growth of the green algae, but also prevents the inhibition of cyanobacterial growth and its associated N_2 fixation. However, Fernández-Valiente et al. (1997) found that the highest level of ammonium fertilizer (140 kg N ha^{-1}) deep-placed in rice-fields near Valencia, Spain, did lead to a significant reduction in nitrogenase activity. The partial reduction in activity increased over the cultivation cycle, being highest at the end.

It is difficult to assess the impact of P fertilization on cyanobacteria in rice-fields, because other fertilizers are almost always added at the same time, and, even in experimental studies with potassium phosphate, controls are seldom included to consider the possible effects of K. However, the highly significant increases in cyanobacterial biomass resulting from phosphate addition found by Bisoyi and Singh (1988) in plots enriched with *Aulosira*, *Aphanothece* or *Gloeotrichia* was almost certainly due to added P.

Many studies have been reported on the use of dried cyanobacteria to inoculate soils as a means of aiding fertility (Roger and Kulasooriya, 1980: Metting, 1988). The term 'algalization' was initially applied to the use of defined mixtures of species (Venkataraman (1972), but has since come to refer to any planned addition of material. The effects of such addition on rice yield were first reported for Japan by Watanabe et al. (1951), when there was a 25% increase in yield after inoculation of poorly drained paddy with *Tolypothrix tenuis*. Studies on the use of inocula have been discontinued in Japan, but numerous studies have been made elsewhere, especially in India (Kaushik, 1998) and more recently in Egypt (e.g.Yanni, 1992 a, b). Unfortunately there are no recent detailed accounts for India in the literature. It has been widely accepted (Agarwal, 1979; Venkataraman, 1981) that algalization can lead to gains in yield, though some of the evidence is equivocal (Whitton and Roger, 1989; Roger et al., 1993). A summary of the results from all studies to 1980 (Roger and Kulasooriya, 1980) showed that there was a mean increase in grain yield due to algalization of 28% in pot experiments and 15% in field experiments. The authors pointed out that most experiments had been designed simply to obtain data about grain yield and little was known about N elsewhere in the ecosystem.

The methodology to prepare inocula was described by Venkataraman (1981). Cyanobacteria are grown from March to May (in India) in open-air shallow tanks, into which farm soil, superphosphate, starter inocula and sometimes also insectides are added. If necessary, lime is added to adjust the soil pH to 7.0 to 7.5. A thick cyanobacterial mat develops withing 15-20 days and the contents of the trays are then allowed to dry, producig flakes which are harvested for distribution. A tray with a surface area of 1.6 m^2 produces sufficient material each season to inoculate one hectare. Typically the inocula are added one week after the rice has been transplanted (Sharma and Gupta, 1983). Roger and Watanabe (1986) reported that algalization was adopted in only two states of India and that the inoculated fields comprised only a few percent of the total area under rice.

Roger (1991) made a statistical analysis of the data on algalization and concluded that the effects of inoculation are erratic and limited. He also suggested that farmers had probably not bothered to report unsuccessful results. The best results appear to have been obtained where mixed inocula are produced from local stocks (Venkataraman, 1981) and where the inocula are added with reduced or no nitrogenous fertilizer (Kaushik, 1998, quoting results from the 1981 All India Coordinated Rice Improvement Project). Experiments which appear to have been well designed and the results of which reported increased straw or grain yield and/or N content include those of Sharma and Gupta (1983), Ahmed and Ahmedunnisa (1984) and Singh and Singh (1985, 1987). Among more recent studies, Ghosh and Saha (1997) reported that inoculation of soil during the period when the rice was growing rapidly with a soil-based mixture of four heterocystous species led to a significant increase in soil N and total N uptake by the crop. Although there was only a small increase in grain yield, N uptake by the grain increased by 30%.

It is difficult to find data to support the claim by Agarwal (1979) that the cyanobacteria introduced as a result of algalization can establish themselves permanently if inoculation is done repeatedly for 3-4 cropping seasons. The one detailed quantitative study on what happened to strains following inoculation is that Reddy and Roger (1988). The fate of five laboratory-grown heterocystous strains representing 75% of the inoculum was studied in 1-m^2 plots of five different soils. During the month following inoculation , the strains multiplied to some extent in all soils, but rarely dominated the indigenous cyanobacteria and did so only when the growth of the indigenous species was poor. The material was then dried, the natural insecticide, neem, added to control grazers, and used for reinoculation. The population changes were followed for a further two months, when it was clear that the inoculated strains still played only a minor role.

A number of studies have to be reported on the selection of natural or mutant strains with properties which would in theory maximize the amount of combined nitrogen available for rice or other plants. These are either strains showing especially high rates of nitrogenase activity in laboratory studies or ones which release much of the nitrogen fixed extracellularly (Spiller et al., 1986). Although mutant strains of cyanobacteria releasing ammonia can lead to transfer of ^{15}N and hence improved growth of rice in pot experiments (Albrecht et al., 1991; Kamuru et al., 1997, 1998: see review by Vaishampayan et al., 1998), it is open to question whether they can compete effectively with other soil strains under field conditions. However, one possible approach is to give the special strains a competitive advantage under flood conditions by immobilizing them on polyurethane foam (Kannaiyan et al., 1997). This study included a field experiment, the results of which indicated that foam-immobilized *Anabaena azollae* excreted considerable amounts of ammonia into the flood water and that this led to increased rice grain and straw yields. (Although the strain was isolated from *Azolla*, evidence was not included to confirm that it was in fact the endosymbiont.) *Anabaena* strains were also immobilized on sugarcane waste (bagasse), suggesting that it might be possible to make the method economically worthwhile.

Algalization seems likely to be most useful where there are marked seasonal changes in land use, such as when the ground is ploughed many times before planting a winter crop, so that the natural soil inoculum is much reduced by the time of the new rice season (Whitton and Roger, 1989). However, few studies have been made on the fate of cyanobacteria as rice paddies dry or, subsequently, during the part of the year when the land is cultivated for other crops. Kaushik (1998) states (without references) that in paddy fields the death of algal biomass is most frequently associated with soil desiccation at the end of the cultivation cycle. As most heterocystous species form akinetes, it is presumably in this form that these species persist until the next period favourable for growth, which may not occur until the next paddy crop. Large-scale mortality of species of Oscillatoriaceae as a result of desiccation would contrast with the ability of cyanobacteria in many other alternately wet-dry habitats to survive the dry periods well (Section II B).

A problem requiring further study is suggested by the observations of Rice et al. (1980) that decaying rice straw inhibited cyanobacterial growth and N_2 fixation, apparently due to phenolic compounds. In view of the current interest in barley straw in controlling cyanobacterial blooms in lakes (Welch et al., 1990; Gibson et al., 1990; Martin and Ridge, 1999), it is strange that no-one appears to have taken this study on the effects of rice straw further. On the one hand it suggests that rice straw might also be used as an aid for controlling cyanobacterial blooms, while on the other it suggests that ploughing barley straw into soils might have an adverse effect on soil cyanobacterial populations.

The use of *Azolla* with its symbiotic nitrogen-fixing *Anabaena* (Chapter 14) has been investigated in many studies (Lumpkin and Plucknett, 1980; Whitton and Roger, 1989; Kannaiyan, 1992; Whitton, 1993). The *Azolla* is used in a variety of ways. One approach is to add a large inoculum with irrigation water to fields at an early stage in the rice crop, leaving the *Azolla* to grow for a few weeks for a few weeks, then partially draining the fields and ploughing the *Azolla* into the soil. Another approach is to grow *Azolla* on its own and then compost the plants and plough the decaying material into the soil. Either way, most of the nitrogen fixed by the symbiont eventually reaches the soil or the flood water and thus some becomes available for the rice (Kannaiyan, 1993; Roger, 1996a).

V. Practical Methods

A. Measurement of abundance

Cyanobacterial abundance in soils has mostly been determined by one or more of three methods - direct observation, plating techniques and measurement of pigments (Roger and Kulasooriya, 1980). Direct counting of soil cyanobacteria has been most successful where the organisms are frequent. Saito and Watanabe (1978) reported a quantitative study where rice-field flood water carrying suspended soil particles was filtered on membrane filters and a gelatinized suspension of the particles was smeared on glass slides. In most circumstances where quantitative data are required, it is important to combine direct microscopy with epifluorescence microscopy, as fluorescence of chlorophyll *a* makes it easy to distinguish the smaller cyanobacteria from bacteria. Combinations of excitation and suppression filters may be chosen to distinguish the accessory pigments phycocyanin and phycoerythrin (Wynn-Williams, 1988), while the fluorochrome Auramine O can be used to differentiate living from dead cells. Hawes and Davey (1989) reported that viable cells of *Phormidium* fluoresced yellow, while non-viable cells retained only the red fluorescence. Cultured material stained much better than field material, suggesting that this method requires further development before being put to routine use.

Plating techniques for counting cyanobacteria depend on the suitability of the chosen medium for growth of cyanobacteria (Reynaud and Roger, 1977) and the reliability of the dilution method used. A general guide on choice of methods for quantifying soil cyanobacteria is given by Whitton (1996). It seems probable that most soil cyanobacteria grow relatively easily if plated on suitable nutrient agar. The absence of cyanobacteria reported in the literature for soils of several regions (e.g. parts of Japan: Watanabe, 1959) was perhaps due to the use of a culture medium with a higher pH than that of the soils tested. It would be well worth making an extensive study of low pH tropical soils using plating techniques and taking care not only to control the pH, but also the buffering capacity of the test medium.

Plating techniques may also overlook or underestimate the abundance of particular species at a site unless a wide enough range of temperatures is tested. The need for this is shown by a study of a single soil sample (total of 0.7 g dry wt) from a dry rice field in the Iraqi marshes (Al-Mousawi and Whitton, 1983). Use of a temperature gradient plate permitted enrichment cultures to be grown over the range 20° to 50°C. Forty-two species were found, with 28 cyanobacteria and 14 eukaryotic algae, but at no single temperature were more than two-thirds of the species observed. There were marked differences in the temperature at which particular species predominated in the mixed community and these values sometimes differed from the optimum for unialgal or axenic cultures of the same species. Substantial growth of a mixed cyanobacterial community occurred at 45°C, but none at 48°C. Similar care should be taken to employ a range of light conditions during plating studies.

To enhance the reproducibility of counts, it may be necessary to homogenize the soil sample, though this leads to its own problems. Some species exist typically as aggregates and/or colonies tightly held together with mucilage, making them difficult to disrupt without damaging many cells. Filamentous forms may be broken into smaller units and, in the case of some moniliform filaments, quite easily into individual cells (Roger and Kulasooriya, 1980), though probably with some loss of cell viability. Other filamentous forms, such as most Oscillatoriaceae, are very difficult to separate into individual cells. As a result the extent to which the colonies developing on agar reflect individual colonies, aggregations, filaments or cells in nature is influenced heavily by the homogenization and dilution methodology. If accurate cell counts are required, it is often essential to combine direct counts for some species with plating techniques for others. However, P. A. Roger and co-workers have shown (e.g. Reddy and Roger, 1988) that useful comparisons can be made based on counts of colony-forming units (CFU) estimated by plating techniques. A survey of the literature (Roger et al., 1987) showed that the CFU of nitrogen-fixing cyanobacteria in rice-field soils range from 10 - 10^7 g^{-1} dry weight of soil, with a mean of 2.5 x 10^5.

A further problem in interpreting the data concerning CFUs on an agar surface results from the fact that many cyanobacteria rapidly undergo cell or filament division on transfer to nutrient-rich media (Whitton, 1992). The release of motile hormogonia is frequent. In some cases (e.g. *Calothrix*) these show a tendency to aggregate, whereas hormogonia from *Nostoc* move some distance from the original source and each one develops a new colony. Nevertheless, a number of interesting studies have been reported based on CFUs. Kaushik (1991, reproduced in Kaushik, 1998) compared the numbers of units of potential nitrogen-fixing cyanobacteria in soils from 11 sites in Sri Lanka. The values ranged over three orders of magnitude, the highest being almost 10^7 CFUs g^{-1} dry soil. Roger and Reynaud (1976, 1977) used plating techniques to quantify biomass in Senegal rice-field soils by determining the mean volume of the 'count unit' (colony, filament or cell) for each species. This involved direct examination of the first dilution and then multiplying the CFU counts by the corresponding ' volume unit'. Quantitative data on CFUs are especially valuable when trying to assess changes taking place as a result of algalization (Section IV), as done by Reddy and Roger (1988) in the comparison of changes in populations of inoculated and indigenous cyanobacteria in rice-field microplots. Pigment analysis is useful for quantifying surface growths dominated by cyanobacteria (Davey, 1988; Rother and Whitton, 1989), but is less suited for sparse populations dispersed throughout the upper part of the soil column. Cores or other types of sample have usually been frozen until required for analysis in order to avoid pigment degradation; this also maximizes extraction efficiency (Hansson, 1988); useful practical details are given by Stal (1988). The use of television image analysis with epifluorescence microscopy for cyanobacteria and other microorganisms in Antartic fellfields has been described in detail by Wynn-Williams (1988, 1990).

If it is important to relate a process such as N_2 fixation to biomass, chlorophyll *a* (Garcia-Pichel and Castenholz, 1991) is usually the only variable easy to quantify in mixed samples, though it may be possible to obtain simultaneous measurements of chlorophyll *a* and dry weight for a few samples. Otherwise, some indication of dry weight may be obtained by using values from laboratory studies in the literature. Islam and Whitton (1992a) reported that the chlorophyll *a* content of a rice-field *Calothrix* in batch culture ranged from 0.25 - 2.8 % dry weight. Values above 1% only occurred for a short period when cellular P was at its maximum (Islam and Whitton, 1992b). In the absence of other data a value of 0.5% dry weight can be used as an approximate estimate for healthy field material, though this is open to considerable error.

Satellite sensor images provide a completely different approach to quantifying surface crusts, as was found during collection of data for a 'normalized difference vegetation index' in semiarid areas of Israel (Karnieli et al., 1996). This led to the use of the unique spectral features of the phycocyanin in the cyanobacteria present in the crusts to develop a spectral crust index for semiarid regions and the suggestion that the approach could be extended to agricultural lands (Karnieli, 1997).

B. Nitrogen Fixation

Nitrogen fixation has been quantified in many cyanobacterial dominated soils and subaerial communities (Sections IIC, IID). An overview for rice fields was provided by Roger and Ladha (1992) and the application of standard methods to cyanobacterial studies has been described by Mague (1978) and Stal (1988). In the case of $^{15}N_2$ measurements, Mague recommended that assays should be made for 30 - 120 min, depending on the rates of N_2 fixation expected; most studies have used 7-ml serum bottles or similar for the incubation studies. For measurements of nitrogenase activity using the acetylene reduction assay methodology, a variety of transparent containers have been employed for measurement of soil nitrogenase activity, such as plastic bags (Yoshida, 1984), perspex cylinders (Reddy and Roger, 1988), glass jars, universal bottles or serum bottles incubated at an angle (Rother and Whitton, 1989). Larger containers are needed if it is essential to maintain the structure of the upper part of the soil column. An assay period of 1 h was found suitable for soil communities in the subtropics (Rother and Whitton, 1989).

Because of the greater nitrogenase activity shown by most species in the light than in the dark, quantitative studies related to field conditions should ideally be carried out *in situ* over 24-h periods (Rother et al., 1988), although this has been done in relatively few studies. Measurement over a 24-h period requires a series of experiments, each carried out over short periods. Because of the heterogeneity of the cyanobacterial communities associated with many soils (Fig. 1), quantitative data often needs a high number of replicates, so biomass needs to be quantified by a simple method. Chlorophyll *a* is the usual variable chosen for this purpose (see above).

C. Assays for Fertility and Toxicity

A methodology for using algae to assess the nutrient status of soils was developed by Tchan (1959; Tchan et al., 1961) long before a similar approach was used for water. This made use of a mixed inoculum, though apparently mostly green algae. However, *Nostoc commune* was selected by Pederson and Shubert (1992) for assaying the toxicity of substances likely to contaminate soil. This involves growing the organism on membrane filters superimposed on the surface of nutrient agar containing various concentrations of the substance under test; this technique maintains the air-solid interface important for many soil surface cyanobacteria. Presumably the method could be modified such that the membrane with the *Nostoc* colony is placed directly on the surface of a soil sample.

VI. Concluding Comments

This review has covered much less than half the literature in mainstream journals and widely available reports and no doubt important topics have been overlooked. More than most aspects of the ecology of cyanobacteria, there is still much of interest in the older literature, though this needs to be read critically in the light of modern studies. The real insight into the subject, however, has come largely from the few research groups that have both treated taxonomic identification as a serious matter and have adapted well-known research techniques to quantify organisms and processes occurring in or on soils. Reliable identification has been hindered by the lack of availability of the one book in English (Desikachary, 1959), which might have benefited people at the many institutes and colleges with limited access to the literature. The production of a good and cheap identification manual for soil and rice-field studies in the tropics and subtropics would greatly improve the situation.

Most quantitative studies have dealt with only one process, nitrogen fixation, combined with some measure of biomass, usually chlorophyll *a*. This is partly because of the possible or known importance of the contribution by cyanobacterial nitrogen fixation in many ecosystems, but also because of the relative ease of measuring nitrogenase activity using the acetylene reduction assay methodology. Unfortunately, most studies on nitrogen fixation have at best been fragmentary and seldom attempted to quantify the process over a whole season. Only a few studies have combined measurements of carbon dioxide and nitrogen fixation. Nevertheless, many of the techniques used for measuring processes and the microenvironment in marine mats (Chapter 4) can be applied to the more distinct soil cyanobacterial communities, as shown by Garcia-Pichel and Belnap (1996) for desert crusts.

Some of the equipment needed for measuring processes and the microenvironment is relatively specialized and unlikely to become available to everyone working on soil cyanobacteria. However, the techniques for counting colony-forming units developed by P.A. Roger, initially with ORSTOM in West Africa and later at IRRI (International Rice Research Institute), are less demanding and can be applied anywhere there is an autoclave and a good microscope. They can produce results which provide considerable insight on the influence of environmental factors and seasonal changes in populations, as shown by the research group at IRRI during the 1980s and since then by some of the people who have moved from there, such as S.A. Kulasooriya in Sri Lanka. The lack of quantitative studies on population dynamics of cyanobacteria in rice-field and other soils, however, still applies in most situations. This is remarkable in view of the resources which have been allocated to algalization studies and how such studies might have been aided with a better understanding of the population dynamics of native and inoculated strains in the soil.

References

Agarwal A (1979) Blue-green algae to fertilize Indian rice paddies. Nature 279: 181.

Ahmed SI and Ahmedunnisa (1984) Utilization of blue-green algae as biofertilizer for paddy cultivation. Pakistan J Sci 27: 355-358.

Albrecht SL, Lattore C, Baker JT and Shanmugan KT (1991) Genetically altered cyanobacteria as nitrogen fertilizer supplier for growth of rice. In: Datta SK and Sloger C (eds) Biological Nitrogen Fixation Associated with Rice Production,. 163-175. Oxford and IBH Pulk vo Pvt Ltd, Bew Delhi.

Ali S and Sandhu GR (1972) Blue-green algae of the saline soils of the Punjab. Oikos 23: 268-272.

Al-Kaisi KA (1976) Contributions to the algal flora of the rice fields of Southeastern Iraq. Nova Hedwigia 27: 813-27.

Al-Mousawi AHA and Whitton BA (1983) Influence of environmental factors on algae in rice-field soil from the Iraqi marshes. Arab Gulf J Scientific Research 1: 237-253.

Anand N (1990) Handbook of Blue-green Algae (of Rice Fields of South India). Bishen Singh Mahendra Pal Singh, 23-A Connaught Place, Dehra Dun, India.

Anand N and Hopper RSS (1987) Blue-green algae from rice fields in Kerala State, India. Hydrobiologia 144: 223-232.

Anderson DC and Rushforth SR (1976) The cryptogam flora of desert soil crusts in southern Utah, U.S.A. Nova Hedwigia 28: 691-714.

Ashley J, Rushforth SR and Johansen JR (1985) Soil algae of cryptogamic crusts from the Utah Basin, Utah, U.S.A. Great Basin Naturalist 45: 432-441.

Atlas RM, Schofield EA, Morelli FA and Cameron RE (1976) Effects of petroleum pollutants on Arctic microbial populations. Environ Pollut 10: 35-43.

Bailey D, Mazurak AP and Rosowski JR (1973) Aggregation of soil particles by algae. J Phycol 9: 99-101.

Balakrishnan MS and Gunale VR (1980) Cyanophycean air pollutants - a possible cause of inhalant allergy. Indian J Air Pollut Control 3: 9-17.

Baldwin NA and Whitton BA (1992) Cyanobacteria and eukaryotic algae occurring in sports turf and amenity grasslands: a review. J Appl Phycol 4: 39-47.

Barhate VP and Tarar JL (1982) Algae on coal-tar. Phykos 21: 141-142.

Bell RA, Athey PV and Sommerfeld MR (1986) Cryptoendolithic algal communities of the Colorado Plateau. J Phycol 22: 429-435.

Belnap J (1995) Soil surface disturbances: their role in accelerating desertification. Env Mon Assess 37: 17-39.

Belnap J (1996) Soil surface disturbances in cold deserts: effects on nitrogenase activity in cyanobacterial-lichen soil crusts. Biol Fertil Soils 23: 362-367.

Belnap J and Gardner JS (1993) Soil microstructure on the Colorado Plateau: The role of the cyanobacterium *Microcoleus vaginatus*. Great Basin Naturalist 53: 40-47.

Belnap J and Gillette DA (1997) Disturbance of biological soil crusts: Impacts on potential wind erodibility of sandy desert soils in Southeastern Utah. Land Degradation and Development 8: 355-362.

Belnap J and Gillette DA (1998) Vulnerability of desert biological soil crusts to wind erosion: the influence of crust development, soil texture, and disturbance. J Arid Environments 39: 133-142.

Belnap J and Harper KT (1995) Influence of cryptobiotic soils crusts on element content of two desert seed plants. Arid Soil Research and Rehabilitation 9: 107-115.

Belnap J, Harper KT and Warren SD (1994) Surface disturbance of cryptobiotic soil crusts: Nitrogenase activity, chlorophyll content, and chlorophyll degradation. Arid Soil Research and Rehabilitation 8: 1-8.

Belnap J, Sanford RL and Lungu L (1996) Biological soil crusts: Ecological roles and response to fire in Miombo Woodlands of Zimbabwe. Trans Zimbabwe Sci Ass 70: 14-20.

Bisiach M (1970) Primi dati sulle infestazioni algali da Cianoficee nelle risiae italiane. Riso 19: 129-134.

Bose P, Nagpal US, Venkataraman GS and Goyal SK (1971) Solubilization of tricalcium phosphate by blue-green algae. Curr Sci 7: 165-166.

Bisoyi RN and Singh PK (1988) Effect of phosphorus fertilization on blue-green algal inoculum production and nitrogen yield under field conditions. Biol Fertil Soils 5: 338-343.

Bristol BM (1919) On the retention of vitality by algae from old stored soils. New Phytol 18: 92-107.

Bristol BM (1920) On the alga-flora of some desiccated English soils: an important factor in soil biology. Annals of Botany 34: 35-81.

Broady PA (1979) The terrestrial algae of Signy Island, South Orkney Islands. Br. Antartic. Surv. Bull. 9: 21-40.

Brock TD (1973) Primary colonization of Surtsey, with special reference to the blue-green algae. Oikos 24: 239-243.

Brown RM, Jr, Larson DA and Bold HC (1964) Airborne algae: their abundance and heterogeneity. Science 143: 583-585.

Buttars SM, St Clair LL, Johansen JR, Sray JC, Payne MC, Webb BL, Terry RE, Pendleton BK and Warren SD (1998) Pelletized cyanobacterial soil amendments: Laboratory testing for survival, escapability, and nitrogen fixation. Arid Soil Res Rehab 12: 165-178.

Cameron HJ and Julian GR (1988) Utilization of hydroxyapatite by cyanobacteria as their sole source of phosphate and calcium. Plant and Soil 109: 123-124.

Carson JL and Brown M, Jr (1978) Studies of Hawaiian freshwater and soil algae II. Algal colonization and succession on a dated volcanic substrate. J Phycol 12: 171-178.

Castenholz RW (1982) Motility and taxes. In: Carr NG and Whitton BA (eds) The Biology of Cyanobacteria, 413-439. Blackwell, Oxford, and Univ. California Press, Berkeley.

Castenholz RW and Waterbury JB (1989) Cyanobacteria. In: Staley JT, Bryant MP, Pfennig N and Holt JG (eds) Bergey's Manual of Systematic Bacteriology Volume 3. Williams and Wilkins, Baltimore: 1710-27.

Das MK and Adhikary SP (1996) Toxicity of three pesticides to several rice-field cyanobacteria. Tropical Agriculture 73: 155-157.

Das SC, Mandal B and Mandal LN (1991) Effect of growth and subsequent decomposition of blue-green algae on the transformation of iron and manganese in submerged soils. Plant and Soil 138: 75-84.

Davey MC (1988) Ecology of terrestrial algae of the fellfield ecosystems of Signy Island, South Orkney Islands. Br Antarctic Surv Bull 81: 69-74.

Davey MC (1991) The seasonal periodicity of algae on Antarctic fellfield soils. Holarctic Ecology 14: 112-120.

Davey MC and Clarke KJ (1991) The spatial distribution of microalgae on Antarctic fellfield soils. Antarctic Science 4: 257-263

Davey MC, Davidson HPB, Richard KJ and Wynn-Williams DD (1991) Attachment and growth of Antarctic soil cyanobacteria and algae on natural and artificial substrata. Soil Biol Biochem. 23: 185-191.

De PK (1939) The role of blue-green algae in nitrogen fixation in rice fields. Proc Roy Soc London, Series B 127: 121-139.

de Caire GZ, de Cano MS, de Mulé MCZ, Palma RM and Colombo K (1997) Exopolysaccharide of *Nostoc muscorum* (cyanobacteria) in the aggregation of soil particles. J Appl Phycol 9: 249-253.

de Halperin DR, de Cano MS, de Mulé MCZ and de Caire GZ (1992) Diazotrophic cyanobacteria from Argentine paddy fields. Phyton 53: 135-142.

Desikachary TV (1959) Cyanophyta. Indian Council of Agricultural Research, New Delhi.

Desertová B (1974) Some interesting algae from soil. Arch Hydrobiol / Suppl 46: 105-119.

de Winder B, Pluis J, de Reus L and Mur LR (1989) Characterization of a cyanobacterial algal dune crust in the coastal dunes in the Netherlands. In: Cohen Y and Rosenberg E (eds) Microbial Mats: Physiological Ecology of Benthic

Microbial Communities, 77-83. Am Soc Microbiol, Washington DC.
Dooley F and Houghton JA (1973) The nitrogen-fixing capacities and the occurrence of blue-green algae in peat soils. Br Phycol J 8: 289-293.
Economou A, Anagnostidis K and Roussomoustakaki M (1984) Structural aspects of the adaptation of some blue-green algae and diatoms to desiccation. In: Margaris NS, Arianoustou-Faraggitaki M and Oechel WC (eds) Being Alive on Land. Proceedings of the International Symposium on Adaptations to the Terrestrial Environment held in Halkidiki, Greece, 1982, 103-114. W. Junk Publishers, The Hague.
Eldridge DJ and Greene RSNB (1994) Microbiotic soil crusts - a review of their roles in soil and ecological processes in the rangelands of Australia. Austr J Soil Res 32: 389-415.
Ernst WHO (1974) Schwermetallvegetation der Erde. Gustav Fischer Verlag, Stuttgart.
Eskew DL and Ting JP (1978) Nitrogen fixation by legumes and blue-green algal - lichen crusts in a Colorado desert environment. Am J Bot 65: 850-856.
Evans RD and Belnap J (1999) Long-term consequences of disturbance on nitrogen dynamics in an arid ecosystem. Ecology 80: 150-160.
Falchini L, Sparvoli E and Tomaselli L (1996) Effect of *Nostoc* (cyanobacteria) inoculation on the structure and stability of clay soils. Biol Fertil Soils 23: 346-252.
Fernández-Valiente E, Quesada A, Prosperi C, Nieva M, Leganés F and Ucha A (1997) Short- and long-term effects of ammonium on photodependent nitrogen fixation in wetland rice fields of Spain. Biol Fertil Soils 24: 353-357.
Forest HS (1965) The soil algal community. II. Soviet soil studies. J Phycol 1: 164-171.
Fritsch FE (1907a) The subaerial and freshwater algal flora of the tropics. Annals of Botany 30: 235-275.
Fritsch FE (1907b) The rôle of algal growth in the colonization of new ground and in the determination of scenery. Geog J: 531-548.
Fritsch FE (1907c) A general consideration of subaërial and fresh-water algal flora of Ceylon. A contribution to the study of tropical algal ecology. Part I. - Subaërial algae and algae of the inland fresh-waters. Proc Roy Soc London, Series B 11: 197-254.
Fritsch FE (1922) The terrestrial alga. J Ecol 10: 220-236.
Gallon JR, Hashem MA and Chaplin AE (1991) Nitrogen fixation by *Oscillatoria* spp. under autotrophic and photoheterotrophic conditions. J Gen. Microbiol 137: 31-39.
Gantar M, Kerby NW, Rowell P and Obreht Z (1991a) Colonization of wheat (*Triticum vulgare* L.) by N_2-fixing cyanobacteria: I. A survey of soil cyanobacterial isolates forming associations with roots. New Phytol 118: 477-483.
Gantar M, Kerby NW and Rowell P 1991b. Colonization of wheat (*Triticum vulgare* L.) by N_2-fixing cyanobacteria: II. An ultrastructural study. New Phytol 118: 485-492.
Garcia-Pichel F and Belnap J (1996) Microenvironments and microscale productivity of cyanobacterial desert crusts. J. Phycol. 32: 774-782.
Garcia-Pichel F and Castenholz RW (1991) Characterization and biological implications of scytonemin, a cyanobacterial sheath pigment. J Phycol 27: 395-409.
Garty J (1990) Influence of epilithic microorganisms on the surface temperature of building walls. Can J Bot 68: 1349-1353.
Ghosh TK and Saha KC (1997) Effects of inoculation of cyanobacteria on nitrogen status and nutrition of rice (*Oryza sativa* L.) in an Entisol amended with chemical and organic sources of nitrogen. Bio Fertil Soils 24: 123-128.
Gibson MT, Welch IM, Barret PRF and Ridge I (1990) Barley straw as an inhibitor of algal growth II: laboratory studies. J Appl Phycol 2: 241-248.
Granhall U (1975) Nitrogen fixaton by blue-green algae in temperate soils, p. 189-198. In: Stewart WDP (ed.) Nitrogen Fixation by Free-living Microrganisms, 189-198. Cambridge University Press, Cambridge, U.K.
Granhall U and Henriksson E (1969) Nitrogen-fixing blue-green algae in Swedish soils. Oikos 20: 175-178.
Grant IF, Tirol AC, Aziz T and Watanabe I (1983) Regulation of invertebrate grazers as a means to enhance biomass and nitrogen fixation of Cyanophyceae in wetland rice fields. Soil Sci Soc Am J 47: 660-675.
Griggs RF (1933) The colonization of Katmai ash, a new and inorganic 'soil'. Am J Bot 2: 92-113.
Gruia L (1964) Algues des sols du massif des Bucegi. Rev Roum de Biol - Botanique 9(5): 295-305.
Gupta AB (1966) Algal flora and its importance in the economy of rice fields. Hydrobiologia 28: 213-222.
Hansson L (1988) Chlorophyll *a* determination of periphyton on sediments: identification of problems and recommendation of method. Freshwat Biol 20: 347-352.
Hawes I and Davey MC (1989) Use of the fluorochrome Auramine O for determination of cell viability in filamentous and thalloid algae. Phycologia 28: 518-523.
Healey FD (1982) Phosphate. In: Carr NG and Whitton BA (eds) The Biology of Cyanobacteria, 105-124. Blackwell, Oxford, and Univ. California Press, Berkeley.
Hershkovitz N, Oren O and Cohen Y 1991. Accumulation of trehalose and sucrose in cyanobacteria exposed to matric water stress. Appl Environ Microbiol 57: 645-648.
Hodges CF (1987a) Blue-green algae and black layer. Landscape Management 26 (10): 38, 42-44.
Hodges CF (1987b) Blue-green algae and black layer Part II. Landscape Management 26 (11): 30-31.
Hodges CF 1992. Interaction of cyanobacteria and sulfate-reducing bacteria in subsurface black-layer formation in high-sand content golf greens. Soil Biol Biochem 24: 15-20.
Hodgins IW and Rogers RW (1997) Correlations of stocking with the cryptogamic soil crust of a semi-arid rangeland in southwest Queensland. Australian J Ecol 22: 425-431.
Howarth RW, Marino R, Lane J and Cole JJ (1988) Nitrogen fixation in freshwater, estuarine, and marine ecosystems. 1. Rates and importance. Limnol Oceanogr 33: 669-687.
International Rice Research Institute (1987) Azolla Utilization. Proceedings of the Workshop on Azolla Use, Fuzhou, Fujian, China, 31 March - 5 April 1985. International Rice Research Institute, PO Box 933, Manila, Philippines.
Islam MR and Whitton BA (1992a) Cell composition and nitrogen fixation by the deepwater rice-field cyanobacterium (blue-green alga) *Calothrix* D764. Microbios 69: 77-88.
Islam MR and Whitton BA (1992b) Phosphorus content and phosphatase activity of the deepwater rice-field cyanobacterium (blue-green alga) *Calothrix* D764. Microbios 69: 7-16.
Jeffries DL, Klopatek JM, Link SO and Bolton H, Jr (1992) Acetylene reduction by cryptogamic crusts from a blackbrush community as related to resaturation and dehydration. Soil Bio Biochem 24: 1101-1105.
Jha MN, Jha UN, Ahmad N and Mallik MK (1986) Cyanobacterial flora of rice field soils of Pusa and its adjoining areas. Phykos 25: 97-101.

Johansen JR (1993) Cryptogamic crusts of semiarid and arid lands of North America. J Phycol 29: 140-147.

John DM (1988) Algal growths on buildings: a general review and methods of treatment. Biodeterioration Abstracts 2: 81-102.

Kabli SA, AlGarni SM and AlFassi FA (1997) Efficiency of cyanobacteria from soil of Western region in the Kingdom of Saudi Arabia as biofertilizer of wheat. Arab Gulf J Scientific Research 15: 481-503.

Kamuru F, Albrecht SL, Allen LH and Shanmugam KT (1998) Growth and accumulation of ^{15}N in rice inoculated with the parent and a nitrogenase-derepressed mutant strain of *Anabaena variabilis*. Appl Soil Ecol 5: 189-195.

Kamuru F, Albrecht SL, Allen LH and Shanmugam KT (1998) Dry matter and nitrogen accumulation in rice inoculated with a nitrogenase-derepressed mutant of *Anabaena variabilis*. Agronomy J 90: 529-535.

Kannaiyan S (1992) Azolla Biofertilizer Technology for Rice Crop. Tamil Nadu Agricultural University, Combatore, India.

Kannaiyan S (1993) Nitrogen contribution by *Azolla* to rice crop. Proc Indian Nat Sci Acad B59: 309-314.

Kannaiyan S, Aruna SJ, Kumari MP and Hall DO (1997) Immobilized cyanobacteria as a biofertilizer for rice crops. J Appl Phycol 9: 167-174.

Karnieli A (1997) Development and implementation of spectral crust index over dune sands. International J Remote Sensing 18: 1207-1220.

Karnieli A, Shachak M, Tsoar H, Zaady E, Kaufman Y , Danin A and Porter W (1996) The effect of microphytes on the spectral reflectance of vegetation in semiarid regions. Remote Sensing of Environment 57: 88-96.

Katznelson R (1989) Clogging of groundwater recharge basins by cyanobacterial mats. FEMS Microbiol Ecol 62: 232-242.

Kaushik BD (1998) Use of cyanobacterial biofertilizers in rice cultivation: A technology improvement. In: Subramanian G, Kaushik BD and Venkataraman GS (eds) Cyanobacterial Biotechnology. Science Publishers, Inc., Enfield, USA: 211-222.

Kaushik BD and Subhashini D (1985) Amelioration of salt-affected soils with blue-green algae. II. Improvement in soil properties. Proc Indian Natn Acad B51: 386-389.

Khan ZUM, Tahmida Begum ZN, Mandal R and Hossain MZ (1994) Cyanobacteria in rice soils. World J Microbiol Biotechnol 10: 296-298.

Khoja TM and Whitton BA (1971) Heterotrophic growth of blue-green algae. Arch Mikrobiol 79: 280-282.

Kleiner KT and Harper KT (1972) Environment and community organization in the grasslands of the Canyonlands National Park. Ecology 53: 229-309.

Kleiner KT and Harper KT (1977) Soil properties in relation to cryptogamic ground cover in Canyonlands National Park. J Rangeland Management 30: 202-205.

Kozyrovskaya N (1990) (Note in) Cyanonews 6(3): 5. (Newsletter distributed by MSU-DOE Plant Research Laboratory, Michigan State University, East Lansing)

Kulasooriya SA (1991) Biological nitrogen fixation in acid rice soils. In: Defurck P and Ponnamperuma FN (eds) Rice Production in Acid Soils of the Tropics. Institute of Fundamental Studies, Kandy, Sri Lanka.

Kulasooriya SA (1998) Cyanobacteria and *Azolla* as biofertilizer for rice. In: Subramanian G, Kaushik BD and Venkataraman GS (eds) Cyanobacterial Biotechnology. Science Publishers, Inc., Enfield, USA: 201-209.

Kulasooriya SA, Roger PA, Barraquio WL and Watanabe I (1980) Biological nitrogen fixation by epiphytic micorganisms in rice fields. IRRI Research Papers 47: 1-10. (International Rice Research Institute, PO Box 933, Manila, Philippines)

Ladha JK and Reddy PM (1995) Extension of nitrogen fixation to rice - necessity and possibilities. Geo Journal 35: 363-372.

Lange OL, Belnap J and Reichenberger H (1998) Photosynthesis of the cyanobacterial soil-crust lichen *Collema tenax* from arid lands in southern Utah, USA: role of waer content on light and temperature responses of CO_2 exchange. Functional Ecology 12: 195-202.

Lange W (1976) Speculations on a possible essential function of the gelatinous sheath of blue-green algae. Can J Microbiol 22: 1181-185.

Leganés F and Fernandez-Valiente E (1992) Effects of phenoxyacetic herbicides on growth, photosynthesis, and nitrogenase activity in cyanobacteria from rice fields. Arch Environ Contam Toxicol 22: 130-134.

Liengen T and Olsen RA (1997) Seasonal and site-specific variations in nitrogen fixation in a high arctic area, Ny-Alesund, Spitsbergen. Can J Microbiol 43: 759-769.

Lindenback SK and Cullimore DR (1989) Preliminary in vitro observations on the bacteriology of the black plug layer phenomenon associated with biofouling of golf greens. J Appl Bacteriol 67: 11-17.

Lipman CB (1941) The successful revival of *Nostoc commune* from a herbarium specimen eighty-seven years old. Bull Torrey Club 68: 669-666.

Lumpkin TA and Plucknett DL (1980) *Azolla* as a Green Manure: Use and Management in Crop Production. Westview Tropical Agriculture Series, Westview Press, Boulder, CO, USA.

Lund JWG (1947) Observations on soil algae. 2. Notes on groups other than diatoms. New Phytol 46: 35-60.

MacGregor AN and Johnson DE (1971) Capacity of desert algal crusts to fix atmospheric nitrogen. Soil Sci Am Proc 35: 843-844.

Mague TH (1978) Nitrogen fixation, p. 363-378. In: Hellebust JA and Craigie JS (eds) Handbook of Phycological Methods. Physiological and Biochemical Methods. Cambridge University Press, Cambridge, U.K.

Mandal R, Tahmida Begum ZN, Khan ZUM and Hossain MZ (1993) N_2-fixing blue-green algae in rice fields and their relationship with soil fertility. Bangladesh J Bot 22: 73-79.

Marticorena B, Bergametti G, Gillette D and Belnap J (1997) Factors controlling threshold velocity in semiarid and arid areas of the United States. J Geophys Res 102: 23277-23287.

Martin D and Ridge I (1999) The relative sensitivity of cyanobacteria and eukaryotic algae to decomposed barley straw. J Appl Phycol. 11:

Mayland HD, McIntosh TH and Fuller WH (1966) Fixation of isotopic nitrogen in a semi-arid soil by algal crust organisms. Soi Sci America Proc 30: 56-60.

Metting B (1981) The systematics and ecology of soil algae. The Botanical Review 47: 195-312.

Metting B (1988) Micro-algae in agriculture. In: Borowitzka MA and Borowitzka LJ (eds) Micro-Algal Biotechnology, 288-304. Cambridge University Press, Cambridge. U.K.

Mishra B (1985) Sporulation in Blue-Green Algae. PhD Thesis, Central Rice Research Institute, Cuttack, India.

Modi DR, Singh AK, Rao KS, Chakravarty D and Singh HN (1991) Construction of multiple herbicide resistant ammonia excreting strains of cyanobacterium *Nostoc muscorum*. Biotechnol Lett 13: 393-398.

Moore AW (1963) Occurrence of non-symbiotic nitrogen-fixing microorganisms in Nigerian soils. Plant and Soil 19: 385-395.

Mrozinska T (1990) Aerophytic algae from North Korea. Algological Studies 58: 29-47.

Natesan R and Shanmugasundaram S (1989) Extracellular phosphate solubilization by the cyanobacterium *Anabaena* ARM310. J Biosci 14: 203-208.

Neuman CM, Maxwell CD and Boulton JW (1996) Wind transport of sand surfaces crusted with photoautotrophic microorganisms. Catena 27: 229-247.

Novelo-Maldanado E and Gonzalez-Gonzalez J (1981) Una methodologia para la colecta y el estudio de la ficoflora edafica. Phycol Lat-Amer 1: 134-145. (J. Cramer, Braunschweig, Germany)

Padhy R.N (1985) Cyanobacteria and pesticides. Res Rev 95: 1-44.

Palmer RJ, Jr and Friedmann EI (1990) Water relations and photosynthesis in the cryptoendolithic microbial habitat of hot and cold deserts. Microb Ecol 19: 111-118.

Pankratova YM and Vakrushev AS (1971) Field determination of the fixation of the atmospheric nitrogen by blue-green algae using N^{15}. Soviet Soil Sci 3(6): 726-733.

Parker BC, Schanen N and Renner R (1969) Viable soil algae from the herbarium of the Missouri Botanical Garden. Ann Missouri Bot Gard 56(2): 113-119.

Parks JM and Rice EL (1969) Effects of certain plants of old-field succession on the growth of blue-green algae. Bull Torr Bot Club 96: 345-360.

Pearson HW, Howsley R, Kjeldson CK and Walsby AE (1979) Aerobic nitrogenase activity by axenic cultures of the blue-green alga *Microcoleus chthonoplastes*. FEMS Microbiol Lett 5: 163-167.

Pederson CL and Shubert L (1992) Use of membrane filters for soil algal bioassays. J Appl Phycol 4: 49-56.

Pedurand P and Reynaud PA (1987) Do cyanobacteria enhance germination and growth of rice? Plant and Soil 101: 235-240.

Peterson HG, Boutin C, Freemark KE and Martin PA (1997) Toxicity of hexazinone and diquat to green algae, diatoms, cyanobacteria and duckweed. Aquatic Toxicology 39: 111-134.

Potts M and Bowman MA (1985) Sensitivity of *Nostoc commune* UTEX 584 (Cyanobacteria) to water stress. Arch Microbiol 141: 51-56.

Prasad BN and Srivastava PN (1968) Systematic and ecological studies on algae of lakaline (Usar) soils. Phukos 7: 102-111.

Prosperi C, Boluda L, Luna C and Fernandez-Valiente E (1992) Environmental factors affecting *in vitro* nitrogenase activity of cyanobacteria isolated from rice fields. J Appl Phycol 4: 197-204

Quesada A, Leganés F and Fernández-Valiente E (1997) Environmental factors controlling N_2 fixation in Mediterranean rice fields. Microb Ecol 34: 39-48

Quesada A, Nieva M, Leganés E, Ucha A, Prosperi C and Fernández-Valiente E (1998) Acclimation of cyanobacterial communities in rice fields and response of nitrogenase activity to light regime. Microb Ecol 35: 147-155.

Rana BC, Gopal T and Kumar HD (1971) Studies on the biological effects of industrial wastes on the growth of algae. Environ Health 13: 138-143.

Rao DLN and Burns RG (1990) The effect of surface growth of blue-green algae and bryophytes on some microbiological, biochemical, and physical soil properties. Biol Fertil Soils 9: 239-244.

Rath B and Adhikary SP (1995) Toxicity of furadan to several N_2-fixing cyanobacteria from the rice fields of coastal Orissa, India. Tropical Agriculture 72: 80-83.

Reddy PM (1983) Control of sporulation in some blue-green algae. Arch. Hydrobiol, Suppl. 63: 443-440.

Reddy PM and Roger PA (1988) Dynamics of algal populations and acetylene-reducing activity in five rice soils inoculated with blue-green algae. Biol Fertil Soils 6: 14-21.

Reynaud PA and Franche C (1986) Isolation and characterization of non-heterocystous tropical cyanobacteria growing on nitrogen-free medium. MIRCEN J 2: 427-443.

Reynaud PA and Roger PA (1977) Milieux sélectifs pour la numération des algues eucaryotes, procaryotes et fixatrices d'azote. Rev Ecol Biol du Sol 14: 421-428.

Rice EL, Lin C-Y and Huang C-Y (1980) Effects of decaying rice straw on growth and nitrogen fixation of a blue green alga. Bot Bull Academia Sinica 21: 111-117.

Rippka R, Deruelles JB, Waterbury JB, Herdman M, Stanier RY (1979) Generic assignments, strain histories and properties of pure cultures of cyanobacteria. J Gen Microbiol 111: 1-61.

Roger PA (1991) Reconsidering the utilization of blue-green algae in wetland rice cultivation. In: Dutta SK and Sloger C (eds) Biological Nitrogen Fixation Associated with Rice Production, 119-141. Oxford and IBH Publishing Co., New Delhi.

Roger PA (1996a) Biology and Management of the Floodwater Ecosystem in Ricefields. International Rice Research Institute, P.O.Box 933, Manila 1099, Philippines.

Roger PA (1996b) The impact of pesticides on ricefield microflora: an analytical review of the literature. In: Pingali PL, Roger PA (eds) Impact of Pesticides on Farmer Health and the Rice Environment, 271-308. International Rice Research Institute and Kluwer Academic Publishers, Dordrecht, The Netherlands, and Norwell, Mass., USA.

Roger PA and Ardales S (1991) The IRRI blue-green algae collection and computerized information on the strains available for distribution. J Appl Phycol 3: 375-376.

Roger PA and Kulasooriya SA (1980) Blue-green Algae and Rice. International Rice Research Institute, PO Box 933, Manila, Philippines.

Roger PA, Kulasooriya SA, Tirol AC and Craswell ET (1980) Deep placement: a method of nitrogen fertilizer application compatible with algal nitrogen fixation in wetland rice soils. Plant and Soil 57: 137-142.

Roger PA and Ladha JK (1992) Biological nitrogen fixation in wetland ricefields: estimation and contribution to nitrogen balance. Plant Soil 141: 41-55.

Roger PA and Reynaud PA (1976) Dynamique de la population algale au cours d'un cycle de culture dans une rizière Sahelienne. Rev Ecol Biol Sol 13: 545-560.

Roger PA and Reynaud PA (1977) La biomasse algale dans les rizières de Sénégal: importance relative des cyanophycées fixatrice de N_2. Rev Ecol Biol Sol 14: 519-530.

Roger PA, Santiago-Ardales S, Reddy PM and Watanabe I (1987) The abundance of heterocystous blue-green algae in rice soils and inocula used for application in rice fields. Biol Fertil Soils 5: 98-105.

Roger PA and Watanabe I (1986) Technologies for utilizing biological nitrogen fixation in lowland rice: potentialities, current usage, and limiting factors. Fertilizer Research 27: 470-477.

Roger PA, Zimmerman WJ and Lumpkin T (1993) Microbiological management of wetland ricefields. In: Metting B (ed.) Soil Microbial Ecology, 417-455. M. Dekker, New York.

Ross RLM and Puritch GS (1981) Identification, abundance, and origin or moss, liverwort, and algal contaminants in

greenhouses of containerized forest nurseries. Can J Bot 11: 356-360.
Rother JA, Aziz A, Hye Karim N and Whitton BA (1988) Ecology of deepwater rice-fields in Bangladesh. 4. Nitrogen fixation by blue-green algal communities. Hydrobiologia 169: 43-56.
Rother JA and Whitton BA (1989) Nitrogenase activity of blue-green algae on seasonally flooded soils in Bangladesh. Plant and Soil 113: 47-52.
Roychoudhury P and Kaushik BD (1989) Solubilization of Mussorie rock phosphate by cyanobacteria. Curr Sci 58: 569-570.
Roychoudhury P, Krishnamurti GSR and Venkataraman GS (1980) Effect of algal inoculation on soil aggregation in rice soils. Phykos 19: 224-227.
Rychert RC and Skujins I (1974) Nitrogen fixation by blue-green algae - lichen crsuts in the Great Basin desert. Soil Sci Am Proc 38: 768-771.
Saito M and Watanabe I (1978) Organic matter production in rice field flood water. Soil Sci. Plant Nutr 28: 427-440.
Scherer S, Ernst A, Chen T-W and Böger P (1984) Rewetting of drought-resistant blue-green algae: time course of water uptake and reappearance of respiration, photosynthesis, and nitrogen fixation. Oecologia, Berl. 62: 418-423.
Schlichting HE, Jr (1961) Viable species of algae and protozoa in the atmosphere. Lloydia 24: 81-88.
Schlichting HE, Jr (1964) Meteorological conditions affecting the dispersal of airborne algae and protozoa. Lloydia 27: 64-78.
Schwabe GH (1963) Blaualgen der phototrophen Grenzschicht (Blaualgen und Lebensraum VII). Pedobiologia 2: 132-152.
Sharma BM and Gupta RS (1983) Effect of algal application on rice yield in Jammu Division. Phykos 22: 176-179.
Shields LM (1957) Algal and lichen floras in relation to nitrogen content of certain volcanic and arid range soils. Ecology 38: 661-663.
Simpson I, Roger PA, Oficial B and Grant IF (1993) Effects of fertilizer and pesticide management on floodwater ecology of a wetland ricefield. I. Experimental design and dynamics of the photosynthetic aquatic biomass. Biol Fert Soils 17: 129-137.
Singh AL and Singh PK (1985) Proceedings of the Indian Academcy of Science (Plant Science) 96: 147-152.
Singh AL and Singh PK (1987) Nitrogen fixation and balance studies of rice soil. Biol Fert Soils 4: 15-19.
Singh AL, Singh PK and Singh PL (1988) Effects of different herbicides on the *Azolla* and blue-green algal biofertilization of rice. J Agric Sci 111: 451-458.
Singh LJ, Tiwari DN and Singh HN (1986) Evidence for genetic control of herbicide resistance in a rice-field isolate of *Gloeocapsa* sp. capable of aerobic diazotrophy under photoautotrophic conditions. J Gen Appl Microbiol 81: 81-88.
Singh NI, Singh NS, Devia GA and Singh SM (1997a) Blue green algae from rice growing areas of Arunachal Pradesh. Phykos 36: 21-26
Singh NI, Dorycanta H, Devi GA, Singh NS and Singh MS (1997b) Blue green algae from rice field soils of Nagaland. Phykos 36: 115-120
Singh RN (1950) Reclamation of usar lands in India through blue-green algae. Nature 165: 325-326.
Singh RN (1961) Role of Blue-green Algae in Nitrogen Economy of Indian Agriculture. Indian Council of Agricultural Research, New Delhi.
Smith GD, Lynch RM, Jacobson G and Barnes CJ (1990) Cyanobacterial nitrogen fixation in arid soils of Central Australia. FEMS Microb. Ecol. 74: 79-90.
Solheim B, Endal A and Vigstad H (1996) Nitrogen fixation in Arctic vegetation and soils from Svalbard, Norway. Polar Biology 16: 35-40.
Spiller H, Lattore C, Hassan ME and Shanmugan KT (1986) Isolation and characterization of nitrogenase depressed mutant strains of cyanobacterium *Anabaena variabilis*. J Biotechnol 165: 412-419.
Stal LJ (1988) Nitrogen fixation in cyanobacterial mats. Methods in Enzymology 167: 474-490.
Steppe TF, Olson JB, Paerl HW, Litaker RW and Belnap J (1996) Consortial N_2 fixation: A strategy for meeting nitrogen requirements of marine and terrestrial cyanobacterial mats. FEMS Microbiol Ecol 21: 149-156.
Stewart WDP (1967) Transfer of biologically fixed nitrogen in a sand dune slack region. Nature 214: 603-604.
Stratton GW (1987) The effects of pesticides and heavy metals towards phototrophic microorganisms. Rev Environ Toxicol 3: 71-147.
Svircev Z, Tamas I, Nenin P and Drobac A (1997) Co-cultivaton of N_2-fixing cyanobacteria and some agriculturally important plants in liquid and sand cultures. Applied Soil Ecology 6: 301-308.
Tarar JL and Shewale TH (1984) Studies on the effects of some fungicides on soil algae of paddy field. Phykos 23: 191-201.
Tchan YT (1959) Study of soil algae. III. Bioassay of soil ferttility by algae. Plant and Soil 10: 220-232.
Tchan YT, Balamm LN, Hawkes R and Draette F (1961) Study of soil algae IV. Estimation of the nutrient status of soil using an algal growth method with special reference to nitrogen and phosphorus. Plant and Soil 14: 147-158.
Tiberg E, Bergman B, Wictorin B and Willén T (1984) Nordic Aerobiology (1984): 24-29 (= 1983, 5th Nord. Symp. Aerobiol., Session II: 24-29).
Tiwari DN, Kumar A and Mishra AK (1991) Use of cyanobacterial diazotrophic technology in rice agriculture. Appl Biochem Biotechnol 28/29: 387-396.
Tiwari GL (1975) A study of the blue-green algae from paddy field soils of India II. Taxonomic considerations of non-heterocystous blue-green algae. Nova Hedwigia 25: 765-798.
Tiwari GL and Pandey RS (1976) A study of the blue-green algae from paddy field soils of India III. Nostocaceae. Nova Hedwigia 27: 701-730.
Trainor FR (1985) Survival of algae in a desiccated soil: a 25 year study. Phycologia 24: 79-82.
Traore TM, Roger PA, Renaud PA and Sasson A (1978) Nitrogen fixation by blue-green algae in paddy field in Mali (French, English Summary). Cah ORSTOM Ser Biol 13: 181-185.
Treub M (1888) Notice sur la nouvelle flora de Krakatau. Ann Jard Bot Buitenzorg 7: 221-223.
Vaishampayan A, Sinha RP and Häder DP (1998) Use of genetically improved nitrogen-fixing cyanobacteria in rice paddy fields: Prospects as a source material for engineering herbicide sensitivity and resistance in plants. Botanica Acta 111(3): 176-190.
Valiente EF, Quesada A, Prosperi C, Nieva M, Leganes F and Ucha A (1997) Short- and long-term effects of ammonium on photodependent nitrogen fixation in wetland rice fields of Spain. Biol Fertil Soils 24: 353-357.
van Liere L and Walsby AE (1982) Interactions of cyanobacteria with light. In: Carr NG and Whitton BA (eds) The Biology of

Cyanobacteria, 9-45. Blackwell, Oxford, and Univ. California Press, Berkeley.

Venkataraman GS (1972) Algal Biofertilizers and Rice Cultivation. Today and Tomorrow Printers, Faridabad, Haryana, India.

Venkataraman GS (1981) Blue-green Algae for Rice Production. A Manual for its Promotion. FAO Bulletin No. 46. FAO, Rome

Watanabe A (1959) Distribution of nitrogen-fixing blue-green algae in various areas of south and east Asia. J Gen Appl Microbiol 5: 21-29.

Watanabe A, Ito R and Konishi C (1951) Effect of nitrogen-fixing blue-green algae on the growth of rice plants. Nature 168: 748-749.

Watanabe I, Roger PA, Ladha JK and Van Hove C (1992) Biofertilizer Germplasm Collections at IRRI. International Rice Research Institute, PO Box 933, Manila, Philippines.

Welch IM, Barrett PRF, Gibson MT and Ridge I (1990) Barley straw as an inhibitor of algal growth I: Studies in the Chesterfield Canal. J Appl Phycol 2: 231-239.

West GS and Fritsch FE (1927) A Treatise on the British Freshwater Algae. Revised Edition. Cambridge University Press, Cambridge.

West NE (1990) Structure and function of microphytic soil crusts in wildland ecosystems of arid to semi-arid regions. Advances in Ecological Research 20: 179-223.

Whitton BA (1980) Zinc and plants in rivers and streams. In: Nriagu JO (ed.) Zinc in the Environment. Part II: Health Effects, 415-437. Wiley-Interscience, New York.

Whitton BA (1987) Survival and dormancy of blue-green algae. In: Henis Y (ed.) Survival and Dormancy of Micro-organisms, 209-266. Wiley, N.Y.

Whitton BA (1990) *Microcoleus*. In: Kumar HD (ed.) Phycotalk, 173-182. Rastogi Publications, Meerut, India.

Whitton BA (1991) Use of phosphatase assays with algae to assess phosphorus status of aquatic environments. In: Jeffrey DW and Madden B (eds) Bioindicators and Environmental Management. Proc. 6th International Bioindicators Symposium, 295-310. Academic Press, London.

Whitton BA (1992) Diversity, ecology and taxonomy of the cyanobacteria. In: Mann NH and Carr NG (eds) Photosynthetic Prokaryotes, 1-51. Plenum Press, N.Y.

Whitton BA (1993) Cyanobacteria and *Azolla* . In: Jones DG (ed.) Exploitation of Microorganisms, 137-167. Chapman and Hall, London.

Whitton BA (1996) Cyanobacteria (blue-green algae). In: Hall DS (ed.) Methods for the Examination of Organismal Diversity in Soils and Sediments, 67-78. CAB International in association with UNESCO and IUBS.

Whitton BA, Aziz A, Kawecka B and Rother JA (1989) Ecology of deepwater rice-fields in Bangladesh. 3. Associated algae and macrophytes. Hydrobiologia 169: 31-42.

Whitton BA, Gale NL and Wixson BG (1981) Chemistry and plant ecology of zinc-rich wastes dominated by blue-green algae. Hydrobiologia 83: 331-341.

Whitton BA, Grainger SLJ and Simon JW (1991) Cell-bound and extracellular phosphatase activities of cyanobacterial isolates. Microb Ecol 21: 85-98.

Whitton BA and Roger PA (1989) Use of blue-green algae and *Azolla* in rice culture. In: Campbell R and Macdonald RM (eds) Microbial Inoculation of Crop Plants. Special Publication of the Society for General Microbiology 25, 89-100. IRL Press at Oxford U.P., Oxford, N.Y., Tokyo.

Wynn-Williams DD (1988) Television image analysis of microbial communities in Antarctic fellfields. Polarforschung 58: 239-249.

Wynn-Williams DD (1990) Microbial colonization processes in Antarctic fellfield soils - an experimental overview. Proc National Institute of Polar Research Symposium on Polar Biology 3: 164-178.

Yanni YG (1992a) The effect of cyanobacteria and *Azolla* on the performance of rice under different levels of fertilizer-N. World J Microbiol Biotechnol: 132-136.

Yanni YG (1992b) Fertilizer responses of rice to nitrogen and cyanobacteria in the presence of insecticides. Soil Biol Biochem 24: 1085-1088.

Yoshida S (1984) Algal nitrogen fixation in grassland. J Japan Grassl Sci 30: 145-150.

Zaady E, Groffman P and Shachak M (1998) Nitrogen fixation in macro- and microphytic patches in the Negev desert. Soil Biol Biochem 30: 449-454.

Zaitseva II (1979) The influience of soil herbicides on the nitrogen-fixing activity of blue-green algae. Byulleten Vesesoyuznogo, Nauchno-Isstedovatel Skogo, Instituta Selskokhozyaistvennoi Mikrobiologii 32: 60-62. (in Russian)

Zimmerman W, Metting B and Rayburn W (1980) The occurrence of blue-green algae in silt loams of Whitman County, Washington. Soil Sci 130: 11-18.

Chapter 9

Limestones

Allan Pentecost[1] and Brian A. Whitton[2]
[1]*Division of Life Sciences, King's College London, Franklin-Wilkins Building, 150 Stamford Street, London SE1 8WA, UK.* [2]*Department of Biological Sciences, University of Durham, Durham DH1 3LE, UK*

Summary

The chapter provides an overview of cyanobacteria in calcareous environments, especially those associated with limestone surfaces, where the organisms may be epilithic or endolithic. The epilithic species of aquatic environments sometimes enhance calcium carbonate deposition, either by trapping particles or by contributing to communities where calcite and aragonite (in freshwaters) or other carbonates are precipitated. Environmental features at a site have an important influence on the process of carbonate precipitation, but photosynthetic activity and also the properties of the sheaths in some species sometimes influence the process. Temporal heterogeneities in water supply and phosphate availability are especially important in determining the species composition. Sites with occasional running water combined with long periods of drought are often dominated by Chroococcales with dark sheaths (e.g. *Gloeocapsa*), while sites with marked temporal differences in ambient phosphate are often dominated by hair-forming genera (e.g. *Rivularia*). Calcium carbonate deposition

B. A. Whitton and M. Potts (eds), The Ecology of Cyanobacteria, 257-279.
© 2000 *Kluwer Academic Publishers. Printed in the Netherlands.*

sometimes leads to a permanent increase in the amount of limestone present, as with travertine in freshwaters or stromatolites in both freshwater and, more especially, marine environments.

Endolithic species occur in a variety of situations, but particularly where the limestone is soft and porous. For example, several genera (e.g. *Hyella*) grow actively into the limestone at sites where removal of surface material tends to occur. Such endolithic communities are best developed on vertical surfaces in the intertidal zone of tropical and subtropical seas, where the more superficial parts of the communities are continually being removed by grazing animals.

I. Introduction

It has long been known that limestone surfaces often have visually conspicuous growths of cyanobacteria and the surfaces as a whole may be rich in species (Jaag, 1945; Golubić, 1967). This chapter provides an overview of the communities present and their component cyanobacteria. Golubić's review in an earlier volume (1973) still provides an excellent introduction to the subject and aspects covered in detail there are mentioned only briefly here.

Some of the diversity of cyanobacteria associated with limestone surfaces can probably be explained by the features of the limestones themselves. These are a diverse group of sediments linked by their high calcium carbonate content (>50% wt). They range from metamorphosed limestones with very low magnesium content to high-magnesian calcites and dolomites, which may reach 50 mole % $MgCO_3$. They vary widely in origin and in their porosity, density and erodability (Tucker and Wright, 1990). They also vary widely in chemical constituents which might be expected to influence cyanobacterial growth, such as phosphate (Davis and Rands, 1982) and organic content.

The cyanobacterial communities of limestone surfaces include those of desert rocks, cliffs of dry and humid regions, various types of emergent structure formed by erosion, the inundated limestones of lakes, rivers and streams and marine examples in the intertidal zone. Highly calcareous freshwaters sometimes, but not always, have conspicuous cyanobacterial communities. Cyanobacteria can also form important communities on limestone artifacts or commercial products where limestone has been incorporated. A wide range of ecosystems will thus be considered in relation to water regime and climatic, tidal and human factors. The organisms, which include the cyanolichens, may be epilithic or endolithic. Some cyanobacteria are noted for their ability to precipitate, trap and bind particles of calcium carbonate to form accumulations of layered travertine or stromatolite. In addition, some are able to etch limestone and grow within the subtrate (Golubić et al., 1970). This ability of a species to deposit or to erode limestone provides it with a potent means of modifying its local environment.

II. Physical and Chemical Features

Limestone underlies an estimated 4% of the global terrestrial surface (Balazs, 1977), much of this being exposed by karstification. However, limestones differ widely in their physical characteristics. In Britain, for example, chalk is a highly porous, soft limestone forming a rolling topography of moderate to low relief contrasting with the tough impermeable Carboniferous limestones forming steep cliffs in areas of high relief. Other limestones, the dolomites, contain a high proportion of magnesium carbonate.

Dense limestones dry quickly and limestone streams are often erratic or ephemeral due to karstification. This permits growth of only the most desiccation-tolerant phototrophs, which include many cyanobacteria (Potts, 1994). Tolerance of high salinity is presumably required by such organisms at the last stages of desiccation, when salinities exceed 100 ppt. However, the high reflectivity of bare limestone surfaces means that less solar radiation is adsorbed than by darker rocks, so the water is cooler and the evaporation rate is retarded in comparison with other rocks. Whether this difference significantly influences colonization and growth is unknown.

Rainwater dissolves limestone and the water becomes alkaline due to the dissolution of calcium carbonate; this may occur directly as a result of contact with the precipitation or indirectly after passage underground. The concentration of dissolved inorganic carbon (DIC), containing carbonate, bicarbonate and free CO_2 is raised, as is the concentration of calcium. In the marine littoral, rainwater is of little importance, but seawater also contains high levels of bicarbonate and calcium (ca 2 mM) and may be supplemented by CO_2 - rich groundwater. Since their cellular calcium requirement is low, cyanobacteria growing on limestone may be responding to elevated levels of DIC, and perhaps also the high pH, rather than Ca.

Part of the Ca requirement which does exist may be extracellular (Somers and Brown, 1978; Smith and Wilkins, 1988).

The effects of pH can be investigated in limestone streams, where its value increases on passing downstream as a response to degassing without any marked change in dissolved inorganic carbon (DIC) (Pentecost, 1992c) In Waterfall Beck, England, the pH of limestone groundwater can be followed from its source for 240 m downstream. The pH starts at 7.62 (average over 1 year) and rises on passing downstream due to CO_2 evasion. In one study the value rose to 8.25, equivalent to a decrease in H^+ of nearly 80%, but with decreases in DIC and HCO_3^- of only 13% and 10%, respectively. Over this stretch of stream the community of *Phormidium* and *Schizothrix* spp. showed little visually obvious change, so that an order of magnitude change in H^+ had no measurable effect.

A study (Davis and Rands, 1982) on the growth of *Hapalosiphon intricatus* in a Florida cave in relation to the mineral composition of the underlying limestone found that the phosphate concentration in the limestone was 2.6 $\mu g\ g^{-1}$ (= 0.85 μg P g^{-1}), while NO_3 ranged from 4 - 27 $\mu g\ g^{-1}$ (= 1 - 6 μg N g^{-1}) and NH_4 from 4 - 13 $\mu g\ g^{-1}$ (= 3 - 10 μg N g^{-1}). In solutions in equilibrium with limestone samples, the phosphate concentration ranged from 5 - 20 $\mu g\ L^{-1}$ (= 1.6 - 6.5 μg P L^{-1}). The authors suggested that the combined nitrogen could have been derived partly from seepage, absorption of ammonia from the resident bat population and nitrogen fixation by the *Hapalosiphon*. A study (Ferris and Lowson, 1997) on vertical dolomitic limestone cliffs in Ontario showed that rock in the vicinity of an endolith community was enriched in some elements (e.g., P, Ba, Zn, Pb) and depleted in others (e.g., Mg, Fe, Cu). The authors interpreted this as evidence that the endolithic microorganisms were playing an active role in biochemical cycling of nutrient and trace elements at the site. However, caution is needed in interpreting the data from this and other analyses, because of the possibility that the organisms might have colonized selectively those parts of the limestone with a particular element composition.

Many of the epilithic forms of terrestrial surfaces and ones wetted only intermittently probably acquire some of their nutrients from rain, though critical studies appear to be lacking. As many industrial, urban and agricultural regions receive a high input of atmospheric N (Bobbink et al., 1998), but a much lower input of atmospheric P, it seems likely that the growth of many epilithic cyanobacterial communities may have become less dependent on dinitrogen fixation for their source of N, but increasingly P-limited.

Features of limestone environments which have a particular influence on cyanobacteria, and which are considered in more detail below, include:

i) Rapid erosion through dissolution;
ii) Water flowing over limestone becomes enriched in dissolved carbon dioxide (mainly bicarbonate) and calcium due to this dissolution;
iii) Iron and phosphate may be removed from solution by the alkaline, calcium-rich water and this may limit growth;
iv) Calcium carbonate crystals can be etched by cyanobacteria growing endolithically;
v) Bare limestone possesses a high reflectivity, thus reducing the amount of solar radiation absorbed at the surface.

III. Environments and their Characteristic Communities

A. Terrestrial

1. Overview

Terrestrial limestones include diverse morphologies such as cliffs, flat pavement and emergent pinnacles. Such surfaces are mostly dominated by cyanobacteria and lichens in varying proportions. The phototroph of the lichens is usually a green alga, but cyanolichens are sometimes important (see below). Cyanobacteria appear to be the dominants of subtropical and tropical limestones, where *Tolypothrix byssoidea* (Fig. 1) is perhaps the most widespread species, while lichens are the typical dominants of drier limestones in temperate regions. However, there are many exceptions to this general picture. For instance, guano deposition favours the presence of lichens (AP, unpublished data). This influence of guano is shown in the striking contrast provided by different atolls in the Indian Ocean. Much of the surface of Aldabra (Whitton, 1971), Astove and Farquhar not covered by higher plants is covered by free-living cyanobacteria, whereas the surface of St. Pierre, which has been mined extensively for guano, is largely covered by lichens (Whitton and Donaldson, 1977). The explanation for this difference is still unclear.

Although it may not explain the differences beween cyanobacterial and lichen dominated communities, a key feature for cyanobacteria is certainly the length of time during the year in which the communities are

Fig. 1. Extensive areas of flat limestone (platin) on Aldabra Atoll largely bare of higher plant vegetation are covered by a cyanobacterial community dominated by *Tolypothrix byssoidea*.

moist enough to carry out metabolic activity. In many situations, such as the communities on cliffs described below, it is possible to compare a range of cyanobacterial communities associated with increasing periods of moist or, in some cases, temporarily submerged conditions.

Many studies have reported that cyanobacteria can be important in the weathering of terrestrial limestone surfaces, often in association with other organisms, such as the combined effects of cyanobacteria and lichens reported for hillsides in the Negev, Israel (Danin and Garty, 1983). Weathering due to cyanobacteria involves a variety of processes, including the growth of endoliths (Chapter 13). Sometimes, the growth of endoliths may be considered opportunistic, making use of cracks and holes already present, such as with the endolithic community dominated by *Gloeocapsa* in the dolomitic limestone studied by Ferris and Lowson (1997: see Section II above). However, boring by cyanobacteria may also play a role (Viles, 1987). It is far from clear why some limestone surfaces are more subject to boring than others, but it seems likely that availability of nutrients, especially phosphate, may be an important factor. Presumably it is only energetically possible for characteristic boring species, such as *Plectonema terebrans*, to carry out the metabolic activity required for etching limestone under especially favourable nutrient conditions.

Deposition may also occur, sometimes in close proximity to erosional processes. For instance, Viles's (1987) study of cyanobacterial boring on terrestrial surfaces of Aldabra found that there was also precipitation of micrite around the filaments, sometimes forming a crust several mm thick. In the case of *Iyengariella endolithica* in a spring-fed lake in N-E. Mexico, the organism was often boring into the rock underneath a zone of active travertine deposition (Seeler and Golubic, 1991). Deposition and erosional processes are treated more fully in Sections V and VI, respectively. Two types of terrestrial environment are described in more detail here.

2. *Tintenstriche*

Colourful vertical streaks of cyanobacteria may be found on steep cliffs worldwide (Plates 26a, 30b) and are associated with temporary water runnels, which may be considered as extreme examples of ephemeral streams. The streaks develop on many lithologies, but hard impervious rocks are especially favoured, as these tend to provide stable cliff faces which permit water to flow, rather than to be absorbed. These tintenstriche (= ink streaks) were first described in detail by Jaag (see his 1945 review for references), who found that the cyanobacterial community differed markedly according to rock lithology. Luttge (1997) provides a recent introduction to the topic. However, these communities have not yet been defined climatically (number of wet days, temperature range, irradiance), nor have they been analysed according the water status (period wet enough to metabolize and period influenced by flowing water with any associated nutrients).

In spite of Jaag's detailed floristic account for the European Alps, the only attempt to describe tintenstriche communities objectively is the limited attempt by Allen (1971). A few researchers have, however, adapted the Braun-Blanquet terminology developed for higher plants to describe cyanobacterial associations, the most comprehensive being Golubić's (1967) account for the limestone of the Dinaric Alps. He recognised several distinct associations which appear to reflect two major variables, water and light. Under high irradiance, with increasing 'wetness', the associations changed from a *Scytonemo gloeocapsetum* to a *Tolypothricetum byssoideae* and finally a *Dichothricetum compactaea.* Characteristic species of the first (most xeric) association include *Scytonema myochrous*, *Gloeocapsa compacta, G. kuetzingiana* and *G. sanguinea*, while the wettest sites had *Dichothrix compacta, Phormidium autumnale* and *Xenococcus kerneri.* Under low illumination, a suite largely dominated by *Schizothrix* spp. occurs. The authors have observed elements of these associations in the United Kingdom and similar communities occur elsewhere (e.g. Fjerdingstad, 1965). The study of tintenstriche adds to the information on cyanobacterial water relations from other types of environment (Chapters 13, 17). For example, *Gloeocapsa calcarea*, after remaining in a desiccator for 6 months, began to fix carbon soon after rewetting (Pentecost, 1992b), while the *Schizothrix* species of permanently wet tintenstriche appear much less tolerant. Those species with the thickest, and often pigmented, sheaths occur at the driest sites, while thin-sheathed species occur in streams. The role of the yellow-brown sheath pigment, scytonemin (Garcia-Pichel and Castenholz, 1991), in protecting against ultraviolet light and the possibly similar role of the pH-sensitive gloeocapsin are described in Chapter 21, Section II A 2.

3. Caves

At the thresholds of limestone caves, cyanobacteria must compete for light with other algae, bryophytes and ferns, but in the deepest recesses of caves they are usually the sole phototrophs. Water relations in caves are as important to growth and colonization as they are in the open air and some caves are surprisingly dry, supporting virtually no phototrophs. However, most caves, at least in Europe, are damp with their threshold walls covered with a green algal felt.

Coccoid cyanobacteria seem to dominate most limestone caves, where they form a crust up to 1 mm thick (AP, unpublished data). In Croatia, the most frequent species have been reported (Golubić, 1967) to be *Aphanocapsa grevillei, Chroococcus turgidus* var. *spelaeus* and *Gloeocapsa kuetzingiana.* Similar associations have been noted in the UK (authors, unpublished data) and China (Tian and He,1996) and probably occur worldwide. However, a study of species diversity in a very old cave in Israel showed that the proportion of coccoid to filamentous forms decreased as irradiance got less (Vinogradova et al., 1998). Twenty genera and 42 species were found in the cave and generic and species diversity both decreased with decreasing irradiance.

Cave *Gloeocapsa* spp. have been found where the light intensity is as low as 1 lux (ca 0.02 μmol photon m^{-2} s^{-1} PAR (Cox et al., 1981). Cell diameters average 1 μm and at this low irradiance would experience a flux of 10^4 photon s^{-1} PAR, which, with an estimated photosythetic efficiency of 30%, would result in a cell division time of 20-25 days. The cells of cave-inhabiting cyanobacteria contain densely packed thylakoids (Cox et al., 1981; Couté, 1982, 1988; Couté and Bury, 1988), a clear adaptation to low light levels. Unlike the low light conditions to be found in deep water, spectral light quality in caves is practically invariant with illumination and chromatic adaptation has not been observed.

Most of these cyanobacteria can also be found on shaded limestone cliffs, but a few appear to be strict cavernicoles. Two species of *Geitleria* are known from caves, where they may be recognised as a delicate ashen felt on cave walls. *G. calcarea* is widely distributed in warm climates (Friedmann, 1979), where it may be found at very low irradiance (Leclerc et al., 1983). Colonies of *Scytonema julianum* form a similar felt, but this species is more widely distributed and is not restricted to cave entrances. In both cases, the unusual pale violet-grey colour is caused by calcification. Calcite is deposited

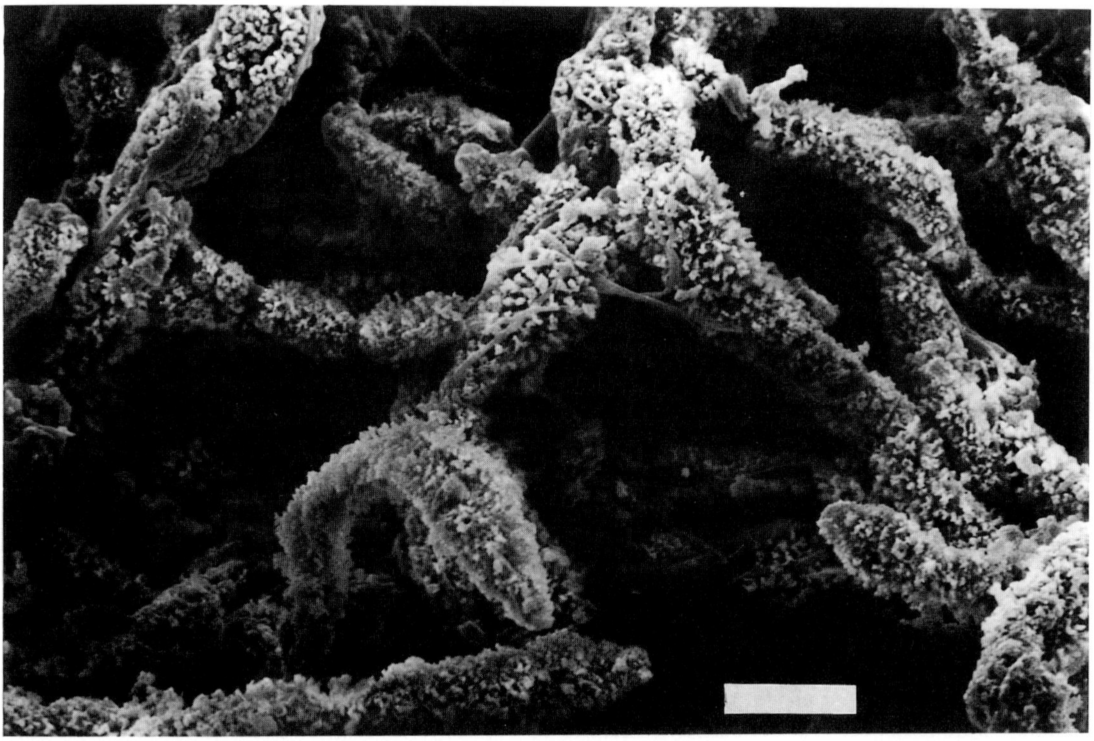

Fig. 2. Scanning electron micrograph of *Scytonema julianum* from Spring Caves, Malham, England. The sheaths are covered with spiky, occasionally dendritic crystals of calcite. Air-dried sample, Au-Pd coated. Bar = 20 μm

on the filaments as dense, radiating microcrystals forming a layer up to 30 μm thick (Fig. 2).

The mineralisation of *Geitleria* appears to be more regular than that of *Scytonema* (Friedmann, 1979; Couté and Bury, 1988) and this has led to the suggestion that it is biologically controlled. These encrustations may however simply be the result of steady evaporation of carbonate-rich water seepages around the filaments. Abiogenic calcite nodules are common at cave entrances and needle-fibre calcites, morphologically similar to these cyanobacterial calcites, are a common feature of cave seep zones (Verrecchia and Verrecchia, 1994). Nevertheless, the regular arrays of microcrystals occasionally found on *Geitleria* make a compelling argument that some form of biologically-induced control is taking place.

Knowledge of cave cyanobacteria is at best rudimentary, yet caves provide an excellent natural laboratory for the investigation of changing illumination under conditions of almost constant temperature. They await further exploitation.

B. Freshwater

The freshwater environment appears the least stressful of calcareous environments, but this is only true where there is permanent water cover. There are a number of records of lakes with an extensive bottom cover of firm stromatolitic structures where cyanobacteria appear to play an important role in their formation. *Rivularia* appears to be the most widely quoted genus (e.g. Loch Borrallie in N-W. Scotland: A. Donaldson, pers. comm.), but coccoid forms are predominant in by far the largest structures known, the 40 m high, tower-like structures in alkaline ($pH > 9.7$) Lake Van, eastern Turkey (Kempe et al., 1991). Three types of community dominated by cyanobacteria are associated with recent high-magnesium stromatolites in landlocked pools on Aldabra Atoll (Braithwaite et al., 1989). Although the cyanobacteria were not calcified, the communities probably played an important role in formation of the stromatolites, with calcification beginning several millimetres below the living cells. The three communities were dominated by:

1) *Phormidium*, *Lyngbya* and *Pleurocapsa*;

2) *Entophysalis*, *Lyngbya* and *Pleurocapsa*;
3) *Pleurocapsa*, *Dichothrix gypsophila* and *Schizothrix*.

Analysis of the processes taking place in the calcifying biofilms of three alkaline lakes (Satonda Crater Lake, Indonesia; Pyramid Lake, USA; Lake Nouertu, China) showed that calcium carbonate formed only rarely around the cyanobacterial filaments (Arp and Reitner, 1998). Extracellular polymers, largely formed by cyanobacteria, apparently retarded precipitation in comparison with surfaces poor in such polymers. Precipitation occurred as the polymers became degraded, perhaps due to the release of Ca^{2+}.

There are also many records of pools and fens fed by calcareous springs and groundwater, where the calcareous deposits, which are often associated with cyanobacteria, remain largely unconsolidated, at least in the short term (Section V C). The Florida Everglades provide a well-known example. Here there are extensive areas of highly calcareous water, with floating or attached masses of phototrophs, whose distribution with respect to phosphate gradients has been investigated in some detail (McCormick and O'Dell, 1996; McCormick et al., 1996; McCormick et al., 1998; Newman et al., 1999). Cyanobacteria (coccoid forms and *Scytonema*) are often abundant, with some of the species becoming calcified, giving rise to a type of stratified marl (Gleason, 1972). What may have been quite similar communities appear to have been responsible for layers in some ancient limestones, such as the botryoidal concretions in the Kirkton Limestone of West Lothian, Scotland (Walkden et al., 1994). A slightly different modern example of extensive deposition of carbonate without its consolidation into a firm layer is the laminated layers known as kopara, which form on the bed of some lakes on Polynesian atolls (Defarge et al., 1994a). The gelatinous sediments consist of alternating thick, red organic-rich layers and thin, white, carbonate-rich layers, which apparently result from alternating periods of saline and fresh water in the lakes, together with the associated communities of cyanobacteria and other microbes.

Cyanobacteria are frequent on ancient limestone surfaces in shallow water, but appear to be infrequent on limestone surfaces in deep water, perhaps because the rocks tend to be covered by fine sediment. Where low irradiance regimes exist close to the surface, occurring within the rock and on the undersides of pebbles, such environments are often dominated by filamentous Oscillatoriaceae (Pentecost, 1978).

Cyanobacteria are often prominent in calcareous situations where the water availability undergoes rapid and unpredictable change such as the littoral of lakes, stream edges, temporary runnels (Jaag, 1945; Golubić, 1967) and pools (Donaldson and Whitton, 1977a, b). Such communities intergrade with the terrestrial ones described above, with their adaptations to desiccation and high insolation, such as thick pigmented sheaths (Chapter 21). Even under permanent flow, abrasion and substrate dissolution can be rapid, making endolithy an advantage (Pentecost, 1992a). Overall, erosion predominates in streams and deposition in lakes, but in streams long periods of deposition may interrupt the erosional cycle, even in upland, fast-flowing sections, where travertine is deposited. Dated travertine sequences have shown that deposition on waterfalls may continue uninterrupted over several centuries (Fabré and Fiche, 1986). Examples of streams which are sufficiently calcareous for travertine deposition to be widespread are described in Section V C.

C. Marine

The marine littoral and sublittoral provide varied environments, which have been described in Chapter 4, together with the processes leading to carbonate trapping and deposition. The following section focusses on endolithic cyanobacteria in limestones and their role in erosion. Typically the organisms are euendoliths - species penetrating the limestone rather than making opportunistic use of cracks. They are common and often abundant (Le Campion-Alsumard, 1991) and their erosional effect is high compared with that of endolithic eukaryotic algae and fungi (Vogel and Glaub, 1998). Bioerosion rates exceeded 200 g m^{-2} yr^{-1} on micritic limstone at two leeward sites on the Bahamas (Vogel and Glaub, 1998). The organisms most frequently reported are *Hyella* spp., *Solentia*, *Plectonema terebrans*, *Mastigocoleus testarum*, *Kyrtuthrix* and *Brachytrichia*. Records for the last two are largely confined to the tropics and subtropics, but it is uncertain whether this is due to the surprising lack of studies on temperate limestones. There is considerable evidence for zonation of communities in the littoral and sublittoral. In the case of sublittoral communities, this zonation is usually attributed to photic conditions (Vogel and Glaub, 1998), but the possibility should be kept in mind that differential effects of grazing and nutrient conditions may also play a role.

The colonization of exposed limestone surfaces is usually rapid, provided that other, already colonized, limestone surfaces occur in the vicinity. Such

Table 1. Frequently recorded cyanobacteria of limestone surfaces in freshwaters, especially those where travertine deposition is taking place. Ecological generalizations are based largely on observations in northern and central Europe. The orders listed are those of Castenholz and Waterbury (1989). M, meteogene; T, thermogene. 1-3, scale indicating frequency in particular category, with 1 = infrequent to 3 = very widespread.

Species	M	T	Notes	References
Chroococcales				
Aphanocapsa grevillei	2	-	Small discrete colonies among filamentous microbes. No calcification within sheath	Grüninger (1965)
Gloeocapsa calcarea	2	-	Widespread; as above	Pentecost (1981)
G. rupestris	1	1	Superficial crusts on sites liable to desiccation	Pentecost (1985b)
Synechococcus elongatus	-	2	Sometimes providing nuclei for aragonite deposition	Pentecost (1990c)
S. minervae	-	1	Distinctive species with elongate cells, not known to calcify	Copeland (1936); Pentecost (1990c)
Pleurocapsales				
Hyella fontana	3	-	Isolated filaments may behave as endoliths in slowly depositing travertine	Golubić (1967)
Pleurocapsa minor	2	-	Probably sometimes endolithic, calcification occasionally reported	Krumbein & Potts (1979) Ordonez & Garcia del Cura (1983)
Oscillatoriales				
Homoeothrix balearica	1	-	Large species forming hard yellowish nodules. Heavily calcified	Pentecost (1988) Winsborough et al. (1994)
H. crustacea	3	-	Well defined yellow-brown colonies 1-3 mm in diameter. Heavily calcified (Plate 16d)	Pentecost (1988)
H. janthina	1	-	Weakly calcified species characterised by the 1-2 µm filaments and dark green colour *en masse*. Several other calcified species are known	Casanova (1994)
Lyngbya martensiana	3	-	Often associated with *Phormidium incrustatum* and distinguished by wider (6-8 µm) trichomes	
L. valderianum	2	1	Usually associated with *P. incrustatum*, but narrower (2- 4 µm) trichomes	Symoens & Mallaise (1967) Pentecost et al. (1997)
L. vandenberghenii			Small rivers with hard water under low flow; under such conditions the combination of the organism and a thin carbonate deposit give the surface a chacteristic milky blue appearance	Symoens & van der Werff (1951) Holmes & Whitton (1981)
Phormidium calcicola	1	-	Similar to *P. incrustatum*, but apparently less common	Kann (1973)

colonization requires the formation of small cells from non-filamentous forms or hormogonia from filamentous forms. Al-Thukair and Golubic (1991) showed that the small cells (baeocytes) of *Hyella immanis* were motile and phototaxic, features which seem likely to be characteristic of the reproductive structures of all euendoliths. Agitation of laboratory cultures of this species enhanced the frequency of borings and the initial boring rate, but had no effect on the continuing boring activity.

Although removal of the superficial parts of the thallus by grazing and erosion is a key problem for most endolithic cyanobacteria, sometimes the organisms grow in situations where calcium carbonate is increasing at the surface, as shown for *Plectonema terebrans* in crustose red algae (Ghiradelli, 1998). Both live and dead algae were included in Ghiradelli's study, but Le Campion-Alsumard et al. (1995) showed that the endolithic communities of live and dead coral on Moorea, French Polynesia, differed. The endoliths of live coral are selected for positive phototropism and fast growth to compete with the accretion rates of the coral skeleton; as long as the coral remains alive, the endoliths are protected from grazing. The main endolith of the coral *Porites lobata* on Moorea was the green alga *Ostreobium quekettii*. Dead and denuded parts of the coral skeleton were colonized at the surface and bored inward by a succession of euendoliths, starting with the green alga *Phaeophila dendroides*, followed by *Plectonema terebrans* and *Mastigocoleus testarum* and leading within 2 yr to a stable community dominated by *Ostreobium quekettii*. The endoliths in dead corals were grazed by molluscs, echinoderms and scarid fish. Such grazing is important for the maintenance of the

Table 1 (continued)

Species	M	T	Notes	References
P. incrustatum	3	-	Preference for shaded sites where it forms 'freshwater stromatolites'	Fritsch (1949) Pentecost (1995b) Freytet & Plaziat (1996)
P. laminosum	-	3	Abundant on travertine terraces where forms smooth green strata which may become calcified	Pentecost et al. (1997)
Plectonema tomasianum	1	-	Superficial growths, rarely encrusted	Glazek (1965)
Schizothrix affinis	2	-	Forms a brownish nodular crust	Caudwell (1987) Pentecost (1987)
S. calcicola	3	-	A 'cluster' species for other calcified *Schizothrix*	Allen (1971) Pentecost (1978)
S. coriacea	2	-	Found as lightly encrusted brownish strata: Plate 16e	Golubić (1967)
S. fasciculata	3	-	Pale brown well calcified nodular colonies often in sites prone to occasional desiccation. There are several other calcified species	Golubić (1967)
S.rubella (syn *S. lardacea)*		2	Forms pink-red thin crusts on intermittently moist limestone rocks (Plate 16c)	
Spirulina labyrinthiformis	-	2	Loose flocs in thermal effluents usually containing sulphide. Usually uncalcified	Pentecost (1995a)
Nostocales				
Calothrix parietina	3	2	Frequent at interface between freshwater aquatic and terrestrial. Sometimes on oncoids, where sheath calcification has been reported.	Pentecost (1989)
Dichothrix gypsophila	2	-	Similar habits to above, tolerant of high salinity. Other species grow on thermogene travertines, usually calcified	Fjerdingstad (1957)
Nostoc sphaeroides	2	-	Superficial, usually uncalcified growths in summer	Parihar & Pant (1982)
Rivularia biasolettiana	2	-	Weakly calcified, merging into species below	Golubić (1967)
R. haematites	3	-	Dark brown heavily calcified colonies, with concentric zones of calcite	Obenlüneschloss (1991)
Scytonema crustaceum	2	-	One of several species forming dark superficial turfs on travertines, usually encrusted below.	Thunmark (1926)
Scytonema myochrous			Forms large and characteristic cushions in moist calcareous environments, trapping particles	Whitton et al. (1986)
Stigonematales				
Fischerella laminosus	-	2	May form a weakly calcified crust. Widely distributed in hot-springs	Pentecost & Terry (1988)

euendolithic community, as otherwise the surface becomes overgrown with an epilithic algal turf.

LeBris et al. (1998) reported on further detailed study on the colonization of blocks of *Porites* in lagoons in French Polynesia. The dynamics of the resulting epilithic and endolithic phototrophic communities differed. The density of the epilithic community and species diversity (cyanobacteria and green algae) increased with the degree of eutrophication and density of borers (sea urchins, parrot-fish); endolithic species constituted 34 - 59% of the community The highly polluted harbour at Faaa provided an extreme case, where the increased density of sea urchins (44 individuals m^{-2}) led to the complete loss of the test blocks within five years. In the oligotrophic waters, endoliths made up 66 - 85% of the community. Cyanobacteria were the first colonizers, followed later by *Ostreobium*. The authors suggested that the extent of epilithic covering or the richness of the endolithic flora could be used as a global indicator of the quality and health of a reef: the higher the ratio of endoliths to epiliths and the higher the diversity of the endoliths, the more healthy the reef.

IV. The Organisms

A. Free-living Species

A relatively wide range of genera and species has been reported from calcareous environments. For instance, over 180 species were reported from terrestrial surfaces of Aldabra Atoll (Whitton, 1971; Whitton and Donaldson, 1977; Potts and Whitton, 1980). In view of the fact that there are extensive areas of limestone in diverse climatic regions for which there have so far been no records of their cyanobacterial flora, let alone a modern attempt to assess their biodiversity, it seems likely that the world list of species will prove to be much higher. For instance, the massive limestone pinnacles found in various parts of the wet tropics, such as Mulu Park in Indonesia, have yet to be studied. Examples from (mostly European) sites with travertine deposition are listed in Table 1. These species may be broadly classified into two groups - those which calcify and eventually become incorporated into the deposit and those which have a more fleeting existence, are normally unmineralised and rarely if ever fossilize.

The possibility that calcification may provide a criterion of taxonomic value (Castenholz, 1989) has been discussed by Pentecost and Riding (1986), Merz (1992) and Merz and Zankl (1993). A conservative taxonomic viewpoint has been adopted in preparing Table 1, with calcifying forms sometimes recognized as separate species. However, molecular studies are required before it can be established whether the difference between a calcifying and an otherwise closely similar non-calcifying form really justifies recognition at the species level.

Most of the epilithic forms have sheaths and form obvious colonies. The heterocystous genera are mostly those with trichomes ending in hairs, an indication that they exist for much of the time under conditions of P limitation (Section VII A). Some of the larger colonies can be very long-lived. For instance, *Rivularia haematites* (Plate 16c), whose laminations (Fig. 6) appear to reflect annual changes, may reach ten years (AP, unpublished data). As the presence of calcite crystals does not appear to inhibit grazing, at least by molluscs, this raises the question whether there is some other factor which might favour these long-lived colonies. One possibility, so far not tested, is that the colonies might produce materials toxic to some of the likely grazers.

There is no evidence to suggest whether organisms of limestone environments are more or less susceptible to other types of biological interaction. However, several members of the Oscillatoriaceae become infected with a pathogen forming circular lesions which can severely affect their overall biomass on travertine (AP, unpublished data). *Phormidium incrustatum* is frequently attacked. The pathogen has not been identified but infected trichomes become bleached and overrun by bacteria.

B. Cyanolichens

These lichens are distinguished by their cyanobacterial photobiont and are a fairly successful group colonizing a variety of terrestrial and freshwater substrata. The lichenized cyanobacteria of limestones have been largely ignored by ecologists, yet, among the limestone macrolichens of the British Isles described in Purvis et al. (1992), almost 50% are cyanolichens, a significantly higher proportion than that found on most other rock types. Large thalli of *Collema* and *Leptogium* are common on hard limestones where there is plenty of moisture, such as on steep slopes facing away from the sun. In surface depressions, where water forms ephemeral pools, the cyanolichens *Lempholemma botryosum* and *Leptogium schraderi* often grow with free-living *Nostoc* (McCarthy, 1983). In more xeric environments, macrolichens tend to be replaced by microlichens with green algal photobionts. One might expect such sites to be favourable for cyanobacteria, but there are surprisingly few encrusting cyanolichens on limestone, except for members of the Pyrenopsidaceae, an uncommon and poorly known family of microlichens and members of the genus *Pyrenocollema* (Purvis et al., 1992). The common encrusting species in northern Europe, *Placynthium nigrum,* is not a member of this family and is in fact characteristic of damp rather than dry limestones (McCarthy, 1983).

The photobiont of all of these lichens apart from *Placynthium* (possibly *Dichothrix*) and *Pyrenocollema* (*Gloeocapsa*) is *Nostoc*. The *Nostoc* apparently always contains obvious heterocysts and thus presumably fixes nitrogen. The cyanobacteria seem to gain structural integrity from the mycobiont, allowing them to colonize otherwise unfavourable surfaces. For example, *Collema cristatum* grows in large circular patches on steep limestone as firmly attached thalli (Fig. 3), colonizing surfaces unsuitable for the free-living cyanobiont. Most of the large cyanolichens form minute, readily detached propagules containing both bionts (isidia) and these are dispersed by invertebrates or rain. A few produce ascospores which must contact free-living cyanobacteria before symbiosis proceeds. With respect to dispersion, limestone cyanolichens resemble most other lichens. The longevity of the lichens perhaps enables them to obtain a large proportion of their P from air-borne dust, which is not normally a significant source for free-living cyanobacteria.

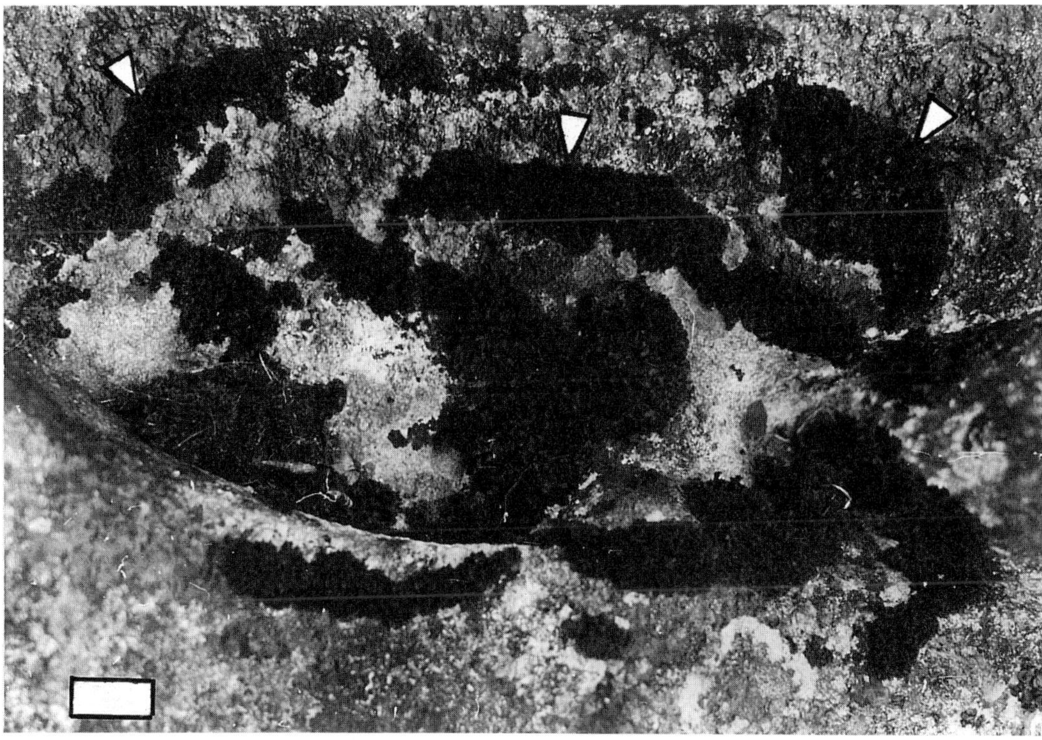

Fig. 3. The foliose cyanolichen, *Collema cristatum*, growing on hard limestone at an upland site in Northern England, where the mean annual temperature is 7°C. The lichen expands radially as a series of dark bands. The outermost band (left arrow) had not progressed during the previous 5 yr, but the fertile patch with apothecia (right arrow) and the inner band (centre arrow) both expanded by 2.1 mm yr^{-1}.

C. Biomass

There are surprizingly few quantitative data on cyanobacterial standing crop from limestone environments, but a few values (expressed as biovolume per unit area) are given in Table 2 to encourage others to record such data. Within this limited dataset, the highest value occurred in a mat of *Phormidium ambiguum*, where the cyanobacteria grew in a thick uncalcified layer exposed to direct sunlight. In contrast, the endolithic and travertine-associated communities had a biomass more than ten times less, presumably because access to light, and, perhaps nutrients was restricted. A similar low biomass is apparent in the uncalcified tintenstriche community, where the algae were influenced markedly by desiccation.

IV. Deposition

A. Processes

Calcium carbonate is deposited by cyanobacteria by two processes:

i) nucleation and precipitation on the cell wall/sheath;

ii) trapping followed by binding (Pentecost and Riding, 1986).

The processes are not mutually exclusive (Pentecost, 1990a), but cell wall precipitation is often referred to as calcification and is distinguished from trapping and binding, which is especially common in marine stromatolites (Chapter 4).

Calcification in the cyanobacteria differs in several ways from that in most other organisms. First, with rare exceptions the mineral phase is precipitated with a low level of organisation when compared with most other organisms, including eukaryotic algae (Lowenstam, 1986). Second, no intracellular structures have been found storing the precursors to

Table 2. Examples of biomass measurements of cyanobacterial communities (as cellular biovolume per unit area). A waterbloom is included for comparison.

Substratum	Location	Dominant Species	Biovolume ($mm^3 cm^{-2}$)	Reference
Meteogene travertine	Waterfall Beck, England	*Schizothrix calcicola* agg *Calothrix parietina*	0.27	Pentecost (1991)
Limestone tintenstriche	Malham Cove, England	*Gloeocapsa alpina* *Gloeocapsa kuetzingiana*	0.32	Pentecost (1982)
Endoliths on submerged limestone cobbles	Malham Cove, England	*Phormidium favosum* *Schizothrix perforans*	0.24	A.P. (unpubl. data)
Mat in warm effluent	Taff Wells, Wales	*Phormidium ambiguum*	4.1	A.P. (unpubl. data)
Waterbloom	L. Lucerne, Switzerland	*Oscillatoria rubescens*	2.4	Zimmermann (1969)

mineralisation, as is the case in, for example, the coccolithophorids (van der Wal et al., 1983). Cyanobacterial calcification is therefore unrefined, with a low level of biological organisation.

The water from which the precipitation occurs has been shown on numerous occasions to be supersaturated with calcite due to degassing (Usdowski et al., 1979; Dandurand et al., 1982; Herman and Lorah, 1987) and apparently unconnected with biological processes. Thus, calcite supersaturation, a prerequisite for calcium carbonate precipitation, need not necessarily be biologically driven, though the cyanobacteria provide a suitable surface for precipitation. Cave travertine (speleothem) often precipitates in the virtual absence of organisms and must be almost entirely abiogenic. On the other hand, freshwater cyanobacteria are often encrusted with calcite, frequently to the virtual exclusion of other phototrophs, and have been shown both to nucleate calcium carbonate *in vitro* (Pentecost and Bauld 1988; Thompson and Ferris, 1990; Heath et al., 1995) and to trap mineral particles (Riding, 1977).

Aqueous chemistry has an important influence on the type of deposit, even where cyanobacteria appear to play a role. A strontium calcite crust formed round many of the cyanobacteria in the discharge zone of groundwater from a serpentinic bedrock (Ferris et al., 1995). The stromatolites foming in the alkaline (pH > 9) Salda Lake, S-W. Turkey, consist of hydromagnesite (Braithwaite and Zedef, 1996); in this case deposition was not congruent with the extensive films of cyanobacteria and diatoms, although the authors considered that the organisms played a role in deposition of the mineral. In marine environments both calcite and aragonite may be formed, but predominantly the latter in the absence of biological activity (Bathurst, 1971).

There may be specific properties of the extracellular sheath which encourage bio-mineralization (Pentecost and Bauld, 1988) and the colony architecture of *Phormidium incrustatum* (Fig. 4; see also Fritsch, 1950) has been shown to contain calcifying microniches, which could be significant in the early stages of calcification (Pentecost, 1995b). It would be a mistake to dismiss biological control: it exists, but at a low level when compared with other biomineralizing systems.

Occasionally the deposits may remain unconsolidated and form a type of stratified marl, as in the Florida Everglades (Section II C). However, the accumulated calcium carbonate often persists, forming substantial build-ups in lakes and streams. The precipitated particles frequently become modified through the process of diagenesis (Bathurst, 1971) to form a dense limestone which may be a travertine (= tufa: precipitation initiated by CO_2 evasion) or calcrete (precipitation by evaporation). Cyanobacteria may play a role not only in travertine formation, but also calcrete formation (Verrecchia, 1994). Some travertine deposits attain a thickness of 200 m and extend over many hundreds of square kilometres. The degassing of CO_2 is assisted by turbulence and warming of bicarbonate-rich groundwaters (Dandurand et al., 1982). Photosynthesis by aquatic plants, especially cyanobacteria and eukaryotic algae, assists deposition by removing further carbon from the system and forcing the relation shown below to the right (Golubić 1973; Spiro and Pentecost, 1991).

$$Ca(HCO_3)_2 = CaCO_3 + H_20 + CO_2$$

Several well defined types of travertine occur, notably the *thermogenes*, which are basically hydrothermal deposits associated with hot springs and the *meteogenes* which are karst water deposits

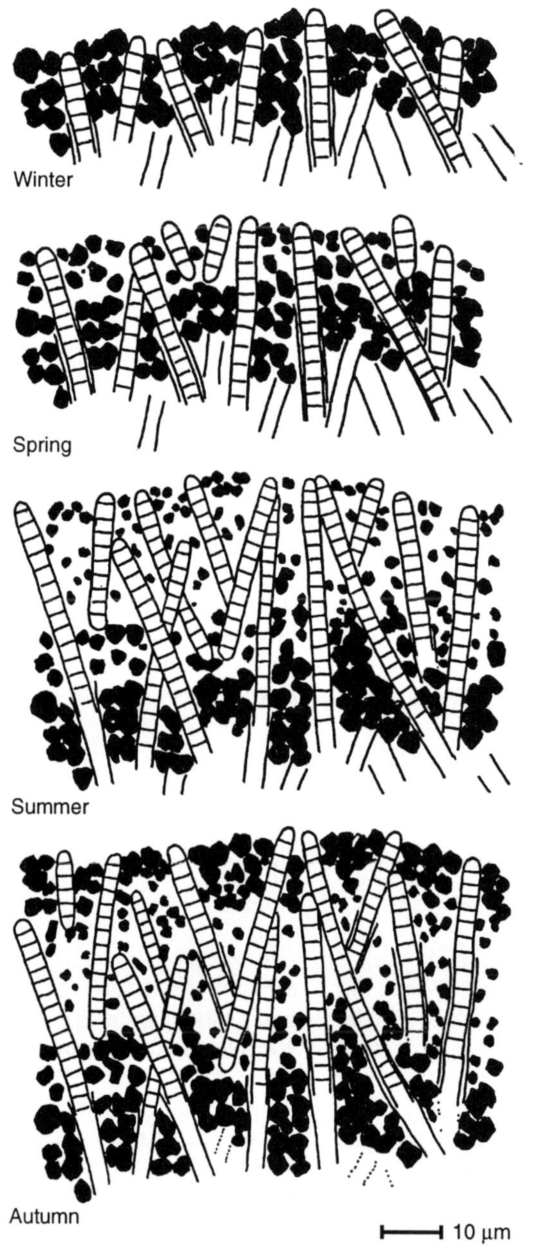

Fig. 4. Development of a laminate *Phormidium incrustatum* travertine over the course of one year in an upland stream in England. Calcite crystals are shown in black.
Winter The trichomes are oriented more or less perpendicular to the deposit surface and grow little or not at all while temperature and irradiance is low. Calcium carbonate continues to precipitate around the trichomes as a result of CO_2 evasion, forming a dense microcrystalline layer.
Spring The trichomes begin to grow upwards and also possibly downwards and small crystals of calcite (micrite) also form. At this time chironomid larvae often build tunnels in the surface layer and feed on the fresh cyanobacterial crop.
Summer This is usually the period of maximum growth with the higher temperatures, but growth may be reduced due to shading by riparian vegetation. Trichomes continue to grow upwards followed by release of hormogonia. Micrite is deposited as a result of both CO_2 evasion and photosynthesis, forming a loose spongy layer.
Autumn Trichome growth slows and another dense later of calcite begins to form and this continues through the winter.

formed from cooler waters. Many cyanobacteria find travertine a suitable environment for growth, and develop both upon and within the deposits (Table 1); almost all active travertines exposed to light are colonized. This close association of cyanobacteria with travertine has been recognized for more than a century (Pia, 1934).

B. Thermogene Travertines

The calcium carbonate in thermogene travertines forms by the reaction of hot, concentrated carbon dioxide solutions with buried carbonates in tectonically and geothermally active regions. As a rule, the deposition rate of thermogene travertines is about an order of magnitude higher than meteogenes due to the higher chemical loads and temperatures. The origins and their relationships with other continental carbonates have been reviewed by Pentecost and Viles (1994).

The thermogene travertines accrete much more rapidly (typically 10-80 mm yr^{-1}) than the meteogene travertines and, where the rate exceeds 100 mm yr^{-1}, cyanobacteria are usually rare (Pentecost, 1990 b,c). With temperatures in the range 30-70°C, one might expect a more rapid rate of cyanobacterial growth, at least over the lower part of this temperature range (Chapter 3). The scarcity of cyanobacteria in these situations may be a direct result of smothering by the accreting carbonate, but other factors could also play a part. Possible reasons include competition from chemautotrophic bacteria in sulphide-rich waters and the fact that a temporal variation in nutrients, which favours at least some species listed in Table 1, probably does not occur.

C. Meteogene Travertines

The calcium bicarbonate in meteogene travertines forms as a result of soil-generated carbon dioxide reacting with subsurface limestones. The commonest cyanobacterium of these travertines appears to be *Phormidium incrustatum*, which is widely distributed in Europe and North America (Pentecost, 1990b). However, although this and other *Phormidium* spp. are widespread and sometimes abundant, there is also often a diverse flora of other cyanobacteria. For instance, of the 36 non-diatom taxa found in two small, fast-flowing rivers in Austria (the calcareous Isar and partially calcareous Gschnitzbach), 29 were cyanobacteria and 10 of these belonged to *Chamaesiphon*).

Water relations are known to influence the distribution of cyanobacteria in travertine-depositing streams (Pentecost, 1982; 1985b), with coccoid forms (*Aphanocapsa, Gloeocapsa*) predominating rather than Oscillatoriaceae, where water availability is less. This also occurs in non-depositing waters flowing over limestone (e.g., tintenstriche) , but on travertine other factors come into play. As the carbonate accretes, photosynthesising organisms need to keep their cells illuminated. Irradiance declines rapidly below the carbonate surface and the photic depth rarely exceeds 3 mm (Pentecost, 1978). Therefore the cells must either migrate upwards or grow towards the light at a rate which is at least equal to the rate of travertine deposition. Does the cyanobacterial growth rate control travertine deposition or *vice versa*? The meteogene travertines of Europe (AP, unpublished data) accrete at 2-5 mm yr^{-1}, which would appear to be sufficiently slow to allow the cyanobacteria to keep pace. In laboratory cultures, trichomes of *P. incrustatum* can grow 5 mm in a few days, so growth should not be limited by carbonate deposition. However, growth rates of calcified colonies measured in the field are similar to those of the travertines themselves.

In view of its importance, the growth of *Phormidium incrustatum* will be described in some detail. The species forms dark green strata often completely covering the deposit surface in shady situations. The trichomes, which are 4-5 μm in diameter, usually occur in fascicles and often form slightly radiating bush-like structures up to 3 mm in diameter (Freytet and Plaziat, 1991, 1996). The fascicles are usually cross-cut with alternating light and dark laminations caused by differences in calcite crystal size (Fig. 4) and *Phormidium* biomass. During the spring and summer when *Phormidium* growth is rapid, small crystals of micrite (calcite crystals <10 μm diameter) are deposited among the trichomes providing a porous, sponge-like structure. During winter, when growth is minimal, a more dense, macrocrystalline layer is formed providing a thinner, lighter coloured lamina when viewed in reflected light. In Europe and North America, the couplets average 2-3 mm in thickness and have been used to measure travertine deposition rates (Pentecost and Viles, in press). During the warmer months, the trichomes of *Phormidium* rise above the travertine surface (Grüninger, 1965) and are presumably lost to grazers, abrasion and, if nutrient conditions are favourable, by hormogonium production. This species is frequently associated with a specialist chironomid fauna (Thienemann, 1934), which along with nematodes, grazes the trichomes.

Several morphologically similar *Phormidium* species are frequently found associated with *P. incrustatum*: *P. martensiana* and *P. valderianum*. They are distinguished largely by trichome diameter, which differs statistically. Kann (1973) attempted a revision of the *P. incrustatum* complex and

Fig. 5. Calcified colonies of *Dichothrix calcarea* from a shallow limestone pool (Thoragill, North Yorkshire, England). Bar = 1 cm

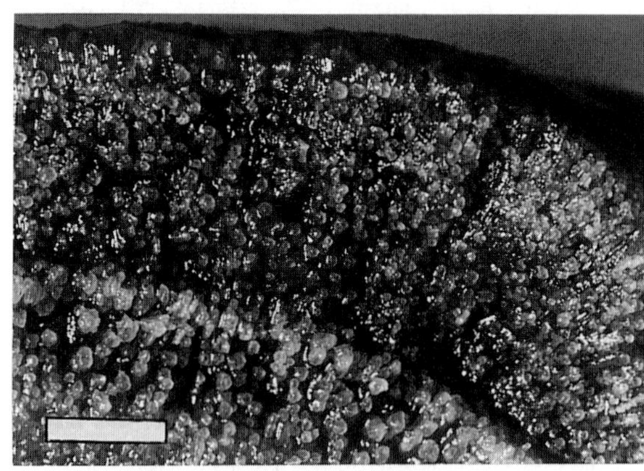

Fig. 6. Section through a colony of *Rivularia haematites* (Waterfall Beck, North Yorkshire, England). Many calcite crystals can be seen embedded in the mucilage. The dense calcified band below was formed in winter and the upper, less mineralized layer was formed during spring and summer. Bar = 1 cm

concluded that there were only two species, *P. incrustatum* and *P. calcicola,* the latter characterised by the presence of a calyptra, but the taxonomic status of this character has yet to be verified. Other forms of *Phormidium* are known from thermogene travertines, where they commonly thrive at water temperatures of 35-55°C. *P. laminosum* is a widely distributed species, forming smooth, coriaceous strata where the travertine is not deposited too rapidly (Pentecost et al., 1997).

Cyanobacteria with sheaths containing two or more trichomes can be abundant on travertines (e.g. Golubić, 1967; Szulc and Smyk, 1994), especially those occurring at lake edges and exposed areas prone to desiccation (Kann, 1959). Here, *Schizothrix* species often predominate, which build yellow-brown calcified sheets (*S. calcicola* and *S. fasciculata*) or nodules (*S. fasciculata* fo. *semiglobosa, S. pulvinata*), the latter often associated with nodules of *Homoeothrix crustacea* (Pentecost, 1988). In the nodular forms (*S. fasciculata, S. pulvinata*) the trichomes form radiating bush-like structure a few millimetres in diameter. These belong to the subgenus Inactis and are often heavily encrusted with calcite. They are particularly common on travertines exposed to dripping or flowing water, or those forming at the edges of lakes, where they may persist for many years. Species forming leathery crusts (*S. coriacea, S. lateritia*) are often found peeling away from rock surfaces and are usually confined to areas of intermittent flow. These belong to the subgenus Hypheothrix and rarely persist. However, current concepts of this genus are dated and await clarification.

Concretionary forms are also encountered in other species of *Homoeothrix*, but more especially in *Dichothrix* (Fig. 5) and *Rivularia* (Fig. 6) Here the individual, approximately hemispherical colonies, may attain several centimetres in diameter and sometimes coalesce to form a calcified sheet. The colonies tend to be most obviously hemispherical in regions of slack water (Fig. 7), but assume a more flattened form where the current speed is high. This general type of life-form reaches its greatest development in fast-flowing travertine-depositing waters, where colony calcification offers clear advantages in providing a firm point of attachment and a hydrodynamically smooth surface.

Accumulation of travertine often occurs by the entrance to caves and this sometimes leads to spectacular asymmetric growths caused by preferential deposition of carbonate on the illuminated side (Golubic, 1973; Cox et al., 1989). Such phenomena are common in S-E. China, where

Fig. 7. Travertine deposition associated with cyanobacterial community (*Dichothrix, Scytonema myochrous*) in trickles adjacent to Al-Tanumah waterfall, near Abha, Saudi Arabia (Whitton et al., 1986).

the cyanobacteria are often associated with bryophytes forming similar structures (Wang et al., 1993). There are two possible reasons for greater precipitation under higher irradiance- biogenic precipitation associated with photosynthetic uptake of CO_2 or greater evaporative loss and hence more deposition. Sheath pigments would enhance radiation absorption and hence evaporation.

D. Oncoids

Oncoids, consisting of rounded and concentrically laminated travertine pebbles, are widespread in calcareous streams and contain a wide range of encrusting cyanobacteria. Most oncoids possess a crust of filamentous species belonging to the genera *Calothrix, Dichothrix, Homoeothrix, Phormidium, Rivularia* or *Schizothrix* (Schafer and Staf, 1978; Pentecost, 1989; Rott, 1991). Because the deposits are continuously moved by water currents they progress slowly downstream, enlarging as the go. They provide an unusual ecological niche where periods of partial burial oscillate with periods of carbonate deposition. At one site in northern England (Pentecost, 1989), growth studies have shown that oncoid travertine accumulation is seasonal and correlated with water temperature. The travertine was colonized throughout the year by *Calothrix parietina*, whose sheaths rapidly became mineralized with calcite. The seasonal variation in growth resulted in concentric laminations which were interrupted by episodes of partial burial. Individual oncoids moved up to 23 m downstream in one year, but did not increase in size above 25 mm. As the size of the oncoids increases at this site, they begin to stabilize and grow laterally, sealing them within the stream bed. The ultimate size attained was dependent upon the local hydrodynamics.

There are probably considerable differences in oncoid development at different sites. It is possible, for instance, that nutrient levels in the water on the lower side of the oncoids in a stream sometimes differ sufficiently from those on the upper side to have an influence on growth. In the case of very large (6-22 cm long) oncoids from the R. Alz, Bavaria, it was shown (Rott, 1991) that differences existed between the outer 4-5 mm (*Homoeothrix crustacea* and *Schizothrix fasciculata*) and the remaining inner part of the oncoids (*Rivularia*). This may reflect environmental changes during the period of oncoid development, such as a possible increase in aqueous combined nitrogen.

VI. Colonization, Succession and Weathering in Terrestrial and Freshwater Environments

Cyanobacteria are among the first phototrophs to colonize clean limestone surfaces. Freshly exposed travertines and substrata such as Iceland spar calcite soon acquire a felt of narrow filamentous cyanobacteria, especially *Phormidium* and *Schizothrix*. Many of these cyanobacteria are endoliths. Golubic et al (1981) succinctly classify endoliths into cryptoendoliths, chasmoliths and euendoliths. The cryptoendoliths and chasmoliths live between mineral grains and within fractures respectively. The euendoliths etch minerals forming a series of deep furrows or tubes, and when intense may lead to fretting and mineral fragmentation. Colonization, however, may take several weeks or many months (Fig. 8) in limestone streams, presumably dependent upon the type of surface (rough or smooth), temperature and light conditions (Pentecost, 1992a).

On drier limestones such as steep cliffs receiving only temporary inundation, the communities consist mainly of coccoid forms and often take a long time to colonize. Garty (1990) found that *Gloeocapsa* took several years to colonize calcareous walls in Canada. At Malham Cove, England, an 80-m high south-facing limestone cliff, coccoid cyanobacteria form dark tintenstriche rarely exposed to water. One streak was cleaned of its cyanobacteria (*Chlorogloea microcystoides*, *Gloeocapsa* spp.) and after ten years of observation the biomass was barely 10% of an adjacent uncleared control area. In the case of a tintenstriche receiving regular wetting (approx 250 days per year), the biomass of a *Schizothrix* community showed complete recovery within one year emphasising the importance of water availability (Pentecost, 1992b). Similarly, substantial colonization of limestone was noted after a period of one year on Aldabra Atoll (Viles, 1988) on frequently wetted sites although some drier sites were still not fully recolonised after 16 years (H.A. Viles, unpublished data).. The significance of water was examined by Viles (1995), who provides a conceptual model of the efficiency of biological weathering by microorganisms as a function of water stress, with epiliths becoming more important than endoliths as water availability increases.

The speed of colonization on submerged limestone by cyanobacteria suggests that the 'climax community' develops quickly. This was certainly the case in the *Schizothrix* tintenstriche noted above,

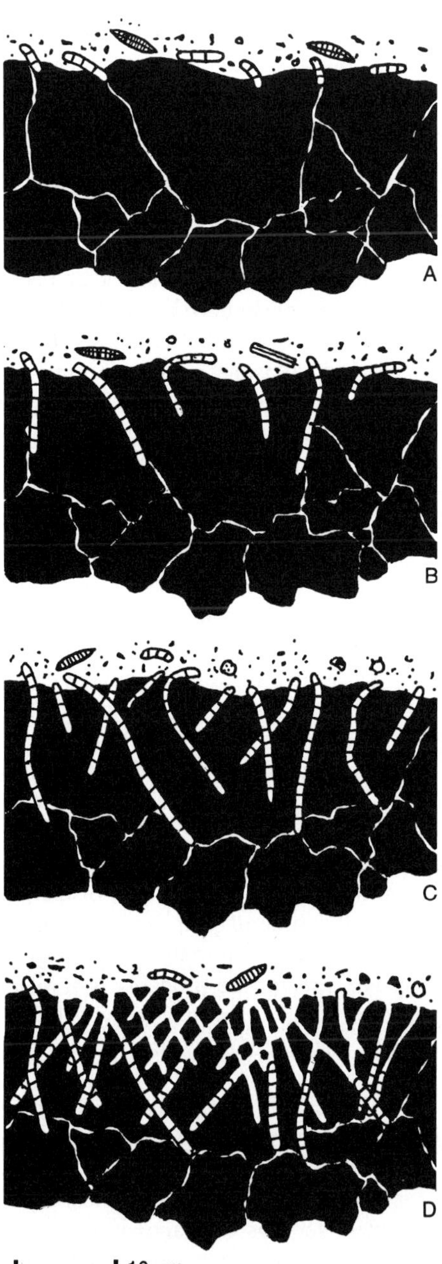

Fig. 8. Colonization and 'micritization' of a dense crystalline limestone as observed in a small hill stream (Waterfall Beck, England). The whole process takes about five years in this upland stream, whose mean water temperature is only about 7°C. Calcite crystals are shown in black.
A. On submersion the limestone acquires a thin aufwuchs containing mucilaginous cyanobacteria and diatoms.
B. *Schizothrix* species begin to etch the calcite crystals, usually following lines of weakness such as crystal boundaries.
C. Further colonization leads to deeper penetration.
D. Intense colonization reveals the cyanobacteria forming a criss-cross pattern guided by crystal planes, weakening the mineral and eventually breaking it up into micrite.

where all but one species (*Stigonema ocellatum*) had recolonized after a year. Since tintenstriche communities rarely possess much structure and zonation, rapid stabilisation is not unexpected. However, Turian (1983) reported an example of succession, beginning with *Gloeocapsa* spp. The surface weathering of submerged limestone by etching is shown diagramatically in Figure 6. Here, intense fretting of calcite crystals in the limestone leads to fragmentation and weakening of the surface. The process is sometimes termed biomicritization, as the larger calcite crystals are reduced to micrite-sized grains (< 10 μm diameter).

In spite of the much slower speed of colonization of terrestrial surfaces, cyanobacterial growths may eventually form aesthetically unattractive streaks and patches on surfaces incorporating limestone or covered with a limestone whitewash. Detailed lists of cyanobacteria on limestone building materials have been recorded (Schlichting, 1975; Strzelczyk, 1981; Anagnostidis et al., 1983). Although such growths can develop on a variety of building materials, the problem appears to be greatest where limestone is employed. The use of ground up coral to whitewash coastal building in the coastal tropics is an important factor contributing to the destruction of coral reefs in some regions such as parts of South Asia. It seems likely that colonization is much more rapid where the limestone wash is relatively rich in P, as might be expected to occur where recent coral is used as source material. There are also many reports of cyanobacteria contributing to degradation of monuments.

While most terrestrial rocks are weathered and denuded by a range of physical and chemical processes, terrestrial limestones are denuded largely by dissolution. Meybeck (1987) calculated rates of chemical weathering on different rock types from dissolved loads of rivers, finding the rate on carbonate rocks to be some 12 times the rate on granite. The weathering of rocks by microorganisms has been recognised for more than a century (Viles, 1995) and measurements of limestone weathering suggest that cyanobacteria are in some places quantitatively important in this process. In Yorkshire, UK, fragments of limestone cliff up to 2 g in weight were dislodged beneath a thick tintenstriche of *Gloeocapsa* and *Scytonema* spp., equivalent to an overall surface weathering rate of about 3mm/100 years (Pentecost, 1992b). In such cases it is probable that the dehydration and rehydration of the sheaths assisted in the loosening of the surface. Large cyanolichens such as *Collema cristatum* show a marked expansion on wetting and are also likely to assist in particle detachment on limestone. In addition to mechanical action, chemical etching is also likely to be significant. Danin et al. (1982) found that limestone and dolomite surfaces in parts of Israel were covered with pits containing cyanobacteria. The pits were 0.1-0.3 mm deep and colonized by *Gloeocapsa* species and lichens. It was suggested that the pits shielded the algae from solar radiation and it is likely that they were formed by dissolution. Viles (1987) also observed a limestone 'biorind' on Aldabra a few millimetres in thickness, where surface precipitation and dissolution processes were apparent.

Cyanobacteria can be harmful to human artifacts not just by the formation of ugly surface streaks, but because of weathering and other types of damage to the surface. Danin and Canova (1990) found that the growth of Pleurocapsales on limestone sculptures and walls in Israel accelerated the weathering rate to about 0.5 mm/100 years. Danin (1993) described in some detail the weathering of the marble Temple of Apollo at Didim, Turkey. Two of the three processes involved were due directly to cyanobacteria: exfoliation associated with Pleurocapsales in fissures and pits formed by cyanobacteria living in circular patches. The third process, the removal of marble crystals leads to the creation of depressions favourable to the growth a microbial community, again dominated by cyanobacteria. The endolithic growth increased the resistance of organisms colonizing the marble fountains of the Alhambra, Granada, to algicides (Bolivar and Sanchez Castillo, 1997).

VII. Influence of Temporal Changes in Nutrients

A. Seasonal

The dominant cyanobacteria of aquatic travertine-depositing sites often show marked seasonal changes in growth and calcite deposition. This is especially obvious in species forming multicellular hairs. Studies on small streams dominated by *Rivularia* in northern England (Plate 26a), and isolates from these streams, provide insight into what may be more general phenomena in limestone streams. The most detailed studies have been made in Upper Teesdale, where stream water is derived partly from limestone springs and partly from peat drainage.

A seasonal study of four streams (Livingstone and Whitton, 1984) showed that phosphate concentrations rose to a high level for a short period in spring, with organic phosphate an important component, but dropped to a low level for the rest of the year, the difference between spring and summer concentrations sometimes approaching three orders of magnitude Subsequent studies over a number of years (Whitton et al., 1998 and unpublished data) have confirmed this general pattern, though the difference between spring and summer phosphate is usually less marked and there are also occasional rises in phosphate concentration for short periods at other times of year, especially autumn. The reasons for the marked seasonal changes in ambient phosphate have not been investigated, but probably relate to the fact that almost all the phosphate input is from the peat rather than the springs and that most phosphate release from the peat occurs in early spring when competition from microbes and roots is weak.

The period of high phosphate concentration in the stream water leads to formation of hormogonia from older *Rivularia* colonies and their aggregation to form new colonies. Surface phosphatase activity of colonies is low at the time of high ambient phosphate, but high for much of the rest of the year (Whitton et al., 1998). Rates of nitrogenase activity are high in spring (BAW, unpublished data), but relatively low for much of the rest of the year. A comparison of nitrogenase activity and CO_2 fixation in summer (Livingstone et al., 1984) suggested that nitrogen fixation contributed only a minor part of the nitrogen requirement at this time of year, with aqueous nitrate probably being more important. Colonies in the Upper Teesdale streams are probably N-limited for a short period in spring, but markedly P-limited for most, if not all, the rest of the year.

The *Rivularia* (cf *R. biasolettiana*) colonies in Upper Teesdale calcify only weakly and relatively little permanent deposit is formed. However, it seems likely that seasonal changes in aqueous P influence growth features in more typical travertine-depositing sites, such as the laminations in oncoids mentioned above and the characteristic laminations of *R. haematites*. Measurements of cyanobacterial growth rates and nutrient budgets in Waterfall Beck, England, also suggested that for most of the year, nitrate-N is adequate, but phosphate may become limiting during summer (Pentecost, 1991; Spiro and Pentecost, 1991). No colony-forming axenic strain of a stream *Rivularia* has yet been brought into culture, but a number of laboratory studies have been made on an isolate of (non-calcifying) *Calothrix parietina* from one of the Upper Teesdale streams, which show that its nitrogenase and phosphatase activities respond similarly to changes in P status as *Rivularia* (Livingstone and Whitton, 1983; Livingstone et al., 1983; Whitton, 1987; Grainger et al., 1989). Changes in field populations of *C. parietina* in these streams probably reflect the smaller phosphate peaks occurring at various times of year and the organism therefore undergoes cycles of morphological changes more frequently than once a year. Oncoids with highly calcified *Rivularia* colonies at several sites in northern England (e.g. Plate 26c) assayed in summer for surface phosphatase also showed high activity, suggesting marked P limitation (J.M. Yelloly, E. Bresnan and BAW, unpublished data).

On a *Scytonema*-dominated xeric tintenstriche in northern England, the heterocyst density in the trichomes was found to be significantly higher than on an adjacent mesic tintenstriche and in nearby streams (Pentecost, 1985a). It was argued that in the xeric sites, nitrate availability was severely restricted due to the intermittent and reduced water flow when compared with the wetter sites. Nitrogen fixation could compensate for reduced nitrate particularly in situations where the tintenstriche remained damp for long periods of time due to very slow or intermittent flow.

B Long-term Effects

Eutrophication of travertine-depositing systems is known to be deleterious. Emig (1917) noted damage in an Oklahoma catchment caused by cattle grazing, and, in England, Edwards and Heywood (1960) found that carbonate encrustations disappeared below a sewage outfall, though the eukaryotic algal flora remained unchanged. Several stream sites are known where *Rivularia* colonies were once present, but now absent (BAW, unpublished data). Phosphate inhibition of calcite nucleation is generally regarded as the cause (Raistrick, 1949; Casanova and Lafont, 1985), the travertine crust being lost together with its associated community. However, enhanced phosphate would also be expected to eliminate organisms adapted to highly variable levels of ambient P, such as the hair-forming genera *Calothrix*, *Dichothrix* and *Rivularia* (Whitton, 1987). The disappearance of oncoids in Little Conestoga Creek, USA, which formed around colonies of *P. incrustatum* was attributed to pollution by acidic industrial effluents (Golubić, 1973; Golubic and Fischer, 1975).

The extent of pollution likely to lead to changes in travertine-deposition seems to vary between sites. In southern England, for instance, most travertine-depositing streams are subject to nutrient enrichment from farms or sewage treatment plants, but deposition remain active, including that associated with cyanobacteria (AP, unpublished data). These observations appear to clash with those of Edwards and Heywood (1960) and the sites in southern England merit more detailed study. Information on seasonal changes in concentrations of aqueous N and P and the N : P ratio is required to establish the limits of eutrophication which communities can withstand before travertine ceases entirely. Such information could be used to evaluate past environmental changes at sites where calcifying genera such as *Rivularia* have disappeared or travertine deposition as a whole has ceased. It is also important for managing land use in the catchments of sites where travertine is still being deposited.

The reverse process, the appearance of calcified (magnesian calcite) cyanobacteria at a site where calcification had previously been absent, was described for a brackish lake on Rangiroa Atoll by Defarge et al. (1994b). In spite of a decrease in the salinity of the lake, the authors argued against interpreting the change as a result of environmental change because a previous dominant species, *Phormidium crossbyanum*, persisted in a non-calcified state. However, it might equally well be argued that the arrival of new, calcified species reflected the fact that there had been a shift in the environment.

VIII. Concluding Comments

There is a considerable literature on deposition and erosion of limestone at terrestrial, freshwater and marine sites. In the case of freshwaters, it is still far from clear why deposition occurs at one site and not another, which, at least superficially, may seem quite similar. More information is needed on the extent to which other factors influence the inhibitory effects of phosphate on deposition. The same applies to individual species such as *Pleurocapsa minor*, which usually do not calcify, but occasionally do so. Almost certainly part of the difficulty in comparing the results of different surveys lies in the temporal variability of the environments of communities which may be growing only very slowly. Under such circumstances, a few spot measurements of nutrients in the water are of little value and may actually give a false impression of the nutrient status of the ecosystem. However, detailed study over long periods requires the development of highly sensitive probes for aqueous N and P fractions, together with recording systems able to operate reliably under the severe environmental conditions likely to be found at many limestone sites.

More information about the influence of nutrients, especially phosphate, is also required for marine environments. In this case it is the influence of eutrophication on dynamics of the whole community of epiliths and endoliths which is important. It would be of considerable value if the suggestion that the relative proportions and species characteristics of this community could be used globally as an indicator of the condition of not only coral reefs, but marine sublittoral limestones in general.

Acknowledgements

Thanks to Dr Heather A. Viles for useful suggestions.

References

Allen TFM (1971) Multivariate approaches to the ecology of algae on terrestrial rock surfaces in North Wales. J Ecol 59: 803-826.

Al-Thukair AA and Golubic S (1991) New endolithic cyanobacteria from the Arabian Gulf. 1. *Hyella immanis* sp. nov. J Phycol 27: 766-780.

Anagnostidis K, Economou-Amilli A and Roussomoustakaki M (1983) Epilithic and chasmolithic microflora (Cyanophyta, Bacillariophyta) from marbles of the Parthenon (Acropolis, Athens, Greece). Nova Hedwigia 38: 227-281.

Arp G and Reitner J (1998) Cyanobacteria in calcifying biofilms of alkaline salt lakes. Abstracts, Cyanobacteria in Action in the Lithosphere and in the Biosphere. 12 International Senckenberg Conference, November 16-19, 1998, Frankfurt, Germany.

Balazs D (1977) The geographic distribution of karst areas. In Ford TD (ed.) Proceedings of the 7th International Speleological Congress, Sheffield, UK, pp. 13-15.

Bathurst RGC (1971) Carbonate Sediments and their Diagenesis. Elsevier, Amsterdam.

Bobbink R, Hornung M and Roelofs JM (1998) The effects of air-borne nitrogen pollutants on species diversity in natural and semi-natural European vegetation. J Ecol 86: 717-738.

Bolivar FC and Sanchez Castillo PM (1997) Biomineralization processes in the fountains of the Alhambra, Granada, Spain. International Biodeterioration & Biodegradation 40: 205-215.

Braithwaite CJR, Casanova J, Frevert T and Whitton BA (1989) Recent stromatolites in landlocked pools on Aldabra, Western Indian Ocean. Palaeogeography, Palaeoclimatology, Palaeoecology 69: 145-165.

Braithwaite CJR and Zedef V (1996) Hydromagnesite stromatolites and sediments in an alkaline lake, Sala Golu, Turkey. J Sed Res 66: 991-1002.

Casanova J (1994) Stromatolites from the East African Rift: a synopsis. In: Bertrand-Sarfati J and Monty C (eds) Phanerozoic Stromatolites II, 193-226. Kluwer, Netherlands.

Casanova J and Lafont R (1985) Les cyanophycées encroutants du Var (France). Verh Int Verein Limnol 22: 2805-2810.

Castenholz RW (1989) Subsection III. Order Oscillatoriales. In Bergey's Manual of Systematic Bacteriology Vol. 3, 1771-1791. Williams and Wilkins, Baltimore, Maryland.

Castenholz RW and Waterbury JB (1989) Taxa of the Cyanobacteria. In Bergey's Manual of Systematic Bacteriology Volume 3, pp. 1712-1728, Williams and Wilkins, Baltimore, Maryland.

Caudwell C (1987) Étude experimentale de la formation de micrite et de sparite dans les stromatolites d'eau à *Rivularia*. Bull Soc Geol Fr 8: 299-306.

Copeland JJ (1936) Yellowstone thermal Myxophyceae. Ann N Y Acad Sci 36: 5-225.

Couté A (1982) Ultrastructure of an aerial calcified cave-dwelling cyanophycean: *Geitleria calcarea*. Hydrobiologia 97: 255-274.

Couté A (1989) *Geitleria calcarea* Friedmann (*Cyanophyceae, Hormogonophycidae, Stigonematales, Stigonemataceae*) : un cas d'adaptation à un milieu extrême. Bull Soc Bot Fr, Actual Bot 1: 113-130.

Couté A and Bury E (1988) Ultrastructure d'une cyanophycée aerienne calcifée cavernicole : *Scytonema julianum* (Frank) Richter (Hormogonophycidae, Nostocales, Scytonemataceae). Hydrobiologia 160: 219-240.

Cox G, Benson D and Dwarte DM (1981) Ultrastructure of a cave-wall cyanophyte, *Gloeocapsa* NS 4. Arch Microbiol 130: 165-174.

Dandurand J L, Gout R, Hoefs J, Menschel G, Schott J and Usdowski E. (1982) Kinetically controlled variations of major components and carbon and oxygen isotopes in a calcite-depositing stream. Chem Geol 36: 299-315.

Danin A (1993) Biogenic weathering of marble monoments in Didim, Turkey, and in Trajan Column, Rome. Water Sci Technol 27: 557-563.

Danin A and Canova G (1990) Deterioration of limestone walls in Jerusalem and marble monuments in Rome caused by cyanobacteria and cyanophilous lichens. Internat Biodeterioration 26: 397-417.

Danin A and Garty J (1983) Distribution of cyanobacteria and lichens on hillsides of the Negev Highlands and their impact on biogenic weathering. Z Geomorphologie NF 27: 423-444.

Danin A, Gerson R, Marton K and Garty J (1982) Patterns of limestone and dolomite weathering by lichens and blue-green algae and their palaeoclimatic significance. Palaeogeogr Palaeoclim Palaeoecol 37: 211-233.

Davis JS and Rands DG (1982) Lime incrusting *Hapalosiphon intricatus* (Cyanophyceae) and phosphate availability in a Florida cave. Schweiz Z Hydol 44: 289-294.

Defarge C, Trichet J, Maurin A and Hucher M (1994a) Kofara in Polynesian atolls - early stages of formation of calcareous stromatolites. Sed Petrol 89: 9-23.

Defarge C, Trichet J and Couté A (1994b) On the appearance of cyanobacterial calcification in modern stromatolites. Sed Geol 94: 11-19.

Donaldson A and Whitton BA (1977a) Chemistry of freshwater pools on Aldabra. Atoll Research Bulletin 213: 1-28.

Donaldson A and Whitton BA (1977b) Algal flora of freshwater habitats on Aldabra. Atoll Research Bulletin 215: 1-26.

Edwards RW and Heywood J (1960) Effect of a sewage effluent discharge on the deposition of calcium carbonate on shells of the snail *Potamopyrgus jenkinsii* (Smith). Nature 186: 492-493.

Emig WH (1917) The travertine deposits of the Arbuckle Mountains, Oklahoma, with reference to the plant agencies concerned in their formation. Bull. Okla Geol Surv 29: 9-75.

Fabré G and Fiche JL (1986) Les concretionnements de l'Aqeduct Romain de Nimes. Mediterranée 1-2: 129-130.

Ferris FG, Fratton CM, Gertis JP, Schultzelam S, Lollar BS (1995) Microbial precipitation of a strontium calcite phase at a groundwater dischange zone near Rock Creek, British Columbia, Canada. Geomicrobiol J 13: 57-67.

Ferris FG and Lowson EA (1997) Ultrastructure and geochemistry of endolithic microorganisms in limestone of the Niagara Escarpment. Can J Microbiol 43: 211-219.

Fjerdingstad E (1957) A lime incrusting algal community of a Danish well. Rev Algol 4: 246-248.

Fjerdingstad E (1965) The algal flora of some 'Tintenstriche' in the 'Alpes Maritimes' (France). Schweiz Z Hydrol 27: 167-171.
Freytet P and Plaziet A (1991) Les formations stromatolitiques (tufs calcaires) recentes de la region de Tournus (Saône et Loire). Geobios 24: 123-139.
Freytet P and Plaziet A (1996) Modern freshwater microbial carbonates: the *Phormidium* stromatolites (tufa-travertine) of Southeastern Burgundy (Paris Basin, France). Facies 34: 219-238.
Friedmann E I (1979) The genus *Geitleria*: distribution of *G. calcarea* and *G. floridana* n. sp. Plant Syst Evol 131: 169-178.
Fritsch FE (1949) The lime-encrusted *Phormidium*-community of British streams. Verh Int Verein Limnol 10: 141-144.
Fritsch FE (1950) *Phormidium incrustatum* (Naeg.) Gom., an important member of the lime-encrusted communities of flowint water. Biologisch Jaarboek 17: 27-39.
Garcia-Pichel F and Castenholz RW (1991) Characterisation and biological implications of scytonemin, a cyanobacterium sheath pigment. J Phycol 27: 395-409.
Garty J (1990) Influence of epilithic microorganisms on the surface temperature of building walls. Can J Bot 68: 1349-1353.
Ghiradelli LA (1998) An endolithic cyanophyte in the cell wall of calcareous algae. Bot Mar 41: 367-373.
Glazek J (1965) Recent oncolites in streams of North Vietnam and the Polish Tatra Mountains. Roczn pol Tow Geol 35: 221-242.
Gleason PJ (1972) The origin, sedimentation and stratigraphy of a calcitic mud located in the southern fresh-water Everglades, Thesis, Department of Geosciences, Pennsylvania State University, University Park, Penn.
Golubić S (1967) Algenvegetation der Felsen. In Elster H-J and Ohle W (eds) Die Binnengewasser 23, E. Schweizerbart'sche Verlagsbuchhandlung, Stuttgart.
Golubić S (1973) The relationship between blue-green algae and carbonate deposits. In: Carr NG and Whitton BA (eds) The Biology of the Blue-green Algae. 434-472. Blackwell, Oxford.
Golubić S, Brent G and LeCampion T (1970) Scanning electron microscopy of endolithic algae and fungi usong a multipurpose casting-embedding technique. Lethaia 3: 203-209.
Golubic S and Fischer AG (1975) Ecology of calcareous nodules forming in Little Conestoga Creek near Lancaster, Pennsylvania. Verh Int Verein Limnol 19: 2315-2323.
Golubić S, Friedmann I and Schneider J (1981) The lithobiontic niche, with special reference to microorganisms. J Sed Petrol 51: 475-478.
Grainger SLJ, Peat A, Tiwari DN and Whitton BA (1989) Phosphomonoesterase activity of the cyanobacterium (blue-green alga) *Calothrix parietina*. Microbios 59: 7-17.
Grüninger W (1965) Rezente Kalktuffbildung im Bereich der Uracher Wasserfälle. Abh Karst u Höhlenkunde E2: 1-113.
Heath CR, Leadbeater BCS, Callow ME (1995) Effects of inhibitors on calcium carbonate deposition mediated by freshwater algae. J Appl Phycol 7: 367-380.
Herman JS and Lorah MM (1987) CO_2 outgassing and calcite precipitation in Falling Spring Creek, Virginia, U.S.A. Chem Geol 62: 251-262.
Holmes NTH and Whitton BA (1981) Phytobenthos of the River Tees and its tributaries. Freshwat Biol 11: 139-168.
Jaag O (1945) Untersuchungen über die Vegetation und Biologie der Algen des näckten Gesteins in den Alpen, im Jura und im schweizerischen Mitteland. Beitr Kryptogamen Flora Schweiz 9: 1-560.
Kann E (1959) Die eulitoral Algenzone im Traunsee. Arch Hydrobiol 55: 129-192.
Kann E (1973) Bemerkungen zur Systematik und Ökologie einiger mit Kalk inkrustierter *Phormidium* Arten. Schweiz Z Hydrol 35: 141-151.
Kempe S, Kazmierczak J, Landmann G, Konuk T. Reimer A and Lipp A (1991) Largest known microbialites discovered in Lake Van, Turkey. Nature 349: 605-608.
Krumbein WE and Potts M (1979) *Girvanella*-like structures formed by *Plectonema gloeophilum* Borzi (Cyanophyta) from the Borrego desert in S. California. Geomicrobiol J 1: 211-218.
LeBris S, Le Campion-Alsumard T and Romano JC (1998) Characteristics of epilithic and endolithic algal turf exposed to different levels of bioerosion in French Polynesian coral reefs. Oceanologica Acta 21: 695-708.
Le Campion-Alsumard T (1991) 3 *Hyella* taxa (endolithic cyanophytes) from tropical environments (Lizard Island, Great Barrier Reef) Arch Hydrobiol 92: 159-166.
Le Campion-Alsumard T, Golubic S and Hutchings P (1995) Microbial endoliths in skeletons of live and dead corals - *Porita lobata* (Moorea, French Polynesia). Mar Ecol Progr Ser 117: 149-157.
Leclerc J C, Couté A and Dupuy P (1983) Annual climatic conditions in two caves from Poitou where pure colonies of shade-loving algae grow. Cryptogam Algol 4: 1-20.
Livingstone D, Khoja TM and Whitton BA (1983) Influence of phosphorus on physiology of a hair-forming blue-green alga (*Calothrix parietina*) from an upland stream. Phycologia 22: 345-350.
Livingstone D, Pentecost A and Whitton BA (1984) Diel variations in nitrogen and carbon dioxide fixation by the blue-green alga *Rivularia* in an upland stream. Phycologia 23: 125-133.
Livingstone D and Whitton BA (1983) Influence of phosphorus on morphology of *Calothrix parietina* (Cyanophyta) in culture. Br Phycol J 18: 29-38.
Livingstone D and Whitton BA (1984) Water chemistry and phosphatase activity of the blue-green alga *Rivularia* in Upper Teesdale streams. J Ecol 72: 405-421.
Lowenstam HA (1986) Mineralisation processes in monerans and protoctists. In: Leadbeater BSC and Riding R (eds) Biomineralisation in Lower Plants and Animals. 1-17. Clarendon Press, Oxford.
Luttge I (1997) Cyanobacterial Tintenstrich communities and their ecology. Naturwissenschaftern 84: 526-534.
McCarthy PM (1983) The composition of some calcicolous lichen communities in the Burren, western Ireland. Lichenologist 15: 231-248.
McCormick PV and O'Dell MB (1996) Quantifying periphyton responses to phosphorus in the Florida Everglades: a synoptic-experimental approach. J N Am Benthol Soc 15: 450-468.
McCormick PV, Rawlik PS, Lurding K, Smith EP and Sklar FH (1996) Periphyton-water quality relationships along a nutrient gradient in the northern Everglades. J N Am Benthol. Soc 15: 433-449.
McCormick PV, Shuford III RBE, Backus JG and Kennedy WC (1998) Spatial and seasonal patterns of periphyton biomass and productivity in the northern Everglades, Florida, U.S.A. Hydrobiologia 362: 185-208.
Merz MUE (1992) The biology of carbonate precipitation by cyanobacteria. Facies 26: 81-102.
Merz MUE and Zankl H (1993) The influence of culture conditions on growth and sheath development of calcifying cyanobacteria. Facies 29: 75-80.
Meybeck M (1987) Global chemical weathering of surficial rocks estimated from river discharge loads. Am Sci 287: 401-428.
Newman S, McCormick PV and Backus JG (1999) Phosphatase activity as an early warning indicator of wetland eutrophication: Problems and prospects. In: Whitton BA and Hernández (eds) Phosphatases in the Environment. Kluwer, Dordrecht.
Obenlüneschloss J (1991) Biologie und Ökologie von drei rezenten Süsswasser-Rivulariaen (Cyanobakterien). – Übertragbarkeit artspezifischer Verkalklangstructuren auf fossile Formen. Göttinger Arb Geol Palaeont 50: 1-86.

Ordonez S and Garcia del Cura MA (1983) Tertiary fluvial carbonates in Central Spain. In: Collinson J D and Lewin J (eds) Modern and Ancient Fluvial Systems. Special Publication of the International Association of Sedimentologists, No 6, pp 485-497.

Parihar NS and Pant G (1982) Bryophytes as rock builders - bryophyte communities associated with travertine formation at Sahasradhara, Dehra Dun. In: Nautiyal DD (ed.) Phyta-Studies on Living and Fossil Plants, pp 277-95, Punjab University, India.

Pentecost A (1978) Blue-green algae and freshwater carbonate deposits. Proc Roy Soc Lond B 200: 43-6.1

Pentecost A (1981) The tufa deposits of the Malham district, North Yorkshire. Fld Stud 5: 365-387.

Pentecost A (1982) A quantitative study of calcareous and Tintenstriche algae from the Malham district, Northern England. Br Phycol J 17: 443-456.

Pentecost A (1985a) Investigation of variation in heterocyst numbers, sheath development and false-branching in natural populations of Scytonemataceae (Cyanobacteria). Arch Hydrobiol 102: 343-353.

Pentecost A (1985b) Association of cyanobacteria with tufa deposits: identity, enumeration and nature of the sheath material revealed by histochemistry. Geomicrobiol J 4: 285-298.

Pentecost A (1987) Growth and calcification of the freshwater cyanobacterium *Rivularia haematites* Proc Roy Soc Lond B 232. 125-136.

Pentecost A (1988) Growth and calcification of the cyanobacterium *Homoeothrix crustacea.* J Gen Microbiol 134: 2665-2671.

Pentecost A (1989) Growth and calcification of *Calothrix*-dominated oncolites from Northern England. In Crick RE (ed.) Origin, Evolution and Modern Aspects of Biomineralization in Plants and Lower Animals, pp. 443-454, Plenum, New York.

Pentecost A (1990a) The identity of suspended particles in a calcite-depositing stream and their significance in trapping and binding phenomena. Limnologica 21: 251-255.

Pentecost A (1990b) The Algal Flora of Travertine. In Herman JS and Hubbard DA (eds) Travertine-Marl: Stream Deposits in Virginia, pp. 117-127. Virginia Division of Mineral Resources Publication 101, Charlottesville, Va.

Pentecost A (1990c) The formation of travertine shrubs: Mammoth Hot Springs, Wyoming. Geol Mag 127 L 159-168.

Pentecost A (1991) Algal and bryophyte flora of a Yorkshire (U.K.) hill stream: a comparative approach using biovolume estimations. Arch Hydrobiol 121: 181-201.

Pentecost A (1992a) A note on the colonization of limestone rocks by cyanobacteria. Arch Hydrobiol 124: 167-172.

Pentecost A (1992b) Growth and distribution of endolithic algae in some North Yorkshire streams. Br Phycol J 27: 145-151.

Pentecost A (1992c) Carbonate chemistry of the surface waters in a temperate karst region : the southern Y, UK. Yorkshire Dales, UK. J. Hydrol. 139: 211-232.

Pentecost A (1995a) The microbial ecology of some Italian hot-spring travertines. Microbios 81: 45-58.

Pentecost A (1995b) Significance of the biomineralizing microniche in a *Lyngbya* (cyanobacterium) travertine. Geomicrobiol J 13: 213-222.

Pentecost A and Bauld J (1988) Nucleation of calcite on the sheaths of cyanobacteria using a simple diffusion cell. Geomicrobiol J 6: 129-135.

Pentecost A, Bayari S and Yesertener C (1997) Phototrophic microorganisms of the Pamukkale travertine, Turkey: their distribution and influence on travertine deposition. Geomicrobiol J 14: 269-283.

Pentecost A and Riding RR (1986) Calcification in Cyanobacteria. In Leadbeater BSC and Riding RR (eds) Biomineralization of Lower Plants and Animals, pp. 73-90, Clarendon Press, Oxford.

Pentecost A and Terry C (1988) Inability to demonstrate calcite deposition by bacterial isolates from travertine. Geomicrobiol J 6: 185-194.

Pentecost A and Viles HA (1994) A review and reassessment of travertine classification. Geogr Phys et Quaternaire 48: 305-314.

Pentecost A and Viles HA (in press) Deposition rates in a travertine from the Schwabian Alb as determined by lamina analysis. Z Geomorphologie, Suppl.

Pia J (1934) Die rezenten Kalksteine. Min Petrog Mitt, Erganzungsband 1: 1-419.

Potts M (1994) Desiccation tolerance of prokaryotes. Microbiol Rev 58: 755-805

Potts M and Whitton BA (1980) Vegetation of the intertidal zone of the lagoon of Aldabra, with particular reference to the photosynthetic prokaryotic communities. Proc Roy Soc 208B: 13-55.

Purvis OW, Coppins BJ, Hawksworth DL, James PW and Moore DM (1992) The Lichen Flora of Great Britain and Ireland. The Natural History Museum, London.

Raistrick B (1949) The stabilisation of the supersaturation of calcium carbonate solutions by anions possessing O-P-O-P-O chains. Faraday Disc Soc 5: 234-237.

Riding RR (1977) Skeletal stromatolites. In Flügel E (ed.) Fossil Algae, Recent Results and Developments, pp. 57-60, Springer, Berlin

Rott E (1991) Oncoids from the summer-warm River Alz (Bavaria) - morphology and dominant cyanophytes. Algological Studies 64: 469-482.

Schafer A and Staf KRG (1978) Permian Saar-Nahe Basin and Recent Lake Constance (Germany): two environments of lacustrine algal carbonates. Spec Publs Int Ass Sedimentol 2: 83-107.

Schlichting HE (1975) Some subaerial algae from Ireland. Br Phycol J 10: 257-261.

Seeler JS and Golubic S (1991) *Iyengariella endolithica* sp. nova, a carbonate boring Stigonematalean cyanobacterium from a warm spring-fed lake - nature to culture. Arch Hydrobiol 92: 399-410.

Smith RJ and Wilkins A (1988) A correlation between intracellular calcium and incident radiation in *Nostoc* 6720. New Phytol 109: 157-162.

Somers GF and Brown M (1978) The affinity of trichomes of blue-green algae for calcium ions. Estuaries 1: 17-28.

Spiro B and Pentecost A (1991) One day in the life of a stream - a diurnal inorganic carbon mass balance for a travertine-depositing stream (Waterfall Beck, Yorkshire). Geomicrobiol J 9: 1-11.

Strzelczyk AM (1981) Stone biodeterioration. In: Rose AH (ed.) Microbial Biodeterioration. Economic Microbiology. Vol. 6. Academic Press, London.

Symoens JJ and Mallaise F (1967) Sur une formation de tuf calcaire observée sur le versant est du plateau des Kundelungu. Bull Acad Roy Sci D'Outre Mer 13: 1148-1151.

Symoens JJ and van der Werff (1951) Note sur des formations de tuf calcaire des environs de Consdorf (Grand-Duché de Luxembourg). Bull Soc Roy Bot Belg 83: 213-218.

Szulc J and Smyk B (1994) Bacterially controlled calcification of freshwater *Schizothrix* stromatolites: an example from the Pieniny Mts, Southern Poland. In Bertand-Sarfati J and Monty CLV (eds) Phanerozoic Stromatolites II, pp. 31-51, Kluwer, Amsterdam.

Thienemann A (1934) Eine gesteinebildende Chironomide *Lithotanytarsus emarginatus* (Goetghebuer) Z Morph Ökol Tiere 28: 480-496.

Thompson JB and Ferris FG (1990) Cyanobacterial precipitation of gypsum, calcite and magnesite from natural alkaline lake water. Geology 18: 995-998.

Thunmark S (1926) Bidrag till Kannendomen om recenta Kalktuffer. Geol Foreningens 48: 541-583.

Tian Y and He F (1996). A preliminary study on the microcommunity ecology of karst Cyanophyta in mat from Fendong Cave, Panxian, Guizhou. Carsalogica Sinica 15 : 233-238.

Tucker ME and Wright VP (1990) Carbonate Sedimentology. Blackwell, London.

Turian G (1983) Cyanophycée "d'orée" (*Chrysocapsa*) pionnière de la colonization bu beton. Saussurea 14: 69-77.

Usdowski E, Hoefs J and Menschel G (1979) Relationship between ^{13}C and ^{18}O fractionation and changes in major element composition in a recent calcite-depositing stream-A model of chemical variations with inorganic $CaCO_3$ precipitation. Earth Planet Sci Lett 42: 267-276.

van der Wal P, de Jong L, Westbroek P, de Bruijn WC and Mulder-Stapel AA (1983) Polysaccharide localization, coccolith formation and golgi dynamics in the coccolithophorid *Hymenomonas carterae*. J Ultrastruct Res 85: 1139-1158.

Verrecchia EP (1994) Microbial and surficial origin of the calcrete laminar horizon. Bull Soc Geol France 165: 583-592.

Verrecchia EP and Verrecchia KE (1994) Needle fiber calcite: a critical review and a proposed classification. J Sed Res A64: 650-664.

Viles HA (1987) Blue-green algae and terrestrial limestone weathering on Aldabra Atoll: an SEM and light microscope study. Earth Surf Process Landforms 12: 319-330.

Viles HA (1988) Cyanobacterial and other biological influences on terrestrial limestone weathering on Aldabra: implications for landform development, Biol Sci Wash Bull 8: 5-13.

Viles HA (1995) Ecological perspectives on rock surface weathering: Towards a conceptual model. Geomorphology 13: 21-35.

Vinogradova ON, Kovalenko OV, Wasser SP and Nevo E (1998) Species diversity gradient to darkness stress in blue-green algae /cyanobacteria: A microscale test in a prehistoric cave, Mount Carmel, Israe. Israel J Plant Sci 46: 229-238.

Vogel K and Glaub I (1998) The role of marine euendolithic cyanobacteria in bioerosion and paleobathymetrical reconstruction of fossil sedimentary basins. Abstracts, Cyanobacteria in Action in the Lithosphere and in the Biosphere. 12 International Senckenberg Conference, November 16-19, 1998, Frankfurt, Germany, 20-21.

Walkden GM, Irwin JER and Fallic AE (1994) Carbonate spherules and botryoids as lake floor cements in the East Kirkton Limestone of West Lothian, Scotland. Trans Roy Soc Edinburgh Earth Sci 84 (3/4): 213-221.

Wang F, Cao J, Huang Ju, Jiang L, Huang Ji and Wang J (1993) Biokarst. Geological Publishing House, Beijing.

Whitton BA (1971) Terrestrial and freshwater algae of Aldabra. Phil Trans Roy Soc Lond B 260: 249-255.

Whitton BA (1987) The biology of Rivulariaceae. In: Fay P and Van Baalen C (eds) The Cyanobacteria - A Comprehensive Review. 543 pp, pp 513-534. Elsevier, Amsterdam.

Whitton BA and Donaldson A (1977) Terrestrial and freshwater algae of three western Indian Ocean islands (Astove, Farquhar and St. Pierre). Atoll Research Bulletin 216: 1-8.

Whitton BA, Khoja TM and Arif IA (1986) Water chemistry and algal vegetation of streams in the Asir Mountains, Saudi Arabia. Hydrobiologia 133: 97-106.

Whitton BA, Yelloly JM, Christmas M and Hernández I (1998) Surface phosphatase activity of benthic algal communities in a stream with highly variable ambient phosphate concentrations. Verh Int Verein Limnol 26: 967-972.

Winsborough BM, Seeler JS, Golubić S, Folk RL and Maguire B (1994) Recent fresh-water lacustrine stromatolites, stromatolitic mats and oncoids from Northeastern Mexico. In: Bertrand-Sarfati J and Monty C (eds) Phanerozoic Stromatolites II pp. 71-100. Kluwer, Netherlands.

Zimmermann U (1969) Ökologische und physiologische Untursuchungen an der planktischen Blaualga *Oscillatoria rubescens* D.C. unter besonderer Berücksichtigung von Licht und Temperatur. Schweiz Z Hydrol 31: 1-58.

Chapter 10

Salts and Brines

Aharon Oren
Division of Microbial and Molecular Ecology, The Alexander Silverman Institute of Life Sciences, and the Moshe Shilo Minerva Center for Marine Biogeochemistry, The Hebrew University of Jerusalem, 91904 Jerusalem, Israel

B. A. Whitton and M. Potts (eds), The Ecology of Cyanobacteria, 281-306.
© 2000 *Kluwer Academic Publishers. Printed in the Netherlands.*

Summary

Cyanobacteria form a major component of the biota of hypersaline environments including salt lakes, solar salterns, hypersaline lagoons, salt flats, and hypersaline sulfur springs. Light-exposed sediments in such environments are often covered with dense communities of filamentous and unicellular cyanobacteria that may be responsible for much of the primary production of these environments. Cyanobacteria are often found in evaporite crusts of gypsum and even halite. A wide range of species, belonging to different taxonomic groups, were reported to live at high salinities, but two major types characteristically dominate hypersaline environments: the filamentous *Microcoleus chthonoplastes*, a cosmopolitan mat-building cyanobacterium found from seawater salinity to salinities exceeding 200‰; and the unicellular *Aphanothece halophytica*. In many hypersaline environments, including the hypolimnion of Solar Lake, Sinai, hypersaline sulfur springs, and benthic mats in salt lakes and salt flats, cyanobacteria are exposed to high sulfide concentrations, either permanently or periodically. Certain species are able to use sulfide as an electron donor in an anoxygenic type of photosynthesis through a process which involves photosystem I only and that produces elemental sulfur or thiosulfate. Some forms are able to grow in the absence of molecular oxygen. *Oscillatoria limnetica*, isolated from Solar Lake, was used as a laboratory model for the study of anoxygenic photosynthesis, and analysis of the phenomenon at the molecular level commenced. To be able to withstand the high osmotic pressure caused by the salt concentrations in their surrounding medium, cyanobacteria living at high salinities possess mechanisms to maintain osmotic equilibrium and cell turgor. Ions (Na^+, K^+, Cl^-) can temporarily enter the cells to counteract rapid increases in medium salinity, however, for the long term, organic solutes are accumulated to provide osmotic balance, as expected in organisms whose enzymatic machinery is inhibited by salt. Different organic osmolytes were found in cyanobacteria: the disaccharides sucrose and trehalose (especially in the less salt tolerant types); glucosylglycerol (in moderately halotolerant species); and glycine betaine (in *A. halophytica* and a few other forms that tolerate very high salt concentrations). Our understanding of the osmoregulatory mechanisms in cyanobacteria increased greatly in recent years, especially with regard to the molecular analysis of glucosylglycerol metabolism in *Synechocystis* PCC 6803. Accumulation of organic osmotic solutes has important implications for the carbon cycle in hypersaline environments where halophilic cyanobacteria provide the greatest contribution to the primary productivity.

I. Introduction

Dense communities of cyanobacteria are often a prominent feature of the biota found in planktonic and benthic environments at high salt concentrations. They form a major component of the biota of hypersaline environments such as salt lakes, solar salterns, hypersaline lagoons, salt flats, and hypersaline sulfur springs (Javor, 1989). Light-exposed sediments in such environments are often covered with conspicuous thick mats of filamentous and unicellular cyanobacteria, which may be responsible for much of the primary production. To a large extent these communities determine the biological properties of such hypersaline environments (Bauld, 1981; Des Marais, 1995). Several types of cyanobacteria thrive at salinities close to NaCl (halite) saturation (Braithwaite and Whitton, 1987; Rothschild et al., 1994).

Abbreviations: DCMU - 3(3,4-dichlorophenyl)1,1-dimethylurea; MAA - mycosporine-like amino acids; RUBISCO - ribulose bisphosphate carboxylase/oxygenase

Cyanobacteria differ greatly in their relationship to salt. Many types are adapted to life at seawater salinities. The diversity of cyanobacterial life in the marine system and the physiology and the ecology of marine cyanobacteria were discussed elsewhere in this volume (Chapter 4). Many of the cyanobacteria that occur in the marine environment (salinities around 35-40‰) are able to grow at higher salinities as well, making it virtually impossible to draw a sharp boundary between species adapted to marine and to hypersaline habitats. Thus, *Microcoleus chthonoplastes*, the main mat building cyanobacterium in the marine littoral and intertidal flats, can be found in hypersaline environments worldwide up to salinities of 200‰ and higher (Dubinin et al., 1992a; Javor, 1989). In all cases a single, cosmopolitan form with the ability to adapt to a wide range of salinities characterizes these environments (Garcia-Pichel et al., 1996). The strain *Synechocystis* PCC 6803, an isolate widely used in the study of the physiology of salt adaptation in cyanobacteria, was isolated from a freshwater

Table 1. Terms used in this chapter to describe salinity and salt composition of brines, and the response of microorganisms to high salt concentrations

Salinity	Total salt content in g salt per kg solution (expressed in ‰)
Thalassohaline	Having an ionic composition resembling that of seawater
Athalassohaline	Having an ionic composition greatly differing from that of seawater
Halophilic	Salt-loving, requiring high salt concentrations for growth
Halotolerant	Being able to grow in the presence of high salt concentrations, but not requiring salt for growth
Stenohaline[1]	Adapted to life within a narrow range of salt concentrations
Euryhaline[1]	Adapted to life within a broad range of salt concentrations
Oligohaline[1]	Growing optimally at low salt concentrations
Mesohaline[1]	Growing optimally at intermediate salt concentrations
Polyhaline[1]	Growing optimally at high salt concentrations

[1]Terms derived from Golubic (1980)

environment, but grows well at salt concentrations two to three times as high as those of seawater (Richardson et al., 1983), and may thus be classified as a halotolerant strain (Table 1). Also true halophiles exist, which do not grow below 10-20 g L^{-1} NaCl (0.2 - 0.4 M). In a comparative study of salt tolerance and salt requirement, Golubic (1980) concluded that cyanobacteria speciate along the salinity gradient, and that separate halophilic taxa occupy environments with relatively constant salinities. He divided the cyanobacteria into stenohaline and euryhaline types (adapted to life within a narrow or a broad range of salt concentrations, respectively), and designated strains as oligo-, meso-, and polyhaline according to the optimal salt concentration for growth (for a definition of these and other related terms see Table 1).

In this chapter the boundary between "marine" and "hypersaline" systems was arbitrarily set at 70‰ salinity; twice the mean value of seawater. The chapter deals mostly with thalassohaline environments that originate from evaporation of seawater and thus reflect its ionic composition. A few cases of athalassohaline lakes (the Dead Sea, hypersaline soda lakes), with salt compositions which differ greatly from that of ocean water, are also discussed.

II. Hypersaline Environments and their Cyanobacterial Communities

Attempts to review the occurrence of cyanobacteria in hypersaline environments are hampered by the poor state of cyanobacterial taxonomy. The species concept is not well defined within the group. Essentially all classification is based on cell, or filament morphology, and different authors often use different designations for what may be the same (morphological) type. On the other hand, what may be referred to by different authors as the same genus and species may encompass a broad range of different organisms (e.g. *Aphanothece halophytica*). In this chapter no attempt was made to adopt a uniform nomenclature system, and most species' designations are those which appeared in the original reports (as cited in the list of references). A non-exhaustive list of genera and species that occur at high salt concentrations (as defined above) includes (as reviewed in: Borowitzka, 1981; Golubic, 1980; Oren, 1993, and Potts, 1980):

— Unicellular types: *Aphanothece* or *Aphanocapsa* (round or ovoid cells surrounded by mucilage); *Synechococcus* (idem without mucilage - note that the presence and thickness of a slime capsule can be highly dependent on the environment making it difficult, at best, to discern some strains of *Aphanothece* from *Synechococcus*; *Synechocystis; Gloeothece* (rod-shaped cells, dividing in a single plane) and *Gloeocapsa* (usually spherical cells which divide in two or three planes; Golubic, 1980). Bergey's Manual of Systematic Bacteriology classifies round to coccoid cells longer than 3 μm as *Cyanothece*, smaller cells as *Synechococcus*, and cells with a well-defined sheath as *Gloeothece* (Waterbury and Rippka, 1989). *Aphanothece halophytica* (also designated *Aphanocapsa halophytica*) is a true halophile (Brock 1976), and is found worldwide at salinities of above 100‰ to values approaching halite saturation. The long, spindle-shaped *Dactylococcopsis salina* is often found as a planktonic species in hypersaline brines (Davis and Giordano, 1996; van Rijn and Cohen, 1983; Walsby et al., 1983).

— Filamentous types: As stated above, *Microcoleus chthonoplastes* is often found as the main component of the cyanobacterial mats at high salt concentrations, as well as in marine systems. Its trichomes are generally encased in multiple-filament sheaths. Filaments that may occur as individual trichomes may easily be mistaken for *Oscillatoria* or *Schizothrix* (Golubic, 1980). *Oscillatoria* forms can often be found, as well as other filamentous non-heterocystous types such as *Phormidium tenue* (Borowitzka, 1981).

Heterocystous cyanobacteria are conspicuously lacking in hypersaline environments and are rarely mentioned in systematic surveys of salt lakes, salterns, environments of evaporites, etc. (Golubic, 1980; Potts, 1980). A single exception may be the report of *Anabaena* in Indian salterns of Sambhar Lake (Seshadri and Buch, 1958). This does not imply that nitrogen fixation does not occur in the presence of high salt concentrations; nitrogen fixation, as measured by the acetylene reduction technique, was detected in hypersaline coastal pools of the Sinai peninsula and in hypersaline mats from Guerrero Negro, Baja California at high salinities (Potts, 1980; Rothschild et al., 1994). Heterocystous cyanobacteria often occur in the intertidal region but are of little or no importance in the open oceans for reasons that are not very clear; most nitrogen fixation in the oceans is due to the non-heterocystous form *Trichodesmium* (Chapter 5).

A. Hypersaline Lagoons and Salt Flats

In marine coastal environments, especially in the tropics and subtropics, shallow areas covered with seawater often become partially isolated from the sea. In sabkhas, such as those along the coast of the Sinai peninsula and elsewhere, a seawater-covered area is isolated from the open sea, but receives a slow exchange of seawater through the separating sand bar (Cohen et al., 1975a, 1977a; Gerdes et al., 1985, 1994; Potts, 1980). Evaporation rates exceed the rate of water inflow and cause an increase in salinity, and a steady-state salinity is reached which exceeds that of seawater.

Three areas in this category were investigated with respect to their cyanobacterial communities: the salt flats of Guerrero Negro, Baja California; the lagoons of Lake Sivash (east Crimea, Ukraine); and the coastal sabkhas of the Sinai Peninsula, Egypt.

Most studies of the Gurrero Negro cyanobacterial mats were performed in areas within the man-made salinas, and these are discussed in Section II.D. Natural salt flats occur outside the salina area, and these were also the subject of study (Rothschild et al., 1994; Stolz, 1990). Evaporite crusts of gypsum and halite are formed, up to 4 cm or more in thickness, and often display distinct colorful layers of unicellular and filamentous cyanobacteria (Rothschild et al., 1994). Evidence was obtained that the cyanobacteria that inhabit these dry crusts remain active for long times, at least for many months after their formation.

Also in the Lake Sivash lagoons (east Crimea), *Microcoleus* mats are present as the main primary producer at salinities between 80 - 160‰ (Zavarzin et al., 1993). Here, however, *Microcoleus* is also present at much higher salinities, even up to 300‰ salinity, at salt concentrations sufficiently high for the development of extremely halophilic archaea (*Haloarcula*, *Halorubrum*) (Zvyagintseva et al., 1995). The close interactions between *Microcoleus* and the halophilic archaea present in these environments is discussed in Section VI.A.

Extensive studies were made of the cyanobacterial communities that inhabit different coastal hypersaline environments on the coast of the Sinai peninsula (Egypt): Solar Lake; a sabkha near Nabq (known also as Sabhka Gavish); and the Ras Muhammad Pool. All receive a supply of seawater through subterranean seepages (Cohen et al., 1975a, 1977a; Ehrlich and Dor, 1985; Gerdes et al., 1985; Javor, 1989; Potts, 1980). In total, 41 species were recorded from the hypersaline coastal pools, including 24 species in Solar Lake and 13 species from the Ras Mohammed Pool (Javor, 1989; Krumbein et al., 1979; Potts, 1980). A clear horizontal zonation of species was reported in the sabkhas. The peripheral zone of the Gavish sabkha and its gypsum crust was dominated by *Pleurocapsa* and *Entophysalis*. The *Pleurocapsa*-inhabited gypsum nodules showed nitrogen fixation (acetylene reduction) activity. Towards the center a red-orange zone of *Aphanothece*, accompanied by *Synechococcus*-type unicellular cyanobacteria was found, while the central area was covered by a green mat of *Microcoleus* together with *Oscillatoria* and *Lyngbya* spp. (Potts, 1980). An additional inhabitant was *Spirulina subsalsa*, found up to a salinity of about 205 ‰. Only two species were found at the highest salinities (250 - 330‰); *Aphanothece halophytica* and *Schizothrix arenaria* (Ehrlich and Dor, 1985; Gerdes et al., 1985).

The varied microbial communities of Solar Lake are discussed in Section IV.A.

B. Hypersaline Lakes

Cyanobacteria are found in many hypersaline lakes, both with a thalassohaline and athalassohaline salt composition. A number of characteristic examples include:

1. Great Salt Lake, Utah

Great Salt Lake, Utah, is a large inland lake. In spite of the fact that no direct connection with the open ocean has existed for tens of thousands of years, the ionic composition of its waters is very similar to that of seawater. The overall salinity of the lake was subject to drastic changes during the past decades. In 1959 the lake was divided by means of a causeway, and this led to the formation of a hypersaline northern arm (total salt concentration up to 250-300 g L^{-1}) and a less saline southern arm (around 120 g L^{-1} in the 1970s; Post, 1977). A large excess of rainfall in the late 1980s - early 1990s in the catchment area of the lake caused a sharp drop in overall salinity.

The biology of the Great Salt Lake was reviewed by Post (1977, 1981). While most of the primary production in the lake is due to eukaryotic unicellular algae (*Dunaliella* spp.), cyanobacteria are a characteristic component of the lake's biota. *Aphanothece halophytica* was found at the highest salinities. In a classic study, Brock (1976) recognized the existence of halophilic cyanobacteria based on the occurrence of *Aphanothece* in the Great Salt Lake and on the properties of the isolated strain in culture. In addition, species of *Phormidium* or *Oscillatoria*, as well as *Microcoleus lyngbyaceus*, *Spirulina major*, and *Nodularia spumigena* were found in the shallow sediments of the lake (Felix and Rushforth, 1979; Post, 1977, 1981; Rushforth and Felix, 1982).

2. The Dead Sea

The Dead Sea is an athalassohaline inland lake on the border between Israel and Jordan. Its extremely saline waters (around 340 g L^{-1} total dissolved salts) have an unusual ionic composition. Presently the lake contains more than 1.8 M Mg^{2+} and 0.4 M Ca^{2+}, in addition to about 1.7 M Na^{+}. The dominant anion is Cl^{-} followed by Br^{-} (about 1% of the anion fraction).

The microorganisms best adapted to this hostile environment are species of halophilic Archaea, and one primary producer, the unicellular green alga *Dunaliella parva*. The biology of the Dead Sea, including information on the dynamics of the communities as influenced by changes in the physical structure of the water column, was reviewed by Oren (1988).

Cyanobacteria do not form a quantitatively important part of the Dead Sea ecosystem. This is expected to some extent from the low pH of the Dead Sea brines (~6); cyanobacteria are not normally abundant at low pH values (Brock, 1973). However, strains of *Aphanothece*, *Microcystis*, *Phormidium* and *Nostoc*, all developing at salt concentrations of up to 180 g L^{-1}, were obtained in enrichment cultures (Volcani, 1944). One obligately halophilic *Aphanocapsa* strain was found to grow at salt concentrations between 60 - 360 g L^{-1}. The isolate was reported to change its morphology response to the salt concentration of the medium. Small cells, singly or in pairs, occurred at the low salinity range and large rounded and vacuolate cells were present at the highest salinities (Volcani, 1944). Though direct microscopic examination of Dead Sea water or sediment samples in recent years never led to the detection of these cyanobacterial types, unicellular cyanobacteria with a similar behavior (this time identified as *Aphanothece halophytica*) showed massive development in artificial solar ponds established at the shore of the Dead Sea (Dor and Hornoff, 1985a; Dor and Paz, 1989, see also Section II.E). Recent attempts to enrich for photosynthetic microorganisms, using Dead Sea sediments from 20-30 m depth as inoculum, and employing media containing 5 to 7 parts of Dead Sea water, gave rise to the development of unicellular *Synechococcus*-type cells (Oren, unpublished results).

3. Solar Lake, Sinai

Solar Lake is a small hypersaline, heliothermal lake on the shore of the Sinai coast. In summer, the lake (maximum depth 4.5-5 m) is mixed, with a uniform salinity of ~180‰. Seepage of seawater from the Gulf of Aqaba (41‰ salinity) and occasional rain floods cause the formation of a less saline layer (as low as 68‰) in autumn - winter, and the lake remains stratified from October until May. During stratification, a chemocline separates an anaerobic hypolimnion (containing up to 39 ppm H_2S) from an oxidized epilimnion. The difference in density of the upper less saline 1 - 2 m of the water column and the deep hypersaline brines prevents mixing, and heliothermal heating causes the hypolimnion to reach maximal temperatures of up to 49-61°C in winter. A detailed account of the limnology of the lake was given by Cohen et al. (1975a, 1977a).

A rich cyanobacterial flora was found both in the water column and in the benthic microbial mats. The list of species found in Solar Lake included *Aphanocapsa concharum*, *Aphanothece stagnina*, *A. microscopica*, *A. halophytica*, *A. litoralis*, *A. pallida*, *Chroococcidiopsis* sp., *Dactylococcopsis salina*, *D. acicularis*, *Entophysalis* sp., *Gloeothece* sp., *Johannesbaptista pellucida*, *Lyngbya aestuarii*, *L. confervoides*, *L. digueti*, *Microcoleus chthonoplastes*, *Oscillatoria redekei*, *O. limnetica*, *O. salina*, *Phormidium hypersalinum*, *P. hypolimneticum*, *Pleurocapsa* sp., *Pseudanabaena catenata*, *Spirulina labyrinthiformis*, and *Synechococcus* sp. (Campbell and Golubic, 1985; Cohen et al., 1975a, 1977b; Jørgensen et al., 1983; Potts, 1980). The mixed water column during the summer months contained *Aphanothece* cells at a low density. *Aphanothece* was also found in the upper water layer during winter stratification. At the border between the aerobic epilimnion and the hot, anaerobic hypolimnion a population of *Dactylococcopsis salina* was found (Potts, 1980; van Rijn and Cohen, 1983, Walsby et al. 1983). *D. salina* (also designated *Myxobactron salinum* by Davis and Giordano, 1996) is gas vacuolate with yellow to orange cells, and grows in culture between 5 - 20 % salt (optimum 7.5 - 15 %) and up to 45°C. The hot, sulfide containing water layer at the bottom of the lake contains a dense population of *Oscillatoria limnetica*, which grows anaerobically, using sulfide as electron donor in an anoxygenic type of photosynthesis (see below). The proper species designation of this, about 3 µm wide filamentous cyanobacterium, is uncertain. The original description of *O. limnetica* relates to a fresh water species with a much smaller filament width than the Solar Lake isolate. Golubic (1980) classified the organism as a *Phormidium* species.

Different types of benthic cyanobacterial mats can be found at Solar Lake. The surface mats (different mat morphologies being described as "shallow flat mat", "deep flat mat" and "blister mat", all at salinities around 80‰ and temperatures between 25 - 30°C) were dominated by *Microcoleus*, *Phormidium*, *Aphanothece*, *Aphanocapsa*, and *Synechococcus*. Nitrogen fixation (as measured by the acetylene reduction assay technique) was shown to occur in the *Microcoleus* mats of Solar Lake (Potts, 1980). The brown-red summer mat was dominated by *Aphanothece halophytica* and *Aphanothece littoralis* (Cohen 1984a; Jørgensen et al., 1983; Krumbein et al., 1977; Revsbech et al., 1985). The peripheral crust around the pool contained *Pseudanabaena*, and the gypsum crust close to water was inhabited by *Entophysalis* (Potts, 1980). *Lyngbya aestuarii* appeared as discontinuous patchy films over the surface of Solar Lake mats (Potts, 1980). An approximately 10 mm thick gelatinous, polysaccharide-rich mat was found in winter in the thermocline area (salinity about 150‰, temperature around 45°C). This mat lacked *Microcoleus*, and was rich in *Phormidium*, *Aphanothece halophytica*, *Aphanocapsa* and *Pleurocapsa*, which imparted a bright orange color to the upper 7 mm of the mat. Below this depth the color changed to light green due to the presence of *Oscillatoria* sp., accompanied by *A. halophytica*.

4. Other Hypersaline Lakes

Microbial mats with cyanobacteria as a predominant part of the biota occur in hypersaline lakes worldwide. An excellent inventory survey of microbial mats in saline lakes was published by Bauld (1981). Following are a few examples, derived from this study and from additional sources:

— Lake Hayward, South-West Australia, is a hypersaline, seasonally stratified lake, with salt concentrations varying from 60 g L^{-1} in the upper water layer to 200 g L^{-1} at the bottom. A 5 – 10 mm thick mucilaginous cyanobacterial mat was found, with *Cyanothece*, *Oscillatoria*, and *Spirulina* species (Burke, 1995).

— A variety of inland lakes are found on the Australian continent, with salinities from 100 g L^{-1} to almost 300 g L^{-1}. Different types of filamentous and unicellular cyanobacteria were found in these environments (Bauld, 1981).

— Hot Lake, Washington; a stratified lake with 100 g L^{-1} total dissolved salts in the surface layer, increasing to 400 g L^{-1} near the bottom, was reported to be inhabited by *Anacystis thermalis*, *Gomphosphaeria aponina*, *Oscillatoria chlorina*, and *Plectonema nostocorum* (Anderson, 1958).

— Lake Bonney in the Dry Valleys of Antarctica has benthic mats dominated by *Lyngbya martensiana* and *Phormidium frigidum*, occurring at a salinity of about 100‰ (Wharton et al., 1983).

— Lake Vanda in Wright Valley, Antarctica, was reported to contain coccoid (*Synechococcus* sp.) and filamentous cyanobacteria (*Phormidium* type) (about 10^6 cells and 10^5 filaments L^{-1}, respectively, at 60 m depth (salinity about 100‰; Goldman et al., 1967).

— Don Juan Pond in Antarctica is a hypersaline pond, with salt concentrations (mainly $CaCl_2$) reported to vary between 200 - 474 g L^{-1}. *Oscillatoria*-like cyanobacteria were reported to occur as a mat near the shore of the lake (Siegel et al., 1979). Few data are available on this interesting ecosystem, and the possibility cannot be ruled out that the growth occurred during inundation by ephemeral freshwater meltstreams, rather than in the concentrated $CaCl_2$ brines (Wright and Burton, 1981).

— Lakes and pools of the Wadi Natrun, Egypt, are characterized by high salinities accompanied by high pH values (10.8 - 11.2). The microbiology of these lakes was described by Imhoff et al. (1978, 1979). Unspecified cyanobacteria were reported to occur in lakes Gaar, Rizunia, Zugm, and Muluk (374, 389, 394, and 159 g L^{-1} salts, respectively). Lake Gabara (91.9 g L^{-1}) contained *Spirulina* sp., a strain of *Synechococcus* was isolated from Lake Hamra (240 g L^{-1} salt).

C. Hypersaline Sulfur Springs

The springs of Hamei Mazor on the western shore of the Dead Sea are characterized by a high salt content (around 170 g L^{-1} total dissolved salts), a high sulfide content (about 2.5 mM), and a pH of 5.2. The temperature of the water at the source is 39°C. The pools and runoff channels formed by the spring were covered by a layer of filamentous cyanobacteria (*Oscillatoria* sp.; Plate 17a), accompanied by patches of purple sulfur bacteria (*Thiocapsa* sp.; Oren, 1989; Plate 17b). The mass occurrence of cyanobacteria in this environment is unusual in view of the low pH of the water, as cyanobacteria are rarely found at pH values below 5.5 to 6 (Brock, 1973; Brock, 1978). Photoassimilation of CO_2 by the cyanobacterial community was not inhibited by DCMU, and the presence of sulfur granules associated with the cyanobacterial filaments suggested that photosynthesis was of the anoxygenic type, with sulfide as electron donor (Oren, 1989; Section IV.B).

D. Salterns

Salterns are man-made shallow ponds used for the production of halite (NaCl) from seawater. Most saltern systems are located in tropical and subtropical areas, and are built as multi-pond systems. Seawater is pumped to the first set of ponds. After evaporation has caused a sufficient rise in salinity, the water is transferred to the next set of ponds, and so on, until a brine saturated with NaCl is obtained, from which halite precipitates in the final set of ponds (the crystallizer ponds). The salinity in each of the ponds is thus kept approximately constant in the course of time, and each pond develops planktonic and benthic microbial communities adapted to the salinity range of the brines found in it. Long before halite crystallizes, other salts, notably gypsum ($CaSO_4.2H_2O$) precipitate, and thick gypsum deposits are characteristically found on the bottom of part of the saltern ponds. A more detailed description of the arrangement and operation of solar salterns was given by Davis (1978) and Javor (1989). Benthic cyanobacterial mats are a conspicuous part of the biota of saltern ponds, from the first seawater evaporation ponds to almost halite saturation. In view of the active role that cyanobacterial mats play in the operation of the salt production process, inventories of cyanobacterial species which occur in these saltern systems were made in solar salt facilities all over the world (Section VII). The types of cyanobacteria found at the different salinities, and the vertical arrangement of these cyanobacteria in the stratified sediments on the bottom of the ponds are probably very similar in saltern facilities worldwide (Plate 17c). Following are a number of case studies. As stated before, use of different nomenclature systems by different authors may result in different species' designations for what may essentially be the same organism.

A general overview of the biota found in salterns worldwide was given by Bauld (1981) and by Davis and Giordano (1996). At intermediate salinities *Aphanothece*, *Oscillatoria*, *Spirulina*, and *Xenococcus* species are found in the benthos where they impart a yellow-orange color to the upper layer, and a green color to the layer below. The presence of cyanobacterial mats enhances solar radiation absorption, and this increases evaporation rates. Another valuable contribution made by cyanobacterial mats in solar evaporation facilities appears to be the prevention of seepage loss through the bottom sediments. In contrast to many natural saline systems, mats in salinas appear to be dominated by unicellular colonial cyanobacteria which produce copious mucilage. In most systems described, filamentous cyanobacteria played a minor role. It was speculated that the unicellular colonial forms were favored by an environment of relatively constant salinity, such as occurs in the evaporation ponds (Bauld, 1981). However, cohesive *Microcoleus* mats occur at salinities from 60 - 120‰ in salterns on the Bretagne coast (France), the Canary islands, and the Balearic islands (Gerdes et al., 1994). Filamentous cyanobacteria also dominate the extensive saltflats of Baja California, Mexico, which serve as the first stage of seawater evaporation in the production of solar salt by the Exportadora de Sal, S.A. (Javor, 1989).In the Guerrero Negro salt flats the main component of the cyanobacterial community is the filamentous *Microcoleus chthonoplastes* that occurs in bundles and forming coherent, highly productive mats at salinities between 60 - 125‰. In these mats, *Microcoleus* was accompanied by *Oscillatoria* sp. The cyanobacterial population changed composition as a function of changes in the salinity of the environment. *Microcoleus* was less abundant at salinities exceeding 150‰. At higher salinities species belonging to the genera *Phormidium*, *Spirulina*, *Aphanothece* and *Synechococcus* predominate (Des Marais, 1995; Javor, 1989; Rothschild, 1991). *Spirulina* was reported to be the major filamentous cyanobacterium in the salterns of Alicante, Spain (Rodriguez-Valera et al., 1985), and *Phormidium* (*Lyngbya*) in de Salins-de-Giraud in the south of France (Caumette et al., 1994). *Aphanothece*, *Aphanocapsa* (*Synechocystis*), *Microcoleus*, *Spirulina*, and *Oscillatoria* were found in a saltern in Bretagne, and in a laboratory model set up to simulate biological processes in the saltern ecosystem (Giani et al., 1989). In an Egyptian saltern, mats composed of *Chroococcus*, *Aphanothece*, *Aphanocapsa*, *Microcoleus*, *Oscillatoria*, *Spirulina*, and *Phormidium* were observed, with a clear stratification: a top 2 – 5 mm thick layer dominated by *Aphanothece*, *Aphanocapsa* and *Chroococcus*; followed by a light green layer (0.2 - 0.9 mm thick) with *Aphanocapsa*, *Aphanothece*, *Microcoleus*, *Spirulina*, and *Oscillatoria*; and a dark green layer (0.5 – 1 mm) with *Spirulina*, *Oscillatoria*, and *Phormidium* (Taher et al., 1995). A checklist of species found in an Indian saltern system contained *Anacystis dimidiatus*, *Coccochloris elabens*, *Gloeocapsa* sp., *Lyngbya majuscula*, *Oscillatoria salina*, *O. formosa*, *Spirulina platensis*, and *Xenococcus acervatus*, all tolerating salinities of up to 174 - 188‰ (Rahaman et al., 1993).

In tropical salinas of the Bahamas, the Caribbean area, Israel, Mexico and South America, the planktonic *Dactylococcopsis* and/or *Synechococcus* may impart a brown color to the brine. *Oscillatoria* may sometimes be present on the bottom of the crystallizer ponds in a layer between the salt and the natural floor materials (Davis and Giordano, 1996). *Dactylococcopsis* and *Synechococcus* were found as the dominant planktonic species in the hypersaline ponds at Yallahs, south Jamaica (Golubic, 1980).

Gypsum crusts that occur in saltern ponds of intermediate salinities are generally densely colonized by different types of cyanobacteria. A great similarity can be found between the biota of the gypsum crusts in the Salins-de-Giraud, France (Caumette, 1993; Caumette et al., 1994), and the salterns of Eilat, Israel (Oren et al. 1995). The biota of the gypsum crusts developing in the Salins-de-Giraud salterns, located on the Mediterranean French coast in the Rhone delta, were well characterized (Caumette, 1993; Caumette et al., 1994). Gypsum crusts were found at salt concentrations between 130 and 200‰. On top of the gypsum crystal layer, a 1 – 2 mm thick layer of *Aphanothece* (*Cyanothece*) was present, embedded in mucoid substance, and imparted a yellow-brown color to the bottom of the ponds during the summer months. Below the 2 mm thick layer of parallel oriented gypsum crystals, an approximately 1 – 2 mm thick green layer was found; inhabited by filamentous cyanobacteria of the genus *Phormidium*. Similar layered gypsum crusts were described from salterns from Alicante, Spain (Rodriguez-Valera et al. 1985), and from Guerrero Negro, Baja California (Rothschild et al., 1994). Thick gypsum deposits with colored layers of unicellular and filamentous cyanobacteria were found in the salterns of Eilat, Israel, at a salinity of 280 – 290 g L^{-1} at the time of sampling (most of this gypsum was actually deposited in a period in which the ponds were kept at a lower salinity (Oren et al., 1995). In this case the cyanobacterial communities (*Aphanothece halophytica* and *Synechococcus*-type unicellular organisms in the upper orange-brown 1 – 2 cm, and *Synechococcus* cells, together with *Phormidium* filaments in the 2 – 4 mm green layer below) were embedded in the gypsum matrix. These resemble the endevaporitic cyanobacterial communities found in gypsum and halite evaporites reported to occur in the Guerrero Negro area (Rothschild et al., 1994). The extent of light penetration to the green layer and to the red-purple layer of photosynthetic purple bacteria found below the cyanobacterial layers was measured by means of optic fiber microprobes. It was found that about 1 % of the photosynthetically active radiation reached the green layer (Oren et al., 1995). The brown color of the upper layer was due to the high carotenoid content of the cyanobacteria (mainly echinenone and myxoxanthophyll). This layer also possessed an extremely high absorption in the near UV range, with a peak around 332 nm and a shoulder around 365 nm.

This absorption was due to a high content of (yet unidentified) mycosporine-like amino acids (MAA's; Oren, 1997). Spectra of the upper cyanobacterial layer in the gypsum crust of the Salins-de-Giraud saltern showed a sharp rise in absorbance from 390 to 350 nm (the lowest wavelength measured; Caumette et al., 1994). These data suggest that UVR-absorbing compounds may be present there in high concentrations. The possible function of these compounds in hypersaline environments is discussed in Section V.B.4.

Aphanothece (*Coccochloris*) mats associated with gypsum crusts were also reported to occur in Australian salterns (Coleman and White, 1993; Jones et al., 1981; Sammy, 1983), and brown-red layers of *Aphanothece* in slime and rich in light-protecting carotenoids were also found in salterns in Egypt (Taher et al., 1995), Brazil (De Medeiros Rocha and Camarra, 1993), and India (Rahaman et al., 1993; Seshadri and Buch, 1958).

As 200‰ salinity is exceeded the number of cyanobacteria decreases, populations of *Dunaliella* increase in abundance, and finally the halophilic archaea predominate in the saturated ponds.

E. Artificial Solar Ponds

Artificial experimental solar ponds were established on the western and northern shore of the Dead Sea. These ponds, intended as large-scale collectors for solar energy, consist of a layer of about 1 m of Dead Sea water, with an approximately 50 cm thick upper layer of fresh water. The latter serves as thermal isolation. Heliothermal heating causes the deeper brines to reach maximum temperatures of 60 - 98°C.

The biology of these stratified artificial lakes with their steep salinity and temperature gradients was studied in detail (Dor and Ehrlich, 1986; Dor and Hornoff, 1985a, 1985b; Dor and Paz, 1989). In the upper convective layer, with salinities between 30 – 140 g L^{-1} and temperatures between 13 - 48°C), different types of cyanobacteria were found; mass occurrences were limited between the boundaries set by a salinity between 30 – 140 g L^{-1}, and a temperature varying from 15 - 33°C. *Aphanothece halophytica* was present at a wide range of temperatures and salinities (16 - 48°C and 40 – 180 g L^{-1} salt). *Phormidium hypolimneticum* developed optimally at 200 g L^{-1} salt and 40°C. Other species found were *Chroococcidiopsis kashaii*, *Chlorogloea fritschii*, *Phormidium hypersalinum*, and *Spirulina subsalsa*. The summer bloom of *Aphanothece halophytica* appeared highly tolerant of high light intensities. In the deeper zone, with temperatures in excess of 35°C, *Aphanothece* formed a soft gelatinous mat (Dor and Paz, 1989). Other photosynthetic microorganisms, such as diatoms, were also present (Dor and Ehrlich, 1986).

Different strains of *Aphanothece halophytica* were isolated from the solar ponds. All proved euryhaline and eurythermal, but they differed in their optimum environmental ranges and in the extent of polymorphism displayed. Cell size generally increased with increasing salinity. Two isolates were studied. In both, the salinity and temperature requirements were shown to be coupled and an increase in temperature was compensated for by an increase in optimum salinity (Dor and Hornoff, 1985a, 1985b).

III. Physiological Properties of the Major Halophilic Cyanobacteria

Though a wide range of cyanobacteria, belonging to different taxonomic groups, were reported to live at high salinities, two major types dominated hypersaline environments: the filamentous *Microcoleus chthonoplastes*, a cosmopolitan mat-building cyanobacterium found from seawater salinity to salinities exceeding 200 g L^{-1}; and the unicellular *Aphanothece halophytica* (also designated as *Aphanocapsa*, *Synechococcus*, *Cyanothece* or *Coccochloris*). These species were the subject of many studies, and more information is known about them than about other halophilic cyanobacteria.

A. Microcoleus chthonoplastes

Microcoleus chthonoplastes is found in a wide range of ecosystems of greatly differing salinities, from marine (Chapter 4) to hypersaline, at salt concentrations exceeding 200‰, and even up to 300‰ (Zavarzin et al., 1993). Strains isolated from habitats all over the world appeared phenotypically and phylogenetically indistinguishable (Garcia-Pichel et al., 1996). However, physiological ecotypes may be distinguished that differ in their optimal salinity for growth and their maximum growth rates, as related to the habitat from which the strains were isolated (Karsten, 1996). A comparative chemosystematic study of the carotenoids and mycosporine-like amino acid compounds in members of the genus *Microcoleus* has also been performed (Karsten and Garcia-Pichel, 1996). Phylogenetic analysis on the basis of 16S rRNA sequences showed *Microcoleus* to be closely related to the *Oscillatoria* group (Garcia-Pichel et al., 1996; Wilmotte, 1994). Due to its mode of growth (thick bundles within a common polysaccharide sheath) *Microcoleus* is one of the main components of cyanobacterial mats worldwide, and other microorganisms find their habitat within the coherent mats built by this form.

The optimal temperature for growth of *Microcoleus* was around 30°C, and optimal rates of photosynthesis were achieved at pH values around 7.5 (best between 7 and 8.5; Dubinin et al., 1992a). Photosynthesis, as measured by oxygen evolution and/or CO_2 photoincorporation, was most rapid at 150‰ salinity, but even at salinities as high as 260‰ photosynthetic activity was detected. However, the most rapid increase in biomass (expressed in terms of synthesis of protein or chlorophyll *a*) was found at 30‰, and the most rapid increase in carbohydrate content at 70‰ salinity (Dubinin et al., 1992a). An important part of the photosynthetically-fixed carbon was probably used for the synthesis of glucosylglycerol; the organic osmotic solute accumulated by *Microcoleus*, and used to achieve osmotic balance with the high salt concentrations of the surrounding medium (Section V.B.2). Rates of photosynthesis in *Microcoleus* mats may be very high, and are limited by the rate of supply of CO_2 (Rothschild and Mancinelli, 1990; Rothschild, 1991).

Microcoleus filaments often are found at or near the interface between the aerobic upper layer of the sediment and the anaerobic deeper layers where dissimilatory sulfate reduction occurs. Filaments often are exposed to sulfide, especially at the end of the night and in early morning, when sulfide produced in the sediment diffuses upward, and sulfide utilization by photosynthetic sulfur bacteria is limited because of lack of available light. Though not as versatile as *Oscillatoria limnetica* from Solar Lake, which is able to grow with high concentrations of sulfide as electron donor (Oren and Padan, 1978; Oren et al., 1985), *Microcoleus* has the capacity to function in the presence of low concentrations of sulfide. It can grow in the presence of 0.15 mM sulfide, and low concentrations of sulfide can actually enhance oxygenic photosynthesis rates (Cohen, 1984b); at concentrations exceeding 1 mM no growth occurs (de Wit et al., 1988). Sulfide was used as electron donor in an anoxygenic type of photosynthesis, in which thiosulfate was the end product of sulfide oxidation (de Wit et al., 1988). Field studies with microelectrodes suggested that oxygenic and anoxygenic photosynthesis occurred simultaneously in *Microcoleus* mats (Jørgensen et al., 1986). *Microcoleus*, like other filamentous cyanobacteria, contains polyunsaturated fatty acids (C14:2, C18:2 and C18:3), the biosynthesis of which is oxygen-dependent. As a consequence no real anaerobic growth will occur. However, in the presence of oxygen, sulfide can drive growth without the participation of photosystem II (de Wit et al., 1988).

As an adaptation to prolonged presence under anaerobic conditions, *Microcoleus chthonoplastes* also possesses a well developed fermentative metabolism that permits fermentation of endogenous carbohydrate reserves, with acetate as the main product. *Microcoleus* is also able to use elemental sulfur as an electron acceptor for anaerobic respiration and this leads to the formation of H_2S in the dark (Dubinin and Gerasimenko, 1993).

B. *Aphanothece halophytica*

Aphanothece halophytica is found worldwide in hypersaline environments. It was first described by Hof and Frémy in 1933, and was extensively studied since Brock (1976) recognized its importance as a model organism for the study of life at high salt concentrations.

The nomenclature of *Aphanothece halophytica* and related organisms is extremely confusing, and it is well possible that morphological types detected in nature and isolated strains described under different names do in fact belong to a single form (as discussed in Borowitzka, 1981). The existing confusion in the nomenclature of these organisms greatly complicates the synthesis of data on physiology and ecology of the different isolates. Thus, Drouet and Daily (1956) incorporated *A. halophytica* in the new taxon *Coccochloris elabens*, and this name was used for the description of an isolate from a salt pond in Puerto Rico (Kao et al., 1973). Brock (1976) used the names *Aphanothece halophytica* and *Aphanocapsa halophytica* in an interchangeable way for his strain from Great Salt Lake, Utah. Spelling variants such as *Aphenotheca* and *Aphenocaspa* (sic) can also be found in the literature (Friedmann et al., 1973). Stanier et al. (1971) placed the organism in typological group 1A of the *Chroococcales* (see also Tindall et al., 1978), and in the classification of Rippka et al. (1979) the species belongs to the genus *Synechococcus*. Thus, the *Synechococcus* strain described from Abu Gabara Lake in Wadi Natrun, Egypt (Imhoff et al., 1978) and from salterns in Queensland, Australia (Coleman and White, 1993) may belong to this group. The genus *Cyanothece*, created by Komárek (1976), includes part of the former genus *Synechococcus*. The strain isolated by De Philippis et al. (1993) from a salt pan in Somaliland, and described as *Cyanothece* sp., was found to be very similar to an isolate from heliothermal salt works in Greece named *Cyanothece halobia* (Roussomoustakaki and Anagnostidis, 1991). In view of the confusing state of the nomenclature, especially in the group of unicellular cyanobacteria, I gladly agree with the statement by Brock (1976) that "it should be emphasized that the species designation has little scientific meaning". An in-depth phylogentic analysis of the unicellular, extremely halotolerant cyanobacteria was published recently (Garcia-Pichel et al., 1998).

To my knowledge few modern taxonomic/ phylogenetic studies, if at all, have been performed with *Aphanothece*. The mol% G+C content of its DNA was reported to be 43 mol% (Waterbury and Rippka, 1989). However, it is still unclear whether *Aphanothece halophytica* Frémy is a single, extremely euryhaline species, adaptable to variety of salinities, or a group of phylogenetically distinct forms.

Significant differences in shape and size were noted for this form, and cell morphology variations complicate the taxonomic picture. The morphological variability was mentioned by Hof and Frémy in 1933, and Yopp et al. (1978a) described their pure culture as consisting of cells variable in size (between 2 and 10 µm) and shape (ellipsoidal, ovoid or cylindrical). Moreover, cell size was found to increase with increasing salinity (Berland et al., 1989; Dor and Hornoff, 1985a; Kao et al., 1973; Yopp et al., 1978a). Morphological mutants have also been isolated (Yopp et al., 1979).

Aphanothece halophytica is mostly found as a benthic species, and its populations cover light-exposed surfaces of sediments and evaporite crusts. Examples of such benthic developments of *Aphanothece* can be found in the Great Salt Lake, Utah (Brock, 1976; Post, 1977), saltern ponds (Caumette, 1993; Caumette et al., 1994; Margheri et al., 1987; Oren et al., 1995), and in other environments, as mentioned in earlier sections. In Solar Lake, Sinai, *A. halophytica* was found to occur as a planktonic species, and formed the dominant component of the phytoplankton during the summer season, when the lake was mixed, and salinity was high (around 180‰; Cohen et al., 1977b). A planktonic strain isolated from Solar Lake was used in studies of the cell wall proteins and their possible relation to cell motility (Simon, 1981); anoxygenic photosynthesis (Garlick et al., 1977); and the properties of RUBISCO (Asami et al., 1983).

Aphanothece halophytica requires a minimum salt concentration of around 30 g L^{-1} and grows up to salinities as high as 300 - 350‰. Most studies indicated optimum concentrations for growth in the range of 60 - 150 ‰ (Borowitzka, 1981; Brock, 1976; Kao et al., 1973; Tindall et al., 1978; Yopp et al., 1978a). Also CO_2 fixation rates were found to be much higher in the presence of 2 M than at 0.25 M NaCl (Takabe et al., 1988). The species displayed a specific requirement for sodium ions, which could not be replaced by potassium or by lithium, and also glycerol does not substitute for salt (Kao et al., 1973). Doubling times of 14.5, 18 and 30 h were measured in the presence of 2, 3, and 4 M NaCl, with an optimal pH of 7 - 7.8 (minimum pH for growth 6.4; Tindall et al., 1978). For an isolate from a solar pond in the Dead Sea area even more rapid growth was reported, with up to four divisions per day. The optimal salt concentration for growth of this strain was found to be temperature-dependent, and increased from 72 - 142 g L^{-1} when the temperature was increased from 30 to 50°C (Dor and Hornoff, 1985a, 1985b).

Aphanothece halophytica used glycine betaine as osmotic solute (Section V.B.3). When exposed to a sudden increase in NaCl concentration from 0.25 to 1 or 2 M, growth temporarily ceased, and resumed after 0.5 - 2 days. After the adaptation period the growth rate was similar to that at the low salt concentration before, but the capacity of the cells for CO_2 fixation when adapted to the presence of 2 M NaCl was 3.7 times higher than of cells growing at 0.25 M NaCl. High NaCl-grown cells also contained increased levels of RUBISCO. The latter perhaps was required for the synthesis of the large amounts of osmotic solute required for cell growth at high salt concentrations (Takabe et al., 1988).

The fact that *A. halophytica* maintains its osmotic balance by the accumulation of organic osmotic solutes, and keeps intracellular ionic concentrations low, is also evident from the fact that its RUBISCO was inhibited by low concentrations (more than 50% inhibition at 0.25 M) of chloride (but not by sodium or potassium ions: Incharoensakdi and Takabe, 1988). In addition, low concentrations of sodium ions (0.25 M) promoted the dissociation of the small subunits of the enzyme (Asami et al., 1983). As in vivo sodium concentrations are expected to be low, the importance of this phenomenon in vivo is not clear. A molecular chaperonin, encoded by the gene *dnaK* was recently characterized. Its genomic transcript increased upon subjection to heat stress and also upon transfer to a hyperosmotic environment, suggesting a role of DnaK in the recovery of *A. halophytica* following hyperosmotic stress (Lee et al., 1997).

Other cellular components of *A. halophytica* that were characterized in detail included C-phycocyanin (Kao et al., 1973), and the plasma membrane lipids. Monogalactosyldiacylglycerol was found to be the most abundant lipid in the plasma membrane of cells grown at 1 M NaCl, while in cells adapted to 3 M NaCl phosphatidylglycerol dominated. The decreasing ratio of uncharged to charged lipids with increasing external salinity was claimed to increase membrane stability in the presence of high salt, and to enable specific modulation of membrane-associated enzyme activity (Ritter and Yopp, 1993).

Aphanothece halophytica communities are often associated with copious amounts of slime. The presence of mucilaginous polysaccharide coatings was reported in many environments (Javor, 1989; Oren et al., 1995). Three independent analyses of the chemical composition of the slime formed by *A. halophytica* gave greatly different results. These data may suggest that three different isolates were involved or that the composition of the slime was highly variable. The "*Cyanothece*" strain isolated from Somaliland had a polysaccharide capsule, the external part of which dissolved in the medium during growth, causing an increase in viscosity. Sugar analyses showed glucose and mannose to be the most abundant sugars, glucose, mannose, xylose, galactose, galacturonic acid, fucose, and glucuronic acid occurring in the ratio of 6.8 : 4.8 : 2.9 : 2.4 : 2 : 1.6 : 1, respectively (De Philippis et al., 1993). Arabinose, ribose, and rhamnose were also detected in the polysaccharides of certain *Cyanothece* strains, and acetyl, pyruvyl and/or sulfate groups may also be present (De Philippis et al., 1998). Uronic acids were absent from the exopolysaccharide formed by an *Aphanocapsa halophytica* strain isolated from an unspecified salt lake, and here fucose was the most abundant sugar (fucose, glucose, mannose, galactose, xylose and rhamnose being found in the ratio 53 : 25 : 15 : 3 : 3 : 2, respectively). Proteins were also present in the slime to about 10% of the total weight, and many of the sugar residues carried sulfate groups (up to 12% by weight). Polysaccharide production increased with increasing nitrate concentration of the medium, and was optimal in the low salinity range (2 - 5% NaCl, with optimal growth at 3 - 12% (Sudo et al., 1995). In the strain investigated by Jones and Yopp (1979), polysaccharide excretion was similar over the whole range of salt concentrations, but was greatly increased when the cultures reached the stationary phase of growth. A similar result was obtained with an isolate designated *Synechococcus* from a gypsum crust from a saltern in Western Australia. Nutrient limitation was probably responsible for the activation of extracellular polysaccharide production (Roux, 1996). Analysis of the slime showed the presence of glucose, fucose, mannose, and galactose in the ratio 1 : 0.6 : 0.32 : 0.23, with 5% minor constituents. Uronic acids or ninhydrin-positive compounds were not detected. The amounts of slime accumulated in *Aphanothece* cultures were also shown to increase with increasing growth temperature (Dor and Hornoff, 1985a). The presence of copious amounts of polysaccharides surrounding the cells makes isolation of *Aphanothece halophytica* from natural samples difficult. Thus, the axenic culture of a strain from a Californian solar evaporation pond was achieved through a protocol involving treatment with different antibiotics, density gradient centrifugation, and ultraviolet radiation (Yopp et al., 1978a).

Aphanothece halophytica can fix nitrogen and has the capacity for anoxygenic photosynthesis with sulfide as electron donor. Nitrogen fixation (acetylene reduction) was detected in evaporite crusts from Guerrero Negro, Baja California, and was attributed to the presence of a "*Synechococcus*"; probably identical to *Aphanothece halophytica* found at similar sites elsewhere. The evaporite crust fixed N_2 in the light both aerobically and anaerobically (up to 2.4 nmol C_2H_4 produced g^{-1} evaporite h^{-1}), but no dark fixation was found (Rothschild et al., 1994). Sulfide-dependent anoxygenic photosynthesis was detected in the Solar Lake strain of *A. halophytica* (Garlick et al., 1977).

IV. Anoxygenic Photosynthesis by Cyanobacteria in Hypersaline Environments

In many hypersaline environments, such as the hypolimnion of Solar Lake, Sinai (during winter stratification), hypersaline sulfur springs, and salt lakes and salt flats, the cyanobacterial populations are exposed to high concentrations of sulfide, either permanently or periodically. The sulfide may be of geothermal origin, as is the case in certain sulfur springs, but in most cases it is derived from dissimilatory sulfate reduction, which is the most important terminal process in anaerobic degradation in marine environments. Benthic cyanobacterial mats may be exposed to high sulfide concentrations in early morning.

Certain species of cyanobacteria are able to use sulfide as the electron donor in an anoxygenic type of photosynthesis, producing elemental sulfur or thiosulfate and using photosystem I only. The physiology and ecology of anoxygenic photosynthesis in cyanobacteria was reviewed by Padan (1979a; 1979b). No direct link exists between halophilism and anoxygenic photosynthesis, and sulfide-dependent anoxygenic photosynthesis was detected in certain freshwater environments (Cohen, 1984b; Padan, 1979b). However, the most comprehensive studies on anoxygenic photosynthesis in cyanobacteria were performed in the hypersaline Solar Lake. The halophilic Solar Lake isolate *Oscillatoria limnetica* became a laboratory model for the study of the biochemistry of anoxygenic photosynthesis in cyanobacteria.

A. Solar Lake

As described above (Section II.B.3), Solar Lake is stratified in winter. The hypolimnion, below a depth of 1 - 2 m, is anaerobic and becomes rich in sulfide (up to 1 mM and even higher) as a result of high rates of dissimilatory sulfate reduction in the bottom sediments. The temperature of the hypolimnion brines (salinity around 180‰) rises during winter stratification to values as high as 61°C due to heliothermic heating (Cohen et al., 1975a; Cohen et al., 1977a). The high transparency of the shallow upper layer allows good light penetration down into the hypolimnion, so that dense communities of sulfide-utilizing anoxygenic phototrophs can develop there (Cohen et al., 1977b).

The filamentous species identified as *Oscillatoria limnetica* (Section II.B.3) forms dense blooms at the lower end of the hypolimnion near the bottom, where sulfide concentrations are the highest. Biomass and primary productivity may be extremely high in this layer (Cohen et al., 1977b). The strain was isolated and was studied extensively during the past twenty years.

Oscillatoria limnetica can shift between two types of photosynthesis: oxygenic photosynthesis, characteristic of all cyanobacteria, in which two photosystems cooperate to split water and allow CO_2 photoassimilation; and anoxygenic photosynthesis, in which sulfide serves as electron donor in a process which does not require participation of photosystem II. Accordingly, DCMU, an inhibitor of photosynthetic electron transport at the acceptor side of photosystem II, does not inhibit sulfide-dependent CO_2 photoassimilation (Cohen et al., 1975b). Photosystem II does not participate in the process, as shown from a comparison of photosynthetic action spectra in the presence and absence of DCMU (Oren et al., 1977). The phenomenon is due to the fact that low concentrations of sulfide inhibit electron flow at the donor side of photosystem II (Oren et al., 1979). It may be noted that in other systems examined, the participation of photosystem II to the photosynthetic processes in the presence of sulfide was significant (Cohen, 1984a). The optimal sulfide concentration for the process was around 2 - 3 mM, and elemental sulfur was shown to be the sole product of sulfide oxidation (Cohen et al., 1975c). The shift from oxygenic to anoxygenic photosynthesis required a lag period of a few hours, and involved synthesis of new proteins (Oren and Padan, 1978), one of which was a sulfide-quinone reductase, the key enzyme required to feed electrons from sulfide into the photosynthetic electron transport chain between the acceptor site of photosystem II and the donor site of photosystem I. This enzyme was partially characterized (Arieli et al., 1994).

Oscillatoria limnetica can fix CO_2 in the light in the presence of high sulfide concentrations, and can also grow anaerobically in the presence of sulfide (Oren and Padan, 1978). Growth in the presence of molecular oxygen requires special adaptations such as the absence of polyunsaturated fatty acids (abundant in the membrane lipids of most filamentous cyanobacteria), and the presence of an oxygen-independent pathway to synthesize monounsaturated fatty acids. *Oscillatoria limnetica* lacks polyenoic acids (Oren et al., 1985), and produces monoenoic acids also under anaerobic conditions: upon anaerobic incubation in the presence of sulfide and DCMU, an increase in abundance of Δ7 and Δ9 C16:1 and Δ9 and Δ11 C18:1 fatty acids was observed; isomers characteristically formed by the bacterial, anaerobic pathway of synthesis of unsaturated fatty acids, rather than by the oxygen-requiring desaturation of the saturated equivalents hexadecenoid and octadecenoic acid (Jahnke et al., 1989).

Another adaptation to anaerobic life found in *Oscillatoria limnetica* is the presence of different modes of energy generation under anaerobic conditions in the dark. One such mechanism involves anaerobic respiration of endogenous storage materials, using elemental sulfur (formed during the day by photosynthetic oxidation of sulfide) as electron acceptor, with the formation of CO_2 and sulfide (Oren and Shilo, 1979). Indications exist that this type of metabolism may be operative in Solar Lake, where sulfide formation at the expense of elemental sulfur was observed below the chemocline during the night (Jørgensen et al., 1979). Another mode of anaerobic growth that reflects the metabolic flexibility of *Oscillatoria limnetica* is fermentation of endogenous carbon reserves, with the formation of lactate as the main product (Oren and Shilo, 1979). Whether lactate is the sole product as suggested remains to be ascertained; acetate and ethanol were found as additional products during anaerobic fermentation of endogenous carbon reserves in a marine *Oscillatoria* sp. (Heyer et al., 1989; Moezelaar et al., 1996).

Screening of the Solar Lake cyanobacterial isolate *Aphanothece halophytica* showed that this strain also had the capacity to use sulfide as electron donor in anoxygenic photosynthesis (Garlick et al., 1977). However, this form was much less sulfide tolerant than *Oscillatoria limnetica*, and the optimal sulfide concentration for the process was 0.7 mM only. Also in the *Aphanothece* strain, elemental sulfur was identified as the product of sulfide oxidation (Garlick et al., 1977). *Microcoleus* in the Solar Lake microbial mat is another versatile form that may use oxygenic and anoxygenic photosynthesis simultaneously in the presence of low sulfide concentrations. In this case photosystem II was partially inhibited, and anoxygenic photosynthesis was partially induced (Cohen, 1984b; Cohen et al., 1986; Jørgensen et al., 1986).

B. Hamei Mazor

While in Solar Lake the high sulfide concentrations are due to biological activity, the sulfide content (2.5 mM) of the Hamei Mazor springs (Section II.*C*) is geothermal. As a result of the low pH of the water (~pH5.2), most of the sulfide presumably exists as the undissociated form (H_2S), which is much more toxic to life than the ionized forms HS^- and S^{2-} present at higher pH values. The *Oscillatoria* mat (Plates 17a, b) that covered the pools and outflow channels of the spring displayed high rates of CO_2 photoassimilation in the presence of sulfide, and little or no inhibition was observed by DCMU (Oren, 1989). Little additional work was done on the strain, and no characterization of fatty acid composition was performed, to assess whether anaerobic growth occurred there.

C. Other Environments

Anoxygenic photosynthesis with sulfide as electron donor was implicated as an ecologically important process in a number of additional hypersaline environments:

— The *Synechococcus* strain isolated from the hypersaline alkaline lakes of Wadi Natrun was shown to be able to perform anoxygenic photosynthesis in the presence of sulfide (Imhoff et al., 1978).

— In the gypsum crusts of the Salins-de-Giraud salterns in the south of France, sulfide from the lower layers was found to reach the *Phormidium* layer at the end of the night (Caumette, 1993; Caumette et al., 1994). The possibility that in the early morning hours the cyanobacterial community oxidizes sulfide photosynthetically is well worth an investigation.

— In the lagoons of Lake Sivash (east Crimea), *Microcoleus* mats also used sulfide as donor, which was oxidized to thiosulfate (Zavarzin et al., 1993).

— In a microbial mat in a hypersaline (74‰ salinity) lagoon on San Salvador Island, Bahamas, DCMU-amended samples had 25% of photosynthetic activity of the control. This anoxygenic photosynthetic activity was possibly due to *Microcoleus* (Pinckney and Pearl., 1997).

V. Osmotic Adaptation of Cyanobacteria Living at High Salt Concentrations

To be able to withstand the high osmotic pressure exerted by the salt concentrations in their surrounding medium, cyanobacteria living at high salinities possess mechanisms to maintain osmotic equilibrium and cell turgor. While ions (K^+, Cl^-) can temporarily enter the cells to counteract rapid increases in medium salinity, in the long term, organic solutes are accumulated to provide osmotic balance, as required in organisms whose enzymatic machinery does not operate properly at high salt concentrations. There were a few reports in the older literature on the occurrence of high intracellular salt concentrations in certain cyanobacteria growing at high salt concentrations (Miller et al., 1976; Yopp et al., 1978b). However, it became increasingly clear that halophilic and halotolerant cyanobacteria attempt to maintain a low salt cytoplasm. Different organic osmolytes occur in cyanobacteria: the disaccharides sucrose and trehalose (especially in the less salt tolerant types); glucosylglycerol (in species with a moderately halotolerance); and glycine betaine (in *Aphanothece halophytica* and a few other types that tolerate high salt concentrations). Our understanding of the mechanisms of osmoregulation in cyanobacteria increased greatly in recent years, especially in the case of glucosylglycerol metabolism in *Synechocystis* strain PCC 6803; the molecular analysis of which progressed rapidly (Hagemann and Zuther, 1992; Hagemann et al., 1996). Two recent review papers give an overview of our present understanding (Hagemann et al., 1998; Joset et al., 1996).

A. Ion Metabolism

Before the discovery of organic compatible solutes, it was suggested that intracellular salt concentrations in halophilic cyanobacteria were quite high. Miller et al. (1976) reported intracellular potassium concentrations as high as 1 M in *Aphanothece halophytica*, with low intracellular sodium concentrations. Blue dextran was used in these studies to estimate the intracellular and extracellular space in cell pellets. Intracellular potassium concentrations were found to increase with the salt concentration of the growth medium, and thus potassium was claimed to be the main osmoticum in *A. halophytica* (Miller et al., 1976; Yopp et al., 1978b).

However, claims of high intracellular salt concentrations in cyanobacteria are problematic in view of the salt sensitivity of intracellular enzymes. Based on a study of the inhibitory effect of salts on RUBISCO activity in *Aphanothece halophytica*, it was concluded that in order for the enzyme to be active in vivo, the intracellular chloride concentration may not exceed 150 mM (Incharoensakdi and Takabe, 1988). Later studies showed lower salt concentrations inside halophilic cyanobacteria. Intracellular chloride concentrations in *Aphanothece halophytica* were found to increase from 35 mM to 150 mM when the medium NaCl concentration increased from 0.5 to 2 M (Incharoensakdi and Takabe, 1988). Reed et al. (1984a) detected only low internal potassium concentrations (180 - 250 mM) in *A. halophytica*. Also, in other species (*Coccochloris elabens*, *Dactylococcopsis salina*, *Synechococcus* sp. DUN52) intracellular potassium was low – between 170 - 310 mM – in cells grown in media with between 0.5 - 4 times the salt concentration of seawater (Reed et al., 1984a). The presence of a relatively large periplasmic space may influence estimates of intracellular salt concentrations. Intracellular salt concentrations were measured in *Synechocystis* PCC 6803 – a strain which can tolerate up to 1.2 M salt – using centrifugation through a layer of silicone oil; a method which does not discriminate between the cytoplasm and periplasmic space. Cells grown at 0.65 M and 1.03 M NaCl had apparent intracellular sodium concentrations of 0.22 and 0.51 M Na, and potassium concentrations of 0.15 and 0.30 M, respectively (Hagemann et al., 1994).

Upon sudden increases in salt concentration in the outside medium, salts may accumulate transiently to achieve a rapid osmotic balance before sufficient amounts of organic osmotic solutes are produced. The latter is a slow process that can take many hours, and it requires adaptation of the photosynthetic apparatus. A multiphasic osmotic adjustment was documented in the euryhaline *Synechocystis* PCC 6714 upon a shift from freshwater to 0.5 M NaCl; a rapid entry of NaCl was observed which permitted a partial recovery of the volume within 2 minutes; followed by an exchange of sodium by potassium in 20 minutes; then accumulation of the organic osmotic solutes glucosylglycerol and sucrose enabled the extrusion of potassium over a period of 24 hours (Reed et al., 1985). Transient accumulation of salt was also documented in *Spirulina subsalsa*, isolated from the Bardawil lagoon in Sinai, Egypt, and growing between 0.25 and 2.5 M NaCl. Following an increase in medium salinity, intracellular sodium and chloride were temporarily greatly increased. Subsequently, sodium was extruded by means of a Na^+/H^+ antiporter and driven by the proton gradient generated by respiration (Gabbay-Azaria and Tel-Or, 1991; Gabbay-Azaria et al., 1992). Respiration rates were enhanced following salt upshock, and cytochrome oxidase was suggested to be involved in the generation of the proton gradient required for Na^+ extrusion via the Na^+/H^+ antiporter (Gabbay Azaria and Tel-Or, 1993). During this adaptation phase, the intracellular concentration of glycine betaine, the osmotic solute found in this organism, increased to achieve a new osmotic equilibrium.

Influx of sodium ions upon salt upshock may have a temporary inhibitory effect on the photosynthetic system, as suggested from studies with *Agmenellum quadruplicatum*, a marine coccoid cyanobacterium which grows from 0 to over 10% NaCl. Upon transfer from 1.85% to 7% NaCl, photosynthesis was severely depressed and recovered only after several hours, during which time excess sodium was pumped out of the cells (Batterton and van Baalen, 1971).

B. Organic Osmotic Solutes

The importance of organic osmotic solutes in the adaptation of cyanobacteria to life at high salt concentrations became clear in the early 1980s as a result of the use of techniques such as HPLC and ^{13}C-NMR spectroscopy. A large-scale survey of the occurrence of osmotic solutes showed that cyanobacteria can be divided into three groups, both according to the types of solutes accumulated, and to salt tolerance: freshwater; marine; and halophilic isolates. Freshwater strains accumulate sucrose and/or trehalose under salt stress. Marine strains, adapted to the presence of seawater salinities, and often tolerating much higher salt concentrations, generally use the heteroside glucosylglycerol (2-*O*-α-D-glucopyranosyl-(1→2)-glycerol) to achieve osmotic equilibrium. Finally, the most salt tolerant strains accumulate high intracellular concentrations of glycine betaine (Mackay et al., 1984; Reed et al., 1984b). Additional osmotic solutes are sometimes found, such as L-glutamate betaine (*N*-trimethyl-L-glutamate), detected in combination with sucrose and/or trehalose in halophilic *Calothrix* isolates (Mackay et al., 1984).

1. Sucrose and Trehalose

The disaccharides sucrose and trehalose are of little importance for the osmotic adaptation of cyanobacteria that live at high salinities because they are not very effective as osmotic solutes, and provide osmotic protection only up to relatively low salinities. Freshwater cyanobacteria may produce sucrose and/or trehalose as a reaction to salt stress (Blumwald and Tel-Or, 1982; Gabbay-Azaria and Tel-Or, 1993), and even then the concentrations accumulated intracellularly may be insufficient to balance the osmotic pressure of the outside medium, as found in the case of sucrose in *Synechococcus* PCC 6311 (Blumwald et al., 1983). When grown at low salt concentrations, also *Microcoleus chthonoplastes* produces trehalose as osmotic solute, replaced by glucosylglycerol at higher salinities (Karsten, 1996).

2. Glucosylglycerol

Glucosylglycerol was first identified by means of natural abundance ^{13}C-NMR in a *Synechococcus* sp. isolated from intertidal rocks in Australia, growing from 30 mM to 1.69 M total salts (Borowitzka et al., 1980). Glucosylglycerol was detected in a wide variety of salt tolerant cyanobacteria. Some of the most widespread halotolerant species accumulate the solute, such as *Microcoleus chthonoplastes* (Kevbrin et al., 1991), different *Synechococcus* and *Synechocystis* isolates (Borowitzka et al., 1980; Richardson et al., 1983), *Spirulina platensis* (Warr et al., 1985), *Agmenellum quadruplicatum* (Tel-Or et al., 1986), and *Microcystis firma* (Erdmann et al., 1992). Large concentrations of glucosylglycerol were detected in natural communities of *Microcoleus chthonoplastes* in Lake Sivash, East Crimea (Zavarzin et al., 1993), and in a hypersaline microbial mat derived from Solar Lake, Sinai (Oren et al., 1994).

Intracellular concentrations of glucosylglycerol can be very high. In *Microcoleus* grown in 140 g L^{-1} salt, glucosylglycerol accounted for 30% of the cell dry weight (Zavarzin et al., 1993). Upon increase of salinity by hyperosmotic shock, additional glucosylglycerol was produced. In the marine strain *Synechococcus* N100, CO_2 was the main carbon source for glucosylglycerol biosynthesis initiated through an increase of medium NaCl concentration from 0.25 to 0.75 M. However, up to 10% may be derived from available intracellular organic carbon reserves. In the dark, no increase in glucosylglycerol was observed (Mackay and Norton, 1987). Synthesis of glucosylglycerol was a slow process, and was only the final phase in a series of events which occur during salt upshock. Multiphasic osmotic adjustment in the euryhaline *Synechocystis* PCC 6714 upon a sudden salinity increase involved rapid entry of NaCl, followed by exchange of sodium by potassium ions, and many hours of adaptation for the completion of glucosylglycerol accumulation to the proper new level required for osmotic balance with the new higher salinity (Reed et al., 1985; Section V.A). Upon salt downshock, *Synechocystis* PCC 6714 released part of the excess glucosylglycerol to the medium (Fulda et al., 1990). In *Synechocystis* up to 90% of the low-molecular weight compounds were excreted after hypoosmotic shock. This hypoosmotically induced exudation was attributed to transient changes in nonspecific membrane permeability (Reed et al., 1986). However, in the marine *Agmenellum quadruplicatum* salt downshock induced increased carbohydrate turnover, and excess glucosylglycerol was probably metabolized to glycogen (Tel-Or et al., 1986).

Biosynthesis of glucosylglycerol was studied in depth, including at the molecular biological level, in *Synechocystis* strain PCC 6803 in a series of papers by the group of Hagemann. These studies provided a detailed picture of the mechanism and the events that determine the regulation of glucosylglycerol biosynthesis. *Synechocystis* PCC 6803 is a euryhaline cyanobacterium and was isolated from a freshwater habitat but it can grow at salinities 2.5 - 3 times those of seawater (up to about 1.2 M NaCl; Richardson et al., 1983; Erdmann et al., 1992). Intracellular glucosylglycerol concentrations were relatively low in cells grown in the presence of 0.65 and 1.03 M NaCl - 45 and 92 mM - respectively, and apparent intracellular sodium and potassium concentrations were high (Hagemann et al., 1994; Section V.A).

Glucosylglycerol is synthesized from ADP-glucose and glycerol-3-phosphate, with glucosylglycerol phosphate as an intermediate, in a two step reaction. A convenient assay for the glucosylglycerol phosphate phosphatase was developed (Schoor et al., 1996), and the *stpA* gene, encoding the glucosylglycerol-phosphate phosphatase, was characterized on the molecular level (Hagemann et al., 1997b).

The enzyme system requires activation; in vivo by hypertonic salt concentrations and in vitro by NaCl addition at the stage of enzyme extraction or assay. Salts such as KCl or $NaNO_3$ were less effective than NaCl as activators, and organic osmolytes were inactive (Hagemann and Erdmann, 1994). The activation of glucosylglycerol biosynthesis took place in presence of chloramphenicol, and thus was not dependent on the synthesis of new proteins (Hagemann et al., 1990). Mutants of *Synechocystis* PCC 6803 were isolated that are unable to grow at elevated salt concentrations. Three of these mutants were obtained by random cartridge mutagenesis, and one of these was shown to be defective in synthesis of glucosylglycerol (Hagemann and Zuther, 1992). The mutant acquired salt resistance upon addition of exogenous glucosylglycerol that was taken up from the medium (Mikkat et al., 1996). This mutant, which tolerated up to 250 mM NaCl only, was found to be defective in the enzyme glucosylglycerol-phosphate phosphatase, and accumulated glucosylglycerol-phosphate intracellularly (Hagemann et al., 1997b).The latter compound was not effective as osmoprotectant and seemed to be toxic (Hagemann et al., 1996).

Synechocystis PCC 6803 has an active transport system for glucosylglycerol that permits uptake of glucosylglycerol from the medium. Uptake activity was enhanced in cells adapted to increasing concentrations of NaCl. Uptake was energy dependent, and was inhibited by uncouplers. The affinity of the uptake system was relatively low (K_m ≅50 μM). Only sucrose and trehalose were found to compete for the transport system. The glucosylglycerol transport protein is encoded by the gene *ggtA*. Transcription of this gene is increased in cells adapted to high salt concentrations. When mutant strains impaired in this gene are grown in high salt media, significant amounts of glucosylglycerol were found in the medium, indicating that the transport system is mainly necessary for recovery of glucosylglycerol that has leaked through the cytoplasmic membrane (Hagemann et al., 1997a; Mikkat et al., 1996, 1997). Transport systems for the osmoprotective compounds trehalose and sucrose were also identified (Mikkat et al., 1997).

At least nine specific proteins were found at increased levels in *Synechococcus* PCC 6803 cells when they were grown at high salinities; the levels of a few other proteins were depressed (Hagemann et al., 1994). Upon salt shock the rate of protein synthesis dropped at first, and was followed by synthesis of specific proteins. Some of these proteins were transiently synthesized during the first hours of the adaptation phase, while others remained to be synthesized at enhanced rates in salt-adapted cells (Hagemann et al., 1991).

3. Glycine Betaine

Glycine betaine is characteristically found as an osmotic solute in the most salt tolerant of the cyanobacteria. Examples of strains that accumulate this compound include *Aphanothece halophytica* (Reed et al., 1984a), *Spirulina subsalsa* isolated from Bardawil Lagoon in Sinai, Egypt, an environment with fluctuating salinity (Gabbay-Azaria et al., 1988), *Dactylococcopsis salina* (Moore et al., 1987), and *Synechocystis* DUN52 (Reed et al., 1984a). The last-named strain was isolated from calcareous stromatolites of intertidal flats in Kuwait. It was identified as a *Synechocystis* rather than *Aphanothece*. Below 70‰ salinity the cells divided in more than one plane. Intracellular glycine betaine concentrations as high as 1.18, 2.43, and 2.98 M were measured in cells grown at 60‰, 100‰ and 200‰ salinity, respectively (Mohammad et al., 1983). Glycine betaine was detected in massive amounts in the *Oscillatoria* mat covering the bottom of the hypersaline sulfur spring of Hamei Mazor (Section II.C), using proton NMR spectroscopy (Oren et al., 1994).

In most cases glycine betaine was found as the sole organic osmotic solute, but in one *Gloeocapsa* strain glycine betaine was reported to occur together with trehalose (Mackay et al., 1984). *Synechocystis* DUN52 contained minor amounts of glucosylglycerol in addition to glycine betaine (Mohammad et al., 1983).

Glycine betaine was very effective in protecting enzymatic activities against the inhibitory action of salt. Glycine betaine protected the structural integrity and activity of Rubisco against the damaging influence of salts in *Aphanothece halophytica* (Incharoensakdi and Takabe, 1988; Incharoensakdi et al., 1986), and restored K^+- induced loss of activity of glucose-6-phosphate dehydrogenase in the same organism (Hawkins et al., 1987). The glucose-6-phosphate dehydrogenase of *Spirulina subsalsa* was markedly inhibited by salt (50% inhibition at 1.25 M NaCl), but in the presence of glycine betaine full activity was obtained at NaCl concentrations as high as 1.5 M (Gabbay-Azaria et al., 1988). In *Synechocystis* DUN52 KCl was stimulatory for glutamine synthetase at concentrations up to 1.3 M, but was inhibitory above 1.4 M. Glycine betaine did not inhibit up to 1.8 - 2 M, and the presence of 1 M betaine alleviated NaCl inhibition by 10 - 30% at NaCl concentrations between 0.8 - 2 M (Warr et al., 1984).

Upon a salt upshock from 0.5 to 1.5 - 2 M NaCl, plasmolysis occurred initially in *Aphanothece halophytica*, and electron micrographs showed that the thylakoid membranes were structurally disorganized. The rate of CO_2 fixation fell immediately, and returned to normal levels within one day after synthesis of sufficient quantities of glycine betaine. The adaptation process was light-dependent, and no betaine was found to accumulate in the dark (Ishitani et al., 1993).

Cyanobacteria that form glycine betaine are also able to accumulate glycine betaine from the outside medium by means of an active transport system. Such transport activities were detected in *Aphanothece halophytica*, *Dactylococcopsis salina*, *Synechococcus* PCC 7418, and *Synechocystis* DUN 52. The transport served as a scavenging system for exogenous glycine betaine, and use of glycine betaine present in the medium may be an effective strategy in environments of fluctuating salinity (Moore et al., 1987). Glycine betaine transport was not found in halophilic cyanobacteria that accumulated sucrose or glucosylglycerol. In *Aphanothece halophytica*, glycine betaine was transported by a light-dependent active transport system which was sensitive to changes in external salt concentration; both transiently and in the long term. Uptake can be rapid. *Aphanothece* accumulated 120 mM glycine betaine intracellularly within 30 minutes when 1 mM was supplied in the medium, and 1 hour of incubation of *Synechocystis* DUN52 with 1 mM glycine betaine resulted in an intracellular concentration of 250 mM. The transport system had a K_m of 2 μM for glycine betaine, a V_{max} of 45 nmol min^{-1} mm^{-3} cell volume, and was optimally active at pH8 - 8.5. Sodium chloride concentrations above 80 mM were required for stimulation of the transport activity. Severe hyperosmotic stress (1 M NaCl) reduced the rate of glycine betaine uptake, but increased the internal betaine concentration (Moore et al., 1987).

Upon salt downshock excess glycine betaine may be excreted from the cells, and released into the surrounding medium (Moore et al., 1987). The release may possibly be due to transient permeability changes of the cell membrane that allow leakage of low molecular weight metabolites (Reed et al., 1986). However, in one study, also involving *Aphanothece halophytica*, a slow decrease (within 10 hours) of intracellular glycine betaine concentrations was observed upon dilution of the medium from 1.5 to 0.5 M NaCl, without any release to the outside medium (Ishitani et al., 1993).

4. Mycosporine-Like Amino Acids as Osmotic Solutes ?

Large concentrations of mycosporine-like amino acids (MAAs) were found in a community of unicellular cyanobacteria that inhabited a gypsum crust developing on the bottom of a hypersaline saltern pond in Eilat, Israel (Oren et al., 1995; Section II.D; Chapter 21). This was the first report of the occurrence of MAAs in a halophilic cyanobacterial community. Two MAAs were detected, one with an absorption maximum at 332 nm, and one at 365 nm. Intracellular MAA concentrations in the cyanobacterial community were estimated to be at least 98 mM, and this already high value was probably an underestimate. With an average molecular weight of around 300, MAAs should thus contribute at least 3% of the cell wet weight. While MAAs were shown to act as sunscreen compounds, protecting the cells against solar UV radiation (Castenholz and Garcia-Pichel, 1997; Garcia-Pichel and Castenholz, 1993), when they occur at such high concentrations they may have an osmotic function as well, and help the cells cope with the high salt concentration of their environment. The possible implications of the accumulation of MAAs in increasing intracellular osmotic pressure was discussed by Garcia-Pichel (1994). When material from the upper layer of the Eilat gypsum crust was subjected to slow dilution with distilled water, MAAs appeared rapidly in the outer medium and the extent of loss of intracellular MAAs was approximately proportional to the extent of the dilution stress applied (Oren, 1997). The mechanism of release may again be related to transient permeability changes of the cell membrane, as was suggested for glycine betaine release (Reed et al., 1986).

VI. Interactions Between Cyanobacteria and Other Microorganisms in Hypersaline Environments

Cyanobacteria are the main primary producers in many hypersaline environments, especially in the upper layers of the sediments, where *Microcoleus*, *Aphanothece*, and other species may form dense mats. Here, cyanobacterial activity determines, to a large extent, the biological properties of the entire ecosystem, and cyanobacteria can thus be expected to interact in different ways with the other members of the biota present. In this section three specific interactions are considered: the role of osmotic solutes produced by cyanobacteria in the carbon cycle of hypersaline environments; the presence of filamentous photosynthetic bacteria associated with bundles of *Microcoleus chthonoplastes*; and the formation and removal of peroxides in *Microcoleus* communities.

A. Cyanobacterial Osmotic Solutes and the Carbon Cycle

The massive accumulation of organic osmotic solutes that results in intracellular concentrations in the molar range has important implications for the carbon cycle in hypersaline environments where primary production is dominated by halophilic cyanobacteria. These osmotic solutes may become available to the heterotrophic communities either when leaking out of the cyanobacterial cells, as a reaction to a decrease in salinity of the environment or upon cell death. Methylated amines were shown to be the main precursor for methane formation in hypersaline environments and glycine betaine, the main compatible solute in the most halotolerant cyanobacterial strains (and also produced by certain anoxygenic phototrophic eubacteria), is the main precursor for the formation of these methylated amines (Oren, 1990). Only recently the first halophilic degrading bacteria that degrade glycine betaine were characterized (*Acetohalobium arabaticum* and related organisms). These may be responsible for the formation of methylated amines from glycine betaine at high salinities (Zhilina and Zavarzin, 1990).

Interesting interactions were shown to occur between the cyanobacterium *Microcoleus chthonoplastes* and halophilic archaea in the microbial mats of the Sivash lagoon (Arabat, Crimea) at salinities between 150 - 300‰ (Zvyagintseva et al., 1995). In these mats, at least 51% of the aerobic heterotrophic bacteria recovered as colonies were halophilic archaea, and this percentage increased to 70 - 80% at the higher salinities examined. The species most frequently isolated were *Haloarcula japonica* and *Halorubrum distributum*. These archaea were thus considered as ecologically significant components of the cyanobacterial mat community. Combined cultivation experiments of *Microcoleus* and halophilic archaea at 150‰ (the upper salt concentration limit at which *Microcoleus* grew), showed the existence of interesting metabolic interactions. *Microcoleus* secreted a number of organic acids into the medium and these were used as carbon sources for the archaea. Paper chromatography of the medium components after incubation of *Microcoleus* in the presence of $^{14}CO_2$ showed the presence of acetate, butyrate, α-ketoglutarate, succinate, malate, fumarate and citrate (Zvyagintseva et al., 1995). The halophilic archaea may further stimulate growth of the *Microcoleus* by removing oxygen and restricting its accumulation to toxic levels (Zvyagintseva et al., 1995).

B. Filamentous Photosynthetic Bacteria associated with Microcoleus chthonoplastes

An interesting association of the aerobic *Microcoleus chthonoplastes* with presumably anaerobic filamentous purple bacteria may exist within *Microcoleus* bundles. Electron microscopic examination of microbial mats of Solar Lake (82‰ salinity) and Guerrero Negro (72 - 91‰ salinity) showed the occurrence of gliding filamentous phototrophic purple bacteria with stacked photosynthetic lamellae between the *Microcoleus* filaments within the common sheath (D'Amelio et al., 1987). One possibility is that these purple bacteria use toxic sulfide to which the cyanobacterial bundles were often exposed while living in sharp gradients of oxygen and sulfide (Section III.A). Alternatively, the bacteria may live on elemental sulfur and thiosulfate formed by *Microcoleus* in the process of anoxygenic photosynthesis (Section IV), or grow photoheterotrophically on organic compounds excreted by the cyanobacterium. The filamentous purple bacterium was not cultured, and a further analysis of the interactions between the halophilic cyanobacterium and the anoxygenic phototrophic bacterium remain to be completed.

C. Formation and Removal of Peroxides in Microcoleus *Communities*

Microcoleus accumulates peroxides during oxygenic photosynthesis. Hydrogen peroxide was produced in the light as a by-product of photosynthetic electron transport at the step of ferredoxin oxidation, and up to 40 % of the electron flow through the electron transport chain was used for peroxide formation (Dubinin et al., 1992b). At least part of the hydrogen peroxide may be derived from dismutation of superoxide radicals by the Fe-containing superoxide dismutase present in the cytoplasm. As *Microcoleus* lacks catalase, peroxide accumulation can easily lead to growth inhibition. The presence of catalase supplied by the heterotrophic components of the mat community prevented autoinhibition of *Microcoleus* by peroxide (Dubinin et al., 1992b).

VII. Biotechnological Aspects of Halophilic Cyanobacteria

The development of massive microbial mats dominated by cyanobacteria in solar saltern ponds is a natural event, but it appears that these mats contribute greatly to the salt production process. Microorganisms and the color they impart to the brine in these ponds are absolutely essential to salt production, since vapor pressures reach such low values as to seriously retard evaporation. Increased solar absorption caused by the algae leads to elevated temperatures, which help to offset the effects of the low vapor pressures, and thus permit adequate evaporation (Davis, 1974).

The presence of a healthy biological system, dominated by cyanobacterial mats at the lower salinities, in solar saltern facilities, is essential for the following reasons: to aid in solar absorption; to release organic matter at the proper levels to enable efficient production of high quality salt; to strip and sequester minerals from the overlying brine (Davis, 1974, 1993); and to control brine leakage (Davis, 1978, 1993; Jones et al., 1981). Fertilization was used in cases where insufficient biota developed (Davis, 1978, 1993). Guidelines for the effective management of saltern ponds with respect to organic matter content were devised (De Medeiros Rocha and Camara, 1993; Sammy, 1983).

Aphanothece releases massive amounts of slime (De Philippis et al., 1993; Jones and Yopp 1979, Sudo et al., 1995; Section III.B). This polysaccharide slime was suggested to have promising industrial uses (De Philippis et al., 1998). The mucilaginous material excreted by *Aphanothece* and other cyanobacteria helps to seal the porous bottom of the evaporation ponds, and thus reduces loss of brine to the ground water (Borowitzka, 1981). However, excess mucilage can impart a high viscosity to the brine, which causes problems in the harvest and the quality of the salt (Borowitzka, 1981; Coleman and White, 1993; De Medeiros Rocha and Camara, 1993; Jones et al., 1981). Fluctuating salinities and high nutrient levels cause *Aphanothece* to produce excessive amounts of organic matter, leading to decreased quantity and quality of the salt harvest. Disturbances (e.g. after input of excessive nutrients, or a steep increase in brine salinity, or brine dilution), may result in excessive reproduction and mucilage release by *Aphanothece halophytica* (Davis, 1993), and parts of the mat may be released from the bottom, to float downstream in the system. Disruptions to one pond have the capacity to cascade through subsequent ponds (Coleman and White, 1993). Excess organic matter causes the formation of hollow or small salt crystals that retain liquid contaminants and contain high levels of magnesium and manganese (Coleman and White, 1993; Davis and Giordano, 1996; De Medeiros Rocha and Camara, 1993). It was recommended that the *Aphanothece* mat should not be fertilized through the input of low salinity water, organic wastes, etc., as its blooms may eventually disturb the formation and precipitation of NaCl (De Medeiros Rocha and Camara, 1993). In Indian saltworks, filamentous cyanobacteria (*Lyngbya majuscula*, *Oscillatoria salina*, *Spirulina platensis*) were reported to form scums on the surface of the ponds which prevented light penetration and hampered the salt crystallization process (Rahaman et al., 1993).

Management practices that maintained biodiversity, and restrained reproduction and organic release by the cyanobacteria, permitted microorganisms to efficiently strip and oxidize nutrients from the brine (Coleman and White, 1993; Davis, 1993; Rahaman et al., 1993). The geometry and size of the ponds are important in order to reduce land runoff. Other management practices include adjustment of brine depth to allow light to reach benthic communities. Reintroduction of *Dactylococcopsis* and *Synechococcus* sp. was used to improve performance of saltern systems (Davis and Giordano, 1996). Introduction of the brine shrimp *Artemia* controlled excessive development of cyanobacteria (De Medeiros Rocha and Camara, 1993). The use of chlorination (using chlorine gas or hypochlorite) also controlled the mucilage-producing cyanobacteria (Seshadri and Buch, 1958; Davis, 1993). However, the most effective strategy is probably to operate stable ecosystems, with a low nutrient input and minimal distribution (Coleman and White, 1993).

Another interesting potential application of halophilic cyanobacteria was described by Cohen and Ideses (1989). Extremely high concentrations of oxycarotenoids, especially canthaxanthin, were found (up to 8.9% of the total organic carbon) in hypersaline mats in the coastal areas of Christmas Island, Central Pacific. These mats were composed mainly of unicellular types (*Synechococcus*, *Synechocystis*, *Pleomorpha*) and thick *Oscillatoria* filaments. Cultivation of these mats in a controlled system was proposed for the production of these valuable carotenoids.

VIII. Conclusions

A great amount of information is available on halophilic cyanobacteria that inhabit hypersaline lakes, salterns, and evaporitic environments. As is evident from the different sections of the present chapter, most of the knowledge on the halophilic cyanobacteria comes from pure culture studies. Physiological, genetic and molecular biological analyses greatly increased our understanding of the ways in which the cyanobacteria cope with high salt concentrations in their environment.

Our understanding of the ecology of the halophilic cyanobacteria lags far behind our insights into their physiology. Studies on the distribution of different types of halophilic cyanobacteria are seriously hampered by the poor state of cyanobacterial systematics and taxonomy (Section II), and little information is available which permits us to understand why certain types are present or absent in different ecosystems. To what extent bacteriophages are important in the regulation of the community density and species composition of halophilic cyanobacteria in nature is not known. Cyanophages (Chapter 20) were described for many non-halophilic species, but there appear to be no reports of phages which attack those species in hypersaline environments. However, the report of the occurrence of three types of restriction endonucleases in *Aphanothece halophytica* (Whitehead and Brown, 1982, 1985) may indicate that defense mechanisms against foreign (bacteriophage) DNA may be important.

Little is known about the *in situ* activities of the cyanobacteria in their environment. The use of oxygen microelectrodes was helpful in the measurement of photosynthetic activities in a few types of hypersaline cyanobacterial mats (e.g. in Solar Lake, Section II.B.3), but the activities of halophilic cyanobacteria on the bottom of saltern ponds, and in such intriguing ecosystems as gypsum or halite evaporites with stratified endevaporitic cyanobacterial communities, remain unstudied (Oren et al., 1995; Rothschild et al., 1994). The extreme salinity conditions in such very aesthetic systems (Plate 17c) make a study of the adaptations of the dominant cyanobacteria a rewarding experience, which may yield interesting results (such as the unexpectedly high concentrations of MAAs detected in the cyanobacteria inhabiting a gypsum crust in a solar saltern, as documented in Section V.B.4). It was recently discovered that a halophilic *Synechococcus* strain isolated from the Baja California saltflats was able to survive a 2-week exposure to high solar UVR and vacuum in the space environment (Mancinelli et al., 1997). Future in-depth studies of halophilic and halotolerant cyanobacteria in their natural environment will undoubtedly present additional surprises.

Acknowledgments

My present research on halophilic cyanobacteria is funded by the Israel Science Foundation administered by the Israel Academy of Sciences and Humanities.

References

Anderson GC (1958) Some limnological features of a shallow saline meromictic lake. Limnol Oceanogr 3: 259-270

Arieli B, Shahak Y, Taglicht D, Hauska G and Padan E (1994) Purification and characterization of sulfide-quinone reductase, a novel enzyme driving anoxygenic photosynthesis in *Oscillatoria limnetica*. J Biol Chem 269: 5705-5711

Asami S, Takabe T, Akazawa T and Codd GA (1983) Ribulose 1,5-bisphosphate carboxylase from the halophilic cyanobacterium *Aphanothece halophytica*. Arch Biochem Biophys 225: 713-721

Batterton JC and Van Baalen C (1971) Growth responses of blue-green algae to sodium chloride concentration. Arch Mikrobiol 76: 151-165

Bauld J (1981) Occurrence of benthic microbial mats in saline lakes. Hydrobiologia 81: 87-111

Berland B, Le Campion T and Campos Baeta Neves MH (1989) Interaction de la salinité et de la temperature sur la morphologie, la croissance et la composition cellulaire d'une Cyanobactérie halotolérante (*Aphanothece* sp.). Bot Mar 32: 317-329

Blumwald E, Mehlhorn RJ and Packer L (1983) Studies of osmoregulation in salt adaptation of cyanobacteria with ESR spin-probe techniques. Proc Natl Acad Sci USA 80: 2599-2602

Blumwald E and Tel-Or E (1982) Osmoregulation and cell composition in salt-adaptation of *Nostoc muscorum*. Arch Microbiol 132: 168-172

Borowitzka LJ (1981) The microflora. Adaptations to life in extremely saline lakes. Hydrobiologia 81: 33-46

Borowitzka LJ, Demmerle S, Mackay MA and Norton RS (1980) Carbon-13 nuclear magnetic resonance study of osmoregulation in a blue-green alga. Science 210: 650-651

Braithewaite CJR and Whitton BA (1987) Gypsum and halite associated with the cyanobacterium *Entophysalis*. Geomicrobiol J 5: 43-55

Brock TD (1973) Lower pH limit for the existence of blue-green algae: evolutionary and ecological implications. Science 179: 480-483

Brock TD (1976) Halophilic blue-green algae. Arch Microbiol 107: 109-111

Brock TD (1978) Thermophilic Microorganisms and Life at High Temperatures. Springer-Verlag, New York

Burke CM (1995) Benthic microbial production of oxygen supersaturates the bottom water of a stratified hypersaline lake. Microb Ecol 29: 163-171

Campbell SE, Golubic S (1985) Benthic cyanophytes (cyanobacteria) of Solar Lake (Sinai). Arch Hydrobiol (Suppl 71) 38/39: 311-329

Caumette P (1993) Ecology and physiology of phototrophic bacteria and sulfate-reducing bacteria in marine salterns. Experientia 49: 473-481

Caumette P, Matheron R, Raymond N and Relexans J-C (1994) Microbial mats in the hypersaline ponds of Mediterranean salterns (Salins-de-Giraud, France). FEMS Microbiol Ecol 13: 273-286

Cohen Y (1984a) The Solar Lake cyanobacterial mats: strategies of photosynthetic life under sulfide. In: Cohen Y, Castenholz RW and Halvorson HO (eds) Microbial Mats: Stromatolites, pp 133-148. Alan R Liss, New York

Cohen Y (1984b) Oxygenic photosynthesis, anoxygenic photosynthesis, and sulfate reduction in cyanobacterial mats. In: Klug MJ and Reddy CA (eds) Current Perspectives in Microbial Ecology, pp 435-441. American Society for Microbiology, Washington

Cohen Y and Ideses R (1989) Hypersaline cyanobacterial mats: environmental manipulations for the production of specific carotenoids and specific polysaccharides. In: Miyachi S, Karube I and Ishida Y (eds) Current Topics in Marine Biotechnology, pp 135-138. Japanese Society for Marine Biotechnology, Tokyo

Cohen Y, Jørgensen BB, Padan E and Shilo M (1975c) Sulphide-dependent anoxygenic photosynthesis in the cyanobacterium *Oscillatoria limnetica*. Nature 257: 489-492

Cohen Y, Jørgensen BB, Revsbech NP and Poplawski R (1986) Adaptation to hydrogen sulfide of oxygenic and anoxygenic photosynthesis among cyanobacteria. Appl Environ Microbiol 51: 398-407

Cohen Y, Krumbein WE, Goldberg M and Shilo M (1977a) Solar Lake (Sinai). 1. Physical and chemical limnology. Limnol Oceanogr 22: 597-608

Cohen Y, Krumbein WE and Shilo M (1975a) The Solar Lake: limnology and microbiology of a hypersaline, monomictic heliothermal heated sea-marginal pond (gulf of Aquaba, Sinai). Rapp Comm Int Mer MÚdit 23: 105-107

Cohen Y, Krumbein WE and Shilo M (1977b) Solar Lake (Sinai). 2. Distribution of photosynthetic microorganisms and primary production. Limnol Oceanogr 22: 609-620

Cohen Y, Padan E and Shilo M (1975b) Facultative anoxygenic photosynthesis in the cyanobacterium *Oscillatoria limnetica*. J Bacteriol 123: 855-861

Coleman MU and White MA (1993) The role of biological disturbances in the production of solar salt. In: Seventh Symposium on Salt, Vol. 1, pp 623-631. Elsevier Science Publishers, Amsterdam

D'Amelio ED, Cohen Y and Des Marais DJ (1987) Association of a new type of gliding, filamentous, purple phototrophic bacterium inside bundles of *Microcoleus chthonoplastes* in hypersaline cyanobacterial mats. Arch Microbiol 147: 213-220

Davis JS (1974) Importance of micro-organisms in solar salt production. In: Fourth Symposium on Salt, Vol. 1, pp 369-372. N. Ohio Geological Society, Cleveland

Davis JS (1978) Biological communities in a nutrient enriched salina. Aquat Bot 4: 23-42

Davis JS (1993) Biological management for problem solving and biological concepts for a new generation of solar saltworks. In: Seventh Symposium on Salt, Vol. 1, pp 611-616. Elsevier Science Publishers, Amsterdam

Davis JS and Giordano M (1996) Biological and physical events involved in the origin, effects, and control of organic matter in solar saltworks. Int J Salt Lake Res 4: 335-347

De Medeiros Rocha R and Camara MR (1993) Prediction, monitoring and management of detrimental algal blooms on solar salt production in north-east Brazil. In: Seventh Symposium on Salt, Vol. 1, pp 657-660. Elsevier Science Publishers, Amsterdam

De Philippis R, Margheri MC, Materassi R and Vincenzini M (1998) Potential of unicellular cyanobacteria from saline environments as exopolysaccharide producers. Appl Environ Microbiol 64: 1130-1132

De Philippis R, Margheri MC, Pelosi E and Ventura S (1993) Exopolysaccharide production by a unicellular cyanobacterium isolated from a hypersaline habitat. J Appl Phycol 5: 387-394

De Wit R, van Boekel WHM and van Gemerden H (1988) Growth of the cyanobacterium *Microcoleus chtonoplastes* on sulfide. FEMS Microbiol Ecol 53: 203-209

Des Marais DJ (1995) The biogeochemistry of hypersaline microbial mats. In: Jones JG (ed) Advances in Microbial Ecology, Vol. 14, pp 251-274. Plenum Press, New York

Dor I and Ehrlich A (1986) Cyanobacteria-diatom competition for the littoral zone of the experimental solar ponds, Dead Sea area, Israel. In: Dubinsky Z and Steinberger Y (eds) Environmental Quality and Ecosystem Stability, Vol. IIIA/B, pp 361-367. Bar-Ilan University Press, Ramat-Gan

Dor I and Hornoff M (1985a) Salinity-temperature relations and morphotypes of a mixed population of coccoid cyanobacteria from a hot, hypersaline pond in Israel. Mar Ecol 6: 13-25

Dor I and Hornoff M (1985b) Studies on *Aphanothece halophytica* Fremy from a solar pond: comparison of two isolates on the basis of cell polymorphism and growth response to salinity, temperature and light conditions. Bot Mar 28: 389-398

Dor I and Paz N (1989) Temporal and spatial distribution of mat microalgae in the experimental solar ponds, Dead Sea area, Israel. In: Cohen Y and Rosenberg E (eds) Microbial Mats.

Physiological Ecology of Benthic Microbial Communities, pp 114-122. American Society for Microbiology, Washington, DC
Drouet F and Daily W (1956) Revision of the coccoid Myxophyceae. Butler Univ Bot Stud 12: 1-218
Dubinin AV and Gerasimenko LM (1993) Dark anaerobic metabolism of halophilic cyanobacterium *Microcoleus chtonoplastes*. Mikrobiologiya (English translation) 62: 391-394
Dubinin AV, Gerasimenko LM and Gusev MV (1992a) Physiological features of a strain of *Microcoleus chthonoplastes* from a hypersaline reservoir. Mikrobiologiya (English translation) 61: 47-51
Dubinin AV, Zastrizhnaya OM and Gusev MV (1992b) Hydrogen peroxide production by the halophilic cyanobacterium *Microcoleus chthonoplastes*. Mikrobiologiya (English translation) 61: -
Ehrlich A and Dor I (1985) Photosynthetic microorganisms of the Gavish sabkha. In: FriedmannGM and Krumbein WE (eds) Hypersaline Ecosystems. The Gavish Sabkha, pp 296-321. Springer-Verlag, New York
Erdmann N, Fulda S and Hagemann M (1992) Glucosylglycerol accumulation during salt acclimation of two unicellular cyanobacteria. J Gen Microbiol 138: 363-368
Felix EA and Rushforth SR (1979) The algal flora of the Great Salt Lake, Utah, USA Nova Hedwigia 31: 163-195
FriedmannGM, Amiel AJ, Braun M and Miller DS (1973) Generation of carbonate particles and laminites in algal mats - Example from a sea-marginal hypersaline pool, Gulf of Aqaba, Red Sea. Am Assoc Pet Geol Bull 57: 541-557
Fulda S, Hagemann M and Libbert E (1990) Release of glucosylglycerol from the cyanobacterium *Synechocystis* spec. SAG 92.79 by hypoosmotic shock. Arch Microbiol 153: 405-408
Gabbay-Azaria R, Schonfeld M, Tel-Or S, Messinger R and Tel-Or E (1992) Respiratory activity in the marine cyanobacterium *Spirulina subsalsa* and its role in salt tolerance. Arch Microbiol 157: 183-190
Gabbay-Azaria R and Tel-Or E (1991) Regulation of intracellular Na^+ content during NaCl upshock in the marine cyanobacterium *Spirulina subsalsa* cells. Biores Technol 38: 215-220
Gabbay-Azaria R and Tel-Or E (1993) Mechanisms of salt tolerance in cyanobacteria. In: Gresshoff PM (ed) Plant Responses to the Environment, pp. 123-132. CRC Press, Boca Raton
Gabbay-Azaria R, Tel-Or E and Schonfeld M (1988) Glycinebetaine as an osmoregulant and compatible solute in the marine cyanobacterium *Spirulina subsalsa*. Arch Biochem Biophys 264: 333-339
Garcia-Pichel F (1994) A model for internal self-shading in planktonic organisms and its implications for the usefulness of ultraviolet sunscreens. Limnol Oceanogr 39: 1704-1717
Garcia-Pichel F and Castenholz RW (1993) Occurrence of UV-absorbing, mycosporine-like compounds among cyanobacterial isolates and an estimate of their screening capacity. Appl Environ Microbiol 59: 163-169
Garcia-Pichel F, Nübel U and Muyzer G (1998) The phylogeny of unicellular, extremely halotolerant cyanobacteria. Arch Microbiol 169: 469-482.
Garcia-Pichel F, Prufert-Bebout L and Muyzer G (1996) Phenotypic and phylogenetic analysis show *Microcoleus chthonoplastes* to be a cosmopolitic cyanobacterium. Appl. Environ. Microbiol. 62: 3284-3291
Garlick S, Oren A and Padan E (1977) Occurrence of facultative anoxygenic photosynthesis among filamentous and unicellular cyanobacteria. J Bacteriol 129: 623-629
Gerdes G, Krumbein WE and Holtkamp S (1985) Salinity and water activity related zonation of microbial communities and potential stromatolites of the Gavish sabkha. In: FriedmannGM and Krumbein WE (ed) Hypersaline Ecosystems. The Gavish Sabkha, pp 169-183. Springer-Verlag, New York
Gerdes G, Krumbein WE and Reineck H-E (1994) Microbial mats as architects of sedimentary surface structures. In: Krumbein WE, Patterson DM and Stal LJ (eds) Biostabilization of Sediments, pp 165-182. BIS, Oldenburg
Giani D, Seeler J, Giani L and Krumbein WE (1989) Microbial mats and physicochemistry in a saltern in the Bretagne (France) and in a laboratory scale saltern model. FEMS Microbiol Ecol 62: 151-162
Goldman CR, Mason DT and Hobbie JE, Jr (1967) Two Antarctic desert lakes. Limnol Oceanogr 12: 295-310
Golubic S (1980) Halophily and halotolerance in cyanophytes. Origins of Life 10: 169-183
Hagemann M and Erdmann N (1994) Activation and pathway of glucosylglycerol synthesis in the cyanobacterium *Synechocystis* sp. PCC 6803. Microbiology 140: 1427-1431
Hagemann M, Fulda S and Schubert H (1994) DNA, RNA, and protein synthesis in the cyanobacterium *Synechocystis* sp. PCC 6803 adapted to different salt concentrations. Curr Microbiol 28: 201-207
Hagemann M, Richter S and Mikkat S (1997a) The *ggtA* gene encodes a subunit of the transport system for the osmoprotective compound glucosylglycerol in *Synechocystis* sp. strain PCC 6803. J Bacteriol 179: 714-720
Hagemann M, Richter S, Zuther E and Schoor A (1996) Characterization of a glucosylglycerol-phosphate-accumulating, salt sensitive mutant of the cyanobacterium *Synechocystis* sp. strain PCC 6803. Arch Microbiol 166: 83-91
Hagemann M, Schoor A, Jeanjean R, Zuther E and Joset F (1997b) The *stpA* gene from *Synechocystis* sp. strain PCC 6803 encodes the glucosylglycerol-phosphate phosphatase involved in cyanobacterial osmotic responds to salt shock. J Bacteriol 179: 1727-1733
Hagemann M, Schoor A, Mikkat S, Effmert U, Zuther E, Marin K, Fulda S, Vinnemeier J, Kunert A, Milkowski C, Probst C and Erdmann N (1998) The biochemistry and genetics of the synthesis of osmoprotective compounds in cyanobacteria. In: Oren A (ed) Microbiology and Biogeochemistry of Hypersaline Environments, pp 177-186. CRC Press, Boca Raton
Hagemann M, Techel D and Rensing L (1991) Comparison of salt- and heat-induced alterations of protein synthesis in the cyanobacterium *Synechocystis* sp. PC 6803. Arch Microbiol 155: 587-592
Hagemann M, Wölfel L and Charger B (1990) Alterations of protein synthesis in the cyanobacterium *Synechocystis* sp. PCC 6803 after a salt shock. J Gen Microbiol 136: 1393-1399
Hagemann M and Zuther E (1992) Selection and characterization of mutants of the cyanobacterium *Synechocystis* sp. PCC 6803 unable to tolerate high salt concentrations. Arch Microbiol 158: 429-434
Hawkins L, Ritter D and Yopp JH (1987) Proposed additional role for glycinebetaine in the adaptation of halophilic cyanobacteria to hypersalinity: protection against pH induced loss of enzymatic activity. Plant Physiol 83 (Suppl.): 85
Heyer H, Stal LJ and Krumbein WE (1989) Simultaneous heterolactic and acetate fermentation in the marine cyanobacterium *Oscillatoria limosa* incubated anaerobically in the dark. Arch Microbiol 151: 558-564
Hof T and Frémy P (1933) On Myxophyceae living in strong brines. Rec Trav Bot Gnarl 30: 140-162
Imhoff JF, Hashwa F and Trüper HG (1978) Isolation of extremely halophilic phototrophic bacteria from the alkaline Wadi Natrun, Egypt. Arch Hydrobiol. 84: 381-388

Imhoff JF, Sahl HG, Soliman GSH and Trüper HG (1979) The Wadi Natrun: chemical composition and microbial mass developments in alkaline brines of eutrophic desert lakes. Geomicrobiol J 1: 219-234

Incharoensakdi A and Takabe T (1988) Determination of intracellular chloride ion concentration in a halotolerant cyanobacterium *Aphanothece halophytica*. Plant Cell Physiol 29: 1073-1075

Incharoensakdi A, Takabe T and Akazawa T (1986) Effect of betaine on enzyme activity and subunit interaction of ribulose-1,5-bisphosphate carboxylase/oxygenase from *Aphanothece halophytica*. Plant Physiol 81: 1044-1049

Ishitani M, Takabe T, Kojima K and Takabe T (1993) Regulation of glycinebetaine accumulation in the halotolerant cyanobacterium *Aphanothece halophytica*. Aust J Plant Physiol 20: 693-703

Jahnke LL, Lee B, Sweeney MJ and Klein HP (1989) Anaerobic biosynthesis of unsaturated fatty acids in the cyanobacterium, *Oscillatoria limnetica*. Arch Microbiol 152: 215-217

Javor B (1989) Hypersaline Environments. Microbiology and Biogeochemistry. Springer-Verlag, Berlin

Jones AG, Ewing CM and Melvin MV (1981) Biotechnology of solar saltfields. Hydrobiologia 82: 391-406

Jones JH and Yopp JH (1979) Cell wall constituents of *Aphanothece halophytica* (Cyanophyta). J Phycol 15: 62-66

Jørgensen BB, Cohen Y and Revsbech NP (1986) Transition from anoxygenic to oxygenic photosynthesis in a *Microcoleus chthonoplastes* cyanobacterial mat. Appl Environ Microbiol 51: 408-417

Jørgensen BB, Kuenen JG and Cohen Y (1979) Microbial transformations of sulfur compounds in a stratified lake (Solar Lake, Sinai). Limnol Oceanogr 24: 799-822

Jørgensen BB, Revsbech NP and Cohen Y (1983) Photosynthesis and structure of benthic microbial mats: microelectrode and SEM studies of four cyanobacterial communities. Limnol Oceanogr 28: 1075-1093

Joset F, Jeanjean R and Hagemann M. 1996. Dynamics of the response of cyanobacteria to salt stress: deciphering of molecular events. Physiol Plant 96: 738-744

Kao OHW, Berns DS and Town WR (1973) The characterization of C-phycocyanin from an extremely halo-tolerant blue-green alga, *Coccochloris elabens*. Biochem J 131: 39-50

Karsten U (1996) Growth and organic osmolytes of geographically different isolates of *Microcoleus chthonoplastes* (Cyanobacteria) from benthic microbial mats: response to salinity change. Phycology 32: 501-506

Karsten U and Garcia-Pichel F (1996) Carotenoids and mycosporine-like amino acid compounds in members of the genus *Microcoleus* (Cyanobacteria): a chemosystematic study. Syst. Appl. Microbiol. 19: 285-294

Kevbrin VV, Dubinin AV and Osipov GA (1991) Osmoregulation in the marine cyanobacterium *Microcoleus chthonoplastes*. Mikrobiologiya (English translation) 60: 407-410

Komárek J (1976) Taxonomic review of the genera *Synechocystis* SAUV. 1892, *Synechococcus* N•G. 1849, and *Cyanothece* gen. nov. (Cyanophyceae). Arch. Protistenk. 118: 119-179

Krumbein WE, Buchholz H, Franke P, Giani D, Giele C and Wonneberger K (1979) O_2 and H_2S coexistence in stromatolites. A model for the origin of mineralogical lamination in stromatolites and banded iron formations. Naturwissenschaften 66: 381-389

Krumbein WE, Cohen Y and Shilo M (1977) Solar Lake (Sinai). 4. Stromatolitic cyanobacterial mats. Limnol Oceanogr 22: 635-656

Lee BH, Hibino T, Jo J, Viale AM and Takabe T (1997) Isolation and characterization of a *dnaK* genomic locus in a halotolerant cyanobacterium *Aphanothece halophytica*. Plant Mol Biol 35: 763-775

Mackay MA and Norton RS (1987) ^{13}C nuclear magnetic resonance study of biosynthesis of glucosylglycerol by a cyanobacterium under osmotic stress. J Gen Microbiol 133: 1535-1542

Mackay MA, Norton RS and Borowitzka LJ (1984) Organic osmoregulatory solutes in cyanobacteria. J Gen Microbiol 130: 2177-2191

Mancinelli RL, White MR and Rothschild LJ (1997) Biopan-survival I: Exposure of the osmophiles *Synechococcus* sp. (Nägeli) and *Haloarcula* sp. to the space environment. Adv Space Res, in press

Margheri MC, Tredici MR, Barsanti L and Balloni W (1987) The photosynthetic community of the Trapani saline lagoons: an alternative option for the exploitation of an extreme environment. Ann Microbiol 37: 203-215

Mikkat S, Effmert U and Hagemann M (1997) Uptake and use of the osmoprotective compounds trehalose, glucosylglycerol, and sucrose by the cyanobacterium *Synechococcus* sp. PCC 6803. Arch Microbiol 167: 112-118

Mikkat S, Hagemann M and Schoor A (1996) Active transport of glucosylglycerol is involved in salt adaptation of the cyanobacterium *Synechocystis* sp. strain PCC 6803. Microbiology 142: 1725-1732

Miller DM, Jones JH, Yopp JH, Tindall RD and Schmid WE (1976) Ion metabolism in a halophilic blue-green alga, *Aphanothece halophytica*. Arch Microbiol 111: 145-149.

Moezelaar R, Bijvank SM and Stal LJ (1996) Fermentation and sulfur reduction in the mat building cyanobacterium *Microcoleus chthonoplastes*. Appl Environ Microbiol 62: 1752-1758

Mohammad FAA, Reed RH and Stewart WDP (1983) The halophilic cyanobacterium *Synechocystis* DUN52 and its osmotic responses. FEMS Microbiol Lett 16: 287-290

Moore, DJ, Reed RH and Stewart WDP (1987) A glycine betaine transport system in *Aphanothece halophytica* and other glycine betaine-synthesizing cyanobacteria. Arch Microbiol 147: 399-405

Oren A (1988) The microbial ecology of the Dead Sea. In: Marshall KC (ed) Advances in Microbial Ecology, Vol. 10, pp 193-229. Plenum Press, New York

Oren A (1989) Photosynthetic and heterotrophic benthic bacterial communities of a hypersaline sulfur spring on the shore of the Dead Sea (Hamei Mazor). In: Cohen Y and Rosenberg E (eds) Microbial Mats. Physiological Ecology of Benthic Microbial Communities, pp 64--76. American Society for Microbiology, Washington, DC

Oren A (1990) Formation and breakdown of glycine betaine and trimethylamine in hypersaline environments. Antonie van Leeuwenhoek 58: 291-298

Oren A (1993) Ecology of extremely halophilic microorganisms. In: Vreeland RH and Hochstein LI (eds) The Biology of Halophilic Bacteria, pp 25-53. CRC Press, Boca Raton

Oren A (1997) Mycosporine-like amino acids as osmotic solutes in a community of halophilic cyanobacteria. Geomicrobiol J, in press

Oren A, Fattom A, Padan E and Tietz A (1985) Unsaturated fatty acid composition and biosynthesis in *Oscillatoria limnetica* and other cyanobacteria. Arch Microbiol 141: 138-142

Oren A, Fischel U, Aizenshtat Z, Krein EB and Reed RH (1994) Osmotic adaptation of microbial communities in hypersaline mats. In: Stal LJ and Caumette P (eds) Microbial Mats. Structure, Development and Environmental Significance, pp 125-130. Springer Verlag, Berlin

Oren A, Kohl M and Karsten U (1995) An endevaporitic microbial mat within a gypsum crust: zonation of phototrophs, photopigments, and light penetration. Mar Ecol Prog Ser 128: 151-159

Oren A and Padan E (1978) Induction of anaerobic, photoautotrophic growth in the cyanobacterium *Oscillatoria limnetica*. J Bacteriol 133: 558-563

Oren A, Padan E and Avron M (1977) Quantum yields for oxygenic and anoxygenic photosynthesis in the cyanobacterium *Oscillatoria limnetica*. Proc Natl Acad Sci USA 74: 2152-2156

Oren A, Padan E and Malkin S (1979) Sulfide inhibition of photosystem I in cyanobacteria (blue-green algae) and tobacco chloroplasts. Biochim Biophys Acta 546: 270-279

Oren A and Shilo M (1979) Anaerobic heterotrophic dark metabolism in the cyanobacterium *Oscillatoria limnetica*: sulfur respiration and lactate fermentation. Arch Microbiol 122: 77-84

Padan E (1979a) Facultative anoxygenic photosynthesis in cyanobacteria. Ann Rev Plant Physiol 30: 27-40

Padan E (1979b) Impact of facultatively anaerobic photoautotrophic metabolism on ecology of cyanobacteria (blue-green algae). In: Alexander M (ed) Advances in Microbial Ecology, Vol. 3, pp 1-48. Plenum Publishing Corporation, New York

Pinckney JL and Pearl HW (1997) Anoxygenic photosynthesis and nitrogen fixation by a microbial mat in a Bahamian hypersaline lagoon. Appl Environ Microbiol 63: 420-426

Post FJ (1977) The microbial ecology of the Great Salt Lake. Microb Ecol 3: 143-165

Post FJ (1981) Microbiology of the Great Salt Lake north arm. Hydrobiologia 81: 59-69

Potts M (1980) Blue-green algae (Cyanophyta) in marine coastal environments of the Sinai Peninsula; distribution, zonation, stratification and taxonomic diversity. Phycologia 19: 60-73

Rahaman AA, Ambikadevi M and Sosamma-Esso (1993) Biological management of Indian solar saltworks. In: Seventh Symposium on Salt, Vol. 1, pp 633-643. Elsevier Science Publishers, Amsterdam

Reed RH, Chudek JA, Foster R and Stewart WDP (1984a) Osmotic adjustment in cyanobacteria from hypersaline environments. Arch Microbiol 138: 333-337

Reed RH, Richardson DL, Warr SRC and Stewart WDP (1984b) Carbohydrate accumulation and osmotic stress in cyanobacteria. J Gen Microbiol 130: 1-4

Reed RH, Warr SRC, Kerby NW, and Stewart WDP (1986) Osmotic shock-induced release of low molecular weight metabolites from free-living and immobilized cyanobacteria. Enzyme Microb Technol 8: 101-104

Reed RH, Warr SRC, Richardson DL, Moore DJ and Stewart WDP (1985) Multiphasic osmotic adjustment in a euryhaline cyanobacterium. FEMS Microbiol Lett 28: 225-229

Revsbech NP, Jørgensen BB, Blackburn TH and Cohen Y (1985) Microelectrode studies of photosynthesis and O_2, H_2S, and pH profiles of a microbial mat. Limnol Oceanogr 28: 1062-1074

Richardson DL, Reed RH and Stewart WDP (1983) *Synechocystis* PCC6803: a euryhaline cyanobacterium. FEMS Microbiol Lett 18: 99-102

Rippka R, Deruelles JB, Waterbury JB, Herdman M and Stanier RY (1979) Generic assignments, strain histories and properties of pure cultures of cyanobacteria. J Gen Microbiol 111: 1-61

Ritter D and Yopp JH (1993) Plasma membrane lipid composition of the halophilic cyanobacterium *Aphanothece halophytica*. Arch Microbiol 159: 435-439

Rodriguez-Valera F, Ventosa A, Juez G and Imhoff JF (1985) Variation of environmental features and microbial populations with salt concentrations in a multi-pond saltern. Microb Ecol 11: 107-115

Rothschild LJ (1991) A model for diurnal patterns of carbon fixation in a Precambrian microbial mat based on a modern analog. BioSystems 25: 13-23

Rothschild LJ, Giver LJ, White MR and Mancinelli RL (1994) Metabolic activity of microorganisms in evaporites. J Phycol 30: 431-438

Rothschild LJ and Mancinelli RL (1990) Model of carbon fixation in microbial mats from 3,500 Myr ago to the present. Nature 345: 710-712

Roux JM (1996) Production of polysaccharide slime by microbial mats in the hypersaline environment of a Western Australian solar saltfield. Int J Salt Lake Res 5: 103-130

Roussomoustakaki M and Anagnostidis K (1991) *Cyanothece halobia*, a new planktic chroococcalean Cyanophyta from hellenic heliothermal saltworks. Arch Hydrobiol 64: 71-95

Rushforth SR and Felix EA (1982) Biotic adjustments to changing salinities in the Great Salt Lake, Utah, USA. Microb Ecol 8: 157-161

Sammy N (1983) Biological systems in north-western Australian solar saltfields. In: Sixth Symposium on Salt, Vol. 1, pp 207-215. The Salt Institute, Alexandria, VA

Schoor A, Hagemann M and Erdmann N (1996) Non-radiometric assay for glucosylglycerol-synthesizing enzymes involved in the cyanobacterial salt adaptation. J Microbiol Meth 27: 139-145

Seshadri K and Buch SD (1958) Elimination of algae in Sambhar Lake brine by chlorination. J Sci Indust Res 17A: 455-457

Siegel BZ, McMurty G, Siegel SM, Chen J and LaRock P (1979) Life in the calcium chloride environment of Don Juan Pond, Antarctica. Nature 280: 828-829

Simon RD (1981) Gliding motility in *Aphanothece halophytica*: analysis of wall proteins in mot mutants. J Bacteriol 148: 315-321

Stanier RY, Kunisawa R, Mandel M and Cohen-Bazire G (1971) Purification and properties of unicellular cyanobacteria (order Chroococcales). Bacteriol Rev 35: 171-205

Stolz JF (1990) Distribution of phototrophic microbes in the flat laminated microbial mat at Laguna Figueroa, Baja California, Mexico. BioSystems 23: 345-357

Sudo H, Burgess JG, Takemasa H, Nakamura N and Matsunaga T (1995) Sulfated exopolysaccharide production by the halophilic cyanobacterium *Aphanocapsa halophytica*. Curr Microbiol 30: 219-222

Taher AG, Abd el Wahab S, Philip G, Krumbein WE and Wali AM (1995) Evaporitic sedimentation and microbial mats in a salina system (Port Fouad, Egypt). Int J Salt Lake Res 4: 95-116

Takabe T, Incharoensakdi A, Arakawa K and Yokota S (1988) CO_2 fixation rate and RuBisCO content increase in the halotolerant cyanobacterium, *Aphanothece halophytica*, grown in high salinities. Plant Physiol 88: 1120-1124

Tel-Or E, Spath S, Packer L and Mehlhorn RJ (1986) Carbon-13 NMR studies of salt shock induced carbohydrate turnover in the marine cyanobacterium *Agmenellum quadruplicatum*. Plant Physiol 82: 646-652

Tindall DR, Yopp JH, Miller DM and Schmid WE (1978) Physico-chemical parameters governing the growth of *Aphanothece halophytica* (Chroococcales) in hypersaline media. Phycologia 17: 179-185

van Rijn J and Cohen Y (1983) Ecophysiology of the cyanobacterium *Dactylococcopsis salina*: effect of light intensity, sulphide and temperature. J Gen Microbiol 129: 1849-1856

Volcani, BE (1944) The microorganisms of the Dead Sea. In: Papers Collected to Commemorate the 70th Anniversary of Dr. Chaim Weizmann. Collective Volume, pp 71-85. Daniel Sieff Research Institute, Rehovoth

Walsby AE, van Rijn J and Cohen Y (1983) The biology of a new gas-vacuolate cyanobacterium, *Dactylococcopsis salina* sp. nov., in Solar Lake. Proc Roy Soc London B 217: 417-447

Waterbury JB and Rippka R (1989) Subsection I. Order *Chroococcales* Wettstein 1924, emend. Rippka et al., 1979. In: Staley JT, Bryant MP, Pfennig N and Holt JG (eds) Bergey's Manual of Systematic Bacteriology, Vol. 3, pp 1728-1746. Williams & Wilkins, Baltimore

Warr SRC, Reed RH, Chudek JA, Foster R and Stewart WDP (1985) Osmotic adjustment in *Spirulina platensis*. Planta 163: 424-429

Warr SRC, Reed RH and Stewart WDP (1984) Osmotic adjustment of cyanobacteria: the effects of NaCl, KCl, sucrose and glycine betaine on glutamine synthetase activity in a marine and a halotolerant strain. J Gen Microbiol 130: 2169-2175

Wharton RA Jr, Parker BC and Simmons GM Jr (1983) Distribution, species composition and morphology of algal mats in Antarctic Dry Valley lakes. Phycologia 22: 355-365

Whitehead PR and Brown NL (1982) *Aha*III: A restriction endonuclease with a recognition sequence coding only A:T basepairs. FEBS Lett 143: 296-300

Whitehead PR and Brown NL (1985) A simple and rapid method for screening bacteria for type II restriction endonucleases - enzymes in *Aphanothece halophytica*. Arch Microbiol 141: 70-74

Wilmotte A (1994) Molecular evolution and taxonomy of cyanobacteria. In: Bryant DA (ed) The Molecular Biology of Cyanobacteria, pp 1-25. Kluwer Academic Publishers, Dordrecht

Wright SW and Burton HR (1981) The biology of Antarctic saline lakes. Hydrobiologia 82: 319-338

Yopp JH, Albright G and Miller DM (1979) Effects of antibiotics and ultraviolet radiation on the halophilic blue-green alga, *Aphanothece halophytica*. Bot Mar 22: 267-272

Yopp JH, Miller DM and Tindall DR (1978b) Regulation of intracellular water potential in the halophilic blue-green alga *Aphanothece halophytica* (Chroococcales). In: Caplan SR and Ginzburg M (eds) Energetics and Structure of Halophilic Microorganisms, pp 619-624. Elsevier/North Holland, Amsterdam

Yopp JH, Tindall DR, Miller DM and Schmid WE (1978a) Isolation, purification and evidence for a halophilic nature of the blue-green alga *Aphanothece halophytica* Frémy (*Chroococcales*). Phycologia 17: 172-178

Zavarzin GA, Gerasimenko LM and Zhilina TN (1993) Cyanobacterial communities in hypersaline lagoons of Lake Sivash. Mikrobiologiya (English translation) 62: 645-652

Zhilina TN and Zavarzin GA (1990) Extremely halophilic, methylotrophic, anaerobic bacteria. FEMS Microbiol Rev 87: 315-322

Zvyagintseva IS, Gerasimenko LM, Kostrikina NA, Bulygina ES and Zavarzin GA (1995) Interaction of halobacteria and cyanobacteria in a halophilic cyanobacterial community. Mikrobiologiya (English translation) 64: 252-258

Chapter 11

Oil Pollution and Cyanobacteria

Samir S. Radwan and Redha H. Al-Hasan
Department of Botany & Microbiology, Kuwait University, P.O. Box 5969, Safat 13060, Kuwait

Summary

There is increasing evidence that ancient cyanobacteria were among the direct biogenic contributors to oil formation. This fact underlines the historical and ecological relations between these photosynthetic microorganisms and petroleum. Evidence that cyanobacteria are capable of hydrocarbon degradation is tentative, but preliminary studies indicate that some strains are capable of oxidizing aromatic and aliphatic oil constituents. Further, in cyanobacterial-dominated mats, the cyanobacteria live in natural association with hydrocarbon-degrading bacteria and fungi that occur in the cyanobacterial polysaccharide layers. Such mat associations flourish in oil-polluted coastal areas of subtropical regions like the Arabian Gulf. The combined activities of the cyanobacteria and the associated oil-degrading organotrophs appear to be crucial and effective in bioremediating such polluted environments.

I. Introduction

Cyanobacteria not only played a key role in the formation of some important petroleum deposits, but they are often found in association with present-day oil seeps and deposits in wet areas (see below). The purpose of this chapter is to assess the possible roles of cyanobacteria in both oil biosynthesis and degradation. If they are important in the latter, clearly they could be useful tools in bioremediating oil-polluted environments, especially coastal regions.

II. Composition of Crude Oil

Crude oil is a complex mixture of thousands of compounds, but each accumulation of oil tends to be unique in composition. Some background about this mixture is needed to understand this chapter. In terms of element composition, oil is composed predominantly of carbon and hydrogen, with minor proportions of oxygen, nitrogen and sulfur, while traces of vanadium, nickel and other elements are also usually present. Petroleum samples, irrespective

B. A. Whitton and M. Potts (eds), The Ecology of Cyanobacteria, 307-319.
© 2000 *Kluwer Academic Publishers. Printed in the Netherlands.*

of their origin, consist of four classes: saturated hydrocarbons, aromatic hydrocarbons, asphaltenes and resins (Colwell and Walker, 1977). The saturated hydrocarbons include the paraffins (alkanes), which have the general formula C_nH_{2n+2}. For values of n less than 5 the compounds are gases at room temperature, for values between 5 and 15 they are liquid and for values more than 15 they are viscous to solid waxes. Alkanes in crude oil may reach in chain length C_{70} or more. Normal straight-chain alkanes usually predominate over the branched (iso-) alkanes. Another major group of the saturated hydrocarbons is the naphthenes, also called cycloalkanes with the general formula C_nH_{2n}; these are liquid at room temperature. The molecular structure of the aromatic hydrocarbons is based on a ring of 6 carbon atoms i.e. a benzene (or substituted benzene) ring. Crude oil contains both simple aromatics (benzene and substituted benzene) and polyaromatics with five or more condensed rings. The asphaltenes and resins are occasionally lumped together under the name "hetero-compounds", on the basis that their molecules contain not only carbon and hydrogen atoms but also other atoms like oxygen, sulfur and nitrogen. Asphaltenes comprise e.g. phenols, fatty acids, ketones, esters and porphyrins. Resins include constituents like pyridines, quinolines, carbazoles, sulfoxides and amides.

Alkanes, cycloalkanes and aromatics are the major constituents of all crude oils, whereas the heterocompounds represent the minor constituents. The dominant compounds are the alkanes, usually with more straight than branched chain alkanes.

III. Biodegradation of Hydrocarbons

A variety of terrestrial and aquatic bacteria and fungi can biodegrade crude oil constituents and the subject has been repeatedly reviewed (Klug and Markovetz, 1971; Levi et al., 1979; Radwan and Sorkhoh, 1993). Oil constituents can potentially be attacked by a great variety of soil and aquatic microorganisms. The above reviews list many tens of genera that oxidise hydrocarbons, both prokaryote and eukaryote. However, most studies have been done on a rather limited number of genera of bacteria, yeasts and filamentous fungi.

The mechanism(s) by which microorganisms take up hydrocarbons is (are) still far from clear, although it is established that such compounds get into the cells as intact molecules. The first contact between the cells and such water insoluble compounds is enhanced by the hydrophobic nature of the cell envelopes (Rapp et al., 1979; Ramsay et al., 1988). It has been proposed that microorganisms take up the dissolved hydrocarbon fraction (Goma et al., 1976; Wodzinski and Coyle, 1974). This may be valid for low molecular weight compounds only; the water solubility decreases dramatically with increasing molecular weight of the hydrocarbon. Many bacteria and yeasts with hydrophobic envelopes have an affinity for hydrocarbon droplets in aqueous media. This process seems to be merely physical, rather than a specific recognition of the substrates by the cells. Many microorganisms not capable of hydrocarbon utilization adhere to droplets of these compounds in aqueous media (e.g. Rosenberg et al., 1980). Further, microorganisms may not be capable of significant adherence to the specific hydrocarbons they efficiently utilize (e.g. Nakhara et al., 1981). Hydrocarbons may also be made available to microorganisms by emulsification into minute droplets (macroemulsions) or solubilization in the aqueous medium (micro-emulsions) (Singer and Finnerty, 1984). These physical forms of hydrocarbons in aqueous media could be established by turbulence and the production of extracellular surfactants.

The biochemical mechanisms of microbial hydrocarbon degradation are well understood (Boulton and Ratledge, 1984; Bühler and Shindler, 1984). An alkane molecule, for example, is oxidized to the fatty alcohol, fatty aldehyde and ultimately to the fatty acid, which is subsequently degraded to acetyl CoA. The latter is further metabolized to yield ATP and cell material. Some of the fatty acids are usually directly esterified in cell lipids, therefore the fatty acid patterns of the cell lipids normally reflect the identity of the alkane oxidised by these cells (Klug and Markovetz, 1967).

Oil constituents differ in their susceptibility to biodegradation, although different authors report different results:

n-alkanes > isoalkanes > low molecular weight aromatics > cycloalkanes (Perry, 1984);

saturates > light aromatics > high molecular weight aromatics > polar compounds

(Jobson et al., 1972; Walker et al., 1976; Fusey and Oudot,1984).

Other authors have found that this pattern is not always strictly valid (e.g. Jones et al., 1983; Cooney et al., 1985). There are even claims that all fractions of crude oil may be biodegraded by bacteria at similar rates (e.g. Horowitz and Atlas, 1977).

IV. Cyanobacteria in the Biogenesis of Oil

Through studies on sedimentary environments and organisms, a wealth of information has become available regarding the survival potential of organic compounds and their species specificity (Schenck and De Leeuw, 1982). There is increasing evidence that ancient cyanobacteria played an especially important role in the biogenic contribution to oil formation. As far as cyanobacterial mats are concerned, it is known that photosynthetically active layers can form on top of older inactive ones and the latter thus become subjected to decomposition and preservation of organic matter (Boon and De Leeuw, 1987). The discovery of cyanobacterial remains in Precambrian stromatolitic rocks (Chapter 2) has enhanced geochemical studies related to this.

Living microbial mats in shallow waters are dominated by cyanobacteria (Chapter 4), but mats in deeper waters are dominated by sulfur-oxidizing bacteria (Williams and Reimers, 1983). Fossilized remains of analogous mat structures from the Miocene Monterey Formation are reportedly associated with oil-producing zones and microbial mats may be a source of the organic matter which became thermally modified to hydrocarbons. Schouten et al. (1997) suggest that S-bound pentakishomohopane in this area was derived from deeper water cyanobacteria and Willingham et al. (1985), in a study of stratiform organic matter in the lower Huronian Supergroup, Canada, concluded that the stratiform kerogens were probably derived from cyanobacterial mats. In siliciclastic, microbial laminated deposits in Puck Bay, Poland, the relative abundance of various hopanoids appears to be an indicator of cyanobacteria and their biodegraded remains (Malinski et al., 1988).

Peniguel et al. (1989) discussed the importance of "microalgae" in stratigraphy and petroleum exploration. The major microalgal groups encountered in petroleum series are not only cyanobacteria, but in later series also Prasinophyceae, Botryococcaceae, *Pediastrum* and diatoms. Earlier research on these groups provided results of only limited use for interpreting stratigraphy and environment, because of their long and widespread occurrence, though they have important implications for petroleum studies and possible industrial developments. However, an increasing number of studies are providing more detailed information on types of environments and the cyanobacterial products that may be associated with them. The analysis of biological marker hydrocarbons allowed Robinson et al. (1989) to suggest that the major sources of sedimentary organic matter in two outcrop samples of Messel oil shale in Germany were dinoflagellates and cyanobacteria. The samples differed principally in their hydrocarbons, with sample 1 containing negligible steroidal hydrocarbons, while sample 2 shows abundant desmethylsterenes and 4-methylsterenes, thus demonstrating the non-homogeneous nature of the Messel deposit. Matsumoto and Watanuki (1990) made a geochemical study of hydrocarbons (n-alkanes, acyclic isoprenoid alkanes, steranes, triterpenes) and fatty acids in sediments of the inland hydrothermal environments of Japan. These compounds could be attributed to various source organisms - bacteria, cyanobacteria, microalgae and vascular plants in and around the hydrothermal environments. Low concentrations of hydrocarbons and fatty acids might reflect the low primary productivity of harsh environments, whereas the abundance of n-alkanoic acids might reflect the fact that fresh organic matter was continually supplied by microbial activity and vascular plants.

A number of other studies have been made on the component lipid molecules in extreme, or at least highly characteristic, modern environments. The major hydrocarbons in soil samples from the McMurdo Dry Valleys of Southern Victoria Land, Antarctica are long-chain n-alkanes, which may result from a mixture of erosion products of sedimentary materials containing vascular plant debris formed during the pre- and inter-glaciation periods of Antarctica (Miocene-Plicene) and microbial debris such as fungi and microalgae (Matsumoto et al., 1990). These authors also found normal alkenes and these may well have been derived from the cyanobacteria and microalgae which occur in parts of the Antarctic cold desert (see Chapter 13) and/or organic debris of wind-transported cyanobacterial mats. Hot spring cyanobacterial mats contain suites of monomethyl alkanes (Shiea et al. 1990), hydrocarbons which are often reported in cultures and natural populations of cyanobacteria. The detection of a series of mid-chain branched alkanes in modern cyanobacterial mats may imply that their occurrence in ancient sediments reflects a direct biogenic contribution, rather than the result of diagenetic processes. In a further study, the same authors (1991) used hot spring microbial mats to distinguish markers for cyanobacterial mats from those for photosynthetic bacterial mats. This involved comparison of the lipids (hydrocarbons, wax esters, alcohols, fatty acids) of two cyanobacterial mats, two anoxygenic photosynthetic bacterial mats and their component photosynthetic bacteria. The results indicated clearly that mat lipids reflect the inputs of the mat-building phototrophs.

Other studies have focussed on particular genera and species. The predominant sources of organic matter and the main diagenetic processes involved have been investigated in *Phormidium valderianum* and *Microcoleus chthonoplastes* mats from evaporite-controlled environments (Grimalt et al., 1992). The changes with depth shown in a detailed study of fatty acids, hydrocarbons, alcohols, ketones and aldehydes were assessed using the results from enrichment cultures of both the cyanobacteria and a range of other organisms in the mats. Rather surprisingly, in spite of the cyanobacteria occurring almost as monocultures in the top layers, they left only minor traces in the extractable lipid sedimentary record. The predominant fatty acids paralleled those of enrichment cultures of purple bacteria and were mixed with acids characteristic of non-phototrophs such as sulfate-reducers. Two calcifying freshwater cyanobacteria (*Scytonema* and *Schizothrix*) in the Everglades, Florida, gave much more characteristic signals (Thiel et al., 1997), with the strong predominance of (among others) n-heptadecane, n-heptadecene, and two monomethl-heptadecanes.

V. Hydrocarbon Degradation Potential of Cyanobacteria

A. Organotrophy in Cyanobacteria

Before dealing specifically with the utilization of hydrocarbons by cyanobacteria a brief account of the more general subject of organotrophy in this group is given. Although cyanobacteria are basically oxygenic photosynthetic organisms, some can also utilize added organic compounds, either by heterotrophy or photoheterotrophy. Reports on this topic are numerous (e.g. Harder, 1917; Kiyohara et al., 1960; Fay and Fogg, 1962; Hoare et al., 1971; Khoja and Whitton, 1971; White and Shilo, 1975; Wolk and Shaffer, 1976; Rippka et al. 1979). In most cases, heterotrophy was tested only on a limited number of organic substrates, namely sugars such as glucose, fructose, ribose, sucrose as well as glycerol.

Rippka et al. (1979) listed the following genera as containing photoheterotrophic isolates: *Dermocarpella*, *Mysosarcina*, *Chroococcidiopsis*, *Nodularia*, *Scytonema*, *Calothrix*, *Fischerella* and *Chlorogloeopsis*. They also listed a number of genera where they did not find evidence for photoheterotrophy, but examples have since been found in at least some of the genera, so detailed study will be needed before any genus can be excluded. Their experiments showed that the tested substrates could be arranged in decreasing order of the isolate numbers they could support as follows: D-glucose > sucrose and D-fructose > D-ribose and glycerol. More et al. (1979) tested nine different cyanobacteria for dark heterotrophic growth on nineteen different organic substrates. The most efficiently utilized substrates were D-glucose, D-fructose, D-ribose and sucrose, whereas D-galactose, maltose, cellobiose, mannitol and D-xylose were only weakly utilized. In all conditions the rate of dark heterotrophic growth was lower than that of photoautotrophic growth.

The possible effects of cultural conditions have mostly not been considered. According to the earlier reports (e.g. Rippka et al., 1979), *Synechococcus* and *Synechocystis* for example can hardly live heterotrophically. Yet, in a more recent study, Anderson and McIntosh (1991) revealed that a glucose-tolerant strain of *Synechocystis* cannot grow on glucose in the dark unless given a daily pulse of white light (5 min at 40 μmol m^{-2} s^{-1}), which is well below that needed for photoautotrophy. The authors found that blue-light pulse, but not pulses of long-wavelength light (i.e. 550-650 nm) were essential for growth. In this context, it is interesting that genetically engineered mutants of *Synechocystis* lacking photosystem II that live photo-heterotrophically, have been produced (Vermass et al., 1986; Jansson et al., 1987). Rippka (1972) found that the growth of *Synechocystis* 6714 on glucose was not enhanced by the addition of other organic compounds to the medium, whereas the addition of amino acids stimulated the growth of *Nostoc* strain MAC (Hoare et al., 1971) and *Plectonema* 73110 (White and Shilo, 1975). At least some cyanobacteria were found to require a period of adaptation before a reproducible rate of growth in the dark on organic substrates is established (Fay, 1965). Interestingly, there are early reports that some cyanobacteria can assimilate acetate: viz. *Nostoc* strains (Allison et al., 1953; Ingram et al., 1973), *Anabaena variabilis* (Pearce and Carr, 1967), *A. flos-aquae*, *Anacystis nidulans* (*Synechococcus* sp.), (Hoare et al., 1967; Pearce and Carr, 1967), *Gloeocapsa alpicola* (Smith et al., 1967) and *Chlorogloeopsis* (Hoare et al., 1967; Miller and Allen, 1972). This result, if confirmed, would contribute to the understanding of modern claims regarding the potential of cyanobacteria to utilize hydrocarbons. As mentioned before, alkanes are utilized via their oxidation to fatty acids that are further oxidized to acetate units.

There are several early studies on the metabolism of exogenous organic compounds by cyanobacteria. It is commonly established that exogenous glucose is metabolized via the oxidative pentose-phosphate cycle (Pelroy et al., 1972; Raboy et al., 1976; Josef-Espardellier et al., 1978; Bottomley and Van Baalen, 1978). In cells of *Synechocystis* 6714 glucose-6-phosphate and 6-phosphogluconate dehydrogenases, the key enzymes of the above cycle, were detected, but little or no phosphofructokinase, keto-deoxyphosphogluconate aldolase or 6-phosphogluconate dehydrase (Pelroy et al., 1972). The two key enzymes were also detected in other cyanobacteria (Raboy et al., 1976; Josef-Espardellier et al., 1978). Evidence for the cyanobacterial glucose metabolism via the oxidative pentose-phosphate cycle has also been collected through studies on the accumulation of glucose analogues (Raboy and Padan, 1978), CO_2 production from glucose (Pelroy et al., 1972), glycogen synthesis (Pelroy and Bassham, 1973) and metabolite labelling by [^{14}C] glucose (Pelroy and Bassham, 1973). According to Smith (1982), fermentation and anaerobic respiration are to be discounted as contributing to the energy generation in the dark heterotrophic cyanobacteria. Instead, aerobic respiration coupled to the catabolism of carbohydrates obtained from the medium provides energy for growth of these organisms in the dark.

The metabolism of exogenous acetate by cyanobacteria has also been studied. Carr and Pearce (1966) calculated that in *Anabaena variabilis* provided with acetate, 18% of the newly assimilated carbon originated from the acetate substrate. The inhibition of non-cyclic photophosphorylation inhibits the photoassimilation of acetate by cyanobacteria (Hoare et al., 1967). This implies that the electron donors and ATP produced via the photochemical reactions are involved in acetate metabolism. Allison et al. (1953) reported that assimilated acetate is used to biosynthesize lipids, but some acetate carbon may also enter into amino acids, e.g. glutamic acid, arginine, proline, leucine and organic acids. The metabolic pathways of such synthetic reactions probably include parts of the interrupted tricarboxylic acid cycle and the glyoxylate cycle (Pearce et al., 1969).

B. Effect of Oil on Photosynthetic Microorganisms

A few studies have been published on the effects of oil and/or oil fractions on growth, oxygen evolution, dark respiration and pigment composition of cyanobacteria and microalgae. In some of these studies a differentiation was made between water-soluble and water insoluble fractions of crude oil.

Singh and Gaur (1988) obtained from five samples of crude oil their paraffinic, asphaltic and aromatic fractions and investigated the effect of each fraction on photosynthetic oxygen evolution and dark respiration by *Anabaena doliolum*. The aromatic and asphaltic fractions inhibited both activities, whereas the paraffinic fractions were stimulatory. Treatment with crude oil exerts reversible as well as irreversible inhibition of microbial photosynthesis (Vandermeulen and Ahern, 1976; Winters et al., 1977; Batterton et al., 1978a, b). Morales-Loo and Goutx (1990) subjected several phytoplankton isolates to the water soluble fraction of the Mexican crude oil "Isthmus Cactus" and measured their growth. The growth of *Nitzschia closterium* (diatom), *Asterionella glacialis* (diatom), *Rhodomonas lens* (Cryptophyceae) and *Dunaliella tertiolecta* (green alga) was inhibited, whereas that of *Skeletonema costatum* (diatom) and *Agmenellum quadruplicatum* (cyanobacterium) was not affected; the growth of *Prorocentrum minimum* (Dinophyceae) was enhanced (Table 1).

There are a number of reports on the stimulation of photosynthesis, dark respiration and growth of photosynthetic microorganisms when they are treated with n-alkanes (Soto et al., 1975; Vandermeulen and Ahern, 1976; Gaur and Kumar, 1981; Schroeder and Rehm, 1981; Gaur and Singh, 1990; Al-Hasan et al., 1994). Nitrogen-fixing cyanobacteria were included by Gaur and Singh (1990). Table 2 provides an example of the response of cyanobacteria to various oil and alkane fractions to two non-axenic cyanobacteria, *Microcoleus chthonoplastes* and *Phormidium corium*, isolated from an oil-rich environment (Al-Hasan et al., 1994). Growth of both cyanobacteria was enhanced by the hydrocarbons, and the chlorophyll to carotene ratio of the cells did not decrease in response to the hydrocarbon treatment. The latter result may imply that crude oil and n-alkanes do not cause any stress to these particular strains. It is clear that photosynthetic microorganisms respond in various ways to crude oil and its constituent fractions.

Table 1. Effects of water-soluble fraction (WSF) of crude oil Isthmus Cactus on growth rate and maximum cell density of marine microalgae (from Morales-Loo and Goutx, 1990)

SPECIES		WSF (%)	Max. specific growth rate (divisions d^{-1})	Stimulation (+) or reduction (-) in growth rate (% of control)	Reduction in max. cell density (%)
Nitzschia closterium	(diatom)	0	3.37	-	-
		50	2.55	- 24.3	- 2.5
		100	1.36	- 59.6	- 65.9
Asterionella glacialis	(diatom)	0	2.35	-	-
		50	1.70	- 27.7	- 50.0
		100	1.07	- 54.5	- 81.3
Rhodomonas lens	(cryptophyte)	0	1.77	-	-
		50	1.67	- 4.0	- 44.0
		100	1.48	- 16.4	- 84.2
Dunaliella tertiolecta	(green alga)	0	2.41	-	-
		50	2.29	- 4.1	- 23.0
		100	2.29	- 4.5	- 24.0
Agmenellum quadruplicatum[1]	(cyanobacterium)	0	6.47	-	-
		50	5.64	+ 3.4	- 4.0
		100	6.14	+ 5.6	+ 7.2
Skeletonema costatum	(diatom)	0	2.44	-	-
		50	2.50	+ 2.4	- 8.0
		100	2.29	- 6.1	- 17.0
Prorocentrum minimum	(dinoflagellate)	0	0.80	-	-
		50	0.85	+ 6.2	+ 43.0
		100	0.89	+11.2	+ 65.0

[1]Specific growth expressed as organic carbon.

Table 2. Effects of crude oil and *n*-alkanes on growth and pigments of non-axenic cyanobacterial cultures (from Al-Hasan et al., 1998)

Hydrocarbons	*Microcoleus chthonoplastes*			*Phormidium corium*		
	Biomass[1]	Biliprotein[2]	Chlorophyll/ carotene[3]	Biomass[1]	Biliprotein[2]	Chlorophyll/ carotene[3]
None (control)	6898	35	2.8	1243	58	1.5
Crude oil	7276	41	2.7	2468	170	3.0
n-Decane (C10)	7619	66	2.8	2007	117	2.1
n-Undecane (C11)	7366	87	2.7	1464	70	2.2
n-Dodecane (C12)	7318	66	2.9	1542	101	2.7
n-Tridecane (C13)	8537	84	2.9	1489	124	2.9
n-Tetradecane (C14)	7141	50	2.9	1472	165	2.1
n-Hexadecane (C16)	7531	72	2.8	1574	173	2.5
n-Nonadecane (C19)	7478	89	2.8	3254	270	2.4

[1]Dry weight in mg L^{-1} [2]In mg L^{-1} [3]Weight/weight

C. Oil Toxicity to Photosynthetic Microorganisms

It is now well established that crude oils include constituents that are particularly inhibitory to photosynthetic organisms even at very low concentrations (see Oil in the Sea, 1985; Narro, 1987). However, the biochemical basis of this toxicity is mostly still unknown. Sikkema et al. (1995) wrote a review on the mechanisms of membrane toxicity of hydrocarbons with emphasis on cyclic ones. Investigators have used both field samples containing mixed populations as well as pure cultures, and the results showed that the toxic effects of different compounds range from being specific to one species to being widely applicable. As mentioned before, the toxic effects were traced by measuring physiological activities such as photosynthesis and growth. In addition, some studies have dealt with a hydrocarbon-induced reduction of nitrogen-fixing activity in cyanobacteria (Gordon and Prouse, 1973; O'Brien and Dixon, 1976; Vandermeulen and Ahern, 1976; Jordan et al., 1978).

In a study on three photosynthetic microorganisms including one cyanobacterium, *Agmenellum quadruplicatum*, Batterton et al. (1978b) found that fuel oils significantly inhibited their growth, but crude oils were much less toxic. The toxicity of the fuel oils was probably due mainly to higher aromatic compounds. Winters et al. (1976) showed that the water-soluble fractions of four tested fuel oils exhibited different toxic effects on two cyanobacteria and four microalgae. The water-soluble fractions of two of the fuel oils killed both the cyanobacteria, *A. quadruplicatum* and *Coccochloris elabens*, and the toxic effect was in part due to the constituent p-toluidine. More than 125 pure aromatic compounds occurring in oil have been tested by several investigators (Winters et al., 1976, 1977; Cerniglia et al., 1981, 1983; Narro, 1985, 1987) for their toxicity to the cyanobacterium *A. quadruplicatum* (data tabulated by Narro, 1987). The following compounds exhibited high toxicity: dimethyl naphthalenes, methyl and ethyl anilines, di- and trimethyl phenols, 2,5-dimethyl pyrrole, indole and its methyl derivatives, quinolines and methyl derivatives, toluidine and methyl and ethyl derivates and p-anisidine. It was also confirmed by Batterton et al. (1978a) that marine cyanobacteria are very sensitive to aniline and p-toluidine, compounds that are believed to be selectively toxic to cyanobacteria.

D. Hydrocarbon Oxidation by Cyanobacteria

There is increasing evidence that photosynthetic microorganisms, particularly cyanobacteria may contribute to the oxidation and degradation of hydrocarbons. However, it is important to emphasize that only in some cases did the authors make clear that the test cyanobacterial cultures were axenic and many studies have been carried out on non-axenic cultures. Such cultures are usually associated with hydrocarbon-utilizers, as we will discuss in more detail. Ridding cyanobacteria from associated organotrophs is sometimes difficult, since in many cases the cyanobacterium ceases to grow with increasing removal of the associated organisms (e.g. Fitzsimons and Smith, 1984). A critical look at the literature has led us to conclude that in many cases there is still no rigorous evidence available for the hydrocarbon-oxidation potential of cyanobacteria. However, the achlorophyllous alga, *Prototheca zopfi* (Walker et al., 1975) does utilize crude oil, n-alkanes, iso-alkanes and aromatic hydrocarbons; most of the later studies with phototrophic microorganisms deal either with aliphatic or aromatic compounds. Therefore, studies on the two chemical groups will be reviewed separately.

1. Aromatic Hydrocarbons.

It has long been known that poly-nuclear aromatic hydrocarbons exhibit serious toxic and carcinogenic effects (e.g. Miller and Miller, 1974; McCann et al., 1875). Therefore, the study of the biotransformation and biodegradability of these aromatic compounds in the environment is not only of basic but also of practical value. It is now established that several bacteria possess the dioxygenase systems catalyzing the incorporation of both atoms of an oxygen molecule into the aromatic ring producing dihydrodiols (see Gibson, 1984). Eukaryotes (e.g. fungi), on the other hand, possess monooxygenases catalyzing the production of arene oxides from aromatic nuclei. These represent the initial and decisive steps of aromatic-hydrocarbon biodegradation.

Among the earliest studies on the potential of photosynthetic microorganisms including cyanobacteria for aromatic hydrocarbon oxidation is that of Ellis (1977). This author investigated the phenol and catechol degradation potential of the microalgae *Chlamydomonas ulvaensis*, *Chlorella pyrenoidosa* and *Scenedesmus brasiliensis*, the phytoflagellate

Euglena gracilis and the cyanobacteria *Anabaena cylindrica* and *Phormidium foveolarum*. The substrates used were uniformly ^{14}C-labelled phenol and catechol, and small proportions of the radioactivity in $^{14}CO_2$ resulting from microbial respiration were recovered. Four of these phototrophs could degrade phenol, but all six degraded catechol. Quite a lot work has been done on the cyanobacteria mediated oxidation of naphthalene. This compound is water-soluble and therefore rather common as a pollutant. There is a wealth of information on the naphtalene biodegradability by bacteria, fungi and higher animals (e.g. Cerniglia and Gibson, 1977). Studies on naphthalene oxidation by cyanobacteria were initiated by studies on *Oscillatoria* sp. and *Agmenellum quadruplicatum*, which were found to oxidize naphthalene under photoautotrophic conditions to 1-naphthol (Cernigila et al., 1979, 1980a). The authors also tentatively identified among the oxidation products cis-1,2-dihydroxy-1,2-dihydronaphthalene and 4-hydroxy-1-tetralonen. They concluded that cyanobacteria probably possess both the monooxygenase and dioxygenase systems that catalyze the very initial step of naphthalene oxidation. Working on *Oscillatoria* sp., Cerniglia et al. (1980c) demonstrated that $^{18}O_2$ is incorporated into 1-naphthol resulting from naphthalene oxidation. Narro (1985) showed that *Oscillatoria* sp. oxidises naphthalene to naphthalene-1,2-oxide, which subsequently isomerizes to a keto-intermediate. Working on nine cyanobacteria and nine microalgae, Cerniglia et al. (1980a) demonstrated that naphthalene oxidation appears to be a widely distributed activity among photosynthetic microorganisms.

Studies have also been done on the oxidation of two other aromatic hydrocarbons, methyl naphthalene, biphenyl phenanthrene and aniline, by photosynthetic microorganisms. Like naphthalene, methyl naphthalenes are water-soluble and toxic (Lee et al., 1974). Cerniglia et al. (1980b) found that phototrophically grown cultures of *Agmenellum quadruplicatum*, *Oscillatoria* sp. and *Anabaena* sp. could oxidize 1- and 2-methyl naphthalene at the methyl group producing respectively, 1- and 2-hydroxymethyl naphthalene. The authors also demonstrated that $^{18}O_2$ was incorporated into 2-methyl naphthalene which produced 2-hydroxymethyl naphthalene. The oxidation of methylnaphthalenes by cyanobacteria is probably similar to that mediated by fungal and mammalian enzyme systems (Cernigila et al. 1984b). It should be emphasized that although cyanobacteria achieve aromatic ring oxidation of naphthalene producing naphthol, they appear to prefer the hydroxylation of the alkyl side-chain when oxidizing methyl naphthalenes.

The oxidation of biphenyl by cyanobacteria has also been investigated. Cerniglia et al. (1980b) showed that *Oscillatoria* sp. could oxidize ^{14}C-biphenyl, and the major metabolite was 4-hydroxybiphenyl with minor proportions of 4,4'-dihydroxybiphenyl. It is thus likely that the oxidation of biphenyl by cyanobacteria is quite similar to that by fungi and mammals (Smith and Rosazza, 1974; Meyer and Scheline, 1976; Wiebkin et al., 1976; Dodge et al., 1979), and differs from that carried out by bacteria which produce dihydroxybiphenyl (Catelani et al., 1971; Gibson et al., 1973).

Studies have also been made on the role of cyanobacteria in the initial oxidative attack on phenanthrene. Narro (1985) demonstrated that *Agmenellum quadruplicatum* cultivated under phototrophic conditions oxidized phenanthrene producing trans-9,10-dihydroxy-9,10-dihydrophenanthrene and 1-methoxy-phenanthrene, in addition to small amounts of phenanthrols. Using $^{18}O_2$, it has been further confirmed that only one atom of the oxygen molecule was incorporated into the trans-9,10-phenanthrene dihydrodiol produced by the cyanobacterium. In other words, a monooxygenase system is involved in the initial oxidative attack by *A. quadrupilicatum* on phenanthrene. This was probably the first report of a prokaryote oxidizing an aromatic hydrocarbon to a dihydrodiol of the trans form.

Another aromatic hydrocarbons group which received the attention of research workers is that of aniline, an aromatic amine, and its derivatives, compounds which are used in industry for the production of dyes, pesticides and pharmaceuticals. Cerniglia et al. (1981) found that two autotrophically grown cyanobacteria, *Agmenellum quadruplicatum* and *Oscillatoria* sp., metabolize aniline producing formanilide, acetanilide and p-aminophenol. It is known that bacteria (Alexander, 1981) and mammals (Gothoskar et al., 1979) also metabolize aromatic amines by N-formylating them.

2. Alkanes

While most of the work on aromatic hydrocarbons was targetted at studying the potential of cyanobacteria for detoxification of these compounds, work on aliphatic hydrocarbons had the major objective of studying the potential of these phototrophs for the complete utilization of these compounds. As in the research on aromatic hydrocarbons, there was also no guarantee for the axenity of the cyanobacterial cultures used in studies on the oxidation of aliphatic compounds. However, the few studies done so far offer interesting results that merit a brief description.

The influence of crude oil and individual alkanes has been tested on the non-axenic strains of *Microcoleus chthonoplastes* and *Phormidium corium* isolated from oil-rich sediments (Al-Hasan et al. (1994). The total lipids from *M. chthonoplastes* grown phototrophically without hydrocarbons contained mainly saturated fatty acids with 16 (hexadecanoic acid) and 18 (octadecanoic acid) carbon atoms. The same was true for lipids from *P. corium*, but in addition, considerable proportions of monoenoic C16 (hexadecanoic acid) and C18 (octadecanoic acid) fatty acids were also present. Growth in the presence of crude oil or n-alkanes led to increased proportions of total unsaturated C16 and C18 fatty acids. This effect is analogous to that occurring in typical hydrocarbon-utilizing bacteria and fungi (Radwan and Sorkhoh, 1993). Further, the two cyanobacteria accumulated fatty acids of equivalent chain lengths to those of the alkane substrates, yet only with the medium-chain alkanes, C_{14} and C_{16}. It was realized early that microorganisms utilize alkanes by initially oxidising them to fatty acids, and that the fatty acid pattern of the cell lipids therefore usually reflects the identity of the alkane utilized (e.g. Yano et al., 1971). Of particular interest is that cyanobacteria grown on n-nonadecane (C_{19}) accumulated unusually high proportions of undecanoic acid (11:0), a result similar to that found for filamentous fungi.

The above results suggested the possibility that the test cyanobacteria might be involved in n-alkane oxidation to fatty acids, so the authors attempted to find out and implement experimental criteria that would rule out the role of organotrophic bacteria, thus allowing an investigation of how far the cyanobacteria may contribute to alkane consumption and oxidation. We will explain the methodology and results in some detail.

1. The cyanobacterial biomass tested was selected such as to contain numbers of organotrophic bacteria well below those needed to exhibit minimal measurable activity when tested alone. In an experiment the cyanobacterial biomass (0.5 g fresh weight) contained about 10^6 organotrophic bacteria. This value for the latter would have been far too low (factor of 10^4) for the measured n-alkane consumption.
2. The incubation period of the biomass with the test alkanes was reduced to only 6 h to avoid excessive bacterial propagation.
3. The distribution of fatty acids resulting from alkane oxidation among the lipid classes was that characteristic of cyanobacteria, but not that of the organotrophic bacteria (Al-Hasan et al., 1998).

Data indicate that the effects of incubating cyanobacterial biomass with n-alkanes on the fatty acid patterns of its total lipids are similar to those reported for organotrophic bacteria and fungi (Radwan and Sorkhoh, 1993). Thus, cyano-bacterial biomass samples incubated with the medium chain alkanes C14 and C16 accumulate fatty acids with the equivalent chain lengths in their total lipids. Biomass samples incubated with very long odd (C_{19}) or even (C_{32}) chain alkanes accumulate respectively rather short odd (11:0 and 13:0) or even (10:0 and 12:0) chain fatty acids, indicating that the long-chain alkanes should have been metabolized through mid-chain oxidation (Radwan et al., 1996). Other important evidence for the potential of cyanobacteria to oxidize alkanes is that fatty acids resulting from that oxidation were found esterified in lipid classes specific of the thylakoids, monogalacto-syldiacylglycerols, digalactosyldiacylglycerols and sulfoquinovosyldiacylglycerols (Al-Hasan et al., 1998).

Chlorinated cyclic aliphatic hydrocarbons have also been reported to be oxidized by cyanobacteria. Kuritz and Wolk (1995) showed that two filamentous cyanobacteria, *Anabaena* sp. and *Nostoc ellipsosporum*, have the potential for degrading the highly chlorinated pesticide lindane. The two cyanobacteria could also be genetically engineered to degrade 4-chlorobenzoate.

E. Cyanobacteria as Immobilizers of Oil-Degraders

There are a few reports indicating that cyanobacteria may play an important, albeit indirect, role in hydrocarbon degradation in nature, by immobilizing oil-degrading bacteria and fungi in their mucilage. Light has been shed on this subject through the discovery of a natural phenomenon associated with extensive oil pollution along the western coast of the Arabian Gulf. The release of oil from a Kuwaiti terminal into the Gulf by the Iraqi forces during the occupation of Kuwait (2 August 1990 - 26 February

1991) caused the greatest single oil spill known. The oil drifted from the Kuwaiti coast and severely polluted stretches of about 770 km of the Saudi Arabian coast. Late in 1991 there developed visually striking blue-green mats over the oil layers, in a region in which all other macroscopically visible forms of life had vanished (Sorkhoh et al., 1992). Such mats covered the regularly inundated lower part of the oiled intertidal zone (Plate 18). The mats with the strongly adhering oil layers are subjected to tearing and dissection into irregular scales. The area appeared as if it had just been subjected to farming management. The mats are associations of photosynthetic and organotrophic microorganisms. Two filamentous cyanobacteria, *Microcoleus chthonoplastes* and *Phormidium corium*, predominate. In the extensive sheath layers of both cyanobacteria up to one million bacterial cells per gram of fresh mat were counted. These bacteria can utilize crude oil and belong predominantly to the genera *Rhodococcus*, *Arthrobacter*, *Pseudomonas* and *Bacillus*. It is apparent that cyanobacteria provide these oil-degraders with a number of advantages in the mat association. Cyanobacteria immobilize the oil-degraders, thus protecting them from being washed out to sea. Cyanobacteria provide oil-degraders with the oxygen needed for their initial attack on the hydrocarbon substrates. Oil degradation may also be stimulated by nitrogen fixed by the cyanobacteria. In other words, the blue-green mats appear to be ideal associations for self-cleaning of the oil-polluted Gulf environment, even if the cyanobacterial role was merely indirect. In this context, it may be mentioned that the associated oil-utilizers could successfully be established in oil polluted sand, which implies that these mats represent a natural source of oil-degrading bacterial cocktails for remediating the oil polluted desert (Sorkhoh et al., 1995).

VI. Concluding Remarks

Although there is an intimate historical relation between crude oil and cyanobacteria, the latter being biogenic sources of the former, a question remains unanswered: why should the photosynthetic cyanobacteria attack complex carbon sources like hydrocarbon? The answer to this question is by no means easy. However, the inhibitory effects of certain crude oil constituents, particularly water-soluble compounds, on photosynthesis may contribute to answering this question. It appears that immediately after an oil spill the photosynthetic potential of cyanobacteria becomes reduced, until toxic constituents volatilize or are washed into the open sea. Further, the fact that cyanobacteria immobilize in their envelope layers oil-degrading bacteria, and probably have done for geological ages, provides the attractive possibility that genetic interaction could possibly have occurred naturally between the organotrophic and phototrophic partners. Through the coding of this gene for catalyzing enzyme systems hydrocarbon degradation may have been transferred from the bacteria to the cyanobacteria. Cobley et al. (1993) reported on the construction of a shuttle plasmid which can be effectively mobilized from *Escherichia coli* into the cyanobacterium *Fremyella diplosiphon*. In this context, we have collected tens of cyanobacteria from the Arabian Gulf coasts and found them all associated with oil-degrading bacteria, even those samples collected from non-polluted areas. Further research is still needed to give a final answer to the question whether or not cyanobacteria may themselves contribute directly to hydrocarbon oxidation. Yet it is clear that at least the indirect role of cyanobacteria in oil-degradation is well established.

Acknowledgments

We thank our co-workers, M. Khanafer, I. El-Nemr, A. El-Desouky and A. Khan. The financial support by Kuwait University through the research projects S0049 and S0071is acknowledged.

References

Al-Hasan RH, Al-Bader DA, Sorkhoh NA, Radwan SS (1998) Evidence for n-alkane consumption and oxidation by filamentous cyanobacteria form oil-contaminated coasts of the Arabian Gulf. Mar Biol 130: 521-527

Al-Hasan RH, Sorkhoh NA, Al-Bader D and Radwan SS (1994) Utilization of hydrocarbons by cyanobacteria from microbial mats on oily coasts of the Gulf. Appl Microbiol Biotechnol 41: 615-619

Alexander M (1981) Biodegradation of chemicals of environmental concern. Science 211: 132-138

Allison RK, Skipper HE, Reid MR, Short WA and Hogan GL (1953) Studies on the photosynthetic reaction I. The assimilation of acetate by *Nostoc muscorum*. J Biol Chem 204: 197-205

Anderson SL and McIntosh L (1991) Light-activated heterotrophic growth of the cyanobacterium *Synechocystis* sp. strain PCC 6803: a blue light-requiring process. J Bacteriol 173: 2761-2767

Batterton J, Winters K and Van Baalen C (1978a) Anilines: selective toxicity to blue-green algae. Science 199: 1068-1070

Batterton J, Winters K and Van Baalen C (1978b) Sensitivity of three microalgae to crude oils and fuel oils. Marine Environ Res 1: 31-41

Boon JJ and De Leeuw JW (1987) Organic geochemical aspects of cyanobacterial mats. In: Fay P and Van Baalen C (eds) The

Cyanobacteria pp 471-492, Elsevier Science Publishers, Amsterdam, New York, Oxford

Bottomley PJ and Van Baalen C (1978) Characteristics of heterotrophic growth in the blue-green alga *Nostoc* sp. strain Mac. J Gen Microbiol 107: 309-318

Boulton CA and Ratledge C (1984) The physiology of hydrocarbon utilizing microorganisms. In: Wiesman A (ed.) Topics in Fermentation and Enzyme Technology, Vol 9, pp 11-77, Ellis Horwood, Chichester, UK

Bühler M and Schindler J (1984) Aliphatic hydrocarbons. In; Rehm H-J and Reed G (eds) Biotechnology - A Comprehensive Treatise in 8 Volumes, Vol 6a, pp 329-385, Verlag Chemie, Weinheim, Germany

Carr NG and Pearce J (1966) Photoheterotrophism in blue-green algae. Biochem J 99: 28-29

Catelani D, Sorlini C and Treccani V (1971) The metabolism of biphenyl by *Pseudomonas putida*. Experientia 27: 1173-1174

Cerniglia CE, Freeman JP, Althaus JR and Van Baalen C (1983) Metabolism and toxicity of 1- and 2-methylnaphthalene and their derivatives in cyanobacteria. Arch Microbiol 136: 177-183

Cerniglia CE, Freeman JP and Van Baalen C (1981) Biotransformation and toxicity of aniline and aniline derivatives in cyanobacteria. Arch Microbiol 130: 272-275

Cerniglia CE and Gibson DT (1977) Metabolism of naphthalene by *Cunninghamella elegans*. Appl Environ Microbiol 34: 363-370

Cerniglia CE, Gibson DT and Van Baalen C (1979) Algal oxidation of aromatic hydrocarbons: formation of 1-naphthol from naphthalene by *Agmenellum quadruplicatum*, strain PR-6. Biochem Biophys Res Comm 88: 50-58

Cerniglia CE, Gibson DT and Van Baalen C (1980a) Oxidation of naphtalene by cyanobacteria and microalgae. J Gen Microbiol 116: 495-500

Cerniglia CE, Lambert KJ, Miller DW and Freeman JP (1984) Transformation of 1- and 2-methylnaphthalene by *Cunninghamella elegans*. Appl Environ Microbiol 47: 111-118

Cerniglia CE, Van Baalen C and Gibson DT (1980b) Oxidation of biphenyl by the cyanobacterium *Oscillatoria* sp. strain JCM. Arch Microbiol 125: 203-207

Cerniglia CE, Van Baalen C and Gibson DT (1980c) Metabolism of naphthalene by the cyanobacterium *Oscillatoria* sp strain JCM. J Gen Microbiol 116: 485-494

Cobley JG, Zerweck E, Reyes R, Mody A, Seludo-Unson JR, Jaeger H, Weerasuriya S and Navankasattusas S (1993) Construction of shuttle plasmids which can be efficiently mobilized from *Escherichia coli* into the chromatically adapting cyanobacterium *Fremyella diplosiphon*. Plasmids 30: 90-105

Colwell RR and Walker JD (1977) Ecological aspects of microbial degradation of petroleum in the marine environment. Crit Rev Microbiol 5: 423-445

Cooney JJ, Silver SA and Beck EA (1985) Factors influencing hydrocarbon degradation in three freshwater lakes. Microb Ecol 11: 127-137

Dodge RH, Cerniglia CE and Gibson DT (1979) Fungal metabolism of biphenyl. Biochem J 178: 223-230

Ellis BE (1977) Degradation of phenolic compounds by fresh-water algae. Plant Sci Lett 8: 213-216

Fay P (1965) Heterotrophy and nitrogen fixation in *Chlorogloea fritschii* Mitra. J Gen Microbiol 39: 11-20

Fay P and Fogg GE (1962) Studies on nitrogen fixation by blue-green algae III. Growth and nitrogen fixation in *Chlorogloea fritschii* Mitra. Arch Mikrobiol 42: 310-321

Fitzsimons AG and Smith RV (1984) The isolation and growth of axenic cultures of planktonic blue-green algae. Br Phycol J 19: 156-162

Fusey P and Oudot J (1984) Relative influence of physical removal and biodegradation in the depuration of petroleum contaminated seashore sediments. Mar Pollut Bull 15: 136-141

Gaur JP and Kumar HD (1981) Growth response of four microalgae to three crude oils and a furnace oil. Environ Pollut Ser A 25: 77-85

Gaur JP and Singh AK (1990) Growth, photosynthesis and nitrogen fixation of *Anabaena doliolum* exposed to Assam-crude extract. Bull Environ Contam Toxicol 44: 494-500

Gibson DT (1984) Microbial degradation of aromatic hydrocarbons. In: Gibson DT (ed.) Microbial Degradation of Organic Compounds, pp 181-251, Marcel Dekker, New York

Gibson DT, Roberts RL, Wells MG and Kobal VM (1973) Oxidation of biphenyl by a *Beijerinckia* species. Biochem Biophys Res Commun 50: 211-219

Goma G, Pareilleux A and Durand G (1976) Specific hydrocarbon solubilization during growth of *Candida lipolytica*. J Ferment Technol 51: 616-621

Gordon DC and Prouse NJ (1973) The effects of three oils on marine phytoplankton photosynthesis. Mar Biol 22: 329-333

Gothoskar SV, Benjamin T, Roller PP and Weisburger EK (1979) N-Formylation of an aromatic amine as a metabolic pathway. Xenobiotica 9: 533-537

Grimalt JO, De Wit R, Teixidor P and Albaiges J (1992) Lipid biogeochemistry of *Phormidium* and *Microcoleus* mats. Organic Geochem 19: 509-530

Harder R (1917) Ernährungsphysiologische Untersuchungen an Cyanophyceen, hauptsächlich dem endophytischen *Nostoc punctiforme*. Z Bot 9: 145-245

Hoare DS, Hoare SL and Moore RB (1967) The photoassimilation of organic compounds by autotrophic blue-green algae. J Gen Microbiol 49: 351-370

Hoare DS, Ingram LO, Thurston EL and Walkup R (1971) Dark heterotrophic growth of an endophytic blue-green alga. Arch Mikrobiol 78: 310-321

Horowitz A and Atlas RM (1977) Continuous open flow-through system as a model for oil degradation in the Arctic Ocean. Appl Environ Microbiol 33: 647-653

Ingram LO, Calder JA, Van Baalen C, Plucker FE and Parker PL (1973) Role of reduced exogenous organic compounds in the physiology of the blue-green bacteria (algae): Photoheterotrophic growth of a heterotrophic blue-green bacterium. J Bact 114: 695-700

Jansson C, Debus RJ, Osiewcz HD, Gurevitz M and McIntosh L (1987) Construction of an obligate photoheterotrophic mutant of the cyanobacterium *Synechocystis* 6803. Plant Physiol 85: 1021-1025

Jobson A, Cook FD and Westlake DWS (1972) Microbial utilization of crude oil. Appl Microbiol 23: 1082-1089

Jones DM, Douglas AG, Parkes RJ, Taylor J, Giger W and Schaffner C (1983) The recognition of biodegraded petroleum-derived aromatic hydrocarbons in recent marine sediments. Mar Pollut Bull 14: 103-108

Jordan DC, McNicol PJ and Marshall MR (1978) Biological nitrogen fixation in the terrestrial environment of a high Arctic ecosystem. Can J Microbiol 24: 643-649

Josef-Espadellier F, Astier C, Evans EH and Carr NG (1978) Cyanobacteria grown under photoautotrophic, photoheterotrophic and heterotrophic regimes: sugar metabolism and carbon dioxide fixation. FEMS Microbiol Lett 4: 261-264

Khoja TM and Whitton BA (1971) Heterotrophic growth of blue-green algae. Arch Mikrobiol 79: 280-282

Kiyohara T, Fujita Y, Hattori A and Watanabe A (1960) Heterotrophic culture of a blue-green alga, *Tolypothrix tenuis*. J Gen Appl Microbiol Tokyo 6: 176-182

Klug MJ and Markovetz AJ (1967) Degradation of hydrocarbons by members of the genus *Candida*. II. Oxidation of n-alkanes and 1-alkanes by *Candida lipolytica*. J Bacteriol 93: 1847-1852
Klug MJ and Markovetz AJ (1971) Utilization of aliphatic hydrocarbons by microorganisms. Adv Microb Physiol 5: 1-43
Kuritz T and Wolk P (1995) Use of filamentous cyanbacteria for biodegradation of organic pollutants. Appl Environ Microbiol 61: 234-238
Lee CC, Craig WK and Smith PJ (1974) Water-soluble hydrocarbons from crude oil. Bull Environ Contam Toxicol 12: 212-216
Levi JD, Shenman JL and Ebbon GP (1979) Biomass from liquid n-alkanes. In: Rose AH (ed.) Economic Microbiology, Vol. 4, Academic Press, London
Malinski E, Witkowski A, Synak E, Szafranck J, Osterroht Ch and Pihlaja K (1988) Hydrocarbon geochemistry of siliciclastic microbial laminated deposits from Puk Bay, Polland. Organic Geochem 12: 81-88
Matsumoto GI, Akiyama M, Watanuki K and Torii T (1990) Unusual distributions of long-chain n-alkanes and n-alkenes in Antarctic soil. Organic Geochem 15: 403-412
Matsumoto GI and Watanuki K (1990) Geochemical features of hydrocarbons and fatty acids in sediments of the inland hydrothermal environments of Japan. Organic Geochem 15: 199-208
McCann J, Choi E, Yamasaki E and Ames BN (1975) Detection of carcinogens as mutagens in the Salmonella microsome test: assay of 300 chemicals. Proc Natl Acad Sci USA 72: 5135-5139
Meyer T and Scheline RR (1976) The metabolism of biphenyl. II. Phenolic metabolites in the rat. Acta Pharacol Toxicol 39: 419-441
Miller EC and Miller JA (1974) Biochemical mechanisms of chemical carcinogenesis. In: Bush H (ed.) Molecular Biology of Cancer, pp 377-402, Academic Press, Inc, New York
Miller JS and Allen MM (1972) Carbon utilization patterns in the heterotrophic blue-green alga *Chlorogloea fritschii*. Arch Microbiol 86: 1-12
Morales-Loo MR and Goutx M (1996) Effects of water soluble fraction of the Mexican crude oil "Isthmus Cactus" on growth, cellular content of chlorophyll a and lipid composition of planktonic microalgae. Marine Biol 104: 503-509
More JE, Williams MM and Smith AJ (1979) Features of organic growth substrate utilization by cyanobacteria In: Nichols JM (ed.) Proceedings of the Third International Symposium on Photosynthetic Prokaryotes, p B49, University of Liverpool, Oxford
Nakahara T, Hisatsuka K and Minoda Y (1981) Effect of hydrocarbon emulsification on growth and respiration of microorganisms in hydrocarbon media. J Ferment Technol 59: 415-418
Narro ML (1985) Oxidation of aromatic hydrocarbons by marine cyanobacteria. Thesis, The University of Texas at Austin, Austin TX 78712
Narro ML (1987) Petroleum toxicity and the oxidation of aromatic hydrocarbons. In: Fay P and Van Baalen C (eds) The Cyanobacteria, pp 493-511, Elsevier Science Publishers, Amsterdam, New York, Oxford
O'Brien PY and Dixon PS (1976) The effects of oil and oil components on algae: a review. Br Phycol J 11: 115-142
Oil in the Sea. Inputs, Fates and Effects (1985) National Research Council, National Academy Press, Washington DC
Pearce J and Carr NG (1967) The metabolism of acetate by the blue-green alga *Anabaena variabilis* and *Anacystis nidulans*. J Gen Microbiol 49: 301-313
Pearce J, Leach CK and Carr NG (1969) The incomplete tricarboxylic acid cycle in the blue-green alga *Anabaena variabilis*. J Gen Microbiol 55: 371-378
Pelroy RA and Bassham JA (1973) Efficiency of energy conversion of aerobic glucose metabolism in *Aphanocapsa* 6714. J Bacteriol 115: 937-942
Pelroy RA, Rippka R and Stanier RY (1972) The metabolism of glucose by unicellular blue-green algae. Arch Mikrobiol 87: 303-322
Peniguel G, Couderc R and Seyve C (1989) Les microalgues actuelles et fossiles-inlerets stratigraphique et petrolier. Bulletin-Centres de Recherche Exploration-Production Elf-Aquitaine 13: 455-482
Perry JJ (1984) Microbial metabolism of cyclic alkanes. In: Atlas RM (ed.) Petroleum Microbiology, pp 61-98, Macmillan Publ. Co. New York.
Raboy B and Padan E (1978) Active transport of glucose and α-methylglucoside in the cyanobacterium *Plectonema boryanum*. J Biol Chem 253: 3287-3291
Raboy B, Padan E and Shilo M (1976) Heterotrophic capacities of *Plectonema boryanum*. Arch Microbiol 110: 77-85
Radwan SS and Sorkhoh NA (1993) Lipids of n-alkane utilizing microorganisms and their application potential. Adv Appl Microbiol 39: 29-90
Radwan SS, Sorkhoh NA, Felzmann H and El-Desouky A (1996) Uptake and utilization of n-octacosane and n-nonacosane by *Arthrobacter nicotianae* KCC B35. J Appl Bacteriol 80: 370-374
Ramsay B, McCarthy I, Guerra-Santos L, Käpelli O and Fiechter A (1988) Biosurfactant production and diauxic growth of *Rhodococcus aurantiacus* when using n-alkanes as the carbon source. Can J Microbiol 34: 1209 -1212
Rapp P, Bock H, Wray W and Wagner F (1979) Formation, isolation and characterization of trehalose dimycolates from *Rhodococcus erythropolis* grown on n-alkanes. J Gen Microbiol 115: 491-503
Rippka R (1972) Photoheterotrophy and chemoheterotrophy among unicellular blue-green algae. Arch Mikrobiol 87: 93-98
Rippka R, Deruelles J, Waterbury JB, Herdman M and Stanier RY (1979) Generic assignments, strain histories and properties of pure cultures of cyanobacteria. J Gen Microbiol 111: 1-61
Robinson N, Eglinton G, Cranwell PA and Zeng YB (1989) Messel oil shale (Western Germany); assessment of depositional palaeoenvironment from the content of biological marker compounds. Chemical Geol 76: 153-173
Rosenberg M, Gutnick D and Rosenberg E (1980) Adherence of bacteria to hydrocarbons: A simple method for measuring cell surface hydrophobicity. FEMS Microbiol Lett 9: 29-33
Schenck PA and De Leeuw JW (1982) Molecular organic geochemistry. In: Hutzinger O (ed.) Handbook of Environmental Chemistry, Vol. 1B, pp 111-129, Springer Verlag, Berlin
Schouten S, Schoell M, Rijpstra WIP, Damste JSS and deLeeuw JW (1997) A molecular stable carbon isotope study of organic matter in immature Miocene Monterey sediments, Pismo basin. Geochim et Cosmochim Acta 61 (10): 2065-2082
Schroeder E and Rehm HJ (1981) Degradation of long chain n-alkanes by Chlorococcales. Eur J Appl Microbiol Biotechnol 12: 36-38
Shiea J, Brassel SC and Ward DM (1990) Mid-chain branched mono- and dimethyl alkanes in hot spring cyanobacterial mats: a direct biogenic source from branched alkanes in ancient sediments? Organic Geochem 15: 223-231
Shiea J, Brassel SC and Ward DM (1991) Comparative analysis of extractable lipids in hot spring cyanobacterial mats and their component photosynthetic bacteria. Organic Geochem 17: 309-319

Sikkema J, De Bont JAM and Poolman B (1995) Mechanisms of membrane toxicity of hydrocarbons. Microbiol Rev 59: 201-222

Singer ME and Finnerty WR (1984) Microbial metabolism of straight chain and branched alkanes. In: Atlas RM (ed.) Petroleum Microbiology pp 1-60, Macmillan Pub. Co, New York

Singh AK and Gaur JP (1988) Effect of Assam crude on photosynthesis and associated electron transport system in Anabaena doliolum. Bull Environ Contam Toxicol 41: 776-780

Smith AJ (1982) Modes of cyanobacterial carbon metabolism. In: Carr NG and Whitton BA (eds) The Biology of Cyanobacteria, Chapter 3, pp 47-85, University of California Press, Berkeley

Smith AJ, London J and Stanier RY (1967) Biochemical basis of obligate autotrophy in blue-green algae and thiobacilli. J Bacteriol 94: 972-983

Smith R and Rosazza J (1974) Microbial models of mammalian metabolism. Aromatic hydroxylation. Arch Biochem Biophys 161: 551-558

Sorkhoh NA, Al-Hasan RH, Khanafer M and Radwan SS (1995) Establishment of oil-degrading bacteria associated with cyanobacteria in oil-polluted soil. J Appl Bacteriol 78: 194-199

Sorkhoh N, Al-Hasan R, Radwan S and Höpner T (1992) Self-cleaning of the Gulf. Nature (London) 359: 109

Soto C, Hellebust JA, Hutchinson TC and Sawa T (1975) Effect of naphthalene and aqueous crude oil extracts on the green flagellate *Chlamydomonas angulosa*. I. Growth. Can J Bot 53: 109-117

Thiel V, MerzPreiss M, Reitner J, Michaelis W (1997) Biomarker studies on microbial carbonates: Extractable lipids of a calcifying cyanobacterial mat (Everglades, USA). Facies 36: 163-172

Vandermeulen JH and Ahern TP (1976) Effect of petroleum hydrocarbons on the algal physiology: review and progress report. In: Lockwood APM (ed.) Effects of Pollution on Aquatic Organisms, pp 107-125, Cambridge University Press, London

Vcrmass WFJ, Williams JGK, Rutherford AW, Mathis P and Arntzen CJ (1986) Genetically engineered mutant of the cyanobacterium *Synechocystis* 6803 lacks the photosystem II chlorophyll binding protein CP-47. Proc Natl Acad Sci USA 83: 9476-9477

Walker JD, Colwell RR and Petrakis L (1976) Biodegradation rates of components of petroleum. Can J Microbiol 22: 1209-1213

Walker JD, Colwell RR, Vaituzis Z and Meyer SA (1975) Petroleum-degrading achlorophyllous alga *Prototheca zopfi*. Nature (London) 254: 423-424

White AW and Shilo M (1975) Heterotrophic growth of the filamentous blue-green alga *Plectonema boryanum*. Arch Mikrobiol 102: 123-127

Wiebkin P, Fry JR, Jones CA, Lowing R and Bridges JW (1976) The metabolism of biphenyl by isolated viable rat hepatocytes. Xenobiotica 6: 725-743

Williams LA and Reimers C (1983) Role of bacterial mats in oxygen-deficient marine basins and coastal upwelling regimes; preliminary report. Geology (Boulder) 11: 267-269

Willingham TO, Nagy B, Nagy LN, Krinsley DH and Mossman DH (1985) Uranium-bearing stratiform organic matter in paleoplacers of the lower Huronian Supergroup, Elliot Lake - Blind River region, Canada. Can J Earth Sci 22: 1930-1944

Winters K, Batterton JC, O'Donnell R and Van Baalen C (1977) Fuel oils: Chemical characterization and toxicity to microalgae. In: Giam CS (ed.) Pollutant Effects on Marine Organisms, pp 167-189, Lexington Books, DC Health, Lexington MA

Winters K, O'Donnell R, Batterton JC and Van Baalen C (1976) Water-soluble components of four fuel oil: chemical characterization and effects on growth of microalgae. Mar Biol 36: 269-276

Wodzinski RS and Coyle JE (1974) Physical state of phenanthrene for utilization by bacteria. Appl Microbiol 27: 1081-1084

Wolk CP and Shaffer PW (1976) Heterotrophic micro- and macrocultures of nitrogen-fixing cyanobacterium. Arch Mikrobiol 110: 145-147

Yano I, Furukawa Y and Kusunose M (1971) Fatty acid composition of *Arthrobacter simplex* grown on hydrocarbons. Occurrence of α-hydroxy-fatty acids. Eur J Biochem 23: 220-228

Chapter 12

Cyanobacterial Dominance in the Polar Regions

Warwick F. Vincent
Département de biologie & Centre d'études nordiques, Université Laval, Sainte-Foy, Québec, Canada G1K 7P4

Summary

Although cyanobacteria are often thought of as warm water organisms, they are the predominant biota in cold polar environments such as ice shelves, glaciers, glacial meltwater streams and ice-capped lakes. Cyanobacteria are the primary colonizers of glacial moraines after the retreat of ice sheets, and they play an important role in the carbon and nitrogen economy of tundra and polar desert soils. Various communities dominated by cyanobacteria inhabit exposed rock surfaces, while others occur within fissures and the interstitial spaces between crystals in certain Arctic and Antarctic rock types (See Chapter 13). Highly pigmented microbial mats dominated by *Nostoc* or oscillatorians (Oscillatoriaceae) are a feature of streams, lakes and ponds in both polar regions, with extreme accumulations up to 90 cm thick and > 40 μg Chl*a* cm^{-2} at some sites. Picocyanobacteria often dominate the phytoplankton of polar and subpolar lakes. In the coastal saline lakes of Antarctica picocyanobacteria achieve some of the highest natural concentrations on record, up to 8 x 10^6 cells mL^{-1}. However, picocyanobacteria are conspicuously absent or rare in the adjacent polar oceans. The ecophysiological characteristics of high-latitude cyanobacteria that contribute to their success and dominance include: an ability to grow over a wide temperature

B. A. Whitton and M. Potts (eds), The Ecology of Cyanobacteria, 321-340.
© 2000 *Kluwer Academic Publishers. Printed in the Netherlands.*

range (but at slow rates); tolerance of desiccation, freezing and salinity stress; a variety of adaptive strategies against high levels of solar radiation (including ultraviolet radiation) in exposed habitats; and acclimation to shade allowing net growth in protected dim light environments. In many polar habitats, the large standing stocks of cyanobacterial biomass are the result of gradual accumulation over many seasons, with only minor losses via biotic and abiotic removal processes. Cyanobacteria are not successful in the polar oceans where slow, temperature-depressed and light-limited growth rates are unable to keep pace with the continuous losses due to grazing, advection and mixing.

I. Introduction

"When I persuaded our botanist, Dr. Berggren, to accompany me in the journey over the ice, I joked with him on the singularity of a botanist making an excursion into a tract, perhaps the only one in the world, that was a perfect desert as regards botany. This expectation was, however, not confirmed. Dr. Berggren's keen eye soon discovered, partly on the surface of the ice, partly in the above mentioned [cryoconite] powder, a brown, polycellular alga, which small as it is, together with the powder and certain other microscopic organisms by which it is accompanied, is the most dangerous enemy to the mass of ice, so many thousand feet in height and hundreds of miles in extent." (p.163 in Leslie, 1879)

These observations by the Swedish explorer Adolf Erik Nordenskiöld describe a remarkable discovery made during his 1870 expedition across the Greenland Ice Sheet. In certain regions of the interior, the surface of the ice sheet is coated by "cryoconite", literally "cold rock dust". These particulates are trapped by black, mucilage-producing cyanobacteria and the resultant dark patches absorb radiation. This in turn influences the local heat balance, producing holes, ponds and streams that accelerate local melting and the degradation of ice. Nordenskiöld proposed several potential origins for the extensive cryoconite powder, but his initial suggestion based on chemical analyses that it might be composed of cosmic dust (micro-meteorites) subsequently proved correct (Maurette et al., 1986). These communities of cyanobacteria in the Greenland interior thus hold the unusual distinction of growing on ice, meltwater and extra-terrestrial substrates.

As explorers began to roam more widely in the polar regions there was increasing evidence, often anecdotal, that cyanobacteria were well established throughout the Arctic and Antarctic. A Swedish expedition to Spitsbergen in 1861 noted cyanobacteria growing in fellfield pools: "and here and there the beautiful *Phalaropus* [northern phalarope] was seen to pluck the alga *Nostoc commune* (See Chapter 17) which is plentiful in these waters" (p. 65 in Leslie, 1879). James Murray, the biologist on Shackleton's 1907-9 expedition to Ross Island, Antarctica, dug through the ice of a frozen lake and found benthic mat material "that on careful thawing released a multitude of living things for study" (Murray, 1910), notably filamentous cyanobacteria and the rotifers which feed on them. These and earlier collections from the Ross Sea region revealed many species of cyanobacteria as well as diatoms and other organisms (Fig. 1). Some two years later, Griffith Taylor, a geographer on the British Antarctic Expedition was sent by Captain Robert Falcon Scott to explore the "Dry Valley", one of a series of largely ice-free valleys at latitude 78°S in the McMurdo Sound region.

As Taylor crossed one of the frozen lakes he looked down and made note of the "extensive water plants" (brightly colored mats of cyanobacteria) beneath him (Taylor, 1916). He also commented on the medicinal taste of one of the ponds, perhaps the first evidence of taste and odor compounds (Persson, 1996) produced by the mat-forming species in this region.

Cyanobacteria are now known to colonize a remarkably diverse range of high latitude environments including rocks, glaciers, ice shelves, streams, ponds and lakes. In polar desert soils these micro-organisms are amongst the primary colonizers, and they continue to play an important role in the overall nitrogen and carbon economy of well-vegetated sites such as tundra and moss banks. At the bottom of certain Antarctic lakes, the biomass accumulation can be spectacular with up to a 90 cm thick layer of biomass and mucilage; some of the thickest microbial biofilms to be found in the natural environment. In the more extreme polar habitats, cyanobacteria constitute not only the dominant phototrophs but also most of the microbial ecosystem biomass. A notable exception to this general trend is the polar marine environment where picocyanobacteria are conspicuously rare or absent.

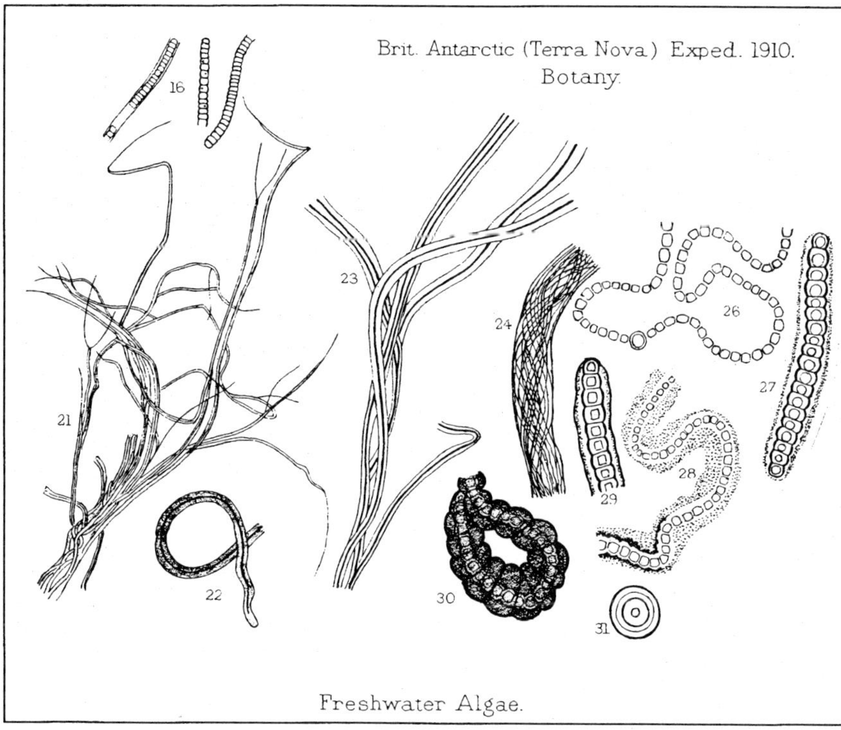

Fig. 1. Drawings of cyanobacteria collected from Ross Island, Antarctica (latitude 77°S) during Scott's Terra Nova expedition. The newly described taxa were *Phormidium priestleyi* (16), *Schizothrix antarctica* (21-24) and *Nostoc fuscescens* var. *mixta* (25-31). From Plate I, Fritsch (1917)

This review first describes the range of cyanobacterial communities and habitats in the polar regions, and considers some of the general issues regarding their distribution and biodiversity. It then examines some of the eco-physiological characteristics that distinguish these assemblages of cyano-bacteria. The final section considers cyanobacterial dominance, and questions why this group of organisms is pre-eminently successful in non-marine high latitude environments yet is so poorly represented in the polar oceans.

II. Habitats and Communities

A. Marine Environments

Chroococcoid forms of cyanobacteria generally ascribed to the genus *Synechococcus* are widely distributed throughout the world oceans, and in many temperate and tropical regions they contribute a major, sometimes dominant, fraction of total phytoplankton biomass and productivity (Waterbury et al., 1986). The polar oceans are a notable exception. In the Arctic as well as Antarctic, concentrations of *Synechococcus* fall to low values, often below 10^2 cells mL^{-1}. Higher concentrations occur in sea ice, although such populations may simply represent cells which were trapped in the ice during freezing, with little *in situ* growth (Walker and Marchant, 1989).

In the Southern Ocean, two types of picocyanobacteria were distinguished on the basis of their cell wall ultrastructure, indicating the likely presence of genetically different strains (Marchant et al., 1987). There was a strong North-South trend in the concentration of these phycoerythrin-containing picocyanobacteria (Fig. 2), with a four order-of-magnitude drop between latitudes 45 and 60°S to a minimum of < 10 cells mL^{-1}. There was also a strong correlation between cell concentration and sea surface temperature, suggesting that low temperatures were the principal determinant of the low abundance in this part of the ocean.

There are many records of picocyanobacteria in the seas of the north polar region, however they are rarely major components of the plankton. There is also

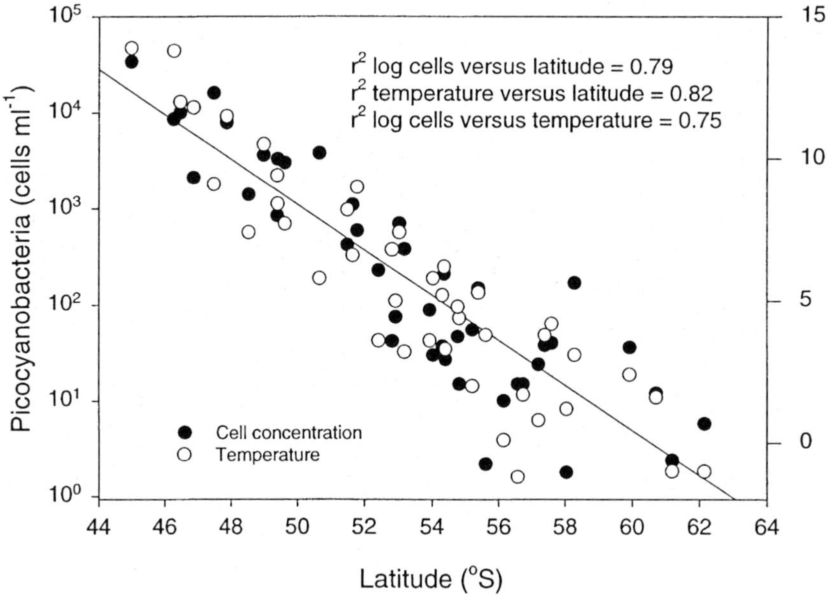

Fig. 2. Relationship between picocyanobacterial abundance (solid circles) and latitude in the Southern Ocean. The data are for 4 transects over the period Sept 1989-Feb 1986, from Table 1 in Marchant et. al. (1987). Also shown are the corresponding water temperature measurements (open circles) and r^2 values for the relationships between variables.

some controversy over to what extent these low density populations represent communities that are actively growing *in situ* versus populations that are transported from lower latitudes by advection. Mishitina et al. (1994) reported that phycoerythrin-rich populations were present at all times during their sampling of the open waters of the Barents, Norwegian and Greenland Seas, implying that there were resident communities of actively growing cells. *Synechococcus* also occurred as an epiphyte on the blades of brown algal macrophytes in the littoral zone of the Barents Sea.

In another study of northern waters, strong seasonal, spatial and depth variations in abundance of picocyanobacteria in the Greenland Sea were measured, with highest concentrations (up to 5470 cells mL^{-1}) in Arctic Intermediate Water. However, they were virtually absent from water collected inside the central Arctic Ocean, leading Gradinger and Lenz (1995) to conclude that picocyanobacteria had little impact on pelagic carbon and energy flux in the Arctic Ocean, and that their periodic appearance in the Greenland Sea was due to a high survival ability during advection from the North Atlantic. Several studies in the northern hemisphere reported a similar trend to that recorded on the Southern Ocean of a decreasing abundance of phycoerythrin-containing cells with increasing latitude; for example in the North Atlantic (Murphy and Haugen, 1985) and the Greenland Sea (Legendre et al., 1993).

B. Ice and Snow

The cryoconite features that were first discovered during Nordenskiöld`s expedition over the Greenland Ice Cap are now known to occur in the lower part of many glaciers in the polar regions as well as on alpine glaciers of the temperate zone (Wharton et al., 1985). The holes, ponds and streams form in the wastage zone and are readily colonized by micro-organisms, particularly cyanobacteria. The Greenland cryoconite communities are dominated by *Calothrix parietina*, a cyanobacterium known from Antarctica as a producer of the black UVR-absorbing pigment scytonemin (See Chapter 21). All of the samples of cryoconite examined by Gendel and Drouet (1960) from the Thule area contained fine mineral particulates loosely bound by the filaments of this cyanobacterium.

Complementary studies on cryoconite communities in the south polar region, specifically on the Canada Glacier in the McMurdo Dry Valleys, showed that the communities were dominated by filamentous cyanobacteria, specifically *Phormidium frigidum*, *Lyngbya martensiana*, *Microcoleus paludosus* var.

acuminatus and *Nodularia harveyana*. Cylindrical cells of the chroococcalean *Synechococcus aeruginosa* were also common. These taxa are found throughout the valleys in other terrestrial and aquatic habitats, and the communities are likely to be derived from wind blown microbial mats (Wharton et al. 1981).

Cyanobacteria appear to be relatively rare in melting snow banks by comparison with other phototrophs, notably the chlorophyte *Chlamydomonas nivalis,* which often dominates snow assemblages in temperate as well as polar latitudes (Vincent, 1988). Cyanobacterial growth rates may be too slow for these transient habitats. However, they are found in a broad range of other ice-dominated systems including ice sheets, glaciers and lake ice. In the ice-covered lakes of the McMurdo Dry Valleys, Antarctica, benthic mats of cyanobacteria can detach and float up to become incorporated in the ice cap during winter freeze-up. These patches of microbial mat then remain within the ice for many years where they continue to be metabolically active in an ice-bubble of gas and melt water. With subsequent freezing at the bottom of the ice cap and ablation at its upper surface the mats gradually move up through the ice over a period of several years and are finally released at the surface. Parker et al. (1982) calculated that this "escape mechanism" of biomass and mineral materials could result in a quantitatively significant loss of nutrients from the lakes each year. Another *Phormidium*-dominated community is also found growing within the liquid water inclusions of the lake ice, and appears to be inoculated from the surface by wind-blown material (Priscu et al. 1998).

One of the most extensive developments of cyanobacteria in the polar regions occurs over an area of 1500 km^2 on the McMurdo Ice Shelf (lat. 78°S). This ablation zone consists of an interconnected complex of meltwater lakes, ponds and streams which contain thick (several mm to cm) microbial mats dominated by oscillatorians or, less commonly, by *Nostoc commune* and *N. microscopicum* with subdominant diatom assemblages (Vincent, 1988; Howard-Williams et al., 1989, 1990; Hawes et al., 1993). The ponds are chemically diverse, ranging from dilute, low conductivity meltwaters (< 200 µS cm^{-1}) to sulfate-rich brines (See Chapter 10) with several times the salinity of sea water (de Mora et al., 1994). Some of the ponds are inundated by a single daily tide (Hawes et al., 1997). The microbial mats are correspondingly varied, but typically consist of an orange or red surface layer which is rich in carotenoid pigments, and a deep blue-green layer which has high concentrations of chlorophyll *a* (Chl*a*) and phycocyanin (Fig. 3).

Studies on the physiological ecology of the ice shelf mats revealed that they act as "compressed euphotic zones" where there are large changes in spectral quality down through the vertical profile (Quesada and Vincent, 1993). Ultraviolet radiation (UVR) and high energy photosynthetically active radiation (PAR) are strongly attenuated by the surface layer of the mat, and most of the photosynthesis takes place in the "deep Chl *a* maximum" layer (DCM) growing towards the base of the profile in an orange or red dim light regime. Some of the microbial mats contain species capable of gliding motility that vertically migrate through the light gradient of the mat in response to changes in ambient solar radiation (Vincent et al., 1993c; Vincent and Quesada, 1994; Quesada and Vincent, 1997).

C. Rock and Soils

Cyanobacteria are the primary colonizers of exposed moraines after the retreat of glaciers and they are widely distributed throughout the soils of both polar regions. Species that fix dinitrogen play an especially important role in the nitrogen economy of these habitats and permit colonization by other micro-organisms and higher plants (Bliss and Gold, 1994). Various types of rock environment provide a favorable habitat for cyanobacteria. The communities form dark surface crusts over the rock (epiliths), and blue-green colored biofilms under translucent stones (subliths or hypoliths). Well-developed communities can also grow within rock fissures (chasmoendoliths) or in the interstitial spaces beneath the surface of porous rocks, particularly sandstones (crypto-endoliths; See Chapter 13; Plates 20 and 21). The severe microclimate of many of these exposed rock habitats contrasts with the more benign conditions experienced by cyanobacteria that grow on (as epiphytes) or in association with moss, lichens and vascular plants.

Dark crusts are found on rock throughout the Arctic and Antarctic and are particularly conspicuous in the polar desert and semi-desert regions (e.g., Aleksandrova, 1988). These occur in habitats that are periodically supplied with snow melt and the assemblages are commonly dominated by *Gloeocapsa* spp. The dark coloration is due to the UVR-screening pigment scytonemin (see Chapter 21). Although this community can withstand bright solar radiation and

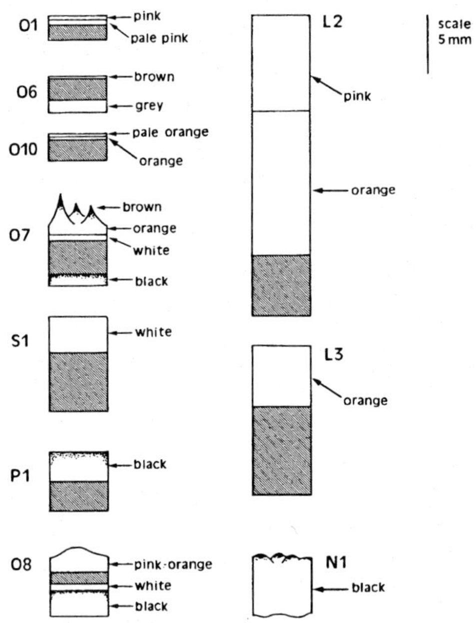

Fig. 3. Schematic diagram of vertical sections through Antarctic microbial mats dominated by cyanobacteria. The shaded area is the blue-green colored DCM. All profiles have been drawn to the same scale, from Vincent et al. (1993b).

desiccation it appears to be intolerant of saline conditions (Broady, 1996). The community is absent, for example, from Signy Island in the maritime Antarctic region, and from coastal areas influenced by sea spray.

Hypolithic communities are commonly found wherever translucent stones occur over the soil surface as they do, for example, in the Vestfold Hills and Schirmacher Oasis regions in the Antarctic continent. At the latter site, the hypolithic environment is colonized exclusively by *Aphanocapsa endolithica.* The Vestfold Hills communities are more diverse with *Chroococcidiopsis*, *Plectonema* and two green algal species (*Desmococcus* and *Prasiococcus*) as the dominant genera (Broady 1996). *Chroococcidiopsis* appears to be completely absent from salt-influenced areas and is limited to mineral, low salinity soils.

Endolithic communities are best described from the south polar region but are also known to occur in the Arctic, for example in fissures within the exposed dolomite sections at Resolute Bay (lat. 74°N), and at 1-5 mm depth within in the sandstones of Ellesmere Island (lat. 81°N). Extensive research on the cryptoendolithic communities of the McMurdo Dry Valleys region, Antarctica, showed that the communities are typically composed of cyanobacteria (in particular the genus *Chroococcidiopsis*), microscopic lichens or the green alga *Hemichloris antarctica* (Nienow and Friedmann, 1993 and refs therein). The chasmoendolithic communities appear to be more diverse and *Chroococcidiopsis* co-exists with a variety of eukaryotic taxa. Vertical zonation is a feature of many of the endolithic communities and cyanobacteria typically form the lowermost stratum. In coastal chasmoendolithic communities the outer green layer of *Desmococcus* or *Prasiococcus* is underlain by an interior blue-green stratum of *Chroococcidiopsis* (Broady, 1981).

Much of what we know about the algal ecology of polar soils comes from more than 25 years of microbiological research on Signy Island (lat. 60°S) in the maritime region of Antarctica. Soil samples from this region contained from 4 to 57 taxa, with most assemblages dominated by cyanobacteria (Broady, 1996). Filamentous cyanobacteria are the primary colonizers of fellfield soils recently exposed by receding ice (Wynn-Williams, 1990); detailed epifluorescence studies showed how these species bind together the soil particles and increase the retention and stability of mineral fines (Wynn-Williams, 1991). There is a high degree of spatial variability between and within frost polygons, with a seasonal succession from chlorophytes in spring to filamentous cyanobacteria in summer (Davey, 1991). The most common form in the Signy Island fellfield soils is *Phormidium autumnale* and it occurs throughout the upper 5 mm of the soil profile. Davey and Clarke (1991) found that *Nostoc* had a median depth of 150 μm while *P. autumnale* had a median depth of 1110 μm. They interpreted this deeper location of *P. autumnale,* as well as the migration behavior of these assemblages, as adaptive responses to avoid desiccation.

Soils on the Antarctic continent that are flushed by meltwater often support visible growths of cyanobacteria (Broady, 1996). Common forms include *Nostoc commune* (see Chapter 17) and members of the Oscillatoriales, with subdominant genera such as *Nodularia*, *Calothrix*, *Scytonema* and *Tolypothrix*. Water availability and salinity appear to be the major determinants that influence distribution of these cyanobacteria. For example, in the Vestfold Hills heterocystous cyanobacteria are absent from soils which are moist but not subject to flowing water, while *Gloeocapsa* becomes locally abundant only where the salinities are low.

Antarctic bryophyte communities often contain an associated epiphytic flora dominated by filamentous and coccoid cyanobacteria. Some species form crusts over the mosses, for example, *Stigonema minutum* at Schirmacher Oasis and *Nostoc* and oscillatorians at the Vestfold Hills; in areas subject to sea spray where birds are prevalent these surface communities are replaced by green crusts of chlorophytes (Broady, 1986).

D. Rivers and Streams

A variety of flowing water environments occur around the margins of Antarctica and these often contain a well developed periphyton dominated by cyanobacteria or, at some locations, by chlorophytes (Vincent et al., 1993a and refs therein). Three groups of stream cyanobacteria were distinguished: surface crusts; cohesive microbial mats; and moss epiphytes; all three types occur over a wide range of substrates including irrigated rock faces, flushed mineral soils and the channels of perennial streams.

The epilithic crusts are typically rich in scytonemin or other UVR-screening pigments and are colored black or brown. The common dominants are *Gloeocapsa*, *Schizothrix* and at some localities (e.g., the Alph River system in southern Victoria Land), the heterocystous genus *Calothrix*. In the latter community, *Gloeocapsa* is subdominant and grows epiphytically over the *Calothrix* as well as directly over the rock substrate.

Two types of microbial mats are found in Antarctic streams. The most common mat is dominated by oscillatorians (particularly *Phormidium*, *Oscillatoria* and *Schizothrix*) in which the filaments are embedded in mucilage where they bind together silt and sand grains. The mats occur up to several mm in thickness (up to 40 μg Chl*a* cm^{-2}) and are often red or orange colored, with a blue-green bottom layer that is rich in phycobilin pigments (e.g., Ellis-Evans and Bayliss 1993; see also Fig. 3). The second group of mats is composed primarily of *Nostoc commune* and form black mucilaginous layers up to 20 mm thick. The *Nostoc* communities tend to occur in more ephemeral environments than do oscillatorian mats such as at the edge of stream beds and in the slow-moving flush areas.

Moss communities are especially common on the bank of streams in the maritime regions of Antarctica as well as on the continent. They generally contain a diverse flora dominated by cyanobacteria, but there are large regional differences in the forms present. For example, at one coastal location in east Antarctica (Strandnebba) the stream-bank moss communities were sometimes completely coated by *Stigonema minutum* and *Plectonema* to the exclusion of other forms, whereas 30 km further to the North the communities were more diverse with five species of diatoms and a desmid as subdominants (Ohtani and Kanda, 1987).

Rivers and streams are also a major component of the Arctic landscape and, as in Antarctica, their periphytic assemblages are often dominated by cyanobacteria. Sheath et al. (1996) recognized two groups of riverine streams in the Arctic tundra: those which flow solely in the Arctic; and those which originate further to the south in the boreal forest. They noted that the latter were typically too deep and turbid to support benthic autotrophs, while the former often contained an extensive, macroscopic growth of periphyton. In their analysis of algal community structure and distribution in 150 segments of tundra streams across North America they recorded 83 taxa, of which 39% were cyanobacteria and 42% were chlorophytes. The most widespread forms were *Rivularia minutula*, *Nostoc commune* and *Tolypothrix tenuis*. Oscillatorian communites also occur in Arctic streams (Croasdale, 1973; Hamilton and Edlund, 1994; Vézina and Vincent, 1997; Sheath and

Müller 1997). The mats are rarely as well developed and extensive as those in Antarctica, perhaps reflecting the increased grazing pressure in the Arctic. Insects, for example, are absent from continental Antarctica but can achieve high population densities in the Arctic. Differences in nutrient supply may also contribute to the contrast between Arctic and Antarctic mat communities.

E. Lakes and Ponds

1. Plankton

Early studies on the limnology of the polar regions drew attention to the apparent lack of cyanobacteria in the plankton despite the eutrophic conditions which would favor blooms of cyanobacteria in water bodies of temperate latitudes (Kalff and Welch, 1974; Kalff et al., 1975). This earlier work on polar phytoplankton tended to emphasize nanoplankton and net plankton species. More recent studies confirmed that the larger colonial and filamentous bloom-forming taxa were relatively rare in polar lakes. However, the advent of fluorescence microscopy revealed that picoplanktonic species of cyanobacteria are abundant and can be the biomass dominants in the phytoplankton community of lakes in both polar regions.

Small, single-celled taxa of planktonic cyanobacteria were recorded in many lakes throughout Antarctica. *Synechococcus* and *Synechocystis* were noted in the plankton of the permanently ice-covered McMurdo Dry Valley lakes during the first investigation of their biological limnology (Goldman et al., 1967). A subsequent study determined that *Synechocystis* was a dominant component of the deep Chl*a* maximum (DCM) in Lake Vanda, the deepest, most oligotrophic of the Dry Valley lake series (Vincent and Vincent, 1982). Further to the north but still within the Antarctic Ross Sea sector, studies on the lakes of the Terra Nova Bay region indicated that up to 50% of the phytoplankton Chl*a* was in the < 2 μm fraction. Cultures of picoplankton samples from these lakes contained *Synechococcus* as well as eukaryotic species and two unidentified taxa that were tentatively identified as prochlorophytes (Andreoli et al., 1992). Picocyanobacteria also appeared to be important elements of lakes of the maritime zone of Antarctica. In five lakes on Signy Island, *Synechococc*us-like cells were the dominant phytoplankton in terms of cell concentrations and Chl*a*, but nanoplankton (mostly flagellates and chlorophytes) dominated the total primary production (Hawes, 1990; Ellis-Evans, 1991).

The most detailed study to date on picocyanobacteria in Antarctic inland waters is from a meromictic, marine-derived lake in the Vestfold Hills (Ace Lake, lat. 68.5°S). *Synechococcus* cells in this lake had dimensions of 0.9 x 1.5 μm and were readily distinguished in flow cytometric analyses through their phycoerythrin fluorescence (Rankin et al., 1997). Cell concentrations were less than 10^4 cells mL^{-1} during winter, but increased rapidly in the DCM to peak concentrations of 8×10^6 cells mL^{-1} at 11 m depth in mid-December. The latter values are almost an order of magnitude higher than the maximum concentrations found in tropical and temperate oceans. The temperature at this depth in the lake remained relatively constant at around 6°C, with a salinity approximately that of sea water, and PAR irradiances less than 1% of those at the surface. Net population growth rates calculated from the cell concentration data for 10 m depth presented by Rankin et al. (1997) averaged 0.03 d^{-1} during spring and early summer, with negative values in the period Jan-July (Fig. 4). Recent studies on isolates from these lakes confirmed that they were rich in phycobilins and devoid of Chl*b*, but also suggested genetic affinities with *Prochlorococcus* (Rankin and Bowman, unpublished data).

Picocyanobacteria are also important components in oligotrophic lakes of the subarctic and Arctic (See Chapters 5 and 7). In the cold, subarctic lakes of northern Québec the < 2 μm fraction typically accounts for 30-60 % of the total planktonic Chl*a* (Bergeron and Vincent, 1997) and is dominated by picocyanobacteria. Similarly, in Arctic lakes picocyanobacteria are often the phytoplankton dominants. A comparison of the growth rates and pigment characteristics of five isolates of *Synechococcus* from Arctic lakes showed differences between strains, and evidence of considerable genetic diversity within this component of the microbiota (Vézina and Vincent, 1997).

Oscillatorian cyanobacteria are often found as a major element of the phytoplankton in polar lakes. Many of these taxa have extremely thin trichomes (ca. 1 μm in diameter), and because of these morphological characteristics their light absorbing properties are as efficient as small coccoid or ovoid cells such as those of *Synechococcus*. These filamentous populations were found in Dry Valley lakes (Vincent and James, 1996) and in Arctic ponds

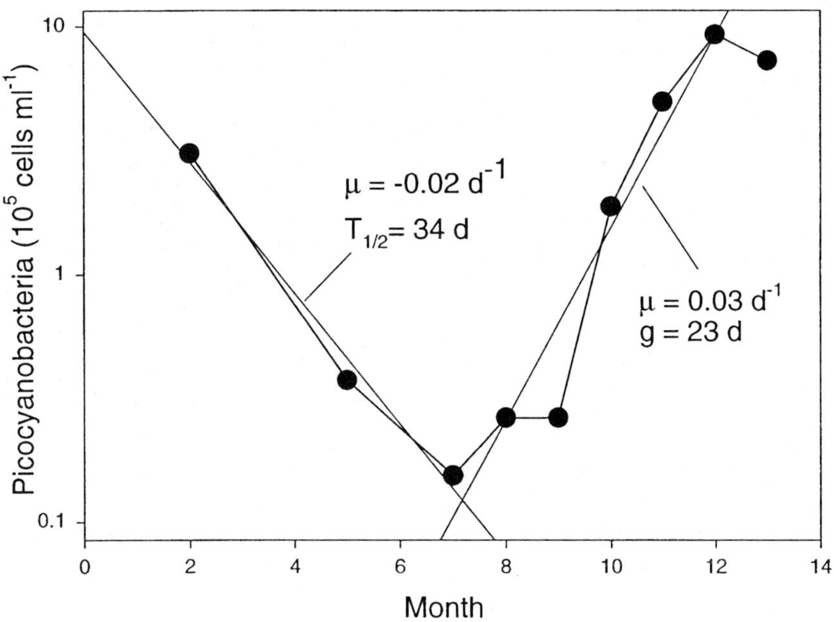

Fig 4. The seasonal distribution of picocyanobacteria at 10 m depth in Ace Lake, Antarctica. The net population loss or growth rates were calculated from natural log regressions of the cell count data versus time. Redrawn from Rankin et. al. (1997).

(Vézina and Vincent,1997). In part, however, these populations may represent filaments washed in from stream beds or the littoral zone where thick mats of cyanobacteria occur.

2. Benthos

The richest biomass accumulations of cyanobacteria in the polar regions occur in the benthic habitats of lakes and ponds. The communities form highly pigmented layers over the bottom substrata and may gradually accumulate as mucilaginous films and mats up to several cm or even several tens of cm in thickness. In the lakes of the McMurdo Dry Valleys region five types of benthic mat were distinguished (Parker and Wharton, 1985):

Moat mats – these form around the edge of the lake where the ice melts each season. They are characterized by thick, spongy layers, and are often pigmented bright orange or brown.

Columnar lift-off mats – these are produced by the trapping of nitrogen and oxygen bubbles within the surface mat, and they grow as upright columnar structures. Portions of the mat may break off and float up under the permanent ice cover of the lake so that they become incorporated into the ice when it freezes (see above). In one lake of the Dry Valleys, Lake Fryxell, the columnar mats precipitate calcium carbonate and remain in place with a hard calcite interior.

Pinnacle mats – these occur to at least 30 m depth within the relatively well illuminated environment of Lake Vanda and form conical structures 2-5 cm high and 3-5 cm wide. These structures incorporate sand grains and calcite crystals and resemble the fossil Precambrian stromatolite *Conophyton* (Chapter 2).

Aerobic prostrate mats – these occur over the surface sediments at depth and precipitate calcite.

Anaerobic prostrate mats – these occur in deep anoxic basins of the lakes.

The mats were typically dominated by filamentous cyanobacteria, usually members of the Oscillatoriaceae. In lakes of the Dry Valleys the dominant form was *Phormidium frigidum* in all mat types, sometimes in association with *Lyngbya martensiana*.

More recent studies in the Larsemann Hills of eastern Antarctica revealed a similarly wide diversity of mat types (Ellis-Evans, 1996, and refs therein):

a) In deeper lakes (> 10 m depth), black *Nostoc* mats several mm in thickness occurred to a depth of c. 1 m.

b) At greater depth *Nostoc* was replaced by orange pigmented mats up to 8 mm thick and dominated by *P. frigidum*. Small surface protrusions of the mat were associated with oscillatorian filaments that overgrew spherical *Nostoc* colonies. These mixed assemblages are morphologically similar to communities reported from elsewhere in Antarctica, specifically at Ablation Point, Alexander Island and in ponds on the McMurdo Ice Shelf which contained *Nostoc microscopicum*. The surface carotenoid pigmentation of the mats diminished with depth in the lake and was absent from communities in water deeper than 6 m, consistent with the role of these pigments as a defense against bright solar radiation (see below).

c) In lakes with a maximum depth of 3-10 m the mats formed circular plates 1-2 cm thick and 10-15 cm in diameter. The plates accumulated in the bottom waters, formed a layer at least 50 cm in thickness, and were dominated by oscillatorians with *Nostoc*, *Gloeocapsa* and *Chroococcus* as subdominants. Similar plate- or disc-shaped growths of cyanobacteria are known to occur in shallow ponds in the southernmost part of Victoria Land (the Pyramid Trough area).

d) In shallow ponds of higher salinity, red pigmented mats covered the bottom and were periodically detached by wind. These mats also contained spherical *Nostoc* colonies on their upper surface, but without the filamentous overgrowth.

Thick benthic mats of cyanobacteria were reported from many other Antarctic locations. For example, divers who examined the bottom of Priyadarshani Lake (maximum depth of 6.5 m) in the Schirmacher Oasis region found 25-90 cm thick cyanobacterial mats that were dominated by oscillatorians as well as *Synechocystis* and *Chroococcus* (Ingole and Parulekar, 1990). Benthic cyanobacteria occur in association with mosses in the maritime lakes of Signy Island. Unlike the lake communities of continental Antarctica these maritime cyanobacterial communities are dominated by *Phormidium* in shallow waters and by *Tolypothrix tenuis* and *Plectonema* (and sometimes the xanthophyte *Tribonema*) at depth (Ellis-Evans, 1996).

III. Biodiversity and Endemism

The biodiversity of polar cyanobacteria is at present a subject of considerable debate and uncertainty. In large part this reflects the inadequacy of current taxonomic criteria for these organisms. Broady (1996) pointed to the lack of consistent taxonomic criteria in analyses of the Antarctic microflora in general, but particularly with respect to the cyanobacteria. Most investigators use the traditional taxonomic scheme of Geitler (1932) and work with samples of freshly collected or preserved field material. Some authors adopted the highly simplified Drouet classification, in which much information was sacrificed. Few studies of polar regions to date have used the criteria of Anagnostidis and Komárek, (1988) that are based on morphological features and attributes of strains in culture. A notable example of this approach in Antarctica was the work of Broady and Kibblewhite (1991); similar studies are only just beginning in the Arctic (Vézina and Vincent, 1997).

The slow rates of speciation by cyanobacteria in general (Castenholz, 1992), in combination with the efficient dispersal abilities of this group of microorganisms, and the relatively young age of ice-free environments in the Arctic and Antarctic, suggest that endemism is likely to be rare amongst polar cyanobacteria. Most of the forms identified to date appear to be cosmopolitan taxa. However, the morphological simplicity of cyanobacteria masks a high level of genetic variability. For example, *Phormidium autumnale* is a commonly encountered species in both polar regions, and recent studies on lakes in the Bylot Island region of the Arctic (Lat. 73°N) demonstrated that several isolates which conformed to the morphological criteria for this taxon differed greatly in their pigment and growth characteristics (Fig. 5).

The application of molecular tools to polar communities (e.g. PCR fingerprinting; sequence analysis of phylogenetically-relevant genes such as 16S rRNA) will be of special interest in determining the genetic relationships between such isolates, and for determining the affinity of the Arctic cyanobacterial flora with that of lower latitudes and Antarctica.

IV. Ecophysiology of Polar Cyanobacteria

A. Temperature Relations

Many observations from temperate lakes and rivers support the tenet that cyanobacteria prefer warm temperatures. For example, bloom-forming species have temperature optima for growth and photosynthesis above 20°C and generally achieve

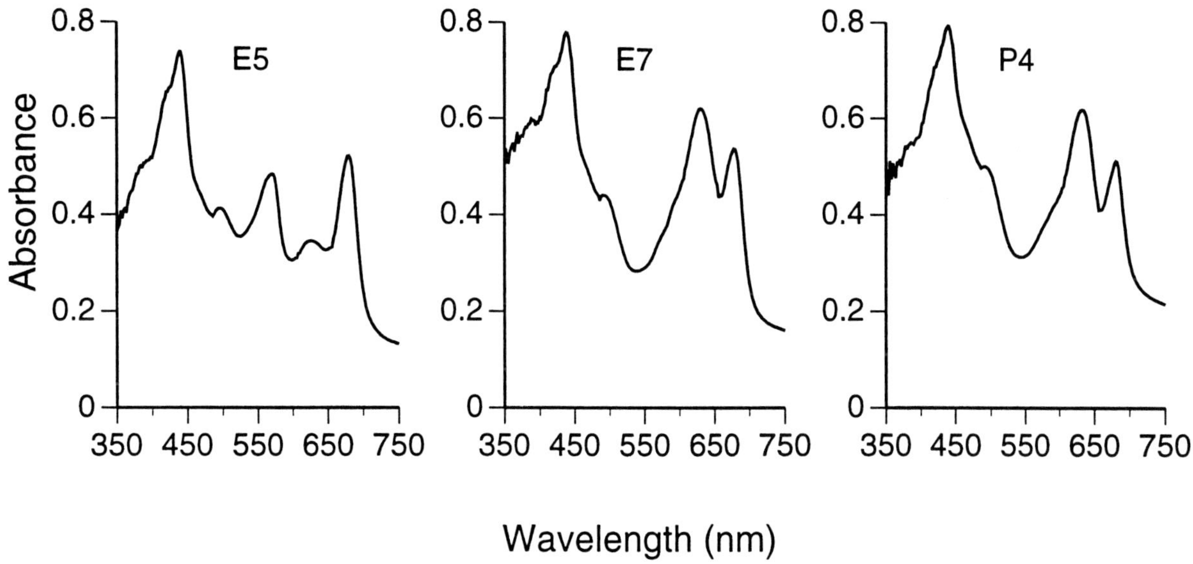

Fig. 5. In vivo absorbance characteristics of three isolates of Arctic cyanobacteria. E5 and E7 are both oscillatorians identified as *Phormidium autumnale*, but E5 contains both phycoerythrin and phycocyanin while E7 contains only phycocyanin. P4 is a picoplanktonic isolate containing phycocyanin but not phycoerythrin (Vézina and Vincent, 1997).

their peak abundance in late summer during the period of highest temperatures (Robarts and Zohary, 1987). Freshwater picocyanobacteria increase in numbers with increasing water temperature in rivers, lakes and the ocean (Bertrand and Vincent, 1994; and refs therein). Although these correlative relationships imply that low temperatures in the polar regions may severely inhibit cyanobacterial growth and abundance, they do not constitute proof of cause and effect, particularly in view of the range of other environmental variables which often co-vary with temperature such as nutrient supply and water column stability.

Field and laboratory measurements on one group of polar cyanobacteria, the mat-formers, indicate that although the low ambient temperatures may not completely inhibit metabolism and growth, they are strongly limiting. In controlled temperature studies on a microbial mat dominated by oscillatorians in a McMurdo Dry Valley stream, photosynthesis increased with increasing temperature, to maximum rates at 25°C. However, respiration rates also increased and in the dark the mats were rapidly destroyed by bacterial decomposition under warmer temperatures (Vincent and Howard-Williams, 1986). Similar assays for photosynthesis in *Phormidium*-dominated mats in the maritime zone (Signy Island) gave an optimum temperature for net photosynthesis of 15°C (Davey, 1989).

In cultures of mat-forming cyanobacteria isolated from the polar regions the temperature optima for growth were consistently above the near-freezing temperatures that typically occur in the natural environment. Several clones of cyanobacteria isolated from southern Victoria Land, Antarctica were unable to grow at temperatures less than or equal to 5°C. This implies that these organisms originated from warmer temperate latitudes (Seaburg et al., 1981). A strain of *Oscillatoria priestleyii* isolated from the McMurdo Ice Shelf had a growth optimum in the range 21-24°C (Castenholz and Schneider,

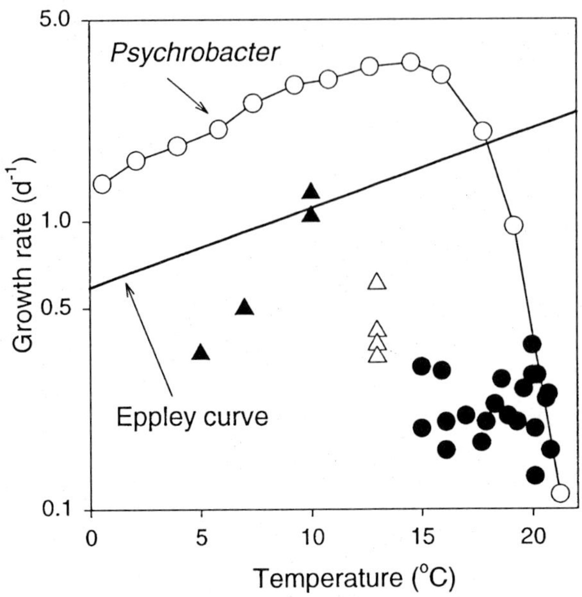

Fig. 6. Growth versus temperature relationships. The black symbols are maximum growth rates (μ_{max}) at optimum temperatures for polar mat-forming cyanobacteria with T_{opt} at or below 20°C (from Tang et. al., 1997). The line is the upper bound for marine phytoplankton given by Eppley (1972). The open symbols are for a marine bacterium isolated from congelation ice in East Antarctica, from Bowman et al. (1997). The open triangles are for picocyanobacterial isolates from Arctic lakes grown at 13°C and optimal light (Vézina & Vincent 1997). The closed triangles are μ_{max} and T_{opt} values for marine diatoms from the Arctic and Antarctic (Jacques, 1983).

1993), well above the summer pond temperatures which were typically in the range 0-8°C.

In the most extensive survey to date, Tang et al. (1997) found that 27 isolates of oscillatorians from lakes, streams and ponds in the Arctic, subarctic and Antarctic were consistently psychrotrophic. Temperature optima for growth were in the range 15-35°C, with some species showing insignificant growth at 5°C. Most isolates, however, grew over a wide temperature range (5-30°C) and, although they were not genetically adapted to low temperatures, they were likely to be tolerant of the variable temperature regime which characterizes their environment. These characteristics contrast with Arctic and Antarctic marine diatoms and sea ice bacteria whose temperature optima for photosynthesis and growth are much lower than those of polar cyanobacteria (Fig. 6). These psychrophilic attributes of the marine communities may reflect the persistent low and stable temperature of these environments, in contrast to the variable thermal regime of non-marine polar habitats (Tang et al., 1997).

B. Desiccation, Freezing and Salinity Tolerance

A primary factor that contributes to the success of polar cyanobacteria is their ability to withstand freezing and the related stresses of high salinity and low water activity. The streams and shallow ponds may dry completely in late summer, or freeze solid and then ablate to leave dry, frozen communities. These later experience elevated salinities as the remaining solutes are re-dissolved and mobilized at the onset of the thaw of the next season. Such freeze-concentration effects may be especially severe for the microbial mats living at the bottom of high latitude ponds that freeze completely in winter. Studies on two Antarctic coastal ponds during the time of freezing revealed the magnitude of salinity variations experienced by their biota (Schmidt et al., 1991). These waters had a relatively low conductivity in summer, but the dissolved salts were largely excluded from the ice during winter freezing, and there was a gradual concentration of solutes in the remaining water. In mid-winter the benthic cyanobacterial mats were covered by a thin layer of concentrated brine with salinities more than six times that of sea water and liquid water temperatures of -12°C (See Chapter 10).

The cyanobacteria of streams in southern Victoria Land are capable of maintaining large populations of viable cells on the dry frozen stream bed throughout winter. These populations provide a large inoculum that may contribute a substantial percentage of the total standing stock during the next growing season. For example, in Fryxell Stream the percent cover of cyanobacteria was 61 (± 19)% prior to summer streamflow and rose slightly to 84 (± 5)% by the end of summer flow. These overwintering assemblages began photosynthesis, respiration and nutrient uptake within 30 minutes to a few hours of rehydration (Vincent and Howard-Williams, 1986).

Studies at two sites in Antarctica confirmed the highly resilient nature of polar cyanobacteria subjected to water stress (Hawes et al., 1992). Mats dominated by *Nostoc commune* placed in the dry atmosphere on the McMurdo Ice Shelf dehydrated rapidly and were completely desiccated within 5 hours (see Chapter 17). Photosynthesis and respiration was measured in the mats within 10 minutes of rewetting. *Phormidium*-dominated mats appeared to be less tolerant of dehydration and took much longer to recover their physiological activity after rewetting.

C. Defenses Against UVR

Many of the polar cyanobacteria inhabit surface or shallow-water environments in which the exposure to continuous UVR in summer may play a role in limiting microbial colonization and growth. Such effects may now be exacerbated by the stratospheric ozone depletion that is occurring over Antarctica as well as increasingly in the high latitudes of the northern hemisphere (Kerr, 1994). Like other phototrophic organisms, cyanobacteria have four lines of defense against UVR, and these strategies are well represented in the polar communities studied to date (Vincent and Quesada, 1994, 1997; Quesada and Vincent 1997):

a) Cyanobacteria can avoid UVR by their choice of habitat such as beneath rock surfaces or deep within microbial mats. The consequence of adopting such a strategy is that only low PAR is available for growth. Examples of polar cyanobacteria in perennial shade environments include sublithic and endolithic communities, phytoplankton and benthos in permanently ice-covered lakes, and the bottom communities of optically-thick microbial mats. Some cyanobacterial species are motile and are capable of migration up and down the mat profile to avoid UVR exposure while ensuring adequate PAR; for example, *Oscillatoria priestleyii* from the McMurdo Ice Shelf (Quesada and Vincent 1997). In some environments, however, this strategy may be more closely linked to water supply and desiccation tolerance than UVR avoidance (Davey and Clarke, 1991).

b) A second type of avoidance strategy is the production of screening compounds that filter out UVR (See Chapter 21). The black or dark gold pigmentation associated with many cyanobacterial communities in the polar regions is due to the pigment scytonemin which absorbs maximally at 390 nm but with a broad absorbance spectrum that extends into the UVB and low energy PAR. Scytonemin is a dimeric molecule (molecular weight of 544) that is probably formed from a condensation of tryptophan and phenylpropanoid derivatives (Proteau et al., 1993). Dark pigmented communities of Arctic cyanobacteria include *Nostoc* colonies and sheets in the shallow waters of lakes and ponds (Fig. 7); stream communities of *Rivularia*, *Stigonema* and *Scytonema* (Sheath et al., 1996); black films and crusts over desert and semi-desert soils (Aleksandrova, 1988) and *Gloeocapsa* crusts on exposed rock faces (Konhauser et al., 1994). Similar communities occur in Antarctica and includc *Nostoc* at the edge of streambeds or in slowly flowing flush environments; *Calothrix* and *Gloeocapsa* at the edge of streams; and black crusts over soils. A dark red pigment occurs in the sheaths of *Gloeocapsa* that forms brown crusts over the rocks in Arctic (Sheath et al., 1996) as well as Antarctic streams; this pigmentation was ascribed to gloeocapsin, but its UVR-absorbing and other molecular properties have yet to be studied. Water soluble pigments that absorb UV-A/B are also well known in cyanobacteria, particularly mycosporine-like amino acids such as asterina-330, shinorine, porphyra-334 and palythene which absorb maximally in the range 320-335 nm (Vincent and Roy, 1993). Aqueous extracts of *Nostoc commune* from Alaska (Fig. 7) and oscillatorian isolates from the McMurdo Ice Shelf (Quesada and Vincent, 1997) have spectra that suggest the presence of this group of sunscreens. These compounds may be responsible for the strong UVR attenuation in the upper few hundred μm of Antarctic *Phormidium* mats (Quesada and Vincent, 1993).

c) UVR damage of biological systems can be the result of direct photochemical degradation of cellular

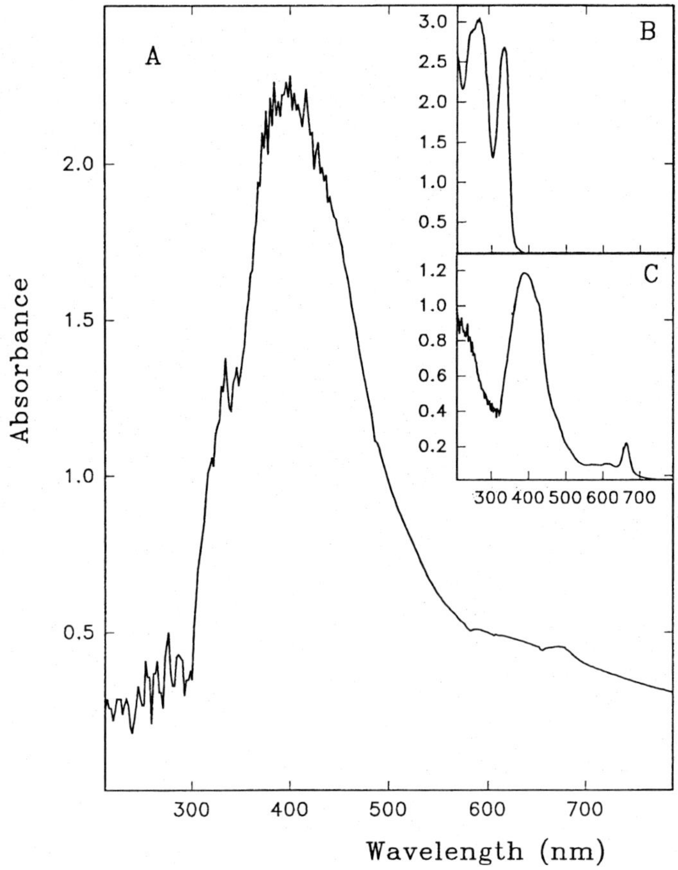

Fig. 7. UV-screening capability of *Nostoc commune* from Toolik Lake, Alaska. Curve A is the *in vivo* absorbance spectrum; curve B is a scan of a 90% methanol-water extract showing the presence of compounds with the absorption characteristics of mycosporine-like amino-acids; curve C is a scan of a 90% acetone-water extract showing the characteristic spectrum of scytonemin (Vincent and Quesada, 1997; Plate 24e).

components, or indirect effects mediated by reactive oxygen species (ROS) (Vincent and Neale, 1999; and refs therein). These latter secondary effects can be substantially offset or eliminated by cellular quenching agents that react with and neutralize ROSs. Carotenoids are well known in this regard, and surface populations of cyanobacteria in the polar regions are often highly pigmented with canthaxanthin, myxoxanthophyll and related compounds. The carotenoid:Chl*a* ratios of polar cyanobacteria in culture are maximal under low temperatures, bright PAR and moderate UVR (Vincent and Quesada, 1997; Tang et al., 1997; Roos and Vincent, unpublished data).

d) Further protection against the long-term effects of UVR exposure is conferred by an ability to identify and repair the photochemical damage to DNA or to the photosynthetic apparatus (See Chapters 15 and 21). These repair mechanisms are stimulated by long wavelength UVR or blue light. Studies on two isolates of oscillatorian cyanobacteria from the McMurdo Ice Shelf showed that there were large differences in UVB sensitivity between species, but for the growth of both strains UVA exposure substantially offset the inhibitory effects of UVB (Quesada et al., 1995).

D. Light Harvesting and Photosynthesis

Arctic and Antarctic cyanobacteria experience extreme variations in PAR supply from continuous light in summer to continuous winter darkness. Many of the communities inhabit subsurface environments that remain highly shaded, even in mid-summer; for example, well beneath the surface of rocks, deep within microbial mats, at the bottom of perennially ice-covered lakes and beneath the surface crust of soils. Other communities occur beneath ice or snow early in the growing season and are then exposed to full sunlight during summer melt. Polar cyanobacteria must therefore be capable of acclimation to a broad range of PAR regimes.

Cultures of Antarctic oscillatorians showed a high level of adaptive flexibility in pigmentation in response to changes in ambient PAR (Quesada and Vincent, 1993). Incubation under dim light (20 μmol photons m^{-2} s^{-1}) resulted in a large increase in cellular Chl*a* and phycobilin content and a corresponding decrease in the carotenoid to Chl*a* ratio. A comparison of incubations under full waveband PAR with those under color filters (which simulated the spectral conditions at the base of the mats) showed that these changing pigment ratios were primarily a response to changes in incident light quantity rather than quality down the mat profile.

The small cell diameter of some polar species may confer a light-capturing advantage under dim PAR conditions by minimizing self-shading effects within the cell. Optically thin cells are a feature not only of the picocyanobacteria but also of members of the Oscillatoriaceae with narrow (1-2 μm) trichomes. Recent studies with isolates of this latter group from Arctic microbial mats indicate that their cellular absorption coefficients (μm^2 $cell^{-1}$) span two orders of magnitude and are a function of cell size and the PAR conditions that they acclimate to during growth. The narrow trichome species from mat communities have similar optical characteristics to the pico-cyanobacteria from Arctic lakes (Vézina and Vincent, unpublished).

One of the intriguing features of the thick microbial mats found in the polar regions is the apparently precarious balance between photosynthesis and respiration. Although the maximum photosynthetic rates can be high per unit area (up to 4 μg C cm^{-2} h^{-1}), the Chl*a* specific carbon uptake is low and reflects the low ambient temperature and the high degree of self-shading within the optically thick mats (Hawes, 1993). Assays of CO_2-exchange performed on cyanobacteria from Antarctic streams indicated that net photosynthesis was generally well below gross photosynthesis, and sometimes near-zero or even negative (Vincent and Howard-Williams, 1986; Hawes and Howard-Williams, 1997). These observations imply that the large accumulations of biomass are the result of many seasons of slow net growth, with relatively little biomass loss through processes such as grazing or sloughing.

E. Nitrogen Fixation

Nitrogen-fixing cyanobacteria, particularly *Nostoc commune*, are contributors of soil nitrogen and carbon in both polar regions, and may play an important role in the primary successional stage of colonization after the retreat of an ice sheet or glacier (Chapin et al., 1992) as well as in mature communities (Henry and Svoboda, 1986). In the Vestfold Hills, Antarctica, and probably other terrestrial sites, nitrogenase activity is strongly regulated by temperature and moisture availability, with maximum activity in December and January when the N_2-fixing mats are well supplied with meltwater and when ground temperatures rise to 8-10°C (Davey and Marchant, 1983). Fixation of dinitrogen by cyanobacteria can be an important contribution to the nitrogen budget of polar lakes, ponds and streams (Alexander et al. 1980, 1989; Hawes et al. 1993).

Nitrogen-fixing cyanobacteria are often found in close association with moss communities. In the maritime Antarctic (Signy Island) nitrogen fixation rates varied from a mean of 46 (dry turf) to 192 (wet carpet) mg N m^{-2} y^{-1}. These rates were comparable to the amount of inorganic nitrogen entering these systems by precipitation; about 65 mg N m^{-2} y^{-1} (Christie, 1987). On Svalbard, at latitude 79°N, the ability to support cyanobacteria, in particular *Nostoc*, varied greatly between bryophyte communities and there were concomitant variations in nitrogen-fixation rates as measured by the acetylene reduction assay technique (Solheim et al., 1996). Nitrogenase activity was stimulated greatly (probably by phosphorus) in regions grazed by geese and the rates in these areas were as much as one order-of-magnitude higher than those measured in cyanobacterial mats in the warmer low latitude mires of the sub- Arctic (Karagatzides et al., 1985) and sub- Antarctic (Smith and Russell, 1982).

V. Why Do Cyanobacteria Dominate (Or Not)?

The large and rapidly expanding literature on high latitude cyanobacteria continues to reinforce a striking dichotomy in their pattern of distribution and abundance. In cold, non-marine habitats such as icecaps, polar desert soils, glacial streams and ice-capped lakes, cyanobacteria are often the dominant phototrophs. In environments such as shallow ponds and rock faces they may contribute most of the total ecosystem standing stock, with spectacular biomass accumulations at some sites. Yet in the marine environment of both polar regions cyanobacteria are rare or conspicuously absent. Any explanation of the pre-eminent success of cyanobacteria in the terrestrial biomes of the Arctic and Antarctica must also consider their weak representation in cold oceanic waters. This concluding section briefly considers five aspects of the eco-physiology of polar cyanobacteria that might account for this fascinating pattern of presence and absence.

A. Temperature Effects

The strong latitude-dependent decrease in marine picocyanobacteria (Fig. 2), and the correlative relationships between cyanobacterial abundance and temperature in the polar regions as well as elsewhere, point to temperature as a likely overall control. The non-marine isolates examined to date have broad tolerances to temperature but lack the adaptive goodness-of-fit to cold that is characteristic of many diatoms and bacteria in the polar ocean. Polar cyanobacteria have extremely slow growth rates at near-zero temperatures relative to psychrophilic species. Even the maximum growth rates of polar cyanobacteria are slow relative to phytoplankton in general with μ_{max} values that fall well below the Eppley curve (Fig. 6). In apparent contradiction to these observations is the dramatic success of cyanobacteria, including picocyanobacteria, in cold non-marine habitats such as cryoconite ponds, polar desert lakes and high latitude streams. The water temperature in some of these non-marine ecosystems fluctuates considerably over the 24 h cycle because of solar heating. The eurythermal capacity of cyanobacteria may provide a competitive advantage in these systems, but not in the polar oceans or in ice-capped lakes where cold temperatures persist throughout the year.

B. Salinity Responses

The greater presence of cyanobacteria in polar freshwaters versus marine habitats might simply reflect their limits of tolerance to salinity. Wright and Burton (1981) noted that the combined extreme of salinity plus low temperatures exerts a severe physiological stress on organisms and could account for the absence of cyanobacteria from some of the hypersaline lake environments in Antarctica. However, this seems an unlikely explanation for the absence of picocyanobacteria in the polar oceans. Picocyanobacteria achieve spectacularly high densities in Ace lake, Antarctica, at salinities similar to seawater and at relatively low temperatures (6°C). On the McMurdo Ice Shelf, mat-forming cyanobacteria occur in ponds that remain cold (< 5°C) throughout the year at salinities at or above seawater (Howard-Williams et al., 1989), and in ponds on Ross Island, microbial mats survive the seasonal extreme of sub-zero liquid water temperatures and salinities up to many times that of seawater (Schmidt et al., 1991).

C. Nutrient Requirements

Bloom-forming and picoplanktonic species of cyanobacteria are extreme in terms of their nutrient requirements. The former group occurs mostly in nutrient-rich environments, while the latter has a high nutrient-scavenging ability and is more typical of oligotrophic conditions. Consistent with this general pattern, picocyanobacteria occur throughout the oligotrophic lakes of both polar regions however, even in enriched lakes of the Arctic and Antarctica, bloom-forming species are rarely present. These latter forms are typically gas vacuolate and can adjust their position in the water column; this strategy would not be favored in the cold unstable waters of polar lakes. A consideration of nutrient-size relationships suggests that picocyanobacteria would have less of a competitive advantage under nutrient-replete conditions than under conditions of nutrient deficiency. Nevertheless, picoplanktonic taxa (but eukaryotes) are often a dominant component of the phytoplankton in the nutrient-rich Arctic Ocean (Gradinger and Lenz, 1995) and Southern Ocean (Vincent, 1988), indicating that picocyanobacteria are not precluded simply on the basis of their small size. In these environments, the success of picoplankton-sized phototrophs is more likely to be the result of selection for organisms which have a superior light-capturing ability and that are less prone to grazing losses than larger cells.

The availability of iron was identified as a potentially important control on phytoplankton production in certain parts of the world ocean, including the seas around Antarctica. Cyanobacteria as a group are known to have a high requirement for iron and therefore may be especially prone to limitation effects. However, this explanation seems unlikely for the absence of picocyanobacteria in the Arctic Ocean where there is a strong freshwater influence from the surrounding land masses, and where terrigenous inputs of iron are probably substantial. Furthermore, picocyanobacteria have a strong scavenging ability for Fe relative to other algal groups.

Polar lakes, ponds and streams which contain mat-forming species of cyanobacteria span a broad range of nutrient conditions, although concentrations within the mats may be orders of magnitude higher than in the overlying water (e.g. up to 1000 μg dissolved reactive P L^{-1} in mats on the McMurdo Ice Shelf; Vincent et al., 1993c). The N_2-fixing ability of *Nostoc* is likely to be a factor that contributes to the widespread distribution of this form in nitrogen-deficient habitats such as glacial moraines and certain meltwater streams.

D. Growth Rates

One of the most striking features of polar cyanobacteria is their slow growth rate in culture as well as in the natural environment. In part this is an effect of low temperature, perhaps compounded by osmotic stress in some habitats. In the polar oceans these temperature-depressed growth rates may be further slowed by light-limitation in the deep mixed layer. However even at the optimum temperature for growth, μ_{max} values for polar cyanobacteria lie well below the Eppley curve (Fig. 6) and in some communities such as the lake and stream microbial mats the net balance between photosynthesis and respiration is close to zero through much of the growing season. This implies that the selection pressure in non-marine polar environments is not for an ability to out-compete other organisms via fast growth. On the other hand, cyanobacteria are able to achieve net growth over a broad range of pH, nutrient concentrations, PAR and UVR fluxes, and temperatures. These broad physiological tolerances are likely to be important in the highly variable non-marine habitats such as soils and ephemeral streams, but may be less useful in the thermally and chemically more stable oceanic environments.

E. Loss Rates

Slow growth can only be one component of an overall strategy for microbial success; it must also be accompanied by an ability to minimize biotic and abiotic processes which remove biomass. In the polar non-marine environment there are many examples of resistance to abiotic loss through processes such as tolerance to freeze-up and desiccation, and the persistence of an overwintering inoculum on dry stream beds or in the bottom waters of lakes. Grazing losses are minimal in these habitats in Antarctica where the only herbivores are microscopic species such as tardigrades, nematodes and rotifers, and these rarely achieve a large population size. Insect grazers are important in Arctic streams and ponds, and may be responsible for the reduced standing stocks in these environments relative to Antarctic communities. In Ace Lake, Antarctica, populations of pico-cyanobacteria begin to decline well before the ambient light conditions begin to deteriorate, perhaps in response to increased populations of protozoan grazers which are known to occur seasonally in these lakes (Laybourne-Parry and Marchant, 1992). This resistance to loss is likely to be much less effective in the polar marine environment where there is a more diverse assemblage of herbivores capable of grazing picocyanobacteria (e.g., choanoflagellates and tunicates), and where continuous advection and mixing may further prevent the accumulation of cells.

In summary, cyanobacteria are able to dominate the microflora at many sites in the polar regions through slow growth, tolerance of severe conditions (e.g., winter freeze-up) and persistence. This strategy is successful in extreme environments in which multi-trophic level communities are poorly developed, and where competition, herbivory and other biological interactions are only weakly expressed. This approach fails, however, in cold oceanic environments where only faster growing, more thermally adapted species are able to keep pace with the continuous removal of biomass via biotic and abiotic loss processes.

Acknowledgments

This paper was written while on sabbatical at Antarctic CRC, University of Tasmania, Hobart, Australia; I thank the director and staff of this institute for their support and hospitality. My polar research group is funded by the Natural Sciences and Engineering Research Council of Canada with logistic support from Centre d'études nordiques and Polar Continental Shelf Project (this is publication PCSP/ÉPCP 01597). I thank John Bowman, Evonne

Tang and Lynne Powell for their review comments on the manuscript.

References

Aleksandrova VD (1988) Vegetation of the Soviet polar deserts. Cambridge University Press, Cambridge

Alexander V, Stanley DM, Daley RJ and McRoy CP (1980) Primary producers. In: Hobbie JE (ed) Limnology of Tundra Ponds, pp179-250. Dowden, Hutchinson and Ross Inc, Stroudsburg, PA

Alexander V, Whalen SC and Klingensmith KM (1989) Nitrogen cycling in Arctic lakes and ponds. Hydrobiologia 172: 165-172

Anagnostidis K and Komárek J (1988) Modern approach to the classification system of cyanophytes, 3.-Oscillatoriales. Archiv für Hydrobiol Supplement band 71 (Algological Studies, 38-39): 291- 302

Andreoli C, Scarabel L, Spini S and Grassi C (1992) Picoplankton in Antarctic lakes of northern Victoria Land during summer 1989-1990. Polar Biol 11: 57-82

Bergeron M and Vincent WF (1997) Microbial food web responses to phosphorus and solar UV radiation in a subarctic lake. Aquatic Microbial Ecol 12: 239-249

Bertrand N and Vincent WF (1994) Structure and dynamics of photosynthetic picoplankton across the saltwater transition zone of the St. Lawrence River. Can J Fish Aquat Sci 51: 161-171

Bliss LC and Gold WG (1994) The patterning of plant communities and edaphic factors along the a high Arctic coastline- implications for succession. Can J Bot 72: 1095-1107

Bowman JP, Nichols DS and McMeekin TA (1997) *Psychrobacter glacincola* sp. nov., a halotolerant, psychrophilic bacterium isolated from Antarctic sea ice. System Appl Microbiol 20: in press

Broady PA (1981) The ecology of chasmolithic algae at coastal locations of Antarctica. Phycologia 20: 259-272

Broady PA (1986) Ecology and taxonomy of the terrestrial algae of the Vestfold Hills. In: Pickard J (ed) Antarctic Oasis, pp 165-202. Academic Press, Sydney

Broady PA (1996) Diversity, distribution and dispersal of Antarctic terrestrial algae. Biodiversity and Conservation 5: 1307-1335

Broady PA and Kibblewhite AL (1991) Morphological characterization of Oscillatoriales (Cyanobacteria) from Ross Island and southern Victoria Land, Antarctica. Ant Sci 3: 35-45

Broady PA and Smith RA (1994) A preliminary investigation of the diversity, survivability and dispersal of algae into Antarctica by human activity. Proceedings of the National Institute of Polar Research (NIPR) Symposium on Polar Biology 7: 185-197

Castenholz RW (1992) Species usage, concept, and evolution in the Cyanobacteria (blue-green algae) J Phycol 28: 737-745

Castenholz RW and Schneider AJ (1993) Cyanobacterial dominance at high and low temperatures: optimal conditions or precarious existence? In: Guerrero R and Perdos-Alio (eds) Trends in Microbial Ecology, pp 19-24. Spanish Society for Microbiology, Barcelona

Chapin FS, Jeffries RL, Reynolds JF, Shaver GR and Svoboda J (1992) Arctic Ecosystems in a Changing Climate: An Ecophysiological Perspective. Academic Press, San Diego, California

Christie P (1987) Nitrogen in two contrasting Antarctic bryophyte communities. J Ecol 75: 173-193

Croasdale H (1973) Freshwater algae of Ellesmere Island, N.W.T. National Museums of Canada Publications in Botany 3: 1-131

Davey MC (1989) The effects of freezing and desiccation on photosynthesis and survival of terrestrial Antarctic algae and cyanobacteria. Polar Biol 10: 29-36

Davey MC (1991) The seasonal periodicity of algae on Antarctic fellfield soils. Holarctic Ecol 14: 112-20

Davey MC and Clarke KJ (1991) The spatial distribution of microalgae on Antarctic fell-field soils. Antarctic Science 3: 257-263

Davey A and Marchant HJ (1983) Seasonal variation in nitrogen fixation by *Nostoc commune* Vaucher at the Vestfold Hills, Antarctica. Phycologia 22: 337-385

de Mora SJ, Whitehead RF and Gregory M (1994) The chemical composition of glacial melt water ponds on the McMurdo Ice Shelf, Antarctica. Ant Sci 6: 17-27

Ellis-Evans JC (1991) Numbers and activity of bacterio- and phytoplankton in contrasting maritime Antarctic lakes. Verh Internat verein Theor Angew Limnol 24: 1149-54

Ellis-Evans JC (1996) Microbial diversity and function in Antarctic freshwater ecosystems. Biodiversity and Conservation 5: 1395-1431

Ellis-Evans JC and Bayliss PR (1993) Biologically active microgradients in cyanobacterial mats of Antarctic lakes and streams. Verh Internat Verein Limnol 25: 948-952

Eppley RW (1972) Temperature and phytoplankton growth in the sea. Fish Bull 70: 1063-1085

Fritsch FE (1917) Freshwater algae. British Antarctic ("Terra Nova") Expedition, 1910, Natural History Report. British Museum (Natural History) p 1-16

Geitler L (1932) Cyanophyceae. In: Kolkwitz R (ed) Dr. L Rabenhorst's Kryptongamenflora von Deutschland, Osterreich und der Schweiz. Akademische Verlagsgesellschaft, Leipzig

Gendel RW and Drouet F (1960) The cyanobacteria of the Thule Area, Greenland. Trans Am Microscop Soc 79: 256-272

Goldman CR, Mason DT and Hobbie JE (1967) Two Antarctic desert lakes. Limnol Oceanogr 12: 295-310

Gradinger R and Lenz J (1995) Seasonal occurrence of picocyanobacteria in the Greenland Sea and central Arctic Ocean. Polar Biol. 15: 447-452

Hamilton PB and Edlund SA (1994) Occurrence of *Prasiola fluviatilis* (Chlorophyta) on Ellesmere Island in the Canadian Arctic. J Phycol 30: 217-221

Hawes I (1990) Eutrophication and vegetation development in maritime Antarctic lakes. In Kerry KR and Hempel G (eds) Antarctic Ecosystems. Ecological Change and Conservation, pp 83-90. Springer-Verlag, Berlin

Hawes I (1993) Photosynthesis in thick cyanobacterial films: a comparison of annual and perennial Antarctic mat communities. Hydrobiologia 252: 203-209

Hawes I and Howard-Williams C (1997) Primary production processes in streams of the McMurdo Dry Valleys, Antarctica. In: Priscu J (ed) Dry Valley a Cold Desert Ecosystem. Antarctic Research Ser, in press

Hawes I, Howard-Williams C and Pridmore RD (1993) Environmental control of microbial biomass in the ponds of the McMurdo Ice Shelf, Antarctica. Arch Hydrobiol 127: 27-287

Hawes I, Howard-Williams C, Schwarz A-MJ and Downes MT (1997) Environment and microbial communities in a tidal lagoon at Bratina Island, McMurdo Ice Shelf, Antarctica. In: Battaglia B, Valencia J and Walton D (eds), Antarctic Communities: Species, Structure and Survival, in press. Cambridge University Press, UK

Hawes I, Howard-Williams C and Vincent WF (1992) Desiccation and recovery of cyano-bacterial mats. Polar Biol 12: 587-594

Henry GHR and Svoboda J (1986) Dinitrogen fixation (acetylene reduction) in high arctic sedge meadow communities. Arctic Alpine Res 18: 181-187

Howard-Williams C, Pridmore RD, Broady PA, and Vincent WF (1990) Environmental and biological variability in the McMurdo Ice Shelf ecosystem. In: Kerry KR and Hempel G (eds) Antarctic Ecosystems. Ecological Change and Conservation, pp 23-31. Springer-Verlag, New York

Howard-Williams C, Pridmore R, Downes MT and Vincent WF (1989) Microbial biomass, photosynthesis and chlorophyll *a* related pigments in the ponds of the McMurdo Ice Shelf, Antarctica. Ant Sci 1: 125-131

Ingole BS and Parulekar AH (1990) Limnology of Priyadarshani Lake, Schirmacher Oasis, Antarctica. Polar Rec 26: 13-17

Jacques G (1983) Some ecophysiological aspects of the Antarctic phytoplankton. Polar Biol 2: 27-33

Kalff J, Kling HJ, Holmgren SH and Welch HE (1975) Phytoplankton, phytoplankton growth and biomass cycles in an unpolluted and in a polluted polar lake. Verh Internat Verein Limnol 19: 487-495

Kalff J and Welch HE (1974) Phytoplankton production in Char Lake, a natural polar lake, and in Meretta Lake, a polluted polar lake, Cornwallis Island, Northwest Territories. J Fish Res Board Can 31: 621-636

Kerr JB (1994) Decreasing ozone causes health concern. Environ Sci Technol 28: 514-518

Konhauser KO, Fyfe WS, Schultze-Lam S, Ferris FG and Beveridge TJ (1994) Iron phosphate precipitation by epilithic microbial films in Arctic Canada. Can J Earth Sci 31: 132-1324

Karagatzides JD, Lewis MC, Schulman HM (1985) Nitrogen fixation in the high arctic tundra at Sarcpa Lake, Northwest territories. Can J Bot 63: 974-79

Laybourne-Parry J and Marchant HJ (1992) The microbial plankton of freshwater lakes in the Vestfold Hills, Antarctica. Polar Biol 12: 405-410

Legendre L, Gosselin, M, Hirsch HJ, Katner G and Rosenberg G (1993) Environmental control and potential fate of size-fractionated phytoplankton production in the Greenland Sea (75 °N). Mar Ecol Prog Ser 98: 297-313

Leslie A (1879) The Arctic Voyages of Adolf Erik Nordenskiöld. MacMillan and Co, London

Marchant HJ, Davidson AT and Wright SW (1987) The distribution and abundance of chroococcoid cyanobacteria in the Southern Ocean. Proceedings of the National Institute of Polar Research (NIPR) Symposium on Polar Biology 1: 1-9

Maurette M, Hammer C, Brownlee DE, Reeh N and Thomsen HH (1986) Placers of cosmic dust in the blue ice lakes of Greenland. Science 233: 869-872

Mishistina IE, Moskvina MI, Rodikova LP and Severina II (1994) Cyanobacteria of the genus *Synechococcus* in Arctic Seas (in Russian). Dokl -Ran 336: 562-565

Murray J (1910) On collecting at Cape Royds. In: Murray J (ed) British Antarctic Expedition 1907-1909 Reports on Scientific Expeditions, Vol. 1 Biology, pp 1-15 Heinemann, London

Murphy LS and Haugen EM (1985) The distribution and abundance of phototrophic ultraplankton in the North Atlantic. Limnol Oceanogr 30: 47-58

Nienow JA and Friedmann EI (1993) Terrestrial lithophytic (rock) communities. In: Friedmann EI (ed) Antarctic Microbiology, pp 343- 412. Wiley-Liss Inc, New York

Ohtani S and Kanda H (1987) Epiphytic algae on the moss community of *Grimmia lawiana* around Syowa Station, Antarctica. Proceedings of the National Institute of Polar Research Symposium on Polar Biology 1: 255-264

Parker BC, Simmons GM, Wharton RA, Seaburg KG and Love FG (1982) Removal of organic and inorganic matter from Antarctic lakes by aerial escape of bluegreen algal mats. J Phycol 18: 72-78

Parker BC and Wharton RA (1985). Physiological ecology of blue-green algal mats (modern stromatolites) in Antarctic oasis lakes. Arch Hydrobiol (Suppl) 71: 331-348

Persson P-E (1996) Cyanobacteria and off-flavours. Phycologia 35: 168-71

Priscu, JC., Fritsen, CH, Adams, EE, Giovannoni, SJ, Paerl, HW, McKay, CP, Doran, PT, Gordon, DA, Lanoil, BD and Pinckney JL (1998) Perennial Antarctic lake ice: an oasis for life in a polar desert. Science 280: 2095-8.

Proteau PJ, Gerwick WH, Garcia-Pichel F and Castenholz R (1993) The structure of scytonemin, an ultraviolet sunscreen pigment from the sheaths of cyanobacteria. Experientia 49: 825-829

Quesada A, Mouget J-L and Vincent WF (1995) Growth of Antarctic cyanobacteria under ultraviolet radiation: UVA counteracts UVB inhibition. J Phycol 31: 242-248

Quesada A and Vincent WF (1993) Adaptation of cyanobacteria to the light regime within Antarctic mats. Verh Internat Verein Limnol 25: 960-965

Quesada A and Vincent WF (1997) Strategies of adaptation by Antarctic cyanobacteria to ultraviolet radiation. Eur J Phycol 32: 335-42

Rankin LM, Franzmann PD, McMeekin TA and Burton HR (1997) Seasonal distribution of picocyanobacteria in Ace lake, a marine-derived Antarctic lake. In: Battaglia B, Valencia J and Walton D (eds), Antarctic Communities: Species, Structure and Survival, pp 178-184. Cambridge University Press, United Kingdom

Robarts RD and Zohary T (1987) Temperature effects on photosynthetic capacity, respiration, and growth rates of bloom-forming cyanobacteria. New Zealand Journal of Marine and Freshwater Research 21: 391-399

Schmidt S, Moskal W, de Mora SJ, Howard-Williams C and Vincent WF (1991) Limnological properties of Antarctic ponds during winter freezing. Ant Sci 3: 379-388

Seaburg KG, Parker BC, Wharton RA and Simmons GM (1981) Temperature growth responses of algal isolates from Antarctic oases. J Phycol 17: 353-60

Sheath RG and Müller KM (1997) Distribution of stream macroalgae in four high Arctic drainage basins. Arctic 50: 355-64.

Sheath RG, Vis ML, Hambrook JA and Cole KM (1996) Tundra stream macro-algae of North America: composition, distribution and physiological adaptations. Hydrobiologia. 336: 67-82

Smith V and Russell (1982) Acetylene reduction by bryophyte-cyanobacteria associations. Polar Biol 1: 153-7

Solheim B, Endal A and Vigstad H (1996) Nitrogen fixation in Arctic vegetation and soils from Svalbard, Norway. Polar Biol 16: 35-40

Tang EPY, Tremblay R and Vincent WF (1997) Cyanobacterial dominance of polar freshwater ecosystems: are high-latitude mat-formers adapted to low temperature? J Phycol 33: 171-181

Taylor G (1916) With Scott: The Silver Lining. Dodd, Mead and Co, New York

Vézina S and Vincent WF (1997) Arctic cyanobacteria and limnological properties of their environment: Bylot Island,

Northwest Territories, Canada (73° N, 80° W) Polar Biol 17: 523-534
Vincent WF (1988) Microbial ecosystems of Antarctica. Cambridge University Press, UK
Vincent WF, Castenholz RW, Downes MT and Howard-Williams C (1993c) Antarctic cyanobacteria: light, nutrients, and photosynthesis in their microbial mat environment. J Phycol 29: 745-755
Vincent WF, Downes MT, Castenholz RW and Howard-Williams C (1993b) Community structure and pigment organisation of cyanobacteria-dominated microbial mats in Antarctica. Eur J Phycol 28: 213-221
Vincent WF and Howard-Williams C (1986) Antarctic stream ecosystems: physiological ecology of a blue-green algal epilithon. Freshwater Biol 16: 219-223
Vincent WF, Howard-Williams C and Broady PA (1993a) Microbial communities and processes in Antarctic flowing waters. In: Friedmann EI (ed) Antarctic Microbiology, pp 543-569, Wiley-Liss Inc, New York
Vincent WF and James MR (1996) Biodiversity in extreme aquatic environments: lakes ponds and streams of the Ross Sea sector, Antarctica. Biodiversity and Conservation 55: 1451-1471
Vincent WF and Neale PJ (1999) Mechanisms of UV damage in aquatic biota. In: de Mora S (ed) UV radiation in the marine environment. Cambridge University Press, in press
Vincent WF and Quesada A (1994) Ultraviolet radiation effects on cyanobacteria: implicat-ions for Antarctic microbial ecosystems. Antarctic Res Ser 62: 111-124
Vincent WF and Quesada A (1997) Microbial niches in the polar environment and the escape from UV radiation in non-marine habitats. In: Battaglia B, Valencia J and Walton D (eds), Antarctic Communities: Species, Structure and Survival, pp 388-395. Cambridge University Press, UK
Vincent WF and Roy S (1993) Solar UV-B effects on aquatic primary production: damage, repair and recovery. Environ Rev 1: 1-12
Vincent WF and Vincent CL (1982) Factors controlling phytoplankton production in Lake Vanda (77°S). Can J. Fish Aquat Sci 39: 1602-9
Walker TD and Marchant HJ (1989) The seasonal occurrence of chroococcoid cyanobacteria at an Antarctic coastal site. Polar Biol 9: 193-196
Waterbury JB, Watson SW, Valois FW and Franks DG (1986) Biological and ecological characterization of the marine unicellular cyanobacterium *Synechococcus*. Can Bull Fish Aquat Sci 214: 71-120
Wharton RA, Parker BC and Simmons GM (1983) Distribution, species composition and morphological algal mats (stromatolites) in Antarctic dry valley lakes. Phycologia 22: 355-365
Wharton RA, McKay CP, Simmons GM and Parker BC (1985) Cryoconite holes on glaciers. Bioscience 35: 499-503
Wharton RA, Vinyard WC, Parker BC, Simmons GM and Seaburg KG (1981) Algae in cryoconite holes on Canada Glacier in southern Victoria Land, Antarctica. Phycologia 20: 208-11
Wright SW and Burton HR (1981) The biology of Antarctic saline lakes. Hydrobiologia 82: 319-338
Wynn-Williams DD (1990) Ecological aspects of Antarctic microbiology. Adv. Microb Ecol 11: 71-146
Wynn-Williams DD (1991) Epifluorescence image analysis of the 3D structure of phototrophic microbial biofilms at Antarctic soil surfaces. Binary Comput Microbiol 4: 53-57

Chapter 13

Cyanobacteria in Deserts - Life at the Limit?

D.D. Wynn-Williams
British Antarctic Survey, Natural Environmental Research Council, High Cross, Madingley Road, Cambridge CB3 0ET, UK

Summary

Cyanobacteria are primitive phototrophic prokaryotes whose long evolutionary history dates back to the Proterozoic era. Their ubiquity on the planet and dominance in hot and cold deserts is a measure of their eco-physiological resilience and adaptability. They have been studied extensively as part of exobiological research into the limits of life in the Solar System. Desert cyanobacterial communities tolerate desiccation that results from acute water deficiency, and they accumulate compatible solutes to counteract osmotic stresses which result from freezing and high salinities. They also accumulate trehalose as a water replacement mechanism to maintain the functional integrity of membranes during anhydrobiosis. Cyanobacteria tolerate high and low extremes of temperature. Their capacity for screening excessive solar radiation (PAR and UVB) by synthesis of "sunscreen" biochemicals whilst retaining a capacity for shade-adaptation, makes them eminently suited for colonisation of

B. A. Whitton and M. Potts (eds), The Ecology of Cyanobacteria, 341-366.
© 2000 *Kluwer Academic Publishers. Printed in the Netherlands.*

diverse lithic habitats. They pioneer the development of microphytic soil crusts which stabilise mobile desert soils. They colonise fissures in rocks as chasmolithic colonists and penetrate the fabric of porous, translucent rocks to provide the primary-producing basis of endolithic communities ranging from the hottest deserts to the cold Dry Valleys of Antarctica. They biodegrade these rocks to create soils which they enrich and inoculate. Their ability to survive at the limits of life on the surface of the Earth is now being studied as an analogue for past life on Mars – the ultimate desert.

I. A Historical Perspective

Cyanobacteria are an ancient group of prokaryotes dating from the earliest evolution of primitive photosynthetic systems in the primordial soup of the Proterozoic era (Schopf and Walter, 1982; Chapter 2). Eleven taxa resembling Oscillatoreacean filamentous cyanobacteria have been described in chert dating from 3.3 - 3.5 Gya in early Archean Apex Basalt in the Warrawoona Group of north-western Australia (Schopf, 1993). The occurrence of these fossil cyanobacteria, which are 1300 million years older than any comparable suite of fossil prokaryotes known at the time, suggests that O_2 – producing cyanobacterial photo-autotrophy was fundamental to life processes on Earth at a very early stage of its evolution. These microbial fossils are not only found as preserved stromatolites in sedimentary desert rocks but also as live analogues in benthic mats of ice-covered lakes in the Antarctic Dry Valleys (Walter and Bauld, 1983; Wharton, 1994; Stal, this volume, chapter 4). Campbell (1979) found that soil accretion, stabilisation and biogenic modification by calcium carbonate-precipitating trichomes of *Microcoleus vaginatus* in microphytic crusts of the Utah desert resembled stromatolitic growth, with a potential for lithification and fossilisation. He suggested that this mechanism may have helped to form mature Precambrian soils. Contemporary cyanobacteria in desert rocks can also be remarkably long-lived. This is exemplified by the carbon dating of cryptoendolithic micro-organisms within Beacon sandstone of the Antarctic cold desert in the order of 1000 years old (Bonani et al., 1988). They are among the few taxa that are capable of tolerating the environmental extremes that limit life in both hot and cold deserts.

The extent of hot and cold deserts is evident from orbital remote imagery. North African deserts are especially extensive, and most of the Antarctic continent can be classified as desert, albeit ice-covered. The McMurdo Dry Valleys region which is the largest ice-free region in Antarctica encompasses an area of 15000km^2, of which 30% is free of snow and ice (Fig. 1). Such regions are commonly dominated by cyanobacteria (Vincent, 1988, this volume) which have the ecophysiological tolerance to cope with the extreme hydrological and thermal stresses.

A major advance in desert cyanobacterial research was made in the 1960s when the NASA-funded Jet Propulsion Laboratories Desert Microflora Program was initiated to develop exobiological expertise (Cameron, 1962, 1966, 1969a; Cameron and Blank, 1966). This approach led to the life detection methodology developed for soils for the Mars Viking landers (Sagan, 1972) and subsequently to research into endolithic cyanobacterial communities in Antarctic rocks (Friedmann and Ocampo, 1976).

Cameron carried out a world-wide survey of desert soil crusts and translucent mineral substrata as potential habitats for microalgae, including cyanobacteria. He defined the critical life-supporting features of the desert habitat, including moisture (availability and purity), solar radiation (quality and intensity), temperature (extremes, mean diurnal and seasonal fluctuations), wind (direction, intensity and frequency) and certain edaphic characteristics such as translucency and stability. He noted that micro-algae (especially filamentous cyanobacteria and coccoid eukaryotic chlorophytes), were the most abundant colonists. He subsequently described cyanobacterial assemblages from a broad geographical diversity of hot and cold deserts. These included polar regions in Antarctica (southern Victoria Land) and the Arctic (Franz Josef Land, New Siberian Islands), high altitude deserts (White Mt. Summit, California and Mt. Aucanquilche, Chile). They also encompassed hot regions ranging from the Middle East, Sahara, USA and South America, and volcanic deserts in California, Alaska and Hawaii (Cameron, 1969b). The predominant strains were reported as the filamentous non-sporeforming Oscillatoriacean genera *Schizothrix, Microcoleus* and *Oscillatoria.* One species, *Schizothrix calcicola,* was frequently found as a single population where no other species was detected by the isolation procedures used.

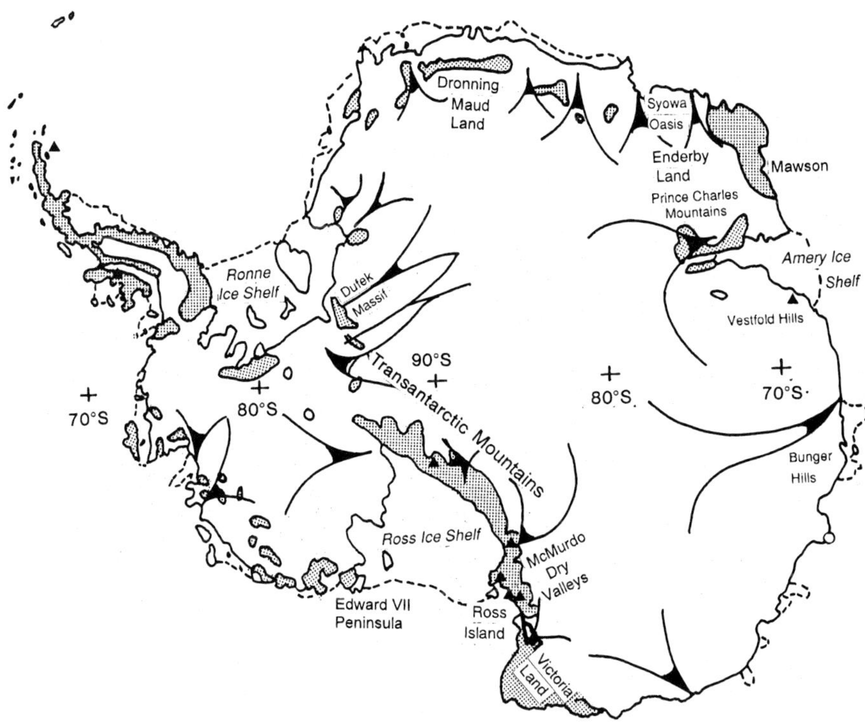

Fig. 1. Antarctica, showing ice-free areas and the catchment and time-averaged flow of desiccating katabatic winds from the polar plateau (after Parish, 1988). Ice-free regions are stippled and areas exceeding 100 km^2 are marked by filled triangles. Scale bar = 1000 km. [after Wynn-Williams, 1991]

However, coccoid genera including *Anacystis* and *Coccochloris* were also found (Cameron, 1969a).

The taxonomy of cyanobacteria described by Cameron (1971c) was summarised relative to other microbial groups (Cameron, 1971a). However, it was of limited value as it was based on the greatly simplified classification proposed by Drouet and Daily (1956) and developed in Drouet (1981) whereby traditional "species" were treated as environmentally induced phenotypic variations of a few genotypes. This approach has been largely superseded by the more informative classification of Anagnostidis and Komarek (1985; 1988) and placed in a modern context by Castenholz (1992). The classification of cyanobacteria by both the International Code of Botanical Nomenclature (as blue-green algae) and the International Code of Bacteriological Nomenclature is incongruous (Whitton, 1992). The taxonomy of Antarctic species has been reviewed by Broady (1996) who has studied specifically the perplexing Oscillatoriales and their extensive synonymy (Broady et al., 1984). These problems may remain unresolved with many ambiguities until the further application of new molecular techniques such as DNA-DNA hybridisation. This has been successfully used by Stulp and others (Nienow and Friedmann, 1993) to demonstrate that endolithic strains of *Chroococcidiopsis* from both hot and cold deserts belong to a single species. Comparison of Tables 1 and 2 for species in hot and cold deserts, respectively, shows many common genera and several common species despite highly disparate temperature regimes. More recently, Turner (1997) has reviewed the molecular systematics of oxygenic photosynthesising bacteria (including cyanobacteria) with special reference to the use of 16S rRNA analysis for *Chroococcidiopsis* and related genera. Billi et al. (1998) have extracted DNA from desert strains of *Chroococcidiopsis* to compare their genomic similarity with special reference to the *jtsZ* gene concerned with cell division.

The eco-physiology of cyanobacteria has been studied in a wide diversity of deserts (Cameron,

1969b; Palmer and Friedmann, 1990). Sites in hot arid regions have included the Sonoran desert in the USA (Cameron, 1962) , the Middle-Eastern Negev (Friedmann, 1971; Lange et al., 1992), west African Namib (Lange et al., 1994b), South African highlands (Wessels and Büdel, 1995; Weber et al. 1996), Scherer and Zhong, 1991) and central Australia (Smith et al. 1990; de Chazal et al., 1992). Studies of cold desert cyanobacteria in Arctic and Antarctic regions have been less extensive than in hot deserts, with early work motivated primarily by exobiology (Cameron, 1969a). In the Antarctic, research was focused mainly on the McMurdo Dry Valleys region.

Soil studies in the Dry Valleys are closely linked to freshwater studies because of their transient nature of water availability in seasonal streams and ponds. Locations of transient terrestrial cyanobacterial desert communities include the periphery of hypersaline Dry Valley lakes (Vincent, 1981), stromatolites (Wharton, 1994), the beds of seasonal melt-water streams (Hawes et al. 1992; Vincent et al., 1993), and the peripheries of ponds (Vincent and James , 1996).

Antarctic cyanobacterial communities of xeric habitats distant from water bodies are constrained to lithobiontic niches by scarce moisture (Golubic et al., 1981). These can be within fissures (chasmolithic) or within the fabric of the rock (endolithic) in extremely dry environments (Friedmann and Ocampo-Friedmann, 1984). Although Friedmann et al. (1967) were familiar with hot desert endolithic communities from the Negev, the relevance of the cold desert endolithic niche as a potential extremely xeric exobiological habitat in Antarctica was overlooked during the development of the Viking programme of 1975–76 (Friedmann and Ocampo, 1976). Ironically, Friedmann (1971) had previously considered cyanobacteria as possible exobiological life-forms because he stated that that the monospecific *Gloeocapsa* community "may... represent an analogous situation to certain extra-terrestrial conditions".

The isolation of Antarctic desert communities from more northerly sources of cyanobacterial inocula is evident from the extent of the circumpolar Southern Ocean and the associated prevailing winds (Wynn-Williams, 1991). However, despite this remoteness, increasing pressure from scientists and tourists in the Dry Valleys has necessitated the development of a management policy (Vincent, 1996). Sites of Special Scientific Interest (SSSI) have been designated for the microbiological protection of cold desert soils in Barwick Valley and endolithic communities at Linnaeus Terrace (British Antarctic Survey, 1997). A Long-term Ecological Research site (near another SSSI) has been established in Taylor Valley at Lake Hoare to promote biological research into desert and lake communities dominated by cyanobacteria (Wharton, 1993).

This review will consider the structure and function of desert cyanobacterial communities and their continued importance as exobiological analogues.

II. Diversity of Desert Ecosystems

A. Hot Desert Ecosystems

Desert biology has traditionally focused on hot deserts (Plate 22a). Fuller (1974) gave a thorough account of desert soils referring to 13 individual world deserts without mentioning the Arctic or Antarctic. Likewise, Davis (1974) described the hydrogeology of arid regions with special reference to the Nubian, Australian and Nevada (USA) deserts without reference to polar regions with their additional complexities of snow-lie, melt-water and permafrost. Within these hot deserts, Friedmann et al. (1967) recognised five major algal habitats (Plate 22b): edaphic (soil colonists), subdivided into endedaphic (within soil), epedaphic (on top of the soil) and hypolithic (under stones), and lithophytic, subdivided into chasmolithic (within rock fissures) and endolithic (within the rock fabric). Epilithic growth on rock surfaces is restricted to seepage zones and areas of moisture accumulation.

Cyanobacteria dominate the microalgal populations of hot deserts, both in soil crusts (de Chazal et al., 1992) and endolithic communities (Friedmann and Ocampo-Friedmann, 1984; Palmer and Friedmann, 1990). They are typically able to withstand high mean summer ground temperatures of 57°C with peaks in excess of 60°C (Mazor et al., 1996). Paradoxically, photosynthesis by lichens of the Negev is adapted to low temperatures and the same seems to be true for cyanobacterial soil crusts (Mazor et al., 1996). *Chroococcidiopsis* dominates certain endolithic communities in both hot and cold deserts (Friedmann, 1971, 1982; Wessels and Büdel, 1995) and is specially remarkable for its heat-tolerance and rate of rehydration-recovery after desiccation (Caiola et al., 1996). Dried spherical microcolonies can remain viable after baking in a hot arid environment at substratum temperatures exceeding 85°C (Potts, 1994). However, despite the stress-tolerance of

Chroococcidiopsis, the most frequently recorded cyanobacterial genera in hot deserts are *Schizothrix, Plectonema, Microcoleus* and *Nostoc* (Whitton, 1987). The ubiquity and survival characteristics of the genus *Nostoc* merit its inclusion as a special topic by Potts in this volume, Chapter 17.

Cyanobacteria comprise the vast majority of the microflora of a hot desert crust (Friedmann and Galun, 1974). Lange et al. (1992) describe the structure and taxonomic composition of a typical crust community 1–3 mm thick on sand dunes in the Nizzana district of the Negev Desert, Israel. It is dominated by four genera of cyanobacteria, although the most common species is *Microcoleus sociatus.* This species, together with sparser *Gloeocapsa* spp. was found within the crust whilst *Nostoc* sp. and *Calothrix parietina* occurred predominantly on the crust surface. This stratified spatial distribution is a characteristic of edaphic and endolithic desert cyanobacterial communities. In this region, fog and dew are abundant, occurring on more than 50% of the mornings. The dew-absorbing crust enables trichomes to remain dormant during the 8-month periods between rainy seasons which provide most of the c. 90 mm precipitation per annum, mainly in winter months. Moisture retained by cyanobacterial exopolysaccharide (EPS) permits photosynthesis in the absence of rain as shown experimentally by the fixation of $^{14}CO_2$ by non-wetted *Microcoleus sociatus* under different levels of humidity (Mazor et al., 1996). Net photosynthesis was not detected at 80% relative humidity (RH), but was significant at 88% RH. However, water-retention by *M. sociatus* was much greater than that by *Nostoc*, so caution must be applied when assessing desiccation protection by cyanobacterial crusts of different species composition. Scherer and Zhong (1991) showed that the proportion of two distinct ecotypes of *Nostoc commune* in desert soils of the Ningxia Province of Inner Mongolia, China changes according to the availability of water and is largely independent of soil type and incident radiation.

Stratification is the result of an ecophysiological compromise between protection from damaging factors such as UVB radiation, and optimal use of life-supporting factors such as photosynthetically-active radiation (PAR) and scarce moisture availability. The interaction between cyanobacterial sheaths and mineral grains stabilises the dune habitat, typically comprising c. 21% sand, 55% silt and 24% clay (Lange et al., 1992). It concurrently increases the water-holding capacity of the porous substratum to create a life-supporting niche (Mazor et al., 1996). Campbell (1979) showed experimentally that re-wetted trichomes of *Microcoleus vaginatus* in a microphytic crust in the Utah desert could glide 500 times their own length in a day, leaving a continuous EPS sheath thread potentially capable of trapping and binding sand grains to create the crust found in the field. From subsequent microscale studies of vertical spatial distribution of cyanobacteria in these Utah crusts, Garcia-Pichel and Belnap (1996) concluded that the motility and morphology of *M. vaginatus* helped to make it the primary coloniser of these desert soils.

In hot desert endolithic niches, cyanobacteria are exposed to more sudden environmental changes than in cold deserts because of daily transitions from hot-dry conditions to warm-humid ones when fog or dew may condense at night (Mazor et al., 1996). Intolerance of such fluctuations limits the species composition of eukaryotic algae in Nizzana sand (Negev Desert) to single members of the genera *Chlorococcum* and *Stichococcus.* Paradoxically, chlorophytes also dominate much colder, drier niches in the Beacon sandstone of southern Victoria Land where dew never occurs (Friedmann, 1982; Nienow and Friedmann, 1993; Palmer and Friedmann, 1990).

B. Cold Desert Ecosystems

The most extensive areas of cold desert soils on the Antarctic continent lie principally within the 4800 km^2 ice-free arid zone (Gunn and Warren, 1962) in southern Victoria Land (Plate 20a)). This area extends from sea level to c. 2000 m altitude. Because of the dominant low-temperature oxidation and salinisation aspects of their pedogenesis, Tedrow and Ugolini (1966) recommended that continental Antarctic soils should be defined as "cold-desert soils" which may or may not be zonal but were always ahumic. The soils and rocks of these and other desert areas of Antarctica have since been comprehensively described and illustrated by Campbell and Claridge (1987). There are other areas of ice-free land in central Victoria land where endolithic cyanobacteria have been found and soil surface cyanobacterial crusts occur near sources of moisture such as melt streams and the periphery of lakes and ponds (Wynn-Williams et al., 1997). The gradient of moisture available in McMurdo Dry Valley (Ross Desert) soils studied by Cameron (1969a) and conditions necessary for life-support in the region are reviewed by Wynn-Williams (1990).

Surprisingly, there is no direct correlation between moisture and microalgal counts because of other factors such as accumulated salinity and the input of propagules. However, extremely arid soils (Victoria and Barwick valleys and Coalsack Bluff) yielded either no algal isolates or only 100 colony-forming units (cfu) g^{-1} soil whilst the benign wet nitrogen-rich ornithogenic soils on nearby Dunlop Island yielded 10^5 cfu g^{-1}.

A comprehensive description by Cameron (1972b) of desert microbial communities in Victoria Valley, southern Victoria Land, summarised the edaphic conditions of the Antarctic desert habitat. He emphasised the very low organic content of soils distant from liquid water sources, usually less than 0.05%, similar to that of the lowest values from hot desert soils of the Sahara, Negev, Atacama and western USA (Cameron, 1969b). Cyanobacteria were represented by cosmopolitan species such as *Anacystis marina, Anacystis montana, Schizothrix calcicola* and *Nostoc commune.* Cameron (1969a) compared Antarctic deserts with those of the Arctic, drawing attention to the temperature-depressing influence of cold waters retained by circumpolar currents at the Antarctic Polar Front which has no Arctic equivalent. This results in a less harsh climate in the North with cryptogamic vegetation at much higher latitudes such as on Spitsbergen (78°N) where the mean temperature for March, the coldest month, is - 15° and for July is + 6.5°C. The resulting shrub vegetation (Coulson et al., 1993) is in marked contrast to the true desert ecosystems found at equivalent southern latitudes such as Linnaeus Terrace (78°S) in Taylor Valley. There the mean air temperature between 1984 and 1986 for the coldest month (July) was - 32.2°C and for the warmest (January) was - 6.6°C (Friedmann et al., 1987). This habitat contains only endolithic lichens or cyanobacterial crusts associated with transient melt-water. At the most southerly extreme Antarctic desert, at 86°S in the La Gorce Mountains, epilithic lichens are very rare (R.N. Weinstein, pers. comm. 1996) and microalgae were limited mainly to *Nostoc commune* and *Schizothrix calcicola* at the periphery of frozen ponds (Cameron, 1972a).

Much of the surface of the Antarctic ice-free desert (Fig. 1) is highly mobile and disturbed by frequent strong katabatic winds flowing downhill from the 2000 m high polar plateau (Parish, 1988). These mainly gravity-driven winds maintain the aridity of ice-free areas by flushing them with cold, dry air (Wynn-Williams, 1991). As in Victoria Valley, this can result in highly unstable desiccated sand dunes (Miotke, 1985) devoid of the biological crusts which are supported by condensation in hot deserts. In a comprehensive world-wide comparison of microflora of hot arid zones and high latitude deserts, Cameron (1971b) found that microalgal dilution counts (Thornton's salt solution at 25°C under 97 μmol m^{-2} s^{-1} PAR) ranged very broadly. Out of 13 diverse hot deserts, counts ranged from 10^6 g^{-1} soil in the Sonoran desert to none detected in one sample from the Atacama desert (Chile). Similarly out of eight cold desert sites, counts ranged from 10^3 g^{-1} soil on High Mt. California, to none detected in several Antarctic Dry Valley soils such as Taylor Valley. Most of the microalgae isolated from both hot and cold deserts were filamentous cyanobacteria whether or not there was any higher vegetation present.

III. The Organisms – Biodiversity

A. Soil Crust Species

Despite the widespread occurrence of desert regions bearing biological crusts and their importance for erosion control and initiating colonisation processes, little is known about the taxonomy and composition of the cyanobacterial microflora of desert soils. Cameron's Desert Microflora Program began with a taxonomic survey of hot deserts in the USA and showed the predominance of cyanobacteria, especially filamentous trichomes (Cameron, 1966). While listing, illustrating and describing his isolates, he drew attention to the confusing problem of synonyms and the difficulty of identifying Oscillatoriaceae such as *Schizothrix calcicola.* The nomenclature of many of the strains referred to here has been revised (Rippka et al., 1979) and will be revolutionised in the future by the application of novel molecular genetics analyses (Hunter and Mann, 1992). Cameron and Blank (1966), Friedmann (1972) and Friedmann and Galun (1974) have reviewed the diversity of desert microalgae with emphasis on hot desert lithosols, chasmoliths and endoliths from Central Asia, the Arctic, the Negev and south-western United States and Mexico. Friedmann and Galun (1974) noted that filamentous cyanobacteria such as *Anabaena, Lyngbya, Microcoleus, Nostoc, Phormidium, Plectonema* and *Scytonema* were the most frequently dominant genera, but cautioned against generalisation because of inter-site variations (Table 1). Coccoid cyanobacteria such as *Synechococcus* and *Synechocystis* were less

Table 1. Cyanobacterial species commonly found in selected hot deserts.

	Hot deserts world-wide Cameron and Blank (1966)[1] Cameron (1971b)	**Atacama desert Chile** Forest and Weston (1966) Schwabe (1963)	**Nizzana, Negev Desert Israel** Lange et al. (1992)	**Namib Desert W. Africa** Büdel and Wessels (1991)	**Ayers Rock Australia** Büdel and Wessels (1991)	**Sonoran Desert Arizona USA** Büdel and Wessels (1991)	**Langjan Nature Reserve S. Africa** Weber et al. (1996)
Cyanobacterial species	**Soil**[2]	**Soil**	**Dunes**	**Cha; End**	**Cha; End**	**Cha: End**	**End.**
Anacystis montana[3]	+	+	-	-	-	-	-
Calothrix parietina	-	-	+	-	-	-	-
Calothrix desertica	-	+	-	-	-	-	-
Chroococcidiopsis sp.	-	-	-	+	+	+	+
Coccochloris peniocystis[3]	-	+	-	-	-	-	-
Chroococcopsis sp.	-	-	-	+	-	-	-
Cyanosarcina sp.	-	-	-	+	+	-	-
Gloeocapsa rupestris	-	-	-	-	+	-	-
Gloeocapsa sp.	-	-	+	+	-	-	-
Gloeothece sp.	-	-	-	-	-	+	-
Microcoleus sociatus	-	-	+	-	-	-	+
Microcoleus chthonoplastes	+	-	-	-	-	-	-
Microcoleus paludosus	+	-	-	-	-	-	-
Microcoleus sp.	+	-	-	-	-	-	-
Myxosarcina sp.	-	-	-	+	+	+	-
Nostoc microscopicum	-	-	-	+	-	+	-
Nostoc muscorum	+	-	-	-	-	-	-
Nostoc sp.	-	-	+	-	-	-	-
Nostochopsis lobatus	-	-	-	-	-	-	+
Oscillatoria sp.	+	-	-	-	-	-	-
Plectonema polymorphum	-	+	-	-	-	-	-
Phormidium angustissimum	-	-	-	-	-	-	-
Phormidium autumnale	+	-	-	-	-	-	-
Porphyrosiphon fuscus	+	-	-	-	-	-	-
Porphyrosiphon notarisii	+	-	-	-	-	-	-
Schizothrix arenaria	+	-	-	-	-	-	-
Schizothrix adunca	-	+	-	-	-	-	-
Schizothrix atacamensis	-	+	-	-	-	-	-
Schizothrix calcicola	+	-	-	-	-	-	-
Schizothrix rubella	+	-	-	-	-	+	-
Scytonema hofmannii	+	-	-	-	-	-	-
Synechococcus elongatus	-	-	-	+	-	-	-
Tolypothrix brevis	-	-	-	+	-	-	-

[1] See Appendix in Cameron and Blank (1966) for many synonyms

[2] Soil = lithosol; A22 Cha = chasmolithic; End = endolithic

[3] Nomenclature of Drouet and Daily (1956)

Table 2. Cyanobacterial species commonly found in Antartic cold deserts.

Cyanobacterial species	Mawson Rock 67°S 63°E Broady, 1981a,c Cha[2]	Vestfold Hills 68°S 78°E Broady, 1981c,1986 Sub; Cha	Davis Station 68°S 78°E Broady, 1981a Cha	Dronning Maud Ld. 71°S, 02°E-03°W Ryan et.al., 1989[1] Epi	McMurdo Dry Valleys 77°S 163°E Seaburg, 1979 Soil[3]; End	Edward VII Pen. 77°-78°S 152°-154°W Broady 1989b Soil; Epi;Cha
Anabaena sp.	-	-	-	-	EIF +[3]	-
Anacystis dimidiata[5]	-	-	-	-	REC +[4]	-
Anacystis marina[5]	-	-	-	-	REC +	-
Anacystis montana[5]	-	-	-	-	REC +	-
Aphanocapsa sp.	-	-	-	-	EIF +	-
Aphanothece sp.	-	+	-	-	-	-
Calothrix parietina	+	+	-	-	-	+
Calothrix gypsophila	-	-	-	-	+	-
Chroococcidiopsis sp.	+	+	+	+	-	+
Coccochloris aeruginosa[5]	-	-	-	-	REC +	-
Coccochloris elabens[5]	-	-	-	-	REC +	-
Coccochloris stagnina[5]	-	-	-	-	REC +	-
Cyanothece aeruginosa	-	-	-	-	-	+
Eucapsis sp.	-	-	-	-	EIF +	-
Gloeocapsa alpina	-	+	-	-	-	-
Gloeocapsa kuetzingiana	-	+	-	-	-	+
Gloeocapsa cf. *punctata*	-	-	-	-	-	+
Gloeocapsa ralfsiana	-	+	-	-	-	+ (S)
Gloeocapsa sp.	-	-	-	+	-	-
Gloeotheca sp.	+	+	+	-	-	-
Hormathonema sp.	-	-	-	-	EIF +	-
Homoeothrix cf. *rivularis*	+	-	-	-	-	+ (E)
Lyngbya sp.	+	-	-	-	EIF +	-
Lyngbya murrayi	-	-	-	+		-
Microchaete sp.	-	-	-	-	EIF +	-
Microcystis stagnalis	-	-	-	-	+	-
Myxosarcina sp.	+	+	-	-	-	-
Nodularia harveyana	-	+	-	-	-	-
Nostoc commune	-	-	-	+	+	-
Nostoc sp.	+	+	-	+	-	+
Oscillatoriaceae	+	+	-	+	-	+
Oscillatoria lutea	-	-	-	-	REC +	-
Oscillatorea submembranacea	-	-	-	-	REC +	-
Phormidium angustissimum	-	-	-	-	+	-
Phormidium autumnale	-	-	-	+	REC +	+
Phormidium frigidum	-	-	-	-	+	-
Phormidium fragile	-	-	-	-	+	-
Phormidium laminosum	-	-	-	+	+	+
Phormidium priestleyi	-	-	-	-	+	-
Phormidium sp.	-	-	-	+	-	-
Plectonema sp.	+	+	+	-	-	-
Pleurocapsa sp.	-	+	-	-	-	-
Porphyrisiphon fuscus	-	-	-	-	-	+ (S)
Porphyrosiphon notarisii	-	-	-	-	REC +	-
Schizothrix calcicola	-	-	-	-	REC +	-
Scytonema sp.	-	-	-	+	-	-
Stigonema minutum	-	-	-	-	-	+
Synechococcus aeruginosa	-	-	-	+	-	-
Tolypothrix tenuis	-	+	-	-	-	-

common. More recently, the taxonomy and function of microphytic crusts has been comprehensively described by West (1990), whilst Lange et al. (1992) focused on the implications of their limited taxonomic biodiversity for photosynthetic activity in sand dunes of the Negev desert. The ability to survive within rock fissures (the chasmolithic habitat) or within the fabric of the rock itself (endolithic) is a characteristic of various cyanobacteria, and eukaryotic algae, whether the desert is hot or cold. Büdel (1991) reported eleven genera of desert cyanobacteria from lithic habitats in Africa, North America Australia and Europe (Table 1). He presented their taxonomic characteristics, including new records for the lithobiontic habitat, and drew attention to UVB-absorbing substances found in the first strain of *Chroococcidiopsis* to be recorded from a terrestrial habitat (hypolithic under quartz pebbles). The protective aspects of lithobiontic habitats will be discussed later.

Knowledge of Antarctic desert cyanobacterial taxonomy and ecology is more comprehensive because of the high profile of microbiological research associated with exobiology and climate change, including global warming and increased UVB radiation during the annual ozone hole (Wynn-Williams, 1996). Cameron (1971c) gave a comprehensive summary of the occurrence, abundance and taxonomic diversity of cyanobacteria (and associated eukaryotic algae) in Antarctic deserts isolated during the Desert Microflora Program. He listed the filamentous species as *Microcoleus vaginatus (Phormidium autumnale), Nostoc commune, Oscillatoria lutea, Oscillatoria submembranacea* and *Schizothrix calcicola.* Coccoid species comprised *Anacystis dimidiata, Anacystis marina, Anacystis montana, Coccochloris aeruginosa, Coccochloris elabens* and *Coccochloris stagnina.* Seaburg et al. (1979) greatly expanded this list for southern Victoria Land, giving alternative names for some of the species reported by Cameron (Table 2). They gave illustrated descriptions of species and details of the habitats from which they were obtained, although the non-aquatic habitats are described broadly as "soils". It is noticeable that most isolates came from littoral habitats which are the most consistent sources of free water for at least part of the year.

More recent studies of Antarctic terrestrial microalgae have been comprehensively summarised by Broady (1996) and Vincent (this volume, chapter 12). Most soil crust communities are from coastal continental sites moistened by marine-influenced precipitation or extensive melt-water (Broady, 1989a; Broady 1986; Broady, 1987) or atypical geothermally-heated and steam-moistened soils (Broady, 1984; Broady et al., 1987; Bargagli et al., 1996), reviewed by Broady (1993). However, a few have included desert microflora that are distant from consistent moisture supplies. Most of these have been studied in southern Victoria Land, but little has been done on soil crust taxonomy in the McMurdo Dry Valleys since Seaburg et al. (1979). Elsewhere, Broady has described colonists from Framnes Mountains near Mawson Station (Broady, 1981a) and Edward VII Peninsula, Marie Byrd Land (Broady, 1989b), and Dronning Maud Land (Engelskjøn, 1986; Ryan et al., 1989). A detailed survey of terrestrial microalgae at Edward VII Peninsula, Marie Byrd Land (77°00' – 78°30'S, 152° – 154°W) yielded 19 species of cyanobacteria (Broady, 1989b). Of these, twelve species were locally abundant. However, most were associated with ponds or other sources of melt-water. Nevertheless, he described thin black crusts over lithosols dominated by the cyanobacterium *Cyanothece aeruginosa* and the chlorophytes *Stichococcus bacillaris* with occasional *Coenocystis* sp. Inland at 1600 – 2470 m a.s.l., on Gjelviksfjella (71°55'S, 02°45'E), Engelskjøn (1986) described encrusted cyanophycean-chlorophycean sub-formations but only associated with seepage trails from melt-water. Dark films of *Gloeocapsa* and *Phormidium* were found on gneiss debris, and extensive sheets of *Nostoc* sp., and *Oscillatoria/Phormidium* were found on locally wetted morainic gravel. However, there was no report of desert biotic crusts. At nearby

FOOTNOTES TO TABLE 2 OPPOSITE

1 including species reported nearby by Engelskjon (1986)

2 Epi (E) = epilithic; chasmo = chasmolithic; endo = endolithic; soil (S) = lithosol

3 EIF = listed as lithobiontic in Friedmann et al. (1988) but not in Seaburg (1979)

4 REC = listed in Cameron (1971a and 1972) but not in Seaburg (1979). Cameron reports *Phormidium autumnale* as Microcoleus vaginatus

5 Nomenclature of Drouet and Daily (1956)

Robertskollen (71°28'S, 03°15'W) 130 km from the sea at 200 – 500 m a.s.l., Ryan et al. (1989) described moist cliff-face sites bearing a film of *Gloeocapsa* sp., *Phormidium* spp. and *Nostoc commune*. No soil crusts were noted, and more prolific cyanobacterial communities were restricted to streams only.

B. Epilithic Species

The survey of Edward VII Land by Broady (1989b) also revealed the importance of epilithic cyanobacterial communities fed by trickles of melt-water over stable rocks This habitat is not subject to the disturbance of dunes by katabatic winds although the communities are vulnerable to sand-blasting and desiccation. Cyanobacterial colonists are completely dependent on transient melt-water, but can develop rapidly in its presence at elevated temperatures due to black-body absorption of solar radiation, and plentiful PAR for photosynthesis during summer months. In these stable habitats, Broady identified 13 epilithic cyanobacterial species in the Edward VII Peninsula region (Broady (1989b), 11 species (mainly *Gloeocapsa* spp. *Homoeothrix* cf. *rivularis* and *Stigonema minutum*) in Princess Elizabeth Land and Mac.Robertson Land (Broady, 1981c). *Gloeocapsa* spp., *Calothrix* spp., and narrow-trichomed Oscillatoriaceae were found, albeit infrequently, as epilithic communities in water percolations on rocks in the Vestfold Hills (Broady, 1986). However, Broady (1981) reported that epilithic algae had not been found in more arid areas of Mawson Rock and the Vestfold Hills where snow lie on the rocks was absent or too transient to initiate growth. The increasing distance inland from warmer, moister air of coastal desert regions to truly arid interior results in a marked decrease in total cyanobacterial species diversity, from 18 genera near the coast at the Vestfold Hills to ten genera in Dronning Maud Land 200 km inland (Engelskjøn, 1986) and down to only five genera 1000 km inland at the Pensacola Mountains near Dufek Massif (Cameron, 1972a; Parker et al., 1977; Claridge, 1987).

C. Endolithic Species

Antarctic endolithic cyanobacteria (Table 2) are more abundant in moister more northerly parts of the McMurdo Dry valleys such as reported at Battleship Promontory (Plate 20b) in the Convoy Range (76°54'S) (Friedmann et al., 1988) and in central Victoria Land at Timber Peak, Priestley Glacier (Plate 21b) (Russell et al., submitted). Although Battleship Promontory also supports lichen-based communities and a sparse occurrence of *Hemichloris antarctica* (green alga), three cyanobacterial communities predominate in contrast to the prevalence of green algae in the drier environment at Linnaeus Terrace (77°35'S) (Nienow and Friedmann, 1993; Friedmann et al., 1993). The most common one at Battleship Promontory is a *Hormathonema-Gloeocapsa* community with lesser prevalence of *Gloeocapsa* and *Chroococcidiopsis* communities, illustrated and described comprehensively by Friedmann et al. (1988). The *Hormathonema – Gloeocapsa* community characteristically occurs in "moist" Beacon sandstone almost devoid of primary iron-staining (0.039% Fe_2O_3 at pH 8.2 compared with 0.132% Fe_2O_3 at pH 5.5 in dry iron-stained sandstone at Linnaeus Terrace) and results in exfoliative weathering. It can also contain *Anabaena* in the upper zone and is characterised by *Aphanocapsa* together with *Gloeocapsa* (up to eight species in the region), *Anabaena* (three species) and *Lyngbya*. The red *Gloeocapsa* community which also contains *Eucapsis* and *Microchaete* is mainly found on vertical faces of large boulders where the horizontal face is usually occupied by lichen-dominated communities. The infrequent *Chroococcidiopsis* community is more characteristic of drier extreme habitats, although *Chroococcidiopsis* can also be found as a phycobiont (cyanobiont) in lichen communities (Büdel and Henssen, 1983).

IV. Desert Niches and Interactions

Microbial life in deserts is poised at the limits of survival. Water is the prime limiting factor, but there are many other physical, chemical and ecological factors, summarised in Table 3, which are influential in maintaining the delicate balance swinging from survival to either establishment or extinction. Climate change such as global warming or elevated UVB due to ozone depletion may swing the balance critically. Cyanobacteria have a remarkable suite of attributes and strategies which enable them to colonise and survive where other organisms cannot do so. They are adept at colonising two very different types of substratum: unstable sand dunes or soil which they actively stabilise, and rocks to which they adhere or penetrate. They survive because they are able to interact with their niche and either exploit its attributes or modify it to make it more suitable.

Table 3. Requirements and strategies for colonisation of deserts by cyanobacteria.

Environmental Factor	Attribute or Strategy	References
Translocation	Aerial transport; animal vectors	Wynn-Williams, 1991
Stable substratum	Rock or static soil	Campbell and Claridge, 1987
	Site orientated N-S across katabatic winds	Parish, 1988
Settlement and attachment	Porous or rough substratum	Campbell and Claridge, 1987
	Sublithic, chasmolithic and endolithic growth	Broady, 1981b; Broady, 1986a; Friedmann, 1982
Sand-blasting by katabatic winds	Endolithic communities	Nienow and Friedmann, 1993
Integration with the microhabitat	Filamentous soil forms	Wynn-Williams, 1997
	Flexible endolithics	Friedmann and Ocampo-Friedmann, 1984
Adhesion to substratum, soil-binding	EPS mucilage production, cementation of grains	Cameron and Devaney, 1970
Scarcity of moisture. Low humidity	Sheltered growth within or under a substratum	Palmer and Friedmann, 1990
	Absorption and storage of snow or melt-water	Friedmann. 1987
	Absorption of water vapour by algae	Vestal, 1988b
	Absorption of condensation by cyanobacteria	Palmer and Friedmann, 1990
	Growth near meltwater source (snow, stream, lake)	Nienow and Friedmann, 1993
Desiccation	Compatible solutes, e.g. trehalose, polyols	Potts and Friedmann, 1988; Potts, 1994
	Anhydrobiosis; Rapid re-hydration	Potts, 1996; Palmer and Friedmann, 1990
	Dormancy	Johnston and Vestal, 1991
Low growth temperatures	Growth on N-facing dark sun-warmed slopes	Nienow et al., 1988b
Freezing	Antifreezes: Compatible solutes, e.g. trehalose	Hershkowitz, 1991
High salinity; evaporite soils	Osmotolerance: Compatible solutes, e.g. trehalose	Csonka and Harrison, 1991
Wet-dry cycles	Crust formation with exopolysaccharides	Mazor et al., 1996
Freeze-thaw cycles	Cryoprotectants, e.g. trehalose, sucrose	Hershkowitz, 1991
Fluctuating temperatures	Thermal buffering of endolithic habitats	Friedmann et al., 1993; McKay and Friedmann, 1985
Rapid response to moisture availability	Membrane permeability; fast metabolic response	Olije et al., 1986
	Fluid membranes at low temperatures	Finegold et al., 1990
Optimization of short growing season	Psychrotolerant enzymes; High Q_{10} at temps >0°C	Ocampo-Friedmann et al., 1988
	Sheltered microhabitats	Palmer and Friedmann, 1990
Light penetration of substratum	Growth in translucent rock e.g. Beacon sandstone	Nienow et al., 1988b
	Sublithic growth under translucent stones	Broady, 1981b
Tolerance of low PAR	Shade adaptation	Nienow et al., 1988
	Spectral adaptation	Hader, 1987
Tolerance of high UVB flux	Sunscreens in EPS sheath, e.g. scytonemin	Garcia-Pichel and Castenholz, 1991
	Intracellular sunscreens, e.g. MAAs	Garcia-Pichel et al., 1993a
Tolerance of high light intensity	Sunscreens such as carotenoids	Downes et al., 1993; Potts et al., 1987
Tolerance of oligotrophic conditions	Pathway flexibility, e.g.re-allocation of C and N	Vestal, 1988
	Independence of fixed C: autotrophy	Friedmann and Ocampo, 1976
	Symbiosis as lichens	Kappen et al., 1981

A. *Sand and Soil*

Miotke (1985) and Campbell and Claridge (1987) have described the instability of cold desert dune systems under the influence of katabatic winds and a similar phenomenon prevails in hot deserts such as the Negev (Mazor et al., 1996). The ability of cyanobacteria to form stabilising crusts on hot desert soils is reviewed in Friedmann and Galun (1974) and Belnap and Gillette (1998). The role of their EPS and filamentous morphology in crust formation is being actively exploited to reclaim desert regions for potential agriculture (Flaibani et al., 1989; Painter, 1993). Cyanobacterial EPS also influences the porosity of the soil so that water is retained longer either by absorption of moisture into the EPS sheath or by obstruction of diffusion. The obstruction of moisture flow is important in habitats such as both hot and cold deserts which have major diurnal fluctuations in temperature. This causes flushing of water vapour out of soil pores by daily expansion and contraction of their atmosphere. Even in cold deserts underlain by permafrost and overlain by transient snow, liquid water released by thawing is partially flushed out of the biotic zone by this process (R. Bargagli, pers. comm).

B. Epilithic, Sublithic and Chasmolithic Habitats

By virtue of their mucilaginous growth form, cyanobacteria are able to adhere to the surface of rocks and retain sparse water resources from dew, intermittent rain or snow, and melt-water in cold regions. They can lodge within the rugosities of rock surfaces in the subaerial epilithic habitat which afford some protection as the meso scale. This growth form has been described by Broady (1981c) at the coastal semi-desert region of Mawson Rock where the microflora was predominantly *Chroococcidiopsis*, *Gloeocapsa*, *Myxosarcina* and *Calothrix*. The epilithic habitat is fully exposed to desiccation stress. However, the loss of water from soil can be obstructed by the presence of translucent stones on the surface to retain sufficient moisture for long enough to sustain an illuminated sublithic habitat (Broady, 1981b).

The sublithic or hypolithic habitat is characteristic of hot deserts (Plate 22c and d) as a moisture-conservation strategy combined with the low light tolerance of cyanobacteria (Friedmann and Galun, 1974). In the cold deserts of Antarctica, visible peripheral growth of cyanobacteria on the underside of quartzitic stones is common in even the driest valleys. Broady (1981b) reported the frequent occurrence of *Chroococcidiopsis* and *Plectonema* and six other species of cyanobacteria growing sublithically under up to 109 mm thickness of quartz in the Vestfold Hills (Fig. 1). Light transmission through stones 13 – 80 mm thick ranged from 0.9 to 2.7% of incident sunlight and the temperature of the moist sublithic soil fluctuated diurnally through a range of up to 13°C. These greenhouse conditions are highly favourable for shade-adapted cyanobacteria. However, sublithic growth only occurs in drainage areas in receipt of snow melt and is therefore restricted to valley floors and approximately horizontal ledges and plateaux.

Fissures, crevices and exfoliations in non-porous but translucent granitic rocks at moister, lower elevations can be colonised on slopes by chasmolithic cyanobacteria (Broady, 1981a). These are usually coccoid forms such as *Chroococcidiopsis* sp. or *Gloeocapsa* sp.. However, they also occur in the Antarctic interior, and a chasmolithic association of *Gloeocapsa* sp. and other cyanobacteria was found in anorthosite in the Dufek Massif (Fig. 1) by Friedmann (1977). He also found an example of a lichen-dominated community with a fungal and algal zone and an innermost layer of cyanobacteria, similar to that in Plate 21a. The characteristics of the chasmolithic niche (Plate 22f) have been described by Friedmann and Ocampo-Friedmann (1984). It is a transition towards the ultimate avoidance strategy of the cryptoendolithic habitat isolated from the rock surface. These transitions are also found in hot deserts. Examples include the habitat diversity of cyanobacteria growing epilithically and cryptoendolithically on Clarens Sandstone cliffs of the Golden Gate Highlands National Park, South Africa (Wessels and Büdel, 1995). Epilithic species in seepage zones (tintenstrich communities) on the sandstone of Moses-cave included *Chamaesiphon* sp., *Chroococcus minutus, Dichothrix baueriana, Gloeocapsa sanguinea, Gloeocapsa muralis, Gloeocapsa biformis, Nostoc* cf. *minutum*, *Oscillatoria* cf. *irrigua* and *Scytonema* sp. However, species diversity of endolithic cyanobacteria within the sandstone Moses-cave was restricted to dominant *Chroococcidiopsis* sp. with sparse *Synechococcus* sp. and *Gloeothece* sp.

C. The Endolithic Habitat

1. Hot Deserts

Light microscopic and SEM studies of the endolithic habitat began in hot deserts (Friedmann et al., 1967; Friedmann, 1971; Friedmann and Ocampo-Friedmann, 1984). This revealed the dominant role of the cyanobacteria *Gloeocapsa*, *Chroococcus turgidus* and *Chroococcidiopsis* (Plate 22e). A wide variety of deserts and translucent rock types were found to contain cryptoendolithic cyanobacteria (Friedmann and Kibler, 1980). These included sandstone from Sinai Desert, Utah and California (USA) and Namibia, and granite from Sinai and Chile. Potts and Friedmann (1981) initiated studies of the ecophysiology of desert communities, showing that strains of dominant *Chroococcidiopsis* (Friedmann, 1980) were less tolerant of water stress than *Chroococcus* strains.

At the beginning of a long-term study of the factors regulating microbial activity and diversity in endolithic communities in diverse parts of the world, Friedmann (1980) showed daily temperature variations from + 22° to + 48°C in granite in the Sonoran Desert. There was little difference between the surface and 20 mm depth, but peak temperatures were much higher than those of the air (+ 38°C at 1 m above the rock). Relative humidity of the air 1 m above the rock fell sharply during the day from 47% to 24%. This suggested that moisture conservation within rocks which generally contained 0.1 – 0.2% water (dw) in this desert region was a critical life-supporting feature for endoliths. Scanning electron microscopy (Friedmann, 1971) showed that the spaces between the outermost layer of rock grains was filled with mineral material (desert varnish). The mineral filling is permeable to liquids and gases but is not penetrable by cells. Micro-organisms were attached to grains underneath.

Vestal (1993) concluded that although hot desert rocks have not been studied as intensively of those of Antarctica, the colonisation potential of endolithic cyanobacteria as an avoidance strategy for survival under osmotic, thermal and UV or light stress was similar to that of cold desert communities. The only major difference was no need for extreme cryoprotection, although compatible solutes produced to deal with osmotic stress would be available for occasions when the night-time temperature fell below freezing point. Physiological studies by Weber et al. (1996) on isolates of *Chroococcidiopsis* sp. and *Nostochopsis lobatus* from a cryptoendolithic community extending 0.4 – 1.9 mm in a sandstone outcrop at Langjan Nature Reserve, Northern Province, South Africa, showed light saturation of photosynthetic electron transport at 200 – 350 μmol m^{-2} s^{-1}. This was very similar to the natural level of 20 – 250 μmol m^{-2} s^{-1} measured in the cyanobacterial zone of the natural rock in full sunlight (1 – 10% of incident light). The zonation observed appeared to be light regulated, as there was negligible difference in temperature between the surface and the cyanobacterial zone, and the pore space for water absorption (8.06% of the rock matrix) was substantial.

Although carbon fixation by photosynthesis may be substantial, there may be limitation by nitrogen in this extremely oligotrophic habitat. Garcia-Pichel and Belnap (1996) found that heterocystous, potentially nitrogen-fixing cyanobacteria (such as *Nostoc* spp.) occurred near the surface of hot desert microphytic crusts but required sunscreens such as scytonemin to survive the intense UVB flux. This is not such a problem for sheltered cyanobacteria in endolithic communities from hot and cold deserts in the Middle East, Mexico, North America, South America, Africa and Antarctica whose nitrogen budgets were studied by Friedmann and Kibler (1980). They concluded that biological nitrogen fixation was not significant in endolithic communities, unlike algal soil crusts in hot deserts, such as Great Basin Desert, which fixed 1 – 10 g N m^{-2} y^{-1} (Rychert and Skujins, 1974). Greenfield (1988) similarly found N_2-fixation to be negligible in sandstone at Beacon Heights whilst its total organic content was 1.56 g m^{-2} (range 1.39 – 1.65 g). Friedmann and Kibler (1980) reported the total organic nitrogen content of rocks from North American and Middle Eastern deserts to range from 1.5 to 8.0 g m^{-2} whilst that of Antarctic Dry Valley rocks from a variety of sites was 1.29 – 7.07 g m^{-2} (Friedmann, 1980; Friedmann et al., 1980) so that nitrogen is probably not rate-limiting for the prevailing very slow C-turnover (Johnston and Vestal, 1991). Nitrogen for desert endoliths is primarily derived abiotically as nitrates, and ammonia from atmospheric electrical discharges and conveyed to the rock by atmospheric precipitation. The incomplete N-cycle also includes decomposition of organic matter to ammonia, re-assimilation of ammonia, ammonia volatilisation, loss of nitrogen during exfoliation, but rarely biological nitrogen fixation (Friedmann and Kibler, 1980).

2. Cold Deserts

The endolithic habitat in Antarctica has received much attention because of the impetus of NASA research into life detection on Mars following the evaluation of the Viking Program in the McMurdo Dry Valleys (Friedmann, 1982; Wynn-Williams, 1990; Nienow and Friedmann, 1993). Characteristics of three examples of this habitat are given in Table 4 and some are illustrated in Plates 20a and 20b. Cyanobacterial communities in moister parts of Victoria Land have been described by Friedmann et al. (1988), but cyanobacteria also occur in lichen-dominated communities further south. There is an outer zone of melanised lichen also containing the eukaryotic algae *Trebouxia* and *Pseudotrebouxia*. There is often also a lower zone containing the cyanobacteria *Gloeocapsa* and *Chroococcidiopsis* in addition to the dominant eukaryotic algae *Hemichloris antarctica, Stichococcus* sp. and *Heterococcus endolithicus* plus hyaline fungi (Plate 21a). The biomass of lichen-dominated communities, derived from analysis of lipid phosphate in sandstone from Linnaeus Terrace (Vestal, 1988a), is locally substantial (2.54 g C m^{-2}, range 1.92 to 3.26). This is a significant potential contribution of carbon available for heterotrophic activity in the oligotrophic desert soil and a source of microbial inoculum released by exfoliation. Distribution is primarily local, but katabatic winds can distribute relatively large pieces of rock over many kilometres. It is important to note that the primary productivity of these eukaryotic and prokaryotic phototrophs supports large and diverse endolithic populations of heterotrophic bacteria (Hirsch et al., 1988; Siebert et al., 1996).

V. Stress Factors in Arid Regions

A. Water Deficiency

Of all the factors associated with desert survival, water availability is paramount. In a comprehensively review of prokaryotic desiccation tolerance, Potts (1994) emphasised the remarkable abilities of cyanobacteria to survive beyond the limits of extinction of other desert taxa. In a more focused study of *Chroococcidiopsis* and *Chroococcus*, Potts and Friedmann (1981) found consistent differences in response to water stress and suggested that this may be a physiological property of cyanobacteria of taxonomic value. They concluded that *Chroococcidiopsis* is probably the most desiccation-resistant cyanobacterium in existence. However, it is absent from the high-altitude (>2400 m) hypolithic desert pavement habitat of the Atacama desert, although the geological-edaphic conditions are similar to other areas of its occurrence. This desert is renowned for its virtual absence of precipitation (< 200 mm y^{-1}, Grosjean et al., 1995) except for a 10-year cycle associated with the El Niño Southern Oscillation oceanic current. It is possible that prolonged drought, combined with the thermal stresses of extreme diurnal ground temperature ranges (c. - 20° to + 40°, (P.O. Montiel, pers. comm.) and a lack of opportunities for dispersal to suitable micro-habitats, is excessive for colonisation by even this hardy organism (Friedmann, 1993).

The extreme tolerance of cyanobacteria to desiccation is exemplified by their ability to tolerate very low water potentials. Water potential (ψ) is the difference in free energy between the system and pure water at the same temperature and is mathematically described as:

$$\psi = 1065.T.\log {}^{p}\!/_{p_0}$$

where:

T = temperature (°K)
p = vapour pressure of the solution
p_0 = vapour pressure of water at the same temperature.

Values of ψ are expressed as negative values in kilo Pascals (kPa) where 100 kPa = 1 bar, and for reference, saturated $CaCl_2$ has a ψ = - 168 MPa. *Crinalium episammum* can survive at a water potential of – 400 MPa and *Nostoc* is typically exposed to desert environments at – 100 MPa (de Winder, 1990; Potts, 1977 unpubl., cited in Potts, 1994). *Chroococcus* and *Chroococcidiopsis* can both fix CO_2 at remarkably low water potentials, below - 3 MPa (Potts and Friedmann, 1981). The mean lowest water potential recorded in Antarctic endolithic habitats is – 66 MPa (Nienow and Friedmann, 1993) whilst the equivalent in hot desert rocks colonised by *Chroococcidiopsis* is - 6.9 MPa (Palmer and Friedmann, 1990). The equivalent stress limits in hot crusts are exemplified by - 2.8 MPa for photosynthesis and - 1.8 MPa for growth in *Microcoleus* (Brock, 1975). However, although *Chroococcidiopsis* could survive low water potentials down to c. – 20 MPa for 24h, it became metabolically inert after 72h at c. – 7 MPa (Potts and Friedmann, 1981).

Water is supplied to desert communities in three forms: precipitation, condensation or water vapour. Limited seasonal precipitation occurs in both hot and cold deserts, either as rain or snow. Between 1962 and 1967, the mean rainfall (± % coefficient of variation, CV) at Avdat in the Negev Desert was 9.45 cm (± 55%) during a mean of 22 (± 37%) rainy days per annum. The corresponding mean annual dew-fall was 3.3 cm (± 15%) during 195 nights (± 9%), giving a mean of 0.17 mm (± 12%) (Friedmann and Galun, 1974). The source of water in Antarctic deserts is melting snow (Friedmann, 1978; Friedmann et al., 1987) but its accumulation depends on slope, exposure and wind-blown drifting. Snowfall occurs three or four times a month on Linnaeus Terrace. It may total 10 cm per annum in the valleys and somewhat more at the higher elevations of Linnaeus Terrace (Thomson et al., 1971). Once melt-water has entered porous rocks, it can be retained for considerable periods of time. Kappen (1981) showed that endolithic rocks at Linnaeus Terrace which usually contained 0.05 to 0.26% water contained 0.45 to 1.15% water five days after snowfall, because the potential evaporation rate is less than 1 – 2 mm^{-2} day^{-1} and the surface crust restricts diffusion. Friedmann et al. (1987) showed a correlation of snowfall with internal rock humidity by continuous nanoclimate recording from 1984–1986. However, this water is not available for metabolism continuously because of diurnal freezing which is less likely to be a constraint in hot deserts.

Lange et al. (1992) have shown that dew-fall is ample for the activation of cyanobacterial desert crusts in the Negev desert. Photosynthesis of crusts continued until the water content of the crusts fell below 0.3 – 0.2 mm, which is close to the mean 0.17 mm derived from data in Friedmann and Galun (1974). Using microprobe technology, Garcia-Pichel and Belnap (1996) showed that both photosynthesis and respiration in cyanobacterial crusts from Utah Desert started within minutes of re-wetting by precipitation or condensation. Benthic mats of *Nostoc* in moraine ponds on the McMurdo Ice Shelf that are seasonally freeze-dried recover respiratory activity rapidly on rehydration, although recovery of nitrogenase activity is much slower than that of photosynthesis (Hawes et al., 1992).

Palmer and Friedmann (1990) found that the Negev Desert cryptoendolithic cyanobacterial community photosynthesises only at matric water potentials approaching that of liquid water. It is therefore significant that photosynthesis by *Chroococcidiopsis*, a dominant cyanobacterium from Nubian sandstone rocks in the Negev desert, ceased at relatively benign water potentials (Potts and Friedmann, 1981). However, this organism was paradoxically tolerant of extreme drought, and recovered very quickly on re-wetting. A rapid response to seasonally periodic water supplies and utilisation of transient dew is a characteristically cyanobacterial response to hot desert hydrology. This may account in part for cyanobacterial dominance in moister desert niches such as Timber Peak and Battleship Promontory relative to Linnaeus Terrace and East Beacon in Victoria Land (Table 4) which are dominated by eukaryotic algae better able to utilise water vapour. Brock (1975) found that *Microcoleus sociatus* from hot desert crusts became inactive under relatively low water stress. Lange et al. (1994a) subsequently confirmed that, unlike *Trebouxia* which is a dominant algal symbiont of endolithic lichens, *M. sociatus* did not show net photosynthesis unless liquid water from dew was available. Water vapour was not sufficient. In contrast, endolithic lichens dominated by eukaryotic algae in Beacon sandstone at Linnaeus Terrace incorporate CO_2 using water vapour alone at a water potential as low as – 46 MPa (70% RH at 8°C) (Palmer and Friedmann, 1990).

B. Desiccation and Wet-Dry Cycles

The physiological effects of desiccation on cyanobacteria are similar to those in diverse prokaryotic and eukaryotic organisms because it affects cell turgidity and the active sites and structural integrity of membranes, proteins, and other vital biomolecules (Potts, 1994). The initial effect of dehydration is to increase osmotic stress so that the concentration of cytoplasmic solutes must increase as a protective response to the decrease in external water potential. Cyanobacteria respond to the stress by producing "compatible" solutes or osmoprotectants which can accumulate to high concentrations without a detrimental effect on enzyme efficiency or metabolic activity. Their colligative properties provide protection against loss of turgidity. Inorganic solutes include K^+, such as in halophilic *Aphanothece halophytica* (Miller et al., 1976). Organic solutes include polyols of low molecular weight such as glycerol, mannitol and sorbitol found in diverse prokaryotic cells (Jennings and Burke, 1990) but not significantly in cyanobacteria. Cyanobacterial cells are probably too permeable for small molecules such as glycerol to be functional in turgor regulation (Walsby, 1980). Larger disaccharide sugars such as trehalose, sucrose and glucosylglycerol are the prevalent compatible solutes in water-stressed terrestrial cyanobacteria (Reed et al., 1984) whilst trehalose has additional functions discussed below. Compatible solutes are especially vital for cyanobacterial survival in saline desert evaporite soils such as those in the Vestfold Hills (Broady, 1986) and elsewhere (Oren, this volume, chapter 10) when they are under the combined stress of desiccation and hypersaline conditions. In these habitats, glycine betaine is the most common compatible solute, as in *Synechocystis* sp. from stromatolites at Al-Khiran (Kuwait) and *Aphanothece halophytica* from Solar Lake, Sinai (Warr et al., 1988).

In his detailed analysis of prokaryotic desiccation tolerance, Potts (1994) describes a "preferential exclusion mechanism", whereby compatible solutes can protect the shell of water round proteins to prevent their denaturation and inactivation under intermediate osmotic stress. However, under extreme water deficit, only polyhydroxyl compounds such as the disaccharides sucrose and trehalose can prevent denaturation of proteins by replacing the shell of water around the macromolecules. This function is described by a "water replacement hypothesis" (Crowe et al., 1992) which is applicable when there is insufficient water available to describe any osmotic interactions (Potts, 1994). Although trehalose is a compatible solute (Csonka and Hanson, 1991), it has more important role in stabilising membranes and depressing their phase transition processes during extreme anhydrobiosis (Crowe et al., 1984, 1992). This results from its ability to replace water lost from the membrane lipid bilayer during desiccation by the formation of hydrogen bonds between its hydroxyl groups the phosphate heads of membrane phospholipids. This situation arises when desert organisms enter a dormant anhydrobiotic state until sufficient water becomes available for re-hydration and resumed metabolic activity (Crowe, 1992; Crowe and Crowe, 1992). The details of these dehydration and rehydration mechanisms, reviewed by Crowe et al. (1992) and Potts (1994), are outside the remit of this review. However, the prevalence of disaccharide sugars (trehalose and sucrose) in desert strains such as *Chroococcidiopsis* sp. (Hershkovitz et al., 1991; Friedmann and Ocampo-Friedmann, 1985) supports their role in desiccation tolerance, especially in cold deserts when they are also needed as cryoprotectants (Montiel and Cowan, 1993).

Accumulations of disaccharide sugars are common in desert cyanobacteria. Under experimental water potential stress of c. – 5 MPa, desiccation resistant strains of *Phormidium autumnale* from the Dan area sewage plant, Israel, and *Chroococcidiopsis* sp. from desert sandstone from Makhtesh Ramon, Israel, accumulated up to 6.2 µg and 3.2 µg trehalose μg^{-1} chlorophyll respectively (Hershkovitz et al., 1991). They also accumulated 6.9 µg and 4.1 µg sucrose μg^{-1} chlorophyll respectively. At higher water potential, the concentration decreased, probably because metabolic activity was being disrupted by the stress itself. Non-drought resistant strains (*Plectonema boryanum* and a *Synechococcus* strain) did not accumulate either sugar under the same conditions.

C. Temperature

The growth and productivity of desert cyanobacteria depends not only on moisture but also on temperature and light. The role of temperature is connected with that of water because of the freezing transition. However, the presence of compatible solutes extends their metabolic window to sub-zero temperatures and Kappen and Friedmann (1983) found that metabolic activity occurred when solar radiation absorbed by the rock warmed it to above - 10°C. Most ecophysiological studies of Antarctic cryptoendolithic ecosystems have concerned lichen communities at Linnaeus Terrace. Based on CO_2 exchange over three years, a gross productivity of 1215 mg C m^{-2} y^{-1} was determined for a horizontal rock, giving a net photosynthetic gain of 606 mg C m^2 y^{-1} (Friedmann et al., 1993). However, the net ecosystem productivity (annual accretion of cellular biomass) amounted to only c. 3 mg C m^{-2} y^{-1}. The difference between the two values was attributed to long-term metabolic costs of the frequent dehydration-rehydration and freeze-thaw cycles or of over-wintering survival costs. The annual gross productivity of cryptoendolithic microbial community of the entire McMurdo Dry Valleys was estimated at 120000 – 180000 kg C, of which 600 – 900 kg C is in microbial biomass. The residue would be soluble organic compounds (such as oxalic acid) which leach into the rocks and/or are lost to the system during exfoliation and transient flushing by melt-water. These data are syntheses from detailed studies of endolithic temperature regimes monitored continuously *in situ* at Linnaeus Terrace and elegantly computer modelled in the laboratory (McKay and Friedmann, 1985; Friedmann et al., 1987; Nienow et al., 1988a). These studies showed the black body absorption effect of rock or soil relative to air temperatures. They also show the buffering effect of the endolithic habitat relative to the extremes of surface temperature which may cross the freezing threshold several times a minute on a day of broken cloud. The warming of the soil or rock prolongs the growing season relative to air temperatures in spring and autumn.

Unlike the psychrophilic strains of algae *Trebouxia* and *Hemichloris antarctica* they obtained from soil and rocks at Linnaeus Terrace, Ocampo-Friedmann et al. (1988) found that the cyanobacterium *Chroococcidiopsis* sp. from a Beacon Valley endolithic community was psychrotolerant. It grew well between 12.5°C and 25.0°C with maximum growth at 17.5°C, and the authors suggested that it may belong to the same species as found in hot desert cryptoendolithic communities. Endolithic carbon turnover time at Battleship Promontory (cyanobacteria-dominated) and Linnaeus Terrace (lichen-dominated) was calculated from the incorporation of C into phospholipids by Vestal (1988b). He noted that photosynthesis in the presence of liquid water was 19-fold greater than in the desiccated state. However, the increase was only

11-fold in the presence of water vapour alone. This is consistent with his detection of cyanobacteria (which require liquid water) in the community as well as eukaryotic phycobionts. Johnston and Vestal (1991) found that lichen communities had psychrophilic temperature responses (maximal rate of 4.5 ng C m^{-2} h^{-1} at 10°C) whilst cyanobacteria communities from Battleship Promontory had maximal rates of 3.0ng Cm^{-2} h^{-1} at 2 – 30°C). From these data, they estimated the Linnaeus Terrace endolithic C turnover time to be 17000 years in lichen communities and 19000 years in endolithic cyanobacterial. However, the lichen rate is likely to be nearer to 10000 years because they did not include photosynthetic activity below 5°C where at least half the lichen photosynthetic gain is produced (Friedmann et al., 1993). The equivalent correction factor for the Battleship Promontory cyanobacteria has not been determined but may well be less because of their higher maximal growth temperature. Nevertheless, the Antarctic cold desert endolithic ecosystem contains probably the slowest growing cyanobacterial community on Earth.

In a hot desert endolithic lichen community in the Negev, Lange et al. (1992) showed that after dew moistening in the field, daily maximum net photosynthesis was 1.11 μmol CO_2 m^{-2} s^{-1} (183.8 mg m^{-2} h^{-1}). This is nearly 400 times the maximum measured rate (ranging from 0.22–0.78 mg CO_2 m^{-2} h^{-1}) for Antarctic endolithic lichens at Linnaeus Terrace (Friedmann et al., 1993). However, much of the hot desert photosynthate is used to support heterotrophic activity in the community at night, which does not occur in Antarctic endolithic communities. For comparison, a *Microcoleus*-dominated soil crust from the same desert region has a mean (± CV) daily maximum net photosynthesis of 0.824 μmol CO_2 $m^{-2}s^{-1}$ (136.4 mg CO_2 m^{-2} h^{-1} ± 20.3%) (Lange et al., 1992). Garcia-Pichel and Belnap (1996) used rapid-response fibre-optic probes and microelectrodes to monitor microscale environmental gradients and associated biomass and primary productivity of two desert soil crusts from Utah, USA. A red crust dominated by *Microcoleus vaginatus* and *Nostoc* sp (sparse) with a dry biomass of 275 mg m^{-2} (6.3 mg Chl_a m^{-2}) had a community net productivity (± SE) of 1.48 ± 0.7 mmol O_2 m^{-2} h^{-1} whilst a denser black crust (biomass 1374 mg m^{-2}, Chl_a 31.6 mg m^{-2}) containing abundant *M. vaginatus*, *Nostoc* sp. and *Scytonema* sp. produced 4.21 ± 0.22 mmol O_2 m^{-2} h^{-1}. Considering the different methodologies used, these values are remarkably similar to those of Lange et al. (1992) who found the equivalent of 2.6 – 4.0 mmol O_2 m^{-2} h^{-1}.

Another assessment of microbial activity and biomass in desert ecosystems is based on the occurrence and preservation microbially-generated ATP, especially in rocks. Although the conversion of ATP to biomass is debatable, calculation of the relative amounts of ATP in different desert habitats can make a valuable comparison of colonisation and activity. Friedmann et al. (1980) found ATP levels of 15.7 and 9.0 mg m^{-2} (mean ± CV = 12.35 mg ± 38%) in two samples of cryptoendolithic cyanobacteria from University Valley, Beacon Heights. The comparable mean for 14 samples of nearby endolithic lichens were 7.18 mg m^{-2} (± 70%), ranging from 0.7 to 14.2 mg. It was evident that high levels of chlorophyll a correlated with high levels of ATP and high estimates of organic matter derived from Kjeldahl nitrogen analyses. In four endolithic cyanobacterial communities, mean values of organic matter were 50.19 g m^{-2} (± 24%), compared with 83.37 mg m^{-2} (± 52%) in lichen communities with a range of 46.0 to 176.5 mg m^{-2}. There was evidently less microbial biomass/activity in cyanobacterial endolithic ecosystems. Subsequent work by Tuovila and LaRock (1987) showed that the ATP assayed in endolithic communities was within living material and not preserved free ATP. In six lichen-dominated endolithic samples, they found somewhat lower levels of 1.48 mg ATP m^{-2} (± 83.5%) with a range of 0.05 to 3.09 mg. They estimated this to be equivalent to 0.369 g organic C m^{-2}.

D. Light and UV-radiation

1. Photosynthetically-Active Radiation

From the earliest ecophysiological studies of desert endolithic communities, light has been considered to be the prime factor for stratification (Friedmann, 1971). Friedmann et al. (1987) showed that light, temperature and moisture conditions were favourable for microbial metabolism for between 375 to 700h y^{-1} at Linnaeus Terrace. Field data and samples from Linnaeus Terrace have been used to validate a light regime model (Nienow et al., 1988b). They concluded that two aspects of light flux regulated stratification: excessive light results in photo-oxidative bleaching; whereas at the lower end, there is insufficient light energy to sustain photosynthetic activity and growth. It is known that dark pigments in the walls of hot desert fungi and in the sheaths of their eukaryotic phycobionts protect against photo-bleaching (Friedmann and Galun, 1974). This is analogous to the decrease in sheath pigmentation with depth in endolithic cyanobacteria as seen in the Negev Desert (Friedmann, 1971). However, *Chroococcidiopsis* forms monospecific endolithic populations near the surface of hot desert rocks (Friedmann and Ocampo-Friedmann, 1976) and is therefore unlikely to be inhibited by photo-oxidation in cold desert endolithic communities.

Estimated annual photon fluxes calculated by Nienow et al. (1988b) from field data (Friedmann et al., 1987) show the steepness of the light gradient penetrating the different zones. Under dry conditions at temperatures above - 10°C, an annual surface receipt of 4000 mol photons m^{-2} decreases to 250 mol at the upper boundary of the pigmented fungal zone, 37 mol at its lower boundary, 2.0 mol at the upper boundary of the *Hemichloris antarctica* zone, and 0.16 mol at its lower boundary where cyanobacteria might be expected. More light penetrates the rock when it is wet, but even then, the photon flux at the bottom of the *Hemichloris* zone only reaches 0.89 mol m^{-2} y^{-1}. The photon flux of 0.05 – 1 µmol m^{-2} s^{-1} (dry or even 2 – 10 µmol m^{-2} s^{-1} when wet) is below the compensation point of most algae. However, cyanobacteria are typically capable of shade-adaptation and *Phormidium frigidum-Lyngbya martensiana* communities can fix carbon significantly (8.20 mg C m^{-2} h^{-1}) at a photon flux of only 1.5 µmol m^{-2} s^{-1} in the benthic prostrate mats of ice-covered Lake Hoare in the McMurdo Dry Valleys (Parker and Wharton, 1985).

Endolithics at Linnaeus Terrace becomes PAR-saturated at 200 – 250 µmol m^{-2} s^{-1} (Vestal, 1988b). This is very similar to the 200 – 300 µmol m^{-2} s^{-1} saturation level for endolithic cyanobacteria in hot Tshipise Sandstone in the Northern Province, South Africa (Weber et al., 1996). However, in microphytic soil crusts of the Negev, saturation does not occur until between 500 and 700 µmol m^{-2} s^{-1} (Lange et al., 1992). Of the light striking the surface of desert sandstone in savannah desert regions of South Africa, only 1.8 – 2.5% reaches the cyanobacteria zone (Büdel and Wessels, 1991). Even less light, only 0.1 – 1.4%, reaches the cyanobacteria in the McMurdo Dry Valleys (Friedmann et al., 1987). This suggests that the hot desert crust may require greater shading from photo-bleaching than an endolithic community. Potts et al. (1987) have suggested that carotenoids, as found in *Nostoc* spp. from diverse geographical locations, may well provide part of this protection. As well as its pigmented inhabitants, the fabric of the rock itself provides protection (Wynn-Williams, 1989). However, in crusts and benthic mats mutual shading by pigmented cyanobacteria is an alternative method of attenuating excessive light (Häder, 1987). This behaviour not only screens out total photon flux but is also selective for certain wavelengths, leaving PAR to penetrate to lower depths. The most important selective effect is the attenuation of UVB radiation which is damaging, not only to photosynthetic systems, but also to functional molecules such as DNA and functional proteins (Karentz, 1994).

2. Ultra-Violet Radiation

Much has been published on the effects of UVB on cyanobacteria ((Wynn-Williams, 1994; Garcia-Pichel, 1998) but little on the effects of UVB on desert communities. However, cyanobacteria related to desert strains have been studied for their production of UV-absorbing pigments. These include scytonemin in the EPS sheath of cyanobacteria such as *Nostoc commune* (Garcia-Pichel and Castenholz, 1991; Ehling-Schultz et al., 1997) and mycosporine-like amino acids (MAAs) within the cells of *Gloeocapsa* spp. (Scherer et al., 1988; Garcia-Pichel et al., 1993, 1993). Scytonemin has also been found in yellowish-brown coloured *Calothrix parietina* in *Microcoleus* crusts on dune soils of the Nizzana district, Negev Desert (Lange et al., 1992). Garcia-Pichel and Belnap (1996) concluded that the investment which surface-bound cyanobacteria make in synthesising scytonemin indicates the ecological importance of colonising of such extreme environments. In polar deserts there will be metabolic competition for biochemical resources to deal with environmental stresses. During the late winter "Ozone Hole", some of the carbon allocated to cryoprotectants may need to be retrieved for synthesis if UVB-protectants such as scytonemin or MAAs. This is an aspect which requires further research.

VI. Bio-Weathering and Nutrient Availability

A feature of rocks colonised by endolithic microbes is the characteristic mosaic effect resulting from exfoliation (Friedmann and Ocampo-Friedmann, 1988). Part of the mechanism of this process depends on the physical effects of heating and cooling and the expansion and contraction of condensed water as it crystallises within the fabric of the rock (Campbell and Claridge, 1987). However, the scarcity of liquid water in endolithic communities makes freeze-thaw disruption an almost negligible factor. The absence of cleavage planes in sandstone makes crumbling into individual grains their main disintegration response. Sandstone layers, as distinct from cleavage planes, are however split apart by the freeze-thaw action of snow melt-water and wind erosion. This is on a different scale from the exfoliation apparently associated with endolithic communities. Thermal disruption is unlikely to be a prime cause of endolithic exfoliation in sandstone since Cooke and Smalley (1968) concluded that expansion and contraction of rock minerals as a result of daily temperature changes in deserts are, by themselves, inadequate causes of rock disintegration. Salt disintegration is likewise unlikely to be influential in the stratified weathering observed in these elevated substrata distant from seepage and salt-accumulation zones. Observations of the exfoliation of endolithics shows that they tend to split along the layer containing the microbial community which suggests that this is a zone of influence and structural weakness.

To investigate the solublising action of minerals by endoliths (chemical weathering), Johnston and Vestal (1989) used dental picks to separate the of zones of endolithic communities dominated by cyanobacteria from Battleship Promontory (BP) and lichens at Linnaeus Terrace (LT) for inorganic analysis. The outer crust and inner bedrock zones at BP contained characteristically much less iron than at LT (c. 200 and 2000 $\mu g\ g^{-1}$ rock respectively), The BP microbial zone contained much more calcium (c. 340 $\mu g\ g^{-1}$) and magnesium (c. 250 $\mu g\ g^{-1}$) than that of LT (c. 140 and 50 $\mu g\ g^{-1}$ respectively). Both contained similar amounts of total phosphorus and potassium (c. 230 and 200 $\mu g\ g^{-1}$ respectively). The BP cyanobacterial zone was much less acid (pH7.95) than that of LT (pH4.85) which had a much lower Fe content (c. 625 $\mu g\ g^{-1}$) than an accumulation zone beneath (c. 2200 $\mu g\ g^{-1}$). They suggested that the Fe had been leached into this zone by the chelative effect of oxalate produced by the endolithic community. To investigate this further, they analysed the distribution of oxalate extracted from different zones of the endolithic community at LT (Johnston and Vestal, 1993). They concluded that oxalate was dissolving the sandstone cement in the lichen-dominated zone and compounding the weathering effect by complexing with Si, Fe and Al.

Russell et al. (1998) used non-destructive FT-Raman spectroscopy (Edwards and Seaward, 1983) to demonstrate the spatial distribution of oxalate in different hydration states within the zones of East Beacon endoliths *in situ* in field-fresh samples. This confirmed oxalate accumulation in the pigmented and hyaline fungal lichen zone with a small amount accumulating in the underlying abiotic zone. Biochemical weathering by oxalate (Weed and Norton, 1991), the physical effect of the growth of microbial tissue (Friedmann and Weed, 1987) and changes in turgor of the cells as they absorb or lose moisture are likely to be major contributors to exfoliation. The expansion of EPS sheaths of cyanobacteria constrained in rock pores during rehydration (Potts, 1994) may exert an additional pressure or lubricating effects on surrounding rock grains. Exfoliation mediated by endolithic cyanobacteria and algae not only biodeteriorates bedrock to create soil but also enriches it with organic nutrients. Concurrently, the process inoculates the region with a microflora which has the potential to develop into a soil community at more favourable sites down-wind.

VII. Thresholds and Extinction

A. Terrestrial Limits of Life

Cyanobacterial life in deserts is poised at the limit of survival. Its existence is threatened by a scarcity of water and its enzymes are stressed their structural and physiological limits. Its membranes have to tolerate wide ranges of fluidity and it is now under increasing threat from elevated UVB flux because of ozone depletion. There are endolithic microbial sites at the edge of the Antarctic ice sheet where survival thresholds have been exceeded and there is only evidence of former endolithic life. This is apparent from "fossil" communities (Friedmann and Weed, 1987) which are revealed by the leaching patterns of their acids which mobilise iron and other minerals (Johnston and Vestal, 1989). To establish the limits of survival (threshold of extinction) for cryptoendolithic communities, Friedmann et al. (1994) established a transect of monitored sites containing Beacon sandstone potentially capable of supporting life in the McMurdo Dry Valleys desert region. They found a sequential decrease in mean air and rock temperatures at Battleship Promontory,

Table 4. Physical and structural descriptions of typical endolithic habitats in southern Victoria Land.

	Linnaeus Terrace[1]	East Beacon	Battleship Promontory
Latitude (S)	77°35'	77°50'	76°54'
Longitude (E)	161°05'	160°53'	160°57'
Altitude (m)	1650	2200	2000
Distance to Polar Ice Sheet (km)	20	15	8
Aspect, facing	north	north	north
Exposure	Sheltered	Sheltered	Sheltered
Grain size[2]			
% at feret dia. < 250 µm	na	37	82
250-500 µm	(200-500 µm)	32	17
<500 µm	na	31	1
Mean circularity	na	0.665	0.644
Mean elongation	na	3.39	3.74
Porosity (%)	8-15	6-7[3]	na
Moisture (range and mean) (%)	0.05-0.26[3]	0.13-1.12 (0.41)[4]	na
Bedrock composition[5] Total (µg g-1) [6]			
Fe	1623-1587	+	164-156
Ca	70-48	+	148-231
K	507-454	++	373-390
Water-soluble (µg g^{-1})			
Fe	8-50	na	<10
Ca	18-1	na	93- <10
K	10-18	na	15-20
NE-facing rock surface temp. (°C)[7]			
Winter mean (June)	-32.1 to 32.4	na	na
Summer mean (January)	-1.4 to -2.2	na	na
Summer maximum (January)	+15.3 to 17.2	na	na
Total hours per annum at >+5°C	210-270	na	na
Mean annual snowfalls, 1984- 85	12.5	na	na
Endolithic %R.H. range Jan-Feb 1984	50-100	na	na
Light (PAR) flux (µmol $m^{-2}s^{-1}$)		na	na
Maximum at surface	1858	c. 1800	c. 1800
Minimum in algal zone	0.05-10	na	na
Annual at surface (mol $m^{-2}s^{-1}$)	3500	na	na
Annual in algae (mol $m^{-2}s^{-1}$)	0.15	na	na

[1] Data are from Johnston & Vestal (1989, 1993) and Nienow & Friedmann (1993)

[2] Based on image analysis of gently crushed rock (author, unpublished)

[3] Kappen et al. (1991)

[4] Data from Greenfield (1988)

[5] Where used: + = present; ++ = abundant

[6] Data from Johnston and Vestal (1989)

[7] Data for 1984-86 from Friedmann et al. (1987)
na = data not available or not applicable

Linnaeus Terrace (see Table 4), Tyrol Valley, (77°35'S, 160°38'E), Mount Fleming (77°33'S, 160°06'E) and Horseshoe Mountain (77°34'S, 159°57'E) which correlated with a gradual extinction of endolithic life. All communities were alive at Battleship Promontory and all were fossilised at Horseshoe Mountain. The threshold of survival-extinction appeared to be at Mount Fleming where the annual mean rock temperature was -24.2°C (max. -4.7°C, min. - 41.8°C) at a mean annual RH of 50% (max. 90%, min. 50%). Data for the cyanobacteria-dominated community at BP were incomplete for technical reasons. However, comparisons of mid-summer data (January 1993) showed Battleship Promontory mean rock temperature to be - 4.4°C (max. + 9.9°C, min. - 14.2°C) at mean RH 54% (max. 96%, min. 25%). Concurrent conditions at Mount Fleming were: mean temp. - 6°C (max. + 5.3°C, min - 19.2°C) at mean RH 52% (max. 90%, min. 27%) and at Horseshoe Mountain were: mean temp. - 10.2°C (max. 7.3°C, min - 22.1°C) at mean RH 55% (max. 85%, min. 27%). Although life processes take place at temperatures above - 10°C (Friedmann et al., 1993), the number of hours per year that the rock temperature was above 0°C was especially influential in defining the viability threshold on the transect.

B. Exobiology

The trends shown in Fig. 2 indicate that the Antarctic desert biome abuts the "Mars biome", and endolithic research relates closely to exobiology. Trace fossils that are 70000 to 2 – 4 million years old in Antarctic sandstone provide evidence of former endolithic activity, despite extinction (Friedmann, 1986; Nienow and Friedmann, 1993). Microscopy of Early Archean Apex Chert of north-western Western Australia Schopf (1993) has shown that trichomic cyanobacterium-like micro-organisms were extant and morphologically diverse at least as early as c. 3.5 million years ago (Mya). When these findings are compared with a model for the history of water on Mars generated by McKay et al. (1992), it seems possible that micro-organisms (potentially

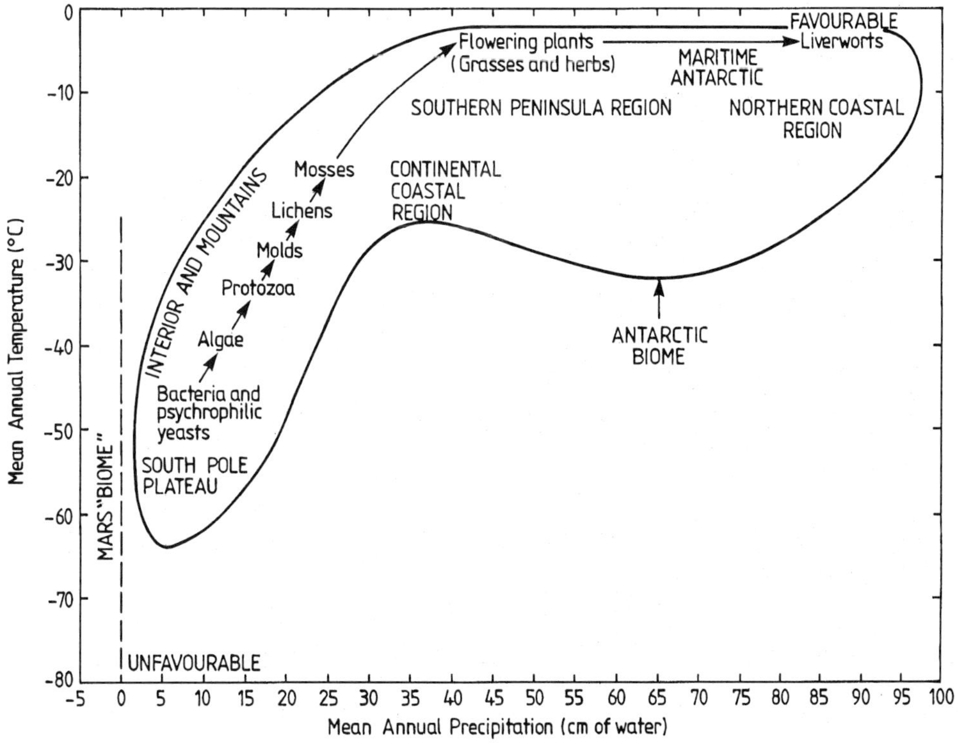

Fig. 2. Diagram of cold desert biomes and more benign habitats resulting from the interaction between mean annual temperature and mean annual precipitation in Antarctica, and projected for Mars. [from Wynn-Williams, 1990, after Cameron et al., 1976, with permission]

cyanobacteria analogous to those surviving at the limits of life in Antarctic deserts) may have developed on Mars.

Based on geomorphological data from the NASA Mariner Mars orbiter (Carr, 1996), the Mars hydrology model indicates four epochs: Epoch I (4.2 – 3.8 Gya) with abundant water (seas and rivers), volcanoes, meteorites and CO_2 at atmospheric pressure (P) ~ 5 atm, temperature (T) > 0°C, with the possibility for the origin of life (including cyanobacteria); Epoch II (3.8 – 3.1 Gya) with ice - covered lakes analogous to those of Antarctic Dry Valleys containing stromatolitic cyanobacterial mats (Wharton, 1994), mean T < 0°C, peak T >0°C with carbonates and dissolved gases; Epoch III (3.1 – 1.5 Gya) with liquid water constrained to the interior of porous rocks, analogous to current Antarctic cold desert endolithic communities at peak T <0°C at P>> 7.4 mb; Epoch IV (1.5 – present) with no liquid water at the surface but with possible microbial life in polar permafrost and in geothermally-warmed deep groundwater, P< 7.4 mb at triple point of water, T << 0°C. As the life-supporting environment deteriorated through to the current extreme desert, putative microbial colonists would have gone through stages of stress resistance until all available biochemical and behavioural strategies failed (Friedmann and Koriem, 1989). They would then have died and degraded, although indicator biomolecules such as porphyrins may have survived if protected from oxidation. These are biochemical fossils.

It is unlikely that viable cyanobacteria have survived, even in an anhydrobiotic state, near the surface of Mars (Crowe and Crowe, 1992). However, sedimentary rocks from Epoch III may contain fossil endolithic communities whose biochemicals may still currently be detectable (Brack et al., 1997), if they are preserved below the zone of degradation by solar radiation and peroxides in Martian soil (Zent and McKay, 1994; Stoker and Bullock, 1997). Since the Viking missions which occurred before the recognition of Epoch III and potential endolithic habitats on Mars (Cameron, 1976; Klein, 1979), the occurrence and viability of endolithic cyanobacteria of Antarctic cold deserts have provided valuable analogues for planning future life-detection experiments for Mars missions (McKay, 1986; McKay, 1997). Friedmann and Ocampo-Friedmann (1994) have even proposed that a *Chroococcidiopsis* would make a suitable pioneer micro-organism for terraforming Mars.

Cyanobacteria are fundamentally important colonists of hot and cold deserts. Their role in xeric ecosystems is a result of their remarkable resistance to desiccation, their tolerance of extreme temperature regimes and their tolerance of extremely high solar radiation whilst being capable of photosynthesis in near darkness. Their production of compatible solutes makes them tolerant of osmotic stresses resulting from desiccation, freezing and extreme salinity. Their synthesis of protective pigments conveys protection against increasing levels of UVB. These, and other survival mechanisms give them the ability to metabolise, albeit slowly, at the limits of survival near the edge of the Antarctic ice sheet which is the closest to a Martian environment on Earth (Wynn-Williams, in press). Cyanobacteria, which are amongst the most primitive organisms on Earth and probable early inhabitants of the "primordial soup", are now being studied for their survival potential in the ultimate desert - Mars.

Acknowledgements

I thank the British Antarctic Survey (BAS), the United States Antarctic Programme (especially Prof. Diana Freckman), US National Science Foundation (especially Dr Polly Penhale) and the Italian Antarctic Programme (ENEA, especially Ing. Mario Zuccelli) for opportunities to work in deserts of Antarctica. I especially thank Dr David Walton (BAS) for his encouragement and support in this line of research. I am grateful to Dr Pedro Montiel and the editors for valuable comments on the paper.

References

Anagnostidis K and Komarek J (1985) Modern approach to the classification of cyanophytes. I - introduction. Arch Hydrobiol (Suppl) 71: 291-302

Anagnostidis K and Komarek J (1988) Modern approach to the classification of cyanophytes. 3 - Oscillatoriales. Arch Hydrobiol (Suppl) 80: 327-472

Bargagli R, Broady PA and Walton DWH (1996) Preliminary investigation of the thermal biosystem of Mount Rittmann fumaroles (northern Victoria Land, Antarctica). Antarct Sci 8: 121-126

Belnap J and Gillette DA (1998) Vulnerability of desert biological soil crusts to wind erosion: the influence of crust development, soil texture and disturbance. J Arid Environ 39: 133-142

Billi D, Caiola MG, Paoluzzi L and Ghelardini P (1998) A method for DNA extraction from the desert cyanobacterium *Chroococcidiopsis* and its application to identification of *ftsZ*. Appl Environ Microbiol 64: 4053-4056

Bonani G, Friedmann EI, Ocampo-Friedmann R, McKay CP and Woelfli W (1988) Preliminary report of radiocarbon dating of cryptoendolithic microorganisms. Polarforsch 58: 199-200

Borcheim KY, Bratbak G and Heldal M (1990) Enumeration and biomass estimation of planktonic bacteria and viruses by transmission electron microscopy. Appl Environ Microbiol 56: 352-356

Brack A, Forterre P, Pillinger CT, Horneck G, Schidlowski M and Wänke H (1997) ESA Exobiology Team Study on the Search for Life in the Solar System: Final Report, Internal Report, European Space Centre, Sassenheim, 154 pp

Braune W and Sanke H (1979) Interferometrische Untersuchungen zur Dynamik von Hydratur und Trockenmassegehalt während der Lichtanhängigen Keimung der Akineten von *Anabaena variabilis*. Z Allg Mikrobiol 19: 535-546

British Antarctic Survey (1997) List of Protected Areas in Antarctica, Foreign and Commonwealth Office, London, 33 pp

Broady PA (1981a) The ecology of chasmolithic algae at coastal locations of Antarctica. Phycologia 20: 259-272

Broady PA (1981b) The ecology of sublithic terrestrial algae at the Vestfold Hills, Antarctica. Br Phycol J 16: 231-240

Broady PA (1981c) Ecological and taxonomic observations on subaerial epilithic algae from Princess Elizabeth Land and MacRobertson Land, Antarctica. Br Phycol J 16:257-266

Broady PA (1984) Taxonomic and ecological investigations of algae on steam-warmed soil on Mt Erebus, Ross Island, Antarctica. Phycologia 23:257-271

Broady PA (1986) Ecology and taxonomy of the terrestrial algae of the Vestfold Hills. In: Pickard J (ed) Antarctic Oasis: Terrestrial Environments and History of the Vestfold Hills, pp 165-202. Academic Press, Sydney

Broady PA (1987) A floristic survey of algae at four locations in northern Victoria Land. NZ Antarct Rec 7:8-19

Broady PA (1989a) Broadscale patterns in the distribution of aquatic and terrestrial vegetation at three ice-free regions on Ross Island, Antarctica. Hydrobiologia 172:77-95

Broady PA (1989b) Survey of algae and other terrestrial biota at Edward VII Peninsula, Marie Byrd Land. Antarct Sci 1:215-224

Broady PA (1993) Soils heated by volcanism. In: Friedmann EI (ed) Antarctic Microbiology, pp 413-432. Wiley-Liss, New York

Broady PA (1996) Diversity, distribution and dispersal of Antarctic terrestrial algae. Biodivers Conserv 5:1307-1335

Broady PA, Garrick R and Anderson G (1984) Culture studies on the morphology of ten strains of Antarctic Oscillatoriaceae(Cyanobacteria). Polar Biol 2:233-244

Broady PA, Given D, Greenfield LG and Thompson K (1987) The biota and environment of fumaroles on Mount Melbourne, northern Victoria Land. Polar Biol 7:97-113

Brock TD (1975) Effect of water potential on a *Micocoleus* (Cyanophyceae) from a desert crust. J Phycol 11: 316-320

Büdel B and Henssen A (1983) *Chroococcidiopsis* (Cyanophyceae), a phycobiont in the lichen family Lichenaceae. Phycologia 22:367-375

Büdel B and Wessels DCJ (1991) Rock inhabiting blue-green algae/cyanobacteria from hot arid regions. Alg Stud 64: 385-398

Caiola MG, Billi D and Friedmann EI (1996) Effect of dessication on envelopes of the cyanobacterium *Chroococcidiopsis* sp. (Chroococcales). Eur J Phycol 31: 97-105

Cameron RE (1962) Species of *Nostoc* Vaucher occurring in the Sonoran Desert in Arizona. Am Microsc Soc Trans 82:379-384

Cameron RE (1966) Soils sampling parameters for extraterrestrial life detection. J Arizona Acad Sci 4: 3-27

Cameron RE (1969a) Cold desert characteristics and problems relevant to other arid lands. In: McGinnies WG and Goldman BF (eds) Arid Lands in Perspective, pp 167-205. American Association of Advanced Science, Washington D.C.

Cameron RE (1969b) (ed) Jet Propulsion Laboratory, Technical Report 32-1378: Abundance of microflora in soils of desert regions, California Institute of Technology, Pasadena, 1-16 pp

Cameron RE (1971a) Antarctic soil microbial and ecological investigations. In: Quam LOand Porter HD (eds) Research in the Antarctic, pp 137-189. American Association for the Advancement of Science, Washington D.C.

Cameron RE (1971b) Application of low-latitude microbial ecology to high-latitude deserts. In: Smiley TL and Zumberge JH (eds) Polar Deserts and Modern Man, pp 71-90. University of Arizona Press, Tucson

Cameron RE (1971c) Ecology of Blue-green algae in Antarctic soils. In: Desikachary TV (ed) International Symposium on Taxonomy and Biology of Blue-green Algae, Madras, 1970, pp 353-384. Bangalore Press, Madras

Cameron RE (1972a) Farthest south algae and associated bacteria. Phycologia 11: 133-139

Cameron RE (1972b) Microbial and ecological investigations in Victoria Dry Valley, Southern Victoria Land, Antarctica. In: Llano GA (ed) Antarctic Terrestrial Biology, pp 195-260. American Geophysical Union, Washington, DC.

Cameron RE and Blank GB (1966) (eds) Jet Propulsion Laboratory, Technical Report 32-971: Desert algae: soil crusts and diaphanous substrata as algal habitats, California Institute of Technology, Pasadena, 1-41 pp

Cameron RE and Devaney JR (1970) Antarctic soil algal crusts. A scanning electron and optical microscope study. Trans Am Microsc Soc 80:264-273

Cameron RE, Honour RC and Morelli FA (1976) Antarctic microbiology-- preparation for Mars life detection, quarantine, and back contamination. In: Heinrich MR (ed) Extreme Environments; Mechanisms of Microbial Adaptation, pp 57-82. Academic Press, New York

Campbell IB and Claridge GGC (1987) Antarctica: Soils, Weathering Processes and Environment, Elsevier, Amsterdam, 368 pp

Campbell SE (1979) Soil stabilization by a prokaryotic desert crust: implications for Precambrian land biota. Orig Life 9:335-348

Carr MH (1996) Water on Mars, Oxford University Press, New York, 229 pp

Castenholz RW (1992) Species usage, concept and evolution in the Cyanobacteria (blue-green algae). J Phycol 28:737-745

Cooke RU and Smalley IJ (1968) Salt weathering in deserts. Nature 220: 1226-1227

Coulson S, Hodkinson ID, Strathdee A, Bale JS, Block W, Worland MR and Webb NR (1993) Simulated climate change: the interaction between vegetation type and microhabitat temperatures at Ny Ålesund, Svalbard. Polar Biol 13:67-70

Crowe LM, Mouradian R, Crowe JH, Jackson SA and Womersley C (1984) Effects of carbohydrates on membrane stability at low water activities. Biochim Biophys Acta 769: 141-150

Crowe JH, Hoekstra FA and Crowe LM (1992) Anhydrobiosis. Ann Rev Physiol 54: 579-599

Crowe LM and Crowe JH (1992) Anhydrobiosis: a strategy for survival. Adv space Res 12: 239-247

Csonka LN and Hanson DA (1991) Prokaryotic osmoregulation: genetics and physiology. Ann Rev Micro 45: 569-606

Davis SN (1974) Hydrogeology of arid regions. In: Brown GW (ed) Desert Biology, pp 1-30. Academic Press, New York de Winder B (1990) Ecophysiological strategies of drought tolerant phototrophic micro-organisms in dune soil. PhD Thesis, University of Amsterdam. Amsterdam, The Netherlands

de Chazal NM, Smaglinski S and Smith GD (1992) Methods involving light variation for isolation of cyanobacteria: characterization of isolates from central Australia. Appl Environ Microbiol 58:3561-3566

Downes MT, Hrstich L and Vincent WF (1993) Extraction of chlorophyll and carotenoid pigments from Antarctic benthic mats for analysis by HPLC. J Applied Phycology 5:623-628

Drouet F (1981) Summary of the classification of blue-green algae. Beih Nova Hedwigia 66:133-209

Drouet F and Daily WA (1956) Revision of the coccoid Myxophyceae. Butler Univ Stud 10:1-218

Edwards HGM and Seaward MRD (1993) Raman microscopy of lichen-substratum interfaces. J Hattori Bot Lab 74:303-316

Ehling-Schultz M, Bilger W and Scherer S (1997) UV-B-induced synthesis of photoprotective pigments and extracellular polysaccharides in the terrestrial cyanobacterium *Nostoc commune*. J Bacteriol 179: 1940-1945

Engelskjøn T (1986) Botany of two Antarctic mountain ranges: Gjelsvikfjella and Muhlig-Hofmannfjella, Dronning Maud Land. I. General ecology and development of the Antarctic cold desert cryptogam formation. Polar Res 4:205-224

Ernst A, Chen TW and Böger P (1987) Carbohydrate formation in rewetted terrestrial cyanobacteria. Oecologia 72:574-576

Finegold L, Singer MA, Federle TW and Vestal JR (1990) Composition and thermal properties of membrane lipids in cryptoendolithic lichen microbiota from Antarctica. Appl Environ Microbiol 56:1191-1194

Flaibani A, Olsen Y and Painter T (1989) Polysaccharides in desert reclamation: Compositions of exocellular proteoglycan complexes produced by filamentous blue-green algae and unicellular green edaphic algae. Carbohydrate Res 190:235-248

Friedmann EI (1971) Light and scanning electron microscopy of the endolithic desert habitat. Phycologia 10:411-428

Friedmann EI (1972) Ecology of lithophytic algal habitats in Middle Eastern and North American deserts. In: Rodin LE (ed) Ecophysiological Foundation of Ecosystems Productivity in Arid Zones, pp 182-185. Nauka USSR, Academy of Sciences, Leningrad

Friedmann EI (1977) Microorganisms in Antarctic desert rocks from Dry Valleys and Dufek Massif. Antarct J U S 12:26-29

Friedmann EI (1978) Melting snow in the Dry Valleys is a source of water for endolithic microorganisms. Antarct J U S 13:162-163

Friedmann EI (1980) Endolithic microbial life in hot and cold deserts. Orig Life 10:233-245

Friedmann EI (1982) Endolithic microorganisms in the Antarctic cold desert. Science 215:1045-1053

Friedmann EI (1986) The Antarctic cold desert and the search for traces of life on Mars. Adv space Res 6:265-268

Friedmann EI (1993) Extreme environments, limits of adaptation and extinction. In: Guerrero R and Pedros-Alio C (eds) Trends in Microbial Ecology, pp 9-12. Spanish Society for Microbiology, Barcelona

Friedmann EI, Druk AY and McKay CP (1994) Limits of life and microbial extinction in the Antarctic desert. Antarct J U S 29:176-179

Friedmann EI and Galun M (1974) Desert algae, lichens and fungi. In: Brown GWJ (ed) Desert Biology, pp 165-212. Academic Press, New York

Friedmann EI, Hua MS and Ocampo-Friedmann R (1988) Cryptoendolithic lichen and cyanobacterial communities of the Ross Desert, Antarctica. Polarforsch 58:251-259

Friedmann EI, Kappen L, Meyer MA and Nienow JA (1993) Long-term productivity in the cryptoendolithic microbial community of the Ross Desert, Antarctica. Microb Ecol 25:51-69

Friedmann EI and Kibler AP (1980) Nitrogen economy of endolithic microbial communities in hot and cold deserts. Microb Ecol 6:95-108

Friedmann EI and Koriem AM (1989) Life on Mars: how it disappeared (if it was ever there). Adv space Res 9:167-172

Friedmann EI, La Rock P and Brunson JO (1980) Adenosine triphosphate(ATP),chlorophyll, and organic nitrogen in endolithic microbial communities and adjacent soils in the Dry Valleys of S. Victoria Land. Antarct J U S 15:164-166

Friedmann EI, Lipkin Y and Ocampo-Paus R (1967) Desert algae of the Negev (Israel). Phycologia 6:185-196

Friedmann EI, McKay CP and Nienow JA (1987) The cryptoendolithic microbial environment in the Ross Desert of Antarctica: Satellite-transmitted continuous nanoclimate data, 1984 to 1986. Polar Biol 7:273-287

Friedmann EI and Ocampo R (1976) Endolithic blue-green algae in the Dry Valleys. Primary producers in the Antarctic desert ecosystem. Science 193:1247-1249

Friedmann EI and Ocampo-Friedmann R (1984) Endolithic microorganisms in extreme dry environments: Analysis of a lithobiontic microbial habitat. In: Klug MJ and Reddy CA (eds) Current Perspectives in Microbiology, pp 177-185. Amer Soc Microbiol., Washington.

Friedmann EI and Ocampo-Friedmann R (1995) A primitive cyanobacterium as pioneer microorganism for terraforming Mars. Adv space Res 15:143-246

Friedmann EI and Weed R (1987) Microbial trace-fossil formation, biogenous, and abiotic weathering in the Antarctic cold desert. Science 236:703-705

Fuller WH (1974) Desert soils. In: Brown GW, Jr (ed) Desert Biology, pp 31-101. Academic Press, New York

Garcia-Pichel F (1998) Solar ultraviolet and the evolutionary history of cyanobacteria. Orig Life Evol Biosph 28: 321-347

Garcia-Pichel F and Belnap J (1996) Microenvironments and microscale productivity of cyanobacterial desert crusts. J Phycol 32:774-782

Garcia-Pichel F and Castenholz RW (1991) Characterization and biological implications of scytonemin, a cyanobacterial sheath pigment. J Phycol 27:395-409

Garcia-Pichel F and Castenholz RW (1993) Occurrence of UV-absorbing, mycosporine-like compounds among cyanobacterial isolates and an estimate of their screening capacity. Appl Environ Microbiol 59:163-169

Garcia-Pichel F, Wingard CE and Castenholz RW (1993) Evidence regarding the UV sunscreen role of a mycosporine-like compound in the cyanobacterium *Gloeocapsa* sp. Appl Environ Microbiol 59:170-176

Hunter CN and Mann NH (1992) Genetic manipulation of photosynthetic prokaryotes. In: Mann NH and Carr NG (eds) Photosynthetic prokaryotes, pp 153-179. Plenum Press, New York

Jennings DH and Burke RM (1990) Compatible solutes - the mycological dimension and their role as physiological buffering agents. New Phytol 116:277-283

Johnston CG and Vestal JR (1989) Distribution of inorganic species in two Antarctic cryptoendolithic microbial communities. Geomicrobiol J 7:137-153

Johnston CG and Vestal JR (1991) Photosynthetic carbon incorporation and turnover in Antarctic cryptoendolithic microbial communities: are they the slowest-growing communities on earth? Appl Environ Microbiol 57:2308-2311

Johnston CG and Vestal JR (1993) Biogeochemistry of oxalate in the Antarctic cryptoendolithic lichen-dominated community. Microb Ecol 25:305-319

Kappen L, Friedmann EI and Garty J (1981) Ecophysiology of lichens in the Dry Valleys of Southern Victoria Land, Antarctica. I. Microclimate of the cryptoendolithic lichen habitat. Flora 171:216-235

Karentz D (1994) Ultraviolet tolerance mechanisms in Antarctic marine organisms. In: Weiler CS and Penhale PA (eds) Ultraviolet Radiation in Antarctica: Measurements and Biological Effects, pp . 93-110. American Geophysical Union, Washington, D.C.

Klein HP (1979) The Viking Mission and the search for life on Mars. Rev Geophys Space Phys 17:1655-1662

Lange OL, Kidron GJ, Büdel B, Meyer A, Kilian E and Abelovich A (1992) Taxonomic composition and photosynthetic characteristics of the "biological crusts" covering sand dunes in the western Negev Desert. Functional Ecology 6:519-527

Lange OL, Meyer A and Büdel B (1994a) Net photosynthesis activation of a desiccated cyanobacterium without liquid water in high air humidity alone. Experiments with *Microcoleus sociatus* isolated from a desert soil crust. Functional Ecology 8:52-57

Lange OL, Meyer A, Zellner H and Heber U (1994b) Photosynthesis and water relations of lichen soil crusts: field measurements in the coastal fog zone of the Namib Desert. Functional Ecology 8:253-264

Mazor G, Kidron GJ, Vonshak A and Abeliovich A (1996) The role of cyanobacterial exopolysaccharides in structuring desert microbial crusts. FEMS Microbiol Ecol 21:121-130

McKay CP (1986) Exobiology and future Mars missions: The search for Mars' earliest biosphere. Adv Space Res 6:269-285

McKay CP (1997) The search for life on Mars. Orig Life Evol Biosph 27:263-289

McKay CP and Friedmann EI (1985) The cryptoendolithlic microbial environment in the Antarctic cold desert: Temperature variations in nature. Polar Biol 4:19-25

McKay CP, Friedmann EI, Wharton RA and Davies W (1992) History of water on Mars: a biological perspective. Adv space Res 12:(4)231-(4)238

Miller DM, Jones JH, Tindall DR and Schmid WE (1976) Ion metabolism in a halophilic blue-green alga, *Aphanothece halophytica*. Arch Mikrob 111:145-149

Miotke F-D (1985) Die Dünen im Victoria Valley, Victoria-Land, Antarktis ein Beitrag zur äolischen Formung im extrem kalten Klima. Polarforsch 55:79-125

Montiel PO and Cowan DA (1993) The possible role of soluble carbohydrates and polyols as cryoprotectants in Antarctic plants. In: Heywood RB (ed) University Research in Antarctica, 1989-92. Proceedings of the British Antarctic Survey Antarctic Special Topic Award Scheme Round 2 Symposium, 30 September - 1 October 1992, pp 119-125. British Antarctic Survey, Cambridge

Nienow JA and Friedmann EI (1993) Terrestrial lithophytic (rock) communities. In: Friedmann EI (ed) Antarctic Microbiology, pp 343-412. Wiley-Liss, New York

Nienow JA, McKay CP and Friedmann EI (1988a) The cryptoendolithic microbial environment in the Ross Desert of Antarctica: Mathematical models of the thermal regime. Microb Ecol 16:253-270

Nienow JA, McKay CP and Friedmann EI (1988b) Cryptoendolithic microbial environment in the Ross Desert of Antarctica: light in photosynthetically active region. Microb Ecol 16:271-289

Ocampo-Friedmann R, Meyer MA, Chen M and Friedmann EI (1988) Temperature response of Antarctic cryptoendolithic photosynthetic microorganisms. Polarforsch 58:121-124

Olie JJ and Potts M (1986) Purification and biochemical analysis of the cytoplasmic membrane from the desiccation-tolerant cyanobacterium *Nostoc commune* UTEX 584. Appl Environ Microbiol 52:706-710

Painter T (1993) Carbohydrate polymers in desert reclamation - the potential of microalgal biofertilizers. Carbohydrate Polym 20:77-86

Palmer RJJ and Friedmann EI (1990) Water relations and photosynthesis in the cryptoendolithic microbial habitat of hot and cold deserts. Microb Ecol 19:111-118

Parish TR (1988) Surface winds over the Antarctic continent: a review. Rev Geophys Space Phys 26: 169-180

Parker BC and Wharton RA (1985) Physiological Ecology of blue green algal mats (modern stromatolites) in Antarctic oasis lakes. Arch Hydrobiol Alg Stud 38/39: 331-348

Parker BC, Ford AB, Allnutt T, Bishop B and Wendt S (1977) Baseline microbiology for soils of the Dufek Massif. Antarct J US 12:24-26

Potts M (1994) Desiccation resistance of prokaryotes. Microbiol Revs 58: 755-805

Potts M (1996) The anhydrobiotic cyanobacterial cell. Physiol Plant 97: 788-794

Potts M and Friedmann EI (1981) Effects of water stress on cryptoendolithic Cyanobacteria from hot desert rocks. Arch Mikrobiol 130:267-271

Potts M, Olie JJ, Nickels JS, Parsons J and White DC (1987) Variation in phospholipid ester-linked fatty acids and carotenoids of desiccated *Nostoc commune* (Cyanobacteria) from different geographic locations. Appl Environ Microbiol 53:4-9

Reed RH, Richardson DL, Warr SRC and Stewart WDP (1984) Carbohydrate accumulation and osmotic stress in cyanobacteria. J Gen Microbiol 130:1-4

Rippka R, Deruelles J, Waterbury JB, Herdman M and Stanuer RY (1979) Generic assignments, strain histories and properties of pure cultures of Cyanobacteria. J Gen Microbiol 111: 1-61

Russell NC, Edwards HGM and Wynn-Williams DD (1998) FT-Raman spectroscopic analysis of endolithic microbial communities from Beacon sandstone in Victoria Land, Antarctica. Antarct Sci 10: 63-74

Ryan PG, Watkins BP, Smith RIL, Dastych H, Eicker A, Foissner W, Heatwole H, Miller WR and Thompson G (1989) Biological survey of Robertskollen, western Dronning Maud Land: area description and preliminary species lists. S Afr J Antarct Res 19: 10-20

Sagan C (1972) The Viking Symposium. Icarus 16: 1-227

Scherer S, Chen TW and Böger P (1988) A new UV-A/B protecting pigment in the terrestrial cyanobacterium *Nostoc commune*. Plant Physiol 88:1055-1057

Scherer S and Zhong ZP (1991) Desiccation independence of terrestrial *Nostoc commune* ecotypes (Cyanobacteria). Microb Ecol 22:271-283

Schopf JW (1993) Microfossils of the early Archean Apex chert: new evidence of the antiquity of life. Science 260: 640-646

Schopf JW and Walter MR (1982) Origin and early evolution of cyanobacteria: the geological evidence. In: Carr NG and

Whitton BW (eds) The Biology of Cyanobacteria, pp 543-564. Blackwell Scientific Publications, Oxford

Seaburg KG, Parker BC, Prescott GW and Whitford LA (1979) The Algae of Southern Victoria land. A Taxonomic and Distributional Study, Bibliotheca Phycologia, Vaduz, 169 pp

Siebert J, Hirsch P, Hoffmann B, Gliesche CG, Peissl K and Jendrach M (1996) Cryptoendolithic microorganisms from Antarctic sandstone of Linnaeus Terrace (Asgard Range): diversity, properties and interactions. Biodivers Conserv 5:1337-1363

Smith G, Lynch R, Jacobson G and Barnes C (1990) Cyanobacterial nitrogen fixation in arid soils of central Australia. FEMS Microbiol Ecol 74:79-90

Stoker CR and Bullock MA (1997) Organic degradation under simulated Martian conditions. J Geophys Res Planets 102: 10881-10888

Tedrow JCF and Ugolini FC (1966) Antarctic soils. In: Tedrow JCF (ed) Antarctic Soils and Soil Forming Process, pp 161-177. American Geophysical Union, Washington, D.C.

Tuovila BJ and LaRock PA (1987) Occurrence and preservation of ATP in Antarctic rocks and its implications for biomass determinations. Geomicrobiol J 5:105-118

Turner S (1997) Molecular systematics of oxygenic photosynthetic bacteria. Plant Syst Evol S11: 13-52

Vestal JR (1988a) Biomass of the cryptoendolithic microbiota from the Antarctic desert. Appl Environ Microbiol 54:957-959

Vestal JR (1988b) Carbon metabolism in the cryptoendolithic microbiota from the Antarctic desert. Appl Environ Microbiol 54:960-965

Vestal JR (1993) Cryptoendolithic communities from hot and cold deserts: speculation on microbial colonization and succession. In: Miles J and Walton DWH (eds) Primary Succession on Land. Special Publication No. 12 of The British Ecological Society, pp 5-16. Blackwell Scientific Publications, Oxford

Vincent WF (1981) Production strategies in Antarctic inland waters: Phytoplankton eco-physiology in a permanently ice-covered lake. Ecology 62:1215-1224

Vincent, WF (1988) Microbial Ecosystems of Antarctica. Cambridge University Press, Cambridge, 304 pp

Vincent WF (1996) (ed) Environmental management of a cold desert ecosystem: The McMurdo Dry Valleys. Report of a National Science Foundation Workshop held at Santa Fe, New Mexico, 14-17 March, 1995, Desert Research Institute, University of Nevada, Reno, NV, 57 pp

Vincent WF, Howard-Williams C and Broady PA (1993) Microbial communities and processes in Antarctic flowing waters. In: Friedmann EI (ed) Antarctic Microbiology, pp 543-569. Wiley-Liss, New York

Vincent WF and James MR (1996) Biodiversity in extreme aquatic environments: lakes, ponds and streams of the Ross Sea sector, Antarctica. Biodivers Conserv 5:1451-1471

Walsby AE (1980) The water-relations of gas-vacuolate prokaryotes. Proc Roy Soc Lond B 208:73-102.

Warr SRC, Reed RH and Stewart WDP (1988) The compatibility of osmotica in cyanobacteria. Plant Cell Environ 11:137-142

Walter MR and Bauld J (1983) The association of sulphate evaporites, stromatolitic carbonates and glacial sediments: examples from the proterozoic of Australia and the Cainozoic of Antarctica. Precambrian Res 21:129-148

Weber B, Wessels DCJ and Büdel B (1996) Biology and ecology of cryptoendolithic cyanobacteria of a sandstone plateau in North-Transvaal, South Africa. Alg Stud 83:565-579

Weed R and Norton SA (1991) Siliceous crusts, quartz rinds and biotic weathering of sandstones in the cold desert of Antarctica. In: Berthelin J (ed) Diversity of environmental biogeochemistry, pp 327-339. Elsevier, Amsterdam

Wessels DCJ and Büdel B (1995) Epilithic and cryptoendolithic cyanobacteria of Clarens sandstone cliffs in the Golden Gate Highlands National Park, South Africa. Botanica Acta 108:220-226

West NE (1990) Structure and function of microphytic soil crusts in wildland ecosystems of arid to semi-arid regions. Adv Ecol Res 20: 179-223

Wharton RAJ (1993) (ed) McMurdo Dry Valleys: a Cold Desert Ecosystem. Report of a National Science Foundation Workshop held at the Institute of Ecosystem Studies the New York Botanical Garden, Millbrook, New York, 5-7 October 1991, Desert Research Institute, Nevada, 51 pp

Wharton RA (1994) Stromatolitic mats in Antarctic lakes. In: Bertrand-Sarfati J and Monty C (eds) Phanerozoic Stromatolites II, pp 53-70. Kluwer Academic Publishers, Dordrecht

Whitton BA (1987) Survival and dormancy of blue-green algae. In: Henis Y (ed) Survival and Dormancy of Microorganisms, pp 109-167. Wiley, New York

Whitton BA (1992) Diversity, ecology and taxonomy of the cyanobacteria. In: Mann NH and Carr NG (eds) Photosynthetic Prokaryotes, pp 1-51. Plenum Press, New York

Wynn-Williams DD (1990) Ecological aspects of Antarctic microbiology. In: Marshall KC (ed) Advances in Microbial Ecology, pp 71-146. Plenum Press, New York

Wynn-Williams DD (1991) Aerobiology and colonization in Antarctica - the BIOTAS Programme. Grana 30:380-393

Wynn-Williams DD (1994) Potential effects of ultraviolet radiation on Antarctic primary terrestrial colonizers: Cyanobacteria, algae and cryptogams. In: Weiler CS and Penhale PA (eds) Ultraviolet Radiation in Antarctica: Measurements and Biological Effects, pp 243-257. American Geophysical Union, Washington, D.C.

Wynn-Williams DD (1996) Antarctic microbial diversity: the basis of polar ecosystem processes. Biodivers Conserv 5:1271-1293

Wynn-Williams DD (in press) Antartica as a model for ancient Mars. J Brit Interplanetry Soc (Spec. Iss.: Hiscox J (ed) The Search for Life on Mars)

Wynn-Williams DD, Russell NC and Edwards HGM (1997) Moisture and habitat structure as regulators for microalgal colonists in diverse Antarctic terrestrial habitats. In: Lyons WB, Howard-Williams C and Hawes I (eds) Ecosystem Processes in Antarctic Ice-free Landscapes. pp 77-88 Balkema, Rotterdam

Xiong F, Komenda J, Kopecky J and Nadbal L (1997 in press) Strategies of ultraviolet-B protection in microscopic algae. Physiol Plant :

Zent AP and McKay CP (1994) The chemical-reactivity of the martian soil and implications for future missions. Icarus 108:146-157

Chapter 14

Detecting the Environment

Nicholas H. Mann
Department of Biological Sciences, University of Warwick, Coventry CV4 7AL, UK

Summary

The cyanobacteria, like all organisms, sense and respond to changes in their environment. This chapter focuses on one aspect, an analysis of the signal transduction pathways that permit the appropriate adaptive response to be set in train. The mechanisms by which prokaryotes respond to their environments largely represent variations and combinations of common themes. The cyanobacteria are no exception, but in many cases there are modifications to the basic signal transduction pathways that are related to the photoautotrophic lifestyle. This is particularly true with respect to the need for integration of the intracellular signals, and the possibility that the thylakoids represent the site for this central process is examined. Whilst transcriptional mechanisms are an important component of the signal transduction processes, signalling elements may operate at any level of the genotype to phenotype pathway and the importance of the post-translational modification of proteins, particularly

B. A. Whitton and M. Potts (eds), The Ecology of Cyanobacteria, 367-395.
© 2000 *Kluwer Academic Publishers. Printed in the Netherlands.*

by phosphorylation, is recognised. The potential to understand cyanobacterial signalling mechanisms has been significantly enhanced by genomic sequence information, and where possible this has been utilised to confirm or extend the present empirical knowledge.

I. Introduction

All organisms are capable of detecting changes in their environment and possess the potential to exhibit responses that enable them to adapt to these changes. The photoautotrophic lifestyle of cyanobacteria implies a primary dependency on physical and chemical factors, such as temperature, and the availability of light and inorganic nutrients. Variations in these factors constitute the primary environmental stimuli to which these organisms must adapt, though cyanobacteria exhibit a wide range of adaptive responses (Tandeau de Marsac and Houmard, 1993). This chapter focuses on the intracellular signals and mechanisms involved in this signal transduction process which lead to adaptation. This requires an examination of the extent to which the signal transduction pathways characterized in other organisms are used and how they are modified to fit the cyanobacterial lifestyle.

The dominant nutritional mode of the cyanobacteria is photoautotrophy. Growth occurs via the light-dependent fixation of CO_2 and the acquisition of simple inorganic nutrients. However, this life style has its own inherent risks, since light above that that used for growth represents a damaging and potentially lethal agent. The light environment may vary in terms of intensity over orders of magnitude with time scales ranging from seconds to days. The availability of the simple inorganic nutrients will also determine both the rate and extent of growth and, indeed, there are complex interactions between nutrient acquisition and light harvesting that may markedly potentiate the damaging effects of light. The interactions between the two limbs of metabolism, photosynthesis and nutrient acquisition, together with the switch to dark metabolism are key points of integration and represent areas where adaptive responses are likely to operate.

Abbreviations: cAMP, cyclic 3′5′-adenosine monophosphate: Ci, inorganic carbon: G6PDH, glucose-6-phosphate dehydrogenase: HPK, histidine protein kinase: Pi, inorganic phosphate: ppGpp, guanosine-5′-diphosphate-3′-diphosphate: PQ, plastoquinone: PQH2, plastoquinol: PSI, photosystem I: PSII, photosystem II: RR, response regulator

In addition to light and the availability of nutrients, bacteria respond to other stimuli that include temperature, osmolarity, toxic agents, surface attachment, population density, and for symbionts and pathogens, the presence of potential host organisms. The list of known mechanisms employed by bacteria to sense and adapt to changes in their environment is continuously being added to and this is particularly the case over the past ten years. This is consistent with a marked change in the conceptual framework that has been applied to the study of bacterial adaptation. The emphasis has moved away from an approach dominated by a conception of transcriptional regulation of gene expression, which stemmed from the analysis of systems such as the *lac* operon and phage lambda, and towards an appreciation that the signalling elements of a bacterial sensory transduction process may operate at any level of the genotype to phenotype pathway.

The environmental stimuli to which bacteria respond may be physical or chemical and external or internal. The signal must interact with a cellular component that either directly or indirectly elicits an adaptive response to complete the transduction pathway. The adaptive response may be an alteration in metabolic capability, the structural organization of the cell, or of behaviour. In the vast majority of cases the cellular component that interacts with the signal is a protein and in many cases, though not all, the adaptive response involves a change in the pattern of gene expression. In some systems the mechanism underlying the adaptive responses may occur at the level of the genome and can involve rearrangement of DNA sequences as happens during heterocyst development (Haselkorn, 1992; Wolk et al., 1994) and in adaptation to heavy metal stress (Chapter 16). There is a range of control processes that elicit a response at the level of transcription. At the most basic level there are the simple systems by which small molecules cause the induction/repression of genes or operons by virtue of their interactions with repressor/inducer proteins. Such transcriptional control mechanisms may become more sophisticated with the transcriptional specificity of the RNA polymerase being modulated by alternative σ-factors as is the case in many global response systems such as heat shock. The activity of RNA polymerase may be altered by interaction with unusual nucleotides as occurs during the stringent response (Condon et al., 1995). Other mechanisms involved in modulating the transcription process include anti-termination, and proteins that influence DNA bending such as integration host factor (IHF).

Adaptive responses may also effect gene expression, but at the translational level, employing mechanisms such as altered mRNA stability, or control of translation initiation. The area that may rank equal in importance to transcriptional control is post-translational control. There is a wide range of mechanisms by which polypeptides may become specifically modified after their synthesis and which may determine their biological activities. Covalent modifications include methylation, ADP-ribosylation, lipoylation etc. and most importantly, phosphorylation. In this context, two particularly important signal transduction mechanisms are the two-component sensory systems and eukaryotic-type protein kinases. Alternatively, the metabolic stability of some polypeptides may be modified in response to environmental signals. Finally some polypeptides may be modified by non-covalent means such as interactions with small molecules, other polypeptides or alterations in their oligomerization state. Frequently, signal transduction pathways will involve multiple elements that operate at different levels. A major development in the study of signal transduction in cyanobacteria has been the sequencing of the complete genome of the unicellular freshwater cyanobacterium *Synechocystis* PCC 6803 (Kaneko et al., 1996; see Kotani and Tabata, 1998) and the availability of this information via the Internet (http://www.cyanobase.kazusa.or.jp/cyano/) (Nakama et al., 1998). Such genomic information may be used to exclude the possibility of some signalling systems or to confirm and extend understanding of others. For example, quorum sensing systems involved in inter-cell signalling have been characterized in a range of bacteria (Fuqua et al., 1994). However, there is no evidence for such a system in cyanobacteria and homologues of the relevant genes are not encoded within the *Synechocystis* PCC 6803 genome (Kaneko et al., 1996), though quorum sensing type molecules (N-acylated homoserine lactones) have been detected in cyanobacterial blooms (Bachofen and Schenk, 1998).

II. Protein Phosphorylation

A. Two-component sensory systems

There has been a remarkable advance in recent years in the understanding of the nature and global importance of two-component sensory systems, which have been discovered in a wide range of prokaryotes (Stock et al., 1989; Parkinson and Kofoid 1992), cyanobacteria and eukaryotes (e.g. Chang and Meyerowitz, 1995). They have also been implicated in an enormous range of bacterial adaptive responses (Stock et al., 1989; Parkinson and Kofoid, 1992; Hoch and Silhavy, 1995). The basic two-component sensory system is composed of two proteins: a histidine protein kinase (HPK) and a response regulator (RR) (Fig. 1). The N-terminal region of the HPK functions as a sensor domain interacting with environmental stimuli and the C-terminal (transmitter) part of the protein is capable of autophosphorylation at a specific histidine residue and subsequent transfer of the phosphate to an aspartyl residue in the RR. Typically, though not always, the HPK has one or more membrane spanning segments associated with the sensor domain that faces towards the periplasm. The transmitter domain extends into the cytoplasm. The RR, which is typically cytoplasmic, also typically consists of two domains. The N-terminal receiver domain contains a phosphorylatable aspartyl residue and the C-terminal output domain, which commonly consists of a DNA-binding feature, is involved in initiating the appropriate adaptive response. The nature of the signal in this sensory transduction pathway is the transfer of the phosphate, usually from ATP, to the HPK and thence to the RR activating it to elicit the adaptive response.

The two classes of protein, HPKs and RRs, constitute two superfamilies which exhibit conserved features at the amino acid sequence level (Fig. 1). In the case of the HPKs these conserved features are apparent within the approximately 240 amino acid transmitter domain, which is in turn composed of two sub-domains separated by a relatively variable linker (Parkinson and Kofoid, 1992). The first sub-domain contains the phosphorylatable histidine residue within a conserved sequence motif known as the H box. The second sub-domain usually contains four conserved sequence features known as the N, D/F and G boxes. The separation between the H and N boxes of the two subdomains is approximately 110 amino acids. The amino acid preferences in the conserved features are detailed by Stock et al. (1995). The conserved

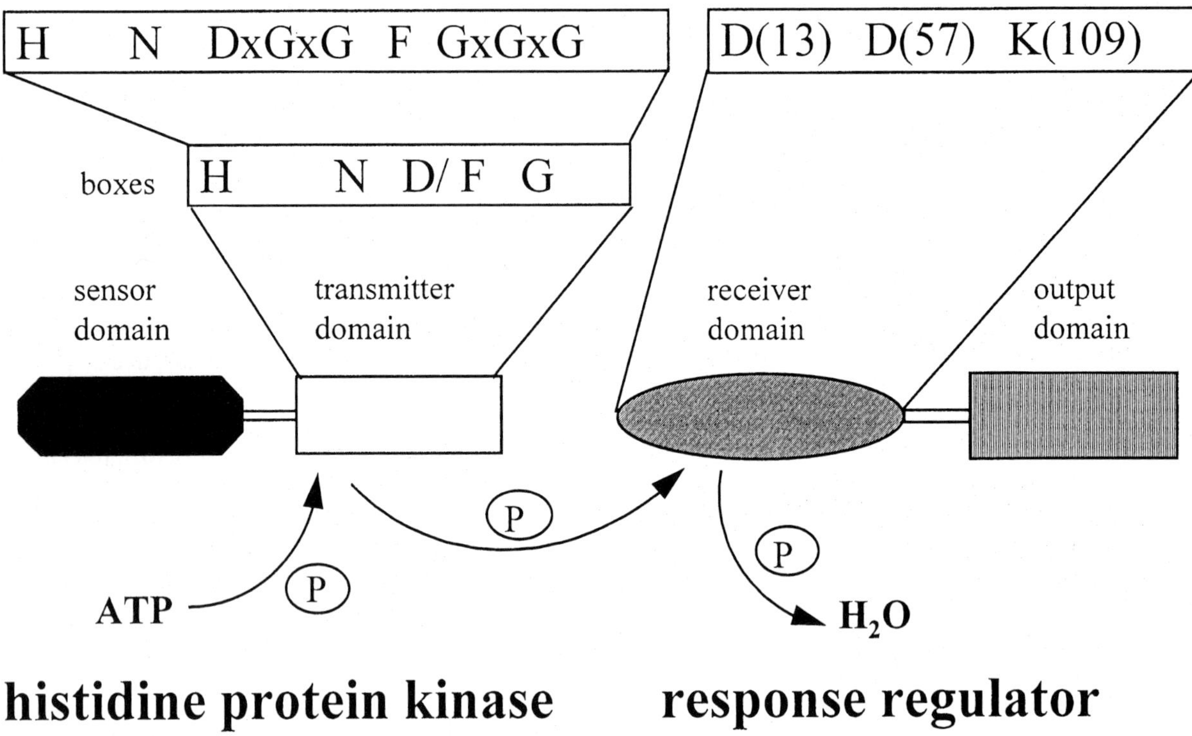

Figure 1. A diagram illustrating the domain organization of the typical histidine protein kinases and response regulators of two-component sensory systems and pattern of phosphotransfer. The H, N, D, F and G boxes of the histidine protein kinases are indicated, together with the conserved amino acid sequence features (D aspartate, F phenylalanine, G glycine, H histidine, K lysine, x any amino acid) for both the histidine protein kinases and the response regulators. The numbering of the conserved residues in the response regulator is that applying to the *Escherichia coli* CheY protein.

receiver domain of RRs is usually approximately 125 amino acids in size. Any two receiver domains usually exhibit 20 - 30% identity and the residues that correspond to Asp-13, Asp-57 and Lys-109 in the well studied and archetypal *E. coli* RR CheY tend to be absolutely conserved (Stock et al., 1989) with Asp-57 being the site of phosphorylation. The receiver domain is usually connected by a variable linker to an output domain. In the cases where the output domain has a DNA-binding activity there are three sub-families typified by OmpR-PhoB, NtrC (NR_I) and FixJ (Stock et al., 1995). In addition to a DNA-binding activity the output domain may exhibit alternative features such as an esterase domain as in the case of the chemotaxis protein CheB. In contrast the CheY family of RRs consist only of the receiver domain.

There are reports of the characterization of some twenty or so proteins from cyanobacteria which fall unambiguously into the classes of HPKs or RRs and analysis of the genome of *Synechocystis* PCC 6803 (Kaneko et al. 1996) indicates at least 44 HPKs and 52 RRs. Cyanobacterial proteins containing either HPK or RR modules have been implicated in a wide range of adaptive responses (Table 1). The RR NblR from *Synechococcus* PCC 7942 is involved in the regulation of phycobilisome degradation in response to nutrient limitation, but also modulates additional functions critical for cell survival during nutrient-limited and high light conditions (Schwarz and Grossman, 1998). A gene *patA* involved in the control of heterocyst pattern formation encodes a 379 amino acid protein consisting of 3 domains of which the C-terminal domain exhibited the sequence features characteristic of a RR (Liang et al., 1992). Transposon mutagenesis of a symbiotic, nitrogen-fixing *Nostoc* strain led to the identification of a gene *devR*, encoding a protein with the conserved features of a RR, similar to the *E. coli* CheY and *Bacillus subtilis* SpoOF, which was required for nitrogen fixation under aerobic conditions, but did not block heterocyst development (Campbell et al., 1996). Similarly, the *rcaC* gene which complements a mutation causing red-light indifference in the

Table 1. *Cyanobacterial adaptive responses for which there is evidence implicating proteins with features characteristic of the modules of two component sensory system histidine protein kinases or response regulators.*

Physiological function	Genes or ORFs	Organism	References
Phycobilisome deradation and survival under nutrient-limited/high light conditions	*nblR*	*Synechococcus* PCC7942	Schwarz and Grossman, 1998
Complementary chromatic adaptation	*rcaC*	*Fremyella diplosiphon*	Chiang et al.,1992
	rcaE	*Fremyella diplosiphon*	Kehoe and Grossman, 1996
	rcaF	*Fremyella diplosiphon*	Kehoe and Grossman, 1997
Heterocyst pattern formation	*patA*	*Anabaena* PCC 7120	Liang et al., 1992
Nitrogen fixation	*devR*	*Nostoc* ATCC 29133	Campbell et al., 1996
CO_2 availability	*hatR*	*Synechocystis* PCC 6803	Bédu et al. 1995
Phytochrome- and ethylene response-like signalling	slr1393, sll1124, slr1969 slr0473	*Synechocystis* PCC 6803	Kaneko et al., 1996
	phy	*Synechocystis* PCC 6803	Hughes et al., 1997
	plpA	*Synechocystis* PCC 6803	Wilde et al., 1997
Herbicide resistance/chemical signalling	*dspA*	*Synechocystis* PCC 6803	Bartsevich and Shestakov, 1995
Phosphate availability	*sphR/sphS*	*Synechococcus* PCC 6803	Aiba et al., 1993
	phoB/phoR	*Synechococcus* WH 7803	Watson et al., 1996

complementary chromatic adapting *Fremyella diplosiphon*, encodes a 632 amino acid protein the N-terminus of which contains two RR receiver domains, one at the N-terminus and the other at the C-terminus (Chiang et al., 1992). More recently, Kehoe and Grossman (1996, 1997) have shown that the *rcaE* and *rcaF* genes of *F. diplosiphon*, which may be involved in the sensing of light quality, encode proteins with the features of an HPK and RR, respectively. Mutations in *Synechocystis* PCC 6803 in a gene designated *dspA*, which lead to cross resistance to the herbicides difunon and diuron as well as the calmodulin antagonists chlorpromazine and trifluoperazine, define a gene encoding an HPK with two typical trans-membrane segments (Bartsevich and Shestakov, 1995). Aiba et al. (1993) characterized two contiguous genes *sphS* and *sphR* from the freshwater strain *Synechococcus* PCC 7942 which encode an HPK and RR respectively. Mutations in these genes prevent the induction of alkaline phosphatase and other polypeptides associated with Pi starvation. Importantly, Nagaya et al. (1994) have shown that SphR is phosphorylated by phosphotransfer from SphS and binds to two distinct sites upstream from the alkaline phosphatase (*phoA*) promoter. In the case of the oceanic strain *Synechococcus* WH7803, two contiguous genes designated *phoB* and *phoR* encode a RR and HPK respectively and transcription of these two genes is induced by Pi starvation (Watson et al., 1996). Several other HPK and RR genes of unclear physiological significance have been identified in cyanobacteria (Mann, 1994) such as those cloned by Nagaya et al. (1993) from *Synechococcus* PCC 7942 on the basis of their ability to complement HPK mutants of *E. coli,* and an incomplete ORF found downstream from the *petH* ($NADP^+$-ferredoxin oxidoreductase) gene of *Synechococcus* PCC 7002 (Schluchter and Bryant, 1992). An adjacent pair of HPK and RR genes have been characterized from *Synechococcus* PCC 7942, but insertional mutagenesis of the RR gene yielded no obvious phenotype (Scanlan and Mann, 1996) and similar observations were made following the disruption of two putative RR genes, designated *srrA* and *srrB*, from *Synechococcus* PCC 7942 (Anandan et al., 1996).

In addition to the prototypical HPKs and RRs, a significant number of proteins have been characterized which possess conserved features of both these families and may further combine these features with domains typical of other signalling systems. Several cyanobacterial proteins fit this description, including the RcaC protein involved in chromatic adaptation (Chapter 15) which has two conserved receiver domains each with a putative aspartate phosphorylation site (Chiang et al., 1992) and also appears to have a histidine autophosphorylation site (Appleby et al., 1996). The 395 amino acid protein encoded by ORF slr2099 in the genome of *Synechocystis* PCC 6803 (Kaneko et al., 1996) has both transmitter and receiver domains, as do several other ORFs. In the case of some of these more complex signalling proteins, it is thought that they may play a part in a four step phosphorelay in which there is a His-Asp-His-Asp sequential phosphotransfer (Appleby et al., 1996). Such a system is typified by the pathway governing the initiation of sporulation in *B. subtilis* and also occurs in the BvgS-BvgA two-component system involved in the regulation of virulence factors in *Bordetella pertussis* (Appleby et al., 1996). Furthermore, with the discovery of RcaE (Kehoe and Grossman, 1996), a sensor kinase that could function upstream from RcaC, such a multiple phosphorelay could be involved in complementary chromatic adaptation (Chapter 15; Appleby et al., 1996). This observation is extended by the discovery of RcaF, which may implicate as many as five phosphoacceptor domains in the signalling pathway (Kehoe and Grossman, 1997; Grossman and Kehoe, 1997). The importance of these multistep relays may be that they offer multiple regulatory checkpoints in a signalling pathway.

Two-component sensory systems, whether of the simple or more complicated forms, obviously play a central role in sensory transduction in cyanobacteria, but information on their physiological functions and possible cross-talk between different signalling pathways is as yet only rudimentary. A detailed analysis of the location of HPK and RR genes in the genome of *Synechocystis* PCC 6803 in relation to other genes of known physiological function may give some clues as to their likely physiological role.

B. Phytochrome- and ethylene response-related proteins

Given the importance of phytochromes in enabling plants to respond to particular changes in their light environment, the question of whether such molecules play a similar role in cyanobacteria has been raised frequently. The phytochromes constitute a family of plant sensors that can exist in two photo-interconvertible isomeric forms and enable plants to respond to particular changes in their light environment by a wide variety of morphogenetic responses. The chromophore is a linear tetrapyrrole that becomes attached to the apoprotein via a thioether bond formed by an autocatalytic self-assembly process (Lagarias and Lagarias, 1989). It will be useful to look briefly at the state of knowledge in *Arabidopsis thaliana*. Five distinct phytochrome genes have been detected (Sharrock and Quail, 1989). Phytochrome A probably mediates the far-red high irradiance response, whereas phytochrome B is responsible for the red/far-red response and the other phytochromes play minor roles (Smith, 1995). Deletion analysis has revealed that there are separate domains within the ~120 kD (1100 amino acids) phytochrome molecule required for structure and biological activity (Cherry et al., 1993). Chromophore attachment is determined within the amino-terminal 400 amino acids, while spectral integrity is influenced by residues up to about position 652. Dimerization is determined within the C-terminal part of the molecule, and key determinants of biological activity appear to reside within sequences close to both the C-terminus and N-terminus. It was first pointed out by Schneider-Poetsch et al., (1991) that there were structural and functional homologies between plant phytochromes and bacterial HPKs covering a domain of about 250 amino acids in the C-terminal part of both groups of proteins and corresponding to the transmitter domain of the HPKs. The apparent significance of these sequence similarities was extended by detailed comparisons of the putative catalytic domains (Thümmler et al., 1995). The functional significance of the sequence similarities between the C-terminus of phytochromes and bacterial HPKs has been called into question by Quail (1994) who has reported that attempts to measure a HPK activity for phytochromes have failed and also points out that in the large majority of phytochromes the highly conserved histidine subject to autophosphorylation in histidine protein kinases is not present.

The ethylene hormone-response pathway in plants affects many aspects of plant growth and development. Studies with *Arabidopsis* mutants displaying dominant ethylene insensitivity facilitated the characterization of the *etr1* gene (Chang et al., 1993). Although the sequence of the amino-terminal half of the deduced ETR1 protein appears to be novel, the carboxyl-terminal half contains both the transmitter and receiver domains of HPKs and RRs respectively (Chang et al., 1993; Chang and Meyerowitz, 1995). A second gene, related to *etr1*, has been characterized and the deduced ERS protein, which exhibits 67% identity with ETR1, has a transmitter domain, but no receiver domain (Hua et al., 1995). It is suggested that both proteins are ethylene sensors and thus there is redundancy in the signalling pathway. The ORF slr1212 in the genome of *Synechocystis* PCC 6803 encodes a polypeptide with marked similarities to ETR1 (Kaneko et al., 1996). Ethylene binding is restricted to land plants and cyanobacteria and a mutant carrying a knockout of slr1212 had lost ethylene binding activity (Esch et al., 1998). It is suggested that the prototypical ethylene-binding domain may have functioned as a copper-binding domain in cyanobacteria and was recruited as an ethylene sensor because it serendipitously created an environment that allowed tight binding of ethylene to copper. However, there is as yet no evidence to exclude a signalling function for ethylene in cyanobacteria.

Kehoe and Grossman (1996) have characterized the *rcaE* gene of *Fremyella diplosiphon* that is involved in complementary chromatic adaptation. The 74 kD RcaE protein has a carboxy-terminal domain with the conserved features of an HPK. Within the amino-terminal half of the protein, however, there is a region of ~140 amino acids that is similar to the chromophore attachment site of phytochromes. RcaE also exhibits similarities with the putative plant ethylene sensors in terms of two motifs, designated T2L and R2L, found in the amino terminal part of the molecule. RcaE is thought to be a member of a large, highly diverged family that occurs in both prokaryotes and eukaryotes. In keeping with this suggestion, several proteins extremely similar to RcaE are encoded within the genome of *Synechocystis* PCC 6803 (Kaneko et al., 1996), although this organism does not undergo complementary chromatic adaptation. The four most similar ORFs (sll1473, slr1393, sll1124 and slr1969) all exhibit the conserved T2L and R2L motifs and all but one (sll1473) also have C-terminal HPK transmitter domains. ORFs sll1124 and sll1473 exhibit some similarity to the phytochrome chromophore attachment region. The gene *plpA* (= sll1124) was disrupted in *Synechocystis* PCC 6803 and the mutant was unable to grow in blue light (Wilde et al., 1997).

In contrast to the other ORFs encoded in the genome of *Synechocystis* PCC 6803 which exhibit limited region of similarity to the plant phytochromes ORF slr0473 displays marked extended similarities (Kaneko et al., 1996). The sequence similarities are not confined to the region aligning with HPK, but extend into the regions involved in chromophore attachment and spectral integrity. Indeed, the cysteine involved in chromophore attachment is present in the cyanobacterial homologue. This putative cyanobacterial phytochrome gene (*phy*) is situated immediately upstream from an ORF that displays the receiver domain of a RR. Expression of the *phy* gene in *E. coli* revealed that the protein was able to autocatalytically react with phycocyanobilin to produce a photochromic holoprotein with absorption maxima at 658 and 702 nm after red and far-red irradiation (Hughes et al., 1997; Lamparter et al., 1997; Yeh et al., 1997). Furthermore, the cyanobacterial phytochrome exhibited ATP-dependent autophosphorylation, apparently on a histidine residue, in the Pr form, that was greatly reduced in the Pfr form (Yeh et al., 1997). The apparent response regulator encoded by the ORF immediately downstream from the phytochrome gene is subject to phosphorylation by the phytochrome (Yeh et al., 1997). Thus, the cyanobacterial phytochrome(s) may provide a model for studying the phytochrome signalling pathway(s) in plants.

C. *Eukaryotic-Type Serine/Threonine and Tyrosine Kinases*

In contrast to the rapid development in our understanding of the significance in prokaryotes of the histidine and aspartate protein phosphorylation there has been comparatively slow progress in establishing the physiological role of serine, threonine and tyrosine protein phosphorylation. This is in stark contrast to the situation in eukaryotes where several hundred protein kinases have been characterized to some extent and the phosphorylation of proteins on hydroxyl-containing amino acids is central to all aspects of cell physiology. It has been known since the late 1970s that monoester phosphorylation of bacterial proteins may occur (Cozzone, 1988), but excluding the case of isocitrate dehydrogenase phosphorylation in *E. coli*, there was limited progress in this area until recently. However, there is extensive unambiguous evidence for the monoester phosphorylation of proteins in a range of cyanobacterial strains using both *in vivo* and *in vitro* labelling techniques, and changes in the pattern of protein phosphorylation have been correlated with a variety of physiological changes (Mann, 1994).

Eukaryotic serine/threonine and tyrosine protein kinases fall into a single superfamily (Hanks et al., 1988). The initial discovery of such "eukaryotic" protein kinases in prokaryotes was based on the design of PCR primers against conserved regions in the catalytic domain of serine/threonine protein kinases and led to the characterization of a serine/threonine protein kinase gene (*pkn1*) from *Myxococcus xanthus* (Muñoz-Dorado et al., 1991) and establishment that this kinase was required for normal differentiation. Subsequent studies indicated that there was a family of such kinases and that they occurred in a range of organisms (Zhang, 1996; Kennelly and Potts, 1996). The first report of an "eukaryotic" kinase in a cyanobacterium employed a similar PCR approach to that used with *M. xanthus* (Zhang, 1993). The *pknA* gene of *Anabaena* PCC 7120 encodes a protein, the N-terminal region of which exhibits significant similarity to the catalytic domains of serine/threonine kinases. Inactivation of *pknA* reduced the frequency of heterocysts in the filaments and colonies of the mutant appeared light green and rough in the absence of combined nitrogen. Southern hybridization experiments indicated that there was a family of such genes in *Anabaena* PCC 7120 (Zhang, 1993). An ORF exhibiting similarity to rat myosin light chain serine/threonine kinase has been reported for *Fremyella diplosiphon* (Grossman et al., 1993). A total of six serine/threonine protein kinase genes have now been characterized from *Anabaena* PCC 7120, and although their analysis is not complete there is evidence that some of them may be required for growth in the absence of combined nitrogen (Zhang, 1996; Zhang and Libs 1998; Zhang et al., 1998a). Phylogenetic analysis based on the catalytic domains of serine/threonine protein kinases from prokaryotes show that five (PknA, PknB, PknC, PknD and PknE) of the six kinases from *Anabaena* PCC 7120 cluster together, while one (HstK) is more closely related to STE7 from *Saccharomyces cerevisiae* (Zhang, 1996). The genomic sequence of *Synechocystis* PCC 6803 reveals seven genes apparently encoding serine/threonine protein kinases and seven serine/threonine and tyrosine phosphatases (Zhang et al., 1998b), all of which obviously must be involved in processes other than growth under dinitrogen-fixing conditions. In addition, there are ORFs encoding polypeptides with marked similarities to the regulatory chain of a cAMP-dependent protein kinase and a protein kinase C inhibitor. In this context, the second ORF of a 16 kb dicistronic message encoding the PSII protein D1 (form II) (Bustos et al., 1990) exhibits extensive similarity to a protein encoded by a maize cDNA which also resembles a protein kinase C inhibitor (Simpson et al., 1994). An important caveat should be borne in mind in that although "eukaryotic"-type protein kinases have been discovered in cyanobacteria and other prokaryotes it may well be the case that much of phosphorylation of proteins as monoesters is carried by kinases that exhibit no homology to the eukaryotic kinases. This is the case for the isocitrate dehydrogenase kinase/phosphatase of *E. coli* (Cortay et al., 1988); thus a new class or classes of prokaryotic kinases may await discovery (Section II.D).

In eukaryotes the number and variety of protein kinases are paralleled by protein phosphatases responsible for dephosphorylating serine/threonine and tyrosine residues. This abundance of phosphatases has not so far been reflected in prokaryotes, except in the case of analysis of eukaryotic-like signalling proteins encoded by the *Synechocystis* PCC 6803 genome (Zhang et al., 1998b). Genes encoding eukaryotic-like phosphatases have also been characterized in *Anabaena* PCC 7120 (Zhang et al., 1998) and *Microcystis aeruginosa* sp. UTEX 2063 (Shi and Carmichael 1997). The most detailed characterization of a cyanobacterial protein phosphatase involves the monoesterase IphP from *Nostoc commune* UTEX 584. The gene (*iphP*), encoding a secreted monoesterase which hydrolysed *p*-nitrophenyl phosphate, was characterized and the predicted sequence of the polypeptide contained the motif:

His-Cys-Xaa-Ala-Gly-Xaa-Xaa-Arg

characteristic of protein-tyrosine phosphatases from eukaryotes (Potts et al., 1993). Indeed, IphP exhibited phosphatase activity towards phosphotyrosine residues, but also towards phosphoserine residues. Further studies on the substrate specificity of IphP showed it to exhibit monoesterase activity toward a broad range of peptide, protein and low molecular weight organophosphate compounds (Howell et al.,

Table 2. *Cyanobacterial adaptive responses in which the monoester phosphorylation of proteins is implicated.*

Environmental Signal	Adaptive response	Organism	References
Light spectral quality	complementary chromatic adaptation	*Fremyella diplosiphon*	Chiang et al., 1992
		Calothrix PCC 7601	Sobczyk et al., 1994
		Fremyella diplosiphon	Kehoe and Grossman, 1997
Light spectral quality	state transition	*Synechococcus* PCC 6301	Allen, 1992
Light intensity	?	*Synechococcus* PCC 7942	Warner and Bullerjahn, 1994
	protein tyrosine phosphorylation	*Anabaena* PCC 7120	McCartney et al., 1997
Ammonium ions/CO_2 fixation	P_{II} phosphorylation	*Synechococcus* PCC 7942	Harrison et al., 1990; Tsinoremas et al., 1991; Forchhammer and Tandeau de Marsac, 1994, 1995ab,
?	β-phycocyanin phosphorylation	*Synechococcus* PCC 6301	Harrison, 1990
Salinity	?	*Synechocystis* PCC 6803	Hagemann et al., 1993
CO_2 availability	Ci transport	*Synechocystis* PCC 6803	Bloye et al., 1992
Availabiliy of fixed nitrogen / N_2 fixation	Heterocyst formation	*Anabaena* PCC 7120	Zhang, 1993, 1996
Ammonium ions	Nitrate transport	*Synechococcus* PCC 6301	Rodriguez et al., 1994

1996). Interestingly, IhpP exhibited MAP kinase phosphatase activity. Introduction of the *iphP* gene ona multicopy plasmid into *Anabaena* PCC 7120 leads to changes in the pattern of protein tyrosine phosphorylation (S. Smith and M. Potts, pers. comm.). However, it is clear from a variety of *in vivo* and *in vitro* studies that turnover of protein monoester phosphates does occur in cyanobacteria and that the activity of these phosphatases may be modulated by redox conditions as well as the concentrations of metabolites such as ribulose 5-phosphate and glucose 6-phosphate (Mann et al., 1991; Allen, 1992).

D. Other Modes of Phosphorylation

Protein phosphorylation involving monoesters has been implicated in a wide range of physiological responses in cyanobacteria (Table 2); however, few of the proteins involved have been identified and kinases other than those of the eukaryotic-type may be involved. P_{II} represents the single example in cyanobacteria where the monoester phosphorylation of a specific identified protein and the physiological significance of the process have been well chacterized. In *E. coli* the P_{II} (GlnB) protein is subject to uridylylation as a response to a decreasing glutamine concentration relative to 2-oxoglutarate, and the modified from of P_{II} activates an adenyltransferase which deadenylates glutamine synthetase causing its activation. The uridylylation of P_{II} also causes an activation of the σ^{54}-dependent transcription of the *glnALG* operon and increased synthesis of glutamine synthetase via a phosphorylation cascade (Ninfa et al., 1995). A P_{II} (GlnB) homologue was identified in *Synechococcus* PCC 7942 and was found to be subject to some form of post-translational modification which was modulated by the ammonium ion concentration and possibly by photosynthetic electron transport (Tsinoremas et al., 1991). Forchhammer and Tandeau de Marsac (1994) showed that, in fact, the P_{II} protein of *Synechococcus* PCC 7942 was subject to phosphorylation of a serine residue rather than uridylylation and it was further demonstrated that the phosphorylation of P_{II} was primarily determined by the N-status of the cell, but also by CO_2 fixation (Forchhammer and Tandeau de Marsac, 1995b). The phosphorylation site of P_{II} was characterized and found to have an RXS motif, also found for eukaryotic cAMP-dependent protein kinases, and the

P_{II} kinase activity was determined by the concentration of 2-oxoglutarate and not by glutamine (Forchhammer and Tandeau de Marsac, 1995a). In fact, P_{II} binds 2-oxoglutarate and ATP in a mutually dependent manner and it has been proposed that the primary function of the *Synechococcus* P_{II} is to sense 2-oxoglutarate, the carbon skeleton required for nitrogen assimilation (Forchhammer and Hedler, 1997). However, there also appears to be a role for P_{II} in the control of nitrate/nitrite transport (Lee et al., 1998). The P_{II} kinase and phosphatase activities are biochemically separable and the latter was capable of dephosphorylating phosphoserine/threonine residues on several proteins, though only PII dephosphorylation was regulated by 2-oxoglutarate and ATP (Irmler et al., 1997).

III. Light/Dark and Redox Sensing

The most obvious environmental factor impacting on the physiology of cyanobacteria is light, because of the daily transitions of night and day, as well as less regular changes in light intensity and spectral quality. Apart from interactions between photosynthetic and respiratory electron transport (Peschek, 1996) and modifications to the structure and function of the light-harvesting apparatus (Chapter 15) there are physiological adaptations in terms of metabolic activity mediated by a variety of different mechanisms, central to which is the redox state of the cell. Furthermore, there is a growing accumulation of evidence from other organisms that redox control of gene expression may be exerted at a variety of levels (Allen, 1993; Allen et al., 1995).

A. Thioredoxin

Thioredoxins are small, disulfide-containing redox proteins and their occurrence in cyanobacteria and role in the control of enzyme activity have recently been extensively reviewed (Gleason, 1994). Cyanobacteria nature are subject to a diurnal light-dark cycle during which the organisms must switch between a photoautotrophic metabolic mode involving the reductive pentose phosphate pathway (RPP) for growth and a dark oxidative mode for cell maintenance. Although enzymes of both the glycolytic and oxidative pentose phosphate (OPP) pathways have been detected in representative species of cyanobacteria, relative specific activities as well as respiratory studies using glucose radioactively labelled in the 1-C and 6-C position have favoured the latter as the most important route of oxidative glucose dissimilation (Carr, 1973; Smith, 1982). The OPP is thought also to be largely responsible for the supply of reductant to nitrogenase in the heterocyst (Apte et al., 1978; Summers et al., 1995b). In higher plants it is well established that several enzymes of photosynthetic and respiratory metabolism are activated by a process involving thioredoxin acting as a protein disulphide reductase (Buchanan, 1991). The switch between "light" and "dark" metabolism involves the activation/inactivation of enzymes via the reduction/oxidation of intramolecular structural disulphide bonds by thioredoxin. Thus the Calvin cycle enzymes:

fructose-1,6-bisphosphatase,
sedoheptulose-1,7-bisphosphatase,
phosphoribulokinase,

and the chloroplast coupling factor CF1 are all activated in the light via the reduction of disulphide bonds and glucose-6-phosphate dehydrogenase (G6PDH), the key enzyme of the respiratory oxidative pentose phosphate pathway, is inactivated. The reduction of thioredoxin itself is coupled to electron transport through the two photosystems via the reduction of ferredoxin and the enzyme ferredoxin-thioredoxin reductase.

Initially, genetic evidence indicated that cyanobacteria appear to contain two thioredoxins, one of which is similar to the plant m-type thioredoxin and a second much less conserved species which may be unique to cyanobacteria (Alam et al., 1989). The genomic sequence of *Synechocystis* PCC 6803 (Kaneko et al., 1996) has revealed four thioredoxin genes, one of which (slr1139) is clearly similar to the atypical *Anabaena* thioredoxin, one of them is m-type (slr0623) and two exhibit less certain affinities (sll1057, slr0233). Attempts have been made to study the light/dark regulation of photosynthetic enzymes with intact cells of the cyanobacterium *Nostoc* MAC (Austin et al., 1992) and evidence was obtained for the reversible light-activation of fructose-1,6-bisphosphatase, sedoheptulose-1,7-bisphosphatase, ribulose-5-phosphate kinase and NADP-linked glyceraldehyde-3-phosphate dehydrogenase. The light-dependent control of activity of ferredoxin:$NADP^+$ reductase, possibly via a thiol/disulphide exchange has been reported for *Anabaena variabilis* (Fillat et al., 1991). The most clear cut evidence from in vitro studies relates to the oxidative pentose phosphate pathway enzyme G6PDH, where thioredoxin control via a thiol/disulphide exchange mechanism has also been proposed both by Cossar et al., (1984) and Udvardy et al., (1984). Inhibition of G6PDH from *Anabaena* PCC 7120 was far more sensitive to thioredoxin-1 than to thioredoxin-2 (Gleason, 1996). This proposed mechanism has received support from the observation that two cysteine residues are absolutely conserved in all the cyanobacterial *zwf* genes so far sequenced, and are absent from the protein from other prokaryotic sources (Newman et al., 1995).

Additionally, there is also in vitro evidence for the thioredoxin-mediated control of fructose-1,6-bisphosphatase activity from *Anacystis nidulans* (=*Synechococcus* PCC 6301), *Anabaena cylindrica, Synechocystis* PCC 6803 and *Anabaena variabilis* (Udvardy et al., 1982; Ip et al 1984; Pacold et al., 1995). In contrast Bishop (1979) reported no thiol activation for the enzyme from *Anacystis nidulans*. In plants there are two distinct fructose-1,6-bisphosphatases, located in the chloroplast and cytosol, which exhibit different regulatory properties. The cytosolic enzyme is regulated allosterically by AMP, whereas the chloroplast enzyme is regulated by thioredoxin (Clancey and Gilbert, 1987). The chloroplast enzyme, compared to the cytosolic form, typically has an insertion of 12-17 amino acids with two adjacent conserved cysteine residues for the thioredoxin-mediated control of activity (Marcus et al., 1988). Comparison of several cyanobacterial fructose 1,6-bisphosphatases (Newman et al., 1995; Summers et al., 1995a) reveals them to lack the region associated with thioredoxin activation of the chloroplast form of the enzyme. Thus if a thiol/disulfide exchange mechanism is involved in the activation of the cyanobacterial enzyme, the residues involved must be completely different from those involved in the chloroplast enzyme.

In the case of phosphoribulokinase there is a report of the activation of the enzyme from *Chlorogloeopsis fritschii* by dithionite and glutathione (Marsten and Codd, 1984) and there are two cysteine residues in the enzyme from *Synechocystis* PCC 6803 which are in a similar position to those involved in thioredoxin-mediated control of the plant enzyme (Su and Bogorad, 1991). The pyridine nucleotide-dependent D-glucose dehydrogenase of *Nostoc* MAC was reportedly deactivated by thioredoxin (Juhász et al., 1986). Finally, glyceraldehyde-3-phosphate dehydrogenase from *Anabaena variabilis* and *Synechocystis* PCC 6803, which both lack one of the conserved cysteines involved in regulation of activity of the plant enzyme, is insensitive to treatment with dithionite, strongly suggesting an absence of thioredoxin control. Thus there is conflicting evidence for the role of thioredoxin in the light-dark control of enzyme activity, with the strongest evidence so far being obtained for the light-dependent activation of fructose 1,6-bisphosphatase and inactivation of G6PDH. However, the importance of the m-type thioredoxin in cyanobacteria is indicated by the fact that, in contrast to the situation with *E. coli*, attempts at insertional mutagenesis of the gene in *Synechococcus* PCC 7942 yielded no viable segregants lacking thioredoxin (Muller and Buchanan, 1989). One last point is worth noting. The gene encoding a thioredoxin-like protein has been characterized from *Synechococcus* PCC 7942 and mutations in the thioredoxin like domain lead to cells which can scarcely grow, whereas mutations in the C-terminus could lead to an increased content of normal phycobilisomes (Collier and Grossman, 1995).

B. Plastoquinone and the Cytochrome b_6f Complex

Given the central position of the inter-photosystem electron transport chain in the energy generating processes of cyanobacteria, and the sensitivity of its redox state to the intensity and quality of light, it is not surprising that there are many reports concerning the signalling role of its components, particularly plastoquinone and the cytochrome b_6f complex . Two major aspects of cyanobacterial physiology which involve signalling from the intersystem electron transport chain are chromatic adaptation, and the response to the spectral quality of the incident light via state transitions (Chapter 15). The possibility that the thylakoids represent a key site of metabolic integration and control in cyanobacteria, particularly with regard to nutrient acquisition, and the role of the redox signalling in these processes are discussed in Section VIII. There is growing evidence from other organisms that the redox state of the cell may be involved in the control of transcription via two-component sensory systems (Allen, 1993), though as yet there is no direct evidence for such effects in cyanobacteria.

IV. Low molecular Weight Signalling Molecules

A. cAMP

cAMP is an important signalling molecule in both prokaryotes and eukaryotes. In enteric bacteria cAMP via its interaction with the CRP protein exerts control over the expression of large number of operons which are sensitive to catabolite repression. CRP complexed with cAMP acts as a transcriptional activator via interaction with the C-terminal domain of the α-subunit of the $\sigma 70$ form of RNA polymerase (Ebright and Busby, 1995). In animal cells cAMP acts as a second messenger in signal transduction primarily via the activation of protein kinase A. A variety of evidence suggests that cAMP may have a significant physiological role in signal transduction in cyanobacteria. cAMP in cyanobacteria was first detected in *Anabaena variabilis* and a correlation between cAMP content and nitrogen starvation was apparent (Hood et al., 1979). Further, evidence linked cAMP concentrations to nitrogen starvation in *A. flos-aquae* (Francko and Wetzel, 1981) and heterocyst pattern formation in *A. variabilis* (Smith and Ownby, 1981). More recently, cAMP concentrations have been shown to change in response to light/dark transitions and pH shifts in *A. cylindrica* (Ohmori et al., 1988; Ohmori, 1989) and exogenously added cAMP affected both respiratory and gliding activities of *Spirulina platensis* (Ohmori et al., 1993). Most recently it has been shown that intracellular cAMP concentrations in *Anabaena cylindrica* change rapidly in response to red/far red light, with red light causing a decrease in cAMP and far red causing an increase (Ohmori et al., 1998) Bianchini et al. (1990) partially characterized an adenylate cyclase activity from *Anabaena* ATCC 29151. The activity, which was in the soluble fraction of the cell and had an apparent molecular weight of about 183400, was activated by Ca^{2+} and bovine brain or spinach calmodulin and inhibited by EGTA. The gene (*cya*) encoding the enzyme adenylate cyclase responsible for the synthesis of cAMP was first characterized from the cyanobacterium *Anabaena cylindrica* (Katayama et al., 1995). The deduced primary structure of the protein revealed it to be a member of the 'universal' class of adenylate cyclases (Danchin, 1993), which includes members from eukaryotes and prokaryotes, and it exhibited marked similarities to the conserved regions of adenylate cyclase from various eukaryotes. In contrast, there was no detectable similarity with the adenylate cyclases from other prokaryotic classes, including that of *E. coli*. Sequence features in the *A. cylindrica* enzyme suggest that it has a signal sequence for protein export and one trans-membrane domain. Subcellular fractionation localized the enzyme to the thylakoid membrane. Four adenylate cyclase genes have been characterized from *Anabaena* PCC 7120 (Katayama and Ohmori, 1997). All four proteins exhibited similarity in their C-terminal domains to the catalytic domains of eukaryotic adenylate cyclases, but also possessed unique features in their N-terminal domains. CyaA contains two putative membrane-spanning domains. CyaB1 and CyaB2 have a region similar to a cGMP-binding module. CyaC contains three distinct features that include, starting from the N-terminus, a response regulator domain, a histidine protein kinase domain, and a second response regulator domain. The transcript of *cyaC* was found to be the most abundant and insertional mutagenesis of *cyaC* caused a marked reduction in cellular cAMP concentration. A similarly complex adenylate cyclase appears to be encoded by the *cyaC* gene of *Spirulina platensis* (Kasahara et al., 1997). The predicted gene product contains, from the N-terminus, a RR receiver domain, a domain similar to part of ETR1 of *Arabidopsis thaliana*, an HPK transmitter domain, a second RR receiver domain and a catalytic domain.

The genome of *Synechocystis* PCC 6803 encodes two proteins which exhibit similarity to adenylate cyclases and also a protein displaying a marked degree of similarity to *E. coli* CRP both in the C-terminal DNA-binding domain and the N-terminal nucleotide-binding domain (Kaneko et al., 1996). *Synechocystis* PCC 6803 was found to exhibit an increase in the intracellular concentration of cAMP in response to a dark/light transition and that blue light (450 nm) was responsible for the effect (Terauchi and Ohmori, 1998). Mutagenesis of the two *cya* genes revealed that that *cya1* (slr1991) was responsible for the effect and was also implicated in motility. Thus, the evidence that cyanobacteria produce cAMP is convincing and a putative cAMP-receptor transcription factor has been identified, though there is much to do on its physiological effects and specific gene targets. Furthermore, that characterization of adenylate cyclase genes which encode proteins which additionally possess RR receiver and HPK transmitter domains suggests that they may be involved in complex signalling pathways that may be similar to, or indeed overlap with, the multistep phosphorelay pathways (Section II.A).

B. Ca^{2+} ions

Like cAMP, Ca^{2+} ions are important as a second messenger in signal transduction in eukaryotes. However, their role in signal transduction in prokaryotes is far less clearly established. There is evidence for an involvement of Ca^{2+} ion signalling in a range of processes in bacteria, including chemotaxis in *Bacillus subtilis* and *E. coli*, and gliding motility in *Myxococcus xanthus* and *Stigmatella aurantiaca* (Norris et al., 1996). In cyanobacteria, Ca^{2+} has been implicated in a variety of cellular processes (Smith, 1988; Onek and Smith, 1992) including, the motility responses of *Spirulina subsalsa* (Abeliovich and Gan, 1982) and *Phormidium uncinatum* (Häder, 1982) as well as heterocyst differentiation in *Nostoc* strain 6720 (Smith et al., 1987). An abundant cell surface protein involved in the swimming motility of marine *Synechococcus* strains has several Ca^{2+}-binding sites (Brahamsha 1996; Section V.B) as does the oscillin protein involved in gliding motility (Hoiczyk and Baumeister, 1997). Estimates of the free intracellular Ca^{2+} concentration in *E. coli* are in the range 0.1 to 1 μM (Gangola and Rosen, 1987), which are several orders of magnitude less than those often found outside the cell. Similarly, in *Anabaena* PCC 7120 the intracellular concentration of Ca^{2+} is in the range 100-200 nM and may alter in response to environmental signals such as cold shock (Torrecilla et al., 1998). In eukaryotes, the intracellular concentration of Ca^{2+} is maintained by the action of Ca^{2+}-ATPases; however, many species of bacteria pump Ca^{2+} out of the cytoplasm by secondary transport involving H^+ or Na^+ antiports (Rosen, 1987). In fact, Ca^{2+}-ATPases do occur in some species of bacteria including cyanobacteria. It was found that *Synechococcus* PCC 7942 contains at least two distinct P-type ATPases: one belongs to the family of typical prokaryotic P-type ATPases and the other markedly resembles eukaryotic P-type ATPases (Kanamaru et al., 1993). Insertion mutants of either of these two ATPase-genes were hypersensitive to osmotic stress. Inactivation of the gene *pacL* encoding the eukaryotic-type Ca^{2+}-ATPase caused a complete loss of ATP-dependent Ca^{2+} transport activity associated with the plasma membrane but, somewhat surprisingly, such a mutant did not exhibit a Ca^{2+}-sensitive phenotype (Berkelman et al., 1994). It may be that the Ca^{2+}-sensitive phenotype was suppressed by the activity of other Ca^{2+}-ATPases, since there appear to be at least three such enzymes encoded by the genome of *Synechocystis* PCC 6803 (Kaneko et., 1996).

In nearly all eukaryotes the protein calmodulin serves as a Ca^{2+} sensor. As calmodulin becomes activated by Ca^{2+} binding to its four EF hand binding sites, it in turn stimulates a variety of enzymes, pumps and other targets including the calmodulin-dependent protein kinase II and the Ca^{2+}-ATPase pump. Thus, an important question regarding the signalling role of Ca^{2+} in cyanobacteria and other prokaryotes is whether a calmodulin homologue is involved or not. Calmodulin has several distinctive characteristics facilitating detection and purification, including heat stability and small size (15-21 kDa). Indirect identification of calmodulin in cell-free extracts has been reported for a variety of bacteria (Onek and Smith, 1992). In cyanobacteria, the first report of a calmodulin-like protein was for *Oscillatoria limnetica*; the evidence was based on the effects of fluphenazine, a Ca^{2+}/calmodulin-inhibitor, on phosphate uptake and also on the radioimmunoassay quantitation of an antigen which cross-reacted with an anti-calmodulin antibody (Kerson et al., 1984). Subsequently, Pettersson and Bergman (1989) described a heat-stable factor in extracts of *Anabaena* strains that exhibited the calmodulin-like property of stimulating cucumber NAD kinase in a Ca^{2+}-dependent reaction. Moreover, the authors were able to detect a 17 kDa protein by western blotting that cross-reacted with a polyclonal antibody against spinach calmodulin. The antigen was detectable in both vegetative cells and heterocysts. Bianchini et al. (1990) reported the partial purification of three heat stable polypeptides from *Anabaena* ATCC 29151 capable of the Ca^{2+}-dependent activation of bovine brain cAMP phosphodiesterase, and they also found that the adenylate cyclase activity from this strain was activated by Ca^{2+} and bovine brain or spinach calmodulin. The requirement of cyanobacterial photosystem II for Ca^{2+} was first demonstrated by England and Evans (1983) and the sensitivity of this Ca^{2+}-dependent activity to anti-calmodulin antibody was reported by Tramontini (1990). Western blot analysis of photosystem II preparations using an anti-calmodulin antibody detected a band of 58-60 kDa and $^{45}Ca^{2+}$ overlay of SDS-PAGE gels of photosystem II preparations revealed bands of approximately 60 and 18 kDa (McColl and Evans, 1990). It should be borne in mind that phycobiliprotein subunits, which are of a similar size to the 18 kDa polypeptide also bind Ca^{2+} in the overlay protocol (Onek et al., 1994). More recently a 21 kDa polypeptide was isolated from *Nostoc* PCC 6720 which exhibited many of the properties associated with calmodulin (Onek et al., 1994). It activated pea NAD kinase in a Ca^{2+}-dependent fashion, reacted with antibodies raised against spinach calmodulin, and exhibited a Ca^{2+}-dependent shift in migration on SDS-PAGE. Unfortunately, no amino acid sequence information was obtained for the protein and thus the true nature of this putative cyanobacterial calmodulin was not established. However, when all the ORFs in the genome of *Synechocystis* PCC 6803 are compared with calmodulin sequences from a variety of sources no significant similarities are detected. Thus, the only safe conclusion from all this evidence is that, if there is a cyanobacterial calmodulin, it is likely to be a functional analogue rather than a true homologue. The real physiological significance of Ca^{2+} ion signalling in cyanobacteria remains to be established.

C. Guanosine tetraphosphate

The synthesis of macromolecular components must obviously be tightly co-ordinated with the growth rate of cell and, given the OFTEN rapidly fluctuating environments, precise mechanisms have evolved to control the synthesis of rRNA and thereby control the protein synthetic capacity of the cell (Condon et al., 1995). In *E. coli* there appear to be at least four mechanisms by which rRNA accumulation is determined and in two of these, the stringent response and growth rate dependent control, the key intracellular signal is the concentration of the unusual nucleotide guanosine 3′-diphosphate 5′-diphosphate (ppGpp). A further role for ppGpp is suggested by the observation that it acts as a positive signal for transcription of the *rpoS* gene encoding the σ^S σ-factor involved in the expression of many stationary phase genes (Gentry et al., 1993). ppGpp was first detected in the unicellular cyanobacterium *Synechococcus* PCC 6301 in response to a shift down in growth rate brought about by a reduction in the incident light intensity (Mann et al., 1975). However, the concentration of ppGpp fell before the accumulation of rRNA restarted suggesting that there were additional mechanisms involved. Smith and Carr (1977) also demonstrated ppGpp synthesis in *Synechococcus* PCC 6301 under conditions of amino acid starvation. Variations in the responses of different cyanobacterial strains were detected by Adams et al (1977) who confirmed the production of ppGpp in *Synechococcus* PCC 6301 when transferred from the light to the dark. Similar effects were observed in a *Nostoc* strain and *Aphanocapsa* 6714, but not in *Anabaena cylindrica* and *Anabaena catenula*. In contrast, there are several reports of ppGpp accumulation in response to nitrogen starvation in *Anabaena cylindrica* (Akinyanju and Smith, 1979, 1982, 1987) and also *Anacystis nidulans* (Friga et al., 1981). The production of ppGpp by *Synechococcus* PCC 6301 has also been confirmed by Borbély et al. (1980), who also established that bacteriophage infection interfered with ppGpp accumulation induced by either energy or nitrogen starvation. Surányi et al. (1987) also showed that there was an inverse correlation between the rate of total RNA synthesis and the pool size of ppGpp in *Synechococcus* PCC 6301, except immediately after a temperature shift when parallel increases were found.

The routes of ppGpp synthesis in *E. coli* are reasonably well characterised. During the stringent response to amino acid starvation, the *relA* gene product, (p)ppGpp synthetase I, catalyses the synthesis of ppGpp and the pentaphosphate derivative pppGpp. pppGpp may be converted to ppGpp by the action of the *gpp* gene product (Condon et al., 1995). Overexpression of RelA causes a marked reduction in the growth rate presumably by inhibiting rRNA accumulation. There is a second pathway for the synthesis of ppGpp employing the product of the *spoT* gene which encodes a bifunctional enzyme, (p)ppGpp synthetase II, which can catalyse both ppGpp synthesis and hydrolysis. This second pathway seems to be associated with ppGpp synthesis in response to fluctuation in intracellular energy source pools. The genome of *Synechocystis* PCC 6803 encodes a polypeptide (slr1325) with marked similarity to the (p)ppGpp 3'-pyrophosphohydrolase encoded by *spoT*. However, no homologue of *relA* can be detected. Thus, it seems probable that ppGpp synthesis (and degradation) in *Synechocystis* PCC 6803 is catalysed only by the SpoT homologue. The true physiological significance of ppGpp in cyanobacteria will only be realized when gene disruption experiments have been carried out in *Synechocystis* PCC 6803 and preferably also a heterocystous strain.

D. Metabolite pools

The concentrations of certain key metabolites are likely to act as signals reflecting the metabolic state of the cell and several metabolites have been implicated as having signalling roles in adaptive responses in cyanobacteria. Intermediates of both the oxidative and reductive pentose phosphate pathways, glucose-6-phosphate and ribulose-5-phosphate, have been shown to exert marked effects on the patterns of protein phosphorylation (Mann et al., 1991), and this is particularly the case for thylakoid protein phosphorylation (unpublished data). Further evidence for a signalling role for glucose-6-phosphate is provided by the fact that glucose, but not analogues that cannot be phosphorylated at the C-6 position, exerted an inhibitory effect on the bicarbonate concentrating system of *Synechocystis* PCC 6803 (Bloye et al., 1992). Also in the area of protein phosphorylation, the activity of the kinase responsible for phosphorylation of the P_{II} protein is stimulated by 2-oxoglutarate (Forchhammer and Tandeau de Marsac, 1995b). In enteric bacteria carbamoyl phosphate and acetyl phosphate can directly phosphorylate some two-component RRs and, indeed, it has been proposed that acetyl phosphate may represent a global signal that there is an imbalance in central metabolism (McCleary et al., 1993). Lastly, although no such roles for carbamoyl phosphate and acetyl phosphate have yet been established in cyanobacteria, cyanate is implicated in the control of expression of genes involved in both carbon and nitrogen metabolism (Suzuki et al., 1996; Chapter 15). It should be noted that cyanate accumulates as a breakdown product of carbamoyl phosphate.

V. Behavioural responses

A. Circadian rhythms

Circadian rhythms which adapt organisms to daily changes in their environments were until recently thought to be exhibited only by eukaryotic organisms. However, there is now unequivocal evidence that circadian rhythms occur in prokaryotes, the bulk of which has been obtained with cyanobacteria (Huang and Grobbelaar, 1995; Johnson et al., 1996; Golden et al., 1997). Given the daily changes in their light environment, it is perhaps not surprising that circadian rhythms should be such a characteristic feature of the physiology of the cyanobacteria. The first reports of circadian rhythms in cyanobacteria documented daily patterns of dinitrogen fixation in cultures of *Synechococcus* and *Oscillatoria* grown in a light/dark cycle (Stal and Krumbein, 1985; Mitsui et al., 1986; Grobbelaar et al., 1986); more recently, they have been shown in the marine *Trichodesmium* IMS101 (Chen et al., 1996; Chen et al., 1998). Importantly these rhythms persist in constant light following entrainment during growth under light/dark cycling, thus fulfilling another requirement of a true circadian rhythm. Indeed, in *Synechocystis* PCC 6803 circadian rhythms persisted in the dark after entrainment (Aoki et al., 1997). The incompatibility of O_2-evolving photosynthesis and O_2-sensitive nitrogenase activity was thus resolved by the two processes exhibiting opposite phases within the circadian rhythm (Mitsui et al., 1986). An important feature of circadian rhythms is that they should be temperature-compensated i.e. their periodicity should be independent of temperature over an appropriate physiological range. Such temperature-compensated circadian behaviour was observed for cell division in the marine phycoerythrin-containing *Synechococcus* WH7803 (Sweeney and Borgese, 1989), and photosynthetic O_2 production, nitrogenase activity and cell division in *Synechococcus* Miami BG 43511 (Mitsui et al., 1993). Similar observations were made with regard to dinitrogen fixation and amino acid transport for *Synechococcus* strain RF-1 (Huang et al., 1990; Chen et al., 1991).

The application of genetic and molecular biological tools is beginning to provide an understanding of the clock mechanism(s) and other features of circadian rhythms in cyanobacteria. Huang et al. (1990) established for *Synechococcus* RF-1 that nitrogenase mRNA abundance was rhythmic under conditions of constant light, implying that transcription itself, or RNA degradation, exhibited periodicity. An important development was the discovery that the genetically amenable *Synechococcus* PCC 7942 exhibited circadian rhythms. The *lux* genes from *Vibrio harveyii* were introduced into this strain under the control of the promoter of the *psbAI* gene (encoding a component of PSII) and the consequent luminescence exhibited a marked circadian rhythm (Kondo et al., 1993). Further studies with this system confirmed entrainment of the luminescence rhythm during a light dark cycle, subsequent persistence of the rhythm under continuous light, and temperature compensation (Kondo et al., 1993; Kondo and Ishiura, 1994). Such studies established that these features reflected transcription of the *psbAI* gene (Liu et al., 1995a). Mutants have been isolated which are apparently affected in their circadian rhythms. Huang et al. (1993) described mutants of *Synechococcus* RF-1 in which either or both the dinitrogen fixation and amino acid uptake rhythms were affected. A large number of mutants of *Synechococcus* PCC 7942 affected in their circadian rhythms have been isolated exhibiting a variety of clock-disrupted phenotypes, such as altered periodicity and arrhythmia; and the majority of these mutations could be complemented by an 8 kb fragment of wild type DNA (Johnson et al., 1996). Random screening of fragments of *Synechococcus* PCC 7942 DNA for circadian promoter activity with the *luxAB* genes (Liu et al., 1995b) revealed that the majority of DNA fragments which exhibited promoter activity displayed rhythmic behaviour and a variety of different phase relationships were apparent. Thus the metabolism of the organism is likely to be dominated by rhythmic behaviour. The *cis*- and *trans*-acting components that effect the circadian control of gene expression have still largely to be defined. However, one gene from *Synechococcus* PCC 7942 which affects the amplitude of the rhythm associated with some promoters has been shown to encode a σ^{70}-like transcription factor. This gene is a member of the family of four principal σ-factor genes in this organism (Tsinoremas et al., 1996) (Section IV.A). The relationship of circadian rhythms to the cell cycle has been investigated and it is apparent that the circadian clock is completely independent of the cell cycle, but the timing of DNA replication and cell division are not independent of the circadian clock (Johnson et al., 1996; Kondo et al., 1997) and this is in keeping with the circadian rhythm of expression of *dnaN* an essential DNA replication gene (Liu and Tsinoremas, 1996). Control of the circadian rhythm is associated with the *kaiABC* gene cluster and there is evidence to suggest that a negative feedback of *kaiC* expression by KaiC generates the circadian oscillation which is sustained by KaiA (Ishiura et al., 1998). The *pex* gene product appears to extend the period of the circadian clock, possibly by interacting with KaiC (Kutsuna et al., 1998). It seems likely that cyanobacterial physiology in nature is dominated by rhythmic cycles and this complicates attempts to understand their ecology based on information from cultures grown under constant illumination. The adaptive significance of a resonating circadian clock has now been demonstrated by the fact that the relative fitness of various strains of cyanobacteria expressing different circadian rhythms was optimal when the circadian period was similar to the light/dark cycle (Yan et al., 1998).

B. Motility

1. Modes of Motility and Taxes

Cyanobacteria exhibit two forms of movement, gliding and swimming, which enable them to respond to changes in their environment. Furthermore, some planktonic species are able to produce gas vesicles which adjust their buoyancy and consequently their position in the water column (Walsby, 1994). Gliding motility has long been recognised as a property of many groups of cyanobacteria and was once thought to be the only active form of motility exhibited by these organisms. Gliding motility involves self-propulsion across a solid or semi-solid surface without the involvement of an organelle (e.g. a flagellum), or a change of shape of the organism. Previously, there have extensive studies on the gliding motility behaviour of cyanobacteria and the ecological significance of these movements (Castenholz, 1982). Much of this earlier work focused on the motility responses to light although chemotactic and aggregative movements were observed. Castenholz (1982) describes three photic motility responses, all of which have been observed with various strains of cyanobacteria. Photokinesis is where the rate of movement is enhanced or inhibited by changes in the light intensity. In a photophobic response a change in light intensity elicits a change in movement, usually a stop, followed by reversal of direction. Phototaxis is a positive or negative movement oriented with respect to the direction of the light.

Few studies have addressed the nature of the chemotactic response associated with gliding motility. Negative chemotactic responses to acids were observed with some Oscillatoriaceae (Fechner, 1915; cited by Richardson and Castenholz, 1989) and sulphide with *O. terebriformis* (Richardson and Castenholz, 1987). Positive light-dependent responses to CO_2, HCO_3^- and O_2 were observed with another *Oscillatoria* (Malin and Walsby, 1985). Richardson and Castenholz (1989) described a transient inhibition of gliding mobility in *O. terebriformis* by fructose. The duration of inhibition was proportional to the concentration of fructose while the rate of movement following restoration of motility was inversely proportional to the concentration. Sulfide exerted a similar effect, but not glucose, acetate or lactate.

Little is known about the mechanism of gliding motility except that it is apparently dependent on both the ΔpH and $\Delta\psi$ components of the proton motive force (Castenholz, 1982). A variety of unusual cellular structures which may or not be associated with gliding motility have previously been observed (Hoiczyk and Baumeister, 1995). Electron microscopy of the envelope structure of four gliding Oscillatoriaceae has shown that the wall is covered with a complex external double layer (Hoiczyk and Baumeister, 1995). The inner of these layers is a tetragonal crystalline S-layer anchored to the inner membrane. The external layer is formed by parallel, helically arranged surface fibrils, which are suggested to be involved in generation of propulsive force, since the orientation of these fibrils is consistent with the direction of filament rotation during gliding. There are reports of actin-like proteins in *Synechocystis* PCC 6803 (Labbe et al., 1996) and *Anabaena* spp. (Guerrero-Barrera et al., 1996). Recently, the surface fibrils of *Phormidium uncinatum* were shown to be composed of a single rod-shaped 66 kDa protein, termed oscillin, which has a N-terminal domain containing 46 repeats of a Ca^{2+}-binding motif (Hoiczyk and Baumeister, 1997).

The first report of a cyanobacterium exhibiting swimming motility was for marine phycoerythrin-containing *Synechococcus* isolates (Waterbury et al., 1985). These did not exhibit photokinesis, photophobic or photactic responses and motility was retained during dark incubation. There was no evidence of flagellar involvement, although the observed rotation of the cells around their longitudinal axis resembled flagellum-mediated motility. This swimming behaviour, which was confined to open ocean isolates, exhibited a marked chemotactic response to nitrogenous compounds including NH_4^+, nitrate, glycine, β-alanine and urea (Willey and Waterbury, 1989). The threshold levels at which chemotactic responses were observed were in the range 10^{-9} - 10^{-10} M, much lower than those reported for other bacteria, and in the range that would be ecologically significant in the ocean. Energization of this motility depends on the sodium motive force in contrast to most other bacteria in which it depends on the proton motive force (Willey et al., 1987). There is still confusion as to the mechanism involved in swimming motility. Self-electrophoresis has been excluded (Pitta and Berg, 1995) and a mechanism involving travelling surface waves has been proposed (Ehlers et al., 1996), but as yet has no experimental support. A molecular biological approach to the problem has identified an abundant cell-surface associated 130 kDa polypeptide that is required for swimming motility (Brahamsha, 1996). Insertional inactivation of the gene *swmA* encoding this protein results in loss of ability to translocate, but the cells still generate torque. The SwmA protein is likely to be glycosylated and contains a number of Ca^{2+}-binding motifs which may be involved in its attachment to the cell surface (Brahamsha, 1996). This is consistent with the observation that calcium is required for swimming motility by *Synechococcus* WH 8113 (Pitta et al., 1997).

The third form of motility exhibited by some species of cyanobacteria is the control of buoyancy to determine position within the water column via the formation of gas vesicles (Walsby, 1994; Chapter 6). The external stimulus triggering gas vesicle formation appears to be low incident light intensity, but nothing is yet known about the nature of the internal signal othern than that it results in a transcriptional response.

2. Signalling and Motility

Both the gliding motility and swimming motility exhibited by cyanobacteria exhibit clear chemotactic components, in contrast to buoyancy control via gas vesicle formation that results in a passive change of position due to buoyancy changes. Thus, the two former processes will involve the processing and transduction of intracellular signals that are likely to have a strong temporal component with regard to the sensing of environmental concentrations of attractants and repellents. Studies on flagellum-mediated chemotaxis in enteric bacteria have defined the components of the sensory transduction pathway (Amsler and Matsumara, 1995) which involves the elements of a two-component sensory system. The detection of chemotactic effectors and transduction of the signal into the cytoplasm involves a group of trans-membrane proteins, termed methyl-accepting chemotaxis proteins (MCPs). The first cytoplasmic component of the sensory transduction pathway is CheA which is a member of the HPK family (section II.A). Interactions between CheA and the MCPs are facilitated by CheW. Autophosphorylation of CheA is followed by phosphotransfer to CheY, a member of the RR family that determines the direction of flagellar rotation. Dephosphorylation of CheY is catalysed by CheZ. The temporal memory of the signalling system resides in the methylation state of the MCPs, with the CheR protein catalysing methylation and CheB, a member of the RR family, acting as a methylesterase after phosphorylation by CheA. The Gram-negative bacterium *Myxococcus xanthus* exhibits gliding motility and social behaviour and the *frz* genes involved in this motility are homologous to the enteric bacterial *che* genes involved in swimming motility, and at least one of the MCP-encoding genes (Shi and Zusman, 1995). Thus there are marked similarities in the signal transduction pathways involved in the swimming and gliding motility of two diverse groups of bacteria. Proteins antigenically related to the MCPs were detected in the membrane fraction of *Synechocystis* PCC 6803, but were not detected in *Oscillatoria tenuis* (Morgan et al., 1993). Analysis of the genomic sequence of *Synechocystis* PCC 6803 reveals several ORFs with marked similarity to chemotaxis proteins from other organisms (Kaneko et al., 1996). There are three well separated discrete groups of apparently chemotaxis-related ORFs, in which the individual member ORFs are contiguous suggesting that they constitute putative operons (Table 3). A non-exhaustive analysis of the polypeptides encoded by these putative operons reveals similarities to the chemotaxis-related proteins of *Myxococcus xanthus* (Shi and Zusman, 1995), *Pseudomonas aeruginosa* (Darzins, 1994), *Bacillus subtilis* (Hanlon et al., 1992) and *Desulfovibrio vulgaris* (Deckers and Voordouw, 1994), which are in turn related to the chemotaxis proteins of *E. coli* (Amsler and Matsumara, 1995). Overall the most frequent similarities were to the *pil* genes of *Ps. aeruginosa* associated with pilus synthesis and twitching motility (Darzins, 1994), and it is somewhat surprising that each of the putative operons begins with two ORFs each exhibiting similarity to the PilGH proteins as does the *pil* operon itself. Consequently, it seems very likely that chemotactic responses in cyanobacteria, involving both gliding and swimming motility, may involve proteins similar to those involved in motility in other bacteria.

VI. Alterations in transcriptional specificity

As with other bacteria, the specificity of the transcriptional apparatus in cyanobacteria may be modified in response to environmental stimuli or during the course of cell differentiation by the interaction of the core RNA polymerase with alternative sigma factors which may modulate the promoter recognition specificity of the enzyme. Alternatively, the action of transcriptional regulatory factors can affect the expression of specific genes or groups of genes under their control (reviewed by Curtis and Martin, 1994).

A. σ - factors

Sequence information for a large number of σ -factors has revealed that they fall into two classes, based on the *E. coli* σ^{70} and σ^{54} subunits, and there is apparently no evolutionary relationship between the two classes (Lonetto et al., 1992). The σ^{70} family has been divided in three groups (Lonetto et al., 1992) with Group 1 representing the essential primary σ-factors in each organism, being responsible for the bulk of transcription and essential for growth. Group 2 σ-factors are very similar in sequence to the group 1 members, but are not essential. Group 3 σ-factors are involved in the transcription of specific regulons and differ considerably in sequence from group 1 members. The σ^{54} family of σ-factors is involved in the transcription of genes involved a diverse range of physiological processes in a wide variety of organisms (Kustu et al., 1989) and is unique in that eukaryotic enhancer-like transcriptional activators are required for effective transcription (North et al., 1993). The first cyanobacterial σ-factor gene to be characterized was that encoding the main σ^{70}-like subunit (SigA) from *Anabaena* PCC 7120 (Brahamsha and Haselkorn, 1991) following the biochemical characterization of the main form of the holoenzyme (Schneider et al., 1987). It is now clear that there are multiple genes encoding σ^{70}-like subunits in several cyanobacterial strains. Two further σ-factor genes *sigB* and *sigC* have been characterized from *Anabaena* PCC 7120 (Brahamsha and Haselkorn, 1992) and four σ-factor genes from *Synechococcus* PCC 7942 (Tanaka et al., 1992a; Tanaka et al., 1992b). Similarly, there are four genes encoding σ^{70}-like subunits in the genome of *Synechocystis* PCC 6803 (Kaneko et al., 1996).

Developmental processes in prokaryotes, such as sporulation in *Bacillus subtilis*, commonly involve sequential gene expression with alternative σ-factors (Haldenwang, 1995). The pattern of transcription of *sigA* in *Anabaena* PCC 7120 is complex with multiple transcripts some of which vary in abundance under nitrogen-fixing growth conditions (Brahamsha and Haselkorn, 1991). Transcription from *sigB* and *sigC* occurs transiently after nitrogen starvation, but neither gene alone is required for nitrogen fixation or heterocyst development (Brahamsha and Haselkorn, 1992). Thus, the role of alternative sigma factors in the differentiation and physiology of heterocysts remains unclear. One clearly established role for one

Table 3. The ORFs constituting three putative chemotaxis operons encoded in the genome of *Synechocystis* sp. PCC 6803 (Kaneko et al., 1996). The homologies were detected using the BLAST algorithm and the SWISSPROT database (unpublished results - this laboratory).

Operon	ORF	Homologue	Organism
1	1 (sll0038)	PilGH	*Pseudomonas aeruginosa*
	2 (sll0039)	PilGH	*Pseudomonas aeruginosa*
	3 (sll0040)	FrzA	*Myxococcus xanthus*
	4 (sll0041)	PilJ	*Pseudomonas aeruginosa*
	5 (sll0042)	DcrH	*Desulfovibrio vulgaris*
	6 (sll0043)	FrzE	*Myxococcus xanthus*
2	1 (sll1291)	PilGH	*Pseudomonas aeruginosa*
	2 (sll1292)	PilGH	*Pseudomonas aeruginosa*
	3 (sll1293)	PilI	*Pseudomonas aeruginosa*
	4 (sll1294)	PilJ	*Pseudomonas aeruginosa*
	5 (sll1296)	FrzE	*Myxococcus xanthus*
3	1 (slr1041)	PilGH	*Pseudomonas aeruginosa*
	2 (slr1042)	PilGH	*Pseudomas aeruginosa*
	3 (slr1043)	CheW	*Bacillus subtilis*
	4 (slr1044)	PilJ	*Pseudomonas aeruginosa*

of the additional sigma factors is for that encoded by the *rpoD2* gene of *Synechococcus* PCC 7942, which is involved in the control of circadian rhythms (Tsinoremas et al., 1996) (Section V.A) and would consequently represent a Group 2 σ-factor. The group 2 σ-factor SigE of *Synechococcus* PCC 7002 is associated with the transcription of post-exponential phase genes (Gruber and Bryant, 1998). *Nostoc punctiforme* ATCC 29133 can establish a symbiotic relationship with several plants including *Anthoceros punctatus* via infection by motile hormogonium filaments and transcription of a group 2 σ-factor gene is transcribed only in the presence of plant exudate (Campbell et al., 1998). The genomic sequence of *Synechocystis* PCC 6803 reveals two σ-factors which probably can be classified as Group 3 since, although they show some relatedness to σ^{70}, they are more closely related to the σ^E and σ^F sporulation-specific σ-factors of *Bacillus subtilis* (Kaneko et al., 1996). Although σ^{54} genes are widely distributed amongst prokaryotes, there is no direct evidence for their occurrence in cyanobacteria. Indeed, there is no ORF encoded by the genome of *Synechocystis* PCC 6803 that exhibits extended similarity with any members of the σ^{54} family. However, there is evidence of proteins with marked similarities to the enhancer-like proteins associated with the activation of transcription of σ^{54}-dependent genes. Analysis of a light-repressed transcript found in *Synechococcus* PCC 7002 revealed an ORF encoding a polypeptide exhibiting similarity to σ^{54} transcriptional modulators from *Klebsiella pneumoniae* and *Azotobacter vinelandii* (Tan et al., 1994). Thus, much remains to be done to clarify the role of alternative σ-factors in cyanobacteria, particularly with regard to the physiological functions of σ^E and σ^F homologues.

B. Regulatory Proteins

Several features of the role of transcriptional regulator proteins are described elsewhere in this volume. Control of expression of the *psbA* genes in response to light intensity and promoter binding factors involved in complementary chromatic adaptation are both discussed in Chapter 15, and the role of the SmtB protein in control of metallothionein expression is covered in Chapter 16. Analysis of the genome of *Synechocystis* PCC 6803 has identified several putative transcriptional regulatory proteins on the basis of sequence homologies (Kaneko et al., 1996), though relatively few examples with good evidence about the signal that these proteins are involved in transducing.

Iron is essential for cyanobacteria (Straus, 1994), as iron is involved in a variety of cellular processes including photosynthetic electron transport. Iron stress results in major physiological and ultrastructural changes, and induces the synthesis of several proteins including a novel chlorophyll-binding protein and flavodoxin which acts as a replacement for the iron-containing ferredoxin. In *E. coli*, expression of genes in response to iron limitation involves the *fur* gene which encodes a DNA-binding repressor protein which is activated by the binding of iron. A homologue of the gene *fur* involved in ferric uptake regulation in *E. coli* was characterized from *Synechococcus* PCC 7942 and the polypeptide product exhibits 41% similarity (36% identity) to the *E. coli* Fur protein and includes the putative Fe-binding site (Ghassemian and Straus, 1996). Insertional mutagenesis of the cyanobacterial *fur* gene caused constitutive synthesis of flavodoxin and hydroxamate siderophores.

Two distinct types of regulatory protein NtcA and NtcB are transcriptional activators that have been implicated in nitrogen metabolism. The *ntcB* gene is apparently involved in the partitioning of fixed N (Chapter 15) and encodes a member of the LysR family of transcriptional activators (Suzuki et al., 1995; Henikoff et al., 1988). The *lrrA* gene product of *Synechococcus* PCC 7942 also belongs to the LysR family (Anandan et al., 1996). NtcA, the global regulator of nitrogen assimilation (Chapter 15) is a member of the CRP-FNR family of transcriptional activators (Vega-Palas et al., 1992; Shaw et al., 1983) and is one of six cyanobacterial proteins which belong to the family. These include the CysR protein involved in the response to S limitation (Laudenbach and Grossman, 1991; Chapter 15) and the PtrA protein that is putatively involved in the response to Pi limitation (Scanlan et al., 1997). Furthermore, a homologue of the CRP protein is encoded by an ORF in the genome of *Synechocystis* PCC 6803 and there are two additional ORFs (sll1924, slr0449) which appear to belong to CRP-FNR family (Kaneko et al., 1996). The structure of the dimeric *E. coli* CRP molecule is known at 2.5 Å resolution (Weber and Steitz 1987). The monomer subunits consist of an amino-terminal domain carrying the cAMP-binding site and the carboxy-terminal domain containing the helix-turn-helix DNA binding domain. The CRP of *E. coli* and presumably the other members of this family are thought to act as transcriptional activators via interactions with the C-terminal domain of the α-subunit of RNA polymerase (Ebright and Busby, 1995). A comparison of sequence features upstream from the promoters of genes involved in sulfur acquisition etc. may reveal the cyanobacterial equivalents of the features recognised by CRP and FNR.

In addition to regulatory proteins that affect transcription, there may be regulatory proteins that exert their influence at a post-transcriptional level. RNA helicases convert dsRNA to ssRNA through an ATP-dependent unwinding of the dsRNA, which may serve to activate the RNA for translation. Two RNA helicases CrhC and CrhL have been characterized from *Anabaena* PCC 7120 and have been implicated in the responses to cold shock and redox state, respectively (Chamot et al., 1998).

VII. Other signalling processes

A. OpcA

Control of activity of the key enzyme of the oxidative pentose phosphate pathway is obviously central to understanding the control of dark metabolism in cyanobacteria and consequently much interest has been focused on the regulatory properties of glucose 6-phosphate dehydrogenase (G6PDH). Apart from the potential control via thioredoxin (Section III.A), metabolites including NADPH (Pelroy et al., 1976; Apte et al., 1978), ribulose -1,5-bisphosphate (Lex and Carr, 1974) and ATP (Grossman and McGowan, 1975), have been implicated in regulation. Schaeffer and Stanier (1978) proposed that cyanobacterial G6PDHs may exist in a range of catalytic/aggregation states and that the transitions between such states determine enzyme activity. Recent studies (Summers et al., 1995b; unpublished data) have shown that a gene, designated *opcA*, immediately downstream from the *zwf* gene (encoding G6PDH) is required for cells to express significant G6PDH activity. Mutations in the *opcA* gene appear to alter the oligomeric state of G6PDH without altering the total amount of G6PDH polypeptide being synthesised (Sundaram *et al.*, 1998). Thus, OpcA may represent another component of the control system regulating G6PDH activity via assembly into the functional oligomeric enzyme.

B. Post-translational covalent modifications

In addition to phosphorylation, proteins are potentially subject to a variety of other modifications, including proteolytic cleavage ADP-ribosylation, uridylylation, methylation, lipoylation etc. There is some limited occurrence of forms of protein modification, other than phosphorylation, in cyanobacteria. Zehr et al., (1993) reported a diel modification, resembling ADP-ribosylation, of the Fe-protein of nitrogenase in natural populations of *Trichodesmium thiebautii.* Similarly, periodic modification of the Fe-protein of nitrogenase in *Gloeothece* ATCC 27152 grown under a light/dark cycle was detected (Du and Gallon, 1993). Silman et al., (1995) reported the ADP-ribosylation-dependent inactivation of glutamine synthetase in *Synechocystis* PCC 6803, though this observation is in conflict with other reports concerning the regulation of glutamine synthetase activity (Chapter 15). However, Joseph and Meeks (1987) have also reported an apparent post-translational modification of glutamine synthetase from a symbiotic *Nostoc* strain.

In all organisms energy-dependent proteases, in addition to their cellular housekeeping role, may mediate the turnover of specific short-lived regulatory proteins, which gives them a central position in regulating the response to many sorts of physiological stress. The Clp ATPases were originally identified as regulatory components of bacterial proteases such as ClpP of *E. coli* (Wawrzynow et al., 1996) and are members of the ubiquitous HSP100 (Schirmer et al., 1996). It was shown that *Synechococcus* PCC 7942 contained a homologue of the chloroplast ClpC protein (Clarke and Eriksson, 1996) and from analysis of the *Synechocystis* PCC 6803 genome it appears that the ClpP protease itself is encoded by a multigene family (Kaneko et al., 1996), in contrast to other eubacteria. In *Synechococcus* PCC 7942 mutation of the *clpP1* gene encoding one of the four ClpP protease isozymes led to restricted growth and light acclimation (Clarke et al., 1998) as well as failure to adapt to UV-B and low temperatures (Porankiewicz et al., 1998). Mutation of the gene encoding one of the Clp ATPases, ClpB, also significantly affected both the development of thermotolerance (Eriksson and Clarke, 1996) and cold acclimation (Porankiwicz and Clarke, 1997) by *Synechococcus* PCC 7942. Thus energy-dependent proteases play an important role in regulating important cyanobacterial adaptive responses.

VII. Cross-talk and Integration of Signalling

Key areas of the integration of metabolism in cyanobacteria are photosynthesis and respiration (Peschek, 1996) and photosynthesis and nutrient acquisition, since limitation for one or more nutrients markedly increases the non-productive excitation of photosynthetic reaction centres leading to photoinhibition. Furthermore, limitation for CO_2 may lead to the production of toxic oxygen species. Apart from effects on the phycobilisome content of the cell (Chapter 15), starvation for nutrients may affect the distribution of light energy between the two photosystems, in a fashion analogous to a light-induced state transition, and a transition from linear to cyclic photosynthetic electron transport. In this context, The NblR response regulator has been shown to play a pivotal role in integrating different environmental signals that link cellular metabolism to light harvesting and the activities of the photosynthetic apparatus. It has been suggested that the cyanobacterial thylakoids are the site of metabolic integration for a range of disparate signals and that protein phosphorylation is involved in this mechanism (Mann, 1994) and much of the evidence described below is consistent with the central role of the intersystem electron transport chain.

Quantitatively, CO_2 is the most important inorganic nutrient for cyanobacteria. Under conditions of low ambient CO_2 availability, active inorganic carbon uptake is driven by PSI-dependent cyclic photophosphorylation (Ogawa et al., 1985; Miller et al., 1988; Ogawa, 1992a; Miyachi et al., 1996). Mutants affected in their ability to transport either CO_2 or HCO_3^- were found to be affected in the genes encoding subunits of the type I NAD(P)H dehydrogenase complex (Ogawa 1991, 1992b). There is conflicting evidence as to whether the type I NAD(P)H dehydrogenase is located in both the thylakoid and cytoplasmic membranes (Schmetterer, 1984), but most observations are consistent with its role in both cyclic photosynthetic and respiratory electron flow (Mi et al., 1992, 1994, 1995; Herbert et al., 1995). When cells are transferred to conditions of low Ci availability there is an inhibition of electron flow from PQ to PSI with a consequent accumulation of plastoquinol (PQH_2) (Badger and Schreiber, 1994). Following a shift from bicarbonate-limited conditions to excess bicarbonate supply *Synechocystis* PCC 6803 exhibited a rapid reduction in bicarbonate uptake rate and this was accompanied by the phosphorylation of a number of polypeptides ranging in size from 15 - 70 kDa (Bloye et al., 1991). Similar changes in the bicarbonate uptake rate and the profile of phosphorylated polypeptides were also observed when bicarbonate-limited cultures were shifted into the dark or were provided with exogenous glucose. All these observations are consistent with the redox state of the PQ pool being the primary signal involved both in the switch to cyclic photosynthetic electron transport and the regulation of the Ci transport system, with the likely involvement of one or more protein kinases.

Other nutrient assimilation processes may interact with the intersystem electron transport chain, particularly that for nitrogen. The phosphorylation state of the P_{II} protein involved in the control of nitrogen assimilation (Chapter 15) is also responsive to carbon availability (Liotenberg et al., 1996) and an imbalance in photosynthetic electron transport (Allen, 1992; Tsinoremas et al., 1991). The DNA-binding activity of the nitrogen-responsive regulatory protein NtcA is subject to redox control (Jiang et al., 1997). The activity of the key enzyme of nitrogen assimilation, glutamine synthetase, is also regulated by the redox state of the PQ pool or the cytochrome *b6f* complex (Reyes et al., 1995) and there is evidence for the bulk of the enzyme being associated with the thylakoid membranes (Sallal, 1996). Under conditions of nitrogen limitation *Synechococcus* PCC 7002 exhibited a marked reduction in the number of active PSII reaction centres (Berges et al., 1996) and changes of nitrogen source effected a state transition-like modification in the distribution of excitation energy in *Phormidium laminosum* (Ochoa de Alda et al., 1996). Such effects may not be confined to nitrogen assimilation, as increases in PSI-driven transport relative to linear electron transport have been observed following P or S deprivation (Fork and Herbert, 1993). The influence of the reduction state of PQ extends to regulation of cellular differentiation, both with regard to hormogonium formation and heterocyst production in *Calothrix* spp. (Campbell et al., 1993). Thus, although interactions between photosynthetic electron transport and nutrient acquisition systems are central to integrating the metabolism of the cell, other aspects of cell growth and differentiation may be determined by the redox poise of the two photosystems.

VIII Conclusions

There is now a clear understanding of the types of signalling processes that enable cyanobacteria to detect and respond to changes in their environment. It is also apparent that these signalling pathways employ elements from every level in the genotype to phenotype pathway. Thus, for example, though differential transcription is of clear importance to cyanobacterial adaptive responses, protein phosphorylation appears to play a central role in modulating transcription via two-component sensory systems. The role of signalling systems employing low molecular weight metabolites or specialised signalling molecules appears firmly established, but much more experimental evidence is required to fully appreciate the significance of such systems. The central importance of the response of cyanobacteria to light, both in terms of intensity and spectral quality, is reflected by the variety of the signal transduction pathways which are affected by this single environmental factor and also by the role of the interphotosystem electron transport chain in integrating so many diverse signals. A surprising aspect to some researchers, though probably not to ecologists, is the potential significance of behavioural responses, both in terms of circadian rhythms and motility, to cyanobacterial adaptation. This is certainly an area where real progress is being made and where cyanobacteria may prove to be the ideal model systems.

Acknowledgements

I thank Dave Scanlan, Noel Carr and John Ormerod for their critical reading of the manuscript and constructive comments.

References

Abeliovich A and Gan A (1982) Site of Ca^{2+} action in triggering motility in the cyanobacterium *Spirulina subsalsa.* Cell Motil 4: 393-403

Adams DG, Phillips DO, Nichols JM and Carr NG (1977) The presence and absence of magic spot nucleotide modulation in cyanobacteria undergoing nutritional shift-down. FEBS Lett 81: 48-52

Aiba H., Nagaya M. and Mizuno T. (1993) Sensor and regulator proteins from the cyanobacterium *Synechococcus* species PCC 7942 that belong to the bacterial signal-transduction protein families: implication in the adaptive response to phosphate. Molec Microbiol 8: 81-91

Akinyanju J and Smith RJ (1979) Accumulation of ppGpp and pppGpp during nitrogen deprivation of the cyanobacterium *Anabaena cylindrica.* FEBS Lett 107: 173-176

Akinyanju J and Smith RJ (1982) Energy deprivation and guanosine 5′-diphosphate 3′-diphosphate synthesis in cyanobacteria. J Bacteriol 149: 681-684

Akinyanju J and Smith RJ (1987) The accumulation of phosphorylated guanine nucleotides in *Anabaena cylindrica.* New Phytol 105: 117-122

Alam J, Curtis S, Gleason FK, Gerami-Nejad M and Fuchs JA (1989) Isolation, sequence and expression in *Escherichia coli* of an unusual thioredoxin gene from the cyanobacterium *Anabaena* PCC 7120 J Bacteriol 171: 162-171

Allen JF (1992) Protein phosphorylation in regulation of photosynthesis. Biochim Biophys Acta 1098: 275-335

Allen JF (1993) Redox control of transcription: sensors, response regulators, activators and repressors. FEBS Lett 332: 203-207

Allen JF, Alexciev K and Håkansson G (1995) Regulation by redox signalling. Curr Biol 5: 869-872

Amsler CD and Matsumara P (1995) Chemotactic signal transduction in *Escherichia coli* and *Salmonella typhimurium*. In: Hoch JA and Silhavy TJ (eds) Two-Component Signal Transduction pp. 89-103 ASM Press, Washington, D.C.

Anandan S, Nalty MS, Cogdell DE and Golden SS (1996) Identification of two classes of transcriptional regulator genes in the cyanobacterium *Synechococcus* PCC 7942 Arch Microbiol 166: 58-63

Aoki S, Kondo T, Wada H, Ishiura M (1997) Circadian rhythm of the cyanobacterium *Synechocystis* PCC 6803 in the dark. J Bacteriol 179: 5751-5755.

Appleby JL, Parkinson JS and Bourret RB (1996) Signal transduction via the multi-step phosphorelay: not necessarily a road less travelled. Cell 86: 845-848

Apte SK, Rowell P and Stewart WDP (1978). Electron donation to ferredoxin in heterocysts of the N_2-fixing alga *Anabaena cylindrica*. Proc Roy Soc B200: 1-25

Austin PA, Ross IS and Mills JD (1992) Light/dark regulation of photosynthetic enzymes within intact cells of the cyanobacterium *Nostoc* sp. MAC. Biochim Biophys Acta 1099: 226-232

Bachofen R and Schenk A (1998) Quorum sensing autoinducers: Do they play a role in natural microbial habitats? Microbiol Res: 153: 61-63.

Badger MR and Schreiber U (1984) Effects of inorganic carbon accumulation on photosynthetic oxygen reduction and cyclic electron flow in the cyanobacterium *Synechococcus* PCC 7942 Photosynth Res 37: 177-191

Bartsevich VV and Shestakov SV (1995) The *dspA* gene product of the cyanobacterium *Synechocystis* PCC 6803 influences sensitivity to chemically different growth inhibitors and has amino acid similarity to histidine protein kinases. Microbiology 141: 2915-2920

Berges JA, Charlebois DO, Mauzerall DC and Falkowski PG (1996) Differential effects of nitrogen limitation on photosynthetic efficiency of photosystems I and II in microalgae. Plant Physiol 110: 689-696

Berkelman T, Garretengele P and Hoffman NE (1994) The *pacL* genes of *Synechococcus* PCC 7942 encodes a Ca^{2+}-transporting ATPase. J Bacteriol 176: 4430-4436

Bianchini GM, Pastini AC, Muschietti JP, Téllez-Iñon MT, Martinetto HT, Torres HN and Flawia MM (1990) Adenylate cyclase activity in cyanobacteria: activation by Ca^{2+}-calmodulin and calmodulin-like activity. Biochim Biophys Acta 1055: 75-81

Bishop RH (1979) Regulatory characteristics of a fructose bisphosphatase from the blue-green bacterium *Anacystis nidulans*. Arch Biochem Biophys 196: 295-300

Bloye SA, Silman NJ, Mann NH and Carr NG (1992) Bicarbonate concentration by *Synechocystis* PCC6803: modulation of protein phosphorylation and inorganic carbon transport by glucose. Plant Physiol 99: 601-606

Borbély G, Kari C, Gulyás A and Farkas GL (1980) Bacteriophage infection interferes with guanosine-3´-diphosphate-5´-diphosphate accumulation induced by energy and nitrogen starvation in the cyanobacterium *Anacystis nidulans*. J Bacteriol 144: 859-864

Brahamsha B (1996) An abundant cell-surface polypeptide is required for swimming by the non-flagellated marine cyanobacterium *Synechococcus*. Proc Natl Acad Sci USA 93: 6504-6509

Brahamsha B and Haselkorn R (1991) Isolation and characterization of the gene encoding the principal sigma factor of the vegetative cell RNA polymerase from the cyanobacterium *Anabaena* PCC 7120. J Bacteriol 173: 2442-2450

Brahamsha B and Haselkorn R (1992) Identification of multiple RNA polymerase sigma factor homologs in the cyanobacterium *Anabaena* PCC 7120. J Bacteriol 174: 7273-7282

Buchanan BB (1991) Regulation of CO_2 metabolism in oxygenic photosynthesis: the ferredoxin/thioredoxin system. Arch Biochem Biophys 288: 1-9

Bustos SA, Schaefer MR and Golden SS (1990) Different and rapid responses of 4 cyanobacterial *psbA* transcripts to changes in light-intensity. J Bacteriol 172: 1998-2004

Campbell DA, Houmard J and Tandeau de Marsac N (1993) Electron transport regulates cellular differentiation in the filamentous cyanobacterium *Calothrix*. Plant Cell 5: 451-463

Campbell EL, Brahamsha B and Meeks JC (1998) Identification and characterization of the *Nostoc punctiforme sigH* and *ctpH* genes. Abstracts of the VI Cyanobacterial Workshop B21.

Campbell EL, Hagen KD, Cohen MF, Summers ML and Meeks JC (1996) The *devR* gene-product is characteristic of receivers of 2-component regulatory systems and is essential for heterocyst development in the filamentous cyanobacterium *Nostoc* ATCC 29133. J Bacteriol 178: 2037-2043

Carr NG (1973). Metabolic control and autotrophic physiology. In: Carr NG and. Whitton BA (eds) The Biology of the Blue-Green Algae, pp 39-65 Blackwell Scientific Publications, Oxford

Castenholz RW (1982) Motility and taxes. In: Carr NG and Whitton BA (eds) The Biology of Cyanobacteria, pp 413-439 Blackwell Scientific Publications, Oxford.

Chamot D, Kujat S, Yu E and Owttrim GW (1998) Cold and redox induced cyanobacterial RNA helicases. Abstracts of the VI Cyanobacterial Workshop A29.

Chang C, Kwok SF, Bleecker AB and Meyerowitz EM (1993) *Arabidopsis* ethylene-response gene ETR1 - similarity to 2-component regulators. Science 262: 539-544

Chang C and Meyerowitz EM (1995) The ethylene response in *Arabidopsis* - a eukaryotic 2-component signalling system. Proc Natl Acad Sci USA 92, 4129-4133

Chen T-H, Chen T-L, Hung L-M and Huang T-C (1991) circadian rhythm in amino acid uptake by *Synechococcus* RF-1. Plant Physiol 97:55-59

Chen Y-B, Dominic B, Mellon MT and Zehr JP (1998) Circadian rhythm of nitrogenase gene expression in the diazotrophic filamentous nonheterocystous cyanobacterium *Trichodesmium* IMS 101. J Bacteriol 180: 3598-3605

Chen Y-B, Zehr JP and Mellon M (1996) Growth and nitrogen fixation of the diazotrophic filamentous nonheterocystous cyanobacterium *Trichodesmium* sp. IMS 101 in defined media: evidence for a circadian rhythm. J Phycol 32: 916-923

Cherry JR, Hondred D, Walker JM, Keller JM, Hershey HP, Vierstra RD (1993) Carboxy-terminal deletion analysis of oat phytochrome A reveals the presence of separate domains required for structure and biological activity. Plant Cell 5: 565-575

Chiang GC, Schaefer MR and Grossman AR (1992) Complementation of a red light-indifferent cyanobacterial mutant. Proc Natl Acad Sci USA 89: 9415-9419

Clancey CJ and Gilbert HF (1987) Thiol/disulfide exchange in the thioredoxin catalysed reductive activation of spinach chloroplast fructose-1,6-bisphosphatase. J Biol Chem 262: 13545-13549

Clarke AK and Eriksson M-J (1996) The cyanobacterium *Synechococcus* sp. PCC 7942 possesses a close homologue to the chloroplast ClpC protein of higher plants. Plant Mol Biol 31: 721-730.

Collier JL and Grossman AR (1995) Disruption of a gene encoding a novel thioredoxin-like protein alters the cyanobacterial photosynthetic apparatus. J Bacteriol 177: 3269-3276

Condon C, Squires C and Squires CL (1995) Control of rRNA transcription in *Escherichia coli*. Microbiol Rev 59: 623-645

Cortay JC, Bleicher F, Rieul C, Reeves HC and Cozzone AJ (1988) Nucleotide sequence and expression of the *aceK* gene coding for isocitrate dehydrogenase kinase/phosphatase in *Escherichia coli*. J Bacteriol 170: 89-97

Cossar JD, Rowell P and Stewart WDP (1984). Thioredoxin as a modulator of glucose-6-phosphate dehydrogenase in a N_2-fixing cyanobacterium. J Gen Microbiol 130: 991-998

Cozzone AJ (1988) Protein phosphorylation in prokaryotes. Ann Rev Microbiol 42: 97-125

Curtis SE and Martin JA (1994) The transcription apparatus and the regulation of transcription initiation .In: Bryant DA (ed) The Molecular Biology of Cyanobacteria, pp 613-639. Kluwer Academic Publishers, Dordrecht, The Netherlands.

Danchin A (1993) Phylogeny and adenylate cyclases. Adv Second Messenger Phosphoprotein Res 27: 109-162

Darzins A (1994) Characterisation of a *Pseudomonas aeruginosa* gene cluster involved in pilus biosynthesis and twitching motility: sequence similarity to the chemotaxis proteins of enterics and the gliding bacterium *Myxococcus xanthus*. Molec Microbiol 11: 137-153

Deckers HM and Voordouw G (1994) Identification of a large family of genes for putative chemoreceptor proteins in an ordered library of the *Desulfovibrio vulgaris* Hildenborough genome. J Bacteriol 176: 351-358

Du C and Gallon JR (1993) Modification of the Fe protein of the nitrogenase of *Gloeothece* (Nageli) sp. ATCC 27152 during growth under alternating light and darkness. New Phytologist 125: 121-129

Ebright RH and Busby S (1995) The *Escherichia coli* RNA polymerase α subunit: structure and function. Curr Opinion Genet Devel 5: 197-203

Ehlers KM, Samuel ADT, Berg HC and Montgomery R (1996) Do cyanobacteria swim using travelling surface waves? Proc Natl Acad Sci USA 93: 8340-8343

England RR and Evans H (1983) A requirement for Ca^{2+} in the extraction of O_2-evolving photosystems 2 preparations from the cyanobacterium *Anacystis nidulans*. Biochem J 210: 473-476

Esch JJ, Rodriguez FI, Binder BM, Hetzel CE and Bleecker AB (1998) Structure, function and evolutionary origin of the copper-based ethylene-binding domain of the ETR1 receptor. Abstracts of the 9th International Conference on *Arabidopsis* Research 393

Fechner R (1915) Die chemotaxis der Oscillatorien und ihre Bewegungserscheinungen uberhaupt. Z Bot 7:289-362

Fillat MF, Edmondson DE and Gomez-Moreno C (1991) Light-dependent de-activation/re-activation of *Anabaena variabilis* ferredoxin:$NADP^+$ reductase. Biochem J 274: 781-786

Forchhammer K and Hedler A (1997) Phosphoprotein P-II from cyanobacteria – Analysis of functional conservation with the P-II signal-transduction protein from *Escherichia coli*. Eur J Biochem 244: 869-875

Forchhammer K and Tandeau de Marsac N (1994) The PII protein in the cyanobacterium *Synechococcus* PCC 7942 is modified by serine phosphorylation and signals the cellular N-status. J Bacteriol 176: 84-91

Forchhammer K and Tandeau de Marsac N (1995a) Functional analysis of the phosphoprotein P_{II} (*glnB* gene product) in the cyanobacterium *Synechococcus* PCC 7942. J Bacteriol 177: 2033-2040

Forchhammer K and Tandeau de Marsac N (1995b) Phosphorylation of the P_{II} protein (*glnB* gene product) in the cyanobacterium *Synechococcus* PCC 7942: analysis of *in vitro* kinase activity. J Bacteriol 177: 5812-5817

Fork DC and Herbert SK (1993) Electron transport and photophosphorylation by photosystem I in vivo in plants and cyanobacteria. Photosyn Res 36: 149-168

Friga G, Borbély G and Farkas GL (1981) Accumulation of guanosine tetraphosphate (ppGpp) under nitrogen starvation in *Anacystis nidulans*, a cyanobacterium. Arch Microbiol 129: 341-343

Francko DA and Wetzel RG (1981) Dynamics of cellular and extracellular cAMP in *Anabaena flos-aquae* (Cyanophyta): intrinsic culture variability and correlation with metabolic variables. J Phycol 17: 129-134

Fuqua WC, Winans SC and Greenberg EP (1994) Quorum sensing in bacteria: the LuxR-LuxI family of cell density-responsive transcriptional regulators. J Bacteriol 176: 269-275

Gangola P and Rosen BP (1987) Maintenance of intracellular calcium in *Escherichia coli*. J Biol Chem 262: 12570-12574

Gentry Dr, Hernandez VJ, Nguyen LH, Jensen DB and Cashel M (1993) Synthesis of the stationary-phase sigma factor σ^S is positively regulated by ppGpp. J Bacteriol 175: 7982-7989

Ghassemian M and Straus NA (1996) Fur regulates the expression of iron-stress genes in the cyanobacterium *Synechococcus* PCC 7942. Microbiology 142: 1469-1476

Gleason FK (1994) Thioredoxins in cyanobacteria: structure and redox regulation of enzyme activity. In: Bryant DA (ed) The Molecular Biology of Cyanobacteria, pp. 715-729 Kluwer academic Publishers, Dordrecht, The Netherlands

Gleason FK (1996) Glucose-6-phosphate dehydrogenase from the cyanobacterium *Anabaena* sp. PCC 7120: purification and kinetics of redox modulation. Arch Biochem Biophys 334: 277-283

Golden SS, Ishiura M, Johnson CH and Kondo T (1997) Cyanobacterial circadian rhythms. Annu Rev Plant Physiol Plant Mol Biol 48: 327-354.

Grobbelaar N, Huang T-C, Lin HY and Chow TJ (1986) Dinitrogen-fixing endogenous rhythm in *Synechococcus* RF-1. FEMS Microbiol Lett 37: 173-177

Grossman AR and Kehoe DM (1997) Phosphorelay control of phycobilisome biogenesis during complementary chromatic adaptation. Photosynth Res 53: 95-108

Grossman A. and McGowan RE (1975) Regulation of the glucose-6-phosphate dehydrogenase in blue-green algae. Plant Physiol 55: 658-662

Grossman AR, Schaefer MR, Chiang GG and Collier JL (1993) The phycobilisome, a light-harvesting complex responsive to environmental conditions. Microbiol Rev 57: 725-749

Gruber TM and Bryant DA (1998) Characterization of the alternative sigma-factors SigD and SigE in *Synechococcus* PCC 7002. SigE is implicated in transcription of post-exponential-phase-specific genes. Arch Microbiol 169: 211-219.

Guerrero-Barrera AL, García-Cuéllar CM, Villalba JD, Segura-Nerito M, Gómez-Lojero C, Reyes ME, Hernández JM, García RM and de la Garza M (1996) Actin-related proteins in

Anabaena spp. and *Escherichia coli*. Microbiology 142: 1133-1140

Häder D-P (1982) Gated ion fluxes involved in photophobic responses of the blue-green alga, *Phormidium uncinatum*. Arch Microbiol 131: 77-80

Hagemann M, Golldack D. Biggins J and Erdmann N (1993) Salt-dependent protein phosphorylation in the cyanobacterium *Synechocystis* sp. PCC 6803. FEMS Microbiol Lett 113: 205-210

Haldenwang WG (1995) The sigma factors of *Bacillus subtilis*. Microbiol Rev 59: 1-30

Hanks KH, Quinn AM and Hunter T (1988) The protein kinase family: conserved features and deduced phylogeny of the catalytic domains. Science 241: 42-52

Hanlon DW, Marquez-Magana LM, Carpenter PB, Chamberlin MJ and Ordal GW (1992) Sequence and characterization of *Bacillus subtilis* CheW. J Biol Chem 267: 12055-12060

Harrison MA (1990) Molecular mechanisms of adaptation in the photosynthetic apparatus. PhD thesis, University of Leeds.

Harrison MA, Keen JN, Findlay JBC and Allen JF (1990) Modification of a GlnB-like gene-product by photosynthetic electron transport in the cyanobacterium *Synechococcus* 6301. FEBS Lett 264: 25-28

Haselkorn R (1992) Developmentally regulated gene rearrangements in prokaryotes. Ann Rev Genet 26: 113-130

Henikoff S, Haughn GW, Calvo JM and Wallace JC (1988) A large family of bacterial activator proteins. Proc Natl Acad Sci USA 85: 6602-6606

Herbert SK, Martin RE and Fork DC (1995) Light adaptation of cyclic electron transport through photosystem I in the cyanobacterium *Synechococcus* sp. PCC 7942. Photosynth Res 46: 277-285

Hoch JA and Silhavy TJ (1995) Two-Component Signal Systems. ASM Press, Washington, D.C.

Hoiczyk E and Baumeister W (1995) Envelope structure of four gliding filamentous cyanobacteria. J Bacteriol 177: 2387-2395

Hoiczyk E and Baumeister W (1997) Oscillin, an extracellular, Ca^{2+}-binding glycoprotein essential for the gliding motility of cyanobacteria. Molec Microbiol 26: 699-708.

Hood E, Armour S, Ownby JD, Handa AK, Bressan RA (1979) Effect of nitrogen starvation on the level of adenosine 3′5′-monophosphate in *Anabaena variabilis*. Biochim Biophys Acta 588: 193-200

Howell LD, Griffiths C, Slade LW, Potts M and Kennelly PJ (1996) Substrate specificity of IphP, a cyanobacterial dual-specificity protein phosphatase with MAP kinase phosphatase activity. Biochemistry 35: 7566-7572

Hua J, Chang C, Qun Q and Meyerowitz EM (1995) Ethylene insensitivity conferred by *Arabidopsis* ERS gene. Science 269: 1712-1714

Huang T-C and Grobbelaar N (1995) The circadian clock in the prokaryote *Synechococcus* RF-1. Microbiology 141: 535-540

Huang T-C, Tu J, Chow T-J and Chen T-H (1990) Circadian rhythm of the prokaryote *Synechococcus* sp. RF-1. Plant Physiol 92: 531-533

Huang T-C, Wang S-T and Grobbelaar N (1993) Circadian rhythm mutants of the prokaryotic *Synechococcus* RF-1. Curr Microbiol 27: 249-254

Hughes J, Lamparter T, Mittmann F, Hartman E, Gärtner W, Wilde A and Börner T (1997). A prokaryotic phytochrome. Nature 386: 663

Ip S-M, Rowell P, Aitken A and Stewart WDP (1984) Purification and characterization of thioredoxin from the N_2-fixing cyanobacterium *Anabaena cylindrica*. Eur J Biochem 141: 497-504.

Irmler A, Sanner S, Dierks H and Forchhammer K (1997) Dephosphorylation of the phosphoprotein P-II in *Synechococcus* PCC 7942: identification of an ATP and 2-oxoglutarate-regulated phosphatase. Molec Microbiol 26: 81-90

Ishiura M, Kutsuna S, Aoki S, Iwasaki H, Andersson CR, Tanabe A, Golden SS, Johnson CH and Kondo T (1998) Expression of a gene cluster *kaiABC* as a circadian feedback process in cyanobacteria. Science 281: 1519-1523

Jiang F, Mannervik B, Bergman B (1997) Evidence for redox regulation of the transcription factor NtcA, acting both as an activator and a repressor, in the cyanobacterium *Anabaena* PCC 7120. Biochem J 327: 513-517

Johnson CH, Golden SS, Ishiura M and Kondo T (1996) Circadian clocks in prokaryotes. Molec Microbiol 21: 5-11

Joseph CM and Meeks JC (1987) Regulation of expression of glutamine synthetase in a symbiotic *Nostoc* strain associated with *Anthoceros punctatus*. J Bacteriol 169: 2471-2475

Juhász A, Csizmadia V, Borbély G, Udvardy J and Farkas GL (1986) The pyridine nucleotide-dependent D-glucose dehydrogenase of *Nostoc* MAC, a cyanobacterium, is subject to thioredoxin modulation. FEBS Lett 194: 121-125

Kanamaru K, Kashiwagi S and Mizuno T (1993) The cyanobacterium, *Synechococcus* sp. PCC7942, possesses two distinct genes encoding cation-transporting P-type ATPases. FEBS Lett 330: 99-104

Kaneko T, Sato S, Kotani H, Tanaka A, Asamizu E, Nakamura Y, Miyajima N, Hirosawa M, Sugiura M, Sasamoto S, Kimura T, Hosouchi T, Matsuno A, Muraki A, Nakazaki N, Naruo K, Okumura S, Shimpo S, Takeuchi C, Wada T, Watanabe A, Yamada M, Yasuda M. and Tabata S (1996) Sequence analysis of the genome of the unicellular Cyanobacterium *Synechocystis* PCC6803 II. Sequence determination of the entire genome and assignment of potential protein-coding regions. DNA Res 3: 109-136

Kasahara M, Yashiro K, Sakamoto T and Ohmori M (1997) The *Spirulina platensis* adenylate cyclase gene, *cyaC*, encodes a novel signal transduction protein. Plant and Cell Physiol 38: 828-836.

Katayama M and Ohmori M (1997) Isolation and characterization of multiple adenylate cyclase genes from the cyanobacterium *Anabaena* PCC 7120. J Bacteriol 179: 3588-3593

Katayama M, Wada Y and Ohmori M (1995) Molecular cloning of the cyanobacterial adenylate cyclase from the filamentous cyanobacterium *Anabaena cylindrica*. J Bacteriol 177: 3873-3878

Kehoe DM and Grossman AR (1996) Similarity of a chromatic adaptation sensor to phytochrome and ethylene receptors. Science 273: 1409-1412

Kehoe DM and Grossman AR (1997) New classes of mutants in complementary chromatic adaptation provide evidence for a novel four-step phosphorelay system. J Bacteriol 179: 3914-3921

Kennelly PJ and Potts M (1996) Fancy meeting you here! a fresh look at "prokaryotic" protein phosphorylation. J Bacteriol 178: 4759-4764

Kerson GW, Miernyk JA and Budd K (1984) Evidence for the occurrence and possible physiological role for cyanobacterial calmodulin. Plant Physiol 75: 222-224

Kondo T and Ishiura M (1994) Circadian rhythms of cyanobacteria: monitoring the biological clocks of individual colonies by bioluminescence. J Bacteriol 176: 1881-1885

Kondo T, Mori T, Lebedeva NV, Aoki S, Ishiura M and Golden SS (1997) Circadian rhythms in rapidly dividing cyanobacteria. Science 275: 224-227

Kondo T, Strayer CA, Kulkarni RD, Taylor W, Ishiura M, Golden SS and Johnson CH (1993) Circadian rhythms in prokaryotes: luciferase as a reporter of circadian gene expression in cyanobacteria. Proc Natl Acad Sci USA 90: 5672-5676

Kotani H and Tabata S (1998) Lessons from the sequencing of the genome of a unicellular cyanobacterium *Synechocystis* sp. PCC6803. Annu Rev Plant Physiol Plant Mol Biol 49: 151-171

Kustu S, Santero E, Keener J, Popham D and Weiss D (1989) Expression of σ^{54} (*ntrA*)-dependent genes is probably united by a common mechanism. Microbiol Rev 53: 367-376

Kutsuna S, Kondo T, Aoki S and Ishiura M (1998) A period-extender gene, pex, that extends the period of the circadian clock in the cyanobacterium *Synechococcus* PCC 7942. J Bacteriol 180: 2167-2174

Labbe JP, Harricane MC, Boyer M, Derancourt J, Roustan C and Benyamin Y (1996) Biochemical evidence for the presence of an unconventional actin protein in a prokaryotic organism. Comp Biochem and Physiol B114: 287-293

Lagarias JC and Lagarias DM (1989) Self-assembly of synthetic phytochrome holoprotein *in vitro*. Proc Natl Acad Sci USA 86: 5778-5780

Lamparter T, Mittmann F, Gärtner W, Börner, Hartmann E and Hughes J (1997) Characterization of recombinant phytochrome from the cyanobacterium *Synechocystis*. Proc Natl Acad Sci USA 94: 11792-11797

Laudenbach DE and Grossman AR (1991) Characterization and mutagenesis of sulfur-regulated genes in a cyanobacterium: evidence for function in sulfate transport. J Bacteriol 173: 2739-2750

Lee HM, Flores E, Herrero A, Houmard J and Tandeau de Marsac N (1998) A role for the signal transduction protein P-II in the control of nitrate/nitrite uptake in a cyanobacterium. FEBS Letters 427: 291-295

Lex M and Carr NG (1974) The metabolism of glucose by heterocysts and vegetative cells of *Anabaena cylindrica*. Arch Microbiol 101: 161-167

Liang J, Scappino L and Haselkorn R (1992) The *patA* gene product, which contains a region similar to CheY of *Escherichia coli*, controls heterocyst pattern formation in the cyanobacterium *Anabaena* 7120. Proc Natl Acad Sci USA 89: 5655-5659

Liotenberg S, Campbell D, Castets A-M, Houmard J and Tandeau de Marsac N (1996) Modification of the P_{II} protein in response to carbon and nitrogen availability in filamentous heterocystous cyanobacteria. FEMS Microbiol Lett 144: 185-190

Liu Y, Golden SS, Kondo T, Ishiura M, and Johnson CH (1995a) Bacterial luciferase as a reporter of circadian gene expression in cyanobacteria. J Bacteriol 177: 2080-2086

Liu Y and Tsinoremas NF (1996) An unusual gene arrangement for the putative chromosome replication origin and circadian expression of *dnaN* in *Synechococcus* PCC-7942 Gene 172: 105-109

Liu Y, Tsinoremas NF, Johnson CH, Lebedeva NV, Golden SS, Ishiura M and Kondo T (1995b) Circadian orchestration of gene expression in cyanobacteria. Genes Dev 9: 1469-1478

Lonetto M, Gribskov M and Gross CA (1992) The σ^{70} family: sequence conservation and evolutionary relationships. J. Bacteriol 174: 3843-3849

Malin G and Walsby AE (1985) Chemotaxis of a cyanobacterium on concentration gradients of carbon dioxide, bicarbonate and oxygen. J Gen Microbiol 130: 983-990

Mann NH (1994) Protein phosphorylation in cyanobacteria. Microbiology 140: 3207-3215

Mann N, Carr NG and Midgley JEM (1975) RNA synthesis and accumulation of guanine nucleotides during growth shift down in the blue green alga *Anacystis nidulans*. Biochim Biophys Acta 402: 41-50

Mann NH, Rippka R and Herdman M (1991) regulation of protein phosphorylation in the cyanobacterium *Anabaena* strain PCC 7120. J Gen Microbiol 137: 331-339

Marcus F, Moberly L and Latshaw SP (1988) Comparative amino acid sequence of fructose 1,6-bisphosphatases: identification of a region unique to the light-regulated chloroplast enzyme. Proc Natl Acad Sci USA 85: 5379-5383

Marsden WJN and Codd GA (1984) Purification and molecular and catalytic properties of phosphoribulokinase from the cyanobacterium *Chlorogloeopsis fritschii*. J Gen Microbiol 130: 999-1006

McCleary WR, Stock JB and Ninfa AJ (1993) Is acetyl phosphate a global signal in *Escherichia coli*? J Bacteriol 175: 2793-2798

McColl SM and Evans EH (1990) The location of a putative EF hand loop Ca^{2+} binding site in the photosystem 2 oxygen evolving complex. Abstracts of the First European Workshop on the Molecular Biology of Cyanobacteria 60-61

Mi HL, Endo T, Schreiber U, Ogawa T and Asada K (1992) Electron donation from cyclic and respiratory electron flows to the photosynthetic intersystem chain is mediated by pyridine-nucleotide dehydrogenase in the cyanobacterium *Synechocystis* PCC-6803. Plant Cell Physiol 33: 1233-1237

Mi H, Endo T, Schreiber U, Ogawa T and Asada K (1994) NAD(P)H dehydrogenase-dependent cyclic electron flow around photosystem-I in the cyanobacterium *Synechocystis* PCC-6803 - a study of dark-starved cells and spheroplasts. Plant Cell Physiol 35: 163-173

Mi H, Endo T, Ogawa T and Asada K (1995) Thylakoid membrane-bound, NADPH-specific pyridine-nucleotide dehydrogenase complex mediates cyclic electron transport in the cyanobacterium *Synechocystis* sp. PCC-6803 Plant Cell Physiol 36: 661-668

Miller AG, Espie GS and Canvin DT (1988) Active transport of CO_2 by the cyanobacterium Synechococcus UTEX 625 Plant Physiol 86: 677-683

Mitsui A, Kumazawa S, Takahashi A, Ikemoto H and Arai T (1986) Strategy by which nitrogen-fixing unicellular cyanobacteria grow photoautotrophically. Nature 323: 720-722

Mitsui A, Suda S, and Hanagata N (1993) Cell cycle events at different temperatures in aerobic nitrogen-fixing marine unicellular cyanobacterium *Synechococcus* sp. Miami BG 043511. J Mar Biotechnol 1: 89-91

Miyachi S, Burger J, Kotzabasis K, Thielmann J and Senger H (1996) Photosynthetic characteristics of 3 strains of cyanobacteria grown under low-CO_2 or high-CO_2 conditions. Zeitschrift fur Naturforschung 51: 40-46

Morgan DG, Baumgartner JW and Hazelbauer GL (1993) Proteins antigenically related to methyl-accepting chemotaxis proteins of *Escherichia coli* detected in a wide range of bacterial species. J Bacteriol 175: 133-140

Muller EGD and Buchanan BB (1989) Thioredoxin is essential for photosynthetic growth. J Biol Chem 264: 4008-4014

Muñoz-Dorado J, Inouye S and Inouye M (1991) A gene encoding a protein serine/threonine kinase is required for normal development of *M. xanthus*, a gram-negative bacterium. Cell 67: 995-1006

Nagaya, M., Aiba, H. and Mizuno, T. (1994) The *sphR* product, a two-component system response regulator protein, regulates phosphate assimilation in *Synechococcus* PCC 7942 by binding to two sites upstream from the *phoA* promoter. J Bacteriol 176: 2210-2215

Nakama Y, Kaneko T, Hirosawa M, Miyajima N and Tabata S (1998) Cyanobase, a WWW database containing the complete nucleotide sequence of the genome of *Synechocystis* PCC6803. Nucleic Acids Res 26: 63-67

Newman J, Karakaya H., Scanlan DJ. and Mann NH (1995) A comparison of gene organization in the *zwf* region of the genomes of the cyanobacteria *Synechococcus* sp. PCC 7942 and *Anabaena* sp. PCC 7120. FEMS Microbiol Lett 133: 187-193

Ninfa AJ, Atkinson MR, Kamberov ES, Feng J and Nifa EG (1995) Control of nitrogen assimilation by the NR_I-NR_{II} two-component system of enteric bacteria. In: Hoch JA and Silhavy TJ (eds) Two-component signal transduction, pp 67-88 ASM Press, Washington, D.C.

Norris V, Grant S, Freestone P, Canvin J, Sheikh FN, Toth I, Trinei M, Modha K and Norman RI (1996) Calcium signalling in bacteria. J Bacteriol 178: 3677-3682

North AK, Klose KE, Stedman KM and Kustu S (1993) Prokaryotic enhancer-binding proteins reflect eukaryote-like modularity: the puzzle of nitrogen regulatory protein C. J Bacteriol 175: 4267-4273

Ochoa de Alda JAG, Tapia MI, Franck F, Llama MJ and Serra JL (1996) Changes in nitrogen source modify distribution of excitation energy in the cyanobacterium *Phormidium laminosum*. Physiol Plant 97: 69-78

Ogawa T (1991) A gene homologous to the subunit-2 gene of NADH dehydrogenase is essential to inorganic carbon transport of *Synechocystis* PCC6803. Proc Natl Acad Sci USA 88: 4275-4279

Ogawa, T (1992a) NADH dehydrogenase - a component of Photosystem-I cyclic electron transport flow driving inorganic carbon transport. Photosyn Res 34: 100

Ogawa T (1992b) Identification and characterization of the *ictA*/*ndhL* gene-product essential to inorganic carbon transport of *Synechocystis* PCC6803. Plant Physiol 99: 1604-1608

Ogawa T, Miyano A and Inouye Y (1985) Photosystem I-driven inorganic carbon transport in the cyanobacterium, *Anacystis nidulans*. Biochim Biophys Acta 808: 77-84

Ohmori M (1989) cAMP in *Anabaena cylindrica*: rapid changes in cellular levels in response to changes in extracellular environments. Plant Cell Physiol 30: 911-914

Ohmori K, Hirose M and Ohmori M (1993) An increase in the intracellular concentration of cAMP triggers formation of an algal mat by the cyanobacterium *Spirulina platensis*. Plant Cell Physiol 34: 169-171

Ohmori M, Katayama M, Kasahara M and Terauchi K (1998) Adenylate cyclase genes responsible for environmental signals. Abstracts of the VI Cyanobacterial Workshop A28.

Ohmori M, Ohmori K and Hasunuma K (1988) Rapid change in cyclic 3,5-AMP concentration triggered by a light-off or light-on signal in *Anabaena cylindrica*. Arch Microbiol 150: 203-204

Onek LA, Lea PJ and Smith RJ (1994) Isolation and characterization of a calmodulin-like protein from the cyanobacterium *Nostoc* sp. PCC 6720. Arch Microbiol 161: 352-358

Onek LA and Smith RJ (1992) Calmodulin and calcium mediated regulation in prokaryotes. J Gen Microbiol 138: 1039-1049

Pacold ME, Stevens FJ, Li D, and Anderson LE (1995) The NADP-linked glyceraldehyde-3-phosphate dehydrogenase of *Anabaena variabilis* and *Synechocystis* PCC 6803, which lack one of the cysteines found in the higher plant enzyme, are not reductively activated. Photosyn Res 43: 125-130

Parkinson JS and Kofoid EC (1992) Communication modules in bacterial signalling proteins. Ann Rev Genet 26: 71-112

Pelroy RA, Kirk MR and Bassham JA (1976) Photosystem II regulation of macromolecule synthesis in the blue-green alga *Aphanocapsa* 6714. J Bacteriol 128: 623-632

Peschek GA (1996) Interaction of photosynthesis and respiration. Biochem Soc Trans 24: 729-733

Pettersson A and Bergman B (1989) Calmodulin in heterocystous cyanobacteria: biochemical and immunological evidence. FEMS Microbiol Lett 60: 95-100

Pitta TP and Berg HC (1995) Self-electrophoresis is not the mechanism for motility in swimming cyanobacteria. J Bacteriol 177: 5701-5703

Pitta TP, Sherwood EE, Kobel AM and Berg HC (1997) Calcium is required for swimming by the non-flagellated cyanobacterium *Synechococcus* strain WH8113. J Bacteriol 179: 2524-2528.

Porankiewicz J and Clarke AK (1997) Induction of the heat shock protein ClpB affects cold acclimation in the cyanobacterium *Synechococcus* PCC 7942. J Bacteriol 179: 5111-5117.

Porankiewicz J, Schelin J and Clarke AK (1998) The ATP-dependent Clp protease is essential for acclimation to UV-B and low temperature in the cyanobacterium *Synechococcus*. Molec Microbiol 29: 275-283.

Potts M, Sun H, Mockaitis K, Kennelly PJ, Reed D and Tonks NK (1993) A protein-tyrosine/serine phosphatase encoded by the genome of the cyanobacterium *Nostoc commune* UTEX 584. J Biol Chem 268: 7632-7635

Quail PH (1994) Photosensory perception and signal transduction in plants. Curr Opinion Genet Devel 4, 652-661

Reyes JC, Crespo JL, García-Domínguez M and Florencio FJ (1995) Electron transport controls glutamine synthetase activity in the facultative heterotrophic cyanobacterium *Synechocystis* sp. PCC 6803. Plant Physiol 109: 899-905

Richardson LL and Castenholz RW (1987) Diel vertical movements of the cyanobacterium *Oscillatoria terebriformis* in a sulfide-rich hot spring microbial mat. Appl Environ Microbiol 53:2142-2150

Richardson LL and Castenholz RW (1989) Chemokinetic motility responses of the cyanobacterium *Oscillatoria terebriformis*. Appl Environ Microbiol 55: 261-263

Rosen BP (1987) Bacterial calcium transport. Biochim Biophys Acta 906: 101-110

Sallal, AKJ (1996) The presence of glutamine synthetase in the thylakoids of nitrogen-fixing cyanobacterium *Nostoc muscorum*. J Basic Microbiol 36: 113-119

Scanlan DJ, Bourne J and Mann NH (1997) A putative transcriptional activator of the Crp/Fnr family from the marine cyanobacterium *Synechococcus* sp. WH7803. J Appl Phycol 8: 565-567

Scanlan DJ and Mann NH (1996) Sequence analysis of a genomic region (Accession No. U70379) from the cyanobacterium *Synechococcus* sp. PCC7942 encoding a NarL homologue and a two-component system pair. Plant Physiol PGR 96-106

Schaeffer F. and Stanier RY. (1978). Glucose-6-phosphate dehydrogenase of *Anabaena* sp. Arch Microbiol 116, 9-19

Schirmer EC, Glover JR, Singer MA and Lindquist S (1996) HSP100/Clp proteins: a common mechanism explains diverse functions. Trends Biol Sciences 21: 289-296.

Schluchter WM and Bryant DA (1992) Molecular characterization of ferredoxin-$NADP^+$ oxidoreductase in cyanobacteria: cloning and sequence of the *petH* gene of *Synechococcus* sp. PCC 7002 and studies on the gene product. Biochemistry 31: 3092-3102

Schmetterer G (1994) Cyanobacterial respiration. In: Bryant DA (ed) The Molecular Biology of Cyanobacteria, pp. 409-435 Kluwer academic Publishers, Dordrecht, The Netherlands

Schneider GJ, Turner NE, Richaud C, Borbely G and Haselkorn R (1987) Purification and characterization of RNA polymerase from the cyanobacterium *Anabaena* 7120. J Biol Chem 262: 14633-14639

Schneider-Poetsch HA, Braun B, Marx S and Schaumburg A (1991) Phytochromes and bacterial sensor proteins are related by structural and functional homologies. FEBS Lett 281: 245-249

Schwarz R and Grossman AR (1998) A response regulator of cyanobacteria integrates diverse environmental signals and is critical for survival under extreme conditions. Proc Natl Acad Sci USA 95: 11008-11013

Sharrock RA and Quail PH (1989) Novel phytochrome sequences in *Arabidopsis thaliana*: structure, evolution and differential expression of a plant regulatory photoreceptor family. Genes Dev 3: 534-544

Shaw DJ, Rice DW and Guest JR (1983) Homology between CAP and Fnr, a regulator of anaerobic respiration in *Escherichia coli.* J Mol Biol 166: 241-247

Shi L and Carmichael WW (1997) ppl-cyano2, a protein serine/threonine phosphatase 1 gene from the cyanobacterium *Microcystis aeruginosa.* Arch Microbiol 168: 528531

Shi W and Zusman DR (1995) The *frz* signal transduction system controls multicellular behaviour in *Myxococcus xanthus*. In: Hoch JA and Silhavy TJ (eds) Two-Component Signal Transduction pp. 419-430

Silman NJ, Carr NG and Mann NH (1995) ADP-ribosylation of glutamine synthetase in the cyanobacterium *Synechocystis* PCC 6803. J Bacteriol 177: 3527-3533

Simpson GG, Clark G and Brown JWS (1994) Isolation of a maize cDNA encoding a protein with extensive similarity to an inhibitor of protein kinase C and a cyanobacterial open reading frame. Biochim Biophys Acta 1222: 306-308

Smith AJ (1982). Modes of cyanobacterial carbon metabolism. In: Carr NG and Whitton BA (eds) The Biology of Cyanobacteria, pp 47-86 Blackwell Scientific Publications, Oxford.

Smith G and Ownby JD (1981) Cyclic AMP interferes with pattern formation in the cyanobacterium *Anabaena variabilis.* FEMS Microbiol. Lett 11: 175-180.

Smith H (1995) Physiological and ecological function within the phytochrome family. Ann Rev Plant Physiol Plant Mol Biol 46: 289-315

Smith RJ (1988) Calcium-mediated regulation in the cyanobacteria? In: Rogers LJ and Gallon JR (eds) Biochemistry of Algae and Cyanobacteria, pp 185-199 Clarendon Press, Oxford.

Smith RJ and Carr NG (1977) The regulation of stable RNA synthesis in the blue-green alga *Anacystis nidulans*: effect of leucine deprivation and 5-methyltryptophan addition. J Gen Microbiol 103: 61-68

Smith RJ, Hobson S and Ellis IR (1987) Evidence for calcium-mediated regulation of heterocyst frequency and nitrogenase activity in *Nostoc* 6720. New Phytol 105: 531-541

Sobczyk A, Bely A, Tandeau de Marsac N and Houmard J (1994) A phosphorylated DNA-binding protein is specific for the red light signal during complementary chromatic adaptation. Molec Microbiol 13, 875-885

Stal LJ and Krumbein WE (1985) Nitrogenase activity in the non-heterocystous cyanobacterium *Oscillatoria* sp. grown under alternating light dark cycles. Arch Microbiol 143: 67-71

Stock JB, Ninfa AJ and Stock AM (1989) Protein phosphorylation and regulation of adaptive responses in bacteria. Microbiol Rev 53: 450-490

Stock JB, Surette MG, Levitt M and Park P (1995) Two-component signal transduction system: structure-function relationships and mechanisms of catalysis. In: Hoch JA and Silhavy TJ (eds) Two-Component Signal Transduction pp. 25-51

Straus NA (1994) Iron deprivation: physiology and gene regulation. In: Bryant DA (ed) The Molecular Biology of Cyanobacteria, pp 731-750 Kluwer Academic Publishers, Dordrecht, The Netherlands

Su X and Bogorad L (1991) A residue substitution in phosphoribulokinase of *Synechocystis* PCC 6803 renders the mutant light-sensitive. J Biol Chem 266: 23698-23705

Summers ML, Meeks JC, Chu S. and Wolf RE Jr. (1995a) Nucleotide sequence of an operon in *Nostoc* ATCC 29133 encoding four genes of the oxidative pentose phosphate cycle. Plant Physiol. 107: 267-268

Summers ML, Wallis JG, Campbell EL and Meeks JC (1995b). Genetic evidence of a major role for glucose-6-phosphate dehydrogenase in nitrogen fixation and dark growth of the cyanobacterium *Nostoc* ATCC 29133 J Bacteriol 177: 6184-6194

Sundaram S, Karakaya H, Scanlan DJ and Mann NH (1998) Multiple oligomeric forms of glucose-6-phosphate dehydrogenase in cyanobacteria and the role of OpcA in the assembly process. Microbiology 144: 1549-1556.

Surányi G, Korcz A, Pálfi Z and Borbély G (1987) Effects of light deprivation on RNA synthesis, accumulation of guanosine 3′(2′)-diphosphate, and protein synthesis in heat-shocked *Synechococcus* PCC 6301, a cyanobacterium. J Bacteriol 169: 632-639

Suzuki I, Horie N, Sugiyama T and Omata T (1995) Identification and characterization of two nitrogen-regulated genes of the cyanobacterium *Synechococcus* PCC7942 required for maximum efficiency of nitrogen assimilation. J Bacteriol 177: 290-296

Suzuki I, Sugiyama T and Omata T (1996) regulation by cyanate of the genes involved in carbon and nitrogen assimilation in the cyanobacterium *Synechococcus* PCC 7942. J Bacteriol 178: 2688-2694

Sweeney BM and Borgese MB (1989) A circadian rhythm in cell division in a prokaryote, the cyanobacterium *Synechococcus* WH7803. J Phycol 25: 183-186

Tan X, Varughese M and Widger WR (1994) A light-repressed transcript found in *Synechococcus* PCC 7002 is similar to a chloroplast-specific small subunit ribosomal protein and to a transcription modulator protein associated with $\sigma 54$. J Biol Chem 269: 20905-20912

Tanaka K, Masuda S and Takahashi H (1992a) The complete nucleotide sequence of the gene (*rpoD1*) encoding the principal sigma factor of the RNA polymerase from the cyanobacterium *Synechococcus* PCC 7942. Biochim Biophys Acta 1132: 94-96

Tanaka K, Masuda S and Takahashi H (1992b) Multiple *rpoD*-related genes of cyanobacteria. Biosci Biotech Biochem 56: 1113-1117

Tandeau de Marsac N and Houmard J (1993) Adaptation of cyanobacteria to environmental stimuli: new steps towards molecular mechanisms. FEMS Microbiol Rev 104: 119-190.

Terauchi K and Ohmori M (1998) An adenylate cyclase, Cya1, mediates the signal of blue light in the cyanobacterium *Synechocystis* sp. PCC 6803. Abstracts of the VI Cyanobacterial Workshop S9.

Thümmler F, Algarra P and Fobo GM (1995) Sequence similarities of phytochromes to protein kinases: implications for the structure, function and evolution of the phytochrome gene family. FEBS Lett 357: 149-155

Torrecilla I, Leganes F, Bonilla I and Fernandez Piñas (1998) Use of recombinant aequorin to monitor Ca^{2+} transients in *Anabaena* sp. PCC 7120 in response to environmental changes. Abstracts of the VI Cyanobacterial Workshop A26.

Tramontini LS (1990) The role of Ca^{2+} in photosystem 2 of *Anacystis nidulans*. PhD thesis, Lancashire Polytechnic, UK.

Tsai J-W (1996) Protein phosphorylation in cyanobacterial light-harvesting complexes. PhD thesis. University of Warwick.

Tsinoremas NF, Castets AM, Harrison MA, Allen JF and Tandeau de Marsac N (1991) Photosynthetic electron transport controls nitrogen assimilation in cyanobacteria by means of posttranslational modification of the *glnB* gene product. Proc Natl Acad Sci USA 88: 4565-4569

Tsinoremas NF, Ishiura M, Kondo T, Anderrson CR, Tanaka K, Takahashi H, Johnson CH and Golden SS (1996) A sigma factor that modifies the circadian expression of a subset of genes in cyanobacteria. EMBO J 15: 2488-2495

Udvardy J, Borbely G, Juhasz A. and Farkas GL (1984). Thioredoxins and the redox modulation of glucose-6-phosphate dehydrogenase in *Anabaena* PCC 7120 vegetative cells and heterocysts. J Bacteriol 157: 681-683

Udvardy J, Godeh MH and Farkas GL (1982) Regulatory properties of a fructose 1,6-bisphosphatase from the cyanobacterium *Anacystis nidulans*. J Bacteriol 151: 203-208

Vega-Palas MA, Flores E and Herrero A (1992) NtcA, a global nitrogen regulator from the cyanobacterium *Synechococcus* that belongs to the Crp family of bacterial regulators. Molec Microbiol 6: 1853-1859

Walsby AE (1994) Gas vesicles. Microbiol Rev: 94-144.

Warner KM and Bullerjahn GS (1994) Light-dependent tyrosine phosphorylation in the cyanobacterium *Prochlorothrix hollandica*. Plant Physiol 105: 629-633

Waterbury JB, Willey JM, Franks DG, Valos FW and Watson SW (1985) A cyanobacterium capable of swimming motility. Science 230: 74-76

Watson GMF, Scanlan DJ and Mann NH (1996) Characterization of the genes encoding a phosphate-regulated two component sensory system in the marine cyanobacterium *Synechococcus* sp. WH7803. FEMS Microbiol Lett: 142: 105-109

Wawrzynow A Banecki B and Zylicz M (1996) The Clp ATPases define a novel class of molecular chaperones. Molec Microbiol 21, 895-899.

Weber IT and Steitz TA (1987) Structure of a complex of catabolite gene activator protein and cyclic AMP refined at 25 Å resolution. J Mol Biol 198: 311-326

Wilde A, Churin Y, Schubert H and Börner T (1997) Disruption of a *Synechocystis* sp. PCC 6803 gene with partial similarity to phytochrome genes alters growth under changing light qualities. FEBS Lett 406: 89-92

Willey JM and Waterbury JB (1989) Chemotaxis towards nitrogenous compounds by swimming strains of marine *Synechococcus* spp. Appl Env Microbiol 55: 1888-1894

Willey, JM, Waterbury JB and Greenberg EP (1987) Sodium-coupled motility in a swimming cyanobacterium. J Bacteriol 169: 3429-3434

Wolk CP, Ernst A and Elhai J (1994) Heterocyst development and metabolism. In: Bryant DA (ed) The Molecular Biology of Cyanobacteria, pp. 769-823. Kluwer academic Publishers, Dordrecht, The Netherlands

Yan OY, Andersson CR, Kondo T, Golden SS and Johnson CH (1998) Resonating circadian clocks enhance fitness in cyanobacteria. Proc Natl Acad Sci USA 95: 8660-8664

Yeh K-C, Wu H-S, Murphy JT and Lagarias JC (1997) A cyanobacterial phytochrome two-component light sensory system. Science 277: 1505-1508

Zehr JP, Wyman M, Miller V, Duguay L and Capone DG (1993) Modification of the Fe protein of nitrogenase in natural populations of *Trichodesmium thiebautii*. Appl Environ Microbiol 59: 669-676

Zhang C-C (1993) A gene encoding a protein related to eukaryotic protein kinases from the filamentous heterocystous cyanobacterium, *Anabaena* PCC 7120. Proc Natl Acad Sci USA 90: 11840-11844

Zhang C-C (1996) Bacterial signalling involving eukaryotic-type protein kinases. Molec Microbiol 20: 9-15

Zhang C-C, Friry A and Peng L (1998a) Molecular and genetic analysis of two closely linked genes that encode respectively, a protein phosphatase 1/2A/2B homolog and a protein kinase homologue in the cyanobacterium *Anabaena* PCC 7120. J Bacteriol 180: 2616-2622

Zhang C-C, Gonzalez L and Phalip V (1998b) Survey, analysis and genetic organization of genes encoding eukaryotic-like signaling proteins on a cyanobacterial genome. Nucl Acids Res 26: 3619-3625.

Zhang C-C and Libs L (1998) Cloning and characterization of the *pknD* gene encoding an eukaryotic-type protein kinase in the cyanobacterium Anabaena sp. PCC7120. Molec Gen Genet 258: 26-33

Chapter 15

Molecular Responses to Environmental Stress

Devaki Bhaya, Rakefet Schwarz[1] and Arthur R. Grossman
The Carnegie Institution of Washington, Department of Plant Biology, 260 Panama Street, Stanford, CA 94305, [1]Department of Botany, The Hebrew University of Jerusalem, 91904, Jerusalem, Israel

B. A. Whitton and M. Potts (eds), The Ecology of Cyanobacteria, 397-442.
© 2000 *Kluwer Academic Publishers. Printed in the Netherlands.*

Summary

Cyanobacteria are remarkable for their ability to flourish in environments with widely fluctuating chemical and physical parameters, such as nutrient availability, light quality and intensity, temperature and osmotic conditions. This chapter focuses on the responses of cyanobacteria to changes in light intensity, light quality and macronutrient availability. Suboptimal light and nutrient conditions result in a number of specific and general responses that strongly influence the physiology of the cell. Some of the changes can be striking, while others are subtle; they can occur in minutes or take place over several days. In recent years knowledge of the molecular responses of cyanobacteria to stress conditions has increased considerably. Capacity to acclimate to suboptimal light levels for photosynthetic carbon fixation is unique among the bacteria. In contrast, an understanding of acclimation to nutrient deprivation has gained much from the information available from analyses of enteric bacteria, although cyanobacteria have a number of interesting deviations in their responses. Perhaps most provocative in studies of acclimation processes in cyanobacteria is the overlap among the responses elicited by different environmental stimuli, which suggests the existence of a hierarchy of responses and the involvement of global regulatory circuits.

I. Introduction

There is a wealth of information on the biochemical and physiological responses of cyanobacteria to environmental stress. The emphasis in this chapter is on the Molecular responses to two key environmental parameters, light and nutrients. Most cyanobacteria are obligate photoautotrophs, making it critical for them to sense and respond to both light intensity and quality (Van Liere and Walsby, 1982). Since light primarily affects photosynthetic activity, many of the responses to a changing light environment modify structural and functional features of the photosynthetic apparatus. Some of the adaptations to light are obvious and striking. For example, certain cyanobacteria can alter the pigment-protein composition of their light-harvesting complexes such that the cells have a different color depending on the wavelengths of light in which they are grown. Other changes are subtle and may involve interaction among several environmental cues. For example, the relative

Abbreviations: A_{max}, absorbance maximum: AP, allophycocyanin: α^{AP}, β^{AP}, the α and β subunits of allophycocyanin, respectively: α^{PC}, β^{PC}, the α and β subunits of phycocyanin, respectively: α^{PE}, β^{PE}, the α and β subunits of phycoerythrin, respectively: C, carbon: CCA, complementary chromatic adaptation: C_i, inorganic carbon: CCM, carbon concentrating mechanism: chl, chlorophyll: DBMIB, 2,5-Dibromo-3-methyl-6-isopropyl-*p*-benzoquinone: DCMU, 3-(3,4-dichlorophenyl)-1,1-dimethylurea: DON,5-diazo-6-oxo-L-norleucine: GL, green light: GOGAT, glutamine 2-oxoglutarate aminotransferase: GS, glutamine synthase: GUS, β-glucuronidase: HCR, high CO_2-requiring: L_C, linker polypeptide in the core of the phycobilisome: L_{CM}, linker polypeptide which serves as an interface between the core of the phycobilisome and the thylakoid membranes and acts as a terminal energy acceptor for the phycobilisome: LHCII, light-harvesting complex of photosystem II: L_R, linker polypeptides in the phycobilisome rods: L_{RC}, linker polypeptide that serves as the interface between the rod and core substructures of the phycobilisome: MSX, L-methionine-D,L-sulfoximine: N, nitrogen: NiR, nitrite reductase: NR, nitrate reductase: P, phosphorus: PBS, phycobilisome: PC, phycocyanin: PC_c, constitutive phycocyanin: PC_i, red light inducible phycocyanin: PE, phycoerythrin: P_i, inorganic phosphate: PQ, plastoquinone: PSI, photosystem I: PSII, photosystem II: RL, red light: RuBPcase, ribulose-1,5-bisphosphate carboxylase: S, sulfur.

partitioning of absorbed excitation energy between photosystem I (PSI) and photosystem II (PSII), depends on the spectral distribution of light in the environment; this ability to redistribute excitation energy is important for balancing electron flow between the photosystems.

The nutrient status of the environment is also critical in modulating the activity and biogenesis of the photosynthetic apparatus. In both terrestrial and aquatic environments cyanobacteria experience prolonged periods during which they are limited for one or more nutrients. Under conditions of limiting nutrient availability, most cyanobacteria increase the efficiency with which they scavenge nutrients from the environment, initiate the mobilization of internal nutrient reserves, and limit metabolic activities which are not essential for survival. Some of the responses to nutrient limitation are specific to the deficiency of a particular nutrient and include elevated synthesis of transport systems and enzymes that may increase the availability of the limiting nutrient to the cell. However, there are also general responses that occur during deficiencies for any of a number of different nutrients (Grossman et al., 1994a). These responses include morphological adaptations, changes in the pools of intracellular reserves and alterations in cellular metabolism, particularly in the photosynthetic apparatus.

Photosynthetic or thylakoid membranes are densely packed within cyanobacterial cells, although they are not stacked as is typical in plant plastids. The protein complexes of the thylakoid membranes that are required for photosynthesis are briefly described below. We refer the reader to several reviews for detailed information concerning the structure and function of the photosynthetic apparatus (Andersson and Styring, 1991; Bryant, 1991; Gantt, 1981; Glazer, 1985; Glazer et al., 1983; Golbeck, 1994; Grossman et al., 1995; Grossman et al., 1993; Pakrasi, 1995; Sidler, 1994; Tandeau de Marsac and Houmard, 1993; Tandeau de Marsac et al., 1990; Vermaas, 1993). There are five major complexes of cyanobacterial thylakoid membranes required for driving the conversion of excitation energy into chemical bond energy. These protein complexes are photosystems (PS) I and II, the cytochrome b_6f complex, the ATP synthase and the phycobilisome (PBS). The PBS is the major light-harvesting complex in cyanobacteria which efficiently gathers light energy and directs most of that energy to the reaction centers of PSII. The antennae pigments of PSI consist mostly of chlorophyll (chl) molecules. When the excitation energy reaches the reaction centers it is converted into chemical energy in the form of a charge separation. In the reaction center of PSII, water molecules are oxidized and the electrons are transferred to the plastoquinone (PQ) pool. Electrons from the PQ pool are passed sequentially to the cytochrome b_6f complex, plastocyanin or cytochrome c_{553} (diffusible electron carriers) and finally to PSI. The electrons from PSI, after further energization, are transferred to ferredoxin and finally to NADPH. Protons accumulate in the thylakoid lumen as a consequence of the splitting of water plus electron transfer through the cytochrome b_6f complex; the proton gradient generated across the thylakoid membranes is coupled to ATP production via the ATP synthase. Both the reductant (NADPH) and energy (ATP) generated by photosynthetic electron transport are central to the metabolism of the cell.

II. Acclimation to Light

In nature, cyanobacteria face light conditions that fluctuate rapidly or are suboptimal for photosynthesis. Both the intensity and quality of the light can vary dramatically during the day or from one habitat to another. In a terrestrial environment, light may be of very high intensity (desert) and potentially damaging to the cell, or of very low intensity (forest floor) and inadequate to support optimal rates of photosynthesis. In the open oceans or in fresh water habitats light is attenuated in the spectral region above 550 nm as a consequence of the presence of particulate matter and the absorptive properties of water (Jerlov, 1976). The light that is available is biased towards the lower wavelengths, which is mostly absorbed by the PBS and used to preferentially excite PSII. Differential excitation could lead to an imbalance in electron flow through the two photosystems, which in turn might reduce photosynthetic efficiency. The range of different light environments experienced by cyanobacteria led to the evolution of mechanisms to detect the features of incident illumination and to modify the photosynthetic apparatus in ways that optimize survival. To accommodate rapid fluctuations in light intensity or quality, the photosynthetic apparatus can be modified within minutes. These short-term modifications are referred to as 'state transitions' and do not require protein synthesis. However, cyanobacteria are also capable of altering (i) the total number of photosystems, (ii) the ratio of PSI to PSII and (iii) PBS structure in response to prolonged changes in the light environment. These long-term acclimation responses require the induction of genes and the *de novo* synthesis of proteins. Furthermore, when the light intensity is very high it can cause 'photoinhibition' or damage to the reaction centers of PSII that results in a decline in the quantum yield of O_2 evolution. Cyanobacteria can modify the protein composition of PSII in high light, which may make the PSII reaction centers less susceptible to photoinhibition. The ways in which their photosynthetic apparatus can adjust to the changing light environment are discussed below.

A. Short-Term Acclimation (State Transitions)

In many environments the photosynthetically active radiation is biased toward certain wavelengths (Van Liere and Walsby, 1982). Since the light-harvesting antennae associated with PSI and PSII have different efficiencies of absorption of excitation energy across the visible spectrum, it is not uncommon for one photosystem to receive more excitation energy than the other. The cells can accommodate the changing spectral environment by adjusting the distribution of the excitation energy between the two photosystems. A process called 'state transitions' has the short-term function of facilitating this redistribution of excitation energy. State transitions can occur in vascular plants, algae and cyanobacteria, are extremely rapid (they take place within minutes of exposure to asymmetric excitation energy) and do not require protein synthesis (Allen, 1992; Biggins and Bruce, 1989; Fujita et al., 1994). Illumination of cyanobacterial cells with light that is absorbed preferentially by the PBS can result in excess excitation of PSII. This triggers a rapid redistribution of the excitation energy such that proportionally more of the absorbed energy excites PSI; under these conditions the cells are in state 2. In contrast, illumination that favors PSI excitation causes relatively more of the harvested light energy to be directed to PSII; under these conditions the cells are in state 1. The biochemical events required for facilitating state transitions were most studied in vascular plants, in which the phosphorylation state of the light-harvesting complex II (LHCII) is related to the shift between state 2 and state 1. A redox-controlled, thylakoid-localized, LHCII protein kinase appears to be required for state transitions (Allen, 1992; Bennett, 1991). When excitation energy is preferentially absorbed by PSII, the PQ pool and cytochrome b_6f become reduced. The reduction of the PQ pool appears to stimulate the LHCII kinase, which leads to increased levels of phosphorylated LHCII. The phosphorylated antennae move from the stacked thylakoid grana to the unstacked thylakoid lamellae where they transfer harvested light energy to PSI (state 2) (Delosme et al., 1996). This redistribution of the excitation energy between the photosystems partially restores balanced linear electron flow. The reverse occurs when the excitation energy preferentially stimulates PSI. A constitutive phosphatase appears to be responsible for dephosphorylating LHCII (Bennett, 1991), which then moves back to the appressed regions of the thylakoid membranes where it functionally associates with PSII (state 1).

The mechanistic aspects of state transitions are controversial in cyanobacteria. Three different models were proposed to explain cyanobacterial state transitions (Allen, 1992; Fujita et al., 1994; Mullineaux, 1992; Mullineaux and Allen, 1990; Salehian and Bruce, 1992; Satoh and Fork, 1983). State transitions in cyanobacteria, like in vascular plants, may require the phosphorylation of components of the light-harvesting antennae complexes (Allen, 1992; Biggins and Bruce, 1989; Sanders et al., 1989). The 'mobile antenna' model hypothesizes that the PBS disconnects from PSII and becomes physically associated with PSI. In the 'spillover' model the PBS do not physically move, however, excess excitation energy accumulated in PSII can spill over, in a somewhat ill-defined manner, into the reaction centers of PSI. In the third model, termed the 'detachment' model, excess excitation of PSII results in PBS detachment; however, the detached PBS does not necessarily reassociate with and donate energy to PSI. None of the models explains all of the experimental data and new approaches, both biophysical and genetic, are being used to elucidate the molecular details of state transitions. Some experiments have suggested a critical role for allophycocyanin α^{AP-B} (encoded by *apcD*) in controlling the distribution of excitation energy between the photosystems (Zhao et al., 1992).

B. Long-Term Acclimations

1. High Light and Photoinhibition

The ways in which photosynthetic organisms adjust to excess excitation is of major ecological and potential commercial importance. Excess excitation of the photosynthetic reaction centers can lead to photoinhibition (Andersson and Styring, 1991; Aro et al., 1993; Kim et al., 1993; Samuelsson et al., 1987) or a decline in the quantum yield of photosynthesis that may reflect damage to reaction centers of PSII. The PSII reaction centers are comprised of four proteins, D1 and D2 which form heterodimers in the membranes, cytochrome b_{559} and the *psbI* gene product. Of these polypeptides, D1 appears to be most easily damaged by excess irradiation. The photooxidation of water by PSII involves the generation of reactive intermediates that can potentially modify proteins. At low light intensities the rate of photosynthetic electron transport is proportional to the photon flux density and damaged D1 polypeptides can be removed from the reaction centers and rapidly replaced by newly synthesized D1 (Guenther and Melis, 1990). At higher light intensities the rate at which the reaction center of PSII is damaged exceeds the rate of repair (De Las Rivas et al., 1992; Öquist et al., 1992; Reuter and Miller, 1993), which results in a decline in the quantum yield of photosynthesis.

In vascular plants and green algae the D1 polypeptide is encoded by a single gene (*psbA*) located on the plastid genome (Andersson and Styring, 1991; Shipton and Barber, 1991). In cyanobacteria such as *Synechococcus* PCC 7942 there are two distinct forms of the D1 protein encoded by a small *psbA* multigene family (Golden, 1994). In contrast, while there are two *psbD* genes for D2 in *Synechococcus* PCC 7942, the polypeptides encoded by these genes are identical (Gingrich et al., 1990). The regulation of the *psbA* and *psbD* genes has been extensively studied in *Synechococcus* PCC 7942 (Golden et al., 1986; Golden et al., 1989; Golden and Stearns, 1988). This organism contains three *psbA* genes, designated *psbAI*, *psbAII*, *psbAIII*. The *psbAI* gene codes for Form I of the D1 polypeptide (also known as D1:1) while both *psbAII* and *psbAIII* code for D1 Form II (D1:2). Of the 25 amino acid differences between Form I and Form II, 12 are clustered within the first 16 amino terminal residues of the polypeptide. Seven amino acid differences are also present in transmembrane helices II and III.

The *psbA* genes exhibit clear differences in their expression patterns with respect to light intensity. At low irradiances, the *psbAI* transcript predominates. If the cells are transferred to higher light intensities, the level of the *psbAI* transcript rapidly declines while the *psbAII* and *psbAIII* transcripts accumulate (Bustos et al., 1990). These changes in transcript levels are also reflected at the polypeptide level; *Synechococcus* PCC 7942 grown at low light contains primarily D1 Form I, which is almost entirely replaced after a short exposure to high light intensities with D1 Form II (Clarke, 1995; Kulkarni and Golden, 1994).

The high light intensity-triggered change in the polypeptide composition of the PSII reaction centers in cyanobacteria may be important for the organism to cope with the potentially damaging effects of excess irradiation. The D1 Form II polypeptide probably helps protect the cells from photodamage. The most convincing data concerning the benefit of D1 Form II during exposure of *Synechococcus* PCC 7942 to high light has come from studies of mutants that do not make either Form I or Form II of the D1 protein. Four different arguments support the view that D1 Form II is critical for acclimation to high light. (i) A mutant unable to synthesize D1 Form II appears to be more susceptible to photoinhibition (Krupa et al., 1991) and less able to recover from photoinhibition than the wild-type strains (Clarke et al., 1993a). (ii) Kulkarni and Golden demonstrated that mutants lacking D1 Form II do not grow as fast as wild-type cells during the transition period following exposure of cells to high light intensities (Kulkarni and Golden, 1995). (iii) The protection afforded by D1 Form II may result, at least in part, from the fact that reaction centers with this form of the D1 polypeptide seem to be more photochemically efficient than reaction centers with D1 Form I (the quantum yield of photosynthesis is about 25% higher for reaction centers with the former) (Clarke et al., 1993b). (iv) It was also suggested that D1, Form II is not as readily damaged in high light as Form I, and does not turn over as rapidly (Kulkarni and Golden, 1994).

Curiously, the level of the D1 Form II protein appears to increase transiently under certain conditions. In experiments by Clarke et al. (1995), when *Synechococcus* PCC 7942 was shifted from 50 to 200 μmol photon m^{-2} s^{-1}, there was a transient rise in the level of D1 Form II in PSII reaction centers; 2 h following the shift to 200 μmol photon m^{-2} s^{-1}, D1 Form II peaked, but decreased to low levels by 4 h after the shift, at which time it was replaced by D1 Form I. This suggests that Form II is only needed during the initial exposure to 200 μmol photon m^{-2} s^{-1} as the cells are acclimating to the new light conditions. However, similarly designed experiments in the laboratory of S. Golden, using different conditions of illumination, yielded significantly different results. In *Synechococcus* PCC 7942 cultures transferred from 130 to 500 μmol photon m^{-2} s^{-1} the D1 Form I was replaced by D1 Form II within the first 2 h of exposure to the higher photon flux density. However, if the cells were maintained at this light intensity, D1 Form II was maintained in the PSII reaction centers (Kulkarni and Golden, 1994). Hence, a shift to moderate photon flux density (200 μmol photon m^{-2} s^{-1}) causes a transient accumulation of D1 Form II while a shift to a relatively high photon flux density (500 μmol photon m^{-2} s^{-1}) sustains the biosynthesis of reaction centers with D1 Form II. This suggests that the reaction centers may be able to acclimate at the moderate intensity light in a manner that obviates the need for by D1 Form II. The advantage of having D1 Form I in the reaction center during growth in low light is not clear.

Light intensity may be critical in determining both transcriptional and post-transcriptional control of the *psbA* mRNA levels (Bustos et al., 1990; Schaefer and Golden, 1989). At low light most of the *psbA* transcript was from *psbAI*. Upon a shift to high light there was a rapid accumulation of the *psbAII* and *psbAIII* transcripts and a concomitant loss of the *psbAI* message. This accumulation of the *psbAII* and *psbAIII* messages is a consequence of activation of these genes (Kulkarni et al., 1992) while the loss of the *psbAI* transcript reflects both a decreased rate of transcription and an increased rate of mRNA turnover (Kulkarni et al., 1992). Based on promoter analyses in which the *psbAII* and *psbAIII* promoters were fused to *lacZ* (encoding β-galactosidase), sequences within the untranslated leader regions of the mRNAs were determined to be responsible for increased expression of the *psbAII* and *psbAIII* genes in high light (Li and Golden, 1993). These sequence elements may function as enhancers that bind proteins that are only present in extracts from high light-grown cells (Li et al., 1995).

After long exposures to high light, the *psbAI* transcript begins to increase while the *psbAII* and *psbAIII* transcripts remain relatively high. In spite of this long-term change in the level of the different *psbA* transcripts, the D1 Form II protein remains associated with PSII, suggesting post-transcription regulation of protein abundance (e.g. a faster turnover of the D1 Form I). Post-transcriptional regulation in this system probably occurs at a number of different levels. The kinetics of degradation of all three of the *psbA* messages are different (Kulkarni et al., 1992). Furthermore, the rate of turnover of the *psbAI* transcript increases upon transfer of the cells to high light. Curiously, the *psbAIII* transcript also shows enhanced degradation in high light while the stability of the *psbAII* transcript is unaffected by light intensity. Recently, Kulkarni and Golden have identified an element that stabilizes *psbA* mRNAs which maps within the region of the transcript coding for the first membrane span of the D1 polypeptide. This stabilization effect is independent of light intensity (Golden and Kulkarni, 1996).

The signal transduction pathway that controls the rates of transcription of the *psbA* gene family has been examined to some extent. The light signal could be transduced directly by a photoreceptor or indirectly through the physiological (e.g. redox) state of the photosynthetic apparatus. Currently there is positive evidence for both models. Tsinoremas et al. (1994) have shown that 10 μmol photon m^{-2} s^{-1} of blue light is enough to elevate message levels, in a manner very similar to that seen with high intensity white light, suggesting the involvement of a blue light photoreceptor. Interestingly, this blue-light effect was red-light reversible. Other blue light-regulated genes exist in cyanobacteria. For example, the *hliA* gene, which encodes a protein associated with the thylakoid membranes and resembles membrane spanning helices of the Lhc polypeptides of vascular plants, appears to be under the control of a blue/UV-A photoreceptor (Dolganov et al., 1995; Chapter 21).

Recently, Öquist's group found that chilling (shifts from 37°C to 25°C) or inhibitors of photosynthetic electron transport such as DCMU (at a level that caused 50% closure of reaction centers) under low light intensity had the same effect as high light treatment in causing the replacement of D1 Form I with D1 Form II. DBMIB did not cause D1 exchange to occur (Campbell et al., 1996; Campbell et al., 1995). These results are not consistent with the idea that reduction of the PQ pool or cytochrome b_6f, as would occur in high light, are important in controlling *psbAII* and *psbAIII*, However, it is possible, as suggested by the authors, that moderate closure of the PSII reaction centers is crucial in regulating the *psbA* gene family. It is not absolutely clear that DCMU is controlling the ratio of D1 Form I to D1 Form II by altering the levels of the *psbA* transcripts. Golden and co-workers found no effect of photosynthetic electron transport inhibitors such as DCMU on the accumulation of the *psbAII* and *psbAIII* transcripts (S, Golden, pers. comm.). These observations can be explained if DCMU, which binds to D1, differentially affects the stability of the two forms of D1 (rather than causing a change in the transcription of the *psbA* genes).

2. Control of Photosystem Stoichiometry

Early studies with cyanobacteria and red algae (which like cyanobacteria contain PBS) demonstrated variability of the quantum yield of photosynthesis for light absorbed by chl *a*. This variability was dependent on the light conditions (e.g. excitation energy preferentially absorbed by PSI or PSII) during growth and correlated with changes in cellular pigment composition (Biggins and Bruce, 1989). Myers et al. (1980) were the first to observe that the PSI:PSII ratio was higher in *Synechococcus* PCC 6301 (*Anacystis nidulans*) cells grown in light absorbed mainly by the PBS than in cells grown in light absorbed mainly by chl *a*. At about the same time, it was demonstrated that the cyanobacterial PSI:PSII ratios were also affected by light intensities; cells grown in weak white light had a higher PSI:PSII ratio than cells grown in strong white light (Kawamura et al., 1979). Following these classic experiments, many data have confirmed that cyanobacteria have the ability to modulate PSI:PSII ratios in response to light intensity and quality (Fujita et al., 1994).

The evolution of mechanisms to modulate the PSI:PSII ratio in response to different wavelengths of light has helped cyanobacteria to optimize photosynthetic efficiency. Based on experimental evidence from several groups, a working model was proposed that rationalizes the modifications of photosystem stoichiometry following growth under different conditions of illumination (Aizawa et al., 1992; Biggins and Bruce, 1989; Fujita et al., 1985; Melis et al., 1989; Murakami and Fujita, 1991). In light that is absorbed mainly by PBS, the PSI:PSII ratio increases. This favors a proportional increase in the absorbance of excitation energy by PSI which, in turn, will result in more balanced, linear electron flow and increased rates of CO_2 fixation. In contrast, cells grown in light that is primarily absorbed by chl *a* (the major light-harvesting pigment for PSI) exhibit elevated levels of PBS relative to chl *a*. This altered ratio, which again tends to favor balanced linear electron flow, is primarily a consequence of a decrease in the number of PSI reaction centers (Aizawa et al., 1992; Fujita and Murakami, 1987; Fujita et al., 1992; Fujita et al., 1988).

In the natural environment cyanobacteria are most frequently exposed to 'white light', which contains all wavelengths of the visible spectrum. Since the PBS absorbs a large proportion of the incident white light, there will be preferential excitation of PSII when cyanobacteria are exposed to nonsaturating white light, which would serve as a signal that prompts elevated synthesis of PSI relative to PSII. As the intensity of white light increases, both photosystems become saturated, and the PSI:PSII ratio falls, approaching that measured in cells grown in chl *a* light. Other light-mediated control mechanisms that are distinct from those that regulate photosystem stoichiometry may also influence photosynthetic activity. For example, prolonged high light intensities may lead to a decrease in the total number of photosystems (Yokoyama et al., 1991).

Experiments performed so far have indicated that changes in PSI:PSII ratios are mediated mainly via alterations in the amount of PSI present in the cell. If the number of PSI and PSII reaction centers are determined on a per cell basis, it appears that the number of PSII centers remain relatively constant while PSI centers can vary by over two fold (Fujita and Murakami, 1987). Dynamic control of PSI levels also appears to occur in vascular plants and in the green alga *Chlamydomonas reinhardtii* (Leong and Anderson, 1986; Melis et al., 1996). The functional advantage of maintaining a relatively invariant pool of PSII and modulating the biosynthesis of PSI is not apparent.

The least well understood aspect of the adjustment of the photosynthetic apparatus to the light environment concerns the transduction of the light signal into altered biosynthetic activity. In an early proposal, the signal for regulating photosystem stoichiometry was suggested to be the ATP:NADPH ratio (Melis et al., 1985). In contrast, Fujita and colleagues (Fujita et al., 1987; Murakami and Fujita, 1991) proposed that the redox state of electron transport components, and in particular of cytochrome b_6, is the signal that controls the level of PSI. The cellular redox level was shown to be important for controlling several genes in *E. coli* (Iuchi and Lin, 1993), and some evidence suggests that the transcription of *Lhc* genes in green algae (Escoubas et al., 1995) is governed by the level of reduction of the PQ pool. Hence, control of both long- and short-term adjustments to the disproportionate absorbance of excitation energy may be mediated by redox-dependent signaling pathways. However, there is little understanding of the signalling molecules that link the redox state of the cell to the rate of synthesis of components of the photosynthetic apparatus. Furthermore, the individual structural polypeptides of the photosystems are often post-translationally modified, associate with one or more prosthetic group, and assemble into functional multisubunit complexes. The coordination of the distinct steps that must be temporally and spatially controlled and that may involve a hierarchy of regulatory events, is still a mystery.

3. Control of Phycobilisome Biosynthesis by Light Quality

a. Phycobilisome Structure

A brief description of the PBS structure is provided to help the reader understand the ways in which light can control PBS biosynthesis. In early studies, electron micrographs of cyanobacteria showed regularly arranged parallel rows of granules on the surface of the thylakoid membrane (Gantt, 1981; Glazer, 1985). These profuse granules were PBS that were extrinsically attached to the membrane surface. PBS may constitute as much as 40% of the protein of the cell and are composed of several different pigmented, water-soluble polypeptides that give cyanobacteria their characteristic blue-green or red color. The PBS functions as a light-harvesting apparatus that absorbs photons and transmits the exciton energy to the photosynthetic reaction centers with an extremely high efficiency (Porter et al., 1978; Searle et al., 1978).

A synthesis of the information generated over a score of years from electron microscopy, biochemical and biophysical studies, X-ray crystallography and molecular analyses of genes encoding PBS polypeptides has led to the development of a detailed model of PBS structure (Bryant, 1991; Bryant et al., 1979; Gantt, 1981; Glazer, 1987; Glazer et al., 1983). A diagrammatic representation of the PBS is shown in Fig. 1 (Section II.3.b for more detail). Each PBS is organized into rod and core substructures. The core, mainly composed of allophycocyanin (AP, A_{max}=650nm) trimers, is in direct contact with thylakoid membranes via a large pigmented polypeptide (L_{CM}) that serves as a terminal energy acceptor of the PBS. There are usually six sets of rods that radiate out from the core in a fan-like array. Rods are composed of hexamers of the chromophorylated phycocyanin (PC, A_{max}=620nm) and sometimes phycoerythrin (PE, A_{max}=565nm) or phycoerythrocyanin (PEC, A_{max}= 570nm). Each of the phycobiliprotein monomers has two different subunits, α and β, with molecular masses of between 15 and 22 kDa each. The chromophores or light-absorbing molecules that are attached to the phycobiliproteins via thioether linkages are linear tetrapyrroles (Glazer, 1987). The numbers and types of chromophores associated with a particular phycobiliprotein subunit are usually invariant (with some exceptions), and the absorbance characteristics of these chromophores are strongly influenced by their conformation and interactions with amino acid residues of the protein moiety of the molecule. The PC and PE hexamers are held together by linker (L) polypeptides, which are mostly non-pigmented (Sidler, 1994). Linker polypeptides serve several functions. These include stabilization of the PBS structure, facilitation of assembly of phycobiliprotein-containing substructures and modulation of the absorption characteristics of the phycobiliproteins to promote unidirectional transfer of energy through the PBS and to the chl *a* of the photosynthetic reaction centers.

b. Complementary Chromatic Adaptation

The construction of a PBS is a dynamic process that is responsive to changing environmental conditions. In cyanobacteria, nutrient levels (Collier and Grossman, 1992; Collier and Grossman, 1994; Yamanaka and Glazer, 1980), light quality and light quantity can all dramatically alter PBS composition (Bogorad, 1975; Grossman, 1990; Grossman et al., 1993; Tandeau de Marsac, 1983; Tandeau de Marsac and Houmard, 1993). The ability of some cyanobacteria to alter the PC/PE composition (PC has an absorbance maximum in the red while PE has an absorbance maximum in the green region of the visible spectrum) of the PBS allows them to efficiently absorb the prevalent wavelengths of light in the environment. Based on the responses of cyanobacteria to light quality, they have been divided into three different groups. Group I cyanobacteria can alter PBS size and number, in response to light cues, but do not markedly alter the absorbance characteristics of their PBS. Group II cyanobacteria can alter the levels of PE in the PBS, and group III organisms can modulate both the PE and PC levels of the PBS via a process termed complementary chromatic adaptation (CCA). The effect of light on the composition of the PBS in a group III organism is shown in Fig. 1. CCA was most extensively examined in the cyanobacterium *Fremyella diplosiphon,* and the closely related *Calothrix* PCC 7601, although the phenomenon was observed in many cyanobacteria (Bryant, 1981; Bryant and Cohen-Bazire, 1981; Tandeau de Marsac, 1977).

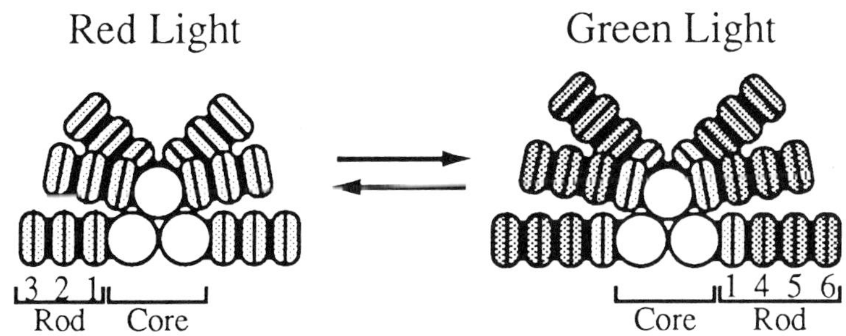

Rod subunit composition	Rod subunit composition
(1) $(\alpha\beta^{PC})_6\ L^{27}_{RC}$	(1) $(\alpha\beta^{PC})_6\ L^{27}_{RC}$
(2) $(\alpha\beta^{PC})_6\ L^{30.5}_{R}$	(4) $(\alpha\beta^{PE})_6\ L^{31.8}_{R}$
(3) $(\alpha\beta^{PC})_6\ L^{32.5}_{R}\ L^{9.7}_{R}$	(5) $(\alpha\beta^{PE})_6\ L^{27.9}_{R}$
	(6) $(\alpha\beta^{PE})_6\ L^{27.6}_{R}$

Fig.1. Chromatic adaptation and the composition of the *F. diplosiphon* PBS from cells maintained in green light (GL) and red light (RL). The composition of the peripheral rods from RL-grown cells are given by 1, 2 and 3, while the composition of the rods in GL-grown cells are given by 1, 4, 5 and 6. The PBS rods from RL grown cells are composed exclusively (or almost exclusively) of PC hexamers, while the rods from GL-grown cells retain only one PC hexamer (nearest to the core) and contain mostly PE hexamers. The rod subunit composition refers to the α and β subunits that comprise the PE hexamer $(\alpha\beta^{PC})_6$ or PC hexamer $(\alpha\beta^{PE})_6$. Each linker (L) polypeptide is associated with a specific set of hexamers and is shown with a superscript that refers to the molecular mass and a subscript that refers to its position within the hexamer (R for rod or RC for rod-core). For further details on core composition and genes encoding phycobiliproteins refer to Grossman et al. (1995).

This process is visually most apparent when the adapting organisms are compared after growth in red light (RL) or green light (GL). In RL *F. diplosiphon* cells are blue, because they accumulate high levels of the blue pigment PC and barely detectable levels of PE. Conversely, in GL, new PBS are synthesized with rods that now contain only a single PC hexamer and as many as three hexamers of PE. The action spectrum for maximal synthesis of PE and minimal synthesis of PC has a peak at 540 nm (GL), while the action spectrum for maximal synthesis of PC and minimal synthesis of PE has a peak at 650 nm (RL) (Haury and Bogorad, 1977; Vogelmann and Scheibe, 1978). Hence, the photoreceptor(s) involved in CCA may show changes in both conformation and activity upon absorbing GL/RL.

Over the last ten years many of the genes encoding the structural components of the PBS were cloned and characterized from a number of different cyanobacterial species (Grossman et al., 1995). In most cases, polycistronic transcripts encoded both the phycobiliprotein subunits and associated L proteins. Furthermore, depending upon the organism, the different phycobiliproteins may be encoded by a multigene family. The PBS genes that are differentially regulated during CCA are those encoding PE, PC and the L polypeptides that associate with the PE and PC hexamers. None of the genes encoding the core components of the PBS are differentially regulated during CCA.

There are three distinct gene sets (designated *cpcB1A1, cpcB2A2 and cpcB3A3*) that each encode α and β subunits of PC in *F. diplosiphon* (Conley et al., 1985; Conley et al., 1988; Conley et al., 1986; Mazel et al., 1988; Mazel and Marliere, 1989). The *cpcB1A1* operon is constitutively expressed and the PC that it encodes is referred to as PC_c (subscript denotes constitutive). It appears to be cotranscribed with *cpcE* and perhaps *cpcF*, which encode a lyase that assists in the attachment of the tetrapyrrole chromophores to the α subunit of PC (Fairchild and Glazer, 1994; Fairchild et al., 1992). A second operon that includes *cpcB2A2* becomes transcriptionally active only in RL (it is inactive in GL) and the PC that it encodes is specifically designated PC_i (subscript denotes inducible) (Conley et al., 1985; Conley et al., 1986). The induction of this operon is critical for CCA. Hexamers of PC_i constitute the majority of the PBS rod substructure when *F. diplosiphon* is grown in RL. Three genes encoding the L polypeptides that associate with PC_i are cotranscribed with *cpcB2A2* and are designated *cpcH2*, *cpcI2* and *cpcD2* (Lomax et al., 1987). The third PC operon contains *cpcB3A3* as well as genes encoding associated L polypeptides. It is only active during sulfur-limited growth (Mazel and Marliere, 1989) and the PC_s (subscript indicating activation during sulfur-limitation), encoded by *cpcB3A3*, has no sulfur-containing amino acids other than the cysteines that are essential for the attachment of the phycobilin chromophores. Hence, when starved for sulfur, *F. diplosiphon* can conserve this nutrient by substituting PC_i and perhaps PC_c with PC_s. PE is encoded by the *cpeBA* operon (Mazel et al., 1986) and the associated linker polypeptides are encoded by a separate operon, *cpeCDE* (Federspiel and Grossman, 1990; Federspiel and Scott, 1992). The *cpeBA* and *cpeCDE* operons are coordinately regulated; transcripts from both of these operons are abundant in GL and barely detectable in RL.

i. Promoter Binding Factors

To study regulatory elements that may bind to the *cpeBA* promoter, soluble proteins from RL- and GL-grown *F. diplosiphon* or *Calothrix* PCC 7601 cells were used in mobility shift assays. Soluble cyanobacterial proteins bind to the *cpeBA* promoter, retarding its mobility during electrophoresis in a polyacrylamide gel. Three proteins appeared to interact with sequences positioned -110 to +81 relative to the transcription start site of the *cpeBA* operon (Sobczyk et al., 1993). RNA polymerase interacted with the sequence from -40 to +15. Two other proteins, designated RcaA and RcaB, appeared to be present only in GL-grown cells. RcaA interacts with the sequence from -67 to -45, which contains a tandem repeat 5'TTGTTA$(N)_4$TTGTTA 3'. Schmidt-Goff and Federspiel (1993) used *in vivo* footprinting, *in vitro* electrophoretic mobility shift assays and dimethyl sulfate and DNase I *in vitro* footprinting, to show that a protein, designated PepB (PE promoter-binding protein), bound to and protected the 5'TTGTTA3' tandem repeat located upstream of the transcription start site. Since RcaA and PepB bind to the same sequences, it is probable that they are identical proteins.

The tandem repeat present upstream of *cpeBA* promoters was also observed in other cyanobacteria that exhibit CCA. However, it was not present in the *cpeCDE* operon, inspite of the fact that this operon was controlled identically to *cpeBA*. Therefore, if the 5'TTGTTA3' motif is important for CCA, additional elements must be involved in coordinating the transcriptional responses of *cpeBA* and *cpeCDE*. The exact position of the binding site for RcaB is not clear.

Proteins were reported to bind to the promoter of the *cpcB2A2* operon, which encodes PC_i (Sobczyk et al., 1994). To further analyze this promoter a chimeric reporter gene was constructed in which the *cpcB2A2* promoter was fused to the *uidA* gene (encoding β-glucuronidase or GUS). The promoter region of the chimeric gene was modified, reintroduced into *F. diplosiphon*, and transformants were assayed for light-responsive GUS activity. A fragment extending from -76 to +25 relative to the *cpcB2A2* transcription start site conferred RL/GL responsiveness to GUS expression. Deletions extending to -37 reduced GUS activity in both RL and GL, suggesting that the sequence between -76 and -37 is essential for the activity of the *cpcB2A2* promoter. Gel mobility shift assays demonstrated that a protein in extracts from RL-grown but not GL-grown cells was able to bind to a -76 to +25 bp fragment. A DNA fragment from -37 to +25 was able to compete for this binding activity (Casey and Grossman, 1994), suggesting that the -37 to +25 region of the *cpcB2A2* promoter interacts with a protein that is either only present or only binds in RL-grown cells. Together, the results suggest that the region *cpcB2A2* operon from -76 to -37 is required for overall promoter activity while the -37 to +25 sequence binds a regulatory protein which, in turn, promotes transcription from *cpcB2A2* in RL. This binding site was shown to contain a direct repeat of the sequence 5'AAATTTGCACAAA3' (Casey and Grossman, 1994).

ii. Mutant Characterization

Over the last several years numerous mutants of *F. diplosiphon* were generated and characterized (Bruns et al., 1989; Chiang et al., 1992; Cobley and Miranda, 1983; Kehoe and Grossman, 1994; Tandeau de Marsac, 1983). These mutants can be placed into at least four phenotypic classes, red (FdR), blue (FdB), green (FdG) and black (FdBk). FdR strains appear red under all conditions of illumination; PE has become constitutive while PC_i is never expressed. The FdB strains are bluer than wild-type cells in RL and require more GL than wild-type cells to suppress PC_i synthesis (Casey et al., 1997). The FdG strains exhibit normal PC_i expression, but the PE genes never become active. In the FdBk mutants there are moderate levels of both PE and PC_i, however, these levels remain the same in RL and GL (Kehoe and Grossman, 1996). Complementation of the mutants back to the wild-type pigmentation phenotype, which can be visually evaluated, is a powerful way of dissecting the regulatory components that control CCA. Recently a number of the different mutants were complemented using a wild-type library from *F. diplosiphon.*

The FdBk mutant was complemented with a gene designated *rcaE* while different FdR mutants were complemented by *rcaF*, *rcaE* (Kehoe and Grossman, 1997) or *rcaC* (Chiang et al., 1992). Significantly, all the components encoded by these genes are proteins that have similarity to components of two component regulatory systems. The 74 kDa protein encoded by *rcaE* has a carboxy-terminal region with similarity to output domains of histidine kinase sensors. Surprisingly, the amino-terminal half of the protein, over a region of approximately 140 amino acids, is similar to the tetrapyrrole chromophore attachment domain of the phytochromes (Kehoe and Grossman, 1996). Recently it was shown, using a combination of antibodies and zinc-enhanced fluorescence, that the RcaE protein is tightly associated with a linear tetrapyrrole (Kehoe and Grossman, unpublished data). Phytochromes are well-characterized photoreceptors that control a wide range of responses in vascular plants (Kendrick and Kronenberg, 1994). RcaE also has sequence similarity to vascular plant ethylene receptors (Chang et al., 1993; Hua et al., 1995; Wilkinson et al., 1995). The ethylene receptors are similar to sensor kinases of bacterial two component regulatory systems. The phenotype of the FdBk mutant (Kehoe and Grossman, 1996), the similarity of RcaE to a vascular plant photoreceptor and sensor histidine kinases, and the association of the RcaE polypeptide with a tetrapyrrole chromophore are all consistent with RcaE being involved in photoperception.

The RcaF protein, encoded by a gene which is initiated 12 bp downstream of the translation termination site of *rcaE*, is predicted to be 124 amino acids and is homologous to response regulators. However, like CheY and Spo0F, response regulators involved in flagellar movement in *E. coli* and sporulation in *Bacillus subtilis*, respectively (Clegg and Koshland, 1984; Parkinson and Kofoid, 1992; Perego and Hoch, 1996; Ravid et al., 1986; Wolfe et al., 1987; Yamaguchi et al., 1986), RcaF does not contain an identifiable output domain. RcaF may act as an intermediate in the phosphorelay pathway controlling CCA and transfer phosphate groups from its cognate sensor (presumably RcaE) to other response regulators such as RcaC (see below).

The *rcaC* gene encodes a protein of 651 amino acids with sequence similarity to response regulators (Appleby et al., 1996; Parkinson and Kofoid, 1992). RcaC has characteristics that make it an unusual response regulator. It is approximately twice as large as most response regulators and has two, instead of one, conserved, aspartate containing receiver domains. One of the receiver domains is at the amino terminus while the other is at the carboxy terminus. Contiguous to the amino terminal receiver domain is a DNA binding motif and between this region and the carboxy terminal receiver domain is a histidine-containing motif that resembles an H block found in some unorthodox sensor proteins (Appleby et al., 1996; Ishige et al., 1994).

The three proteins critical for the control of CCA, RcaE, RcaF and RcaC, are all homologous to elements of bacterial two component regulatory systems. However, the phosphorelay system that controls CCA is unique in that it includes, among these three proteins, at least five potential phosphoacceptor domains. Some systems have been characterized that require four phosphoacceptor domains; these include the phosphorelay systems that control sporulation in *B. subtilis* (Burbulys et al., 1991), transcriptional regulation of virulence factors in *Bordetella pertussis* (Ishige et al., 1994; Uhl and Miller, 1996) and osmoregulation in *Saccharomyces cerevisiae* (Posas et al., 1996).

Based on the phenotypes of the CCA mutants and on studies in which each of the regulatory genes is overexpressed in wild-type cells or CCA mutants, Kehoe and Grossman have hypothesized a linear phosphorelay pathway (Kehoe and Grossman, 1997; Grossman and Kehoe, 1997), which is shown in Fig. 2. In this model, RL stimulates and GL inhibits the transfer of phosphoryl groups along the phosphorelay pathway. RcaE is likely to be associated with the chromophore(s) that absorbs the different wavelengths of light. When RcaE absorbs RL it

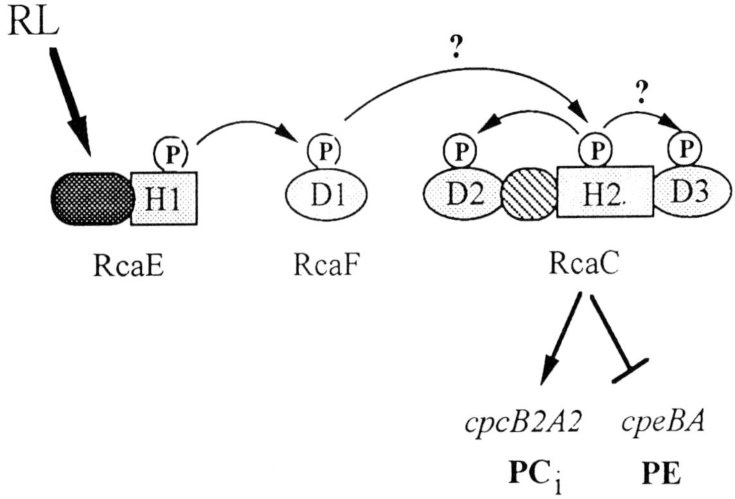

Fig. 2. Model for control of phycobiliprotein genes by red light (RL) during complementary chromatic adaptation. The phosphorelay that is activated by RL and regulates the genes for PC_i (*cpcB2A2*) and PE (*cpeBA*) as shown. RL, probably absorbed by a chromophore attached to the phytochrome-like domain of RcaE (dark oblong) triggers autophosphorylation of the conserved histidine residue of the H block (H1). The phosphoryl group, P, is transferred to the aspartate residue (D1) of RcaF. RcaF may transfer the phosphoryl group to a histidine in the H block (H2) of RcaC. Finally the phosphoryl group may be transferred to the aspartate residue (D2) in the receiver domain at the amino terminus or perhaps to an aspartate residue (D3) at the carboxy terminus. The arrows indicate phosphoryl group transfer reactions; the question marks indicate reactions that might occur. The stippled circle represents a DNA binding domain on RcaC.

undergoes an autophosphorylation on the histidine of the H block. This phosphate group is then transferred to an aspartate residue of the response regulator RcaF. RcaF may also serve as an entrance point for phosphoryl donation from proteins other than RcaE. RcaF may then transfer the phosphoryl group to an H block within RcaC, which can then pass it to either the amino or (perhaps) the carboxy terminal receiver domains. The amino terminal receiver domain of RcaC is critical for CCA (Kehoe and Grossman, 1995). The role of the carboxy terminal receiver domain is unclear, although it may function to fine tune the system with respect to environmental conditions (e.g. light intensity, nutrient availability) or control other processes affected by light quality, such as hormogonia formation (Tandeau de Marsac, 1994). It is unclear whether RcaC binds to the promoter regions of the *cpcB2A2* and *cpeBA* operons to directly regulate transcription, or whether additional regulatory factors are required.

When GL is absorbed by RcaE, the phosphorelay is maintained in a dephosphorylated state. This could be accomplished if RcaE binds to and blocks promiscuous phosphorylation of RcaF (that can occur through interactions with other sensors in the cell or small phosphodonor molecules) or if RcaE acts as a phosphatase, removing any phosphate that is transferred to RcaF.

III. Responses to Nutrient Limitation

The responses of cyanobacteria to nutrient deficiency have been studied extensively (Allen, 1984; Bryant, 1986; Bryant, 1991; Healey, 1982; Reithman et al., 1988; Simon, 1987; Tandeau de Marsac and Houmard, 1993). Some of the responses of these responses are specific, triggered by a deficiency for one particular nutrient, while others are general, and occur during deficiency for any of a number of different nutrients. The former category includes increased synthesis of specific transport systems and enzymes that transform inaccessible forms of a nutrient into those that the cell can metabolize. The latter category includes changes in cell morphology, intracellular nutrient reserves, and alterations in the activity of numerous physiological processes, including photosynthesis (Grossman et al., 1994a; Grossman et al., 1994b). For convenience, *Synechococcus* PCC 7942, *Anacystis nidulans*, *Synechococcus* PCC 6301 and *Synechococcus leopoliensis*, which are all very similar strains, will sometimes be referred to as *Synechococcus* sp.

Likewise, a number of *Anabaena* strains are also treated as a group. The following discussion is limited to the responses of cyanobacteria to macronutrient limitation, which includes carbon (C), nitrogen (N) phosphorus (P) and sulfur (S).

A. Carbon Limitation

Cyanobacteria are able to perform efficient photosynthesis in aquatic environments despite a limited supply of CO_2 and the low affinity that the primary CO_2 fixing enzyme, ribulose-1,5-bisphosphate carboxylase-oxygenase (RuBPcase), has for CO_2. When cyanobacterial cells are transferred from high (1 - 5% CO_2 in air) to low CO_2 concentrations (air level of CO_2 or below), they exhibit a suite of light-dependent responses that include (i) a 10 - 20 fold increase in the apparent photosynthetic affinity for inorganic carbon (Ci), which reflects an increased ability of the cells to concentrate C_i (Aizawa and Miyachi, 1986; Badger, 1987; Badger and Price, 1992; Coleman, 1991; Kaplan et al., 1991; Kaplan et al., 1994; Kaplan et al., 1990; Karagouni et al., 1990; Miller et al., 1990; Pierce et al., 1988; Raven, 1991), (ii) an increase in the RuBPcase level (Mayo et al., 1989), (iii) elevated numbers of carboxysomes (polyhedral bodies that contain most of the RuBPcase) (McKay et al., 1993; Turpin et al., 1984), (iv) the accumulation of several polypeptides in the cytoplasmic membranes, and in particular a polypeptide of 42 kDa (Omata and Ogawa, 1986), (v) thickening of a specific layer of the cell envelope (Kaplan et al., 1990), (vi) modification of purine biosynthesis (Schwarz et al., 1992) and (vii) changes in the pattern of phosphorylation of polypeptides (Bloye et al., 1992).

The induction of the C_i concentrating mechanism (CCM) is a prominent feature of the acclimation process and occurs in most cyanobacteria that have been examined. One study did suggest that the marine cyanobacterium *Synechococcus* WH7803 did not induce a CCM (Karagouni et al., 1990). A re-evaluation of this indicates that this organism does concentrate C_i when maintained in low intensity light (A. Kaplan, pers. comm.); in the initial study (Karagouni et al., 1990) the relatively high photon flux densities used may have caused considerable photodamage.

The CCM promotes the fixation of CO_2 by raising its concentration around RuBPcase, an enzyme with a relatively low affinity for its substrate. CCM activity may also be partially responsible for differences between carbon isotope ($^{13}C/^{12}C$) ratios in the atmosphere and organic matter. While discrimination between the carbon isotopes is primarily a result of RuBPcase activity, which discriminates against the heavier isotope, the CCM may influence the isotopic composition of the internal C_i pool if the C_i that enters the cell cannot readily leak out.

A current model of our understanding of the CCM is presented in Fig. 3. It has two major components, the machinery involved in the uptake of C_i and a structural entity required for the efficient fixation of CO_2, the carboxysome. C_i enters the cells from the surrounding unstirred layer and accumulates in the cytoplasm via an energy-dependent transport system(s). Regardless of whether HCO_3^- or CO_2 is supplied, HCO_3^- is the species that appears in the cytoplasm of the cell; it diffuses into the carboxysomes where carbonic anhydrase catalyzes the rapid equilibrium between HCO_3^- and CO_2. The elevated concentration of CO_2 at the site of carboxylation inhibits photorespiration by competing with O_2 and reducing the rate of oxygenation. Part of the CO_2 may leak from carboxysomes to the cytoplasm, but a plasmalemma-located carbonic anhydrase activity may convert it back to HCO_3^- in an energy-dependent process, thereby minimizing wasteful loss of CO_2 from the cells.

1. Mechanism of C_i Uptake

A ratio of $[C_i]in/[C_i]out$ that reached 1000 has been observed in low-CO_2 adapted cells. Taking into account the ΔpH and the membrane potential, an active process must be responsible for concentrating C_i against its electrochemical potential. The addition of HCO_3^- to cultures of *Synechococcus* PCC 7942 grown on air levels of CO_2 triggers hyperpolarization of the membrane potential, indicating that the transport of C_i involves a primary electrogenic pump (Kaplan et al., 1990). This pump might be directly energized by ATP or it might facilitate C_i transport by generating a trans-membrane electrochemical potential gradient for another ion. For instance, the dependence of HCO_3^- uptake on Na^+ suggests that uptake may be driven by a secondary active Na^+ symport system (Reinhold et al., 1984); energy for uptake would be stored as a transmembrane Na^+ gradient. A light-dependent primary electrogenic Na^+ pump in cyanobacteria (Broun et al., 1990) might be able to establish the required gradient. The involvement of Na^+/HCO_3^- symport in the CCM is supported by characterizations of a High-CO_2-

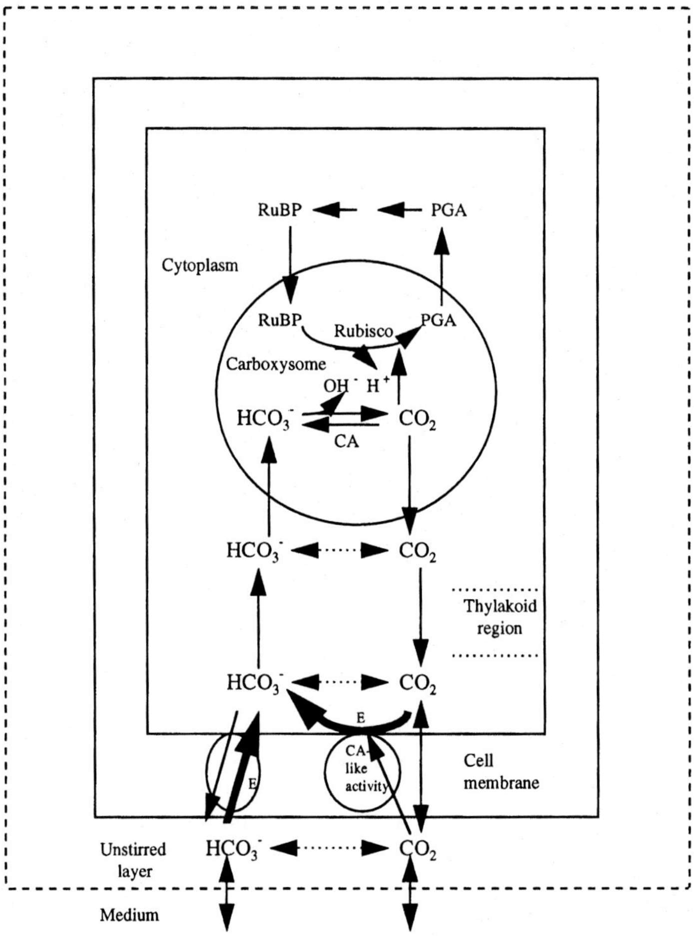

Fig.3. A schematic presentation of C_i fluxes associated with the CCM in cyanobacteria. The cytoplasm, carboxysomal space, thylakoid region, unstirred layer surrounding the cell and the bulk medium are indicated. Energy dependent C_i fluxes are designated (E) and presented as heavy lines. Dashed arrows indicate that the C_i species are not at chemical equilibrium in the cytoplasm and the unstirred layer. As indicated in the text, the CO_2 generated in the carboxysomes by the activity of carbonic anhydrase is mostly fixed by RuBPcase, but some may diffuse out of the carboxysome and into the cytoplasm of the cell.

Requiring (HCR) mutant (IL-2) impaired in HCO_3^- transport that has a lesion in a gene encoding a protein similar to the Na^+/pantothenate symporter of *E. coli* and to a family of multi-drug resistance proteins (Ronen-Tarazi et al., 1995b).

One of the major proteins induced upon exposure of *Synechococcus* PCC 7942 to air levels of CO_2 is a 42 kDa membrane protein. The gene encoding this protein, *cmpA*, is part of the *cmpABCD* gene cluster which encodes proteins homologous to components of ABC transport systems (Suzuki et al., 1992). Based on this homology, it was proposed that the 42 kDa polypeptide was involved in C_i transport. Surprisingly, inactivation of the *cmpA* gene did not affect growth at air levels of CO_2 or C_i transport (Omata et al., 1990). It is possible that the transport components encoded by the *cmpABCD* operon are essential at concentrations of CO_2 lower than that in air.

The transport of HCO_3^- and CO_2 into cyanobacterial cells probably occurs via distinct transport systems. This is supported by the finding that cyanobacterial mutants can be specifically impaired in HCO_3^- (Ronen-Tarazi et al., 1995b) or CO_2 transport (Katoh et al., 1996a). The defective gene in a strain unable to transport CO_2, *cotA*, is similar to *cemA,* which encodes a polypeptide that is localized to the inner membrane of pea chloroplasts (Katoh et al., 1996a). The *cotA* gene product has 440 amino acids and is located in both the cytoplasmic and thylakoid membranes (Ogawa, 1996). Strains in which *cot*A was inactivated exhibited no light-induced proton extrusion, suggesting that impaired CO_2 uptake in the mutant is a secondary effect of the lesion (Katoh et al., 1996b). Differential dependence of CO_2 and HCO_3^- on Na^+ levels also suggests that they enter the cell via different transport systems. HCO_3^- transport requires high concentration of Na^+ (mM range) while much lower concentrations (µM range) support CO_2 transport (Miller and Canvin, 1985). Interestingly, Na^+-independent HCO_3^- transport is present in cells maintained as 'standing cultures' (i.e. not bubbled or shaken) which probably experience extremely low external C_i concentrations.

Both the CO_2 and HCO_3^- transport systems were characterized in *Synechococcus* PCC 7942 after the medium was depleted of C_i. These experiments demonstrated that high CO_2-grown cells were able to take up CO_2 while air-adapted cells transported both CO_2 and HCO_3. The $K_{1/2}$ for the transport of HCO_3^- was 60-100 mM HCO_3^- while the Vmax was 1000 µmol HCO_3^- mg chl^{-1} h^{-1} (Volokita et al., 1984). Using mass spectrometry (Espie et al., 1991), the CO_2 transporter of *Synechococcus* UTEX 625 was shown to have an affinity for CO_2 ($K_{1/2}$ = 0.4 µM) that was approximately 10-fold higher than the affinity measured by previously described technique. Interestingly, high CO_2-grown cells were found to be capable of transporting HCO_3^- when mass spectrometry was used to quantitate C_i uptake during steady state photosynthesis, as opposed to previous measurements made under nonsteady state conditions (Sultemeyer et al., 1995; Yu et al., 1994).

Regardless of which C_i species is supplied, HCO_3^- is the species that accumulates at the inner face of the cytoplasmic membrane, and the C_i species of the cytoplasm are not at chemical equilibrium (Price and Badger, 1989a; Schwarz et al., 1988; Volokita et al., 1984). These findings suggest that a carbonic anhydrase activity located on the cytoplasmic membrane is associated with the CO_2 transporter and is responsible for a vectorial conversion of CO_2 (out) to HCO_3^- (in). A carbonic anhydrase activity that might be associated with C_i transport has been measured in cytoplasmic membrane preparations from *Synechocystis* PCC 6803 (Bèdu et al., 1993). A cytoplasmic membrane-associated carbonic anhydrase may also serve the important function of minimizing the leak of cytoplasmic CO_2 molecules from the cell by converting them back to HCO_3^- (Fridland et al., 1996). Furthermore, it was suggested that the transport of CO_2 might occur by passive diffusion through the membrane followed by conversion to HCO_3^- (Fridland et al., 1996); the saturable kinetics observed for CO_2 transport would reflect the latter step.

Light is required to energize and possibly activate C_i uptake (Kaplan et al., 1987). Energization of the transport process appears to rely on the generation of ATP via PSI-dependent cyclic electron flow (Ogawa et al., 1985) while activation requires PSII activity, albeit at a very low level. The NADH dehydrogenase complex, part of the cyanobacterial respiratory chain, was proposed to be important for cyclic electron flow (Yu et al., 1993). Mutants in the genes *ndhB*, *ndhK* and *ndhL*, encoding subunits of this complex, require high CO_2 for growth and are impaired in C_i accumulation (Marco et al., 1993; Ogawa, 1991), presumably as a consequence of impaired cyclic electron flow. Surprisingly, *Synechocystis* PCC 6803 mutants inactivated for *ndhC* (Ogawa, 1992) and *ndhF* (Schluchter et al., 1993; Yu et al., 1993) could grow under low CO_2 conditions. This suggests that either not all of the subunits of the NADH dehydrogenase complex are equally required for efficient cyclic electron flow and growth on air, or that the *ndhC* and *ndhF* genes are functionally redundant. The importance of photosynthetic electron transport for the HCO_3^- uptake is also suggested by the finding that a mutant in a genomic region containing the *psaI/psaL* genes, (which encode subunits VIII and XI of PSI), is impaired in HCO_3^- transport and unable to grow on air (Ronen-Tarazi et al., 1995b). Finally, a strong correlation between C_i uptake and fluorescence quenching in *Synechococcus* UTEX 625 cells links C_i uptake to photosynthetic electron transport, although the nature of this interaction is not understood (Espie and Kandasamy, 1992; Miller et al., 1988; Miller et al., 1990).

2. Functional Significance of the Carboxysomes

Carboxysomes, polyhedral bodies present in cyanobacteria and certain other bacteria (Codd, 1988), represent a specialized cellular structure that appears to be designed to facilitate CO_2 fixation. Immunogold labeling demonstrated that most of the RuBPcase is present in in the carboxysome (McKay et al., 1993). Some studies with *Anabaena* PCC 7120 suggested that RuBPcase activase and chaperonin 60 are also carboxysome-localized (Jäger and Bergman, 1991), although other studies using *Synechococcus* PCC 7942 have localized RuBPcase activase to the cytoplasm (Friedberg et al., 1993). The enzymes of the reductive pentose phosphate pathway, such as phosphoribulose kinase, appear to be soluble, cytoplasmic proteins (McKay et al., 1993). The spatial separation between RuBPcase and the enzymes of the carbon reduction pathway suggests that large fluxes of metabolites between the carboxysomes and the cytoplasm are critical for efficient CO_2 fixation in cyanobacteria.

Carboxysomes were partially purified from some bacteria and found to contain several polypeptides in addition to the large and small subunit of RuBPcase. In the sulfur bacterium *Thiobacillus neopolitanus* the carboxysomes are delineated by a proteinaceous shell that appears to be devoid of lipid (Holthijzen et al., 1986); this is probably also the case for cyanobacterial carboxysomes. The gene encoding the major carboxysome shell polypeptide of *T. neopolitanus* was cloned and the deduced protein sequence has homology to the proteins encoded by the carboxysome related genes of *Synechococcus* PCC 7942 (English et al., 1994).

Models have been proposed to describe C_i fluxes in actively photosynthesizing cyanobacterial cells in which both the generation and fixation of CO_2 are restricted to the carboxysome (Reinhold et al., 1991; Reinhold et al., 1989). These models assume that there is no carbonic anhydrase to facilitate the conversion of HCO_3^- to CO_2 in the cytoplasm of the cell, and that the C_i species do not reach chemical equilibrium in this compartment. When the accumulated, cytoplasmic HCO_3^- enters the carboxysomes, CO_2 is rapidly generated by carboxysomal carbonic anhydrase. In recent models the carbonic anhydrase is placed at the center of the carboxysome and the carboxysome shell would not have to serve as a diffusion barrier for CO_2 since the CO_2 that is generated within the polyhedral body would be fixed by RuBPcase during diffusion outwards toward the cytoplasm of the cell (Reinhold et al., 1991).

The model described above predicts that (i) intracellular carbonic anhydrase is confined to the carboxysomes since, if it were soluble in the cytoplasm, it would inhibit photosynthesis, and that (ii) the C_i species present in the cytoplasm are not at chemical equilibrium. Experimental evidence supports this model. Most of the intracellular carbonic anhydrase in *Synechococcus* PCC 7942 was found to be associated with carboxysomes (Price et al., 1992). Furthermore, *Synechococcus* PCC 7942 in which human carbonic anhydrase was expressed as a soluble cytoplasmic protein required elevated levels of CO_2 for growth; this phenotype was a consequence of a very high rate of C_i leakage (Price and Badger, 1989b). Finally, mutants of *Synechococcus* PCC 7942 impaired in the *icfA* gene, which is located 20 kbp downstream of the *rbcLS* genes (Fukuzawa et al., 1992) and encodes carboxysomal carbonic anhydrase, required high CO_2 for growth. In one of these mutants the carbonic anhydrase activity facilitated the equilibrium between C_i species at a rate 250 times that of the uncatalyzed reaction (Price et al., 1992) but that was still 30-fold slower than that of wild-type cells. According to the model, this mutant should have sufficient carbonic anhydrase activity for maximal photosynthesis. Hence, other factors such as the correct association of the carbonic anhydrase with RuBPcase and other carboxysomal proteins and the generation of a high enough CO_2 concentration within the carboxysome to activate RuBPase may be critical for CO_2 fixation.

The importance of carboxysome structure for efficient CO_2 fixation became apparent with the analysis of various HCR mutants that are either lacking or contain abnormal carboxysomes (Friedberg et al., 1989; Ogawa et al., 1994a; Ogawa et al., 1994b; Pierce et al., 1988; Price and Badger, 1991; Schwarz et al., 1995; Suzuki et al., 1990). HCR mutants of *Synechococcus* PCC 7942 with altered carboxysome structure were shown to have lesions in any of five distinct genes, designated *ccmK, L M, N, O*, which are clustered on the genome in the region of *rbcLS* (Ronen-Tarazi et al., 1995a). Genes critical for the biogenesis of carboxysomes were isolated from *Synechocystis* PCC 6803. One of these genes, *ccm*A, is unique, while the other is similar to the *ccmM* of *Synechococcus* PCC 7942 (Ogawa et al., 1994a; Ogawa et al., 1994b). All of the mutants with defective carboxysomes accumulated C_i at a rate similar to that of wild-type cells, but were unable to efficiently utilize the internal C_i pool for CO_2 fixation (the apparent photosynthetic affinity for external C_i is 2-3 orders of magnitude lower than that of wild-type cells). This phenotype strongly suggests that efficient CO_2 fixation is tightly linked to carboxysome structure, which in turn, may be required for establishing elevated concentrations of CO_2 near the active site of RuBPcase. Besides being a substrate for RuBPcase, CO_2 serves as an activator of the enzyme; the elevated level of CO_2 in the carboxysome is probably essential for activation. Support for the importance of this aspect of carboxysomal control of C_i fixation comes from studies of an HCR mutant in which RbcS has an extension at its carboxy terminus and has an aberrant carboxysome structure. The in vitro kinetic characteristics of the activated RuBPcase from the mutant are very similar to those of the wild-type enzyme. However, activity measurements in situ suggest that when the mutant cells are grown in low levels of CO_2 their RuBPcase is not activated probably because the modified carboxysome cannot accumulate CO_2 to the levels required for activation. As expected, growth at elevated CO_2 levels compensates for the lesion; presumably because the RuBPcase can now be activated (Schwarz et al., 1995).

3. C_i-Dependent Gene Expression and Signal Transduction

Several cyanobacterial genes are induced upon exposure of cyanobacterial cells to low levels of CO_2. Some of these genes are exclusively transcribed in an atmosphere of low CO_2 (the genes are 'stringently controlled') while others are transcribed in both low and high atmospheric CO_2, but transcript abundance is elevated under the former conditions (the genes are 'upregulated'). The genes that are stringently controlled include *purK* (encoding subunit II of phosphoribosyl aminoimidazole carboxylase), and *cmpA* (encoding a 42 kDa protein integral to the cytoplasmic membranes). Deletion analysis of the *cmpA* promoter indicated that positive and negative cis-acting elements are likely to be involved in the regulation by ambient CO_2 (Kaplan, 1994). The *cmpA* gene is not exclusively controlled by CO_2 levels since it turns out to be identical to the *cbpA* gene which encodes a putative carotenoid-binding protein that is synthesized during exposure of *Synechococcus* PCC 7942 to elevated light intensities (Reddy et al., 1989). High light and low CO_2 may affect the transcription of this gene by altering topology of the genomic DNA; reagents that inhibit gyrase activity appear to prevent activation of the gene in high light (Reddy et al., 1989). Furthermore, if localized changes in DNA topology are important for control, it might explain why many of the genes that are activated at low C_i concentrations are clustered.

The group of genes upregulated in an atmosphere of low CO_2 includes *ndhB* (encoding subunit II of NADH dehydrogenase), *rbcLS* and the *ccmNO* operon; the latter genes, located immediately upstream of *rbcLS*, encode proteins involved in carboxysome biogenesis. The *ccmN* gene product may be a regulatory element since a point mutation in *ccmN* prevented accumulation of the CmpA protein (Schwarz et al., 1988) and its mRNA (Schwarz, unpublished data).

Very little information is available as to the nature of the signal and elements of the signal transduction pathway that control the acclimation of cyanobacteria to changes in ambient C_i levels. Badger (1987) demonstrated that the ability of cyanobacteria to accumulate C_i internally depended on the ambient concentration of C_i rather than that of CO_2. A rapid decline in the ability to accumulate C_i was observed when *Synechocystis* PCC 6803 cells were shifted from photoautotrophic to photoheterotrophic growth by supplying the culture with glucose. A decline following a similar time course was also observed after the addition of HCO_3^- to low-CO_2 grown cells (Bloye et al., 1992). It is not known whether the inhibition of CCM induction by high C_i levels and a shift from autotrophic to photoheterotrophic conditions occurs via the same mechanism.

Based on an analysis of the time course of acclimation at various concentrations of CO_2 and O_2, Kaplan has suggested that the cells may respond to the relative rate at which RuBPcase catalyzes carboxylation and oxygenation reactions (Kaplan et al., 1990). Metabolites that accumulate as a consequence of the elevated oxygenation activity that accompanies the transfer of high CO_2-grown cells to low CO_2 conditions may promote the induction of the CCM. This possibility gains some support from the analysis of a *Chlamydomonas reinhardtii* mutant impaired in the conversion of phosphoglycolate to glycolate, which cannot acclimate to low C_i levels (Marek and Spalding, 1991).

Interestingly, a significant increase in the extent of phosphorylation of several polypeptides was noted when low-CO_2-grown *Synechocystis* PCC 6803 cells were supplied with either glucose or HCO_3^- (Bloye et al., 1992). The nature of the phosphorylated polypeptides and their possible role in suppressing the induction of the CCM has not been explored.

Recently, a gene designated *hatR* was identified in *Synechocystis* PCC 6803 which, when inactivated, eliminated the induction of a high affinity C_i transport system normally observed upon transfer of high CO_2 grown cells to air levels of CO_2 (Bèdu et al., 1995). The *hatR* mutant grew somewhat slower than wild-type cells in air, suggesting that the C_i transport system that cannot be induced in the *hatR* mutant is not essential for growth under these conditions, but may be required at lower C_i concentrations. The deduced amino acid sequence of the *hatR* gene product exhibits a 40% similarity to CheY (*E. coli*) and SpoOF (*Bacillus subtilis*), small response regulators involved in motility (Ravid et al., 1986; Yamaguchi et al., 1986) and sporulation (Perego and Hoch, 1996), respectively.

Mutants impaired only in the ability to adapt to low CO_2 are expected to show physiological characteristics similar to wild-type cells, when they are grown at high concentrations of CO_2. Most of the HCR mutants that have been isolated exhibit an apparent photosynthetic affinity for C_i that is two orders of magnitude lower than that of wild-type cells, even when they are grown at high CO_2 concentrations. HCR mutants that show photosynthetic characteristics similar to that of wild-type cells under high concentrations of CO_2 are altered in a gene immediately downstream of the *rbcLS* operon. This gene, *purK* (Schwarz et al., 1992), encodes subunit II of phosphoribosyl aminoimidazole carboxylase, an enzyme required for the formation of inosine 5`-monophosphate, an intermediate in purine biosynthesis (Ebbole and Zalkin, 1987). The transcript for *purK* is not detected when *Synechococcus* PCC 7942 is grown on high CO_2, but is induced following short exposure to low CO_2. Under high CO_2, subunit I apparently suffices for purine biosynthesis. Based on the phenotype of the mutant, PurK appears to be important for raising the affinity of the phosphoribosyl aminoimidazole carboxylase for CO_2 (Schwarz et al., 1992). Although this lesion is not related to the CCM or its regulation, it does imply that other carboxylation reactions in the cell are extremely sensitive to CO_2 levels and their regulation is critical for survival in a low CO_2 environment. It also raises the possibility that a drop in the purine levels in the cell, which occurs shortly after exposure to low Ci, may participate in signaling during acclimation.

B. Nitrogen Limitation

The energy and reductant that are generated from photosynthesis are in large part required for the assimilation of nitrogen (N) and the reduction of CO_2 (Flores et al., 1983; Ho and Krogman, 1982; Manzano et al., 1976). Cyanobacteria can use a variety of inorganic and organic sources of N to fulfil their nutritional requirements. The most abundant inorganic sources are nitrate, ammonium and N_2, while common organic sources are urea and certain amino acids. However, all these sources must first be converted to ammonium before they can be utilized for biosynthetic processes. Ammonium is the preferred N source, permitting the fastest growth rate and is therefore, a key molecule in controlling global aspects of N metabolism. Ammonium is incorporated into carbon skeletons by the sequential action of glutamine synthetase (GS) and glutamate synthase (GOGAT; glutamine 2-oxoglutarate amino transferase). GS catalyses the formation of glutamine by the addition of ammonium to 2-oxoglutarate, in a reaction that requires ATP. GOGAT regenerates two molecules of glutamate by catalyzing a transamination between glutamine and 2-oxoglutarate. Thus GS/GOGAT activity controls the levels of glutamate, which is the major N donor in the cell contributing to the synthesis of amino acids and nucleic acids via the action of different aminotransferases (Flores and Herrero, 1994). In *E. coli*, where N assimilation has been studied in great detail, GS levels are known to be controlled very precisely and at a number of different levels (Ninfa et al., 1995). Further, there is an NTR regulon which consists of genes and operons that are regulated by N availability and which is controlled by a two-component system consisting of a sensor (NtrB or NRII) and a regulator (NtrC or NRI). The parallels and differences between N assimilation in cyanobacteria and enteric bacteria will form a framework for this section. The following discussion first considers utilization of organic sources of N, then inorganic sources (N_2, ammonium and nitrate), and ends with a review of global N regulation.

1. Organic Nitrogen Sources

a. Urea

It is well documented that many species of cyanobacteria, both marine and freshwater, can use urea as a sole N source (Flores and Herrero, 1994; Neilson and Doudoroff, 1973). The concentrations of urea are variable in different environments, but in sea water the concentrations may be 0.1 - 0.5 μM and constitute a significant N source. Urea can be transported into cyanobacterial cells and metabolized to produce one molecule of CO_2 and two molecules of ammonium by the enzyme urease which has been isolated from some cyanobacterial species (Argall et al., 1992; Singh, 1990). There are conflicting reports about the role of ammonium in inhibiting urease activity and it is not clear whether urease is present in all cyanobacteria (Collier and Palenik, 1996; Flores and Herrero, 1994; Singh, 1990). Recently, Collier and Palenik (1996) characterized the urease from the marine *Synechococcus* WH 7805 and isolated, for the first time, the genes encoding urease. This urease has a K_m of 320 μM and a pH optimum of 8.6. The structural genes (*ureABC*) and accessory genes, encoding proteins required to synthesize an active urease (*ureDEFG*), are similar to those of other bacteria. However, little is still known about how urea uptake and assimilation are regulated, although the high K_m of urease for its substrate suggests that there might be a urea concentrating mechanism. It has recently been shown that *Synechococcus* sp.strain PCC 7002cells grown in the presence of 50mM urea and elevated CO_2 levels undergo rapid chlorosis and cell death during stationary phase (however, a *ureC* mutant, which has no urease activity, does not show this phenotype). This phenomenon could be mimicked by the addition of polyunsaturated fatty acids or paraquat to the medium, both of which induce severe oxidative stress (Sakamoto et al., 1998). Further work is required to understand the exact sequence of events that trigger extensive chlorosis and cell-death under certain growth conditions.

b. Amino Acids

The amino acids arginine, glutamine and asparagine can serve as N sources for many cyanobacteria (Flores and Herrero, 1994; Flores and Muro-Pastor, 1988, Flores and Muro-Pastor, 1990). Arginine is taken up by a transport system specific for basic amino acids (Herrero and Flores, 1990). It is converted to urea and ornithine by the enzyme arginase, or to citrulline and ammonium by arginine deiminase. While some cyanobacteria like *Synechocystis* PCC 6803 grow as well on arginine as on nitrate, others, like *Anabaena* PCC 7120 and *Pseudanabaena* PCC 6903 grow more slowly on arginine than on nitrate (Flores and Muro-Pastor, 1990; Herrero and Flores, 1990; Neilson and Doudoroff, 1973).

In a detailed study, the ability of nine cyanobacterial strains (both unicellular and filamentous, as well as N_2 fixers and non-fixers) to take up various amino acids was examined (Montesinos et al., 1997). All of the strains had a transport system for neutral amino acids, while some also had systems specific for basic amino acids, confirming earlier work that was done on amino acid transport (Flores and Muro-Pastor, 1988; Labarre et al., 1987). This is in striking contrast to the situation observed for the enteric bacteria where there are often transport systems tailored for a specific amino acid. This may reflect a fundamental difference in the availability of amino acids in the enteric milieu versus that in soil or aquatic environments. In the heterotrophic, unicellular *Synechocystis* PCC 6803, three amino acid transport systems have been identified (Labarre et al., 1987). One of the systems is specific for basic amino acids and glutamine, a second is specific for neutral amino acids (excluding glutamine) and a third is specific for glutamate and glutamine. In the heterocyst forming *Anabaena* PCC 7120, at least five amino acid transport systems were identified (Montesinos et al., 1995). Three of these systems were active transport systems; one system was relatively specific for basic amino acids, while the two other systems had overlapping specificities for neutral amino acids. The remaining two systems were low affinity transporters; one was passive for basic amino acids and a putative one was for acidic amino acids. In the same study, Montesinos et al. isolated mutants (based on resistance to toxic analogues of amino acids) that were severely impaired in amino acid transport. Interestingly, some of these mutants grew slowly in medium lacking combined nitrogen. This was used to argue that some amino acid transport systems may have a secondary, though not unimportant, function to recapture amino acids that leak out of the cell, or to assist in the transfer of amino acids along the cyanobacterial filament (from the heterocysts to vegetative cells) during growth on N_2.

The first report identifying genes encoding components of amino acid transport systems was presented by Flores and co-workers (Montesinos et al., 1997). Two genes for polypeptides of amino acid transport systems were identified as encoding putative transport polypeptides based on the analysis of the entire genome of *Synechocystis* PCC 6803 (Kaneko et al., 1996). One of the genes, *natA* (ORF slr0467) encodes a protein with homology to an ABC-type permease (nucleotide binding protein), while the second, *natB* (ORF slr0559), encodes a protein similar to a periplasmic binding protein. Insertional mutants of *natA* and *natB* were unable to transport neutral amino acids (except for glutamine), indicating that these genes encode components of neutral amino acid uptake systems. Homologues of *natA* and *natB* were found in all the cyanobacteria examined. This work demonstrates how the information available from the complete sequence of the *Synechocystis*

PCC 6803 genome offers the possibility of combining a strong molecular approach with classical genetics and physiology to identify the functions of members of specific gene families.

2. Inorganic Nitrogen Sources

a. Nitrogen Fixation

i. Strategies Among Cyanobacteria

Many unicellular and filamentous cyanobacteria fix N_2. Some of the latest findings in this area are considered here; the reader is referred to recent reviews for more detail (Fay, 1992; Haselkorn and Buikema, 1992; Wolk, 1996; Wolk et al., 1994). Nitrogenase is the conserved enzyme complex that catalyzes the ATP dependent transfer of six electrons from ferredoxin or flavodoxin to N_2 (Howard, 1994). Hence, the requirement for both reductant (as reduced ferredoxin) and energy (ATP) link N_2 fixation to photosynthesis. Nitrogenase is irreversibly inactivated by O_2, therefore, all N_2-fixing organisms have evolved methods to protect the enzyme from the damaging O_2-rich atmosphere. This problem is exacerbated in cyanobacteria since they generate O_2 intracellularly during photosynthesis. Cyanobacteria have evolved a number of strategies to overcome the problem of O_2 toxicity to the nitrogenase (Fay, 1992). Filamentous cyanobacteria such as *Anabaena* PCC 7120 form specialized cells called heterocysts that develop at regular intervals along the filament and contain the N_2-fixing machinery. A unique cell wall that serves as a barrier to atmospheric O_2 and the elimination of O_2 evolving PSII activity help prevent the inactivation of nitrogenase in the heterocyst (Haselkorn, 1992; Wolk, 1996; Wolk et al., 1994). In some non-heterocystous filamentous cyanobacteria, such as *Trichodesmium*, N_2 fixation is dependent on the presence of light (Ohki and Fujita, 1988; Chapter 5), implying the necessity for photosynthetically-derived reductant and energy, and exhibits a diurnal rhythm (Wyman et al., 1996). The partial elimination of O_2 from the site of N_2 fixation may be a consequence of colony morphology and/or an ability of the cells to efficiently scavenge O_2 via elevated respiratory activity. The diffusion of O_2 and the penetration of light to cells near the center of a *Trichodesmium* colony may be extremely limited, which may help preserve nitrogenase activity (Carpenter and Price, 1976). However, not all non-heterocystous N_2 fixers can fix N_2 in aerobic or microaerobic conditions; for instance, *Plectonema boryanum* fixes N_2 only under conditions of anaerobiosis (Stewart and Lex, 1970).

Unicellular cyanobacteria from various environmental niches have been shown to fix N_2 (Carpenter and Price, 1976; Huang and Chow, 1988; Reddy et al., 1993; Wyatt and Silvey, 1969). These organisms are ubiquitous, and unique in their ability to fix N_2 within a cell that maintains a functional O_2 evolving photosynthetic apparatus. In certain strains of *Gloeothece* and *Synechococcus* grown under a light/dark cycle there is a diurnal pattern of N_2 fixation that peaks in the dark (Huang and Chow, 1988; Rai et al., 1992). The temporal separation of photosynthesis and N_2 fixation ensures that photosynthetic O_2 evolution is not concurrent with N_2 fixation (Mitsui et al., 1986). However, even though these cells do not generate O_2 during the dark period, the intracellular levels of O_2 may become high enough during the light period to cause some inactivation of the nitrogenase enzyme, and the enzyme must be continuously replenished. This requirement for *de novo* synthesis of nitrogenase is also reflected in the diurnal rhythm of nitrogenase mRNA production: more nitrogenase transcript accumulates during the dark phase of growth (Huang and Chow, 1990). Energy and reductant for N_2-fixation come from dark respiration of stored carbohydrates that are synthesized during the light period.

The genes encoding the nitrogenase enzyme and related proteins were characterized and sequenced from a number of different species (Haselkorn and Buikema, 1992; Wolk et al., 1994). The identification of a region of the cyanobacterial genome that contained a cluster of genes involved in N_2 fixation, including *nifHDK*, which encode structural components of the nitrogenase, was made several years ago (Mazur et al., 1980). One of the most striking features of this region of the *Anabaena* PCC 7120 genome was the existence of a large insertion (11 kbp) in the coding region of *nifD* in vegetative cells (Golden et al., 1985; Haselkorn, 1992; Wolk et al., 1994). During heterocyst development the 11 kbp insertion is spliced from the genome creating a functional *nifHDK* operon. The gene encoding the excisase (*xis*A), which removes the 11 kbp insert, was cloned and sequenced (Lammers et al., 1986). Subsequently, other examples of genomic rearrangements that occur during heterocyst formation have been characterized in *Anabaena* PCC 7120, *Anabaena cylindrica* and *Nostoc* MAC

(Haselkorn, 1992; Carrasco et al, 1995; Carrasco and Golden, 1995). Interestingly, the 11 kbp insertion is not present in non-heterocystous, N_2-fixing cyanobacteria (Fujita et al., 1991). Recently, it was shown that in addition to the major nitrogenase, *Anabaena*, ATCC 29413 synthesizes vanadium and Fe nitrogenase enzymes that are similar to those found in the soil bacterium *Azotobacter vinelandii* (Kentemich et al., 1988; Kentemich et al., 1991; Thiel, 1993).

Like other diazotrophs, cyanobacteria prefer fixed N sources; N_2 fixation is a last resort since it requires high levels of both energy and reductant. Preferred N sources such as ammonium or nitrate inhibit the expression of the N_2 fixation system, primarily at the level of transcription. Ammonium can independently inhibit the development of both heterocysts (in *Anabaena* PCC 7120) and nitrogenase synthesis (in heterocyst- and non-heterocyst forming N_2 fixers) (Mulligan and Haselkorn, 1989; Mullineaux et al., 1981; Rai et al., 1992). The repressive effect of ammonium and nitrate probably operate through the same pathway (i.e. GS activity is required for repression) (Fujita et al., 1991; Huang and Chow, 1990; Martin-Nieto et al., 1991). The regulation of the *nif* genes in *Azotobacter vinelandii* and *Klebsiella pneumoniae* is well understood, and is under the control of the *nifLA* gene products (Merrick, 1992). So far, no homologues of *nifLA* have been reported in cyanobacteria, although a gene encoding a polypeptide with 41% identity to *nifL* from *Klebsiella pneumoniae* was noted by Kaneko et al. (1996) in the *Synechocystis* PCC 6803 genome.

ii. Protein Turnover

Protein turnover, an important aspect of the cyanobacterial responses to nutrient deficiency, has been examined most thoroughly in N deficient cells. A large proportion of cellular protein is degraded in *Anabaena variabilis* during the differentiation of heterocysts (Ownby et al., 1979; Thiel, 1990). Both phycobiliproteins and other cellular proteins are degraded in N deprived *Synechococcus* and *Spirulina* (Boussiba et al., 1984; Boussiba and Richmond, 1980; Wyman et al., 1985). The amino acids released from the degradation of proteins that are not essential during N limitation may provide substrates for *de novo* synthesis of proteins that are required for the acclimation process. There are reports of general proteases and those that specifically degrade PC (Boussiba and Richmond, 1980; Elmorjani and Herdman, 1987; Foulds and Carr, 1977; Wood and Haselkorn, 1979; Wood and Haselkorn, 1980; Yamanaka and Glazer, 1980). A gene encoding a general protease that does not preferentially degrade phycobiliproteins has been cloned from *Anabaena variabilis* (Lockau et al., 1988; Maldener et al., 1991). It appears to be a serine protease, but shows no significant similarity to other known serine proteases. This protease is identical to a Ca^{++}-requiring protease that thought to be induced during N deficiency and heterocyst development in *Anabaena* sp. (Wood and Haselkorn, 1979; Wood and Haselkorn, 1980). The activity of this protease, however, was shown to be identical in nitrate-grown and N_2-fixing *Anabaena* (Lockau et al., 1988; Maldener et al., 1991). A transient increase in the activity of the protease may occur during the mass differentiation of heterocysts after transfer from growth on fixed N to N_2-fixing conditions, but this increase is not maintained during the steady state growth of N_2-fixing cultures. Furthermore, while inactivation of this gene eliminated the Ca^{++}-dependent protease activity, there was no effect on heterocyst differentiation, suggesting that it is not critical for heterocyst development or that its function can be substituted by other proteases (Section III.D.4.b).

b. Ammonium Assimilation

Ammonium levels in the environment may be very low, but ammonium can still be taken up rapidly from the medium by cyanobacteria using a high affinity permease system (Boussiba, 1989; Boussiba et al., 1984; Flores et al., 1980; Ohmori et al., 1977). Once ammonium enters the cell it can be used either for the reductive amination of 2-oxoglutarate by glutamate dehydrogenase (GDH) to form glutamate, or for the formation of glutamine by GS. Glutamine and 2-oxoglutarate are the substrates for the enzyme GOGAT. This enzyme catalyzes the formation of two molecules of glutamate, regenerating glutamate from glutamine. Glutamate is a central metabolite in N metabolism since it can be converted to the amino acids glutamine, arginine, proline and 5-aminolevulinic acid. The potential importance of the GDH and the GS/GOGAT systems in several cyanobacterial species have been studied in a variety of ways. The use of radiolabelled ^{13}N-ammonium, in combination with inhibitors that block either GS (L-methionine-D,L- sulfoximine or MSX) or GOGAT (azaserine) suggests that the majority of ammonium assimilation in cyanobacteria is via the GS/GOGAT system (Meeks et al., 1978).

i. GS/GOGAT

The GS enzyme, encoded by *glnA*, is an abundant protein in the cyanobacterial cell and its structure is similar to that of GS in other prokaryotes. It requires Mg^{++} for activity, although other divalent cations can substitute for Mg^{++} (Mèrida et al., 1990). In *E. coli*, multilevel controls dictate GS activity. First, it is controlled at the transcriptional level. Second, the enzyme activity is regulated by reversible covalent modification via adenylylation such that the dodecameric enzyme is less active when N is in excess. Third, there is allosteric feedback inhibition exerted by several different small molecules (including alanine, glycine, tryptophan, CTP, AMP, carbamoyl phosphate and glucosamine 6-phosphate). The regulation of *glnA* expression is discussed in the next section (III.B.2.b.ii). So far, there is no evidence that GS is adenylylated in cyanobacteria, but there are observations that certain amino acids and nucleotides may bind GS and inhibit its activity (Blanco et al., 1989; Mèrida et al., 1990). Ammonium rapidly inhibits GS in *Synechocystis* PCC 6803 but appears to have no effect on the GS of *Anabaena* PCC 7120. However, when the GS of *Anabaena* PCC 7120 is heterologously expressed in *Synechocystis* PCC 6803, it is inactivated by ammonium (Mèrida et al., 1991; Mèrida et al., 1992). The mechanism by which ammonium inhibits the cyanobacterial GS is not yet understood. Reyes and Florencio (1995b) have suggested that GS inactivation occurs via the noncovalent binding of a phosphorylated protein in cells growing in medium containing ammonium. Based on cross-linking studies this protein was calculated to have a molecular mass of 14 kDa, but no further identification has yet been made.

A second gene encoding GS has been found in *Synechocystis* PCC 6803 (Reyes and Florencio, 1994). The gene, designated *glnN*, encodes a protein that is very different from GSI (50 kDa) and GSII (40 kDa), that have been previously identified in prokaryotes, and is considered to constitute a new gene family, GSIII (75 kDa). A *glnA* mutant of *Synechocystis* PCC 6803 was found to be viable, suggesting that GlnN encodes a functional GS, even though it accounts for only 3% of the total GS activity in wild-type cells. The presence of a second GS may also explain the results of Wagner et al. (1993), who found that a *glnA* mutant of *Agmenellum quadruplicatum* (*Synechococcus* PCC 7002) was viable, although it did not grow as well as wild-type cells. This suggests that GSIII activity cannot be fully compensated for by GSI. Interestingly, in *Synechocystis* PCC 6803, *glnN* was strongly induced in N-free medium and the *glnN* gene appeared to be present only in non-N_2 fixing cyanobacteria (Reyes and Florencio, 1994). More information is needed to understand the role of the different GS proteins in cyanobacteria and the nature of the modifications that may alter their activities.

GOGAT has been purified from *Synechococcus* PCC 6301 (Marquès et al., 1992) and appears more similar to the plant enzyme than to the bacterial counterpart in that it can accept electrons from reduced ferredoxin. It is a single polypeptide containing one molecule of FMN and may contain nonheme iron. Two different genes, *gltB* and *gltS*, encoding ferredoxin-dependent GOGAT have been isolated in *Synechocystis* PCC 6803 (Navarro et al., 1995). The deduced amino acid sequence of *gltB* shows 55% identity to the alfalfa NADH-GOGAT and about 43% identity to the plant ferredoxin-GOGAT and to the bacterial GOGAT. GltB appears to have a 3Fe-4S cluster and a FMN-binding domain. Insertional inactivation of either *gltB* or *gltS* demonstrated that neither gene alone was essential for viability; however, a double mutant of *gltB, gltS* was non-viable.

ii. Regulation of glnA

Expression of *glnA* is regulated by the N source in the medium such that it is lower in ammonium- than in nitrate- or N_2-grown cells (Cohen-Kupiec et al., 1993; Mérida et al., 1990; Wagner et al., 1993). In *Synechococcus* PCC 7942 a single *glnA* transcript is induced when ammonium is removed from the medium; induction does not appear to require nitrate. Moreover, the presence of a typical NtcA binding site in the region of the promoter suggests that this transcriptional activator is required for *glnA* expression (Section III.B.3.a). In *Anabaena* PCC 7120, four mRNAs with different 5' ends are transcribed from the *glnA* gene (Tumer et al., 1983). Three of these transcripts are synthesized in ammonium-grown cells while one is induced during N-deficient conditions. While NtcA is known to bind to the *glnA* promoter, further analysis is required to establish how the specificity of the transcription start sites is established. The pattern of *glnA* transcripts in *Calothrix* PCC 7601 also appears to depend on the source of N present in the growth medium (Elmorjani et al., 1992).

In *Synechocystis* PCC 6803 the expression of *glnA* may be under redox control (Reyes and Florencio, 1995a). There was a dramatic decrease in *glnA* mRNA in cells transferred from light to dark or treated with inhibitors of photosynthetic electron transport (DCMU or DBMIB). This control seems to be exerted at the level of transcription since the half-life of the *glnA* transcript was the same in the light and in the dark. GS levels, however, did not fluctuate, so the significance of these results is not clear.

c. Nitrate and Nitrite Assimilation

Nitrate, and to a lesser extent nitrite, are abundant, readily available sources of N for cyanobacteria. Both of these molecules can be taken up from the medium by a high affinity transport system. The major rate-limiting step of nitrate assimilation is transport into the cell; ammonium can rapidly inhibit this transport activity, although the mechanism of inhibition is not understood (Lara et al., 1987). The uptake of nitrate and nitrite has been extensively characterized in both filamentous and unicellular cyanobacteria (Flores et al., 1983; Meeks et al., 1983; Rodriguez et al., 1992). Competition experiments suggest that the import of these compounds can be mediated by the same high affinity system (Luque et al., 1994b; Madueño et al., 1987). Recent evidence from Omata's group indicates that there may also be a distinct nitrite transporter (Maeda and Omata, 1997), although it is still uncharacterized. The identification of transporters for nitrate and nitrite involved the isolation and characterization of mutants impaired in various aspects of N metabolism (Madueño et al., 1988; Omata, 1991; Omata et al., 1989; Luque et al., 1992; Omata, 1991). The analyses of these mutants also led to the identification of genes encoding polypeptides involved in N-metabolism (Luque et al., 1992; Omata, 1996; Omata et al., 1993; Suzuki et al., 1993). To date, eight genes from this cluster have been characterized. Six of the genes are in the *nirA* operon, which includes *nirA, nrtA, nrtB, nrtC, nrtD* and *narB*. A second operon, oriented in the opposite direction to the *nirA* operon, contains *nirB* and *ntcB* (Suzuki et al., 1995a). The most complete information is available about the genes encoding the nitrate reductase (*narB*) and the nitrite reductase (*nirA*) (Luque et al., 1993; Omata et al., 1993).

i. Nitrate and Nitrite Reductase

Nitrite reductase (NiR), encoded by *nirA*, the first gene of the *nirA* operon, is a polypeptide of 56 kDa which has strong similarity to plant nitrite reductase (Luque et al., 1993; Suzuki et al., 1993). NiR is proposed to contain a siroheme prosthetic group, and a Fe_4S_4 cluster. In *Plectonema boryanum*, which is a nonheterocystous, N_2-fixing cyanobacterium (Suzuki et al., 1995b), the *nirA* gene encodes a novel nitrite reductase in which there is a C-terminal extension that is absent in the *Synechococcus* PCC 7942 and plant NiRs. Intriguingly, this extra domain has strong homology to ferredoxin. Additional studies are required to determine the function of this domain. Suzuki et al. suggested that the ferredoxin-like domain may be involved in transferring electrons to the NiR catalytic domain, where the electrons could be used to reduce nitrite. It is also possible that the ferredoxin-like domain controls the activity of the enzyme by sensing the redox state of the cell. This is an attractive idea since it would allow for a mechanism by which the cell rapidly coordinates NiR activity with the rate of photosynthesis and CO_2 fixation.

The cyanobacterial nitrate reductase (NR) is a polypeptide of about 75 kDa and contains molybdenum and [Fe-S] clusters, which make it more similar to the bacterial type of NR than to the eukaryotic NR, which contains FAD and heme as prosthetic groups. Reduced ferredoxin was shown to transfer electrons to NR. This is further evidence that N-assimilation is coupled to photosynthetic electron transport (Flores and Herrero, 1994). Andriesse et al. (1990) sequenced a gene, designated *narB*, that was suggested to encode the NR of *Synechococcus* PCC 7942 (based on its similarity to other bacterial NRs). When *narB* was expressed in *E. coli* under the control of an inducible promoter, protein extracts exhibited the concurrent accumulation of a 76 kDa protein and NR activity (Rubio et al., 1996). The characteristics of this heterologously expressed NR were similar to those of the *Synechococcus* PCC 7942 enzyme (both in terms of the K_m value and requirement for reduced ferredoxin). Finally, the *narB* gene was shown to be the terminal gene of the *nirA* operon (Luque et al., 1992; Suzuki et al., 1993).

Recently, Suzuki et al. (1995a) characterized a second operon, containing the genes *nirB* and *ntcB*, that is close to, but in the opposite orientation of the *nirA* operon. NirB is a protein of 345 amino acids with no known homology to any proteins in the database. Insertional inactivation of the *nirB* gene causes a 60% decrease in NiR activity, while the activity of NR and GS/GOGAT was only slightly higher than in wild-type cells. The authors suggested that NirB is required for maximal NiR activity as a consequence of i) activation of genes required for the assembly of NiR or its cofactors, or ii) stabilization of the NiR protein. NtcB was recently shown to be involved in the positive regulation of the *nirA* operon (Section B.2.c.iii) (Aichi and Omata, 1997).

ii. Transport Systems

The cluster of genes downstream of *nirA* in *Synechococcus* PCC 7942 includes *nrtABCD*, which encode components of the nitrate and nitrite transport systems. A *nrtA* mutant was unable to grow on nitrate as a sole N source (Omata, 1988, Omata, 1991). The *nrtA* gene product is 48 kDa and appears to be a membrane-bound substrate binding protein. It binds both nitrite and nitrate with a Kd value of approximately 0.3 μM for both substrates (Maeda and Omata, 1997). Most substrate-binding proteins of Gram negative bacteria are soluble periplasmic proteins with a high affinity for a particular substrate. Interestingly, NrtA may be a lipoprotein that is anchored to the cell membrane (Maeda and Omata, 1997). Globomycin, which specifically blocks cleavage of the signal peptide of lipoprotein precursors causes the intracellular accumulation of the NrtA precursor. Furthermore, many of the putative periplasmic substrate binding proteins of *Synechocystis* sp strain PCC 6803 (derived from analysis of the complete genome sequence) have typical lipoprotein consensus sequences. Maeda and Omata have suggested that anchored substrate binding proteins may be a general feature of transport systems in certain cyanobacteria. Substrate-binding lipoproteins are rare in Gram negative bacteria, although in Gram positive bacteria they are critical for anchoring proteins to the cytoplasmic membrane. The necessity for anchoring these proteins to the membrane in *Synechococcus* PCC 7942 is puzzling since the binding proteins of cyanobacteria would be confined to the periplasm by a cell wall and the periplasmic substrate binding proteins of other Gram negative bacteria, such as *E. coli*, function efficiently without a membrane anchor.

Other components of the nitrate/nitrite transport system are not as well characterized as NrtA. The NrtB protein has similarity to the integral membrane components of ABC transporters, while NrtC and NrtD exhibit similarity to the ATP binding proteins of the transporter. The NrtC polypeptide is unique in that it has an amino terminal domain that is very similar to NrtD and a carboxyl terminal domain with 30% identity to NrtA (Kobayashi et al., 1997). A mutant lacking the carboxyl terminal domain of NrtC was, unlike wild-type cells, able to transport nitrate in the presence of ammonium. This suggests that the carboxyl terminal domain may be involved in ammonium-regulated inhibition of the transporter. The mechanism of this inhibition is not yet known, although it may involve a phosphorylation event, as an ammonium-sensitive protein kinase activity has been demonstrated in *Synechococcus* PCC 7942 (Rodriguez et al., 1994).

iii. Regulation of Genes Involved in Nitrogen Assimilation

Ammonium inhibits nitrate/nitrite uptake in all cyanobacteria examined so far (Herrero and Guerrero, 1986; Lara et al., 1987; Martin-Nieto et al., 1991), which is partly a consequence of repression of the *nirA* operon (Suzuki et al., 1993). Transcription initiation from *nirA* is almost immediate following the shift to ammonium-free, nitrate-containing medium and requires the activator NtcA (Section III.B.3.a) which binds to the *nirA* promoter. The transcripts that are synthesized (>7.5 kb) cover the six genes of the operon, accumulate to a maximum after approximately 20 min (but are unstable), and 90 min after the initial increase they drop to 10% of the maximal value. This suggests that transcription of the operon is controlled by feed-back regulation through the accumulation of a metabolite(s) generated by N-assimilation (Suzuki et al., 1993). The *nir*A transcripts also accumulate when cells are transferred from ammonium-containing to N-free medium, implying that ammonium alone can repress transcription and that nitrate is not required for induction. However, Kikuchi et al. (1996) examined the effect of nitrate and nitrite in MSX treated cells in which there is no negative feedback by the ammonium generated internally via the reduction of nitrate. Under these conditions, nitrite activated transcription of the *nirA* operon. This activation has been shown to work through ntcB, a protein of the LysR family encoded by the nirB-ntcB operon (Aichi and Omata, 1997). These results suggest that positive regulation of the operon by nitrite is masked by ammonium repression. Thus, there are likely to be a number of factors that lead to finely tuned expression of the *nirA* operon.

Ammonium does not cause repression of the *nirA* operon in the presence of MSX, an inhibitor of GS activity, demonstrating that a metabolite downstream of ammonium in the N-assimilation pathway is responsible for repression (Suzuki et al., 1993). Glutamine is not likely to be the regulator since treatment of the cells with 5-diazo-6-oxo-L-norleucine (DON), which inhibits glutamine amido transferase causing the level of intracellular glutamine to rise, does not repress the *nirA* operon. Hence, a metabolite of glutamine must signal repression and efforts have been made to identify this molecule (Suzuki et al., 1996). The most direct way to identify controlling molecules is to monitor the repressive effects of metabolites on the level of the *nirA* transcript. As mentioned above, when *nirA* is induced in the absence of ammonium, the transcript level increases to a peak and then drops rapidly to a very low level, making it difficult to evaluate any inhibitory effect that a metabolite might have on the steady state level of the *nirA* trancript. To overcome this problem, Suzuki et al. (1996) used a mutant that was defective in active nitrate transport and therefore was in a state of apparent N deficiency. As a consequence it accumulated high levels of the *nirA* transcript. This mutant was used to test various metabolites of N metabolism for their ability to repress *nirA*, and at the same time, to activate *rbcLS*, which was shown to be upregulated by the addition of ammonium.

Initially, of the metabolites tested, carbamoyl phosphate (CP), and its decomposition product cyanate, were found to prevent accumulation of both *nirA* and *glnA* transcripts. These metabolites also triggered accumulation of the *rbcLS* transcript and transcripts from *ccmKLMNO* and *icfA*, which encode carboxysome-related proteins and carbonic anhydrase, respectively (Omata, 1996). Evidence that cyanate is the metabolite responsible for repression of the *nirA* operon rather than ammonium which accumulates in cells as a consequence of cyanate breakdown, was based on experiments that demonstrated that even when ammonium reassimilation was blocked by the addition of MSX, cyanate still repressed *nirA* and activated *rbcLS*.

The presence of cyanase activity was demonstrated in *Synechococcus* UTEX 625, *Synechococcus* sp strain PCC 7942 and *Synechocystis* PCC 6803 and the gene encoding cyanase, *cynS*, was shown to be repressed by ammonium (Miller and Espie, 1994; Harano et al., 1997). Studies on the regulation of *cynS* showed that cellular cyanase levels are 6 to 12 fold higher in nitrate-grown cells than in ammonium-grown cells. The transcriptional activator NtcA appears to be involved in this regulation (Harano et al., 1997). The regulation of cyanase activity seems to fit the hypothesis that cyanate levels may reflect the nitrogen status of the cell; the low cyanase activity in ammonium-grown cells would allow cyanate accumulation while the high cyanase activity in nitrogen-limited cells would keep intracellular cyanate levels low. These results raise the interesting possibility that cyanate acts as a global regulator of genes involved in both N and C assimilation (Section III.B.3.a). A model showing the effect of cyanate on genes involved in N and C assimilation is presented in Fig. 4. However, recent work from the group of Omata (pers. comm.) showed that in the *cynS* mutant where cyanate is present at elevated levels there is little inhibition of the accumulation of *nirA* transcripts although activation of of *rbcLS*, *ccmKLMNO* and *icfA* is still observed. These data place some doubt on the details of the model and suggest that a nitrogen metabolite other than cyanate may be involved in the control of *nirA* expression.

3. Global Nitrogen Regulation

The control of N metabolism is critical to the cell and must be responsive to (i) the intracellular status of N and (ii) the availability of external N sources. As already mentioned, ammonium metabolism and GS/GOGAT activity are central to N assimilation. The discovery that ammonium may influence the transcription of genes via cyanate was discussed in the previous section. In this section we focus on the recent developments regarding the role of NtcA and PII, central control molecules in enteric bacteria, in the regulation of N metabolism is cyanobacteria. We also discuss a newly discovered protein, NtcB, which may be a second global regulator that links C and N metabolism.

a. NtcA

The existence of a global controller of N assimilation was hinted at by the isolation of several pleiotropic mutants that were unable to grow on nitrate and also lacked expression of all known N-regulated genes (Luque et al., 1994a; Vega-Palas et al., 1992; Vega-Palas et al., 1990). The *ntcA* gene, isolated by complementation of these pleiotropic mutants (Luque

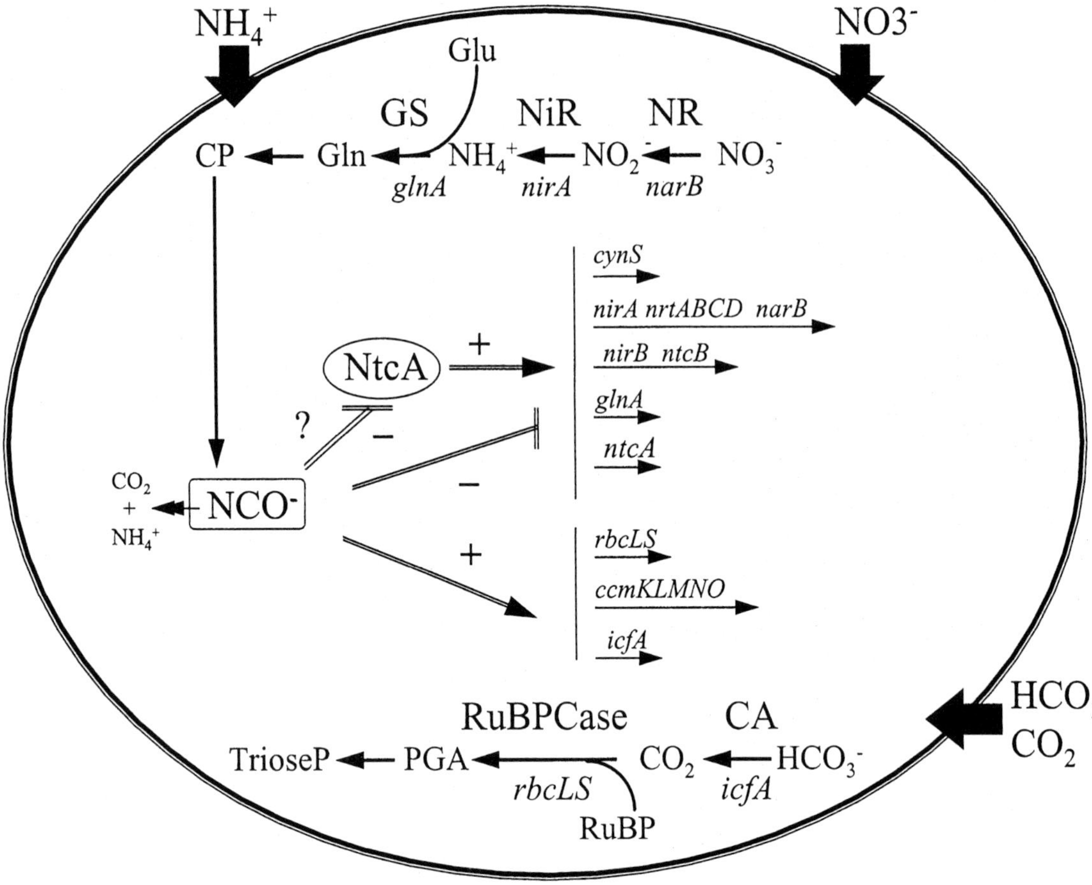

Fig. 4. Model for the regulation of carbon and nitrogen assimilation genes. The transport of ammonium (NH_4^+), nitrate (NO_3^-), bicarbonate (HCO_3^-) and CO_2 into the cell through specific transporters is shown. The metabolic conversion of HCO_3^- to phosphoglycerate (PGA) and triose phosphate (triose-P) via the activity of CA *(icfA)* and RuBPcase (*rbcLS*) is shown. The N assimilation pathway from nitrate (NO_3^-) to cyanate (NCO^-), with the intermediates is shown. The enzymes involved are nitrate reductase (NR), encoded by *narB*, nitrite reductase (NiR), encoded by *nirA*. Ammonium plus glutamate (Glu) is converted to glutamine (Gln) by glutamine synthetase (GS) encoded by *glnA*. Gln is converted to carbamoyl phosphate, which may decompose to cyanate which, in turn, is converted to CO_2 and NH_4 by cyanase (*cynS*). The positive regulation of the *nirA* and *nirB* operons, *glnA, ntcA* and *cynS* genes by NtcA and their possible repression by NCO^- is shown. The positive effect of NCO^- on the C assimilation operons, *rbcLS, ccmKLMNO* and *icfA* is also shown. The question mark indicates that it is not known whether activation/repression by NCO^- directly involves NtcA.. See text for qualification of the model.

et al., 1994a; Vega-Palas et al., 1990), has a typical helix-turn-helix DNA-binding motif at its C-terminus and is similar to transcription activators of the CRP-family. The gene encoding NtcA is present in all cyanobacteria that have been analyzed so far. Binding of NtcA to the promoter regions of several N-regulated genes was demonstrated and a consensus NtcA binding sequence ($GTAN_8TAC$) was identified. Since the binding site is a palindrome, NtcA probably interacts with the promoter as a dimer (or higher order aggregate), typical for the CRP family of trancription activators (Busby and Ebright, 1994; Johnson, 1995; Kolb et al., 1993). The NtcA binding site was located upstream of the *nirA*, *glnA* and *nirB* transcription initiation sites and substitutes for the canonical -35 motif. There are three potential NtcA binding sites (nirI, nirII and nirIII) symmetrically placed in the 286 bp sequence between the *nirA* and *nirB* coding regions (*nirA* and *nirB* are oriented in opposite direction to each other). Of these, nirI and nirIII are likely to be the NtcA binding sites associated with the *nirA* and *nirB* operons, respectively. The significance of nirII, located between the two other sites, is not known. However, the presence of multiple binding sites suggests that transcription from *nirA* and *nirB* may be under complex, N-modulated control.

A transcription factor, designated BifA (or VFI) was identified on the basis of its interactions with the *nifH* and *xisA* promoters; both genes are only expressed during heterocyst development. This protein was also shown to bind upstream of the *rbcLS* operon (Ramasubramanian et al., 1994; Wei et al., 1993). BifA is now known to be the same protein as NtcA. Thus, NtcA seems to regulate genes encoding proteins critical for N-assimilation, C-assimilation and heterocyst development (Frias et al.,1994).

The *ntcA* gene was shown to be more active in *Synechococcus* PCC 7942 cells grown in the absence of ammonium (Luque et al., 1994a). It has a typical NtcA binding site centered at -35 relative to transcription initiation, suggesting that it can regulate its own expression. The *ntcA* gene in *Anabaena* PCC 7120 appears to have multiple transcription initiation sites (Ramasubramanian et al., 1996), one of which is responsible for a major transcript that initiates at -49 (relative to the translation start site) and is synthesized regardless of the source of combined N This major transcript was more abundant in cells grown in medium lacking combined N. Two additional transcripts present in heterocysts have 5' ends that map to positions -180 and -190 relative to the translation start site. The relative abundance of these transcripts fluctuated during heterocyst development; the -49 and -180 transcripts were most abundant 12 h after N was removed from the culture medium while the -190 transcript was expressed at later times. A NtcA binding site was located upstream of the transcript that initiated at -49, but was not associated with the sites at -180/190. Not enough work has been done to establish whether multiple promoters are characteristic of heterocyst development and what factors regulate expression from these promoters.

The *ntcB* gene, which is co-transcribed with the *nirB* gene, encodes a protein of 309 amino acids with similarity to LysR transcriptional activators. An *ntcB* mutant exhibited an intriguing phenotype in that it grew slower than wild-type cells on nitrate or nitrite, but as fast as wild-type cells on ammonium (Suzuki et al., 1995a). This suggests that the mutant has a reduced ability to assimilate nitrate and nitrite. Furthermore, the *ntcB* mutant showed signs (based on its bleached appearance) of being N-limited even when maintained on ammonium. Hence, NtcB may play an important role in regulating the partitioning of fixed N by controlling certain biosynthetic activities under both N-sufficient and deficient conditions.

b. PII protein

Early studies (Allen et al., 1985; Sanders et al., 1989) on the potential role of phosphorylation of thylakoid membrane proteins in *Synechococcus* PCC 6301 revealed the presence of two soluble proteins that were phosphorylated under conditions of illumination that preferentially excited PSII. One of these proteins (molecular mass 13 kDa) was shown to have an N-terminal sequence with similarity to the PII protein of *E. coli*. Tandeau de Marsac's group isolated a gene from *Synechococcus* PCC 7942, designated *glnB*, that encodes a PII homologue. This protein contains 112 amino acids, is 64% identical to the proteobacterial PII protein and was present in all cyanobacterial species examined (Tsinoremas et al., 1991).

PII plays a complex and central role in N-control in enteric bacteria (Ninfa et al., 1995). A bifunctional uridylyl transferase (encoded by *glnD*) catalyzes the reversible uridylylation of PII at a specific tyrosine residue; in the uridylated form PII signals N limitation. The uridylyl transferase is sensitive to changes in the ratio of glutamine and 2-oxoglutarate, which reflects the N status of the cell. Thus, PII functions as a second messenger and interacts with NtrB/NtrC, the sensor/regulator pair that controls the NTR regulon and directly regulates GS activity by altering its adenylylation status (via the adenyltransferase encoded by *glnE*). PII homologues were found in several groups of bacteria, but its role in regulation is not always the same as that of PII in the enteric bacteria. For instance, in the free-living N_2-fixing bacterium *Azospirillum brasilense,* PII is critical for nitrogen-fixation, but seems to have no interaction with the NTR system (de Zamaroczy et al., 1993).

In cyanobacteria the PII polypeptide is not uridylylated, but is phosphorylated at a serine residue (Ser 49) in approximately the same position as the tyrosine residue that is uridylated in enteric bacteria (Forchhammer and Tandeau de Marsac, 1994; Forchhammer and Tandeau de Marsac, 1995a; Forchhammer and Tandeau de Marsac, 1995b). This is interesting for a number of reasons. First, phosphorylation at a serine residue suggests that a serine kinase, often used for signaling in eukaryotic organisms (and atypical of prokaryotic organisms), is functioning in this control system. Serine kinases have been detected in cyanobacteria (Warner and

Bullerjahn, 1994; Zhang, 1993) and the occurrence of protein kinases and their potential roles in the control of cyanobacterial metabolism have been reviewed (Kennelly and Potts, 1996; Mann, 1994; Chapter 17). It is important to identify the specific protein kinase and phosphatase that act on PII and to elucidate the ways in which cellular metabolites alter their activities. Second, the results suggest that PII functions to coordinate C and N assimilation. Phosphorylation of PII is stimulated by an increase in the 2-oxoglutarate/glutamine ratio. The cellular levels of 2-oxoglutarate would not only reflect the rate of N-assimilation, but would also depend upon the rate of CO_2 fixation. The critical questions of whether and how PII communicates with the transcriptional apparatus of the cell need clarification. At this point, no homologues of NtrB/NtrC, the regulators that govern the activity of the NTR regulon in enteric bacteria, have been found in cyanobacteria. Theoretically, PII could act directly as a transcriptional regulator or modify other regulators in the cell. The former possibility is unlikely since it lacks a DNA binding domain. The possibility that PII acts via NtcA was evaluated by Forchhammer and Tandeau de Marsac (1995a) using a *glnB* (PII) mutant. This mutant exhibited normal accumulation of transcripts encoding NiR and NR when it was transferred from ammonium to nitrate containing medium, unlike an *ntcA* mutant which shows low level constitutive accumulation of these transcripts. Thus, the role of PII in regulating N and C assimilation awaits further analysis.

C. Phosphorus Limitation

There is considerable evidence that inorganic phosphate (Pi) is frequently limiting in freshwater and marine environments (Healey, 1982; Heath and Cooke, 1975; Hecky and Kilham, 1988; Vaulot et al., 1996; Wynne, 1977). In terrestrial environments a substantial portion of P is sequestered in the form of poorly soluble mineral Pi (usually calcium phosphates) or organic matter of high molecular weight (Goldstein, 1994). Most cyanobacteria cannot access either of these forms of phosphate, which must first be converted to soluble Pi or low molecular weight organic phosphate. Cyanobacteria have evolved a number of strategies to deal with P deprivation (Bhaya, 1996; Grossman et al., 1994a; Healey, 1982). To overcome short periods of starvation they are able to access stored polyphosphate bodies. However, when faced with longer term P deprivation they need to increase their capacity to take up Pi and scavenge P from any available source, which may require the induction of several genes. In the last few years there has been a growing interest in some of the molecular responses to P stress in both marine and freshwater cyanobacteria (Carr and Mann, 1994). The responses of *E. coli* to P stress have been studied in great detail and provide a useful paradigm for exploring cyanobacterial responses. Therefore a brief account of the acclimation of *E. coli* to P limitation is included here.

If *E. coli* is placed in medium lacking P, genes required for the utilization of alternate P sources and encoding a high-affinity Pi-specific transporter may be induced by more than 100-fold. These genes are part of a PHO regulon that comprises 31 genes organized into eight different transcriptional units (Wanner, 1993). The concept of a regulon implies the presence of a master regulator that controls several genes that are coordinately activated in response to a single external cue. The PHO regulon includes (i) the *phn* operon, which encodes proteins that facilitate the breakdown and uptake of phosphonates; (ii) *phoA*, which encodes a periplasmic alkaline phosphatase that cleaves Pi from a variety of organic phosphates; (iii) *phoBR*, which encodes the sensor/regulator pair that control expression from the operons that comprise the PHO regulon; (iv) *phoE*, which encodes a polyanionic porin; (v) the *pstSCAB-phoU* operon of which *pstSCAB* encodes an ABC-transport system for Pi uptake and *phoU* which encodes a polypeptide involved in repression of the PHO regulon; (vi) *ugpBAECQ*, which encodes proteins required for the uptake of *sn*-glycerol-3-phosphate and phosphodiesterase (*ugpQ*) activity and (vii) *phoH* and (viii) *psiE* of unknown function.

The major control of the PHO regulon involves sensing extracellular levels of Pi by the Pi-specific transport system (encoded by *pstSCAB*) and transmission of this signal to the sensor protein PhoR. PhoR is assumed to exist in two forms, only one of which is involved in the transcriptional activation of genes. Activation of PhoR probably occurs via autophosphorylation at a conserved histidine residue when the cell senses Pi deprivation. PhoR then transfers the phosphoryl group to a conserved aspartate residue in the receiver domain of PhoB. Phosphorylated PhoB can specifically recognize promoters of the PHO regulon and activate their transcription (Makino et al., 1994). In addition to this Pi-dependent control, the PHO regulon may be subject to cross regulation by two other cell-signaling pathways (Wanner, 1994a; Wanner, 1994b; Wanner, 1995). These Pi-independent controls are coupled to later steps in P-metabolism and may constitute a form of cross-talk that links central P metabolic pathways to Pi uptake.

1. Phosphorus Acquisition

a. Phosphate Uptake Systems

Synechococcus sp. grown in P-deficient medium exhibits a 50-fold increase in V_{max} for Pi transport although the $K_{1/2}$ remains unchanged, implying that there is upregulation of a single Pi transport system (Grillo and Gibson, 1979). More than one transport system may be operative in *Synechococcus* sp. but characterizations of these putative transporters have not been performed (Healey, 1982). In *E. coli* the Pi transport system is comprised of four polypeptides, PstS, the periplasmic Pi-binding protein, PstC and PstA, the integral membrane proteins, and PstB, the permease (nucleotide binding protein). This is typical of the family of ABC transporters (Ames, 1986). In cyanobacteria only the gene encoding the phosphate-binding protein was identified and characterized. PstS from marine *Synechococcus* WH7803 has a molecular weight of about 33 kDa and is 35% identitical to the *E. coli* PstS protein. *pstS* transcripts accumulated under Pi-depleted conditions (Carr and Mann, 1994; Scanlan et al., 1993). In the freshwater strain *Synechococcus* PCC 7942, Aiba and Mizuno (1994) characterized a novel protein designated SphX (33 kDa) which was particularly abundant in P-limited cells. Although they did not identify this as a Pi- binding protein, subsequent comparative analyses have shown that SphX is most likely to be the homologue of PstS (Aiba and Mizuno, 1994; Mann and Scanlan 1994; Wagner et al., 1994). Analysis of the entire *Synechocystis* PCC 6803 genome sequence has revealed an operon that might encode members of the Pi transport system (Kaneko et al., 1996). This operon contains two *pstS*-like genes, one which encodes a polypeptide with greater homology to the marine-type PstS and the other more similar to the fresh water-type PstS (N. H. Mann, pers. comm.). The significance of this finding awaits biochemical characterization of the putative PstS polypeptides.

b. Phosphatases

Most cyanobacteria are unable to take up large organic P-containing molecules but can cleave the Pi moiety from a variety of substrates using extracellular phosphatases with broad substrate specificities. So far, three distinct phosphatases have been identified in different cyanobacterial species. Of these the best characterized is an atypical alkaline phosphatase (PhoA) of *Synechococcus* PCC 7942. This enzyme accumulates in P-deficient cultures of *Synechococcus* PCC 7942 and is a periplasmic protein with a molecular mass of approximately 150 kDa (Block and Grossman, 1988). The *phoA* gene encoding the alkaline phosphatase appears to be transcriptionally controlled, and mRNA encoding PhoA only accumulates in P-deficient cells. Characterization of PhoA demonstrated that it is unlike other previously described phosphatases (Ray et al., 1991). The first one third of the protein has 29% identity with mammalian and bacterial 5'-nucleotidases (Zimmermann, 1992). The 5'-nucleotidases have a broad substrate specificity which includes 5'-purine and pyrimidine mononucleotides as well as more complex nucleotides such as FAD and UDP-glucose. A region of about 50 amino acids shows strong homology to the UshA protein, a UDP sugar hydrolase of *E. coli*. The last half of the protein has sequences similar to P-loops, which are motifs found in kinases that are involved in binding nucleotide triphosphates via the terminal phosphate ester. The presence of multiple P-loops in the alkaline phosphatase might allow for the binding of phosphate groups attached to a variety of different compounds, thereby expanding the substrate specificity of the enzyme. This would be a desirable trait for an enzyme utilizing diverse phosphorylated compounds in its natural environment. In this context it is noteworthy that a strain in which the *phoA* gene was insertionally inactivated was able to grow as fast as wild-type cells in medium lacking P. This suggests that although PhoA is not essential for survival under P-starvation conditions, it may give the cells a particular advantage in certain environmental milieus. Since P is most often the limiting nutrient in freshwater ecosystems (Hecky and Kilham, 1988), the phosphatases of cyanobacteria probably serve a vital role in natural cyanobacterial populations.

A second alkaline phosphatase, designated PhoV, was recently identified in *Synechococcus* PCC 7942 (Wagner et al., 1995). This enzyme, like PhoA, is periplasmically located and has both a broad pH optimum (in the alkaline range) and substrate specificity for phosphomonoesters. It has a molecular mass of 61 kDa, requires Zn^{++} for activity, and exhibits 34% identity to the alkaline phosphatase, PhoA, from *Zymomonas mobilis*. Preliminary evidence suggests that the *phoV* gene is not regulated by P levels and may be a constitutive periplasmic phosphatase. A more recent comparison of PhoV with proteins in the database shows its striking similarity to a well-defined Ca^{++} ATPase from *Flavobacterium odoratum* (Peiffer et al., 1996). The significance of this similarity is unclear.

Potts and co-workers identified and cloned a gene (*ihpP*) from *Nostoc commune* UTEX 584 encoding an enzyme with phosphomonoesterase activity (Potts et al., 1993; Xie et al., 1989; See Chapter 17). This enzyme, designated IphP, shows no obvious similarity to known prokaryotic phosphatases, but does have a motif (His-Cys-X_5-Arg) characteristic of the active site of eukaryotic protein-tyrosine phosphatases and dual-specificity protein phosphatases. In eukaryotes tyrosine phosphatases are often involved in signal transduction (Howell et al., 1996; Keyse, 1995; Shenolikar, 1994). The IphP polypeptide could dephosphorylate phosphotyrosine, phosphoserine and phosphothreonine of a variety of phosphoproteins, making it the only dual-specificity phosphatase so far identified in cyanobacteria. Whether IphP has any role in signal transduction or P acquisition in cyanobacteria is not clear (Kennelly and Potts, 1996; Mann, 1994; Wanner, 1994a; Zhang, 1993). IphP does have a signal peptide, suggesting that it might be in the periplasm. It also exhibits broad phosphomonoesterase activity towards peptides, proteins and low molecular weight organophosphates and has measureable pyrophosphatase activity. Thus, there is growing evidence that several phosphatases are present as periplasmic or extracellular proteins in cyanobacteria. Depending on the substrate specificity of these phosphatases, there may be efficient scavenging of P from a variety of available sources.

2. Gene Regulation During Phosphorus Limitation

The regulation of genes that are induced during P deprivation has been studied in the freshwater species *Synechococcus* PCC 7942 and the marine unicellular cyanobacterium *Synechococcus* WH 7803. In both species homologues of the two-component sensor/regulator pair, PhoR/PhoB, were identified. Heterologous complementation of an *E. coli* strain lacking PhoR with a library of *Synechococcus* PCC 7942 DNA was used to isolate *sphS* and *sphR*, whose gene products have significant homology to PhoR and PhoB, respectively (Aiba et al., 1993; Watson et al., 1996). SphS and SphR are almost certainly the sensor/regulator pair that controls the response of *Synechococcus* PCC 7942 to P limitation. This is supported by the finding that a mutant of *Synechococcus* PCC 7942 in which the *sphRS* operon was inactivated could not induce periplasmic alkaline phosphatase activity in response to P limitation (nor was there the induction of the abundant, periplasmic Pi binding protein). This raises the question of whether the equivalent of a PHO regulon exists in cyanobacteria, and whether it is under the control of the SphR/SphS sensor/regulator pair. In *E. coli*, all PHO regulon promoters (except one) have an 18 base consensus sequence (comprised of two 7 base direct repeats separated by 4 bases) called the 'pho box' that replaces the typical -35 sequence. Phosphorylated PhoB binds to this sequence (Wanner, 1993). The phosphorylated form of SphR binds two specific regions upstream of the *phoA* promoter (at positions -167 to -130 and -66 to -18) and to the promoter of the *sphX* gene (encoding the putative Pi-binding protein) (Aiba and Mizuno, 1994; Nagaya et al., 1994). Further work on other P-regulated promoters will be required to establish the existence of a possible pho box in *Synechococcus* PCC 7942. Chimeric genes in which the *phoA* promoter from -66 to +190 (relative to the transcription start site) were fused to the *uidA* gene to create a translational fusion showed inducible β-glucuronidase activity (GUS) when the cells were grown in medium lacking P. In site-directed mutants of this construct in which 3 bases in the positions -34, -35, -36 were altered, the inducible GUS activity was completely lost (D. Bhaya and A. Grossman, unpublished data), suggesting that there is possible sequence-specific binding of SphR to the *phoA* promoter.

3. Polyphosphates

Polyphosphate granules are used by cyanobacteria (and other organisms) as intracellular P-reserves. In addition, polyphosphates may serve a number of functions in microorganisms. They may act as i) chelators, ii) buffers against adverse conditions such as alkaline environments, heat and oxidative stress, iii) regulators of transcription and iv) molecules that allow survival during stationary phase (Kornberg, 1994; Chapter 17).

Polyphosphate is generally present in exponentially growing cells, and may accumulate to high levels in N- or S-deficient cells (except in heterocysts) and be degraded in Fe- and C-deficient cells, as well as in P-deficient cells (Tang et al., 1995). Interestingly, the activity of polyphosphate synthetase is detectable in extracts of P-deficient cyanobacteria, even though the level of polyphosphates may be low (Grillo and Gibson, 1979; Mackerras et al., 1990). The persistence of this activity involved in the synthesis of these storage granules may explain the overaccumulation of reserve material that occurs when Pi is resupplied to P-deficient cells. This 'overplus' phenomenon results in an initial, massive accumulation of polyphosphate granules, which are subsequently degraded as the cells return to nutrient-replete exponential growth. The ability to overaccumulate polyphosphate may allow cyanobacteria to take advantage of pulses of available nutrients that are common in patchy, nutrient-poor environments. It has also been suggested that the buoyancy of marine *Trichodesmium* may be controlled, in part, by diurnal changes in polyphosphate content (Romans et al., 1994; See Chapter 5). The identification of putative genes in *Synechocystis* PCC 6803 encoding a polyphosphate kinase and a polyphosphatase (Kaneko et al., 1996) should lead to further analysis of the role of polyphosphates in the physiology and metabolism of the cell.

D. Sulfur Limitation

Cyanobacteria that experience sulfur limitation cease growth, alter photosynthetic electron transport, degrade their PBS and increase the efficiency with which they transport sulfate and other S-containing molecules. Indeed, *Synechococcus* PCC 7942 is able to use a number of organic and inorganic S compounds including sulfate, thiosulfate, cysteine, cystine, reduced glutathione, and thiocyanate (Laudenbach and Grossman, 1991; Lawry and Jensen, 1986; Schmidt et al., 1982). In this section the transport of sulfate into cyanobacterial cells is considered since both the change in photosynthetic electron transport and the degradation of the PBS are discussed elsewhere in this review (Section III.D.4).

1. Sulfate Uptake

Sulfate uptake is an active process, dependent upon light, and varies with both temperature and pH, is sensitive to osmotic shock and is inhibited by thiosulfate and sulfite (Jeanjean and Broda, 1977; Utkilen et al., 1976). Utkilen and co-workers (1976) reported a $K_{1/2}$ of 0.75 μM and a V_{max} of 0.7 pmol/(10^6cells x min) for sulfate transport at 42°C in S starved *Synechococcus.* Similar physiological studies (Green and Grossman, 1988) demonstrated that while the V_{max} for sulfate transport increased between 10- and 20-fold during S-deprivation, the $K_{1/2}$ was approximately 1 μM in both S sufficient and S deficient medium. This suggested that S deficiency caused elevated accumulation of a single sulfate transport system. Once sulfate is transported into *Synechococcus* cells it is reduced via the PAPS sulfotransferase pathway (Schmidt and Christen, 1978). Other cyanobacteria can reduce sulfate via the APS sulfotransferase pathway (Tsang and Schiff, 1975).

2. Gene Regulation Upon Sulfur Limitation

A region of the *Synechococcus* PCC 7942 genome (8.3kb) was identified that contained genes involved in sulfate acquisition (Green et al., 1989; Laudenbach et al., 1991; Laudenbach and Grossman, 1991). A large increase in the level of mRNA transcribed from these genes occurs when the cells are deprived of S. Sequence characterization of this region of the cyanobacterial genome revealed the presence of genes encoding four components of a typical periplasmic transport system (*cysA, cysT, cysW* and *sbpA*) (Ames, 1986). One of the components, localized to the periplasmic space, is a polypeptide that binds to the substrate (SbpA, sulfate binding protein). Interactions of the substrate with the binding protein cause a conformational change in the protein that results in trapping of the substrate molecule. The protein ligand complex then interacts with two hydrophobic proteins (CysT and CysW) that are thought to span the cytoplasmic membrane and form a pore. This interaction causes the release of the substrate, which then traverses the pore and enters the cell against a concentration gradient using energy derived from ATP hydrolysis (which is a function of the nucleotide binding protein CysA). Inactivation of the genes encoding CysT, CysW or CysA resulted in cells that were unable to utilize sulfate and required an alternate S source for growth. In contrast, inactivation of the gene encoding the periplasmic sulfate binding protein (*sbpA*) did not prevent the cells from utilizing sulfate as the sole S source. These results suggested that either SbpA was not absolutely required for growth on sulfate, or that a second gene existed whose gene product could substitute for SbpA.

The sulfate permease may also be utilized for the uptake of thiosulfate. Thiosulfate transport may also be augmented by the binding of thiosulfate to a specific substrate-binding protein designated CysP in *E. coli* (Hryniewicz et al., 1990). Based on homology to the *E. coli* gene, the *cysP* gene in *Synechococcus* PCC 7942 is immediately downstream of *cysW*. The putative *cysP* gene is followed by two genes that have homologies to *cysT* and *cysW*. These may be part of a transport system that has a higher specificity for thiosulfate than for sulfate. There does not appear to be a separate nucleotide binding protein associated with this system so CysA may function for both transport systems.

Downstream of the putative thiosulfate transport genes is a gene designated *rhdA*, which encodes a protein with some similarity to the enzyme rhodanese (which transfers the thiol group of thiosulfate to thiophilic acceptor molecules) (Laudenbach et al., 1991). This protein accumulates to high levels in the periplasmic space upon S starvation. However inactivation of *rhdA* does not prevent utilization of various inorganic S sources. RhdA may have a catalytic role in acclimation or may simply bind certain S-containing compounds and prevent their leakage from the cell.

3. CysR

A gene located in the middle of the sulfate permease operon, designated *cysR*, encodes a protein that resembles transcriptional regulators such as FixK and CRP (Laudenbach and Grossman, 1991). When the *cysR* gene is interrupted, thiocyanate can no longer serve as a sole S source, although other S sources can support growth of the *cysR* mutant. Recently, it was found that CysR was involved in the regulation of several genes on the pANL plasmid of *Synechococcus*. These genes encode a 36 kDa periplasmic protein (SrpA) which is similar to catalase, enzymes involved in cysteine biosynthesis (SrpG and SrpH) and a protein (ChrA) with remarkable homology to the chromate-inducible *chrA* gene product of *Alcaligenes eutrophus*. The *chrA* gene product may be involved in transport or efflux of heavy metals such as chromate. This provides the first indication of the role of the hitherto cryptic plasmids of cyanobacteria (Nicholson et al., 1995; Nicholson and Laudenbach, 1995) and suggests temporal control of gene regulation during S limitation.

Following sulfur deprivation the genes encoding the sulfate and thiosulfate transport systems and *cysR* are activated (the activator of the operons encoding the sulfate and thiosulfate transport systems have not been identified). Increased synthesis of the CysR protein causes the activation of numerous genes on the large *Synechococcus* PCC 7942 plasmid. For some of the plasmid localized genes the function was established, however, others encode proteins with intriguing similarities to proteins that have defined catalytic function, but the role that they play in the response of cyanobacteria to sulfur limitation is not clear.

4. Changes in the Photosynthetic Apparatus

a. Photosynthetic Activity and Gene Regulation

N and S limitation lead to a decline in photosynthetic electron transport and a general loss of pigmentation (Ochoa de Alda et al., 1996a, 1996b). O_2 evolution and PSII activity often dramatically decline in parallel with the loss of the PC and AP content of cell. In cultures deprived of N or S, the PC content can decline to virtually undetectable levels, while chl levels are less affected. Similar changes are observed in non-N_2-fixing strains depleted of N, and during the transient N deficiency in N_2-fixing strains. A decrease in PSII activity during nutrient-deprivation may be advantageous since, in the absence of anabolic processes, the formation of excited chl molecules that are unable to transfer energy to the photosynthetic reaction centers might generate toxic oxygen species. In this regard, it is interesting to note that both the chl and β-carotene levels in N, S, or P deprived cells decline relative to the levels of xanthophylls, and in particular relative to zeaxanthin (Fresnedo et al., 1991; Gombos and Vigh, 1986; Grossman et al., 1992). Maintenance of high zeaxanthin levels may be beneficial to nutrient-limited cells since zeaxanthin may function in dissipating the excitation energy absorbed by chl molecules (Demmig-Adams, 1990). In contrast to the decline in PSII activity, both S and N starved cells maintain PSI activity and the production of ATP (Collier et al., 1994) via cyclic photophosphorylation. The continued production of ATP in this manner would provide energy to drive metabolic processes that must be maintained in nutrient deficient cells. This energy would also be required for transporting the limiting nutrient into the cell when it does become available.

The levels of transcripts encoding specific components of the photosynthetic apparatus were shown to decline during nutrient limitation. In *Synechococcus* PCC 7002, transcripts from the *cpcBACDEF* operon were essentially undetectable 3 - 5 h after the initiation of N deprivation, (Bryant, 1991; de Lorimier et al., 1984). Data obtained by Gasparich et al. (1987), using a *cpcB-lacZ* chimeric gene suggested that N depletion caused a marked decrease in transcription of the *cpcBA* operon. In *Synechococcus* PCC 7942 the levels of mRNA encoding both PC and AP declined rapidly during either N or S deprivation and less rapidly (and not to as low a level) during P-deprivation. However, all of these mRNA species remained at 5 - 10% of the level found in nutrient-replete cells (Collier and Grossman, 1992), even 48 h after cells are transferred to medium devoid of N or S. Change in the steady state levels of phycobiliprotein mRNAs may be a consequence of both altered rates of transcription and mRNA turnover. The low levels of PC and AP mRNA that remain in nutrient-deprived cells may not be translated (Grossman, unpublished data), suggesting that, in addition to transcriptional regulation, post-transcriptional processes are involved in inhibiting phycobiliprotein synthesis during nutrient deprivation. A decrease in the level of mRNA encoding PC and AP also occurs in the N-fixing cyanobacterium *Anabaena* during the initial phase of heterocyst differentiation (some bleaching occurs at this time) (Johnson et al., 1988; Wealand et al., 1989). The decrease in PC and AP mRNA is temporary, and normal levels of both are re-established in vegetative cells (Belknap and Haselkorn, 1987), but not heterocysts, after N_2 fixation begins.

The photosynthetic apparatus that remains in heterocysts may be similar to that found in S-deprived, and in N_2 deprived non-N_2-fixing, cyanobacteria. Heterocysts do not have a functional PSII, but maintain an active PSI for the production of ATP via cyclic electron flow. The phycobiliproteins that remain after heterocyst differentiation, consist almost entirely of PC, and are not assembled into PBS but can still transfer harvested light energy to PSI (Peterson et al., 1981; Tyagi et al., 1981; Yamanaka and Glazer, 1983). The generation of ATP by cyclic electron flow in heterocysts would help support the energetically expensive process of N_2 fixation.

b. Phycobilisome Degradation

Cyanobacterial cells that are starved for any of a number of different nutrients lose their pigmentation in a process known as chlorosis or bleaching. In some cases chlorosis involves the active degradation of pigments, while in other cases there is no degradation, but pigment synthesis stops although the cell continues to divide with a consequent dilution of the pigments. Both types of bleaching are observed, and the distinction may be important for understanding the modulation of photosynthetic activity and the function of the photosynthetic apparatus during various stress conditions.

A decline in the level of PBS, as measured by a net loss of PC, was most extensively studied in N- and S-deprived cells. Both conditions cause a rapid and near complete degradation of the PBS (Collier and Grossman, 1992; Lau et al., 1977; Yamanaka and Glazer, 1980). The degradation process is light dependent and is inhibited if DCMU is added to the cultures upon imposition of the nutrient limitation conditions (Schwarz and Bhaya, unpublished data). This suggests that either the generation of energy (ATP or an electrochemical gradient) or regulation of the redox state of the cell (reduction of electron carriers) via photosynthetic electron transport is absolutely required for PBS degradation. In contrast to the situation for N and S deprived cells, P depleted *Synechococcus* PCC 7942 cells have a low PBS content because the PBS are diluted during growth on medium lacking P; little PBS degradation occurs (Collier and Grossman, 1992). Degradation of the PBS (as well as other cellular proteins) during N deficiency could provide amino acids for the synthesis of proteins or other cellular constituents important for the acclimation process. The use of phycobiliproteins as amino acid storage molecules may be especially important for marine cyanobacteria (Wyman et al., 1985; Yeh et al., 1986), since N is frequently limiting in marine environments. PBS are a poor source of S-containing amino acids, but are nevertheless degraded in S deprived cells. In fact, *Fremyella diplosiphon* has a set of PC genes that are devoid of all but the essential S-containing amino acids and are only expressed when the cells are maintained on low-S medium (Mazel and Marliere, 1989). PBS destruction during nutrient-deprivation may also help protect the cells from phototoxicity (discussed above), although there may be other mechanisms that dramatically reduce the transfer of energy from the PBS to chl.

The degradation of the PBS of *Synechococcus* PCC 7942 is virtually identical during N and S deprivation. It is an ordered process in which degradation begins at the periphery of the complex, eliminating the core distal PC hexamers, with subsequent elimination of hexamers closer to the core (Collier and Grossman, 1992; Duke et al., 1989; Yamanaka and Glazer, 1980). The degradation of PBS polypeptides results in a decrease in the PBS size (sedimentation coefficient) and a reduction in the ratio of PC to AP. The smaller PBS can still function in harvesting light energy. Continued nutrient deprivation results in the complete degradation of the remaining PBS structure. Upon adding the limiting nutrient back to deprived cultures, new PBS are rapidly synthesized.

The existence of a protease synthesized *de novo* in response to N deprivation and involved in PBS degradation was suggested from a studies with *Anabaena*, *Spirulina* and *Synechococcus* (Boussiba and Richmond, 1980; Elmorjani and Herdman, 1987; Foulds and Carr, 1977; Wood and Haselkorn, 1979; Wood and Haselkorn, 1980; Yamanaka and Glazer, 1980). Mutants of *Synechococcus* PCC 7942 were isolated that are unable to degrade their PBS during either S or N deprivation, although no new PBS were synthesized during the period of stress. Surprisingly, these cells grew at a similar rate and exhibited a similar susceptibility to high light during nutrient deficiency as wild-type cells (Collier and Grossman, 1994). Hence, even though the cells had a considerable number of PBS when N or S deprived, they appeared to be no more photosensitive than normally acclimating cells. Additional experiments suggested that the PBS present in the mutant organisms deprived of N or S were not able to efficiently transfer harvested light energy to the reaction center of PSII; this uncoupling of the light-harvesting complex from the primary photochemical reactions of PSII may explain the lack of photosensitivity.

The 'non-bleaching' mutants were complemented to the bleaching, wild-type phenotype with a gene that was designated *nblA*. The *nblA* gene contains an open reading frame of 59 amino acids and a transcript covering this gene accumulates at high levels in cells deprived of N or S. Insertional inactivation of *nblA* resulted in a nonbleaching phenotype, confirming that NblA was required for bleaching. The *nblA* gene is present on the genomes of other cyanobacteria (*Synechocystis* PCC 6803, *Synechococcus* PCC 7942, *F. diplosiphon*) and red algae (unpublished data). Analysis of the genome of *Synechocystis* PCC 6803 revealed two *nblA*-like sequences that are contiguous on the genome (Kaneko et al., 1996). These genes are only 24% identical, suggesting that the duplication that gave rise to the tandem repeat was not recent and that the gene products may have diverged sufficiently to have distinct biological functions. No homology was observed between NblA and any protein listed in the databases. There are a number of possibilities for the way in which NblA might function. i) It may activate a protease, such as the one studied by Wood and Haselkorn (1980), to specifically degrade PBS polypeptides. ii) It may interact with the polypeptides of the PBS rendering them susceptible to cellular proteases. Such interactions may involve covalent attachment, similar to the way in which ubiquitin binds to proteins in eukaryotes (Hershko, 1988), or the disruption hydrophobic and/or ionic interactions among various constituents of the PBS. In fact, the presence of ubiquitin has recently been demonstrated in *Anabaena* (Durner and Boeger, 1995). iii) NblA may activate other genes that are directly involved in degrading the PBS.

Two other distinct non-bleaching mutants of *Synechococcus* PCC 7942 have recently been characterized. One of these mutants is in a gene, designated *nblB*. The polypeptide product of the nblB gene resembles the lyases (Fairchild et al., 1992; Fairchild and Glazer, 1994; Jung et al., 1995; Swanson et al., 1992; Kahn et al., 1997; Zhou et al., 1992) that attach chromophores to apophycobiliprotein subunits. The nblB gene product may be involved in removing chromophores from the phycobiliprotein subunits prior to proteolytic attack (Dolganov and Grossman, 1998).

Another non-bleaching mutant was altered in a gene, *nblR*, encoding a response regulator (Schwarz and Grossman, 1998). NblR is required for i) modulating PBS levels during nutrient-replete growth, ii) activating *nblA* and eliciting PBS degradation during S and N limitation, iii) modulating PBS levels upon illumination with high light, and iv) survival during S or N starvation and during exposure to high light. The requirement of NblR for survival during adverse conditions may reflect a role of this regulator in altering photosynthetic electron transport activity, which is absolutely essential for the survival. How NblR senses different environmental signals is not known, although the redox state of the cell would be a common link among the cues sensed by this response regulator.

IV. Concluding Remarks

The responses of cyanobacteria were discussed in regard to suboptimal light and to nutrient limitation: two environmental parameters that impinge strongly on the photosynthetic capacity of the organism and on their survival. The development and use of genetic and molecular tools has enabled cyanobacteriologists to begin to dissect some of the major responses to stress and to identify molecules that control these responses. The approaches are becoming more satisfying as researchers are adopting sophisticated molecular techniques to examine gene regulation. These tools can be used in conjunction with our knowledge of the complete sequence of the *Synechocystis* PCC 6803 genome.

As more is learnt about acclimation responses in cyanobacteria we are likely to get clues about how more complex photosynthetic organisms cope with their environment and the evolution of acclimation processes. At the very least, we feel that this will be true with respect to the ways in which cyanobacteria control photosynthetic electron transport. Furthermore, models that were developed to describe how cyanobacteria respond to diverse environments can now be evaluated in natural populations, which could be of significant ecological value. The field is also poised to contribute to our understanding of global regulators, which might provide unique insights into the remarkable abilities of cyanobacteria (like those who study them) to survive inhospitable environs.

Acknowledgments

We are very grateful to Aaron Kaplan, David Kehoe and Tatsuo Omata for allowing us to use and modify the figures presented here. We would like to thank E. Flores, S. S. Golden, A. Kaplan, N. H. Mann, T. Omata, S. Tabata for sharing unpublished data, preprints and helpful discussions. Kathi Bump was invaluable in typing and formatting this chapter. We acknowledge support from NIH, USDA NSF and the Human Frontiers Program for more than two decades of support. This is CIW publication No. 1336.

References

Aiba H and Mizuno T (1994) A novel gene whose expression is regulated by the response-regulator, SphR, in response to phosphate limitation in *Synechococcus* species PCC 7942. Mol Microbiol 13: 25-34

Aiba H, Nagaya M and Mizuno T (1993) Sensor and regulator proteins from the cyanobacterium *Synechococcus* species PCC 7942 that belong to the bacterial signal-transduction protein families: implication in the adaptive response to phosphate limitation. Mol Microbiol 8: 81-91

Aichi M and Omata T (1997) Involvement of NtcB, a LysR family transcription factor, in nitrite activation of the nitrate assimilation operon in the cyanobacterium *Synechococcus* sp. strain PCC 7942. J Bacteriol 179:4671-4675

Aizawa K and Miyachi S (1986) Carbonic anhydrase and CO_2 concentrating mechanism in microalgae and cyanobacteria. FEMS Microbiol Rev 39: 215-233

Aizawa K, Shimizu T, Hiyama T, Satoh K, Nakamura Y and Fujita Y (1992) Changes in composition of membrane proteins accompanying the regulation of PSI/PSII stoichiometry observed with *Synechocystis* PCC 6803. Photosyn Res 32: 131-138

Allen JF (1992) Protein phosphorylation in regulation of photosynthesis. Biochim Biophys Acta 1098: 275-335

Allen JF, Sanders CE and Holmes NG (1985) Correlation of membrane protein phosphorylation with excitation energy distribution in the cyanobacterium *Synechococcus* 6301. FEBS Lett 193: 271-275

Allen M M (1984) Cyanobacterial cell inclusions. Annu Rev Microbiol 38: 1-25

Ames GF-L (1986) Bacterial periplasmic transport systems: structure, mechanism, and evolution. Annu Rev Biochem 55: 397-425

Andersson B and Styring S (1991) Photosystem II: Molecular organization, function, and acclimation. Curr Top Bioenergetics 16: 1-81

Andriesse X, van Arkel G and Weisbeek P (1990). Cloning of the nitrate reductase gene of *Synechococcus* PCC 7942. In: Third International Symposium on NO_3-Assimilation: Molecular and Genetic Aspects. Bombanes

Appleby JL, Parkinson JS and Bourret RB (1996) The multi-step phosphorelay: Not necessarily a road less traveled. Cell 86: 845-848

Argall ME, Smith GD, Stamford NPJ and Youens BN (1992) Purification and properties of urease from the cyanobacterium *Anabaena cylindrica*. Biochem International 6: 1027-1036

Aro E-M, Virgin I and Andersson B (1993) Photoinhibition of photosystem II. Inactivation, protein damage and turnover. Biochim Biophys Acta 1143: 113-134

Badger MR (1987) The CO_2-concentrating mechanism in aquatic phototrophs. In: Hatch MD & Boardman NK (eds) The Biochemistry of Plants: A Comprehensive Treatise. Vol. 10: Photosynthesis pp 219-274 Academic Press, New York

Badger MR and Price GD (1992) The CO_2 concentrating mechanism of cyanobacteria and green algae. Physiol Plant 84: 606-615

Bèdu S, Beuf L and Joset F (1993) Membrane and soluble carbonic anhydrase activities in a cyanobacterium, *Synechocystis* PCC 6803. In: Murata N (ed) Research in Photosynthesis. Vol. III pp. 18.189-18.882 Kluwer, Dordrecht, The Netherlands

Bèdu S, Pozuelos P, Cami B and Joset F (1995) Uptake of inorganic carbon in the cyanobacterium *Synechocystis* PCC

6803: physiological and genetic evidence for a high-affinity uptake system. Mol Microbiol 18: 559-568

Belknap WR and Haselkorn R (1987) Cloning and light regulation of expression of the phycocyanin operon of the cyanobacterium *Anabaena*. EMBO J 6: 871-884

Bennett J (1991) Protein phosphorylation in green plant chloroplasts. Annu Rev Plant Physiol 42: 281-311

Bhaya D (1996) Molecular responses of cyanobacteria to macronutrient limitation. J Sci Ind Res 55: 630-637

Biggins J and Bruce D (1989) Regulation of excitation energy transfer in organisms containing phycobilins. Photosyn Res 20: 1-34

Blanco F, Alana A, Llama MJ and Serra JL (1989) Purification and properties of glutamine synthetase from the non-N_2-fixing cyanobacterium *Anabaena laminosum*. J Bacteriol 171: 1158-1165

Block MA and Grossman AR (1988) Identification and purification of a derepressible alkaline phosphatase from *Anacystis nidulans* R2. Plant Physiol 86: 1179-1184

Bloye SA, Silman NJ, Mann NH and Carr NG (1992) Bicarbonate concentration by *Synechocystis* PCC 6803. Plant Physiol 99: 601-606

Bogorad L (1975) Phycobiliproteins and complementary chromatic adaptation. Annu Rev Plant Physiol 26: 369-401

Boussiba S (1989) Ammonia uptake in the alkalophilic cyanobacterium *Spirulina platensis*. Plant Cell Physiol 30: 303-308

Boussiba S, Resch CM and Gibson J (1984) Ammonia uptake and retention in some cyanobacteria. Arch Microbiol 138: 287-292

Boussiba S and Richmond AE (1980) C-phycocyanin as a storage protein in the blue-green alga *Spirulina platensis*. Arch Microbiol 125: 143-147

Broun II, Fadeyev SI, Kirik II, Severina II and Skulachev VP (1990) Light-dependent delta-μNa generation and utilization in the marine cyanobacterium *Oscillatoria brevis*. FEBS Lett 270: 203-206

Bruns B, Briggs WR and Grossman AR (1989) Molecular characterization of phycobilisome regulatory mutants in *Fremyella diplosiphon*. J Bacteriol 171: 901-908

Bryant DA (1981) The photoregulated expression of multiple phycocyanin species: general mechanism for control of phycocyanin synthesis in chromatically adapting cyanobacteria. Eur J Biochem 119: 425-429

Bryant DA (1986) The cyanobacterial photosynthetic apparatus: comparisons to those of higher plants and photosynthetic bacteria. Can Bull Fish Aquat Sci 214: 423-500

Bryant DA (1991) Cyanobacterial phycobilisomes: progress towards a complete structural and functional analysis via molecular genetics. In: Bogorad L & Vasil IK (eds) The Photosynthetic Apparatus: Molecular Biology and Operation. Vol. 7 pp. 257-300 Academic Press, New York

Bryant DA and Cohen-Bazire G (1981) Effects of chromatic illumination on cyanobacterial phycobilisomes: Evidence for the specific induction of a second pair of phycocyanin subunits in *Pseudanabaena* 7409 grown in red light. Eur J Biochem 119: 415-424

Bryant DA, Guglielmi G, Tandeau de Marsac N, Castets A-M and Cohen-Bazire G (1979) The structure of cyanobacterial phycobilisomes: a model. Arch Microbiol 123: 113-127

Burbulys D, Trach K and Hoch JA (1991) Initiation of sporulation in *B. subtilis* is controlled by a multicomponent phosphorelay. Cell 64: 545-552

Busby S and Ebright RH (1994) Promoter structure, promoter recognition, and transcription activation in prokaryotes. Cell 79: 743-746

Bustos SA, Schaefer MR and Golden SS (1990) Different and rapid responses of four cyanobacterial *psbA* transcripts to changes in light intensity. J Bacteriol 172: 1998-2004

Campbell D, Clarke AK, Gustafsson P and Öquist G (1996) D1 exchange and the Photosystem II repair cycle in the cyanobacterium *Synechococcus*. Plant Science 115: 183-190

Campbell D, Zhou G, Gustafsson P, Öquist G and Clarke AK (1995) Electron transport regulates exchange of two forms of photosystem II D1 protein in the cyanobacterium *Synechococcus*. EMBO J 14: 5457-5466

Carpenter E and Price CC (1976) Marine *Oscillatoria (Trichodesmium)*: Explanation for aerobic nitrogen fixation without heterocyts. Science 191: 1278-1280

Carrasco CD and Golden JW (1995) Two heterocyst-specific DNA rearrangements of *nif* operons in *Anabaena cylindrica* and *Nostoc* sp. strain MAC. Microbiol 141: 2479-2487

Carrasco CD, Buettner JA and Golden JW (1995) Programmed DNA rearrangement of a cyanobacterial *hupL* gene in heterocysts. Proc Natl Acad Sci USA 92: 791-795

Carr NG and Mann NH (1994) Oceanic cyanobacterial picoplankton. In: Bryant DA (ed) The Molecular Biology of Cyanobacteria. pp. 27-48 Kluwer, Dordrecht

Casey ES and Grossman AR (1994) In vivo and in vitro characterization of the light-regulated *cpcB2A2* promoter of *Fremyella diplosiphon*. J Bacteriol 176: 6362-6374

Casey ES, Kehoe DM and Grossman AR (1997) Suppression of mutants aberrant in light intensity responses of complementary chromatic adaptation. J Bacteriol 179: 4599-4606

Chang C, Kwok SF, Bleecker AB and Meyerowitz EM (1993) *Arabidopsis* ethylene-response gene *ETR*1: Similarity of product to two-component regulators. Science 262: 539-544

Chiang GG, Schaefer MR and Grossman AR (1992) Complementation of a red-light indifferent cyanobacterial mutant. Proc Natl Acad Sci USA 89: 9415-9419

Clarke A, Soitamo A, Gustafsson P and Öquist G (1993a) Rapid interchange between two distinct forms of cyanobacterial photosystem II reaction-center protein D1 in response to photoinhibition. Proc Natl Acad Sci USA 30: 9973-9977

Clarke AK, Campbell D, Gustafsson P and Öquist G (1995) Dynamic responses of photosystem II and phycobilisomes to changing light in the cyanobacterium *Synechococcus* sp. PCC 7942. Planta 197: 553-562

Clarke AK, Hurry VM, Gustafsson P and Öquist G (1993b) Two functionally distinct forms of the photosystem II reaction-center protein D1 in the cyanobacterium *Synechococcus* sp. PCC 7942. Proc Natl Acad Sci USA 90: 11985-11989

Clegg D and Koshland D (1984) The role of a signaling protein in bacterial sensing: Behavioral effects of increased gene expression. Proc Natl Acad Sci USA 81: 5056-5060

Cobley JG and Miranda RD (1983) Mutations affecting chromatic adaptation in the cyanobacterium *Fremyella diplosiphon*. J Bacteriol 153: 1486-1492

Codd GA (1988) Carboxysomes and ribulose bisphosphate carboxylase/oxygenase. In: Ross AH & Tempest DW (eds) Advances in microbial physiology. Vol. 29 pp. 115-164 Academic Press, London

Cohen-Kupiec R, Gurevitz M and Zilberstein A (1993) Expression of *glnA* in the cyanobacterium *Synechococcus* sp. strain PCC 7942 is initiated from a single *nif*-like promoter under various nitrogen conditions. J Bacteriol 175: 7727-7731

Coleman JR (1991) The molecular and biochemical analyses of CO_2-concentrating mechanisms of cyanobacteria and green algae. Plant Cell & Environment 14: 861-867

Collier JL and Grossman AR (1992) Chlorosis induced by nutrient deprivation in *Synechococcus* sp. strain PCC 7942: not all bleaching is the same. J Bacteriol 174: 4718-4726

Collier JL and Grossman AR (1994) A small polypeptide triggers complete degradation of light-harvesting phycobiliproteins in nutrient-deprived cyanobacteria. EMBO J 13: 1039-1047

Collier JL, Herbert SK, Fork DC and Grossman AR (1994) Changes in the cyanobacterial photosynthetic apparatus in response to macronutrient deprivation. Photosyn Res 42: 173-183

Collier JL and Palenik BP (1996) Urea utilization by marine *Synechococcus*. J Phycol 32 Suppl: 12.

Conley PB, Lemaux PG and Grossman AR (1985) Cyanobacterial light-harvesting complex subunits encoded in two red light-induced transcripts. Science 230: 550-553

Conley PB, Lemaux PG and Grossman AR (1988) Molecular characterization and evolution of sequences encoding light harvesting components in the chromatically adapting cyanobacterium *Fremyella diplosiphon*. J Mol Biol 199: 447-465

Conley PB, Lemaux PG, Lomax TL and Grossman AR (1986) Genes encoding major light-harvesting polypeptides are clustered on the genome of the cyanobacterium *Fremyella diplosiphon*. Proc Natl Acad Sci USA 83: 3924-3928

De Las Rivas J, Andersson B and Barber J (1992) Two sites of primary degradation of the D1-protein induced by acceptor or donor side photo-inhibition in photosystem II core complexes. FEBS Lett 301: 246-252

de Lorimier R, Bryant DA, Porter RD, Liu W-Y, Jay E and Stevens SE (1984) Genes for the α and β subunits of phycocyanin. Proc Natl Acad Sci USA 81: 7946-7950

Delosme R, Olive J and Wollman F-A (1996) Changes in light energy distribution upon state transitions: an in vivo photoacoustic study of the wild type and photosynthesis mutants from *Chlamydomonas reinhardtii*. Biochim Biophys Acta 1273: 150-158

de Zamaroczy M, Paquelin A and Elmerich C (1993) Functional organization of the *glnB-glnA* cluster of *Azospirillum brasilense*. J Bacteriol 175: 2507-2515

Demmig-Adams B (1990) Carotenoids and photoprotection in plants: A role for the xanthophyll zeaxanthin. Biochim Biophys Acta 1020: 1-24

Dolganov NAM, Bhaya D and Grossman AR (1995) Cyanobacterial protein with similarity to the chlorophyll *a/b*-binding proteins of higher plants: Evolution and regulation. Proc Natl Acad Sci USA 92: 636-640

Dolganov NAM and Grossman AR (1998) A polypeptide with similarity to phycocyanin α subunit phycocyanobilin lyase involved in the degradation of the phycobilisomes. J Bacteriol, Submitted

Duke CS, Cezeaux A and Allen MM (1989) Changes in polypeptide composition of *Synechocystis* sp. strain 6803 phycobilisomes induced by nitrogen starvation. J Bacteriol 171: 1960-1966

Durner JB and Boeger P (1995) Ubiquitin in the prokaryote *Anabaena variabilis*. J Biol Chem 270: 3720-3725

Ebbole D and Zalkin H (1987) Cloning and characterization of a 12-gene cluster from *Bacillus subtilis* encoding nine enzymes for *de novo* purine nucleotide synthesis. J Biol Chem 262: 8274-8287

Elmorjani K and Herdman M (1987) Metabolic control of phycocyanin degradation in the cyanobacterium *Synechocystis* PCC 6803: A glucose effect. J Gen Microbiol 133: 1685-1694

Elmorjani K, Liotenberg S, Houmard J and Tandeau de Marsac N (1992) Molecular characterization of the gene encoding glutamine synthetase in the cyanobacterium *Calothrix* sp. PCC 7601. Biochem Biophys Res Commun 189: 1269-1302

English RS, Lorbach SD, Qin X and Shively JM (1994) Isolation and characterization of a carboxysome shell gene from *Thiobacillus neapolitanus*. Mol Microbiol 12: 647-654

Escoubas J-M, Lomas M, LaRoche J and Falkowski PG (1995) Light intensity regulation of *cab* gene transcription is signaled by the redox state of the plastoquinone pool. Proc Natl Acad Sci USA 92: 10237-10241

Espie GS and Kandasamy RA (1992) Na^+-independent HCO_3^--transport and accumulation in the cyanobacterium *Synechococcus* UTEX 625. Plant Physiol 98: 560-568

Espie GS, Miller AG and Canvin DT (1991) High affinity transport of CO_2 in the cyanobacterium *Synechococcus* UTEX 625. Plant Physiol 97: 943-953.

Fairchild CD and Glazer AN (1994) Oligomeric structure, enzyme kinetics, and substrate specificity of the phycocyanin α subunit phycocyanobilin lyase. J Biol Chem 269: 8686-8694

Fairchild CD, Zhao J, Zhou J, Colson SE, Bryant DA and Glazer AN (1992) Phycocyanin α-subunit phycocyanobilin lyase. Proc Natl Acad Sci USA 89: 7017-7021

Fay P (1992) Oxygen relations of nitrogen fixation in cyanobacteria. Microbiol Rev 56: 340-373

Federspiel NA and Grossman AR (1990) Characterization of the light-regulated operon encoding the phycoerythrin-associated linker proteins from the cyanobacterium *Fremyella diplosiphon*. J Bacteriol 172: 4072-4081

Federspiel NA and Scott L (1992) Characterization of a light-regulated gene encoding a new phycoerythrin-associated linker protein from the cyanobacterium *Fremyella diplosiphon*. J Bacteriol 179: 5994-5998

Flores E, Guerrero MG and Losada M (1980) Short-term ammonium inhibition of nitrate utilization by *Anacystis nidulans* and other cyanobacteria. Arch Microbiol 128: 137-144

Flores E, Guerrero MG and Losada M (1983) Photosynthetic nature of nitrate uptake and reduction in the cyanobacterium *Anacystis nidulans*. Biochim Biophys Acta 722: 408-416

Flores E and Herrero A (1994) Assimilatory nitrogen metabolism and its regulation. In: Bryant DA (ed) The Molecular Biology of Cyanobacteria. pp. 487-517 Kluwer, Dordrecht

Flores E and Muro-Pastor MI (1988) Uptake of glutamine and glutamate by the dinitrogen-fixing cyanobacterium *Anabaena* sp. PCC 7120. FEMS Microbiol Lett 56: 127-130

Flores E and Muro-Pastor MI (1990) Mutational and kinetic analysis of basic amino acid transport in the cyanobacterium *Synechocystis* sp. PCC 6803. Arch Microbiol 154: 521-527

Forchhammer K and Tandeau de Marsac N (1994) The P_{II} protein in the cyanobacterium *Synechococcus* sp. strain PCC 7942 is modified by serine phosphorylation and signals the cellular N-status. J Bacteriol 176: 84-91

Forchhammer K and Tandeau de Marsac N (1995a) Functional analysis of the phosphoprotein P_{II} (*glnB* gene product) in the cyanobacterium *Synechococcus* sp. strain PCC 7942. J Bacteriol 177: 2033-2040

Forchhammer K and Tandeau de Marsac N (1995b) Phosphorylation of the P_{II} protein (*glnB* gene product) in the cyanobacterium *Synechococcus* sp. strain PCC 7942: Analysis of in vitro kinase activity. J Bacteriol 177: 5812-5817

Foulds IJ and Carr NG (1977) A proteolytic enzyme degrading phycocyanin in the cyanobacterium *Anabaena cylindrica*. FEMS Microbiol Lett 2: 117-119

Frias JE, Flores E and Herrero A (1994) Requirement of the regulatory protein NtcA for the expression of nitrogen assimilation and heterocyst development genes in the cyanobacterium *Anabaena* sp. PCC 7120. Mol Microbiol 14: 823-832

Fresnedo O, Gomez R and Serra JL (1991) Carotenoid composition in the cyanobacterium *Phormidium laminosum*: Effect of nitrogen starvation. FEBS Lett 282: 300-304

Fridland L, Kaplan A and Reinhold L (1996) Quantitative evaluation of the role of a putative CO_2-scavenging entity in the cyanobacterial CO_2-concentrating mechanism. Biosystem 37: 229-238

Friedberg D, Jager KM, Kessel M, Silman NJ and Bergman B (1993) Rubisco but not Rubisco activase is clustered in the carboxysomes of the cyanobacterium *Synechococcus* sp. PCC 7942: Mud-induced carboxysomeless mutants. Mol Microbiol 9: 1193-1201

Friedberg D, Kaplan A, Ariel R, Kessel M and Seijffers J (1989) The 5' flanking region of the gene encoding the large subunit of ribulose-1,5-bisphosphate carboxylase/oxygenase is crucial for growth of the cyanobacterium *Synechococcus* PCC 7942 in air level of CO_2. J Bacteriol 171: 6069-6076

Fujita Y and Murakami A (1987) Regulation of electron transport composition in cyanobacterial photosynthetic system: Stoichiometry among photosystem I and II complexes and their light-harvesting antennae and cytochrome b_6f complex. Plant Cell Physiol 28: 1547-1553

Fujita Y, Murakami A and Aizawa K (1992) Light acclimation of thylakoid system in cyanophytes: Regulation of PSI formation in response to light regime. In: Murata N (ed) Research in Photosynthesis. Vol. IV (pp. 301-308) Kluwer, Dordrecht

Fujita Y, Murakami A, Aizawa K and Ohki K (1994) Short-term and long-term adaptation of the photosynthetic apparatus: Homeostatic properties of thylakoids. In: Bryant DA (ed) The Molecular Biology of Cyanobacteria. pp. 677-692 Kluwer, Dordrecht

Fujita Y, Murakami A and Ohki K (1987) Regulation of photosystem composition in the cyanobacterial photosynthetic system: The regulation occurs in response to the redox state of the electron pool located between the two photosystems. Plant Cell Physiol 28: 283-292

Fujita Y, Murakami A, Ohki K and Hagiwara N (1988) Regulation of photosystem composition in cyanobacterial photosynthetic system: Evidence indicating that photosystem I formation is controlled in response to the electron transport state. Plant Cell Physiol 29: 557-564

Fujita Y, Ohki K and Murakami A (1985) Chromatic regulation of photosystem composition in the photosynthetic system of red and blue-green algae. Plant Cell Physiol 26: 1541-1548

Fujita Y, Takahashi Y, Shonai F, Ogura Y and Matsubara H (1991) Cloning, nucleotide sequences and differential expression of the *nifH* and *nifH*-like (*frxC*) genes from the filamentous nitrogen-fixing cyanobacterium *Plectonema boryanum*. Plant Cell Physiol 32: 1093-1106

Fukuzawa H, Suzuki E, Komukai Y and Miyachi S (1992) A gene homologous to chloroplast carbonic anhydrase (*icfA*) is essential to photosynthetic carbon fixation in the cyanobacterium *Synechococcus* PCC 7942. Proc Natl Acad Sci USA 89: 4437-4441

Gantt E (1981) Phycobilisomes. Annu Rev Plant Physiol 32: 327-347

Gasparich GE, Buzby JS, Bryant DA, Porter RD and Stevens Jr SE (1987) The effects of light intensity and nitrogen starvation on the phycocyanin promoter in the cyanobacterium *Synechococcus* sp. PCC 7002. In: Biggins J (ed) Progress in Photosynthesis Research. Vol. IV pp. 761-764 Martinus-Nijhoff Publishers, Dordrecht

Gingrich JC, Gasparich GE, Sauer K and Bryant DA (1990) Nucleotide sequence and expression of the two genes encoding D2 protein and the single gene encoding the CP43 protein of photosytem II in the cyanobacterium *Synechococcus* sp. PCC 7002. Photosyn Res 24: 137-150

Glazer AN (1985) Light harvesting by phycobilisomes. Annu Rev Biophys Biophys Chem 14: 47-77

Glazer AN (1987) Phycobilisomes: assembly and attachment. In: Fay P & Van Baalen C (eds) The Cyanobacteria. pp. 69-92 Elsevier Biomedical, Amsterdam

Glazer AN, Lundell DJ, Yamanaka G and Williams RC (1983) The structure of a "simple" phycobilisome. Annals Institut Pasteur/Microbiology 134B: 159-180

Golbeck JH (1994) Photosystem I in cyanobacteria. In: Bryant DA (ed) The Molecular Biology of Cyanobacteria. pp. 319-360 Kluwer, Dordrecht

Golden JW, Robinson SJ and Haselkorn R (1985) Rearrangement of nitrogen fixing genes during heterocyst differentiation in *Anabaena*. Nature 314: 419-423

Golden SS (1994) Light-responsive gene expression and the biochemistry of the photosystem II reaction center. In: Bryant DA (ed) The Molecular Biology of Cyanobacteria. pp. 693-714 Kluwer, Dordrecht

Golden SS, Brusslan J and Haselkorn R (1986) Expression of a family of *psbA* genes encoding a photosystem II polypeptide in the cyanobacterium *Anacystis nidulans* R2. EMBO J 5: 2789-2798

Golden SS, Cho DC and Nalty MS (1989) Two functional *psbD* genes in the cyanobacterium *Synechococcus* sp. strain PCC 7942. J Bacteriol 171: 4707-4713

Golden SS and Kulkarni RD (1996) Stability determinants of *Synechococcus psbA* transcripts. In US-Japan Workshop on Advances in the Molecular Biology of Photosynthesis. 1996. Grand Canyon National Park, USA

Golden SS and Stearns GW (1988) Nucleotide sequence and transcript analysis of three photosystem II genes from the cyanobacterium *Synechococcus* sp. PCC 7942. Gene 67: 85-96

Goldstein AH (1994) Involvement of the quinoprotein glucose dehydrogenase in the solubilization of exogenous mineral phosphates by gram-negative bacteria. In: Torriani-Gorini A, Yagil E & Silver S (eds) Phosphate in Microorganisms, pp 197-203. American Society for Microbiology, Washington, DC

Gombos Z and Vigh L (1986) Primary role of the cytoplasmic membrane in thermal acclimation evidenced in nitrate-starved cells of the blue-green alga, *Anacystis nidulans*. Plant Physiol 80: 415-419

Green LS and Grossman AR (1988) Changes in sulfate transport characteristics and protein composition of *Anacystis nidulans* R2 during sulfur deprivation. J Bacteriol 170: 583-587

Green LS, Laudenbach DE and Grossman AR (1989) A region of a cyanobacterial genome required for sulfate transport. Proc Natl Acad Sci USA 86: 1949-1953

Grillo JF and Gibson J (1979) Regulation of phosphate accumulation in the cyanobacterium *Synechococcus*. J Bacteriol 140: 508-517

Grossman AR (1990) Chromatic adaptation and the events involved in phycobilisome biosynthesis. Plant Cell Environ 13: 651-666

Grossman AR, Bhaya D, Apt KE and Kehoe DM (1995) Light-harvesting complexes in oxygenic photosynthesis: Diversity, control and evolution. Annu Rev Genet 29: 231-287

Grossman AR, Bhaya D and Collier JL (1994a) Specific and general responses of cyanobacteria to macronutrient deprivation. In: Torriani-Gorini A, Yagil E & Silver S (eds) Phosphate in Microorganisms, pp 112-118. ASM Press, Washington, DC

Grossman AR, Collier JL and Laudenbach DE (1992) The responses of *Synechococcus* sp. strain PCC 7942 to sulfur-limited growth. In: Murata N (ed) Proceedings of the IXth International Congress on Photosynthesis. Vol. IV, pp 19.3-19.10. Kluwer, Dordrecht

Grossman AR, Kehoe DM (1997) Phosphorelay control of phycobilisome biogenesis during complementary chromatic adaptation. Photosyn Res 53: 95-108

Grossman AR, Schaefer M, Chiang G and Collier J (1994b) The responses of cyanobacteria to environmental conditions: Light and nutrients. In: Bryant D (ed) The Molecular Biology of Cyanobacteria, pp 641-675 Kluwer, Dordrecht

Grossman AR, Schaefer MR, Chiang GG and Collier JL (1993) The phycobilisome: A light harvesting complex responsive to environmental conditions. Microbiol Rev 57: 725-749

Guenther JE and Melis A (1990) The physiological significance of photosystem heterogeneity in chloroplasts. Photosyn Res 23: 195-203.

Harano Y, Suzuki I, Maeda S, Kaneko T, Tabata S, Omata T (1997) Identification and nitrogen regulation of the cyanase gene from the cyanobacteria *Synechocystis* sp.strain PCC 6803 and *Synechococcus* sp. strain PCC 7942. J. Bacteriol 179: 5744-5750

Haselkorn R (1992) Developmentally regulated gene rearrangements in prokaryotes. Annu Rev Genet 26: 111-128

Haselkorn R and Buikema WJ (1992) Nitrogen fixation in cyanobacteria. In: Stacey G, Burris RH & Evans HJ (eds) Biological Nitrogen Fixation. pp. 166-190 Chapman & Hall, New York

Haury JF and Bogorad L (1977) Action spectra for phycobiliprotein synthesis in a chromatically adapting cyanophyte, *Fremyella diplosiphon*. Plant Physiol 60: 835-839

Healey FP (1982) Phosphate. In: Carr NG & Whitton BA (eds) The Biology of Cyanobacteria. Botanical Monographs. Vol. 19, pp.105-124. University of California Press, Berkeley

Heath RT and Cooke GD (1975) The significance of alkaline phosphatase in a eutropic lake. Verh Int Verein Theor Angew Limnol 19: 959-965.

Hecky RE and Kilham P (1988) Nutrient limitation of phytoplankton in freshwater and marine environments: A review of recent evidence on the effects of enrichment. Limnol Oceanogr 33: 796-822

Herrero A and Flores E (1990) Transport of basic amino acids by the dinitrogen-fixing cyanobacterium *Anabaena* PCC 7120. J Biol Chem 265: 3931-3935

Herrero A and Guerrero MG (1986) Regulation of nitrate reductase in the cyanobacterium *Anacystis nidulans*. J Gen Microbiol 132: 2463-2468

Hershko A (1988) Ubiquitin-mediated protein degradation. J Biol Chem 263: 15237-15240

Ho KK and Krogman DW (1982) Photosynthesis. In: Carr NG & Whitton B, A. (eds) The Biology of Cyanobacteria, pp 191-215. University of California Press, Berkeley

Holthijzen YA, van Breemen JFL, Kuenen JG and Konings L (1986) Protein composition of the carboxysomes of *Thiobacillus neopolitanus*. Arch Microbiol 144: 398-404

Howard JB, and Rees DC (1994) Nitrogenase: A nucleotide-dependent molecular switch. Annu Rev Biochem 63: 235-264

Howell LD, Griffiths C, Slade LW, Potts M and Kennelly PJ (1996) Substrate specificity of IphP, a cyanobacterial dual-specificity protein phosphatase with MAP kinase phosphatase activity. Biochem 35: 7566-7572

Hryniewicz M, Sirko A, Palucha A, Böck A and Hulanicka D (1990) Sulfate and thiosulfate transport in *Escherichia coli* K-12: identification of a gene encoding a novel protein involved in thiosulfate binding. J Bacteriol 172: 3358-3366

Hua J, Chang C, Sun Q and Meyerowitz EM (1995) Ethylene insensitivity conferred by *Arabidopsis ETS* gene. Science 269: 1712-1714

Huang RC and Chow TJ (1990) Characterization of the rhythmic nitrogen-fixing activity of *Synechococcus* sp. RF-1 at the transcription level. Curr Microbiol 20: 23-26

Huang TC and Chow TJ (1988) Comparative studies of some nitrogen-fixing unicellular cyanobacteria isolated from rice fields. J Gen Microbiol 134: 3089-3097

Ishige K, Nagasawa S, Tokishita S-I and Mizuno T (1994) A novel device of bacterial signal transducers. EMBO J 13: 5195-5202

Iuchi S and Lin ECC (1993) Adaptation of *Escherichia coli* to redox environments by gene expression. Mol Microbiol 9: 9-15

Jager KM and Bergman B (1991) Localization of a multifunctional chaperonin (GroEL protein) in nitrogen-fixing *Anabaena* PCC 7120: presence in vegetative cells and heterocysts. Planta 183: 120-125

Jeanjean R and Broda E (1977) Dependence of sulfate uptake by *Anacystis nidulans* on energy, osmotic shock, and on sulfate starvation. Arch Microbiol 114: 19-23

Jerlov NG (1976) Marine Optics. Elsevier, Amsterdam

Johnson AD (1995) The price of repression. Cell 81: 655-658

Johnson TR, Haynes II JI, Wealand JL, Yarbrough LR and Hirschberg R (1988) Structure and regulation of genes encoding phycocyanin and allophycocyanin from *Anabaena variabilis* ATCC 29413. J Bacteriol 170: 1858-1865

Jung LJ, Chan CF and Glazer AN (1995) Candidate genes for the phycoerythrocyanin α subunit lyase. J Biol Chem 270: 12877-12884

Kahn K, Mazel D, Houmard J, Tandeau de Marsac N, Schaefer M (1997) A role for CpeYZ in cyanobacterial phycoerythrin biosynthesis. J Bacteriol 179: 998-1006

Kaneko T, Sato S, Kotani H, Tanaka A, Asamizu E, Nakamura T, Miyajima N, Hirosawa M, Sugiura M, Sasamoto S, Kimura T, Hosouchi T, Matsuno A, Muraki A, Nakazaki N, Naruo K, Okumura S, Shimpo S, Takeuchi C, Wada T, Watanabe A, Yamada M, Yasuda M and Tabata S (1996) Sequence analysis of the genome of the unicellular cyanobacterium *Synechocystis* sp. strain PCC 6803. II. Sequence determination of the entire genome and assignment of potential protein-coding regions. DNA Research 3: 109-136

Kaplan A, Schwarz R, Ariel R and Reinhold L (1990) The "CO_2 concentrating mechanism" of cyanobacteria: Physiological, molecular and theoretical studies. Botan Mag Tokyo 2: 53-72

Kaplan A, Schwarz R, Lieman-Hurwitz J and Reinhold L (1991) Physiological and molecular aspects of the inorganic carbon-concentrating mechanism in cyanobacteria. Plant Physiol 97: 851-855

Kaplan A, Schwarz R, Lieman-Hurwitz J, Ronen-Tarazi M and Reinhold L (1994) Physiological and Molecular Studies on the Response of Cyanobacteria to Changes in the Ambient Inorganic Carbon Concentration. In: Bryant DA (ed) The Molecular Biology of Cyanobacteria, pp 469-485. Kluwer, Dordrecht

Kaplan A, Zenvirth D, Marcus Y, Omata T and Ogawa T (1987) Energization and activation of inorganic carbon uptake by light in cyanobacteria. Plant Physiol 84: 210-213

Karagouni AD, Bloyce SA and Carr NG (1990) The presence and absence of inorganic carbon concentrating systems in unicellular cyanobacteria. FEMS Microbiol Lett 68: 137-142

Katoh A, Lee K-S, Fukuzawa H, Ohyama K and Ogawa T (1996a) *cemA* homologue essential to CO_2 transport in the cyanobacterium *Synechocystis* PCC 6803. Proc Natl Acad Sci USA 93: 4006-4010

Katoh A, Sonoda M, Katoh H and Ogawa T (1996b) Absence of light-induced proton extrusion in a *cotA*-less mutant of *Synechocystis* sp. strain PCC 6803. J Bacteriol 178: 5452-5455

Kawamura M, Mimuro M and Fujita Y (1979) Quantitative relationship between two reaction centers in the photosynthetic system of blue-green algae. Plant Cell Physiol 20: 697-705

Kehoe DM and Grossman AR (1994) Complementary chromatic adaptation: Photoperception to gene regulation. Seminars in Cell Biology 5: 303-313

Kehoe DM and Grossman AR (1995) The use of site directed mutagenesis in the analysis of complementary chromatic adaptation. P. Mathis (ed) in Proceedings from the Xth International Photosynthesis Congress: Photosynthesis: from light to biosphere. Vol. III, pp 501-504. Kluwer Academic Publishers, Dordecht

Kehoe DM and Grossman AR (1996) Similarity of a chromatic adaptation sensor to phytochrome and ethylene receptors. Science 273: 1409-1412

Kehoe DM and Grossman AR (1997) Analysis of mutants to delineate the unique four-step phosphorelay pathway controlling light harvesting structure. J Bacteriol 179: 3914-3921

Kendrick RE and Kronenberg GHM (1994) Photomorphogenesis in Plants. Kluwer, Dordrecht

Kennelly PJ and Potts M (1996) Fancy meeting you here! a fresh look at "Prokaryotic" protein phosphorylation. J Bacteriol 178: 4759-4764

Kentemich R, Danneberg G, Hundeshagen B and Bothe H (1988) Evidence for the occurrence of the alternative, vanadium-containing nitrogenase in the cyanobacterium *Anabaena variabilis*. FEMS Microbiol Lett 51: 19-24

Kentemich T, Haverkamp G and Bothe H (1991) The expression of a third nitrogenase in the cyanobacterium *Anabaena variabilis*. Z Naturoforsch 46c: 217-222

Keyse SM (1995) An emerging family of dual specificity MAP kinase phosphatases. Biochim Biophys Acta 1265: 152-160

Kikuchi H, Aichi M, Suzuki I and Omata T (1996) Positive regulation by nitrite of the nitrate assimilation operon in the cyanobacteria *Synechococcus* sp. strain PCC 7942 and *Plectonema boryanum*. J Bacteriol 178: 5822-5825

Kim JH, Glick RE and Melis A (1993) Dynamics of photosystem stoichiometry adjustment by light quality in chloroplasts. Plant Physiol 102: 181-190

Kobayashi M, Rodriguez R, Lara C and Omata T (1997) Involvement of the C-terminal domain of an ATP-binding subunit in the regulation of the ABC-type nitrate/nitrite transporter of the cyanobacterium *Synechococcus* sp. strain PCC 7942. J Biol Chem 272: 27197-27201

Kolb A, Busby S, Buc H, Garges S and Adhya S (1993) Transcriptional regulation by cyclic AMP and its receptor protein. Annu Rev Biochem 62: 749-79

Kornberg A (1994) Inorganic polyphosphate: A molecular fossil come to life. In: Torriani-Gorini A, Yagil E & Silver S (eds) Phosphate in Microorganisms, pp 204-209. ASM Press, Washington, D C

Krupa Z, öquist G and Gustafsson P (1991) Photoinhibition of photosynthesis and growth responses at different light levels in *psbA* gene mutants of the cyanobacterium *Synechococcus*. Physiol Plant 82: 1-8

Kulkarni RD and Golden SS (1994) Adaptation to high light intensity in *Synechococcus* sp. strain PCC 7942: Regulation of three *psbA* genes and two forms of the D1 protein. J Bacteriol 176: 959-965

Kulkarni RD and Golden SS (1995) Form II of D1 is important during transition from standard to high light intensity in *Synechococcus* sp. strain PCC 7942. Photosyn Res 46: 435-443

Kulkarni RD, Schaefer MR and Golden SS (1992) Transcriptional and posttranscriptional components of *psbA* response to high light intensity in *Synechococcus* sp. strain PCC 7942. J Bacteriol 174: 3775-3781

Labarre J, Thuriaux P and Chauvat F (1987) Genetic analysis of amino acid transport in the facultatively heterotrophic cyanobacterium *Synechocystis* sp. strain 6803. J Bacteriol 169: 4668-4673

Lammers PJ, Golden JW and Haselkorn R (1986) Identification and sequence of a gene required for a developmentally regulated DNA excision in *Anabaena*. Cell 44: 905-911

Lara C, Romero JM and Guerrero MG (1987) Regulated nitrate transport in the cyanobacterium *Anacystis nidulans*. J Bacteriol 169: 4376-4378

Lau RJ, MacKenzie MM and Doolittle WF (1977) Phycocyanin synthesis and degradation in the blue-green bacterium *Anacystis nidulans*. J Bacteriol 132: 771-778

Laudenbach DE, Ehrhardt D, Green L and Grossman AR (1991) Isolation and characterization of a sulfur-regulated gene encoding a periplasmically localized protein with sequence similarity to rhodanese. J Bacteriol 173: 2751-2760

Laudenbach DE and Grossman AR (1991) Characterization and mutagenesis of sulfur-regulated genes in a cyanobacterium: Evidence for function in sulfate transport. J Bacteriol 173: 2739-2750

Lawry NH and Jensen TE (1986) Condensed phosphate deposition, sulfur amino acid use, and unidirectional transulfuration in *Synechococcus leopoliensis*. Arch Microbiol 144: 317-323

Leong TY and Andersonn JM (1986) Light-quality and irradiance adaptation of the composition and function of pea thylakoid membranes. Biochim Biophys Acta 850: 57-63

Li R, Dickerson NS, Mueller UW and Golden SS (1995) Specific binding of *Synechococcus* sp. strain PCC 7942 proteins to the enhancer element of *psbAII* required for high-light-induced expression. J Bacteriol 177: 508-516

Li R and Golden SS (1993) Enhancer activity of light-responsive regulatory elements in the untranslated leader regions of cyanobacterial *psbA* genes. Proc Natl Acad Sci USA 90: 11678-11682

Lockau W, Massalsky B and Dirmeier A (1988) Purification and partial characterization of a calcium-stimulated protease from the cyanobacterium *Anabaena variabilis*. Eur J Biochem 172: 433-438

Lomax TL, Conley PB, Schilling J and Grossman AR (1987) Isolation and characterization of light-regulated phycobilisome linker polypeptide genes and their transcription as a polycistronic mRNA. J Bacteriol 169: 2675-2684

Luque I, Flores E and Herrero A (1993) Nitrate reductase gene from *Synechococcus* sp. PCC 7942: Homology between cyanobacterial and higher-plant nitrate reductase. Plant Mol Biol 21: 1201-1205

Luque I, Flores E and Herrero A (1994a) Molecular mechanism for the operation of nitrogen control in cyanobacteria. EMBO J 13: 2862-2869

Luque I, Flores E and Herrero A (1994b) Nitrate and nitrite transport in the cyanobacterium *Synechococcus* sp. PCC 7942 are mediated by the same permease. Biochim Biophys Acta 1184: 296-298

Luque I, Herrera A, Flores E and Madueño F (1992) Clustering of genes involved in nitrate assimilation in the cyanobacterium *Synechococcus*. Mol Gen Genet 232: 7-11

Mackerras AH, de Chazal NM and Smith GD (1990) Transient accumulations of cyanophycin in *Anabaena cylindrica* and *Synechocystis* 6308. J Gen Microbiol 136: 2057-2065

Madueño F, Borrias WE, van Arkel GA and Guerrero MG (1988) Isolation and characterization of *Anacystis nidulans* R2 mutants affected in nitrate assimilation: Establishment of two new mutant types. Mol Gen Genet 213: 223-228

Madueño F, Flores E and Guerrero MG (1987) Competition between nitrate and nitrite uptake in the cyanobacterium *Anacystis nidulans*. Biochim Biophys Acta 896: 109-112

Maeda S-I and Omata T (1997) Substrate-binding lipoprotein of the cyanobacterium *Synechococcus* sp. strain PCC 7942 involved in the transport of nitrate and nitrite. J Biol Chem 272: 3036-3041

Makino K, Amemura M, Kim S-K, Nakata A and Shinagawa H (1994) Mechanism of transcriptional activation of the phosphate regulon in *Escherichia coli*. In: Torriani Gorini A, Yagil E & Silver S (eds) Phosphate in Microorganisms. Cellular and Molecular Biology, pp 5-12. ASM Press, Washington, DC

Maldener I, Lockau W, Cai Y and Wolk CP (1991) Calcium-dependent protease of the cyanobacterium *Anabaena*: Molecular cloning and expression of the gene in *Escherichia coli*, sequencing and site-directed mutagenesis. Mol Gen Genet 225: 113-120

Mann NH (1994) Protein phosphorylation in cyanobacteria. Microbiology 140: 3207-3215

Mann NH and Scanlan DJ (1994) The SphX protein of *Synechococcus* species PCC 7942 belongs to a family of phosphate binding proteins. Mol Microbiol 14:595-596

Manzano C, Candau P, Gomez-Moreno C, Relimpio AM and Losada M (1976) Ferredoxin-dependent photosynthetic reduction of nitrate and nitrite by particles of *Anacystis nidulans*. Mol Cell Biol 10: 161-169

Marco E, Ohad N, Schwarz R, Lieman-Hurwitz J, Gabay C and Kaplan A (1993) High CO_2 concentration alleviates the block in photosynthetic electron transport in an *ndhB*-inactivated mutant of *Synechococcus* sp. PCC 7942. Plant Physiol 101: 1047-1053

Marek LF and Spalding MH (1991) Changes in photorespiratory enzyme activity in response to limiting CO_2 in *Chlamydomonas reinhardtii*. Plant Physiol 97: 420-425

Marquès S, Florencio FJ and Candau P (1992) Purification and characterization of the ferredoxin-glutamine synthetase activity in the unicellular cyanobacterium *Synechococcus* sp. PCC 6301. Eur J Biochem 206: 69-77

Martin-Nieto J, Herrero A and Flores E (1991) Control of nitrogenase mRNA levels by products of nitrate assimilation in the cyanobacterium *Anabaena* sp. strain PCC 7120. Plant Physiol 97: 825-828

Mayo MP, Elrifi IR and Turpin DH (1989) The relationship between ribulose bisphosphate concentration, dissolved inorganic carbon (DIC) transport and DIC-limited photosynthesis in the cyanobacterium *Synechococcus leopoliensis* grown at different concentrations of inorganic carbon. Plant Physiol 90: 720-727

Mazel D, Guglielmi G, Houmard H, Sidler W, Bryant DA and Tandeau de Marsac N (1986) Green light induces transcription of the phycoerythrin operon in the cyanobacterium *Calothrix* 7601. Nucleic Acids Res 14: 8279-8290

Mazel D, Houmard J and Tandeau de Marsac N (1988) A multigene family in *Calothrix* sp. PCC 7601 encodes phycocyanin, the major component of the cyanobacterial light-harvesting antenna. Mol Gen Genet 211: 296-304

Mazel D and Marliere P (1989) Adaptive eradication of methionine and cysteine from cyanobacterial light-harvesting proteins. Nature 341: 245-248

Mazur BJ, Rice D and Haselkorn R (1980) Identification of blue-green algal nitrogen fixation genes by using heterologous DNA hybridization probes. Proc Natl Acad Sci USA 77: 186-190

McKay RML, Gibbls SP and Espie GS (1993) Effect of dissolved inorganic carbon on the expression of carboxysomes, localization of rubisco and the mode of inorganic carbon transport in cells of the cyanobacterium *Synechococcus* UTEX 625. Arch Microbiol 159: 21-29

Meeks JC, Wolk CP, Lockau W, Schilling N, Shaffer PW and Chien WS (1978) Pathways of assimilation of $[^{13}N]N_2$ and $^{13}NH_4^+$ by cyanobacteria with and without heterocysts. J Bacteriol 134: 125-130

Meeks JC, Wykoff KL, Chapman JS and Enderlin CS (1983) Regulation of expression of nitrate and dinitrogen assimilation by *Anabaena* species. Appl Environ Microbiol 45: 1351-1359

Mells A, Manodori A, Glick RE, Ghirardi ML, McCauley SW and Neale PJ (1985) The mechanism of photosynthetic membrane adaptation to environmental stress conditions: A hypothesis on the role of electron-transport capacity and of ATP/NADPH pool in the regulation of thylakoid membrane organization and function. Physiol Veg 23: 757-765

Melis A, Manodori A, Glick RE, Ghirardi ML, McCauley SW and Neale PJ (1989) Acclimation of the photosynthetic apparatus to photosystem I or photosystem II light: Evidence from quantum yield measurements and fluorescence spectroscopy of cyanobacterial cells. Z Naturforsch 44c: 109-118

Melis A, Murakami A, Nemson JA, Aizawa K and Ohki K (1996) Chromatic regulation in *Chlamydomonas reinhardtii* alters photosystem stoichiometry and improves the quantum efficiency of photosynthesis. Photosyn Res 49: 253-265

Mèrida A, Flores E and Florencio FJ (1991) Regulation of glutamine synthetase activity in the unicellular cyanobacterium *Synechocystis* sp. strain PCC 6803 by the nitrogen source: Effect of ammonium. J Bacteriol 173: 4095-4100

Mèrida A, Flores E and Florencio FJ (1992) Regulation of *Anabaena* sp. strain PCC 7120 glutamine synthetase activity in a *Synechocystis* sp. strain PCC 6803 derivative strain bearing the *Anabaena glnA* gene and a mutated host *glnA* gene. J Bacteriol 174: 650-654

Mèrida A, Leurentop L, Candau P and Florencio FJ (1990) Purification and properties of glutamine synthetases from the cyanobacteria *Synechocystis* sp. strain PCC 6803 and *Calothrix* sp. strain PCC 7601. J Bacteriol 172: 4732-4735

Merrick MJ (1992) Regulation of nitrogen fixation genes in free-living and symbiotic bacteria. In: Stacey G, Burris RH & Evans HJ (eds) Biological nitrogen fixation, pp 835-876. Chapman & Hall, New York

Miller AG and Canvin DT (1985) Distinction between HCO_3^-- and CO_2- dependent photosynthesis in the cyanobacterium *Synechococcus leopoliensis* based on the selective response of HCO_3^-- transport to Na^+. FEBS Lett 187: 29-32

Miller AG, Espie GS (1994) Photosynthetic metabolism of cyanate by the cyanobacterium *Synechococcus* UTEX 625. Arch Microbiol 162:151-157

Miller AG, Espie GS and Canvin DT (1988) Chlorophyll *a* fluorescence yield as a monitor of both active CO_2 and HCO_3^--transport by the cyanobacterium *Synechococcus* UTEX 625. Plant Physiol 86: 655-658

Miller AG, Espie GS and Canvin DT (1990) Physiological aspects of CO_2 and HCO_3^- transport by cyanobacteria: A review. Can J Bot 68: 1291-1302

Mitsui A, Kumazawa S, Takahashi A, Ikemoto H, S. C and Arai T (1986) Strategy by which nitrogen-fixing unicellular cyanobacteria grow photoautotropically. Nature 323: 720-722

Montesinos ML, Herrero A and Flores E (1995) Amino acid transport systems required for diazotropic growth in the cyanobacterium *Anabaena* sp. strain PCC 7120. J Bacteriol 177: 3150-3157

Montesinos ML, Herrero A and Flores E (1997) Amino acid transport in taxonomically diverse cyanobacteria and identification of two genes encoding elements of a neutral amino acid permease putatively involved in recapture of leaked hydrophobic amino acids. J Bacteriol 179: 853-862

Mulligan ME and Haselkorn WJ (1989) Nitrogen fixation *(nif)* genes of the cyanobacterium *Anabaena* species strain PCC 7120. The *nifB-fdxN-nifS-nifU* operon. J Biol Chem 264: 19200-19207

Mullineaux CW (1992) Excitation energy transfer from phycobilisomes to photosystem I in cyanobacteria. Biochim Biophys Acta 1100: 285-292

Mullineaux CW and Allen JF (1990) State-1-state 2 transitions in the cyanobacterium *Synechococcus* 6301 are controlled by the redox state of electron carriers between photosystem I and photosystem II. Photosyn Res 23: 297-311

Mullineaux PM, Gallon JR and Chaplin AE (1981) Acetylene reduction (nitrogen fixation) by cyanobacteria grown under alternating light-dark cycles. FEMS Microbiol Lett 10: 245-247

Murakami A and Fujita Y (1991) Regulation of photosystem stoichiometry in the photosynthetic system of the cyanophyte *Synechocystis* PCC 6714 in response to light intensity. Plant Cell Physiol 32: 223-230

Myers J, Graham J-R and Wang RT (1980) Light harvesting in *Anacystis nidulans* studied in pigment mutants. Plant Physiol 66: 1144-1149

Nagaya M, Aiba H and Mizuno T (1994) The *sphR* product, a two-component system response regulator protein, regulates phosphate assimilation in *Synechococcus* sp. strain PCC 7942 by binding to two sites upstream from the *phoA* promoter. J Bacteriol 176: 2210-2215

Navarro F, Chavez S, Candau P and Florencio FJ (1995) Existence of two ferredoxin-glutamate synthases in the cyanobacterium *Synechocystis* sp. PCC 6803: Isolation and insertional inactivation of *gltB* and *gltS* genes. Plant Mol Biol 27: 753-767

Neilson AH and Doudoroff M (1973) Ammonia assimilation in blue-green algae. Arch Microbiol 89: 15-22

Nicholson ML, Gaasenbeek M and Laudenbach DE (1995) Two enzymes together capable of cysteine biosynthesis are encoded on a cyanobacterial plasmid. Mol Gen Genet 247: 623-632

Nicholson ML and Laudenbach DE (1995) Genes encoded on a cyanobacterial plasmid are transcriptionally regulated by sulfur availability and CysR. J Bacteriol 177: 2143-2150.

Ninfa AJ, Atkinson MR, Kamberov ES, Feng J and Ninfa EG (1995) Control of nitrogen assimilation by the NR_I-NR_{II} two-component system of enteric bacteria. In: Hoch JA & Silhavy TJ (eds) Two-Component Signal Transduction, pp 67-89. ASM Press, Washington, DC

Ochoa de Alda J, Tapia MI, Franck F, Llama MJ and Serra JL (1996a) Changes in nitrogen source modify distribution of excitation energy in the cyanobacterium *Phormidium laminosum.* Physiol Plant 97: 69-78

Ochoa de Alda JAG, Tapia MI, Llama MJ and Serra JL (1996b) Changes in photosynthetic yield, amino acids, and organic acids are induced by ammonium addition to cells of *Phormidium Laminosum* (cyanophyceae). J Phycol 32: 602-608

Ogawa T (1991) A gene homologous to the subunit-2 gene of NADH dehydrogenase is essential to inorganic carbon transport of *Synechocystis* PCC 6803. Proc Natl Acad Sci USA 88: 4275-4279

Ogawa T (1992) Identification and characterization of the *ictA/ndhL* gene product essential to inorganic carbon transport of *Synechocystis* PCC 6803. Plant Physiol 99: 1604-1608

Ogawa T (1996) *cemA* homologue in cyanobacteria: its function and localization. In: US-Japan Workshop on Advances in the Molecular Biology of Photosynthesis. 1996. Grand Canyon National Park, USA

Ogawa T, Amichay D and Gurevitz M (1994a) Isolation and characterization of the *ccmM* gene required by the cyanobacterium *Synechocystis* PCC 6803 for inorganic carbon utilization. Photosyn Res 39: 183-190

Ogawa T, Marco E and Orus MI (1994b) A gene (*ccmA*) required for carboxysome formation in the cyanobacterium *Synechocystis* sp. strain PCC 6803. J Bacteriol 176: 2374-2378

Ogawa T, Miyano A and Inoue Y (1985) Photosystem-I-driven inorganic carbon transport in the cyanobacterium *Anacystis nidulans*. Biochim Biophys Acta 808: 77-84

Ohki K and Fujita Y (1988) Aerobic nitrogenase activity measured as acetylene reduction in the marine non-heterocystous cyanobacterium *Trichodesmium* sp. grown under artificial conditions. Marine Biology 98: 111-114

Ohmori M, Ohmori K and Strotmann H (1977) Inhibition of nitrate uptake by ammonia in a blue-green alga, *Anabaena cylindrica*. Arch Microbiol 114: 225-229

Omata T (1991) Cloning and characterization of the *nrtA* gene that encodes a 45 kDa protein involved in nitrate transport in the cyanobacterium *Synechococcus* PCC 7942. Plant Cell Physiol 32: 151-157

Omata T (1996) Nitrogen regulation of carbon and nitrogen assimilation genes in the cyanobacterium *Synechococcus* sp. strain PCC 7942. In: US-Japan Workshop on Advances in the Molecular Biology of Photosynthesis. 1996. Grand Canyon National Park, USA

Omata T, Andriesse X and Hirano A (1993) Identification and characterization of a gene cluster involved in nitrate transport of the cyanobacterium *Synechococcus* sp. PCC 7942. Molec Gen Genet 236: 193-202

Omata T, Carlson TJ, Ogawa T and Pierce J (1990) Sequencing and modification of the gene encoding the 42-kilodalton protein in the cytoplasmic membrane of *Synechococcus* PCC 7942. Plant Physiol 93: 305-311

Omata T and Ogawa T (1986) Biosynthesis of a 42-kD polypeptide in the cytoplasmic membrane of the cyanobacterium *Anacystis nidulans* strain R2 during adaptation to low CO_2 concentration. Plant Physiol 80: 525-530

Omata T, Ohmori M, Arai N and Ogawa T (1989) Genetically engineered mutant of the cyanobacterium *Synechococcus* PCC 7942 defective in nitrate transport. Proc Natl Acad Sci USA 86: 6612-6616

Öquist G, Chow WS and Anderson JM (1992) Photoinhibition of photosynthesis represents a mechanism for the long-term regulation of photosystem II. Planta 186: 450-460

Ownby JD, Shannahan M and Hood E (1979) Protein synthesis and degradation in *Anabaena* during nitrogen starvation. J Gen Microbiol 110: 255-261

Pakrasi HB (1995) Genetic analysis of the form and function of photosystem I and photosystem II. Annu Rev Genet 29: 755-776

Parkinson JS and Kofoid EC (1992) Communication modules in bacterial signaling proteins. Annu Rev Genet 26: 71-112

Peiffer WE, Desrosiers MG and Menick DR (1996) Cloning and expression of the unique Ca^{2+}-ATPase from *Flavobacterium odoratum*. J Biol Chem 271: 5095-5100

Perego M and Hoch JA (1996) Protein aspartate phosphatases control the output of two-component signal transduction systems. Trends Genet 12: 97-101

Peterson RB, Dolan E, Calvert HE and Ke B (1981) Energy transfer from phycobiliproteins to photosystem I in vegetative cells and heterocysts of *Anabaena variabilis*. Biochim Biophys Acta 634: 237-248

Pierce J, Carlson TJ and Williams JGK (1988) Anomalous oxygen sensitivity in a cyanobacterial mutant requiring the expression of ribulose bisphosphate carboxylase from a photosynthetic anaerobe. Proc Natl Acad Sci USA 86: 5753-5757

Porter G, Tredwell CJ, Searle GFW and Barber J (1978) Picosecond time-resolved energy transfer in *Porphyridium cruentum*. Part I. In the intact alga. Biochim Biophys Acta 501: 232-245

Posas R, Wurgler-Murphy SM, Maeda T, Witten EA, Thai TC and Saito H (1996) Yeast HOG1 MAP kinase cascade is regulated by a multistep phosphorelay mechanism in the SLN1-YPD1-SSK1 "two-component" osmosensor. Cell 86: 865-875

Potts M, Sun H, Mockaitis K, Kennelly PJ, Reed D and Tonks NK (1993) A protein-tyrosine/serine phosphatase encoded by the genome of the cyanobacterium *Nostoc commune* UTEX 584. J Biol Chem 268: 7632-7635

Price GD and Badger M (1989a) Expression of human carbonic anhydrase in the cyanobacterium *Synechococcus* PCC 7942. Plant Physiol 91: 505-513.

Price GD and Badger MR (1989b) Expression of human carbonic anhydrase in the cyanobacterium *Synechococcus* PCC 7942 creates a high CO_2 requiring phenotype. Plant Physiol 91: 505-513

Price GD and Badger MR (1991) Evidence for the role of carboxysomes on cyanobacterial CO_2-concentrating mechanism. Can J Bot 69: 963-973

Price GD, Coleman JR and Badger MR (1992) Association of carbonic anhydrase activity with carboxysomes from the cyanobacterium *Synechococcus* PCC 7942. Plant Physiol 100: 784-793

Rai AN, Borthakur M and Bergman B (1992) Nitrogenase depression, its regulation and metabolic changes associated with diazotrophy in the non-heterocystous cyanobacterium *Plectonema boryanum* PCC 73110. J Gen Microbiol 138: 481-491

Ramasubramanian TS, Wei T-F and Golden JW (1994) Two *Anabaena* sp. strain PCC 7120 DNA-binding factors interact with vegetative cell- and heterocyst-specific genes. J Bacteriol 176: 1214-1223

Ramasubramanian TS, Wei T-F, Oldham AK and Golden JW (1996) Transcription of the *Anabaena* sp. strain 7120 *ntcA* gene: Multiple transcripts and NtcA binding. J Bacteriol 178: 922-926

Raven JA (1991) Implications of inorganic C utilization: Ecology, evolution and geochemistry. Can J Bot 69: 908-924

Ravid S, Matsumura P and Eisenbach M (1986) Restoration of flagellar clockwise rotation in bacterial envelopes by insertion of the chemotaxis protein CheY. Proc Natl Acad Sci USA 83: 7157-7161

Ray JM, Bhaya D, Block MA and Grossman AR (1991) Isolation, transcription, and inactivation of the gene for an atypical alkaline phosphatase of *Synechococcus* sp. strain PCC 7942. J Bacteriol 173: 4297-4309

Reddy KJ, Haskell JB, Sherman DM and Sherman LM (1993) Unicellular, aerobic nitrogen-fixing cyanobacteria of the genus of *Cyanothece*. J Bacteriol 175: 1284-1292

Reddy KJ, Masamoto K, Sherman DM and Sherman LA (1989) DNA sequence and regulation of the gene (*cbpA*) encoding the 42-kilodalton cytoplasmic membrane carotenoprotein of the cyanobacterium *Synechococcus* sp. strain PCC 7942. J Bacteriol 171: 3486-3493

Reinhold L, Kosloff R and Kaplan A (1991) A model for inorganic carbon fluxes and photosynthesis in cyanobacterial carboxysomes. Can J Bot 69: 984-988

Reinhold L, Volokita M, Zenvirth D and Kaplan A (1984) Is HCO_3^- transport in *Anabaena* a Na^+-symport? Plant Physiol 76: 190-192

Reinhold L, Zviman M and Kaplan A (1989) Quantitative model for inorganic carbon fluxes and photosynthesis in cyanobacteria. Plant Physiol Biochem 27: 945-954

Reithman H, Bullerjahn G, Reddy KJ and Sherman LA (1988) Regulation of cyanobacterial pigment-protein composition and organization by environmental factors. Photosyn Res 18: 133-161

Reuter W and Mueller C (1993) Adaptation of the photosynthetic apparatus of cyanobacteria to light and carbon dioxide. J Photochem Photobiol 21: 3-27

Reyes JC and Florencio FJ (1994) A new type of glutamine synthetase in cyanobacteria: The protein encoded by the *glnN* gene supports nitrogen assimilation in *Synechocystis* sp. strain PCC 6803. J Bacteriol 176: 1260-1267

Reyes JC and Florencio FJ (1995a) Electron transport controls transcription of the glutamine synthetase gene (*glnA*) from the cyanobacterium *Synechocystis* sp. PCC 6803. Plant Mol Biol 27: 789-799

Reyes JC and Florencio FJ (1995b) A novel mechanism of glutamine synthetase inactivation by ammonium in the cyanobacterium *Synechocystis* sp. PCC 6803. Involvement of an inactivating protein. FEBS Lett 367: 45-48

Rodriguez R, Garcia-Gonzalez M, Guerrero MG and Lara C (1994) Ammonium-sensitive protein kinase activity in plasma membranes of the cyanobacterium *Anacystis nidulans*. FEBS Lett 350: 19-23

Rodriguez R, Lara C and Guerrero MG (1992) Nitrate transport in the cyanobacterium *Anacystis nidulans* R2. Kinetic and energetic aspects. Biochem J 282: 639-643

Romans K, Carpenter EJ and Bergman B (1994) Bouyancy regulation in the colonial diazotrophic cyanobacterium *Trichodesmium tenue*: Ultrastructure and storage of carbohydrate, polyphosphate, and nitrogen. J Phycol 30: 935-942

Ronen-Tarazi M, Lieman-Hurwitz J, Gabay C, Orus M and Kaplan A (1995a) The genomic region of *rbcLS* of *Synechococcus* sp. strain PCC 7942 contains genes involved in the ability to grow under low CO_2 concentrations and in chlorophyll biosynthesis. Plant Physiol 108: 1461-1469

Ronen-Tarazi M, Schwarz R, Bouevitch A, Lieman-Hurwitz J, Erez J and Kaplan A (1995b) Response of photosynthetic microorganisms to changing response of ambient concentration of CO_2. In: Joint I (ed) Molecular Ecology of Aquatic Microbes. NATO ASI Series, Vol G38: 323-334

Rubio LM, Herrero A and Flores E (1996) A cyanobacterial *narB* gene encodes a ferredoxin-dependent nitrate reductase. Plant Mol Biol 30: 845-850

Sakamoto T, Delgaizo VB and Bryant DA (1998) Growth on urea can trigger death and peroxidation of the cyanobacterium *Synechococcus* sp. strain PCC 7002. Appl Environ Microbiol 64: 2361-2366

Salehian O and Bruce D (1992) Distribution of excitation energy in photosynthesis: Quantification of fluorescence yields from intact cyanobacteria. Journal Lumin 51: 91-98

Samuelsson G, Lonneborg A, Gustafsson P and Öquist G (1987) The susceptibility of photosynthesis to photoinhibition and the capacity of recovery in high and low light grown cyanobacteria *Anacystis nidulans*. Plant Physiol 83: 438-441

Sanders CE, Melis A and Allen JF (1989) In vivo phosphorylation of proteins in the cyanobacterium *Synechococcus* 6301 after chromatic acclimation to photosystem I or photosystem II light. Biochim Biophys Acta 976: 168-172

Satoh K and Fork DC (1983) A new mechanism for adaptation to changes in light intensity and quality in the red alga, *Porphyra perforata*. I. Relation to state1-state 2 transitions. Biochim Biophys Acta 722: 190-196

Scanlan DJ, Mann NH and Carr NG (1993) The response of the picoplanktonic marine cyanobacterium *Synechococcus sp.* WH7803 to phosphate starvation involves a protein homologous to the periplasmic phosphate-binding protein of *Escherichia coli*. Mol Microbiol 10: 181-191

Schaefer MR and Golden SS (1989) Differential expression of members of a cyanobacterial *psb*A gene family in response to light. J Bacteriol 171: 3973-3981

Schluchter W, Zhao J and Bryant DA (1993) Isolation and characterization of the *ndhF* gene of *Synechococcus* sp. strain PCC 7002 and initial characterization of an interposon mutant. J Bacteriol 175: 3343-3352

Schmidt A and Christen U (1978) A factor-dependent sulfotransferase specific for 3'-phosphoadenosine-5'-phosphosulfate (PAPS) in the cyanobacterium *Synechococcus* 6301. Planta 140: 239-244

Schmidt A, Erdle I and Kost H-P (1982) Changes of c-phycocyanin in *Synechococcus* 6301 in relation to growth on various sulfur compounds. Z Naturforsch 37c: 870-876

Schmidt-Goff CM and Federspiel NA (1993) In vivo and in vitro footprinting of a light-regulated promoter in the cyanobacterium *Fremyella diplosiphon*. J Bacteriol 175: 1806-1813

Schwarz R, Friedberg D, Reinhold L and Kaplan A (1988) Is there a role for the 42kDa polypeptide in inorganic carbon uptake by cyanobacteria. Plant Physiol 88: 284-288

Schwarz R and Grossman AR (1998) A response regulator of cyanobacteria integrates diverse environmental signals and is critical for survival under extreme conditions. Proc Natl Acad Sci USA, In press

Schwarz R, Lieman-Hurwitz J, Hassidim M and Kaplan A (1992) Phenotypic complementation of high-CO_2-requiring mutants of the cyanobacterium *Synechococcus* sp. strain PCC 7942 by inosine 5'-monophosphate. Plant Physiol 100: 1987-1993

Schwarz R, Reinhold L and Kaplan A (1995) Low activation state of ribulose 1,5-bisphosphate carboxylase/oxygenase in a carboxysome-defective *Synechococcus* mutant. Plant Physiol 108: 183-190

Searle GFW, Barber J, Porter G and Tredwell CJ (1978) Picosecond time-resolved energy transfer in *Porphyridium cruentum*. Part II. In the isolated light-harvesting complex (phycobilisomes). Biochim Biophys Acta 501: 246-256

Shenolikar S (1994) Protein serine/threonine phosphatases - New avenues for cell regulation. Annu Rev Cell Biol 10: 55-86

Shipton CA and Barber J (1991) Photoinduced degradation of the D1 polypeptide in isolated reaction centers of photosystem II: Evidence for an autoproteolytic process triggered by the oxidizing side of the photosystem. Proc Natl Acad Sci USA 88: 6691-6695

Sidler WA (1994) Phycobilisome and phycobiliprotein structures. In: Bryant DA (ed) The Molecular Biology of Cyanobacteria. (pp. 139-216) Kluwer, Dordrecht

Simon RD (1987) Inclusion bodies in the cyanobacteria: cyanophycin, polyphosphate, polyhedral bodies. In: Fay P & Van Baalen C (eds) The Cyanobacteria, pp 199-225. Elsevier Science Publishers, Amsterdam.

Singh S (1990) Urea uptake in cyanobacteria *Anabaena doliolum* and *Anacystis nidulans*. Indian J Exp Biol 28: 378-379

Sobczyk A, Bely A, Tandeau de Marsac N and Houmard J (1994) A phosphorylated DNA binding protein is specific for the red-light signal during complementary chromatic adaptation in cyanobacteria. Mol Microbiol 13: 875-885

Sobczyk A, Schyns G, Tandeau de Marsac N and Houmard J (1993) Transduction of the light signal during complementary chromatic adaptation in the cyanobacterium *Calothrix* sp. PCC 7601: DNA-binding proteins and modulation by phosphorylation. EMBO J 12: 997-1004

Stewart WDP and Lex M (1970) Nitrogenase activity in the blue-green alga *Plectonema boryanum* strain 594. Arch Microbiol 73: 250-26.

Sultemeyer D, Price GD, Yu J-W and Badger MR (1995) Characterization of carbon dioxide and bicarbonate transport during steady-state photosynthesis in the marine cyanobacterium *Synechococcus* strain PCC 7002. Planta 197: 597-607

Suzuki E, Fukuzawa H, Abe T and Miyachi S (1990) Identification of the genomic region which complements a temperature-sensitive, high CO_2-requiring mutant of the cyanobacterium *Synechococcus* PCC 7942. Molec Gen Genet 226: 401-408

Suzuki I, Horie N, Sugiyama T and Omata T (1995a) Identification and characterization of two nitrogen-regulated genes of the cyanobacterium *Synechococcus* sp. strain PCC 7942 required for maximum efficiency of nitrogen assimilation. J Bacteriol 177: 290-296

Suzuki I, Kikuchi H, Nakanishi S, Fujita Y, Sugiyama T and Omata T (1995b) A novel nitrate reductase gene from the cyanobacterium *Plectonema boryanum*. J Bacteriol 177: 6137-6143

Suzuki I, Omata T and Sugiyama T (1992) Gene expression and regulation of nitrate assimilating enzymes in *Synechococcus* PCC 7942. In: Murata N (ed) Research in Photosynthesis. Proceedings of the IXth International Congress on Photosynthesis. Vol. IV, pp 19.75-19.78. Kluwer, Dordrecht

Suzuki I, Sugiyama T and Omata T (1993) Primary structure and transcriptional regulation of the gene for nitrate reductase from the cyanobacterium *Synechococcus* PCC 7942. Plant Cell Physiol 34: 1311-1320

Suzuki I, Sugiyama T and Omata T (1996) Regulation by cyanate of the genes involved in carbon and nitrogen assimilation in the cyanobacterium *Synechococcus* sp. strain PCC 7942. J Bacteriol 178: 2688-2694

Swanson RV, Zhou J, Leary JA, Williams T, de Lorimier R, Bryant DA and Glazer AN (1992) Characterization of phycocyanin produced by *cpcE* and *cpcF* mutants and

identification of an intergenic suppressor of the defect in bilin attachment. J Biol Chem 267: 16146-16154.
Tandeau de Marsac N (1977) Occurrence and nature of chromatic adaptation in cyanobacteria. J Bacteriol 130: 82-91
Tandeau de Marsac N (1983) Phycobilisomes and complementary adaptation in cyanobacteria. Bulletin de L'Institut Pasteur 81: 201-254
Tandeau de Marsac N (1994) Differentiation of hormogonia and relationships with other biological processes. In: Bryant DA (ed) The Molecular Biology of Cyanobacteria, pp 825-842. Kluwer, Dordrecht
Tandeau de Marsac N and Houmard J (1993) Adaptation of cyanobacteria to environmental stimuli: new steps towards molecular mechanisms. FEMS Microbiol Rev 104: 119-190
Tandeau de Marsac N, Mazel D, Capuano V, Damerval T and Houmard J (1990) Genetic analysis of the cyanobacterial light-harvesting antenna complex. In: Drews G & Dawes EA (eds) The Molecular Biology of Membrane-bound Complexes in Phototrophic Bacteria. pp. 143-153 Plenum Press, New York
Tang M, Jensen T and Corpe WA (1995) The occurrence of polyphosphate bodies in polyhedral bodies (carboxysome-s) in *Synechococcus leopoliensis* (Cyanophyceae). Microbios 81: 59-66
Thiel T (1990) Protein turnover and heterocyst differentiation in the cyanobacterium *Anabaena variabilis*. J Phycol 26: 50-54
Thiel T (1993) Characterization of genes for an alternative nitrogenase in the cyanobacterium *Anabaena variabilis*. J Bacteriol 175: 6276-6286
Tsang ML-S and Schiff JA (1975) Studies of sulfate utilization by algae 14. Distribution of adenosine 5'-phosphosulfate (APS) and adenosine 3'-phosphate 5'-phosphosulfate (PAPS) sulfotransferase in assimilatory sulfate reducers. Plant Sci Lett 4: 301-307
Tsinoremas NF, Castets AM, Harrison MA, Allen JF and Tandeau de Marsac N (1991) Photosynthetic electron transport controls nitrogen assimilation in cyanobacteria by means of posttranslational modification of the *glnB* gene product. Proc Natl Acad Sci USA 88: 4565-4569
Tsinoremas NF, Schaefer MR and Golden SS (1994) Blue and red light reversibly control *psbA* expression in the cyanobacterium *Synechococcus* sp. strain PCC 7942. J Biol Chem 269: 16143-16147
Tumer NE, Robinson SJ and Haselkorn R (1983) Different promoters for the *Anabaena* glutamine synthetase gene during growth using molecular or fixed nitrogen. Nature 306: 337-342
Turpin DH, Miller AG and Canvin DT (1984) Carboxysome content of *Synechococcus leopoliensis* (Cyanophyta) in response to inorganic carbon. J Phycol 20: 249-253
Tyagi VVS, Ray TB, Mayne BC and Peters GA (1981) The *Azolla-Anabaena azollae* relationship. Plant Physiol 68: 1479-1484
Uhl MA and Miller JF (1996) Integration of multiple domains in a two-component sensor protein: The *Bordetella pertussis* BvgAS phosphorelay. EMBO J 15: 1028-1036
Utkilen HC, Heldal M and Knutsen G (1976) Characterization of sulfate uptake in *Anacystis nidulans*. Physiol Plant 38: 217-220
Van Liere L and Walsby AE (1982) Interactions of cyanobacteria with light. In: Carr NG & Whitton B A (eds) The Biology of Cyanobacteria, pp 9-47. University of California Press, Berkeley
Vaulot D, LeBot N, Marie D and Fukai E (1996) Effect of phosphorus on the *Synechococcus* cell cycle in surface Mediterranean waters during summer. Appl Environ Microbiol 62: 2527-2533
Vega-Palas MA, Flores E and Herrero A (1992) NtcA, a global nitrogen regulator from the cyanobacterium *Synechococcus* that belongs to the Crp family of bacterial regulators. Mol Microbiol 6: 1853-1859
Vega-Palas MA, Madueño F, Herrero A and Flores E (1990) Identification and cloning of a regulatory gene for nitrogen assimilation in the cyanobacterium *Synechococcus* sp. strain PCC 7942. J Bacteriol 172: 643-647
Vermaas W (1993) Approaches to analyze photosystem 2 structure and function. Annu Rev of Plant Physiol Plant Molec Biol. 44: 457-481
Vogelmann TC and Scheibe J (1978) Action spectrum for chromatic adaptation in the blue-green alga *Fremyella diplosiphon*. Planta 143: 233-239
Volokita M, Zenvirth D, Kaplan A and Reinhold L (1984) Nature of the inorganic carbon species actively taken up by the cyanobacterium *Anabaena variabilis*. Plant Physiol 76: 599-602
Wagner F, Mario G, Ahorn H, Peschek GA and Falkner G (1994) Isolation and functional reconstitution of a phosphate binding protein of the cyanobacterium *Anacystis nidulans* induced during phosphate-limited growth. J Biol Chem 269: 5509-5511
Wagner K-U, Masepohl B and Pistorius EK (1995) The cyanobacterium *Synechococcus* sp. strain PCC 7942 contains a second alkaline phosphatase encoded by *phoV*. Microbiol 141: 3049-3058
Wagner SJ, Thomas SP, Kaufman RI, Nixon BT and Stevens SE (1993) The *glnA* gene of the cyanobacterium *Agmenellum quadruplicatum* PR-6 is nonessential for ammonium assimilation. J Bacteriol 175: 604-612.
Wanner BL (1993) Gene regulation by phosphate in enteric bacteria. J Cell Biochem 51: 47-54
Wanner BL (1994a) Multiple controls of the *Escherichia coli* Pho regulon by the P_i sensor PhoR, the catabolite regulatory sensor CreC, and acetyl phosphate. In: Torriani-Gorini A, Yagil E & Silver S (eds) Phosphate in Microorganisms, pp 13-21. ASM Press, Washington, D C
Wanner BL (1994b) Phosphate-regulated genes for the utilization of phosphonates in members of the family *Enterobacteriaceae*. In: Torriani-Gorini A, Yagil E & Silver S (eds) Phosphate in Microorganisms, pp 215-222. ASM Press, Washington, DC
Wanner BL (1995) Signal transduction and cross regulation in the *Escherichia coli* phosphate regulon by PhoR, CreC, and acetyl phosphate. In: Hoch JA & Silhavy TJ (eds) Two-Component Signal Transduction, pp 203-223. ASM Press, Washington, D. C.
Warner KM and Bullerjahn GS (1994) Light-dependent tyrosine phosphorylation in the cyanobacterium *Prochlorothrix hollandica*. Plant Physiol 105: 629-633
Watson GMF, Scanlan DJ and Mann NH (1996) Characterization of the genes encoding a phosphate-regulated two component sensory system in the marine cyanobacterium *Synechococcus* sp. WH7803. FEMS Microbiol Lett 142: 105-109
Wealand JL, Myers JA and Hirschberg R (1989) Changes in gene expression during nitrogen starvation in *Anabaena variabilis* ATCC 29413. J Bacteriol 171: 1309-1313
Wei T-F, Ramasubramanian TS, Pu F and Golden JW (1993) *Anabaena* sp. strain PCC 7120 *bifA* gene encoding a sequence-specific DNA binding protein cloned by in vivo transcriptional interference selection. J Bacteriol 175: 4025-4035
Wilkinson JQ, Lanahan MB, Yen H-C, Giovannoni JJ and Klee HJ (1995) An ethylene-inducible component of signal transduction encoded by *Never-ripe*. Science 270: 1807-1809

Wolfe AJ, Conley P, Kramer TJ and Berg HC (1987) Reconstitution of signaling in bacterial chemotaxis. J Bacteriol 169: 1878-1885.

Wolk CP (1996) Heterocyst formation. Annu Rev Genet 30: 59-78

Wolk CP, Ernst A and Elhai J (1994) Heterocyst metabolism and development. In: Bryant DA (ed) The Molecular Biology of Cyanobacteria, pp 769-823. Kluwer, Dordrecht

Wood NB and Haselkorn R (1979) Proteinase activity during heterocyst differentiation in nitrogen-fixing cyanobacteria. In: Cohen GN & Holzer H (eds) Limited Proteolysis in Microorganisms, pp 159-166. US DHEW Publication No. (NIH) 79-1591, Bethesda, MD

Wood NB and Haselkorn R (1980) Control of phycobiliprotein proteolysis and heterocyst differentiation in *Anabaena.* J Bacteriol 141: 1375-1385

Wyatt JT and Silvey JKG (1969) Nitrogen fixation by *Gloeocapsa.* Science 165: 908-909

Wyman M, Gregory RPF and Carr NG (1985) Novel role for phycoerythrin in a marine cyanobacterium, *Synechococcus* strain DC2. Science 230: 818-820

Wyman M, Zehr JP and Capone DG (1996) Temporal variability in nitrogenase gene expression in natural populations of the marine cyanobacterium *Trichodesmium thiebautii.* Appl Environ Microbiol 62: 1073-1075

Wynne D (1977) Alterations in activity of phosphatases during the *Peridinium* bloom in Lake Kinneret. Physiol Plant 40: 219-224

Xie W-Q, Whitton BA, Simon JW, Jager K, Reed D and Potts M (1989) *Nostoc commune* UTEX 584 gene expressing indole phosphate hydrolase activity in *Escherichia coli.* J Bacteriol 171: 708-713

Yamaguchi S, Aizawa S-I, Kihara M, Isomura M, Jones CJ and Macnab RM (1986) Genetic evidence for a switching and energy-transducing complex in the flagellar motor of *Salmonella typhimurium.* J Bacteriol 168: 1172-1179

Yamanaka G and Glazer AN (1980) Dynamic aspects of phycobilisome structure. Phycobilisome turnover during nitrogen starvation in *Synechococcus* sp. Arch Microbiol 124: 39-47

Yamanaka G and Glazer AN (1983) Phycobiliproteins in *Anabaena* 7119 heterocysts. In: Papageorgiou GC & Packer L (eds) Photosynthetic Prokaryotes: Cell Differentiation and Function, pp 69-90. Elsevier Science Publishing, Amsterdam

Yeh SW, Ong LJ and Glazer AN (1986) Role of phycoerythrin in marine picoplankton *Synechococcus* sp. Science 234: 1422-1423

Yokoyama E, Murakami A, Sakurai H and Fujita Y (1991) Effect of supra-high irradiation on the photosynthetic system of the cyanophyte *Synechocystis* PCC 6714. Plant Cell Physiol 32: 827-834

Yu J-W, Price GD and Badger MR (1994) Characterization of CO_2 and HCO_3^--uptake during steady-state photosynthesis in the cyanobacterium *Synechococcus* PCC 7942. Aust J Plant Physiol 21: 185-195

Yu L, Zhao J, M‚hlenhoff U, Bryant DA and Golbeck JH (1993) PsaE is required for in vivo cyclic electron flow around photosystem I in the cyanobacterium *Synechococcus* sp. PCC 7002. Physiol Plant 103: 171-180

Zhang C-C (1993) A gene encoding a protein related to eukaryotic protein kinases from the filamentous heterocystous cyanobacterium *Anabaena* PCC 7120. Proc Natl Acad Sci USA 90: 11840-11844

Zhao J, Zhou J and Bryant DA (1992) Energy transfer processes in phycobilisomes as deduced from analyses of mutants of *Synechococcus* sp. PCC 7002. In: Murata N (ed) Research in Photosynthesis. Vol. I, pp 25-32. Kluwer, Dordrecht

Zhou J, Gasparich GE, Stirewalt V L, de Lorimier R and Bryant DA (1992) The *cpcE* and *cpcF* genes of *Synechococcus* sp. PCC 7002. Construction and phenotypic characterization of interposon mutants. J Biol Chem 267: 16138-16145

Zimmermann H (1992) 5'-nucleotidase: molecular structure and functional aspects. Biochem J 285: 345-365

Chapter 16

Metal Metabolism and Toxicity: Repetitive DNA

Nigel J. Robinson, Julian C. Rutherford, Mathew R. Pocock and Jennifer S. Cavet
Department of Biochemistry & Genetics, The Medical School, University of Newcastle, NE2 4HH, UK

Summary

Significant transitions in the availability, toxicity and subsequent use of metals are predicted to have coincided with the release of dioxygen in quantity and hence coincided with the first appearance of cyanobacteria. During the rapid adaptive radiation of the cyanobacteria there will have been (i) selection for the evolution of new metal resistance determinants and (ii) greater opportunity for more available metals to be recruited for use in newly evolved proteins. Today, the availability of metals in many environments fluctuates both spatially and temporally not only as a result of 'natural' processes, but also as a consequence of anthropogenic factors. The first part of this chapter considers the metabolism and toxicity of metals in cyanobacteria. The second part of this chapter considers a separate topic, the abundance of short sequences which are now known to be repeated throughout the

B. A. Whitton and M. Potts (eds), The Ecology of Cyanobacteria, 443-463.
© 2000 *Kluwer Academic Publishers. Printed in the Netherlands.*

DNA of many cyanobacteria. It is predicted that the study of cyanobacteria will also give unique insights into the evolution of repetitive DNA.

I. Metal Metabolism and Toxicity

A. Introduction and Scope

In this chapter we will discuss the availability and metabolism of metals of the copper and zinc triads, copper and zinc being exploited in biology while the remaining metals (cadmium, mercury, silver and gold) are considered to be exclusively toxic. In excess, copper and zinc are also toxic, copper being the more potent. It is clear that the availability of iron is an important determinant of cyanobacterial productivity. However, this subject has been discussed elsewhere (Tandeau de Marsac and Houmard, 1993; Carr and Mann, 1994; Grossman et al., 1994) and is excluded from this chapter. Iron is only mentioned where it is relevant to considerations of the use of copper.

The entire catalogue of all known enzymes which require copper is relatively small (many of these are extracellular and in higher organisms they are typically involved in the biochemistry of connective tissues) (Vallee and Falchuk, 1993), and few copper requiring proteins are known in cyanobacteria. By contrast, the catalogue of known enzymes that require zinc for catalytic activity exceeds 300 (Vallee, 1991). Zinc is the only element encountered at the active sites of enzymes which participate in all six categories of reaction recognised by IUB nomenclature (i.e. oxidoreductases, transferases, hydrolases, lyases, isomerases and ligases) (Vallee and Falchuk, 1993). In addition, zinc is a structural component of many proteins (Berg and Shi, 1996). However, the extent to which zinc 'pervades' metabolism in different cyanobacteria remains to be established (Section D.4).

The physical and chemical properties of zinc make it ideally suited to certain types of catalysis (Vallee and Falchuk, 1993). By contrast it has been argued that many of the functional surfaces created by structures that contain zinc could have evolved without any metal dependency (O'Halloran, 1993). Many of these surfaces, for example in zinc-finger proteins, allow interactions with other macromolecules, proteins or nucleic acids (Berg and Shi, 1996). These interactions are thus zinc-dependent. Such a dependency could be a relic of a former regulatory role for zinc or a clue to an existing role. In the following section it is argued that if zinc occupancy of such structures is/was regulatory, then such a role could have first emerged in 'some' strains of cyanobacteria. The proposal that zinc is a messenger in some cells has been widely disseminated (for example, Berg and Shi, 1996) but to date only supported by circumstantial evidence. Fraústo da Silva and Williams (1991) suggest that in eukaryotes zinc can act as a 'master hormone' with rapid cell division depending upon metallothionein mediated storage/transfer of zinc. Again, the possibility that such a role may, at least in part, have emerged during the diversification of cyanobacteria is considered (Section D.1).

B. Changes in Metal Availability through Time

Our prognosis that the study of cyanobacteria will provide unique insight into metal metabolism and toxicity has been substantially influenced by 'The Biological Chemistry of the Elements' by Fraústo da Silva and Williams (1991). In part, these arguments are based upon knowledge of the propensities of different metal ions to form oxides and sulphides, their relative solubilities and predictions about changes in water chemistry caused by the appearance of dioxygen.

Central to these discussions is the concept that the biological recruitment of elements is influenced by their chemical state and availability (Fraústo da Silva and Williams, 1991). During early evolution it is deduced that metals such as zinc and copper would have been locked up in insoluble sulphides. For iron, cobalt and nickel it is suggested that at a predicted pH close to the solubility product of the precipitates, small metal-sulphide clusters would exist in the seawater solution. It is thus believed that 'biology could take advantage of the sulphide clusters of these three metals' while by contrast copper is so insoluble that it would have been totally excluded. The appearance of dioxygen would have led to the removal of sulphide from the sea and an increase in the availability of metal ions such as zinc and copper(II). There would however have been a coincident decrease in the availability of iron.

We propose that the cyanobacteria command a major evolutionary transition in the toxicity and use of metal ions. As discussed later, molecular phylogenies indicate that cyanobacteria underwent a rapid evolutionary radiation (Giovannoni et al., 1988). Thus, at a time when the availability of many metal ions was (relatively) swiftly changing, new species of cyanobacteria were rapidly evolving. Exposure to elevated concentrations of metals such as copper and zinc would have initially selected for resistance determinants, including zinc and copper efflux systems. However, the exploitation of these newly available ions then became an option for subsequent evolution. The use of these metals would in turn create a further selection for more precise homoeostasis, including zinc and copper uptake systems and perhaps metallothioneins. By contrast, the reduced availability of iron would select for novel iron scavenging systems and/or alternative biochemistry which reduced the need for iron by replacing it with a different but chemically suited metal ion, such as copper. It is our general thesis that these evolutionary transitions are reflected in cyanobacterial diversity (discussed in the following sections). For example, while some cyanobacteria can exploit copper as an alternative to iron in photosynthetic electron transport, others do not possess this capability (Section D.6). *Synechococcus* PCC 7942 has precise zinc homoeostasis mediated by metallothionein and in this organism zinc is exploited in zinc-fingers, for example in DNA primase (Section D.1). By contrast DNA primase in *Synechocystis* PCC 6803 does not contain a carboxy-terminal zinc-finger and no metallothionein gene has been identified within the *Synechocystis* PCC 6803 genome, at least based upon sequence similarity. Whether or not *Synechocystis* PCC 6803 exploits zinc, and zinc-fingers, to any lesser extent than *Synechococcus* PCC 7942 is presently unknown (Section F).

C. Spatial Distribution of Metals and Metal Resistance in Cyanobacteria

1. Metal Pollution and Variation in Metal Levels

Pollution of aquatic environments with metals of the copper and zinc triads became common knowledge in the 1950s following two incidents, both in Japan (cited in Mason, 1991; Clark, 1992). The first resulted from the use of mercuric oxide as a catalyst in the production of acetaldehyde by a factory at Minimata Bay. Mercury discharged into the bay led to the subsequent accumulation of these metal ions in fish and the appearance of nervous and other disorders in humans (seafood being a substantial component of the local diet). The second resulted from cadmium containing mining effluent being released into the Jintsu river which was used to irrigate downstream paddy fields. Contamination of many lakes, rivers and estuaries with cadmium, mercury, zinc and/or copper, from mine drainage or industrial processes has subsequently been documented in numerous surveys (Say and Whitton, 1981; Förstner, 1983). Copper pollution of aquatic environments has also resulted from its use in antifouling paint for ships' hulls (subsequently replaced by tin compounds), as an algicide and as a wood preservative. It was estimated that 180 t copper entered the coastal waters of California between Santa Barbara and San Diego from antifouling paints in one year. Municipal wastes often contain elevated levels of metals, for example the municipal waste from Los Angeles was, at one time, estimated to contribute 510 t copper to the sea annually (cited in Clark, 1992).

Cadmium is generally depleted in the surface layers of sea water (Boyle et al., 1976; Bruland et al., 1978) but this can be reversed by upwellings particularly in coastal regions where cadmium has been introduced into the environment via anthropogenic processes. There is one example of the apparent substitution of cadmium for zinc in the marine diatom *Thalassiosira weissflogii* growing in highly zinc depleted waters (Price and Morel, 1990). However, as previously noted, cadmium is generally thought to be exclusively toxic and not used in biology, and indeed there are no documented cases of the use of cadmium in cyanobacteria. Exposure to elevated cadmium is generally thought to have 'only' occurred as a result of man's influences on environments. Thus, where cadmium sensitive phenotypes have been observed in gene deletion mutants (for example Turner et al., 1993) this has invariably been attributed to the 'gratuitous' handling of cadmium by the products of genes which have evolved in response to some selection imposed by zinc (for example Palmiter, 1998). The availability of different metals can be greatly influenced by localised anoxic conditions and by differences in pH; for example copper is poorly available in highly alkaline lakes (Section D.6). The presence of higher concentrations of other ions, such as calcium and phosphate, can dramatically reduce zinc toxicity to cyanobacteria (Say and Whitton, 1977; Shehata and Whitton, 1982). In soils, the availability of metals can fluctuate both spatially and temporally and these changes can operate over relatively short distances (adjacent to, or distant from, particles) and timescales (for example with rapid changes in soil moisture content). It is, of course, the potential for movement to micro niches where metal availability changes and the potential for swift fluctuations in metal availability that select for metal perception systems (Section E).

2. Toxicity and Tolerance in the Responses of Cyanobacteria to Elevated Concentrations of Metals

Numerous studies have documented growth, morphological, physiological and biochemical responses of different cyanobacteria to supra-optimal concentrations of metals of the copper and zinc triads (reviewed in Whitton, 1980; Rai et al., 1981; Vymazal et al., 1985; Reed and Gadd, 1990). Morphological effects of cadmium and elevated concentrations of zinc include inhibition of daughter cell separation, production of filaments in normally unicellular strains (such as *Synechococcus* PCC 6301) and cell elongation (Chintamani and Mohanty, 1989; Whitton and Shehata, 1982; Shehata and Whitton, 1982). In contrast, elevated copper has been shown to cause production of sub-spherical rather than coccoid *Synechococcus* PCC 6301 (Whitton and Shehata, 1982). Elevated concentrations of metals influence pigment composition, for example with increased carotenoid to chlorophyll ratios and overall decreases in phycocyanin, chlorophyll *a* and total protein in *Synechococcus* PCC 6301 exposed to elevated zinc (Chintamani and Mohanty, 1989). An analysis of the effects of elevated concentrations of copper on *Nostoc calcicola* revealed that photosynthesis was readily inhibited (Verma and Singh, 1991). Polyphosphate bodies are implicated in the sequestration of metals of the copper and zinc triads (Jensen et al., 1982) and these increase in response to elevated concentrations of such metals.

At some metal polluted aquatic sites, for example zinc enriched waters of high pH, cyanobacteria are abundant and can be the dominant organisms (Whitton, 1980; Plate 23, this volume). Some strains of cyanobacteria isolated from metal polluted sites are resistant to elevated metal concentrations in subsequent axenic culture (Shehata and Whitton, 1982; Takamura et al., 1989). Strains have also been selected in laboratory studies by step-wise adaptation of cultures to gradually increasing metal concentrations (Singh and Pandey, 1982; Gupta et al., 1993). Enhanced resistance in such a copper resistant line of *Nostoc calcicola* was correlated with enhanced copper efflux while enhanced cadmium resistance in *Synechococcus* PCC 6301 was correlated with enhanced endogenous metal chelation mediated by increased expression of metallothionein (Section D).

D. Genes involved in Copper and Zinc Homoeostasis and Metabolism

1. Zinc and Metallothionein: SmtA, MtnA

Metallothioneins (MTs) bind and sequester metals of the copper and zinc triads (Kägi and Schäffer, 1988, for a review). They are, by definition, rich in cysteine and bind the metal ions in metal-thiolate clusters. Cyanobacterial MTs are unusual in also containing histidine residues which have been implicated in the coordination of metals of the copper and zinc triads, at least in other proteins. Recent NMR studies, coupled with site directed mutagenesis, have demonstrated that cyanobacterial MT coordinates three zinc ions, two exclusively to cysteine-thiolate groups and the third to a site composed of two cysteine-thiolate and two histidine-imidazole groups (Daniels et al., 1998). MT genes have been isolated from a wide range of eukaryotes, but cyanobacterial MT genes are the only prokaryotic MT genes which have (to date) been characterised. A cyanobacterial MT-like protein was first isolated from *Anacystis nidulans* (*Synechococcus* PCC 7942 or 6301) exposed to cadmium (Maclean et al., 1972). Subsequently MTs were purified from a marine cyanobacterium *Synechococcus* RRIMP NI (Olafson et al., 1979), and freshwater strains designated *Synechococcus* UTEX-625 and *Synechococcus* TX-20 (again both strains are thought to be synonymous with, or at least very closely related to, *Synechococcus* PCC 7942 and 6301). The protein from *Synechococcus* TX-20 was subsequently sequenced and spectroscopic studies indicated the presence of a single metal cluster involving coordination of metal ions to thiol groups of cysteine residues (Olafson et al., 1988). Fragments of the corresponding genes were obtained from DNA purified from *Synechococcus* PCC 7942 and *Synechococcus* PCC 6301 via PCR using degenerate primers (Robinson et al., 1990). These PCR fragments were used as probes to isolate the corresponding genomic region and the gene designated *smtA* (Huckle et al., 1993). A related gene, designated *mtnA*, has also been identified in DNA sequences from *Synechococcus vulcanus* (Shimizu et al., 1992). To date, no homologue of *smtA* has been detected in the fully sequenced genome of *Synechocystis* PCC 6803. However, the small size of the anticipated open reading frame (ORF), and the relative lack of sequence homology among class II MTs, mean that such an ORF could be easily overlooked.

Olafson and co-workers (1988) noted that MT accumulated in *Synechococcus* TX-20 following exposure to cadmium and zinc, but not copper ions. After purification the protein was associated with either cadmium or zinc ions with copper as only a minor component. Takatera and Watanabe (1992), following their purification protocols, made similar observations concerning the metal-binding specificities of cyanobacterial MT. The metal-binding specificities of known cyanobacterial MTs are therefore unlike other known microbial MTs (from fungi). The latter are synthesised in response to copper ions, are associated with copper ions *in vivo* and mutants deficient in the corresponding genes (for example *CUP1* in *Saccharomyces cerevisiae*) are hypersensitive to excess copper but not zinc. The *smtA* gene has been over-expressed in *Escherichia coli* to generate a recombinant fusion protein (Shi et al., 1992). SmtA associated with all of the analysed metal ions (zinc, cadmium, copper and mercury) following purification from *E. coli* grown in metal supplemented media. Relative (to equine renal MT) affinities for these metals were estimated by displacing metal ions with protons at low pH (Shi et al., 1992). More protons were required to displace zinc from cyanobacterial MT than from equine renal MT suggesting that the former has a particularly high affinity for zinc. Evidence that cyanobacteria containing *smtA* accumulate more zinc than mutants deficient in *smtA* also supports the proposal that SmtA binds zinc within cyanobacterial cells (Turner et al., 1995).

Mutants of *Synechococcus* PCC 7942 which are deficient in a functional MT gene have been created by homologous recombination mediated gene insertion (Turner et al., 1993). These cells are approximately 5-fold more sensitive to excess zinc, show some reduced tolerance to cadmium, but have 'normal' levels of tolerance to excess copper. The *smtA* gene has been used as a selectable marker to select for transformants of MT deficient cells on the basis of their restored resistance to zinc (Turner et al., 1993; 1995).

Two different types of mutations which affect the *smtA* gene have been characterised in cadmium tolerant cyanobacteria. In both cases, these were detected in cultured cells selected to grow in media sequentially enriched with increasing concentrations of cadmium. First, an increase in the number of copies of the *smtA* gene (Gupta et al., 1992) and second, the deletion of a repressor (see following sections) (Gupta et al., 1993). Both mutations are assumed to confer an increase in the expression of *smtA* and hence the amount of metal sequestered. It will be of interest to examine whether similar mutations have occurred in cyanobacteria selected for resistance to cadmium or zinc in metal polluted 'natural' environments.

Upstream of *smtA* is the DNA primase gene *dnaG* (Bird et al., 1998). DNA primase is required for the synthesis of short RNA primers that are essential for both leading strand synthesis at the origin of replication and for repeated reinitiation of lagging strand DNA synthesis at replication forks. In many other bacteria *dnaG* is at the centre of a macromolecular synthesis operon which is generally composed of *rpsU-dnaG-rpoD* encoding proteins required for the initiation of synthesis of protein, DNA and RNA respectively. This gene architecture is widely conserved and it has been suggested that the coordinated expression, within an operon, of genes required for the synthesis of the key informational macromolecules must be advantageous (Versalovic et al., 1993). The absence of such an operon in *Synechococcus* PCC 7942 could suggest that alternative mechanisms are involved in the coordination of macromolecular synthesis in this cyanobacterium, compared to other analysed bacteria. At the amino-terminal end of all DNA primases, including the one from *Synechococcus* PCC 7942, is a conserved zinc-binding region (Stamford et al., 1992) and the purified primase from *E. coli* has been shown to contain tightly bound zinc (Griep and Lokey, 1996). However, at the carboxy-terminal end of the *Synechococcus* PCC 7942 primase is an additional eukaryotic style zinc-finger which is absent from all other reported primases (Bird et al., 1998). In *E. coli*, the extreme carboxy-terminus of primase (the last 8 amino acids) interacts with the replicative helicase and it is the strength of this interaction which 'sets the replication fork clock' (Tougu and Marians, 1996). DNA replication and cell division in *Synechococcus* PCC 7942 do not fit the prokaryotic paradigm (Binder and Chisholm, 1990; 1995) with asynchronous initiation of chromosome replication.

It is tempting to speculate that atypical coordination of DNA synthesis in this cyanobacterium is influenced by atypical regulation of primase-helicase interaction which is, in turn, influenced by zinc occupancy of the carboxy-terminal zinc-finger in DNA primase.

In addition to metal regulated expression, animal MT shows programmed changes in expression during development, and in its location during the cell cycle. For example, in cultured animal cells MT accumulates in the nucleus at early S-phase, at the onset of DNA synthesis (Tsujikawa et al., 1991). Exchange of zinc has also been demonstrated between animal MT and a number of zinc requiring proteins *in vitro* (Zeng et al., 1991a, b; Cano-Gauci and Sarker, 1996). A global regulatory role for animal MT, modifying the availability of zinc, has been suggested. Could such a role for MT and indeed a regulatory role for zinc, have first emerged in cyanobacteria? By analogy to proposed roles for MTs in higher eukaryotes, changes in the expression of MT in *Synechococcus* PCC 7942 could modify the availability of zinc and hence regulate zinc occupancy of the zinc-finger in DNA primase.

It is notable that a gene encoding a sub-unit of phenylanyl tRNA synthetase, *pheT*, is co-transcribed with *dnaG* in *Synechococcus* PCC 7942 (Bird et al., 1998). Phenylalanyl tRNA synthetase, in common with many other aminoacyl tRNA synthetases, not only catalyses the association of an amino acid with its cognate tRNA molecule, but also catalyses the production of diadenosine tetraphosphate, Ap4A, in a zinc dependent manner (Plateau et al., 1981). Changes in zinc availability mediated by SmtA are therefore likely to modulate production of Ap4A, a known regulator of the cell cycle (Nishimura, 1998), and hence rates of cell proliferation.

2. Copper-transporters: PacS, CtaA

Genes encoding P-type ATPases are known from a variety of organisms, encoding products which transport different cations in different directions (in, out or cation exchange) (Silver et al., 1993, for a synopsis). It is not possible to predict, with any degree of certainty, which cations (protons, sodium, potassium, calcium, magnesium, cadmium, copper or possibly zinc) are specifically transported by a given P-type ATPase, merely from sequence information. However, an identifiable subset of these defined as P1-type (Lutsenko and Kaplan, 1996) or CPx-type (representing a diagnostic motif) (Solioz and Vulpe, 1996) are known to specifically transport transition metals.

Energy dependent copper efflux was proposed as a mechanism of copper resistance in *Synechococcus* sp. (Olafson, 1986) and in *Nostoc calcicola* (Verma and Singh, 1991). Subsequently, genes encoding copper-transporting P-type ATPases, designated *pacS* (Kanamura et al., 1994), and *ctaA* (Phung et al., 1994) have been isolated, suggesting roles for this class of membrane proteins in copper acquisition, distribution and homoeostasis in, at least some, cyanobacteria. Kanamura and coworkers (1993) cloned two genes encoding deduced P-type ATPases, named *pacS* and *pacL*. PacL was shown to be a calcium-transporting ATPase (Berkelman et al., 1994) while PacS was shown to transport copper (Kanamura et al., 1994). The *pacS* transcripts increase in abundance in response to elevated concentrations of copper and silver ions and Western blots confirm that the PacS protein similarly accumulates in copper treated *Synechococcus* PCC 7942. An insertional mutant of *pacS* is hypersensitive to elevated concentrations of copper and, unexpectedly, the PacS protein was found to be localised to the thylakoid membrane, not the plasmalemma (Kanamura et al., 1994). PacS thus appears to be involved in copper homoeostasis within intracellular compartments. This is unusual since most bacterial P-type ATPases are, of course, thought to be involved in the movement of metal ions between the extracellular environment and the cell. Indeed this is thought to be the role for the other characterised copper-transporting ATPase from *Synechococcus* PCC 7942, CtaA (Phung et al., 1994). Mutants deficient in functional *ctaA* show increased tolerance to copper compared to wild type cells suggesting a role for CtaA in the import of copper ions.

3. Zinc Exporter ZiaA and other Homologues from Synechocystis *PCC 6803*

Within the genome of *Synechocystis* PCC 6803 (Kaneko et al., 1996) is a cluster of ORFs encoding deduced proteins with similarity to known proteins involved in copper and zinc homoeostasis in other bacteria (Fig. 1). This includes sequences with similarity to *czcA* and *czcB*. These two genes were originally isolated from *Alcaligenes eutrophus* CH34. This organism was recovered from a zinc decantation tank (Mergeay et al., 1978) and the plasmid encoded *czc* (for cobalt, zinc and cadmium) confers resistance to these three metal ions. The *czc* determinant is composed of three *czcCBA* (Nies and Silver, 1995; Diels et al., 1995) structural genes. The CzcCBA proteins combine in a 1:1:1 ratio to form a cation efflux complex (Nies and Silver, 1995). Deletion of *czcC* leads to loss of cadmium and cobalt resistance but the CzcBA complex alone still confers high efficiency zinc efflux (Nies et al., 1989). The presence of homologues of *czcBA* but not *czcC* in *Synechocystis* PCC 6803 could therefore suggest an exclusive role in zinc efflux, although this remains to be tested. In addition, there are two divergently transcribed (from the *czc*-like) genes designated *czcR* and *czcS* (Fig. 1) which show similarity to two component adaptive response regulators (reviewed in Parkinson and Kofoid, 1992). Sequences encoding a histidine kinase and a response regulator are also present in sequences flanking *czc* from *Alcaligenes eutrophus* and these are known to regulate the expression of the *czc* genes (van der Lelie et al., 1997). Elsewhere in the *Synechocystis* PCC 6803 genome is an ORF encoding a protein with similarity to CzcD. CzcD is involved in the metal responsive expression of the *czc* determinant (Neis, 1992) and has similarity to ZRC1 and COT1 which confer zinc/cadmium and cobalt resistance respectively in *Saccharomyces cerevisiae* (Kamizomo et al., 1989; Conklin et al., 1992).

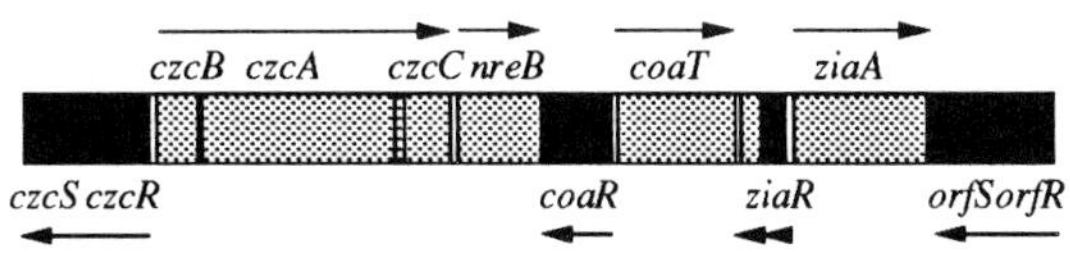

Fig. 1. A region of the *Synechocystis* PCC 6803 genome containing a cluster of genes encoding deduced proteins involved in the transport and perception of metal ions. In addition to designated ORFs described in the text, this region contains a second pair of genes encoding a deduced two component sensor (*orfS* and *orfR*) which could be involved in metal perception.

Two ORFs within this cluster encode proteins with sequence features of transition metal-transporting CPx-type ATPases. One of these was initially designated PacS due to its similarity to the copper transporter, but an analysis of deletion mutants has revealed sensitivity to zinc, reduced zinc export to the periplasm, and the gene has been redesignated *ziaA* (Thelwell et al., 1998). The gene confers enhanced resistance to zinc and reduced zinc accumulation when introduced into *smt* deficient *Synechococcus* PCC 7942 (Thelwell et al., 1998). Divergently transcribed from *ziaA* is an ORF encoding a protein with similarity to SmtB (see following sections). This protein has been shown to be a sensor of internal zinc concentrations and redesignated ZiaR (Thelwell et al., 1998) (Fig. 1). This suggests that the cluster of genes could include two distinct zinc transporters, with *czc* responding to periplasmic zinc levels mediated by a two component sensor and the ATPase responding to internal zinc levels mediated by an SmtB-like metal sensor ZiaR.

Elsewhere in the *Synechocystis* PCC 6803 genome are other deduced metal transporters (including the *czcD*-like sequence). It is now necessary to delete these genes, and those in Fig. 1, to establish the effects on metal tolerance and accumulation and resolve which metals are transported.

4. The Use of Zinc

Despite the plethora of known zinc requiring proteins in other organisms, remarkably few cyanobacterial proteins have, to date, been unequivocally demonstrated to bind to, and to require this metal ion. Alkaline phosphatase is known to be a zinc 'requiring' enzyme in other prokaryotes (Simpson and Vallee, 1968). The *Synechococcus* PCC 7942 enzyme is larger than any other characterised alkaline phosphatase (Ray et al., 1991). Most importantly, the protein from *Synechococcus* PCC 7942 is irreversibly 'inhibited' by zinc (Block and Grossman, 1988). However, data are eclectic, there having been no extensive and systematic survey of zinc requirements in cyanobacteria. It would be premature to conclude that the use of zinc is any less common in cyanobacteria than in other organisms.

Zinc certainly is needed for optimal growth of at least some species and is required for some biochemical pathways. For example, zinc is needed for toxin production in *Microcystis aeruginosa* while other metals tested, including copper, manganese and nickel, are not required (Lukac and Aegerter, 1993). Of course, MT from *Synechococcus* PCC 7942 binds and detoxifies excess zinc, but it remains to be established whether or not it also has a role in storing, or otherwise modifying the availability of this metal ion for zinc requiring proteins.

Sequences of several cyanobacterial proteins are known which contain deduced zinc-binding motifs. One which has been mentioned in a previous section is DNA primase. In addition to the predicted zinc-binding site common to all primases, the *Synechococcus* PCC 7942 protein also contains the additional zinc-finger and this structure is required for normal levels of DNA synthesis in *Synechococcus* PCC 7942 (Bird et al., 1998). Zinc-finger motifs have also been described in other deduced cyanobacterial proteins, for example unknown ORFs from *Synechocystis* PCC 6803 (Ogura et al., 1991), but binding of zinc to these other sites remains to be tested. RNA polymerase is known to be a zinc containing enzyme and the β' subunit of *E. coli* RNA polymerase contains a zinc-binding pocket which is also present in the related sequence of the *rpoC1* gene product of *Anabaena* PCC 7120 (Bergsland and Haselkorn, 1991). Indeed, this four cysteine containing motif is preserved throughout RpoC1 sequences from eight different cyanobacteria (including the *Cyanophora* cyanelle, *Prochlorococcus* sp. and *Prochlorothrix hollandica*) (Palenik and Haselkorn, 1992) suggesting that zinc is likely to be an essential requirement for transcription throughout cyanobacterial diversity. However, the association of zinc with this deduced binding pocket remains to be tested. The deduced sequence of carbonic anhydrase from *Synechococcus* PCC 7942, encoded by the *icfA* gene, contains five amino acids (Cys-39, Glu-82, His-98, Cys-101 and Glu-153) which are candidates for zinc-binding and are conserved in all prokaryotic carbonic anhydrases (Fukuzawa et al., 1992). However, once again, association of zinc with the cyanobacterial carbonic anhydrase has not formally been demonstrated. The deduced sequence of 5-aminolevulinic acid dehydratase (ALAD) from *Synechococcus* PCC 7942 contains a conserved zinc-binding domain found in ALAD from mammals, yeast and other eubacteria (Jones et al., 1994). Unlike ALAD from higher plants the cyanobacterial protein was shown not to require magnesium or manganese ions, and it was proposed to contain tightly bound zinc, analogous to the mammalian, yeast and other known eubacterial enzymes.

5 Zinc Import: An ABC Importer

In the *Synechocystis* PCC 6803 genome there are two genes (slr2045 and slr2044) encoding components of a deduced ATP-binding cassette type importer which, from the local gene architecture, appear likely to be co-transcribed with a gene (slr2043) encoding a deduced periplasmic zinc-binding protein. The latter, herein designated ItzA for inward transport of zinc, has sequence similarity to extra-plasmamembrane zinc-binding proteins Pzp1 from *Haemophilus influenzae* (Lu et al., 1997), AdcA from *Streptococcus pneumoniae* (Dintilhac et al., 1997) and ZnuA from *E. coli* (Patzer and Hantke, 1998). These proteins are all proposed to act in the periplasmic- (or at least the extra-plasmamembrane in the Gram positive *Streptococcus pneumonia*) binding of zinc and subsequent donation of the metal ions to the ABC importer. Presumably, zinc is exploited by all of the cyanobacteria which contain such an import system. A survey of the species, and habitat, distribution of these genes within The Cyanobacteria would be of interest.

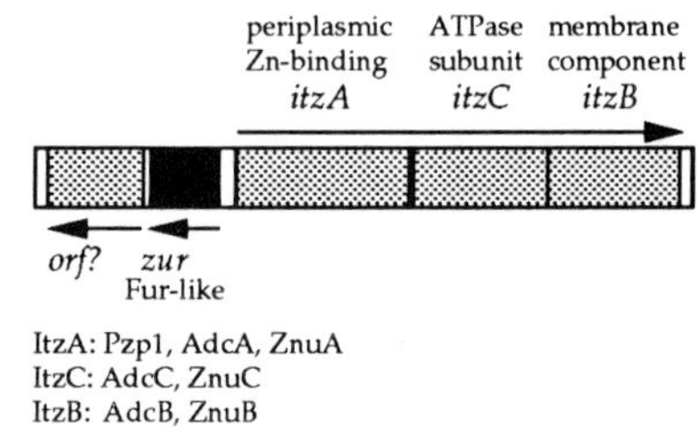

Fig. 2. A divergon which encodes a deduced zinc sensor, now designated Zur (Section E.3), which regulates a deduced ABC zinc import system, now designated Itz. The related gene products from other bacteria are shown.

6. The Copper-iron Connection in Photosynthetic Electron Transport

There are relatively few known uses for copper in cyanobacteria, but one which has been studied in some detail is in plastocyanin. Two proteins can function as electron carriers between photosystem II and photosystem I, the iron-haem protein cytochrome c_{553} and copper containing plastocyanin. Some cyanobacteria, and indeed some eukaryotic algae, contain genes for both carriers while others contain only cytochrome c_{553} (Bohner and Böger, 1978; Wood, 1978; Merchant and Bogorad, 1986; Sandmann, 1986). In *Anabaena variabilis*, which contains both genes, their expression is complementary. Plastocyanin (encoded by the *petE* gene) expression increases with added copper and cytochrome c_{553} shows a coincident decrease in expression (Bovy et al., 1992). This change in expression is regulated at the level of transcription in

Anabaena variabilis. *Synechocystis* PCC 6803 also contains both genes, but in this cyanobacterium expression of *petE* in response to copper is controlled at the translational or post-translational level. *Synechococcus* PCC 7942 lacks plastocyanin and expression of cytochrome c_{553} is not regulated by copper. An ability to switch between the use of these alternative redox-active metals (copper and iron) in at least some cyanobacteria presumably reflects critical fluctuations in availability during evolution, at least in some habitats. Morand et al. (1994) have noted that the absence of plastocyanin in some strains correlates with known environmental factors, for example *Spirulina maxima* growing in Lake Texcoco (Mexico) in which copper is insoluble due to alkalinity (greater than pH 10). By contrast, these authors suggest that the exclusive presence of plastocyanin in higher plants could reflect the enhanced availability of copper to these organisms due to the evolution of mechanisms for scavenging insoluble copper. This is made possible by direct contact between higher plants and insoluble elements in solid substrates, a situation which is uncommon in aquatic cyanobacteria. An analogous argument (to that proposed for higher plants) could be applied to cyanobacteria which grow on solid substrates, while endosymbionts may have a ready supply of available copper from the host. It could be predicted that plastocyanin would be more commonly used in cyanobacteria which occupy these latter environments.

E. Perception of Metals of the Copper and Zinc Triads in the Environment

1. SmtB

In response to elevated concentrations of a number of trace metal ions (including cadmium and zinc), there is an increase in the abundance of *smtA* transcripts. Heat shock does not elicit an equivalent response (Huckle et al., 1993). Inhibitor studies have shown that there is no effect of cadmium on *smtA* transcript stability, indicating control at the level of transcriptional induction. Expression of a reporter gene (*lacZ*) driven by the *smtA* operator-promoter was similarly shown to increase in response to a number of trace metal ions. However, at maximum permissive concentrations only zinc, and to a lesser extent cadmium and copper ions, gave substantial increases in reporter gene expression (Huckle et al., 1993).

Immediately upstream (100 bp) from the *smtA* gene is a divergently transcribed gene encoding a protein designated SmtB (Huckle et al., 1993). The SmtB polypeptide contains a deduced helix-turn-helix DNA-binding motif. It also shows sequence similarity to numerous bacterial proteins a number of which are known transcriptional repressors involved in the perception of metals (Huckle et al., 1993; Turner et al., 1996). These include plasmid encoded ArsR from *E. coli* (San Francisco et al., 1990), *Staphylococcus xylosus* (Rosenstein et al., 1992), *Staphylococcus aureus* (Ji and Silver, 1992) and chromosomally encoded ArsR from *E. coli* (Xu et al., 1996), all of which are metal-oxyanion responsive repressors of *ars* (arsenic resistance) operons. Other homologues are CadC proteins from *S. aureus* (Endo and Silver, 1995) and *Listeria monocytogenes* (Lebrun et al., 1994), which are cadmium responsive repressors of genes encoding cadmium efflux ATPases, and ZntR from *S. aureus* which appears to regulate expression of a cobalt/zinc exporter (Xiong and Jayaswal, 1998). Finally, MerR from *Streptomyces lividans* which is thought to be a mercury responsive repressor of genes encoding mercury detoxification systems (including mercury reductase), however, it is unlike the more well known MerR transcriptional switches which have been well characterised in other bacteria.

In mutants of *Synechococcus* PCC 7942 deficient in the *smt* divergon (*smtA* and *smtB*) there is highly elevated basal expression (of *lacZ*) associated with the *smt* operator-promoter (Huckle et al., 1993). Repression, and metal-dependent expression was restored in cells containing plasmid borne and/or chromosomal *smtB* (Huckle et al., 1993) implying that SmtB is a metal responsive repressor of *smtA* expression. Electrophoretic mobility shift assays detected multiple complexes forming with the *smtA* operator-promoter, at least one of which was attributed to SmtB, since this complex was absent in cells devoid of *smtB*. Furthermore, this complex dissociated from DNA in response to zinc, and bound again following treatment of extracts with chelating agents to remove zinc (Morby et al., 1993). These observations have been confirmed using recombinant SmtB (Erbe et al., 1995). In addition, site directed mutations within *smtB* have established that a pair of histidine residues (H105 and H106) are essential for the perception of zinc by SmtB while cysteine residues are not (Turner et al., 1996). This is of interest since cysteine residues have been proposed, and in two cases demonstrated, to be important for the perception of metal ions by other members of this family of repressors (Shi et al., 1994; Thelwell et al., 1998). This indicates the intriguing possibility of a family of metal-sensors possessing distinct metal-binding sites which give diversity in the spectra of metals sensed; a preference for cadmium by CadC, arsenic by ArsR, zinc by SmtB and mercury by MerR. The possibility remains that other members of this family of metal sensors with different metal specificities may also be present in, at least some, cyanobacterial genomes (see following sections). A partial structure of SmtB is now known (Cook et al., 1998) and this has allowed the zinc-sensing sites to be modelled. It was suggested from these structural data that apo-SmtB may associate with DNA that is bent. Metallo-regulation by SmtB could involve metal-dependent loss of DNA bending, perhaps at some intermediate level of zinc, in addition to dissociation at highly elevated concentrations of inducer.

2. ZiaR and other Homologues of SmtB: A Family of Metal Sensors

Upstream of *mtnA* from *Synechococcus vulcanus* (Shimizu et al., 1992) is a partly sequenced, divergently transcribed ORF, which encodes a protein with similarity to SmtB, designated MtnB (Turner and Robinson, 1995). This is presumed to be an analogous zinc-sensing repressor. In addition, within the *Synechocystis* PCC 6803 genome are two further ORFs encoding deduced proteins with similarity to SmtB. One of these, *ziaR*, has already been noted. Reporter gene contructs show that ZiaR acts as a repressor required for zinc-dependent expression from the *ziaA* operator-promoter (Thelwell et al., 1998). Gel retardation assays detected ZiaR binding to the *ziaA* operator-promoter and ZiaR-DNA binding was enhanced by treatment with the metal chelator 1,10 phenanthroline. The analysis of site directed mutants of ZiaR established that a pair of cysteine-residues adjacent to the helix-turn-helix DNA-binding site are essential for zinc perception as is a histidine residue (Thelwell et al., 1998) which aligns with a histidine residue (H106) in SmtB which forms a second pair of sites located at the SmtB dimer interface (Cook et al., 1998).

A second ORF within the *Synechocystis* PCC 6803 genome with similarity to *smtB* has been designated *arsA* and is most similar to the oxyanion sensor gene *arsR*, although again it remains to be established which (if any) metals are sensed. The location of *arsA* provides few clues about the likely action of its product. It is near to a gene encoding putative aquaporin. The identification of a family of related metal sensors which differ in the spectra of metals they perceive may be exploited to identify the determinants of metal specificity.

3. A Putative Low Zinc Sensor: Zur

Divergently transcribed from the *itz* genes (see section D.5) is a gene (sll1937) encoding a Fur-like protein (Fig. 2). Another gene in *Synechocystis* PCC 6803 encodes a Fur-like protein which is the more similar in sequence to Fur from *Synechococcus* PCC 7942, the latter having been demonstrated to act in response to elevated concentrations of iron to repress expression of genes only required under low iron conditions (Ghassemian and Straus, 1996). The function of the product of sll1937 is therefore unknown. However, in *E. coli* it has recently been demonstrated that a Fur homologue, redesignated Zur, is in fact a zinc sensor which represses genes that are only required under low zinc conditions (Patzer and Hantke, 1998). We therefore suggest that sll1937 is a Zur-like sensor which regulates the divergently transcribed, deduced zinc import system. We look forward to the characterisation of the full complement of sensors that respond to fluctuations in zinc, and the full complement of homoeostatic systems that they regulate in *Synechocystis* PCC 6803 (Fig. 3).

4. Further Sequences from the *Synechocystis* PCC 6803 Genome

Several putative metal sensors have already been described in the context of discussions of the associated structural genes and descriptions of SmtB, ZiaR and Zur in the preceding sections. In addition, there are two ORFs encoding proteins with similarity to the mercury ion sensor, MerR (not SmtB-like). One of these is located with the region shown in Fig. 1 and partly aligns with MerR and partly with precorrin isomerase (encoded by *cobH*). Precorrin isomerase is involved in the biosynthesis of vitamin B-12, the sole cobalt containing cofactor. The recruitment of DNA fragments encoding domains of such an enzyme suggests that this regulator may be involved in the perception of cobalt and indeed our data support this hypothesis and the sensor has been provisionally designated CoaR (Rutherford et al., unpublished). There are uncharacterised two component sensors in the complete *Synechocystis* PCC 6803 genome and it is, of course, possible that

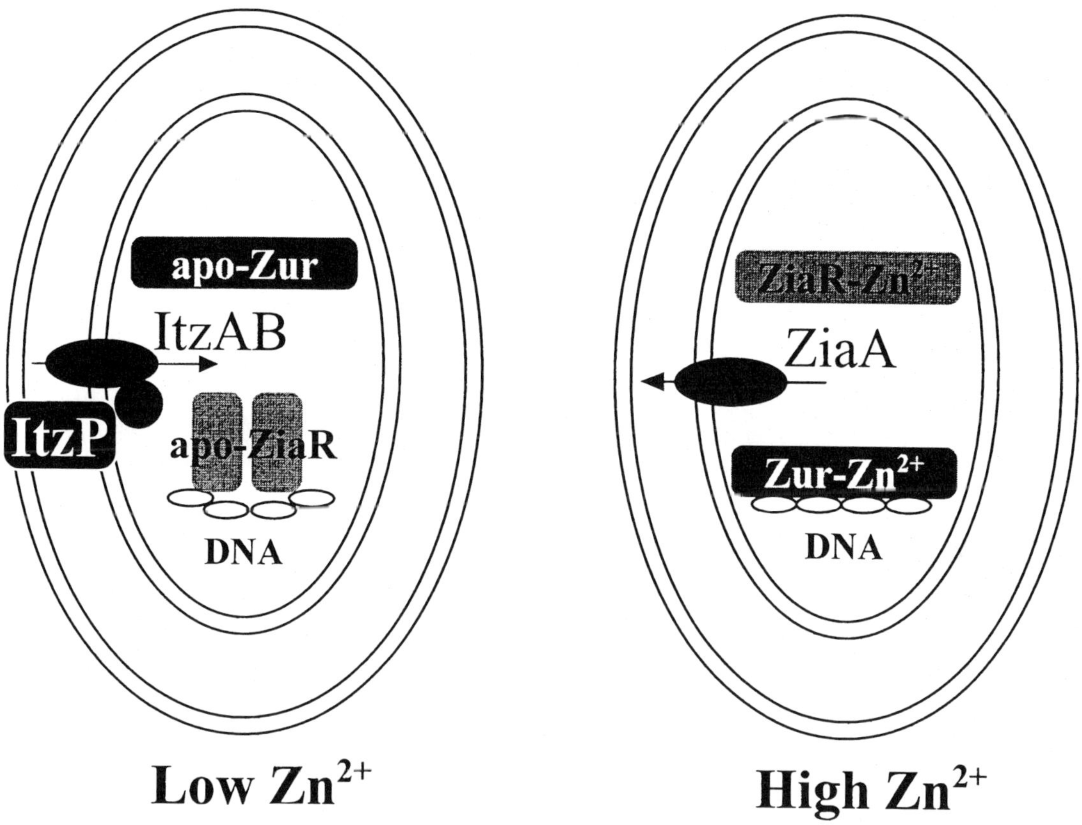

Fig 3. A model showing two proposed 'high and low' zinc sensors in *Synechocystis* PCC 6803 and the proposed modes of action of the products of the genes they regulate.

some of these could be implicated in the perception of metals of the copper and zinc triads.

Once the complement of sensors, homoeostatic systems and metabolic requirements for specific metals are known it may become possible to establish how responses to changes in the levels of a metal in the environment are modulated coincident with changes in endogenous demand. A lack of complete specificity or precise correspondence in the spectra of metals sensed, transported or sequestered by the components of each operon/regulon is also likely to cause 'cross-talk' between the homoeostatic systems for different metals.

F. Contrasting Responses of Two Cyanobacteria to elevated Zinc

If the detoxification, and subsequently the more widespread use, of zinc first emerged during the adaptive radiation of the cyanobacteria, then a diversity of responses to elevated zinc might be anticipated. Indeed, *Synechocystis* PCC 6803 and *Synechococcus* PCC 7942 possess closely related zinc sensors, ZiaR and SmtB, but they regulate very different structural proteins with very different consequences for the cell biology of zinc. The former triggers expulsion of zinc via ZiaA mediated efflux, while the latter triggers internal sequestration by SmtA (Huckle et al., 1993; Thelwell et al., 1998) (Fig. 4).

Is zinc homoeostasis in *Synechocystis* PCC 6803 purely regulated by metal ion influx/efflux, while in *Synechococcus* PCC 7942 it is regulated by sequestration in the cytoplasm? Is the periplasm used as an alternative zinc store in *Synechocystis* PCC 6803 or does this organism have a lesser requirement for zinc storage? Does this reflect more widespread use of zinc in *Synechococcus* PCC 7942 and merely a requirement for the detoxification of unnecessary zinc in *Synechocystis* PCC 6803, or did the former evolve

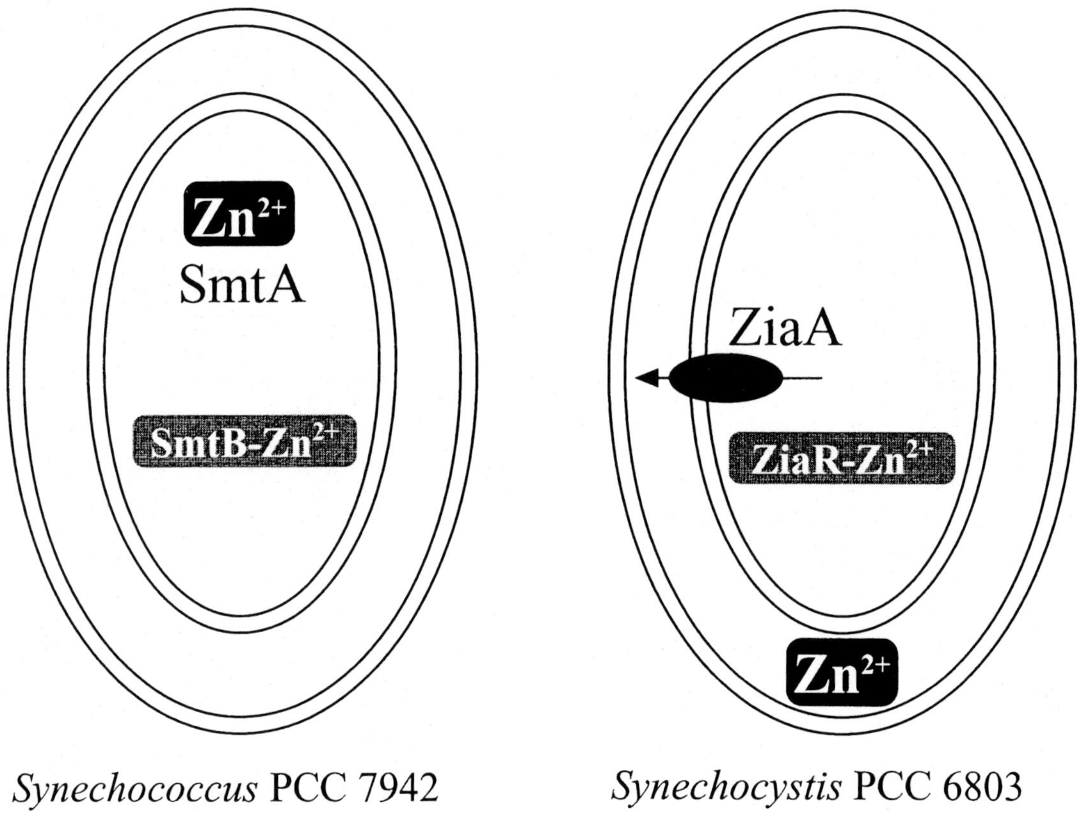

Fig. 4. These two cyanobacteria have closely related sensors of elevated cytoplasmic zinc but they regulate very different structural proteins with different consequences for the cell biology of zinc.

in more zinc deficient environments which have selected for more efficient sequestration and storage of endogenous zinc?

II. Repetitive DNA in Cyanobacteria

A. The Known Repeats

A number of repeated DNA sequences have been described in the genomes of different cyanobacteria. Three types of short tandemly repeated repetitive (STRR) sequences were first observed in *Calothrix* PCC 7601. A repeated heptanucleotide forms the core STRR sequence, and this is known to also be present in *Nostoc* PCC 7906 and *Anabaena* PCC 7120 (Mazel et al., 1990). STRR elements are estimated to occur approximately 100 times within the 12 Mb genome of *Calothrix* PCC 7601. STRR elements are even more abundant in *Anabaena azollae* than *Anabaena* PCC 7120, but an examination of site location suggested that the elements have a different distribution even in these closely related strains (Jackman and Mulligan, 1995). Six new STRR sequences were also identified in *Anabaena azollae*. In addition to STRR sequences, *Anabaena* PCC 7120, and several other cyanobacteria, also contain long(er) tandemly repeated repetitive (LTRR) sequences which are 37 bp in length, found in clusters and also conserved in sequence in several strains (Masepohl et al., 1996). Two repeats, designated REP-A and REP-B have been described in *Microcystis* sp. which appear to be absent from *Synechococcus* PCC 7942 and *Synechocystis* PCC 6803 (Asayama et al., 1996).

The octameric sequence 5' GCGATCGC 3', designated HIP1 (for highly iterated palindrome) was first identified at the junctions of a gene (the *smtB* repressor) deletion event in *Synechococcus* PCC 6301 (Gupta et al., 1993). The sequence was noted to occur many times within the *smt* region and subsequent analyses of database entries indicated that HIP1 was over-represented in many cyanobacterial

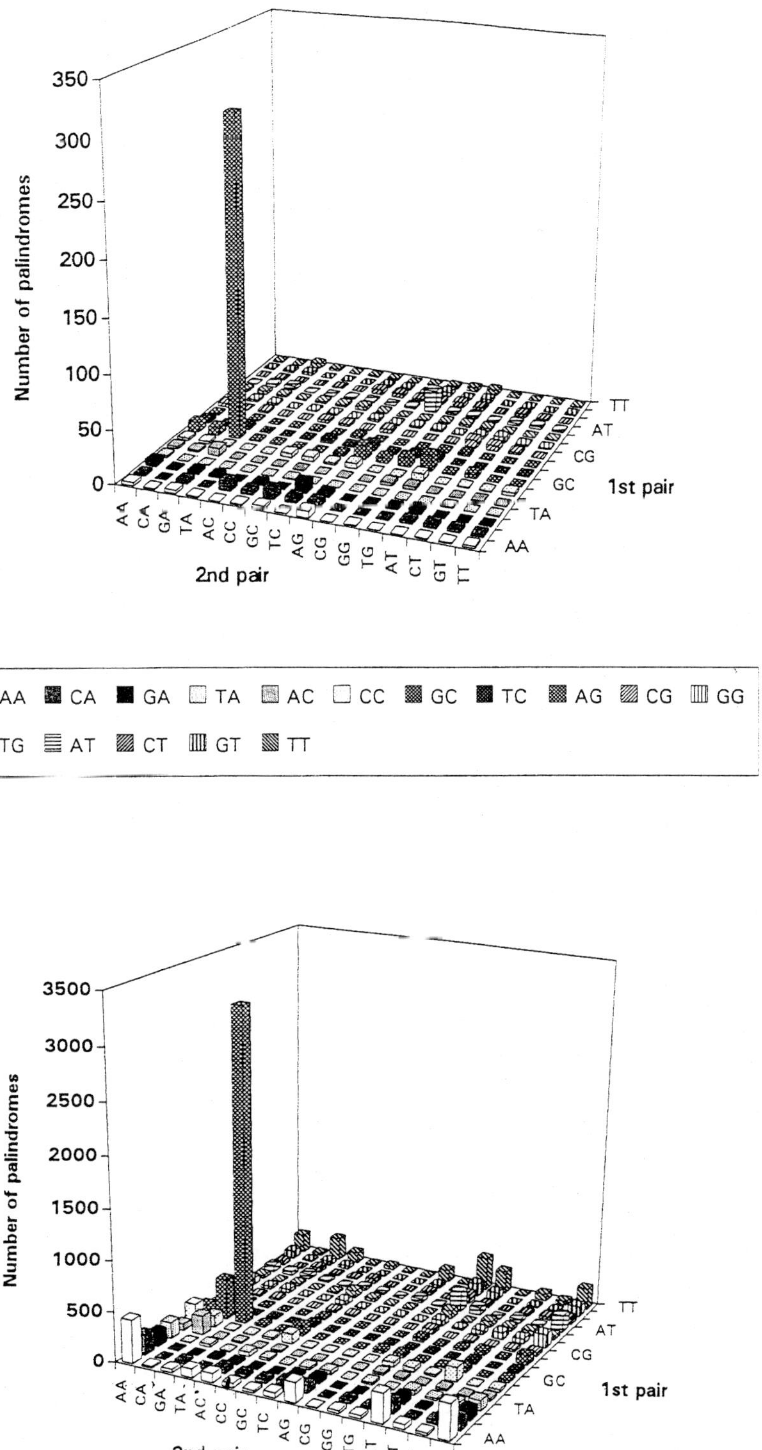

Fig. 5. Singular over-representation of HIP1 in currently known sequence (142, 954 nucleotides) from *Synechococcus* PCC 7942 (upper panel) and in the complete genome of *Synechocystis* PCC 6803 (bottom panel). The first and second nucleotide pairs are listed on the axes (from left to right and from front to back) in the order shown on the key. The patterns on the key correspond to the first nucleotide pair.

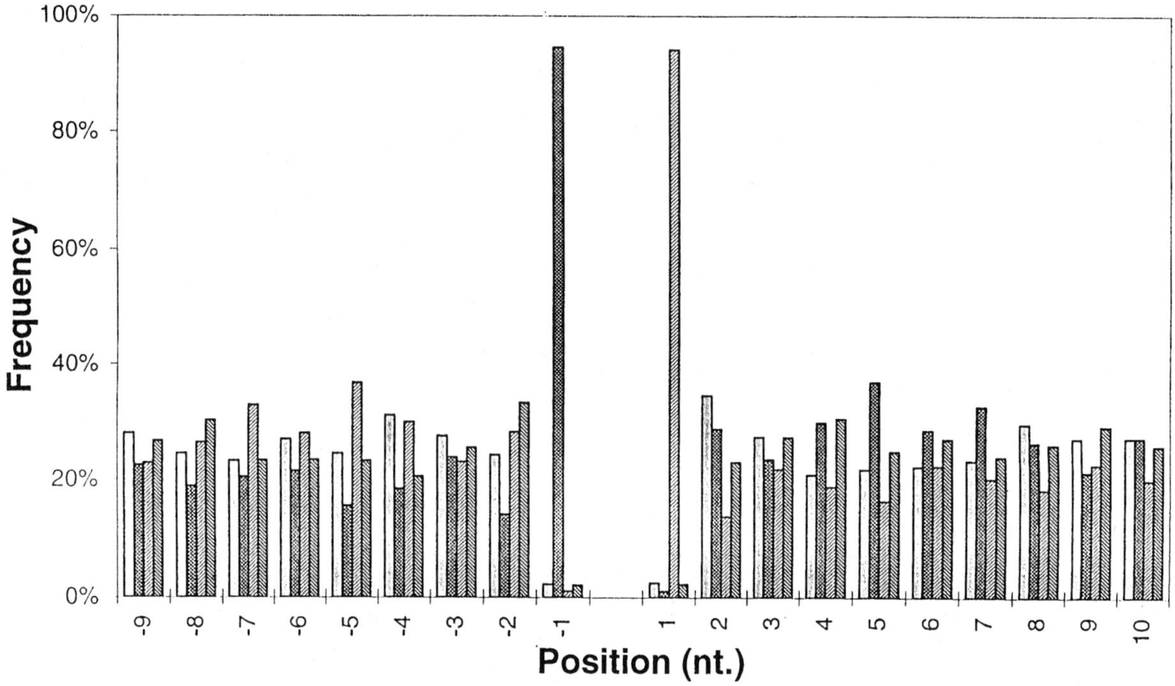

Fig. 6. Analysis of nucleotides flanking octameric HIP1 sites in *Synechocystis* PCC 6803 reveals that the majority of sites are flanked by G and C thus identifying a novel over-represented decameric sequence in this organism (designated $HIP1_{n=10}$). The frequency of each nucleotide 5' (negative nucleotide positions) and 3' (positive nucleotide positions) are represented in the order AGCT respectively.

genomes (Robinson et al., 1995). An analysis of the frequency of occurrence of all 256 possible octameric palindromes revealed that there was singular over-representation of HIP1 in several strains (Fig. 5). This sequence constituted 2.5% of novel sequences obtained from clones selected at random from a library prepared from *Synechococcus* PCC 6301. This is equivalent to a site occurring on average once every 320 nucleotides within this organism. However, analysis of equivalent "random" sequences from one strain, *Calothrix* D253, indicated that HIP1 is not abundant in all cyanobacterial genomes (Robinson et al., 1995).

An oligonucleotide including HIP1 gave multiple products with template DNA from cyanobacteria known to contain abundant HIP1, but not from strains in which HIP1 is rare, or absent (Robinson et al., 1995). Furthermore characteristic HIP1-PCR banding patterns were obtained from some strains suggesting possible applications in strain identification and/or the diagnostic testing of environmental samples for the presence of cyanobacteria.

Examination of the sequences flanking all of the HIP1 sites in the complete genome of *Synechocystis* PCC 6803 reveals a remarkable consensus (Fig. 6) with an extended HIP1 site 5' GGCGATCGCC 3' being present in this organism. This decameric version of HIP1 (here designated $HIP1_{n=10}$) is distributed throughout the genome (Fig. 7) and is not clustered into one region (for example proximal or distal to the origin of replication). $HIP1_{n=10}$ is found both outside, and within, protein coding regions and occurs in all three reading frames.

However, the extended HIP1 decamer does not account for the majority of the HIP1 sites in *Synechococcus* PCC 7942, where only the octamer appears to be over-represented. In known sequences from *Anabaena* PCC 7120 a different decameric variant of HIP1, 5' AGCGATCGCT 3' appears to be over represented (unpublished observations).

Fig. 7. The decameric variant of HIP1 ($HIP1_{n=10}$) is distributed throughout the complete genome of *Synechocystis* PCC 6803 and occurs both inside and outside open reading frames. The upper lines represent the locations of the decamers and the lower bars show the positions of open reading frames. Less than ten copies in total of any decameric sequence would be expected to occur by random chance within this entire genome.

B. Evolutionary Considerations

In common with many other repetitive DNA sequences, the manner by which HIP1 propagates in cyanobacterial DNA is unknown. Its proliferation could merely reflect rates of propagation and drift to fixation in excess of rates of loss by negative purifying selection and drift to extinction. If this were the case, then HIP1 could be considered to be essentially parasitic. Alternatively, the proliferation of HIP1 could be advantageous to the organism and hence the sequence conserved. This would imply that HIP1 is functional. Functions attributed to other repeats in other groups of organisms include roles in the termination, stability and regulation of translation of certain transcripts (Gilson et al., 1986; Newbury et al., 1987a, b; Stern et al., 1988), the promotion of genomic rearrangements (Gilson et al., 1984; Shyamala et al., 1990), the chromosomal integration of DNA and the organisation of bacterial chromosome structure (Gilson et al., 1984; Gilson et al.,1987; Stern et al., 1988). At present, no clear function has been attributed to HIP1 but there is evidence to suggest that it is more likely to be functional rather than parasitic.

One feature which distinguishes HIP1 from many, although not all, other repetitive DNA elements is its prevalence within protein coding regions (Robinson et al., 1995). This provides an opportunity to investigate the manner by which HIP1 propagates since it is possible to align regions of DNA that encode related proteins with some confidence. Such comparisons of HIP1 containing cyanobacterial sequences with HIP1 free sequences from other bacteria (for example from *E. coli*), reveal that HIP1 does not introduce gaps into such alignments. HIP1 aligns with partial HIP1 sites in homologous sequences from non-cyanobacteria consistent with its propagation via nucleotide substitutions rather than insertion (Robinson et al., 1997). In most cases only

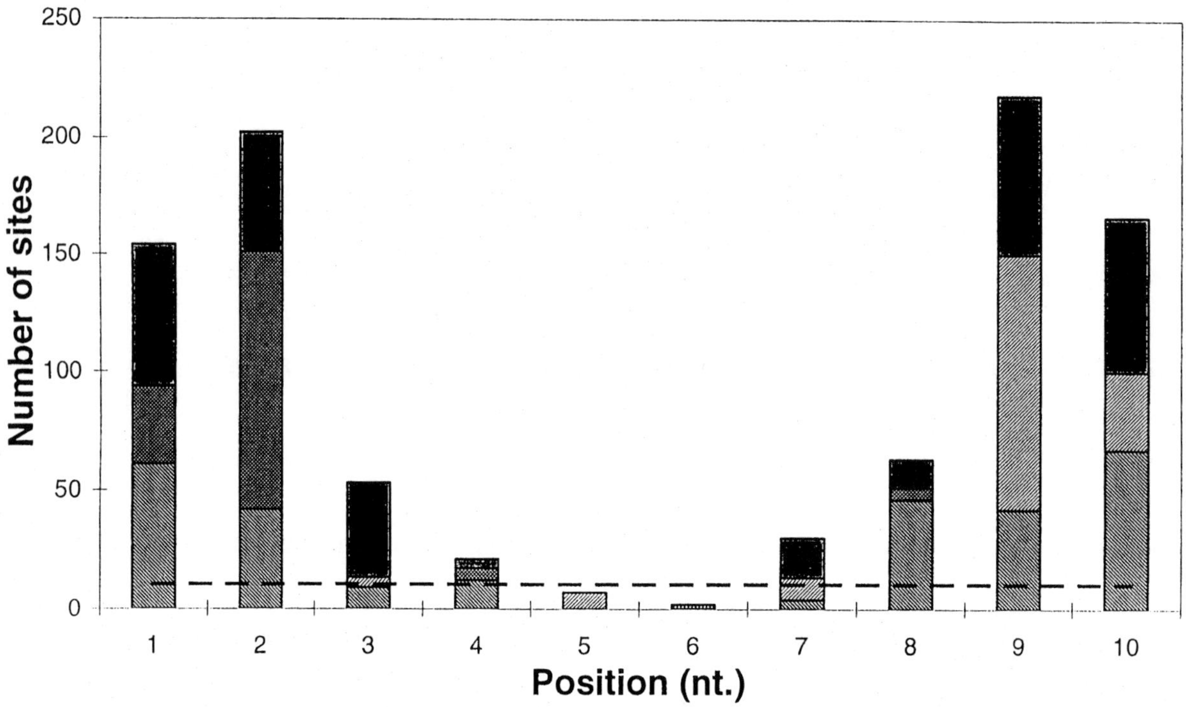

Fig. 8. The number of times that each near-HIP1$_{n=10}$ site occurs within the complete genome of *Synechocystis* PCC 6803. The near-HIP1$_{n=10}$ sites are defined as those in which 9 out of 10 nucleotides match the consensus sequence. It is most common for these sites to have a nucleotide incorrect within the outer four nucleotides. The dashed line shows the predicted number of times any given decamer should occur by chance.

a small number of nucleotide substitutions would be required to create HIP1 sites at equivalent locations in the homologous non-cyanobacterial sequences. Many of the nucleotide differences are either synonymous, or result in only conservative amino acid changes. These observations suggest that either a common ancestor contained partial HIP1 sites which have been lost in other bacteria (such as *E. coli*) or that a common ancestor contained partial HIP1 sites which subsequently accumulated nucleotide substitutions within some cyanobacteria to generate perfect HIP1 sites. Both explanations are consistent with the presence of selection within cyanobacteria to retain HIP1 sites that do not otherwise impair gene function.

Fig. 8 reveals the number of each different 'near' HIP1$_{n=10}$ site, in which 9 out of 10 nucleotides match the consensus sequence, within *Synechocystis* PCC 6803. Of course, it might be anticipated that any sequence closely related to HIP1$_{n=10}$ could be over-represented due to the probability of it arising via mutation of the abundant HIP1$_{n=10}$ sites. Generation of these sites by this mechanism would not be expected to impose any bias on the position of the miss-matched nucleotide. However, the data (Fig. 8) show that while miss-matches within the outer four nucleotides are common and do indeed occur at a frequency greater than that anticipated for any given decamer by random chance, miss-matches within the central nucleotides are less common. The approximately symmetrical appearance of this graph presumably reflects the palindromic nature of the sequence, and indicates that the observed differences in occurrence are significant. One possible interpretation of these observations is that there is also selection for near-HIP1$_{n=10}$ sites provided only one of the outer nucleotides is incorrect. One obvious explanation is that this reflects the specificity of a putative binding protein. However, attempts to detect such a protein by gel retardation assays have not, to date, identified any complexes which form specifically and exclusively with HIP1 (Robinson et al., 1997).

Since HIP1 was first identified at the borders of a gene deletion it has been hypothesised that it could promote widespread gene rearrangement, generating a pool of genetic diversity and thereby enhancing adaptation to environmental change. However, frequencies of deletion between a defined pair of HIP1 sites were shown to be similar in *E. coli* and in *Synechococcus* PCC 7942 (Robinson et al., 1997) indicating that there is not a mechanism of HIP1-mediated gene deletion which is peculiar to cyanobacteria. Data indicate deletion by a copy-choice mechanism, most probably replication slippage (Robinson et al., 1997). Nonetheless, it remains formally possible that an innate propensity of HIP1 to promote intramolecular recombination by replication slippage could be advantageous and cause selection for its proliferation in some cyanobacteria. It is also feasible that the properties of this sequence make it a suitable substrate for intramolecular recombination. Such a function has been proposed for repeated sequences in the radiation resistant bacterium *Deinococcus radiodurans* (Fujitani et al., 1991). In common with many cyanobacteria (Mann and Carr, 1974) *D. radiodurans* contains multiple chromosomes per cell (cited in Minton, 1994). Its remarkable resistance to double strand DNA breaks caused by ionising radiation is thought to involve extensive recombination amongst these chromosomes (Minton, 1994). It has been hypothesised that such proficiency at DNA-repair allows the deinobacteria to survive extensive oxidative DNA-damage accumulated during prolonged desiccation in its natural habitats (Minton, 1994). The presence of multiple chromosome copies in many cyanobacteria is similarly thought to be an important factor in increasing desiccation tolerance in some strains (Potts, 1994). It is therefore hypothesised that HIP1 may promote recombination repair following desiccation and/or exposure to other factors that cause DNA-damage.

Phylogenetic analysis, based upon DNA sequences encoding 16S rRNA, of the species distribution of HIP1 over-representation indicates that this trait does not define a clade (Robinson et al., 1997). HIP1-rich strains are interspersed with more closely related HIP1 free strains. This could be consistent with the lateral transfer of an inserted DNA element between some species of cyanobacteria, however, as already discussed, sequence alignments have indicated that HIP1 is unlikely to be an inserted DNA sequence. The polyphyletic distribution of HIP1 therefore indicates either (i) a common origin of HIP1 over-representation which pre-dates speciation events followed by loss of this trait from several strains or (ii) multiple independent origins of HIP1 over-representation within the cyanobacteria. There are examples of the use of phylogenetic analyses to investigate whether unusual genetic phenomena are primitive but subsequently lost from some lineages, or more recent and only acquired in some lineages (Landweber and Gilbert, 1994). It is not possible to resolve which applies to HIP1 since neither the presence or absence of HIP1 can be confined to the more recently diverging branches. This is because it is often difficult to interpret early branching order in the cyanobacteria with any confidence. It is thought that many modern cyanobacterial lineages arose in a rapid evolutionary radiation and the resulting similar depths of these early branchings underlie uncertainty in the branching order of these organisms (Giovannoni et al., 1988).

Likely explanations for the observed species distribution of HIP1 over-representation are that it has either been lost from species in which it has ceased to confer a selective advantage or only ever acquired by those in which it is advantageous. A rigorous investigation of the habitat distribution of HIP1-rich species may therefore identify correlates which reveal why it has been selected. For example, *Calothrix desertica* was isolated from fine desert sand in Chile (Rippka et al., 1979), while the closely related *Calothrix* D253 was isolated from a mangrove in Cuba. It will be of interest to establish whether the former, unlike the latter, has massive over-representation of HIP1 within its nucleic acids. An extensive analysis of the HIP1 content of multiple strains of cyanobacteria from habitats which are prone to desiccation, or in which this is unlikely (for example marine habitats), could prove to be informative. Furthermore, a detailed inspection of the genes in *Synechocystis* PCC 6803 which do, or do not, contain $HIP1_{n=10}$ may aid the identification of factors which have encouraged its proliferation.

III. Concluding Remarks

With the development of PCR-based methods for rapidly analysing DNA sequences from organisms in different environments, the technology of molecular genetics is increasingly being used to address ecological, environmental and evolutionary questions.

Most of the chapters provide elegant examples of such applications of molecular ecology. In the context of both the study of repetitive DNA and of genes involved in metal-metabolism and toxicity, this chapter provides examples of the converse; namely the use of environmental knowledge to enhance understanding of biochemistry and genetics. The complete sequencing of microbial genomes has revealed a substantial number of functionally unknown, 'FUN', genes (at the time of editing this article in the region of 30% of genes within several fully sequenced bacterial genomes fall into this category). Others show sequence similarities to known genes but their functions are only tentatively assigned. The systematic deletion of genes has often failed to reveal obvious phenotypes in culture, while some of the phenotypes which have been observed are likely to be gratuitous. Researchers engaged in the quest for gene function are increasingly eager to understand the challenges imposed by natural environments.

Acknowledgements

Research on metal-gene-interactions and HIP1 is supported by the Biotechnology and Biological Sciences Research Council and the Natural Environmental Research Council.

References

Asayama M, Kabasawa M, Takahashi I, Aida T and Shirai M (1996) Highly repetitive sequences and characteristics of genomic DNA in unicellular cyanobacterial strains. FEMS Microbiol Lett 137: 175–181

Berg JM and Shi Y (1996) The galvanization of biology: A growing appreciation for the roles of zinc. Science 271: 1081–1085

Bergsland KJ and Haselkorn R (1991) Evolutionary relationships among eubacteria, cyanobacteria and chloroplasts: evidence from the *rpoC1* gene of *Anabaena* sp. strain PCC 7120. J Bacteriol 173: 3446–3455

Berkelman T, Garret-Engele P and Hoffman NE (1994) The *pacL* gene of *Synechococcus* sp. strain PCC 7942 encodes a Ca^{2+} -transporting ATPase. J Bacteriol 176: 4430–4436

Binder BJ and Chisholm SW (1990) Relationship between DNA-cycle and growth-rate in *Synechococcus* sp. strain PCC 6301. J Bacteriol 172: 2313–2319

Binder BJ and Chisholm SW (1995) Cell-cycle regulation in marine *Synechococcus* sp. strains. Appl Eviron Microbiol 61: 708–717

Bird AJ, Turner-Cavet JS, Lakey JH and Robinson NJ (1998) A carboxyl-terminal Cys_2/His_2-type zinc-finger motif in DNA primase influences DNA content in *Synechococcus* PCC 7942. J Biol Chem 273: 21246–21252

Block MA and Grossman AR (1988) Identification and purification of a derepressible alkaline phosphatase from *Anacystis nidulans* R2. Plant Physiol 86: 1179–184

Bohner H and Böger P (1978) Reciprocal formation of cytochrome c-553 and plastocyanin in *Scenedesmus*. FEBS Lett 85: 337–339

Bovy A, de Vrieze G, Borrias M and Weisbeek P (1992) Transcriptional regulation of the plastocyanin and cytochrome c_{553} genes from the cyanobacterium *Anabaena* sp. PCC 7937. Mol Microbiol 6: 1507–1513

Boyle E, Sciater FR and Edmond JM (1976) On the marine geochemistry of cadmium. Nature (London) 263: 42–44

Bruland KW, Knauer GA and Martin JH (1978) Cadmium in northeast Pacific waters. Limnol Oceanog 23: 618–626

Cano-Gauci DF and Sarkar B (1996) Reversible zinc exchange between metallothionein and the estrogen receptor zinc finger. FEBS Lett 386: 1–4

Carr NG and Mann NH (1994) The oceanic cyanobacterial picoplankton. In: Bryant DA (ed) The Molecular Biology of Cyanobacteria, pp 27–48. Kluwer Academic Publishers, Dordrecht

Chintamani A and Mohanty P (1989) Effect of elevated levels of zinc on growth of *Synechococcus* PCC 6301. Zentralblatt für Mikrobiologie 144: 531–536

Clark RB (1992) Marine Pollution. Oxford University Press, Oxford

Conklin DS, McMaster JA, Culbertson MR and Kung C (1992) *COT1*, a gene involved in cobalt accumulation in *Saccharomyces cerevisiae*. Mol Cell Biol 12: 3678–3688

Cook WJ, Kar SR, Taylor KB and Hall LM (1998) Crystal structure of the cyanobacterial metallothionein repressor SmtB: A model for metalloregulatory proteins. J Mol Biol 275: 337–346

Daniels MJ, Turner-Cavet JS, Selkirk R, Sun H, Parkinson JA, Sadler PJ and Robinson NJ (1998) Coordination of Zn^{2+} (and Cd^{2+}) by prokaryotic metallothionein: Involvement of His-imidazole. J Biol Chem 273: 22957–22961

Diels L, Dong Q, van der Lelie D, Baeyens W and Mergeay M (1995) The *czc* operon of *Alcaligenes eutrophus* CH34: from resistance mechanism to the removal of heavy metals. J Ind Microbiol 14: 142–153

Dintilhac A, Alloing G, Granadel C and Claverys JP (1997) Competence and virulence of *Streptococcus pneumoniae*: Adc and PsaA mutants exhibit a requirement for Zn and Mn resulting from inactivation of putative ABC metal permeases. Mol Microbiol 25: 727–739

Endo G and Silver S (1995) CadC, the transcriptional regulatory protein of the cadmium resistance system of *Staphylococcus aureus* plasmid pI258. J Bacteriol 177: 4437–4441

Erbe JL, Taylor KB and Hall LM (1995) Metalloregulation of the cyanobacterial *smt* locus: identification of SmtB-binding sites and direct interaction with metals. Nucl Acids Res 23: 2472–2478

Förstner U (1983) Binding forms of heavy-metals in sediments and sludges - sorption, mobilisation, chemical-extraction and bioavailability. Fresensius Zeitschrift Für Analytische Chemie 316: 604–611

Fraústo da Silva JJR and Williams RJP (1991) Zinc: Lewis acid catalysis and regulation. In: The Biological Chemistry of the Elements, pp 299– 318. Clarendon Press, Oxford

Fujitani S, Komano T and Inouye S (1991) A unique repetitive DNA sequence in the *Myxococcus xanthus* genome. J Bacteriol 173: 2125–2127

Fukuzawa H, Suzuki E, Komukai Y and Miyachi S (1992) A gene homologous to chloroplast carbonic anhydrase (*icfA*) is essential

to photosynthetic carbon dioxide fixation by *Synechococcus* PCC 7942. Proc Natl Acad Sci USA 89: 4437–4441
Ghassemian M and Straus NA (1996) Fur regulates the expression of iron-stress genes in the cyanobacterium *Synechococcus* sp. strain PCC 7942. Microbiology 142: 1469–1476
Gilson E, Clément J-M, Brutlag D and Hofnung M (1984) A family of dispersed repetitive extragenic palindromic sequences in *E. coli* EMBO J 3: 1417–1421
Gilson E, Clément J-M, Perrin D and Hofnung M (1987) Palindromic units: a case of highly repetitive DNA sequences in bacteria. Trends Genet 3: 226–230
Gilson E, Rousset JP, Clément J-M and Hofnung M (1986) A subfamily of *E. coli* palindromic units implicated in transcription termination? Ann Inst Pasteur/Microbiol 137B: 259–270
Giovannoni SJ, Turner S, Olsen GJ, Barns S, Lane DJ and Pace NR (1988) Evolutionary relationships among cyanobacteria and green chloroplasts. J Bacteriol 170: 3584–3592
Griep MA and Lokey ER (1996) The role of zinc and the reactivity of cysteines in *E. coli* primase. Biochemistry 35: 8260–8267
Grossman AR, Schaefer MR, Chiang GG and Collier JL (1994) The responses of cyanobacteria to environmental conditions: Light and nutrients. In: Bryant DA (ed) The Molecular Biology of Cyanobacteria, pp 641–675. Kluwer Academic Publishers, Dordrecht
Gupta A, Morby AP, Turner JS, Whitton BA and Robinson NJ (1993) Deletion within the metallothionein locus of Cd-tolerant *Synechococcus* PCC 6301 involving a highly iterated palindrome (HIP1). Mol Microbiol 7: 189–195
Gupta A, Whitton BA, Morby AP, Huckle JW and Robinson NJ (1992) Amplification and rearrangement of a prokaryotic metallothionein locus *smt* in *Synechococcus* PCC 6301 selected for tolerance to cadmium. Proc Roy Soc (London) B 248: 273–281
Huckle JW, Morby AP, Turner JS and Robinson NJ (1993) Isolation of a prokaryotic metallothionein locus and analysis of transcriptional control by trace metal ions. Mol Microbiol 7: 177–187
Jackman DM and Mulligan ME (1995) Characterization of a nitrogen-fixation (*nif*) gene-cluster from *Anabaena azollae* 1a shows that closely related cyanobacteria have highly variable but structured intergenic regions. Microbiol 141: 2235–2244
Jensen TE, Baxter M, Rachlin JW and Jani V (1982) Uptake of heavy metals by *Plectonema boryanum* (Cyanophyceae) into cellular components, especially polyphosphate bodies: An x-ray energy dispersive study. Eviron Pollut (Series A) 27: 119–127
Ji G and Silver S (1992) Regulation and expression of the arsenic resistance operon from *Staphylococcus aureus* plasmid pI258. J Bacteriol 174: 3684–3694
Jones MC, Jenkins JM, Smith AG and Howe CJ (1994) Cloning and characterisation of genes for tetrapyrrole biosynthesis from the cyanobacterium *Anacystis nidulans* R2. Plant Mol Biol 24: 435–448
Kägi JHR and Schäffer A (1988) Biochemistry of metallothionein. Biochemistry 27: 8509–8515
Kamizomo A, Nishizawa M, Teranishi Y, Murata K and Kimura A (1989) Identification of a gene conferring resistance to zinc and cadmium ions in the yeast *Saccharomyces cerevisiae*. Mol Gen Genet 219: 161–167
Kanamura K, Kashiwagi S and Mizuno T (1993) The cyanobacterium, *Synechococcus* species PCC7942, possesses two distinct genes encoding cation-transporting P-type ATPases. FEBS Lett 330: 99–104
Kanamura K, Kashiwagi S and Mizuno T (1994) A copper-transporting P-type ATPase found in the thylakoid membrane of the cyanobacterium *Synechococcus* species PCC7942. Mol Microbiol 13: 369–377
Kaneko T, Sato S, Kotani H, Tanaka A, Asimizu E, Nakamura N, Miyajima N, Hirosawa M, Sugiura M, Sasamoto M, Kimura T, Hosouchi T, Matsuno A, Muraki A, Nakazaki N, Naruo K, Okumura S, Shimpo S, Takeuchi C, Wada T, Watanabe A, Yamada M, Yasuda C and Tabata S (1996) Sequence analysis of the genome of the unicellular cyanobacterium *Synechocystis* sp. strain PCC6803. II. Sequence determination of the entire genome and assignment of potential protein-coding regions. DNA Res 3: 109–136 & Suppl 185–209
Landweber LF and Gilbert W (1994) Phylogenetic analysis of RNA editing: A primitive genetic phenomenon. Proc Natl Acad Sci USA 91: 918–921
Lebrun M, Audurier A and Cossart P (1994) Plasmid-borne cadmium resistance genes in *Listeria monocytogenes* are similar to *cadA* and *cadC* of *Staphylococcus aureus* and are induced by cadmium. J Bacteriol 176: 3040–3048
Lukac M and Aegerter R (1993) The influence of trace metals on growth and toxin production of *Microcystis aeruginosa.* Toxicon 31: 293–305
Lutsenko S and Kaplan JH (1996) P-type ATPases. Trends Biochem Sci 21: 467
Maclean FI, Lucis OJ, Shakh ZA and Jansz ER (1972) The uptake and subcellular distribution of Cd and Zn in microorganisms. Fed Proc 31: 699
Mann NH and Carr NG (1974) Control of macromolecular composition and cell division in the blue-green alga *Anacystis nidulans.* J Gen Microbiol 83: 399–405
Masepohl B, Görlitz K and Böhme H (1996) Long tandemly repeated repetitive (LTRR) sequences in the filamentous cyanobacterium *Anabaena* sp. PCC 7120. Biochim Biophys Acta 1307: 26–30
Mason CF (1991) Biology of freshwater pollution. Longman Scientific and Technical, Harlow
Mazel D, Houmard J, Castets AM and Tandeau de Marsac N (1990) Highly repetitive DNA sequences in cyanobacterial genomes. J Bacteriol 172: 2755–2761
Merchant S and Bogorad L (1986) Regulation by copper of the expression of plastocyanin and cytochrome c-552 in *Chlamydomonas reinhardtii*. Mol Cell Biol 6: 462–469
Mergeay M, Houba C and Gerits J (1978) Extrachromosomal inheritance controlling resistance to cadmium, cobalt and zinc ions: evidence from curing in a *Pseudomonas*. Arch Int Physiol Biochim 86: 440–441
Minton KW (1994) DNA repair in the extremely radioresistant bacterium *Deinococcus radiodurans*. Mol Microbiol 13: 9–15
Morand LZ, Cheng RH, Krogmann DW and Ho KK (1994) Soluble electron transfer catalysts of cyanobacteria. In: Bryant DA (ed) The Molecular Biology of Cyanobacteria, pp 381–407. Kluwer Academic Publishers, Dordrecht
Morby AP, Turner JS, Huckle JW and Robinson NJ (1993) SmtB is a metal regulated repressor of the cyanobacterial metallothionein gene *smtA*: identification of a Zn inhibited DNA-protein complex. Nucl Acids Res 21: 921–925
Newbury SF, Smith, NH and Higgins CF (1987a) Differential mRNA stability controls relative gene expression within a polycistronic operon. Cell 51: 1131–1143
Newbury SF, Smith NH, Robinson EC, Hiles ID and Higgins CF (1987b) Stabilization of translationally active mRNA by prokaryotic REP sequences. Cell 48: 297–310

Nies DH (1992) CzcR and CzcD, gene products affecting regulation of cobalt, zinc and cadmium resistance (*czc*) in *Alcaligenes eutrophus*. J Bacteriol 174: 8102–8110

Nies DH and Silver S (1995) Ion efflux systems involved in bacterial metal resistances. J Ind Microbiol 14: 186–199

Nies DH, Nies A, Chu L and Silver S (1989) Expression and nucleotide sequence of plasmid determined divalent cation efflux system from *Alcaligenes eutrophus*. Proc Natl Acad Sci USA 86: 7351–7355

Nishimura A (1998) The timing of cell division: Ap4A as a signal. Trends Biochem Sci 23: 157–159

Ogura Y, Yoshida T, Nakamura Y, Takemura M, Oda K and Ohyama K (1991) Gene encoding a putative zinc finger protein in *Synechocystis* PCC 6803. Agric Biol Chem 55: 2259–2264

O'Halloran TV (1993) Transition metals in control of gene expression. Science 261: 715–725

Olafson RW (1986) Physiological and chemical characterization of cyanobacterial metallothioneins. Environ Health Perspect 65: 71–75

Olafson RW, Abel K and Sim RG (1979) Prokaryotic metallothionein: primary characterization of a blue green alga heavy metal binding protein. Biochem Biophys Res Commun 89: 36–43

Olafson RW, McCubbin WD and Kay CM (1988) Primary- and secondary-structural analysis of a unique prokaryotic metallothionein from a *Synechococcus* sp. cyanobacterium. Biochem J 251: 691–699

Palenik B and Haselkorn R (1992) Multiple evolutionary origins of prochlorophytes, the chlorophyll *b*-containing prokaryotes. Nature (London) 355: 265–267

Palmiter RD (1998) The elusive function of metallothioneins. Proc Natl Acad Sci USA 95: 8428–8430

Parkinson JS and Kofoid EC (1992) Communication modules in bacterial signalling proteins. Annu Rev Genetic 26: 71—112

Patzer SI and Hantke K (1998) The ZnuABC high-affinity zinc uptake system and its regulator Zur in *Escherichia coli*. Mol Microbiol 28: 1199–1210

Phung LT, Ajlani G and Haselkorn R (1994) P-type ATPase from the cyanobacterium *Synechococcus* 7942 related to the human Menkes and Wilson disease gene products. Proc Natl Acad Sci USA 91: 9651–9654

Plateau P, Mayaux JF and Blanquet S (1981) Zinc(II)-dependent synthesis of diadenosine 5', 5''' -P(1) ,P(4) -tetraphosphate by *Escherichia coli* and yeast phenylalanyl transfer ribonucleic acid sythetases. Biochemistry 20: 4654–4662

Potts M (1994) Desiccation tolerance of prokaryotes. Microbiol Rev 58: 755–805

Price NM and Morel FMM (1990) Cadmium and cobalt substitution for zinc in a marine diatom. Nature (London) 344: 658–660

Rai LC, Gaur JP and Kumar HD (1981) Phycology and heavy-metal pollution. Biol Rev 56: 99–151

Ray JM, Bhaya D, Block MA and Grossman AR (1991) Isolation, transcription and inactivation of the gene for an atypical alkaline phosphatase of *Synechococcus* sp. strain PCC 7942. J Bacteriol 173: 4297–4309

Reed RH and Gadd GM (1990) Metal tolerance in eukaryotic and prokaryotic algae. In: Shaw AJ (ed) Heavy Metal Tolerance in Plants: Evolutionary Aspects, pp105–118. CRC Press, Boca Raton

Rippka R, Deruelles J, Waterbury JB, Herdman M and Stanier RY (1979) Generic assignments, strain histories and properties of pure cultures of cyanobacteria. J Gen Microbiol 111: 1–61

Robinson PJ, Cranenburgh RM, Head IM and Robinson NJ (1997) HIP1 propagates in cyanobacterial DNA via nucleotide substitutions but promotes excisions at similar frequencies in *E. coli* and *Synechococcus* PCC 7942. Mol Microbiol. 24: 181–189

Robinson NJ, Gupta A, Fordham-Skelton AP, Croy RRD, Whitton BA and Huckle JW (1990) Prokaryotic metallothionein gene characterisation and expression: chromosome crawling by ligation mediated PCR. Proc Roy Soc (London) B 242: 241–247

Robinson NJ, Robinson PJ, Gupta A, Bleasby AJ, Whitton BA and Morby AP (1995) Singular over-representation of an octameric palindrome, HIP1, in DNA from many cyanobacteria. Nucl Acids Res 23: 729–735

Rosenstein R, Nikoleit K and Götz F (1992) Binding of ArsR, the repressor of the *Staphylococcus xylosus* (pSX267) arsenic resistance operon to a sequence with dyad symmetry within the *ars* promoter. Mol Gen Genet 242: 566–572

Sandmann G (1986) Formation of plastocyanin and cytochrome c_{553} in different species of blue-green algae. Arch Microbiol 145: 76–79

San Francisco MJD, Hope CL, Owolabi JB, Tisa LS and Rosen BP (1990) Identification of the metalloregulatory element of the plasmid-encoded arsenical resistance operon. Nucl Acids Res 18: 619–624

Say PJ and Whitton BA (1977) Influence of zinc on lotic plants. II. Environmental effects on toxicity of zinc to *Hormidium rivulare*. Freshwater Biol 7: 377–384

Say PJ and Whitton BA (1981) Changes in flora down a stream showing a zinc gradient. Hydrobiologia 76: 255–262

Shehata FHA and Whitton BA (1982) Zinc tolerance in strains of the blue-green alga *Anacystis nidulans*. Br Phycol J 17: 5–12

Shi J, Lindsay WP, Huckle JW, Morby AP and Robinson NJ (1992) Cyanobacterial metallothionein gene expressed in *E. coli*-metal binding properties of the expressed protein. FEBS Lett 303: 159–163

Shi W, Wu J and Rosen BP (1994) Identification of a putative metal binding site in a new family of metalloregulatory proteins. J Biol Chem 269: 19826–19829

Shimizu TT, Hiyam T, Ikeuchi M and Inoue Y (1992) Nucleotide sequence of a metallothionein gene of the thermophilic cyanobacterium *Synechococcus vulcanus*. Plant Mol Biol 20: 565–567

Shyamala V, Schneider E and Ames GF-L (1990) Tandem chromosomal duplications: role of REP sequences in the recombination event at the join point. EMBO J 9: 939–946

Silver S , Nucifora G and Phung LT (1993) Human Menkes X-chromosome disease and the staphylococcal cadmium-resistance ATPase: a remarkable similarity in protein sequence. Mol Microbiol 10: 7–12

Simpson RT and Vallee BL (1968) Two differential classes of metal atoms in alkaline phosphatase of *E. coli*. Biochem 7: 4343–4350

Singh SP and Pandey AK (1982) Cadmium-mediated resistance to metals and antibiotics in a cyanobacterium. Mol Gen Genet 187: 240–243

Solioz M and Vulpe C (1996) CPx-type ATPases: a class of P-type ATPases that pump heavy metals. Trends Biochem Sci 21: 237–241

Stamford NPJ, Lilley PE and Dixon NE (1992) Enriched sources of *E. coli* replication proteins. The *dnaG* primase is a zinc metalloprotein. Biochim Biophys Acta 1132: 17–25

Stern MJ, Prossnitz E and Ames GF-L (1988) Role of the intercistronic region in post-transcriptional control of the gene expression in the histidine transport operon of *Salmonella typhimurium*: involvement of REP sequences. Mol Microbiol 2: 41–152

Takamura N, Kasai F and Watanabe MM (1989) Effects of Cu, Cd and Zn on photosynthesis of freshwater benthic algae. J Appl Phycol 1: 39–52

Takatera K and Watanabe T (1992) Application of high performance liquid chromatography/inductively coupled plasma mass spectrometry to the speciation of cadmium-binding metallothionein-like protein in a cyanobacterium. Anal Sci: 469–474

Tandeau de Marsac N and Houmard J (1993) Adaptation of cyanobacteria to environmental stimuli: new steps towards molecular mechanisms. FEMS Microbiol Rev 104: 119–190

Thelwell C, Robinson NJ and Turner-Cavet JS (1998) An SmtB-like repressor from Synechocystis PCC 6803 regulates a zinc exporter. Proc Natl Acad Sci USA 95: 10728–10733

Tougu K and Marians KJ (1996) The interaction between helicase and primase sets the replication fork clock. J Biol Chem 271: 21398–21405

Tsujikawa K, Imai T, Kakutani M, Kayamori Y, Mimura T, Otaki N, Kimura M, Fukuyama R and Shimizu N (1991) Localization of metallothionein in nuclei of growing primary cultured adult rat hepatocytes. FEBS Lett 283: 239–242

Turner JS, Glands PD, Samson ACR and Robinson NJ (1996) Zn^{2+}-sensing by the cyanobacterial metallothionein repressor SmtB: different motifs mediate metal-induced protein-DNA dissociation. Nucl Acids Res 24: 3714–3721

Turner JS, Morby AP, Whitton BA, Gupta A and Robinson NJ (1993) Construction and characterisation of Zn^{2+}/Cd^{2+} hypersensitive cyanobacterial mutants lacking a functional metallothionein locus. J Biol Chem 268: 4494–4498

Turner JS and Robinson NJ (1995) Cyanobacterial metallothioneins: Biochemistry and molecular genetics. J Ind Microbiol 14: 119–125

Turner JS, Robinson NJ and Gupta A (1995) Construction of Zn^{2+}/Cd^{2+}-tolerant cyanobacteria with a modified metallothionein divergon: Further analysis of the function and regulation of *smt*. J Ind Microbiol 14: 259–264

Vallee BL (1991) Introduction to metallothionein. Meth Enzymol 205: 3–7

Vallee BL and Falchuk KH (1993) The biochemical basis of zinc physiology. Physiol Rev 73: 79–118

Van der Lelie D, Schwuchow T, Schwidetzky U, Wuertz S, Baeyens W, Mergeay M and Nies H (1997) Two-component regulatory system involved in transcriptional control of heavy-metal homoeostasis in *Alcaligenes eutrophus*. Mol Microbiol 23: 493–503

Verma SK and Singh HN (1991) Evidence for energy-dependent copper efflux as a mechanism of Cu^{2+} resistance in the cyanobacterium *Nostoc calcicola*. FEMS Microbiol Lett 84: 291–294

Versalovic J, Koeuth T, Britton R , Geszvain K and Lupski JR (1993) Conservation and evolution of the *rps*U-*dnaG*-*rpo*D macromolecular synthesis operon in bacteria. Mol Microbiol 8: 343–355

Vymazal J (1985) Occurrence and chemistry of zinc in freshwaters - its toxicity and bioaccumulation with respect to algae: A review. Part 1: Occurrence and chemistry of zinc in freshwaters. Acta Hydrochimica and Hydrobiologia 13: 627–654

Whitton BA (1980) Zinc and plants in rivers and streams. In: Shubert LE (ed) Algae as Ecological Indicators, pp 257–280. Academic Press, London

Whitton BA and Shehata FHA (1982) Influence of cobalt, nickel, copper and cadmium on the blue-green alga *Anacystis nidulans*. Environ Pollut (Series A) 27: 275–281

Wood RM (1978) Interchangeable copper and iron proteins in algal photosynthesis. Eur J Biochem 87: 9–19

Xiong A and Jayaswal RK (1998) Molecular characterization of a chromosomal determinant conferring resistance to zinc and cobalt ions in *Staphylococcus aureus*. J Bacteriol 180: 4024–4029

Xu C, Shi W and Rosen BP (1996) The chromosomal *arsR* gene of *E. coli* encodes a *trans*-acting metalloregulatory protein. J Biol Chem 271: 2427–2432

Zeng J, Heuchel R, Schaffner W and Kägi JHR (1991a) Thionein (apometallothionein) can modulate DNA binding and transcription activation by zinc finger containing factor Sp1. FEBS Lett 279: 310–312

Zeng J, Vallee BL and Kägi JHR (1991b) Zinc transfer from transcription factor IIIA fingers to thionein clusters. Proc Natl Acad Sci USA 88: 9984–9988

Chapter 17

Nostoc

Malcolm Potts
Department of Biochemistry, Virginia Polytechnic Institute and State University, Blacksburg, VA 24061, USA

B. A. Whitton and M. Potts (eds), The Ecology of Cyanobacteria, 465-504.
© 2000 *Kluwer Academic Publishers. Printed in the Netherlands.*

Summary

Species of *Nostoc* are found on every continent on Earth in a wide range of terrestrial and aquatic ecosystems where their growths become both conspicuous and abundant. Environments where water is scarce, or absent for protracted periods, remain some of the most extreme on the Earth yet they typically support populations of the terrestrial desiccation-tolerant form *N. commune.* The nitrogen-fixing activities of *Nostoc* spp. contribute to the quality of nutrient-poor soils, especially in karst regions, and several forms contribute to the productivity of rice paddies throughout an expansive region of Asia (Chapter 8). Some *Nostoc* spp. enter into a range of associations with higher and lower plants and these include the only known symbiosis with a flowering plant (Chapter 19). The only known endocyanosis of a fungus, *Geosiphon pyriforme,* involves *N. punctiforme* and may represent a type of photoautotrophic association which was one step in the evolution of all land plants. Growths of *Nostoc* elaborate a range of natural products that range from the innocuous, which contribute no more than a mildly offensive odor to potable water supplies, to potent toxins which cause sickness, cell transformation and the death of wildlife and humans (Chapter 22). Despite the fact that *Nostoc* spp., collectively, represent a unique biological resource, there is no adequate species concept for the group; the origins of *Nostoc* and other heterocystous cyanobacteria are unclear; and the physiological and molecular traits which could provide diagnostic features of a *Nostoc* remain elusive (Chapter 1). *Nostoc* is collected and cultivated as a source of food by the indigenous populations of many countries and in at least one of these, China, *Nostoc* occupies a position in human social behavior; it has done so for at least 1500 years. This chapter considers in a broad sense what is known and what needs to be known about *Nostoc* spp.

I. Introduction

There is a considerable accumulation of literature on *Nostoc* spp. and an extensive number of studies are in the Chinese literature. The studies of Mollenhauer (1970, 1985a, 1985b, 1986a, 1986b), recent accounts by Dodds et al. (1995) and Gao (1998), the revision of Komárek and Anagnostidis (1989), and two reviews (Potts 1994, 1996) focused on aspects of the eco-physiology, biochemistry and cell structure of *Nostoc.* A comprehensive consideration of the metabolism and development of heterocysts, and a description of the use of *Nostoc* spp. as genetic tools with which to understand the molecular basis for heterocyst formation, appeared recently (Wolk et al., 1994; Wolk, 1996). The present and limited account is not a review. It is a selective consideration that emphasizes aspects of molecular ecology in order to place concepts raised by other contributors to this volume in perspective.

Abbreviations: ABA, abscisic acid; DSB's, double-strand breaks; EPS, extrapolysaccharide; HIP, highly-iterated palindromic repeat sequence; PAS, periodic acid-Schiff reagent; STRR, short tandem repeat repetitive sequence; UVR, ultra-violet radiation; Wsp, water-stress proteins

II. Natural Communities

A. Niches

1. Karst and Tintenstrich

Species of *Nostoc* enjoy a range that is predominantly terrestrial and descriptions from aquatic systems are almost exclusively to do with freshwater forms. In fact, Komárek and Anagnostidis (1989) concluded that all of the data about *Nostoc* species from submersed and marine biotypes concern other taxa.

One environment that clearly does not appear to be colonized by *Nostoc* is the interior of rocks. This must be due to some physiological constraint because other cyanobacteria, including heterocystous forms

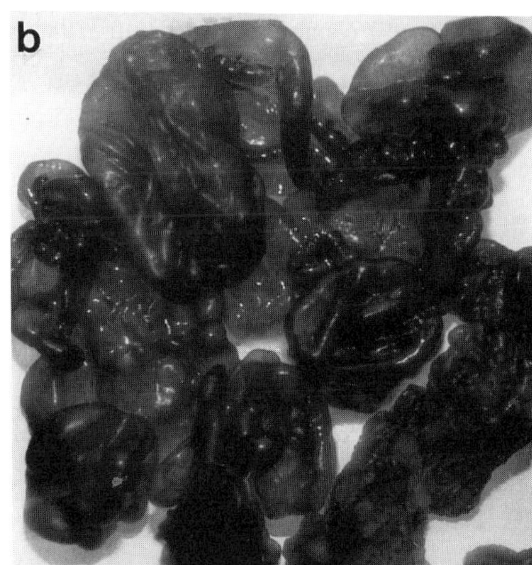

Fig. 1. Nostoc commune.
a) Karst scenery. "Sugar" limestone in the area around Malham Tarn, Ingleborough, England. Communities of *Nostoc commune* accumulate in the shallow depressions of the rock that periodically fill with rainwater. b) Fully hydrated colonies that have the consistency of a semi-rigid gel.

such as *Mastigocoleus testarum*, *Kyrtuthrix dalmatica* and *Scytonema endolithicum,* do infest the rocks and sediments of freshwater, terrestrial and marine localities (Chapters 8,13).

A cursory inspection of nutrient-poor soils, especially those in areas where limestone is prevalent, often reveals the parchment-like colonies of *N. commune* (Scherer, 1991). When dry, the colonies form darkened crusts with the consistency and appearance of burned bacon rinds. These crusts are readily detached and can accumulate in rock hollows that periodically fill with rainwater (Potts, 1994). The characteristic strands of *N. commune* var. *flagelliforme* cover exposed rocks in semi arid areas such as the Gobi Desert in Inner Mongolia (Scherer and Zhong, 1991).

Nostoc commune is especially prevalent in limestone or karst regions such as those of the Burren, Co. Clare, Eire; Malham Tarn, England; and the Adriatic coast of the former Jugoslavia (Fig. 1a). The

Caher River drains the limestone-rich Burren and provides numerous habitats where macroscopic colonies of *Nostoc* and *Rivularia* can be found (Potts, unpublished data; Fig. 1b). Limestone pebbles that are used as a ballast to cover roofs of university buildings, schoolhouses and the like in the United States, as well as gravel paths ("Kiesweg" in German), and flat coral limestone pavement ("platin" in French Creole) in the Tropics (Plate 24c), provide the perfect conditions of water retention and drainage for *Nostoc* growth.

Tintenstrich communities (German for "ink strip") receive their name because of the streaks such cyanobacterial growths display, particularly in most spectacular form on a background of vertical white limestones and dolomites (Lüttge, 1997; Plate 30b). *Nostoc* may occur in such communities but does so typically as more mature crusts that can also form on granitic and sandstone substrates (Lüttge. 1997).

The pH requirements of *Nostoc* spp. from limestone regions suggest a preference for alkaline conditions (Büdel et al., 1993) although a strain of *N. microscopicum* isolated from a rice field in the Philippines showed optimum growth in tap water at a pH between 6 and 7 (Emralino and Rodulfo, 1987).

2. Soils and Water

There is no shortage of microbial inocula anywhere on the planet and any apparent restricted distribution of a given form must be either a real one, with some distinct ecological basis or, the reflection of insufficient documentation of sporadic growths. For example, the soils in certain parts of the Dry Valleys, Antarctica, are truly sterile, and this is not for want of inocula. Mollenhauer (1970) provided an exhaustive listing of sites where *N. punctiforme* was recorded and concluded that the apparent lack of this form in Australia and Antarctica was probably real. However, Seaburg et al. (1979) listed seven different *Nostoc* species, including *N. punctiforme*, from a littoral benthic mat in Lake Bonney, Taylor Valley, Southern Victoria Land, Antarctica. Curiously, the seven species were not recorded from the other Lakes in the valley (Lakes Fryxell, Hoare, Chad and Choice) despite their relative proximity to one another. In contrast *N. commune* was present throughout the area in a range of habitats including the plankton, soils and glacial meltstreams (Chapter 12). These marked differences in distribution patterns of different forms appear convincing; so why should such differences have arisen? For the most part the distribution of *N. commune* seems to be determined, not surprisingly, by a persistence of liquid water. In The Vestfold Hills, Antarctica, mucilaginous to leathery and irregularly-lobed colonies of this strain have a sparse and scattered distribution and desiccate rapidly under full sunlight in soils which may attain temperatures of 25°C (Broady, 1986). Here, spherical colonies of 1 cm diameter or more also erupt from the surfaces of moss cushions which may provide a means to retain additional water as well as insulation.

When a source of liquid water is available some remarkable habitats can be colonized by *Nostoc* spp. and other cyanobacteria. *Nostoc*, although scarce, was present in the remarkable consortia which develop within solar-generated water pockets in the permanent ice covers of Antarctic lakes (Priscu et al., 1998).

An abundant growth of *N. microscopicum* was detected at a depth of 25m within a cave in Clarens-sandstone of the Drakensberg Mountains, South Africa where seepage water (pH9 to 9.2) was the only source of moisture (Büdel et al., 1993). Brown granules of *Nostoc* were reported from within a water supply tank where illumination was scarce (Hu et al., 1994), while more extensive growths occur in the hypertrophic coastal lagoon in Albufera, Valencia E. Spain (Moreno et al., 1994). Studies of the growth of *Nostoc parmelioides* in streams indicated a clear preference for rough as opposed to smooth substrates (Dudley and D'Antonio, 1991). *N. parmelioides* was little affected by grazing of the caddisfly *Agapetus celatus* but greatly affected by disturbance and only basal crusts were able to withstand disturbances in the absence of severe scour.

The widespread use of phosphate and nitrate-based fertilizers has all but eradicated *Nostoc* from many soils, including residential grass lawns, where it would otherwise proliferate. In China, the collecting of *N. commune* var. *flagelliforme* for food and pasturing of cattle are diminishing this resource (Gau, 1998).

Reports of *Nostoc* from some 1000m depth in the Indian Ocean and the Mediterranean are questionable, unconfirmed and therefore remain intriguing (Bernard 1963; Bernard and Lecal, 1960).

3. Aldabra Atoll

Aldabra Atoll, in the extreme Southwest corner of the Indian Ocean, offers an unsurpassed setting in which to study the ecology of both freshwater (Donaldson, 1976; Whitton Donaldson and Potts, 1979) and marine (Potts 1977; Potts and Whitton, 1980) cyanobacteria. *Nostoc commune* is common among the grass *Sporobolus virginicus*, especially in groves of *Cocos nucifera* (coconut), on the atoll where daily temperatures may reach the mid 40s centigrade. Shallow water-filled depressions in limestone ("platin") support dense growths of the characteristic spherical colonies of *N. sphaericum*; in drier areas the colonies tend to be leathery, irregularly-lobed or even hair-like in appearance. It seems certain that physical conditions play a major role in determining the appearance of these communities but can one be certain that these forms are genotypically equivalent; or is there a continuum of genetic diversity (Chapter 1)?

The rates of nitrogen fixation per unit chlorophyll recorded for some terrestrial *Nostoc* populations on Aldabra are among the highest recorded in the literature for in situ studies of cyanobacteria (Whitton et al., 1979; Chapter 8). During the dry season when the southeast trade winds blow (May through November) colonies are desiccated for long periods. During the wet season caused by the westnorthwest monsoon (December through April) the communities, on a daily basis, undergo a classic pattern of wetting and drying. Desiccated colonies of *N. commune* from Aldabra were found to take up water significantly faster than those of either *N. sphaericum* or a *Nostoc-Phormidium* community; rates of uptake were also faster than by *N. commune* from a temperate locality (Tarn Moor, England). In fact about half the maximal rate of nitrogen fixation was attained after rewetting of the colonies for only one hour (Whitton et al., 1979).

Free-living *Nostoc* spp., as well as *Tolypothrix byssoidea* and some lichenized communities on Aldabra, contribute nitrogen to the scarce, shallow and impoverished soils. The impact of nitrogen-fixing communities, including cyanobacteria, is apparent. These soils support a grass and shrub vegetation that in turn sustains a population of more than 100000 giant land tortoises (*Geochelone gigantea*); many members of which have weights in excess of 50kg.

B. The Colony

1. Physical Properties

Nostoc spp. develop colonies that have a range of characteristic shapes, sizes, smells, textures and colors and many become both abundant and visually conspicuous. If there is a persistent source of precipitation these colonies may assume dramatic proportions. Colonies of *N. pruniforme*, as large as 25 cm in diameter, grow in Mare's Egg Spring, Oregon, USA (Dodds and Castenholz, 1987; Plate 24a). Depending on the frequency of wetting *N. commune* may form spheres ("pearls"), from tens of mm to 3cm in diameter, or discoid and flattened crusts (Martinez and Querijero, 1986). The formation of spherical macroscopic colonies deserves mention. Other cyanobacteria as well as some eukaryotic algae that do not attach to substrata can form spherical colonies (aegagropiles and "Seeknüdel") of appreciable size, when they grow in sheltered waters where there is only periodic agitation. This is a passive physical phenomenon (e.g. *Scytonema*; Potts and Whitton, 1980). Spherical colonies of *Nostoc* do arise in both submerged and exposed situations but their formation is certainly not passive; it represents a distinct part of their life cycle. The characteristic"hair-like" colonies of *N. commune* var *flagelliforme* show different characteristics of water loss and retention than *N. commune* (Gao et al., 1998).

A considerable layer of extrapolysaccharide (EPS; glycan; sheath material) separates cells from the atmosphere and thus constitutes a barrier that appears to remain unbroken even during the budding process and despite the associated populations of epilithic microorganisms. Cyanophages or other bacteriophage -like structures were not noted within these colonies. A distinct pellicle encompasses the outer periphery of the thallus and this layer has a microstructure that includes fibrils perpendicular to the surface of the colony. This layer is enriched in silicon (Hill et al., 1994b). A feature of desiccated materials of *N. commune* collected *in situ*, and of "pearls" (Plate 24d) which bud from the thallus (Potts, 1994), is that the interior of the colony is devoid of any other microorganisms save for *N. commune.* Young colonies grown under laboratory conditions are packed with filaments. In contrast the interior of colonies that grow in situ are often depleted of *Nostoc* filaments and mostly consist of the gel-like EPS (Martinez and Querijero, 1986; Dodds and Castenholz, 1987).

Colonies are pigmented and appear in a range of colors from dark green to black, yellow-green to red-brown. Sections through colonies may reveal a yellow-brown coloration due to scytonemin (Chapter 21) that is most intense at the exposed surface of the colony (Plate 24e). A brown *Nostoc* species recorded from Java was found to be resistant to high light intensity and UVR (Chazal and Smith, 1994). The

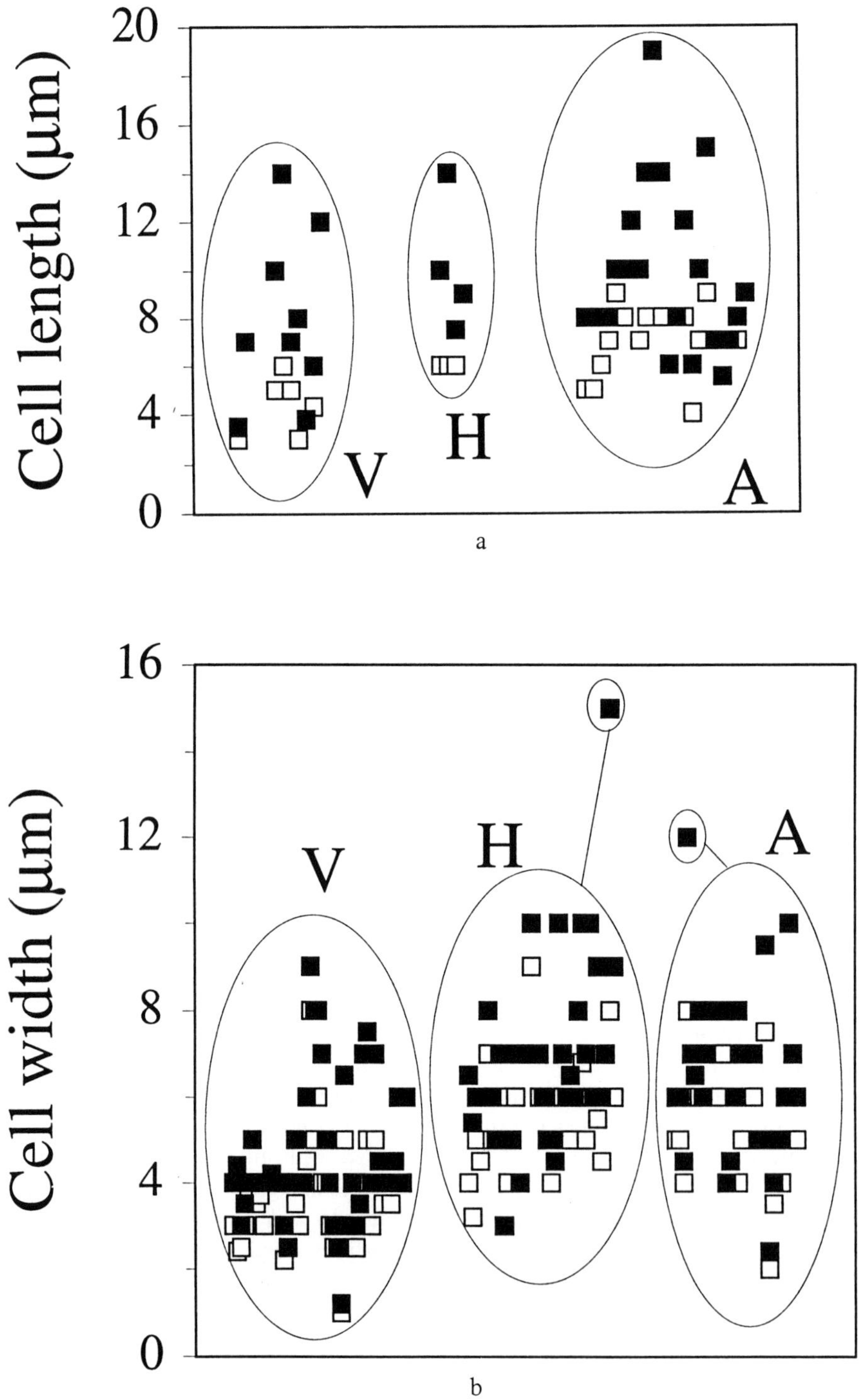

Fig. 2. Distributions of cell lengths and cell widths in vegetative cells (V), heterocysts (H) and akinetes (A) of all *Nostoc* species described by Geitler (1932). a. Cell length distribution b. Cell width distributiion. □ minimum measurement for a given species, ■ maximum measurement for a given species

disappearance of peripheral granules from vegetative cells of this strain during exponential growth coincided with the release of a brown pigment. Curiously the brown pigment, and a number of UVR-absorbing pigments were not observed under low oxygen conditions nor were the pigments produced when cells were grown under a 4 h : 20 h light : dark cycle.

2. Macromolecular Components

"The purification of the mucilage was difficult and tedious, and the hydrolysis products so complex as to render analysis difficult": these comments were made by Hough et al. (1952) following attempts to purify the EPS of gelatinous nodules of a *Nostoc* sp. collected from around the Freshwater Biological Association, Ambleside in 1948. EPS was extracted from the culture liquid of *Nostoc* sp. strain D and found to contain arabinose and galacturonic acid, ribose, xylose, glucose and two unidentified compounds. SEM revealed the mucilage had velum-like and net-like structures that were also seen in the isolated polysaccharide (Cupac and Gantar, 1992). There are reports that the sheath material of *N. commune* contains a granular component associated with the fibrillar network, possessing some morphological and physiological similarities to the matrix granule of animal cartilage, collagen, (Fjerdingstad et al., 1979; Abdelahad et al., 1984, Komárek and Anagnostidis, 1989). Synthesis of the structural protein collagen that is universally distributed among metazoans only became possible when the oxygen concentration in primitive Earth rose from less than 3-10 to 100% of present atmospheric levels (Canfield and Teske, 1996). Abundant proteins (see below) are present in the EPS of *N. commune* and may contribute to the cartilage-like properties of the EPS matrix. The fibrous structure typically seen in thin sections of the EPS (Potts, 1994) is generally indicative of gels.

Kumar et al. (1996) demonstrated the efficient UVR/B shielding role of $FeCl_3$ and pigmented products from *N. spongiaeforme* when they were incorporated in 13mm-thick agarose gels. It's a pity these experiments were not also conducted with the same gels that had been dried to a thin film to simulate the form of a dried *Nostoc* colony. Other extracellular pigments included a violet pigment, nostocine A (1) (6-methyl,3,6,7,8 penta-azabicyclo[4.3.0]nonan 1,3,5(7) trien-9-one) with a broad spectrum of growth inhibitory activity (Hirata et al., 1996).

All of the *Nostoc commune* proteins identified to date in the EPS were carbohydrate-modifying proteins and their biochemical properties were studied in detail (Scherer and Potts, 1989; Hill et al., 1994a; Potts, 1994, 1996). The EPS of *Nostoc commune* has unique rheological properties that may be determined, in part, by these secreted proteins (Potts 1994; Helm and Potts, unpublished data).

C. Cells

1. Organization

Cells of *Nostoc* spp. are spherical, barrel-shaped or oval and their contiguous arrangement leads to the formation of unbranched filaments which if straight (small writhe, large twist) may exceed 1 mm in length. The width of cells in the 40 or so descriptions annotated by Geitler (1932) vary over approximately a one order of magnitude from a minimum of around 1 mm to a maximum of around 9 mm (Fig. 2a). Heterocysts and akinetes (Section III.C) tend to be broader and longer than vegetative cells (Fig. 2b).

The smallest *Nostoc* recorded has a cell width that is one fiftieth that of *Chroococcus turgidus* UTEX 123; a representative of a group which may represent the largest of the cyanobacteria (Potts et al., 1983). These are purely empirical observations and without any biophysical measurements it is hard to make any generalizations as to their significance (Koch, 1996).

The filaments of *N. commune* are embedded within, and distributed throughout, a dense EPS. The overall arrangement of filaments in the EPS and their structural and biochemical properties were described in detail (Hill et al., 1994b; Potts, 1994, Potts, 1996). One striking feature of the distribution of filaments in "pearls" is that no one filament touches another; there appears to be a very ordered and regular 3-dimensional distribution of the filaments. In the desiccated thallus the filaments, when examined in SEM, appear to reside in narrow "tunnels" which permeate the glycan. These "tunnels" have ribbed extensions when observed in transverse section in the SEM, are clearly cylindrical when observed in a plane parallel to the tunnel and appear to be empty. In some materials these "tunnels" were less distinct or almost absent. Here, the glycan sheath at the periphery of the filaments appeared heavily pitted with a microporous structure that followed the contour of the "tunnels" observed in other materials.

Magnesium, calcium, silicon, phosphorus, and sulfur predominated the EDX spectrum of all materials that were studied. Relative to the glycan, cells were enriched in sulfur and phosphorus. The most obvious result of these analyses was the identification of a conspicuous accumulation of silicon and calcium in the external layer.

Light microscopy and TEM resolved the immediate environment of each cell in desiccated colonies of field material of *N. commune* as a homogeneous envelope that was stained only weakly (Hill et al., 1994b; Potts, 1994). This envelope layer, S_1, followed the contours of the individual cells and was constricted at their crosswalls. The envelope layer was appressed to the EPS where a dense staining membrane-like interface layer was discerned. The EPS, but not the envelope layer, was heavily labeled with antibodies prepared using a purified extracellular preparation (Hill et al., 1994b).

The pattern of labeling of the EPS suggested that the apparently homogenous fibrous structure was immunologically heterogeneous. When rehydrated cells of *N. commune* CHEN pearls or *N. commune* DRH1 were prepared under similar conditions of fixation, dehydration, infiltration, and critical point drying, the envelope layer and glycan showed a different ultrastructure in comparison to the desiccated materials. The most apparent differences were as follows: following short term rehydration (30 min) portions of the envelope layer had the capacity to be post-stained (S_2). These areas appeared to consist of fibrils parallel to both the stained membrane interface layer and the contiguous fibrils of the glycan. After longer periods of rehydration (60 min), the developing layer, S_2, became more reticulate and stained with a greater intensity. In fully rehydrated material an obvious translucent envelope layer could not be discerned. Although the "tunnel"-like structures (described above) were absent, filaments were encased in a layer which had different structural properties to the bulk EPS. Analysis through light microscopy also indicated different staining properties of this layer. Filaments tended to occupy the outer portions of the colony where they formed smaller "pearls" and packets, while non-encased filaments were present in the central portions of each "pearl".

During growth, minute "pearls" bud off from the parent pearl at the periphery of the latter (Plate 24d). These buds originate as encased filaments immediately below the surface of the pearl and, at this stage, an external layer is already apparent at the periphery of these small packets of filaments. The staining characteristics of the material in and surrounding these packets differed from those of the parent glycan (sheath) matrix. For example, with Alcian blue at pH2.5 the small packets stained an intense dark blue in comparison to the light blue-green color of the surrounding glycan (Hill et al., 1994b).

2. *Ultrastructure*

Ultrastructural studies of *Nostoc* spp., like those of cyanobacteria in general, tended to focus on the form and occurrence of cell inclusions (Allen, 1984; Jensen 1984; Simon 1987). Twenty nine *Nostoc* isolates contained 12 different types of inclusions. Other than a seeming ubiquitous occurrence of membrane-limited crystalline bodies there was no obvious trend in association of certain structures with particular forms of *Nostoc*.

Cytochemical analysis revealed that a high salt environment activated an oubain-sensitive Na^+, K^+ ATPase, presumably involved in efflux of Na^+ ions (Iwano, 1995). Exposure of *Nostoc* sp. UAM 205 and other cyanobacteria to the insecticide trichlorfon lead to destabilization of the heterocyst envelope. This may be selective for communities in situ as the effects of the insecticide were largely prevented under anaerobic conditions (Orus and Marco, 1991). Many studies focused on the ultrastructure of *Nostoc* spp. in interactions with plants (Gantar et al. 1991; Johansson and Bergman, 1992), and effects of heavy metals (Gupta and Singhal 1996; Fernandezpinas et al., 1995).

D. *Structure and Function*

1. *Exopolysaccharide*

A number of different attributes were ascribed to the extracellular EPS of bacteria (Potts 1994; Hill et al., 1997). The alginate of *Pseudomonas aeruginosa* restricts diffusion of oxygen, a feature which is shared by cyanobacterial EPSs (Chang, 1980). Experiments using light microscopy identified a conspicuous reaction of heterocysts in desiccated material with PAS, a stain used in glycoprotein detection (Hill et al., 1991b). In fully hydrated pearls a more general and conspicuous reaction with PAS was observed. Two polypeptides, that showed a positive glycan reaction, were detected in aqueous extracts of desiccated samples from six different geographic locations.

One principal function of the EPS is that it provides a repository for water. The EPS represents a mixed system, where water and the polysaccharide tend to mix as thoroughly as they can for thermodynamic reasons (Potts 1994). Work is required to remove water from the gel and this can be lost through the application of pressure or temperature, and through evaporation. The outer layer of the colonies may act as a membrane which is under pressure from the bulk glycan. One explanation for the striking form of the colonies of *N. commune* DRH1 is that a sphere represents the minimum surface area for a given volume which would clearly provide a reduction, and uniformity, in the net rate of evaporation of water. Of course for the same considerations the spherical glycan surface has a reduced capacity for gas uptake. Cyanobacterial sheath materials were found to retard gas exchange (Chang, 1980). Spherical colonies are formed by many *Nostoc* spp. growing in situ, including those that grow submerged, suggesting functions in addition to a retardation of water loss (Martinez and Querijero, 1986; Dodds and Castenholz, 1987; Plate 24a).

Previous studies of cyanobacterial envelopes and sheath structures suggested that these may serve to concentrate metals and thus may aid in excluding the colonies from predation by gastropods, insects, etc. (Tease and Walker, 1987). Although microorganisms are certainly present at the surfaces of the *N. commune* colonies; the outer silicon-rich layer represents an impenetrable barrier for them. The silicon-rich layer must be made through physico-chemical precipitation as it is hard to account for a concerted synthesis of this layer on behalf of the cells. More likely the layer is the product of some oxygen/drying-dependent effect on the peripheral sheath, although in liquid cultures of *N. commune* strain DRH1 a discrete pellicular structure was also seen (Hill et al., 1994b).

The EPS represents the bulk of the colony and constitutes a considerable diversion of the carbon and nitrogen budget (Ernst et al., 1987). The UVR pigments represent another sizable fraction of the dry weight of a desiccated colony. In addition, Wsp is the most abundant protein in the EPS. These three components are secreted from the cells and Wsp and UV-pigment synthesis may be related (Hill et al., 1994a,b). UV-A/B irradiation of N. commune DRH1 lead to increased amounts of water-soluble UV-absorbing pigments as well as the extracellular glycan (Ehling-Shulz et al., 1997). It remains to be determined whether *N. commune* has scavenging mechanisms for any or all of these extracellular components or their degradation products, if any.

There is a uniformity in the EPS isolated from materials collected from a range of different climatic environments (Hill et al., 1994b; Huang et al., submitted). The latter are characterized by extended periods of desiccation and often rapid, and intermittent, periods of rehydration. The extracellular glycan appears to represent a buffer zone between the atmosphere and the cells. The prodigious investments made in sheath synthesis and those components found within the sheath, and our interpretation of structure and composition suggest a principal role for the glycan in desiccation tolerance.

2. *UVR-Absorbing Pigments*

At least two distinct UVR-absorbing pigments are present in the EPS of *N. commune.* One, pigmented yellow, has the spectral properties of scytonemin, a 544 molecular weight dimeric molecule of indolic and phenolic subunits, known only from the extracellular sheath materials of certain cyanobacteria (Chapter 21; Plate 24e). The other is a complex mixture consisting of mycosporine amino acids with chromophores, linked to galactose, glucose, xylose, and glucosamine, which have absorption maxima at 335 and 312 nm (Böhm et al. 1995). Either alone, or in combination, scytonemin and mycosporines afford protection from incident solar irradiation (Chapter 21). The mycosporine compounds may constitute up to 10% by dry weight of desiccated colonies and their release upon rehydration constitutes a significant loss of cellular carbon and nitrogen (Potts, 1994). In contrast, scytonemin is not lost upon aqueous extraction. The banding pattern of scytonemin noted in colonies may represent the vestiges of the old surface of the colony as the colonies grow, or some phenomenon associated with swelling and shrinkage of the colony (Plate 24e). The strategy of synthesizing both water-soluble and lipid-soluble components is clear. The former can saturate the glycan compartment rapidly upon rehydration but at the expense of a high loss of the pigment; scytonemin provides a more localized screening with the advantage that the pigment is retained by the colonies. Scytonemin may likely have a much more important role than the aqueous UVR-absorbing pigments during the protection of cells upon dispersal of colony fragments.

E. Interactions with Other Organisms

Nostoc enters into a range of associations with higher and lower plants as well as with the fungus *Geosiphon pyriforme* (Chapter 19; Schübler et al., 1995, 1996; Plate 28). In addition there are a number of interesting observations on indirect interactions between *Nostoc*, fungi, insects, plants and bacteria. A *Nostoc* sp., in association with *Oscillatoria* spp. and *Desulfovibrio* sp., forms black subsurface globules and layers, often indicative of death of turf, on high sand golf greens (Hodges, 1992a). However, the presence of these cyanobacteria in black-layered sands decreased the root and shoot dry weight loss of the grass *Agrostis palustris* when it was infected by the fungal pathogen *Pythium torulosum* (Hodges, 1992b). The isolated *Nostoc* sp. impeded water flow in artificial sand columns but did not lead to blackening of the substrate. Blackening occurred only when the columns were inoculated with *Nostoc,* or the *Oscillatoria* strains, and cultures of *Desulfovibrio* sp.

The use of a fresh inoculum of cyanobacteria that included *Nostoc muscorum* decreased leaf blast, but not neck blast, of Japonica rice (line Giza 171) caused by *Pyricularia oryzae* (Yanni and Sehly, 1991). Induced morphological changes were described in a *Nostoc parmelioides-Cricotopus fuscatus* symbiosis in a stream (Kleinhaus and Kaiser 1988).

Myxococcus fulvus BGO2 is a cyanobacteriolytic bacterium. Experiments suggest that the absence of myxobacteria in cyanobacterial blooms and absence of cyanobacterial population control by related lytic bacteria which are found in blooms, may reflect insufficient inorganic nutrient fertility to support sufficient populations of these lytic bacteria (Fraleigh and Burnham, 1988).

III. Evolution

A. The Rise of Nostoc

1. Origin

To consider the evolution of *Nostoc* spp., two time periods are of relevance; the period leading up to the time when ancestral *Nostoc* acquired its characteristic morphology; and the period which has elapsed since then.

Life may have emerged 3.85 billion years ago (Mojzsis et al., 1996). An availability of combined nitrogen is not thought to have posed a serious limitation for the evolution of early life because atmospheric chemistry is such that lightning may have provided as much as 10^6 tons of nitrogen per year to the primitive ocean (Young and McElroy, 1979). An availability of nitrogen is also unlikely to have been a determinant in the subsequent evolution of *Nostoc* and other heterocystous cyanobacteria.

The attaching of absolute time scales to phylogenetic trees is troublesome, and the inferring of evolutionary relationships and times of divergence from such trees remains intractable (Pennisi, 1998). From comparative amino acid sequence analysis it was argued, with some criticism (e.g. Hasegawa and Fitch, 1996), that prokaryotes and eukaryotes last shared a common ancestor approximately 2 My ago (Doolittle et al., 1996). If this is true then there is serious doubt whether 3.5 My old microfossils (Chapter 2; Plate 4) really correspond to extant cyanobacteria (Feng et al., 1997). That is, fossil cyanobacteria and modern day counterparts are analogues, rather than homologues. Such a conclusion has important consequences for understanding the tempo of speciation in cyanobacteria (Section III.A.2). However, it was argued recently that the exchange of genetic material for neutrally evolving genes should be of serious concern when dating divergence times using molecular clocks (Rudi et al. 1998). In any event, the evolution of heterocystous cyanobacteria in general, and *Nostoc* in particular, must be considered cautiously in light of these data.

Of the 2000 or so microfossils that were ascribed to cyanobacteria, some 15% are believed to represent ancestors of those extant forms which can be assigned to group IV *sensu* Rippka et al. (1979) using morphological criteria (Fig. 3). Of these, *Nostocomorpha*, *Palaeonostoc* and *Veteronostocale* are Precambrian genera that show similarity to extant forms of *Nostoc* (Schopf, 1994). In contrast, the oldest microfossils thought to be of prokaryotic origin are those described from the Early Archaean Apex Basalt of northwestern Western Australia but of these, none were ascribed to cyanobacteria with any certainty and none have morphologies comparable with present-day *Nostoc* (Schopf, 1993). Either *Nostoc*-like forms were not as widespread at this time or they appeared later.

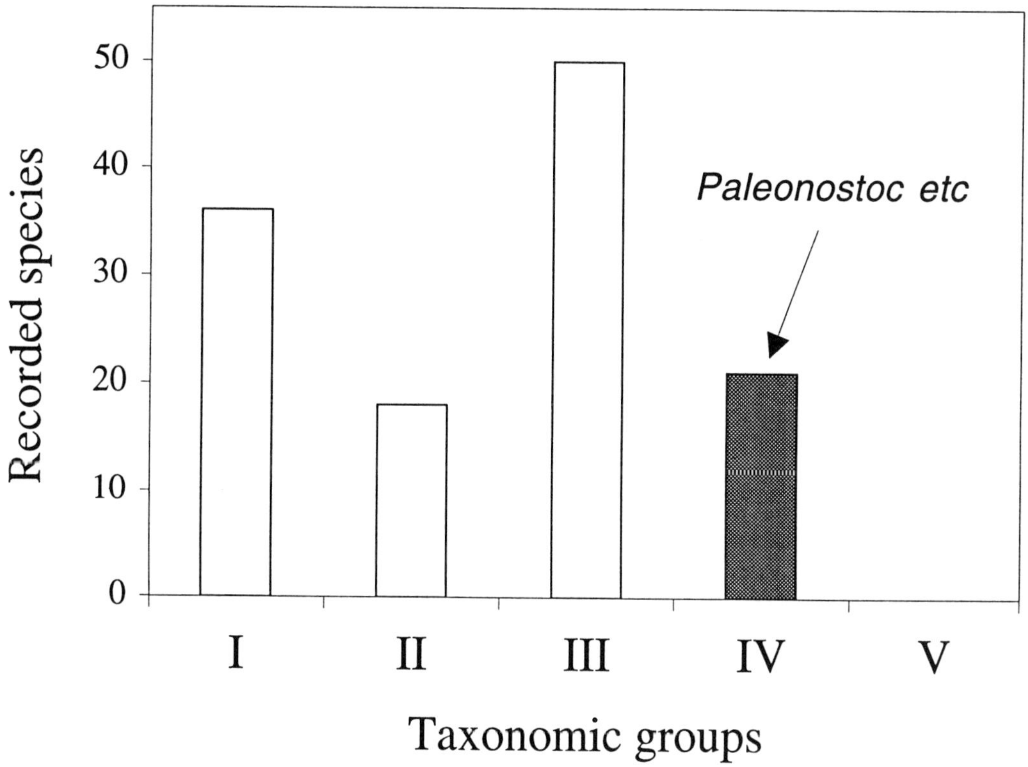

Fig. 3. Records of microfossils thought to be of cyanobacterial origin. Records are those listed in Schopf (1994). Taxonomic groups correspond to those in the scheme of Rippka et al. (1979).

It was concluded that *Nostoc* and other heterocystous forms arose later than their non-heterocystous counterparts (Doolittle, 1982; Giovannoni et al., 1988). Even as a cautionary one the observation is not especially apparent from a scrutiny of published phylogenetic trees (Nelissen et al., 1996; Wilmotte, 1994). There are far too few sequences available for heterocystous cyanobacteria at the time of writing to permit any meaningful conclusions to be made about the origin of this group. Current bacterial classification at the genus level and above is based predominantly upon 16S rRNA sequence analysis, and unless a similar evolutionary relationship is supported by other gene sequences, the inferences derived from it may be in question (Gupta et al., 1997). 16S rRNA sequences are unlikely to provide indications of intra species variability in view of their high level of conservation (Ward et al., 1992). In fact because 16S rRNA genes, and other genes used by systematists such as *rpoB*, perform the same “housekeeping” duties in every taxon, they may be interchangeable across taxa (Cohan, 1996). The genetic diversity between morphologically similar, or diverse, cyanobacteria is poorly understood which suggests that current phylogenies based upon 16SrRNA sequences must be viewed most cautiously (Wilmotte, 1994; Chapter 1). A comparison of DNA-DNA hybridization percentages and 16S rRNA similarities provided contradictory results (Fox et al., 1992; Ward et al., 1992). More importantly, a theoretical basis for the empirical correspondence between 16S rRNA sequence clusters and ecological populations has yet to be found (Cohan 1996).

The patterns of sequence divergence observed by systematists involve substitutions that are of no fitness consequence and they are in genes that are not involved in population-specific adaptations. Forms of *Nostoc* spp., and *N. commune* in particular, are especially well suited for the resolution of some of these problems by virtue of their prevalence as conspicuous, morphologically-similar communities in a broad range of habitats from the Tropics to the polar regions.

Despite the caution emphasized by the preceding discussion there is one feature of the phylogenetic tree shown by Nelissen et al. (1996) that bears mention. There is a hint of trichotomy within the grouping of the ten heterocystous forms. One cluster includes *Nostoc*, *Anabaena*, *Cylindrospermum* and *Nodularia* strains (Order Nostocales *sensu* Komárek and Anagnostidis, 1989; Section IV *sensu* Rippka et al., 1979). A second includes *Fischerella* and *Chlorogloeopsis* strains (Order Stigonematales, *sensu* Anagnostidis and Komárek, 1990; Section V, *sensu* Rippka et al., 1979) and a third comprises the sequences from *Scytonema* and *Calothrix* strains (Order Nostocales *sensu* Komárek and Anagnostidis, 1989; Section IV *sensu* Rippka et al. 1979).

Developing an understanding of the phylogeny of heterocystous cyanobacteria requires more exhaustive molecular surveys of new isolates from well-documented field locations. Rudi et al. (1997) focused on the comparative sequence analysis of DNA encoding the variable regions V6, V7 and V8 of 16S rRNA in 10 strains of *Nostoc*, *Anabaena* and *Aphanizomenon*. It was concluded that the strain category they represented was monophyletic but consisted of three evolutionary different lineages (two *Nostoc* and one *Anabaena/Aphanizomenon*). Despite the very limited data set the heterogeneity within the *Nostoc* category, particularly the divergence of *Nostoc* sp. NIVA-CYA 246 (= *Anabaena* sp. PCC 7120) from other named *Nostoc* strains, collected from field sites in Norway and Antarctica, was obvious. In a subsequent study the authors added two strains, *N. commune* and *N. commune* var. *flagelliforme* (both isolated from field sites in China) to the data set (Rudi et al., 1998). Comparison of the phylogeny based on 16S rDNA and the translated (amino acid) sequence of *rbcX* revealed several discrepancies but both had the same clustering of five independent *Nostoc* isolates (*Nostoc* lineage II). Both phylogenies emphasized the clear divergences between lineage II, lineage I (*Anabaena/Aphanizomenon*) and lineage III (*Anabaena* sp. PCC 7120).

2. Tempo of Speciation

What of the time since *Nostoc* acquired its morphology? Stulp (1983) described three possible scenarios to account for the antiquity of cyanobacteria and their high degree of genetic relatedness. Either speciation began in the Middle Precambrian with a rate of base substitution three orders of magnitude less than that estimated for eukaryotes or, speciation occurred very recently (some 25 to 50 million years ago) with a rate of base substitution comparable to that measured in eukaryotes or, speciation occurred early, in a comparatively short period, and then was arrested. If one assumes that the second scenario is unlikely then the other two, collectively, reflect one process of very slow genetic change (Chapter 2; "stasis" see Gould and Eldredge, 1993). Schopf proposed that the morphology (and physiology) of cyanobacteria evolved little or not at all over thousands of millions of years (Schopf, 1994). Hyperbradytely refers to a slow or suspended tempo of evolution. The proposal (Section III.A.1) that the last common ancestor of prokaryotes and eukaryotes occurred no earlier than 2 billion years ago contradicts such hyperbradytely. Was the genetic complexity (and physiological capacity) of ancient *Nostoc* similar if not equivalent to that of extant forms such as *N. commune* or have extant forms retained their morphological appearance despite a highly evolved genome? Rudi et al. (1998), in defense of hyperbradytely, suggested that only the highly conserved genes used in the estimations of Doolittle and colleagues could share approximately 2 billion year old common ancestors. They proposed that the divergence between different lineages of heterocystous cyanobacteria such as for the *rbcLX* locus was caused mainly by neutral mutations, suggesting homogenization within these groups through neutral gene exchange.

If hyperbradytely did prevail then a mechanism must be proposed to account for such a sluggish evolutionary drift. Gould and Eldredge (1993) have emphasized that morphological stasis is caused by developmental constraints that can be broken only by a restructuring of the genome through genetic drift in small, sexually reproductive populations. This view is, of course, tailored for eukaryotic systems. For cyanobacteria, a slow rate of evolutionary change could reflect a slow rate of accumulation of mutations. If this indeed is the case then there must be a plausible explanation. Several possibilities can be considered including one or all of the following: an inordinately high fidelity of DNA replication and minimal to no mutagenesis, minimal to no recombination, or highly efficient DNA repair. The first mechanism is unlikely in view of the marked conservation of the DNA replication machinery in the

Archaea, Bacteria and Eukarya. The other two possibilities warrant further comment.

The relevance of recombination in the evolution of bacteria in general and in cyanobacteria in particular, and its role, if any, in populations of the latter when they grow in situ, is exceedingly hard to evaluate. Recombination does occur in strains of cyanobacteria manipulated under laboratory conditions, cyanobacteria do contain *rec* genes, evidence for interchromosomal recombination was obtained, and the expectation that recombination will be observed in natural populations was voiced (Castenholz, 1992; Curtis and Martin, 1994; Gurevitz et al., 1991;Wolk, 1997). But what is the true significance of these observations? Genetic exchange in bacteria is thought to be exceedingly rare; the measured or estimated rates of recombination fall in the range of 10^{-7} to 10^{-8} per gene segment per genome per generation (Cohan, 1996). Does genetic exchange play the same role in preventing divergence in bacteria as it does in higher eukaryotes? Some important considerations to weigh when considering answers to this question are as follows: bacteria can undergo homologous recombination with related species that are up to 25% (or more) divergent in the sequences of homologous genes; bacteria can accept and express new genes on plasmids from diverse sources and there are some well characterized cases of lateral gene transfer (Groisman et al., 1992; Cohan, 1996).

The rates of recombination measured in bacteria are sufficiently low that they may be unable to preserve genetic diversity by countering the process of "periodic selection" (Cohan, 1996). Simply stated, when an adaptive mutation is selected for in an asexual population the entire genome of the individual that carries the mutation is also selected for because of the low recombination frequency. As a consequence the diversity at all loci in the genomes of all the individuals of the population is eventually lost.

If the conclusions of Rudi et al. (1998) prove to be correct then the picture that emerges for cyanobacteria is that forms diverged at their adaptive genes. This resulted in the range of morphologies and phenotypes seen in the fossil record and is reflected in the appearances of present-day forms. Mechanisms such as conjugation, transduction, simultaneous competence, codon usage etc. provided a means for genetic exchange and homologous recombination but only between phyletically closely related forms (e.g. *Nostoc* forms). The latter process contributed to the homogenization in neutrally-evolving genes whose analysis in extant forms may be problematic when attempting to determine divergence times.

With respect to morphologically-complex cyanobacteria, such as *Nostoc*, one must question how morphological characters which may have arisen early (e.g. sheath formation and colony structure) evolved in response to environmental selection pressure - if indeed they did.

B. Genomics and Genetics

1. Complexity and Organization

The structure of DNA, and thus the conformation of the nucleoid of cells is affected by the stability of base stacking (dinucleotides), methylation, repair, replication, transcription, context-dependent mutation pressures, and constraints on helicity (Karlin et al., 1997). DNA structure is also influenced by the availability of water molecules, UVR, osmotic pressure, temperature and protonation (Potts, 1994). The interplay between these factors makes it likely that there are features of an organism's genome which mirror its environment. In the case of *Nostoc* spp., and other terrestrial or aerophytic forms whose genomes are regularly desiccated, there may be characteristic genome "signatures." Karlin et al. (1997) compared compositional biases, distributions of short oligonucleotides, and dinucleotide relative abundance differences in 15 prokaryotes for which substantial (in many cases complete) genome sequence data were available. These analyses demonstrated that the cyanobacteria *Synechocystis* PCC 6803, *Synechococcus* sp. and *Anabaena* sp. did not form a coherent group and were as far from each other as general gram-negative sequences are from general gram-positive sequences. (Karlin et al., 1997). It is not unreasonable to suggest that the sequencing of the complete genome of *N. commune* could reveal much about the capacity of this organism for desiccation tolerance.

The appearance of the chromosomal DNA of *Nostoc commune* collected from natural populations is variable and depends in large part upon the degree of hydration of the nucleoplasm (Fig. 4). The nucleoid of *Anabaena* sp. PCC 7118 had a structure that included protrusions, lobes and distinct isolated elements (Pinevich and Grigoryeva, 1994). The figures provided by these authors are somewhat reminiscent of the characteristic "doughnut"-shaped nucleoids present in *Dinococcus radiodurans* (J. Battista, pers. comm.).

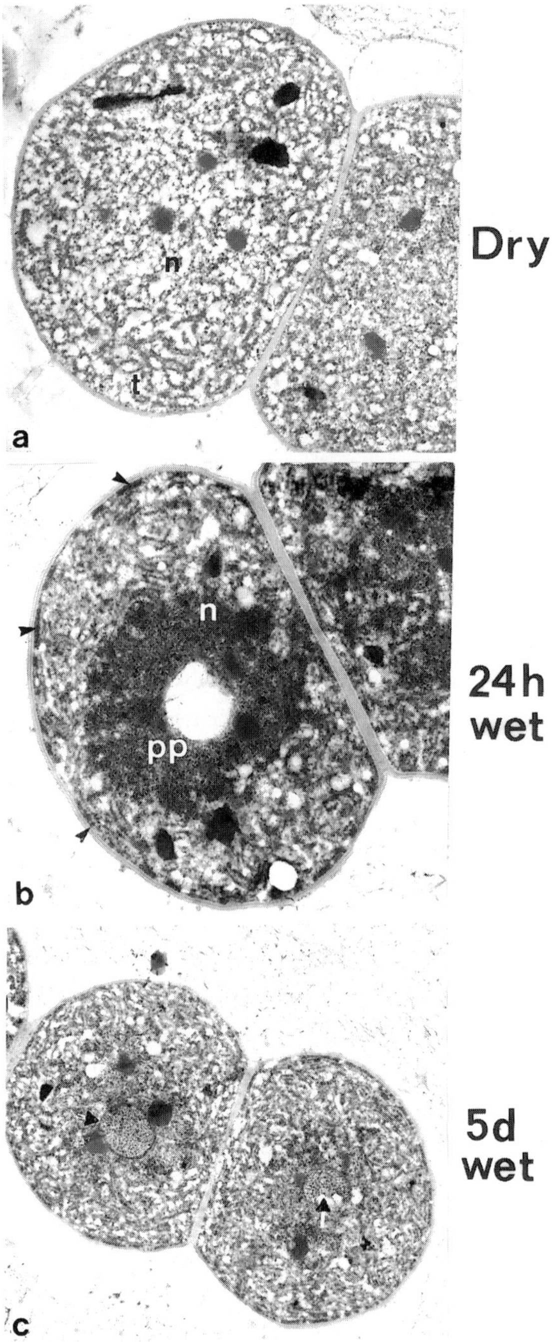

Fig. 4. Appearance of the nucleoplasm in *N. commune* is influenced by degree of hydration. Electron micrographs of cells of field materials of *N. commune*: a.. Desiccated for 12 years. t = thylakoid. b. Rehydrated 24 h pp = polyphosphate, n = nucleoplasm; arrows denote membrane-bound ribosomes. c. Rehydrated 5 days; arrows denote atypical discontinuity (spherical) in the nucleoplasm, possibly polyphosphate? Magnification approx. x 10,000. Courtesy of Alan Peat.

Through analogy with the genome of *A. variabilis* it is assumed that the genomes of *Nostoc* spp. are circular (Herdman, 1982). The genome sizes of *Nostoc* strains deposited in the Pasteur Culture Collection were measured as 4 to 6.4×10^9 dalton or $6.1 - 9.7 \times 10^6$ bp (Herdman et al., 1979). Rapidly renaturing fractions in *Nostoc* sp. PCC 6719, of unknown sequence or identity, had a kinetic complexity of 8.1×10^6 dalton (equivalent to a repeat structure of approximately 12 kbp). This fraction represented around 8.8% of the total genome with a copy number of 43 relative to the major chromosomal fraction.

The functional significance of multiple genome copies in cyanobacterial cells remain unknown. Multiple copies of the genome may guarantee mandatory condensation of the nucleoid at low water potentials (Zimmerman and Murphy, 1996; Woldringh et al., 1995). For terrestrial cyanobacteria such as *Nostoc* spp. this response may be of physiological and structural significance. In this regard it would be of interest to determine the effects of cell dehydration in *Anabaena* strains AB22 and CPB5437 that have lost the capacity to synthesize the histone-like protein HU; a protein thought to play a role in specific recombination events (Khudyakov and Wolk, 1996). Importantly, *E. coli hupAB* mutants which lacked HU were sensitive to UV radiation because of the role of HU in homologous recombination (Li and Waters, 1998).

Nostoc genomes, like those of other cyanobacteria, are often heavily methylated (Jäger and Potts, 1988; Fig. 5). Such modification is known to influence the sensitivity of cells to UV irradiation and photoproducts such as H_2O_2 (Yallaly and Eisenstark, 1990). Little is known in regard to the overall structural properties this modification confers on the whole chromosome with the possible exception of enhancing the rate of depurination (Potts, 1994).

The chromosomes of *Nostoc* spp. and other heterocystous forms are larger than those of non-heterocystous cyanobacteria. Herdmann and co-workers (1979,1982) suggested that redundant DNA created from duplications of an ancestral genome may have provided templates for mutation which could have lead to the development of new genes. This gene duplication theory is one of two models that were proposed to account for the evolution of new genes. The second is the overprinting theory, and one which has some experimental support, where new sequences are produced from extant unused reading frames on sense or antisense strands (Ikehara et al.,

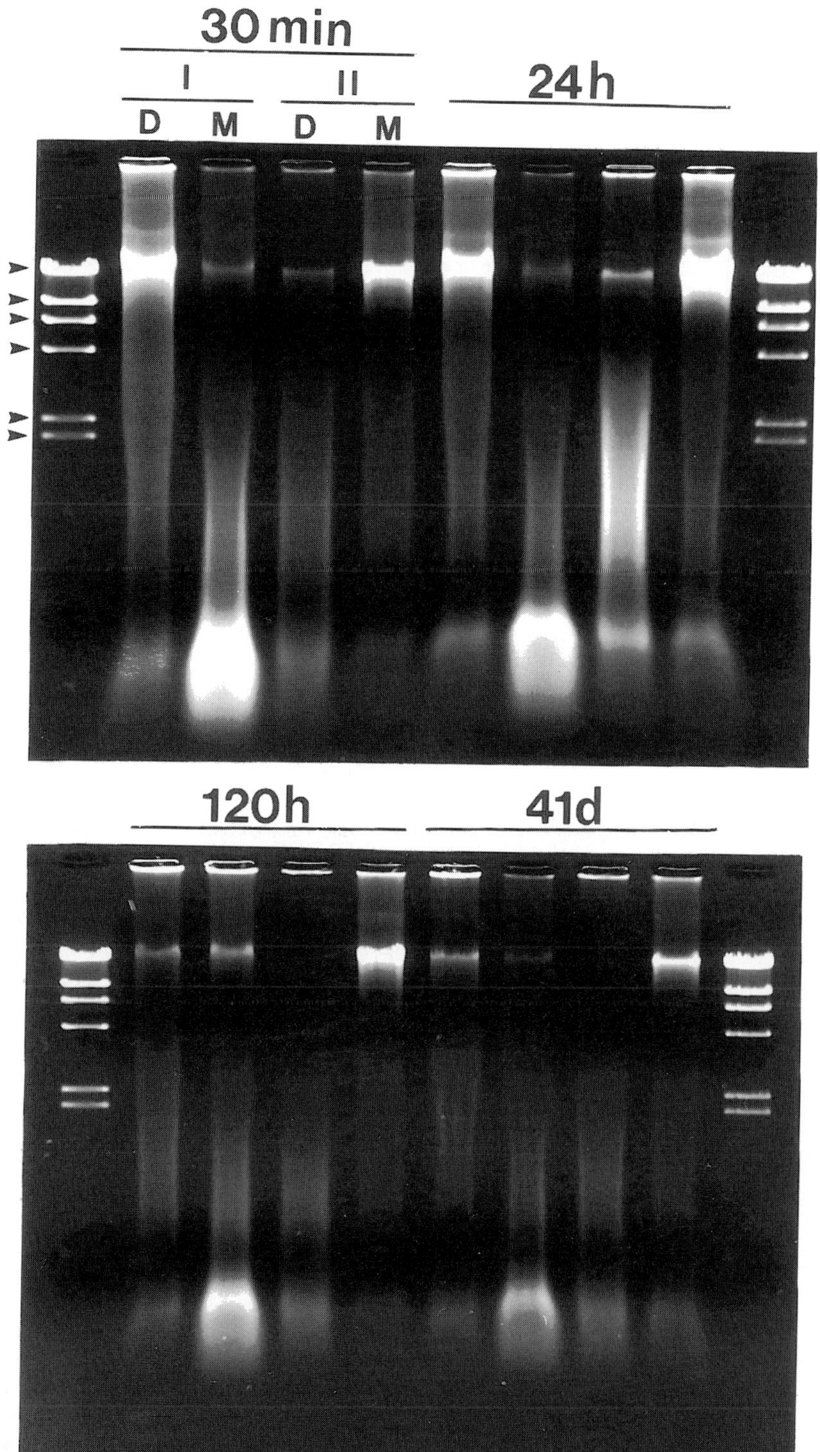

Fig. 5. Genomic DNA of *N. commune* consists of hypo- (I) and hypermethylated (II) fractions (Jäger and Potts 1988). Desiccated colonies were rehydrated for either 30 min, 24h, 120 h or 41 days before purification of DNA fractions. Fractions were treated with either *Dpn*I (D) or *Mbo*I (M) For each set of rehydrations samples were loaded as follows: Fraction I : *Dpn*I, *Mbo*I; Fraction II: *Dpn*I, *Mbo*I. Lambda *Hin*dIII markers in lanes 1 and 10 of each gel. Courtesy Karin Jäger.

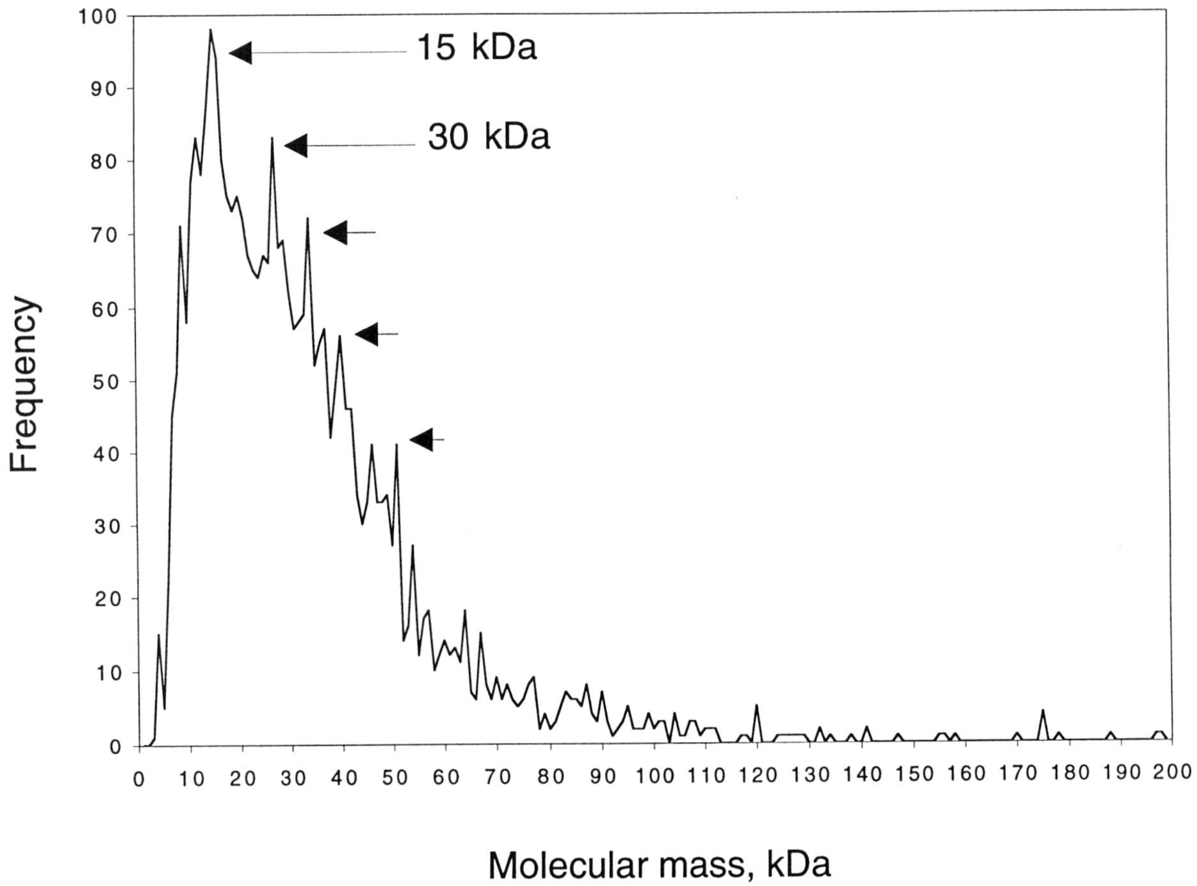

Fig. 6. Distinct size bias and periodicity in the sizes of cyanobacterial polypeptides. All the potential 3169 polypeptides encoded by the genome of *Synechocystis* PCC 6803 were included in the calculation (Kaneko et al., 1996). Arrows indicate multiples (n) of the basic repeating unit of 15 kDa.

1996). As the genomic sequences of *Nostoc* spp. become available in the future it will be possible to assess the validity of the latter theory, as well as the possible relevance of lateral transfer through comparisons of the mol % G+C contents of selected genes.

Some 3169 putative ORFs were identified following DNA sequence analysis of the chromosome of *Synechocystis* sp. PCC 6803 (Kaneko et al., 1996). The chromosomes of *Nostoc* spp. are approximately two to three times larger than that of *Synechocystis* sp. PCC 6803. These data may suggest that the *Nostoc* genome could encode as many as 10000 polypeptides. Genes associated specifically with the differentiation of heterocysts represent one class of genes that do not appear to be present in non-heterocystous forms such as *Synechocystis* - are there others? *Nostoc* spp. are capable of a number of developmental options and perhaps all of these require novel genes (Cohen et al., 1994). Certainly one that bears consideration is the capacity for *Nostoc* spp. to invade lower and higher plants and to establish stable relationships with them (Chapter 19). Otherwise do *Nostoc* genomes simply carry a considerable fraction of redundant DNA? At this time of writing, however, the sequences of only 70 or so *Nostoc* spp. proteins were deposited in data banks and some of these reflect the dual naming of *Anabaena* sp. PCC 7120 as *Nostoc* sp. PCC 7120.

While there are no specific data available at this time to assess the numbers of proteins encoded by the genomes of *Nostoc* spp. it is useful to consider certain data in the context of the evolution of cyanobacterial genomes. Some 46% of the 1000 sampled proteins in

E. coli have molecular masses of n × 14000 ± 2500 dalton (n = 1, 2, etc.; Savageau, 1986). This was interpreted as the consequence of polypeptides arising from a fundamental structural unit. With the availability of the complete DNA sequence of *Synechocystis* sp. PCC 6803 (Kaneko et al., 1996) it was possible to analyze all of the potential polypeptide sequences of this cyanobacterium. The 3169 polypeptide products identified in *Synechocystis* sp. PCC 6803 appear to be designed on a standard unit of 15 kDa, essentially an identical size to that identified in *E. coli*, with obvious peaks of n=2, n=3 (Fig. 6). Will this prove to be the case for other cyanobacteria? Did incremental changes to the basic unit facilitate microevolution, and did quantum changes to the next stable size of protein accommodate macroevolution?

2. Nucleic Acid Stability Under Stress

Could a low rate of evolutionary change be the price paid, or rather the consequence, for being able to colonize extreme environments where DNA damage may be accelerated? UVR, free radical generation, photoxidation and desiccation all lead to DNA damage and presumably they imposed a considerable restraint upon the colonization of habitats by early life forms including terrestrial *Nostoc* spp. (Potts, 1994). A consideration of the consequences of DNA damage may shed light upon selection pressures imposed upon populations of cyanobacteria. One process that causes considerable damage to cellular DNA is exposure to ionizing radiation. ^{60}Co irradiation introduces damage to DNA principally through the introduction of double-stranded breaks (DSB's). Cyanobacteria are known to be highly resistant to ^{60}Co and ^{60}Co irradiation has sometimes been used to render cultures of cyanobacteria axenic in his laboratory. The practice was not widespread, presumably due to concerns about enhancement of mutation frequencies. Were those fears founded ? The following suggests perhaps they were not. Species of the bacterial genus *Deinococcus,* especially *D. radiodurans,* have a startling resistance to ionizing and ultraviolet radiation, as well as mutagens that form adducts and cross links in DNA. The means by which they do so has been studied in detail (Agostini et al., 1996) and a consideration of the data is important to allow speculation about certain physiological responses of *Nostoc* spp. For example, following exposure to 1 Mrad (10000 Gy), it was determined that the genome of *D. radiodurans* sustains about 100 DSBs which are repaired within 29 h without lethality, mutagenesis or rearrangements (Daly and Minton, 1996; J. Battista, pers. comm.). Following an exposure to 1.75 Mrad, which degrades the four chromosomes of a cell into about 500 fragments, as many as 175 crossovers per chromosome (700 crossovers per nucleoid) underwent repair (Daly and Minton, 1995a, b). A model to explain how *D. radiodurans* accomplishes such a resurrection was proposed by Minton and colleagues (Fig. 7; Daly and Minton, 1995b). The essential feature of the model is that interchromosomal recombination is made possible through large numbers of Holiday junctions. The latter ensure that a DSB is not lethal because an identical undamaged DNA duplex is available nearby (Daly and Minton, 1995a, b; Minton and Daly, 1995; Minton, 1996). It was considered that RecA played a crucial role in these pre-aligned repair reactions by mediating strand invasion, although RecA was not thought to be necessary to maintain the Holiday junctions as it was sufficiently toxic to be tolerated by the host only during DNA repair (Minton and Daly, 1995). Recent work demonstrated that the initial repair, i.e. in the first few hours following irradiation, was *recA*-independent and was hypothesized to reflect a single-strand annealing reaction which required repeat sequences (Daly and Minton, 1996).

One important feature of the model and these proposals is the use to which redundant DNA is put. After many years of study of this phenomenon it was proposed recently that the extreme resistance of *D. radiodurans* to levels of ionizing radiation that do not and have never existed on Earth, is simply a

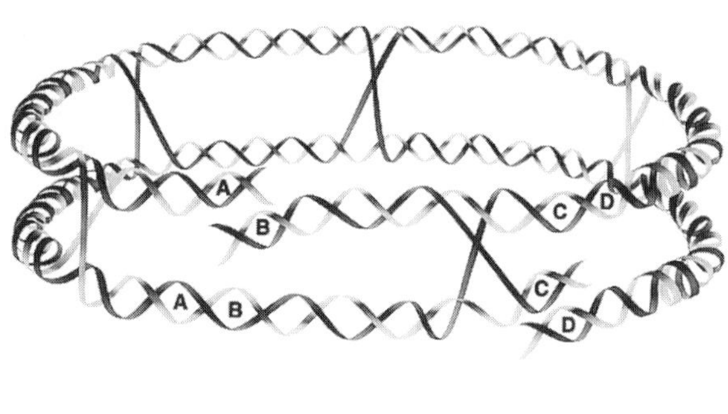

Fig. 7. Model of how the *Deinococcus* genome is repaired following damage from ionizing radiation or desiccation. Hypothetical double chromosome structure showing double-strand DNA breaks (at loci A-B and C-D) held in alignment through persistent Holiday junctions (Daly and Minton, 1995). Courtesy of the authors with permission of the publisher.

reflection of the capacity of the strain for desiccation tolerance (Mattimore and Battista, 1996). The capacity of *N. commune* for desiccation tolerance must reflect, in part, its capacity to perform efficient DNA repair; in this regard more information is needed on the role of redundant DNA sequences in cyanobacteria, such as multiple genomes, STRR's, HIP sequences (Chapter 16), and the role of chromosome structure and DNA methylation in these organisms.

3. Genes for Carbohydrate Metabolism

Pathways of carbohydrate metabolism and their constituent enzymes are thought to have arisen early during evolution (Marsh and Lebherz 1992). A consideration of the genes encoding carbohydrate-modifying enzymes in *Nostoc* is especially relevant in view of the obvious importance that extracellular carbohydrates play in colony structure. The evolution of fructose-1,6-bisphosphate aldolase (fructose-1,6-bisphosphate D-glyceraldehyde-3-phosphate lyase; aldolase; EC 4.1.2.13) is of particular interest as this enzyme occupies a pivotal position at the convergence of the pathways of glycolysis (catabolism) and photosynthesis (anabolism). Two mechanistically-distinct forms of fructose bisphosphate aldolase exist, both are believed to be ancient and both are considered to be the products of convergent evolution (Marsh and Lebherz, 1992; Rutter, 1964). The class I aldolases appear to be the predominant forms found in eukaryotic cells while comparatively little is known about the class II enzymes (Marsh and Lebherz, 1992; Gibson and Tabita, 1988, Marsh and Lebherz, 1992, Schäferjohann et al., 1995).

A class II aldolase activity was demonstrated in cell-free extracts of several cyanobacteria including *Nostoc muscorum, Anabaena variabilis, Plectonema* sp. and *Anabaenopsis* sp. and the activity was found to be used strictly for hexose synthesis during the light-independent reactions of photosynthesis (Rutter, 1964;Willard and Gibbs, 1968). The organization of the *fda* region in *N. commune* 584 is atypical (Joardar et al., unpublished data). In the Bacteria, genes encoding Calvin Cycle enzymes are sometimes duplicated, they are often clustered, linked tightly and orientated in the same direction on the chromosome (Alefounder et al., 1989; Chen et al., 1991; Gibson and Tabita, 1988; Gibson et al., 1991; Schäferjohann et al., 1995; Singer et al., 1992; van den Bergh et al., 1996). Recently, it was proposed that there may be a dichotomy in the evolution of the bacterial class II aldolases (Schäferjohann et al., 1995). Our analysis and comparisons of Fda sequences, including that of *N. commune* UTEX 584, suggest four clusters of related sequences among the class II aldolases (Joardar et al. unpublished data). The sequences of those aldolases used only in the Calvin cycle constitute cluster 1. Such a grouping strengthens the suggestion that "...photosynthetic aldolase(s)" - and also those of the other autotrophs included in cluster 1 - "(are) specifically adapted in functional and regulatory characteristics to the needs of the Calvin cycle..." (Gibson et al., 1991). It can be noted, however, that eukaryotic algae and plants relied upon class I aldolases for photosynthesis rather than improvising on a gene acquired from a cyanobacterial ancestor (Marsh and Lebherz, 1992).

The first rigorous evidence of the presence of sucrose phosphate synthase in a prokaryotic organism was presented for *Anabaena* sp. strain 7119 (Porchia and Salerno, 1996). Unlike the equivalent enzymes of plants and algae the cyanobacterial enzymes are monomeric, small (45 to 47 kDa versus 260 to 520 kDa), and have a specific requirement for Mn^{2+} or Mg^{2+} for their catalytic activities. This is an important finding and it raises some very interesting questions. Presumably the plant and algal enzymes derived from an ancestral cyanobacterial enzyme. Why should cyanobacteria be the only prokaryotes to have evolved the capacity to synthesize sucrose? It is thought that sucrose is the principal carbon which moves from the vegetative cells to heterocysts (Wolk et al., 1994). Recently, inactivation of *sucA*, encoding sucrose synthase and thought to be involved in sucrose breakdown, lead to a Fix^- phenotype in *Anabaena* sp. PCC 7120 (Buikema et al., 1998). It seems unlikely, however, that forms such as *Nostoc* acquired the ability to make sucrose solely to support nitrogen fixation. In fact a gene corresponding to that which encodes the sucrose phosphate synthase of *Zea mays* was identified in *Synechocystis* sp. PCC 6803, and its putative product of 720 amino acids shares 47.7% sequence identity with the plant enzyme (Kaneko et al., 1996). The capacity to synthesize sucrose may have arisen early during the evolution of cyanobacteria and possibly for the following reason: sucrose is a non-reducing sugar that acts, with an effectiveness equivalent to that of trehalose, to prevent leakage from desiccated membranes upon their rehydration (Potts, 1994). For example, sucrose and trehalose account for the very low mid point temperatures of the membrane phase transitions in desiccated and rehydrated cells of *N. commune* CHEN (T_{mdry} = 8°C and T_{mwet} = 6°C, respectively (Hill et al., 1997). There is reason to suppose that sucrose synthesis in cyanobacteria arose through the need to protect cells during desiccation, probably the principle stress imposed upon ancient terrestrial and aerophytic forms.

4. DNA Transfer

Knowledge of the genetics of heterocystous cyanobacteria was made possible, in large part, through the development of a conjugation system for filamentous cyanobacteria and its application to *Anabaena* sp. PCC 7120 (Wolk et al., 1994; Haselkorn, 1992). The development of methods used to manipulate this strain were made possible through isolation of the replication origin of a plasmid that was endigenous to *Nostoc* sp. strain PCC 7524 (Schmetterer and Wolk, 1988).

An auxotrophic valine-requiring strain of *N. muscorum* was reported to be transformed with DNA in a process that was insensitive to ribonuclease and protease and dependent upon DNA concentration (Trehan and Sinha, 1982). Environmental factors such as UV irradiation and desiccation may have direct effects upon the genomic integrity of *Nostoc* spp. (and other cyanobacteria; Chapter 21) growing in situ but the effects of more indirect factors under these conditions, such as cyanophages and exogenous DNA, remain largely unknown (Chapter 20).

C. Aquisition of the Heterocyst

1. Selective Pressure

Structural, physiological and molecular properties of heterocysts are understood in detail (Wolk et al., 1994) but what selective advantage was imposed upon *Nostoc* spp. and other forms which achieved the capacity to develop them? How and why did the capacity arise? Within which lineage did the capacity arise and how many times did the capacity arise within the Cyanobacteria?

Many cyanobacteria fix nitrogen efficiently under ambient oxygen concentrations and they do so without the recourse of heterocyst differentiation. Paradoxically, oxygen evolution, which is widely regarded as the selective pressure which lead to the development of heterocysts, was made possible through the evolution of a component of nitrogenase itself. Ancient duplications and divergence of a type I (containing Mo and Fe) NifH protein lead to the formation of a protochlorophyllide reductase that subsequently diverged to permit reduction of chlorin and then formation of chlorophyll; an event that occurred later during the early evolution of cyanobacteria (Burke et al., 1993). The selective pressure of oxygen on the acquisition of heterocysts is unclear. Based upon sulfur-isotope studies it was suggested that between 0.64 and 1.05 billion years ago atmospheric oxygen levels rose from 5% to 18% of the present (100%) level (Habicht and Canfield, 1996; Canfield and Teske, 1996). These data suggest that forms corresponding in morphology with present day *Nostoc* and other heterocyst-forming species arose before the oxygen content of the atmosphere was especially high. The oxygen concentration inside actively-photosynthesizing cells is another matter however; and one that probably was more instrumental in the need to partition nitrogenase and oxygenic photosynthesis; although this is not certain and it cannot explain adequately why only some nitrogen-fixing forms of cyanobacteria differentiated heterocysts.

The capacity for heterocyst differentiation is thought to have arisen only once in cyanobacterial evolution (Doolittle, 1982). However, a second Nif system, Nif2, was discovered recently in *Anabaena variabilis* ATCC 29413 (Thiel et al., 1995; Schrautemeier et al., 1995). The Nif2 system functions both in heterocysts and vegetative cells but only under anaerobic conditions. The Nif2 system was absent from a number of strains classified as *Anabaena* spp., including *Anabaena* sp. PCC 7120, and also absent from a number of *Nostoc* strains including free-living isolates from *Azolla* and *Macrozamia*. The system was detected in five independent isolates classified as *Anabaena* from *Azolla*. These data imply that the presence of the Nif2 system reflects a restricted physiological capacity; a more thorough analysis of its distribution is needed before it can be determined whether possession of this system is a useful taxonomic marker within the *Anabaena/Nostoc* group. This finding of a second Nif system in vegetative cells of some heterocystous forming strains raised the question of whether Nif1 and Nif2 arose independently; a question which is not yet resolved with certainty. The restricted distribution of the Nif2 system may suggest a lateral transfer event.

2. Relationship to Akinetes

There are examples of genes that function in the development of both akinetes and heterocysts (e.g. Leganes, 1994) and it was suggested that both cell types share part of a common pathway. It was suggested that heterocyst differentiation may have built upon akinete formation, rather than vice versa, because the need to sporulate would have preceded, in evolutionary time, the need to fix nitrogen aerobically (Wolk et al., 1994). If akinete production was such a pressing need for primitive cyanobacteria then we must question why no other group of extant forms makes them. What could be the selective advantage of being able to make akinetes? The akinetes of *Nostoc* sp. PCC 7524 and *A. cylindrica* were found to be more resistant to drying and freezing than vegetative cells (see Herdmann, 1988). In fact electron microscope autoradiography using ^{3}H-thymidine indicated that akinetes were not quiescent and had a mode of DNA synthesis that matched that in germinating akinetes and vegetative cells (Favali and Caiola, 1986). Geitler (1932) commented that he "spotted" akinetes in *N. commune* on only one occasion and questioned whether other reports of akinetes in this cyanobacterium concerned other forms. In fact he attributed the rare development of akinetes in *N. commune* to the resistance of the colony (i.e. the vegetative cells) to desiccation and freezing. By doing so he implied a function for akinetes. However, it does not seem that the function of akinetes is specifically for desiccation tolerance or freezing tolerance (Potts, 1994). Akinetes were absent from all of the field materials of *N. commune* I collected during the past 23 years, and they do not occur in laboratory cultures of *N. commune* DRH1 or UTEX 584 (unpublished observations). The vegetative cells of *N. commune* DRH1 do show electron translucent wall layers that resemble those of akinetes (Potts, 1994) and other *Nostoc* strains do produce akinetes (Mollenhauer, 1986b). Akinetes are resting stages that may confer some selective advantage. But non-heterocystous cyanobacteria are as equally distributed and seemingly as successful as heterocystous cyanobacteria yet they had no recourse to develop akinetes. Why?

3. Gene Expression in the Differentiated Heterocyst

Nearly 45% of the *Anabaena variabilis* ATCC 29413 DNA sense strand (1900 different mRNA species) is transcribed by terminally-differentiated heterocysts and more than half of the transcribed mRNAs (approximately 1000 genes) are heterocyst specific (Lynn et. al., 1986; Buikema and Haselkorn, 1993). The ability to differentiate a heterocyst thus requires an amount of DNA that is greater than the total genome size of a small prokaryote e.g. *Mycoplasma genitalium* (580 kbp). In view of the proposed inordinately slow rate of evolution of cyanobacteria (Chapter 2), the comparatively recent appearance of forms capable of diverting some 15 – 20% of the coding capacity of their genome to a single physiological function is incongruous. Did the capacity to differentiate a heterocyst evolve slowly through gradual rearrangements and modifications of existing genes and pathways or did the event occur rapidly following acquisition of DNA by lateral transfer?

The genome of *Anabaena* sp. PCC 7120 is approximately 6.42 Mb and was partially mapped using homologous and heterologous gene probes (Wolk, 1997). Only a relatively small number of genes were mapped but the positions of a sufficient number of those required for heterocyst differentiation were located. From these analyses it appears that genes encoding components of the heterocyst differentiation machinery are scattered throughout the genome. For example, with respect to position O^{+} on a circular map, *hetR* (Buikema and Haselkorn 1991) is located at position 2.96 Mb, genes of the *nifB* and *nifH* operons (Mulligan and Haselkorn, 1989) are clustered at 1.7 Mb, and *hepA* is located at position 3.47 Mb (Holland and Wolk, 1990). Can we expect a similar pattern of distribution of heterocyst-specific genes in *Nostoc* spp. and other heterocystous forms?

Unlike non-heterocystous forms, *Nostoc, Anabaena* and other heterocystous cyanobacteria contain multiple copies of genes which encode products similar to eukaryotic RNA-binding proteins (Mulligan et al., 1994). Rather than the presence of heterocysts this may reflect the larger genomes of the latter and more complex physiology such as the presence of life cycles.

4. *Significance in Plant Symbioses*

Why do symbioses between plants and cyanobacteria involve heterocystous and not non-heterocystous forms, and why is it that *Nostoc* spp. are the predominant heterocystous forms in such associations? Symbiotic cyanobacteria do have the capacity to form motile hormogonia and this property is thought to be important for the establishment of an association with plants (Tandeau de Marsac, 1994; Chapter 19). In fact specific genes in *Nostoc punctiforme* ATCC 29133, *hrmUA*, may regulate hormogonium formation through metabolism of a metabolite from the symbiont *Anthoceros punctatus* (Cohen and Meeks, 1997). However, although numerous cyanobacteria produce motile hormogonia the ability to enter into symbioses with plants is restricted to a few genera. Interesting data were provided recently by Meeks and colleagues. *Nostoc* sp. strain UCD311 is a transposon-induced mutant of *Nostoc* sp. strain ATCC 29133 that is unable to fix nitrogen in air but it does so under anoxic conditions. The mutant phenotype results from disruption of *devR* that is essential to the development of mature heterocysts but not akinetes (Campbell et al., 1996). Therefore, the ability of the mutant to invade and establish an association with *Anthoceros punctatus* suggests that protection against oxygen inactivation is provided within the tissues of the host. A range of other ancillary factors must be taken into account with respect to oxygen sensitivity. For example, *Nostoc cordubensis* Prosperi can form cultures with or without mucilage much of which is produced by heterocysts (Prosperi, 1994). Mucilaginous heterocysts retained nitrogen-fixing capability at high oxygen concentrations while heterocysts lacking mucilage were unable to fix nitrogen at oxygen concentrations higher than 20%. At this time there appears to be no adequate explanation for the prevalence of *Nostoc* spp. in plant associations.

IV. Physiology and Growth

A. *Life Cycles*

The life cycles of *Nostoc* spp. represent physiological (and presumably genetic) capacities that are unknown in other genera (Komárek and Anagnostidis, 1989). These characteristics include, in simple terms, the formation of motile hormogonia with or without gas vesicles, the formation of seriate (filamentous) and aseriate (ball-like) phases, formation of micro and macro colonies, reproduction through budding and fragmentation of colonies, or development of hormogonia (Mollenhauer1986b; Mollenhauer, 1970; Dodds et al., 1995; Tandeau de Marsac, 1994; Lazaroff, 1973; Potts and Bowman, 1985). So distinct are these life cycles that it is not surprising they figured prominently in classical descriptions of *Nostoc* and were thus used extensively in taxonomic schemes (Geitler, 1932). While the validity of the more elaborate of the latter are questionable there is no doubt that the genetic determinants of these life cycles will be found to have solid taxonomic utility for the discrimination of certain *Nostoc* spp.

Studies with *Nostoc commune* strain UTEX 584, *Nostoc muscorum* A and *Nostoc commune* 45 suggested that the spectral quality of incident radiation as well as the presence of excreted metabolites of unknown identity influenced morphological changes including the production of hormogonia (Isono and Fujita, 1981; Lazaroff, 1973; Robinson and Miller, 1970; Martinez et al., 1985; Tandeau de Marsac, 1994). However, many details are lacking and in some cases the data of different authors are contradictory. Further advances in this area may benefit from the increasing awareness of, and interest in, the components of protein phosphorylation and dephosphorylation networks in bacterial physiology and their roles in sensing (Kennelly and Potts, 1996; McCartney et al., 1997). For example, cAMP stimulated significantly the aggregation and gliding movement of filaments of *Spirulina platensis*, and rapid changes in levels of cAMP in *Anabaena cylindrica* accompanied responses to environmental factors including changes in pH (Ohmori, 1989; Ohmori et al.,1992); the gene encoding adenylate cyclase in (cya) in *Anabaena cylindrica* is now available (Katayama et al., 1995).

The morphological features of the life cycles of several *Nostoc* spp. are documented in detail (Mollenhauer, 1970; Lazaroff, 1973; Martinez et al., 1985) and photographs of some representative stages were provided by Rippka et al. (1979). The following describes only the essential features of the life cycles of *Nostoc commune* strains UTEX 584 and DRH1. These strains differ in aspects of their life cycle, in morphology, and in a number of different genetic markers. It is unclear whether the original isolate of *N. commune* UTEX 584 derived from classic field material of *N. commune*.

– *Nostoc* sp. Gibson 4 (T. Gibson, Scotland?) → M. Dyar (U. Washington) → M.B. Allen (U. California-Berkeley → deposited in UTEX Culture

Collection as *N. commune* UTEX 584 → M. Potts; contaminated culture rendered axenic by Nancy S. Morrison (Florida State University).

– *Nostoc commune* collected in Hunan Province, Henyong District, China → S. Scherer (U. Konstanz, Germany) → M. Potts; *N. commune* strain DRH1 isolated by Donna R. Hill (Virginia Tech).

The following features apply to *N. commune* UTEX 584 and are described, in part, in Potts and Bowman (1985). Stationary phase liquid cultures of *N. commune* UTEX 584 tend to occur solely in the dark-green aseriate stage and they remain so following months of incubation (Fig. 8). Long-term growth on solid media leads to the development of a spreading, pustulate, solid, gelatinous thallus. Inoculation of fresh media (in the presence or absence of combined nitrogen) leads to a dispersion of the aseriate masses (Fig. 8d). Although this probably occurs rapidly it is not readily apparent and first becomes manifest as a film of hormogonia on the surface of liquid media within several hours or a halo of lighter colored material around colonies transferred to fresh solid media (Fig. 9a). Within approximately 12hr of transfer the motility of the hormogonia ceases, and intercalary heterocysts become apparent within 24 hr (Fig. 9b). The appearance of these stages is highly reproducible. The following are not as readily synchronized as the initial events in the cycle: the extracellular polysaccharide of single filaments becomes prominent, except around heterocysts; and the single filament tends to contort which gives the appearance of a change in diameter (Fig. 9c). The degree of contortion of the filament, and its apparent increase in apparent diameter, continues. These stages which are intermediate between the seriate and mature aseriate phases may have either terminal or intercalary heterocysts. The mature aseriate stage tends to be spherical and represents the full contortion of the interheterocyst portion of a filament. Because heterocysts do not divide or grow they remain as the connection points for strings of aseriate spheres much like beads on a string (Fig. 8b). These tend to occur as two spheres held by a single heterocyst, or in some cases three spheres but rarely more (Fig. 8a).

The cell walls of the hormogonia presumably contain hydrophobic components that may aid dispersal. The specific cause of the contortion of the filaments is not known but surely reflects a change in the rheological properties the EPS. The EPS imposes a mechanical restraint upon the filament that as it grows is restricted to a smaller and smaller available space. The filament therefore is likely to have acquired energy rather like a coiled spring from the pressure imposed from the sheath. Examination of the EPS of aseriate masses following subculture and release of cells shows it to be rigid and pronounced. The change between the sol and gel properties of the EPS was studied at the ultrastructural level by Bazzichelli et al., 1985, 1986). It seems certain that the only way to effect such changes is through the secretion of one or more carbohydrate-modifying enzymes.

Sessile colonies and motile hormogonia were recognized as the two main developmental stages in the life cycle of a *Nostoc* sp. isolated from the endocytobiosis with *Geosiphon pyriforme* (Bilger et al. 1994). Hormogonia displayed an enhanced capacity for nonradiative dissipation of absorbed light energy.

The mature thallus of *N. commune* in nature has a characteristic structure (Fig.1b). Prolonged incubation of colonies on semi-solid media generates small spherical colonies, "pearls" ("Perlen des Rozenkranzes" Mollenhauer, 1986a) from 0.5 to 4 mm in diameter which bud from the mature thallus. The mass of small colonies has the appearance of caviar (Fig. 18d in Potts 1994; Plate 24d). A strain was isolated from the inner portion of one of these colonies and designated *N. commune* strain DRH1. Some features of the life cycle of *Nostoc commune* and *N. commune* strain DRH1 were presented in Potts (1994; Fig. 18) and Hill et al. (1994b). *N. commune* strain DRH1 grows in liquid culture as contorted filaments in loose aggregates that form spherical masses. In the initial stages of their formation in liquid culture these colonies appear as small balls and can be resolved with a dissecting microscope. Homogenization of liquid cultures and transfer of aliquots of the cell suspension to semi-solid media leads to the development of macroscopic spherical colonies. "Pearls" placed on sterile agar at one side of a Petri plate ultimately generate pearls at some distance from the inoculum presumably as a result of the formation of motile hormogonia.

A distinguishing characteristic of strains of *N. commune* is the massive amount of EPS produced in liquid and identification as a carbohydrate-modifying enzyme(s) (Hill et al. 1994a, 199b); it seems likely that the role for Wsp is the modification/modulation of the glycan structure perhaps through limited hydrolysis of cross links.

Staining of "pearls" with carbohydrate-specific dyes at different pHs demonstrated how the layers of

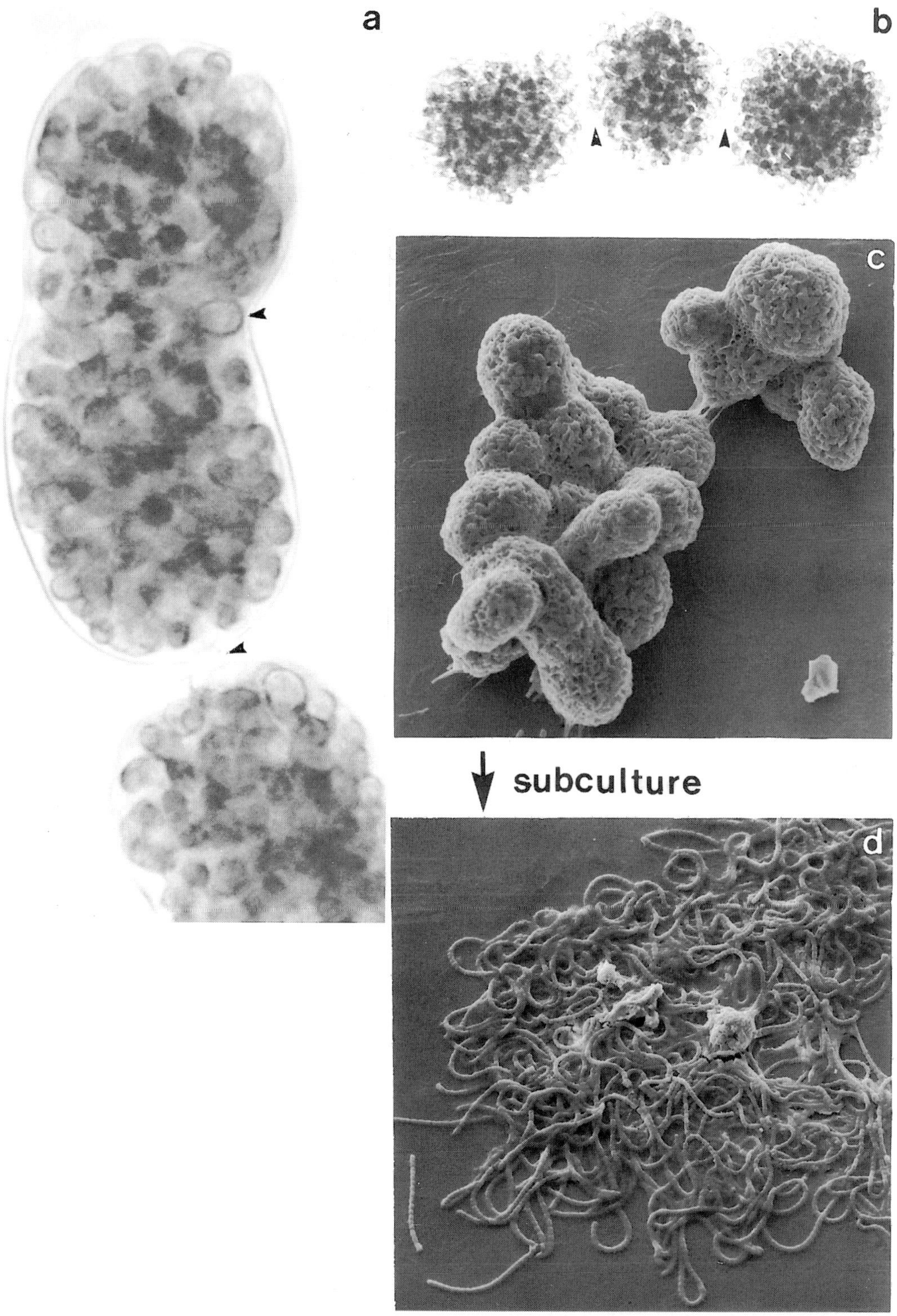

Fig. 8. Life cycle of *Nostoc commune* UTEX 584. a. Late aseriate stage. Note very prominent sheath of the cell packets and lack of sheath around the heterocyst (arrow) that joins the packets; bar = approx. 15 mm. b. String of mature aseriate packets that derived from a single filament with two heterocysts. c. aseriate packets showing fine structure of the surface sheath (SEM). d. Hormogonia after transfer of aseriate stages to fresh media. (SEM; compare with Fig. 9.).

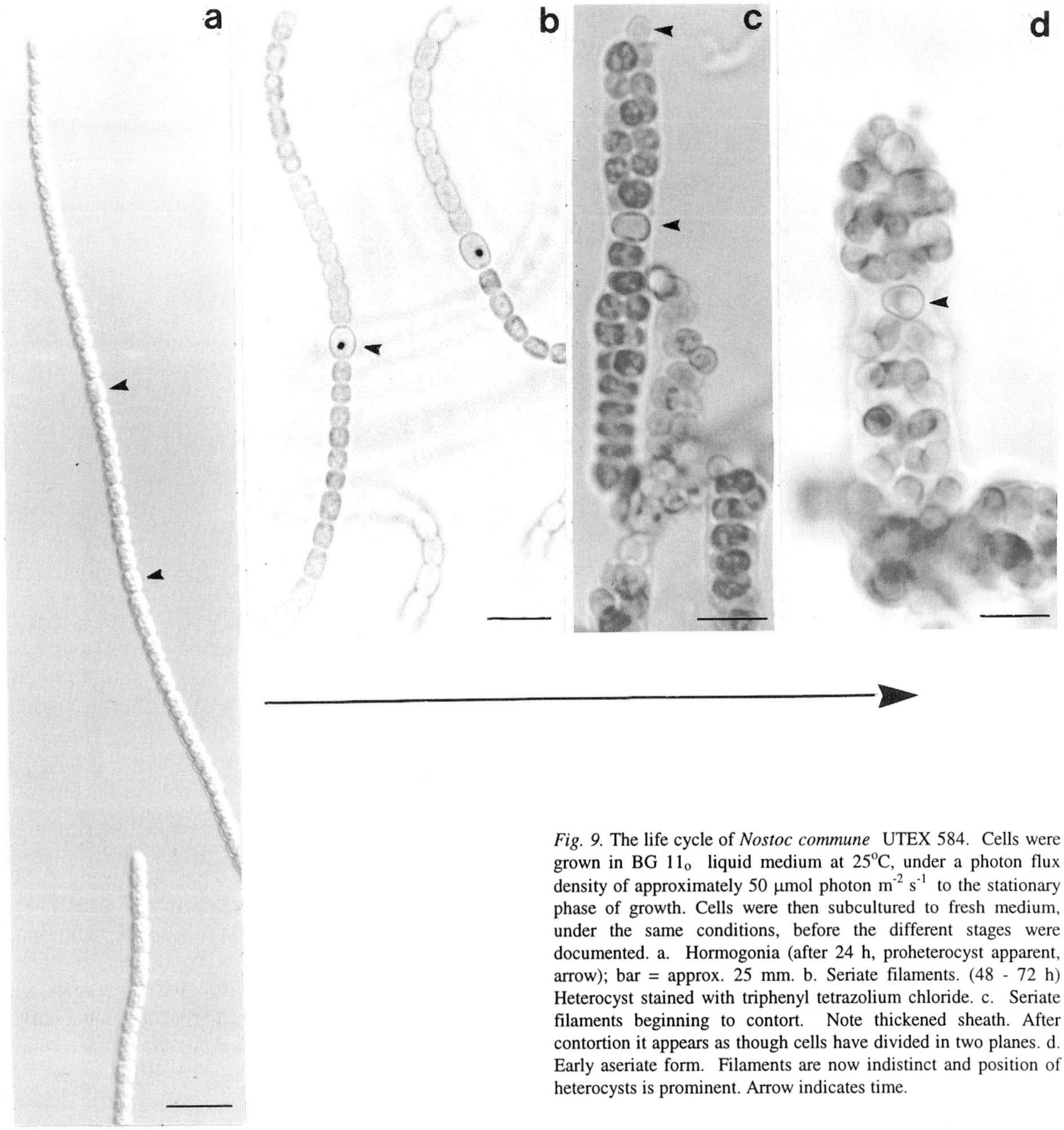

Fig. 9. The life cycle of *Nostoc commune* UTEX 584. Cells were grown in BG 11_0 liquid medium at 25°C, under a photon flux density of approximately 50 µmol photon m^{-2} s^{-1} to the stationary phase of growth. Cells were then subcultured to fresh medium, under the same conditions, before the different stages were documented. a. Hormogonia (after 24 h, proheterocyst apparent, arrow); bar = approx. 25 mm. b. Seriate filaments. (48 - 72 h) Heterocyst stained with triphenyl tetrazolium chloride. c. Seriate filaments beginning to contort. Note thickened sheath. After contortion it appears as though cells have divided in two planes. d. Early aseriate form. Filaments are now indistinct and position of heterocysts is prominent. Arrow indicates time.

carbohydrate differ markedly in different sections of the thallus (Hill et al. 1994b). In material of thalli collected from Aldabra Atoll it is possible to resolve single filaments totally calcified. These are released from fragmented colonies and may, if viable, serve as a source of inocula for dispersal (Hill and Potts, unpublished data).

The characteristic morphologies of *Nostoc commune*, *N. commune* var. *flagelliforme* and *N. parmelioides* are maintained in all those habitats within which these forms accumulate. These characteristic morphologies, found in colonies from the Tropics to polar regions, are striking and must be determined only in part by environmental variables.

B. Protein Phosphorylation

Studies suggest that protein phosphorylation and two-component regulatory systems may play an important role in physiological and developmental responses of cyanobacteria, including their responses to osmotic shock, high light, ammonia, chromatic adaptation and presence of intermediary metabolites at physiological concentrations (Grossman et al 1994; Hagemann et al. 1993; Howell et al., 1996; Mann et al., 1991; Mann, 1994; Potts et al., 1993; Rodríguez et al., 1994; Warner and Bullerjahn, 1994; Zhang 1993; McCartney et al., 1997; Chapters 14 and 15). The study of the role of protein phosphorylation in cyanobacterial development is in its infancy but the indications are that phosphorylated proteins may intervene in developmental pathways (Campbell et al., 1996; Hagen and Meeks, 1998).

Although two component regulatory systems appear to be involved in several specific aspects of cyanobacterial physiology it is probable that protein tyrosine phosphorylation in particular, plays a central role in the overall integration of metabolism as it does in eukaryotic cells. The phosphorylation and dephosphorylation of proteins on tyrosine constitute a pivotal mechanism for the regulation of enzyme activity and thus cellular functions. It was assumed, until only recently, that such protein modifications occurred exclusively within the confines of the eukaryotic cell (Kennelly and Potts, 1996). This assumption was based, in part, on the extensive documentation of phosphohistidine and phospho-carboxyl amino acids in bacterial proteins, and the involvement of the latter in a diverse set of bacterial signal transduction mechanisms.

The finding of a dual specificity protein serine/threonine tyrosine phosphatase (IphP) in a cyanobacterium, *Nostoc commune* UTEX 584, raised important questions as to the origin of tyrosine phosphorylation, and the evolution of its function (Potts et al., 1993;McCartney et al., 1997). A gene library of *N. commune* UTEX 584 was constructed in λgt 10 and was plated in the presence of 5-bromo-4-chloro-3-indolyl phosphate (BCIP) – a non-specific substrate for phosphomonoesterases which generates indigo-carmine (a colored product) upon hydrolysis of the phosphoester bond in the presence of oxygen. One blue plaque was recovered from a plating of around 20000 plaques. An *Eco*RI fragment was subsequently recovered, subcloned in a plasmid vector, and it was found that this fragment directed the synthesis of a secreted *i*ndole *p*hosphate *h*ydrolase (IphP; Xie et al., 1989). In retrospect this finding has all the elements of serendipity: λgt 10 is not an expression vector, and the fact that the gene product was correctly processed, secreted beyond the outer membrane of *E. coli,* and was enzymatically-active, is unusual. Upon DNA sequence analysis of *iphP* it was noted that a portion of the derived amino-acid sequence of IphP bore a striking resemblance to the consensus active site domain of the eukaryotic protein tyrosine phosphatases (PTPase's) – a group of enzymes implicated in the control of cell transformation and oncogenesis in humans. Upon purification and characterization of IphP it was found that the enzyme did indeed possess PTPase activity, as well as a secondary protein serine phosphatase activity; such dual specificity is a property that is shared with the VH1 gene product of vaccinia virus (Potts et al., 1993).

The question of the target (substrate) of IphP is an important one. In a characterization of IphP the presence of an aromatic ring either as part of the leaving group alcohol or immediately adjacent to it, as in 5'-AMP, was a strong positive determinant for hydrolysis. Among peptide and protein substrates a rough, imperfect, correlation between charge character and hydrolysis was noted in which proteins and phosphorylation sites of an acidic nature seemed favored. What could be the natural substrate of IphP - a secreted phosphoprotein perhaps? A secreted protein tyrosine phosphatase with a modular effector domain was described in the bacterial pathogen *Salmonella typhimurium.* (Kaniga et al., 1996). It seems not unreasonable that *in vivo* the target of IphP is phosphoprotein either on *N. commune* UTEX 584 itself or on other organisms with which it comes into contact.

As is evident from the preceding discussion the components of signal transduction networks modulated through protein phosphorylation are being uncovered in cyanobacteria. The finding of a calmodulin-like polypeptide in *Anabaena* strains (Pettersson and Bergman, 1989) is therefore of some interest because in eukaryotic cells protein phosphorylation-dephosphorylation networks are attenuated considerably by calmodulin. Added to this are the reports of abscisic acid (ABA), a potential calcium agonist in cyanobacteria (Hirsch et al., 1989; (Marsálek et al., 1992). In *N. muscorum* imposition of salt stress lead to the appearance of ABA in the extracellular medium and it was considered that it may become immobilized in the EPS (Marsálek et al., 1992). One of the most obvious effects of adding exogenous ABA to *Nostoc* 6720 was a 20% increase in intracellular calcium and a doubling of the heterocyst frequency (Smith et al., 1987; Marsálek et al., 1992; Huddart et al., 1986). These data raise important questions to do with Ca^{2+} fluxes that may have the potential to induce physiological changes in field populations and the role(s), if any, of ABA. On Aldabra Atoll, a typical value for Ca^{2+} in freshwater pools which support *Nostoc* growths is around 100 mg L^{-1} (2.6 mM; Potts, 1977); a value slightly higher than that at which the biphasic uptake of Ca^{2+} was saturated in *Nostoc* MAC (Pandey et al., 1996). The mean content of Ca^{2+} in trichomes of *Nostoc calcicola* was measured as 64 to 322 fg cell-1 (Fagerbakke et al., 1991); Ca^{2+} constituted one of four elements that represented the more dissolvable fraction in cells. The conclusion of the study was that there may be a transport system for elements within the trichome to facilitate an equal distribution. Either cell division was nearly synchronous, or a rapid redistribution of cell material takes place immediately prior to division. The common occurrence of cyanobacteria in limestone areas where calcium is present at high concentrations in solution makes it hard to infer any obvious role for calcium through the drawing of analogies with well-characterized eukaryotic systems. Field communities of *Nostoc* forms often grow in close proximity to other organisms including higher and lower plants; it necessary to evaluate any reports of abscisic acid either in field colonies, or "axenic" cultures, most cautiously.

1. Secondary Products

A wide range of secondary products are produced by *Nostoc* spp. (Golakoti et al. 1995; Kaya et al. 1996; See Chapter 22). Bioactive compounds present in methanolic extracts and extracellular products from *Nostoc muscorum* evoked a significant inhibition of growth of the phytopathogen *Sclerotinia sclerotiorum* (Mule et al., 1991). The cyclic depsipeptides nospeptins A and B were isolated from *N. minutum* (NIES 26) and were found to be potent inhibitors of elastase and chymotrypsin (Okino et al. 1997). Two bacteriohopanoids with acyclic pentol sidechains were purified from *Nostoc* sp. strain PCC 6720 (Zhao et al., 1996). A Boeseken complex of boric acid was reported from the marine form *Nostoc linkia* (Hemscheidt et al., 1994). In this strain the structural skeletons of 7.7 paracyclophanes were polyketides that formed from the dimerization of acetate-derived nonaketides (Bobzin and Moore 1993). The synthesis of heterocyst glycolipids is thought to originate with polyketides, and a gene encoding a polyketide synthase like enzyme was characterized in *N. punctiforme* strain ATCC 29133 (Campbell et al., 1997)

C. Cyanoglobins

The gene encoding cyanoglobin, *glbN*, was identified through sequence analysis of the *nifU-nifH* region of *Nostoc commune* UTEX 584 (Potts et al., 1992). GlbN is a monomeric hemoglobin that binds oxygen reversibly, with high affinity, and with non-cooperativity (Thorsteinsson et al., 1996).

Cyanoglobin is synthesized only in cells starved of combined nitrogen when they are subjected to microoxic or anoxic conditions (Hill et al., 1996). The available data suggest that GlbN may function as part of a terminal cytochrome oxidase complex that is needed under microxic conditions when ATP supply for nitrogen fixation may be limited (Hill et al., 1996).'s The on-and-off rates of cyanoglobin for oxygen are almost 4-fold, and 6-fold higher, respectively, than the equivalent rates for leghemoglobin (Thorsteinsson et al., unpublished data); a property which is indicative of an involvement in rapid oxygen flux and one which is consistent with the proposed role for this hemoprotein.

The list of nonvertebrate oxygen-binding proteins that includes GlbN is growing (Andersson et al., 1996; Hardison., 1996). In a recent study Moens et al. (1996) proposed that all globins evolved from a family of ancestral, approximately 17-kDa hemoproteins, which displayed the globin fold and functioned as redox proteins. Alignments of available sequences defined three separate clusters. It was proposed that the horizontal transfer of a globin gene occurred from an ancestor common to the ciliates (*Paramecium* and *Tetrahymena)* and the green alga *Chlamydomonas,* to a cyanobacterial ancestor. The alignment of a restricted number of sequences is provided here and includes in addition those of a putative globin of *Synechocystis* PCC 6803; a representative of a new class of non-leghemoglobin sequences from plants (Andersson et al., 1996); and the sequence of the cytochrome oxidase of *Pseudomonas perfectomarina (stutzeri)* (Jüngst et al., 1991) whose N-terminal region shares sequence similarity with *Nostoc* GlbN (Hill et al., 1996). Evidence to support a horizontal transfer from an

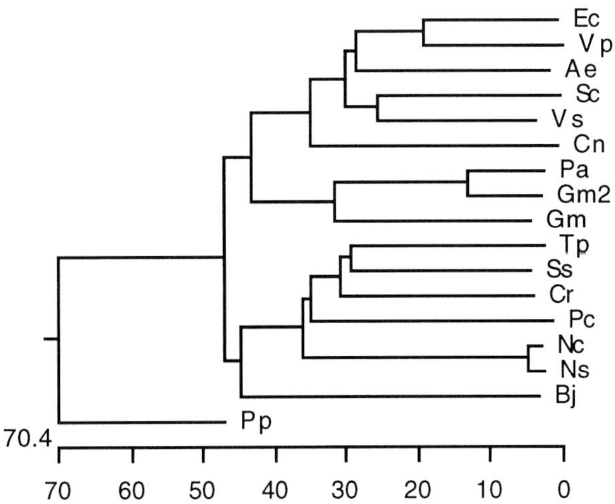

Fig. 10. Alignment of selected hemoproteins and cyanoglobins. Sequences were acquired from the Genbank, PIR and Cyanobase databases. Abbreviations:Tp, *Tetrahymena pyriformis*; Ss, *Synechocystis* PCC 6803; Cr, *Chlamydomonas reinhardtii*; Pc, *Paramecium caudatum*; Nc, *Nostoc commune* UTEX 584; Ns, *Nostoc* MUN 8820; Sc, *Saccharomyces cerevisiae*; Vs, *Vitreoscilla* sp.; Ec, *Escherichia coli*; Ae, *Alcaligenes eutrophus*; Cn, *Candida norvegensis*; Pa, *Parasponia andersonii*; Gm, *Glycine max;* Gm2, *Glycine max* non-leghemoglobin; Pp, *Pseudomonas perfectomarina*; *Vp, Vibrio parahaemolyticus;* Bj, *Bradyrhizobium japonicum.* Sequences were aligned using the Clustal algorithm. The multiple alignment parameters were: gap penalty 10, gap length 10, with a pairwise alignment parameter of Ktuple 1 and a percent accepted mutation (PAM) of 2.5 per residue. Ancestral residues were determined using the Hartigan-Fitch-Sankoff algorithm that provides the most parsimonious set of residues from two residue sets and the residue weight table; here scoring for both identical and chemically-similar residues. In the cladogram branch distances represent the actual length of sequence divergence that was calculated from comparing sequence pairs in relation to the phylogenetic tree. The x ordinate measures distance between sequences.

ancestral ciliate to an ancestor of *Nostoc* is not readily apparent (Fig. 10) although the restricted distribution of *glbN* may in itself suggest it was acquired through lateral transfer. The sequence similarity between the protozoan sequences and cyanobacterial globins is apparent. In view of the fact that protozoa often occur in association with *Nostoc* colonies, and the range of symbiotic associations between protozoa and cyanobacteria (Carr and Whitton, 1973, 1983), the suggestion of the transfer of an ancestral gene between members of these two groups appears to be not unreasonable. Given the identification of proteins of 18 kDa in a range of cyanobacteria that are immunologically-related to GlbN (Hill et al., 1996), and the identification of a putative hemoprotein in *Synechocystis* sp. strain PCC 6803 which shows obvious sequence similarity to protozoan globins, it is possible that "cyanoglobins" may be widespread in cyanobacteria; they may have quite different roles to fit different ecological contexts. The gene that may encode the globin in *Synechocystis* PCC 6803 is positioned within a cluster of genes that may encode sensory transduction histidine kinases - a situation which has obvious parallels in other bacterial systems (Gilles-Gonzalez et al., 1991).

The restricted distribution of *glbN* and GlbN to a small subset of *Nostoc* strains suggested that the presence of *glbN* may provide a useful genotypic marker (Hill et al., 1996). The marker is probably indicative of a shared physiological capacity by these strains, and one that may be of ecological relevance. In this context it is of interest that *N. commune* strain DRH1 – the derivative of field material of *N. commune* – seems to lack GlbN (Hill et al. 1996).

The sequence similarity between the regions both upstream and downstream of *glbN* in *N. commune* 584 and *Nostoc* sp. 8820 is high with the exception that *Nostoc* sp. 8820 lacks the NtcA-binding site identified in the *nifU-glbN* intergenic region of *N. commune* 584 (Hill et al., 1996; Jackmann and Mulligan, 1995). Curiously the NtcA binding region in *N. commune* 584 is flanked by short tandem repeats (AATTACG) and inverted repeats provide the potential for this region to fold into a stable secondary structure (Hill et al., 1996). The functions of such elements is unresolved (Masepohl et al., 1996). Perhaps the NtcA binding site of *N. commune* 584 *glbN* was added to the promoter through a separate recombination event after acquisition of the ancestral gene. If, as seems to be the case, *glbN* has its own promoter, one must still question why the gene is positioned between the *nif* genes of two contiguous *nif* operons.

Nostoc MUN 8820 contains the *glbN* gene positioned between *nifU* and *nifH* but lacks the AATTACG STRR sequences present in the *nifU-glbN* region of *N. commune* UTEX 584 (Hill et al., 1996). Yet, *Nostoc* PCC 6720, which lacks a *glbN* gene at this position contains 6 copies of STRR TTAGTCA and 6 copies of GGACTAT (Beesley et al., 1994).

D. Desiccation Tolerance

1. Significance

The pivotal role of water in the emergence of life is readily acknowledged. In contrast, the importance of water in cell structure and function is usually taken for granted, it is persistently understated or, more than often, it is ignored (Potts, 1994). In 1702 van Leeuwenhoek described the springing to life of dried objects – in this case rotifers – upon their rehydration. These observations generated little attention. In paraphrase of Albert Szent-Györgyi, Clegg (1986) noted: "biology has forgotten the dried cell – or has never discovered it." A detailed review of desiccation tolerance in prokaryotes (Potts, 1994), and a brief overview in the context of cyanobacteria (Potts, 1996), appeared recently.

Despite its important environmental consequences none of the books or reviews on cyanobacteria that appeared during the past 26 years make more than a passing reference to desiccation tolerance, if they do so at all. In the early 1970s, any appreciation of desiccation tolerance which one derived from the current literature tended to rely upon a good deal of phenomenological and anecdotal observations of microbial communities growing in situ – in many instances they still do.

There are numerous cases where intriguing cellular phenomena, first observed in communities which constitute diverse components of natural ecosystems, were analyzed at the ecological level, subsequently at the physical and biochemical level, and then finally at the molecular level. But, as pointed out and emphasized by Walsby (1994) there are precious few instances where the information was used to redesign physiological studies for application in the natural habitat where the communities occurred. The latter approach permits a fine analysis of natural selection and can, in the long run, permit the attenuation and manipulation of the activities of communities growing in situ. The widespread distribution of cyanobacteria, and their pervasive roles in microbial ecology, make them especially suited to such studies (Walsby, 1994).

Most cyanobacteria show a degree of tolerance to the effects of air-drying; one exception is the ultrasensitive *Pseudoanabaena galeata* (Geitler, 1982). Lyophilization of 13 cultures of cyanobacteria belonging to seven genera was measured in a variety of suspending substances. There was variation in the survival depending on the medium. With the exception of *Nostoc muscorum*, none of the strains survived lyophilization in skim milk, egg albumin medium or Jansen salts medium (Corbett and Parker, 1976).

Table 1 provides data that emphasize the low level of hydration of the cytoplasm in *N. commune* and other anhydrobiotes.

2. Natural Communities

The capacity of field populations of *Nostoc* to withstand cycles of desiccation and rehydration was known for a considerable amount of time (Section V). It took 5 hours for the thalli of desiccated *N. commune* to become fully hydrated in field tests in Schirmacher Oasis, Antarctica (Chapter 12). Similarly, upon dehydration water loss was complete within 5 hours. Under laboratory conditions water gain proceeded at a rate 6-fold faster than in the field (Gupta and Kashyap, 1994). *Nostoc commune* mats in McMurdo Sound Antarctica recovered to pre-desiccation rates of photosynthesis and respiration within as little as 10 min of rewetting while recovery of acetylene reduction was slower (> 24 h). *Phormidium* mats were less tolerant of desiccation and required 10 days for recovery. The partial hydration of *N. commune* during aerial exposure tended to slow recovery upon full rehydration (Hawes et al., 1992).

The first description of the structural properties of desiccated cyanobacteria was for those isolated from hot-desert rocks (Potts et al. 1983). Subsequent studies with *Nostoc commune* described techniques to permit the fixation and preparation of desiccated cells that avoided artifacts and permitted immunocytochemistry (Peat and Potts, 1987; Peat et al., 1988). Application of the latter technique confirmed that the water stress proteins (Wsp) of *N. commune* were secreted proteins, intimately associated with the extracellular glycan. The finding that Fe protein of nitrogenase was stable in cells desiccated for more than a decade complemented studies directed at the analysis of protein stability (Peat et al.,1988).

3. Mechanisms

Following the transfer of cells of *N. commune* UTEX 584 from liquid culture to the air the cells loose the capacity to fix nitrogen (within approximately 19 minutes) but maintain a pool of intracellular ATP, equivalent to that in liquid growing cells, for several hours. Upon rehydration desiccated cells of *N. commune* recover the capacities for respiration, photosynthesis and nitrogen

Table 1. Proteins in desiccated *N. commune* lack a monolayer coverage of water

Protein hydration[a] ($g\ H_20\ g\ protein^{-1}$)	Event	Consequence for protein	Cell hydration ($g\ H_20\ g\ cell\ solid^{-1}$)
0.0 to 0.1	charged group hydration polar group hydration disorder/order transition	proton redistribution disulfides reorder protein "loosens up"	anhydrobiotic cell e.g. *N. commune* **0.02 to 0.05**
0.1 to 0.2	acids saturate clusters mobilize side chain polar saturation	side chain and backbone conformational shifts	↑
0.2 to 0.3	peptide NH saturates water structure – condensation and rearrangment	**onset of enzyme activity** ↓	**decreasing H_2O** ↑ *Bacillus* spore **0.21 to 0.58**
0.3 to 0.4	polar group monolayer coverage finally complete monolayer coverage	dynamic icrease to solution value with increase in activity	at **0.3** sensitive bacteria die
0.5 to 0.7	**water content of protein crystals**		

[a]Some data adapted from refs in Potts (1994)

fixation in that order (for refs see Potts, 1994). Mel'nikova (1985) noted that one of the early events to occur upon dehydration of *Nostoc* was an increase in an ability of the nuclear material to bind acridine orange; young growing cells appeared more resistant than mature cells about to divide

Characterization of desiccated colonies of *Nostoc commune* identified a novel class of acidic polypeptides – these were termed water stress proteins (Wsp) and they constituted some 70% of the total soluble protein of desiccated cells. Wsp polypeptides are highly stable in desiccated cells, and they are induced in liquid grown cultures if those cells are subjected to multiple cycles of drying (Scherer and Potts, 1989). Initially, it was not appreciated that Wsp proteins are secreted by *N. commune*, but it was noted that the accumulations of Wsp in laboratory grown cultures never achieved those present in field materials. Scherer and Potts (1989) provide a preliminary characterization of Wsp. Recently, it was shown that Wsp polypeptides are secreted; an unexpected finding given the concentrations of these proteins; Wsp associate with a 1,4-β-D-xylanxylanohydrolase activity that is also secreted from the cells; they form complexes with secreted UVR-absorbing pigments through salt-dependent ionic interactions; they are immunologically-related to several carbohydrate-modifying enzymes including β-endogalactosidase (Hill et al., 1994a). At this point it can be noted that there are conspicuously few characterizations of proteins that are secreted by cells of cyanobacteria (see also the discussion of IphP above). The solving of the structure of the E_{335} carbohydrate-containing chromophore of the UVR-absorbing pigment, completed in the laboratory of S. Scherer, showed that the pigment contained xylose. These data, and those provided in Hill et al. (1994a) indicate that the role of Wsp may be to do with the synthesis and/or modification of carbohydrate-containing components which are secreted from the cells. Further evidence was obtained to substantiate this idea. An 8-kb *Eco* RI fragment was isolated recently from a library of *N. commune* UTEX 584 DNA in pTrc99A. The fragment was contained in a clone that showed intense cross-reaction with Wsp antibodies. Of the 5 ORFs that have been identified, one shows conspicuous derived amino acid sequence similarity to proteins involved in the secretion of cyclic glucans by *Agrobacterium tumefaciens* (as well as HepA from *Anabaena* PCC 7120 - a protein involved in the development of the heterocyst carbohydrate wall layer(Holland and Wolk, 1990). The other ORFs all show correspondence with carbohydrate-modifying enzymes (unpublished data).

It appears that the capacity to maintain proteins in a functionally-stable state is a central feature of the desiccation tolerance of *N. commune* (Potts, 1994). For example, the onset of the synthesis of all classes of lipid in desiccated cells resumes upon rehydration, instantaneously (Taranto et al., 1993). Some of the potential roles of proteins in responses of cells to water-deficit are summarized in Table 2.

Production of EPS, as described in Potts (1994), may also represent a mechanism for desiccation tolerance. The EPS of *Nostoc commune* is a complex, high molecular weight glycan that accumulates to more than 60% of the dry weight of colonies. The EPS retarded water loss from phosphatidylcholine membrane vesicles dried in vitro at 0% relative humidity (-400 MPa). A mixture of trehalose:sucrose, in a molar ratio which was determined to be present in cells in vivo, and low concentrations of the EPS, had a strong synergistic preventative effect on the leakage of a trapped solute (carboxyfluoroscein) when membrane vesicles were either desiccated or freeze-dried (Hill et al., 1977). The mechanism of protection by the EPS was determined to involve the prevention of membrane fusion. Freeze-fracture electron microscopy resolved the infrastructure at the contact points between the EPS and the outer membrane of desiccated cells in vivo. The freeze-fracture pattern of the outer membrane was not described previously in biological materials. The pattern was indicative of the P_b gel state phase where fatty acid moieties are tilted with respect to the bilayer normal. The capacity of the EPS to prevent membrane leakage, the atypical structure of the outer membrane, and the marked depression in the membrane phase transition temperature of desiccated cells which was measured *in vitro* through Fourier transform infrared spectroscopy, constitute protective mechanisms which prevent membrane damage during desiccation (or freeze-drying) and rehydration (Hill et al.,1997).

V. Patents and Applications

A small number of patents described methods that employ whole cells or components of *Nostoc* spp. The life cycle of *Nostoc muscorum* A was manipulated as the basis for a patent to screen for anti-adhesin antibiotics (Lazaroff, 1992). Plating of heterocysts (apparently with the capacity to germinate) and isolation of microcolonies derived from single heterocysts allowed the isolation of strains which differed in the properties of hormogonial motility and aggregation. Antibiotics that prevent bacterial attachment, but which are not bacteriostatic or bactericidal, can be identified through their interference with hormogonial motility.

Stevens and Murphy (1996) listed several strains of *Nostoc* and *Anabaena* as host cells within which to synthesize the *Bacillus thuringiensis* ssp. *israeliensis cryIVD* product. The latter is active as a pesticide against Diptera.

Genes encoding the Nsp7524V restriction-modification enzymes from *Nostoc* PCC 7524 were isolated and a method was devised to express the active enzymes in *E. coli* (Ueno et al., 1995). *Nostoc* and other cyanobacteria form the basis of a substrate designed to produce vegetation on bare terrain

Table 2. Potential roles for proteins in the desiccation response of *N. commune*

1. specific peptidase	protein targeting e.g. ubiquitin
2. special sigma factors	to induce expression of stress genes
3. control derepression of catabolic enzymes	to escape stress
4. catabolic enzymes	to provide energy
5. proteins for surviving stress	to maintain homoeostasis
	to maintain pH etc
	to stabilize DNA and RNA
	to uncouple viability from cell division
	to regulate recruitment of mRNA
	to regulate carbohydrate metabolism
6. sensory proteins	to transduce the stress
7. structural (protective) proteins	to achieve tolerance (protection ?)

(Chiaffredo and Figureau 1995). Cell extracts of *Nostoc* and other cyanobacteria were used in a flocculation process to trap minerals (Leslie et al. 1984) and also to produce a dried, flowable, water suspendible composition for the treatment of soils (Schaefer and Boyum, 1988). Merck and Co. Inc. developed several patents for natural products of *Nostoc* spp. that are effective against *Cryptococcus* sp. – a causal agent of secondary fungal infections in patients with AIDS (Hirsch et al.1990).

Nostoc punctiforme was found to be effective in treating effluent from the polyfibre industry that contained domestic sewage (Hosetti and Patil 1989). In view of its growth characteristics in a new growth medium (BW) a *Nostoc* sp. was considered to be promising as a source of phycobiliproteins for the production of natural dyes (Silva et al 1989).

The wide scope of these techniques suggest considerable applications for whole cells of *Nostoc* spp. as well as their natural products.

VI. A Chronology of *Nostoc*

The discovery and subsequent study of *Nostoc* constitutes a fascinating and a humorous story. Added to this are a number of peculiarities of its interactions with humans that warrant discussion.

A. The Hair-Vegetable

Because *Nostoc commune* grows in the wild, without roots, stems, or leaves, and floats on water, the Chinese compared it to the "immortals" who are said to travel the world at will. *Nostoc commune* became known as Tian-Xian-Mi (Rice of Heavenly Immortals; Chinese names and places were translated according to the Pin-yin system proposed by the People's Republic of China) or Tian-Xian-Cai (Vegetable of Heavenly Immortals). According to legend, Hung Ge, an alchemist, physician and Taoist theoretician of the Eastern Jin Dynasty (317-420 AD) used *N. commune* as a staple food during periods of famine when he was a hermit in Southern China. Hung Ge attributed his health to the consumption of *N. commune* and he later introduced *N. commune* to the Emperor following an invitation to Court. The ailing Emperor is said to have regained his health after eating *Nostoc* and was so grateful that he granted the name Ge-Xian-Mi (Rice of Immortal Ge) for the cyanobacterium in honor of Hung Ge; the name is retained to this day (J-S Chen, pers. comm; T-C Huang, pers. comm.; Y-T Wong, pers. comm.; C.T Teng, pers. comm.).

The Ben-Cao-Gang-Mu (or Pen-Tsao-Kang-Mu; General Outlines and Divisions of Herbal Medicine) by Shi-Zhen Li (1517-1593?) of the Ming Dynasty (Li, 1596) has 52 volumes and lists at least 1892 medicines, among which are *Nostoc commune* and *N. commune* var. *flagelliforme*. The latter is referred to as "Fa-Cai" in mandarin Chinese (Fa means hair of human head; Cai means vegetable). The derivation of the word is readily explained from the hair-like morphology of desiccated colonies (Aruga, 1992). Earliest descriptions of Fa-Cai date back to the Sung Dynasty (960-1279 AD). The pronunciation of Fa-Cai (or Fa-Tsai) is similar to the pronunciation of a phrase that means "acquiring wealth" (or "make money"). A different word, but with the same pronunciation, is also part of the greeting "gongxi facai" which means "wish you prosperity." These different plays on words account for the traditional eating of *N. commune* var. *flagelliforme* as a delicacy at the time of the Chinese New Year - a custom that is thought to bring good luck (Fig. 11); a prestigious position indeed for such a lowly side dish (Plate 1i).

Fig. 11. Good health and wealth! The cover of a box of *N. commune* var. *flagelliforme* sold at the time of the Chinese New Year. Courtesy of S. Scherer; Plate 1i.

The following was translated from a popular Chinese newspaper: *"In Ninxia, Northwest China, ...(he) pointed to a barren hill beyond the ditch and indicated to us that this is where Fa-Cai is grown. The place shown is surrounded by barbwire and guarded by a vicious dog and a guard hut. I walked up the hill and saw for the first time the cultivated Fa-Cai...which is also called black gold" (for its price; Suen, 1994). Its price is based on gram* *quantities. With the price skyrocketing, it has been collected aggressively to near extinction. A one-year medium-scale test showed that from one Mu (a Chinese unit of land measurement equivalent to about 14.8 acres), the equivalent to about US$140 (in 1994) worth of Fa-Cai could be produced."*

There seems to be quite a business in *Nostoc*. In the market of Ilocos Norte, Laoag City, Philippines *N. commune* is sold in dried form – curiously in the fish section rather than the vegetable section of the market – and stored dry in 50-kg jute sacks (Martinez, 1988)!

The finding of *Nostoc* growths is taken with the utmost seriousness in China. According to a report in the overseas edition of *The People's Daily: "good news came recently from the Shi-Wu-Tou Village, Xin-Hua County, Shen-Nueng-Jia Forest District, Hubei Province (in Southeast China). The rare, precious Ge-Xian-Mi has been found to grow in about ten Mu (about 148 acres) of water-filled paddy in that village. The annual yield reached five kilograms. According to the literature, the only other places that also produce Ge-Xian-Mi are: six Mu (about 90 acres) in Africa and 0.7 Mu (about 10 acres) of water-filled paddy or swamp in Xiang-Fan of Hubei Province. At present, there is no whole-sale quantity of Ge-Xian-Mi on the domestic or overseas market, so the price for it is very high."*

Like all things in China the eating of *Nostoc* is likely to have gone on for a considerable period of time; something that Geitler appeared to be unaware of when he implied that natives should put *N. commune* to culinary use (Geitler, 1932).

Nostoc is a common dietary supplement for the endigenous populations of Thailand, Peru, China, Ecuador, Fiji, Java, Japan, Mexico, Mongolia and Siberia who have long appreciated the plant-like nutritional properties of a number of different forms including *N. commune* var. *flagelliforme*, *N. parmelioides* (*edule*), *N. ellipsosporum*, *N. verrucosum* and *N. pruniforme* (Jassby, 1988; Plate 1i). Does it not seem reasonable to question whether primitive man also sampled such delicacies?

2. *The Philosopher's Stone*

Descriptions of *Nostoc* from Europe appeared some 500 years ago, well before the scourge of the Black Death in the 1600s and more than 200 years before Antonie van Leeuwenhoek described animalcules under his prototype microscope in 1702 (Potts, 1997). It is clear why *Nostoc* came to prominence at this time; colonies of *N. commune* in soils appear as blackened, brittle, nondescript crusts when they are dried, but become conspicuous and swollen as dark blue-green or olive-green masses, with the consistency of firm gelatin, when they are wetted (Fig. 1b). The swelling colonies is rapid, there is a considerable increase in their bulk, and often rehydration induces the release of the characteristic "earthy" smell of *trans*-1,10-dimethyl-*trans*-2-decalol, or geosmin (Naes, 1988). Anything that is conspicuous and that has an unusual smell tends to catch people's attention. The rapid appearance of growths of *N. commune* after thundershowers lead to the common belief during the Middle Ages that such colonies fell from the sky and they were referred to as "Sternschnuppen" – shooting stars (Mollenhauer, 1986a). *"Nostoch is that which we call a falling star, a kind of gelly or slime found oftentimes in the Summer in fields, and meadowes"* (French, 1650). The different names used to describe the colonies of this cyanobacterium (Table 3) attest not only to the lively nature of human superstition and imagination, they emphasize that *Nostoc* spp. may have had a widespread occurrence in soils prior to the introduction and extensive use of chemical fertilizers.

The 15th century scientist, philosopher and alchemist Aureolis Phillipus Theophrastus Bombastus von Hohenheim (Paracelsus) invented the name "Nostoch" (Potts, 1997).

"Nostoch...Pollution of some plethoricall and wanton Star, or rather excrement blown from the nostrills of some rheumatick planet" (Paracelsus in Charlton, 1650).

This observation by Paracelsus is not the outrageous statement it may seem, rather it hid, until recently, the clue to the source of a word which defied etymological analysis for 500 years. Nostoch is a play on two words by a very clever man; an English word and a German word, which both describe a part of the human anatomy intimately associated with extracellular polysaccharide. *Nost*rhyl and Nasen*loch* – nostril (Potts 1997). In fact one can readily appreciate the probable origin of the Old Dutch word "snot,"; a vulgar euphemism for nasal secretion. As quite correctly pointed out by Wolfgang Krumbein, Paracelsus was the first to recognize and study biofilms (Krumbein, 1993).

It is perhaps ironic that some 500 years after being classified as a plant in ancient China *Nostoc,* in "enlightened" Europe, was relegated to the realm of the supernatural and people were lead to believe that *"Nostoch understandeth the nocturnall"* (Paracelsus; Charlton translation, 1650). Europeans were most

Table 3. National origins of words used to describe *Nostoc commune*

GERMANY	FRANCE	UNITED KINGDOM	SCANDINAVIA	PHILIPPINES
Sternschnuppen	archee celest	fairies' butter	troldsmør	batkil
Stiernschott	beurre magique	fallen stars	skyfallsalgen	bakate
Erdblume	crachat d'un diable	lock-lubberty		bal-baladyok
Erdgallert	crachat de la lune	scoom		kulintaba
Glasgallerte	crachat de Mai	shot star		taba-taba
Kuckucksspeichel	ecume printiare	spittle of the stars		
Himmelsblatt	feuille du ciel	star-jelly		
Himmelsblume	fleur du soleil	star shot		
Schleimling	merde du coucou	star-slime		
Zitteralge	salive de coucou	star slough		
Zittertang		star-slubber		
		will-o'-the-wisp		
		witches butter		

certainly collecting *Nostoc* for study at this time – although we have to wonder what they were doing with it – as a reference in a diary confirms; *"a bottle of Nostock"* (De La Pryme, 1701). It took a protracted amount of time for *Nostoc* to again be accepted as a member of the plant world.

3. From Blue-Green Alga to Cyanobacterium

The humbling realization that rainwater lead to the mysterious appearance of *N. commune* is attributed to Rene Antoine Ferchault de Reamur in 1727, at a time when classical botany was enjoying its vogue (Mollenhauer 1986a). It is hard to believe, nevertheless, that those people who were collecting *N. commune* much earlier than this date didn't catch on to the rehydration phenomenon. The form of the *N. commune* macroscopic colony still, however, posed problems for taxonomists in the 1700s; it still is today. In 1741 *Nostoc* was considered to be a form of *"Tremella,"* a fungal genus whose name derives from the latin *tremere*, meaning to tremble or quiver in the same fashion that the turgid *Nostoc* colony does if moved (Mollenhauer, 1986a). Such superficial similarities to fungi were subsequently discounted and *Nostoc*'s welfare was taken over first by Albrecht (Wilhelm) Roth (1757-1834), who first provided a definition of the term "algae," and then more avidly by a clergyman and botanist, Jean Pierre Etienne Vaucher (1763-1841), who is credited with the first clear-cut definition of the genus *Nostoc* (Mollenhauer and Kovácik, 1988).

Gustave-Adolphe Thuret (1817-1875) and (Jean-Baptiste) Édouard Bornet (1828-1911) provided the starting point for a new phase in phycology with the publication from 1886 through 1888 of their "Revision des Nostocacées hétérocystées (Mollenhauer and Kovácik, 1988; Bornet and Flahault, 1884-1888). The Revision forms the later starting point for the taxonomy of the group as was ratified by the International Botanical Congress in 1910. Thuret and Bornet clearly shared a passion for their work and this pair of phycologists for many years lived and worked together at Thuret's Mediterranean villa at Antibes. The figures and drawings of *Nostoc* spp. and other cyanobacteria that appear in their publications are outstanding and are of the highest quality imaginable. Drouet (1978) claimed because the Revision was so totally accepted by botanists of the time that phycology, as a profitable career, attracted no dedicated researchers during the following 60 years. *Nostoc* at this time then was firmly entrenched within the plant world as a genus of the blue-green algae.

At the risk of appearing superficial I end this consideration by recognizing the subsequent treatments of the blue-green algae by Geitler (1932), Desikachary (1959), Starmach (1966) and others; the growing realization of the dichotomy between prokaryote and eukaryote (Stanier, 1980); the explosion of names including cyanophytes, myxophytes, bilibacteria, blue-green bacteria and, more recently, cyanoprokaryotes; the recognition but the mixed and lukewarm reception of the Cyanobacteria; and the extant problems and issues to do with taxonomy (Chapter 1).

VII. Epilogue

It is fitting, by way of a summary, to confirm that an appreciation of the ecological significance and success of *Nostoc* did not escape those outside the scientific community. The avant-garde musical group, Zoviet-France, used the title "Gathering nostoc" for a piece of music on their album Loh Land recorded in April 1985 in Amsterdam (Zoviet-France, 1985). Ben Ponton, the co-founder of the group, explained that the use of the word nostoc was deliberate; it *"included a sense of closeness to the earth, of belonging to and living within the biosphere, a sense that has in contemporary times become atrophied and displaced. Nostoc is a fine example of (an organism) living literally hand to mouth"* (Ben Ponton, pers. comm.).

Acknowledgments

This work was made possible by the National Science Foundation. I thank all collaborators who contributed to studies on *Nostoc.* Dieter Mollenhauer, Ben Ponton, Jiann-Shin Chen, Tan-Chi Huang, Young-Tsu Wong and Chao-Tsi Teng provided valuable information. I am grateful to Kenneth Minton, Richard Castenholz, Donna Hill, Alan Peat and Karin Jäger for slides and figures.

References

Abdelahad N, Bazzichelli G, and Ventola F (1984) An electron microscopic investigation on granules observed in the sheath of *Nostoc commune* VAUCH. Arch Hydrobiol Suppl 67 35: 205-211

Agostini HJ, Carroll JD and Minton KW (1996) Identification and characterization of *uvrA*, a DNA repair gene of *Deinococcus radiodurans*. J. Bacteriol 178:6759-6765

Alefounder PR, SA Baldwin, RN Perham and NJ Short (1989) Cloning, sequence analysis and over-expression of the gene for the class II fructose-1,6-bisphosphatealdolase of *Escherichia coli*. Biochem J 257: 529-534

Allen MM (1984) Cyanobacterial cell inclusions. Ann Rev Microbiol 38: 1-25

Anagnostidis and Komárek (1990) Modern approach to the classification system of cyanophytes. 5-Stigonematales. Arch Hydrobiol Suppl 86, Algological Studies 59: 1-73

Andersson CR, Jensen EO, Llewellyn DJ, Dennis ES and Peacock WJ (1996) A new hemoglobin gene from soybean: a role for hemoglobin in all plants. Proc Natl Acad Sci USA 93: 5682-5687

Aruga Y (1992) Habitat and distribution of "facai," *Nostoc flagelliforme* (Cyanophyta). Jap J Phycol (Sorui) 40: 307-309

Bazzichelli G, Abdelahad N and Ventola F (1985) Structural modifications in the extracellular investment of *Nostoc commune* Vauch. during the life cycle. I Motile and nonmotile hormogonium, biseriate stage. J Ultrastruc Res 91: 174-181

Bazzichelli G, Abdelahad N and Ventola F (1986) Structural modifications in the extracellular investment of *Nostoc commune* Vauch. during the life cycle. II Microcolonies and "balls" of filaments. J Ultrastruc Mol Struc Res 94: 114-120

Beesley CE, Smith RJ, Temple SJ and Lea PJ (1994) Cloning and nucleotide sequence of the gene encoding dinitrogenase reductase (*nifH*) from the cyanobacterium *Nostoc* 6720. Biochem Biophys Acta 1219: 548-550

Bernard F (1963) Density of flagellates and Myxophyceae in the heterotrophic layers related to environment. In: Openheimer C (ed.) Symposium on Marine Microbiology, pp 215-228. Charles C Thomas, Springfield, Illinois

Bernard F and Lecal J (1960) Plancton unicellulaire recolté dans l'Ocean Indien par le Charcot (1950) et le Norsel (1955-56). Bull Inst Oceanogr Monaco 1166:1-59

Bilger W, Büdel B, Mollenhauer R and Mollenhauer D (1994) Photosynthetic activity of two developmental stages of a *Nostoc* strain (Cyanobacteria) isolated from *Geosiphon pyriforme* (Mycota). J Phycol 30: 2225-230

Bobzin SC and Moore RE (1993) Biosynthetic origin of 7.7 cyclophanes from cyanobacteria. Tetrahedron Lett 49: 7615-7626

Böhm G A, Pfleiderer W, Böger P and Scherer S (1995) Structure of a novel oligosaccharide-mycosporine-amino acid ultraviolet A/B sunscreen pigment from the terrestrial cyanobacterium *Nostoc commune*. J Biol Chem 270: 8536-8539

Bornet É and Flahault C (1884-1888) Revision des Nostocacées hétérocystées contenues dans les principaux herbiers de France. Ann Sci Nat Bot (7) 3 (6): 323 381; 4 (6): 343-373; 5 (2, 3): 51-129; 7 (3-5): 177-262

Broady PA (1986) Ecology and taxonomy of the terrestrial algae of the Vestfold Hills. In Pickard J (ed.) Antarctic Oasis, pp 165-202. Academic Press

Büdel B, Wessels DCJ and Mollenhauer D (1993) Massenentwicklung von *Nostoc* cf. *microscopicum* (Carmichael) Harvey ex Bornet & Flahault in alkalischem, salzhaltigem Milieu einer Höhle in den Drakensbergen Sudafrikas (Golden Gate Highlands National Park). Arch Protistenkd 143: 229-35

Buikema WJ, Guo Y, Zuther E, Hagemann M and Haselkorn R (1998) Sucrose synthase in *Anabaena* 7120 is required for heterocyst function. Abs VIth Cyanobacterial Workshop, Asilomar, B6.

Buikema WJ and Haselkorn R (1991) Characterization of a gene controlling heterocyst differentiation in the cyanobacterium *Anabaena* 7120. Genes Devel 5: 321-330

Buikema WJ and Haselkorn R (1993) Molecular genetics of cyanobacterial development. Annu Rev Plant Physiol Plant Mol Biol 44: 33-52

Burke DH, Hearst JE and Sidow A (1993) Early evolution of photosynthesis: clues from nitrogenase and chlorophyll iron proteins. Proc Natl Acad Sci USA 90: 7134-7138

Campbell EL, Cohen MF and Meeks JC (1997) A polyketide synthase like gene is involved in the synthesis of heterocyst glycolipids in *Nostoc punctiforme* strain ATCC 29133. Arch Microbiol 167: 251-258

Campbell EL, Hagen KD, Cohen MF, Summers ML and Meeks JC (1996) The *devR* gene product is characteristic of receivers of two component regulatory systems and is essential for heterocyst development in the filamentous cyanobacterium *Nostoc* sp. strain ATCC 29133. J Bacteriol 178:2037-2043

Canfield DE and A Teske (1996) Late Proterozoic rise in atmospheric oxygen concentration inferred from phylogenetic and sulphur-isotope studies. Nature 382: 127-132

Carr NG and Whitton BA (1973) The biology of blue-green algae. Botanical Monographs No 9. Blackwell Scientific Publications, Oxford, London, Edinburgh, Boston, Melbourne.

Carr NG and Whitton BA (1983) The biology of cyanobacteria. Botanical Monographs No 19. Blackwell Scientific Publications, Oxford, London, Edinburgh, Boston, Melbourne.

Castenholz RW (1992) Species usage, concept, and evolution in the Cyanobacteria (Blue Green Algae). J Phycol 28: 737-745

Chang T-P (1980) Mucilage sheath as a barrier to carbon uptake in a cyanophyte, *Oscillatoria rubescens* DC Arch Hydrobiol 88:128-133

Charlton (1650) Van Helmont's paradoxes. Transl Suppl 98

Chazal DE and Smith GD (1994) Characterization of a brown *Nostoc* species from Java that is resistant to high light intensity and UV. Microbiology 140: 3183-3189

Chen J-H, Gibson JL McCue LA and Tabita FR (1991) Identification, expression and deduced primary structure of transketolase and other enzymes encoded within the form II CO_2 fixation operon of *Rhodobacter sphaeroides*. J Biol Chem 266: 20447-20452

Chiaffredo M and Figureau C (1995) Substrate containing Cyanophyceae and Bryophyte protonemas for producing vegetation on bare terrain. United States Patent 5441877.

Clegg JS (1986) The physical properties and metabolic status of *Artemia* cysts at low water contents: The "water replacement hypothesis." In: Leopold AC (ed.) Membranes, metabolism, and dry organisms; pp 169-187. Comstock Publishing Associates, Ithaca and London.

Cohan FM (1996) The role of genetic exchange in bacterial evolution. ASM News 62: 631-636

Cohen MF and Meeks JC (1997) A hormogonium regulating locus, *hrmUA*, of the cyanobacterium *Nostoc punctiforme* strain ATCC 29133 and its response to an extract of a symbiotic plant partner Anthoceros punctatus. Mol Plant-Microbe Int 10: 280-289

Cohen MF, Wallis JG, Campbell EL and Meeks JC (1994) Transposon mutagenesis of *Nostoc* sp. strain ATCC 29133, a filamentous cyanobacterium with multiple cellular differentiation alternatives. Microbiology 140: 3233-3240

Corbett LL and Parker DL (1976) Viability of lyophilized cyanobacteria (blue-green algae). Appl Environ Microbiol 32: 777-780

Cupac S and Gantar M (1992) Chemical composition and morphological structure of the mucilaginous sheath of the cyanobacterium *Nostoc* D. Biomed Lett 47: 133-138

Curtis SE and Martin JA (1994) The transcription apparatus and the regulation of transcriptional initiation. In: Bryant DA (ed.) The Molecular Biology of Cyanobacteria, pp 613-639. Kluwer Academic Publishers, The Netherlands

Daly MJ and Minton KW (1995a) Interchromosomal recombination in the extremely radioresistant bacterium *Deinococcus radiodurans*. J Bacteriol 177:5495-5505

Daly MJ and Minton KW (1995b) Resistance to radiation. Science 270: 1318

Daly MJ and Minton KW (1996) An alternative pathway of recombination of chromosomal fragments precedes *recA*-dependent recombination in the radioresistant bacterium *Deinococcus radiodurans*. J Bacteriol 178: 4461-4471

De La Pryme (1701-2) Let. In: Diary (Surtees) p 247

Desikachary TV (1959) Cyanophyta. Indian Council of Agricultural Research, New Delhi

Dodds WK and Castenholz RW (1987) Effects of grazing and light on the growth of *Nostoc pruniforme* (Cyanobacteria). Br Phycol J 23:219-227

Dodds WK, Gudder DA and Mollenhauer D (1995) The ecology of *Nostoc*. J Phycol 31:2-18

Donaldson AD (1976) The freshwater blue-green algae of Aldabra Atoll. PhD Thesis, Durham University, England

Doolittle RF, Feng D-F, Tsang S, Cho G and Little E (1996) Determining divergence times of the major kingdomes of living organisms with a protein clock. Science 271: 470-477

Doolittle WF (1982) Molecular evolution. In: Carr NG and Whitton BA (eds) The Biology of Cyanobacteria, pp 307-331. Botanical Monographs No 19, Blackwell Scientific Publications, Oxford, London, Edinburgh, Boston, Melbourne.

Drouet F (1978) Revision of the Nostocaceae with Constricted Trichomes. Beih Nova Hedwigia 57: 1-258

Dudley TL and D'Antonio (1991) The effects of substrate texture grazing and disturbance on macroalgal establishment in streams. Ecology 72: 297-309

Ehling-Shulz M, Bilger W and Scherer S (1997) UV-B-induced synthesis of photoprotective pigments and extracellular polysaccharides in the terrestial cyanobacterium *Nostoc commune*. J Bacteriol 179: 1940-1945

Emralino GA and Rodulfo BR (1987) Cultural studies on blue-green algae. III. *Nostoc microscopicum* Carmichael. Phillip J Sci 116: 67-72

Ernst A, Chen T-W and Böger P (1987) Carbohydrate formation in rewetted terrestrial cyanobacteria. Oecologia 72: 574-576

Fagerbakke KM, Heldal M and Norland S (1991) Variation in elemental content among and within trichomes in *Nostoc calcicola* 79WA01 measured by x-ray micro-analysis. FEMS Microbiol Lett 81: 227-232

Favali MA and Caiola MG (1986) Autoradiography and ultrastructural study on *Nostoc* species labeled with ^{3}H-thymidine. Microbios 45: 33-39

Feng D-F, Chop G and Doolittle RF (1997) Determining divergence times with a protein clock:update and reeavluation. Proc Natl Acad Sci USA 94: 13028-13033

Fernandezpinas F, Mateo P and Bonilla I (1995) Ultrastructural changes induced by selected cadmium concentrations in the cyanobacterium *Nostoc* UAM208. J Plant Physiol 147: 452-456

Fjerdingstad E, Fjerdingstad E and Fjerdingstad EJ (1979) Evidence for the presence of collagen in the sheaths of a blue-green alga: an electron microscopical and histochemical study of *Scytonema myochrous* status petalonematoides. Arch Hydrobiol 24: 307-323

Fox GE, Wisotzkey JD and Jurtshuk P (1992) How close is close: 16S rRNA sequence identity may not be sufficient to guarantee species identity. Int J Syst Bacteriol 42: 166-170

Fraleigh PC and Burnham JC (1988) Myxococcal predation on cyanobacterial populations: nutrient effects. Limnol Oceanogr 33: 476-483

French J (1650) Chymn Dict

Gantar M, Kerby NW and Rowell P (1991) Colonization of wheat (*Triticum vulgare* L.) by N_2-fixing cyanobacteria. II An ultrastructural study. New Phytol 118: 485-492

Gao K (1998) Chinese studies on the edible blue-green alga, *Nostoc flagelliforme*: a review. J Appl Phycol 10: 37-49

Gao K, Qiu B, Xia J, Yu A and Li Y (1998) Effect of wind speed on loss of water from *Nostoc flagelliforme* colonies. J Appl Phycol 10: 55-58

Geitler L (1932) Cyanophyceae. In: Kolkwitz (ed.) Dr L Rabenhorst's Kryptogamenflora von Deutschland, Österreich und der Schweiz. Akademische Verlagsgesellschaft, Leipzig

Geitler L (1982) Zur Kenntnis der Oscillatoriacee *Pseudanabaena galeata* (Cyanophyta). Plant Syst and Evol 141: 169-175

Gibson JL, Falcone DL and Tabita FR (1991) Nucleotide sequence, transcriptional analysis, and expression of genes encoded within the form I CO_2 fixation operon of *Rhodobacter sphaeroides*. J Biol Chem 266: 14646-14653.

Gibson JL and Tabita FR (1988) Localization and mapping of CO_2 fixation genes within two gene clusters in *Rhodobacter sphaeroides*. J. Bacteriol 170: 2153-2158

Gilles-Gonzalez MA, Ditta GS and Helinski DR (1991) A haemoprotein with kinase activity encoded by the oxygen sensor of *Rhizobium meliloti*. Nature 350: 170-172

Giovannoni SJ, Turner S, Olsen GJ, Barns S, Lane DJ and Pace NR (1988) Evolutionary relationships among cyanobacteria and green chloroplasts. J Bacteriol 170: 3584-3592

Golakoti T, Ogino J, Heltzel CE, Husebo TL, Jensen CM, Larsen LK, Patterson GML, Moore RE, Mooberry SL, Corbett TH and Valeriote FA (1995) Structure determination, conformational analysis, chemical stability studies, and antitumor evaluation of the cryptophycins. Isolation of 18 new analogs from *Nostoc* sp. strain GSV 224. J Am Chem Soc117: 12030-12049

Gould SJ and Eldredge N (1993) Punctuated equilibrium comes of age. Nature 366: 223-227

Groisman EA, Saier Jr MH and Ochman H (1992) Horizontal transfer of a phosphatase gene as evidence for mosaic structure of the *Salmonella* genome. EMBO J 11:1309-1316

Grossman AR, Schaefer MR, Chiang GG and Collier JL(1994) The responses of cyanobacteria to environmental conditions: light and nutrients. In: Bryant DA (ed.) The Molecular Biology of Cyanobacteria, pp 641-675. Kluwer Academic Publishers, Netherlands.

Gupta A and Singhal GS (1996) Changes in ultrastructure and inhibition of electron transport activity induced by Thallium in *Nostoc muscorum*. Photosynthetica 32: 393-398

Gupta RK and Kashyap AK (1994) Uptake and loss of water by Antarctic cyanobacterium *Nostoc commune* under field and laboratory conditions. Phykos 33: 25-30

Gupta RS, Bustard K, Falah M and Singht D (1997) Sequencing of heat shock protein 70 (DnaK) homologs from *Deinococcus proteolyticus* and *Thermomicrobium roseum* and their integration in a protein-based phylogeny of prokaryotes. J Bacteriol 179: 345-357

Gurevitz M, Osiewacz HD and Keren Y (1991) Molecular evidence for interchromosomal recombination in the cyanobacterium *Synechocystis* sp. PCC 6803. Plant Sci 78: 217-224

Habicht KS and Canfield DE (1996) Sulphur isotope fractionation in modern microbial mats and the evolution of the sulphur cycle. Nature 382: 342-343

Hagemann M, Golldack D, Biggins J and Erdmann N (1993) Salt-dependent protein phosphorylation in the cyanobacterium *Synechocystis* PCC 6803. FEMS Microbiol Lett 113: 205-210

Hagen KD and Meeks JC (1998) Genetic evidence for the involvement of a phosphorelay system in heterocyst maturation in *Nostoc punctiforme*. Abs VIth Cyanobacterial Workshop, Asilomar, B4.

Hardison RC (1996) A brief history of hemoglobins: Plant, animal, protist, and bacteria. Proc Natl Acad Sci USA 93:5675-5679

Hasegawa M and Fitch WM (1996) Dating the cenancester of organisms. Science 274:1750-1751.

Haselkorn R (1992) Developmentally regulated gene rearrangements in prokaryotes. Annu Rev Genet 26: 111-128

Hawes I, Howard-Williams C and Vincent WF (1992) Desiccation and recovery of Antarctic cyanobacterial mats. Polar Biol 12: 587-594

Hemscheidt T, Puglisi MP, Larsen LK, Patterson GML, Moore RE, Rios JL and Clardy J (1994) Structure and biosynthesis of borophycin, a new Boeseken complex of boric acid from a marine strain of the blue-green alga *Nostoc linckia*. J Org Chem 59: 3467-3471

Herdman M (1982) Genomes and Genetics. In: Carr NG and Whitton BA (eds) The Biology of Cyanobacteria, pp 263-305. Botanical Monographs No 19. Blackwell Scientific Publications, Oxford, London, Edinburgh, Boston, Melbourne

Herdman M (1988) Cellular differentiation: akinetes. Meth Enzymol 167: 222-232

Herdmann M, Janvier M, Rippka R and Stanier RY (1979) Genome size of cyanobacteria. J Gen Microbiol 111: 73-85

Hill DR, Belbin TJ, Thorsteinsson MV, Bassam D, Brass S, Ernst A, Böger P, Paerl H, Mulligan ME and Potts M (1996) GlbN (Cyanoglobin) is a peripheral membrane protein that is restricted to certain *Nostoc* spp. J Bacteriol 178: 6587-6598

Hill DR, Hladun H, Scherer S and Potts M (1994a) Water stress proteins of *Nostoc commune* (Cyanobacteria) are secreted with UV-A/B-absorbing pigments and associate with 1,4-b-D-xylanxylanohydrolase activity. J Biol Chem 269:7726-7734

Hill DR, Keenan TW, Helm RF, Potts M, L M Crowe and JH Crowe (1997 Extracellular polysaccharide of *Nostoc commune* (Cyanobacteria) inhibits fusion of membrane vesicles during desiccation. J Appl Phycol 9: 237-248

Hill DR, Peat A and Potts M (1994b) Biochemistry and structure of the glycan secreted by desiccation-tolerant *Nostoc commune* (Cyanobacteria). Protoplasma 182: 126-148

Hirata K, Nakagami H, Takashina J, Mahmud T, Kobayashi M, In Y, Ishida T and Miyamoto K (1996) A novel violet pigment, nostocine A, an extracellular metabolite from cyanobacterium *Nostoc spongiaeforme*. Heterocycles 43: 1513-1519

Hirsch CF, Liesch JM, Salvatore MJ, Schwartz RE and Sesin DF (1990) Antifungal fermentation product and method. United States Patent 4946835

Hirsch R, Hartung W and Gimmler H (1989) Abscisic acid content of algae under stress. Bot Acta 102: 326-334

Hodges CF (1992a) Interaction of cyanobacteria and sulfate-reducing bacteria in subsurface black-layer formation in high-sand content golf greens. Soil Biol Biochem 24: 15-20

Hodges CF (1992b) Pathogenicity of *Pythium torulosum* to roots of *Agrostis palustris* in black-layered sand produced by the interaction of the cyanobacteria species *Lyngbya*, *Phormidium*, and *Nostoc* with *Desulfovibrio desulfuricans*. Can J Bot 70: 2193-2197

Holland D and Wolk CP (1990) Identification and characterization of *hetA*, a gene that acts early in the process of morphological differentiation of heterocysts. J Bacteriol 172: 3131-3137

Hosetti BB and Patil HS (1989) Impact of growth of *Nostoc punctiforme* and *Escherichia coli* on polyfibre wastewater. J Environ Biol 10: 379-381

Hough L, Jones LKN and Wadman WH (1952) An investigation of the polysaccharide of components of certain fresh-water algae. J. Chem Soc 652: 3393-3399

Howell LD, Griffiths C, Slade LW, Potts M and Kennelly PJ (1996) Substrate specificity of IphP, a cyanobacterial; dual-specificity protein phosphatase with MAP kinase phosphatase activity. Biochemistry 35: 7566-7572

Hu TL, Kean CK and Leong OS (1994) Odorous problem associated with *Nostoc* sp. in water supply system of central Taiwan. Int Conf Asian Water Tech 100-110.

Huang Z, Liu Y, Paulsen SP and Klaveness D (submitted) Studies on polysaccharides from three species of edible *Nostoc* (Cyanophyta, Cyanobacteria) with different colony morphologies. 1. Comparison of monosaccharide composition and viscosity between field colonies and suspension cultures.

Huddart H, Smith RJ, Langton PD, Hetherington AM and Mansfield TA (1986) Is abscisic acid a universally active calcium agonist? New Phytol 104: 161-173

Ikehara K, Amada F, Yoshida S, Mikata Y and Tanaka A (1996) A possible origin of newly-born bacterial genes: significance of GC-rich nonstop frame on antisense strand. Nucl Acids Res 24: 4249-4255

Isono T and Fujita Y (1981) Studies on morphological changes of the blue-green alga *Nostoc muscorum* A with special reference to the role of light. Plant Cell Physiol 22: 185-195

Iwano M (1995) Selective effect of salt stress on the activity of two ATPases in the cell membrane of *Nostoc muscorum*. Plant Cell Physiol 36:1297-1301

Jackman DM and Mulligan ME (1995) Characterization of a nitrogen-fixation (*nif*) gene cluster from *Anabaena azollae* 1a shows that closely related cyanobacteria have highly variable but structured intergenic regions. Microbiology 141: 2235-2244

Jäger K and Potts M (1988) Distinct fractions of genomic DNA from cyanobacterium *Nostoc commune* that differ in the degree of methylation. Gene 74: 197-201

Jassby A (1988) *Spirulina*: a model for microalgae as human food. In: Lembi C A and Waaland J R (eds) Algae and Human Affairs, pp 149-179. Cambridge University Press, Cambridge

Jensen TE (1984) Cyanobacterial cell inclusions of irregular occurrence: systematic and evolutionary implications. Cytobios 39: 35-62

Johansson C and Bergman B (1992) Early events during the establishment of the *Gunnera*/*Nostoc* symbiosis. Planta (Berl) 188: 403-413

Jüngst X, Wakabayashi A, Matsubara H and Zumft WG (1991) The *nirSTBM* region coding for cytochrome cd_1-dependent nitrite respiration of *Pseudomonas stutzeri* consists of a cluster of mono-, di- and tetraheme proteins. FEBS Lett 279: 205-209

Kaneko T, Sato S, Kotani H, Tanaka A, Asamizu E, Nakamura Y, Miyajima N, Hirosawa M, Sugiura M, Sasamoto S, Kimura T, Hosouchi T, Matsuno A, Muraki A, Nakazaki N, Naruo K, Okumura S, Shimpo S, Takeuchi C, Wada T, Watanabe A, Yamada M, Yasuda M and Tabata S (1996) Sequence analysis of the genome of the unicellular cyanobacterium *Synechocystis* sp. strain PCC 6803. II. Sequence determination of the entire genome and assignment of potential protein-coding regions. DNA Res 3: 109-136

Kaniga K, Uralil J, Bliska JB and Galan JE (1996) A secreted protein tyrosine phosphatase with modular effector domains in the bacterial pathogen *Salmonella typhimurium*. Mol Microbiol 21: 633-641

Karlin S, Mrázek J and Campbell AM (1997) Compositional biases of bacterial genomes and evolutionary implications. J Bacteriol 179: 3899-3913

Katayama M, Wada Y and Ohmori M (1995) Molecular cloning of the cyanobacterial adenylate cyclase gene from the filamentous cyanobacterium *Anabaena cylindrica*. J Bacteriol 177: 3873-3878

Kaya K, Sano T, Beattie KA and Codd GA (1996) Nostocyclin, a novel 3-amino-6-hydroxy-2-piperidone-containing cyclic depsipeptide from the cyanobacterium *Nostoc* sp. Tetra Lett 37: 6725-6728

Kennelly PJ and Potts M (1996) Fancy meeting you here! a fresh look at "prokaryotic" protein phosphorylation. J Bacteriol 178: 4759-4764

Khudyakov I and Wolk CP (1996) Evidence that the *hanA* gene coding for HU protein is essential for heterocyst differentiation in, and cyanophage A-4(L) sensitivity of, *Anabaena* sp. strain PCC 7120 J Bacteriol 178:3572-3577

Kleinhaus S and Kaiser AD (1988) Ecology and biomechanical consequences of living together. Induced morphological change in a *Nostoc* and *Cricotopus* symbiosis. Am Zool 28: 34A

Koch A L (1996) What size should a bacterium be? A question of scale. Annu Rev Microbiol 50: 317-348

Komárek J and Anagnostidis K (1989) Modern approach to the classification system of cyanophytes. 4-Nostocales. Arch Hydrobio Suppl 823, Algological Studies 56: 247-345

Krumbein WE (1993) Paracelsus und die mucilaginischen Substanzen 500 Jahre EPS-Forschung. Deutsche Gesellschaft fur Meeresforschung -Mitteilungen 3: 8-14.

Kumar A, Tyagi MB, Srinivas G, Singh N, Kumar HD, Sinha RP and Hader DP (1996) UVB shielding role of $FeCl_3$ and certain cyanobacterial pigments. Photochem Photobiol 64:321-325

Lazaroff N (1973) Photomorphogenesis and Nostocacean Development. In: Carr NG and Whitton BA (eds) The Biology of Blue-Green Algae, pp 279-319. Botanical Monographs No 9. Blackwell Scientific Publications, Oxford, London, Edinburgh, Boston, Melbourne

Lazaroff N (1992) Screening for anti-adhesin antibiotics by employing photoinduced cyanobacteria. United States Patent 5,112,742

Leganés F (1994) Genetic evidence that *hep*A gene is involved in the normal deposition of the envelope of both heterocysts and akinetes in *Anabaena variabilis* ATCC 29413. FEMS Microbiol Lett 123: 63-68

Leslie JF, Greasham RL and Hulbert MH (1984) Biological flocculation of mineral slimes. United State Patent 4456532

Li S and Waters R (1998) *Escherichia coli* strains lacking protein HU are UV sensitive due to a role for HU in homologous recombination. J Bacteriol 180: 3750-3756

Li SC (1596) Pen Ts'ao Kang Mu (Chinese Pharmaceutical Compendium); reprinted (1977) by People's Medical Publishing House, Beijing

Lüttge U (1997) Cyanobacterial *Tintenstrich* communities and their ecology. Naturwissenschaften 84: 526-534

Lynn ME, Bantle JA and Ownby JD (1986) Estimation of gene expression in heterocysts of *Anabaena variabilis* by using DNA-RNA hybridization. J Bacteriol 167: 940-946

Mann NH (1994) Protein phosphorylation in cyanobacteria. Microbiology 140: 3207-3215

Mann NH, Rippka R and Herdmann M (1991) Regulation of protein phosphorylation in the cyanobacterium *Anabaena* strain PCC 7120. JGen Microbiol 137: 331-339

Marsálek B, Zahradnícková and Hronková M (1992) Extracellular abscisic acid produced by cyanobacteria under salt stress. J Plant Physiol 139: 506-508

Marsh JJ and Lebherz HG (1992) Fructose-bisphosphate aldolases: an evolutionary history. TIBS 17:110-113

Martinez M (1988) *Nostoc commune* Vauch., a nitrogen-fixing blue-green alga, as source of food in the Philippines. Phil Agricul 71: 295-307

Martinez MR and Querijero NMB (1986) Macrocolony formation in the nitrogen-fixing blue-green alga, *Nostoc commune* VAUCH. Phil Agri 69: 547-565

Martinez MR, Querijero NMB and Villareal-Castillo L (1985) Photocontrolled growth, nitrogenfixation and morphological developmental life *Nostoc commune* 45. Phil J Sci Monograph 16: 85-103

Masepohl B, Gorlitz K and Böhm H (1996) Long tandemly repeated repetitive (LTRR) sequences in the filamentous cyanobacterium *Anabaena* sp PCC 7120. Biochim Biophys Acta Gene-Structure and Expression 1307 (1) : 26-30.

Mattimore V and Battista JR (1996) Radioresistance of *Deinococcus radiodurans* functions necessary to survive ionizing radiation are also necessary to survive prolonged desiccation. J Bacteriol l178: 633-637

McArtney B., Howell D., Kennelly PJ and Potts M (1997) Protein tyrosine phosphorylation in *Anabaena* sp. PCC 7120 (Cyanobacteria). J Bacteriol 179: 2314-2318

Mel'nikova VV (1985) Response of poikiloxerophilic blue-green water plants *Nostoc commune* Vauch. to dehydration. Tstitologiya 27: 440-444

Minton KW (1996) Repair of ionizing-radiation damage in the radiation-resistant bacterium *Deinococcus radiodurans*. Mut Res 363: 1-7

Minton KW and Daly MJ (1995) A model for repair of radiation-induced DNA double-strand breaks in the extreme radiophile *Deinococcus radiodurans*. BioEssays 17: 457-464

Moens L, Vanfleteren J, Van de Peer Y, Peeters K, Kapp O, Czeluzniak J, Goodman M, Blaxter M and Vinogradov S (1996) Globins in nonvertebrate species: Dispersal by horizonal gene transfer and evolution of the structure-function relationships. Mol Biol Evol 13: 324-333

Mojzsis SJ, Arrhenius G, McKeegan KD, Harrison TM, Nutman AP and Friend CRL (1996) Evidence for life on Earth before 3,800 million years ago. Nature 384: 55-59

Mollenhauer D (1970) Beitrage zur Kenntnis der Gattung *Nostoc*. Abh senckenb naturforsch Ge 524: 1-80

Mollenhauer D (1985a) Blaualgen der Gattung*Nostoc* - ihre Rolle in Forschung und Wissenschaftsgeschichte I. Natur und Museum 115: 305-319

Mollenhauer D (1985b) Blaualgen der Gattung *Nostoc* - ihre Rolle in Forschung und Wissenschaftsgeschichte II. Natur und Museum 115:369-379

Mollenhauer D (1986a) Blaualgen der Gattung *Nostoc* - ihre Rolle in Forschung und Wissenschaftsgeschichte III. Natur und Museum 115:43-59

Mollenhauer D (1986b) Blaualgen der Gattung *Nostoc* - ihre Rolle in Forschung und Wissenschaftsgeschichte IV. Natur und Museum 115:104-120

Mollenhauer D and Kovácik L (1988) Who was who in Cyanophyte research. I. Arch Hydrobiol Suppl 80: 19-33

Moreno J, Rodriguez H, Vargas MA, Rivas J, Guerrero MG and Lee Y-K (1994) Nitrogen-fixing cyanobacteria as source of phycobiliprotein pigments. Composition and growth performance of ten filamentous heterocystous strains. J Appl Phycol 7: 1-2

de Mule MCZ, de Caire GZ, de Cano MS and Halperin DR (1991) Bioactive compounds from *Nostoc muscorum* (Cyanobacteria). Cytobios 66: 266-267

Mulligan ME and Haselkorn R (1989) Nitrogen fixation (nif) genes of the cyanobacterium *Anabaena* species strain PCC 7120. The *nifB-fdxN-nifS-nifU* operon. J Biol Chem 264:19200-19207

Mulligan ME, Jackman DM and Murphy ST (1994) Heterocyst-forming filamentous cyanobacteria encode proteins that resemble eukaryotic RNA-binding proteins of the RNP family. J. Mol. Biol. 235: 1162-1170

Naes H (1988) Geosmin production. Meth Enzymol 167: 605-608

Nelissen B, De Baere R, Wilmotte A and De Wachter R (1996) Phylogenetic relationships of nonaxenic filamentous cyanobacterial strains based on 16S rRNA sequence analysis. J Mol Evol 42: 194-200

Ohmori K, Hirose M and Ohmori M (1992) Function of cAMP as a mat-forming factor in the cyanobacterium *Spirulina platensis*. Plant Cell Physiol 33: 21-25

Ohmori M (1989) cAMP in *Anabaena cylindrica*: rapid changes in cellular levels in response to changes in extracellular environments. Plant Cell Physiol 30: 911-914

Okino T, Qi S, Matsuda H, Murakami M and Yamaguchi K (1997) Nostopeptins A and B, elastase inhibitors from the cyanobacterium *Nostoc minutum* 60: 158-161

Orus MI and Marco E (1991) Heterocyst structure alteration and O_2-mediated inhibition of dinitrogen fixation by trichlorfon in *Anabaena* PCC 7119. J Exp Bot 42:1595-1600

Pandey PK, Singh BB, Mishra R and Bisen PS (1996) Ca^{2+} uptake and its regulation in the cyanobacterium *Nostoc* MAC. Curr Microbiol 32: 332-335

Peat A and Potts M (1987) The ultrastructure of immobilized desiccated cells of the cyanobacterium *Nostoc commune* UTEX 584. FEMS Microbiol Lett 43: 223-227

Peat A, Powell N and Potts M (1988) Ultrastructural analysis of the rehydration of desiccated *Nostoc commune* HUN (Cyanobacteria) with particular reference to the immunolabeling of NifH. Protoplasma 146: 72-80

Pennisi E (1998) Genome data shake tree of life. Science 280: 672-6764

Petterston A and Bergman B (1989) Calmodulin in heterocystous cyanobacteria: biochemical and immunological evidence. FEMS Microbiol Lett 60: 95-100

Pinevich A and Grigoryeva I (1994) Epifluorescence microscopy of *Anabaena* sp.: nucleoid configurations and evidence for inclusion-associated DNA. Experientia 50: 44-48

Porchia AC and Salerno GL (1996) Sucrose biosynthesis in a prokaryotic organism: Presence of two sucrose-phosphate synthases in *Anabaena* with remarkable differences compared with the plant enzymes. Proc Natl Acad Sci USA 93: 13600-13604

Potts M (1977) Studies on blue-green algae and photosynthetic bacteria in the lagoon of Aldabra Atoll. PhD Thesis, University of Durham, England

Potts M (1994) Desiccation tolerance of prokaryotes. Microbiol Rev 58: 755-805

Potts M (1996) The anhydrobiotic cyanobacterial cell. Physiol Plant 97:788-794

Potts M (1997) Etymological basis for the genus name *Nostoc* (Cyanobacteria). Int J Syst Bacteriol 47: 584

Potts M, Angeloni SV, Ebel RE and Bassam D (1992) Myoglobin in a cyanobacterium. Science 256: 1690-1692

Potts M and Bowman MA (1985) Sensitivity of *Nostoc commune* UTEX 584 to water stress. Arch Microbiol 141: 51-56

Potts M, Ocampo-Friedmann R, Bowman MA and Tüzün B (1983) *Chroococcus* S24 and *Chroococcus* N41 (cyanobacteria): morphological, biochemical and genetic characterization and effects of water stress on ultrastructure. Arch Microbiol 135: 81-90

Potts M, Sun H, Mockaitis K, Kennelly P, Reed D and Tonks NK (1993) A protein\ tyrosine/serine phosphatase encoded by the genome of the cyanobacterium *Nostoc commune* UTEX 584. J Biol Chem 268:7632-7635

Potts M and Whitton BA (1980) Vegetation of the intertidal zone of the lagoon of Aldabra, with particular reference to the photosynthetic prokaryotic communities. Proc Roy Soc Lond B 208: 13-55

Priscu JC, Fritsen CH, Adams EE, Giovannoni SJ, Paerl HW, McKay CP, Doran PT, Gordon DA, Lanoil BD and Pinckney JL (1998) Perennial Antarctic Lake ice: an oasis for life in a polar desert. Science 280: 2095-2098

Prosperi CH (1994) A cyanophyte capable of fixing nitrogen under high levels of oxygen. J Phycol 30: 222-224

Rippka R, Deruelles J, Waterbury JB, Herdman M and Stanier RY (1979) Generic assignments, strain histories and properties of pure cultures of cyanobacteria. J Gen Microbiol 111: 1-61

Robinson BL and Miller JH (1970) Photomorphogenesis in the blue-green alga *Nostoc commune* 584. Physiol Plant 23: 461-472

Rodríguez R, García-González M,. Guerrero MG and Lara C (1994) Ammonium-sensitive protein kinase activity in plasma membranes of the cyanobacterium *Anacystis nidulans*. FEBS Lett 350: 19-23

Rudi K, Skulberg OM and Jakobsen KS (1998) Evolution of cyanobacteria by exchange of genetic material among phyletically related strains. J Bacteriol 180: 3453-3461

Rudi K, Skulberg OM, Larsen F and Jakobsen KS (1997) Strain characterization and classification of oxyphotobacteria in clone cultures on the basis of 16S rRNA sequences from the variable regions V6, V7, and V8. Appl Environ Microbiol 63: 2593-2599

Rutter WJ (1964) Evolution of aldolases. Fed Proc Fed Am Soc Exp Biol 23: 1248-1257

Savageau MA (1986) Proteins of *Escherichia coli* come in sizes that are multiples of 14kDa: domain concepts and evolutionary implications. Proc Natl Acad Sci USA 83: 1198-1202

Schaefer Jr JW and Boyum KW (1988) Microbial compositions and methods for treating soil. United States Patent 4774186

Schäferjohann J, Yoo JG and Bowien B (1995) Analysis of the genes forming the distal parts of the two *cbb* CO_2 fixation operons from *Alcaligenes eutrophus*. Arch Microbiol 163: 291-299

Scherer S (1991) Mikroorganismen in extremen Lebensraümen. Cyanobakterien in Wüstengebieten. Anpassungen an UV-B und Trockenstreb. Biol in unserer Zeit 4: 220-222

Scherer, S and Potts M (1989) Novel water stress protein from a desiccation-tolerant cyanobacterium: purification and partial characterization. J Biol Chem 264: 12546-12553

Scherer S and Zhong Z-P (1991) Desiccation independence of terrestrial *Nostoc commune* ecotypes (Cyanobacteria). Microb Ecol 22: 271-283

Schmetterer G and Wolk CP (1988) Identification of the region of cyanobacterial plasmid pDU1 necessary for replication in *Anabaena* sp. strain M-131. Gene 62: 101-109

Schopf JW (1993) Microfossils of the Early Archaean Apex Chert: New evidence of the antiquity of life. Science 260: 640-646

Schopf JW (1994) Disparate rates, differing fates: Tempo and mode of evolution changed from the Precambrian to the Phanerozoic. Proc Natl Acad Sci 91: 6735-6742

Schrautemeier B, Neveling U and Schmitz S (1995) Distinct and differently regulated Mo-dependent nitrogen-fixing systems evolved for heterocysts and vegetative cells of *Anabaena variabilis* ATCC 29413: characterization of the *fdxH1/2* gene regions as part of the *nif1/2* gene clusters. Mol Microbiol 18: 357-369

Schübler A, Bonfante P, Schnepf E, Mollenhauer D and Kluhe M (1996) Characterization of the *Geosiphon pyriforme* symbiosome by affinity techniques: confocal laser scanning microscopy (CLSM) and electron microscopy. Protoplasma 190: 53-67

Schübler A, Schnepf E, Mollenhauer D and Kluge M (1995) The fungal bladders of the endocyanosis *Geosiphon pyriforme*, a *Glomus*-related fungus: cell wall permeability indicates a limiting pore radius of only 0.5 nm. Protoplasma 185: 131-139

Seaburg KG, Parker BC, Prescott GW and Whitford LA (1979) The algae of Sothern Victorialand, Antarctica. Bibliothecia Phycologica Band 46. Trauss and Cramer GmbH, Hirschberg, Germany

Silva HJ, Cortinas TI and Ertola RJ (1989) Effect of nutritional factors on the culture of *Nostoc* sp. as a source of phycobiliproteins. Appl Microbiol Biotechnol 31: 293-297

Simon RD (1987) Inclusion bodies in the cyanobacteria: cyanophycin, polyphosphate, polyhedral bodies. In: Fay P and Van Baalen C (eds) The Cyanobacteria pp 199-225. Elsevier Science Publishers BV (Biomedical Division)

Singer MP, Rossmiessl BM, Liebke CH and Gross CA (1992) The *Escherichia coli ts8* mutation is an allele of *fda*, the gene encoding fructose-1,6-diphosphate aldolase. J Bacteriol 173: 6242-6248

Smith RJ, Hobson S and Ellis IR (1987) The effect of abscisic acid on calcium-mediated regulation of heterocyst frequency and nitrogenase activity in *Nostoc* 6720. New Phytol 105: 543-549

Stanier RY (1980) The journey, not the arrival, matters. Ann Rev Microbiol 34: 1-48

Starmach K (1966) Cyanophyta Sinice, Glaucophyta-Glaukofity. In: Starmach K (Hrsg) Flora slodkowodna Polski 2.(Panstwow Wydawnictwo Naukowe) Warszawa. 806 + I S

Stevens Jr SE and Murphy RC (1996) Recombinant biopesticide and method of use thereof. United State Patent 5518897

Stulp BK (1983) Morphological and molecular approaches to the taxonomy of the genus *Anabaena* (Cyanophyceae, Cyanobacteria). Ph D Thesis, University of Groningen

Suen M-H (1994) English translation. On-site news. Sci Tech Daily (China) 1.

Tandeau de Marsac N (1994) Differentiation of hormogonia and relationships with other biological processes. In: Bryant DA (ed.) The Molecular Biology of Cyanobacteria, pp 825-842. Kluwer Academic Publishers, The Netherlands

Taranto P, Keenan TW and Potts M (1993) Rehydration induces a rapid onset of lipid biosynthesis in desiccated *Nostoc commune* (Cyanobacteria). Biochim Biophys Acta 1168: 228-237

Tease BE and Walker RW (1987) Comparative composition of the sheath of the cyanobacterium *Gloeothece* ATCC 27152 cultured with and without combined nitrogen. J Gen Microbiol 133: 3331-3339

Thiel T, Lyons EM, Erkner JC and Ernst A (1995) A second nitrogenase in vegetative cells of a heterocyst-forming cyanobacterium. Proc Natl Acad Sci USA 92: 9358-9362

Thorsteinsson MV, Bevan DR, Ebel RE, Weber RE and Potts M (1996) Spectroscopical and functional characterization of the hemoglobin of *Nostoc commune* UTEX 584 (Cyanobacteria). Biochim Biophys Acta 1292: 133-139

Trehan K and Sinha U (1982) DNA-mediated transformation in *Nostoc muscorum*, a nitrogen-fixing cyanobacterium. Aust J Biol Sci 35: 573- 577

Ueno T, Ito H, Kotani H and Nakajima K (1995) NSP7524V restriction-modification genes. United States Patent 5441881

van den Bergh ERE, Baker SC, Raggers RJ, Terpstra P, Woudstra EC, Dijkhuizen L and Meijer WG (1996) Primary structure and phylogeny of the calvin cycle enzymes transketolase and fructose bisphosphate aldolase of *Xanthobacter flavus*. J Bacteriol 178: 888-893

Walsby AE (1994) Gas vesicles. Microbiol Rev 58:94-144

Ward DM, Bateson MM, Weller R and Ruff-Roberts AL (1992) Ribosomal RNA analysis of microorganisms as they occur in nature. Adv Microb Ecol 12: 219-286

Warner KM and GS Bullerjahn (1994) Light-dependent tyrosine phosphorylation in the cyanobacterium *Prochlorothrix hollandica*. Plant Physiol 105: 629-633

Whitton BA, Donaldson A and Potts M (1979) Nitrogen fixation by *Nostoc* colonies in terrestrial environments of Aldabra Atoll, Indian ocean. Phycologia 18: 278-287

Willard JM and Gibbs M (1968) Fructose-diphosphate aldolase from blue-green algae. Meth Enzymol 42: 228-234

Wilmotte A (1994) Molecular evolution and taxonomy of the cyanobacteria. In: Bryant DA (ed.) The Molecular Biology of Cyanobacteria, pp 1-25. Kluwer Academic Publishers, The Netherlands

Woldringh CL, Jensen PR and Westerhoff HV (1995) Structure and partitioning of bacterial DNA: determined by a balance of compaction and expansion forces? FEMS Microbiol Lett 131: 235-242

Wolk CP (1996) Heterocyst formation. Annu Rev Genet 30: 59-78

Wolk CP (1997) *Anabaena* sp. strain PCC 7120. In: de Bruijn FJ, Lapslei JR and Weinstock G (eds) Bacterial Genomes: Structure and Analysis, Chapman and Hall

Wolk CP, Ernst A and Elhai J (1994) Heterocyst metabolism and development In: Bryant DA (ed.) The Molecular Biology of Cyanobacteria, pp 769-823. Kluwer Academic Publishers, The Netherlands

Xie W-Q, Whitton BA, Simon JW, Jäger K, Reed D and. Potts M (1989) *Nostoc commune* UTEX 584 gene expressing indole phosphate hydrolase activity in *Escherichia coli.* J Bacteriol 171: 708-713

Yallaly P and Eisenstark A (1990) Influence of DNA adenine methylase on the sensitivity of *Escherichia coli* to near-ultraviolet radiation and hydrogen peroxide. Bioch Biophys Res Comm 169: 64-69

Yanni YG and Sehly MR (1991) Rice performance and natural infection with blast (*Pyricularia oryzae* Cav.) under different algalization techniques and rates of fertilizer nitrogen. World J Microbiol Biotechnol 7: 43-47

Young YL and McElroy MB (1979) Fixation of nitrogen in the prebiotic atmosphere. Science 203: 1002-1004

Zhang C-C (1993). A gene encoding a protein related to eukaryotic protein kinases from the filamentous heterocystous cyanobacterium *Anabaena* PCC 7120. Proc Natl Acad Sci USA 90: 11840-11844

Zhao N, Berova N, Nakanishi K, Rohmer M, Mougenot P and Jürgens UJ (1996) Structures of two bacteriohopanoids with acyclic pentol side-chains from the cyanobacterium *Nostoc* PCC 6720. Tetrahedron Lett 52: 2777-2788

Zimmerman SB and Murphy LD (1996) Macromolecular crowding and the mandatory condensation of DNA in bacteria. FEBS Lett 390: 245-248

Zoviet-France (1985) Loh Land. 1st Ed. Staaltape, Amsterdam

Chapter 18

Arthrospira (*Spirulina*): Systematics and Ecophysiology

Avigad Vonshak[1] and Luisa Tomaselli[2]
[1]*Microalgal Biotechnology Laboratory, the Jacob Blaustein Institute for Desert Research, Ben-Gurion University of the Negev, Sede Boker Campus 84990, Israel*
[2]*Centro di Studio dei Microrganismi Autotrofi del Consiglio Nazionale delle Ricerche, Piazzale delle Cascine 27, 50144 Firenze, Italy*

Summary

Notwithstanding the official recognition in 1989 of the cyanobacteria *Arthrospira* and *Spirulina* as distinct genera, the term "*Spirulina*" has been used to indicate the species of both genera indifferently, particularly the species "*S. platensis*" and "*S. maxima*" cultivated and sold as food, feed and a specialty product source. The successful commercial exploitation of *Arthrospira,* because of its high nutritional value, chemical composition

B. A. Whitton and M. Potts (eds), The Ecology of Cyanobacteria, 505-522.
© 2000 *Kluwer Academic Publishers. Printed in the Netherlands.*

and the safety of its biomass, has made it one of the most important industrially cultivated microalgae. This chapter describes the ecology of *Arthrospira*, together with morphological and ultrastructural features relevant for supporting the systematic position of this organism. While the confused taxonomy of *Arthrospira* and its relationship with *Spirulina* have been resolved by 16S rRNA sequence analysis, the long debated problem of species definition is still ongoing. One study of ten strains suggested that the morphological criteria used to identify the species (*A. maxima, A. platensis*) corresponded to the molecular data obtained by total DNA restriction profile analyses. However, another study based on a wider range of classical species and forty different strains failed to show a clear correspondance with molecular data obtained for one part of the genome. This emphasises the need to address the problem of the definition of a species by using a polyphasic approach. Knowledge of the ecophysiology of *Arthrospira*, essential for understanding the growth requirements of this alkaliphilic organism in the natural environment, has been used in developing suitable technologies for mass cultivation. The relationships between environmental and cultural factors, which govern productivity in outdoor cultures, are discussed in connection with growth yield and efficiency. The response of *Arthrospira* and its modification under stress are described, together with the strategy of osmotic adjustment and the mechanism of internal pH regulation to alkalinity. The metabolic plasticity of the response of this cyanobacterium to disparate environmental stimuli is demonstrated in the natural environment, but is also well-expressed in the maintenance of highly productive monoculture in intensive outdoor cultivation systems.

I. Introduction

Ever since *Arthrospira* was first reported in 1852 by Stizenberger, many species of this genus of helically coiled cyanobacteria have been described and isolated. However, the classification of this genus has long been a source of confusion. Geitler (1932) invalidated the genus *Arthrospira* in his revision of the Cyanophyceae and included all regularly helically coiled oscillatorian organisms in the previously described genus *Spirulina* Turpin 1829. Since then, the term *Spirulina* has mostly been used to indicate species of either genus indifferently. As described later, it is now recognized that there are two very distinct genera (Castenholz, 1989). Therefore, many species that currently bear the generic epithet, *Spirulina*, should be re-included in the original genus *Arthrospira*. However, the term Spirulina is widely used to indicate the economically important species, "*S. platensis*" and "*S. maxima*", so we use it in parts of this chapter to avoid confusion with previous studies.

Arthrospira has been reported to exist in environments varying in their osmoticum, temperature and salt concentrations, most of them being of high alkalinity (Iltis 1969a, b; Busson, 1971). Filaments of the genus *Spirulina* in the strict sense also occur in many of these environments.

There is great interest in past use of the use of *Arthrospira* as a traditional food. Dangéard (1940), Brandily (1959), Léonard (1966) and Léonard and Compère (1967) all described how African tribes living along Lake Chad collect this microscopic "alga" (*A. platensis*). It is collected from water bodies near the lake and sun-dried on the shores to produce a hardened dark cake called "dihé", which is broken into small pieces and used in different forms by the local populations as part of their daily diet. At about the same time, and on the other side of the globe, *Arthrospira* was also recorded in the water of Lake Texcoco, Mexico. Here, it had been used as food by the natives living in the area (Clément, 1968). Travelers to Mexico during the 16th century described how the Aztecs used a soft a blue-green material, harvested with fine nets from the lake, for making a kind of bread called "tecuitlatl" (Ciferri, 1983). It is clear that these two different cultures, living far apart, discovered the nutritional value of *Arthrospira* independently.

Later attention was re-focused on "*Spirulina*" by the pioneering work done at the Institute Française du Pétrole, with studies on the cyanobacterial blooms occurring in the evaporation ponds of the industrial soda production facility at Lake Texcoco near Mexico City. This led to the first detailed study of the growth requirements and physiology of *Arthrospira*. The Ph.D. study by Zarrouk (1966) was the basis for establishing the first large-scale production plant of *Arthrospira*. The work was followed up by several groups in Italy, France and Israel and was summarized in a detailed review paper by Ciferri (1983). The extensive research on cell biology, biochemistry and biotechnology carried out since then is reviewed in the book edited by Vonshak (1997a).

This chapter provides a perspective on the systematics and the ecophysiology of "*Spirulina*", now correctly re-assigned to *Arthrospira* Stizenberger 1852.

II. Morphology

A. The Helical Shape of the Filament

The main morphological feature of *Arthrospira* is the patterned arrangement of its multicellular cylindrical trichomes in an open helix. The trichomes are composed of cylindrical cells that undergo binary fission in a single plane, perpendicular to the main axis. Trichome elongation occurs through multiple intercalary cell division along the entire filament. Multiplication occurs only by fragmentation. Trichome breakage is transcellular and involves destruction of an intercalary cell. The mechanism as it occurs in "*S. maxima*" and in "*S. platensis*" was described in detail by Tomaselli et al. (1981). In *Spirulina*, trichome breakage is intercellular (Rippka et al., 1979; Anagnostidis and Komárek, 1988).

The cell width ranges from about 3 to 12 μm, though occasionally reaches 16 μm. The helix pitch typically ranges from 10 to 70 μm and its diameter from 20 to 100 μm. These two parameters which define the shape of the helix architecture are highly dependent on growth and environmental conditions (Plate 27d).

The effect of temperature on filament structure was studied by Van Eykelenburg (1979), who subsequently described the rapid reversible change from the helix to the spiral shape on solid media (Van Eykelenburg and Fuchs, 1980). A detailed study of the effect of physical and chemical conditions on the helix geometry of *Arthrospira* was done by Jeeji Bai and Seshadri (1980) and Jeeji Bai (1985). They reported the occurrence of straight or nearly straight spontaneous culture variants also repeatedly observed by Lewin (1980) and by other authors. Though there are considerable variations in the degree of coiling between different strains of the same species and within the same strain, the helical shape of the trichome is regarded as a peculiar property of the genus. Once a strain has converted to the straight form, either naturally or due to physical or chemical treatments, it does not usually revert to the helical form. Jeeji Bai (1985) suggested that this may be due to a mutation affecting some trichomes under certain growth conditions. The common occurrence of straight trichomes in cultures of *Arthrospira* also suggests that the helical character may be carried on plasmids, but no one has yet demonstrated the existence of plasmids in *Arthrospira* or *Spirulina* strains. Fox (1996) successfully reverted several straight filaments to the typical coiled shape by exposing the filaments to high light irradiance. He believes the shape to be associated with helical protection against photolysis, which occurs frequently in natural water bodies. Perhaps because laboratory cultures grown at low light intensity, or well-mixed, dense outdoor cultures do not need to protect themselves against photolysis, straight filaments are frequently observed under both sets of conditions.

The ability to reverse filament configuration is further confirmed by the observation that it is possible to cause a straight filament to revert to the helical shape by growing it without intensive mixing at optimal light and temperature conditions, after inducing filament breakage with sonication (A. Vonshak, unpublished). We have also observed that under certain stress conditions (high growth temperature and very high alkaline medium) filaments have a more tightly coiled appearance.

In general, it seems that artificial laboratory conditions favor the development of straight forms. Trichomes of several *Spirulina* species also straighten after they have been streaked on plates during isolation procedures or after repeated transfers onto solid media (L. Tomaselli, unpublished).

B. Ultrastructure

The cell organization of *Arthrospira* is typical of prokaryotes, with a lack of membrane-bound organelles (Van Eykelenburg, 1979; Tomaselli et al., 1976; Tomaselli et al., 1993b; Fig. 1). The gram-negative multilayered cell wall is surrounded by a diffluent mucilaginous polysaccharide envelope. The thin cell wall has four layers, with an easily detectable electron-dense layer corresponding to the peptidoglycan layer (Van Eykelenburg, 1977). The regularly spaced cross-walls divide the trichome into cells connected by plasmadesmata. Cross-walls are formed by centripetal growth and extensions of both the peptidoglycan and the more internal layer of the cell wall toward the center of the cell. The cell has a number of inclusions mostly corresponding to the typical ones for cyanobacteria described in previous volumes (Fogg et al., 1973; Carr and Whitton, 1973; 1982). Thylakoid membranes with phycobilisomes are arranged in parallel bundles. The other main subcellular inclusions are (in addition to carboxysomes, ribosomes and DNA fibrils) gas vacuoles, polyglucan granules (especially near the cross-walls), polyphosphate granules and large

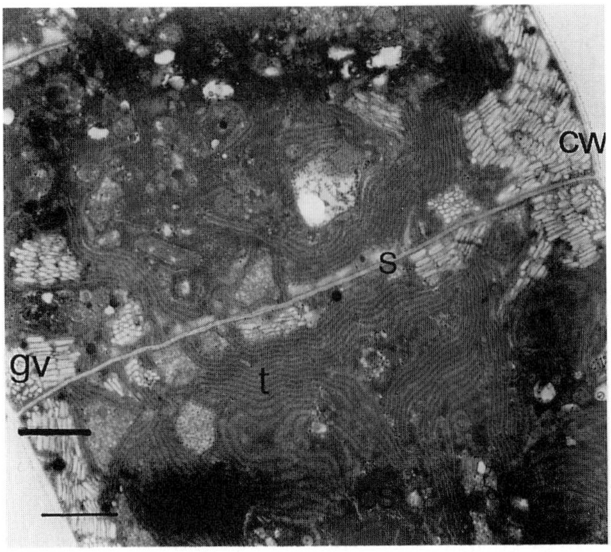

Fig. 1. Thin section showing the fine structure of *Arthrospira maxima*. Note: cell-wall (cw), septum (s) and thylakoids (t). Bar marker = 1 μm (photo courtesy of M.R. Palandri)

cyanophycin granules. Structures like the cylindrical bodies reported for some other members of the Oscillatoriaceae are often present. These inclusions,whose explanation is still unknown, are lacking in *Spirulina* (Tomaselli et al., 1996).

III. Systematics

The first report of a taxon which can be assigned to *Arthrospira* Stizenberger 1852 dates back to more than 150 years ago, when Wittrock and Nordstedt (1844) reported the occurrence near Montevideo of a blue-green alga with helical shaped filaments, which they described as "*Spirulina jenneri* f. *platensis*", even though it possessed septa. At that time the genus *Spirulina* was considered to be unicellular. Some years later, Stizenberger (1852), who had observed septa in some helically coiled oscillatoriacean forms, proposed that the forms, with a multicellular structure, be included in a new genus, *Arthrospira*. This distinction was confirmed by Gomont (1892) in his taxonomic study on the Oscillatoriaceae. He left the apparently aseptate forms in *Spirulina*, and placed the septate forms in *Arthrospira* Stizenberger 1852. However, Geitler's revision (1932) of the Cyanophyceae re-unified the members of these two genera in *Spirulina* Turpin 1829. He based classification on the close similarity in morphology and ignored the presence of septa, visible or otherwise, and thus made no distinction between the forms previously recognized as separate genera. Geitler actually split the genus into two subgeneric taxa (Section I. Arthrospira and Section II. Euspirulina) on the basis of the criterion originally used by Stizenberger (1852) to separate the genera *Spirulina* and *Arthrospira*, i.e. visible or non-visible cross-walls. The forms with septa that were easily observable under a light microscope were classified in the Section Arthrospira and those with unlikely or only artificially observable septa (after trypsin treatment or staining with neutral red) in the Section Euspirulina.

In Bergey's Manual of Systematic Bacteriology, Castenholz (1989) attempted to resolve the matter by restoring *Arthrospira* to its original rank of genus. He suggested using three features to separate *Spirulina* and *Arthrospira* in Subsection III of the Order Oscillatoriales, which shares the property of helically coiled trichomes. These are:

1) degree of inclination of the pitch of trichome helix (from transverse axis)
2) aspect and visibility (light microscope) of cross-walls between the cells in the filament
3) distribution (electron microscopy) of junctional pores in the cell wall.

Thus *Arthrospira* can be differentiated from *Spirulina* on the basis of the degree of inclination of the pitch of the trichome helix, which forms an angle > 45°

Table 1. Main features separating the genera *Arthrospira* Stizenberger 1852 and *Spirulina* Turpin 1829.

Criteria	**Arthrospira**	**Spirulina**
Trichome diameter	2.5-16 µm	0.5-5 µm
Type of helicity	loosely coiled	tightly coiled
Cross-walls	visible under light microscope	invisible under light microscope
Cell wall pore pattern	single row around the trichome	several rows on concave side of the coil
Mode of trichome fragmentation	intracellular (necridium)	intercellular
Cylindrical bodies	present	absent
Anoxygenic photosynthesis	absent	present in some strains
C-phycoerythrin	not found	present in some strains
γ-linolenic acid	present	absent

from the transverse axis, the presence of easy visible septa and the distribution of junctional pores in one circular row around the cross-walls.

Tomaselli et al. (1996) suggest that there are other meaningful criteria to distinguish between the two, including 16S rRNA sequence data which clearly show that *Spirulina* and *Arthrospira* are phylogenetically distant (Giovannoni et al., 1988; Nelissen et al., 1994). Another useful criterion is that, unlike *Spirulina*, *Arthrospira* contains the unsaturated fatty acid γ-linolenic acid) (Cohen and Vonshak, 1991). *Spirulina* and *Arthrospira* are both aquatic, but, unlike *Spirulina*, *Arthrospira* has gas-vacuolated cells. The main criteria for distinguishing the two genera are listed in Table 1.

Although *Arthrospira* is helically coiled in natural populations, the occurrence of straight or slightly straight filamentous forms, similar to *Oscillatoria*, have been reported repeatedly in culture, making simple morphological observation invalid in the identification of *Arthrospira* (Lewin, 1980). The presence of straight forms of *Arthrospira* and their similarity to some members of the tribe "Oscillatoriae" (Castenholz, 1989) due to a similar junctional pore distribution (Guglielmi and Cohen-Bazire, 1982) emphasize the need to find additional phenotypic features to distinguish *Arthrospira*.

According to the Botanical Code, *A. jenneri* (Hass.) Stizenberger is the type species and *A. platensis* PCC 7345, which was isolated from a saline marsh (Del Mar Slough, California, USA), is the reference strain for the Bacteriological Code (Rippka and Herdman, 1992) (Table 2).

Hindák (1985) assigned all the planktonic forms of *Arthrospira* (including those previously described as "*S. platensis*) in alkaline saline lakes to "*S. fusiformis*" Voronichin 1934 (synonyms *A. maxima* Setchell et Gardner 1917 and *A. geitleri* De Toni 1935), since, in agreement with Fott and Karim (1973), he considered "*S. platensis*" to be a benthic species. On the basis of this assumption, most of the African populations were included under *A. fusiformis*. This opinion, also shared by Komárek and Lund (1990), has been debated by Tomaselli (1997). Komárek and Lund (1990), in their revision of the taxonomy and nomenclature of the three related species "*S. platensis*", "*S. maxima*" and "*S. fusiformis*", suggested that within the planktonic forms there are two taxa ("*maxima*" and "*fusiformis*"), and furthermore that these two taxa, which could represent two species (*A. fusiformis* Komárek and *A. maxima*), have different geographical distributions. Hence they consider *A. maxima* to be pantropical, *A. fusiformis* to be limited to Africa and tropical and central Asia, and consider the 'benthic' *A. platensis* to be essentially restricted to South America.

More recently Desikachary and Jeeji Bai (1992) proposed that all the calyptrate forms of *A. fusiformis* be included in the new species *A. indica* Desikachary

Table 2. Species of the genus *Arthrospira* Stizenberger 1852.

Species	**First description**	**References**
A. fusiformis Komárek 1990	Siberian steppe, Russia, Lake Tunatan	Voronichin, 1934
A. gomontiana Setchell 1895	North America, stagnant water	Geitler, 1932
A. indica Desikachary & Jeeji Bai 1992	Madurai, India, natural pond	Desikachary & Jeeji Bai, 1992
A. jenneri Stizenberger 1852	Europe, stagnant water	Gomont, 1892
A. khannae Drouet & Strickland 1942	Rangoon, Myanmar, natural pond	Desikachary, 1959
A. massartii Kufferath 1914	Luxemburg, spring water	Geitler, 1932
A. maxima Setchell & Gardner 1917	Oakland, California, warm, saline pond	Gardner, 1917
A. platensis Gomont 1892	Montevideo, Uruguay, stagnant water	Wittrock & Nordstedt, 1844
A. spirulinoides Ghose 1923	Lahore, Pakistan, standing rain water	Geitler, 1932
A. tenuis Brühl & Biswas 1922	Bengal, India, artificial basin	Geitler, 1932

and Jeeji Bai 1992; this implies that the thickening of the apical cell wall (calyptra) is a significant taxonomic feature. The proposal is based on the fact that the calyptra was never reported in *A. fusiformis*. If the existence of the calyptra can be demonstrated in the original material, *A. indica* will became a later synonym of *A. fusiformis* (Desikachary and Jeeji Bai, 1996). Nevertheless, the proposal for a new species *A. indica* is questionable since other authors (Fott and Karim, 1973) consider "*S. geitleri*" which is a synonym of the calyptrate "*S. maxima*" to be identical to "*S. fusiformis* ".

The traditional separation of the common species into *A. maxima* and *A. platensis* should be maintained until more detailed studies have been completed (Tomaselli, 1997). Considering the great variations in size and shape of trichome geometry in natural and cultured *Arthrospira* populations, taxa based on morphological features should be confirmed through use of molecular biology techniques. Total DNA restriction profile analysis, performed on ten *Arthrospira* strains, led to the recognition of two well-separated genotypic groups (Viti et al., 1997) coinciding with the botanical species *A. maxima* and *A. platensis*, to which the strains had previously been assigned (Tomaselli et al., 1993a).

A further phylogenetic study, using ARDRA (Amplified Ribosomal DNA Restriction Analysis) of the ITS (Internal Transcribed Spacer) as a molecular taxonomic marker, was made on more than 40 strains of *Arthrospira*, which differed (with one exception) from the previous ones (Scheldeman et al., 1999). The results of this study led the strains being grouped into only two main genotypic clusters with no clear correlation to their geographic origin or phenotypic features. This is in contrast to the previous study by Viti et al. (1997), which showed a large consensus between the genotypic clusters and the other features of the strains. The lack of a close concordance between the results of these studies emphasizes the need to address the long debated problem of the definition of species through a polyphasic approach using phenotypic and different types of genotypic information.

The question of the proper designation of *Arthrospira* extends to the commercial products and human health, since *Arthrospira* biomass is sold as a health food. The widespread use of the term Spirulina for such material clearly causes problems in accepting the correct terminology, but it is highly important to be aware of the type of organism being cultivated. Some *Spirulina* spp. (e.g. *S. subsalsa*) are in fact cultivated for commercial purposes in mistaken belief that they have the nutritional quality and toxicological safety of the traditional edible species of *Arthrospira* (Shimada et al., 1989; Brieba and Merinos, 1993).

IV. Occurrence and Distribution

A. General

Species of *Arthrospira* have been isolated mainly from alkaline, brackish and saline waters in tropical and semitropical regions (Castenholz, 1989). *A. jenneri* and *A. platensis* were described by Geitler

(1932) as cosmopolitan. Though most of the *Arthrospira* species have been found in alkaline water, where they form massive blooms, some other species have been reported to occur in fresh waters, but without such predominance. Other species with more restricted distributions occur in running fresh waters (*A. okensis*) and springs (*A. massartii*). *A. spirulinoides*, *A. gomontiana* and *A. tenuis* are reported for stagnant fresh waters. Desikachary (1959) considers *A. jenneri* and *A. platensis* as typical of tropical, saline environments.

Among the various species included in the genus *Arthrospira*, the most widely distributed, *A. platensis*, is mainly in Africa (Chad, Kenya, Egypt, Ethiopia, Sudan, Libya, Algeria, Congo, Zaire, Zambia), Asia (Pakistan, India, Sri Lanka, Myanmar, Thailand) and South America (Uruguay, Peru). *A. maxima* (syn. *A. geitleri*) appears to be confined to Central America (Mexico, California-USA). This species is the main component of the phytoplankton of the solar evaporation channels in Lake Texcoco (Mexico), while *A. platensis* dominates the alkaline saline lakes of the semi-desertic Sudan-Sahel area (Chad) and of the Rift Valleys (Kenya). The abundant occurrence of the species *A. platensis* and *A. maxima* in saline, alkaline, tropical and sub-tropical waters has led to recognition of these environments as typical for the entire genus.

The geographical distribution of the main populations of *Arthrospira* is given in Table 3. *Arthrospira* species have apparently not been reported from marine environments. However, when an adequate supply of bicarbonate, N and P, together with suitable pH and salinity values are provided, *Arthrospira* species can be highly productive in sea water (Tredici et al., 1986).

The most detailed studies on the ecology of *A. platensis* are those of Iltis (1968; 1969a, b; 1970a, b; 1971a, b; 1972) for tropical African lakes, and that of Busson (1971), who described past and recent use of "*S. geitleri*" (syn. *A. maxima*) as a food source in Mexico, in addition to its ecology. Other studies include those of Jeeji Bai and Seshadri (1980), Hindák (1985), Komárek and Lund (1990) and Desikachary and Jeeji Bai (1992). The last deals with the distribution of Indian isolates and revises the taxonomy of some culture collection strains, assigning them to the calyptrate species, *A. indica*.

B. Water Basins of the Sudan-Sahel Zone

The extensive survey of the phytoplankton of the permanent and temporary lakes near Lake Chad performed by Iltis for ORSTOM (Office de la Recherche Scientifique et Technique Outre-mers) has contributed much to our knowledge of *A. platensis*. Its main distribution is in the water basin of the Sudan-Sahel zone in Central Africa, north-east of Lake Chad towards the region of Ounianga Kebir, and the basins of the East African Rift Valleys. It also occurs in Ethiopia (Lake Chitu and Lake Aranguadi) and Zaire (Lake Kivu) (Table 3).

Most of the water bodies in which *Arthrospira* thrives are permanent or temporary saline alkaline lakes (natron lakes); these are scattered in the hollows of the fossil dune system of the Sahara and the Sudan-Sahel areas and fed by aquifers and the scanty rain (about 300 mm per year). The salinity of these lakes is highly variable; when evaporation exceeds inflow the salinity rises and leads to an accumulation of salt deposits on the shores (Iltis, 1971b). The mean water temperature is ~25°C, with oscillations of about ± 10°C. During the warm season, the water temperature in shallow lakes may reach 38°C. Mean length of daylight is about 12 hours and solar radiation is over 2000 µmol photon m^{-2} s^{-1}. The disposition and slope of the impermeable clay and water-bearing sandy layers favor subterranean seepage of water away from Lake Chad (Maglione, 1969). The relationships between the superficial and the subterranean waters affect both the salinity and the ionic composition. The latter is characterized by prevalence of monovalent ions, by Na^+ much higher than K^+, and by high concentrations of carbonate and bicarbonate, which lead to pH values from 9.5 to 11.0. Ca^{2+} and Mg^{2+} concentrations are low. SO_4^{2-} content is usually high, while that of Cl^- is very low. The water of the lakes is often blue-green in color due to the massive development of phytoplankton, which can attain very high densities.

A. platensis is present as nearly monospecific populations in the highly saline lakes (Iltis 1969a), only a few other phototrophs being found. In contrast, in the mesohaline lakes, *Arthrospira* coexists with a number of other phototrophs (especially diatoms and dinoflagellates) in addition to the dominant cyanobacteria. *Anabaenopsis arnoldii* and many *Oscillatoria* spp. are often important components of the cyanobacterial population. Several *Spirulina* species may be present, especially *S. laxissima*, *S. major* and *S. subsalsa*. A smaller form of *A. platensis*, var. *minor* Rich, was also present (Iltis, 1971b). In the studies performed on samples of two African natron lakes, Lake Mombolo and Lake Rombou, of the region of Kanem lying north-east of Lake Chad, Margheri et al. (1975) found

Table 3. Geographical distribution of the main populations of *Arthrospira* spp.

COUNTRY	ENVIRONMENT	SPECIES	REFERENCES
AFRICA			
Chad	natron lakes (Bodou, Mombolo, Rombou, Yoan) and pools (Latir, Iseirom, Latir, Liva), Kanem region	*A. platensis, A.platensis* f. *minor*	Léonard & Compère, 1967; Iltis, 1972, Rich, 1931
Kenya	natron lakes (Bogoria, Crater, Elmenteita, Nakuru), Rift Valley	*A. platensis*	Rich, 1931; Vareschi, 1982
	Lake Bogoria	*A. platensis, A.platensis* f. *minor, A. fusiformis*	Tuite, 1981; Hindák, 1985
	Lake Simbi	*A. platensis, A. platensis* = *A. fusiformis*	Melack, 1979 Kebede & Ahlgren, 1996
Ethiopia	Lake Aranguadi, Lake Chitu, Green Lake	*A. platensis, A. platensis* = *A. fusiformis*	Melack, 1979 Kebede & Ahlgren, 1996
Egypt	Lake Maryut	*A. platensis*	El-Bestawy et al., 1996
Sudan	Lake Dariba, Jebel Marra	*A. geitleri*	Fott & Karim, 1973
Algeria	pond Tamanrasset	*A. platensis*	Fox, 1966
Congo	Laka Mougounga	*Arthrospira* sp.	Fox, 1966
Zaire	Lake Kivu	*Arthrospira* sp.	Fox, 1996
Zambia	Lake Bangweolou	*Arthrospira* sp.	Fox, 1966
ASIA			
India	ponds	*A. maxima*	Desikachary & Jeeji Bai, 1996
	Lonar Lake; ponds; tank (Madurai, MCRC isolate)	*A. indica*	Desikachary & Jeeji Bai, 1992
Myanmar	Crater lakes	*Arthrospira* sp.	Min Thein, 1993
Pakistan	fish pond, Lahore	*Arthrospira* sp.	Fox, 1996
Sri Lanka	Lake Beria	*Arthrospira* sp.	Fox, 1996
China	fish ponds, Nanking	*A. platensis*	Tsen & Chang, 1990
Thailand	tapioca factory effluent lakes, Bangkok	*Arthrospira* sp.	Fox, 1996
Russia	Tunatan lake, Siberian steppe	*A. fusiformis*	Voronichin, 1934
Pakistan	pond, Lahore	*Arthrospira* sp.	Fox, 1996
AMERICA			
Mexico	Lake Texcoco solar evaporator	*A. maxima*	Busson, 1971
California	pond, Oakland; coastal lagoon, Del Mar	*A. maxima, A. platensis*°	Gardner, 1917; Lewin, 1980
Peru	Lake Huachachina	*A. platensis, A. maxima*	Busson, 1971; Desikachary & Jeeji Bai, 1992, 1996
Uruguay	Montevideo	*A. platensis*°°	Gomont, 1892
EUROPE			
Spain	Lake Santa Olalla	*Arthrospira* sp.	Rippka & Herdman, 1992
France	tiny lake, Camargue	*Arthrospira* sp.	Fox, 1996
Hungary	Adasztevel-Oroshaz	*Arthrospira* sp.	Busson, 1971
Azerbaijan	water basin, Khumbasha	*Arthrospira* sp.	Fox, 1996

° reference strain PCC 7345.

°° holotype

the above-mentioned *Spirulina* species and several *Arthrospira* spp. The overwhelming majority of the isolated strains of *Arthrospira* were identified as *A. platensis*; one strain was identified as *A. platensis* var. *minor* and one other as *A. maxima* (Tomaselli et al., 1993a). Within the photosynthesizing population of the permanent and temporary lakes, Iltis and Riou-Duwat (1971c) observed a very high, albeit seasonally variable, density of rotifers (mainly *Brachionus* spp.), up to several hundred individuals per liter. In some lakes *Arthrospira* constitutes an excellent feed for the cichlid fish *Tilapia*. Another consumer of *Arthrospira* and other large filamentous cyanobacteria such as *Anabaenopsis* is the flamingo (*Phoenicopterus minor*). The pink color of these birds comes from the carotenoid pigments originating in cyanobacteria.

The detailed studies performed by Iltis over the course of several years have established a direct correlation between the biomass density of *Arthrospira* and the salinity of the water. The massive *Arthrospira* blooms occur at salinity levels from 22 to 60 g L^{-1}, high carbonate-bicarbonate concentrations (pH8.5 - 11.0), and temperatures from 25 - 40°C. Generally, the waters populated by *Arthrospira* have a mean salinity of 37 g L^{-1}. Nevertheless, *Arthrospira* has been found to occur at salinity levels ranging from 8.5 to 200 g L^{-1} (in some cases, up to 270 g L^{-1}: Iltis, 1968). Biomass density can exceed 1 g L^{-1}.

Tuite (1981) found the density of *Arthrospira* to be highly variable. The decline in the *Arthrospira* population was preceded by a period of increased water conductivity. In many of the natron lakes, *Arthrospira* was absent or present only at low densities. Marked fluctuations were recorded in other lakes. In some cases *Arthrospira* maintained high densities for periods up to several years, and appeared to be the only significant cyanobacterial species in the phytoplankton.

Arthrospira biomass density was lower in the temporary lakes than in the permanent lakes, but reached very high levels (up to 5 g L^{-1}) in the former immediately prior to their desiccation. The phytoplankton of the temporary water bodies is usually dominated by cyanobacteria and diatoms. *Arthrospira* blooms tend to develop after a bloom of green flagellates and other microalgae following the refilling of a lake by rainwater (Iltis, 1969b; 1970a). *Arthrospira* flourishes all year round, with a small decline during the short and scarce rainy season in the permanent lakes, but lasts no more than two months in the temporary water bodies. In these lakes the density of the populations depends not only on the mean salinity, but also on the time period during which the basin is flooded. *Arthrospira* blooms were not observed in water bodies flooded for less than four months.

C. Water Basins of the East African Rift Valley

In the volcanic regions of the East African Rift Valley there are many highly-alkaline saline lakes (HCO_3^- and CO_3^{-2} more than 1000 meq L^{-1}); these lakes are developed along the valley floor, the geology of which is dominated by alkaline, trachyte lavas rich in Na^+. The high alkalinity is the result of geological and climatic features:

i) presence of carbonates and bicarbonates derived from leaching of carbonatic volcanic rocks;

ii) absence of significant amounts of Ca^{+2} and Mg^{+2} in the surrounding land;

iii) concentration of ions due to evaporation.

Underground seepage through soluble volcanic rocks may also contribute to the increase in the salinity of the waters (Grant et al., 1990). Due to the generally hot and dry climate and the high solar irradiance, evaporation proceeds through brine saturation to crystallization. Therefore, the salinity and the ionic composition of the water are both subject to marked changes, due to the different solubility of the ions and the origins of the soluble salts. The saltiness of these extremely alkaline lakes in the East African Rift Valley ranges from 5 to 30%, according to seasonal weather conditions; the pH is almost consistently above 10 (Beadle, 1974). The waters are low in SO_4^{-2} and high in Na^+. In many of these lakes, *A. platensis* is dominant, as it is in the Chad area. In Lake Elmenteita and Lake Nakuru, Grant and Tindall (1986) found *A. platensis* to constitute the major bloom, although they also observed significant numbers of other cyanobacteria as well as diatoms, which became dominant as the cyanobacterial blooms decreased. In his review on "*Spirulina*", Ciferri (1983) summarized most of the studies on the distribution of this cyanobacterium in Africa, with particular emphasis on the alkaline lakes of the Rift Valley.

Arthrospira is also found in some lakes of the northern Algerian Sahara having saline waters (1 - 8%) and pH levels ranging from 7 to 9. These waters, which have lower alkalinity (about 20 meq L^{-1} HCO_{3-1} plus CO_3^{-2}), but Na^+ levels similar to those of the highly alkaline saline waters of the East African Rift Valley, may be considered to be sulfate-chloride rich waters (with Cl^- the major anion). Fott and Karim (1973) found that the phytoplankton population of a saline alkaline lake (pH about 10) in Jebel Marra, Sudan, comprised only one cyanobacterium, which they named as *Spirulina geitleri*. Later, Fott (1977) identified this species as *S. fusiformis* (today *A. fusiformis*). Hindák (1985) assigned the *Arthrospira* populations occurring in Lake Bogoria (Kenya) to be *S. fusiformis.*

D. Water Basins of the Solar Evaporator at Lake Texcoco, Mexico

As previously mentioned, a massive population of *Arthrospira* was found in some of the external sectors of the gigantic spiral-shaped solar evaporator used to extract salts from the saline carbonate-bicarbonate rich waters of Lake Texcoco, Mexico. The evaporator lies 2200 m above sea level and is about 1 m in depth. The alkaline (pH 10) saline water is slowly moved from the outer part of the evaporator towards the center (Gallegos, 1993). The water is high in HCO_3^-, CO_3^{-2}, Cl^- and Na^+; the salt concentration ranges from 11 to 39 g L^{-1} (Busson 1971). In the past, the company that operated the plant, Sosa Texcoco (and later Spirulina Mexicana), had its Spirulina production facilities in the outside rim of the solar evaporator. In the alkaline saline waters, the species *A. maxima* is present in almost monospecific populations, being associated with only a few other cyanobacteria such as *Synechocystis aquatilis*.

V. Physiology of *Arthrospira*

Ecological studies performed during the last three decades have led to understanding of the conditions that can favor its abundant growth in phytoplankton populations. This information has been useful for developing the technology for mass cultivation of *Arthrospira* (Richmond, 1988).

One of the basic principles of this technology is the re-creation of determinate environmental and cultivation conditions such as to permit high productivity in a monoalgal culture. The main environmental factors which govern the productivity of a phototrophic microorganism are light intensity and temperature. Nutrient concentration, pH, salinity, cell density, mixing, etc., are strictly cultural factors (Vonshak, 1987a).

In nature, the organism uses its gas vacuoles to regulate its position along the underwater light gradient and follow the daily and seasonal light changes. Yet in the shallow culture ponds, light availability has to be optimized by modifying operational parameters such as optimal cell density, the degree of mixing and culture depth in order to obtain the highest productivity (Vonshak et al., 1982).

Photoinhibition and photooxidative damage due to prolonged exposure to high solar irradiance of the outdoor culture can be prevented to a certain degree by constant mixing of the culture and by maintenance of high cell densities. This also provides a more efficient light utilization (Richmond, 1996). The light regime imposed by mixing in the *Arthrospira* ponds at the very least prevents excessive exposure to high irradiance by repeatedly shifting the cells from the lighted into the dark layers and vice-versa. Intensive mixing also ensures an even distribution of nutrients and at the same time prevents filament aggregation and sinking (Grobbelaar, 1996).

The fact that *Arthrospira* requires high alkalinity for growth ensures selective conditions in the growth medium and for this reason *Arthrospira* culture growing outdoors in open ponds remains in quasi-monoculture (Vonshak et al., 1983). Moreover, when other essential growth elements are present in sufficient amounts, the elevated concentration of bicarbonate-carbonate favors the high productivity observed in the natural habitat.

A. Growth: Yield and Efficiency

The growth of *Arthrospira* follows the pattern common to many other micro-organisms which undergo simple cell division involving no differentiation or sexual steps. Thus, the specific growth rate (μ) is described by the following equation:

$$\mu = \frac{1}{x} . dx / dt$$

where x is the initial biomass concentration.

Ogawa and Terui (1970) made the first assessment of quantum yield (Y_{kcal}) (measured as the amount of dry algal biomass harvested per kcal light energy absorbed) for *Arthrospira.* Calculated values of Y_{kcal} range from 0.01 to 0.02 g cell $kcal^{-1}$. These values correspond to an efficiency of 6 - 12%. In a later study, Ogawa and Aiba (1978) estimated the quantum requirement of *Arthrospira* cultures grown at steady state conditions; the Q_{CO2} value was about 20 mol quanta mol^{-1} CO_2, which corresponds to an efficiency of 10%.

The relationship between the specific growth rate and the specific light energy absorption rate was used to establish a mathematical equation describing the growth of *Arthrospira* in a batch culture (Iehana, 1987). The equation indicates that in cultures with a high cell concentrations the specific growth rate increases linearly with the increase in the specific absorption rate of light energy. Two other groups, Lee et al. (1987) and Cornet et al. (1992a, b) have published detailed studies on attempts to establish a mathematical model for the growth of *Arthrospira* under a variety of growth conditions. It seems that all of the models fit the experimental growth data well under normal steady state conditions, where light is either limiting or at its saturation level. These models require modification if stress conditions, such as photoinhibition or environmental stress (i.e. temperature or salinity) are introduced.

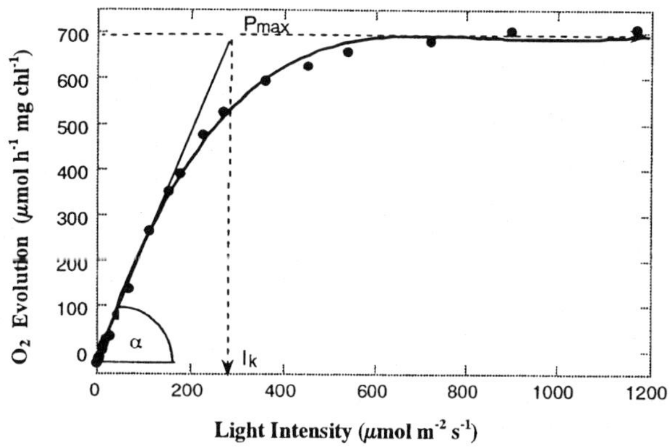

Fig. 2. Typical curve of specific photosynthetic rate (P) as a function of light intensity (I), illustrating the maximum photosynthetic rate (Pmax), the saturation onset parameter (Ik) and the slope (α) of the linear part of the curve (P-I curve).

B. Response to Environmental Factors

1. Effects of Light

a. Effects on Growth and Photosynthesis

Zarrouk (1966) was the first to study the response of "*Spirulina maxima*" to light; he reached the conclusion that growth of this organism saturates at levels of 25 - 30 Klux. Yet it is very difficult to compare these results to more recent studies because the method used to measure the light intensity and the light path in the vessel cultures is not described. From data obtained in our laboratory, growth of *Arthrospira platensis* is saturated at a range of 150 - 200 μmol photon m^{-2} s^{-1}. This is about 10 to 15% of the total solar irradiance in the 400 - 700 nm range. This value is highly dependent on growth conditions and correlates with the chlorophyll to biomass ratio. Using a turbidostatic culture, we have estimated that the μ_{max} of some *Arthrospira* strains corresponds to a doubling time of 8 - 10 h. For the strain *A. maxima*, 6Mx the value of μ_{max} was 0.0692 h^{-1}, corresponding to a doubling time of 10 hours (Tomaselli et al., 1997). This strain was able to alter its light harvesting capacity during acclimation to sudden irradiance changes. This is an important feature in selecting strains for outdoor cultivation.

The most common way to study the photosynthetic response of algal cultures to light is to measure the photosynthetic activity (P) versus irradiance (I) (i.e. the P-I curve). A typical P-I curve is shown in Fig. 2. In the dark, the rate of oxygen evolution or carbon fixation will be negative, due to respiration. As irradiance is increased, a point is reached at which the photosynthetic rate is just balanced by the rate of respiration. This is the compensation point (I_c). As irradiance is further increased, the rate of photosynthesis increases linearly. Following further increase in irradiance, the curve eventually levels off, as photosynthesis becomes saturated and reaches a maximum, P_{max}. The initial slope of the P-I curve is a useful indicator of quantum yield, i.e. photosynthetic efficiency.

The irradiance at onset of light saturation is defined by the value of I_k, which represents the point at which the extrapolation of the initial slope of the P-I curve crosses P_{max}. Exposing *Arthrospira* cultures to high photon flux densities above the saturation point may result in a reduced rate of photosynthesis; this phenomenon is defined as photoinhibition. The classical view that photoinhibition is observed only at high irradiance values seems to be a very simplistic one. Photoinhibition may be observed even at relatively low irradiance levels when other environmental stresses are involved. This will be discussed later.

The P_{max} and I_k levels are highly dependent on growth conditions. *Arthrospira* cultures grown at high or low light intensities will have different P_{max} and I_k values. Changes in these values may represent the organism's ability to photoadapt, or acclimate, to different light environments, as demonstrated in another cyanobacterium *S. subsalsa* grown at two photon flux densities (Tomaselli et al., 1995). Table 4 summarizes experiments carried out by Vonshak et al. (1996a) indicating different values of measured photosynthetic parameters in three different *Arthrospira* strains. Although the strains were grown under the same temperature and light conditions, they have different α and I_k values. Similar observations were reported by Tomaselli et al. (1993a).

b. Light Stress – Photoinhibition

Photoinhibition, as mentioned earlier, is defined as a loss of photosynthetic capacity due to damage caused by photon flux densities (PFDs) in excess of those required to saturate photosynthesis. The phenomenon of photoinhibition has been studied extensively and is well documented in algae and higher plants (Powles, 1984; Kyle and Ohad, 1986). Photoinhibition in laboratory *Arthrospira* cultures was first studied by Kaplan (1981), who observed a reduction in the photosynthetic activity when the cells were exposed to strong light under CO_2-depleted conditions. It was suggested that the reduction in photosynthetic activity was due to the accumulation of H_2O_2.

Different strains of *Arthrospira* may differ in their sensitivity to light stress (Vonshak et al., 1988b). In at least one strain such a difference was probably due to a difference in the turnover rate of a specific protein, D1, which is part of PSII. The different responses of different *Arthrospira* strains to photoinhibitory stress may be genotypic characteristics as well being dependent on growth conditions. Cultures grown at high light intensities exhibit a higher resistance to photoinhibition (Vonshak et al., 1996a). It should be emphasized that photoinhibition affects not only the P_{max} level, but actually has a stronger effect on light-limited photosynthetic activity. Thus a significant reduction in α is observed in photoinhibited cultures.

Table 4. Specific growth rate (μ) and photosynthetic parameters of three *Arthrospira* strains. $\mu = h^{-1}$; I_k = irradiance at onset of light saturation (μmol photon m^{-2} s^{-1}); Pmax = maximal light-saturated rate of O_2 evolution (μmol O_2 h^{-1} mg Chl a^{-1}); α = light-limited slope of the P-I curve (μmol O_2 mg Chl a^{-1}/μmol photon m^{-2}). Values are mean ± SE (*n*=3).

PARAMETERS	STRAINS		
	BP	P4P	Z19/2
μ	0.047	0.044	0.045
I_k	258 ± 20	185 ± 15	325 ± 40
Pmax	536 ± 4	585 ± 10	585 ± 12
α	2.3 ± 0.3	3.85 ± 0.9	2.1 ± 0.5

2. Effects of Temperature

a. Effect of Temperature on Growth

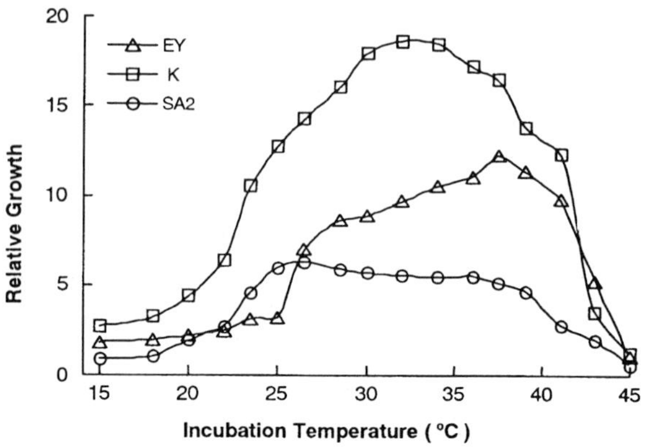

Fig. 3. Temperature response of 3 *Spirulina* isolates measured as the increase in Chlorophyll concentration after incubating the cells for 4 days on a temperature block under constant light (100 μmol m^{-2} s^{-1})

In nature, *Arthrospira* is found in permanent or temporary water bodies at relatively high temperatures. The optimal temperature for laboratory cultivation of this organism ranges from 30 - 38°C. However, many *Arthrospira* strains differ in their optimal growth temperature as well as in their sensitivity to extreme values. For instance, in the comparison of strains shown in Fig. 3, strain SA2 has a relatively low optimal growth temperature (24 - 28°C), while the EY strain grows well up to 40 - 42°C. The isolate marked K has a relatively wide range of optimal growth temperatures, 28 - 36°C.

Growth temperature also affects the biochemical composition of the cells (Tomaselli et al., 1988). When "*S. platensis*" M2 is grown under light-limited turbidostatic conditions at its maximum growth temperature (42°C), a marked decrease in the photosynthetic pigments and proteins level is observed. Under these conditions, an increase in carbohydrate cell content associated with changes in the degree of fatty acid saturation (i.e. reduction of γ-linolenic synthesis in favor of linoleic acid accumulation) is also observed.

b. Effect of Temperature on Photosynthesis and Respiration

The net productivity of an algal culture is directly correlated with the gross rate of CO_2 fixation, or O_2 evolution (photosynthesis), and with the rate of respiration. These two metabolic activities are temperature dependent, while only CO_2 fixation, or O_2 evolution, is both light- and temperature-dependent. A detailed study of the response of an *Arthrospira* strain marked M2, performed by Torzillo and Vonshak (1994), determined the optimal temperature for photosynthesis (35°C) and also measured the effect of temperature on the dark respiration rate of *Arthrospira* by following the O_2 uptake rate in the dark. A temperature-dependent exponential relationship was obtained, with the respiration rate increasing as temperature increased. The temperature-dependent dark respiration rate is given by:

$$R = 0.771.e^{(0.0636.T)}$$

where R is the respiration rate (μmol O_2 mg^{-1} chl h^{-1}) and T is the temperature (°C). At 50°C and 15°C, dark respiration rates dropped almost to zero. An Arrhenius plot for respiration showed an activation energy of 48.8 kJ mol^{-1} for *Arthrospira*. The temperature coefficient (Q_{10}) calculated for the 20 - 40°C range was 1.85. The respiration to photosynthesis ratio in *Arthrospira* was 1% at 20°C and 4.6% at 45°C (Torzillo and Vonshak 1994). These low values confirm the general assumption that cyanobacteria have low respiration rates (van Liere and Mur, 1979). The respiration-to-photosynthesis rates measured in these experiments were much lower than those reported for outdoor cultures of *Arthrospira*, where up to 34% of the biomass produced during daylight may be lost through respiration at night (Guterman et al., 1989; Torzillo et al., 1991). In the *Arthrospira* strain M2, the optimum temperature for respiratory activity was much higher than for photosynthetic activity. Nevertheless, the photosynthetic activity of the cells was more resistant to temperature extremes than was dark respiration, at the minimum and maximum temperatures tested.

c. Interaction with Light

Deviations from optimal growth temperatures have an inhibitory effect on photosynthetic capacity. This reduction in activity represents a limitation that is immediately overcome after a shift back to the optimal temperature, if no other damage was occurred. The kinetics of recovery from low temperature incubation indicate that some repair process must take place before the original level of photosynthetic activity is reached. This observation concerns solely cultures incubated at a low temperature in the light. It is thus suggested that *Arthrospira* cultures grown at less than the optimal temperature are more sensitive to photoinhibition than those grown at the optimal temperature. The latter will be able to better handle excessive light energy, since they have a higher rate of electron transport, an active repair mechanism and more efficient means of energy dissipation (Vonshak, 1997b).

3. Response to Salinity

Cyanobacteria inhabit environments characterized by drastically differing saline levels. In the past 15 years many studies have been published on the response of cyanobacteria to different saline environments. The studies mainly focus on the specific role of organic compounds as osmoregulants (Borowitzka, 1986), the modifications occurring in photosynthetic and respiratory activities (Vonshak and Richmond, 1981), and the variations in the protein synthesis pattern (Hagemann et al., 1991). Different *Arthrospira* species have been isolated from a variety of alkaline environments. We will describe the work done using strains isolated from alkaline and brackish waters.

a. Effect of Salinity on Growth

The occurrence of *Arthrospira* spp. in marine habitats has not been documented, at least not as dominant populations like those found in the natron lakes. This is most likely due more to the very low content of bicarbonate than to the high content of NaCl. *Arthrospira*, unlike marine organisms, does not require high NaCl concentration for growth, rather it only tolerates the salt. The exposure of *Arthrospira* cultures to high NaCl concentrations at first results in an immediate cessation of growth; after a lag period, a new steady state of growth is established (Vonshak et al., 1988b). The length of the time lag is exponentially correlated to the degree of salinity stress imposed on the cells. In many cases, this lag is associated with a decline in chlorophyll and biomass concentrations in the culture (Vonshak et al., 1988b). The response of *Arthrospira* to salinity in terms of the degree to which its growth is inhibited, its adaptation to salt levels and the rate of such adaptation varies widely, depending on the strain used in the study. A decrease in the growth rate due to salt stress has also been demonstrated in other cyanobacteria such as *Anacystis* (Vonshak and Richmond, 1981) and *Nostoc* (Blumwald and Tel-Or, 1982).

b. Effect of Salinity on Photosynthesis and Respiration

It has been suggested that exposure to high salinity is accompanied by a higher demand for energy by the stressed cells (Blumwald and Tel-Or, 1982). Changes in the photosynthetic and respiratory activities of an *Arthrospira* strain were measured over a period of 48h beginning 30 min. after exposure to 0.5 and 1.0 M NaCl. A marked decrease in the photosynthetic oxygen evolution rate was observed 30 min. after exposure to the salt. This decline was followed by a recovery period, characterized by a lower steady state rate of photosynthesis (Vonshak, 1988b).

The immediate inhibition of the photosynthetic and respiratory systems after exposure to salt stress have been explained by Ehrenfeld and Cousin (1984) and Reed et al. (1985), respectively in *Dunaliella* and in *Synechocystis*. They showed that a short-term increase in cellular sodium concentration was due to a transient increase in the permeability of the plasma membrane during the first seconds of exposure to high salt concentration. It has been suggested that the inhibition of photosynthesis due to the rapid entry of sodium, might be the result of the detachment of phycobilisomes from the thylakoid membranes (Blumwald et al., 1984). High rates of dark respiration in cyanobacteria due to salinity stress had been reported previously (Vonshak and Richmond, 1981; Fry et al., 1986; Molitor et al., 1986). This high activity may be associated with the increased level of maintenance energy required for pumping out the excess of sodium ions.

c. Osmoregulation and Strain-Specific Response of Arthrospira to Salinity

During the course of adaptation to salinity, an osmotic adjustment is required. In *Arthrospira*, low molecular weight carbohydrates accumulate. These have been identified as a nine carbon heteroside called glucosyl-glycerol, plus a trehalose (Martel et al., 1992). When the biomass compositions of two *Arthrospira* strains grown under different salt stress conditions were compared, significant changes, mainly an increase in carbohydrates and a decrease in the protein level, were observed. These changes correlated with the degree of stress imposed (i.e., a higher level of carbohydrates at a higher salt concentration). The difference in the level of carbohydrates accumulated by the two strains may also reflect a difference in their ability to adapt to salt stress. Changes in lipid synthesis have also been demonstrated; in salt-stressed cells, an increase in lipid content and in the degree of fatty acid saturation is observed with specific modifications in the long-chain fatty acids (C18). In *Arthrospira* strain M2 the oleic acid content doubled when the organism was grown in presence of 0.5 M NaCl (Tomaselli et al., 1993a).

C. The Alkaliphilic Nature of Arthrospira

As previously indicated, most *Arthrospira* species were isolated from alkaline and saline or brackish waters characterized by high levels of carbonate-bicarbonate and high pH levels. The physiological characteristics which enable a micro-organism to thrive in such an extreme environment are still not fully understood. It should be pointed out that *Arthrospira* not only survives at high pH values, but actually thrives in such conditions. Belkin and Boussiba (1991), who compared the growth pH optima for the two cyanobacteria *Anabaena* and *Arthrospira*, demonstrated that while the optimal pH for *Anabaena* was in the range of 6.8 - 7.2, the maximal growth rate for *Arthrospira* was obtained in the 9.5 - 9.8 range. When incubated at pH7.0, the growth rate of *Arthrospira* was severely inhibited and was only 20% of that obtained under the optimal conditions. This high pH (> 8) requirement clearly defines *Arthrospira* as an obligatory alkaliphile (Grant et al., 1990).

One of the major problems faced by cells in an alkaliphilic environment is that of regulating their internal pH. Belkin and Boussiba (1991) were the first to measure the ability of *Arthrospira* to maintain a pH gradient across its cytoplasmatic membrane. It was demonstrated that at external pH values of 10.0 and 11.5 the intracellular pH values were only 8.0 and 8.5 respectively.

Many alkaliphiles require sodium in order to survive in extreme alkaline environments (Horikoshi and Akiba, 1982). Padan et al. (1981) demonstrated that an active sodium-proton antiporter is required in order to maintain a low internal pH. In a more recent study, Schlesinger et al. (1996) demonstrated that "*Spirulina platensis*" requires sodium in order to maintain optimal growth at pH10. The authors demonstrate that under sodium deprivation the pH gradient collapses and the cells undergo fast lysis. Depriving the cells of sodium causes inactivation of the sodium-proton antiporter that is responsible for the inward pumping of protons and the outward pumping of sodium. This mechanism is one way to maintain lower internal pH levels in cells under alkaliphilic conditions. Further studies on the energy requirement and involvement of light in the regulatory process are yet to be carried out.

D. How Does Arthrospira *Compete in Culture ?*

Studies performed on different *Arthrospira* isolates from various sources demonstrate highly variable responses to different ecological factors (Tomaselli et al., 1987; Vonshak 1987b, Vonshak et al., 1996a, b). It is on the basis of such variability that strains suitable for biotechnological applications are selected, such as photo-, thermo- or osmotolerant ones (Tomaselli et al., 1993a, Vonshak and Guy, 1992; Vonshak et al 1988b). Detailed studies aimed at measuring the ecological specificity of the different species, particularly those used for commercial purposes, are scarce (Kebede and Ahlgren., 1996). Nevertheless, full knowledge of the ecological demands of the chosen species and the use of more productive selected strains is important if high productivity is to be achieved in intensive outdoor cultivation systems. The use by commercial producers of multistrain inocula which respond positively to changes in abiotic factors (due to their heterogeneity) represents another way of compensating for the present lack of well-characterized strains.

Like other components of the phytoplankton, *Arthrospira* plays a role in cycling nutritional elements and in their passage into different food chains; moreover, it provides an important direct food source for zooplankton, fishes, birds and humans. The same factors that in the original habitat regulate the occurrence of *Arthrospira* in monospecific populations are those required for both successfully maintaining a monoculture in intensive production ponds outdoors and for achieving high productivity. Factors such as high alkalinity and salinity are the key features that selectively limit the growth of other organisms and predators.

Even if high pH and salt concentration reduce contamination from both air and water, *Arthrospira* cultures may, when effective pond management is lacking, be greatly contaminated by other organisms such as heterotrophic bacteria, fungi, protozoa, rotifers, other cyanobacteria and microalgae. Contamination of outdoor *Arthrospira* cultures by other microalgae is of frequent occurrence; these contaminants are usually green algae (*Chlorella*) and diatoms. As the diatoms live tightly attached to surfaces (such as paddle wheels), they are difficult to wash out of the culture system. More problematic is the presence of other filamentous *Oscillatoria*-like cyanobacteria, which have similar ecological demands and may invade the culture. An increase in the organic load subsequent to deterioration processes (cell lysis) also stimulates the development of fungi and bacteria. Nevertheless, once the optimal conditions for the growth of *Arthrospira* are maintained in ponds, this organism successfully out-competes other occasionally occurring phototrophs for the primary resources and may then reach high population densities. This is attributable to its fast rates of growth and photoacclimation, and its high tolerance to environmental stress (irradiance, alkalinity, salinity, temperature).

VI. Concluding Remarks

Arthrospira is a cosmopolitan cyanobacterium, yet occupies very selective habitats. The main ecological demands are stable alkaline, saline waters, warm growth temperatures, high irradiance, and eutrophic conditions, (i.e. abundance of bicarbonate and other macronutrients). Consequently, primary productivity of *Arthrospira* in some tropical and subtropical lakes can be very high.

The use of *Arthrospira* as food by indigenous populations in different parts of the world is well documented. In recent years, this cyanobacterium has gained a considerable reputation as a health food additive and is marketed commercially. In many cases the term "Spirulina" has become a practical synonym for cultivated microalgae. At present, *Arthrospira* represents the second most important commercial microalga for the production of biomass as a health food and an animal feed (after *Chlorella*). With the current rapid increase in production and a demand reaching 1500 t in 1996 (predicted to reach 2000 t by the beginning of the new millennium), there is no doubt that *Arthrospira* will be the most important mass-produced microalga for commercial purposes in the near future.

The accumulated knowledge of the ecology, physiology and biochemistry of *Arthrospira* has made such results feasible, yet much more work is required, in terms of strain selection and better understanding of the factors which govern growth and light utilization in dense cultures, to support this growing agro-industry.

References

Anagnostidis K and Komárek J (1988) Modern approach to the classification system of cyanophytes. 3. Oscillatoriales. Arch Hydrobiol, Suppl 80, Algological Studies 50/53: 327-329

Beadle LC (1974) The Inland Waters of Tropical Africa. An Introduction to Tropical Limnology. Longman, London

Belkin S and Boussiba S (1991) Resistance of *Spirulina platensis* to ammonia at high pH values. Plant Cell Physiol 32: 953-958

Blumwald E, Mehlhorn RJ and Packer L (1984) Salt adaptation mechanisms in the cyanobacterium *Synechococcus* 6311. In: Sybesma S (ed.) Advances in Photosynthesis Research, pp 627-630, Martinus Nijhoff /Dr W Junk Publs, The Hague

Blumwald E and Tel-Or E (1982) Osmoregulation and cell composition in salt adaptation of *Nostoc muscorum*. Arch Microbiol 132: 168-172

Borowitzka LJ (1986) Osmoregulation in blue-green algae. In: Round FE and Chapman D (eds) Progress in Phycological Research, Vol. 4, pp. 243-256, Biopress Ltd, Bristol

Brandily MY (1959) Depuis des lustres une tribu primitive du Tchad exploite la nourriture de l'an 2000. Sci Avenir 152: 516-519

Brieba C and Merinos C (1993) Some advances in the *Spirulina* culture in seawater. In: Progress in Biotechnology of Photoautotrophic Microorganisms. Abstr 6th ICAA, p 93, Ceské Budejovice, Czech Republic

Busson F (1971) *Spirulina platensis* (Gom.) Geitler et *Spirulina geitleri* J. de Toni, Cyanophycées Alimentaires. Service de Santé, Marseille

Carr NG and Whitton BA (eds) (1973) The Biology of Blue-Green Algae. Blackwell Scientific, Oxford

Carr NG and Whitton BA (1982) The Biology of Cyanobacteria. Blackwell Scientific, Oxford

Castenholz, RW (1989) Subsection III, Order Oscillatoriales. In: Staley JT, Bryant MP, Pfenning N and Holt JG (eds) Bergey's Manual of Systematic Bacteriology, Vol. 3, pp 1771-1780, Williams and Wilkins Co, Baltimore

Ciferri O (1983) *Spirulina*, the edible micro-organism. Microbiol Rev 47: 551-578

Clément G (1968) A new type of food algae. In: Mateles RJ and Tannenbaum SR (eds) Single-cell Protein. MIT Press, Cambridge, Mass

Cohen Z and Vonshak A (1991) Fatty acid composition of *Spirulina* and *Spirulina*-like strains in relation to their chemotaxonomy. Phytochemistry 30: 205-208

Cornet JF, Dussap CG, Cluzel P and Dubertret G (1992b) A structured model for simulation of cultures of the cyanobacterium *Spirulina platensis* in photobioreactors: II. Identification of kinetic parameters under light and mineral limitations. Biotechnol Bioengin 40: 826-835

Cornet JF, Dussap CG and Dubertret G (1992a) A structured model for simulation of cultures of the cyanobacterium *Spirulina platensis* in photobioreactors: I. Coupling between light transfer and growth kinetics. Biotechnol Bioengin 40, 817-827

Dangeard P (1940) Sur une algue bleue alimentaire pour l'homme: *Arthrospira platensis* (Nordst.) Gomont. Actes Soc Linn Bordeaux, 91:39-41

De Toni GB (1935) Noterelle di nomenclatura algologica 3: 1-2

Desikachary TV (1959) Cyanophyta. Indian Council for Agricultural Research, pp 187-198, New Delhi.

Desikachary TV and Jeeji Bai N (1992) Taxonomic studies in *Spirulina* . In: Seshradi CV and Jeeji Bai N (Eds) *Spirulina* ETTA National Symposium, pp 12-21, MCRC, Madras

Desikachary TV and Jeeji Bai N (1996) Taxonomic studies in *Spirulina* II. The identification of *Arthrospira* ("*Spirulina*") strains and natural samples of different geographical origins. Arch Hydrobiol, Suppl 116, Algological Studies 83: 163-168

Ehrenfeld J and Cousin JL (1984) Ionic regulation of the unicellular green alga *Dunaliella tertiolecta*: response to hypertonic shock. J Membr Biol 77: 45-55

El-Bestawy E, Bellinger EG and Sigee DC (1996) Elemental composition of phytoplankton in a subtropical lake: X-ray microanalytical studies on the dominant algae *Spirulina platensis* (Cyanophyta) and *Cyclotella meneghiniana* (Bacillariophyceae) Eur J Phycol 31: 157-166

Fogg GE, Stewart W D P, Fay P and Walsby A E (1973) The Blue-Green Algae. Academic Press, London

Fott B (1977) Taxonomy of planktonic Spirulinas in saline tropical lakes. Schweiz Z Hydrobiol 39:139-140

Fott B and Karim GAA (1973) *Spirulina* plankton community in a Lake in Jebel Marra, Sudan. Arch Protistenk 115: 408-418

Fox R D (1996) *Spirulina* Production and Potential. Edisud, Aix-en-Provence, France

Fry IV, Huflejt M, Erber WWA, Peschek GA and Packer L (1986) The role of respiration during adaptation of the freshwater cyanobacterium *Synechococcus* 6311 to salinity. Arch Biochem Biophys 244: 686-690

Gallegos AJ (1993) The past, present and future of algae in Mexico. Bull Inst océanograph, Monaco, 12: 133-139

Gardner NL (1917) New Pacific coast marine algae I. Univ Calif Publ Bot 6: 377-418

Geitler L (1932) Cyanophyceae. In: Rabenhorst's Kryptogamenflora von Deutschland, Österreich und der Schweiz, Vol. 14, pp 916- 931, Akad Verlag, Leipzig

Giovannoni SJ, Turner S, Olsen GJ, Barns S, Lane DJ and Pace NR (1988) Evolutionary relationships among cyanobacteria and green chloroplasts. J Bacteriol 170: 3584-3592

Gomont M (1892) Monographie des Oscillariées. Ann Sci Nat Bot, Sér 7, 15: 263-368; 16: 91-264

Grant WD, Mwatha WE and Jones BE (1990) Alkaliphiles: ecology, diversity and applications. FEMS Microbiol Rev 75: 255-270

Grant WD and Tindall BJ (1986) The alkaline saline environment. In: Herbert RA and Codd GA (eds) Microbes in Extreme Environments, pp 22-54, Academic Press, London

Grobbelaar JU, Nedbal L and Ticghy V (1996) Influence of high frequency light/dark fluctuation on photosynthetic characteristics of microalgae photoacclimated to different light intensities and implication for mass algal cultivation. J Appl Phycol 8: 335-342

Guglielmi G and Cohen-Bazire G (1982) Structure et distribution des pores et des perforations de l'enveloppe de peptidoglycane chez quelques cyanobactéries. Protistologica 18: 151-165

Guterman H, Vonshak A and Ben-Yaakov S (1989) Automatic on-line growth estimation method for outdoor algal biomass production. Biotechnol Bioengin 34: 143-152

Hagemann M, Techel D and Rensing L (1991) Comparison of salt- and heat-induced alterations of protein synthesis in the cyanobacterium *Synechocystis* sp. PCC 6803. Arch Microbiol 155: 587-592

Hindák F (1985) Morphology of trichomes in *Spirulina fusiformis* Voronichin from Lake Bogoria, Kenya. Arch Hydrobiol, Suppl 71, Algological Studies 38/39: 209-218

Horikoshi K and Aiba T (1982) Alkalophilic microorganisms. Springer-Verlag, Berlin

Iehana M (1987) Kinetic analysis of the growth of *Spirulina* sp. in batch culture. J Ferment Technol 65: 267-272

Iltis A (1968) Tolérance de salinité de *Spirulina platensis* (Gom.) Geitler, (*Cyanophyta*) dans les mares natronées du Kanem (Tchad). Cah ORSTOM, sér Hydrobiol, 2: 119-125

Iltis A (1969a) Phytoplancton des eaux natronées du Kanem (Tchad). I. Les lacs permanents a Spirulines. Cah ORSTOM, sér Hydrobiol, 3: 29-44

Iltis A (1969b) Phytoplancton des eaux natronées du Kanem (Tchad). II. Les mares temporaires. Cah ORSTOM, sér. Hydrobiol, 3: 3-19

Iltis A (1970a) Phytoplancton des eaux natronées du Kanem (Tchad). III. Variations annuelles du plancton d'une mare temporaire. Cah ORSTOM, sér. Hydrobiol, 4: 53-59

Iltis A (1970b) Phytoplancton des eaux natronées du Kanem (Tchad). IV. Note sur les espèces du genre *Oscillatoria*, sous-genre *Spirulina* (Cyanophyta). Cah ORSTOM, sér. Hydrobiol, 4: 129-134

Iltis A (1971a) Note sur *Oscillatoria* (sous-genre *Spirulina*) *platensis* (Nordst.) Bourrelly (Cyanophyta) au Tchad. Cah ORSTOM, sér. Hydrobiol, 5: 53-72

Iltis A (1971b) Phytoplancton des eaux natronées du Kanem (Tchad). V. Les lacs mésohalins. Cah ORSTOM, sér Hydrobiol, 5: 73-84

Iltis A (1972) Algues des eaux natronées du Kanem (Tchad). I. Première partie. Cah ORSTOM, sér Hydrobiol, 4: 129-134

Iltis A and Riou-Duwat S (1971c) Variations saisonnières du peuplement en rotifères des eaux natronées du Kanem (Tchad). Cah ORSTOM, sér Hydrobiol, 5: 101-112

Jeeji Bai N (1985) Competitive exclusion or morphological transformation? A case study with *Spirulina fusiformis*. Arch Hydrobiol, Suppl 71, Algological Studies 38/39: 191-199

Jeeji Bai N and Seshadri CV (1980) On coiling and uncoiling of trichomes in the genus *Spirulina*. Arch Hydrobiol, Suppl 60, Algological Studies 26: 32-47

Kaplan A (1981) Photoinhibition in *Spirulina platensis*: response of photosynthesis and HCO_3^- uptake capability to CO_2^- depleted conditions. J Exper Bot 32: 669-673

Kebede E and Ahlgren G (1996) Optimum growth conditions and light utilisation efficiency of *Spirulina platensis* (equal *Arthrospira fusiformis*) (Cyanophyta) from Lake Chitu, Ethiopia. Hydrobiologia 332: 99-109

Komárek J. and Lund JWG (1990) What is "*Spirulina platensis*" in fact? Arch Hydrobiol, Suppl 85, Algological Studies 58: 1-6

Kyle DJ and Ohad I (1986) The mechanism of photoinhibition in higher plants and green algae. In: Staehelin LA and Arntzen CJ (eds) Encyclopedia of Plant Physiology, New Series Vol. 19 Photosynthesis III, pp 468-475, Springer-Verlag, Berlin

Lee HY, Erickson LE and Yang SS (1987) Kinetics and bioenergetics of light-limited photoautotrophic growth of *Spirulina platensis*. Biotechnol Bioengin 29: 832-838

Léonard J (1966) The 1964-65 Belgian trans-saharan expedition. Nature, London, 209: 126-128

Léonard J and Compère P (1967) *Spirulina platensis* (Gom.) Geitler, algue bleue de grande valeur alimentaire par sa richesse en protéines. Bull Jard bot Nat Belg 37: 1-23

Lewin RA (1980) Uncoiled variants of *Spirulina platensis* (Cyanophyceae: Oscillatoriaceae). Arch Hydrobiol, Suppl 60, Algological Studies 26: 48-52

Maglione G. (1969) Premières données sur le regime hydrogéochimique des lacs permanents du Kanem (Tchad). Cah ORSTOM, sér Hydrobiol, 3: 121-141

Margheri MC, Tomaselli Feroci L e Materassi R (1975) Contributo sperimentale alla sistematica del genere *Spirulina* Turpin. Atti XVII Congr Naz Soc. Ital Microbiol, Vol. II, pp 891-901, Padova

Martel A, Yu S, Garcia-Reina G, Lindblad P and Pedersén M (1992) Osmotic-adjustment in the cyanobacterium *Spirulina platensis*: Presence of an α-glucosidase. Plant Physiol Biochem 30: 573-578

Melack JM (1979) Photosynthesis and growth of *Spirulina platensis* (Cyanophyta) in an equatorial lake (Lake Simbi, Kenya). Limnol Oceanogr 24: 753-760

Min Thein (1993) Production of Spirulina in Myanmar. Bull Inst océanograph, Monaco, 12: 175-178

Molitor V, Erber W and Peschek GA (1986) Increased levels of cytochrome oxidase and sodium proton antiporter in the plasma membrane of *Anacystis nidulans* after growth in sodium enriched media. FEBS Lett 204: 251-254

Nelissen B, Wilmotte A, Neefs J-M and De Wachter R (1994) Phylogenetic relationship among filamentous helical cyanobacteria investigated on the basis of 16S ribosomal RNA gene sequence analysis. System Appl Microbiol 17: 206-210

Ogawa T and Aiba S (1978) CO_2 assimilation and growth of a blue-green alga, *Spirulina platensis*, in continuous culture. J. Appl. Chem. Biotechnol. 28: 5151-5157

Ogawa T and Terui G (1970) Studies on the growth of *Spirulina platensis* (I) On the pure culture of *Spirulina platensis*. J Ferment Technol 48: 361-364

Padan E, Zilberstein D and Schuldiner S (1981) pH Homeostasis in bacteria. Biochem Biophys Acta 650: 151-166

Powles SB (1984) Photoinhibition of photosynthesis induced by visible light. Annu Rev Plant Physiol 35: 15-37

Reed RH, Richardson DL and Stewart WDP (1985) Na^+ uptake and extrusion in the cyanobacterium *Synechocystis* PCC 6714 in response to hyper-saline treatment. Evidence for transient changes in plasmalemma Na^+ permeability. Biochim Biophys Acta 814: 347-353

Rich F (1931) Notes on *Arthrospira platensis*. Rev Algol 6: 75-82

Richmond A (1988) *Spirulina*. In: Borowitzka MA and Borowitzka LY (eds) Micro-algal Biotechnology, pp 85-101, Cambridge, Cambridge Univ Press, UK

Richmond A (1996) Efficient utilization of high irradiance for production of photoautotrophic cell mass: a survey. J Appl Phycol 8: 381-386

Rippka R, Deruelles J, Waterbury JB, Herdman M and Stanier RY (1979) Generic assignments, strain histories and properties of pure cultures of cyanobacteria. J Gen Microbiol 111: 1-61

Rippka R and Herdman M (1992) Catalogue of strains. Pasteur Culture Collection of cyanobacterial strains in axenic culture, 103 pp. Institut Pasteur, Paris

Scheldeman P, Baurain D, Bouhy R, Scott M, Mühling M, Whitton BA, Belay A, Wilmotte A (1999) *Arthrospira* ('*Spirulina*') strains from four continents are resolved into two only two clusters, based on ARDRA (Amplified Ribosomal DNA Restriction Analysis) of the ITS (Internally Transcribed Spacer). FEMS Microbiol Lett 172: 213-222

Schlesinger P, Belkin S and Boussiba S (1996) Sodium deprivation under alkaline conditions causes rapid death of the filamentous cyanobacterium *Spirulina platensis*. J Phycol 32: 608-613

Shimada A, Oguchi M, Otsubo K, Nitta K, Koyano T and Miki K (1989) Application of tubular photo-bioreactor system to culture *Spirulina* for food production and gas exchange. In: Miyachi S, Karube I and Ishida Y (eds) Current Topics in Marine Biotechnology, pp 147-161, Jap Soc Mar Biotechnol, Tokyo

Stizenberger E. (1852) *Spirulina* und *Arthrospira* (nov. gen.). Hedwigia 1: 32-41

Tomaselli L (1997) Systematic and Ecology. In: Vonshak A (ed.) *Spirulina platensis* (*Arthrospira*): Physiology, Cell-biology and Biotechnology. pp 1-19, Taylor & Francis Ltd, London, UK

Tomaselli L, Boldrini G and Margheri MC (1997) Physiological behaviour of *Arthrospira* (*Spirulina*) *maxima* during acclimation to changes in irradiance. J Appl Phycol 9: 37-43

Tomaselli Feroci L, Margheri MC and Pelosi E (1976) Die Ultrastruktur von *Spirulina* im Vergleich zu *Oscillatoria*. Zbl Bakt Abt II, 131: 592-599

Tomaselli L, Giovannetti L and Margheri MC (1981) On the mechanism of trichome breakage in *Spirulina platensis* and *S. maxima*. Ann Microbiol 31: 27-31

Tomaselli L, Giovannetti L, Sacchi A and Bocci F (1988) Effects of temperature on growth and biochemical composition in Spirulina platensis strain M2. In: Staedler T, Mollion J, Verdus MC, Karamanos Y, Morvan H, Christiaen D (eds) Algal Biotechnology, pp 305-314, Elsevier Applied Sciences, London and New York

Tomaselli L, Giovannetti L and Torzillo G (1993a) Physiology of stress response in *Spirulina* spp. Bull Inst océanograph, Monaco, 12: 65- 75

Tomaselli L, Margheri MC and Sacchi A (1995) Effects of light on photosynthetic activity in a phycoerythrin-rich strain of *Spirulina subsalsa*. Aquat microb Ecol 9: 27-31

Tomaselli L, Palandri MR and Tani G (1993b) Advances in preparative techniques for observation of the fine structure of *Arthrospira maxima* Setch. et Gardner (syn. *Spirulina maxima* Geitler). Arch Hydrobiol, Suppl 100, Algological Studies 71: 43-50

Tomaselli L, Palandri MR and Tredici MR (1996) On the correct use of the *Spirulina* designation. Arch Hydrobiol, Supp 116, Algological Studies 83: 539-548

Tomaselli L, Torzillo G, Giovannetti L, Pushparaj B, Bocci F, Tredici M, Papuzzo T, Balloni W and Materassi R (1987) Recent research on *Spirulina* in Italy. Hydrobiologia 151/152: 79-82

Torzillo G, Sacchi A, Materassi R and Richmond A (1991) Effect of temperature on yield and night biomass loss in *Spirulina platensis* grown outdoors in tubular photobioreactors. J Appl Phycol 3: 103-108

Torzillo G and Vonshak A (1994) Effect of light and temperature on the photosynthetic activity of the cyanobacterium *Spirulina platensis*. Biomass Bioeng 6: 399-404

Tredici MR, Papuzzo T and Tomaselli L (1986) Outdoor mass culture of *Spirulina maxima* in sea-water. Appl Microbiol Biotechnol 24: 42-50

Tsen CT and Chang TP (1990) Ultrastrukturen von vier «Spirulina»-Arten (Cyanophyceae). Arch Hydrobiol, Suppl 87, Algological Studies 60: 33-41

Tuite CH (1981) Standing crop densities and distribution of *Spirulina* and bentic diatoms in East African alkaline saline lakes. Freshwater Biol 11: 345- 360

Turpin PJF (1829) *Spirulina oscillarioide.* In: Dictionnaire des Sciences Naturelles, Vol 50, pp 309-310, De Lévrault, Paris

Van Eykelenburg C (1977) On the morphology and ultrastructure of the cell wall of *Spirulina platensis.* A Leeuwenhoek J Microbiol 43: 89-94

Van Eykelenburg C (1979) The ultrastructure of *Spirulina platensis* in relation to temperature and light intensity. A Leeuwenhoek J Microbiol 45: 369-375

Van Eykelenburg C and Fuchs A (1980) Rapid reversible macromorphological changes in *Spirulina platensis.* Naturwissenschaften 67: 200-204

Van Liere L and Mur L (1979) Growth kinetics of *Oscillatoria agardhii* Gomont in continuous culture limited in its growth by the light energy supply. J Gen Microbiol 115: 153-159

Vareschi E (1982) The ecology of lake Nakuru (Kenya). III: Abiotic factors and primary production. Oecologia 55: 81-101

Viti C, Ventura S, Lotti F, Capolino E, Tomaselli L and Giovannetti L (1997) Genotypic diversity and typing of cyanobacterial strains of the genus *Arthrospira* by very sensitive total DNA restriction profile analysis. Res Microbiol 148: 605-611

Vonshak A (1987a) Biological limitations in developing the biotechnology for algal mass cultivation. Science de L'eau. (J. of Water Sci.) 6: 99-103

Vonshak A (1987b) Strain selection of *Spirulina* suitable for mass production. Hydrobiologia 151/152: 79-82

Vonshak A (ed.) (1997a) *Spirulina platensis (Arthrospira)*: Physiology, Cell-Biology and Biotechnology. Taylor & Francis Ltd, London, UK

Vonshak A (1997b) Spirulina: growth, physiology and biochemistry. In: Vonshak A (ed.) *Spirulina platensis (Arthrospira)*: Physiology, Cell-Biology and Biotechnology, pp 43-65, Taylor & Francis Ltd, London, UK

Vonshak A, Abeliovich A, Boussiba S, Arad S and Richmond A (1982) Production of *Spirulina* biomass: effects of environmental factors and population density. Biomass 2: 175-185

Vonshak A, Boussiba S, Abeliovich A and Richmond A (1983) Production of *Spirulina* biomass: maintenance of pure culture outdoors. Biotechnol Bioeng 25: 341-349

Vonshak A, Chanawongse L, Bunnang B and Tanticharoen M (1996a) Light acclimation and photoinhibition in three *Spirulina platensis* (cyanobacteria) isolates. J Appl Phycol 8: 35-40

Vonshak A and Guy R (1992) Photoadaptation, photoinhibition and productivity in outdoor grown *Spirulina platensis* strains. Plant Cell Environm 15: 613-616

Vonshak A, Guy R and Guy M (1988a) The response of the filamentous cyanobacterium *Spirulina platensis* to salt stress. Arch Microbiol 150: 417-420

Vonshak A, Guy R, Poplawsky R and Ohad I (1988b) Photoinhibition and its recovery in two different strains of *Spirulina*. Plant Cell Physiol. 29: 721-726

Vonshak A, Kancharaksa N, Bunnang B and Tanticharoen M (1996b) Role of light and photosynthesis on the acclimation process of the cyanobacterium *Spirulina platensis* to salinity stress. J Appl Phycol 8: 119-124

Vonshak A and Richmond A (1981) Photosynthetic and respiratory activity in *Anacystis nidulans* adapted to osmotic stress. Plant Physiol 68: 504-509

Vonshak A and Richmond A (1988) Mass production of *Spirulina* - an overview. Biomass 15: 233-241

Voronichin NN (1934) Biologie des bassins mineral des steppes Koulondine. Expedition de l'Academie des Sciences d'URSS 1931-1933 I. Leningrad. Trav Soviet pour serv l'étude ressour natur, Sér Siberie 8: 182-183

Wayne LG, Brenner DJ, Colwell RR, Grimont PAD, Kandler O, Krichevsky MI, Moore LH, Moore WEC, Murray RGE, Stackebrandt E, Starr MP & Trüper HG (1987) Report of the *ad hoc* committee on reconciliation of approaches to bacterial systematics. Int J System Bacteriol 37: 463-464

Wittrock and Nordstedt (1844) Algae aquae dulcia exiccata, fascicule XIV n° 679. In: Descriptiones systematice dispositae, p 59

Zarrouk C (1966) Contribution à l'étude d'une Cyanophycée. Influence de divers facteurs physiques et chimiques sur la croissance et la photosynthèse de *Spirulina maxima* (Setchell et Gardner) Geitler. Ph D Thesis, University of Paris, France

Chapter 19

Symbiotic Interactions

David G. Adams
Department of Microbiology, University of Leeds, Leeds LS2 9JT, UK

Summary

Cyanobacteria are unique in their capacity to form symbiotic associations with a remarkable range of eukaryotic hosts including plants, fungi, sponges and protists. In most cases the benefit to the host is clear, and is invariably metabolic provision, mostly in the form of combined nitrogen, or occasionally carbon. Being photoautotrophs, and in many cases facultative heterotrophs and nitrogen fixers, cyanobacteria can provide non-photosynthetic hosts with both nitrogen and carbon. The benefit to the cyanobacteria is less clear. Certainly, they often receive

B. A. Whitton and M. Potts (eds), The Ecology of Cyanobacteria, 523-561.
© 2000 *Kluwer Academic Publishers. Printed in the Netherlands.*

carbon from photosynthetic hosts, but they are capable of carbon fixation themselves. The enclosed environment provided by the host may protect the cyanobacteria from environmental extremes such as high light intensity and desiccation, and possibly from predation.

Many cyanobacterial symbionts have characteristics in common; they are filamentous, fix nitrogen in specialised cells known as heterocysts, and they develop motile filaments known as hormogonia that serve as the infective agents in many of the symbioses. The best studied cyanobacterial associations are those with plants, and these offer great potential as experimental systems. Unlike the symbionts in rhizobial and actinorhizal symbioses, cyanobacteria are not restricted to the roots of plants, but can infect thalli, stems and leaves. Some plants enhance their chances of infection by producing chemical signals that induce hormogonia formation and serve as chemoattractants to guide the hormogonia into the plant tissue. Within the host additional plant signals repress further hormogonia formation and stimulate heterocyst development. The cyanobacteria then undergo considerable morphological and physiological changes, again in response to plant signals, usually resulting in reduced growth rate and carbon dioxide fixation, enhanced nitrogen fixation and loss, to the plant, of much of the nitrogen fixed. With recent advances in molecular genetic techniques for symbiotically competent cyanobacteria, the molecular basis of these changes is now being elucidated, and the next few years promise a rapid, and exciting expansion of our understanding of these associations.

I. Introduction

Cyanobacteria form symbiotic associations with a wide range of eukaryotic hosts, and in the majority of cases the cyanobacterial symbionts (cyanobionts) are capable of nitrogen fixation, the products of which become available to the host. As a consequence of their oxygenic photosynthesis cyanobacteria have evolved several mechanisms for protecting nitrogenase from oxygen inactivation (Fay, 1992; Gallon, 1992). The most remarkable of these is the heterocyst (Adams, 1992, 1997; Wolk et al., 1994; Wolk, 1996), which develops at regular intervals within filaments of vegetative cells, and provides an anaerobic environment suitable for nitrogen fixation (Fay, 1992; Gallon, 1992). The specialised nature of the heterocyst means that it must rely on neighbouring vegetative cells for the provision of fixed carbon, while in turn supplying these cells with fixed nitrogen (Wolk et al., 1994). This interdependence provides an interesting parallel with the cyanobacteria-plant symbioses, in which the cyanobiont provides fixed nitrogen to the host, which provides fixed carbon in return.

The infective agents in many cyanobacterial symbioses, particularly those involving plants, are specialised filaments known as hormogonia, produced by heterocystous cyanobacteria in the Nostocaceae and Stigonemataceae. Hormogonia lack heterocysts but possess gliding motility (Tandeau de Marsac, 1994; Adams, 1997), a crucial factor in their ability to establish symbioses, because the long, immotile parent filaments are unable to gain entry to the plant structures that house the symbiotic colonies (Section IV.A). Hormogonia formation and hormogonia themselves are responsive to plant extracellular signals (Section IV.B), and it may be for these reasons that plant cyanobionts are almost always members of the genus *Nostoc* (Section III).

In most of the symbioses described in this review the symbionts are recognisible as cyanobacteria and are capable of independent growth, although in some cases they have defied laboratory cultivation. However, there are two exceptions: the prochlorophytes and the cyanelles. The former are a group of oxygenic photosynthetic prokaryotes that closely resemble cyanobacteria, but differ in the possession of both chlorophylls *a* and *b* and the lack of phycobilin pigments (Section II.E). A description of one prochlorophyte, *Prochloron*, is included because it forms a symbiosis with ascidians. The second exception is the cyanelle, which is one of a range of plastids found in the cells of photosynthetic eukaryotes, and was once thought to be a cyanobacterial endosymbiont (Section II.F). Cyanelles indeed possess many similarities with cyanobacteria, yet with genomes of only 130 to 140 kbp, they are incapable of independent growth. A brief description of cyanelles is included here because they almost certainly evolved from a cyanobacterial ancestor, and represent the extreme of adaptation to a symbiotic lifestyle in which the symbiont has lost much of its own genetic capacity and become dependent on the host.

This chapter concentrates on the literature from 1985 onwards, work prior to this being well covered in the many review articles and books referred to in the text. Despite the wide-ranging nature of cyanobacterial symbioses, most research has focussed on the experimentally amenable plant associations, and the literature reflects this. However, this review will also address recent advances in knowledge of the many other cyanobacterial associations and their environmental significance.

II. The Symbioses and Their Environmental Impact

A. Plants

1. Loose Associations

Loose associations between plants and cyanobacteria are likely to be widespread. For example, it has long been known that the presence of both free-living (including epiphytic) and symbiotic (*Azolla*-associated; Section II.A.5) cyanobacteria in rice fields improves both growth and crop yield of the rice (see Chapter 8). Epiphytic growth of *Gloeotrichia pisum* on submerged roots and stems of deepwater rice has also been observed (Whitton et al., 1988: Plate 15c, d) and is associated with nitrogen-fixing activity (Rother et al., 1988). Similarly, colonies of the heterocystous genera *Nostoc*, *Gloeotrichia*, *Anabaena*, *Calothrix* and *Cylindrospermum* have been found growing epiphytically on the lower epidermis and the reproductive pockets of duckweed (*Lemna*) leaves, and are presumably responsible for the nitrogen-fixing activity associated with duckweed blooms (Duong and Tiedje, 1985).

Although knowledge of the effects of cyanobacteria on crop plants other than rice is very limited, the importance of wheat as a temperate food crop has prompted investigations in this area, and has provided evidence that cyanobacteria associated with wheat roots can stimulate plant growth. For example, an ammonia-excreting mutant of *Anabaena variabilis* has been shown to enhance growth of wheat when the cyanobacterium is directly associated with the plant roots (Spiller and Gunasekaran, 1990; Spiller et al., 1993). Similarly, the roots of hydroponically-grown wheat seedlings can become colonized by certain N_2-fixing cyanobacteria (Fig. 1; Gantar et al., 1991a, 1991b), stimulating root length and enhancing the nitrogen content of the plants (Obreht et al., 1993). Indeed, $^{15}N/^{14}N$ fractionation experiments with sand-grown wheat plants have shown that some plant nitrogen is acquired via cyanobacterial nitrogen fixation (Gantar et al., 1995b). This effect is dependent on the specific combination of plant cultivar and cyanobacterial strain, but close association of the two is not essential. The nitrogenase activity of some of the cyanobacteria can be sustained in the dark when they are associated with the roots of wheat plants, presumably supported by carbon compounds present in root exudates (Gantar et al., 1995b).

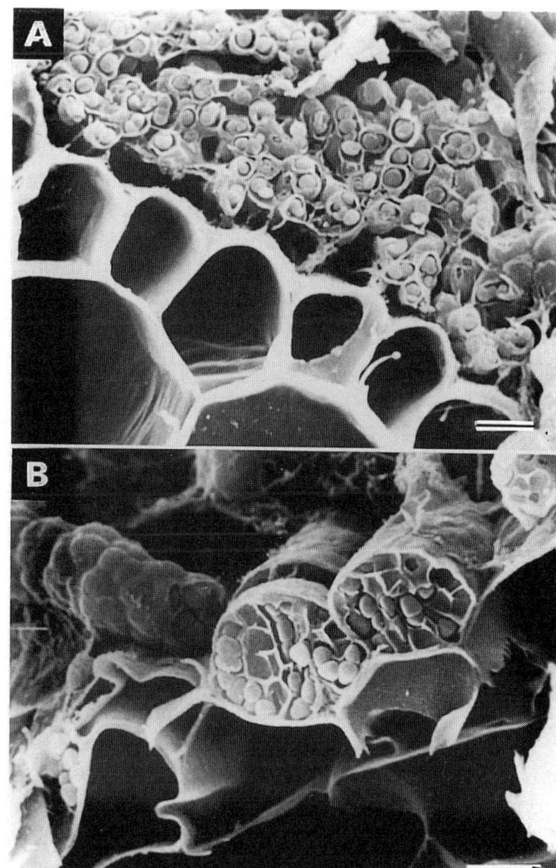

Fig. 1. Scanning electron micrographs of wheat roots. (A) Thick layer of cyanobacteria on the root surface. (B) Packages of cyanobacterial filaments tightly associated with the root surface after removal of loosely associated cyanobacteria by washing. Bars = 10 μm. (Reproduced with permission from Gantar et al., New Phytol. 118: 485--492, 1991, © Cambridge University Press).

2. Bryophytes (Mosses, Hornworts and Liverworts)

Among bryophytes the Hepaticae (liverworts), Anthocerotae (hornworts), and Musci (mosses), all show examples of epiphytic or endophytic associations with cyanobacteria, primarily *Nostoc*. Associations with mosses are mostly epiphytic (Meeks, 1990), the exceptions being species of *Sphagnum* in which the *Nostoc* occupies hyaline cells, the function of which is unknown. *Hapalosiphon* sometimes also grows closely associated with *Sphagnum*, but critical studies on how far this can be endophytic appear to be lacking. It has been suggested that buffering within hyaline cells helps *Nostoc* to grow at pH values < 5 (Meeks, 1990; Peters, 1991). However species of the N_2-fixing genera *Hapalosiphon* and *Tolypothrix* are sometimes frequent in pools at pH values as low as 4.1 (B.A. Whitton, pers. comm.).

These endophytic associations are largely limited to plants growing above the water line and have nitrogen fixation rates below those of epiphytic associations, which are usually confined to submerged plants. There is some evidence for transfer to the *Sphagnum* of nitrogen fixed by the cyanobacteria, and for the elevation of nitrogen fixation rates of epiphytically associated *Nostoc* spp. above the same free-living strains (Meeks, 1990). The nitrogen-fixing epiphytic and endophytic cyanobacterial associations with mosses such as *Sphagnum* and *Drepanocladus* may contribute to the nitrogen economy of bog and mire habitats, particularly in subarctic regions (Meeks, 1990). Epiphytic associations also occur with grassland and forest mosses such as *Ceratodon* and *Funaria*, although the rates of nitrogen fixation are much lower than those of *Sphagnum* associations.

Cyanobacterial associations with liverworts are very rare, with only four of over 340 genera known to be involved, two forming epiphytic and two endophytic associations. Four of the six hornwort genera form endophytic associations. The epiphytic associations are poorly understood, although they may be more common than once thought (Dalton and Chatfield, 1985; Brasell et al., 1986). The endophytic associations have been studied in much more detail because of their adaptability to laboratory experimentation. Hornworts such as *Anthoceros* and *Phaeoceros*, and liverworts such *Blasia*, can be grown axenically in shaken liquid culture, with or without their symbiotic cyanobacterial partners (Enderlin and Meeks, 1983; Meeks, 1988; Kimura and Nakano, 1990; Meeks, 1990). The symbiont-free plants can be readily re-infected with axenic cyanobacteria (Plate 29e), making this an excellent model system for the study of cyanobacteria-plant symbiosis. In their natural habitat these liverworts and hornworts grow as a prostrate gametophyte thallus a few centimeters in length, being attached to the substratum by rhizoids. Symbiotic colonies are visible as dark spots approximately 0.5 mm in diameter within the plant tissue (d 28D; Section V).

3. Gymnosperms (Cycads)

The cycads are the most primitive group of seed plants (gymnosperms) alive today, and comprise approximately 150 species found in tropical and subtropical areas of the world, almost exclusively in the southern hemisphere (Peters, 1991; Lindblad and Bergman, 1990; Peters et al., 1986). They are palm-like in appearance with a trunk ranging in height from a few centimetres to 20 metres (Plate 28b), and a conspicuous tap root from which may develop three additional root types: lateral, coralloid and apogeotropic (Ahern and Staff, 1994). From the standpoint of symbiosis the most important of these is the coralloid roots (so-called because of their resemblance to coral; Fig. 2A) because these are frequently infected with nitrogen-fixing cyanobacteria (Peters et al., 1986; Lindblad and Bergman, 1990; Peters, 1990; Ahern and Staff, 1994), forming a dark band between the inner and outer cortex of the root (Fig. 2B; Section V). In their limited natural habitats, such as the *Eucalyptus* forests of Australia, the cycads can make significant contributions to the local nitrogen economy (Lindblad and Bergman, 1990).

4. Angiosperms (Gunnera)

Of all the angiosperms (flowering plants) only the genus *Gunnera* forms symbiotic associations with cyanobacteria. Consisting of approximately 50 species, all of which can become infected with cyanobacteria, the genus is widely distributed in the tropical/subtropical regions of the southern hemisphere where they prefer wet areas, often at altitude (Peters et al., 1986; Bonnett, 1990; Bergman et al., 1992b). The plants vary in size from small creeping forms with leaves 1-10 centimetres across, to very large rhizomatous plants with rhubarb-like leaves up to two meters across (Plate 28a; Peters et al., 1986; Bonnett, 1990; Bergman et al., 1992b; Osborne et al., 1992). *Gunnera* is the only one of the

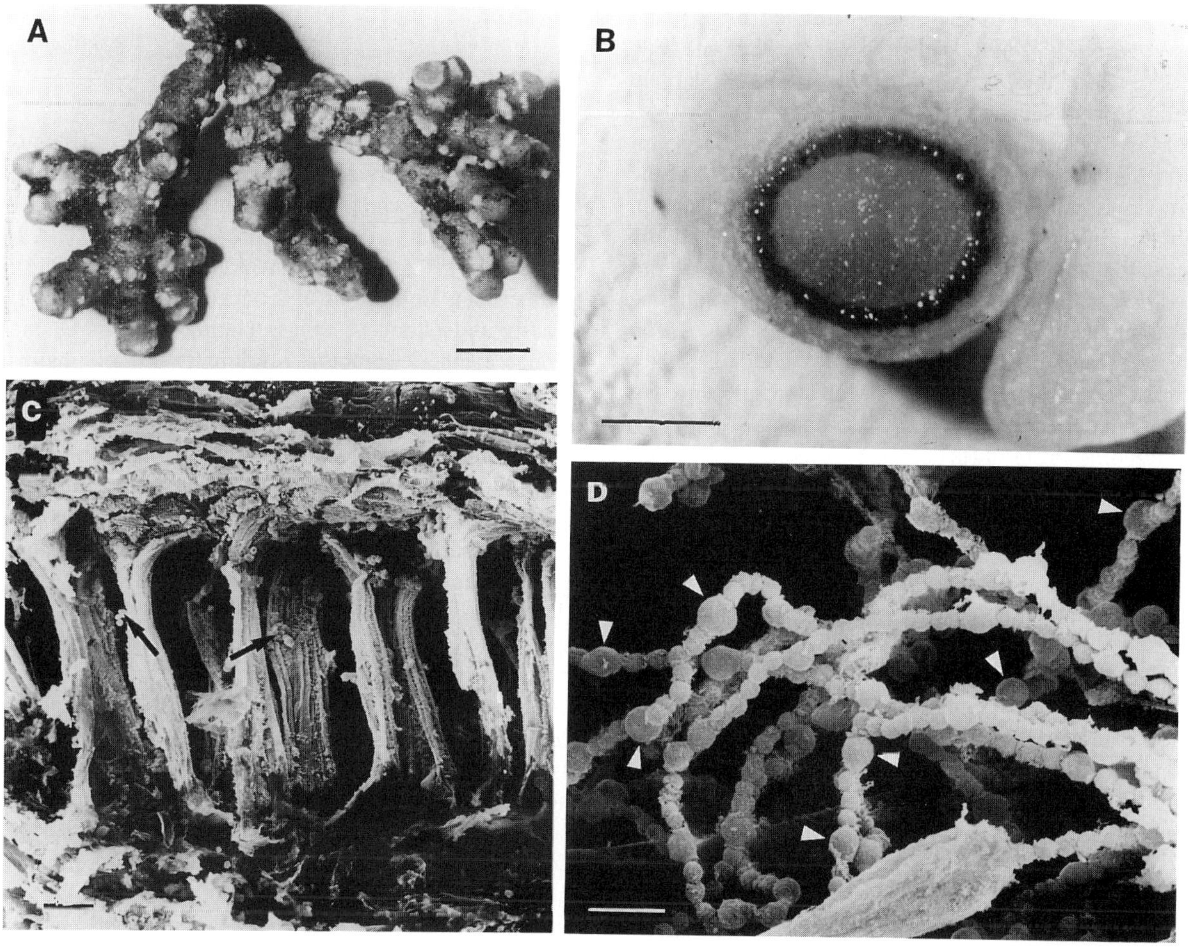

Fig. 2. Cyanobacterial symbiosis with the cycad *Zamia skinneri*. (A) Coralloid root collected from 10 cm below the surface. (B) Transverse section of the root showing the dark cyanobacterial zone. (C) Scanning electron micrograph of the cyanobacterial zone. Most of the cyanobacteria and mucilage have been removed by washing, to reveal the elongated *Zamia* cells traversing the zone. A few cyanobacterial cells can still be seen (arrows). (D) Scanning electron micrograph of the cyanobacterial zone showing cyanobacterial filaments with heterocysts (a few of which are indicated by arrowheads). Bars = 2 mm (A), 1 mm (B), 20 μm (C) and 10 μm (D). (Reproduced with permission from Lindblad et al., New Phytol. 101: 707--716, 1985, © Cambridge University Press).

plant-cyanobacteria symbioses in which the cyanobiont (*Nostoc* sp.) is found intracellularly, within specific mucus-secreting glands on the plant stem at the base of each petiole (Fig. 8; Section V). Even the largest *Gunnera* plants colonised by *Nostoc* are able to fix sufficient nitrogen for their own needs, and this becomes available to the local nitrogen economy (Bonnett, 1990; Osborne et al., 1991). *Gunnera* is only angiosperm forming symbioses with cyanobacteria, and the cyanobionts occur intracellularly. Since all major crop plants are angiosperms, an improved understanding of the *Nostoc-Gunnera* system may improve the chances of developing novel nitrogen-fixing symbioses with economically important plants.

5. Pteridophytes (Azolla)

Within the Pteridophyta only *Azolla* is known to form symbioses with cyanobacteria. The rhizome of the fern forms extensive branches bearing alternate leaves consisting of two clearly defined lobes, each approximately 1 mm in length (Plate 28C and Fig. 3a). The thin, convex ventral lobe is largely achlorophyllous and floats on or just below the water surface supporting the fleshy, chlorophyllous dorsal lobe above the water. During the development of the dorsal lobe an infolding of the epidermis produces an extracellular cavity that is normally infected by cyanobacteria (Fig. 3C), usually referred to as *Anabaena azollae*, although their taxonomic assignment is controversial (Section III.A). *Azolla* can be freed of its cyanobionts, but no longer fixes nitrogen and becomes reliant on the presence of combined nitrogen (Lin and Watanabe, 1988).

Records from 11th century Vietnam and 14th century China describe the use of *Azolla* as a green manure in rice cultivation, and it is still in use throughout the world (Chapter 8). The fern can be grown on its own and incorporated into the soil prior to transplanting the rice, or as an intercrop alongside the rice (Nierzwicki-Bauer, 1990). Incorporation of the *Azolla*, usually by mechanical turning of the soil, or by induced chemical or natural decomposition, results in the release of combined nitrogen, usually as ammonia (Nierzwicki-Bauer, 1990). The use of *Azolla* in this way can increase rice grain yields by up to 40% and improve the protein content of the grain (Nierzwicki-Bauer, 1990). However, there are limitations its use: its application is labour intensive, growth is adversely affected by wind, wave action and high temperature, and is often limited by the availability of phosphorus, and it is susceptible to fungal and insect pests, as well as a number of the pesticides employed in rice culture (Peters et al., 1986; Nierzwicki-Bauer, 1990; Braun-Howland and Nierzwicki-Bauer, 1990; Watanabe and Liu, 1992; Roger, 1995; Kundu and Ladha, 1996; Cassman et al., 1996).

In addition to rice cultivation *Azolla* is used as an animal fodder, as a weed suppressor when in dual cultivation with rice, and has been suggested as a means of improving water quality by the removal of excess quantities of nitrate and phosphorus (Nierzwicki-Bauer, 1990; Watanabe and Liu, 1992). The water fern can also be used as food for fish reared alongside rice in flooded rice fields, and it may help reduce loss of nitrogen from the fields through ammonia volatilization (Watanabe and Liu, 1992). *Azolla* has also been used as a replacement for inorganic nitrogen for the growth of tomato plants (Milicia and Favilli, 1993).

B. Fungi

1. Lichens

Lichens are associations between a fungus (mycobiont, usually an ascomycete) and an autotroph (photobiont), which is a green alga or a cyanobacterium (Rai, 1990b; Werner, 1992; Honegger, 1992, 1993; Ahmadjian, 1993; Hill, 1994). Of the many definitions of a lichen, perhaps the most readily understood is "an association of a fungus and a photosynthetic symbiont resulting in a stable thallus of specific structure" (Ahmadjian, 1993). There are many primitive fungal-photobiont associations producing morphologically simple and inconspicuous thalli, often refered to as microlichens to differentiate them from the smaller number of macrolichens forming thalli that rise above the substratum and that very clearly differ from the aposymbiotic state (Honegger, 1993). Some 8-15% of the 15000-20000 species of lichen contain a cyanobacterium as phycobiont, usually *Nostoc*, although species of *Calothrix*, *Scytonema* and *Fischerella* are also common (Peters et al., 1986; Rai, 1990b; Büdel, 1992; Ahmadjian, 1993; Hill, 1994). The only lichens capable of nitrogen fixation are those containing heterocystous cyanobacteria. Bipartite lichens consist of two partners, a fungus and either a green alga or a cyanobacterium, whereas tripartite lichens contain a fungus, a green alga (which occupies most of the thallus) and a cyanobacterium (contained within specialised structures known as cephalodia, either within or on the surface of the thallus; Plate 28e). On rare occasions the bipartite and tripartite morphologies can be combined in a single chimeroid thallus known as a photosymbiodeme (Ott, 1988).

The 8% of terrestrial ecosystems dominated by lichens constitute regions where vascular plants struggle to grow, and the ability of lichens to survive cold, heat and desiccation, in a dormant state, gives them a significant ecological advantage (Honegger, 1991, 1993; Ahmadjian, 1993; Honegger et al., 1996). Although the growth rate of lichens is slow, cyanolichens can contribute significantly to the nitrogen budget of specific ecosystems, particularly in forests where this occurs by leaching of organic

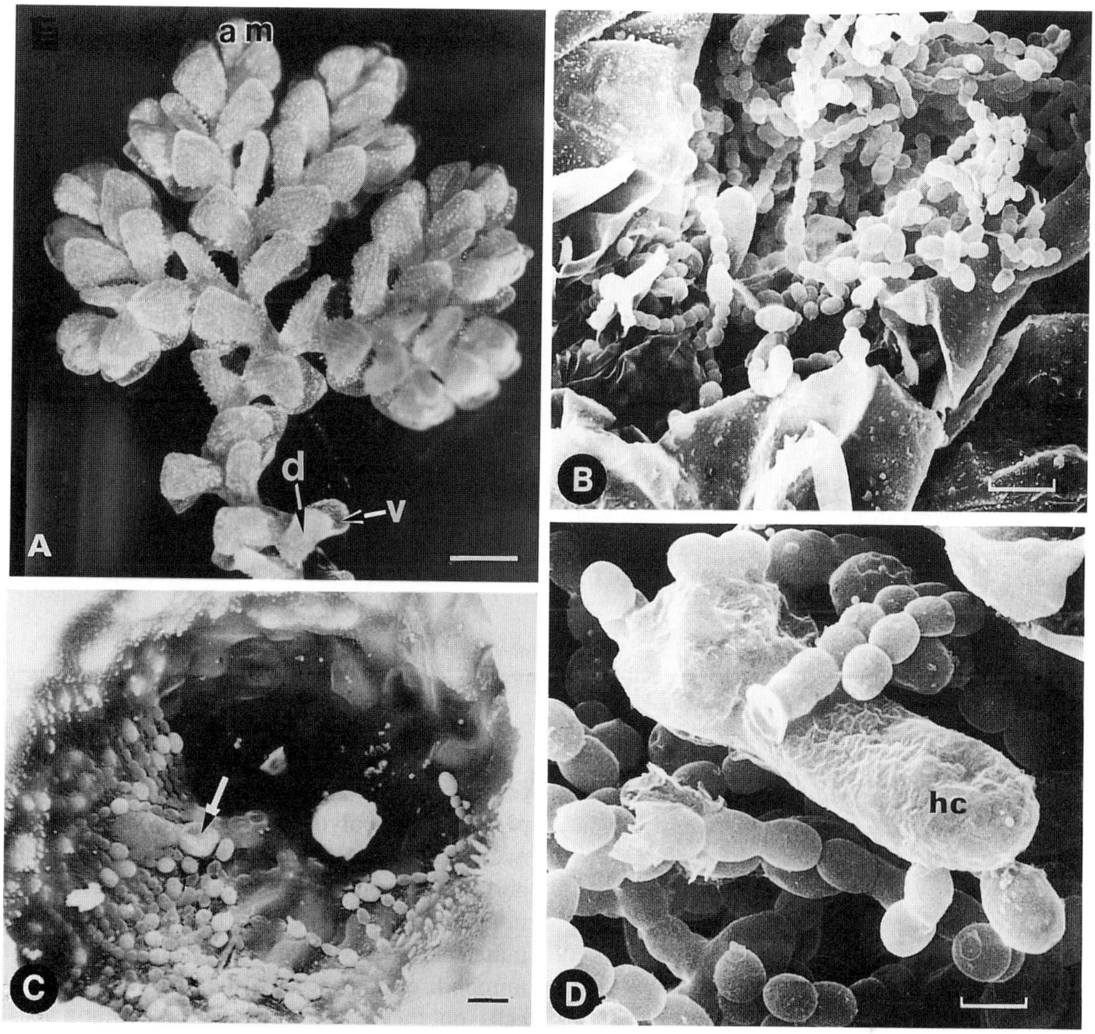

Fig. 3. The water fern *Azolla*. (A) *Azolla* sporophyte viewed from above showing the apical meristem (am) and the dorsal (d) and ventral (v) leaf lobes; (B) - (D) Scanning electron micrographs of leaf cavities of *Azolla* spp. (B) Densely packed cyanobacterial filaments within a leaf cavity. (C) *A. filiculoides* showing the distribution of cyanobacterial filaments containing heterocysts (the larger cells) around the periphery of the leaf cavity. An *Azolla* hair cell (arrow) is also visible. (D) Cyanobacterial filaments associated with a plant hair cell (hc). Bars = 1.5 mm (A), 20 μm (B and C), and 5 μm (D). (Reproduced with permission from: (B) and (D) Shi et al., Planta 172: 298--308, 1987, © Springer-Verlag GmbH & Co. KG; (C) Reprinted from FEMS Microbiol Lett 34, Robins et al., 155-160, © 1996 with kind permission from Elsevier Science - NL, Sara Burgerhartstraat 25, 1055 KV Amsterdam, The Netherlands; (A) photograph courtesy of S. Nierzwicki-Bauer).

nitrogen from thalli, or by the decay of older thalli that fall to the ground (Ahmadjian, 1982, 1993). Cyanolichens can also be important providers of fixed nitrogen in the subarctic and arctic/antarctic regions where they fix nitrogen at a surface temperature as low as 0°C (Peters et al., 1986), and photosynthesise at -10°C (Lange et al., 1996). For example, the nitrogen-fixing biota of the sub-antarctic MacQuarie Island is dominated by cyanobacteria growing both epiphytically and symbiotically with plants such as liverworts and mosses, and lichens such as *Peltigera* sp. (Line, 1992).

Much of the carbon fixed by the photobiont may be used (in the form of polyols), not as a nutrient, but as a physiological buffer, helping the lichen to survive environmental stresses (Farrar, 1988; Ahmadjian, 1993), the major one being the daily cycle of drying and wetting experienced by many lichens (Ahmadjian, 1993). Indeed, lichens show rapid recovery of nitrogen fixation after periods of desiccation (Ahmadjian, 1982; Peters et al., 1988; Hill, 1994). Rapid recovery of photosynthesis also occurs, although green algal photobionts become turgid and show considerable photosynthetic activity when a dry thallus is exposed to high relative humidity, whereas cyanobacterial photobionts require liquid water to become turgid and to reactivate photosynthesis (Section VI.D; Büdel and Lange, 1991; Lange et al., 1993). The ability of green algal photobionts to absorb water from the air is probably a result of their high cell content of polyols (Hill, 1994). This difference in response to water availability may explain why lichens with cyanobacterial symbionts are confined to the maritime regions of Antarctica, and are absent from the extreme environments of continental Antarctica with very low temperatures and limited water availability (Schroeter et al., 1994).

Non-lichenized associations of algae and fungi are also common, and this is not surprising given the huge numbers of fungi (45,000 species, excluding lichen fungi) and algae (21,000 species), and their frequent occurrence in the same environments (Ahmadjian, 1982, 1993). However, little is known of equivalent cyanobacterial associations, although there are many examples of fungal parasitism of cyanobacteria, particularly *Scytonema*, that fall short of the definition of a lichen given above (Ahmadjian, 1982; 1993). *Geosiphon pyriforme*, an association of *Nostoc sphaericum* with a phycomycete, is unique because the cyanobacterium lives endosymbiotically within vesicles inside the fungal hyphae (Sections II.B.2 and V).

2. *Geosiphon pyriforme*

In common with the lichens, *Geosiphom pyriforme* is a symbiotic association between a fungus and a cyanobacterium (Mollenhauer, 1992; Kluge et al., 1991; 1992, 1994; Mollenhauer and Kluge, 1994; Schüßler et al., 1996; Mollenhauer et al., 1996). However, in many other respects these two forms of symbiosis are very different. The fungus is a phycomycete, thought to be related to *Glomus* (Schüßler et al., 1994; Gehrig et al., 1996), whereas the cyanobacterium is *Nostoc*, thought to be *N. punctiforme* or *N. sphaericum* (Kluge et al., 1991; Bilger et al., 1994), although other *Nostoc* spp. can be incorporated by the fungus (Kluge et al., 1994). The cyanobionts live intracellularly within specialised bladders produced by the fungal hyphae (Plate 28f and Fig. 4; Section V).

Geosiphon grows on the surface and in the uppermost layers of moist loamy soils, but appears to be extremely rare, there being only five reports of it being found in nature (Kluge et al., 1991). This rarity, and difficulty in finding the very small organism even in locations where it is known to occur, have hampered studies of this fascinating symbiosis. Although the cyanobiont can be readily grown free-living, it was not until comparatively recently that the intact association was grown in the laboratory (Mollenhauer and Mollenhauer, 1988). This has facilitated physiological studies, demonstrating that the association can fix both nitrogen (Kluge et al., 1992) and CO_2 (Kluge et al., 1991; Section VI.D). It is not known if the fungus can exist in nature without its cyanobacterial partner, but laboratory conditions suitable for growth of the intact symbiosis will not support growth of the fungus lacking its endosymbiont (Kluge et al., 1991). The cyanobiont, however, is thought to survive in nature when akinetes liberated from decaying *Geosiphon* bladders germinate and continue to live in the soil (Mollenhauer, 1992). In the field *Geosiphon* is found with the liverwort *Blasia* and the hornwort *Anthoceros*, and there is evidence that the *Nostoc* cyanobionts of *Geosiphon* can infect the bryophytes, and vice versa (Mollenhauer, 1992; Kluge et al., 1994).

C. Diatoms

The marine pennate diatom *Rhizosolenia* is abundant in the North Pacific and frequently forms a symbiotic association with the filamentous, heterocystous cyanobacterium *Richelia intracellularis,* which fixes both CO_2 and N_2 for the host (Rai, 1990a). Each cyanobacterial filament, of which there can be several in each diatom cell, usually contains a single basal heterocyst, although occasionally one is present at each end. The apex of *Rhizosolenia clevei* contains cytoplasm only in the region occupied by the cyanobacterium, although the cyanobiont is always external to the host cytoplasm, and its cell wall remains intact (Janson et al., 1995b). The level of glutamine synthetase protein is very low in the cyanobiont, and this may result in release of fixed nitrogen to the host (Janson et al., 1995b). However, the presence of *Richelia* is not essential to the growth and survival of *Rhizosolenia*, because the diatom can be found without the cyanobiont (Villareal, 1987).

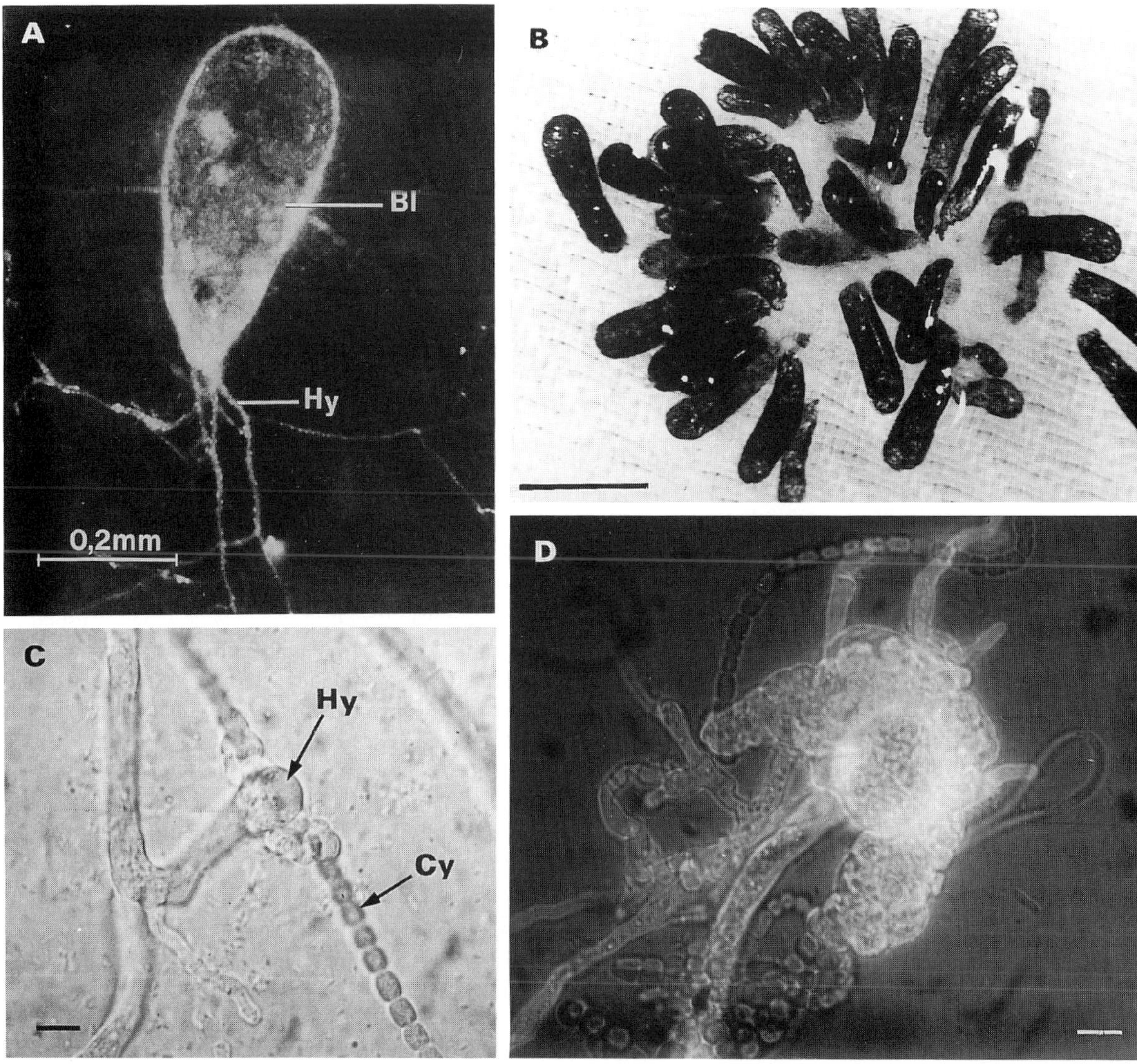

Fig. 4. Geosiphon pyriforme. (A) Mature bladder (Bl) showing adjacent fungal hyphae (Hy). (B) Growth on plankton gauze, showing the characteristic bladders containing the cyanobacterium *Nostoc*. (C) Early stage in the incorporation of *Nostoc* by the fungus. A fungal hypha (Hy) has transformed into a dumb-bell shaped structure in which a blunt projection of fungal cytoplasm surrounds the cyanobacterium (Cy). (D) A later stage of incorporation in which the fungal wall has softened and additional cytoplasmic bulges surround the cyanobacterium. Bars = 0.2 mm (A), 1 mm (B), and 10 μm (C and D). (Reproduced with permission from: (A) Bilger et al., J. Phycol. 30: 225--230, 1994, © Phycological Society of America; (B) Kluge et al., Bot. Acta. 105: 343--344, 1992, © Georg Thieme Verlag; (C) and (D) Mollenhauer et al., Protoplasma 193: 3--9, 1996, © Springer-Verlag Wien).

Richelia intracellularis is also found in diatoms of the genus *Hemiaulus*, which was once thought to be a relatively uncommon and unstable symbiosis (Rai, 1990a). However, it has recently been shown that 91-100% of three *Hemiaulus* spp. in the Bahama Islands and the Caribbean Sea contain *Richelia intracellularis*, this symbiosis being up to 254 times as common here as the *Rhizosolenia-Richelia* association (Villareal, 1994). The *Hemiaulus* symbioses have been found as far ranging as 40°N in the Atlantic and 31°S in the Pacific. Epifluorescent microscopy is required for reliable visualisation of the cyanobiont, which is usually not visible in living material using transmitted light, possibly explaining the previous underestimates for the occurrence of this symbiosis (Villareal, 1991, 1994).

Two diatoms of the Epithemiaceae, *Rhopalodia gibba* and *Epithemia turgida*, possess unicellular cyanobacterial symbionts in their cytoplasm (Rai, 1990a; DeYoe et al., 1992). The cyanobionts have not been isolated or identified, but appear to be capable of nitrogen fixation. Their number per diatom cell ranges from 1-8, and increases in response to decreasing external nitrate concentration (DeYoe et al., 1992). However, this increase does not occur under phosphorus limitation, possibly due to reduced cyanobiont growth rate. This diatom-cyanobacterium symbiosis is clearly at an advantage in nitrogen-limited habitats, but probably at a disadvantage when nitrogen supply is adequate. Although the number of cyanobionts is then reduced, at least one remains in each diatom cell, presumably at a metabolic cost to the host (DeYoe et al., 1992).

D. Dinoflagellates

Cyanobacterial symbionts known as phaeosomes (Taylor, 1990) are found in a variety of tropical marine dinoflagellates (Lucas, 1991; Gordon et al., 1994; Janson et al., 1995a). In the subtropical waters of the Gulf of Aqaba, heterotrophic dinoflagellates of genera *Ornithocercus*, *Histioneis* and *Citharistes* have been shown to contain cyanobacterial symbionts, thought to be members of the unicellular genera *Synechococcus* and *Synechocystis* (Gordon et al., 1994). The cyanobacteria are located between the upper and lower lists of the horizontal groove of cells of *Ornithocercus* and *Histioneis*, and within special chambers inside the cells of *Citharistes* (Fig. 5). The cyanobacteria were thought to benefit by the provision of a suitable microaerobic environment for the efficient functioning of nitrogenase (Gordon et al., 1994). However, recent immunolocalisation studies with the phaeosomes of *Ornithocercus magnificus* and *O. steinii*, which are cyanobacteria of the unicellular genus *Synechococcus*, failed to detect nitrogenase, even in samples collected at night, when some unicellular cyanobacteria are known to fix nitrogen to avoid the inihibitory effects of photosynthetically generated oxygen during the day (Janson et al., 1995a). The cyanobionts seem to provide fixed carbon, rather than nitrogen, for their non-photosynthetic host, although they may also be phagotrophically consumed when they senesce.

E. Prochlorophytes

The prochlorophytes are a group of oxygenic photosynthetic prokaryotes that resemble cyanobacteria in many ways, but differ in the possession of both chlorophylls *a* and *b* and the lack of phycobilin pigments (Burger-Wiersma and Matthijs, 1990; Mur and Burger-Wiersma, 1992; Bullerjahn and Post, 1993; Post and Bullerjahn, 1994; Matthijs et al., 1994). Only three prochlorophytes are known and these have been placed in three different genera: *Prochloron didemni*, *Prochlorothrix hollandica* and *Prochlorococcus marina*. The phylogenetic position of these prokaryotes, and their relationship to the evolution of plastids, is still controversial (Bullerjahn and Post, 1993; Post and Bullerjahn, 1994; Matthijs et al., 1994; Bhattacharya and Medlin, 1995), but recent evidence indicates that *Prochloron* is closely related to *Prochlorothrix* and the cyanobacteria, whereas *Prochlorococcus* is more closely related to the γ-purple bacteria (Shimada et al., 1995). The first of the prochlorophytes to be discovered was *Prochloron didemni* which is found in a symbiotic relationship with marine colonial ascidians (chiefly didemnids) in (sub)tropical coastal waters. The other two are free-living strains that will not be discussed here, but have been extensively reviewed in recent years (Burger-Wiersma and Matthijs, 1990; Mur and Burger-Wiersma, 1992; Bullerjahn and Post, 1993; Post and Bullerjahn, 1994; Matthijs et al., 1994).

Prochloron is unicellular, with coccoid cells 6 to 30 μm in diameter, and is found as a symbiont in ascidians of the Didemnidae, where it usually occurs extracellularly either on the surface, or within the cloacal cavity of the host (Lewin, 1989; Lewin and Cheng, 1989; Griffiths and Thinh, 1989; Swift and Robertson, 1991). However, *Prochloron* has been found in the ascidian *Lissoclinum punctatum* both extracellularly and intracellularly in the tunic cells (Hirose et al., 1996). The intracellular *Prochloron* are morphologically similar to those found extracellularly, and show no signs of rejection or degeneration, implying a stable symbiosis. Cells of *Prochloron* contain carboxysomes and are bounded by a cytoplasmic membrane, periplasmic space, outer membrane and outer sheath (Swift and Leser, 1989; Bullerjahn and Post, 1993; Post and Bullerjahn, 1994), and those cells from some didemnids contain one or more characteristic vacuoles that occupy much of the cell volume (Swift, 1989).

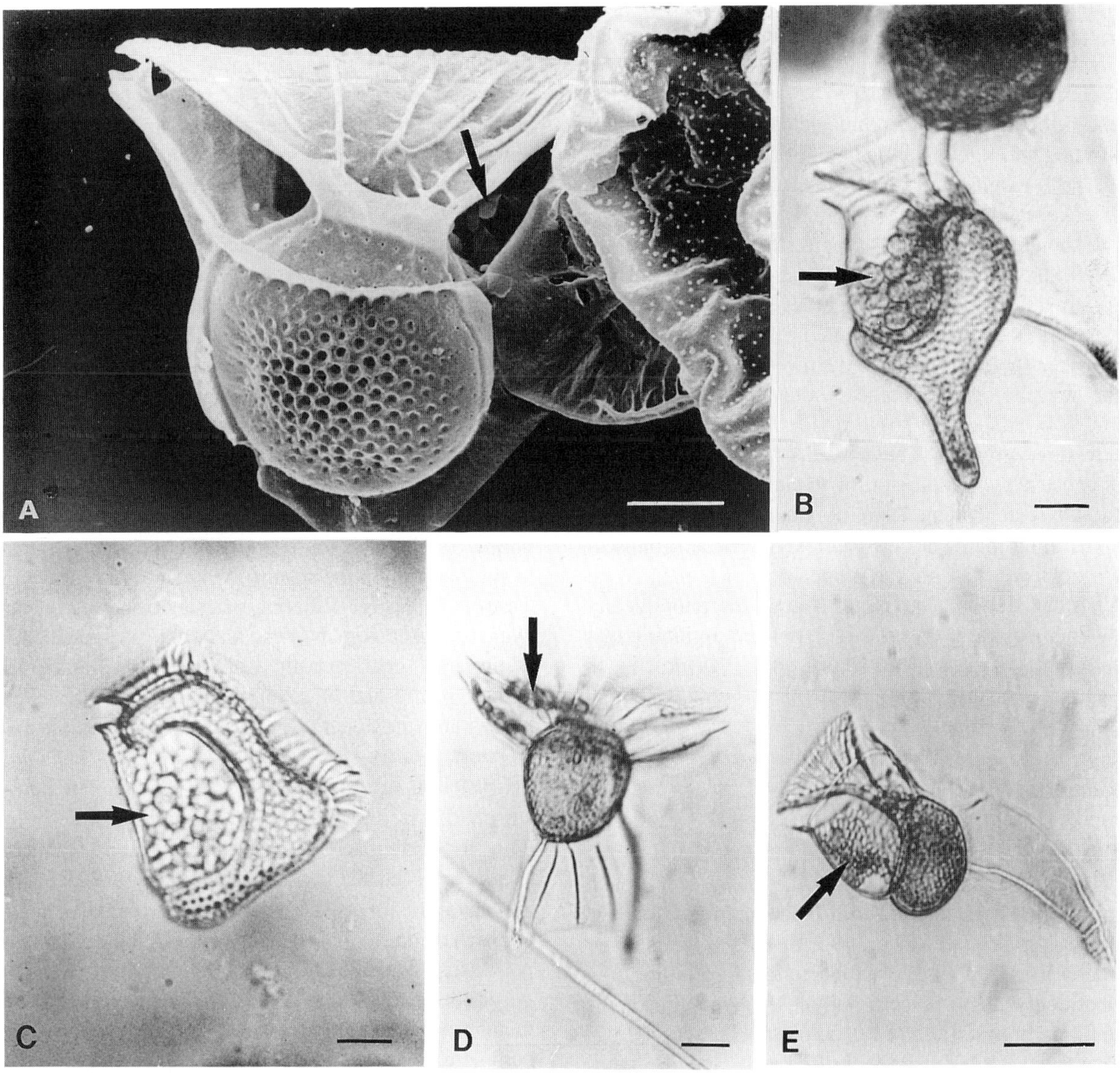

Fig. 5. Cyanobacterial symbiosis with dinoflagellates. (A) Scanning electron micrograph of *Ornithocercus magnificus* daughter cell shortly after division showing cyanobacteria in the cingulum (arrow); (B) *Histioneis highleyi*; (C) *Citharistes apsteinii*; (D) *Ornithocercus quadratus*; (E) *Histioneis elongata*. Arrows in (B) - (D) indicate the location of cyanobacteria. Bars = 10 μm. (Reproduced with permission from Lucas, Ophelia 33: 213--224, 1991, © Ophelia Publications).

Biochemical and physiological studies with *Prochloron* (Whatley and Alberte, 1989) have been hampered by the inability to grow it in culture. However, it is known to fix CO_2 for its host (Section VI.D) and there has been one report of acetylene reduction, indicating nitrogen fixation, in one association (Paerl, 1984), although the N_2-fixing activity of some *Prochloron*-ascidian associations in the Seychelles has been shown not to be associated with *Prochloron* itself (Odintsov, 1991). Chlorophyll fluorescence quenching analysis has recently been used to assess the photosynthetic performance of *Prochloron* within the host *Lissoclinum patella* (Schreiber et al., 1997). The symbiont displayed exceptionally high quantum yields under both light-limiting and light-saturated conditions.

The results of oligonucleotide cataloguing (Stackebrandt et al., 1982) and DNA-DNA reassociation studies (Stam et al., 1985; Holton et al., 1990) have implied that *Prochloron* isolated from different hosts at one site, or even hosts at widely different locations, appear to be closely related, or possibly the same species. However, an analysis of the morphological characteristics of both symbiont and host led Cox (1986) to divide *Prochloron* isolates into three groups. These results have been partially supported by comparison of nucleotide sequences of the DNA-dependent RNA polymerase (*rpoC*) genes, confirming that at least two closely related groups of *Prochloron* exist in the same area of Palau, West Caroline Island, although some morphologically distinct isolates are very closely related, implying that different physiological states of *Prochloron* in different hosts may affect their morphological appearance (Palenik and Swift, 1996).

F. Cyanelles

Cyanelles are one of a range of plastids found in the cells of photosynthetic eukaryotes, and were once thought to be cyanobacterial endosymbionts of their eukaryotic hosts (Pascher, 1929). They possess many similarities with cyanobacteria, including a peptidoglycan wall, chlorophyll *a*, phycobiliproteins and carboxysome-like structures (Trench, 1982; Wasmann et al., 1987; Löffelhardt and Bohnert, 1994a, b; Pfanzagl et al., 1996a, b), yet their genomes are only 130 to 140 kbp, similar in size to those of chloroplasts, but only about 5% of those of free-living cyanobacteria. The phylogenetic relationship between cyanobacteria and plastids has received considerable attention in recent years, and remains controversial, although the emerging view seems to be that all plastids, including cyanelles, arose by almost simultaneous divergences from a single common ancestor, itself the result of an endosymbiosis between a phagotrophic host and a cyanobacterium (Douglas, 1992, 1994; Löffelhardt and Bohnert, 1994a, b; Helmchen et al., 1995; Bhattacharya et al., 1995). Cyanelles have been found in the cells of very different eukaryotes that show neither systematic nor evolutionary relatedness, although the taxonomic position of these eukaryotes is often confused (see Wasmann et al., 1987, for a discussion of this). Determining the ancestry of the secondary plastids present in many photosynthetic eukaryotes is complex because many are likely to have acquired these by injestion of another eukaryote already containing a plastid (Delwiche and Palmer, 1997).

The obligately photoautotrophic alga *Cyanophora paradoxa* can be grown easily, and is consequently well studied (Wasmann et al., 1987; Löffelhardt and Bohnert, 1994a, b), the complete nucleotide sequence of the cyanelle genome having been published (Stirewalt et al., 1995). The centre of *Cyanophora* cyanelles contains a characteristic protein inclusion body resembling the cyanobacterial carboxysome (Blank, 1985) and containing most of the cyanelle ribulose bisphosphate carboxylase (Mangeney and Gibbs, 1987), whereas phosphoribulokinase is found in soluble form in the stroma (Mangeney et al., 1987). This resembles the situation in cyanobacteria, yet the cyanelle inclusion bodies lack the proteinaceous membrane characteristic of the cyanobacterial carboxysome, and it is unlikely that the inclusion bodies are functional carboxysomes (Löffelhardt and Bohnert, 1994a).

Cyanelles are unique among plastids in being surrounded by a peptidoglycan wall of the same A1α-type as cyanobacteria (Löffelhardt and Bohnert, 1994a, b; Pfanzagl et al., 1996a, b), providing an explanation for the lethal effect of β-lactam antibiotics on the obligately autotrophic organisms possessing cyanelles (Berenguer et al., 1987). The genes for the enzymes of peptidoglycan synthesis and metabolism have not been identified within the complete sequence of the *Cyanophora paradoxa* genome, and are presumably encoded by the nucleus (Stirewalt et al., 1995). The peptidoglycan is located between the cyanelle membrane and an outer envelope membrane (Löffelhardt and Bohnert, 1994a, b). The origin of the latter is unknown, but may be a remnant of the outer membrane of the cyanobacterial ancestor of the cyanelle.

G. Animals: Sponges, Ascidians, Echiuroid Worms and Midge Larvae

Almost 40 genera of marine sponge are known to contain symbiotic cyanobacteria, found both intra- and intercellularly, and distributed either throughout the sponge or just in the superficial tissues (Fig. 6; Rai, 1990a; Wilkinson, 1992). The cyanobionts are usually members of the unicellular genus *Aphanocapsa*, or the filamentous genus *Phormidium* (Rai, 1990a), although the cyanobiont of the tropical marine sponge *Dysidea herbacea* has been identified as *Oscillatoria spongelia* (Unson and Faulkner, 1993; Hinde et al., 1994; Unson et al., 1994). All of these genera lack heterocysts, although the cyanobionts of at least two sponges have been shown to fix nitrogen (Rai, 1990a). Cyanobionts are photosynthetically active, only being found in sponges in the photic zone, and transfer up to 12% of the carbon they fix to the host, probably in the form of glycerol (Rai, 1990a).

The degree of metabolic reliance of some sponges on the presence of the cyanobionts is illustrated by *Chondrilla nucula* which cannot colonise dark sites, and, if transferred to the dark, looses its cyanobionts and dies (Arillo et al., 1993). The benefits to the host have been demonstrated in Jamaican coral reef sponges in which those containing cyanobacterial symbionts grow more rapidly (Wilkinson and Cheshire, 1988). The cyanobionts are also thought to help protect the sponge from damage by high light intensity (Rai, 1990a). On rare occasions the relationship between growth of host and cyanobiont can become unbalanced. An example occurs in the sponge *Geodia papyraceae* from a mangrove island off Belize, that contains intercellular coccoid cyanobacteria equal in volume to almost half its own cellular tissue (Ruetzler, 1988). Under favourable growth conditions the cyanobiont cells multiply more rapidly than the sponge can eliminate them, causing host tissue damage, so-called mangrove sponge disease.

Sponges, such as *Theonella swinhoei* and *Siphonochalina tabernacula* (Fig. 6), that contain nitrogen-fixing cyanobionts, most likely benefit from the provision of combined nitrogen (Rai, 1990a) and this presumably benefits the local nitrogen economy. Indeed, the sponge *Chondrilla nucula*, which is ubiquitous in coral reefs off Puerto Rico, contains N_2-fixing cyanobacterial symbionts, and releases sufficient nitrate to account for 50-120% of the nitrogen required to sustain reef productivity (Corredor et al., 1988). Many sponges also produce bioactive natural products, at least some of which may help protect against potential predators and bacterial invasion (Faulkner et al., 1994). *Oscillatoria spongelia*, the cyanobacterial symbiont of the marine sponge *Dysidea herbacea*, produces a group of polychlorinated (Unson and Faulkner, 1993; Faulkner et al., 1994; Handayani et al., 1997) and polybrominated (Unson et al., 1994) secondary metabolites. One of the latter is inhibitory to the growth of a range of Gram-positive and Gram-negative bacteria, and deters feeding by generalist fishes (Faulkner et al., 1994). A novel ultraviolet-absorbing mycosporine-like amino acid is also produced by the cyanobiont of *Dysidea herbacea* (Bandaranayake et al., 1996; Bandaranayake, 1998). The production of such secondary metabolites was once thought to be limited to the sponge tissue, or to associated heterotrophic bacteria (Unson et al., 1994; Faulkner et al., 1994). Indeed, the sponge *Theonella swinhoei* produces a range of bioactive natural products thought to be produced by cyanobacterial symbionts, but recently several of these have been shown to be present in associated heterotrophic bacteria rather than the cyanobionts (Bewley et al., 1996).

Two of the least studied animal associations are with ascidians and echiuroid worms. Two American Pacific coast ascidians contain numerous tunic pockets filled with a variety of algae and cyanobacteria, that probably enter during the incorporation of sand grains (Lambert et al., 1996). The cyanobacteria include *Dermocarpa*, *Xenococcus* and *Synechococcus*. Representatives of the additional cyanobacterial genera *Oscillatoria*, *Spirulina* and *Anabaena* have been found in the inter-tunic cavity of the double-tunic ascidians *Pyura cancellata* and *P. carnea* (Lambert et al., 1996). Cyanobacteria have been reported in the subepidermal connective tissue of two species of echiuroid worm, *Ikedosoma gogoshimense* and *Bonellia fuliginosa*, but little is known of these associations (Rai, 1990a).

An apparently mutualistic association occurs between larvae of the chironomid midge *Cricotopus nostocicola* living inside colonies of *Nostoc parmelioides* in moderately fast-flowing streams (Sheath and Cole, 1992; Dodds et al., 1995). The larvae eat the *Nostoc* and are protected from predation, while *Nostoc* colonies containing

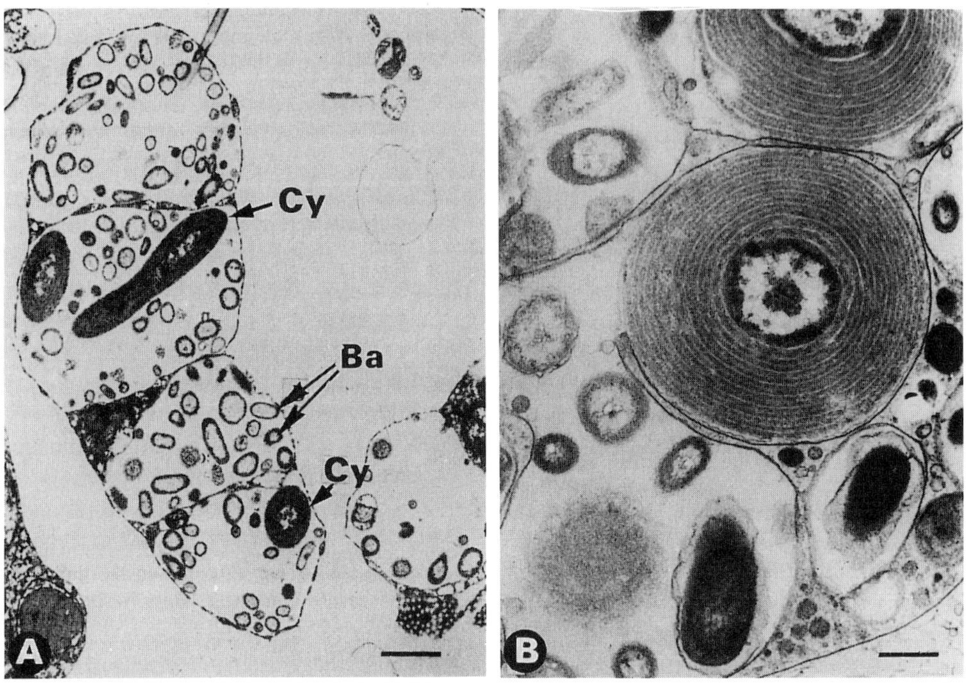

Fig. 6. Symbiotic cyanobacteria in the sponge *Siphonochalina tabernacula.* (A) Cyanobacteria (Cy) and bacteria (Ba) within sponge cell vacuoles. (B) Section through sponge vacuole showing cyanobacterium with spiral thylakoids, and adjacent smaller bacteria. Bars = 5 µm (A) and 1 µm (B). (Reproduced with permission from Wilkinson and Fay, Nature 279: 527--529, 1979, © Macmillan Magazines Limited).

the larvae photosynthesise more rapidly than unoccupied colonies (Ward et al., 1985), possibly because the former colonies grow further into the water flow, enhancing outward diffusion of O_2, which is inhibitory to nitrogen fixation, and promoting the inward diffusion of CO_2, which is limiting for growth (Dodds, 1989). The colony may also benefit by being attached to the rocks by silk from the larvae, while the larvae are thought to trigger hormogonia formation, promoting the expansion of the colony, and providing a habitat for later generations of larvae (Dodds and Marra, 1989).

III. The Symbionts

A. *Cyanobacteria*

In the plant-cyanobacteria symbioses the cyanobiont is almost always *Nostoc* (Table 1). The ability to form hormogonia is the key to success as a symbiont, because hormogonia are the infective agents in most if not all of these symbioses. The immotile parent filaments would be unable to gain entry to the plant structures that house the symbiotic colonies. It is clear, however, that a wide range of *Nostoc* species and strains are symbiotically competent in a wide

Table 1. Identity and location of the symbionts in plant-cyanobacteria symbioses.

Plant host	Plant structure infected[a]	Cyanobiont	Cyanobiont location
Cycads	Coralloid roots (Fig. 2)	*Nostoc* (+ *Calothrix*)[b]	Intercellular
Gunnera	Stem glands (Fig. 8)	*Nostoc*	Intracellular[d]
Liverworts	Auricles (Plate 29a-e and Fig. 7)	*Nostoc*	Intercellular
Hornworts	Slime cavities	*Nostoc* (+ *Calothrix*)[b]	Intercellular
Azolla	Dorsal leaf cavities (Fig. 3)	*Anabaena* (*Nostoc* ?)[c]	Intercellular

[a] The figures in the text that illustrate each structure are given in parentheses.
[b] The symbionts are *Nostoc* spp. in almost all cases; there have been rare reports of *Calothrix* spp. as symbionts (Section III.A).
[c] The *Azolla* symbiont is usually referred to as *Anabaena* sp., but the alternative genera *Nostoc* and *Trichormus* have been suggested (Section III.A).
[d] *Gunnera* is the only plant symbiosis in which the cyanobacteria are found intracellularly (see Section V).

range of plants (Leizerovitch et al., 1990; Zimmerman and Culley, 1991; Bergman et al., 1992b; Lotti et al., 1996). Members of genera other than *Nostoc* can also form symbioses with plants. For example, the cycad *Encephalartos* has been reported to contain *Calothrix* spp. as cyanobionts (Grobbelaar et al., 1987), as has the hornwort *Phaeoceros* (West and Adams, 1997). In common with *Nostoc*, *Calothrix* is heterocystous and capable of hormogonia formation.

Notwithstanding the previous example, cycad symbionts are usually *Nostoc* spp. (Plate 29f), and morphological comparisons (Grobbelaar et al., 1987) and restriction fragment polymorphism (RFLP) analysis of cyanobiont DNA (Lindblad et al., 1989), have shown that a variety of strains can be involved, although it seems that a single *Nostoc* strain is predominant in a particular root. Similar conclusions have been drawn from comparisons of lectin binding, electrophoretic protein profiles and zymograms of *Nostoc* spp. from cycads at a single site (Zimmerman and Rosen, 1992). As with *Gunnera*, although several *Nostoc* spp. can infect the plant (Zimmerman and Bergman, 1990; Söderbäck, 1992; Bergman et al., 1992a), it is likely that only one or a few are present in an individual plant (Bergman et al., 1992b).

The cyanobionts of lichens show much greater diversity, with both filamentous and unicellular genera reported. However, the picture is less than clear because most identifications of cyanobionts have relied on direct observation, rather than culturing, and cyanobacteria often display very different morphologies when symbiotically-associated rather than free-living (Rai, 1990b; Section VI.B). Members of at least four genera of filamentous, heterocystous cyanobacteria are known to be involved, *Nostoc*, *Calothrix*, *Scytonema* and *Fischerella* (Rai, 1990b). Of the unicellular cyanobacteria, the most common are *Gloeocapsa*, *Gloeothece* and *Synechocystis*, although their exact taxonomic status is often uncertain (Rai, 1990b). A recent comparsion of PCR amplified $tRNA^{Leu}$ (UAA) introns from the *Nostoc* symbionts of the lichens *Nephroma arcticum*, *Peltigera aphthosa*, *P. membranacea* and *P. canina* revealed that a single symbiont was present in an individual thallus (Paulsrud and Lindblad, 1998). Closely related cyanobionts were also present in the same lichen collected from different locations (in Sweden and in Finland), implying a high degree of specificity in the choice of *Nostoc* symbiont (Paulsrud et al., 1998). The same technique has been used to show that the same *Nostoc* symbiont occurs in the bipartite and tripartite lobes of a photosymbiodeme of *P. aphthosa* (Paulsrud et al., 1998).

In the case of the unique fungal association *Geosiphon pyriforme*, only *Nostoc* spp. seem to be cyanobionts. It is interesting that in nature *Geosiphon* occurs at the same sites as *Blasia* and *Anthoceros*, and there is evidence that the *Geosiphon* cyanobionts can infect the bryophytes, and vice versa (Mollenhauer, 1992; Kluge et al., 1994).

The cyanobiont of *Azolla* has been traditionally referred to as *Anabaena azollae*, although this taxonomic asignment has been controversial, as the discussion below reveals. *Azolla* is unusual in carrying a colony of the cyanobiont within the sporocarp during sexual reproduction, making re-infection unnecessary (Peters et al., 1986; Peters and Meeks, 1989; Plazinski, 1990; Bergman et al., 1992b), and removing the need for the cyanobiont to produce hormogonia for infection to occur. Members of a non-hormogonium-forming genus such as *Anabaena* could therefore be considered potential symbionts in this system. However, undifferentiated filaments, resembling hormogonia, are associated with the apical meristem of each *Azolla* stem, providing the inoculum for new leaf cavities (Peters and Meeks, 1989).

The *Azolla* cyanobiont was first described as *Nostoc* by Strasburger (1873), who later assigned it to the genus *Anabaena* and re-named it *Anabaena azollae* (Strasburger, 1884). This classification has often been questioned, and the genus *Nostoc* has been proposed, based on comparisons of the arrangement of *nif* genes (Meeks et al., 1988) and the numbers of superoxide dismutase isoenzymes (Canini et al., 1992b) in symbiotic and free-living *Anabaena* and *Nostoc* strains. The results of RFLP analysis of the genomic DNA of cyanobionts from seven *Azolla* species also imply that they are more closely related to *Nostoc* than to *Anabaena* (Plazinski et al., 1990). This assertion is supported by the results of polymerase chain reaction (PCR) amplification of *Azolla* cyanobiont DNA (Kim et al., 1997). However, comparisons of cellular fatty acids have implied that the *Azolla* symbionts are equally distinct from both *Anabaena* and *Nostoc* (Caudales et al., 1992). The alternative name *Trichormus azollae* has been suggested (Grilli-Caiola pers. comm., quoted in Bergman et al., 1992), and supported by subsequent observations of the characteristics of akinetes present in the cyanobiont of *Azolla filiculoides* (Grilli-Caiola et al., 1993).

The true identity of the *Azolla* symbiont has remained elusive, because there appears to be more than one cyanobiont associated with the fern (Peters and Meeks, 1989). The minor symbiont is readily isolated (Gebhardt and Nierzwicki-Bauer, 1991), whereas the major symbiont has proved impossible to cultivate free-living, although there has been one report of limited growth of a symbiont reacting positively to a monoclonal antibody raised to the major symbiont (Tang et al., 1990). Attempts to identify the major symbiont have therefore to rely on the isolation of sufficient quantities of cells for immediate analysis.

Immunological (Arad et al., 1985) and lectin haemagglutination studies (McCowen et al., 1987; Zimmerman et al., 1989), RFLP analyses (Franche and Cohen-Bazire, 1985, 1987; Meeks et al., 1988; Plazinski et al., 1988; Gebhardt and Nierzwicki-Bauer, 1991), and enzymatic comparisons (Zimmerman et al., 1989), have confirmed that the readily culturable cyanobionts of *Azolla* are not the same as the freshly-isolated symbiont. PCR-based techniques have been used successfully to differentiate between *Azolla* symbionts from different plant species (Eskew et al., 1993; Kim et al., 1997). The genetic diversity of *Anabaena azollae* strains from different ferns has also been assessed by RFLP analysis using *nif* gene probes (Van Coppenolle et al., 1993a), and the results suggest a taxonomy of *A. azollae* parallel with *Azolla* itself, determined by DNA amplification using arbitrary primers (Van Coppenolle et al., 1993b). Similarly, analysis of the fatty acid composition of cyanobionts from different American *Azolla* species implies that the cyanobionts form several clusters that correlate well with the taxonomic grouping of the plants, implying that the cyanobionts have coevolved with the plant (Caudales et al., 1995), presumably as a result of the presence of *Anabaena azollae* throughout the life cycle of *Azolla*, avoiding the need for constant re-infection.

B. Bacteria

In addition to the cyanobacteria, the leaf cavities of *Azolla* contain heterotrophic bacteria (Grilli Caiola et al., 1988; Nierzwicki-Bauer and Aulfinger, 1990; Plazinski et al., 1990b), including abundant *Arthrobacter* spp. (Wallace and Gates, 1986; Petro and Gates, 1987; Forni et al., 1989; Nierzwicki-Bauer and Aulfinger, 1990; Braun-Howland and Nierzwicki-Bauer, 1990; Forni and Grill-Caiola, 1992; Leonardi et al., 1993). Immunogold labelling and transmission electron microscopy have shown that some of these bacteria contain nitrogenase and are presumably capable of nitrogen fixation (Lindblad et al., 1991). The role of the bacteria in the symbiosis is unknown, but they have been considered a third member of the association (Petro and Gates, 1987; Forni et al., 1989; Carrapiço and Tavares, 1989; Carrapiço, 1991). Some of the bacterial strains produce the plant hormone indole-3-acetic acid (Forni and Grilli Caiola, 1988; Forni et al., 1992a), although it remains to be seen what significance this may have for the symbiosis. The bacteria also produce capsular and exopolysaccharides which may contribute to the mucilage produced by the cyanobionts (Forni et al., 1992b). As with the cyanobiont, the bacteria appear to remain associated with *Azolla* throughout its life cycle (Carrapiço and Tavares, 1989; Braun-Howland and Nierzwicki-Bauer, 1990; Carrapiço, 1991). Indeed, bacteria are found together with cyanobacterial akinetes beneath the indusium of *Azolla* megasporocarps, and they commence cell division at the same time as the akinetes germinate (Aulfinger et al., 1991).

Little is known of the involvement of bacteria in cyanobacterial symbioses apart from *Azolla*. In the case of *Gunnera*, although bacteria are present with the cyanobacteria in the early stages of gland formation, they do not enter the host cells along with the cyanobacteria (Bergman et al., 1992b).

IV. Host-Cyanobacteria Interactions Prior to Infection

A. What Makes a Successful Cyanobiont?

The two most essential characteristics for a cyanobacterium to form symbioses with plants are the ability to develop both heterocysts and hormogonia (Campbell and Meeks, 1989; Meeks, 1990; Johansson and Bergman, 1994; Bergman et al., 1996; Meeks, 1998). Heterocysts are essential to provide fixed nitrogen for both partners, whereas the hormogonia have the necessary motility to enable the otherwise immotile filaments to gain entry to the plant structures that they will occupy (Section V). In symbioses with non-photosynthetic hosts, the cyanobacteria may benefit the host by carbon dioxide fixation alone, and in these cases the ability to produce heterocysts is not essential. For example in lichen and sponge symbioses non-heterocystous cyanobacteria can sometimes be found, although, in the former at least, the heterocystous forms are still the most common (Section III.A). However, the importance of heterocysts probably extends beyond their ability to fix nitrogen, because they are associated with the formation of a particular type of hormogonia, restricted to cyanobacteria of the families Nostocaceae and Stigonemataceae. In these cyanobacteria hormogonia form by rapid cell division without growth, resulting in a decrease in cell size (Tandeau de Marsac, 1994; Adams, 1997). In other hormogonia-forming cyanobacteria, such as *Oscillatoria*, the hormogonia are formed by simple fragmentation of the parent trichome and, apart from being shorter in length, resemble the original trichome in all respects, including being motile or immotile. It may be that the characteristic small-celled hormogonia formed by reductive cell division, exemplified by *Nostoc*, are the only sort which can infect plants, possibly because efficient infection requires the possession of motility together with chemotaxis, and possibly the ability to initiate hormogonia formation in response to plant extracellular compounds (Section IV.B).

However, the ability to form hormogonia does not guarantee symbiotic competence, because many hormogonia-forming strains fail to establish symbioses with suitable host plants (Meeks, 1990; Johansson and Bergman, 1994; Bergman et al., 1996). The ability to metabolise organic carbon sources is likely to be essential for many of the cyanobionts of plants that receive little or no light, and many cyanobionts are indeed facultative heterotrophs. This capacity also enables them to commit more vegetative cells to become heterocysts, thereby reducing their ability to fix CO_2, but enhancing the capacity to fix N_2, and this elevated heterocyst frequency is commonly seen in the cyanobionts of higher plants, and in those of tripartite lichens in which the green algal symbiont can provide fixed carbon for the other two partners (Section VI.B).

B. Signalling between Potential Partners

Because hormogonia are the primary infective agents in most if not all plant symbioses, it is no surprise that plant hosts produce extracellular signals that trigger hormogonia formation in suitable cyanobacteria. *Anthoceros punctatus* releases a low molecular mass, heat-labile product that stimulates hormogonia formation in *Nostoc* strains (Campbell and Meeks, 1989). This does not occur when the hornwort is cultured in medium containing excess NH_4^+, implying that the compound is produced when the plant is nitrogen starved. The importance of this plant signal is illustrated by a transposon mutant of the symbiotic strain *Nostoc* 29133, which shows increased sensitivity to the *Anthoceros* hormogonia-inducing factor and has a much higher initial frequency of infection of the hornwort than the wild-type (Cohen et al., 1994).

The acidic mucilage of the *Gunnera* stem gland also serves as a hormogonia-inducing signal, the active component of which seems to be a heat-labile protein of less than 12 kDa, present only in the mucilage and not in extracts of leaf, stem, or root, nor in seed rinse (Rasmussen et al., 1994). The mucilage also enhances the growth of compatible cyanobacterial strains, stimulates the synthesis of a 65 kDa protein, and induces the production of a new 40 kDa protein, again only in compatible strains. The mucilage contains sugars and amino acids characteristic of arabinogalactan proteins (AGPs), and reacts with an AGP-specific monoclonal antibody (Rasmussen et al., 1996). A mucilage AGP may therefore influence differentiation of compatible cyanobacteria prior to infection of the plant. Phenolic compounds found in the vicinity of *Gunnera* symbiotic tissue are also potential signals in this system, with possible roles in attraction, defence or the transcriptional activation of cyanobacterial genes involved in symbiosis (Bonnett, 1990; Bergman et al., 1992a). Flavonoids of legumes induce symbiosis-specific gene expression in rhizobia (van Rhijn and Vanderleyden, 1995), and there is evidence of a degree of homology between some of these flavonoid-responsive genes and certain gene sequences of the cyanobacteria infecting *Azolla* (Plazinski et al., 1991) and *Gunnera* (Rasmussen et al., 1996).

Signals that function in the opposite direction, from cyanobacterium to *Gunnera*, have recently been postulated, based on several lines of evidence (Bergman et al., 1996). Firstly, there is some homology between the rhizobial genes *nodEF*, *nodMN* and the *nod* box, and genomic regions of cyanobacteria that infect *Gunnera* (Rasmussen et al., 1996) and *Azolla* (Plazinski et al., 1991). In rhizobia these *nod* genes code for the biosynthesis of lipooligosaccharides that serve as nodule-inducing factors, produced in response to plant flavonoids (van Rhijn and Vanderleyden, 1995). Secondly, target *Gunnera* cells are thought to undergo enhanced mitotic division in response to the presence of symbiotic cyanobacteria, resulting in the formation of a symbiotic "organ" within the stem gland, bearing functional similarities to the root nodules of legumes (Johansson, 1994; Bergman et al., 1996). The growth of the finger-like transfer cells into symbiotic colonies in bryophytes such as *Anthoceros* and *Blasia* (Section V) may also result from the presence of signals derived from the cyanobacteria. Finally, cell division during embryogenesis in carrot is stimulated by extracts of certain cyanobacteria (Wake et al., 1991, 1992). A possible source of the signals that induce such changes in the plant host may be the arabinogalactan proteins released by many cyanobacteria (Bergman et al., 1996); such AGPs are thought to have important roles in plant growth and development (Pennell, 1992).

Interactions between plant and cyanobacterium are not limited to the endophytic associations. Hydroponically grown wheat roots also release a hormogonia inducing factor that is produced at a higher level in nitrogen-free medium (Gantar et al., 1993). The initial colonization of the roots by *Nostoc* 2S9B occurs by motile hormogonia, even when the initial cyanobacterial inoculum contains no hormogonia (Gantar et al., 1993). This is because the hormogonia inducing factor triggers hormogonia formation, thereby overcoming an apparent autoinhibitor of hormogonia development present in the aseriate packages of filaments that are characteristic of *Nostoc* 2S9B.

For efficient infection of most plants the mere presence of hormogonia is unlikely to be sufficient, because of the restricted accessibility of the plant structures involved. The hormogonia must be guided into the plant by chemoattraction. Although the mucilage of *Gunnera* stem glands can induce hormogonia formation, it does not appear to contain a chemoattractant (Rasmussen et al., 1994). However, when starved of combined nitrogen the liverwort *Blasia* releases extracellular signals that not only trigger hormogonia formation (Babic, 1996), but also serve as very effective chemoattractants (Knight and Adams, 1996). Germinating wheat seeds also release hormogonia chemoattractants (Knight and Adams, 1996).

C. Other Important Factors

Little is known about the recognition processes involved in the infection of hosts by cyanobacteria. In the lichen, bryophyte and *Azolla* symbioses, the host plant is known to produce lectins that recognise specific sugar residues on the cyanobacterial cell surface (Braun-Howland and Nierzwicki-Bauer, 1990; Kardish et al., 1991; Bergman et al., 1992b; Honegger, 1993). The use of fluorescein isothiocyanate-conjugated lectins and confocal laser scanning microscopy has recently revealed changes in the extracellular slime of the *Nostoc* cyanobiont of *Geosiphon pyriforme* during its life cycle, with α-D-mannosyl or α-D-glucosyl residues (or glycoconjugates containing them) being characteristic of the post-hormogonia stage that is capable of forming symbiosis with the fungus (Schüßler et al., 1997).

The formation of both endophytic and epiphytic associations requires the attachment of cyanobacteria to the host surface, and both fimbriae (pili) and extracellular polysaccharides have been implicated in the process. Fimbriae have been observed in the *Nostoc* cyanobiont of *Peltigera canina* and may play a role in specificity (Dick and Stewart, 1980). Plant and cyanobacterial polysaccharides are important for the attachment of cyanobacteria to plant cells and inert surfaces (Robins et al., 1986). The involvement of fimbriae and polysaccharides in the colonization of wheat roots by *Nostoc* 2S9B involves two stages: first hormogonia become weakly attached by means of fimbriae; second, the hormogonia develop into heterocystous filaments that become firmly attached by the production of a mucilaginous sheath (Gantar et al., 1993). The characteristics of the sheath seem to be important for the attachment to wheat roots, because the isolated sheath of *Nostoc* 2S9B, which forms a tight association with the roots, is capable of tight attachment to the root surface, whereas that of *Anabaena* C5, which forms only a loose association with the roots, is only loosely attached (Gantar et al., 1995a).

In the *Gunnera-Nostoc* symbiosis the importance of fimbriae has been questioned because they are found in both infective and non-infective *Nostoc* strains, are most abundant in two non-infecting strains, and appear to be absent from some infective strains (Johansson and Bergman, 1994). Nevertheless, fimbriae may serve a non-specific function in initial adhesion of potential symbionts to the *Gunnera* surface, prior to the involvement of further, as yet unknown recognition signals that determine the ultimate success or failure of the infection process. That the plant host plays an important role in determining the fate of a potential symbiont is illustrated by the observation that some *Nostoc* strains that successfully infect both *Gunnera* and bryophytes, become intracellular in the former but not in the latter (Johansson and Bergman, 1994).

V. Host Structures and Their Infection

Unlike the bacterial symbionts in rhizobial and actinorhizal symbioses, cyanobacteria occupy existing plant structures. They are not limited to areas of the plant receiving light, but can be found growing photo- or chemoheterotrophically in, for example, the coralloid roots of cycads, which receive little or no light.

The cyanobacterial symbionts of the liverwort *Blasia* occupy auricles, which are almost spherical structures on the ventral surface of the thallus (Plate 29a-e and Fig. 7), whereas in the much thicker thallus of the hornworts *Anthoceros* and *Phaeoceros* the cyanobacteria are found within the thallus, in slime cavities that open to the ventral surface via narrow slit-like pores (Plate 29A). In both cases the mature symbiotic colonies are up to 0.5 mm in diameter. Oxygen microelectrode measurements have shown that in *Anthoceros* the cavity is microaerobic and, as a result, can support nitrogenase activity in a mutant of the symbiotic *Nostoc* 29133 that is unable to fix nitrogen under aerobic conditions because of a defective heterocyst envelope (Campbell and Meeks, 1992). The cyanobacteria enter *Blasia* auricles, and presumably hornwort slime cavities, as hormogonia, whereupon they lose motility and differentiate heterocysts (Kimura and Nakano, 1990).

In *Azolla*, cyanobacteria occupy a cavity formed in the dorsal leaf lobe by an infolding of the epidermis (Peters et al., 1986; Peters and Meeks, 1989; Braun-Howland and Nierzwicki-Bauer, 1990). This cavity is extracellular and initially open to the outside, becoming closed as the leaf matures. The mature cavity is ellipsoidal inshape and approximately 0.3 mm in length. The cyanobacteria are surrounded by an envelope of plant origin (Uheda, 1986; Braun-Howland and Nierzwicki-Bauer, 1990; Uheda and Kitoh, 1991) and do not fill the cavity, but occupy the periphery (Fig. 3C), the centre being gaseous (Robins et al., 1986; Peters and Meeks, 1989; Peters, 1991; Klein et al., 1992).

The cyanobiont remains intimately associated with the fern throughout the plant's life cycle, which can involve both sexual reproduction, through sporocarp formation, or asexual vegetative fragmentation of the sporophyte. The apical meristem of each *Azolla* stem carries a colony of undifferentiated cyanobiont filaments, resembling hormogonia, protected by the newly developing bi-lobed leaves. As new leaves grow, their developing leaf cavities become infected from this apical *Anabaena azollae* colony (Peters and Meeks, 1989; Braun-Howland and Nierzwicki-Bauer, 1990; Perkins and Peters, 1993; Peters and Perkins, 1993). Continuity of the symbiosis is maintained through the sexual cycle by the packaging of *Anabaena azollae* into the developing megasporocarps, where they differentiate intoakinetes (Peters and Meeks, 1989; Braun-Howland and Nierzwicki-Bauer, 1990; Peters, 1991; Aulfinger et al., 1991; Perkins and Peters, 1993; Peters and

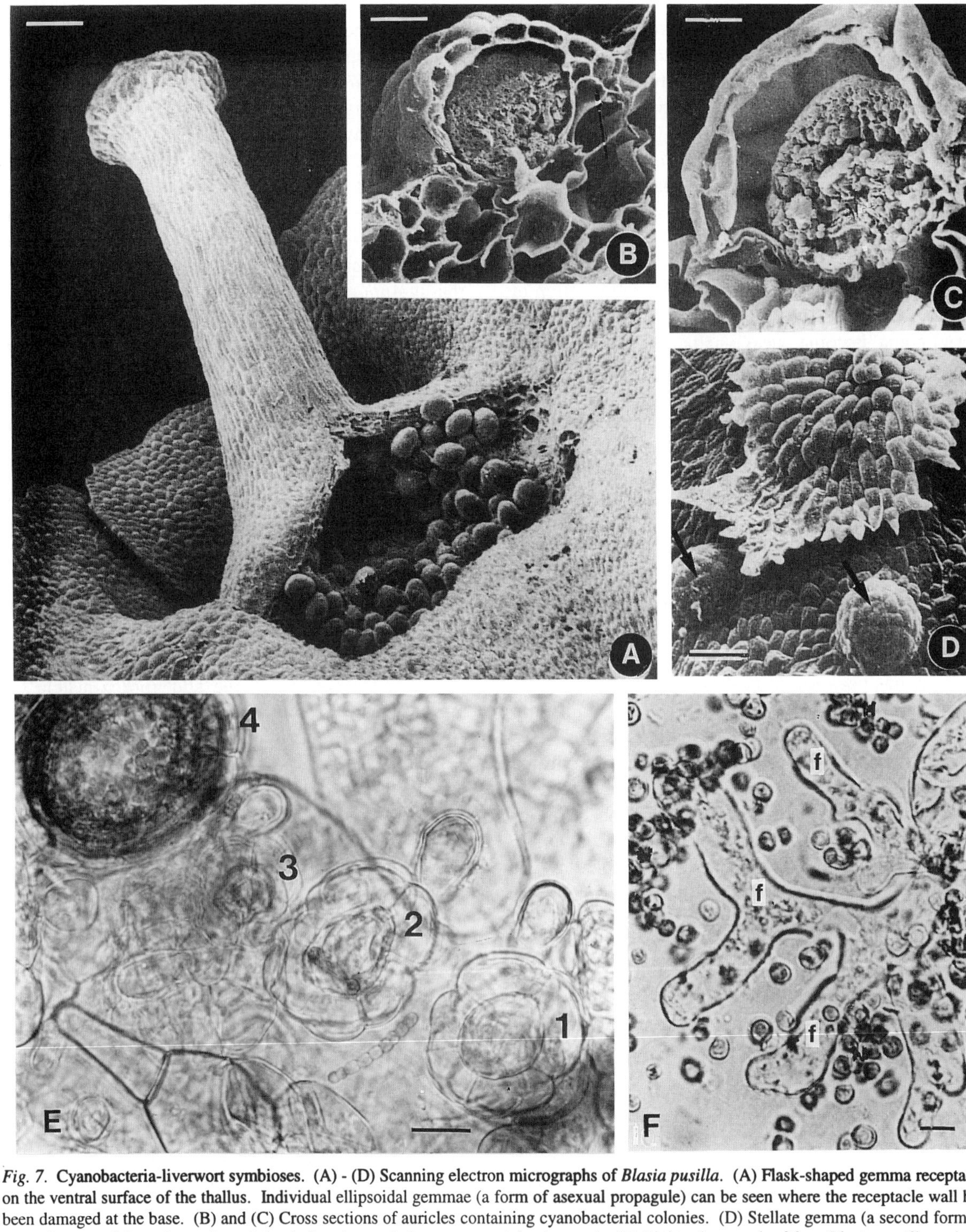

Fig. 7. **Cyanobacteria-liverwort symbioses. (A) - (D) Scanning electron micrographs of *Blasia pusilla*. (A) Flask-shaped gemma receptacle on the ventral surface of the thallus. Individual ellipsoidal gemmae (a form of asexual propagule) can be seen where the receptacle wall has** been damaged at the base. (B) and (C) Cross sections of auricles containing cyanobacterial colonies. (D) Stellate gemma (a second form of asexual propagule) on the surface of the thallus, in proximity to two auricles (arrows). (E) Light micrograph of four *Blasia pusilla* auricles (1-4). Auricle 1 is uninfected, 2 and 3 have recently become infected and contain single cyanobacterial filaments, and 4 contains a developing symbiotic colony. (F) Filamentous protrusions (f) extruded from an infected auricle, surrounded by cyanobacterial cells showing the lack of the filamentous habit that characterises the free-living state. Bars = 100 μm (A), 20 μm (B and C), 50 μm (D), and 10 μm (E and F). (Reproduced with permission from: (A) - (D) Honnegger, Flora 170: 290--302, 1980, © Gustav Fischer Verlag; (F) Rodgers and Stewart, New Phytol. 78: 441--458, 1977, © Cambridge University Press; (E) photograph courtesy S. Babic).

Perkins, 1993; Grilli Caiola et al., 1993), which are a spore-like resting stage. When the megasporocarp germinates the akinetes also germinate and serve as the inoculum for the embryonic sporophyte.

Cycad coralloid roots are usually found beneath the soil surface, although they may extend above the surface in older plants (Lindblad et al., 1985; Lindblad and Bergman, 1990; Peters, 1991; Ahern and Staff, 1994). The cyanobionts occupy the well-defined open zone between the inner and outer cortical layers; a region traversed by elongated root cells (Fig. 2). Coralloid roots (Fig. 2A) begin as precoralloids that are produced by axenically grown plants and are therefore not a response to the presence of cyanobacteria. However, invasion by cyanobacteria stimulates irreversible changes in the precoralloids, transforming them into typical coralloid roots (Ahern and Staff, 1994). The mode of entry of the cyanobacteria into the precoralloid root has long been a source of speculation, although a break in the dermal layer of the roots seems to be the most likely site of entry (Lindblad and Bergman, 1990; Peters, 1991; Ahern and Staff, 1994). Once inside the root the cyanobacteria follow a continuous channel from the surface into the cortex (Lindblad and Bergman, 1990; Staff and Ahern, 1993; Ahern, 1993; Ahern and Staff, 1994).

In the *Gunnera-Nostoc* association the cyanobacteria invade mucus secreting glands at the base of each petiole (Fig. 8; Towata, 1985; Peters et al., 1986; Bonnett, 1990; Johansson and Bergman, 1990; Bergman et al., 1992b). This is the only plant-cyanobacterium symbiosis in which the cyanobacterium penetrates the host cell walls and becomes intracellular. The glands consist of a number of papillae, separated by intercellular channels that aid in infection, and during gland development the epidermal layer is cast aside, permitting entry of the cyanobacteria (Bonnett, 1990). Ultrastructural studies of the infection process have revealed six discrete stages (Johansson and Bergman, 1990). The first stage is the formation of the gland, accompanied by the excretion of a carbohydrate-rich acidic mucus that covers the surface and gathers a population of *Nostoc*, fungi and bacteria. This mucilage contains at least three signalling compounds that induce hormogonia formation, and influence cyanobacterial gene expression and the growth of compatible *Nostoc* strains (Section IV.B). During stage two of the infection process cyanobacteria that have accumulated on the gland are induced to differentiate hormogonia, and during stage three these migrate towards the interior of the gland via narrow intercellular channels formed by the dissolution of cell walls during the early stages of gland formation (Johansson and Bergman, 1990; Bergman et al., 1992b). In the interior of the gland the *Gunnera* cell walls dissolve and the cyanobacteria enter the cells without penetrating the host plasmalemma (stages four and five). It is not clear whether the dissolution of host cell walls is caused by the plant, the cyanobacteria or the other bacteria present, but only the cyanobacteria enter the host cells (Bergman et al., 1992b). In the final stage the *Nostoc* hormogonia differentiate a high frequency of heterocysts (Section VI.B) and begin fixing nitrogen, while the gland channels disappear, rendering further infection impossible.

Although in all of the above examples the cyanobacteria invade existing plant structures, the presence of the cyanobionts does induce some changes in the plant tissues. For example, in the cycad *Macrozamia communis* most uninfected precoralloid roots senesce, whereas infected precoralloids are stimulated to differentiate into coralloids (Ahern and Staff, 1994). The filamentous protrusions (Fig. 7F) that grow into the cyanobacterial colony in *Anthoceros* are stimulated by the presence of the cyanobiont (Rogers and Stewart, 1977), and there is evidence that cyanobacteria stimulate cell division of specific *Gunnera* cells, resulting in the formation of a symbiotic "organ" within the stem gland (Bergman et al., 1996). Even in a non-host plant such as the carrot (*Daucus carota*) extracts of some cyanobacteria stimulate cell division during embryogenesis (Wake et al., 1991, 1992). However, in contrast to these relatively minor changes in the host plant, the lichens are unique in possessing an overall morphology completely different to the free-living partners, and the areas occupied by the cyanobacterial or algal partners do not exist in the un-lichenised fungus (Section II.B.1). The phycobiont is either randomly distributed within the thallus, or occupies distinct zones (Fig. 9B). In the tripartite lichens, containing both a green alga (the primary photobiont) and a cyanobacterium (the secondary photobiont), the two phycobionts are always separated, occupying either distinct layers or specialised structures known as cephalodia, either on the surface or within the thallus (Honegger, 1991, 1992, 1993; Ahmadjian, 1993; Hill, 1994).

Geosiphon pyriforme is unusual because the cyanobiont is intracellular, entering the fungal hyphae by endocytosis (Mollenhauer, 1991; Kluge et al.,

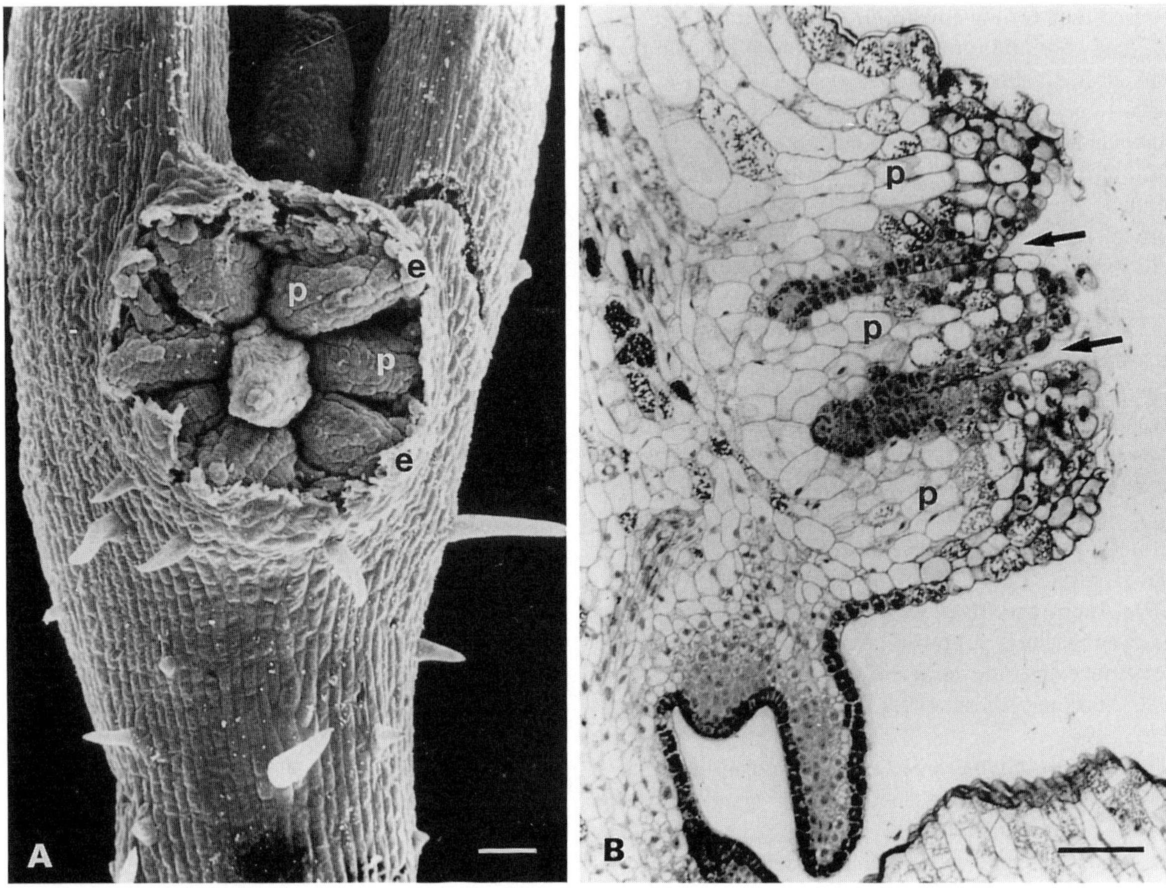

Fig. 8. **The angiosperm *Gunnera chilensis*. (A) Scanning electron micrograph showing a stem gland located at the base of two leaf petioles. The gland is just breaking through the plant epidermis (e) and the individual papillae (p) can be seen, separated by channels leading into the** stem tissue. (B) Longitudinal section of a stem gland. The channels (arrows) separating the papillae (p) are lined by a layer of electron dense cells. (Reproduced with permission from Bergman et al., New Phytol. 122: 379--400, 1992, © Cambridge University Press). Bars = 100 μm.

1994; Mollenhauer et al., 1996). Only immotile "primordia", that develop from motile hormogonia, are incorporated by the host (Mollenhauer et al., 1996; Schüßler et al., 1997). When the tips of the fungal hyphae encounter suitable *Nostoc* filaments the fungal wall softens and the cyanobacterial vegetative cells, but not any existing heterocysts, are engulfed by the fungal plasmalemma, by a process of phagocytosis (Fig. 4C, D; Mollenhauer et al., 1996; Mollenhauer and Mollenhauer, 1996). New heterocysts develop from the engulfed vegetative cells. Once infected the hyphae swell to form pear-shaped multinucleate bladders, up to 0.5 mm in diameter and 2 mm in length, containing the cyanobacteria (Fig. 4A and B; f 28F). The bladder wall contains fibrillar chitin (Schüßler et al., 1996) and possesses an electron dense outer layer, which may explain the unusually small limiting pore size (pore radii approximately 0.5 nm) that results in the bladder wall acting as an osmotic barrier impermeable to glucose and virtually impermeable to sucrose (Schüßler et al., 1995). The bladders presumably develop as a result of infection, because bladders lacking endosymbionts are never found, andthe endosymbionts never occur elsewhere in the fungal hyphae (Kluge et al., 1991).

The *Nostoc* filaments occupy the inner surface of the bladders, in a cup-shaped compartment known as

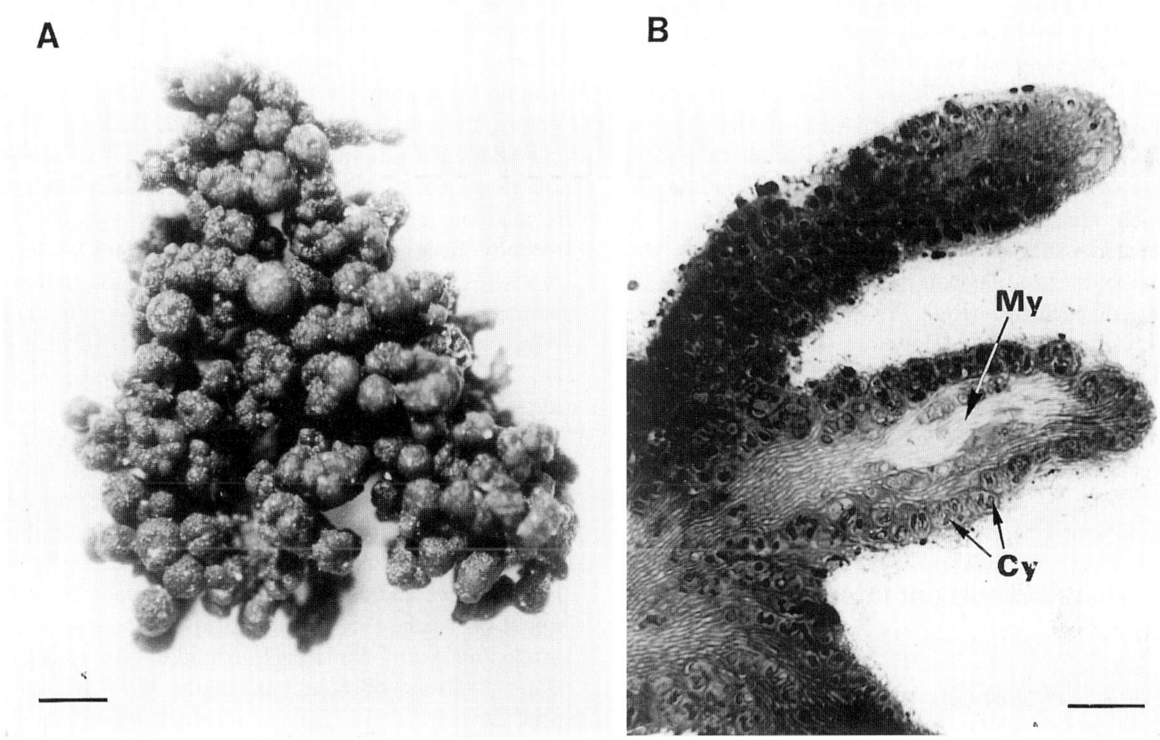

Fig. 9. The lichen *Lichina confinis.* (A) View of the thallus. (B) Longitudinal section of part of the thallus showing the cyanobiont (Cy) and mycobiont (My). Bars = 0.5 mm. (A) and 50 μm (B). (Reproduced with permission from Janson et al., New Phytol. 124: 149--160, 1993, © Cambridge University Press).

the symbiosome, bordered by a host membrane (Schüßler et al., 1996). The region between the symbiosome membrane and the *Nostoc* cell wall, known as the symbiosome space, is filled with an amorphous material, as are other areas of the compartment not occupied by the cyanobiont. The precise nature of the amorphous material is not known, but it appears to contain components characteristic of the fungal cell wall, including polysaccharides and fibrillar chitin, and is probably produced by the fungal (symbiosome) plasma membrane (Schüßler et al., 1996). The symbiosome space, which is effectively the interface between the *Nostoc* cell wall and the fungal plasma membrane, has similarities with the symbiotic interface between the fungal wall and the plant plasma membrane in arbuscular mycorrhizas (Schüßler et al., 1996). Indeed, the fungus *Geosiphon pyriforme* is closely related to the arbuscular mycorrhiza-forming fungi of the genus *Glomus* (Schüßler et al., 1994).

In all plant- and fungus-cyanobacteria symbioses the host benefits from nutrients, primarily combined nitrogen, provided by the cyanobiont. In the plant symbioses there is transfer in the opposite direction, with the cyanobiont receiving mostly fixed carbon from the plant. Efficient transfer of nutrients is therefore essential, and in many cases there are obvious structures that aid in this. In the liverwort and hornwort symbioses the host cells lining the symbiotic cavity produce branched, filamentous protrusions extending into the symbiotic colony (Fig. 7F), and their transfer cell morphology implies an involvement in the exchange of metabolites (Stewart and Rogers, 1977). In the *Azolla* leaf cavity, branched and simple hairs (Fig. 3D) possess transfer cell characteristics and are thought to play a role in exchange of nutrients between the cyanobiont and the plant (Calvert et al., 1985; Braun-Howland and Nierzwicki-Bauer, 1990). Branched hairs are thought to be involved in nitrogen exchange, and simple hairs in carbon exchange (Calvert et al., 1985). A similar transfer cell morphology is seen in the elongated cells that traverse the open area, occupied by cyanobacteria, between the inner and outer cortex of the cycad coralloid roots (Fig. 2C; Lindblad et al., 1985). In lichens there is close physical contact between the cyanobiont and the fungal hyphae, without the latter penetrating the cyanobacterial cell wall (Rai, 1990b). In addition, fungal haustoria, which are specialised peg-like projections of the hyphae, can be found closely associated with cyanobacterial filaments and may play a role in nutrient exchange (Rai, 1990; Janson et al., 1993; Ahmadjian, 1993). However, there is confusion about the frequency and function of haustoria, and the extent to which they can penetrate the cyanobiont cell walls (Ahmadjian, 1993). In contrast, no additional structures are needed in *Gunnera* and *Geosiphon pyriforme* because the intracellular location of the cyanobiont ensures efficient exchange of metabolites.

VI. Host-Cyanobiont Interactions Post-Infection

A. Cell division Control and Hormogonia Formation

The establishment of a stable symbiosis between cyanobacterium and host demands that the latter exerts considerable control over the free-living characteristics of the former. For example, the growth rate of the cyanobacterium must match that of the host, yet free-living cyanobacteria have doubling times considerably greater than the same symbiotically associated strains (Peters and Meeks, 1989; Hill, 1989; Braun-Howland and Nierzwicki-Bauer, 1990). The rate of cyanobiont cell division is also linked to the varied rate of growth of different regions of host tissue. For example, in *Azolla* the rate of cyanobiont cell division is slower in leaf cavities than at the shoot apex where host cell growth is rapid (Hill, 1989). The mechanism of this growth control is unknown, but for reasons discussed in Section C below, it would seem not to be limitation of nitrogen availability, even though the host takes most of the nitrogen fixed by its partner. Indeed, it may be that the primary control is exerted on cell division of the symbiont, resulting in reduced demand for nutrients, the excess becoming available for use by the host (Hill, 1989).

The production of hormogonia-inducing signals by the host plant (Section IV.B) creates a potential conflict with the requirement for symbiotically associated cyanobacteria to develop heterocysts. Hormogonia do not form heterocysts, and a viable symbiotic colony would not be produced if, immediately upon infection, the cyanobacteria were induced to form hormogonia once more. There must therefore be a means of repressing hormogonia formation, and evidence for such a system has recently been obtained (Cohen and Meeks, 1997; Meeks, 1998) by analysis of a transposon-induced mutant of *Nostoc punctiforme* that is 50-fold more infective than the wild-type in *Anthoceros* (Cohen et al., 1994). Two genes, *hrmA* and *hrmU*, were identified as part of an operon, and expression of both was induced by an aqueous extract of *Anthoceros* tissue, but not by the hormogonia inducing factor (HIF) of the hornwort. The aqueous extract contains a hormogonia repressing factor (HRF) that inhibits HIF-induced hormogonia formation in wild-type *N. punctiforme*, and it seems that the mutant is more infective because it continues to respond to HIF, producing hormogonia and extending its infective state. These observations imply that the gene products of the *hrmUA* operon block hormogonium formation, perhaps by the production of an inhibitor or by catabolism of an activator (Cohen and Meeks, 1997).

B. Morphological Modifications

The gross cellular morphology of cyanobacteria is frequently altered in symbiosis. In liverworts and hornworts cyanobiont vegetative cells become enlarged and spherical, and the cell to cell junctions weaken, such that even gentle pressure will cause the cells to part company (Fig. 7F; Meeks, 1990). In cyanolichens filamentous cyanobionts such as *Scytonema* and *Calothrix* can become unicellular (Rai, 1990), and *Nostoc* cell size increases (Ahmadjian, 1993). In mature *Azolla* leaf cavities vegetative cell size increases significantly, and although consistent within each filament, cell size and shape can be highly variable between filaments (Braun-Howland and Nierzwicki-Bauer, 1990; Peters, 1991). The *Nostoc* cyanobiont of *Cycas revoluta* increases in cell size when symbiotically associated (Sharma et al., 1992), as does the *Nostoc* symbiont of *Geosiphon pyriforme* after phagocytosis by the fungal host (Kluge et al., 1992, 1994; Mollenhauer et al., 1996).

Table 2. Heterocyst frequency and glutamine synthetase characteristics of symbiotically associated cyanobacteria.

Host	Heterocyst frequency[a]	Glutamine synthetase		Form of combined nitrogen released
		Amount of protein[b]	Specific activity[b]	
Cycads	40-45 %	100 %	100 %	citrulline/glutamine
Gunnera	60-80 %	100 %	70 %	NH_4^+
Hornworts and liverworts	30-50 %	~86 %	~38 %	NH_4^+
Azolla	30-40 %	5-10 %	5-10 %	NH_4^+
Lichens: bipartite	10-55 %	<10%	<10%	NH_4^+
tripartite	4-8 %	<10%	<10%	NH_4^+

[a] Heterocyst frequencies are expressed as a percentage of total cells. Typical values for free-living cyanobacteria are 4-10%.
[b] Values are for the symbiont as a percentage of the same cyanobacterium in the free-living state.
Table compiled from Bergman et al. (1992b) and Rai (1990a, 1990b).

In addition to these morphological changes, many symbiotically associated cyanobacteria commonly exhibit heterocyst frequencies considerably higher than in the free-living state (Table 2). This is true of all cases where the host is photosynthetic, permitting the cyanobacteria to enhance their nitrogen-fixing capabilities by increasing heterocyst numbers at the expense of vegetative cells. The resulting loss of CO_2-fixing capacity can be compensated by the supply of carbon skeletons by the host, or in the case of tripartite lichens, by the green algal partner.

Heterocyst frequency is usually lowest in new symbiotic colonies and highest in older colonies. For example, in the newest *Azolla* leaf cavities, close to the apical meristem of each stem, the heterocyst frequency is zero or very low, but increases as leaves age, until senescence begins and nitrogenase activity declines (Kaplan et al., 1986; Peters et al., 1986; Grilli Caiola et al., 1989; Canini et al., 1990; Bergman et al., 1992b). Indeed, in basal (oldest) leaf cavities heterocyst frequency is still high, whereas nitrogenase activity is lowered (Canini et al., 1990), possibly as a result of the degeneration of heterocysts evident by electron microscopy (Braun-Howland et al., 1988). However, the results of DAPI staining of DNA in the *Azolla* cyanobiont have been taken to indicate that heterocyts are still functional in basal leaf cavities (Canini et al., 1992a).

In cycad coralloid roots the frequency of heterocysts is lowest at the root tip and highest in the basal, older parts, where high frequencies of two, three or even four adjacent heterocysts are found (Lindblad et al., 1985a, b; Lindblad and Bergman, 1990; Peters, 1991). Similarly, heterocyst frequency in *Gunnera* is lowest at the growing tip, becoming higher in older parts of the plant (Lindblad et al., 1985a, b; Söderbäck et al., 1990; Bergman et al., 1992b; Stock and Silvester, 1994). In two-membered lichens cyanobiont heterocyst frequency is usually little changed from the free-living state (Table 2), presumably because the cyanobiont must supply both fixed carbon and nitrogen for the symbiosis. However, heterocyst frequencies in three-membered lichens are considerably elevated (Table 2), although they are often lowest in cephalodia in the apical (newer) regions of the thallus, and increase in the older tissues (Rai, 1988, 1990; Janson et al., 1993).

Increases in the heterocyst frequency of free-living cyanobacteria can be induced by the manipulation of environmental conditions, such as the provision of fructose (Rozen et al., 1986, 1988), immobilisation of cells in polyurethane and polyvinyl foams and other hollow matrices (Shi et al., 1987; Shi and Hall, 1988; Mahesh and Kannaiyan, 1993), and exposure to increased light intensity or the amino acid analogue 7-azatryptophan (Adams, 1992). However, the frequencies achieved are never as high as those in symbiotic colonies, and there are reasons to believe that the mechanisms that regulate heterocyst development and frequency in symbiosis are different to those operating in the free-living state. For example, in free-living cyanobacteria the stimulus to develop heterocysts is nitrogen starvation (Wolk et al., 1994), yet in symbiosis with plants the cyanobacteria show no sign of nitrogen starvation (see the end of Section VI.C for a discussion of this).

Further evidence comes from studies of a *Nostoc* mutant defective in nitrate assimilation (Campbell and Meeks, 1992). Nitrate fails to repress nitrogen fixation and heterocyst development in the free-living mutant strain, but represses both when the strain is symbiotically associated with the hornwort *Anthoceros*. Ammonium is an effective repressor in both wild-type and mutant whether free-living or symbiotically associated. The repressive effect of nitrate in the symbiotically associated mutant does not result from the accumulation of NH_4^+, generated by the reduction of NO_3^- to NH_4^+ in *Anthoceros*, because NH_4^+ pools are the same in tissue from hornwort grown in the presence and absence of NO_3^-. The implication of these observations is that nitrogenase expression and heterocyst development in symbiotically associated *Nostoc* are controlled by plant signals. A greater understanding of the factors involved in the regulation of heterocyst development in symbiosis may come from the application of molecular genetic techniques to cyanobionts such as *Nostoc* 29133 (Campbell et al., 1996).

The true significance of the elevated heterocyst frequencies found in many cyanobacterial symbioses is difficult to assess for several reasons. Distinguishing heterocysts within symbiotic colonies can be very difficult, because they often loose their characteristic regular shape and thickened walls (Meeks, 1990). The loss of the latter may occur as a result of the anaerobic nature of some of the symbiotic cavities (Lindblad et al., 1991b; Campbell and Meeks, 1992), because the formation of the additional heterocyst wall layers is inhibited under anaerobiosis (Rippka and Stanier, 1978). Even when heterocysts can be recognised with certainty by light or electron microscopy, their full metabolic capabilities cannot be known, and at least in bryophytes it seems likely that some of the heterocysts are senescent or dead (Meeks, 1990). A similar conclusion may be drawn from the cycads *Zamia* and *Cycas* in which heterocyst frequency is highest in the basal region, yet nitrogenase activity is highest at the coralloid root tip (Lindblad and Bergman, 1990). Notwithstanding these observations, there is a general correlation between the increase in heterocyst frequency of many symbiotically associated cyanobacteria and elevated rates of nitrogen fixation.

C. N_2 Fixation and Transfer of Fixed Nitrogen

In plant-cyanobacteria symbioses the primary function of the cyanobiont is to provide the plant with combined nitrogen, and all the known cyanobionts in these cases are indeed capable of nitrogen fixation (Rai, 1990). In these plant symbioses (if not all cyanobacterial symbioses) nitrogenase protein is confined to the heterocysts (Bergman et al., 1986; Braun-Howland et al., 1988; Bergman and Rai, 1989; Rai et al., 1989; Söderbäck et al., 1990), that provide the necessary anaerobic conditions for nitrogenase. This may be important in the aerobic leaf cavities of *Azolla* (Grill Caiola et al., 1989; Mutuskin and Kolesnikov, 1991), in which additional protection from superoxide radicals may be provided by iron superoxide dismutase, shown by immunogold labelling to be highest in cyanobionts in leaves with the highest nitrogenase activity (Canini et al., 1992c). In contrast, the symbiotic colonies in the cycad *Macrozamia riedlei* are thought to be microaerobic, and the freshly isolated cyanobionts exhibit increased sensitivity to oxygen (Lindblad et al., 1991b), possibly because of the incomplete development of heterocyst wall layers in the absence of oxygen (Rippka and Stanier, 1978).

In many cyanobacteria-plant symbioses the cyanobacteria are capable of supplying the entire nitrogen requirements of the host. In even the largest *Gunnera* plants, with leaves two meters across, this is achieved by a cyanobiont mass that may be as little as 1 % of the total plant weight (Osborne et al., 1992). Indeed, in the case of *Gunnera*, the plant may have little capacity to use alternative sources of combined nitrogen, and consequently be almost entirely dependent on the cyanobionts for its nitrogen (Osborne, 1989; Osborne et al., 1992). Compared with the same free-living strains, symbiotically associated cyanobacteria exhibit increased rates of nitrogen fixation, this increase generally being least in the newest colonies and greatest in older colonies, and this correlates with increases in heterocyst frequency in symbiosis (Section VI.B). The best correlation seems to be with the frequency of single rather than total heterocysts, possibly because the many multiple (adjacent) heterocysts present in for example *Gunnera* may function inefficiently as a result of poor transfer of photosynthate from vegetative cells (Bergman et al., 1986, 1992). In *Gunnera* at least, the lowered rate of nitrogenase activity in older tissues is not due to reduced levels of nitrogenase protein, but is most likely due to a reduction in the supply of photosynthates (Söderbäck et al., 1990).

Dinitrogen fixed by the cyanobiont is released to the host as ammonia in the case of *Anthoceros*, *Blasia*, *Azolla*, *Gunnera* (Meeks et al., 1985a, b; Bergman et al., 1992; Silvester et al., 1996) and many lichens (Rai et al., 1990), and in an unknown organic form (possibly citrulline or glutamine) in cycads (Table 2). The proportion of combined nitrogen retained by the symbiont can be as little as 5% in some lichens (Rai, 1990a; Ahmadjian, 1993), 10% in *Gunnera* (Silvester et al., 1996), 20% in the hornwort *Anthoceros* (Meeks et al., 1985a), and is thought to be 60% in *Azolla* (Meeks et al., 1985b, 1988; Peters and Meeks, 1989). In the case of *Azolla* this leads to ammonium concentrations of 0.8 mM in leaf cavities at the apex, to 6 mM in the basal (oldest) leaf cavities (Canini et al., 1990). Ammonia release by the cyanobiont is a consequence of decreased activity of glutamine synthetase, the first enzyme in the glutamine synthetase-glutamate synthase (GS-GOGAT) pathway, which is the primary route of ammonia assimilation in cyanobacteria and in bryophytes and *Azolla* (Meeks, 1990; Bergman et al., 1992), but not in lichens which employ glutamate dehydrogenase (GDH; Rai, 1990b).

There are several means by which the decrease in GS activity is achieved (Table 2). In *Azolla* it results from a decrease in the amount of enzyme in the cyanobiont (Lee et al., 1988; Peters and Meeks, 1989), which correlates with a ten-fold decrease in the level of GS (*glnA*) mRNA (Nierzwicki-Bauer and Haselkorn, 1986). Similarly, in cyanolichens both GS and GOGAT activity is reduced by over 95% due to a reduction in the amount of GS protein (Rai, 1990a). In contrast, in the hornwort *Anthoceros* an undetermined post-translational modification of the enzyme is thought to be responsible for the decreased GS activity, because the amount of GS protein in free-living and symbiotically associated *Nostoc* is little different (Table 2; Joseph and Meeks, 1987; Lee et al., 1988; Rai et al., 1989; Meeks, 1990). In the *Gunnera* symbiont GS specific activity is 70% of that in the free-living state althought the amount of protein is unchanged (Table 2; Bergman et al., 1992b; Silvester et al., 1996). In contrast, GS activities and protein levels in cycads are little changed in the symbiotically associated *Nostoc* (Lindblad and Bergman, 1986), and this is consistent with the release of organic nitrogen, rather than NH_4^+, to the plant (Table 2; Pate et al., 1988).

In the cyanolichens (Hällbom et al., 1986), bryophytes (Rai et al, 1989) and *Azolla* (Bergman and Rai, 1989) the level of GS protein in heterocysts is similar to that of vegetative cells. However, in the apical regions of the lichen *Lichina confinis* thallus, GS protein cannot be detected in heterocysts of the *Calothrix* symbiont, although it remains relatively abundant in vegetative cells (Janson et al., 1993). This contrasts with the situation in free-living cyanobacteria in which heterocysts have approximately twice the GS activity and protein levels of vegetative cells (Bergman et al., 1985; Rai et al., 1989; Renström-Kellner et al., 1990; Bergman et al., 1992b).

Initial uptake of the ammonia released by the cyanobionts in bryophytes and *Azolla* occurs via the GS-GOGAT pathway of the host (Meeks et al., 1985b; Peters et al., 1985; Meeks, 1990; Rai, 1990), whereas in cyanolichens it occurs via fungal glutamate dehydrogenase (Rai, 1990). The form in which nitrogen is transported from the symbiotic tissue to the rest of the host is known in only a few cases. In *Azolla* the major compounds moving from the leaf cavity to the stem apex are thought to be glutamate, glutamine, ammonia and possibly a glutamate derivative (Peters et al., 1985; Braun-Howland and Nierzwicki-Bauer, 1990; Bergman et al., 1992), whereas in cycads glutamate and citrulline move from the roots to the stem (Pate et al., 1988; Bergman et al., 1992), and in tripartite cyanolichens alanine is thought to be form of nitrogen transferred from the cephalodia to the main thallus (Rai, 1990). In *Gunnera monoica* the phloem may play an important role in the transport of recently fixed nitrogen from mature to apical regions of the plant, and to the leaves (Stock and Silvester, 1994). This is important because the apex is a region of intense meristematic activity but has the lowest specific nitrogenase activity. The phloem is also involved in the transport of carbon into the symbiotic tissues, implying bidirectional transport (Stock and Silvester, 1994).

In plant-cyanobacteria symbioses the transfer to the host of much of the nitrogen fixed by the cyanobiont does not seem to lead to nitrogen starvation of the cyanobiont. There are several reasons for this conclusion. Firstly, the phycobiliproteins, which serve as both photosynthetic accessory pigments and nitrogen reserves in cyanobacteria, are still present in the cyanobionts of bryophytes (Rai et al, 1989; Meeks, 1990), *Azolla* (Kaplan et al., 1986; Braun-Howland and Nierzwicki-Bauer, 1990), cycads (Lindblad and Bergman, 1989) and *Gunnera* (Söderbäck and Bergman, 1992). Secondly, the unique cyanobacterial nitrogen reserve, cyanophycin, a co-polymer of arginine and aspartic acid (Simon, 1987), is also present in the cyanobionts of bryophytes (Rai et al., 1989), *Gunnera* (Söderbäck et al., 1990; Jager et al., 1997), cycads (Lindblad et al., 1985a) and *Azolla* (Bergman et al., 1992b). Nitrogen starvation would be expected to result in breakdown of nitrogen reserves and hence a reduction in the cellular contents of both phycobiliproteins and cyanophycin.

D. CO_2 Assimilation and Transfer of Carbon

The primary route of CO_2 fixation in free-living and symbiotically-associated cyanobacteria is the Calvin cycle, with ribulose-1, 5-bisphosphate carboxylase/oxygenase (Rubisco) as the primary carboxylating enzyme (Tabita, 1994). The rates of light-dependent O_2 evolution and CO_2 fixation in the cyanobionts of bryophytes (Meeks, 1990), cycads (Lindblad and Bergman, 1990; Lindblad et al., 1991b) and *Gunnera* (Bonnett, 1990; Söderbäck and Bergman, 1993) are either greatly reduced, or undetectible, immediately after their separation from the host plant (Table 3). In these cases the cyanobiont grows photo- or chemoheterotrophically, receiving fixed carbon from its photosynthetic host, probably in the form of sucrose (Steinberg and Meeks, 1991; Bergman et al., 1992; Söderbäck and Bergman, 1993), as is the case with *Azolla* (Peters and Meeks, 1989; Braun-Howland and Nierzwicki-Bauer, 1990; Peters, 1991). The cyanobiont of *Azolla* can fix CO_2, but this probably accounts for no more than 5% of the carbon fixation of the symbiosis (Kaplan and Peters, 1988; Peters and Meeks, 1989) and the level of Rubisco mRNA is at least five-fold lower than in free-living *Anabaena* (Nierzwicki-Bauer and Haselkorn, 1986). Most CO_2 fixation is thought to occur in vegetative cells at the apical meristem of *Azolla*, with little in the older leaves (Braun-Howland and Nierzwicki-Bauer, 1990). Surprisingly, immediately after isolation from *Azolla*, *Anabaena azollae* has a rate of light-dependent CO_2 fixation similar to that of the free-living organism (Table 3).

Immediately after its separation from symbiosis, the *Nostoc* symbiont of *Anthoceros* has a rate of light-dependent CO_2 fixation eight-fold lower than the same cyanobacterium grown in the free-living state (Table 3; Steinberg and Meeks, 1989; Meeks, 1990). The level of Rubisco protein is little different in the two cases (Steinberg and Meeks, 1989; Rai et al., 1989; Meeks, 1990), implying that activity is regulated by post translational modification of the enzyme by an unknown mechanism (Steinberg and Meeks, 1989; Meeks, 1990). In contrast, the freshly isolated cyanobiont of *Gunnera chilensis* shows Rubisco activity higher than the free-living *Nostoc*, even though the symbiotically associated symbiont has lowered CO_2-fixing activity (Table 3; Söderbäck and Bergman, 1993). This rules out a post-translational control such as that in *Anthoceros*. It seems that the photosynthetic machinery of the symbiont is largely intact, with chlorophyll *a* (Söderbäck et al., 1990), phycobiliproteins and Rubisco (Söderbäck and Bergman, 1992) being present at levels equivalent to the free-living symbiont, and that a reduction in photosynthetic electron transport is responsible for the reduced *in vivo* CO_2-fixing activity of the cyanobiont. A similar situation may arise in cycads such as *Cycas revoluta*, in which, despite the lack of *in vivo* CO_2 fixation, considerable Rubisco activity can be measured in crude extracts of the freshly isolated cyanobiont (Table 3; Lindblad et al., 1987; Lindblad and Bergman, 1990). An exception may be the cyanobiont from *Cycas circinalis* which, immediately after isolation, shows high rates of photosynthetic O_2 evolution, possibly as a result of the coralloid roots, and therefore the cyanobionts, being close to the surface and thus exposed to light (Perraju et al., 1986).

The cyanobionts of both bipartite and tripartite lichens are capable of photosynthetic CO_2 fixation, and Rubisco protein can be detected by immunogold labelling (Table 3; Bergman and Rai, 1989; Janson et al., 1993). They transfer some of this carbon, in the form of glucose, to the fungal host (Rai, 1990b; Feige and Jensen, 1992; Bogner et al., 1993; Ahmadjian, 1993; Schroeter, 1994; Hill, 1994), where it is thought to be rapidly converted to mannitol (Ahmadjian, 1993), and in bipartite cyanolichens the mycobiont is entirely dependent on this for its carbon requirements. Maximum photosynthetic rates in lichens are attained when the thallus is damp rather

Table 3. Characteristics of light-dependent CO_2 fixation and ribulose bisphosphate carboxylase (Rubisco) in symbiotically associated cyanobacteria.

Host	Light-dependent CO_2 fixation[a]	Rubisco	
		Amount of protein[a]	Specific activity[a]
Azolla	100%[c]	n.d.[b]	n.d.
Hornwort (*A. punctatus*)	12%[d]	100%[d]	12%[d]
Cycads	0[e]	n.d.	87%[e]
Gunnera	<2%[f]	100%[g]	160%[f]
Lichens: bipartite	n.d.	~100%[h]	n.d.
tripartite	~8%[I]	~100%[h]	~8%[i]

[a] Values are expressed as a percentage of those for the free-living cyanobcateria.
[b] The amount of protein has not been determined but Rubisco mRNA is <20% of that in free-living *Anabaena* (Nierzwicki-Bauer and Haselkorn, 1986; see Section VI.D).
Table compiled from the following references: [c] Peters and Meeks (1989); [d] Steinberg and Meeks (1989); [e] Lindblad et al. (1987); [f] Söderbäck and Bergman (1993); [g] Söderbäck and Bergman (1992); [h] Bergman and Rai (1989); [i] Rai et al. (1981).
n.d. = not determined

than wet, because CO_2 can diffuse rapidly through the air spaces in a damp thallus, but diffusion is reduced in a wet thallus in which the air is replaced by water, and in which the fungal hyphae swell and block the gaseous pores (Lange et al., 1988; Cowan et al., 1992; Hill, 1994). However, in wet conditions cyanolichens are at an advantage because the cyanobionts possess a CO_2-concentrating mechanism to ensure that photosynthesis is unaffected by the presence of liquid water (Palmqvist, 1993). This mechanism operates under conditions of low CO_2 availability, and is thought to help avoid limitation of photosynthesis by slow diffusion of CO_2 through the lichen tissue (Badger et al., 1993; Palmqvist, 1993, 1995; Palmqvist et al., 1994; Máguas and Brugnoli, 1996). The CO_2-concentrating mechanism present in some green algal symbionts is far less efficient (Badger et al., 1993; Palmqvist, 1993).

In tripartite lichens the cyanobiont may contribute little fixed carbon to the partnership, but, with its elevated heterocyst frequency, concentrate on nitrogen fixation (Palmqvist, 1993). However, separate measurements of photosynthetic activity in the green algal and cyanobacterial partners in the tripartite lichen *Placopsis contortuplicata* imply that both act as photobionts (Schroeter, 1994). In *Geosiphon pyriforme* the photosynthetic activity of the cyanobiont may be higher in symbiosis than in its free-living state (Bilger et al., 1994), and the low permeability of the fungal bladder wall enclosing the symbiotic colony (Section V) implies that the fungus is largely reliant on the cyanobiont for carbohydrate provision (Schüßler et al., 1995).

Little is known about carbon fixation and transfer in the many other cyanobacterial symbioses. Cells of *Prochloron* (Section II.E) contain carboxysomes (Swift and Leser, 1989; Bullerjahn and Post, 1993; Post and Bullerjahn, 1994) and a partial sequence of Rubisco has been determined (Shimada et al., 1995). Carbon fixed by *Prochloron* is probably derived from respiration of the host (Alberte et al., 1987), and released from bicarbonate in seawater by the effect of carbonic anhydrase present in *Prochloron* (Dionisio-Sese et al., 1993). The ascidian-*Prochloron* symbiosis is dependent on carbon fixed by the latter and passed to the host (Alberte et al., 1986, 1987; Olson, 1986), possibly as glucose or maltose (Lewin, 1994), although the transfer of carbon is comparatively slow (Griffiths and Thinh, 1987). In ascidians such as *Didemnum molle* this carbon is insufficient to meet the animal's needs and must be supplemented externally, whereas in *Lissoclinum voeltzkowi*, the host's entire organic carbon requirement is met by *Prochloron* (Koike et al., 1993).

The cyanobionts of sponges are photosynthetically active, transfering up to 12% of the carbon they fix to the host, probably in the form of glycerol (Rai,

1990a). *Richelia intracellularis*, the cyanobiont of *Rhizosolenia*, fixes both CO_2 and N_2 for its diatom host (Rai, 1990a). In contrast, the unicellular cyanobionts of the dinoflagellates *Ornithocerus magnificus* and *Ornithocerus steinii*, seem to provide fixed carbon, rather than nitrogen, for their non-photosynthetic host, although they may also be phagotrophically consumed when they senesce (Janson et al., 1995a).

VII. Reconstitution of the Symbioses

Reconstruction of symbioses by introducing axenic cultures of the two symbiotic partners has been successfully achieved with all of the plant symbioses except *Azolla*, this being the only case in which it has not been possible to grow the major cyanobiont in free culture. However, the artificial reconstruction of the *Azolla* symbiosis has been achieved by removal of a megasporocarp apical cap containing the cyanobiont and transferring this to the decapitated megasporocarp of another *Azolla* plant (Lin et al., 1988; Lin and Watanabe, 1988; Watanabe et al., 1989; Braun-Howland and Nierzwicki-Bauer, 1990). In contrast, the bryophyte symbioses have proved particularly easy to reconstitute (Plate 29e), not just with the original cyanobiont, but with cyanobionts from *Gunnera*, cycads, and even free-living strains (Meeks, 1988, 1990; Kimura and Nakano, 1990). All of the symbiotically-competent cyanobacteria are hormogonia-forming strains of the genus *Nostoc*. Similar observations have been made with *Gunnera* which can be infected by both symbiotic and free-living *Nostoc* strains (Bonnett, 1990; Bergman et al., 1992b; Johansson and Bergman, 1994), but not by hormogonia-forming cyanobacteria of other genera such as *Calothrix*, *Fischerella* and *Chlorogloeopsis* (Johansson and Bergman, 1994).

Reconstitution experiments such as these are very helpful in studying the infection process and the establishment of symbiotic colonies, and can provide information on the specificity of the association. They confirm the broad competence of *Nostoc* spp. as symbionts, but are likely to give a false impression of the breadth of strains capable of infecting plants in the field, because such experiments are artificial in their use of a single axenic plant and a single axenic cyanobacterium. Even competition experiments using several cyanobacteria at once (Meeks, 1990) do not mimic the natural environment in which many cyanobacterial strains will be present, not to mention the unknown influences of the many other bacteria and eukaryotes. For example, *Calothrix* and *Chlorogloeopsis*, both capable of forming hormogonia, can infect *Phaeoceros* in the laboratory, yet the latter has not been found in field samples of any bryophyte, and the former only very rarely, possibly reflecting both the relative abundance of these strains in the soil and their inability to compete with the natural populations of *Nostoc* (West and Adams, 1997).

Although it is often possible to grow separately the photobiont and mycobiont partners of lichens, resynthesis of the symbiosis has proved extremely difficult. This is in part because of the very slow growth rates of the mycobionts in particular. However, there are an increasing number of examples of partial or complete resyntheses of both bipartite and tripartite lichens (Ahmadjian, 1989, 1993; Stocker-Wörgötter and Türk, 1991, 1994; Hill, 1994; Stocker-Wörgötter, 1995). In the case of *Geosiphon*, *Nostoc* spp. other than *N. punctiforme* can be incorporated by the fungus as endosymbionts (Kluge et al., 1994). Indeed, it is thought that in nature *Geosiphon* and the bryophytes *Blasia* and *Anthoceros* can exchange cyanobionts (Mollenhauer, 1992; Kluge et al., 1994).

Attempts to construct artificial associations between symbiotic and free-living cyanobacteria and plants have involved the introduction of cyanobacteria into higher plant protoplasts, or the growth of cyanobacteria with plant tissue cell cultures and plant regenerates (Gusev and Korzhenevskaya, 1990 and Korzhenevskaya et al., 1993).

VIII. Concluding Remarks

There is little doubt that many cyanobacterial associations remain to be discovered; even our understanding of the cyanobacterial associations described in this chapter is mostly rudimentary. The best studied are the cyanobacteria-plant symbioses, and these have considerable potential as experimental systems, because of their flexibility and simplicity in comparison with the rhizobia-legume associations. One of the most interesting aspects of these cyanobacteria-plant associations is the signalling between symbiont and host. The identity of the chemical signals and of the genes they regulate are presently unknown, but with the development of molecular genetic techniques for symbiotic cyanobacteria, progress is now being made towards their elucidation. An understanding of the regulatory systems involved in the establishment and maintenance of plant-cyanobacteria associations may open the possibility of enhancing existing symbioses, or even establishing novel ones with for example crop plants. Cyanobacteria such as *Nostoc* spp. have considerable potential as nitrogen-fixing plant symbionts. They can respond to plant signals for hormogonia induction and chemotaxis; they have in the heterocyst, a highly evolved oxygen protection system that permits nitrogen fixation under aerobic conditions; they are all photoautotrophs, but many are also facultative photo- and chemoheterotrophs, enabling them to adapt to any light regime; finally, cyanobacteria are extremely catholic in their choice of hosts, establishing symbioses with a wide range of plants without seeming to elicit a defense response. All this offers hope that the development of novel associations with agriculturally important plants may be feasible.

Acknowledgements

Many thanks to Birgitta Bergman, the late David Hall, Peter Lindblad, Ian Lucas, Sandra Nierzwicki-Bauer, Dieter Mollenhauer, and Arthur Schüßler for providing photographs and for communicating data prior to publication.

References

Adams DG (1992) Multicellularity in cyanobacteria. In: Mohan S, Dow C and Cole JA (eds) Prokaryotic Structure and Function: A New Perspective. Symp Soc Gen Microbiol 47: 341-384

Adams DG (1997) Cyanobacteria. In: Shapiro J and Dworkin M (eds) Bacteria as Multicellular Organisms, pp 109-148. Oxford University Press, New York

Ahern CP (1993) Establishment of the cycad/cyanobacteria symbiosis. J Computer-Aided Microsc 5: 43-46

Ahern CP and Staff IA (1994) Symbiosis in cycads: the origin and development of coralloid roots in *Macrozamia communis* (Cycadaceae). Am J Bot 81: 1559-1570

Ahmadjian, V (1982) Algal/fungal symbioses. In: Round FE and Chapman DJ (eds) Progress in Phycological Research, vol 1, pp 179-233. Elsevier Biomedical Press, Amsterdam

Ahmadjian, V (1989) Studies on the isolation and synthesis of bionts of the cyanolichen *Peltigera canina* (Peltigeraceae). Plant Systematics and Evolution 165: 29-38

Ahmadjian, V (1993) The Lichen Symbiosis. John Wiley and Sons, New York

Alberte RS, Cheng L and Lewin RA (1986) Photosynthetic characteristics of *Prochloron*-ascidian symbioses. I. Light and temperature responses of the algal symbiont of *Lissoclinum patella*. Marine Biol (Berlin) 90: 575-587

Alberte RS, Cheng L and Lewin RA (1987) Characteristics of *Prochloron*-ascidian symbioses. 2. Photosynthesis-irradiance relationships and carbon balance of associations from Palau, Micronesia. Symbiosis 4: 147-170

Arad H, Keysari A, Tel-Or E and Kobiler D (1985) A comparison between cell antigens in different isolates of *Anabaena azollae*. Symbiosis 1: 195-204

Arillo A, Bavestrello G, Burlando B and Sara M (1993) Metabolic integration between symbiotic cyanobacteria and sponges - a possible mechanism. Marine Biol 117: 159-162

Aulfinger H, Braun-Howland EB, Kannaiyan S and Nierzwicki-Bauer SA (1991) Ultrastructural changes of the endosymbionts of *Azolla microphylla* during megaspore germination and early plantlet development. Can J Bot 69: 2489-2496

Babic S (1996) Hormogonia formation and the establishment of symbiotic associations between cyanobacteria and the bryophytes *Blasia* and *Phaeoceros*. Ph.D. thesis, University of Leeds, England

Badger MR, Pfanz H, Büdel B, Heber U and Lange OL (1993) Evidence for the functioning of photosynthetic CO_2-concentrating mechanisms in lichens containing green algal and cyanobacterial photobionts. Planta 191: 59-72

Bandaranayake WM (1998) Mycosporines: are they nature's sunscreens? Natural Product Reports 15: 159-172

Bandaranayake WM, Bemis JE and Bourne DJ (1996) Ultraviolet absorbing pigments from the marine sponge *Dysidea herbacea*: Isolation and structure of a new mycosporine. Comp Biochem Physiol C: Pharmacology, toxicology and endocrinology 115: 281-286

Bar E and Tel-Or E (1994) Effect of light and oxygen on nitrogenase activity and dinitrogenase reductase (Fe-protein) content in *Azolla-Anabaena* association. J Plant Physiol 144: 438-443

Berenguer J, Rojo F, dePedro MA, Pfanzagl B and Löffelhardt W (1987) Penicillin-binding proteins in the cyanelles of *Cyanophora paradoxa*, a eukaryotic photoautotroph sensitive to beta-lactam antibiotics. FEBS Lett 224: 401-405

Bergman B, Johansson C and Söderbäck E (1992a) Tansley Review No. 42. The *Nostoc-Gunnera* symbiosis. New Phytol 122: 379-400

Bergman B, Lindblad P, Pettersson A, Renström E and Tiberg E (1985) Immunogold localization of glutamine synthetase in a nitrogen-fixing cyanobacterium, *Anabaena cylindrica*. Planta 166: 329-334

Bergman B, Lindblad P and Rai AN (1986) Nitrogenase in free-living and symbiotic cyanobacteria: immunoelectronmicroscopic localization. FEMS Microbiol Lett 35: 75-78

Bergman B, Matveyev A and Rasmussen U (1996) Chemical signalling in cyanobacterial-plant symbioses. Trends Plant Sci 1: 191-197

Bergman B and Rai AN (1989) The *Nostoc-Nephroma* symbiosis: localization, distribution pattern and levels of key proteins involved in nitrogen and carbon metabolism of the cyanobiont. Physiol Plant 77: 216-224

Bergman B, Rai AN, Johansson C and Söderbäck E (1992b) Cyanobacterial-plant symbioses. Symbiosis 14: 61-81

Bewley CA, Holland ND and Faulkner DJ (1996) Two classes of metabolites from *Theonella swinhoei* are localized in distinct populations of bacterial symbionts. Experimentia 52: 716-722

Bhattacharya D, Helmchen T, Bibeau C and Melkonian M (1995) Comparisons of nuclear encoded small subunit ribosomal RNAs reveal the evolutionary position of the Glaucocystophyta. Mol Biol Evol 12: 415-420

Bhattacharya D and Medlin L (1995) The phylogeny of plastids: a review based on comparisons of small-subunit ribosomal RNA coding regions. J Phycol 31: 489-498

Bilger W, Büdel B, Mollenhauer R and Mollenhauer D (1994) Photosynthetic activity of two developmental stages of a *Nostoc* strain (cyanobacteria) isolated from *Geosiphon pyriforme* (mycota). J Phycol 30: 225-230

Blank RJ (1985) Is the central inclusion body of the cyanelle of *Cyanophora paradoxa* a carboxysome? Endocytobiosis Cell Res 2: 113-118

Bogner E, Wastlhuber R, Schlegl I and Loos E (1993) Glycogen, amylase and alpha-glucosidase as possible components in the glucose-release system of the cyanobiont of *Peltigera horizontalis* - partial-purification and characterization. Symbiosis 14: 485-494

Bonnett HT (1990) The *Nostoc-Gunnera* association. In: Rai AN (ed) CRC Handbook of Symbiotic Cyanobacteria, pp 161-171. CRC Press, Boca Raton, Florida

Brasell, HM, Davies, SK and Mattay, JP (1986) Nitrogen fixation associated with bryophytes colonizing burnt sites in Southern Tasmania, Australia. J Bryol 14: 139-149

Braun-Howland EB and Nierzwicki-Bauer SA (1990) *Azolla-Anabaena* symbiosis: biochemistry, physiology, ultrastructure, and molecular biology. In: Rai AN (ed) CRC Handbook of Symbiotic Cyanobacteria, pp 65-117. CRC Press, Boca Raton, Florida

Braun-Howland EB, Lindblad P, Nierzwicki-Bauer SA and Bergman B (1988) Dinitrogenase reductase (Fe-protein) of nitrogenase in the cyanobacterial symbionts of three *Azolla* species: Localization and sequence of appearance during heterocyst differentiation. Planta 176: 319-332

Büdel B (1992) Taxonomy of lichenized procaryotic blue-green algae. In: Reisser W (ed) Algae and Symbioses: Plants, Animals, Fungi, Viruses. Interactions Explored, pp 301-324. Biopress Ltd., Bristol

Büdel B and Lange OL (1991) Water status of green and blue-green phycobionts in lichen thalli after hydration by water vapour uptake: Do they become turgid? Bot Acta 104: 361-366

Büdel B and Rhiel E (1987) Studies on the ultrastructure of some cyanolichen haustoria. Protoplasma 139: 145-152

Bullerjahn GS and Post AF (1993) The prochlorophytes: are they more than just chorophyll a/b-containing cyanobacteria? Crit Rev Microbiol 19: 43-59

Burger-Wiersma T and Matthijs HCP (1990) The biology of the Prochlorales. In: Codd GA, Dijkhuizen L and Tabita FR (eds) Advances in Autotrophic Microbiology and One-Carbon Metabolism, Vol 1, pp 1-24. Kluwer, Dordrecht

Calvert HE, Pence MK and Peters GA (1985) Ultrastructural ontogeny of leaf cavity trichomes in *Azolla* implies a functional role in metabolic exchange. Protoplasma 129, 10-27

Calvert HE, Perkins SK and Peters GA (1983) Sporocarp structure in the heterosporous water fern *Azolla mexicana* Presl. Scanning Electron Microsc 3: 1499-1510

Campbell EL, Hagen KD, Cohen MF, Summers ML and Meeks JC (1996) The *devR* gene product is characteristic of receivers of two-component regulatory systems and is essential for heterocyst development in the filamentous cyanobacterium *Nostoc* sp. strain ATCC 29133. J Bacteriol 178: 2037-2043

Campbell EL and Meeks JC (1989) Characteristics of hormogonia formation by symbiotic *Nostoc* spp. in response to the presence of *Anthoceros punctatus* or its extracellular products. Appl Environ Microbiol 55: 125-131

Campbell EL and Meeks JC (1992) Evidence for plant-mediated regulation of nitrogenase expression in the *Anthoceros-Nostoc* symbiotic association. J Gen Microbiol 138: 473-480

Canini A, Bergman B, Civitareale P, Rotilio G and Grilli Caiola M (1992c) Localization of iron-superoxide dismutase in the cyanobiont of *Azolla filiculoides* Lam. Protoplasma 169: 1-8

Canini A, Grilli Caiola M, Civitareale P and Galiazzo F (1992b) Superoxide dismutase in symbiotic, free-living and wild *Anabaena* and *Nostoc* (Nostocales, Cyanophyta). Phycologia 31: 225-230

Canini A, Grilli Caiola M and Ferrusci L (1992a) Quantitiative and qualitative DNA variations in *Anabaena azollae* Strasb. living in *Azolla filliculoides* Lam. Cytometry 13: 299-306

Canini A, Grilli Caiola M and Mascini M (1990) Ammonium content, nitrogenase activity and heterocyst frequency within the leaf cavities of *Azolla filiculoides* Lam. FEMS Microbiol Lett 71: 205-210

Carrapiço F (1991) Are bacteria the third partner of the *Azolla-Anabaena* symbiosis? Plant Soil 137: 157-160

Carrapiço F and Tavares R (1989) New data on the *Azolla-Anabaena* symbiosis. I. Morphological and histochemical aspects. In: Skinner FA, Boddey RM and Fendrik I (eds) Nitrogen Fixation With Non-Legumes, pp 89-94. Kluwer Academic Publishers, Dordrecht

Cassman KG, De Datta SK, Amarante ST, Liboon SP, Samson MI and Dizon MA (1996) Long-term comparison of the agronomic efficiency and residual benefits of organic and inorganic nitrogen sources for tropical lowland rice. Expl Agric 32: 427-444

Caudales R, Moreau RA, Wells JM and Antoine AD (1993) Cellular lipid and fatty acid composition of cyanobionts from *Azolla caroliniana*. Symbiosis 14: 191-200

Caudales R, Wells JM and Antoine AD (1992) Cellular fatty acid composition of symbiotic cyanobacteria isolated from the aquatic fern *Azolla*. J Gen Microbiol 138: 1489-1494

Caudales R, Wells JM, Antoine AD and Butterfield JE (1995) Fatty acid composition of symbiotic cyanobacteria from different host plant (*Azolla*) species. Evidence for coevolution of host and symbiont. Int J System Bacteriol 45: 364-370

Cohen MF and Meeks JC (1997) A hormogonium regulating locus, *hrmUA*, of the cyanobacterium *Nostoc punctiforme* strain ATCC 29133 and its response to an extract of a symbiotic plant partner *Anthoceros punctatus*. Mol. Plant- Microbe Interactions 10: 280-289

Cohen MF, Wallis JG, Campbell EL and Meeks JC (1994) Transposon mutagenesis of *Nostoc* sp. strain ATCC 29133, a filamentous cyanobacterium with multiple cellular differentiation alternatives. Microbiol 140: 3233-3240

Corredor JE, Wilkinson CR, Vicente VP, Morell JM and Otero E (1988) Nitrate release by Caribbean reef sponges. Limnol Oceanogr 33: 114-120

Cowan IR, Lange OL and Green TGA (1992) Carbon-dioxide exchange in lichens: determination of transport and carboxylation characteristics. Planta 187: 282-294

Cox G (1986) Comparison of *Prochloron* from different hosts. I. Structural and ultrastructural characteristics. New Phytol 104: 429-445

Dalton, DA and Chatfield, JM (1985) A new nitrogen-fixing cyanophyte-hepatic association: *Nostoc* and *Porella*. Am J Bot 72: 781-784

Delwiche CF and Palmer JD (1997) The origin of plastids and their spread via secondary symbiosis. Plant System Evol S11: 53-86

DeYoe HR, Lowe RL and Marks JC (1992) Effects of nitrogen and phosphorus on the endosymbiont load of *Rhopalodia gibba* and *Epithemia turgida* (Bacillariophycaea). J Phycol 28: 773-777

Dick, H and Stewart, WDP (1980) The occurrence of fimbriae on a N_2-fixing cyanobacterium which occurs in lichen symbiosis. Arch Microbiol 124: 107-109

Dionisio-Sese ML, Shimada A, Maruyama T and Miyachi S (1993) Carbonic anhydrase activity of *Prochloron* sp. isolated from an ascidian host. Arch Microbiol 159: 1-5

Dodds WK (1989) Photosynthesis of two morphologies of *Nostoc parmelioides* (Cyanobacteria) as related to current velocities and diffusion patterns. J Phycol 25: 258-262

Dodds WK and Marra J (1989) Behaviors of the midge *Cricotopus* (Diptera:Chironomidae) related to mutualism with *Nostoc parmelioides* (Cyanobacteria). Aquat Insects 11: 201-208

Dodds WK, Gudder DA and Mollenhauer D (1995) The ecology of *Nostoc*. J Phycol 31: 2-18

Douglas SE (1992) Eukaryote-eukaryote endosymbioses: insights from studies of a cryptomonad alga. BioSystems 28: 57-68

Douglas SE (1994) Chloroplast origins and evolution. In: Bryant DA (ed) The Molecular Biology of Cyanobacteria, pp 91-118. Kluwer Academic Publishers, Dordrecht

Dunham DG and Fowler K (1987) Taxonomy and species recognition in *Azolla* Lam. In: *Azolla* Utilization, pp 7-16. International Rice Research Institute, Los Baños, Philippines

Duong TP and Tiedje JM (1985) Nitrogen fixation by naturally occurring duckweed-cyanobacterial associations. Can J Microbiol 31: 327-330

Eskew DL, Caetano-Anollés G, Bassam BJ and Gresshoff PM (1993) DNA amplification fingerprinting of the *Azolla-Anabaena* symbiosis. Plant Mol Biol 21: 363-373

Enderlin CS and Meeks JC (1983) Pure culture and reconstitution of the *Anthoceros-Nostoc* symbiotic association. Planta 158: 157-165

Faulkner JD, Unson MD and Bewley CA (1994) The chemistry of some sponges and their symbionts. Pure Appl Chem 66: 1983-1990

Fay P (1992). Oxygen relations of nitrogen fixation in cyanobacteria. Microbiol Rev 56: 340-373

Feige GB and Jensen M (1992) Basic carbon and nitrogen metabolism of lichens. In: Reisser W (ed) Algae and Symbioses: Plants, Animals, Fungi, Viruses. Interactions Explored, pp 277-299. Biopress Ltd., Bristol

Forni C, and Grilli Caiola M (1988) Physiological characterization of the bacteria, isolated from the aquatic fern *Azolla*. In: Bothe H, de Bruijn FJ and Newton WE (eds) Nitrogen Fixation: Hundred Years After. Proceedings of the 7th International Congress on Nitrogen Fixation, p 235. Gustav Fisher, Stuttgart

Forni C, and Grilli Caiola M (1992) The morphogenesis of *Arthrobacter* species isolated from the fern *Azolla filiculoides* Lam. Microbiologica 15: 271-279

Forni C, Grilli Caiola M and Gentili S (1989) Bacteria in the *Azolla-Anabaena* symbiosis. In: Skinner FA, Boddey RM and Fendrik I (eds) Nitrogen Fixation With Non-Legumes, pp 83-88. Kluwer Academic Publishers, Dordrecht

Forni C, Riov J, Grilli Caiola M and Tel-Or E (1992a) Indole-3-acetic acid (IAA) production by *Arthrobacter* species isolated from *Azolla*. J Gen Microbiol 138: 377-381

Forni C, Haegi A, del Gallo M and Grilli Caiola M (1992b) Production of polysaccharides by *Arthrobacter globiformis* associated with *Anabaena azollae* in *Azolla* leaf cavity. FEMS Microbiol Lett 93: 269-274

Franche C and Cohen-Bazire G (1985) The structural *nif* genes of four symbiotic *Anabaena azollae* show a highly conserved physical arrangement. Plant Sci 39: 125-131

Franche C and Cohen-Bazire G (1987) Evolutionary divergence in the *nifHDK* gene region among nine symbiotic *Anabaena azollae* and between *Anabaena azollae* and some free-living heterocystous cyanobacteria. Symbiosis 3: 159-178

Fuchs B, Suttner P, Sterner S, Wastlhuber R and Loos E (1994) Disproportionating transglycosylase (D-enzyme) in green algae and cyanobacteria: partial purification and characterization. Z Naturforschung C-A Journal of Biosciences 49: 163-170

Gallon JR (1992) Tansley Review No. 44. Reconciling the incompatible: N_2 fixation and O_2. New Phytol 122: 571-609

Gantar M, Kerby NW and Rowell P (1991a) Colonization of wheat (*Triticum vulgare* L.) by N_2-fixing cyanobacteria: II. An ultrastructural study. New Phytol 118: 485-492.

Gantar M, Kerby NW and Rowell P (1993) Colonization of wheat (*Triticum vulgare* L.) by N_2-fixing cyanobacteria: III. The role of a hormogonia-promoting factor. New Phytol 124: 505-513

Gantar M, Kerby NW, Rowell P and Obreht Z (1991b) Colonization of wheat (*Triticum vulgare* L.) by N_2-fixing cyanobacteria. I. A survey of soil cyanobacterial isolates forming associations with roots. New Phytol 118: 477-483

Gantar M, Kerby NW, Rowell P, Obreht Z and Scrimgeour C (1995b) Colonization of wheat (*Triticum vulgare* L.) by N_2-fixing cyanobacteria: IV. Dark nitrogenase activity and effects of cyanobacteria on natural ^{15}N abundance in the plants. New Phytol 129: 337-343

Gantar M, Rowell P, Kerby NW and Sutherland IW (1995a) Role of extracellular polysaccharide in the colonization of wheat (*Triticum vulgare* L.) roots by N_2-fixing cyanobacteria. Biol Fertil Soils 19: 41-48

Gebhardt JS and Nierzwicki-Bauer SA (1991) Identification of a common cyanobacterial symbiont associated with *Azolla* spp. through molecular and morphological characterization of free-living and symbiotic cyanobacteria. Appl Environ Microbiol 57: 2141-2146

Gehrig H, Schüßler A and Kluge M (1996) *Geosiphon pyriforme*, a fungus forming endocytobiosis with *Nostoc* (cyanobacteria), is an ancestral member of the Glomales - evidence by SSU ribosomal RNA analysis. J Mol Evol 43: 71-81

Gordon N, Angel DL, Neori A, Kress N and Kimor B (1994) Heterotrophic dinoflagellates with symbiotic cyanobacteria and nitrogen limitation in the Gulf of Aqaba. Marine Ecology - Progress Series 107: 83-88

Griffiths DJ and Thinh L-V (1987) Photosynthesis by *in situ* and isolated *Prochloron* (Prochlorophyta) associated with the didemnid ascidians. Symbiosis 3: 109-122

Griffiths DJ and Thinh L-V (1989) Current status of the taxonomy of the genus *Prochloron*. In: Staley JT (ed) Bergey's Manual of Systematic Bacteriology, vol 3, pp 1802-1805. Williams and Wilkins, Baltimore

Grilli Caiola M, Canini A and Moscone D (1989) Oxygen concentration, nitrogenase activity and heterocyst frequency in the leaf cavities of *Azolla filiculoides* Lam. FEMS Microbiol Lett 59: 283-288

Grilli Caiola M, Forni C and Castagnola M (1988) Bacteria in the *Azolla-Anabaena* association. Symbiosis 5: 185-198

Grilli Caiola M, Forni C and Castagnola M (1993) *Anabaena azollae* akinetes in the sporocarps of *Azolla filiculoides* Lam. Symbiosis 14: 247-264

Grobbelaar N, Scott WE, Hattingh W and Marshall J (1987) The identification of the coralloid root endophytes of the southern African cycads and the ability of the isolates to fix dinitrogen. South African J Bot 53: 111-118

Gusev MV and Korzhenevskaya TG (1990) Artificial associations. In: Rai AN (ed) CRC Handbook of Symbiotic Cyanobacteria, pp 173-230. CRC Press, Boca Raton, Florida

Hällbom L, Bergman B and Rai AN (1986) Immunogold localization of glutamine synthetase in the cyanobionts of the lichens *Peltigera aphthosa* and *Peltigera canina*. Lichen Physiol Biochem 1: 27-34

Handayani D, Edrada RA, Proksch P, Wray V, Witte L, VanSoest RWM, Kunzmann A, Soedarsono (1997) Four new bioactive polybrominated diphenyl ethers of the sponge *Dysidea herbacea* from west Sumatra, Indonesia. J Natural Products 60: 1313-1316

Helmchen TA, Bhattacharya D and Melkonian M (1995) Analyses of ribosomal RNA sequences from glaucocystophyte cyanelles provide new insights into the evolutionary relationships of plastids. J Mol Evol 41: 203-210

Hill DJ (1989) The control of the cell cycle in microbial symbionts. New Phytol 112: 175-184

Hill DJ (1994) The nature of the symbiotic relationship in lichens. Endeavour, New Series 18: 96-103

Hinde R, Pironet F and Borowitzka MA (1994) Isolation of *Oscillatoria spongeliae*, the filamentous cyanobacterial symbiont of the marine sponge *Dysidea herbacea*. Marine Biol 119: 99-104

Hirose E, Maruyama T, Cheng L and Lewin RA (1996) Intracellular symbiosis of a photosynthetic prokaryote, *Prochloron* sp., in a colonial ascidian. Invertebrate Biol 115: 343-348

Holton RW, Stam WT and Boele-Bos SA (1990) DNA-DNA reassociation studies with DNA from *Prochloron* (Prochlorophyta) samples of Indo-West Pacific origin. J Phycol 26: 358-361

Honegger R (1991) Functional aspects of the lichen symbiosis. Annu Rev Pl Physiol 42: 553-578

Honegger R (1992) Lichens: mycobiont-photobiont relationships. In: Reisser W (ed) Algae and Symbioses: Plants, Animals, Fungi, Viruses, Interactions Explored, pp 256-275. Biopress, Bristol

Honegger R (1993) Tansley Review No. 60. Developmental biology of lichens. New Phytol 125: 659-677

Honegger R, Peter M and Scherrer S (1996) Drought-induced structural alterations at the mycobiont-photobiont interface in a range of foliose macrolichens. Protoplasma 190: 221-232

Jäger KM, Johansson C, Kunz U and Lehmann H (1997). Sub-cellular element analysis of a cyanobacterium (*Nostoc* sp.) in symbiosis with *Gunnera manicata* by ESI and EELS. Bot Acta 110: 151-157

Janson S, Carpenter EJ and Bergman B (1995a) Immunolabelling of phycoerythrin, ribulose 1,5-bisphosphate carboxylase/oxygenase and nitrogenase in the unicellular cyanobionts of *Ornithocercus* spp. (Dinophyceae). Phycologia 34: 171-176

Janson S, Rai AN and Bergman B (1993) The marine lichen *Lichina confinis* (O. F. Müll) c. Ag.: ultrastructure and localization of nitrogenase, glutamine synthetase, phycoerythrin and ribulose 1, 5-bisphosphate carboxylase/oxygenase in the cyanobiont. New Phytol 124: 149-160

Janson S, Rai AN and Bergman B (1995b) Intracellular cyanobiont *Richelia intracellularis* - Ultrastructure and immuno-localization of phycoerythrin, nitrogenase, Rubisco and glutamine synthetase. Marine Biol 124: 1-8

Johansson C (1994) Establishment of the *Gunnera-Nostoc* symbiosis. PhD thesis, Stockholm University, Sweden

Johansson C and Bergman B (1992) Early events during the establishment of the *Gunnera/Nostoc* symbiosis. Planta 188: 403-413

Johansson C and Bergman B (1994) Reconstruction of the symbiosis of *Gunnera manicata* Linden: cyanobacterial specificity. New Phytol 126: 643-652

Joseph CM and Meeks JC (1987) Regulation of expression of glutamine synthetase in a symbiotic *Nostoc* strain associated with *Anthoceros punctatus*. J Bacteriol 169: 2471-2475

Kaplan D, Calvert HE and Peters GA (1986) The *Azolla-Anabaena azollae* relationship. XII. Nitrogenase activity and phycobiliproteins of the endophyte as a function of leaf age and cell type. Plant Physiol 80: 884-890

Kaplan D and Peters GA (1988) Interaction of carbon metabolism in the *Azolla-Anabaena* symbiosis. Symbiosis 6: 53-68

Kardish N, Silverstein L, Fleminger G and Galun M (1991) Lectin from the lichen *Nephroma leavigatum* Ach. localization and function. Symbiosis 11: 47-62

Kies L (1984) Cytological aspects of blue-green algal endosymbiosis. In: Weissner W, Robinson DG and Starr RC (eds) Compartments in Algal Cells and Their Interaction, pp 191-199. Springer-Verlag, Berlin, Heidelberg

Kies L (1992) Glaucocystophyceae and other protists harbouring procaryotic endocytobionts. In: Reisser W (ed) Algae and Symbiosis: Plants, Animals, Fungi, Viruses. Interactions Explored, pp 353-377. Biopress Ltd, Bristol

Kim JJ-H, Krawczyk K, Lorentz WP and Zimmerman WJ (1997) Fingerprinting cyanobionts and hosts of the *Azolla* symbiosis by DNA amplification. World J Microbiol Biotechnol 13: 97-101

Kimura J and Nakano T (1990) Reconstitution of a *Blasia-Nostoc* symbiotic association under axenic conditions. Nova Hedwigia 50: 191-200

Klein E, Bar E, Forni C, Malkin S and Tel-Or E (1992) The application of cryo-SEM techniques to the study of the symbiotic association in the *Azolla* leaf cavity. J Microscopy 167: 273-278

Kluge M, Mollenhauer D and Mollenhauer R (1991) Photosynthetic carbon assimilation in *Geosiphon pyriforme* (Kützing) F. v. Wettstein, an endosymbiotic association of fungus and cyanobacterium. Planta 185: 311-315

Kluge M, Mollenhauer D and Mollenhauer, R (1994) *Geosiphon pyriforme* (Kützing) von Wettstein, a promising system for studying endocyanoses. Prog Bot 55: 130-141

Kluge M, Mollenhauer D, Mollenhauer, R and Kape R (1992) *Geosiphon pyriforme*, an endosymbiotic consortium of a fungus and a cyanobacterium (*Nostoc*), fixes nitrogen. Bot Acta 105: 343-344

Knight CD and Adams DG (1996) A method for studying chemotaxis in nitrogen-fixing cyanobacterium-plant symbioses. Physiol Molec Plant Pathol 49: 73-77

Koike I, Yamamuro M and Pollard PC (1993) Carbon and nitrogen budgets of two ascidians and their symbiont, *Prochloron*, in a tropical seagrass meadow. Aust J Mar Freshwater Res 44: 173-182

Korzhenevskaya TG, Baulina OI, Gorelova OA, Lobacova, ES, Butenko, RG and Gusev, MV (1993) Artificial syncyanoses - the potential for modeling and analysis of natural symbioses. Symbiosis 15: 77-103

Kundu DK and Ladha JK (1996) Enhancing soil nitrogen use and biological nitrogen fixation in wetland rice. Expl Agric 31: 261-277

Lambert G, Lambert CC and Waaland JR (1996) Algal symbionts in the tunics of six New Zealand Ascidians (Chordata, Ascidiacea). Invertebrate Biol 115: 67-78

Lange OL, Budel B, Meyer A and Kilian E (1993) Further evidence that activation of net photosynthesis by dry cyanobacterial lichens requires liquid water. Lichenologist 25:175-189

Lange OL, Green TGA and Ziegler H (1988) Water related photosynthesis and carbon isotope discrimination in species of lichen genus *Pseudocyphellaria* with green or blue-green photobionts in photosymbiodemes. Oecologia 75: 494-501

Lange OL, Hahn SC, Müller G, Meyer A and Tenhunen JD (1996) Upland tundra in the foothills of the Brooks Range, Alaska: influence of light, water content and temperature on CO_2 exchange of characteristic lichen species. Flora 191: 67-83

Lee KY, Joseph CM and Meeks JC (1988) Glutamine synthetase specific activity and protein concentration in symbiotic *Anabaena* associated with *Azolla caroliniana*. Antonie van Leeuwenhoek 54: 345-355

Leizerovitch I, Kardish N and Galun M (1990) Comparison between eight symbiotic, cultured *Nostoc* isolates and a free-living *Nostoc* by recombinant DNA. Symbiosis 8: 75-85

Leonardi D, Canini A and Forni C (1993) Immunological comparison between *Arthrobacter* isolates and bacteria living in *Azolla filiculoides* Lam. Symbiosis 15: 269-279

Lewin RA (1989) Order *Prochlorales*. In: Staley JT (ed) Bergey's Manual of Systematic Bacteriology, vol 3, pp 1799-1802. Williams and Wilkins, Baltimore

Lewin RA (1994) The cell sap of *Prochloron* (Prochlorophyta). Phycologia 33: 71-76

Lewin RA and Cheng L (eds) (1989) *Prochloron* a Microbial Enigma. Chapman and Hall, New York

Lin C and Watanabe I (1988) A new method for obtaining *Anabaena*-free *Azolla*. New Phytol 108: 341-344

Lin C, Watanabe I, Liu CC, Zheng DV and Tang LF (1988) Re-establishment of symbiosis to *Anabaena*-free *Azolla*. In: Bothe H, de Bruijn FJ and Newton WE (eds) Nitrogen Fixation: Hundred Years After, pp 223-227. Gustav Fisher, Stuttgart

Line MA (1992) Nitrogen fixation in the sub-Antarctic MacQuarie Island. Polar Biol 11: 601-606

Lindblad P, Atkins CA and Pate JS (1991b) N_2-fixation by freshly isolated *Nostoc* from coralloid roots of the cycad *Macrozamia riedlei* (Fisch. ex Gaud.) Gardn. Plant Physiol 95: 753-759

Lindblad P and Bergman B (1986) Glutamine synthetase: activity and localization in cyanobacteria of the cycads *Cycas revoluta* and *Zamia skinneri*. Planta 169: 1-7

Lindblad P and Bergman B (1989) Occurrence and localization of phycoerythrin in symbiotic *Nostoc* of *Cycas revoluta* and in the free-living isolate *Nostoc* 7422. Plant Physiol 89: 783-785

Lindblad P and Bergman B (1990) The cycad-cyanobacterial symbiosis. In: Rai AN (ed) CRC Handbook of Symbiotic Cyanobacteria, pp 137-159. CRC Press, Boca Raton, Florida

Lindblad P, Bergman B, Hofsten AV, Hällbom L and Nylund JE (1985b) The cyanobacterium-*Zamia* symbiosis: an ultrastructural study. New Phytol 101: 707-716

Lindblad P, Bergman B and Nierzwicki-Bauer, SA (1991a) Immunocytochemical localization of nitrogenase in bacteria symbiotically associated with *Azolla* spp. Appl Environ Microbiol 57: 3637-3640

Lindblad P, Hällbom L and Bergman B (1985a) The cyanobacterium-*Zamia* symbiosis: C_2H_2/reduction and heterocyst frequency. Symbiosis 1:19-28

Lindblad P, Haselkorn R, Bergman B and Nierzwicki-Bauer, SA (1989) Comparison of DNA restriction fragment length polymorphisms of *Nostoc* strains in and from cycads. Arch Microbiol 152: 20-24

Lindblad P, Rai AN and Bergman B (1987) The *Cycas revoluta-Nostoc* symbiosis: enzyme activities of nitrogen and carbon metabolism in the cyanobiont. J Gen Microbiol 133: 1695-1699

Löffelhardt W and Bohnert HJ (1994a) Structure and function of the cyanelle genome. Int Rev Cytol 151: 29-65

Löffelhardt W and Bohnert HJ (1994b) Molecular biology of cyanelles. In: Bryant DA (ed) The Molecular Biology of Cyanobacteria, pp 65-89. Kluwer Academic Publishers, Netherlands

Lotti F, Giovannetti L, Margheri MC, Ventura S and Materassi R (1996) Diversity of DNA methylation pattern and total DNA restriction pattern in symbiotic *Nostoc*. World J Microbiol and Biotechnol 12: 38-42

Lucas IAN (1991) Symbionts of the tropical Dinophysiales (Dinophyceae). Ophelia 33: 213-224

Madan AP and Nierzwicki-Bauer SA (1993) In situ detection of transcripts for ribulose-1, 5-bisphosphate carboxylase in cyanobacterial heterocysts. J Bacteriol 175. 7301-7306

Máguas C and Brugnoli E (1996) Spatial variation in carbon isotope discrimination across the thalli of several lichen species. Plant, Cell and Environment 19: 437-446

Mahesh G and Kannaiyan S (1993) Effect of immobilization of cyanobacteria in solid matrix on ammonia excretion and nitrogen-fixing activity. Biotechnol Lett 15: 975-978

Mangeney E and Gibbs SP (1987) Immunocytochemical localization of ribulose-1, 5-bisphosphate carboxylase/oxygenase in the cyanelles of *Cyanophora paradoxa* and *Glaucocystis nostochinearum*. Eur J Cell Biol 43: 65-70

Mangeney E, Hawthornthwaite AM, Codd GA and Gibbs SP (1987) Immunocytochemical localization of phosphoribulose kinase in the cyanelles of *Cyanophora paradoxa* and *Glaucocystis nostochinearum*. Plant Physiol 84: 1028-1032

Matthijs HCP, Van der Staay GWM and Mur LR (1994) Prochlorophytes: the 'other' cyanobacteria? In: Bryant DA (ed) The Molecular Biology of Cyanobacteria, pp 49-64. Kluwer Academic Publishers, Netherlands

McCowen SM, MacArthur L and Gates JE (1987) *Azolla* fern lectins that specifically recognize endosymbiotic cyanobacteria. Curr Microbiol 14: 329-333

Meeks JC (1988) Symbiotic associations. In: Packer L and Glazer AN (eds) Meth Enzymol 167, pp 113-121. Academic Press Inc., London

Meeks JC (1990) Cyanobacterial-bryophyte associations. In: Rai AN (ed) CRC Handbook of Symbiotic Cyanobacteria, pp 43-63. CRC Press, Boca Raton, Florida

Meeks JC, Enderlin CS, Joesph CM, Chapman JS and Lollar MWL (1985a) Fixation of $[^{13}N]N_2$ and transfer of fixed nitrogen in the *Anthoceros-Nostoc* symbiotic association. Planta 164: 406-414

Meeks JC, Joseph, CM and Haselkorn, R (1988) Organization of the *nif* genes in cyanobacteria in symbiotic association with *Azolla* and *Anthoceros*. Arch Microbiol 150: 61-71

Meeks JC, Steinberg N, Joseph CM, Enderlin CS, Jorgensen PA and Peters GA (1985b) Assimilation of exogenous and dinitrogen-derived $^{13}NH_4^+$ by *Anabaena azollae* separated from *Azolla caroliniana* Willd. Arch Microbiol 142: 229-233

Meeks JC, Steinberg NA, Enderlin CS, Joseph CM and Peters GA (1987) *Azolla-Anabaena* relationship. XIII. Fixation of $[^{13}N]N_2$. Plant Physiol 84: 883-886

Milicia F and Favilli F (1993) *Azolla* symbiotic systems application as biofertilizer for green garden crops. Symbiosis 14: 495-500

Mollenhauer D (1992) *Geosiphon pyriforme*. In: Reisser W (ed) Algae and Symbioses: Plants, Animals, Fungi, Viruses. Interactions Explored, pp 339-351. Biopress Ltd., Bristol.

Mollenhauer D and Kluge M (1994) *Geosiphon pyriforme*. Endocytobios Cell Res 10: 29-34

Mollenhauer D and Mollenhauer R (1988) *Geosiphon* cultures ahead. Endocytosis Res 5: 69-73

Mollenhauer D and Mollenhauer R (1996) *Nostoc* in symbiosis - Taxonomic implications. Algological Studies 83: 435-446

Mollenhauer D, Mollenhauer R and Kluge M (1996) Studies on initiation and development of the partner association in *Geosiphon pyriforme* (Kütz.) v. Wettstein, a unique endocytobiotic system of a fungus (Glomales) and the cyanobacterium *Nostoc punctiforme* (Kütz.) Hariot. Protoplasma 193: 3-9

Mur LR and Burger-Wiersma T (1992) The order Prochlorales. In: Balows A, Trüper HG, Dworkin M, Harder W and Schleifer K-H (eds) The Prokaryotes, Vol II, pp 2105-2110. Springer-Verlag, Heidelberg

Mutuskin AA and Kolesnikov PA (1991) Symbiotic *Azolla-Anabaena* association: some photosynthetic characteristics, nitrogenase activity and oxidative systems. Photosynthetica 25: 639-644

Nierzwicki-Bauer SA (1990) *Azolla-Anabaena* symbiosis: Use in agriculture. In: Rai AN (ed) CRC Handbook of Symbiotic Cyanobacteria, pp 119-136. CRC Press, Boca Raton, Florida

Nierzwicki-Bauer SA and Aulfinger H (1990) Ultrastructural characterization of eubacteria residing within leaf cavities of symbiotic and cyanobiont-free *Azolla mexicana*. Current Microbiol 21: 123-129

Nierzwicki-Bauer SA and Aulfinger H (1991) Occurrence and ultrastructural characterization of bacteria in association with and isolated from *Azolla caroliniana*. Appl Environ Microbiol 57: 3629-3636

Nierzwicki-Bauer SA and Haselkorn R (1986) Differences in mRNA levels in *Anabaena* living freely or in symbiotic association with *Azolla*. EMBO J 5: 29-35

Obreht Z, Kerby NW, Gantar M and Rowell P (1993) Effects of root-associated N_2-fixing cyanobacteria on the growth and nitrogen content of wheat (*Triticum vulgare* L.) seedlings. Biol Fertil Soils 15: 68-72

Odintsov VS (1991) Nitrogen fixation in *Prochloron* (Prochlorophyta)-ascidian associations - is *Prochloron* responsible. Endocytobiosis and Cell Research 7: 253-258

Olson RR (1986) Light-enhanced growth of the ascidian *Didemenum mollelProchloron* sp. symbiosis. Marine Biol (Berlin) 93: 437-442

Osborne BA (1989) Comparisons of photosynthesis and productivity of *Gunnera tinctoria* (Molina) Mirbel with and without the phycobiont *Nostoc punctiforme* L. Plant, Cell and Environ 12: 941-946

Osborne BA, Cullen A, Jones PW and Campbell GJ (1992) Use of nitrogen by the *Nostoc-Gunnera tinctoria* (Molina) Mirbel symbiosis. New Phytol 120: 481-487

Osborne BA, Doris F, Cullen A, McDonald R, Campbell G and Steer MW (1991) *Gunnera tinctoria*: an unusual nitrogen-fixing invader. BioScience 41: 224-234

Ott S (1988) Photosymbiodemes and their development in *Peltigera venosa*. Lichenologist 20: 361-368

Paerl HW (1984) N_2 fixation (nitrogenase activity) attributed to a specific *Prochloron* (Prochlorophyta)-ascidian association on Palaau, Micronesia. Marine Biol (Berlin) 81: 251-254

Palenik B and Swift H (1996) Cyanobacterial evolution and prochlorophyte diversity as seen in DNA-dependent RNA polymerase gene sequences. J Phycol 32: 638-646

Palmqvist K (1993) Photosynthetic CO_2-use efficiency in lichens and their isolated photobionts: the possible role of a CO_2-concentrating mechanism. Planta 191: 48-56

Palmqvist K (1995) Uptake and fixation of CO_2 in lichen photobionts. Symbiosis 18: 95-109

Palmqvist K, Máguas C, Badger MR and Griffiths H (1994) Assimilation, accumulation and isotope discrimination of inorganic carbon in cyanobacterial lichens: evidence for the operation of a CO_2 concentrating mechanism in cyanobacterial lichens. Crypto Bot 4: 218-226

Pascher A (1929) Studien über Symbiosen. I. Über einige Endosymbiosen von Blaualgen in Einzellern. Jahrbuch Wiss Bot 71: 386-462

Pate JS, Lindblad P and Atkins CA (1988) Pathways of assimilation and transfer of fixed nitrogen in coralloid roots of cycad-*Nostoc* symbioses. Planta 176: 461-471

Paulsrud P and Lindblad P (1998) Sequence variation of the $tRNA^{leu}$ intron as a marker for genetic diversity and specificity of symbiotic cyanobacteria in some lichens. Appl Environ Microbiol 64: 310-315

Paulsrud P, Rikkinen J and Lindblad P (1998) Cyanobiont specificity in some *Nostoc*-containing lichens and in a *Peltigera aphthosa* photosymbiodeme. New Phytol 139: 517-524

Pennell RI (1992) Cell surface arabinogalactan proteins, arabinogalactans and plant development. In: Collow JA and Green JR (eds) Perspectives in Plant Cell Recognition, pp 105-121. Cambridge University Press, Cambridge

Perkins SK and Peters GA (1993) The *Azolla-Anabaena* symbiosis:- endophyte continuity in the *Azolla* life-cycle is facilitated by epidermal trichomes. I. Partitioning of the endophytic *Anabaena* into developing sporocarps. New Phytol 123: 53-64

Perraju BTVV, Rai AN, Kumar AP and Singh HN (1986) *Cycas circinalis-Anabaena cycadeae* symbiosis: photosynthesis and the enzymes of nitrogen and hydrogen metabolism in symbiotic and cultured *Anabaena cycadeae*. Symbiosis 1: 239-250

Peters GA (1991) *Azolla* and other plant-cyanobacteria symbioses: Aspects of form and function. Plant Soil 137: 25-36

Peters GA and Meeks JC (1989) The *Azolla-Anabaena* symbiosis: basic biology. Annu Rev Plant Physiol 40: 193-210

Peters GA, Kaplan D, Meeks JC, Buzby KM, Marsh BH and Corbin JL (1985) Aspects of nitrogen and carbon interchange in the *Azolla-Anabaena* symbiosis. In: Ludden PW and Burris JE (eds) Nitrogen Fixation and CO_2 Metabolism, pp 213-222. Elsevier, New York

Peters GA and Perkins SK (1993) The *Azolla-Anabaena* symbiosis - endophyte continuity in the *Azolla* life-cycle is facilitated by epidermal trichomes. II. Re-establishment of the symbiosis following gametogenesis and embryogenesis. New Phytol 123: 65-75

Peters GA, Toia Jr RE, Calvert HE and Marsh BH (1986) Lichens to *Gunnera* - with emphasis on *Azolla*. Plant and Soil 90: 17-34

Petro MJ and Gates JE (1987) Distribution of *Arthrobacter* sp. in the leaf cavities of four species of the N-fixing *Azolla* fern. Symbiosis 3: 41-48

Pfanzagl B, Allmaier G, Schmid ER, De Pedro MA and Löffelhardt W (1996b) *N*-acetylputrescine as a characteristic constituent of cyanelle peptidoglycan in glaucocystophyte algae. J Bacteriol 178: 6994-6997

Pfanzagl B, Zenker A, Pittenhauer E, Allmaier G, Martinez-Torrecuadrada J, Schmid ER, De Pedro MA and Löffelhardt W (1996a) Primary structure of cyanelle peptidoglycan of *Cyanophora paradoxa*: a prokaryotic cell wall as part of an organelle envelope. J Bacteriol 178: 332-339

Plazinski J (1990) The *Azolla-Anabaena* symbiosis. In: Gressshoff PM (ed) The Molecular Biology of Symbiotic Nitrogen Fixation, pp 51-75. CRC Press, Boca Raton, Florida

Plazinski J, Croft L, Taylor R, Zheng Q, Rolfe BG and Gunning BES (1991) Indigenous plasmids in *Anabaena azollae*: their taxonomic distribution and existence of regions of homology with symbiotic genes of *Rhizobium*. Can J Microbiol 37: 171-181

Plazinski J, Franche C, Liu C-C, Lin T, Shaw W, Gunning BES and Rolfe BG (1988) Taxonomic status of *Anabaena azollae*: An overview. Plant Soil 108: 185-190

Plazinski J, Taylor R, Shaw W, Croft L, Rolfe BG and Gunning BES (1990b) Isolation of *Agrobacterium* sp. strain from the *Azolla* leaf cavity. FEMS Microbiol Lett 70: 55-59

Plazinski J, Zheng Q, Taylor R, Croft L, Rolfe BG and Gunning BES (1990a) DNA probes show genetic variation in cyanobacterial symbionts of the *Azolla* fern and a closer relationship to free-living *Nostoc* strains than to free-living *Anabaena* strains. Appl Environ Microbiol 56: 1263-1270

Post AF and Bullerjahn GS (1994) The photosynthetic machinery in Prochlorophytes: structural properties and ecological significance. FEMS Microbiol Rev 13: 393-414

Rai, AN (1988) Nitrogen metabolism. In: Galun M (ed) CRC Handbook of Lichenology vol I, pp 201-237. CRC Press, Boca Raton, Florida

Rai, AN (1990a) Cyanobacteria in Symbiosis. In: Rai, AN (ed) CRC Handbook of Symbiotic Cyanobacteria, pp 1-7. CRC Press, Boca Raton, Florida

Rai AN (1990b) Cyanobacterial-fungal symbioses: the cyanolichens. In: Rai AN (ed) CRC Handbook of Symbiotic Cyanobacteria, pp 9-41. CRC Press, Boca Raton, Florida

Rai AN, Borthakur M and Paul D (1996) Symbiotic cyanobacteria - biotechnological applications. J Scient Industl Res 55: 742-752

Rai AN, Borthakur M, Singh S and Bergman B (1989) *Anthoceros-Nostoc* symbiosis: immunoelectronmicroscopic localization of nitrogenase, glutamine synthetase, phycoerythrin and ribulose-1,5-bisphosphate carboxylase/oxygenase in the cyanobiont and the cultured (free-living) isolate *Nostoc* 7801. J Gen Microbiol 135: 385-395

Rai AN, Borthakur M, Söderbäck E and Bergman B (1992) Immunogold localization of hydrogenase in the cyanobacterial-plant symbioses *Peltigera canina*, *Anthoceros punctatus* and *Gunnera megallanica*. Symbiosis 12: 131-144

Rai AN, Rowell P and Stewart WDP (1981) Nitrogenase activity and dark CO_2 fixation in the lichen *Peltigera aphthosa* Willd. Planta 151: 256-264

Rasmussen U, Johansson C, and Bergman, B (1994) Early communication in the *Gunnera-Nostoc* symbiosis: Plant-induced cell differentiation and protein synthesis in the cyanobacterium. Molecular Plant-Microbe Interactions 7, 696-702

Rasmussen U, Johansson C, Renglin A, Petersson C and Bergman, B (1996) A molecular characterization of the *Gunnera-Nostoc* symbiosis: comparison with *Rhizobium*- and *Agrobacterium*-plant interactions. New Phytol 133: 391-398

Renström-Kellner E, Rai AN and Bergman B (1990) Correlation between nitrogenase and glutamine synthetase expression in the cyanobacterium *Anabaena cylindrica*. Physiol Plant 80: 12-19

Rippka R and Stanier RY (1978) The effects of anaerobiosis on nitrogenase synthesis and heterocyst development by Nostocacean cyanobacteria. J Gen Microbiol 105: 83-94

Robins RJ, Hall DO, Shi D-J, Turner RJ and Rhodes, MJC (1986) Mucilage acts to adhere cyanobacteria and cultured plant cells to biological and inert surfaces. FEMS Microbiol Lett 34: 155-160

Roger PA (1995) Biological N_2 fixation and its management in wetland rice cultivation. Fertiliser Research 42: 261-276

Rogers GA and Stewart WDP (1977) The cyanophyte-hepatic symbiosis. I. Morphology and physiology. New Phytol 78: 441-458

Rother JA, Aziz A, Karim NH and Whitton BA (1988) Ecology of deepwater rice-fields in Bangladesh. 4. Nitrogen fixation by blue-green algal communities. Hydrobiologia 169: 43-56

Rozen A, Arad H, Schönfeld M and Tel-Or E (1986) Fructose supports glycogen accumulation, heterocyst differentiation, N_2 fixation and growth of the isolated cyanobiont *Anabaena azollae*. Arch Microbiol 145: 187-190

Rozen A, Schönfeld M and Tel-Or E (1988) Fructose-enhanced development and growth of the N_2-fixing cyanobiont *Anabaena azollae*. Z Naturforsch 43c: 408-412

Ruetzler K (1988) Mangrove sponge disease induced by cyanobacterial symbionts: Failure of a primitive immune system? Dis Aquat Org 5: 143-149

Schreiber U, Gademann R, Ralph PJ and Larkum AWD (1997) Assessment of photosynthetic performance of *Prochloron* in *Lissoclinum patella* in hospite by chlorophyll fluorescence measurements. Plant Cell Physiol 38: 945-951

Schroeter B (1994) In situ photosynthetic differentiation of the green algal and the cyanobacterial photobiont in the crustose lichen *Placopsis contortuplicata*. Oecologia 98: 212-220

Schroeter B, Green TGA, Kappen L and Seppelt RD (1994) Carbon dioxide exchange at subzero temperatures. Field measurements on *Umbilicaria aprina* in Antarctica. Cryptogam Bot 4: 233-241

Schüßler A, Bonfante P, Schnepf E, Mollenhauer D and Kluge M (1996) Characterization of the *Geosiphon pyriforme* symbiosome by affinity techniques: confocal laser scanning microscopy (CLSM) and electron microscopy. Protoplasma 190: 53-67

Schüßler A, Meyer T, Gehrig H and Kluge M (1997) Variations of lectin binding sites in extracellular glycoconjugates during the life cycle of *Nostoc punctiforme*, a potentially endosymbiotic cyanobacterium. Eur J Phycol: 32: 233-240

Schüßler A, Mollenhauer D, Schnepf E and Kluge M (1994) *Geosiphon pyriforme*, an endosymbiotic association of fungus and cyanobacteria: the spore structure resembles that of arbuscular mycorrhizal (AM) fungi. Bot Acta 107: 36-45

Schüßler A, Schnepf E, Mollenhauer D and Kluge M (1995) The fungal bladders of the endocyanosis *Geosiphon pyriforme*, a *Glomus*-related fungus: cell wall permeability indicates a limited pore radius of only 0.5 nm. Protoplasma 185: 131-139

Sharma A, Mishra DP and Kumar A (1992) Characterization of a symbiotic, heterocystous, N_2-fixing cyanobacterium from *Cycas* coralloid roots. World J Microbiol Biotechnol 8: 529-531

Sheath RG and Cole KM (1992) Biogeography of stream macroalgae in North America. J Phycol 28: 448-460

Shi D-J, Brouers M, Hall DO and Robins RJ (1987) The effects of immobilization on the biochemical, physiological and morphological features of *Anabaena azollae*. Planta 172: 298-308

Shimada A, Kanai S and Maruyama T (1995) Partial sequence of ribulose-1,5-bisphosphate carboxylase/oxygenase and the phylogeny of *Prochloron* and *Prochlorococcus* (Prochlorales). J Mol Evol 40: 671-677

Simon RD (1987) Inclusion bodies in the cyanobacteria. In: Fay P and Van Baalen C (eds) The Cyanobacteria, pp 199-225

Silvester WB, Parsons R and Watt PW (1996) Direct measurement of release and assimilation of ammonia in the *Gunnera-Nostoc* symbiosis. New Phytol 132: 617-625

Söderbäck E (1992) Developmental patterns in the *Nostoc-Gunnera* symbiosis. PhD thesis, Stockholm University, Sweden. Akademitryck AB

Söderbäck E and Bergman B (1992) The *Nostoc-Gunnera magellanica* symbiosis: phycobiliproteins, carboxysomes and Rubisco in the cyanobiont. Physiol Plant 84: 425-432

Söderbäck E and Bergman B (1993) The *Nostoc-Gunnera* symbiosis: carbon fixation and translocation. Physiol Plant 89: 125-132

Söderbäck E, Lindblad, P and Bergman B (1990) Developmental patterns related to nitrogen fixation in the *Nostoc-Gunnera magellanica* Lam., symbiosis. Planta 182: 355-362

Spiller H and Gunasekaran M (1990) Ammonia-excreting mutant of the cyanobacterium *Anabaena variabilis* supports growth of wheat. Appl Microbiol Biotechnol 33: 477-480

Spiller H, Stallings Jr. W, Woods T and Gunasekaran M (1993) Requirement for direct association of ammonia-excreting *Anabaena variabilis* mutant (SA-1) with roots for maximal growth and yield of wheat. Appl Microbiol Biotechnol 40: 557-566

Stackebrandt E, Seewaldt E, Fowler, VJ and Schleifer K-H (1982) The relatedness of *Prochloron* sp. isolated from different didemnid ascidian hosts. Arch Microbiol 132: 216-217

Staff IA and Ahern CP (1993) Symbiosis in cycads with special reference to *Macrozamia communis*. In: Stevenson DW and Norstog JK (eds), The Biology, Structure and Systematics of the Cycadales. Proceedings of CYCAD 90, the second international conference on cycad biology, pp 200-210. Palm and Cycad Societies of Australia Ltd., Queensland, Australia

Stam WT, Boele-Bos SA and Stulp BK (1985) Genotypic relationships between *Prochloron* samples from different localities and hosts as determined by DNA-DNA reassociations. Arch Microbiol 142: 340-341

Steinberg NA and Meeks JC (1989) Photosynthetic CO_2 fixation and ribulose bisphosphate carboxylase/oxygenase activity of *Nostoc* sp. strain UCD 7801 in symbiotic association with *Anthoceros punctatus*. J Bacteriol 171: 6227-6233

Steinberg NA and Meeks JC (1991) Physiological sources of reductant for nitrogen fixation activity in *Nostoc* sp. strain UCD 7801 in symbiotic association with *Anthoceros punctatus*. J Bacteriol 173: 7324-7329

Stergianou KK and Fowler K (1990) Chromosome numbers and taxonomic implications in the fern genus *Azolla* (Azollaceae). Plant Syst Evol 173: 223-239

Stewart WDP and Rogers GA (1977) The cyanophyte-hepatic symbiosis. II. Nitrogen fixation and the interchange of nitrogen and carbon. New Phytol 78, 459-471

Stirewalt VL, Michalowski CB, Loffelhardt W, Bohnert HJ and Bryant DA (1995) Nucleotide sequence of the cyanelle genome from *Cyanophora paradoxa*. Plant Mol Biol Reporter 13: 327-332

Stock PA and Silvester WB (1994) Phloem transport of recently-fixed nitrogen in the *Gunnera-Nostoc* symbiosis. New Phytol 126: 259-266

Stocker-Wörgötter E (1995) Experimental cultivation of lichens and lichen symbionts. Can J Bot 73 (supplement 1): S579-S589

Stocker-Wörgötter E and Türk R (1990) Thallus formation of the cyanobacterial lichen *Peltigera didactyla* from soredia under laboratory conditions. Bot Acta 103: 315-321

Stocker-Wörgötter E and Türk R (1991) Artificial resynthesis of thalli of the cyanobacterial lichen *Peltigera praetextata* under laboratory conditions. Lichenologist 23: 127-138

Stocker-Wörgötter E and Türk R (1994) Artificial resynthesis of the photosymbiodeme *Peltigera leucophlebia* under laboratory conditions. Cryptogam Bot 4: 300-308

Strasburger E (1873) Über *Azolla*, pp 76-86. Herman Dabis Verlag, Jena

Strasburger E (1884) Das Botanische Practicum. Gustav Fischer, Jena and Leipzig

Swift H (1989) The cytology of *Prochloron*. In: Lewin RA and Cheng L (eds) *Prochloron*, a Microbial Enigma, pp 71-81. Chapman and Hall, New York

Swift H and Leser GP (1989) Cytochemical studies on prochlorophytes: localization of DNA and ribulose 1,5 bisphosphate carboxylase-oxygenase. J Phycol 25: 751-761

Swift H and Robertson DL (1991) Structural aspects of *Prochloron*-tunicate symbiosis. Symbiosis 10: 95-113

Tabita FR (1994) The biochemistry and molecular regulation of carbon dioxide metabolism in cyanobacteria. In: Bryant DA (ed) The Molecular Biology of Cyanobacteria, pp 437-467. Kluwer Academic Publishers, Dordrecht

Tan BC, Payawal P, Watanabe I, Lacdan N and Ramirez C (1986) Modern taxonomy of *Azolla*: a review. Phil Agric 69: 4491-4516

Tang LF, Watanabe I and Liu CC (1990) Limited multiplication of symbiotic cyanobacteria of *Azolla* spp. on artificial media. Appl Environ Microbiol 56: 3623-3626

Taylor FJR (1990) Symbiosis in marine protozoa. In: Capriulo GM (ed), Ecology of Marine Protozoa, pp 323-340. Oxford University Press, Oxford

Towata EM (1985) Mucilage glands and cyanobacterial colonization in *Gunnera kaalensis* (Haloragaceae). Bot Gaz 146: 56-62

Trench RK (1982) Physiology, biochemistry, and ultrastructure of cyanellae. In: Round FE and Chapman DJ (eds) Progress in Phycological Research vol 1, pp 257-288. Elsevier Biomedical Press, Amsterdam

Uheda E (1986) Isolation of empty packets from *Anabaena*-free *Azolla*. Plant Cell Physiol 27: 1187-1190

Uheda E and Kitoh S (1991) Electron microscopic observations of the envelopes of isolated algal packets of *Azolla*. Can J Bot 69: 1418-1419

Unson MD and Faulkner DJ (1993) Cyanobacterial symbiont biosynthesis of chlorinated metabolites from *Dysidea herbacea*. Experimentia 49: 349-353

Unson MD, Holland ND and Faulkner DJ (1994) A brominated secondary metabolite synthesised by the cyanobacterial symbiont of a marine sponge and accumulation of the crystalline metabolite in the sponge tissue. Marine Biol 119: 1-11

Vagnoli L, Margheri MC, Allotta G and Materassi R (1992) Morphological and physiological properties of symbiotic cyanobacteria. New Phytol 120: 243-249

Van Coppenolle B, Watanabe I, Van Hove C, Second G, Huang N and McCouch SR (1993a) Genetic diversity and phylogeny analysis of *Azolla* based on DNA amplification by arbitrary primers. Genome 36: 686-693

Van Coppenolle B, McCouch SR, Watanabe I, Huang N and Van Hove C (1993b) Genetic diversity and phylogeny analysis of *Anabaena azollae* based on RFLPs detected in *Azolla-Anabaena azollae* DNA complexes using *nif* gene probes. Theoret Appl Genet 91: 589-597

van Rhijn P and Vanderleyden J (1995) The *Rhizobium*-plant symbiosis. Microbiol Rev 59: 124-142

Villareal TA (1991) Nitrogen-fixation by the cyanobacterial symbiont of the diatom genus *Hemiaulus*. Marine Ecology Progress Series 76: 201-204

Villareal TA (1994) Widespread occurrence of the *Hemiaulus*-cyanobacterial symbiosis in the southwest north Atlantic ocean. Bulletin of Marine Science 54: 1-7

Wake H, Akasaka A, Umetsu H, Ozeki Y, Shimomura K and Matsunaga T (1992) Promotion of plantlet formation from somatic embryos of carrot treated with a high molecular weight extract from a marine cyanobacterium. Plant Cell Rep 11, 62-65

Wake H, Umetsu H, Ozeki Y, Shimomura K and Matsunaga T (1991) Extracts of marine cyanobacteria stimulated somatic embryogenesis of *Daucus carota* L. Plant Cell Rep 9: 655-658

Wallace WH and Gates JE (1986) Identification of eubacteria isolated from leaf cavities of four species of the N-fixing *Azolla* fern as *Arthrobacter* Conn and Dimmick. Appl Environ Microbiol 52: 425-429

Ward AK, Dahm CN and Cummings KW (1989) *Nostoc* (Cyanophyta) productivity in Oregon stream ecosystems: invertebrate influences and differences between morphological types. J Phycol 21: 223-227

Wasmann CC, Löeffelhardt W and Bohnert HJ (1987) Cyanelles: organization and molecular biology. In: Fay P and Van Baalen C (eds) The Cyanobacteria, pp 303-324. Elsevier, Amsterdam

Watanabe I, Lin C, Ramirez C, Lapis MT, Santiago-Ventura T and Liu CC (1989) Physiology and agronomy of *Azolla-Anabaena* symbiosis. In: Skinner FA, Boddey RM and Fendrik I (eds) Nitrogen Fixation with Non-Legumes, pp 57-62. Kluwer Academic Publishers, Dordrecht

Watanabe I and Liu CC (1992) Improving nitrogen-fixing systems and integrating them into sustainable rice farming. Plant Soil 141: 57-67

Werner D (1992) Symbiosis of Plants and Microbes, 389 pp. Chapman and Hall, London

West N and Adams, DG (1997) Phenotypic and genotypic comparison of symbiotic and free-living cyanobacteria from a single field site. Appl Environ Microbiol 63: 4479-4484

Whatley FR and Alberte RS (1989) Biochemical features of *Prochloron*. In: Lewin RA and Cheng L (eds) *Prochloron*, a Microbial Enigma, pp 53-64. Chapman and Hall, New York

Whitton BA, Aziz A, Kawecka B and Rother JA (1988) Ecology of deepwater rice-fields in Bangladesh. 3. Associated algae and macrophytes. Hydrobiologia 169: 31-42

Wilkinson CR (1992) Symbiotic interactions between marine sponges and algae. In: Reisser W (ed) Algae and Symbioses: Plants, Animals, Fungi, Viruses. Interactions Explored, pp 112-151. Biopress Ltd., Bristol

Wilkinson CR and Cheshire AC (1988) Growth rate of Jamaican coral reef sponges after Hurricane Allen. Biol Bull Mar Biol Lab Woods Hole 175: 175-179

Wilkinson CR and Fay P (1979) Nitrogen fixation in coral reef sponges with symbiotic cyanobacteria. Nature 279: 527-529

Wolk CP (1996).Heterocyst formation. Annu Rev Genet 30, 59-78

Wolk CP, Ernst A and Elhai J (1994) Heterocyst metabolism and development. In: Bryant DA (ed) The Molecular Biology of Cyanobacteria, pp 769-823. Kluwer Academic Publishers, Dordrecht

Zimmerman WJ and Bergman B (1990) The *Gunnera* symbiosis: DNA restriction fragment length polymorphism and protein comparisons of *Nostoc* symbionts. Microb Ecol 19: 291-302

Zimmerman WJ and Cullen DE (1991) Genetic variation at the *apcAB*, *cpcAB*, *gvpA1*, and *nifH* loci and in DNA methylation among N_2-fixing cyanobacteria designated *Nostoc punctiforme*. Microb Ecol 21: 199-209

Zimmerman WJ and Rosen BH (1992) Cyanobiont diversity within and among cycads of one field site. Can J Microbiol 38: 1324-1328

Zimmerman WJ, Rosen BH and Lumpkin TA (1989) Enzymatic, lectin, and morphological characterization and classification of presumptive cyanobionts from *Azolla* Lam. New Phytol 113: 497-503

Chapter 20

Cyanophages and Their Role in the Ecology of Cyanobacteria

Curtis A. Suttle
Departments of Earth and Ocean Sciences, Botany, and Microbiology and Immunology, The University of British Columbia, 6270 University Boulevard, Vancouver, British Columbia, Canada V6T 1Z4

B. A. Whitton and M. Potts (eds), The Ecology of Cyanobacteria, 563-589.
© 2000 *Kluwer Academic Publishers. Printed in the Netherlands.*

Summary

Cyanophages belong to three recognized families of double-stranded DNA viruses; Myoviridae (contractile tails); Styloviridae (long non-contractile tails); and Podoviridae (short tails). They have a complex pattern of host ranges, are widely distributed, and can be readily isolated from marine and fresh waters. Although cyanophages are related to other bacteriophages, it is likely that they evolved more than 3 billion years ago when cyanobacteria diverged from other prokaryotes. In marine waters, genetically-diverse Myoviridae which infect *Synechococcus* spp. are the most abundant cyanophages; Styloviridae and Podoviridae are most commonly isolated from fresh waters. Morphological evidence also suggests that freshwater and marine myoviruses are more closely related to each other than they are to other bacteriophages. Cyanophages that infect phycoerythrin-rich *Synechococcus* spp. can be extremely abundant in coastal marine environments where they can occur at titers in excess of 10^6 mL^{-1} and 10^5 g^{-1} of sediment. In surface waters abundance varies over orders of magnitude on a seasonal basis. Abundance follows that of *Synechococcus*, with evidence for a threshold in *Synechococcus* of ca. 10^3 to 10^4 mL^{-1} beyond which cyanophage abundance increases greatly. In nearshore waters the high concentrations of cyanophages and *Synechococcus* result in high encounter frequencies and selection for *Synechococcus* communities that are largely resistant to infection. Encounters are much less frequent offshore and this leads to the appearance of a community that appears to have low resistance to infection. In freshwaters, viruses which infect filamentous cyanobacteria appear to be most abundant and also show strong seasonal dynamics however; even in the most eutrophic environments titers are orders of magnitude less than in productive coastal waters. Little effort was made to screen freshwaters for cyanophages that infect phycoerythrin-rich *Synechococcus.* In marine surface waters turnover times for cyanophage populations range from hours to days. Solar radiation has a major affect on cyanophage infectivity and results in the selection of cyanophage communities that are more resistant to destruction by sunlight during summer. In contrast to surface waters, infectious cyanophages can persist in sediments for at least 100 years. Although the effect of cyanophages on the mortality of cyanobacterial communities is likely to be variable, current estimates suggest that cyanophages are responsible for the removal of approximately 3% of marine *Synechococcus* on a daily basis. In addition to lytic infection, lysogenic associations were clearly demonstrated in filamentous and unicellular cyanobacteria, but the ecological implications of lysogeny remain unexplored. Environmental factors and the physiological state of cyanobacteria clearly affect cyanophage-cyanobacterial interactions but remain poorly understood.

I. Introduction

Safferman and Morris (1963) first reported the isolation of a virus that infected filamentous cyanobacteria. Over the next two decades numerous cyanophage systems were isolated and a great deal of work was done in characterizing and understanding the biological interactions between these viruses and their hosts (reviewed by: Brown, 1972; Padan and Shilo, 1973; Safferman, 1973; Stewart and Daft, 1977; Sherman and Brown, 1978; Martin and Benson, 1988). Considerable emphasis was also directed towards understanding the effect of cyanophages on natural populations and communities of freshwater cyanobacteria. The impetus for these studies was largely the potential importance of cyanophages as controlling factors for cyanobacterial blooms. When it became apparent that cyanophages were unlikely to be useful as biological control agents ecological investigations largely ceased.

In the last few years there was a resurgence of work focused on the ecology of cyanophages, particularly with respect to those which infect marine *Synechococcus* spp. (Carr and Mann, 1994). These efforts stemmed from observations that virus-like particles were typically present in natural waters at abundances in excess of 10^7 mL^{-1} (Bergh et al., 1989); a significant proportion of cyanobacteria in marine phytoplankton communities appeared to contain visible virus particles (Proctor and Fuhrman, 1990); and viruses that infect cyanobacteria were readily isolated from marine waters (Suttle et al., 1990, 1992). Subsequently, it was discovered that infectious cyanophages can be extremely abundant in seawater, with concentrations frequently in excess of 10^4 to 10^5 mL^{-1} (Suttle and Chan, 1993; Suttle et al., 1993; Waterbury and Valois, 1993). Given the global significance of cyanobacteria as primary producers and nitrogen fixers, as well as the potential of cyanophages to serve as mortality agents, vectors of genetic information and regulators of community structure, it is timely to consider what transpired in the decade since the last review of cyanophages was published. Unlike previous reviews that typically provided extensive information on cyanophage biology, the primary emphasis of this contribution is to focus on aspects that are of ecological and environmental importance.

II. Taxonomy, Morphology and Evolution of Cyanophages

A. Taxonomy and Morphology

The taxonomy of viruses that infect cyanobacteria was extensively reviewed by Safferman et al. (1983), and briefly commented upon by Ackermann and DuBow (1987a). Cyanophages belong to three families of tailed phages that contain double-stranded DNA and infect bacteria and aarchaea (Ackermann and DuBow, 1987b; Murphy et al., 1995). The Myoviridae posses a contractile tail that is separated from the capsid by a neck; Siphoviridae (also referred to as Styloviridae) have a long, non-contractile tail that can appear flexible in micrographs; Podoviridae have short non-contractile tails. Representatives from each of these families infect both unicellular and filamentous cyanobacteria. Safferman et al. (1983) proposed that the taxonomy of cyanophages should not be confused with the scheme of coliphage type species that was established by the *International Committee on Taxonomy of Viruses* (ICTV). This decision was based on observations that the host range of cyanophages extends across genera, whereas, the infectivity of coliphages is typically restricted to a genus. In part, the broad host range reported for some cyanophages may reflect difficulties with the taxonomy of cyanobacteria (discussed further in Section III.B). In the following sections cyanophages are described according to the genera proposed by Safferman et al. (1983) although these assignments were not adopted in the most recent overview of viral taxonomy by the ICTV (Murphy et al., 1995).

1. Cyanomyovirus

These are cyanophages belonging to the Myoviridae (Fig. 1a), and were the first viruses described which infected cyanobacteria (Safferman and Morris, 1963). Viruses that belong to this group are the most frequent isolates from marine waters (Suttle and Chan, 1993; Waterbury and Valois, 1993; Wilson et al., 1993), and can also be readily isolated from fresh waters (Gromov, 1983). The proposed type species is Cyanophage AS-1. Cyanophage AS-1 was isolated from a waste stabilization pond and infects unicellular freshwater cyanobacteria originally assigned to the genera *Anacystis* and *Synechococcus* (Safferman et al., 1972). Morphological characteristics include an isometric head (diameter 90 nm), a contractile tail of 244 x 23 nm that contracts to 93 nm. Representatives from this group also infect filamentous cyanobacteria assigned to the genus *Anabaena* (e.g. Gromov, 1983). There is considerable variation in size and morphology among isolates within this group, both in the size of the capsid and in the structure of the tail (Gromov, 1983; Suttle and Chan, 1993; Waterbury and Valois, 1993), which may suggest that the proposed genus consists of a number of species. For example, Waterbury and Valois (1993) isolated four distinct morphotypes from a single water sample. This variation was also observed at the molecular level; the number of capsid proteins reported varied from 14 for two marine isolates (Wilson et al., 1993) to 30 for the type species; the mol% G+C content was determined to vary from ca. 37 to 55%, and the genome molecular weight from 24×10^6 to 57×10^6 daltons (Adolph and Haselkorn, 1971; Safferman et al., 1972; Sherman and Connelly, 1976). The latter correspond to genomic complexities of ca. 37 to 88 kb (based on 649 base pairs per dalton for lambda). Based on restriction mapping (Bancroft and Smith, 1989) the genome size of two closely related cyanophages, AN-13 and AN-23 (Hu et al., 1981) was 46 kb while AN-10 and A4-L (Kozyakov 1977) were ca. 67 kb. Wilson et al. (1993) reported genome sizes for several marine isolates to range from ca. 80 to 100 kb. Note that at least one of the cyanophages identified as a cyanostylovirus in Wilson et al. (1993) was subsequently reclassified as a cyanomyovirus (Wilson et al. 1996). The ICTV classified phages within this group as unassigned species in the Myoviridae (Murphy et al., 1995).

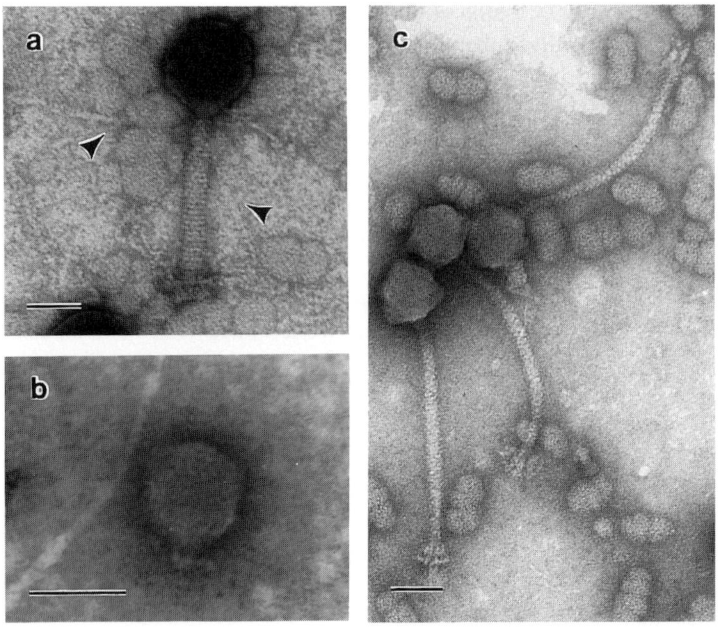

Fig. 1. Representatives of viruses that infect cyanobacteria. Scale bar equals 50 nm. a) Cyanomyovirus (*Myoviridae*) isolate S-PWM4. Cyanomyoviruses possess a contractile tail that is separated from the capsid by a neck. Note the presence of neck hairs (arrows), an unusual feature that is found on some cyanomyovirus isolates from fresh and marine waters. b) Cyanopodovirus (*Podoviridae*) isolate S-BBP1. Cyanopodoviruses are distinguished by short non-contractile tails. c) Cyanostylovirus (*Siphoviridae*) isolate S-BBS1. Cyanostyloviruses have a long non-contractile tail that can appear flexible in micrographs.

2. *Cyanopodovirus*

Cyanopodoviruses (Fig. 1b) belong within the Podoviridae but no genus was assigned by the ICTV. They are present in marine and fresh waters and are commonly isolated from waste stabilization ponds (e.g. Hu et al., 1981). The isometric capsids are typically similar in diameter to cyanostyloviruses, but the tails are much shorter. The type species is Cyanophage LPP-1 that infects algae assigned to the genera *Lyngbya, Plectonema and Phormidium* (Schneider et al., 1964; Sherman and Haselkorn, 1970). The isometric head is 59 nm in diameter and the tail 15 to 20 nm long; there are about 10 structural proteins and the genome is 27 x 10^6 daltons (ca. 42 kb). The reports of mol% G + C contents range from the low 50s in LPP-1 (Luftig and Haselkorn, 1967) to the mid 60s in SM-1 (MacKenzie and Haselkorn, 1972a). Bancroft and Smith (1989) performed restriction mapping of Cyanopodovirus A-4L, which infects *Anabaena* PCC 7120 (Kozyakov, 1977), and concluded that the genome was approximately 40 kb in size.

3. *Cyanostylovirus*

Viruses within the proposed genus *Cyanostylovirus* belong to the family Siphoviridae (Fig 1c). The type species, Cyanophage S-1, infects the genus *Synechococcus,* has an isometric head, is 50 nm in diameter, and has a rigid tail of 140 nm (Adolph and Haselkorn, 1973a). Other viral isolates within this group (e.g. S-2L and SM-2) posses similar morphology (Fox et al., 1976; Khudyakov, 1977), although the tails can be considerably longer reaching lengths of 200 to 300 nm (e.g. Gromov, 1983, Suttle and Chan, 1993; Sode et al., 1994). Reports of genome sizes range widely from ca. 26 x 10^6 daltons (ca. 40 kb) to ca. 90 to 100 kb; reports of the number of structural proteins also varies widely from 13 to 23 (e.g. Adolph and Haselkorn, 1973a; Wilson et al., 1993). The mol% G + C contents reported for these viruses were typically higher (ca. 60 to 74%) than reported for cyanomyoviruses (Adolph and Haselkorn, 1973a; Khudyakov, 1977; Benson and Martin, 1984). ICTV placed this group of viruses in the family Siphoviridae, but did not assign them to a genus.

4. Nomenclature of Cyanophage Isolates

Historically, the nomenclature of cyanophages was based entirely on the taxonomy of the organisms which were infected and did not reflect the relatedness among the viruses. Consequently, it is not possible to tell anything about the taxonomic grouping of a cyanophage isolate from its nomenclature. For example, N-1 and AN-10 are cyanomyoviruses; whereas, N-2 and AN-9 are cyanopodoviruses. The confusion is exacerbated by the fact that even the host organism is not clear from the nomenclature of the cyanophages. For instance, cyanophages belonging to AN infect cyanobacteria originally assigned to *Anabaena* and *Nostoc*; the AS group infects cyanobacteria classified as *Anacystis nidulans* and *Synechococcus cedrorum*; and the SM group infects cyanobacteria identified as *S. elongatus* and *Microcystis aeruginosa*. Matters are complicated further because the validity of many of these species has been called into question (Section III.B and Chapter 1).

More recently, there was an effort to include information about the taxonomic affiliation of cyanophages in their strain designations (Suttle and Chan, 1993; Wilson et al., 1993) and it is recommended that this procedure be followed in the future. The suggested nomenclature for new cyanophage isolates is Cyanophage *Xx*-YYZaa where, *Xx* is the first letters of the genus and species names of the host on which the virus was isolated, YY is a descriptor for the origin of the isolate, Z is the virus family (i.e. M = Myoviridae, P = Podoviridae, S = Siphoviridae), and aa is a reference number for the virus isolate. Although not perfect, this nomenclature includes taxonomic information about the host for consistency with previous cyanophage strain designations, and indicates the family to which the virus was assigned. The designation also provides information on the location where the viruses were isolated.

B. Evolution

Although the origin of cyanophages is unknown, it is possible to make inferences regarding their evolutionary history. The vast array of genetic architecture that is found in viruses indicates that they are polyphyletic and do not share a common ancestor. The most common explanation for their origin is that they represent "escaped" genetic material from organisms. Cyanophages are unlikely to have arisen from cyanobacteria, however, and they do not appear to be closely related to the organisms they infect (Luftig and Haselkorn, 1967; Cowie and Prager, 1969) although there can be considerable homology among highly conserved genes such as DNA polymerases (Braithwaite and Ito, 1993). Nonetheless, taxonomic groups of viruses can share common ancestry and each of the three families of viruses that are represented in the cyanophages are thought to be monophyletic (Ackermann and DuBow, 1987b). If so, the origin of cyanophages probably occurred early in evolutionary time, prior to the divergence of cyanobacteria from other bacteria. Cyanobacteria represent an ancient branch in bacterial evolution (Woese, 1987), and it is thought they were well established by about 3.5 billion years ago (Schopf and Packer, 1987; Chapter 2); the implication is that cyanophages are even older. Because cyanobacteria may have evolved long before eukaryotes it is possible cyanophages were the original "predators" of cyanobacteria.

Although it seems evident that the origin of cyanophages can be traced to that of cyanobacteria, the possibility remains that the Myoviridae, Styloviridae and Podoviridae arose more recently and that similarities among all prokaryotes allowed these viruses to radiate throughout members of the Bacteria and even the Archaea. This seems unlikely given the relatively narrow host range of most phages. Perhaps additional data on the morphology, biochemistry and molecular biology of cyanophages will ultimately allow definitive statements to be made regarding their origin.

Morphological evidence suggests that cyanomyoviruses isolated from marine and fresh waters are more closely related to each other than to other myoviruses. For example, cyanomyoviruses which share morphological features such as neck hairs (Fig. 1a) were isolated from freshwater and marine environments (Adolph and Haselkorn, 1973b; Padan and Shilo, 1973; Suttle and Chan, 1993); these structures are rare in other Myoviridae (Ackermann and DuBow, 1987a).

III. Diversity

A. Genetic Diversity

The morphology and molecular characteristics of the proposed genera of cyanophages suggest that genetic diversity may be high, although there are few data which specifically examine diversity within cyanophages. Muradov et al. (1990) examined five isolates of viruses in the NP-1T group which infect cyanobacteria assigned to the genera *Plectonema* and *Nostoc*. Based on their description it appears that these isolates were *Podoviridae*. They found a different pattern in the number of restriction fragments obtained with *Eco*RV and *Sau*3A restriction endonucleases indicating that each of the isolates was genetically different. Wilson et al. (1993) also examined restriction fragment length polymorphisms (RFLP) in several cyanophages that infect marine *Synechococcus* sp. strain DC2 (synonym WH7803). Although originally characterized as cyanomyoviruses (S-BM1, S-PM1, S-WHM1) and cyanostyloviruses (S-BS1, S-PS1) at least one of the styloviruses was reassigned to the *Myoviridae* (Wilson et al., 1996). Regardless of their taxonomy the viruses fell into three groups based on RFLP (restriction fragment length polymorphism) analyses. In a similar study (unpublished), Chan and Suttle examined three cyanomyoviruses (S-PWM1, S-PWM3, S-PWM4). Although all of these viruses were also originally isolated using *Synechococcus* sp. strain DC2 as the host organism, they differed in the extent of their host range (Suttle and Chan, 1993). The variation in host range was reflected in genetic differences because RFLP patterns were different for each virus. Despite the fact that there is clearly considerable genetic variation within natural communities of cyanophages, this has yet to be quantified by sequence analysis.

One promising approach for quantifying genetic diversity among cyanophages and within cyanophage communities may be to examine sequence variation in DNA polymerase genes. DNA polymerase genes occur in all living cells and are also widely distributed in viruses which contain double-stranded DNA (Ito and Braithwaite, 1991, Braithwaite and Ito, 1993). Sequence differences in this gene were shown to reflect genetic differences among algal viruses (Chen and Suttle, 1996; Chen et al., 1996); it is reasonable to assume that DNA polymerase gene sequences also reflect genetic differences within groups of cyanophages (Short and Suttle, 1999).

B. Host Range

1. LPP Group

There is considerable confusion in the literature regarding the host specificity of viruses that infect cyanobacteria. A good deal of the confusion undoubtedly stems from the disarray that has surrounded cyanobacterial taxonomy and species designations in the past (Chapter 1). Although great progress is being made on the molecular taxonomy of cyanobacteria (e.g. Wilmotte, 1994; Turner, 1997), much of the requisite work on cyanobacterial taxonomy that is required to precisely define cyanophage host range has yet to be published.

The original cyanophage isolate (LPP-1) and a group of related phages were isolated from waste stabilization ponds scattered throughout the United States (Safferman et al., 1969a) as well as from the Ukraine (Mendzhul et al., 1974, Gromov, 1983). The morphology of the LPP group of cyanophages was variously described; with non-contractile (Padan and Shilo, 1973) or contractile (Martin and Benson, 1988) tails. It is fairly clear that LPP cyanophages possess short non-contractile tails similar to that of Bacteriophage T7 (Luftig and Haselkorn, 1968a; 1968b) and that they should be grouped within the Podoviridae, as was done by Safferman et al. (1983). The LPP cyanophages were described as causing lysis of several species within three genera of cyanobacteria, *Lyngbya, Plectonema* and *Phormidium* (Safferman and Morris, 1963). Padan and Shilo (1973) suggested that the very similar mol% G + C contents (46%) and buoyant densities (1.705 g cm^{-3}) of the DNA isolated from susceptible isolates of these genera (Edelman et al., 1967), indicated that these strains were closely related. It was suggested, based on the genetic and morphological similarities among the LPP group of cyanobacteria, that they be included in a single species, *Schizothrix calcicola* (Drouet, 1963). Similarly, Stam (1980) recommended the merger of the genera, while Rippka and Herdman (1992) placed cyanobacteria susceptible to infection by LPP-1 (PCC 6306, 6402, 7310, 7410, 7505, 7602) into a closely related genus (*Leptolyngbya*, Cluster 1) within the Order Oscillatoriales. However, recent work (Turner, 1997) indicated that some members assigned to this genus belonged in other taxonomic groups, and that *Leptolyngbya* was probably not a valid genus and should be further subdivided.

The most extensive study on the host range of LPP-1 was conducted by Johnson and Potts (1985). They examined the sensitivities of 33 strains and variants of cyanobacteria to lysis by the LPP-1 archaetype, as well as several LPP-1 serotypes, LPP-2 serotypes and new LPP isolates. They also concluded that the host range of LPP cyanophages was restricted to members of the *Lyngbya, Plectonema* and *Phormidium* subgroup of cyanobacteria. However, they emphasized that different LPP cyanophage isolates varied in host range and that caution should be used when assigning cyanobacteria to specific taxa based on phage susceptibility. Interpretations of host range studies are further complicated by the fact that there can be changes within the LPP cyanobacterial subgroup in terms of sensitivity to cyanophage infection (Johnson and Borovosky, 1987).

It must be emphasized that considerable taxonomic confusion remains within cyanobacteria that were assigned to the "LPP" genera. Although different groups of genetically related cyanobacteria are beginning to be quite well defined (Wilmotte, 1994; Turner, 1997), there are numerous examples where cyanobacteria were incorrectly assigned to a given genus. Consequently, the LPP group of cyanobacteria should not be considered a taxonomically valid group, and it is likely that the cyanobacteria which are infected by the LPP group of cyanophages will ultimately be placed in a single genus.

2. A, AN, N and NP Groups

Cyanophages also infect and cause lysis of filamentous cyanobacteria assigned to *Nostoc* spp. and *Anabaena* spp. and include both podoviruses (A-1, A-4(L), N-2, AN series except AN 10 and AN-15) and myoviruses (N-1, A-2, AN-10 and AN-15, A-(L) series except A-4(L); Adolph and Haselkorn, 1971; Kozyakov et al., 1972; Kozyakov, 1977; Hu et al., 1981; Gromov, 1983). Although the viruses represent two families, they possess, with few exceptions, remarkably similar host ranges and infect both *Nostoc* spp. and *Anabaena* spp. (Hu et al., 1981). In further studies with cyanophage N-1, Rippka and Herdman (1992) confirmed lysis of cyanobacteria originally assigned to *Anabaena variabilis* (PCC 7118 = ATCC 27892), *Anabaena* sp. (PCC 6411 = ATCC 27898), and *Nostoc muscorum* and *Anabaena* sp. (PCC 7119 = ATCC 29151, PCC 7112 = ATCC 27347 and 27893). Based on a variety of criteria they concluded that all of these cyanobacterial isolates belonged within a very closely related taxonomic group of *Nostoc* (Chapter 17). Consequently, the original conclusion that these viruses possessed a broad host range was an artifact of cyanobacterial nomenclature. It is also important to note that host range mutations also occur spontaneously in this group of viruses (Sarma and Kaur, 1993); hence, caution must be exercised when using sensitivity to cyanophage infection as a taxonomic criterion.

Little can be said about the NP group of cyanophages (NP-1T strains) that were reported to infect a variety of cyanobacterial isolates assigned to the genera *Nostoc* and *Plectonema* (Muradov et al., 1990). Based on the authors' description the viruses represented a homogeneous group of particles ca. 78 nm in diameter with no obvious tails and an estimated genome size of 43.2 kb (based on the measured length of the uncoiled DNA in electron micrographs). The five isolates of the viruses that were examined apparently had the same host range, yet were different from each other based on RFLP analysis, burst size, adsorption rate constants, latent period and length of the lytic cycle. It is not clear if the relatively broad host range that was reported was the result of problems in the taxonomic assignments of the host, or whether these cyanophages actually infected two genera of cyanobacteria. The authors indicated that these viruses were lysogenic, although no evidence for this was provided in their publication other than the observation of turbid plaques.

3. AS and SM Groups

Similar difficulties exist with host range studies conducted on cyanophage isolates that infect unicellular cyanobacteria. The myovirus AS-1 was originally reported to infect *Anacystis nidulans* (UTEX 625 = PCC 6301), and *Synechococcus cedrorum* (UTEX 1191 = PCC 6908) (Safferman et al., 1972). These cyanobacteria, as well as three others (PCC 6311, PCC 7942, PCC 7943) were infected by AS-1 (Rippka and Herdman, 1992); moreover, PCC 6301, PCC 7942 and PCC 7943 showed a high degree of DNA/DNA homology (Wilmotte and Stam, 1984). It is likely that all five strains are independent isolates of a single species (Wilmotte and Stam, 1984, Waterbury and Rippka, 1989).

Another example of cyanophages that were reported to infect different genera can also be attributed to taxonomic confusion. The podovirus SM-1 was originally described as infecting *Synechococcus elongatus* (UTEX 563 = PCC 6907) and *Microcystis aeruginosa* (NRC-1; Safferman et al., 1969b), while the myovirus AS-1 was described as infecting *M. aeruginosa* (NRC-1), as well as *Synechococcus cedrorum* (UTEX 1191; Safferman et al., 1972). Subsequently, it was shown that SM-1 infected a number of close relatives to PCC 6907

(PCC 6307, PCC 6713, PCC 6904, PCC 6911) which belong within a closely related cluster of *Synechococcus* species including *S. cedrorum* (PCC 6908) (Waterbury and Rippka, 1989, Rippka and Herdman, 1992). Yet, *M. aeruginosa* (NRC-1 = PCC 7941) has characteristics which clearly distinguish it from *Synechococcus* isolates which can be infected by SM-1 and AS-1, including gas vacuoles, toxin production, β-cyclocitral production, as well as a much lower mol% G + C content (42.5%) than the *Synechococcus* isolates (56 to 71.4%) (Waterbury and Rippka, 1989, Rippka and Herdman, 1992). However, MacKenzie and Haselkorn (1972a) reported that the mol% G + C content of Strain NRC-1 was 66%, which places it in the range of the *Synechococcus* isolates, and indicates that the original NRC-1 which was assigned to the genus *Synechococcus* by Stanier (MacKenzie and Haselkorn, 1972a) was not PCC 7941. In fact, Rippka (personal communication) determined that the isolate originally assigned to *M. aeruginosa* (NRC-1), and which is sensitive to phage SM-1, corresponded to *Synechococcus* sp. (PCC 6911) not *M. aeruginosa* (NRC-1 = PCC 7941). Therefore, the cyanobacteria which can be infected by SM-1 and AS-1 once again appear to be restricted to a closely related taxonomic group. Taxonomic confusion is the same reason that the siphovirus SM-2 were reported to cause lysis of the genera *Synechococcus* and *Microcystis* (Fox et al., 1976; Leach et al., 1980).

All of the previously mentioned host-range studies were performed with cyanobacteria and cyanophages isolated from fresh waters. In recent years there was increasing interest in the role of cyanophages in the marine environment. The majority of cyanophages in seawater infect representative cyanobacteria within *Synechococcus* Marine-cluster A such as DC2 (= WH 7803), SYN48 (= WH 6501), and WH 8012 (Suttle and Chan, 1993, Waterbury and Valois, 1993). *Synechococcus* isolates in this group contain phycoerythrin as a photosynthetic pigment, have an obligate salt requirement, and are dominant members of marine cyanobacterial communities (Waterbury and Rippka, 1989). Interestingly, some cyanophages which infect Marine-cluster A are also able to infect isolates of *Synechococcus* outside of this taxonomic group. Waterbury and Valois (1993) isolated two myoviruses that were able to infect several *Synechococcus* isolates from Marine-cluster A as well as an isolate from cluster B (WH 8101). Cells from Cluster B did not possess phycoerythrin. Similarly, Suttle and Chan (1993) isolated a myovirus that infected several phycoerythrin- and phycocyanin-dominant strains of cyanobacteria. There is also evidence for a broader host range for marine cyanophages than those isolated from freshwater in a study by Moisa et al. (1981). They isolated 31 strains of cyanophages from marine and fresh waters, and for 9 isolates obtained from the Black Sea communities of Eforie, Mangalia and Constanta they reported host ranges which spanned both filamentous and unicellular cyanobacteria. Unfortunately, little information was provided on the host strains or phage purification procedures, although the data certainly suggested that the host range for marine cyanophages may be much broader than for freshwater cyanophages.

IV. Distribution, Abundance and Seasonal Dynamics

Although first discovered in freshwater, cyanophages are also widely distributed in the marine environment and they are probably ubiquitous in aquatic systems. Following their initial discovery, most work on the distribution and abundance of cyanophages focused on understanding their role in the control of cyanobacterial populations in eutrophic freshwater environments. These studies were completed with the hope that cyanophages might be useful as biological control agents for cyanobacterial blooms. The recent realization that viruses are extremely abundant in the sea led to further investigations on the presence, abundance, and ecological role of cyanophages in the marine environment.

A. Freshwater

The existence of cyanophages was demonstrated by the purification of a virus which caused the lysis of several isolates of filamentous cyanobacteria belonging to a group of closely related cyanobacteria that were classified into three genera; *Lyngbya, Plectonema* and *Phormidium* (Safferman and Morris, 1963). Further investigation demonstrated that LPP cyanophages were found in waste stabilization ponds in the United States (Safferman and Morris, 1967; Safferman et al., 1969a; Safferman, 1973), India (Singh and Singh, 1967) and Scotland (Daft et al., 1970), as well as in fishponds in Israel (Padan and Shilo, 1969) and rice fields in India (Singh, 1967). Despite their widespread occurrence, the titers of LPP cyanophages were at most a few hundred infectious units per mL in waste stabilization ponds throughout the United States. Similarly, LPP cyanophages were detected in 90% of 40 water samples collected from fishponds in Israel, but the highest titers were only 13 mL^{-1}. Although high titers of infectious LPP cyanophages in natural waters were not observed, there was good evidence for strong seasonal variations in their abundance. In waste stabilization ponds in Arkansas, USA, and fish-rearing ponds in Israel, titers of LPP cyanophages were at times more than 10-fold greater during October and November, and from May to July (Safferman and Morris, 1967; Padan and Shilo, 1969; Safferman, 1973), than during February and March. In similar studies Fallon and Brock (1979) observed seasonal changes in the titer of lytic pathogens of cyanobacteria that coincided with changes in cyanobacterial abundance, however; the pathogens were much more abundant than reported for LPP cyanophages and reached peak titers of > 1000 mL^{-1}. Clearly, seasonal changes in cyanophage populations are related to environmental changes which also affect host-cell abundance, although the precise cause of these variations have yet to be demonstrated.

Viruses that infect cyanobacteria in the *Anabaena-Nostoc* group are also widespread and were found in lakes, reservoirs and sewage settling ponds (Adolph and Haselkorn, 1971; Mendzhul et al., 1973; Kozyakov, 1977; Hu et al., 1981; Moisa et al., 1981; Phlips et al., 1990). They are less frequently isolated than the LPP cyanophages, however, and also appear to be much less abundant; typically they are only detectable following amplification in enrichment culture.

A third group of cyanophages was reported to infect cyanobacteria assigned to the genera *Nostoc* and *Plectonema* (Muradov et al., 1990). These viruses were reported to be distributed widely in the former Soviet Union where they were isolated from fish farms and waste stabilization ponds. High titers of these viruses were found during summer with abundances ranging from 4500 to 9500 infectious units mL^{-1} (Muradov et al., 1990).

The range of occurrence of viruses that infect the *Anacystis-Synechococcus* group of cyanobacteria is as widespread as that of other freshwater cyanophages (Safferman et al., 1972; Sherman and Connelly, 1976; Gromov, 1983; Kim and Choi, 1994) although titers were not reported routinely. Interestingly, although phycoerythrin-rich *Synechococcus* spp. can be dominant in many freshwater systems (e.g. Caron et al., 1985; Stockner and Shortreed, 1991), there were few attempts to isolate viruses which infect these cyanobacteria. On several occasions natural virus communities from Lake Constance in Germany were screened, and viruses which infected several phycoerythrin-rich *Synechococcus* strains were isolated (Chan and Suttle, unpublished data). More efforts should be made to screen freshwaters for these cyanophages because they are abundant in marine systems where they represent the only group that can infect phycoerythrin-rich isolates of *Synechococcus* (see below).

B. Marine

The first report of cyanophages in the marine environment was for Romanian coastal waters of the Black Sea (Moisa et al., 1981). It is surprising that this observation went largely unnoticed given the fact that cyanobacteria, and the closely related prochlorophytes, are recognized as being the most important primary producers in much of the world's oceans (Joint and Pomroy, 1983; Li et al., 1983; 1992; Waterbury et al., 1986; Olson et al., 1990; See Chapter 5). The impetus for further study followed observations that high concentrations of virus-like particles appeared to be ubiquitous in the sea (Proctor et al., 1988; Bergh et al., 1989). Subsequent investigations demonstrated that a significant proportion of heterotrophic bacteria and *Synechococcus* spp. in marine waters contained visible virus-like particles (Proctor and Fuhrman, 1990) and that viruses infecting major marine primary producers, including *Synechococcus* spp., could be readily isolated from seawater (Suttle et al., 1990). Most of the cyanophages that were isolated were Myoviridae, although Styloviridae and Podoviridae were also isolated (Suttle and Chan, 1993; Suttle et al., 1993; Waterbury and Valois, 1993).

It was soon apparent that very high concentrations of cyanophages that infect phycoerythrin-rich strains of marine *Synechococcus* spp. were present in a variety of environments. In coastal waters of the Gulf of Mexico viruses that infect *Synechococcus* Strains DC2 and SYN48 occurred routinely at concentrations of 10^5 mL^{-1} and greater (Suttle and Chan, 1993; 1994). Similarly, Waterbury and Valois (1993) found infectious cyanophage concentrations > 10^4 mL^{-1} in Woods Hole Harbor, while Garza and Suttle (unpublished data) recorded abundances in fjords along the coast of British Columbia ranging between 10 and 5000 mL^{-1} in August. The highest concentration of infectious cyanophages reported for a natural environment was during a bloom of *Synechococcus* spp. at a station overlying the continental shelf in the western Gulf of Mexico (Suttle et al., 1996). The abundance of cyanophages that infected *Synechococcus* Strain DC2 was ca. 10^6 mL^{-1}, which represented about 10% of the entire viral community, as determined from direct counts of viral particles. In nearshore waters of the Gulf of Mexico (Port Aransas, Texas) and Woods Hole, infectious cyanophage concentrations reached their highest abundance during the summer and fell when water temperatures were highest and *Synechococcus* spp. was most abundant (Waterbury and Valois, 1993; Suttle and Chan, 1994).

The abundance of cyanophages that infect many non-phycoerythrin containing *Synechococcus* spp. is typically undetectable. However, for at least one strain (SNC1 = UTEX 2624) for which infectious cyanophages were detectable, the pattern of seasonal changes in abundance was similar to viruses which infected the phycoerythrin-rich strains DC2 (= WH 7803) and SYN48 (= WH 6501); the maximum abundance was only about a tenth as much (Suttle and Chan, 1994). It is important to note that there is considerable variation among the phycoerythrin-rich strains in terms of the abundance of infectious cyanophages that are present in a given environment. For example, in Woods Hole Harbor the abundance of cyanophages that infected *Synechococcus* strains WH 8012 and WH 8018 were frequently similar, but at other times were more than one order of magnitude different (Waterbury and Valois, 1993). Similar observations were made for coastal waters of the Gulf of Mexico where the concentration of viruses that infect some phycoerythrin-rich strains of *Synechococcus* (838BG and SYN48) were less than 10% of those infecting *Synechococcus* Strain DC2 (Suttle and Chan, 1993; 1994).

It is clear that the abundance of infectious cyanophages is tied to that of *Synechococcus* spp., and not directly to temperature. A regression analysis on seasonal data collected during a three-year period on the abundance of cyanophages infecting several strains of *Synechococcus* spp. versus temperature only explained 53 to 70% of the variance (Suttle and Chan, 1993). This conclusion is consistent with data collected on a seaward transect from the coast of Texas. In surface waters both the abundance of cyanophages that infect *Synechococcus* Strain DC2, and the concentration of *Synechococcus* cells, dropped more than ten-fold; whereas, the temperature increased by about 1°C (Suttle and Chan, 1994). At the seaward end of the transect, in waters which overlaid the continental shelf, cyanophage abundance dropped from about 10^4 mL^{-1} at the surface to ca. 100 mL^{-1} at 97 m depth, while water temperature dropped to about 21°C. Rodda (1996) completed a seasonal study in the same environment and came to a similar conclusion. Principle Component Analysis demonstrated that the abundance of infectious cyanophages was related first, and most strongly, to the abundance of *Synechococcus* and secondly, to temperature. At stations in the central Gulf of Mexico, infectious cyanophages were undetectable (unpublished data).

For marine environments there is evidence for a threshold in the abundance of *Synechococcus* spp. beyond which infectious cyanophage concentration increases greatly. A sharp increase in infectious cyanophage concentrations from ca. 10^2 mL^{-1} to 10^5 mL^{-1} was observed to be associated with *Synechococcus* populations in excess of ca. 10^3 cells mL^{-1} (Suttle and Chan 1994). A larger data set revealed a similar relationship although the threshold was at about 10^4 cells mL^{-1} (Rodda 1996) These data are consistent with the idea that when host cell abundance exceeds a certain level, host-virus contact rates exceed a critical level and infectious viruses propagate rapidly (Wiggins and Alexander, 1985).

A surprising finding was that infectious cyanophages can occur at considerable depth in marine sediments. The occurrence and abundance of cyanophages that infect *Synechococcus* sp. strain DC2 were measured in the sediment and overlaying water at a site 45 km offshore in the Gulf of Mexico in May, August, and December of 1995 (Rodda et al., in review). Cyanophage titers ranged from 1.4 x 10^5 mL^{-1} at the sea surface to 1.0 x 10^4 mL^{-1} at 46.6 m, just above the sediment surface. In the sediment the titer of infectious cyanophages decreased exponentially with depth from 9.4 x 10^4 mL^{-1} at the water/sediment interface to 3.0 x 10^2 mL^{-1} at 30 cm below the sediment surface (Fig. 2). These sediments were anaerobic below 2 cm. The source of infectious cyanophages at 30 cm depth in sediments lying below the photic zone is an intriguing question. The concentration of cyanophages in the water immediately above the sediment surface was about an order of magnitude less that at the sediment/water interface. This suggests that the viruses probably attach to particles either directly or via infected cells, and sink to the sediment surface where they are subsequently buried. Based on absolute dating estimates sedimentation rates in this region of the

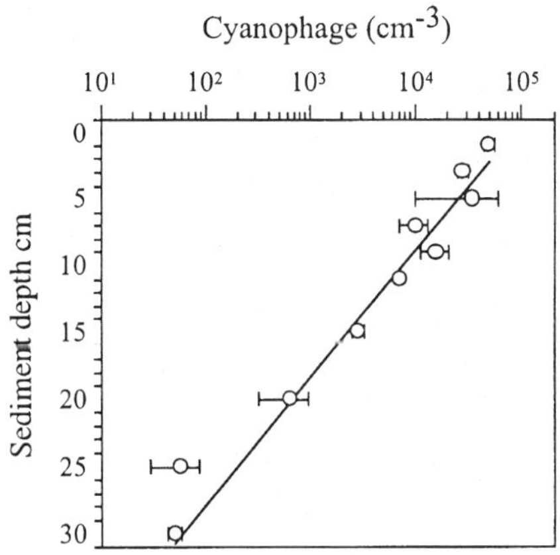

Fig. 2. Abundance of cyanophages infecting marine *Synechococcus* Strain DC2 in sediments 45 km offshore in the Gulf of Mexico. Based on absolute dating estimates and sedimentation rates the viruses at 30 cm depth are in excess of 100 years old.

Gulf of Mexico are ca. 0.33 cm y^{-1} (Snedden, 1985; Berryhill, 1986); therefore, viruses at 30 cm are approximately 100 years old. These estimates are supported by little evidence for physical disturbance or reworking of the sediments (Snedden, 1985), and calculated rates of diffusion which would require > 300 years for a virus to diffuse 30 cm (Rodda et al., in review). These data suggest that sediments may be an important reservoir of infectious cyanophages.

Unlike freshwater environments where viruses that infect filamentous cyanobacteria are commonly isolated, viruses in marine waters which infect unicellular cyanobacteria appear to be by far the most abundant. The only example of a marine filamentous cyanobacterium that was demonstrated to be infected by a virus was *Phormidium persicinum* (Ohki and Fujita, 1996). Unlike the podoviruses that have short non-contractile tails and infect other filamentous cyanobacteria, the virus which infected *P. persicinum* possessed a long flexible tail and was morphologically similar to siphoviruses. At present, there is no information on the distribution or abundance of these cyanophages in nature.

To date, viruses which infect other marine filamentous cyanobacteria were not isolated. It would be particularly interesting if viruses can be found that infect the oscillatorian, *Trichodesmium*; a major primary producer and nitrogen fixer in tropical waters (Carpenter, 1983; Carpenter and Romans, 1991; Chapter 5). Given the prevalence of other viruses which infect filamentous cyanobacteria it seems likely that they exist.

V. Fate of Cyanophages in the Natural Environment

The major loss processes for viruses in aquatic environments are destruction by solar radiation, attachment to particles, attachment to host cells and consumption by protozoan grazers (Suttle and Chen, 1992). Of these processes, destruction by solar radiation in near-surface waters is the most significant and can result in the destruction of the entire infectious viral community in a matter of hours (Suttle and Chen, 1992; Wommack et al., 1996; Nobel and Fuhrman, 1997; Wilhelm et al., 1998a). Nevertheless, photoreactivation can restore infectivity to a significant portion of the damaged viruses (Weinbauer et al., 1997; Wuet al., 1967). Destruction rates of cyanophages exposed to solar radiation are in the range of those measured for marine bacteriophages (Suttle and Chan, 1994); therefore, sunlight-dependent removal rates of cyanophage infectivity in surface waters should be similar to removal rates measured for other viruses. This presents a paradox (Suttle et al., 1993). If the infectivity of cyanophages is very sensitive to solar radiation how can very high concentrations of infectious cyanophages persist in surface waters when UVB flux is high?

In situ measurements showed that decay rates of infectivity in natural cyanophage communities in full sunlight ranged from < 0.1 to ca. 0.4 h^{-1} at the surface and decreased strongly with depth. Most of the decrease I infectivity was explained by the attenuation of UVB (Garza, 1996; Garza and Suttle, in review; Chapter 21). The relationship between depth and the removal rate of cyanophage infectivity that resulted from solar radiation was similar to that predicted for other marine bacteriophages (Suttle and Chen, 1992). There appeared to be a minor light-dependent component in the decay of cyanophage infectivity that was not the result of UVB (Garza and Suttle, in review). When integrated over the surface mixed layer and averaged over 24 h, sunlight was responsible for removal rates of cyanophage infectivity of 0.53 and 0.75 d^{-1} at two stations on the continental shelf of the Gulf of Mexico.

Although cyanophage infectivity decays as the result of exposure to UVB radiation, natural cyanophage communities are more resistant to damaging solar radiation in the summer, than in winter and spring. In a series of experiments, the decay rates of infectivity of natural cyanophage communities exposed to sunlight were compared to the decay rates of cyanophage isolates under the same conditions (Garza and Suttle, in review). During the spring and winter, decay rates of cyanophage isolates and natural cyanophage communities were generally similar, but during summer, decay rates of isolates were as much as two-fold higher than in the natural communities. As the sensitivity of the isolates to damaging radiation did not change, these results indicate that the natural viral community became more resistant to damage caused by sunlight. The mechanism whereby resistance to damaging radiation can be increased is unknown, but potentially could involve changes to the DNA itself.

The presence of high concentrations of infectious cyanophages in surface waters is attributable, in part, to increased resistance of natural communities to sunlight during periods of highest UVB flux, and repair of damaged DNA by host-cell and viral encoded mechanisms (Miller and Kokjohn, 1990; Friedberg et al., 1995; Chapter 21) during darkness or when the viruses and host cells are mixed below the surface. Mixing below the surface will not only result in a decrease in exposure to UVB, but will also result in a shift in spectral quality to longer wavelengths, allowing the potential repair of the damaged DNA by the light-dependent enzyme, photolyase. The loss of viral infectivity following exposure to sunlight is largely the result of damage to DNA from the formation of pyrimidine dimers; exposure to sunlight generally will not interfere with the attachment of cyanophages to host cells and the subsequent injection of viral nucleic acids. The damaged viral DNA can then be repaired by host-cell mechanisms such as photoreactivation, (i.e. the restoration of infectivity to UV-irradiated viruses upon exposure to visible light in the presence of host bacteria; Dulbecco, 1949, 1950). Both filamentous and unicellular cyanobacteria are known to have very efficient light-dependent mechanisms that can repair UV-damaged cyanophage DNA (Wu et al., 1967; Singh, 1975; Asato, 1976; Amla, 1979; Hwang-Lee et al., 1985; Levine and Thiel, 1987). Moreover, it was demonstrated recently that photoreactivation can be very important in maintaining the infectivity of natural phage communities in the sea (Weinbauer et al. 1997; Wilhelm et al., in press; in review)

The significance of sunlight, relative to other factors responsible for the decay of cyanophage infectivity in the upper mixed layer of the water column, depends on a number of things including insolation at the surface, the optical clarity of the water, and the depth of the mixed layer. In contrast to solar radiation, light independent factors which contribute to the destruction of cyanophage infectivity are assumed to be relatively constant with depth in the mixed layer. The observation that the virucidal properties of seawater can be eliminated or reduced by heating or filtration was recognized for many years (e.g. Lycke et al., 1965; Matossian and Garabedian, 1967; Mitchell and Jannasch,1969; Mitchell, 1971; Shuval et al., 1971). Although the heat-labile particulate material that is responsible for much of the light-independent decay of viral infectivity can include bacteria (e.g. Moebus, 1992), it is apparent that much of the virucidal material can be in the > 1.0 µm size fraction, and is not directly attributable to the predominant bacteria (Suttle and Chen, 1992). The explanation offered is that heat-labile colloidal material is responsible for the removal of viral infectivity. There is also evidence for colloidal heat-labile virucidal material in the < 0.2-µm size fraction (Nobel and Fuhrman, 1997). The other processes that were shown to remove infectious viruses from seawater include flagellate grazing and attachment to host cells, although these processes were relatively minor and were estimated to account for < 1% of measured viral removal rates (Suttle and Chen, 1992; Gonz·lez and Suttle, 1993).

Decay rates of natural cyanophage communities were also temperature dependent. In the dark, decay rates varied from ca 0.05 d^{-1} between 4 and 15°C, to 0.07 and 0.13 d^{-1} at 20 and 25°C, respectively (Garza and Suttle, in review). These rates were similar to those that were measured for natural communities of algal viruses in seawater (Cottrell and Suttle, 1995); they are still much less than rates which were measured in sunlight. Consequently, even in warm waters, solar radiation remains a major factor responsible for the decay of cyanophage infectivity in the natural environment.

VI. Effect of Cyanophages on Cyanobacterial Populations and Communities

Clearly, the fact that infectious cyanophages exist in aquatic systems, and the fact that they have a relatively short residence time (see above), implies that they are responsible for some portion of cyanobacterial mortality. Given that this mortality is selective, it can affect community structure of planktonic communities. There were two primary foci for studies on the impact of cyanophages on the mortality of cyanobacteria. The first, and earliest work, dealt with freshwaters and was primarily concerned with the potential use of cyanophages as biological control agents to eliminate, or reduce, nuisance and toxic blooms of cyanobacteria. The impetus for this work was the major ecological and economic effects of such blooms. The second focus was to try and determine the effect of cyanophages on the mortality of *Synechococcus* spp. in marine environments. This research was driven largely by the importance of *Synechococcus* spp. as major primary producers in the marine environment and the high abundance of co-occurring cyanophages.

A. Mortality

A number of methods were used to examine the effect of viruses on mortality in aquatic communities of heterotrophs and photosynthetic microbes (Suttle, 1994). These included the use of transmission electron microscopy to determine the percentage of infected cells (Proctor and Fuhrman, 1990); inferring viral production rates from rates of viral destruction (Heldal and Bratbak, 1991; Suttle and Chen, 1992); and calculating contact rates between viruses and potential host cells (Murray and Jackson, 1992; Waterbury and Valois, 1993; Suttle and Chan, 1993). These approaches were used to infer the effect of cyanophages on the mortality of marine *Synechococcus* spp. Unfortunately, all of the methods require specific assumptions which weaken the conclusions. On the other hand, it is encouraging that entirely independent methods lead to similar estimates of cyanobacterial mortality as a consequence of viral infection (Plate 10c).

1. Frequency of Infected Cells

The first attempt to deduce the effects of viruses on the mortality of *Synechococcus* spp. is also perhaps the most elegant. The approach was based on the principle that if the proportion of the latent period (the time between when the cells are infected and when lysis begins to occur) during which viral particles are visible within cells is known, and if the proportion of cells containing visible viral particles is also known, then the proportion of the population which is infected can be calculated. Proctor and Fuhrman (1990) examined samples from five locations in the western Atlantic and Caribbean, and determined that on average 0.8 to 2.8% of cyanobacterial cells contained visible phage particles. To estimate the proportion of the cyanobacterial community that was infected they used data for a marine bacterium, in which viral particles were visible for the last 10% of the latent period. Hence, the actual proportion of cyanobacteria which were infected at any given time would be 10 times the proportion which contained visible viral particles. This would mean that, on average, 8 to 28% of the cyanobacterial populations were infected by cyanophages. However, Waterbury and Valois (1993) argued that these are probably overestimates as viruses are probably only visible for about half of the latent period within infected cyanobacteria (Padan and Shilo, 1973; Sherman et al., 1976). Consequently, the percentage of visibly infected cells should be multiplied by two, rather than 10, in order to estimate the proportion of the population that is infected.

In reality, there are very few time-course data on the appearance of viral particles within infected cyanobacteria. For the filamentous cyanobacterium *Plectonema boryanum* (UTEX 594) infectious viruses appeared approximately half way through the latent period (Padan et al., 1970). However, there are many differences in the replication of cyanophages in filamentous and unicellular cyanobacteria, and it is likely inappropriate to infer details of the lytic cycle in marine *Synechococcus* spp. from observations made on freshwater filamentous cyanobacteria. Data from a freshwater *Synechococcus* sp. (PCC 6911) indicated that virus particles were visible within infected cells about 75% through the latent period (MacKenzie and Haselkorn, 1972b). A further complication in the estimation of the percent of the cyanobacterial population that is infected from the proportion of visibly infected cells is that the rise period (the time from the beginning of cell lysis to complete lysis of a culture) can be comparable in length to the latent period (MacKenzie and Haselkorn, 1972b; Benson and Martin, 1981). In a reevaluation of the parameters used to infer bacterial mortality from visibly infected bacteria, Proctor et al. (1993) acknowledged that the current model, where the proportion of cyanobacteria which were infected in a population was calculated on the basis of the proportion of cells in which viral particles were visible, was inadequate.

The available data cannot be used to define precisely the relationship between the proportion of cyanobacteria in a population that is infected by viruses, and the proportion of visibly infected cells. Nonetheless, as a first approximation it is probably reasonable to assume that viruses are visible for about

25% of the lytic cycle, based on the data of MacKenzie and Haselkorn (1972b). This would mean that if 0.8 to 2.8% of *Synechococcus* cells contained visible virus particles, then 3 to 12% of the total population was infected. As the lytic cycle of cyanophages in unicellular cyanobacteria ranges from ca. 5 to 40 hours (MacKenzie and Haselkorn, 1972b; Safferman et al., 1972; Suttle and Chan, 1993; Kim and Choi, 1994) one cannot state with certainty what this represents in terms of daily mortality, although it seems likely that several percent of the *Synechococcus* population is lysed on a daily basis. In a steady-state system in which 50% of the cells live to reproduce and 50% die, then something that removes 50% of the cells will account for 100% of the mortality (Proctor and Fuhrman, 1990). Therefore, if 3 to 12% of the population is infected and the lytic cycle is less than the doubling time of the cells, then 6 to 24% of the total mortality may be the result of viral infection if all infected cells are killed by viruses.

2. Decay Rates

Another approach that was used to infer the effect of viruses on cyanobacterial mortality was to assume that the rates of removal and production of infectious cyanophages were balanced. If the number of viruses produced per lytic event is known then the number of cells in which lysis occurred can be estimated from the number of viruses that were lost from the system. This approach was first used to estimate mortality rates in marine bacteria caused by phages (Heldal and Bratbak, 1991; Suttle and Chen, 1992). The same approach can be used for cyanobacteria.

Infectious cyanophages are continuously lost from aquatic environments because of a variety of factors including solar radiation, and attachment to particles. The first attempt to estimate mortality rates imposed on natural communities of *Synechococcus* spp. by cyanophages was based on average decay rates of infectivity for cyanophages and other viruses as a function of solar radiation, and attenuation coefficients for damaging radiation in seawater (Suttle and Chan, 1994). It was estimated that the decay rate averaged over 24 hours and integrated over the surface mixed layer (7.5 m) was 2 d^{-1} for optically transparent waters off the coast of Texas. As the average concentration of cyanophages infecting DC2 was 15300 mL^{-1}, the rate of viral production required to balance the daily removal of viruses was 2 x 15300 or 30600 viruses mL^{-1} d^{-1}. A burst size of 250 viruses produced per cell lysed (Suttle and Chan, 1993) resulted in an estimate of 122 cells mL^{-1} d^{-1} in which lysis would have to occur if viral production was to balance the removal resulting from exposure to solar radiation. The concentration of *Synechococcus* at this location was 1,878 cells mL^{-1}; therefore, in a steady-state system it was estimated that about 6.5% of the cells would have to be lysed on a daily basis in order to balance the removal resulting from solar radiation. As more data became available (see below) it was apparent that the average decay rate and burst size in natural communities were in error a magnitude of about three-fold; as one error was an overestimate and the other an underestimate the effects of these processes on the estimated mortality of *Synechococcus* spp. should be minimal.

More recently, in situ decay rates of natural communities of cyanophages were used to estimate the effect of viruses on the mortality of *Synechococcus* spp. (Garza, 1996; Garza and Suttle, in review). In two stations off the coast of Texas the decay rates of infectious cyanophages were estimated to be 0.53 and 0.75 d^{-1} when averaged over 24 hours and integrated over the surface mixed layer (28.2 and 10.4 m, respectively). The concentrations of cyanophages that infected *Synechococcus* strain DC2 in these samples were 1.68 x 10^5 and 1.05 x 10^6 mL^{-1}, respectively. Consequently, balancing the removal rates of infectious cyanophages would require average production rates of 8.9 x 10^4 and 7.9 x 10^5 cyanophages mL^{-1} d^{-1}, respectively. Based on TEM observations of these samples, the average burst size was estimated to be 81 viruses per *Synechococcus* cell lysed (Suttle et al., 1996). Therefore, the numbers of *Synechococcus* cells in which lysis must have occurred, in order to balance the measured loss rates of infectious cyanophages, were 1099 and 9722, which represents approximately 1 and 8% of the *Synechococcus* population, respectively. This is a minimum estimate as it only considered viruses that caused lysis of *Synechococcus* strain DC2. In addition, it assumed that all *Synechococcus* cells were responsible for production of infectious cyanophages. In reality, there maybe subsets of the *Synechococcus* population which are responsible for most of the viral production.

It is encouraging that results from two independent approaches – cyanophage decay rate experiments and observations of the proportion of infected *Synechococcus* cells in natural communities – were consistent. These estimates suggest that the mortality caused by cyanophages, although not large, is significant, and results in the infection and lysis of several percent of the cells in natural *Synechococcus* communities on a daily basis.

3. Contact Rates

A third approach that can be used to make inferences about the maximum effect of viruses on the mortality of cyanobacteria is to examine the contact rates between viruses and potential host cells (Murray and Jackson, 1992). If the abundances of infectious viruses and host cells are known the contact rate between the two can be calculated (Suttle & Chan, 1994) as $(S_h\ 2\pi\ d\ D_V)\ V\ P$, where S_h is the Sherwood number (dimensionless) for *Synechococcus* (1.01), d is the cell diameter (1.5×10^{-4} cm), D_V is the diffusivity of the viruses ($3.456 \times 10^{-3}\ cm^2\ d^{-1}$), V is the concentration of infectious viruses (cm^{-3}) and P is the concentration of *Synechococcus* (cm^{-3}). An essentially equivalent calculation is to infer infection rates from theoretical rates of phage adsorption and measured concentrations of *Synechococcus* and cyanophages (Waterbury and Valois, 1993). In both approaches the potential infection rate is directly proportional to the product of the viral and host cell abundances. As the infection rate cannot exceed the rate at which viruses encounter or adsorb to host cells, these rates provide maximum estimates of the viral infection rate or the potential for viral infection (Suttle, 1994).

Based on abundance data for cyanophages and *Synechococcus* cells in Woods Hole Harbor and waters farther offshore, Waterbury and Valois (1993) argued that cyanophages could only be responsible for the mortality of at most a few percent of *Synechococcus* cells on a daily basis. However, in these waters the highest encounter rate occurred when the abundances of cyanophages and *Synechococcus* were 1.14×10^4 and $3.8 \times 10^5\ mL^{-1}$, respectively. In contrast, in the coastal waters of Texas cyanophage abundances and hence encounter rates can be much higher, and result in most *Synechococcus* cells being contacted by infectious cyanophage on a daily basis (Suttle and Chan, 1994). These calculations assume that the concentration of cyanophages that infect specific hosts such as *Synechococcus* Strains DC2 or WH 8012 are indicative of the total number of infectious cyanophages, and that all infectious cyanophages have the potential to infect all *Synechococcus* cells. It is probably reasonable to assume that the abundance of cyanophages that infect these strains of *Synechococcus* is representative of the total number of infectious cyanophages; however, the proportion of the *Synechococcus* population that can be infected by the most abundant cyanophages likely varies (see below).

4. Photosynthetic Rates

An approach that was used to infer the potential for viruses to affect primary productivity was to enrich natural phytoplankton communities with elevated levels of the co-occurring viral assemblages and follow the changes in the rates of photosynthesis. Experiments of this type showed that photosynthetic rates often decreased in response to elevated viral concentrations, then gradually reached an asymptote where additional increases in viral abundance had no further effect on photosynthesis (Suttle et al., 1990; Suttle, 1992). The y intercept of a regression through the initial slope of the rate of photosynthesis versus viral abundance provides an estimate of the photosynthetic rate in the absence of viruses. This use of this type of approach suggests that phytoplankton primary productivity would be about 3% higher if viruses were absent (Suttle, 1994). Unfortunately, because the lytic cycle of unicellular cyanobacteria can take 10 hours or more, and because the photosynthetic rates of infected cells do not decrease until near the point of cell lysis, interpretation of these types of data can be especially difficult.

B. Community Composition and Biological Control

The role of cyanophages in the control and maintenance of community composition was of interest ever since the first cyanophages were isolated. The primary impetus for studies in this area was the enormous environmental and economic damage caused by blooms of freshwater cyanobacteria. In recent years the discovery of very high abundances of cyanophages in seawater resulted in a shift in emphasis towards the obtaining of an understanding of the effect of phages on the mortality of cyanobacteria in marine environments. In part, a comprehending of the effect of cyanophages in nature is predicated on an understanding of the development and maintenance of host-cell resistance in natural cyanobacterial communities.

1. Host-cell Resistance

Not only can contact rates be used to establish the maximum number of cells that can be infected by cyanophages on a daily basis, but if independent estimates of mortality as the result of infection are available, then contact rates can also be used to infer the proportion of a *Synechococcus* population that is resistant to infection. For example, in the surface mixed layer of the coastal waters of Texas the percentage of *Synechococcus* cells in which lysis occurred due to viral infection was estimated to range from 0.2% per day in nearshore waters to 6.6% per day 83 km offshore (Suttle and Chan, 1994). This calculation assumed that solar radiation was responsible for most of the destruction of infectivity and that the burst size was about 250 (Suttle and Chan, 1993). In fact, the destruction rates and burst sizes were probably overestimated by about three-fold (see above). Calculated contact rates in the surface mixed layer at the offshore station indicated that 5% of *Synechococcus* were contacted by infectious cyanophages on a daily basis. This estimate was similar to the estimated mortality rate of 6.6% per day, and indicated that most contacts resulted in infection; hence, most *Synechococcus* were not resistant to most cyanophages. In contrast, in nearshore waters, the estimated mortality rate of *Synechococcus* as a result of infection was only 0.5% per day; yet, 83% of *Synechococcus* were contacted by cyanophages each day. Consequently, few contacts resulted in infection, and this suggests that either most *Synechococcus* were resistant or, that the efficiency of infection was very low.

A different approach used by Waterbury and Valois (1993) provided further evidence that when *Synechococcus* and cyanophages are abundant, most *Synechococcus* are resistant to the most abundant cyanophages. Clonal isolates of *Synechococcus* and cyanophages were obtained by serial dilution of water collected from Woods Hole Harbor. The cyanobacteria isolated from the highest dilutions (i.e. those that were most abundant) were resistant to the most abundant cyanophages; whereas, cyanobacteria that were isolated from lower dilutions were sensitive to many of the cyanophages. The authors argued that the observation that most *Synechococcus* were resistant to infection by their co-occurring phages was consistent with results from continuous cultures of heterotrophic bacteria (Lenski and Levin, 1985; Lenski, 1988). In the latter there is a the culmination in stable phage and host populations with phage about an order of magnitude less abundant than bacteria.

The scenario in the Gulf of Mexico is different because the abundance of cyanophages is frequently 10-fold greater than that of *Synechococcus*. As a consequence contact rates are sufficiently high that small populations of sensitive cells could not persist unless successful contacts were very low. Low contact success was also proposed as an explanation for the persistence of natural populations of the marine phytoflagellate *Micromonas pusilla* in the presence of high titers of infectious viruses (Cottrell and Suttle, 1995).

Further evidence that low contact success might explain the persistence of *Synechococcus* in the presence of high cyanophage titers was obtained using fluorescent-labeled viruses as probes to identify cells susceptible to infection (Hennes et al., 1995). In cultures of a coastal clone of *Synechococcus* it was found that a cyanophage isolate attached to only about 10% of the host population at any given time (Hennes and Suttle, unpublished data); yet, when viruses were observed to attach, they attached over the entire cell surface. Therefore, at any given time only a portion of the cells were susceptible to infection suggesting that viral receptors were not present throughout the cell cycle. From an evolutionary perspective this is rational; the probability of infection is reduced by minimizing the number, or time of exposure, of receptors to cyanophages. Although the nature of the viral receptor is unknown, work on Cyanophage AS-1 indicated that proteins are involved (Samimi and Drews, 1978). In other phage-host systems receptors are known to involve essential proteins involved in nutrient acquisition (e.g. Killmann et al., 1995). The observation that receptors were not present on all cells at a given time is consistent with the idea that proteins involved with nutrient acquisition were only present on the cell surface at a particular stage of the cell cycle. Moreover, in natural populations of *Synechococcus* viral receptors were observed to be very rare, or localized in a small region of the cell surface. Observations that only a portion of cells in a population were susceptible to infection at any one time are consistent with the data and model of Lenski (1988). The model suggests that one way to maintain stability in populations of host cells and viruses is to have a large fraction of the host population be protected from phage infection at any given time, but not to have this protection be heritable as is the case for resistance.

For other isolates of *Synechococcus*, viruses were observed to attach to all cells in a clonal population (Hennes et al., 1995). Perhaps continuous expression of viral receptors is more prevalent in offshore isolates where contact frequencies are lower, where there is less selection to avoid infection (Suttle and Chan, 1994), and where low nutrient concentrations result in selection for high levels of nutrient transport proteins. Alternatively, some cyanobacteria may become less resistant to infection after being maintained in culture for a number of years (Waterbury and Valois, 1993).

In many ways the relationship between cyanobacteria and cyanophages in freshwater appears similar to that in marine systems, although the biomass of nitrogen-fixing filamentous and colonial cyanobacteria are frequently much more abundant in freshwater. Like their marine counterparts, freshwater cyanobacteria which are sensitive to infection are never dominant when phages which infect them are abundant (Safferman and Morris, 1967; Padan and Shilo, 1973). This observation implies selection for cyanobacterial communities that are resistant to infection by the dominant phage. In continuous-culture experiments the abundances of filamentous cyanobacteria and infectious cyanophages underwent a series of increasingly-dampened, out-of-phase oscillations, eventually resulting in a stable coexistence between the phage and the host (Cowlishaw and Mrsa, 1975; Cannon et al., 1976; Barnet et al., 1981). During the experiments Barnet et al. (1981) found that there was selection for both mutant cyanophage and cyanobacteria, and this resulted in a range of adsorption and infection efficiencies. The results were similar to those obtained from the study of chemostats of *Escherichia coli* and Phage T4 (Levin et al., 1977; Lenski and Levin, 1985), and suggested that a combination of differing growth rates and phage susceptibilities were responsible for the coexistence of cyanophages and cyanobacteria.

In addition to the host range of specific cyanophages it is necessary to also consider host-cell resistance. Wide variations in the extent to which specific strains of cyanobacteria are susceptible to infection by cyanophages was demonstrated most clearly for viruses that infected marine *Synechococcus* spp. Titers of viruses that infect some strains of *Synechococcus* can reach 10^4 to 10^6 mL^{-1}, whereas titers of viruses which infect other strains vary from undetectable to a few mL^{-1} (Suttle and Chan 1993, 1994; Waterbury and Valois, 1993). Exclusively, the isolates which are most sensitive to infection by a wide range of cyanophages belong to the phycoerythrin-rich group of *Synechococcus* spp. that were termed Marine-cluster A (Waterbury and Rippka, 1989). In contrast, phycocyanin-dominant isolates were resistant to infection by most cyanophages in natural waters.

Phycoerythrin-rich *Synechococcus* are by far the most abundant unicellular cyanobacteria in oligotrophic oceanic waters (Waterbury et al., 1986) as well as in many oligotrophic large lakes (Caron et al., 1985; Stockner and Shortreed, 1991). Phycocyanin-dominant *Synechococcus* are restricted to productive waters. Phycoerythrin-rich cells are photosynthetically much more efficient than those with phycocyanin as their dominant pigment under the shorter wavelengths characteristic of oligotrophic water (Lewis et al., 1986). Also, the abundances of *Synechococcus* and infectious cyanophages decrease as one moves offshore (Suttle and Chan, 1994); therefore, the contact frequency between *Synechococcus* and cyanophages also decreases. The lower contact frequency means that the potential mortality from viral infection is decreased relative to other factors such as grazing, and results in less selective pressure for cyanobacteria to be resistant to infection.

2. Community Composition and Biological Control

The potential for viruses to control the composition of communities is intimately tied to the development of host resistance. The isolation of a cyanophage from a waste stabilization pond (Safferman and Morris, 1963) quickly lead to speculation and research on the potential for cyanophages to be used as biological control agents for blooms of cyanobacteria (Safferman and Morris, 1964a; Safferman, 1968; Shilo, 1971). Observations that cyanobacteria never co-occur in high abundance with cyanophages that cause their lysis (Safferman and Morris, 1967; Padan and Shilo, 1973) provided evidence that viruses controlled the distribution of some cyanobacteria. Evidence that viruses can affect the distribution and abundance of cyanobacteria was obtained from attempts to control experimental blooms of *Plectonema boryanum* (UTEX 594) in 8000 liter ponds by the addition of Cyanophage LPP-1 (Desjardins and Olson, 1983). The experiments showed that LPP-1 could control *P. boryanum*, but that water quality did not necessarily improve as other phytoplankton, including cyanobacteria, ultimately dominated the ponds. In other ponds the presence of cyanophages appeared to preclude the establishment of a bloom.

Experiments with naturally occurring cyanobacterial blooms also demonstrated that the abundance of cyanobacteria can be reduced by the addition of cyanophages and bacterial pathogens (Martin and Benson, 1988). In 80 liter enclosure experiments viruses and other pathogens of cyanobacteria were added to naturally occurring blooms and this resulted in a substantial decrease in cyanobacterial biomass. Even the addition of cyanophages alone was reported to cause a considerable decrease in cyanobacteria. The presence of virus-like particles within *Aphanizomenon* cells at the end of a bloom also suggested that viruses may be involved in bloom termination (Coulombe and Robinson, 1981). The promise of biological control remains an attractive proposition and led to continued efforts to isolate viruses that infect bloom-forming cyanobacteria (e.g. Phlips et al., 1990; Monegue and Phlips, 1991).

Cyanophages are important species-specific mortality agents of cyanobacteria, and as such undoubtedly play a significant role in the control of the structure of cyanobacterial communities. The relatively narrow host range of cyanophage isolates, coupled with the ability of cyanobacteria to develop resistance, makes it unlikely that cyanophages alone will be an effective biological control agent. It may be possible, however, that through the use of cyanophages as one of an array of pathogens, the biomass in blooms may be reduced. In nature there are a myriad of complex interactions that occur, and although we may recognize only a few different cyanobacteria based on morphological criteria, there is much more diversity at the molecular level in terms of host-cyanophage interaction.

VII. Environmental and Physiological Effects on Cyanophages

There are a number of environmental and physiological considerations that affect the stability, attachment and replication of cyanophages, and which are consequently important for their persistence in nature. Viruses vary greatly in terms of their stability, which makes it difficult to make generalizations. Nonetheless, many cyanophages were shown to retain their infectivity over a wide range of environmental conditions, although most of these studies were conducted with cyanophage isolates rather than natural cyanophage communities. Moreover, most of these studies were conducted on time scales of minutes to hours, rather than days to weeks, making extrapolation to the natural environment difficult.

A. Environmental Stability of Infectious Cyanophages

Temperature is a key environmental variable that effects infectivity. Studies on cyanophage isolates revealed that viruses can generally be stored for a month or more at 4°C with little loss of titer (Safferman and Morris, 1964b; Safferman et al., 1969b) although freezing can have a large effect (Desjardins and Olson, 1983). In seawater samples (with particulate material > 0.2 μm removed) cyanophage titers were stable for several months when stored at 4°C (Rodda and Suttle, unpublished data), and titers in excess of 10^5 mL^{-1} were measured in 0.2 μm filtered samples stored a year or more (Suttle and Chan, 1993). At higher temperatures, titers of some cyanophage isolates were stable an hour or more at 40°C (Safferman and Morris, 1964b; Safferman et al., 1969b) while other thermosensitive strains were reported to be unable to form plaques at 35°C (Padan et al., 1971). At temperatures above 50°C infectivity was typically rapidly lost (Safferman and Morris, 1964b; Padan et al., 1971; Sarma and Singh, 1995). The only data on natural cyanophage communities indicated that decay rates of infectivity in untreated seawater varied from ca 0.05 d^{-1} between 4 and 15°C, to 0.07 and 0.13 d^{-1} at 20 and 25°C, respectively (Garza, 1996; Garza and Suttle, in review).

The infectivity of natural cyanophage communities is also very sensitive to solar radiation, and this can be the major loss process in natural waters (Suttle and Chan, 1994; Garza, 1996; Garza and Suttle, in review). For natural cyanophage communities at two stations on the continental shelf of the Gulf of Mexico, sunlight was responsible for removal rates of cyanophage infectivity of 0.53 and 0.75 d^{-1}, when integrated over the surface mixed layer and averaged over 24 h (Section IV). Clearly, the effect of solar radiation on the survival of cyanophages in natural waters depends strongly on other environmental factors such as water transparency, solar angle and insolation.

Similar to other phages, there is no correlation between cyanophage morphology and the effect of pH on stability, although many freshwater cyanophages are infectious over a much broader range of pH than other bacteriophages. For example, LPP-1, AS-1 and SM-1 were stable over the range from around pH 5 to 11 (Safferman and Morris, 1964b; Safferman et al., 1969b; 1972); whereas, many bacteriophages loose infectivity when the pH exceeds 8 or 9 (Adams, 1959; Ackermann and DuBow, 1987a). These observations have environmental significance because some cyanobacteria show optimal growth at pH values above 8 while others are known to occur in lakes with a pH as high as 11 (Padan and Shilo, 1973). The broad range in the pH tolerance of many cyanophages suggests that environmental changes in pH are unlikely to affect their distribution in nature. There are no comparative studies on the pH tolerances of marine cyanophages.

Cyanophage infectivity can also depend on the concentration of cations. For example, LPP-1 (Safferman and Morris, 1964b; Schneider et al.,1964) and SM-2 (Martin and Benson, 1988) are much more stable in the presence of Mg^{+2} than when resuspended in deionized water. Also, AS-1 was reported to be more stable in the presence of cations (Amla, 1981; Desjardins and Olson, 1983), although it could be dialyzed against distilled water without loss of infectivity (Safferman et al., 1972). Interestingly, LPP-1, SM-2 and AS-1 belong to three different families of viruses; hence, a requirement for cations cannot be associated with a particular taxonomic group. In contrast, some cyanophages such as the podovirus SM-1 had no cation requirement and remained infectious when resuspended in distilled water (Safferman et al. 1969b). Other cations such as Mn^{2+} or Na^{+} also stabilized the infectivity of some cyanophage particles (e.g. Amla, 1981; Benson and Martin, 1988). In the case of LPP-1, Mg^{2+} was shown to be required to maintain capsid integrity; at Mg^{2+} concentrations of 1 mM or above LPP-1 remained infectious while at a concentration of 0.1 mM, infectivity was lost (Goldstein et al., 1967). This observation is potentially significant since many lakes have Mg^{2+} concentrations in the 0.1 to 1 mM range, while some lakes have even lower concentrations (Horne and Goldman, 1994). The requirement for cations is not universal, and Cyanophage AN-23 was shown not to tolerate NaCl concentrations greater than 0.05 M (Bancroft and Smith, 1992), which is much lower than is found in seawater.

B. Adsorption

Although some cyanobacteria have strategies to reduce the frequency with which contact with cyanophages result in infection (see above), in general most contacts between bacteriophages and host cells are thought to result in adsorption (Anderson, 1949; Schwartz, 1976). This is also true for many cyanophages. For example, when 9×10^4 mL^{-1} of Cyanophage S-BBS1 were added to 9×10^6 mL^{-1} of exponentially growing *Synechococcus* sp. Strain BBC1, the viral removal rate was 0.035 min^{-1}; hence, ca. 88% of the viruses adsorbed after 1 hour (Suttle and Chan, 1993). Contact rates during the same experiment, calculated using transport theory (Murray and Jackson, 1992) and assuming a cell diameter of 1.5 μm and diffusivity of the viruses of 6.32×10^{-8} cm^2 s^{-1} (calculated based on the diffusivity of a spherical virus with a diameter of 50 nm; Berg, 1983), yielded an adsorption rate constant of 0.0325 min^{-1}. These data suggest that approximately 86% of the viruses should have contacted a host cell during a 1 hour incubation. Therefore, the adsorption rate was near the theoretical maximum based on diffusion theory, and most contacts resulted in infection. This is significant since it is stated frequently that cyanophage adsorption kinetics are much slower than those for other bacteriophages. Although slow rates of adsorption can occur in some systems (e.g. Safferman et al., 1972) this is not a general rule. The likely reason that measured rates of viral removal are often slower than observed for bacteriophage is that the abundance of cyanobacteria used in adsorption experiments is typically much lower than in experiments with heterotrophic bacteria.

However, for some viruses, including certain cyanophages, the frequency with which adsorption occurs following contact between a virus and host cell is affected by environmental factors or the physiological status of the host. It is not unusual, for example, for phages to require a cofactor such as a divalent cation for adsorption. Environmentally important cations such as Ca^{2+} or Mg^{2+} in mM concentrations are most frequently required, although other cations can sometimes substitute (Ackermann and DuBow, 1987b). Although it was not demonstrated in nature, given the high concentration of cations required for adsorption of some phage, it is conceivable that the distribution of some freshwater cyanophages may be limited by the availability of cations. Nutrient effects on cyanophage adsorption rates were not reported. The only reported data are for Cyanophage S-PM2, for which adsorption kinetics were the same on phosphate-replete or -deplete *Synechococcus* Strain DC2 (Wilson et al. 1996).

Light can also affect rates of cyanophage adsorption. Cyanophage AS-1 was observed to adsorb to host cells approximately two-fold faster in the light than in the dark (Cseke and Farkas, 1979). The data are unusual in that the adsorption kinetics were not first order. Almost all of the adsorption occurred within the first 10 min, and a large proportion of the phage remained unbound even after 90 min. In addition, at least some of the binding was reversible. The mechanism suggested to account for this observation was that light caused charge neutralization at the cell surface via photochemical reactions or through light-induced changes in the ionic composition adjacent to the cell surface. Martin and Benson (1988) claimed to have evidence of a similar phenomenon with Cyanophage SM-2. Interestingly, these effects were eliminated or reduced by increasing the Na^+ concentration from 12 mM to 110 mM, also possibly by charge neutralization. If light effects on cyanophage adsorption are widespread this could be of considerable ecological importance. Clearly, the effect of light on cyanophage adsorption in an area that is in need of further investigation.

C. Replication

1. Light

Intuitively, replication should be sensitive to environmental effects because anything that affects host-cell physiology potentially affects viral replication. For example, ATP is required for viral replication (Padan et al., 1970; Sherman and Haselkorn, 1971; Adolph and Haselkorn, 1972), and consequently cyanophage production is curtailed in darkness. If *Plectonema boryanum* (UTEX 594) is kept in the dark throughout the period of cyanophage growth, there is a lag in the appearance of free viral particles and the burst size is greatly reduced (Padan et al., 1970; Sherman and Haselkorn, 1971). If infected cells are returned to the light, even after many hours of darkness, viral replication proceeds normally although the final burst size can be even larger. In contrast, in the nitrogen-fixing cyanobacterium *Nostoc muscorum* (Adolph and Haselkorn, 1972) and in unicellular cyanobacteria (MacKenzie and Haselkorn, 1972c; Allen and Hutchison, 1976) viral replication essentially stopped in darkness.

Not only does the physiological state of cyanobacteria affect viral replication but, depending on the host, viral infection can have very different effects on photosynthetic metabolism. In filamentous cyanobacteria photosynthetic rates are often inhibited strongly following infection (Ginzburg et al., 1968; Adolph and Haselkorn, 1972; Sherman, 1976; Amla et al., 1987) although exceptions do occur (Wu and Shugarman, 1967; Ginzburg et al., 1976; Bisen et al., 1988). There is good evidence that in systems where viral replication can continue in the dark, or where photosynthesis is strongly suppressed following infection, some of the energy for viral synthesis is obtained via oxidative phosphorylation (Padan et al., 1970; Sherman and Haselkorn, 1971; Bisen et al., 1988; Singh et al., 1994). Furthermore, studies which used DCMU (Padan et al., 1970; Sherman and Haselkorn, 1971; Adolph and Haselkorn, 1972) or which measured enzyme activities (Kashyap and Singh, 1989) demonstrated that energy from cyclic photophosphorylation can be used for viral synthesis.

In contrast to filamentous cyanobacteria, in which photosynthesis is shut down as viral replication occurs, in unicellular cyanobacteria photosynthetic rates of infected cells typically remain unchanged until near the point of cell lysis (MacKenzie and Haselkorn, 1972c; Sherman, 1976; Suttle and Chan, 1993). The addition of DCMU either completely blocked phage replication (MacKenzie and Haselkorn, 1972c; Sherman, 1976) or strongly reduced it (Allen and Hutchison, 1976). Even though infection of unicellular cyanobacteria generally does not lead to a decrease in photosynthetic rates, there is evidence that non-cyclic photophosphorylation is inhibited (Teklemariam et al., 1990).

The interactions that are observed between cyanophage infection and host cell photosynthesis make sense from an evolutionary perspective. Given the difference in cell size, the carbon reserves that are available for respiratory energy production will likely be greater in filamentous cyanobacteria than in unicellular forms. Also, many of the filamentous cyanobacteria that were studied with respect to viral infection were frequently associated with surface water or other relatively high-light environments where photosynthesis was unlikely to be light limited. On the other hand, the unicellular genera for which viruses were isolated were not typically associated with surface blooms; note that *Synechococcus* sp. was misidentified as *Microcystis aeruginosa* in a number of earlier studies (see *Host Range*). It is also interesting that ultrastructural changes were found to occur in unicellular and filamentous cyanobacteria after infection. In filamentous cyanobacteria displacement of the photosynthetic lamellae was associated with the production of viral particles (Smith et al., 1966); whereas, in the unicellular forms there were no obvious changes in the thylakoids (Sherman et al., 1976).

2. Nutrients

Obviously, protein and DNA synthesis are an integral part of viral replication, so it is reasonable to anticipate that under the conditions of nitrogen or phosphorus limitation, as exist in many natural waters, viral replication will be affected. Yet, surprisingly few studies rigorously examined the effect of nutrient limitation on cyanophage replication. The studies that were done generally focused on metabolic changes in the infected cell, rather than on more ecologically-significant factors such as length of the lytic cycle and burst size.

One study which examined the response of nitrogen- or energy-starved *Anacystis nidulans* to infection by AS-1 (BorbÈly et al., 1980) demonstrated how viral infection prevented a typical starvation response by the host. These data lead the authors to suggest that infection may release cellular materials that mask the nutritional status of the cell. Other studies focused on the nitrogen metabolism of infected cells and demonstrated that nitrate reductase activity increased (Bisen et al., 1986a); the dynamics of amino acid pools were different in infected and non-infected cells (Kadyrova et al. 1995); and nitrogen fixation continued until shortly before cell lysis (Stewart and Daft, 1977). However, basic work on the effect of the nitrogen status of cyanobacteria on cyanophage replication remains to be done. The only study that provided some information was work done on two myovirus isolates (S-PM2 and S-WHM1) which were amplified on marine *Synechococcus* Strain DC 2 grown in nutrient replete condition or under nitrate limitation (Wilson et al., 1996). It was reported that the titer of S-WHM1 was reduced by 25% while no difference was detected in S-PM2.

Studies on the effect of phosphorus limitation on cyanophage replication are similarly scant. The exception is a study by Wilson et al. (1996) who examined the effect of phosphate depletion on the kinetics of cyanophage replication in *Synechococcus* Strain DC 2. The reduction in titer for three cyanomyoviruses (S-PM2, S-WHM1 and S-BM1) amplified on phosphorus-depleted cells ranged from 59 to 91% of that in nutrient-replete controls. The amplification of S-PM2 was examined in detail and it was found that although the burst size was reduced the length of the lytic cycle remained about 10 hours under phosphate-deplete conditions. However, only about 55% of the cells to which viruses adsorbed produced viral progeny when phosphate was in short supply.

The effects of nutrient-limited growth on the interactions between cyanophages and the cells they infect remains a key area for investigation. It is only through a teasing apart of these relationships that we will fully understand the role that cyanophages play in the environment.

VIII. Lysogeny

A. Filamentous Cyanobacteria

By the early 1970's the existence of lysogenic associations between cyanophages and filamentous cyanobacteria were well established (e.g. Cannon et al., 1971; Padan et al., 1972; Singh and Singh, 1972; Khudyakov and Gromov, 1973; Rimon and Oppenheim, 1975), although whether lysogeny was truly demonstrated in some of the early studies (Sherman and Brown, 1978) is a matter of debate. Much of the earlier work focused on lysogens of *Plectonema boryanum* and was directed towards developing tools to understand the molecular genetics of cyanobacteria. This literature was reviewed extensively by Sherman and Brown (1978) and Martin and Benson (1988). Since then new lysogens of filamentous cyanobacteria were isolated sporadically (Bisen et al., 1986b; Franche, 1987; Ohki and Fujita, 1996), but the primary focus for work with lysogenic cyanophages remains the molecular biology of cyanobacteria (e.g. Bagchi et al., 1987; Schneider and Haselkorn, 1988; Khudyakov and Wolk, 1996), a discussion of which is beyond the scope of this chapter.

B. Unicellular Cyanobacteria

In contrast to the wealth of studies on temperate cyanophages of filamentous cyanobacteria, data on lysogens of unicellular cyanobacteria were slow to materialize. In fact, with the exception of some preliminary evidence for lysogeny in *Anacystis nidulans* (Bisen et al., 1985), lysogens which infect unicellular cyanobacteria remained undocumented until the isolation of a temperate cyanophage which infected marine *Synechococcus* sp. strain NKBG 042902 (Sode et al., 1994).

C. Significance

Consideration of the ecological and evolutionary significance of lysogeny in cyanophages was not addressed. However, given the increased recognition of the importance of cyanobacteria (including prochlorophytes) in the world's oceans and lakes, and the renewed interest in the role of cyanophages in the environment, it is an area that is ripe for investigation. In particular, lysogenic cyanophages may be important as vectors of genetic information within cyanobacterial communities. Although the movement of cyanobacterial DNA by viruses (transduction) was not demonstrated there is clear evidence that lysogenic cyanophages integrate into the host genome (e.g. Franche, 1987). The triggers that induce prophages in cyanobacteria to enter the lytic cycle vary. For some lysogenized cyanobacteria environmental factors, including UVR (Ohki and Fujita, 1996), temperature (Rimon and Oppenheim, 1975) or exposure to high levels of copper (Sode et al., 1997), trigger induction, while in other cases attempts at induction were unsuccessful, even with mutagens such as mitomycin C (e.g. Padan et al., 1972).

At the present time there are no estimates of the proportion of cyanobacteria that are lysogenized in nature; however, estimates indicate that a relatively small proportion of the entire marine bacterial community may be lysogenized. Jiang and Paul (1996) were unable to detect lysogenic viral production in all but a few of 11 samples collected in oligotrophic offshore environments, but obtained evidence of induction at 6 of 11 eutrophic stations. In a similar study, Weinbauer and Suttle (1996) used a more sensitive approach for the detection of phage production in 9 samples from the Gulf of Mexico. They estimated that 0.07 to 4.4% (average 1.5%) of the natural bacterial community could be induced by the addition of Mitomycin C. These studies showed that a small proportion of marine bacterial communities as a whole were induced by mitomycin C or UVC radiation, so it is reasonable to suppose that cyanobacteria respond in a similar way. The greatest difficulty with interpreting these experiments, in the context of natural microbial communities, is the effectiveness of the inducing agents in stimulating phage production in lysogens. Mitomycin C or UVC radiation are not environmentally relevant and may not be effective inducing agents.

In the natural environment it is not known what factors are important for the establishment of a lysogenic association in cyanobacteria, nor what triggers lysogens to enter the lytic pathway. One possibility is that nutrients may be involved. Wilson et al. (1996) reported that lysis occurred in only 9% of the cells when infectious cyanophages were added to *Synechococcus* grown under low phosphate conditions. In contrast, lysis occurred in all cells that were grown under phosphate replete conditions. One interpretation of this result is that low phosphate conditions resulted in the establishment of a lysogenic association; however, this conclusion assumes that low phosphate concentrations do not affect cell viability. From an evolutionary perspective it makes sense that the nutritional status of the host may affect the establishment of lysogeny.

IX. Conclusion

The three taxonomic groups of viruses that infect cyanobacteria are representatives of the same well-known families of double-stranded DNA viruses that include T4 (Myoviridae), T7 (Podoviridae) and lambda (Styloviridae). Consequently, it is not surprising that a fairly comprehensive understanding of the biology of cyanophages is beginning to emerge given that there are many parallels between the biology and that of some of the most thoroughly studied viruses. In contrast, very little is known about the role of phages in the ecology of cyanobacteria; the implications of which are extremely different than they are for enteric bacteria. Cyanobacteria are among the most important primary producers on the planet. They exist in environmental extremes ranging from polar lakes to hydrothermal springs and are responsible for a significant proportion of the primary production in the world's oceans and lakes. Cyanophages also appear to be ubiquitous in aquatic environments, and can occur at abundances in excess of 10^6 mL^{-1}. Despite their abundance, and widespread distribution, our understanding of the role of cyanophages in the regulation of cyanobacterial productivity, abundance, distribution and diversity, is in its infancy. Yet, there is great potential for the achievement of a comprehensive understanding of their significance. The methods for culturing many types of cyanobacteria are well established, and there were numerous advances in recent years for techniques to isolate and culture cyanophages from the environment. All of these factors suggest that the ecological implications of cyanophage infection should be a fruitful area for research in the future.

Acknowledgments

I am indebted to the individuals who over the years made my laboratory an exciting environment to study and think about viruses and their role in aquatic environments. In particular, the hard work and insights of Feng Chen, Matthew Cottrell, Hudson DeYoe, Randy Garza, Kilian Hennes, Kristen Rodda, Steven Short, Steven Wilhelm, Markus Weinbauer and Yanping Zhang are acknowledged. Special thanks are reserved for Amy Chan whose efforts contributed in innumerable ways to the research efforts in the laboratory, and for Lita Proctor who introduced me to the world of viruses in the sea. The comments of Steven Wilhelm improved the manuscript. The support of the US Office of Naval Research, US National Science Foundation and Natural Science and Engineering Research Council of Canada is gratefully acknowledged.

References

Ackermann H-W and DuBow MS (1987a) Viruses of Prokaryotes (Volume 2): Natural Groups of Bacteriophages. CRC Press, Boca Raton, Florida

Ackermann H-W and DuBow MS (1987b) Viruses of Prokaryotes (Volume 1): General Properties of Bacteriophages. CRC Press, Boca Raton, Florida

Adams MH (1959) Bacteriophages. John Wiley and Sons, NY

Adolph KW and Haselkorn R (1971) Isolation and characterization of a virus infecting the blue-green alga *Nostoc muscorum*. Virology 46: 200-208

Adolph KW, and Haselkorn R (1972) Photosynthesis and the development of the blue-green algal virus N-1. Virology 47: 370-374

Adolph KW and Haselkorn R (1973a) Isolation and characterization of a virus infecting a blue-green alga of the genus *Synechococcus*. Virology 54: 230-236

Adolph KW and Haselkorn R (1973b) Blue-green algal virus N-1: Physical properties and disassembly into structural parts. Virology 53: 427-440

Allen MM and Hutchinson F (1976) Effect of some environmental factors on cyanophage AS-1 development in *Anacystis nidulans*. Arch Microbiol 110: 55-60

Anderson TF (1949) The reactions of viruses with their host cells. Bot Rev 15: 464-505

Amla DV (1979) Photoreactivation of ultraviolet irradiated blue-green alga *Anacystis nidulans* and Cyanophage AS-1. Arch Virol 59: 173-179

Amla DV (1981) Chelating agent shock of cyanophage AS-1 infecting unicellular blue-green algae. Indian J Exp Biol 19: 209-211

Amla DV, Rowell P and Stewart WDP (1987) Metabolic changes associated with cyanophage N-1 infection of the cyanobacterium *Nostoc muscorum*. Arch Microbiol 148: 321-327

Asato, Y (1976) Ultraviolet light inactivation and photoreactivation of AS-1 cyanophage in *Anacysitis nidulans*. J Bacteriol 126: 550-552

Bagchi SN, Kaloya P and Bisen PS (1987) Effect of cyanophage N-1 infection on the synthesis and stability of *Nostoc muscorum* nitrate reductase. Curr Microbiol 15: 61-65

Bancroft I and Smith RJ (1989) Restriction mapping of genomic DNA from 5 cyanophages infecting the heterocystous cyanobacteria *Nostoc* and *Anabaena*. New Phytol 113: 161-166

Bancroft I and Smith RJ (1992) Oriented degradation of the genome of cyanophage AN-23 after exposure to NaCl. New Phytol 121: 571-575

Barnet YM, Daft MJ and Stewart WDP (1981) Cyanobacteria-cyanophage interactions in continuous culture. J Appl Bacteriol 51: 541-552

Benson R and Martin E (1981) Effects of photosynthetic inhibitors and light- dark regimes on the replication of cyanophage SM-2. Arch Microbiol 129: 165-167

Benson R and Martin E (1984) Physiochemical characterization of cyanophage SM-2. Arch Microbiol 140: 212-214

Berg HC (1983) Random Walks in Biology. Princeton University Press, Princeton

Bergh O, Borsheim KY, Bratbak G and Heldal M (1989) High abundance of viruses found in aquatic environments. Nature 340: 467-468

Berryhill HL Jr (1986) Late Quaternary facies and structure, northern Gulf of Mexico : interpretations from seismic data. American Association of Petroleum Geologists, Tulsa, OK, USA

Bisen PS, Audholia S and Bhatnagar AK (1985) Mutation to resistance for virus AS-1 in the cyanobacterium *Anacystis nidulans*. Microbios Lett 29: 7-13

Bisen PS, Audholia S, Bhatnagar AK and Bagchi SN (1986b) Evidence for lysogeny and viral resistance in the cyanobacterium *Phormidium uncinatum*. Curr Microbiol 13: 1-5

Bisen PS, Audholia S, Shukla HD, Gupta A and Singh DP (1988) Evidence for photosynthetic independence of viral multiplication in cyanophage LPP-1 infected cyanobacterium *Phormidium uncinatum*. FEMS Microbiol Lett 52: 225-228

Bisen PS, Bagchi SN and Audholia S (1986a) Nitrate reductase activity of a cyanobacterium *Phormidium uncinatum* after cyanophage LPP-1 infection. FEMS Microbiol Lett 33: 69-72

Borbéy G, Kari C, Gulyas A and Farkas GL (1980) Bacteriophage infection interferes with quanosine 3'-diphosphate-5'-diphosphate accumulation induced by energy and nitrogen starvation in cyanobacterium *Anacystis nidulans*. J Bacteriol 144: 859-864

Braithwaite DK and Ito J (1993) Compilation, alignment, and phylogenetic relationships of DNA polymerases. Nucleic Acids Res 21: 787-802

Brown RM, Jr (1972) Algal Viruses. Adv Virus Res 17: 243-277

Cannon RE, Shane MS and Bush VN (1971) Lysogeny of a blue-green alga, *Plectonema boryanum*. Virology 45: 149-153

Cannon RE, Shane MS and Whitaker JM (1976) Interaction of *Plectonema boryanum* (Cyanophyceae) and the LPP-cyanophages in continuous culture. J Phycol 12: 418-421

Caron DA, Pick FR and Lean DRS (1985) Chroococcoid cyanobacteria in Lake Ontario: vertical and seasonal distributions during 1982. J Phycol 21: 171-175

Carpenter EJ (1983) Physiology and ecology of marine planktonic *Oscillatoria* (*Trichodesmium*). Mar Biol Lett 4: 69-85

Carpenter EJ and Romans K (1991) Major role of the cyanobacterium *Trichodesmium* in nutrient cycling in the north Atlantic Ocean. Science 254: 1356-1358

Carr NG and Mann NH (1994) The oceanic cyanobacterial picoplankton. In: Bryant DA (ed) The Molecular Biology of Cyanobacteria, pp 27-48. Kluwer Academic Publications, Dordrecht

Chen F and Suttle CA (1996) Evolutionary relationships among large double-stranded DNA viruses that infect microalgae and other organisms as inferred from DNA polymerase genes. Virology 219: 170-178

Chen F, Suttle CA and Short SM (1996) Genetic diversity in marine algal virus communities as revealed by sequence analysis of DNA polymerase genes. Appl Environ Microbiol 62: 2869-2874

Cottrell MT and Suttle CA (1995) Dynamics of a lytic virus infecting the photosynthetic marine picoflagellate, *Micromonas pusilla*. Limnol Oceanogr 40: 730-739

Coulombe AM and Robinson GGC (1981) Collapsing *Aphanizomenon flos-aquae* blooms: Possible contributions of photo-oxidation, O_2 toxicity, and cyanophages. Can J Bot 59: 1277-1284

Cowie DB and Prager L (1969) Cyanophyta and their viruses. Carnegie Institution of Washington Yearbook, 1968. Carnegie Institution of Washington, Washington, DC, p 391-397

Cowlishaw J and Mrsa, M. (1975) Co-evolution of a virus-alga system. Appl Microbiol 29: 234-239

Cseke CS and Farkas GL (1979) Effect of light on the attachment of cyanophage AS-1 to *Anacystis nidulans*. J Bacteriol 137: 667-669

Daft MJ, Begg J and Stewart WDP (1970) A virus of blue-green algae from freshwater habitats in Scotland. New Phytol 69: 1029-1038

Desjardins PR and Olson GB (1983) Viral control of nuisance cyanobacteria (blue-green algae). II. Cyanophage strains, stability studies of phages and hosts, and effects of environmental factors on phage-host interactions. California Water Resources Center, Contribution No. 185, University of California, Davis

Drouet F (1963) Ecophenes of *Schizothrix calcicola* (Oscillatoriaceae). Proc Acad Nat Sci Phil 115: 261-281

Dulbecco R (1949) Reactivation of ultra-violet-inactivated bacteriophage by visible light. Nature 163: 949-950

Dulbecco R (1950) Experiments on photoreactivation of bacteriophages inactivated with ultraviolet radiation. J Bacteriol 59: 329-347

Edelman M, Swinton D, Schiff JA, Epstein HT and Zeldin B (1967) Deoxyribonucleic acid of the blue-green algae (Cyanophyta). Bacteriol Rev 31: 315-331

Fallon RD and Brock TD (1979) Lytic organisms and photooxidative effects: Influence on blue-green algae (cyanobacteria) in Lake Mendota, Wisconsin. Appl Environ Microbiol 38: 499-505

Fox JA, Booth SJ and Martin EL (1976) Cyanophage SM-2: A new blue-green algal virus. Virology 73: 557-560

Franche C (1987) Isolation and characterization of a temperate cyanophage for a tropical Anabaena strain. Arch Microbiol 148: 172-177

Friedberg EC, Walker GC and Siede W. (1995) DNA repair and mutagenesis. Washington, DC: ASM Press.

Garza DR (1996) In situ production and decay of marine viruses. MA Thesis Marine Science, University of Texas at Austin

Garza DR and Suttle CA (in review) The effect of cyanophages on the mortality of *Synechococcus* spp. and seasonal changes in the resistance of natural viral communities to UV radiation.

Ginzberg D, Padan E and Shilo M (1968) Effect of cyanophage infection on CO_2 photoassimilation in *Plectonema boryanum*. J Virol 2: 695-701

Ginzberg D, Padan E and Shilo M (1976) Metabolic aspects of LPP cyanophage replication in the cyanobacterium *Plectonema boryanum*. Biochim Biophys Acta 423: 440-449

Goldstein DA, Bendet IJ, Lauffer MA and Smith KM (1967) Some biological and physicochemical properties of blue-green algal virus LPP-1. Virology 32: 601-613

González JM and Suttle CA (1993) Grazing by marine nanoflagellates on viruses and virus-sized particles: Ingestion and digestion. Mar Ecol Progr Ser 94: 1-10

Gromov BV (1983) Cyanophages. Ann Microbiol 134: 43-59

Heldal M and Bratbak G (1991) Production and decay of viruses in aquatic environments. Mar Ecol Progr Ser 72: 205-212

Hennes KP, Chan AM and Suttle CA (1995) Fluorescently labeled virus probes show that natural virus populations can control the structure of marine microbial communities. Appl Environ Microbiol 61:3623-3627

Horne AJ and Goldman CR (1994) Limnology. McGraw-Hill, New York

Hu N-T, Thiel T, Gidding TH and Wolk CP (1981) New *Anabaena* and *Nostoc* cyanophages from sewage settling ponds. Virology 114: 236-246

Hwang-Lee L, Cohn G, Cosowsky L, McGowan RE and Blamire J (1985) DNA metabolism during infection of *Anacystis nidulans* by cyanophage AS-1. VII. UV-induced alterations of the AS-1/*A. nidulans* lytic cycle. Microbios 43: 277-295

Ito J and Braithwaite DK (1991) Compilation and alignment of DNA polymerase sequences. Nucleic Acids Res 19: 4045-4057

Jiang SC and Paul JH (1996) Occurrence of lysogenic bacteria in marine microbial communities as determined by prophage induction. Mar Ecol Progr Ser 142: 27-48

Johnson DW and Borovsky D (1987) Changes in sensitivity to cyanophage infection in axenic LPP cyanobacteria. Microbios Lett. 35: 105-112

Johnson DW and Potts M (1985) Host range of LPP cyanophages. Int J Syst Bacteriol 35: 76-78

Joint IR and Pomroy AJ (1983) Production of picoplankton and small nanoplankton in the Celtic sea. Mar Biol 77: 19-27

Kadyrova GKh, Muradov MM, Khalmuradov AG, Koltukova NV and Mendzhul MI (1995) Dynamics of the amino acid pool in the cyanobacterium *Anabaena variabilis* in the early period of cyanophage A-1 development. Mikrobiologiia 64: 469-472

Kashyap AK and Singh S (1989) Changes in photoelectron transport activity in cyanophage N-1-infected cells of *Nostoc muscorum*. Curr Microbiol 18: 151-155

Khudyakov IY (1977) Characteristics of a new cyanophage S-2L lysing the unicellular cyanobacterium belonging to the *Synechococcus* genus. Mikrobiologiya 46: 547-553

Khudyakov IY and Gromov BV (1973) The temperate cyanophage A-4 (L) of the blue-green alga *Anabaena variabilis*. Mikrobiologiia : 904-907

Khudyakov IY and Wolk CP (1996) Evidence that the han A gene coding for hu protein is essential for heterocyst differentiation in, and cyanophage A-4(L) sensitivity of, *Anabaena* sp. Strain PCC 7120. J Bacteriol 178: 3572-3577

Killmann H, Videnov G, Jung G, Schwarz H and Braun V (1995) Identification of receptor binding sites by competitive peptide mapping: Phages T1, T5, F80 and Colicin M bind to the gating loop of FhuA. J Bacteriol 177: 694-698

Kim M and Choi YK (1994) A new *Synechococcus* cyanophage from a reservoir in Korea. Virology 204:338-342

Kozyakov SYa (1977) Cyanophages of the series A(L) specific for blue-green alga *Anabaena variabilis*. In: Gromov BV (ed) Experimental Algology, pp 151-171. Leningrad State University, Leningrad

Kozyakov SYa, Gromov BV and Khudyakov IYa (1972) A-1(L) cyanophage of the blue-green alga *Anabaena variabilis*. Mikrobiologiya 41: 555-559

Leach JE, Lee KW, Benson RL and Martin EL (1980) Ultrastructure of the infection cycle of cyanophage SM-2 in *Synechococcus elongatus* (Cyanophyceae). J Phycol: 307-310

Lenski RE (1988) Dynamics of interactions between bacteria and virulent bacteriophage. Adv Microb Ecol 10: 1-39

Lenski RE and Levin BR (1985) Constraints on the coevolution of bacteria and virulent phage: a model, some experiments and predictions for natural communities. Am Nat 125: 585-602

Levin BR, Stewart FM and Chao L (1977) Resource-limited growth, competition, and predation: a model and experimental studies with bacteria and bacteriophage. Am Nat 111: 3-24

Levine E and Thiel T (1987) UV-inducible DNA repair in the cyanobacteria *Anabaena* spp. J Bacteriol 169: 3988-3993

Lewis MR, Warnock RE and Platt T (1986) Photosynthetic response of marine picoplankton at low photon flux. In: Platt T and Li WKW (eds) Photosynthetic Picoplankton, Can Bull Fish Aquat Sci 214: 235-250

Li WKW, Dickie PM, Irwin BD and Wood AM (1992) Biomass of bacteria, cyanobacteria, prochlorophytes and photosynthetic eukaryotes in the Sargasso Sea. Deep Sea Res 39: 501-519

Li WKW, Subba Rao DV and Harrison WG (1983) Autotrophic picoplankton in the tropical ocean. Science 219: 292-295

Luftig R and Haselkorn R (1967) Morphology of a virus of blue-green algae virus and properties of its deoxyribonucleic acid. J Virol 1: 344-361

Luftig R and Haselkorn R (1968a) Studies on the structure of blue-green algal virus LPP-1. Virology 34: 644-674

Luftig R and Haselkorn R (1968b) Comparison of blue-green algae virus LPP-1 and morphologically related viruses G111 and Coliphage T7. Virology 34: 675-678

Lycke E, Magnusson S and Lund E (1965) Studies on the nature of the virus inactivating capacity of sea water. Archiv Gesamte Viruschung 17: 409-413

MacKenzie JJ and Haselkorn R (1972a) Physical properties of blue-green algal virus SM-1 and its DNA. Virology 49: 497-504

MacKenzie JJ and Haselkorn R (1972b) An electron microscope study of infection by the blue-green algal virus SM-1. Virology 49: 505-516

MacKenzie JJ and Haselkorn R (1972c) Photosynthesis and the development of blue-green algal virus SM-1. Virology 49: 517-521

Martin EL and Benson R (1988) Phages of cyanobacteria. In: Calendar R (ed) The Bacteriophages, Vol 2, pp 607-645. Plenum Publishing Corp.

Matossian AM and Garabedian GA (1967) Virucidal action of sea water. Am J Epidemiol 85: 1-8

Mendzhul MI, Bobrovnik SA and Lysenko TG (1974) Study of cyanophage LPP-1 adsorption onto cells of cyanophyceae (*Plectonema boryanum*). Vop Virus 1: 31-36

Mendzhul MI, Lysenko TG, Bobrovnik SA and Spivak MYa (1973) Detection of A-1 virus of blue-green alga *Anabaena variabilis* in the Kremenchug artificial reservoir. Mikrobiol Zh 35: 747-751

Miller RV and Kokjohn TA (1990) General microbiology of *recA*: environmental and evolutionary significance. Annu Rev Microbiol 44:365-94

Mitchell R (1971) Destruction of bacteria and viruses in seawater. J San Eng Div, Proc Am Soc Civil Eng 97: 425-432

Mitchell R and Jannasch HW (1969) Processes controlling virus inactivation in seawater. Environ Sci Technol 3: 941-943

Moebus K (1992) Laboratory investigations on the survival of marine bacteriophages in raw and treated seawater. Helgol Meeresunters 46: 251-273

Moisa I, Sotropa E and Velehorschi V (1981) Investigations on the presence of cyanophages in fresh and sea waters of Romania. Rev Roum Med - Virol 32: 127-132

Monegue RL and Phlips EJ (1991) The effect of cyanophages on the growth and survival of *Lyngbya wollei*, *Anabaena flos-aquae* and *Anabaena circinalis*. J Aquat Plant Manage 29: 88-93

Muradov MM, Cherkasova GV, Akhmedova DU, Kamilova FD, Mukhamedov RS, Abdukarimov AA and Khalmuradov AG (1990) Comparative study of NP-IT cyanophages, which lysogenize nitrogen-fixing bacteria of the genera *Nostoc* and *Plectonema*. Microbiology 59: 558-563 (Translation of Mikrobiologiya 59: 819-826

Murphy FA, Fauquet CM, Bishop DHL, Ghabrial SA, Jarvis AW, Martelli GP, Mayo, MA and Summers MD (1995) The Classification and Nomenclature of Viruses, Sixth Report of the International Committee on Taxonomy of Viruses, Archiv Virol (Supplement 10) Springer, Vienna

Murray AG and Jackson GA (1992) Viral dynamics: A model of the effects size, shape, motion and abundance of single-celled planktonic organisms and other particles. Mar Ecol Progr Ser 89: 103-116

Noble RT and Fuhrman JA (1997) Virus decay and its causes in coastal waters. Appl Environ Microbiol 63: 77-83

Ohki K and Fujita Y (1996) Occurrence of a temperate cyanophage lysogenizing the marine cyanophyte *Phormidium persicinum*. J Phycol 32: 365-370

Olson RJ, Chisholm SW, Zettler ER, Altabet MA and Dusenberry JA (1990) Spatial and temporal distributions of prochlorophyte picoplankton in the North Atlantic Ocean. Deep Sea Res 37: 1033-1051

Padan E, Ginzburg D and Shilo M (1970) The reproductive cycle of cyanophage LPP1-G in *Plectonema boryanum* and its dependence on photosynthetic and respiratory systems. Virology 40: 514-521

Padan E, Rimon A, Ginzberg D and Shilo M (1971) A thermosensitive cyanophage (LPP1-G) attacking the blue-green alga *Plectonema boryanum*. Virology 45: 773-776

Padan E and Shilo M (1969) Distribution of cyanophages in natural habitats. Verh Int Verein Limnol 17: 747-751

Padan E and Shilo M (1973) Cyanophages-viruses attacking blue-green algae. Bacteriol Rev 37: 343-370

Padan E, Shilo M and Oppenheim AB (1972) Lysogeny of the blue-green alga *Plectonema boryanum* by LPP2-SPI cyanophage. Virology 47: 525-526

Phlips EJ, Monegue RL and Aldridge FJ (1990) Cyanophages which impact bloom-forming cyanobacteria. J Aquat Plant Manage 28: 92-97

Proctor LM and Fuhrman JA (1990) Viral mortality of marine bacteria and cyanobacteria. Nature 343: 60-62

Proctor LM, Fuhrman JA and Ledbetter, MC (1988) Marine bacteriophages and bacterial mortality. Eos 69: 1111-1112

Proctor LM, Okubo A and Fuhrman JA (1993) Calibrating estimates of phage-induced mortality in marine bacteria:

Ultrastructural studies of marine bacteriophage development from one-step growth experiments. Microb Ecol 25:161-182

Rimon A and Oppenheim AB (1975) Heat induction of the blue-green alga *Plectonema boryanum* lysogenic for the cyanophage SPIcts1. Virology 64: 454-463

Rippka R and Herdman M (1992) Pasteur Culture Collection of Cyanobacteria, Catalogue and Taxonomic Handbook, Volume I: Catalogue of Strains, Institut Pasteur, Paris

Rodda KM (1996) Temporal and Spatial Dynamics of *Synechococcus* spp. and *Micromonas pusilla* host-viral systems. MA Thesis Marine Science, University of Texas at Austin

Rodda KM, Suttle CA, Clark LL and Ingall E (in review) Infectious viruses persist in marine sediments for at least 100 years

Safferman RS (1968) Virus diseases in blue-green algae. In: Jackson DF (ed) Algae, Man and the Environment, pp 429-439. Syracuse University Press, New York

Safferman RS (1973) Phycoviruses. In: Carr NG, Whitton BA (eds) The Biology of Blue-Green Algae, pp 214-237. Blackwell Scientific Publications, Oxford

Safferman RS, Cannon RE, Desjardins PR, Gromov BV, Haselkorn R, Sherman LA and Shilo M (1983) Classification and nomenclature of viruses of cyanobacteria. Intervirology 19: 61-66

Safferman RS, Diener TO, Desjardins PR and Morris ME (1972) Isolation and characterization of AS-1, a phycovirus infecting the blue-green algae, *Anacystis nidulans* and *Synechococcus cedrorum*. Virology 47: 105-113

Safferman RS and Morris ME (1963) Algal virus: isolation. Science 140: 679-680.

Safferman RS and Morris ME (1964a) Control of algae with viruses. J Am Water Works Assoc 56: 1217-1224

Safferman RS and Morris ME (1964b) Growth characteristics of the blue-green algal virus LPP-1. J Bacteriol 88: 771-775

Safferman RS and Morris ME (1967) Observations on the occurrence, distribution and seasonal incidence of blue-green algal viruses. Appl Microbiol 15: 1219-1222

Safferman RS, Morris ME, Sherman LA and Haselkorn R (1969a) Serological and electron microscopic characterization of a new group of blue-green algae viruses (LPP-2). Virology 39: 775-781

Safferman RS, Schneider IR, Steere RL, Morris ME and Diener TO (1969b) Phycovirus SM-1: A virus infecting unicellular blue-green algae. Virology 37: 386-395

Samimi B and Drews G (1978) Adsorption of cyanophage AS-1 to unicellular cyanobacteria and isolation of receptor material from *Anacystis nidulans*. J Virol 25: 164-174

Sarma TA and Kaur B (1993) Spontaneous and induced host range mutants of cyanophage N-1. Arch. Virol. 130: 195-200

Sarma TA and Singh R (1995) Characterization of TS-mutants of cyanophage N-1 by their inactivation by physical and chemical agents. Acta Virol 39: 65-68

Schopf JW and Packer BM (1987) Early Archaen (3.3 billion to 3.5 billion years old) microfossils from Warrawoona group, Australia. Science 137: 70-73

Schneider IR, Diener TO and Safferman RS (1964) Blue-green algal virus LPP-1: purification and partial characterization. Science 144: 1127-1130

Schneider GJ and Haselkorn R (1988) Characterization of two early promoters of cyanophage N-1. Virology 167: 150-155

Schwartz M (1976) The adsorption of coliphage Lambda to its host: effect of variation in the surface density of receptor and in phage-receptor affinity. J Molec Biol 103: 521-536

Sherman LA (1976) Infection of *Synechococcus cedrorum* by the cyanophage AS-1M. III. Cellular metabolism and phage development. Virology 71: 199-206

Sherman LA and Brown RM Jr (1978) Cyanophages and viruses of eukaryotic algae. In: Fraenkel-Conrat H, and Wagner RR (eds) Comprehensive Virology 12, pp 145-234. Plenum Press, New York

Sherman LA and Connelly M (1976) Isolation and characterization of a cyanophage infecting the unicellular blue-green algae *A. nidulans* and *S. cedrorum*. Virology 72: 540-554

Sherman LA, Connelly M and Sherman DM (1976) Infection of *Synechococcus cedrorum* by the cyanophage AS-1M. I. Ultrastructure of infection and phage assembly. Virology 71: 1-16

Sherman LA and Haselkorn R (1970) LPP-1 infection of the blue-green alga *Plectonema boryanum*. J Virol 6: 820-846

Sherman LA and Haselkorn R (1971) Growth of the blue-green algae virus LPP-1 under conditions which impair photosynthesis. Virology 45: 739-746

Shilo M (1971) Biological agents which cause lysis of blue-green algae. Mitt Internat Verein Limnol 19: 206-213

Short SM and Suttle CA (1999) Use of the polymerase chain reaction and denaturing gel electrophoresis to study diversity in natural virus communities. Hydrobiologia 40: 19-32

Shuval HI, Asce M, Thompson A, Fattal B, Cymbalista S and Wiener Y (1971) Natural virus inactivation processes in seawater. J San Eng Div Proc Am Soc Civil Eng 97: 587-600

Singh PK (1967) Occurrence and distribution of cyanophages in ponds, sewage and rice fields. Arch Mikrobiol 89: 169-172

Singh P (1975) Photoreactivation of UV-irradiated blue-green algae and algal virus LPP-1. Arch Mikrobiol 103: 297-302

Singh RN and Singh PK (1967) Isolation of cyanophages from India. Nature 216: 1020-1021

Singh RN and Singh PK (1972) Transduction and lysogeny in blue-green algae. In: Desikachary TV (ed) Taxonomy and Biology of Blue-Green Algae, pp 258-262. University of Madras Press, Madras, India

Singh S, Bhatnagar A and Kashyap AK (1994) Energetics of cyanophage N-1 multiplication in the diazotrophic cyanobacterium *Nostoc muscorum*. Microbios 78: 259-265

Smith KM, Brown RM Jr and Walne PL (1966) Electron microscopy of the infection process of the blue-green algal virus. Virology 30: 182-192

Snedden JW (1985) Origin and Sedimentary Characteristics of Discrete Sand Beds in Modern Sediments of the Central Texas Continental Shelf, PhD Thesis, Louisiana State University

Sode K, Oonari R and Oozeki M (1997) Induction of a temperature marine cyanophage by heavy metal. J Mar Biotechnol

Sode K, Oozeki M, Asakawa K, Burgess JG and Matsunaga T (1994) Isolation of a marine cyanophage infecting the marine unicellular cyanobacterium, *Synechococcus* sp. NKBG 042902. J Mar Biotechnol 1: 189-192

Stam WT (1980) Relationships between a number of filamentous blue-green algal strains (Cyanophyceae) revealed by DNA-DNA hybridization. Arch Hydrobiol Suppl 56, Algological Studies 25: 351-374

Stewart WDP and Daft MJ (1977) Microbial pathogens of cyanophycean blooms. In: Droop MR, Jannasch HW (eds) Adv Aquat Microbiol, pp 177-218. Academic Press, New York

Stockner JG and Shortreed KRS (1991) Autotrophic picoplankton: community composition, abundance and distribution across a gradient of oligotrophic British Columbia and Yukon Territory lakes. Int Rev ges Hydrobiol 76: 581-601

Suttle CA (1992) Inhibition of photosynthesis in phytoplankton by the submicron size fraction concentrated from seawater. Mar Ecol Progr Ser 87:105-112

Suttle CA (1994) The significance of viruses to mortality in aquatic microbial communities. Microb Ecol 28: 237-243

Suttle CA and Chan AM (1993) Marine cyanophages infecting oceanic and coastal strains of *Synechococcus*: Abundance, morphology, cross-infectivity and growth characteristics. Mar Ecol Progr Ser 92: 99-109

Suttle CA and Chan AM (1994) Dynamics and distribution of cyanophages and their effect on marine *Synechococcus* spp. Appl Environ Microbiol 60: 3167-3174

Suttle CA, Chan AM, Chen F and Garza DR (1993) Cyanophages and sunlight: A paradox. In: Guerrero R and Pedros-Alio C (eds) Trends in Microbial Ecology, pp 303-307. Span Soc Microbiol, Barcelona

Suttle CA, Chan AM and Cottrell MT (1990) Infection of phytoplankton by viruses and reduction of primary productivity. Nature 347: 467-469

Suttle CA, Chan AM, Rodda KM, Short SM, Weinbauer MG, Garza DR and Wilhelm SW (1996) The effect of cyanophages on *Synechococcus* spp. during a bloom in the western Gulf of Mexico. Eos 76 (suppl): OS207-OS208

Suttle CA and Chen F (1992) Mechanisms and rates of decay of marine viruses in seawater. Appl Environ Microbiol 58: 3721-3729

Suttle CA, Chen F and Chan AM (1992) Marine viruses: decay rates, diversity and ecological implications. In: Nash CC (ed) International Marine Biotechnology Conference "IMBC-91", pp 153-163. W. Brown, Dubuque

Teklemariam TA, Demeter S, De·k Z, Sur·nyi G and BorbÈly G (1990) AS-1 cyanophage infection inhibits the photosynthetic electron flow of photosystem II in *Synechococcus* sp. PCC 6301, a cyanobacterium. FEBS Lett 270: 211-215

Turner S (1997) Molecular systematics of oxygenic photosynthetic bacteria. Plant Systematics and Evolution S11: 13-52

Waterbury JB and Rippka R (1989) Subsection I. Order Chroococcales Wettstein 1924, emend. Rippka et al., 1979. In: Staley JT, Bryant MP, Pfennig N, Holt JB (eds) Bergey's Manual of Systematic Bacteriology, pp 1728-1746. Williams and Wilkins, Baltimore

Waterbury JB and Valois FW (1993) Resistance to co-occurring phages enables marine *Synechococcus* communities to coexist with cyanophages abundant in seawater. Appl Environ Microbiol 59: 3393-3399

Waterbury JB, Watson SW, Valois FW and Franks DG (1986) Biological and ecological characterization of the marine unicellular cyanobacterium *Synechococcus*. In: Platt T and Li WKW (eds) Photosynthetic Picoplankton, Can Bull Fish Aquat Sci 214: 71-120

Weinbauer MG and Suttle CA (1996) Potential significance of lysogeny to bacteriophage production and bacterial mortality in coastal waters of the Gulf of Mexico. Appl Environ Microbiol 62: 4374-4380

Weinbauer MG, Wilhelm SW, Suttle CA and Garza DR (1997) Photoreactivation compensates for UV damage and restores infectivity to natural marine viral communities. Appl Environ Microbiol 63: 2200-2205

Wiggins BA and Alexander M (1985) Minimum bacterial density for bacteriophage replication: implications for significance of bacteriophages in natural ecosystems. Appl Environ Microbiol 49: 19-23

Wilhelm SW, Weinbauer MG, Suttle CA and Jeffrey WH (1998a) The role of sunlight in the removal and repair of viruses in the sea. Limnol Oceanogr 43: 586-592

Wilhelm SW, Weinbauer MG, Suttle CA, Pledger RJ and Mitchell DL (1998b) Measurements of DNA damage amd photoreactivation imply that most viruses in marine surface waters are infective. Aquat Microb Ecol 14: 215-222

Wilmotte A (1994) Molecular evolution and taxonomy of the cyanobacteria. In: Bryant DA (ed) The Molecular Biology of Cyanobacteria, pp 1-25. Kluwer Academic Publishers, Amsterdam

Wilmotte AMR and Stam WT (1984) Genetic relationship among cyanobacterial strains originally designated as *Anacystis nidulans* and some other *Synechococcus* strains. J Gen Microbiol 130: 2737-2740

Wilson WH, Carr NG and Mann NH (1996) The effect of phosphate status on the kinetics of cyanophage infection in the oceanic cyanobacterium *Synechococcus* sp.WH7803. J Phycol 32: 506

Wilson WH, Joint IR, Carr NG and Mann NH (1993) Isolation and molecular characterization of five marine cyanophages propagated on *Synechococcus* sp. strain WH7803. Appl Environ Microbiol 59: 3736-3743

Woese CR (1987) Bacterial evolution. Microbiol Rev 51: 221-271

Wommack KE, Hill RT, Muller TA and Colwell RR (1996) Effects of sunlight on bacteriophage viability and structure. Appl Environ Microbiol 62: 1336-1341

Wu JH, Lewin RA and Werbin H (1967) Photoreactivation of UV-irradiated blue-green alga virus LPP-1. Virology 31: 657-664

Wu JH, Shugarman and PM (1967) Effect of virus infection rate on photosynthesis and respiration of a blue-green alga, *Plectonema boryanum*. Virology 32: 166-167

Chapter 21

Cyanobacterial Responses to UV-Radiation

Richard W. Castenholz[1] and Ferran Garcia-Pichel[2]
[1]*Dept. of Biology, University of Oregon, Eugene, Oregon 97403, USA.* [2]*Department of Microbiology, Arizona State University, Tempe, Arizona 85287-2701, USA*

Summary

The influence of ultraviolet radiation (UVR) on microbial populations is the subject of intense investigation. One reason is the awareness of how decreasing regional ozone levels in the stratosphere result in an increase of the UVB flux that reaches the Earth's surface, and the fact that microbial populations and species may show a more immediate and greater measurable sensitivity to small increases in UVR than the larger macrophytes and metazoa. Some cyanobacteria, representing probably the oldest oxygenic inhabitants of the planet, evolved various methods (or complex strategies) for coping with present levels of UVR, and possibly with the higher levels of UVR which occurred in the early Precambrian and which may occur in the future. Cyanobacteria

B. A. Whitton and M. Potts (eds), The Ecology of Cyanobacteria, 591-611.
© 2000 *Kluwer Academic Publishers. Printed in the Netherlands.*

invaded a large number of extreme environments (or remain as relicts of similar Precambrian habitats), including shallow waters and exposed terrestrial surfaces. As a consequence they must often cope with high solar irradiance in which UVR can be the most inhibitory region of the spectrum. Cyanobacteria evolved different strategies to counter the effects of UVR. The best understood of these include the development of sunscreen pigments that envelope the cell and function even when cells are at rest, the synthesis of compounds such as mycosporine-like amino acids (the true value of which is still not certain), development of efficient systems for repair of damaged DNA and for replacement of UVR-damaged compounds, and implementation of directed gliding motility for escaping the diurnally high intensities of solar irradiance when soft microbial mats or sediments are present. This review considers the various specific effects of UVR on cyanobacteria and their compensating responses, with particular emphasis on the reactions which occur, or are likely to occur, in natural habitats today.

I. Introduction

A. The Detrimental Effects of UVR on Microorganisms

Because of its potential to directly affect molecules of central biological importance, ultraviolet radiation (UVR) is considered detrimental to all life forms. But because UVR is part of the total solar spectrum at the Earth's surface avoidance is difficult. Although they are only a small proportion of the total energy or photon flux, the wavelengths which are termed UVR (UVA: 320-400 nm; UVB: 280-320 nm; UVC: 190-280 nm) become progressively more detrimental as wavelength decreases, and exponential increases in potential damage to cells would be expected. Virtually all UVC and most of the UVB do not currently reach the Earth's surface, although these bands presumably penetrated the atmosphere in significant amounts (virtually unattenuated) during the Archean and early Proterozoic Eons, before ozone accumulated in the stratosphere (Kasting et al., 1992). Indeed it is possible that UVR played a significant role during the evolution of cyanobacteria. A plausible scenario for this role was examined recently (Garcia-Pichel, 1998). Under very clear skies at temperate to equatorial latitudes, total UVB today may be as high as 7-8 W m^{-2} and UVA as high as 45-50 W m^{-2}, as compared to values up to ~1100 W m^{-2} for visible wavelengths (~400 - 700 nm). Except for a small portion of near UVA, UVR is not photosynthetically active. Therefore, UVR is considered to be something to tolerate, counteract, or avoid. However, both UVA and violet/blue radiation are positively effective in pigment synthesis and in photoreactivation of UVB-induced dimerization between adjacent pyrimidine bases of DNA. Other possible beneficial effects may be the use of UVA and/or UVB as a cue in photo-motility responses, such as photophobic reactions and phototaxis (see later). Specific UVR "photoreceptors" are unknown.

The negative effects of UVR exposure are many and diverse. At the phenomenon level, photosynthesis, growth, motility, photomovements, cell differentiation and other processes are affected (Table 1; Braune and Dohler, 1996; Young et al., 1993; Holm-Hansen et al., 1993; Vincent and Roy, 1993; Häder et al., 1986; Jagger, 1985). UVA (320-400 nm) is primarily associated with production of reactive oxygen species such as $^1O_2$. The latter causes lipid peroxidation, chlorophyll photobleaching, phycobiliprotein degradation (Loa and Glazer, 1996; Wingard et al., 1997; Jeffrey and Mitchell, 1997) and inhibition of growth, but it also directly affects the D1/D2 protein matrix of PS II (Renger et al., 1989). UVB (280-320 nm) results in similar damage and inhibits various activities (e.g. RUBISCO, ATP-synthase, synthesis of chlorophyll *a*, energy transfer from phycobilisome to chlorophyll, nitrogen fixation, etc.), but it also directly targets DNA and results in dimeric photoproducts between adjacent pyrimidimes and other mutagenic damage, some of which can be corrected by photoreactivation or excision repair (Holm-Hansen et al., 1993; Vincent and Roy, 1993; Mitchell and Karentz, 1993).

Abbreviations: UVR = ultraviolet radiation; UVB = ultraviolet B; UVA = ultraviolet A; MAA = mycosporine-like amino acid; PS II = photosystem II; Gyr = 10^9 years.

Table 1. Typical adverse effects of UVR on metabolic processes in cyanobacteria. Abbreviations: FR, fraction of control metabolic activity remaining after given treatment; PSII, photosystem II; RUBISCO, ribulose-bis-phosphate carboxylase/oxygenase; 1ary ET, primary electron transport; Chl a, chlorophyll a; PBP, phycobiliproteins; PBS, phycobilisome; LPP, PBS linker-polypeptide, PC: phycocyanin; GAAS, glutamate synthase; PPR, photophobic response. *A., Anacystis (=Synechococcus); Anab, Anabaena; O., Oscillatoria ; N., Nostoc.*

Process	Possible target[a]	Wavelength	Dose rate / Time (FR) [b] Wm^{-2}/ min (%)	Organism	Reference
Photosynthesis					
Carbon fixation	PSII/RUBISCO	UVB	2.1 / 360 (60) [c]	*Anab. flos-aquae*	Döhler et al. 1986
			0.2/constant (60)	*O. amphigranulata*	Wingard et al. 1997
Oxygen evolution	1ary E T	UVA[d]	8/30	*A. nidulans* ACIAMT M6	Hirosawa and Miyachi 1983
			5/45 (60)	*A. nidulans*	Kulandaivelu et al. 1989
	PSII	UVB	5/18 (60)	*A. nidulans*	Kulandaivelu et al. 1989
	PSII	UVC	5/9 (60)	*A. nidulans*	Kulandaivelu et al. 1989
PBS excitation transfer	LPP	UVB/A	30/180	*Anab.* sp. and *N. carmium*	Sinha et al. 1995
Tetrapyrrol synthesis	ALS	UVA	8/constant	*A. nidulans* ACIAMT M6	Hirosawa and Miyachi, 1983
Chromatic adaptation	?	UVA	2/10	*Fremyella diplosiphon* M100	Ohki et al. 1982
Pigment loss	PC,	UVB	4.1 /600 (35)	*S. leopoliensis* SAG 1401-1	Döhler et al. 1986
	Chl a	UVB	4.1 /600 (80)	*S. leopoliensis* SAG 1401-1	Döhler et al. 1986
	all PBP	UVB	0.2/ 4 days (60)	*O. amphigranulata*	Wingard et al. 1997
	Chl a	UVB	0.2/ 4 days (65)	*O. amphigranulata*	Wingard et al. 1997
Growth					
Delay	multiplc	UVA	8/constant	*A. nidulans* ACIAMT	Hirosawa and Miyachi, 1983
			9/constant	*Oscillatoria* BG091600	Wachi et al 1995
			4/constant	*Chlorogloeopsis* O-89-Cgs.(1)	Garcia-Pichel et al. 1992
Rate repression	multiple	UVA	1/constant (0)	*Synechococcus* NKBG042902	Wachi et al. 1995
			12/ constant (0)	*Oscillatoria* BG091600	Wachi et al. 1995
			5/constant (80)	*Phormidium murrayi*	Vincent and Quesada, 1994
	multiple	UVB	4.1/ 600[c]	*Synechococcus leopoliensis*	Döhler et al.1986
			4.1/ 600[c]	*Anabaena cylindrica*	Döhler et al. 1986
			1/ constant (40)	*Nostoc commune* DRH1	Ehling-Schulz et al. 1997
			5/constant (38)	*Phormidium murrayi*	Vincent and Quesada, 1994
			4.3/ constant (25)	*Gloeocapsa* C-90-Cal-G(2)	Garcia-Pichel et al 1993
	multiple	UVC	5/constant (7)	*Phormidium murrayi*	Vincent and Quesada, 1994

[over

Table 1 continued

Process	Possible target[a]	Wavelength	Dose rate / Time (FR) [b] Wm^{-2}/ min (%)	Organism	Reference
Survival	multiple	UVB	5.5/ 60-100 (33)	***N. spongiaeforme***	Tyagi et al. 1992
			5/ 75 (33)	*Anab.* sp. and *N. carmium*	Sinha et al. 1995
			5/ 105 (33)	*N. commune* and *Scytonema* sp.	Sinha et al. 1995
		UVC	2.2/ 0.75 (33)	*Aphanothece halophytica*	Yopp et al. 1979
			17.5/ 6.6 (33)	*Agmenellum quadruplicatum*	Van Baalen, 1969
Nitrogen metabolism					
Nitrogen fixation	nitrogenase ?	UVB	2.1/180 (60)	***Anabaena flos-aquae***	Newton et al. 1979
			5.5/ 5 (60)	*Anabaena flos-aquae*	Newton et al. 1979
			5.5/ 15 (60)	*N. spongiaeforme*	Tyagi et al. 1992
			5/30 (0)	*N. calcicola*	Kumar et al. 1996
nitrate uptake	membrane proteins?	UVB	4.1/600[c] (75)	*Anab. cylindrica*	Döhler et al. 1986
ammonium uptake	membrane proteins?	UVB	0.95/30	*Anab. variabilis* [e]	Braune and Döhler, 1996
amino acid metabolism	GAAS	UVB		*Anab. cylindrica*	Döhler et al 1986
Motility and Taxes					
step-down PPR	?	UVB	0.2 120 (0)	*Phormidium uncinatum*	Häder, 1984
cell motility	?	UVB	4/290 (0)	*Phormidium uncinatum*	Häder et al, 1986
Cell differentiation					
heterocyst formation		UVB	37 KJm^{-2} (0) [f]	***Anabaena aequalis***	Blakefield and Harris, 1994
akinete induction		UVB	37 KJm^{-2} (0) [f]	*Anabaena aequalis*	Blakefield and Harris, 1994

a) The assignment of a possible target is in most cases ours.
b) When possible, the irradiation conditions have been recalculated for consistance in units. If exposure time exceeded 2 days, it has been reported as "constant".
c) daily exposures of 300 min, twice
d) short-term effect, acclimated cells show no or much dampened depressions.
e) germinating akinetes were irradiated. No effect was seen if irradiated just before germination.
f) Only absolute dose received was given, excission repair had been artificially inhibited with caffein.

B. Exposure of Cyanobacteria to UVR in Natural Habitats

1. Physical Exposure: Incident Irradiance and Optical Modifications in the Environment

The almost exclusive photoautotrophic mode of growth of cyanobacteria demands that they thrive in habitats exposed directly, or indirectly, to solar radiation. More often than not, this requirement results in their exposure not only to the visible part of the sun´s spectrum but also to its shorter wavelengths.

Some geographical and environmental factors that affect overall solar insolation will in the same manner affect the fluxes of UVR incident at ground level. Latitude determines a large portion of both the maximal instantaneous and yearly average incident UV irradiance, as well as the amplitude of its seasonal variation. Large seasonal variation and low maximal values are associated with higher latitudes, whereas strong, weakly seasonal irradiances correspond to lower latitudes. High altitude, by decreasing the path length for atmospheric attenuation, also results in stronger incident fluxes. The altitude gradients for UVR are much steeper than those for visible radiation, and thus the ratio of UVR to visible increases with altitude. Some other meteorological factors may modify specifically the incident UV irradiance, such as cloud cover (Gautier et al., 1994), pollution (Liu et al., 1991) or the optical thickness of the stratosphere (Roy et al., 1994). Although direct measurements of UVR are preferred, excellent tools are available in the form of graphs and tables (Robinson, 1966) or computer programs (Musil and Bhagwandin, 1992; Bjorn and Murphy, 1985) for the calculation of predicted values of natural UVR flux under given seasonal and geographical parameters. Data can, in some cases, be retrieved from large databases in several meterological institutions.

The degree of physical exposure to UVR, however, is strongly dependent on the physico-chemical characteristics of the local environment, to the extent that knowledge of the UV irradiance reaching ground (or sea) level may be indicative of, but usually insufficient to understand, the physical exposure in the natural environment. For a given cell, the relevant parameter which describes appropriately the instantaneous physical exposure to any wavelength is the scalar irradiance, E_0 (or fluence rate, y); a measure of the radiation incident on a point from all directions as measured in the cell´s immediate vicinity (see Jagger, 1967; Kirk, 1983). Absorptive and scattering processes within the medium may make E_0 differ significantly from the incident irradiance, but these modifications are not always intuitively evident or easy to measure. (Fig. 1).

Given the diversity of cyanobacterial habitats, it may be convenient to consider three "optical" types separately: planktonic, terrestrial, and sedimentary habitats. In planktonic habitats the populations are suspended in water (i.e. in the ocean, lakes or ponds). Terrestrial habitats refer to the solid surfaces exposed directly to the atmosphere such as rocks, tree barks, and soil surfaces in which cyanobacteria are often well represented if not dominant. In sedimentary habitats the populations are immersed in a matrix of solid or semi-solid mineral or organic matter. This type includes shallow water sediments, microbial mats, soils, and porous rocks (Revsbech et al., 1983). In both planktonic and sedimentary habitats one can find a heterogeneity of microhabitats, where deeper zones usually correspond to refuges from UVR.

Exposure in planktonic habitats is determined by the extinction rate of UVR within the water column and by the hydrodynamic mixing regime of the water masses. Typically, cells are subjected to complex, alternating high and low exposure regimes as they are brought up and down by the turbulent mixing processes (Cullen and Neale, 1994; Chapter 5). The instantaneous physical exposure for a given cell, E_0, can be conveniently described by three parameters: downwelling incident irradiance at the water surface (E_0, $Z = 0$), UVR attenuation coefficient in the water (a), and the depth of the cell in the water column (z) [assuming that a is constant with z, and that upwelling UV irradiance is so small that it can be neglected]. Both E_0D and a have a strong wavelength dependence. Pure water (and oceanic seawater) is quite transparent to UVR, so that in oligotrophic waters UVR can penetrate to nearly the maximum depth of the euphotic zone, particularly in the UVA range (Smith and Baker, 1981; Helbling et al., 1994; Prezelin et al., 1994). In turbid coastal waters the penetration of both UVA and UVB is restricted by dissolved and suspended materials (Vosjan and Pauptit, 1992; Piazena and Häder, 1994). In most coastal, estuarine, and inland waters terrestrial-derived "gelbstoff" is the paramount factor that determines the restricted penetration of UVR in the water column (Kirk, 1994; Scully et al., 1996; Laurion et al., 1997). A time integration of E_0-cell yields the dose received (D). This, however, is virtually impossible to calculate because the function [z(t)] for a cell in mixed waters is a result of complex hydrodynamic mixing processes. Additionally, the formation of thermoclines within the euphotic zone may temporarily or permanently trap cells at depths where they become exposed to high irradiance (Milot-Roy and Vincent, 1994). Deeper thermoclines in lakes and in the open ocean may provide stable environments for planktonic populations where they take refuge from UVR. If one assumes that a population of cells is homogeneously distributed and well mixed within the water column, a space-averaged exposure or a space-average dose can be more easily calculated. This practice stems from photobiological studies in the laboratory (Morowitz, 1950; van Liere and Walsby, 1982) but has found its application in ecological research (Helbling et al., 1994; Garcia-Pichel and Bebout, 1996), at least for comparative purposes. A comparison of several natural marine waters showed that the doses averaged for the euphotic zone ranged between 3 and 9% of those incident at the surface (Garcia-Pichel, 1996).

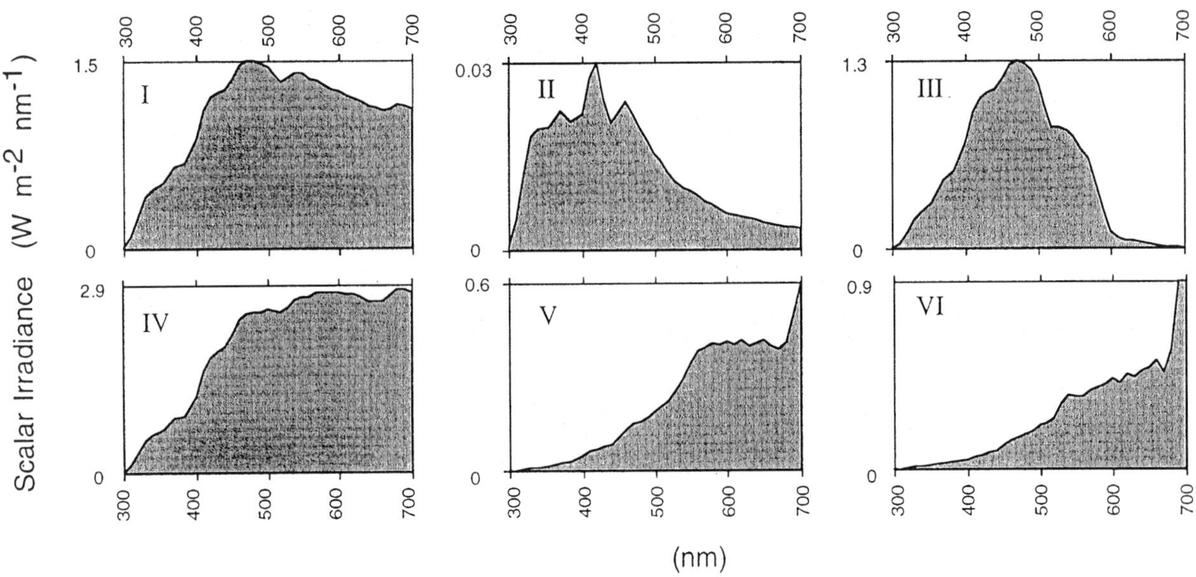

Fig. 1: Effects of the habitat on the physical exposure of cyanobacteria. I. The spectral scalar irradiance (sun and sky radiation) incident at ground level at noon in a clear midsummer day at 41°N Lat. The rest of the plates depict the in situ scalar irradiance experienced by cyanobacterial cells thriving in several habitats exposed to the incident fluxes of I (note different scales). II: a "strong shade" habitat (N-facing surface illuminated by extremely diffuse sky radiation only), where scalar irradiance is very low but the relative importance of UV is enhanced. III: a planktonic habitat (under 1 m of clear open-ocean water), where all fluxes remain fairly high and UVR-B and visible are more strongly attenuated than UVA. IV: the surface of beach (quartz, feldspar) sand, where all UVB, UVA, and visible are higher than incident (by 120, 150, and 205%, respectively) due to light trapping effects. V: 300 μm deep in a wet topsoil, where UVB and UVA have been attenuated below 5% of incident but ca. 20% of the visible light remains. VI: scalar irradiance within the thallus of the terrestrial cyanobacterial lichen *Collema* sp. (See Color Plate 32c, d and e). The spectra were calculated from the following sources: F. Garcia-Pichel (unpublished); Garcia-Pichel, 1995; Büdel et al., 1997; Smith and Baker, 1981.

In terrestrial habitats, two factors locally modify the incident UVR; the orientation of the surface with respect to the solar vertical; and the albedo of the supporting surface. Instantaneous exposure and dose are highest in surfaces oriented orthogonal to the Sun´s vertical, and lowest in those in the optical shadow of solids. However, due to the preferential atmospheric (Rayleigh) scatter of the shorter wavelengths in the atmosphere, the spectral composition of the diffuse light which penetrates the shade under solids is highly enriched in UVR (Robinson, 1966; Frederick et al., 1989). At wavelengths of 300 nm, approximately one half of the incident radiation is diffuse and one-half comes directly from the solar disc. It follows that populations thriving on shaded surfaces will receive lower absolute UVR doses, but higher UVR doses relative to the visible than populations on sunny surfaces (Fig. 1). This implies that UVR may play a particularly important role in the biology of cyanobacteria exposed to diffuse light. The radiation reflected from the surface may change the absolute and spectral characteristics of E_0-cell, with respect to the incident fluxes. Substrates such as white sandstone and carbonaceous materials (limestone, coral sand, concrete) reflect UVR strongly (Koller, 1965; Diffey et al., 1995).

In contrast, the optics of sedimentary environments is characterized by the strong attenuation of both visible and UVR due to absorption and multiple (Mie) scattering by the matrix and the organisms themselves, so that phototrophic growth is restricted to thin (millimeter to centimeter) surface layers (Friedmann and Ocampo, 1976; Pierson et al., 1987; Jørgensen and Marais, 1988; Garcia-Pichel and Castenholz,1994; Kühl and Jørgensen, 1994; Garcia-Pichel and Bebout, 1996). Close to the surface of these layers, light trapping phenomena result in localized irradiance maxima which are higher than incident irradiance. Typically, there is an onset of quasi-exponential attenuation of light below the light-trapping zones. Thus, in sedimentary habitats one finds extremely steep gradients of exposure ranging from zones of increased exposure (Fig. 1, IV) to zones which are virtual refuges from UVR (Fig. 1, V). Because light-trapping effects are greatest at longer wavelengths, and absorption by particulates is most pronounced at shortest wavelengths, the ratio of UVR to visible light invariably decreases with depth. A comparison of data from several sedimentary microenvironments showed that the space-averaged UVR dosage rates within the euphotic zone ranged between 15 and 33% of the incident, a much higher UVR exposure than those in the euphotic zones of natural waters (Garcia-Pichel and Bebout, 1996).

2. *Biologically Effective Exposure: Biological Weighting Functions*

If one strives to predict the biological impact of exposure to UVR on the basis of the spectral irradiance (or doses) received, and given that all biological effects of UVR present strong wavelength dependence, a function which relates both parameters is needed. These biological weighting functions (BWF) yield, upon multiplication by the spectral irradiance, an estimate of the biologically effective exposure. Biological weighting functions are, in principle, equivalent to the action spectra used in photobiology to elucidate the spectral characteristics of photoreceptors or target molecules (Clayton, 1970). They differ from them in that absolute units are required and the use of polychromatic radiation is preferred to monochromatic bands in order to include any synergistic effects of either visible light or other UV wavelengths (i.e. Cullen et al., 1992). Unfortunately no rigorous cyanobacterial BWF for any target process is yet available even though there is ample evidence of the effect of UVR-induced growth depression, and damage to DNA, photosynthesis, dinitrogen fixation and several other metabolic processes (Table 1) as well as strong wavelength dependence. Even though the use of BWF as a predictive tool is not without problems, the determination of cyanobacterial BWFs for several target processes should be a priority task if one wishes to understand the importance of UVR on the abundance and distribution of cyanobacteria in nature, and if one wishes to predict the possible affect of a reduction in stratospheric ozone on cyanobacterial populations.

3. *Metabolically-Timed Exposure*

Cyanobacteria seem to be successful colonizers of harsh habitats where metabolic activity is restricted by environmental conditions, such as lack of water or nutrients, long-term freezing or low temperatures. Under conditions of restricted or interrupted metabolism, the biological impact of exposure to solar radiation may seriously exceed that predicted by simply expressing exposure in terms of physical time. One might account for these differences by calculating either dose modification factors or by using metabolically-timed exposures, in that the exposure (either as cumulative dose or as instantaneous irradiance) is divided by the fraction of the total time in which the cells are active. This may be relatively easy in some cases, where the time partition is clear, or quite complicated if the metabolic activity slows rather than halts. An extreme example may be the surface-dwelling soil cyanobacteria of the Colorado Plateau (*Scytonema* sp., *Nostoc* sp.). These populations may experience ca. 100 hours of wetness (metabolic activity) during the winter season and additional scattered periods when moisture is available in the summer (Garcia-Pichel and Belnap 1996; Chapter 17). Approximately one half of the periods of rewetting occur during daytime. However, the colonies are exposed to UVR year round. Thus, the damage accumulated during a year's exposure needs to be taken care of within some 120 hours when cells are potentially active. This represents a metabolically-timed dosage of about 70 times the actual incident dose. The metabolically-timed irradiance is likely to be a better predictive factor in the terrestrial habitat, in polar regions and in alpine situations, where desiccation and cold are important ecological factors (Chapter 12).

4. *Cyanobacterial Habitats With High UVR Exposure*

Cyanobacteria are conspicuous in numerous habitats that are clearly exposed to high solar irradiance. It is thought that the Proterozoic Eon (2.5 - 0.57 Gyr) was abundantly populated by cyanobacteria, or their immediate ancestors, in shallow seas and bounding land masses, presumably exposed to high solar irradiance with a higher UVR flux than at present (Kasting et al., 1992). Today there is certainly a greater variety of specialized aquatic and terrestrial habitats than in the Precambrian, and some of these are also exposed to high irradiance and are dominated by cyanobacteria; others are shaded and protected.

a. Hot Springs

Although hot springs may occur today in numerous sites associated with volcanic or tectonic activity, most terrestrial springs are exposed to full sun simply because most vascular plants in the immediate vicinity of geothermal activity are killed. In addition, some of the major regions of hot springs, such as those of Yellowstone National Park, USA, emerge at high elevation (> 2000 m) and thus intercept even higher intensities of UVR than those closer to sea level. In general, hot springs are shallow runoffs of clear water and flow over the cyanobacterial mats or biofilms which form on substrates bathed by water at temperatures as high as 73-74°C in some regions (e.g. western North America; Chapter 3). Although many mechanisms for UVR resistance or avoidance exist, the conspicuous screening pigments, such as scytonemin in sheathed cyanobacteria, are lacking above about 55°C, the point at which sheathed forms such as *Calothrix* sp and *Pleurocapsa* sp. first appear (e.g. Wickstrom and Castenholz, 1978; Plate 30a).

b. Intertidal and Other Shallow Marine Habitats

Intertidal marine habitats of "normal" salinity, such as mud flats and marshes, are mostly devoid of extensive or perennial cyanobacteria-dominated microbial mats, principally because of a variety of eukaryotic competitors and herbivores. Nevertheless, some intertidal flats are inundated by water so infrequently (e.g. Laguna Guerrero Negro, Mexico; Chapter 10), that most eukaryotes are absent over large areas, and cyanobacterial mats (often dominated by scytonemin-rich *Lyngbya* cf. *aestuarii* and *Calothrix* sp.), cover vast areas (Javor and Castenholz, 1981, 1984; Plate 31a). Other intertidal sedimentary habitats, such as Great Sippewissett Marsh, Massachusetts, are so rich in biogenic sulfide that many eukaryotic grazers are absent or restricted and highly exposed cyanobacterial mats develop annually (Nicholson et al., 1987). The benthos of shallow, natural or man-made, hypersaline pools and lagoons also provide high insolation habitats where cyanobacteria usually dominate (Des Marais, 1995; Chapter 10).

In the tropics or semi-tropics on carbonaceous and other hard substrates, cyanobacteria compete well with other micro- and macro- phototrophs, particularly during warmer water periods. Many of these form crusts or small cushions in the upper- or supra-littoral zone. Others are subtidal in the generally clear shallow waters of unpolluted tropical and oceanic waters and are therefore exposed to high UVR levels. Many of the latter form wisp-like ephemeral covers or tougher gelatinous tufts on various substrates.

c. Benthic Freshwater Habitats

Clear water habitats occur in oligotrophic lakes throughout the Earth, and these include a large proportion of alpine/subalpine lakes. Although most of these habitats are not dominated by cyanobacteria, some ultra-oligotrophic lakes, such as Waldo Lake, Oregon, have dominant benthic covers of scytonemin-rich *Stigonema* and *Scytonema* to depths of several meters (Castenholz, unpublished data). Some of these oligotrophic fresh waters are dominated by cyanobacteria by virtue of an extreme deficiency of combined nitrogen, but not necessarily of other nutrients. In these environments heterocystous cyanobacteria may be conspicuous, such as the huge slow-growing, scytonemin-containing colonies of *Nostoc* cf. *pruniforme* (Dodds and Castenholz, 1988; Chapter 17; Plate 24a).

Unexpected cyanobacterial dominance also occurs in benthic mats in shallow cold water ponds and lakes in the Antarctic and Arctic (Vincent, 1988; Vincent et al., 1993; Tang et al., 1997). A high cyanobacterial biomass develops in these regions probably as a result of a lack of efficient herbivores, but perhaps also as a consequence of the tolerance of many cyanobacteria to freezing (Chapter 12). Many of the freshwater ponds in these regions freeze throughout the water column in winter. Again, these populations may be exposed to UVR for long continuous stretches during the polar summers, although at lower fluence rates than at middle latitudes with higher sun angles. Some of the cyanobacteria are scytonemin-containing; many are not. However, some nitrogen-deficient, shallow melt water streams and ponds in Antarctica are dominated by scytonemin-containing species of *Nostoc* (Plate 30c).

d. Terrestrial Habitats

In exposed terrestrial habitats that are extreme by virtue of periodic or long-term desiccation, cyanobacteria thrive by virtue of their tolerance to desiccation but presumably also because of a high tolerance to UVR to which they are so often exposed (Potts, 1994). The terrestrial habitats with conspicuous cyanobacterial mats or crusts include many of the warm and cold deserts of the Earth, where the cyanobacteria may form a ground cover, often between and around desert shrubs (see Garcia-Pichel and Belnap, 1996; Mazor et al., 1996; Plate 31b). The harder terrestrial substrates, such as cliff faces, are often covered by periodically-wetted cyanobacterial crusts which are often dark in color due to scytonemin or "gloeocapsin" (Plates 2d and 24c). These dark areas are often referred to as "tintenstriche" (ink streaks; Plate 30b). Cyanolichens, common in a large variety of terrestrial habitats, may also be exposed to high solar irradiance, and as a consequence contain scytonemin within the thallus (Büdel et al., 1997; Plate 32c, d and e). Even more extreme terrestrial habitats, such as the outer shells of hot and cold desert sandstones harbor cryptoendolithic cyanobacteria (e.g. *Chroococcidiopsis* spp; Plates 20, 21 and 22). It is possible that these too may be exposed to high UVR fluence rates (Nienow and Friedmann, 1993), although specific UVR intensities, in these situations, were not measured.

e. Marine and Freshwater Plankton

The cyanobacteria of the plankton are exposed, at least periodically, to high solar irradiance when turbulence brings them near the surface under conditions of high light (Chapter 5, 6). Picoplanktonic cyanobacteria, such as marine species of *Synechococcus* and *Prochlorococcus*, fall into this category, and these species are particularly common in clear tropical waters. Nevertheless, there are periods of vertical stratification when the cells may remain deep in the water column. UVR, however, may penetrate to considerable depth; in some cases UVB to over 30 m and UVA in excess of 60 m (Booth and Morrow, 1997; Jeffrey et al., 1996; Holm-Hansen et al., 1993). Freshwater unicellular cyanobacteria of picoplanktonic size also occur in many lakes, including clear alpine and subalpine lakes (See Chapter 7; Weisse, 1993; Eguchi et al., 1996; Postius et al., 1996). Gas vacuolate cyanobacteria in marine waters are represented mainly by species of *Trichodesmium* in warmer waters, and by species of *Anabaena, Aphanizomenon, Oscillatoria, Limnothrix, Microcystis, Coelosphaerium* and others in fresh and brackish waters. Although regulation of vertical position in the water column occurs in some of the gas vacuolate forms, some forms are simply buoyant (Walsby, 1994). These cyanobacteria may simply accumulate on the surface during calm periods of low turbulence, with high mortality as a result of exposure to high solar irradiance. Many of the larger planktonic cyanobacteria, such as *Trichodesmium, Aphanizomenon, Dactylococcopsis*, and others are known to accumulate large quantities of putative intracellular sunscreens of the MAA type, but MAAs do not seem to occur in picoplanktonic forms (Garcia-Pichel and Appel, unpublished data).

II. Strategies of UVR Tolerance by Cyanobacteria

A. UV-Screening Compounds

1. MAAs and Other Compounds

The use of UVR-absorbing compounds as sunscreens is known in many organisms. A sunscreen represents a passive mechanism for protection; UVR radiation is, to some extent, absorbed by the sunscreen and cannot interact with potential cellular targets. The melanins of tegumental animals are perhaps the best known example of such sunscreens (Kollias et al., 1991). In higher plants phenylpropanoids and related substances (anthocyanins, flavone glycosides) provide protection against UVR (Takahashi et al., 1991). In microorganisms, sunscreen roles were suggested for, or ascribed to, a variety of substances including melanins and usnic acid in fungi, prodigiosine in *Serratia marcensis* and violacein in *Chromobacterium violaceum.* However, the use of sunscreens in microorganisms presents special problems because of their small cell size. Since absorption depends on both the concentration of a compound and on the pathlength offered to the incident radiation, the concentrations of absorbing substances required by a microorganism must be high, which requires a significant metabolic investment. Modeling suggests that cell wall or membrane bound sunscreens are not optically feasible for bacteria; internally accumulated sunscreens can only be effective if cells are larger than several μm in diameter; only cells (or colonies) larger than 10 μm can make use of internal sunscreens with low cost/benefit ratios (Garcia-Pichel, 1994; Garcia-Pichel, 1996). Most cyanobacteria are thus in the range of cellular/colonial sizes for which the use of sunscreens is possible, but caution should be exercised in assigning sunscreen functions to putative substances.

UVR absorption is a necessary, but certainly not a sufficient, criterion to determine that a compound may function as a sunscreen. An interesting example is that of the biopterin glucoside reported from a UVA-tolerant marine *Oscillatoria* (Matsunaga et al., 1993; Wachi et al., 1995). Because the cellular content of this compound can be raised by exposure of the cells to UVA, in a manner proportional to its intensity, the compound was considered to be photoprotective. This and other biopterin glycosides were subsequently identified in other cyanobacteria (e.g. *Synechococcus leopoliensis*, *Microcoleus chthonoplastes*) but UV induction could only be demonstrated in some cases (Grether et al., 1996). The maximum content of biopterin glycoside measured in cyanobacteria under any condition did not exceed 1 mg per g dry weight, an amount which would screen no more than about 1% of incident photons. Thus, while the involvement of pterins in the UV photobiology of cyanobacteria seems important (see also Shibata et al., 1996), their role as sunscreens is unlikely.

In cyanobacteria, the extracellular sheath pigments (scytonemin, "gloeocapsin") observed in some terrestrial forms by Nägeli (1849) were thought to be used for protection against high light (Lemmerman, 1910). Shibata (1969) discovered that colorless UV-absorbing compounds accumulated in large quantities in cyanobacterial cells. These compounds are now known to belong to the family of compounds known as mycosporine-like amino acid derivatives (MAAs; Garcia-Pichel and Castenholz, 1993). Several other cyanobacterial compounds and structures were suggested to have possible screening functions and these include; external calcite deposits and intracellular gas vesicles (van Liere and Walsby, 1982); carotenoids in general (Paerl et al., 1983); cellular membrane-bound carotenoids (Carr and Wyman,1986); $FeCl_3$, and a brown pigment released into the medium (Kumar et al., 1996). However, little if any, supporting evidence for a sunscreen role was presented, and these compounds are not discussed further here.

The family of substances known as mycosporines (mycosporine-like amino acid derivatives or MAAs) encompasses a series of colorless, low molecular weight, water soluble compounds which have a single absorption maximum within the solar UVR range. Their name derives from their original discovery in fungal spores (Favre-Bomvin et al., 1976). Structurally they are condensation derivatives of a cyclohexenone ring and amino acid (or imino alcohol) residues (Fig. 2). Their synthesis probably originates from the first part of the shikimate pathway (Favre-Bomvin et al., 1987). Monosubstituted MAAs absorb maximally around 310 nm, whereas bisubstituted MAAs absorb maximally at longer wavelengths (between 320 and 360 nm). They occur in large concentrations in fungi (Favre-Bomvin et al., 1976; Arpin et al., 1977), eukaryotic micro- and macro-algae (Carreto et al., 1989; Sivalingam et al., 1974; Karentz et al., 1991) and lichens (Büdel et al., 1997). They also occur in a variety of marine invertebrates (Karentz et al., 1991; Dunlap and Chalker, 1986) in which they are generally thought to be obtained from the diet (Dunlap and Shick, 1998). In a single survey of 20 cyanobacterial isolates from high insolation environments, 13 contained one or several MAAs (13 different MAAs in all; Garcia-Pichel and Castenholz, 1993). Isolates from marine, freshwater, terrestrial or planktonic origin contain MAAs. Although some have been identified, most of the cyanobacterial MAAs still await chemical characterization. Shinorine, asterina-330 and porphyra-334, mycosporine-gly (Karsten and Garcia-Pichel, 1996; Garcia-Pichel and Appel, unpublished data) all occur in cyanobacteria (Fig.2). The UV-absorbing compound reported from *Nostoc commune* (Scherer et al., 1988) was shown to be a mixture of oligosaccharide-MAAs (OS-MAA; Böhm et al., 1995). The glycosylated MAAs of *N. commune* represent the only known example of MAAs that are actively excreted and accumulated extracellularly (Chapter 17); in all other cases studied MAAs were accumulated intracellularly.

The evidence presented for the sunscreen role of intracellular MAAs is inconclusive in some cases, and it is possible that they serve more than one role in the cellular metabolism of all or some organisms (Chapter 10). This should not be too surprising given the diversity of the compounds and organisms. Whereas most MAAs are stable compounds, mycosporine-gly is rather unstable (Ito and Hirata, 1977) and mycosporine-gln undergoes facile photosensitized hydrolysis (Bernillon et al., 1990). These traits are not desirable for a sunscreen; mycosporine-gly may play an antioxidant role instead (Dunlap and Yamamoto, 1995). Perhaps the most elegant demonstration of the sunscreen role of MAAs yet is with echinoderm eggs (Adams and Schick, 1996), but their probable role as sunscreens in cyanobacteria is also well documented. Using a set of stringent optical and physiological criteria the hypothesis that MAAs are sunscreens in *Gloeocapsa*

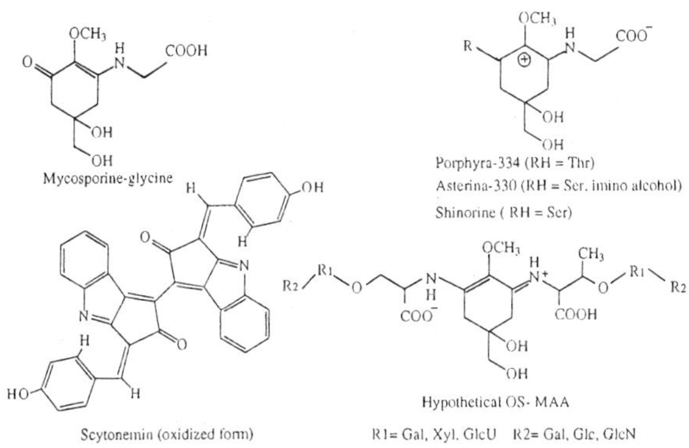

Fig.2. Chemical structure of some sunscreen compounds identified in cyanobacteria. The structures are scytonemin (Proteau et al.,1993), OS-MAA (Böhm et al., 1995), mycosporine-glycine (Ito and Hirata, 1977), porphyra-334 (Takano et al., 1979), shinorine (Tsujino et al., 1980) and asterina 330 (Nakamura et al.,1981).

sp. could not be discounted (Garcia-Pichel et al., 1993). In cyanobacteria, a constitutive level of MAAs appears to be present under all growth conditions, but the specific content can be significantly raised by exposure to UV-B/A radiation with a maximum at 320 nm (Garcia-Pichel and Castenholz, 1993). In *Gloeocapsa* sp. the levels were shown to be a direct function of the incident UV-320 irradiance, and longer wavelengths were incapable of eliciting the response. Similar evidence was gathered for the OS-MAAs from terrestrial *Nostoc commune* (Böhm et al., 1995; Ehling-Schulz et al., 1997).

The single-cell sunscreen effect of intracellular MAAs in cyanobacteria is modest (10-30% of incident photons were intercepted in a fairly large-celled *Gloeocapsa* sp.), although colony-formation and mat-type growth may substantially increase the efficiency. The concentration of the sunscreen is not too high (below 1% of the dry weight, in most cases) and it seems that the accumulation of intracellular MAA may provide some benefit for a small investment. But, why is the intracellular content of MAA not raised higher so that more significant screening would occur? There may be physiological limitations to the accumulation of osmotically active compounds such as MAAs within cells (Garcia-Pichel 1994). It is possible that the maximal specific content of MAA in the cells is regulated by osmotic mechanisms. This is consistent both with the fact that field populations of halotolerant cyanobacteria have unusually high MAA contents (Oren, 1997, Chapter 10), and with the finding that an increase in the salinity of the medium raised both the basal and the UV-induced specific content of MAAs in strains of *Microcoleus chthonoplastes* (Karsten and Garcia-Pichel, unpublished data). The adaptive advantage gained by avoiding osmotic limitations and by enhancing the optical efficiency may have driven the evolution of extracellular MAAs (OS-MAA). Beyond anecdotal observations, field data or correlations between cyanobacterial MAAs and other environmental parameters are still lacking.

2. Sheath Pigments

Cyanobacterial sheaths are often colored. Yellow to brown colors are the most common but red, blue or black also occur (Plate 32a, b and c). The yellow to brown coloration is due to the presence of scytonemin, a UV absorbing pigment (Plate 24e). The synthesis and excretion in the cells and sheath is not constitutive but induced most effectively by exposure to UVR (Garcia-Pichel and Castenholz, 1991). Scytonemin is a dimeric indole alkaloid with no related compounds known among natural products (Fig. 2), and it is most probably synthesized from aromatic amino acid residues (Proteau et al., 1993). The screening by scytonemin is most efficient at the longest wavelengths of the UV (max ~384 nm), although it absorbs throughout the solar UV range, and it is interesting that the screening complements that of MAA absorption. Scytonemin absorption is also strong in the violet/blue region but diminishes greatly at longer wavelengths. Its role as a sunscreen was clearly demonstrated in an isolate of terrestrial *Chlorogloeopsis* sp. by compliance with a set of stringent optical and physiological criteria (Garcia-Pichel et al. 1992). Strong evidence for the role of scytonemin derived from studies with isolates and collected materials (Garcia-Pichel and Castenholz, 1991; Garcia-Pichel and Castenholz, 1993). It was shown recently that scytonemin was responsible for UVR screening in a variety of cyanobacterial lichens

(Büdel et al., 1997; Plate 32c, d and e). UVR-induced cultures may accumulate as much as 5% of the cellular dry weight as scytonemin; field samples may reach an even higher specific content. Scytonemin may accumulate to such an extent in the sheaths of some exposed cyanobacteria, that paracrystalline granules of the pigment become visible under the microscope. UVA was effective in inducing scytonemin synthesis in all cases studied but detailed action spectra are lacking. UVB was relatively ineffective in initiating scytonemin production in *Nostoc commune* (Ehling-Schulz et al., 1997). Scytonemin-producing cyanobacteria are unknown from the planktonic environment (Garcia-Pichel and Castenholz, 1991) but are common in terrestrial soils and on rock surfaces, upper intertidal surfaces, and in ponds of the Arctic, Antarctic, subalpine and alpine regions. Many of these habitats may be exposed to high incident UV irradiance, but not all (Pentecost, 1993).

The common denominator for presence of scytonemin seems to be periodic exposure under conditions of restricted metabolism, a point embraced by many (Whitton and Potts, 1982; Potts, 1994; Garcia-Pichel and Castenholz, 1991; Pentecost, 1993). Pentecost (1993) provided field evidence for this in a study of shaded terrestrial *Scytonema* populations. Scytonemin content failed to correlate directly with incident UV irradiance, and only when water availability and cell division rates were taken into account could the data set be explained. It is conceivable that the large investments required for effective use of scytonemin pay off only when exposure is linked to periods of metabolic inactivity. Competition with other cyanobacterial forms that rely on active repair mechanisms would then be successful. Our observations, at this point, indicate that sheathed cyanobacteria that occupy even shaded, but periodically desiccated habitats, may produce a high scytonemin content. The presence and distribution in time and space of cyanobacteria that produce scytonemin in nature implies that the pigment is an important factor for fitness in environments of high "metabolically-timed exposure" (Section B.3).

Little is known about the red/violet pigments ("gloeocapsin") other than their *in situ* reaction to acid/base (Nägeli and Schwenderer, 1877), and their common occurrence in natural populations of cyanobacteria (e.g. *Gloeocapsa* sp.; Plate 2c) on vertical faces of cliffs in many mountainous regions ("Tintenstriche" or ink streaks: Jaag, 1945; Plate 30b) and in some cyanobacterial soil crusts and filamentous tropical forms (e.g. *Porphyrosiphon* sp.). Apparently, no gloeocapsin-producing strains are in culture (Plate 32a, b and c). The peripheral pigment distribution in colonial forms parallels that of scytonemin, with the implication that exposure to solar irradiance causes its accumulation. Optical evidence with natural samples suggests that the sunscreen effect due to gloeocapsin is substantial (Garcia-Pichel, unpublished data).

B. Avoidance Responses by Motile Cyanobacteria

1. Benthic, Microbial Mat Populations

There are numerous reports of the downward movement of motile "oscillatorian" cyanobacteria from microbial mat surfaces into the mat matrix or into soft sediments during periods of high solar irradiance (Plate 32d; e.g. see Ramsing and Prufert-Bebout, 1994). Although the downward movements usually amount to only about 1 mm or less, the attenuation of visible and UVR radiation within this short distance is extreme (see Garcia-Pichel and Bebout, 1996). Vertical movements of this type were described in shallow hot or warm spring mats (Castenholz, 1968; Richardson and Castenholz, 1987; Castenholz et al., 1991; Chapter 3); in shallow subtidal, intertidal marine and hypersaline mats or sediments (Whale and Walsby, 1984; Castenholz, 1994; Bebout and Garcia-Pichel, 1995); and in some freshwater mud flats (Pentecost, 1984). Although there is a positive correlation between increasing irradiance and downward positioning, there are not many quantitative data available. The distance of downward descent (up to about 0.5 mm) in soft microbial mats of hypersaline salinas of Guerrero Negro, Mexico, correlated well with the increase in surface irradiance. The fluence rate of visible irradiance at the depth of the main cyanobacterial band during the daylight hours (except for early morning and evening) corresponded to that of photosynthetic saturation and not of photoinhibition (Garcia-Pichel and Castenholz, 1994). The band of motile cyanobacteria in the study consisted primarily of *Oscillatoria* sp. and *Spirulina* cf. *subsalsa.* Thus, it is possible that optimization of photosynthetic irradiance was achieved and was not simply the avoidance of harmful intensities. Nevertheless, the involvement of UV irradiance was not investigated in this study.

Garcia-Pichel and Castenholz, (1994) suggested that UV irradiance was a primary cue for the vertical movements of cyanobacteria. In an indirect measurement of vertical movement, primarily of *Microcoleus* cf. *chthonoplastes* in a mat from Solar Lake, Egypt, UVB was particularly effective (Bebout and Garcia-Pichel, 1995). Artificial illumination was used at intensities that corresponded to strong natural fluxes of the UVB component, but UVA and visible light were much lower than natural. In a more recent outdoor study of motile *Spirulina* and *Oscillatoria* in the hypersaline mats of Guerrerro Negro, it was shown that upward movements were most sensitive to the natural levels of incident UVA above ~ 1.5 W m^{-2} and UVB an order of magnitude lower (Kruschel and Castenholz, 1998). That is, upward movements in bright daylight were effectively prevented by UVB or UVA in contrast to broad visible irradiance alone. Downward movements, beginning from low light occurred in response to rather high incident intensities of UVA (>10 W m^{-2}) and broad visible irradiance over ~ 400 W m^{-2}), but not to UVB alone. The spectral/irradiance responses were complex and depended on the species and the mat type (Kruschel and Castenholz, 1998). In addition, it was shown that natural UVB and UVA were the spectral regions most inhibitory to photosynthesis in both of the predominant motile cyanobacteria of these mats. A similar migratory response and inhibition pattern was measured in an *Oscillatoria* sp. in an Antarctic saline pond (Nadeau and Castenholz, in preparation).

Several studies suggest that low and natural intensities of UVR (especially UV-B) inhibit photosynthesis, motility, phototactic orientation, and photophobic responses, but often only after 30 min to 1-2 hours after exposure (Häder 1984, Häder et al. 1986). However, in some cases inhibition occurred within 5 min (Donkor et al., 1993). In the studies of migrations within natural mats in Guerrero Negro, downward movements were observed within 30 min after sudden exposure to high solar irradiance, presumably before significant inhibition of the motility responses took place. In addition, it should be realized that natural diel vertical movements usually occur in response to gradual increases and decreases in solar irradiance, not the sudden changes used in experiments. However, on partly cloudy days sudden changes in solar irradiance may occur, but usually with the susceptible cyanobacteria safely below the surface. In the Antarctic saline pond sensitivity to different wavelengths and fluence rates was similar to the behavior of the "oscillatorians" in Mexico, but response rates were about 4 times slower, presumably because of the low temperature (< 8 - 10° C) (T-L. Nadeau and R.W. Castenholz, in prep.). The avoidance of high solar irradiance by downward migration in mats or sediments definitely appears to be a strategy of cyanobacteria that results in their tolerance of habitats exposed to high UVR and visible radiation without the problem of long term direct exposure and the need for physiological acclimation. Indeed, in the few cases investigated, the generally high cell content of major light-harvesting pigments (phycobilins and chlorophyll-a) was retained, allowing the cells to take advantage of low photon fluence rate, such as in early morning, late afternoon, and periods of overcast.

2. Planktonic Avoidance Responses

Since some planktonic, gas vacuolate, cyanobacteria are known to be capable of vertical positioning in the water column (Walsby, 1994), it is possible that a subtle regulation in some cases may be in response to UV wavelengths which penetrate the water. There is ample evidence that marine phytoplankters, including cyanobacteria, are negatively affected by UVR at levels comparable to those in nature (Holm-Hansen et al., 1993), but there is no information at present on specific avoidance responses related to UV irradiance.

C. Tolerance to UVR Requiring Active Repair or De Novo Synthesis

The acclimation of cells to UVR may involve active repair mechanisms, in which damaged targets are partially substituted or the damage is repaired without the need for *de novo* synthesis of all the components. This is exemplified by the DNA repair mechanisms. Alternatively, increased expression of damaged target proteins, or the possible synthesis of UVR-resistant forms of target proteins may also be involved in acclimation to UVR. Some cyanobacteria can thrive under strong fluxes of UVR, even without sunscreen components. This implies that they must have adapted by possessing efficient repair/synthetic mechanisms. But these may not be constitutively expressed, a fact which becomes clear through observation of the time-course of growth after sudden exposure to UVR in which an initial depression in growth is followed by recovery in a matter of hours/days (Hirosawa and Miyachi, 1983; Garcia-Pichel et al., 1993; Wachi et al., 1995). The physiological basis of this phenomenon was only explored superficially. Exposure of cyanobacteria to UVR elicited *de novo* synthesis of some known (RecA, chaperones) and many unidentified peptides (Owttrim and Coleman, 1989; Chitnis and Nelson, 1991). That extrachromosomal genetic elements may be involved in cyanobacterial UV resistance is an interesting possibility (Kumar and Kumar, 1990).

1. DNA-Damage Repair

Exposure of DNA to UVB and UVC radiation can cause several DNA lesions, the most common being the photodimerization of adjacent pyrimidine bases, although adducts and cross-links may also be formed (Harm, 1980; Jagger, 1985; Häder and Tevini, 1987). DNA-repair mechanisms are universal; all cyanobacteria studied had some type of excision repair and photoreactivation activity (Levine and Thiel, 1987). In photoreactivation UVA/blue light is used by the enzyme DNA photolyase to split pyrimidine dimers caused by exposure to UVB/C radiation. Photoreactivation was demonstrated in all cyanobacteria tested (Van Baalen, 1969; Singh, 1975; Asato, 1972; O'Brien and Houghton, 1982; Levine and Thiel, 1987). Cyanobacterial photolyases contain two chromophores. The light harvesting chromophore that transfers excitation energy to the active FADH group is a diazaflavin (maximum absorbance at 437 nm), which expands the photoreactivating capabilities from the UVA into the blue (Eker et al., 1990). Cyanobacterial photoreactivation appears to be very efficient in comparison with the process in *Escherichia coli* (Fig. 2.4 in Jagger, 1985). While there is direct evidence for the presence of dark excision repair in cyanobacteria (Levine and Thiel, 1987; O'Brien and Houghton, 1982), cyanobacteria seem to rely heavily on photoreactivation for DNA repair (Levine and Thiel, 1987; Blakefield and Harris, 1994).

2. Repair of the Photosynthetic Machinery

After acute exposures to UVR, if subsequent treatments are moderate, at least a partial recovery may occur; photosystem II activity resumes and specific contents of bleached photopigments rise. This response might simply be due to an increased (or renewed) expression of genes that encode replacements for the damaged molecules. However, a specific, light regulated repair system for damage to PSII (which does not involve protein or RNA synthesis) was claimed for *A. nidulans* (Bhattacharjee and David, 1977; Bhattacharjee et al., 1987). On the other hand, it was shown that the genes in *psbA* multigene family (coding for 3 different forms of the protein D1 of PSII) in *Synechococcus* PCC 7942 were expressed differentially according to the light intensity (Golden, 1994). Analysis of the performance of mutants defective in some of the D1 forms showed that one of them allowed increased resistance to photoinhibition caused by visible light (Golden, 1994). Similarly, mutagenesis of the *psbA* gene of *Synechocystis* 6803 resulted in photo-tolerant mutants (Satoh, 1998). Since the main target of UV damage to PSII seemed to be the D1 protein, it is possible that specific D1 polypeptides resistant to UV damage evolved in cyanobacteria (Garcia-Pichel, 1992). The presence of a repair mechanism for the photosynthetic apparatus, other than the *de novo* synthesis of affected polypeptides/chromophores, remains to be explored experimentally. *De novo* synthesis of the D1 and D2 reaction center subunits is known as a key step of the repair process in *Synechocystis* sp. PCC 6803 (Sass et al., 1997).

3. Protection by Carotenoids

Carotenoid absorption maxima occur mainly in the visible spectral range (i.e. > 400 nm), but often extend minimally into the UVA region (< 400 nm) (Goodwin, 1980; Koyama and Hashimoto, 1993), and some have a small absorption peak in the UV-B region. Their function as UVR-screening pigments is minimal, although in some eukaryotes (e.g. the green algae: *Haematococcus, Trentepohlia,* and *Dunaliella*) high cytoplasmic accumulations of secondary carotenoids (e.g. astaxanthin, ß-carotene, etc.) provide almost complete screens of violet/blue portions of the spectrum (e.g. Ben-Amotz et al., 1989; Hagen et al., 1994). In cyanobacteria carotenoids occur in the outer cellular membrane as well as in the thylakoids, and during long term exposure to high natural or artificial irradiance, very high ratios of combined carotenoids to chlorophyll*a* and phycobilins occur, usually as a result of a lowering of the contents of the latter two pigments. Cyanobacterial cells with high carotenoid to chlorophyll/phycobilin ratios are better able to tolerate higher light intensities, particularly at suboptimal growth temperatures (Castenholz, 1972), but UVR was not identified as a factor. Accentuated UV-inhibition would normally be expected at sub-optimal temperatures because of temperature-dependent biosynthetic repair processes (see Roos and Vincent, 1998). In a strain of *Chroococcidiopsis* (Garcia-Pichel and Dor, unpublished data) exposure to UVB (4 h per day at 1.2 W m^{-2}) led to an absolute, 2-fold increase in carotenoid content. Myxoxanthophylls and xanthines increased while β-carotene decreased and echinenone remained the same. A one-day exposure of *Nostoc commune* to UVB radiation resulted in a small increase in myxoxanthophyll and echinenone content (Ehling-Schultz et al., 1997). In one case a relatively UVR-resistant strain of *Gloeocapsa alpicola* possessed a higher carotenoid content than the wild-type (Buckley and Houghton, 1976).

In many cases the change to high carotenoid:chlorophyll/phycobilin ratios in cyanobacteria may be in response to low availability of several nutrients, including nitrogen (Collier and Grossman, 1992). It is not known if such cells have an increased resistance to UVR. In some non-photosynthetic microorganisms such as *Neurospora crassa,* carotenoid-less mutants or wild-type cells grown in the presence of inhibitors of carotenoid synthesis were more sensitive to UVB than wild-type cells which possessed carotenoids but the nature of the carotenoid protection was unknown (Morris and Subden, 1974). The UVB protective effect of various carotenoids was recently evaluated in *E. coli* transformed with several carotenogenic genes (Sandmann et al., 1998).

In cyanobacteria also, it appears that the role of carotenoids in UVR protection is indirect; functioning as anti-oxidants, quenchers of photosensitization products such as triplet chlorophyll, singlet O_2, peroxy radicals, and also as inhibitors of free radical reactions (Krinsky, 1979; Britton, 1995). High intensity visible wavelengths (especially blue) and UVA are especially effective in producing the photoproducts that carotenoids have the ability to "quench," but the relative importance of UVA remains unevaluated. It is common to find anecdotal allusions in the scientific literature as to the UV protective role of carotenoids in cyanobacteria.

4. Other UVR Stress Responses

Radiation in the 295-390 nm range (UVB and UVA), at an intensity which resulted in 30-40% cell survival, changed the pattern of protein synthesis in the unicellular cyanobacterium *Anacystis nidulans* R-2 (= *Synechococcus* sp.; Shibata et al., 1991). Sixteen "UV-shock" proteins were synthesized. Heat shock without UVR induced the synthesis of seven proteins that were similar, in terms of SDS-PAGE mobility, to seven of the "UV-shock" proteins. Six distinct proteins with similarity to the "UV-shock" proteins were also induced simply by oxidative stress. More recently, Shibata et al. (1996) demonstrated that a form of pro-oxidant lumazine from cyanobacterial cells generated superoxide anion under "near" UVR. Shibata et al. (1991) suggested that exposure to UVR shifted synthetic metabolism to the production of enzymes that scavenge reactive oxygen moieties, thereby limiting UVR damage. P. Nicholson et al. (1987, 1991) also observed the synthesis of some, apparently unique, "stress" proteins which may have been involved in DNA repair after exposure to UVC radiation. Stress proteins, especially some of the heat shock proteins, were extensively characterized (e.g. Craig et al., 1993; Sanders, 1993, Satoh and Murata, 1998). Unique stress proteins may also be involved in responses to desiccation and high salt concentrations (Potts, 1994; Close and Lammers, 1993). Although little is known about the specific function of most of the proteins associated with UVR stress, a well known response to high irradiance (not specifically UVR) is the increased synthesis of superoxide dismutase (Mn SOD in some species of cyanobacteria) above constitutive levels of Fe SOD (Whitelam and Codd, 1986; Van Liere and Walsby, 1982; Fridovich, 1975).

III. Effects of UVR in Nature and Whole Community Responses

Although intuitively the effect of UV radiation on microbial communities in which cyanobacteria or microalgae are the predominant primary producers would be expected to be negative, a few studies indicated that the results are complex and positive for certain species and negative for others. Obviously, diverse photosynthetic and non-photosynthetic microorganisms exhibit differential tolerances to UVR and to different portions of the UV spectrum.

Earlier in this chapter the responses of vertically migrating cyanobacteria in relation to UVR were discussed (Section II.B). Although the entire microbial mat community was not investigated, the studies on photoresponses did involve natural populations. Recent studies of the high temperature (65-73°C) *Synechococcus cf. lividus* biofilms at Octopus Spring in Yellowstone National Park indicate that high natural daytime levels of UVR can result in a substantial (estimates range from 20 to

48%) inhibition of short-term photosynthesis particularly at non-optimal temperatures (Miller et al., 1998; Chapter 3). However, these populations of *Synechococcus* persisted even at supraoptimal temperatures in summer under full solar irradiance, although they were without oxygenic competitors above about 65°C. It is likely that cell repair processes (possibly mainly at night) are required for survival.

A natural warm spring stream with a cyanobacterial biofilm community composed almost entirely of *Calothrix*, a species rich in the sheath pigment scytonemin, was the subject of three-month summer experiments in Yellowstone National Park which aimed to determine how production and maintenance of scytonemin and light harvesting pigments were influenced by visible and UV components of natural solar irradiance. In addition, the sensitivity of photosynthesis to UV radiation after long term maintenance and growth in the presence or absence of UV irradiance was evaluated (Brenowitz and Castenholz, 1997). Intact mats and cleared substrates were treated under filters transparent to UV radiation and visible light, those that exclude only UVR as well as under neutral density screens. After three months under all treatments, the predominant organism in intact mats and on newly colonized substrate was still the same morphotype of *Calothrix*. Intact *Calothrix* mats produced high levels of scytonemin, except under low visible irradiance (i.e. ~12% T) with UVR removed (< 1% UVR T). Intact *Calothrix* mat and newly colonized mat with substantial scytonemin, after three months, did not show significant UVR inhibition of photosynthesis. Long-term new growth with < 1% UVA exposure produced cells with reduced scytonemin levels. These showed greater UVR photosynthetic inhibition than cells grown in the presence of higher levels of UVR (Brenowitz and Castenholz, 1997). Thus, cells with substantial scytonemin were well protected from specific inhibition by UVR, although the scytonemin content varied considerably. It is not known to what extent cell metabolism was required, in addition to scytonemin, for UVR protection, since all tests were conducted on natural populations at the common field temperature of *Calothrix* (i.e. ~ 25°C).

These results, obtained from virtually monospecific natural populations of cyanobacteria, contrast with those of Bothwell et al. (1993) who observed large differences in species composition in a microbial community which developed on artificial substrate over several weeks with and without added UVR. An inhibition of diatom accumulation under UVR occurred for the first two to three weeks, but after three to five weeks the total diatom abundance was two to four times greater under UVR than in the UVR minus controls. In addition a change in dominant diatom species also occurred. These experiments used materials from a stream that harbored a diverse community of benthic diatoms and herbivores, and some of the results may be explained by the greater sensitivity of diatom consumers (especially chironomid larvae) to UVR than the diatoms themselves (Bothwell et al., 1994). In other studies, with diverse natural assemblages in alpine lakes, biomass accrual (including cyanobacteria) was inhibited by natural levels of UVR (Vinebrook and Leavitt, 1996). In a more detailed study in a ultraoligotrophic lake Vinebrook and Leavitt (personal communication) found that the effects of natural UV were mostly direct, as herbivores were largely insensitive to UVR. Growth of eukaryotic periphyton (mainly diatoms) was suppressed by UVR (particularly UVB), whereas growth of cyanobacterial picoplankton (*Synechococcus* sp.) was largely unaffected. Surprisingly, the growth of cyanobacterial periphyton (*Merismopedia* sp. and *Phormidium* sp.) was strongly stimulated by the inclusion of UVA!

There is recent evidence that the UV component of irradiance is necessary to maintain the normal three dimensional structure of cyanobacterial mats, at least aerial mats on black mangrove in the tropics (R.P. Sheridan, personal communication). Sheridan has evidence that species more resistant to UVR normally occupy the mat surface, giving protection to more sensitive species below. Without UVR, opportunistic species of cyanobacteria overgrew the laboratory-maintained mats, and with this disorganization, nitrogen fixation was severely reduced. The addition of UVR in the laboratory maintained a condition more similar to that of the field.

IV. Conclusions

It is obvious that an increasing number of research scientists are considering the effects of UVR on various photosynthetic bacteria and microalgae. Not all have been cited here. Many are reductionists and fail to consider the variable modifying effects of factors other than UVR that are intimately involved and environmentally relevant. However, most studies progressed in recent years, from simply measuring the effects of UVB or UVA alone in the laboratory without some background of visible irradiance, to at least trying to simulate natural conditions. In some cases the effects of various levels of natural solar irradiance are measured with and without the UVR components that may be excluded by various filters. But even in these experiments usually only short-term effects, such as those of hour-long photosynthetic rates, were considered. In their defense, short-term studies may necessarily have to precede longer, more complex experiments. Long-term studies on essentially monospecific natural populations or culture clones and on whole communities are rare and should be pursued. It is also important in future work to evaluate the importance of the diel dark period in recuperation and in repair of UVR damage that occurs during daylight.

In addition to the detrimental effects of UVR and the strategies of tolerance or avoidance exhibited by cyanobacteria and other photosynthetic microorganisms, it is now beginning to be realized that UVR can also be used as a signal or cue for various biochemical or motility responses. In these cases the UV receptor and transduction systems are still unknown.

Acknowledgments

RWC gratefully acknowledges the US National Science Foundation for support of research by both authors reported here. F G-P also acknowledges the support of the Max Planck Society and of the Biotechnology Program of the European Commission. We thank M. Ehling-Schulz, S. Scherer, D. Vinebrooke and P. Leavitt for making unpublished data available to us.

References

Adams N L and Schick JM (1996) Mycosporine-like amino acids provide protection against ultraviolet radiation in eggs of the green sea urchin *Strongylocentrotus droebachiensis*. Photochem Photobiol 64: 149-158

Arpin N, Favre-Bomvin J and Thivend S (1977) Structure la mycosporine-2, nouvelle molecule, isolĕé de *Botrytis cynerea*. Tetrahedron Lett 10: 819-820

Asato Y (1972) Isolation and characterization of ultraviolet light-sensitive mutants of the blue-green alga *Anacystis nidulans*. J Bacteriol 110: 1058-1064

Bebout BM and Garcia-Pichel F (1995) UVB-induced vertical migrations of cyanobacteria in a microbial mat. Appl Environ Microbiol 61: 4215-4222

Ben-Amotz A, Shaish A and Avron M (1989) Mode of action of the massively accumulated ß-carotene of *Dunaliella bardawil* in protecting the alga against damage by excess irradiation. Plant Physiol 91: 1040-1043

Bernillon J, Parusssine E, Letoublon R, Favre-Bonvin J and Arpin N (1990) Flavin-mediated photolysis of mycosporines. Phytochem 29: 81-84

Bhattacharjee SK and David KAV (1977) Unusual resistance to ultraviolet light in dark phase of blue-green bacterium *Anacystis nidulans.* Nature 265: 183-184

Bhattacharjee SK, Mathur M, Rane SS and David KAV (1987) UV-sensitivity of cyanobacterium *Anacystis nidulans* Part I. Evidence for photosystem II (PSII) as a lethal target and constitutive nature of a dark repair system against damage to PSII. Indian J Exper Biol 25: 832-836

Bjorn LO and Murphy TM (1985) Computer calculation of solar ultraviolet radiation at ground level. Physiol Veg 23: 555-561

Blakefield MK and Harris DO (1994) Delay of cell differentiation in Anabaena aequalis caused by UV-B radiation and the role of photoreactivation and excision repair. Photochem Photobiol 59: 204-208

Böhm GA, Pfleiderer W, Böger P and Scherer S (1995) Structure of a novel oligosaccharide-mycosporine-amino-acid ultraviolet A/B sunscreen pigment from the terrestrial cyanobacterium *Nostoc commune*. J Biol Chem 270: 8536-8539

Booth CR and Morrow JH (1997) The penetration of UV into natural waters. Photochem Photobiol 65: 254-257

Bothwell ML, Sherbot DMJ and Pollock CM (1994) Ecosystem response to solar ultraviolet-B radiation: influence of trophic level interactions. Science 265: 97-100

Bothwell ML, Sherbot D, Roberge AC and Daley RJ (1993) Influence of natural ultraviolet radiation on lotic periphytic diatom community growth, biomass accrual, and species composition: short-term versus long-term effects. J Phycol 29: 24-35

Braune W and Döhler G (1996) ^{15}N-uptake, influenced by UV-B radiation, and pattern of amino acid pools during akinete germination in *Anabaena variabilis* (Cyanobacteria). J Basic Microbiol 36: 2199-227

Brenowitz S and Castenholz RW (1997) Long-term effects of UV and visible irradiance on natural populations of a scytonemin-containing cyanobacterium (*Calothrix* sp.) FEMS Microbiol Ecol 24: 343-352

Britton G (1995) Structure and properties of carotenoids in relation to function. FASEB J 9: 1551-1558

Buckley CE and Houghton JA (1976) A study of the effects of near UV radiation on the pigmentation of the blue-green alga *Gloeocapsa alpicola.* Arch Microbiol 107: 93-97

Büdel B, Karsten U and Garcia-Pichel F (1997) Ultraviolet-absorbing scytonemin and mycosporine-like amino acid derivatives in exposed, rock-inhabiting cyanobacterial lichens. Oecologia 112: 165-172

Carr NG and Wyman M (1986) Cyanobacteria: their biology in relation to oceanic picoplankton. In: Platt T and Li WKW (eds) Photosynthetic picoplankton, pp 159-204. Can Bull Fish Aquat Sci 214

Carreto J I, DeMarco SG and Lutz VA (1989) UV-absorbing pigments in the dinoflagellates *Alexandrium excavatum* and *Prorocentrum micans:* Effects of light intensity. In: Okaichi T, Anderson DM and Nemoto T (eds) Red Tides: Biology, Environmental Science and Toxicology, pp 333-336. New York, Elsevier

Castenholz RW (1968) The behavior of *Oscillatoria terebriformis* in hot springs. J Phycol 4: 132-139

Castenholz RW (1972) Low temperature acclimation and survival in thermophilic *Oscillatoria terebriformis.* In: Desikachary TV (ed) Taxonomy and Biology of Blue-green Algae, pp 406-418. University of Madras, India

Castenholz RW (1994) Microbial mat research: the recent past and new perspectives. In: Stal LJ and Caumette P (eds) Microbial Mats, pp 3-18, NATO ASI Series, Vol. 35, Springer-Verlag, Berlin

Castenholz RW, Jørgensen BB, D'Amelio E and Bauld J (1991) Photosynthetic and behavioral versatility of the cyanobacterium *Oscillatoria boryana* in a sulfide-rich microbial mat. FEMS Microbiol Ecol 86: 43-58

Chitnis PR and Nelson N (1991) Molecular cloning of the genes encoding two chaperone proteins of the cyanobacterium *Synechocystis* sp PCC 6803. J Biol Chem 266: 58-65

Clayton RK (1970) Light and Living Matter. McGraw-Hill, New York

Close TJ and Lammers PJ (1993) An osmotic stress protein of cyanobacteria is immunologically related to plant dehydrins. Plant Physiol 101: 773-779

Collier JL and Grossman AR (1992) Chlorosis induced by nutrient deprivation in *Synechococcus* sp. strain PCC 7942: not all bleaching is the same. J Bacteriol 174: 4718-4726

Craig E, Gambill B and Nelson R (1993) Heat shock proteins: molecular chaperones of protein biogenesis. Microbiol Rev 57: 402-414

Cullen JJ and Neale PJ (1994) Ultraviolet radiation, ozone depletion and marine photosynthesis. Photosynth Res 39: 303-320

Cullen JJ, Neale PJ and Lesser MP (1992) Biological weighting function for the inhibition of phytoplankton photosynthesis by ultraviolet radiation. Science 258: 646-650

Des Marais DJ (1995) The biogeochemistry of hypersaline microbial mats. Adv Microbiol Ecol 14: 251-274

Diffey BL, Green AT, Loftus MJ,Johnson GJ and Lee PS (1995) A portable instrument for measuring ground reflectance in the ultraviolet. Photochem Photobiol 61: 68-70

Dodds WK and Castenholz RW (1988) The biological effects of nitrate fertilization and water replacement in an oligotrophic cold water pond. Hydrobiologia 162: 141-146

Döhler G, Bierman I and Zink J (1986) Impact of UV-B radiation on photosynthetic assimilation of ^{14}C bicarbonate and inorganic ^{15}N-compounds by cyanobacteria. Zeit Naturforsch 41: 426-432

Donkor VA, Amewowor DHAK and Häder D-P (1993) Effects of tropical solar radiation on the velocity and photophobic behavior of filamentous gliding cyanobacteria. Acta Protozoologica 32: 67-72

Dunlap WC and Chalker BE (1986) Identification and quantification of near UV absorbing compounds (S-320) in a hermatipic scleractinian. Coral Reefs 5: 155-159

Dunlap WC and Shick JM (1998) Ultraviolet radiation-absorbing mycosporine-like amino acids in coral reef organisms: a biochemical and environmental perspective. J Phycol 34: 418-430

Dunlap WC and Yamamoto Y (1995) Small-molecule antioxidants in marine organisms: antioxidant activity of mycosporine-glycine. Comp Biochem Physiol B 112: 105-114

Eguchi M, Oketa T, Miyamoto N, Maeda H and Kawai A (1996) Occurrence of viable photoautotrophic picoplankton in the aphotic zone of Lake Biwa, Japan. J Plankton Res 18: 539-550

Ehling-Schulz M, Bilger W and Scherer S (1997) UV-B-induced synthesis of photoprotective pigments and extracellular polysaccharides in the terrestrial cyanobacterium *Nostoc commune*. J Bacteriol 179: 1940-1945

Eker EPM, Kooiman P, Hessels JKC and Yasui A(1990) DNA photoreactivating enzyme from the cyanobacterium *Anacystis nidulans*. J Biol Chem 265: 8009-8015

Favre-Bomvin J, Arpin N and Brevard C (1976) Structure de la mycosporine (P310). Can J Chem 54: 1105-1113

Favre-Bomvin J, Bernillon J, Salin N and Arpin N (1987) Biosynthesis of mycosporines mycosporine glutaminol in *Trichotecium roseum*. Phytochemistry 26: 2509-2514

Frederick JE, Snell HE and Haywood EK (1989) Solar ultraviolet radiation at the earth's surface. Photochem. Photobiol. 50: 443-450

Fridovich I (1975) Superoxide dismutases. Ann Rev Biochem 44: 147-159

Friedmann EI and Ocampo R (1976) Endolithic blue-green algae in the dry valleys: primary producers in the Anctarctic desert ecosystem. Science 193: 1247-1249

Garcia-Pichel F (1992) Ultraviolet-absorbing substances in cyanobacteria and their role as sunscreens. PhD Thesis, Department of Biology, University of Oregon, Eugene

Garcia-Pichel F (1994) A model for internal shelf-shading in planktonic microorganism and its implications for the usefulness of sunscreens. Limnol Oceanogr 39: 1704-1717

Garcia-Pichel F (1996) The absorption of ultraviolet radiation by microalgae: simple optics and photobiological implications. Scientia Marina 60: 73-79

Garcia-Pichel F (1998) Solar ultraviolet and the evolutionary history of cyanobacteria. Origins of Life Evol Biosphere 28: 321-347

Garcia-Pichel F and Bebout BM (1996) The penetration of ultraviolet radiation into shallow water sediments: high exposure for photosynthetic communities. Mar Ecol Progress Ser 131: 257-262

Garcia-Pichel F and Belnap J (1996) Microenvironments and microscale productivity of cyanobacterial desert crusts. J Phycol 32: 774-782

Garcia-Pichel F and Castenholz RW (1991) Characterization and biological implications of scytonemin, a cyanobacterial sheath pigment. J Phycol 27: 395-409

Garcia-Pichel F and Castenholz RW (1993) Occurrence of UV-absorbing, mycosporine-like compounds among cyanobacterial isolates and an estimate of their screening capacity. Appl Environ Microbiol 59: 163-169

Garcia-Pichel F and Castenholz RW (1994) On the significance of solar ultraviolet radiation for the ecology of microbial mats. In: Stal LJ and Caumette P (eds) Microbial Mats. Structure, Development and Environmental Significance, pp 77-84, NATO ASI Series, Vol. 35, Springer-Verlag, Berlin

Garcia-Pichel F, Sherry ND and Castenholz RW (1992) Evidence for a UV sunscreen role of the extracellular pigment scytonemin in the terrestrial cyanobacterium *Chlorogloeopsis* spp. Photochem Photobiol 56: 17-23

Garcia-Pichel F, Wingard CE and Castenholz RW (1993) Evidence regarding the UV-sunscreen role of a mycosporine-like compound in the cyanobacterium *Gloeocapsa* sp. Appl Environ Microbiol 59: 170-176

Gautier C, He G and Yang S (1994) Role of clouds and ozone on spectral ultraviolet B radiation and biologically active UV dose over Antarctica. In: Weiler CS and Penhale PA (eds) Ultraviolet Radiation in Antarctica: Measurements and Biological Effects. Vol 62, pp 83-91, American Geophysical Union. Washington, DC

Golden SS (1994) Light-responsive gene expression and the biochemistry of the photosystem II reaction center. In: Bryant DA (ed) The Molecular Biology of Cyanobacteria, pp 693-714. Kluwer Academic Publishers, Dordrecht

Goodwin TW (1980) The Biochemistry of the Carotenoids, Vol. I Plants. (2nd ed) Chapman and Hall, London

Grether T, Maier J, Karsten U, Garcia-Pichel F and Ninnemann H (1996) UV-light induction of biosynthesis of pteridine glycosides of cyanobacteria from microbial mats (Abstract). UV/Blue Light: Perception and Responses in Plants and Microorganism, Marburg (Germany)

Häder D-P (1984) Effects of UV-B on motility and photoorientation in the cyanobacterium, *Phormidium uncinatum.* Arch Microbiol 140: 34-39

Häder D-P and Tevini M(1987) General Photobiology. Pergamon Press, Oxford

Häder D-P, Watanabe M and Furuya M (1986) Inhibition of motility in the cyanobacterium, *Phormidium uncinatum,* by solar and monochromatic UV irradiation. Plant Cell Physiol 27: 887-894

Häder D-P, Worrest RC, Kumar HD and Smith RC(1995) Effects of increased solar ultraviolet radiation on aquatic ecosystems. Ambio 24: 174-180

Hagen C, Braune W and Bjørn LO (1994) Functional aspects of secondary carotenoids in *Haematococcus lacustris* (Volvocales). III. Action as a "sunshade". J Phycol 30: 241-248.

Harm W (1980) Biological Effects of Ultraviolet Radiation. Cambridge University Press Cambridge

Helbling EW, Villafañe V and Holm-Hansen O (1994) Effects of ultraviolet radiation on Antarctic marine phytoplankton photosynthesis with particular attention to the influence of mixing. In: Weiler CS and Penhale PA (eds) Ultraviolet Radiation in Antarctica: Measurements and Biological Effects, Vol. 62, pp 207-227. American Geophysical Union, Washington, DC

Hirosawa T and Miyachi S (1983) Effect of long-wavelength ultraviolet (UV-A) radiation on the growth of *Anacystis nidulans.* Plant Sci Lett 28: 291-298

Holm-Hansen O, Lubin D and Heibling EW (1993) Ultraviolet radiation and its effects on organisms in aquatic environments. In: Young AR, Björn LO, Moan J, and Nultsch W (eds) Environmental UV Photobiology, pp 379-425. Plenum Press, New York

Ito S and Hirata Y(1977) Isolation and identification of a mycosporine from the zooanthid Palythoa tuberculosa. Tetrahedron Lett 28: 2429-2430

Jaag O (1945) Untersuchungen über die Vegetation und Biologie der Algen des nackten Gesteins in den Alpen, im Jura und im schweizerischen Mittelland. Beitr. Kryptogamenflora Schweiz (Mat Flore Cryptog Suisse) 9(3)

Jagger J (1967) Introduction to Research in Ultraviolet Photobiology. Prentice Hall, Englewood Cliffs, New Jersey

Jagger J (1985) Solar-UV Actions on Living Cells. Praeger, New York

Javor BJ and Castenholz RW (1981) Laminated microbial mats, Laguna Guerrero Negro, Mexico. Geomicrobiol J 2: 237-273

Javor BJ and Castenholz RW (1984) Productivity studies of microbial mats, Laguna Guerrero Negro, Mexico. In: Cohen Y, Halvorson H and Castenholz RW (eds) Microbial Mats: Stromatolites, pp 149-170, Alan R. Liss, Inc, New York

Jeffrey WH and Mitchell DL (1997) Mechanisms of UV-induced DNA damage and response in marine microorganisms. Photochem Photobiol 65: 260-263

Jeffrey WH, Pledger RJ, Aas P, Hager S, Coffin RB, Von Haven R and Mitchell DL (1996) Diel and depth profiles of DNA photodamage in bacterioplankton exposed to ambient solar ultraviolet radiation. Mar Ecol Prog Ser 137: 283-291

Jørgensen BB and Des Marais DJ (1988) Optical properties of benthic photosynthetic communities: fiber-optic studies of cyanobacterial mats. Limnol Oceanogr 33: 99-113

Karentz D, McEuen FS, Land MC and Dunlap WC (1991) Survey of mycosporine-like amino acid compounds in Antarctic marine organisms: potential protection from ultraviolet exposure. Mar Biol 108: 157-166

Karsten U and Garcia-Pichel F (1996) Carotenoids and mycosporine-like amino acids in members of the genus *Microcoleus* (Cyanobacteria): a chemosystematic study. System Appl Microbiol 19: 285-294

Kasting JF, Holland HD and Kump LR (1992) Atmospheric evolution: the rise of oxygen. In: Schopf JW and Klein C (eds) The Proterozoic Biosphere, a Multidisciplinary Study, pp 159-168. Cambridge Univ Press, Cambridge

Kirk JT (1994) Optics of UV-B radiation in natural waters. Arch Hydrobiol Beih Ergebn Limnol 43: 1-16

Kirk JTO (1983) Light and Photosynthesis in Aquatic Ecosystems. Cambridge University Press, Cambridge

Koller LR (1965) Ultraviolet Radiation. J Wiley and Sons, New York

Kollias N, Sayre RM, Zeise L and Chedekel MR (1991) Photoprotection by melanin. J Photochem Photobiol B: Biol 9: 135-160

Koyama Y and Hashimoto H (1993) Spectroscopic studies of carotenoids in photosynthetic systems. In: Young A and Britton G (eds) Carotenoids in Photosynthesis, pp 327-408. Chapman and Hall, London

Krinsky NI (1979) Carotenoid protection against oxidation. Pure and Appl Chem 51: 649-660

Kruschel C and Castenholz RW (1998) The effect of solar UV and visible irradiance on the vertical movements of cyanobacteria in microbial mats of hypersaline waters. FEMS Microbiol Ecol 27: 53-72

Kulandaivelu G, Gheeta V and Periyanan S (1989) Inhibition of energy transfer reactions in cyanobacteria by different ultraviolet radiation. In: Singhal GS (ed) Photosynthesis, Molecular Biology and Bioenergetics, pp 305-313, Springer-Verlag, New York

Kühl M and Jørgensen BB (1994) The light field of microbenthic communities: radiance distribution and microscale optics of sandy coastal sediments. Limnol Oceanogr 39: 1368-1398

Kumar A, Tyagi M, Srinivas G, Singh N, Kumar HD, Sinha RP and Häder DP (1996) UVB shielding role of $FeCl_3$ and certain cyanobacterial pigments. Photochem Photobiol 64: 321-325

Kumar D and Kumar HD (1990) Nature of UV resistance in the cyanobacterium *Nostoc linckia.* Curr Science 59: 412-414

Laurion I, Vincent WF and Lean DRS (1997) Underwater ultraviolet radiation: development of spectral models for northern high latitude lakes. Photochem Photobiol 65: 107-114

Lemmerman E (1910) Algen I. Kryptogamenflora der Mark Brandenburg. Leipzig: 1-259

Levine E and Thiel T (1987) UV-inducible DNA repair in the cyanobacteria *Anabaena* spp. J Bacteriol 169: 3988-3993

Liu SC, McKeen SA and Madronich S (1991) Effects of anthropogenic aerosols on biologically active ultraviolet radiation. Geophys Res Lett 99: 2265-2268

Loa, K and Glazer, AN (1996) Ultraviolet-B photodestruction of a light-harvesting complex. Proc Natl Acad Sci USA 93: 5258-5263

Matsunaga T, Burgess JG, Yamada N, Komatsu K, Yoshida S and Wachi Y (1993) An ultraviolet (UV-A) compound biopterin glycoside from the marine planktonik cyanobacterium *Oscillatoria* sp. Appl Microbiol and Biotechnol 39: 250-253

Mazor G, Kidron GJ, Vonshak A and Abeliovich A (1996) The role of cyanobacterial exopolysaccharides in structuring desert microbial crusts. FEMS Microbiol Ecol 21: 121-130

Miller SR, Wingard CE and Castenholz RW (1999) Effects of visible light and UV radiation on photosynthesis in a population of a hot spring cyanobacterium, a *Synechococcus* sp., subjected to high temperature stress. Appl Environ Microbiol 64: 3893-3899

Milot-Roy V and Vincent WF (1994) UV radiation effects on photosynthesis: the importance of near-surface thermoclines in a subarctic lake. Arch Hydrobiol Beih Ergbn Limnol 43: 171-184

Mitchell DL and Karentz D (1993) The induction and repair of DNA photodamage in the environment. In: Young, AR, Björn LO, Moan J and Nultsch W (eds) Environmental UV PhotoBiology, pp 345-377. Plenum Press, New York

Morowitz HJ (1950) Absorption effects in volume irradiation of microorganism. Science 111: 229-230

Morris SAC and Subden RE (1974) Effects of ultraviolet radiation on carotenoid containing and albino strains of *Neurospora crassa.* Mutation Res 22: 105-109

Musil CF and Bhagwandin N (1992) The SUN program for computation of solar ultraviolet spectral irradiances: solar exposure limits in South Africa. South African J Sci 88: 406-410

Nägeli C (1849) Gattungen einzelliger Algen, physiologisch und systematisch bearbeitet. Neue Denkschrift Allg Schweiz Natur Ges 10: 1-138

Nägeli C and Schwenderer S (1877) Das Mikroskop. Willhelm Engelmann Verlag, Leipzig

Nakamura H, Kobayashi J and Hirata Y (1981) Isolation and structure of a 330 nm UV-absorbing substance, Asterina-330 from the starfish *Asterina pectinifera.* Chem Lett: 1413-1414

Neale PJ, Lesser MP and Cullen JJ (1994) Effects of ultraviolet radiation on the photosynthesis of phytoplankton in the vicinity of McMurdo Station, Antarctica. In: Weiler, CS and Penhale PA (eds) Ultraviolet Radiation in Antarctica: Measurements and Biological Effects. Vol 62, pp 125-142, American Geophysical Union, Washington, DC

Newton JW, Tyler DD and Slodki MD (1979) Affect of ultraviolet-B radiation on blue-green algae (Cyanobacteria), possible biological indicators of stratospheric ozone depletion. Appl Environ Microbiol 37: 1137-1141

Nicholson JM, Stolz JF and Pierson BK (1987) Structure of a microbial mat at Great Sippewissett Marsh, Cape Cod, Massachusetts. FEMS Microbiol Ecol 45: 343-364

Nicholson P, Osborn RW and Howe CJ (1987) Induction of protein synthesis in response to ultraviolet light, nalidixic acid and heat shock in the cyanobacterium *Phormidium laminosum.* FEBS Lett 221: 110-114

Nicholson P, Varley JPA and Howe CJ (1991) A comparison of stress responses in the cyanobacterium *Phormidium laminosum.* FEMS Microbiol Lett 78: 109-114

Nienow, JA and Friedmann, EI (1993) Terrestrial lithophytic (rock) communities. In: Friedmann, EI (ed) Antarctic Microbiology, pp 343-412, Wiley-Liss, New York

O'Brien PA and Houghton JA (1982) Photoreactivation and excision repair on UV induced pyrimidine dimers in the unicellular cyanobacterium *Gloeocapsa alpicola* (*Synechocystis* PCC 6803). Photochem Photobiol 35: 359-364

Ohki K, Watanabe M and Fujita Y (1982) Action of near UV and blue light on the photocontrol of phycobilin formation; a complementary chromatic adaptation. Plant Cell Physiol 23: 651-656

Oren A (1997) Mycosporine-like amino acids as osmotic solutes in a community of halophilic cyanobacteria. Geomicrobiol J 14: 231-240

Owttrim GW and Coleman JR (1989) Regulation of expression and nucleotide sequence of the *Anabaena variabilis recA* gene. J Bacteriol 171: 5713-5719

Paerl HW, Tucker J and Bland PT (1983) Carotenoid enhancement and its role in maintaining blue-green algal (*Microcystis aeruginosa*) surface blooms. Limnol Ocean 28: 847-857

Pentecost A (1984) Effects of sedimentation and light intensity on mat-forming Oscillatoriaceae with particular reference to *Microcoleus lyngbyaceus* Gomont. J Gen Microbiol 130: 983-990

Pentecost A (1993) Field relationships between scytonemin density, growth and irradiance in cyanobacteria occurring in low illumination regimes. Microb Ecol 26: 101-110

Piazena H and Häder D-P (1994) Penetration of solar UV radiation in coastal lagoons of the Southern Baltic sea and its effect on phytoplankton communities. Photochem Photobiol 60: 463-469

Pierson BK, Oesterle A and Murphy GL (1987) Pigments, light penetration and photosynthetic activity in the multi-layered microbial mats of Great Sippewissett Salt Marsh, Massachusetts. FEMS Microbiol Ecol 45: 365-376

Postius C, Ernst A, Kenter U and Böger P (1996) Persistence and genetic diversity among strains of phycoerythrin-rich cyanobacteria from the picoplankton of Lake Constance. J Plankton Res 18: 1159-1166

Potts M (1994) Desiccation tolerance of prokaryotes. Microbiol Rev 58(4): 755-805

Prezelin, B. B., N. P. Boucher and R. C. Smith (1994). Marine primary production under the influence of the Antarctic ozone hole: Icecolors'90. In: Weiler CS and Penhale PA (eds) Ultraviolet Radiation in Antarctica: Measurement and Biological Effects. Vol 62, pp 159-186. American Geophysical Union, Washington, DC

Proteau PJ, Gerwick WH, Garcia-Pichel F and Castenholz RW(1993) The structure of scytonemin, an ultraviolet sunscreen pigment from the sheaths of cyanobacteria. Experientia 49: 825-829

Ramsing NB and Prufert-Bebout L (1994) Motility of *Microcoleus chthonoplastes* subjected to different light intensities quantified by digital image analysis. In: Stal LJ and Caumette P (eds) Microbial Mats. Structure, Development and Environmental Significance, pp 183-191, NATO ASI Series, Vol 35, Springer-Verlag, Berlin

Renger G, Völker M, Eckert HJ, Fromme R, Hohm-Veit S and Gräber P (1989) On the mechanism of photosystem II deterioration by UV-B irradiation. Photochem Photobiol 49: 97-105

Revsbech NP, Jørgensen BB, Blackburn TH and Cohen Y (1983) Microelectrode studies of the photosynthesis and O_2, H_2S and pH profiles of a microbial mat. Limnol Oceanogr 28: 1062-1074

Richardson LL and Castenholz RW (1987) Diel vertical movements of the cyanobacterium *Oscillatoria terebriformis* in a sulfide-rich hot spring mat. Appl Environ Microbiol 53: 2142-2150

Robinson, N. (1966). Solar Radiation. Elsevier, Amsterdam

Roos JC and Vincent WF (1998) Temperature dependence of UV radiation effects on Antarctic cyanobacteria. J.

Roy, CR, Gies HP, Tomlinson DW and Lugg DL (1994) Effects of ozone depletion on the ultraviolet radiation environment at the Australian stations in Antarctica. In: Weiler, CS and Penhale PA (eds) Ultraviolet Radiation in Antarctica: Measurements and

Biological Effects Vol 62, pp 1-15. American Geophysical Union, Washington, DC

Sanders B (1993) Stress proteins in aquatic organisms: an environmental perspective. Crit Rev Toxicol 23: 49-75

Sandmann G, Kuhn S, and Böger P (1998) Evaluation of structurally different carotenoids in *Escherichia coli* transformants as protectants against UV-B radiation. Appl Environ Microbiol 64: 1972-1974.

Sass L, Spetea C, Máté Z, Nagy F and Vass I (1997) Repair of UV-B induced damage of photosystem II via *de novo* synthesis of the D1 and D2 reaction centre subunits in *Synechocystis* sp. PCC 6803. Photosyn Res 54: 55-62

Satoh K (1998) Creation of photo-tolerant mutants of a cyanobacterium, *Synechocystis* sp. PCC 6803, by *in vitro* random mutagenesis of the *psbA* gene. In Satoh K and Murata N (eds) Stress Responses of Photosynthetic Organisms, pp 3-14. Elsevier Science BV, Amsterdam

Satoh K and Murata N (eds) (1998) Stress Responses of Photosynthetic Organisms. Elsevier Science BV, Amsterdam

Scherer S, Chen TW and Böger P (1988) A new UV-A/B protecting pigment in the terrestrial cyanobacterium *Nostoc commune*. Plant Physiol 88: 1055-1057

Scully NM, McQueen DJ, Lean DRS and Cooper WJ (1996) Hydrogen peroxide formation: the interaction of ultraviolet radiation and dissolved organic carbon in lake waters along a 43-75°N gradient. Limnol Oceanogr 41: 540-548

Shibata H, Katsuya B and Ochiai H (1991) Near-UV irradiation induces shock proteins in *Anacystis nidulans* R-2; possible role of active oxygen. Plant Cell Physiol 32: 771-776

Shibata H, Noda T, Ogura Y, Suginaga K, Matsui Y, Ozoe Y, Sawa Y and Kona Y (1996) A soluble form of pro-oxidant lumazine isolated from cyanobacterial cells generates superoxide anion under near-UV irradiation. Biochim Biophys Acta 1274: 129-134

Shibata K (1969) Pigments and a UV-absorbing substance in corals and a blue-green alga living in the Great Barrier Reef. Plant Cell Physiol 10: 325-335

Singh PK (1975) Photoreactivation of UV-irradiated blue-green algae and algal virus LPP-1. Arch Microbiol 103: 297-302

Sinha RP, Lebert M, Kumar A, Kumar HD and Häder D-P (1995) Spectroscopic and biochemical analyses of UV effects on phycobiliproteins of *Anabaena* sp. and *Nostoc carmium.* Botanica Acta 108: 87-92

Sivalingam PM, Ikawa T, Yokohama Y and Nisizawa K (1974) Distribution of a 334 UV-absorbing substance in algae with special regard of its possible physiological roles. Bot Mar 19: 9-29

Smith RC and Baker KS (1981) Optical properties of the clearest natural seawaters. Appl Optics 20: 177-184

Takahashi A, Takeda K and Ohnishi T (1991) Light-induced anthocyanin reduces the extent of damage to DNA in UV-irradiated *Centaurea cyanus* cells in culture. Plant Cell Physiol 32: 541-547

Takano S, Nakahashi A, Uemura D and Hirata Y (1979) Isolation and structure of a 334 nm absorbing substance, porphyra-334, from the red alga *Porphyra tenera* Kjellman. Chem Lett: 419-420

Tang EPY, Tremblay R and Vincent WF (1997) Cyanobacterial dominance of polar freshwater ecosystems: are high-latitude mat formers adapted to low temperature? J Phycol 33: 171-181

Tsujino I, Yabe K and Sekikawa I (1980) Isolation and structure of a new amino acid, shinorine, from the red alga *Chondrus yendoi.* Bot Mar 23: 65-68

Tyagi R, Srinivas G, Vyas D, Kumar A and Kumar HD (1992) Differential effect of ultraviolet-B radiation on certain metabolic processes in a chromatically adapting *Nostoc.* Photochem Photobiol 55: 401-407

Van Baalen C (1969) The effects of ultraviolet radiation on a coccoid blue-green alga: survival, photosynthesis and photoreactivation. Plant Physiol 43: 1689-1695

Van Liere L and Walsby AE (1982) Interactions of cyanobacteria with light. In: Carr NG and Whitton BA (eds), The Biology of Cyanobacteria, pp 9-45. Blackwell Scientific Publications, Oxford

Vincent WF (1988) Microbial Ecosystems of Antarctica. Cambridge University Press, Cambridge

Vincent WF and Quesada A (1994) Ultraviolet radiation effects on cyanobacteria: implications for Antarctic microbial ecosystems. In: Weiler CS and Penhale PA (eds) Ultraviolet Radiation in Antarctica: Measurements and Biological Effects, pp 111-124. Antarctic Res Series Vol 62, Am Geophys Union, Washington DC

Vincent WF and Roy S (1993) Solar ultraviolet-B radiation and aquatic primary production: damage, protection and recovery. Environ Rev 1: 1-12

Vincent WF, Castenholz RW, Downes MT and Howard-Williams C (1993) Antarctic cyanobacteria: light, nutrients, and photosynthesis in the microbial mat environment. J Phycol 29: 745-755

Vinebrooke RD and Leavitt PR (1996) Effects of ultraviolet radiation on periphyton in an alpine lake. Limnol Oceanogr 41: 1035-1040

Vosjan JH and Pauptit E (1992) Penetration of photosynthetically available light (PAR), UV-a and UV-b in Admiralty bay, King George Is., Antarctica. Circumpolar J 1-2: 50-58

Wachi Y, Burgess JG, Iwamoto K, Yamada N, Nakamura N and Matsunaga T (1995) Effect of ultraviolet-A (UV-A) on growth, photosynthetic activity and production of biopterin glucoside by the marine UV-A resistant cyanobacterium *Oscillatoria* sp. Biochim Biophys Acta 1244: 165-168

Walsby AE (1994) Gas vesicles. Microbiol Rev 58: 94-144

Weisse T (1993) Dynamics of autotrophic picoplankton in marine and freshwater ecosystems. Adv Microbial Ecol 13: 327-370

Whale GF and Walsby AE (1984) Motility of the cyanobacterium *Microcoleus chthonoplastes* in mud. Brit Phycol J 19: 117-123

Whitelam GC and Codd GA (1986) Damaging effects of light on microorganisms. In: Herbert RA and Codd GA (eds) Microbes in Extreme Environments, pp 129-169. Academic Press, London

Whitton, BA and M Potts (1982) Marine littoral. In: Carr NG and Whitton BA (eds) The Biology of Cyanobacteria pp 512-542. Blackwell Scientific Publishers, Oxford

Wickstrom CE and Castenholz RW (1978) Association of *Pleurocapsa* and *Calothrix* (Cyanophyta) in a thermal stream. J Phycol 14: 84-88

Wingard CE, Schiller JR and Castenholz RW (1997) Evidence regarding the possible role of c-phycoerythrin in ultraviolet-B tolerance in a thermophilic cyanobacterium. Photochem Photobiol 65: 833-842

Yopp JH, Albright G and Miller DM (1979) Effects of antibiotics and ultravioletradiation on the halophilicblue-green alga, *Aphanothece halophytica.* Botanica Marina 22: 167-272

Young AR, Björn LO, Moan J and Nultsch W (eds) (1993) Environmental UV Photobiology. Plenum Press, New York

Chapter 22

Cyanotoxins

Crawford S Dow and Uthaya K Swoboda
Biological Sciences, University of Warwick, Coventry, UK

B. A. Whitton and M. Potts (eds), The Ecology of Cyanobacteria, 613-632.
© 2000 *Kluwer Academic Publishers. Printed in the Netherlands.*

Summary

The toxicity of some cyanobacteria to mammals has been known for over a century, but only recently has the diversity of toxic organisms and the toxins produced by them become clear. Most studies have been done on freshwater and marine bloom-forming species, but benthic and floating forms are sometimes also toxic. The toxins include neurotoxins (e.g. anatoxin-a and saxitoxin), hepatotoxins (over 60 variants of microcystins) and cytotoxins (scytophycin, cyanobacterin, hapalindole, acutiphycin, lyngbyatoxin). Mouse bioassays of toxicity are still conducted in many countries, but increasingly other methods are being adopted; these include simpler organisms or tissues for the bioassays, immunological procedures, various chromatographic techniques and mass spectrometry. Although the toxins appear in general to be more toxic to mammals, including humans, than aquatic organisms, there is experimental evidence for their toxicity to various animals, including the young stages of some fish species. In addition, the toxicity of many lake populations of cyanobacteria is known to have an effect on some potential zooplankton grazers and the toxins may thus have an important effect on the food web in a lake.

I. Introduction

Cyanobacteria are a highly successful group of oxygenic photoautotrophs with a long evolutionary history. They occupy a wide range of ecological niches, show considerable diversity of cell expression, yet have simple metabolic requirements. It is against these features that this chapter considers what is increasingly seen as one of the economically most important features of cyanobacteria, the toxicity of many strains to other organisms. For instance, the dense freshwater blooms discussed in Chapter 6 not only affect the water's taste, odour and appearance, but more significantly, they are frequently highly toxic to wildlife, domestic livestock and humans, causing a range of allergic and gastroenteric responses. In the case of humans, the primary cyanobacterial toxicoses include acute liver diseases (hepatotoxicoses), peracute neurotoxicoses and gastrointestinal disturbances (Kuiper-Goodman et al., 1999).

In spite of their aquatic origin, most cyanotoxins appear to be less toxic to aquatic biota than to humans and other terrestrial mammals (Sivonen and Jones, 1999). Microcystin, for instance, showed low toxicity to mussels and the crayfish *Procambarus clarkii* (Vasconcelos, 1998). Nevertheless, there is evidence for the influence of cyanotoxins on a range of aquatic organisms and some examples are described in this chapter. The ecological and economic impacts of toxins seem likely to increase as blooms increase in many countries as a result of nutrient eutrophication (Carmichael, 1997; Codd and Beattie, 1991; Hunter, 1995b; Chorus and Bartram, 1999).

II. Poisoning Incidents

Toxic cyanobacterial blooms have been reported for over a century, one of the first being a *Nodularia* bloom described by Francis in *Nature* in 1878 (Carmichael, 1994a; Lukkainen et al., 1994). Cyanotoxins cause acute and possibly chronic health problems in humans and fatal poisonings in other animals, fish and birds (Carmichael, 1992, 1994a, b; Hunter, 1995a, b; Hunter and Roberts, 1991). However, diagnosis of cyanotoxicosis is difficult, primarily because many blooms are not hazardous even when dominated by species known to produce toxins at other times, so diagnosis of toxicosis following ingestion of cyanobacteria requires confirmation. This remains difficult because of a lack of concise information on appropriate diagnostic procedures (Beasley et al., 1989).

In recent years, the presence and frequency of cyanobacterial blooms in amenity water bodies has greatly increased the interest shown by various groups, such as the water management organizations and environmental protection agencies (Codd and Beattie, 1991; Keevil, 1991). This interest can be traced not only to illness and death of domestic animals and wildlife, as well as to human illnesses which have been attributed to contact and/or ingestion of toxic cyanobacterial blooms, but to the potential of the microcystins as potent tumour promoters (Falconer, 1991). There is still insufficient information to evaluate the extent of the problem caused by the latter, since it is not yet considered in most water quality guidelines, such as the World Health Organization's recommended limit of 1 μg microcystin LR L^{-1} drinking water (WHO, 1997).

There are numerous reported instances of cyanobacterial toxicoses. Until recently, however, there have been no definitive reports of human deaths due to cyanotoxins. However, world attention was drawn to this problem in 1996, when two Brazilian haemodialysis units, which had obtained water from a reservoir contaminated with cyanobacteria, caused toxic hepatitis in 126 patients, 60 of whom died (Pouria et al., 1998). Analysis of the water source and the dialysis fluid revealed the presence of cyanotoxins (microcystin LR and related toxins), whilst immunoassays confirmed that acute or lethal concentrations of cyanotoxin were present in the patients' livers.

During August and September 1989, 20 sheep and 15 dogs died after ingesting a *Microcystis* scum from Rutland Water, England. In addition, wind-surfers, who had come into contact with this scum, complained of skin rashes, mouth blisters and severe thirst (Codd and Beattie, 1991; Hunter and Roberts, 1991; Pearson et al., 1990). Similarly, soldiers, who had been on a canoeing exercise on Rudyard Lake, England, and had come into contact with a *M. aeruginosa* scum, subsequently showed signs of cyanotoxicosis - sore throat, headaches, blistered mouth, diarrhoea and vomiting. Two were admitted to hospital with fever and atypical pneumonia (Turner et al., 1990).

In 1990 at Loch Insh, Scotland, several dogs died after ingesting a neurotoxic *Oscillatoria* scum (Codd and Beattie, 1991; Gunn et al., 1992). Two of these, Scottish terriers returning from a walk on the loch shore, died within ten minutes of the onset of clinical symptoms. A few days later an adult springer spaniel died after drinking from the lake. The dog showed cyanosis, rigors, limb twitching and hypersalivation and died within thirty minutes of the onset of illness. The following summer a collie bitch died within fifteen minutes of having swum in the loch. Numerous incidents of animal deaths, particularly cattle, sheep and dogs, have been recorded in many other parts of the world - Australia, mainland Europe, South Africa and North America (Carmichael, 1992; Chengappa et al, 1989; Lukac and Aegerter, 1993; Luukainen et al., 1993, 1994). However, reports on the impact of cyanotoxins upon fish, amphibian or insect populations are less abundant and not always easy to evaluate critically.

Mammals in temperate regions are usually poisoned in the summer and early autumn, when four factors converge:

i) Calm conditions or light wind;
ii) High water temperature (15 - 30 °C);
iii) pH between 6.0 to 9.0;
iv) abundant nutrients (phosphate and often also nitrate).

Under these conditions cyanobacteria outgrow eukaryotic microalgae. The floating masses formed by a number of species (Chapter 6) can be concentrated by currents and light winds to form a thick scum on the shores of lakes and reservoirs, making the ingestion of a fatal dose a ready possibility (Carmichael, 1994a, b). Most poisoning incidents are, therefore, associated with bloom-forming planktonic species. However, the Loch Insh incidents were caused by a neurotoxic, benthic *Oscillatoria*. Flecks of benthic material detached from the sediment were driven on-shore by wind and wave action. Unlike the plankton forms, toxic hazards from the benthos are only likely to become apparent when they accumulate on the shore.

The lethal dose of water contaminated with cyanobacteria is dependent on various factors such as the cellular concentration and type of toxin(s), the biomass concentration, mode of exposure and the susceptibility of the victim, notably, age, sex, weight and species (Carmichael, 1994a). Unfortunately, it would appear that the foul odour and taste of the contaminated water does little to deter susceptible animals (Carmichael, 1994a).

Relatively little is known of the influence of cyanotoxins on the natural grazers of cyanobacteria and other aquatic biota. There is evidence to show that zooplankton may exhibit both physiological and behavioural adaptations, which enhance their ability to co-exist with toxic cyanobacteria (DeMott et al., 1991). Copepods discriminate between toxic and non-toxic taxa of cyanobacteria and this may be an adaptive behaviour favouring the ecological success and evolution of toxic strains (DeMott and Moxter, 1991). Indeed, field studies have shown that zooplakton avoid depths where toxic species are abundant (Forsyth et al., 1990) and a positive correlation has been shown between grazing pressure by *Daphnia* species and the toxicity of *Microcystis* dominated blooms. The latter provides further support for the hypothesis that cyanotoxins have evolved as a defence against grazers (Carmichael, 1992; DeMott et al., 1991; Lukac and Aegerter, 1993; Sivonen et al., 1992; Hanazato, 1996). Direct proof that microcystin is responsible for acute and long-term toxicosis to *Daphnia* was shown by comparison of the effects of a microcystin-containing strain and a mutant lacking the microcystin responsible for its formation (Dittmann et al., 1998). The *Daphnia* showed the same digestion rates with the wild-type and mutant strains.

Marine cyanobacteria from intertidal and sublittoral zones (*Lyngbya*, *Schizothrix*, *Oscillatoria*)

Table 1. Confirmed toxin-producing species of cyanobacteria.

SPECIES	
Anabaena circinalis Rabenh.	*Microcystis wesenbergii* (Komárek in Kondrateva, 1968)
A. flos-aquae (Lyngb) Breb.	*Microcystis*: other forms of uncertain status (*M. botrys, M. viridis*)
A. hassallii (Kütz.) Wittr.	*Nodularia spumigena* Mertens
A. lemmermanni Richter	*Nostoc linckia* (Roth) Born. et Flah.
A. spiroides var. contracta Kleb.	*N. paludosum* Kütz.
A. variabilis Kütz.	*N. rivulare* Kütz.
Anabaenopsis milleri Woron.	*N. zetterstedtii* Areschoug
Aphanizomenon flos-aquae (L.) Ralfs	*Oscillatoria acutissima* Kuff.
Coelosphaerium kuetzingianum Näg.	*O. agardhii/rubescens* group (= *Planktothrix*)
Cylindrospermum sp.	*O. formosa* Bory
Cylindrospermopsis raciborskii (Wolos.) Seenaya et Subba Raju	*O. nigroviridis* Thwaites
Fischerella epiphytica Ghose	*Oscillatoria* sp.
Gloeotrichia echinulata Richter	*Pseudanabaena catenata* Lauterb.
Gomphosphaeria lacustris Chod.	*Schizothrix calcicola* (Ag.) Gom
G. naegliana (Unger) Lemmerm.	*Scytonema pseudohofmanni* Bharadw.
Hormothamnion enteromorphoides Grun.	*Synechococcus* strains Miami BCII 6S & ATCC 18800
Lyngbya majuscula Harvey	*Synechocystis* sp.
Lyngbya wollei	*Tolypothrix byssoidea* (Hass.) Kirchn.
Microcystis aeruginosa Kütz.	*Trichodesmium erythraeum* Ehrb.
	Umezakia natans Harada

have yielded a range of toxic compounds associated with contact dermatitis (Moore, 1984), but the extent to which oceanic planktonic cyanobacteria produce toxins is still unclear. However, blooms of *Trichodesmium* have been implicated in the deaths of fish, oysters and crabs and the avoidance of such blooms by tuna has been observed (Chidambaram and Unny, 1944; Nagabhushanam, 1967; Chellam and Algarswami, 1978).

Evaluation of the ecological impact of the toxins is limited. However, it is known that they can impact upon a wide range of aquatic life forms – zooplankton (DeMott et al., 1991), *Daphnia* (Lightner, 1978), *Paramecium* (Lawton and Codd, 1988), *Artemia salina* (Kiviranta et al., 1991) and freshwater snails (Mohammed et al., 1992).

Cyanotoxicosis is recognised as being a world-wide problem, toxic biomass having been reported from almost all regions where it has been sought. There are few published studies on the frequency of cyanobacterial toxicity in environmental samples. An investigation of cyanobacterial blooms in Finnish fresh and coastal waters found that 44% were lethally toxic (Sivonen et al., 1990). Approximately 67% of the cyanobacterial blooms collected from over one hundred freshwater locations in the United Kingdom were toxic (Codd and Beattie, 1991). Accepting the variability of toxin expression (discussed later), laboratory and field data support the view that the majority of cyanobacterial species produce toxic compounds. The toxicology of these, their mode of synthesis, functional significance at the cellular and ecological level and the risk to human health in particular are all questions awaiting answers.

III. Which Taxa are Toxic?

Those genera reported as having toxic species are listed in Table 1.

Atoxigenic and toxigenic strains have been isolated from the same bloom, but there is little information as to the factors that regulate expression i.e. a toxic strain may become a non-toxic strain or *vice versa*. Toxicity not only varies between strains, but between clones of the same isolate (Carmichael, 1992; Lukac and Aegerter, 1993; Utkilen and Gjölme, 1992). In addition, some strains produce three or more toxins, with the relative proportions being influenced by environmental factors such as light, pH, nitrate, phosphate, metal ions and temperature (Carmichael, 1992; Keevil, 1991; Lukac and Aegerter, 1993; Sivonen, 1990; Watanabe and Oishi, 1985; Wicks and Thiel, 1990; Utkilen and Gjölme, 1992).

Field data on the influence of environmental factors on toxin formation in particular species have to be assessed with care, because any observed effect may be due to varying influences on different species. For instance, a comparison of 72 Finnish lakes (Vaitomaa et al., 1998) indicated that higher concentrations of aqueous P favoured hepatotoxic *Microcystis* and anatoxin-a containing *Anabaena* blooms, whereas *Anabaena* blooms with unknown neurotoxicity were associated with low P and high nitrate (N).

IV. Types of Cyanobacterial Toxin

Cyanobacteria are a rich source of biologically active secondary metabolites i.e. compounds that are not utilised by the organism's primary metabolism (Carmichael, 1992; Lukac and Aegerter, 1993), many of which may have pharmaceutical potential. The majority of cyanotoxins can probably be regarded as as secondary metabolites. They may be classified into two general groups:

(a) Those causing acute lethal poisoning - the *neurotoxins* and *hepatotoxins*;

(b) Those which are not highly lethal to animals, but show more selective bioactivity - *cytotoxins* (Table 2).

A. Neurotoxins

Neurotoxins are produced by several genera e.g. *Anabaena*, *Aphanizomenon*, *Oscillatoria* and *Trichodesmium* (Carmichael, 1992, 1994a, b; Keevil, 1991). The major neurotoxins are anatoxin-a, homoanatoxin-a, anatoxin-a(s), saxitoxin and neosaxitoxin (aphanotoxins I and II).

1. Anatoxin-a

Anatoxin-a is produced by some strains of *Anabaena.* The compound is a secondary amine i.e. 2-acetyl-9-azabicyclo-[4.2.1]non-2-ene, and is a structural analogue of cocaine and the neurotransmitter acetylcholine. The enzyme acetylcholinesterase breaks down this substance, thereby preventing muscle over-stimulation. Anatoxin-a is lethal, as acetylcholinesterase cannot degrade it and consequently muscle over-stimulation occurs i.e. the blocking of postsynaptic cholinergic transmission results in muscle fasciculations leading to fatigue and paralysis. Signs of poisoning are staggering, muscle fasciculations, gasping, convulsions and in birds, opisthotonos. Death by respiratory failure occurs within minutes to a few hours depending on species and dose (Carmichael, 1992, 1994a, b; Hunter, 1995a; Keevil, 1991; Repavich et al., 1990). The intraperitoneal LD_{50} for mice is 200 $\mu g\ kg^{-1}$ body wt, with a survival time of 4-7 minutes (Carmichael and Gorham, 1978; Carmichael et al., 1977; Carmichael and Biggs, 1978) .

Detection is primarily by mouse bioassay. However the use of three analytical detection methods has been reported - high performance liquid chromatography (Astrachan and Archer, 1981; Wong and Hindi, 1982), gas chromatography-mass spectrometry (Smith and Lewis, 1987) and gas chromatography-electron capture detection (Stevens and Krieger, 1988).

2. Homoanatoxin-a

This toxin has been purified from *Oscillatoria formosa* and characterised as a secondary amine alkaloid, methylene-anatoxin-a. Similar to anatoxin -a it is a potent neuromuscular blocking agent with an intraperitoneal LD_{50} in mice of 250 $\mu g\ kg^{-1}$ body weight (Skulberg et al., 1992). Toxicosis in the lethal dose range leads to severe body paralysis, convulsions and death by respiratory failure.

3. Anatoxin-a(s)

Anatoxin-a(s), the s denoting salivation in vertebrates, is produced by strains of *Anabaena flos-aquae* (Mahmood and Carmichael, 1986, Matsunaga et al., 1989). Symptoms are similar to those of anatoxin-a with the addition of ataxia, diarrhoea, hypersalivation, and tremors. However, this toxin is structurally and physiologically different from anatoxin-a. It is a naturally occurring organophosphate i.e. a *N*-hydroxyguanidine methyl phosphate ester and functions as an acetyl cholinesterase inhibitor similarly to the organophosphate pesticides e.g. malathion and parathion (Carmichael, 1992, 1994a, b). The intraperitoneal LD_{50} for mice is 20 $\mu g\ kg^{-1}$ body weight, ten times more lethal than anatoxin-a. Anatoxin-a(s) is stable in acid, but unstable in basic conditions.

4. Aphanotoxins I and II (Saxitoxin and Neosaxitoxin)

The aphanotoxins are neurotoxic alkaloids from *Aphanizomenon flos-aquae* (Sawyer et al., 1968). Chromatographic and pharmacological evidence has

Table 2. Cyanobacterial toxins.

SPECIES	TOXIN	STRUCTURE	LD_{50}*
Neurotoxins			
Anabaena flos-aquae			
NRC-44-1 (Canada)	Anatoxin-A	Secondary amine alkaloid (MW 165)	200
NRC-525-17 (Canada)	Anatoxin-A(S)	N-hydroxy guanadine methyl-phosphate ester (MW 252)	20
Aphanizomenon flos-aquae			
NH-1 (USA)	Aphantoxin (neosaxitoxin)	Purine alkaloid (MW 315)	10
NH-5 (USA)	AphantoxinII (saxitoxin)	(MW 299)	
Hepatoxins			
Anabaena flos-aquae			
S-23-g-1 (Canada)	Microcystins	Heptapeptides (MW 994)	50
Microcystis aeruginosa			
WR-70 (South Africa)	Cyanoginosins	Heptapeptides (MW 909-1044)	50
Waterbloom (Australia)	Cyanoginosin	Heptapeptide (MW 1035)	50
Waterbloom (USA)	Microcystin	Heptapeptide (MW 994)	50
NRC-1(SS-17) (Canada)	Microcystin	Heptapeptide (MW 994)	50
Strain 7820 (Scotland)	Microcystin	Heptapeptide (MW 994)	50
Waterbloom (Norway)	Microcystin	Heptapeptide (MW 994)	50
M-228 (Japan)	Microcystin	Heptapeptide (MW 994 and 1044)	50
Microcystis viridis	Cyanoviridin	Heptapeptide (MW 1039)	
Nodularia spumigena	Nodularin	Pentapeptide (MW 824)	30-50
Oscillatoria agardhii var. *isothrix* (Norway)	Microcystins	Heptapeptides (MW 1009)	300-500
Oscillatoria agardhii (Norway)	Microcystins	Heptapeptides (MW 1023)	500-1000
Cylindrospermopsis raciborskii	Cylindrospermopsin		
Aphanizomenon ovalisporum	Cylindrospermopsin		
Hormothamnion enteromorphoides	Hormothamnin A	Cyclic undecapeptide	
Anabaena BQ-16-1		Cyclic decapeptide containing chlorine	
Anabaena laxa	Laxaphycin A	Cyclic peptide (MW 1350-1400)	
	Laxaphycin B	Cyclic peptide (MW 1150-1290)	
Cytotoxins			
Scytonema pseudohofmanni BC-1-2 (Hawaii)	Scytophycin	Methylformamide A (MW 821)	
		Methylformamide B (MW 819)	650
Scytonema hofmanni UTEX-1581 (Texas)	Cyanobacterin	Chlorinated diaryllactone	
Hapalosiphon fontinalis V-3-1 (Marshall Islands)	Hapalindole A	Substituted indole alkaloid	
Tolypothrix byssoidea H-6-2	Tubercidin	Pyrrolopyrimidine	
Oscillatoria acutissima B-1 (Hawaii)	Acutiphycin	Macrolide	
Fischerella muscicola UTEX 1829	Fischerellin	Cyanobacterial inhibitor	
Scytonema mirabile BY-8-1		Antimicrobial	
Microcystis aeruginosa	volatile sulfur compounds		
Microcystis wesenbergii	volatile sulfur compounds		

*$\mu g\ kg^{-1}$ body weight

established that the aphanotoxins strongly resemble saxitoxin and neosaxitoxin. The aphanotoxins interfere with neurotransmission, i.e. nerve impulses, by blocking the neurone sodium channels across the axon membrane. Saxitoxin has also been recorded from *Lyngbya wollei* (Carmichael et al., 1997), an organism which forms nuisance growths on ponds in south-east USA, and a bloom-forming strain of *Oscillatoria* isolated from a bloom in Lake Varese, Italy (Pomati et al., 1998). These toxins are also associated with a number of marine dinoflagellates e.g. *Alexandrium tamarense*, which are responsible for paralytic shellfish poisoning in humans and massive deaths of marine animals associated with "red tides" in coastal waters world-wide. These toxins cause symptoms such as irregular breathing, loss of co-ordination, twitching and death by respiratory failure (Carmichael 1992, 1994a; Hunter, 1995a; Keevil, 1991). The intraperitoneal LD_{50} for mice using batch-cultured cells is approximately 5 mg kg^{-1} (each gram of lyophilized cells yields 1.3 mg aphantoxin I and 0.1 mg aphanatoxin II) (Mahmood and Carmichael, 1986).

B. Hepatotoxins

Cyclic hepatotoxic peptides are the most common offenders world-wide in cases of waterborne disease caused by toxic cyanobacteria. The major cyanotoxins causing death and illness in animals are the peptide hepatotoxins (liver toxins) (Carmichael 1992, 1994a; Keevil, 1991; Luukkainen et al., 1993, 1994). They are classified into the cyclic heptapeptides i.e. the microcystins (Fig. 1), which have been isolated and characterised from *Microcystis* and the filamentous genera *Anabaena (A. flos-aquae)*, *Nostoc* (*N. rivulare*) and *Oscillatoria* (*O. agardhii*) (Carmichael, 1992, 1994a; Codd and Beattie, 1991; Keevil, 1991; Luukkainen et al., 1993, 1994) and the cyclic pentapeptide nodularin (Fig. 2) isolated from the brackish water cyanobacterium *Nodularia spumigena* (Codd and Beattie, 1991; Hunter, 1995a, Keevil, 1991; Luukkainen et al., 1993). A detailed account of the occurrence of microcystins is given by Park and Watanabe (1996), Watanabe (1996) and other chapters in the book *Toxic Microcystis* edited by Watanabe et al. (1996).

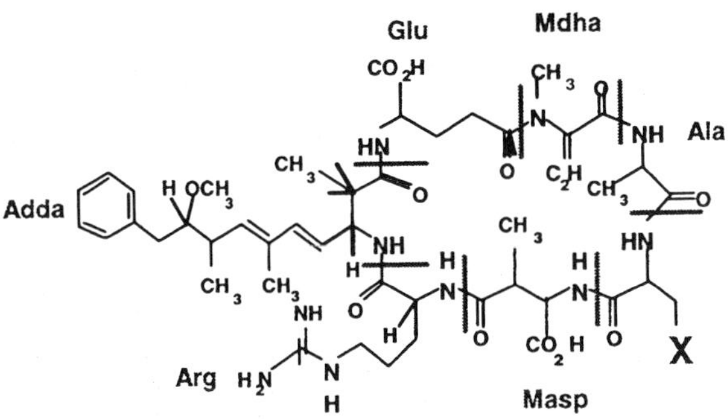

Fig 1. The general structure of microcystins (MCYST) is cyclo (-D-Ala-*X*-D-MeAsp-*Z*-Adda-D-Glu-Mdha) where *X* and *Z* (=Arg as above) represent the variable L amino acids at positions 2 and 4. *X* is usually leucine (L), arginine (R), tyrosine (Y) and phenylalanine (F); while *Z* is typically arginine (R), or alanine (A) and methionine (M). Microcystins are usually given two letter suffixes, such as MCYST-LR (MW = 994Da), containing leucine and arginine and MCYST-RR (MW = 1037), containing two arginines (Carmichael, 1992, 1994a; Luukkainen et al., 1993, 1994). Adda is the novel C_{20} *B*-amino acid (2S, 3S, 8S, 9S)-3-amino-9-methoxy-2,6,8-trimethyl-10-phenyldeca-4,6-dienoic acid and is the key component for toxicity of these compounds (Carmichael, 1992; Hunter, 1995a; Luukkainen et al., 1993, 1994). Mdha is *N*-methyl-dehydroalanine and D-MeAsp the D-erythro-*B*-methyl-aspartic acid (Carmichael, 1992, 1994a; Harada et al., 1991; Keevil, 1991; Kiviranta et al., 1992; Lanaras et al., 1991; Luukkainen et al., 1993, 1994). D-MeAsp and D-Glu are connected by an isolinkage and also D-MeAsp and Mdha exist in demethylated forms [i.e. D-Asp and (dehydroalanine) Dha; Lanaras et al., 1991; Luukkainen et al., 1993]. Moreover, L-serine has been found instead of Mdha (Namikoshi et al. 1992).

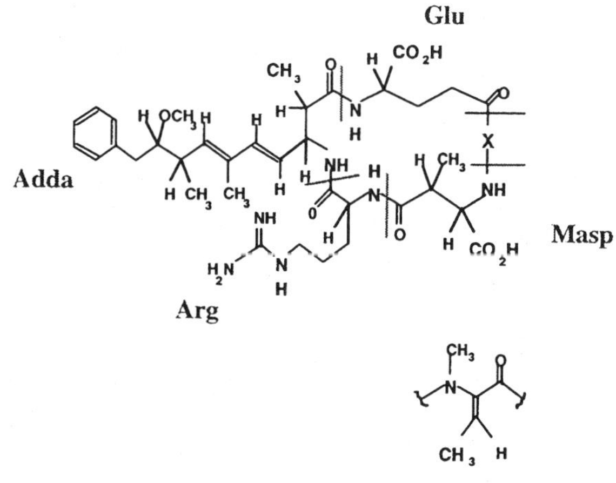

Fig 2. The structure of nodularin. The largest variation in structure for the cyclic hepatopeptides has been studied in the brackish filamentous cyanobacterium, *Nodularia spumigena* (Carmichael, 1992; Hunter, 1995a). Rinehart et al. (1988) and Sivonen et al. (1989, 1990) have confirmed the structure for this *Nodularia* peptide as a cyclic pentapeptide (five membered ring; MW of 824Da) and named it nodularin (NODLN). The general structure of nodularin produced by this filamentous organism is: cyclo (-D-MeAsp-L-Arg-Adda-D-Glu-*N*-Me-Z-dehydrobutyrine).

Another peptide toxin, cylindrospermopsin, has been isolated in Australia from *Cylindrospermopsis raciborskii* and also from a recently characterised organism from Japan, *Umezakia natans* (Harada et al., 1994), and *Aphanizomenon ovalisporum* in Israel (Banker et al., 1997) and Australia (Smith et al., 1998). This toxin causes similar symptoms to those of other hepatotoxins, but details of their biological activity have yet to be reported (Harada et al., 1994).

Acute hepatotoxicosis is the main type of poisoning caused by cyanobacteria (Carmichael, 1992, 1994a, b; Codd and Beattie, 1991; Fitzgeorge et al., 1994; Keevil, 1991; Sivonen, 1990). The symptoms of poisoning in laboratory animals (mice, rats and rabbits) include anorexia, diarrhoea, pallor mucous membranes, vomiting, weakness and death (Carmichael, 1992, 1994a; Hunter, 1995a; Keevil, 1991). Death (within 1-2 h) is due to intrahepatic haemorrhage - liver necrosis and the disintegration of the architecture of the liver i.e. hepatic parenchyma and hypovolaemic shock (Carmichael, 1992, 1994a; Fitzgeorge et al., 1994; Hunter, 1995a; Keevil, 1991; Repavich et al., 1990). There is a marked increase (up to 100%) in liver weight as a consequence of internal haemorrhaging (Carmichael, 1992, 1994a; Keevil, 1991). These toxins cause striking ultrastructural changes in isolated hepatocytes (Runnegar and Falconer, 1986; Eriksson et al., 1989), including a decrease in the polymerization of actin; that is, these toxins appear to interact with the cell's cytoskeletal system. Why there is specificity of these toxins for liver cells is not clear.

Microcystins act similarly to another group of cyclic peptides, the hepatotoxic heptapeptides "phallotoxins" (e.g. phalloidin) of the toadstool *Amanita phalloides,* the green death cap (Faulstich and Wieland, 1992; Lukac and Aegerter, 1993; Keevil, 1991; Wieland and Faulstich, 1978).

Microcystins have the general structure *cyclo*(-D-Ala-L-X-erythro-β-methyl-D-isoAsp-L-Z-Adda-D-isoGlu-N-methyldehydro-Ala) where X and Z represent certain L-amino acids (Fig. 1) (Botes et al., 1984, 1985; Carmichael et al., 1988; Lanaras et al., 1991; Rinehart et al., 1994). The presence of Adda [(2*S*,3*S*,8*S*,9*S*)-3-amino-9-methoxy-2,6,8-trimethyl-10-phenyldeca-4,6-dienoic acid] is the most unusual structural feature. Adda plays an important role in hepatotoxicity since hydrogenation or ozonolysis of the diene system in this unit gives an inactive product (Harada et al., 1990a; Harada et al., 1990b).

Over 60 variants of microcystins have been characterised (Table 3), some of the most recent being isolated from *Oscillatoria agardhii* - [Dha^7] MCYST-FR (Luukkainen et al., 1994) and [$D\text{-}Asp^3$, Dhb^7] MCYST-RR (Sano and Kaya, 1995). Dimethyl microcystins appear to be unique to Scandinavian cyanobacteria with MCYST-YR and its dimethyl forms having been found so far only in *Microcystis* spp. (Luukkainen et al., 1994). Differences between the microcystins are primarily in the type of L-amino acid present and in the presence or absence of a methyl group on one of the D-amino acids or on a dehydroamino acid or both. Studies on the biosynthesis of microcystins suggest a non-ribosomal mechanism of synthesis. Tillett et al. (1998) reported that a 20-kb gene cluster cloned and characterised from *Microcystis* had significant similarity to other peptide synthetases and has a specific role in microcystin biosynthesis. The basic difference between hepatotoxic and non-toxic strains of *M. aeruginosa* PCC7806 is the presence of one more genes coding for microcystin synthetases (Börner et al., 1998). Hybridization experiments with DNA fragments of a 29-kb peptide synthetase operon of hepatotoxic *Anabaena* showed that peptide synthetase genes are common in cyanobacteria (Rouhiainen et al., 1998), so it seems likely that they will prove to be involved in production of many cyanotoxins.

The hepatotoxins are strong inhibitors of type 1 and 2A serine protein phosphatases (PP1c and PP2Ac) (MacKintosh et al., 1990). These enzymes are vital to various cellular processes such as cell growth and tumour suppression and therefore these toxins are possible potent cancer promoters (Carmichael, 1992, 1994a; Luukkainen et al., 1993, 1994; MacKintosh et al., 1990; Runnegar et al., 1995). Recent research has indicated that microcystin-LR is an extremely potent tumour promoter in laboratory animals (Nishiwaki-Matsushima et al., 1992) and is the most potent liver carcinogen yet characterised. The occurrence of these toxins in potable water may therefore present a serious health hazard to humans if very low concentrations of the peptide toxins are consumed over a long period of time, contributing to chronic liver illnesses, such as liver tumours or necrosis (Falconer, 1991; Falconer et al. 1988).

C. Cytotoxins

Although the neurotoxins and hepatotoxins are the high profile cyanotoxins because of their lethality and ubiquity, there are several less well studied cyanotoxins, the majority of which are secondary metabolites. Of interest is the observation that many have antialgal, antimycotic or antibacterial activity, whilst others are active against cell tissue lines and have moderate antitumour activities (Gerwick et al., 1994; Patterson et al., 1994).

1. Scytophycins

These are lipophilic toxins, produced by *Scytonema pseudohofmanni,* which are moderately toxic to mice (LD_{50} of 650 μg kg^{-1}). They also have strong cytotoxic activity against cell cultures e.g. human epidermoid carcinoma and mouse fibroblasts. In addition they are active against intraperitoneally implanted lymphocytic leukemia and lung carcinoma (Moore et al., 1986; Carmichael et al., 1990).

Table 3. Examples of structural variants of microcystin (adapted from Carmichael, 1992).

MICROCYSTIN	ORGANISM	MW (Da)	FORMULA
MCYST-LA	*Microcystis aeruginosa*[b]	909	$C_{46}H_{67}N_7O_{12}$
MCYST-Laba	*Microcystis aeruginosa*[b]	923	$C_{47}H_{69}N_7O_{12}$
MCYST-AR	*Microcystis* spp.[c]	952	$C_{46}H_{68}N_{10}O_{12}$
MCYST-YA	*Microcystis aeruginosa*[b]	959	$C_{49}H_{65}N_7O_{12}$
MCYST-LR	*Microcystis aeruginosa*[b] *Anabaena flos-aquae*[b]	994	$C_{49}H_{74}N_{10}O_{12}$
MCYST-LY	*Microcystis aeruginosa*[b]	1001	$C_{52}H_{71}N_7O_{13}$
[D-Asp^3,Dha^7] MCYST-RR	*Oscillatoria agardhii*[c]	1009	$C_{47}H_{71}N_{13}O_{12}$
[Dha^7] MCYST-FR	*Microcystis* spp.[c]	1014	$C_{51}H_{70}N_{10}O_{12}$
[D-Asp^3, ADMAdda^5] MCYST-Lhar	*Nostoc* sp.[b]	1022	$C_{50}H_{74}N_{10}O_{13}$
[D-Asp^3] MCYST-RR	*Oscillatoria agardhii*[c]	1023	$C_{48}H_{73}N_{13}O_{12}$
[Dha^7] MCYST-RR	*Microcystis aeruginosa*[b]	1023	$C_{48}H_{73}N_{13}O_{12}$
[Dhb^7] MCYST-RR	*Oscillatoria agardhii*[c]	1023	$C_{48}H_{73}N_{13}O_{12}$
MCYST-FR	*Microcystis* spp.[c]	1028	$C_{52}H_{72}N_{10}O_{12}$
MCYST-(O)R	*Microcystis* spp.[c]	1028	$C_{48}H_{72}N_{10}O_{13}S$
MCYST-YM(O)	*Microcystis aeruginosa*[c]	1035	$C_{51}H_{69}N_7O_{14}S$
[ADMAdda^5] MCYST-Lhar	*Nostoc* sp.[b]	1036	$C_{51}H_{76}N_{10}O_{13}$
MCYST-RR	*Microcystis aeruginosa*[b] *Microcystis viridis*[b]	1037	$C_{49}H_{75}N_{13}O_{12}$
MCYST-YR	*Microcystis aeruginosa*[b] *Microcystis viridis*[b]	1044	$C_{42}H_{72}N_{10}O_{13}$
[D-Asp^3] MCYST-HtyR	*Anabaena flos-aquae*[b]	1044	$C_{52}H_{72}N_{10}O_{13}$
MCYST-HtyR	*Anabaena flos-aquae*[b]	1058	$C_{53}H_{74}N_{10}O_{13}$
MCYST-WR	*Microcystis* spp.[c]	1067	$C_{54}H_{73}N_{11}O_{12}$
MICROCYSTIN-LR			
MCYST-LR	*Microcystis aeruginosa*[b] *Anabaena flos-aquae*[b]	994	$C_{49}H_{74}N_{10}O_{12}$
[D-Asp^3,Dha^7] MCYST-LR	*Microcystis aeruginosa*[b]	966	$C_{47}H_{70}N_{10}O_{12}$
[D-Asp^3] MCYST-LR	*Anabaena flos-aquae*[b] *Microcystis viridis*[c]	980	$C_{48}H_{72}N_{10}O_{12}$
[Dha^7] MCYST-LR	*Microcystis aeruginosa*[b]	980	$C_{48}H_{72}N_{10}O_{12}$
[DMAdda^5] MCYST-LR	*Microcystis* spp.[c] *Nostoc* sp.[b]	980	$C_{48}H_{72}N_{10}O_{12}$
[D-Asp^3, ADMAdda^5] MCYST-LR	*Nostoc* sp.[b]	1008	$C_{49}H_{72}N_{10}O_{13}$
[Mser^7] MCYST-LR	*Microcystis* spp.[c]	1012	$C_{49}H_{76}N_{10}O_{13}$
[ADMAdda^5] MCYST-LR	*Nostoc* sp.[b]	1022	$C_{50}H_{74}N_{10}O_{13}$
[Dser^1, ADMAdda^5] MCYST-LR	*Nostoc* sp.[b]	1038	$C_{50}H_{74}N_{10}O_{14}$
[ADMAdda^5, Mser^7] MCYST-LR	*Nostoc* sp.[b]	1040	$C_{50}H_{76}N_{10}O_{14}$
[D-Glu(OC^{3}H^{7}O)^6] MCYST-LR	*Microcystis* spp.[c]	1052	$C_{52}H_{80}N_{10}O_{13}$

Key: [a]Aba, aminoisobutyric acid; Dha, dehydroalanine; DMAdda, *O*-dimethylAdda; ADMAdda, *O*-acetyl-*O*-dimethylAdda; Mser, *N*-methylserine; Har, homoarganine; M(O), methionine-S-oxide; Hty, homotyrosine.

[b]Toxins isolated from strains of culture collections.

[c]Toxins isolated from blooms.

2. *Cyanobacterin*

This product of *Scytonema hofmanni* is a chlorine containing diaryl-lactone which has anti-cyanobacterial activity; as a consequence it has been proposed as a possible algicide (Mason et al., 1982; Gleason and Paulson, 1984).

3. *Hapalindole A*

A cytotoxic alkaloid has been isolated from strains of *Hapalosiphon fontinalis*. This lipophilic compound has a broad range of antialgal and antimycotic activity (Moore et al., 1984).

4. *Acutiphycin*

Two novel macrolides, acutiphycin and 20,21-didehydroacutiphycin, have been isolated from *Oscillatoria acutissima* (Barchi et al., 1984). They are cytotoxic to tissue cells and have antitumour properties (lung carcinoma).

5. *Lyngbyatoxin*

Lyngbyatoxin is a potent tumour promoter (Nishizuka, 1984) and activates calcium-activated, phospholipid-dependent protein kinase C causing contraction of vascular smooth muscle (Robinson et al., 1991). It resembles the phorbol esters (tumour promotion and activation of protein kinase C) and has been isolated from *Lyngbya majuscula* (Fujiki et al., 1983), a species forming extensive floating masses in shallow, sheltered waters in warmer regions. Lyngbyatoxin is also a potent irritant causing swimmer's itch (Cardellina et al., 1979; Moore, 1981).

6. *Endotoxin (Lipopolysaccharide)*

Cyanonobacterial lipopolysaccharides, in addition to their endotoxic activity, have been implicated as being contact irritants. In common with Gram-negative bacteria, cyanobacteria form lipopolysaccharide as an important part of their outer cell envelope, but unlike the former, cyanobacterial lipopolysaccharide lacks any phosphate in the lipid A core (Keleti and Sykora, 1982). Compared with enterobacterial endotoxins, such as that from *Salmonella* spp., cyanobacterial lipopolysaccharide was found to be ten times less toxic in bioassays (Codd, 1984). Nevertheless, it has been implicated in human illnesses (Billings, 1981; Codd and Bell, 1985). Toxicological studies in the laboratory have been contradictory - some have shown no effect (Weise et al., 1970; Keleti et al., 1979), whilst others have been lethal (Keleti et al., 1979; Keleti and Sykora, 1982).

D. *Other "Toxins"*

Cyanobacteria also produce other secondary metabolites falling within the broad category of cyanotoxins. These are compounds which are toxigenic, but whose function and ecological importance are ill-defined. Doubtless the list of toxigenic substances identified from cyanobacteria will continue to grow; however their contribution to, or role in cyanobacterial physiology, growth and survival is at present sketchy.

V. Toxin Analysis

It is not possible to determine whether a cyanobacterial bloom is toxic by its appearance. Consequently a major difficulty in the study of cyanotoxins lies in the methods available for the detection and assessment of toxigenicity. The mouse bioassay remains the primary means of assessing function and potency of the cyanotoxins. However, it has severe limitations in routine laboratory and environmental studies. Many water authorities or companies do not have animal house facilities. The method is also unsuitable for measuring the low concentrations of cyanotoxins that usually prevail in populations not forming a scum (Sivonen and Jones, 1999). Further, there is increasing opposition in many countries to the use of animals for any form of toxicity testing. Consequently a range of other assay procedures have been investigated.

1. *Mouse Bioassay*

The mouse bioassay is the recognised standard in terms of establishing the LD_{50}, symptoms and effects of cyanotoxins. The toxicity of cyanobacterial biomass may be assessed by mouse bioassay (Chaivimol et al., 1994a, b; Dow et al., 1994; Swoboda and Dow, 1994; Swoboda et al., (1994). Routinely, adult male and female mice are injected intraperitoneally with the sample to be assayed. The mice are observed frequently (every 15 min) for toxic symptoms and toxicity break points are normally scored within 24 h. Toxic symptoms vary depending upon material and concentration, but are in general distinct for a category of toxin. It is difficult to differentiate between closely related toxins.

2. Alternative Bioassays

The amount of zooplankton is known to decrease under cyanobacterial bloom conditions (Lightner, 1978); consequently, bioassays have been devised using *Daphnia* sp. and *Artemia salina* (Kiviranta et al., 1991). A few studies have been done with other crustaceans and rotifers (Gilbert, 1994), but not taken to the extent where they can be used for bioassays. However, a bioassay has been developed for the fruitfly *Drosophila melanogaster* (Swododa and Dow, 1994) and microinjection into mosquitoes has also been suggested (Turell and Middlebrook, 1988).

Major difficulties in the use of alternative bioassays such as those based on *Artemia salina* are highlighted by Kiviranta et al. (1991). Moderate to high concentrations of toxin (neuro- and hepatotoxins) can be detected in bloom samples using this organism, but detection of the lower concentrations found easily by mouse bioassay is less reliable. In contrast, a laboratory-grown strain of *Oscillatoria agardhii* was toxic to larvae of the mosquito *Aedes aegyptii* (Kiviranta and Abdelhameed, 1994), *Artemia salina* and *Daphnia pulex*, but non-toxic to mice (Reinikainen et al., 1995). Also toxicity differs for the larval and adult stages of the shrimp, particularly for the neurotoxins. Pure anatoxin-a-hydrochloride is not toxic to larvae, but when added to non-toxic cyanobacterial biomass the death percentage of the larvae increases significantly. This may be a consequence of a synergistic effect with other compounds or of the fact that the pure toxin is not absorbed by the larvae. It is possible that some compounds may enable the toxin to enter the larvae by affecting the biochemistry of the crustacean or by hydrogen bonding to chelates, which are more easily absorbed. The use of alternative bioassays must therefore by approached with caution.

3. Tissue Culture Cytotoxicity

It is well documented that hepatotoxins cause deformation and ultrastructural changes of the cytoskeleton of mouse liver hepatocytes primarily due to the inhibition of phosphatase activity (Dabholkar and Carmichael , 1987; Eriksson et al., 1990). Aune and Berg (1986) proposed the use of freshly prepared rat hepatocytes to study/assess the toxicity of cyanobacterial blooms. They have shown that the *in vitro* toxicity to cells can be correlated with control animal experiments when crude algal biomass is used. Similar effects have been observed with permanent cell lines and erythrocytes (Grabow et al., 1982). However, despite the remarkable toxic potential of hepatotoxins *in vivo*, no cell lysis, liberation of lactate dehydrogenase or haemolysis has been observed after application of pure hepatotoxin to primary or permanent cell lines (Runnegar and Falconer, 1982, 1986; Eriksson et al., 1987).

The ambiguity of the data from bioassays may be increased due to substances other than the recognised cyanotoxins having cytotoxic effects on cultured cell lines. Henning et al. (1992) found that the cytotoxic effects of crude aqueous extracts from *Microcystis aeruginosa* PCC 7806 are a consequence of a 35-kDa, heat-sensitive substance, not microcystin (994 Da, heat-resistant). This substance is effective against Chang liver, Chimpanzee liver cells, HeLa cells and African green monkey cells.

4. Chromatographic Analysis

Initially thin-layer chromatography (TLC) was used to purify the hepatotoxins, but lacked sensitivity and specificity. This was superceded by high performance liquid chromatography (HPLC) with which it has been possible to isolate hepatotoxins from several cyanobacterial species and environmental biomass with high sensitivity (Botes et al., 1982; Brooks and Codd, 1986; Sivonen, 1990; Harada, 1996). Resolution of the toxic fraction from *Microcystis aeruginosa* PCC7806 into closely related peptides has been achieved by reversed-phase HPLC (Cremer and Henning, 1991; Chaivimol et al., 1994a, b; Dow et al., 1994; Swoboda and Dow, 1994; Swoboda et al., 1994). Integration of diode array spectroscopy with RP-HPLC has provided a rapid and sensitive method of assessing the nature and concentration of toxic peptides from cyanobacterial blooms. All of these procedures involve solvent extraction and clean-up prior to analysis, though reversed-phase fast protein liquid chromatography (RP-FPLC) has been used to separate distinct toxic peptides in a single preparative scale run (Cremer and Henning, 1991). However, it is RP-HPLC which has been applied most frequently to the characterisation of field samples of cyanobacteria.

5. Mass Spectrometry

The structural characteristics of the hepatotoxins have been ascertained primarily by fast atom bombardment mass spectrometry (FAB/MS) and proton nuclear magnetic resonance (NMR) (Meriluoto et al., 1989; Sandstrom et al., 1990; Skulberg, et al., 1992; Kusumi, 1996). More recently, matrix assisted laser desorption ionisation (MALDI) mass spectrometry has permitted procedurally easier analysis of cyanobacterial biomass. This presents the possibility of rapid, sensitive analysis of environmental biomass for the presence of cyanotoxins (Chaivimol et al., 1994a; Dow et al., 1994; Swoboda et al., 1994). A recent (1998) practical guide to the methodology is given by the (UK) Environment Agency.

From the available structural data (mass spectrometry and amino acid sequences) Lanaras et al. (1991) have ascertained the 3-dimensional structure of microcystin-LR and nodularin.

6. Protein Phosphatase Inhibition Assay

Confirmation that the hepatotoxic cyanobacterial toxins produce their toxic effects through the inhibition of protein phosphatases 1 and 2A has laid the foundation for toxin assays based on the inhibition of these enzymes (Sim and Mudge, 1994; Swoboda et al, 1994). This assay is an effective and reliable means of detecting all hepatotoxic cyanobacterial toxins and has been enhanced by the development of a colorimetric procedure (Ward et al., 1997).

7. Immunological Procedures

With the development of immunoassays for a number of low molecular weight toxins, particularly mycotoxins and marine toxins, antibodies against the microcystins have been sought. Both monoclonal (Kfir et al., 1986) and polyclonal antibodies (Brooks and Codd, 1988) have been successfully raised, but initially with low sensitivity in immunoassays. This was subsequently improved by Chu et al. (1989) using microcystin-LR-polylysine and microcystin-LR-ethylenediamine-modified bovine serum albumin to immunize rabbits. The latter proved to be the better immunogen. The raised antibodies had good cross-reactivities with microcystin-RR, microcystin-YR and nodularin and supported the hypothesis that the configuration of ADDA in these toxins is similar. In addition, it is the arginine residue between the unusual amino acids erythro-β-methyl-*d*-iso-aspartate and the β-amino acid residue of ADDA which plays a dominant role in expressing antibody specificity.

These antibodies did not cross-react with toxic peptide extracts from *Oscillatoria* or *Anabaena* blooms, but gave positive responses with toxic *Microcystis aeruginosa* blooms (Brooks and Codd, 1988).

Nagata et al. (1995) have produced monoclonal antibodies against microcystin-LR which showed cross-reactivity with microcystin-RR, -Yr, -LA and several other derivatives. Although the epitope of this antibody is not clear the importance of the Adda moiety for antibody binding has been indicated. Unfortunately this antibody also reacts to the non-toxic monomethyl ester of microcystin-LR giving a toxicological false positive. However, a competitive ELISA method has been developed with detection limits of 0.05 $\mu g\ L^{-1}$ (Ueno et al., 1996).

VI. Ecological Implications

A. Human Health

1. Toxicology

The Tolerable Daily Intake (TDI), i.e. the dose level in humans that is considered to be without adverse effects when taken daily over a life-time, must be determined before safe levels of toxicants or contaminants can be advised. In establishing these values there are many obstacles: inadequate or non-existant human studies; the difference in sensitivity between animals and humans - the former being used in laboratory studies and challenged with diverse preparations of toxin(s); the routes of exposure - many toxins are more toxic when administered via the intravenous or intraperitoneal route than by the oral route, the latter being the main route of human exposure. The data currently available are insufficient for the determination of TDI values for the great majority of cyanotoxins. A further complication is introduced when the structural variants of the hepatotoxins are considered, which usually show very different toxicities in laboratory studies.

Evidence for the adverse effects of cyanotoxins on human health comes from three sources: epidemiology, poisoning events - human and animal, and toxicological studies.

Several studies of human populations have produced epidemiological evidence linking poisoning and or injury to the presence of cyanotoxins in drinking or recreational water. However, most of these studies have been retrospective and complete epidemiological data, particularly regarding exposure, are rare. Nevertheless, epidemiological evidence is of special importance in demonstrating the link between toxin exposure and human health outcomes which cannot be derived from animal experiments.

Although animal toxicity challenges are not performed under conditions of natural exposure and have involved several different species, they are of importance since they are performed under defined laboratory conditions. These studies provide the primary evidence for the role(s) of cyanotoxins in human and animal poisonings.

Of all the cyanotoxins the cyclic peptides represent the greatest threat to human health. Acute exposure to high doses may cause death from liver haemorrhage or from liver failure, whilst chronic low-dose exposure may promote the growth of tumours. The risk of long-term exposure to comparatively low concentrations of these toxins in drinking water is of particular concern

The alkaloid neurotoxins have shown only acute effects in mammals. In relation to these toxins an important exposure route is via the consumption of fish, shellfish and animals which have been exposed to water bodies contaminated with toxic cyanobacteria. In effect they act as a means of concentrating the toxin resulting in acute exposure on consumption.

The histopathology of cylindrospermopsin which causes major kidney damage, has been well characterised (Hawkins et al., 1997; Falconer et al., 1999; Seawright et al., 1999). However, the effects, if any, of long term exposure to low concentrations are unknown.

Cyanobacterial lipopolysaccharide can elicit both allergic and toxic responses in humans but very little is known about their acute or chronic effects. The lack of axenic cyanobacterial strains has hindered studies on the structure and toxicity of cyanobacterial lipopolysaccharide.

2. Short-term Effects

Gastrointestinal and hepatic illness attributed to cyanobacterial toxins in water supplies have all been coincident with either the breakdown of a natural cyanobacterial bloom (Tisdale, 1931; Zilberg, 1966; Teixera et al., 1993) or with the artificial lysis of the population e.g. by the application of copper sulphate (Lippy and Erb, 1976; Byth, 1980; Jochimsen et al., 1998). In these cases cyanotoxin is released from the cells and whereas treatment procedures will remove the bulk of the cellular biomass they are much less efficient in the removal of free cyanotoxin.

3. Long-term Effects

Short exposure to toxins may result in long-term injury, however, it is the effect of chronic, low concentration exposure which is of greatest concern. Continual oral exposure to low doses of microcystins have shown chronic liver injury, but more important is the possibility of carcinogenesis and tumour growth promotion. The incidence of human hepato-cellular carcinoma in China is one of the highest in the world. Cancer mortality rates are lowest when water is drawn from deep wells rather than from surface sources such as ponds or ditches which sustain abundant cyanobacterial populations, most frequently of *Oscillatoria* species (Yu, 1989, 1995).

Experimental evidence is at best, contradictory. Intraperitoneal administration of microcystin-LR to mice (20 μg kg^{-1} body wt, 100 times over 28 weeks) (Ito et al, 1997) induced neoplastic liver nodules. However, oral administration of a higher dose over the same time period (80 μg kg^{-1} body wt, 100 times over 28 weeks) showed no evidence of liver injury or nodule formation, highlighting the importance of the challenge route. [Note: microcystin-LR does not readily cross cell membranes, hence does not enter most tissues. After oral uptake it is transported across the ileum into the bloodstream via a bile acid type transporter present in hepatocytes and cells lining the small intestine. There is no evidence of hydrolysis of microcystin by peptidases in the stomach and it is apparent that a significant amount of microcystin-LR passes the intestinal barrier and is absorbed. It is concentrated in the liver as a consequence of active uptake by hepatocytes (Runnegar et al., 1981). Some microcystin variants are more hydrophobic and may cross cell membranes by other mechanisms].

Microcystins have been reported to promote the appearance of cancer once initiation has occurred (Yamasaki, 1988; Fitzgerald and Yamasaki, 1990; Falconer, 1991). Microcystin-LR is a potent inhibitor of eukaryotic protein serine/threonine phosphatases 1 and 2A both *in vitro* (Honkanen et al., 1990; MacKintosh et al., 1990) and *in vivo* (Runnegar et al., 1993). Substances which inhibit these protein phosphatases are considered to be tumour promoters e.g. okadaic acid, tautomycin and calyculin (and nodularin) (Fujiki and Suganuma, 1993). Inhibition of protein phosphatase inhibition results in higher phosphorylation of target proteins, such as tumour suppresser proteins. This post-translational modification can result in excessive signalling and may lead to cell proliferation, cell transformation and tumour promotion. The implications of protein phosphatase inhibition in humans due to low level chronic exposure to microcystins are not known.

At present, the human evidence for microcystin carcinogenicity is inadequate and the animal evidence is limited. However, the WHO (1998) has published a provisional TDI of 0.067 $\mu g\ kg^{-1}$ body wt per day and a provisional guideline value of 0.04 $\mu g\ kg^{-1}$ body wt per day for microcystin-LR.

4. Recreational Exposure

Contact with cyanobacteria in bathing or recreational waters can lead to skin irritation - "swimmer's itch" - and increased likelihood of gastrointestinal symptoms (Moikeha and Chu, 1971; Hashimoto et al., 1976; Pilotto et al., 1997). Severe skin reactions have been reported involving the marine cyanobacterium, *Lyngbya majuscula*, which causes blistering when trapped beneath a bathing costume (Grauer, 1961). This organism expresses powerful dermal toxins aplysiatoxins and debromoaplysiatoxin (Mynderse et al., 1977), which are potent tumour promoters and protein kinase C activators (Fujiki et al., 1990).

Lyngbya majuscula may grow epiphytically on edible algae e.g. *Acanthophora spicifera.* In addition cyanobacteria have also been suspected to be a source of toxins in the ciguatera food chain leading to human poisoning (Hahn and Capra, 1992; Endean et al., 1993) and *Lynbya* might therefore be the agent involved here.

B. Toxicity to Fish

When fish are challenged with cyanotoxins by force-feeding or by intraperitoneal injection, they develop symptoms similar to laboratory mammals. There is evidence to indicate that in the natural environment cyanotoxins affect healthy fish. Histopathological investigations of fish deaths during cyanobacterial blooms in the UK indicated that the cause of death was due to damage to the gills, digestive tract and liver (Rodger et al., 1994). Gill damage by dissolved microcystin-LR has been shown experimentally in *Tilapia* and trout (Garcia, 1989; Gaete et al., 1994; Bury et al., 1996). However, gill damage may also be caused by the high pH values associated with cyanobacterial photosynthesis and the high ammonia concentrations arising from the decomposition of cyanobacterial cells. Irrespectively of this, gill damage almost certainly enhances microcystin uptake leading to liver necrosis.

The effects of microcystins on the European carp, *Cyprinus carpio*, were assessed by comparison of fish from Lake Wellington (no evidence of significant growth of *Microcystis aeruginosa*) with fish from Lake Mokoan (visible *M. aeruginosa* scum with detectable concentrations of microcystin for four consecutive months). The Lake Mokoan fish displayed atrophy of hepatocytes, pinpoint necrosis on the gills, epithelial damage, elevated aspartate aminotransferase activity and serum bilirubin concentrations, all consistant with impaired hepatocyte function (Carbis et al., 1997).

In other instances, it has been shown that immersion of adults or juveniles in contaminated water had no pathological consequences (Tencalla et al., 1994). However, the most definitive effect on fish concerns Atlantic salmon reared in net pens in coastal waters of British Columbia and Washington State, USA (Anderson et al., 1993). An unidentified microcystin-producing organsism caused progressive degeneration of the liver of salmon smolts, a disease referred to as Net Pen Liver Disease, with significant economic consequences.

Recent studies have addressed the effects of cyanobacterial toxins on developing fish embryos. In general no acute toxic effects were observed after exposure to microcystin (-LR, -RR and -YR). Saxitoxin delayed hatching and led to malformations and mortalities. Of greater interest is the observation that far more pronounced effects were evident following challenge with various aqueous crude extracts of environmental cyanobacterial biomass and batch cultures than with purified toxin (Oberemm et al., 1999).

Also of considerable ecological interest are the reports of grazing carp which are able to differentiate between toxic and non-toxic cyanobacterial blooms.

C. Toxicity to Insects

There are no published data on the effect of cyanobacterial blooms on insect populations in the natural environment. However, the sensitivity of *Daphnia* (Baird et al., 1989; Laurén-Määttä et al., 1997), *Locusta migratoria* (Hiripi et al., 1998), *Schistocerca gregaria* (Mcelhiney et al., 1998) and other insects to a range of cyanobacterial toxins points to ecological consequences. Insect challenges are by ingestion of cyanobacterial cells, with mortality showing a dose response curve (Kotak et al., 1996). The implications with respect to the aquatic food chain have yet to be addressed.

VII. Concluding Comments

Despite cyanobacteria having been recognised for some considerable time it is only over the past twenty years that there have been significant advances in characterisation of the toxic species and their toxins *per se*. Early investigations suggested that only a few species were toxic and that they expressed only a few toxins which were produced intermittently. The perception of risk to human health and possible environmental impacts were considered minimal and at best received scant attention. This picture has changed dramatically. The number of toxic species of cyanobacteria, or more correctly species which may produce toxins, is continually increasing as is the number and variety of toxins.

With the recognition of the potency of cyanotoxins in general, scientific investigation has focused on the risks of acute and chronic injury to human health - particularly those posed by drinking water and recreational use of waterbodies. Consequently the isolation and characterisation of many of the toxins has advanced in that their structure and modes of action have been elucidated. However, major questions remain about the synthesis pathways, particularly those of the hepatotoxins. How is toxin produced within the cell, how is it regulated and by what? Several ecological parameters have been implicated in the regulation of toxin synthesis e.g. phosphate and nitrate concentrations, light intensity and temperature, but no definitive evidence has been forthcoming. There are only a few indisputable facts - toxic species are numerous, they may, depending on environmental or culture regimes, produce a diverse range of potent toxins, and cyanobacterial blooms of both fresh and marine waters are increasing in frequency with the increase in eutrophication. To control such events it is essential that the parameters impinging on toxin production are identified.

Instances of the successful management of cyanobacterial blooms which have already formed are rare. More often than not the treatment e.g. lysis of the biomass or phosphate removal has been at best neutral, while sometimes the net effect has been to enhance the problem.

Analytical techniques for the detection of cyanotoxins have improved markedly in both sensitivity and specificity, considerably extending the list of toxic species and their toxic products. Fortunately the most serious risk to human health - the hepatotoxins and their role in tumour promotion (a role still to be categorically substantiated) - is reduced due, not to treatment processes in the water industry, but to the degradative action of microbes. Although hepatotoxins are stable toxins, they are readily degraded in the natural environment, presumably to non-toxic entities.

The effects of cyanotoxins upon natural ecosystems are largely unknown. However, extrapolation from laboratory studies indicates that there must be pronounced effects on several eukaryotic components. Whether by grazing or diminution of a food source, it is inconceivable that toxic cyanobacterial blooms do not significantly compromise the immediate ecosytems. However, the extent to which these impact on food webs and population dynamics is only just starting to be investigated critically. At present environmental questions still have to be answered largely by informed deduction.

Much has been published on the adverse effects of the cyanotoxins, but considerably less on their potential value. As secondary metabolites cyanotoxins may offer a diverse source of important compounds, particularly in medicine.

The knowledge base concerning toxic cyanobacteria has increased tremendously over the past two decades but as greater insight is gained into the cyanobacteria and their toxins the number and complexity of the questions and dilemmas increase.

References

Anderson RJ, Luu HA, Chen DZX, Holmes CFB, Kent M, LeBlanc M, Taylor FRJ and Williams DE (1993) Chemical and biological evidence links microcystins to salmon "Netpen liver Disease". Toxicon 31: 1315-1323

Astrachan NB and Archer BG (1981) Simplified monitoring of anatoxin-a by reverse-phase high performance liquid chromatography and the sub-acute effects of anatoxin-a in rats. In: Carmichael WW (ed.) The Water Environment: Algal Toxins and Health. 437-446. Plenum Press, New York

Aune T. and Berg K. (1986). Use of freshly prepared rat hepatocytes to study toxicity of blooms of the blue-green algae *Microcystis aeruginosa* and *Oscillatoria agardhii*. J Toxicol Environ Health 19: 325-336

Baird DJ, Soares AMVM, Girling AE, Barber I, Bradley MC and Calow P (1989) The long-term maintenance of *Daphnia magna* for use in ecotoxicity tests: Problems and prospects. In: Lokke H, Tyle H and Bro Rasmussen F (eds) Proceedings of the First Conference on Ecotoxology. 144-148. Lyngby, Denmark

Banker PD, Carmeli S, Hadas O, Teilsch B. Porat R and Sukenik A (1997) Identification of cylindrospermopsin in *Aphanizomenon ovalisporum* (Cyanophyceae) isolated from Lake Kinneret, Israel. J Phycol 33: 613-616

Barchi JJ Jr, Moore RE and Patterson GML (1984) Acutiphycin and 20,21-didehydroacutiphycin, new antineoplastic agents from the cyanophyte *Oscillatoria acutissima*. J Am Chem Soc 106: 8193-8197

Beasley VR, Cook WO, Dahlem AM, Hooser SB, Lovell RA and Valentine WM (1989) Algae intoxication in livestock and waterfowl. Clin Toxicol 5: 336-345

Billings WH (1981) Water-associated human illness in northeast Pennsylvania and its suspected association with blue-green algae blooms. In: Carmichael WW (ed.) The Water Environment: Algal Toxins and Health. 243-255. Plenum Press, New York

Börner T, Dittmann E, Neilan BA, Erhard M and von Döhren H (1998) Microcystin synthetase: genes and enzymes. In: Carmichael WW and Paerl HW (eds) Abstracts, 4th International Conference on Toxic Cyanobacteria, Beaufort, North Carolina: 58

Botes DP, Kruger H and Viljoen CC (1982) Isolation and characterisation of 4 toxins from the blue-green-alga *Microcystis aeruginosa*. Toxicon 20: 945-954

Botes DP, Tuinman AA, Wessels PL, Viljoen CC, and Kruger H (1984) The structure of cyanoginosin-LA, a cyclic heptapeptide toxin from the cyanobacterium *Microcystis aeruginosa*. J Chem Soc Perkin Transactions 1: 2311-2318

Brooks WP and Codd GA (1986) Extraction and purification of toxic peptides from natural blooms and laboratory isolates of the cyanobacterium *Microcystis aeruginosa*. Lett Appl Microbiol 2: 1-3

Brooks WP and Codd GA (1988) Immunoassay of hepatotoxic cultures and water blooms of cyanobacteria using *Microcystis aeruginosa* peptide toxin polyclonal antibodies. Environ Technol Lett 9: 1343-1348

Bury NR, Flik G, Eddy FB and Codd GA (1996) The effects of cyanobacteria and the cyanobacterial toxin microcystin-LR on Ca^{2+} transport and Na^+/K^+-ATPase in Tilapia gills. J Exp Biol 199: 1319-1326

Carbis CR, Rawlin GT, Grant P, Mitchell GF, Anderson JW and McCauley I (1997) A study of feral carp, *Cyprinus carpio* L, exposed to *Microcystis aeruginosa* at Lake Mokoan, Australia, and possible implications for fish health. J Fish Dis 20: 81-91

Cardellina JA II, Marner FJ and Moore RE (1979) Seaweed dermatitis: structure of lyngbyatoxin A. Science 204: 193-195

Carmichael WW (1992) Cyanobacteria secondary metabolities – the cyanotoxins. J Appl Bacteriol 72: 445-459

Carmichael WW (1994a) Toxins of cyanobacteria. Sci Am 270: 64-70

Carmichael WW (1994b) In: Codd GA, Jeffries TM, Keevil CW, and Potter E (eds) Detection Methods for Cyanobacterial Toxins. Proceedings of the First International Symposium on Detection Methods for Cyanobacterial Toxins, September 1993. The Royal Society of Chemistry, Cambridge

Carmichael WW (1997) The cyanotoxins. Adv Bot Res 27: 211-256

Carmichael WW and Biggs DF (1978) Muscle sensitivity differences in two avian species to anatoxin-a produced by the freshwater cyanophyte *Anabaena flos-aquae* NRC-44-1. Can J Zool 56: 510-512

Carmichael WW, Eschedor JT, Patterson GML and Moore RE (1988) Toxicity and partial structure of a hepatotoxic peptide produced by the cyanobacterium *Nodularia spumigena* Mertens. Appl Env Microbiol 54: 2257-2263

Carmichael WW, Evans WR, Yin QQ, Bell P and Mocauklowski E (1997) Evidence for paralytic shellfish poisons in the freshwater cyanobacterium *Lyngbya wollei* (Farlow ex Gomont) comb. nov. Appl Environ Microbiol 63: 3104-3110

Carmichael WW and Gorham P (1978) Anatoxins from clones of *Anabaena flos-aquae* isolated from lakes of Western Canada. Mitt Int Ver Limnol 21: 285-295

Carmichael WW, Mahmood NA and Hyde EG (1990) Natural toxins from cyanobacteria (blue-green algae). In: Hall S and Strichartz G (eds) Marine Toxins, Origin, Structure and Molecular Pharmacology 418: 87-106. American Chemical Society, Washington DC

Chaivimol J, Swoboda UK and Dow CS (1994a) Characterisation of hepatotoxic freshwater *Oscillatoria* spp.: variation in toxin and temporal expression. In: Codd GA, Jeffries TM, Keevil CW, and Potter E (eds) Detection Methods for Cyanobacterial Toxins. Proceedings of the First International Symposium on Detection Methods for Cyanobacterial Toxins, September 1993. 161-163. The Royal Society of Chemistry, Cambridge

Chaivimol J, Swoboda UK and Dow CS (1994b) Phosphatase assay as a determinant of hepatotoxicity. In: Codd GA, Jeffries TM, Keevil CW, and Potter E (eds) Detection Methods for Cyanobacterial Toxins. Proceedings of the First International Symposium on Detection Methods for Cyanobacterial Toxins, September 1993. 172-174. The Royal Society of Chemistry, Cambridge

Chellum A and Alagarswami K (1978) Blooms of *Trichodesmium thiebautii* and their effect on experimental pearl culture at Veppalodai. Ind J Fish 25: 237-239

Chengappa MM, Pace LW and Mclaughlin BG (1989) Blue-green algae (*Anabaena spiroides*) toxicosis in pigs. J Am Vet Med Ass 194(12): 1724-1725

Chidambaram K and Unny MM (1944) Note on the swarming of the planktonic alga *Trichodesmium erythraeum* in the Pambau area and its effect on the fauna. Current Science 10: 263

Chorus I and Bartram J (eds) (1999) Toxic Cyanobacteria in Water. A Guide to their Public Health Consequences, Monitoring and Management. E & FN Spon, London & New York

Chu FS, Huang X, Wei RD and Carmichael WW (1989) Production and characterisation of antibodies against microcystins. Appl Environ Microbiol 55: 1928-1933

Codd GA (1984) Toxins of freshwater cyanobacteria. Microbiol Sci 1: 48-52

Codd GA and Beattie KA (1991) Cyanobacteria (blue-green alga) and their toxins: awareness and action in the United Kingdom. PHLS Microbiology Digest Supplement, 8: 82-86. London

Codd GA and Bell SG (1985) Eutrophication and toxic cyanobacteria in water. Sci Technol 21: 1-13

Cremer J and Henning K (1991) Application of reversed-phase medium-pressure liquid chromatography to the isolation, separation and amino acid analysis of two closely related peptide toxins of the cyanobacterium *Microcystis aeruginosa* strain PCC7806. J Chromatography 587: 71-80

Dabholkar AS and Carmichael WW (1987) Ultrastructural changes in the mouse liver induced by hepatotoxin from the freshwater cyanobacterium *Microcystis aeruginosa* strain 7820. Toxicon 25: 285-292

DeMott WR and Moxter F (1991) Foraging on cyanobacteria by copepods: Responses to chemical defenses and resource abundance. Ecology 72: 1820-1834

DeMott WR, Zhang Q-X and Carmichael WW (1991) Effect of toxic cyanobacteria and purified toxins on the survival and feeding of a coopepod and three species of *Daphnia*. Limnol Oceanogr 36: 1346-1357

Dittmann E, Erhard M, Rohrlack T, von Döhren H, Kohl JG and Börner T (1998) Characterization of a microcystin synthetase deficient mutant of *Microcystis aeruginosa* PCC 7806. In: Carmichael WW and Paerl HW (eds) Abstracts, 4th International Conference on Toxic Cyanobacteria, Beaufort, North Carolina: 27

Dow CS, Swoboda UK and Firth P (1994) Release and degradation of microcystin during a *Microcystis aeruginosa* bloom in a fresh water reservoir. In: Codd GA, Jeffries TM, Keevil CW, and Potter E (eds) Detection Methods for Cyanobacterial Toxins. Proceedings of the First International Symposium on Detection Methods for Cyanobacterial Toxins,

September 1993. 158-160. The Royal Society of Chemistry, Cambridge

Endean R, Monks SA, Griffith JK and Llewellyn LE (1993) Apparent relationship between toxins elaborated by the cyanobacterium *Trichodesmium erythraeum* and those present in the flesh of the narrow-barred spanish mackerel *Scomberomorus commersoni*. Toxicon 31: 1155-1165

Environment Agency (1998) The Determination of Microcystin Algal Toxins in Raw and Treated Waters by High Performance Liquid Chromatography 1998. Methods for the Examination of Waters and Associated Materials. Environment Agency of England and Wales, Bristol BS32 4UD, 12 pp.

Eriksson JE, Hagerstrand H and Isomaa B (1987) Cell selective cytotoxicity of a peptide from the cyanobacterium *Microcystis aeruginosa*. Biochem Biophys 930: 304-310

Eriksson JE, Meriluoto JA and Lindholm T (1989) Accumulation of peptide toxin from the cyanobacterium *Oscillatoria agardhii* in the fresh water mussel *Anadonta cygnea*. Hydrobiologia 183: 211-216

Eriksson JE, Toivola D, Meriluoto JAO, Karaki H, Han Y-G and Hartshorne D (1990) Hepatocyte deformation induced by cyanobacterial toxins reflects inhibition of protein phosphatases. Biochem Biophys Res Comm 173: 1347-1353

Falconer IR (1991) Tumour promotion and liver-injury caused by oral consumption of cyanobacteria. J Environ Toxicol Water Qual 6: 511-574

Falconer IR, Hardy SJ, Humpage AR, Froscio SM, Tozer GJ and Hawkins PR (1999) Hepatic and renal toxicity of the blue-green alga (cyanobacterium) *Cylindrospermopsis raciborskii* in male Swiss Albino mice. J Environ Toxicol Water Qual 14:1 (in press)

Falconer IR, Smith JV, Jackson ARB, Jones A and Runnegar MTC (1988) Oral toxicity of a bloom of the cyanobacterium *Microcystis aeruginosa* administered to mice over periods up to 1 year. J Toxicol Environ Health 24: 291-305

Faulstich H and Wieland T (1992) Mushroom poisons. In: Tu AT (ed.) Food poisonings, Handbook of Natural Toxins 7; 207-235. New York, Marcel Dekker, Inc

Fitzgeorge RB, Clark SA and Keevil C.W. (1994) Routes of intoxication. In: Codd GA, Jeffries TM, Keevil CW, and Potter E (eds) Detection Methods for Cyanobacterial Toxins. Proceedings of the First International Symposium on Detection Methods for Cyanobacterial Toxins, September 1993. 69-74. The Royal Society of Chemistry, Cambridge

Fitzgerald DJ and Yamasaki H (1990) Tumor promotion: models and assay systems. Teratogen Carcinogen Mutagen 10: 89-102

Forsyth DJ, James MR and Cryer M (1990) Alteration of seasonal and diel patterns in vertical migeration of zooplankton by *Anabaena* and planktivorous fish. Arch Hydrobiol 117: 385-404

Francis G (1878) Poisonous Australian lake. Nature 18: 11-12

Fujiki H and Suganuma M (1993) Tumor promotion by inhibitors of protein phosphatases 1 and 2A: the okadaic acid class of compounds. Adv Cancer Res 61: 143-194

Fujiki H, Suganuma M, Suguri H, Yoshizawa S, Takagi K, Nakayasu M, Ojika M, Yamada K, Yasumoto T, Moore RE and Sugimura T (1990) New tumor promoters from marine natural products. In: Hall S and Strichartz G (eds) Marine Toxins: Origin, Structure and Molecular Pharmacology. 232-240. American Chemical Society, Washington DC

Fujiki H, Sugimura T and Moore RE (1983) New classes of environmental tumor promoters: indole alkaloids and polyacetates. Environ Health Perspect 50: 85-90

Gaete V, Canelo E, Lagos N and Zambrano F (1994) Inhibitory effects of *Microcystis aeruginosa* toxin on ion pumps of the gill of freshwater fish. Toxicon 82: 121-127

Garcia GO (1989) Toxicity of the cyanobacterium *Microcystis aeruginosa* strain 7820 to trout and tilapia: a clinical and histopathological study. MSc Thesis, University of Stirling, UK

Gerwick WH, Roberts PJ, Proteau PJ and Chen J-L (1994) Screening cultured marine microalgae for anticancer-type activity. J Appl Phycol 6: 143-149

Gilbert JJ (1994) Susceptibility of planktonic rotifers to a toxic strain of *Anabaena flos-aquae*. Limnol Oceanogr 39: 1286-1297

Gleason FK and Paulson JP (1984) Site of action of the natural algicide, cyanobacterin in the blue-green alga, *Synchococcus* sp. Arch. Microbiol. 138: 273-277

Grabow WOK, DuRandt WC, Prozesky OW and Scott WE (1982) *Microcystis aeruginosa* toxin: cell culture toxicity, hemolysis and mutagenicity assays. Appl Environ Microbiol 43: 1425-1433

Grauer F (1961) Seaweed dermatitis. Arch Dermatol 84: 720-732

Gunn GJ, Rafferty AG,Rafferty GC, Cockburn N, Edwards C, Beattie KA and Codd GA (1992) Fatal canine neurotoxicosis attributed to blue-green algae (cyanobacteria). Vet Rec 4: 301-302

Hahn ST and Capra MF (1992) The cyanobacterium *Oscillatoria erythraea* – a potential source of toxin in the ciguatera food-chain. Food Add Contam 9: 351-355

Hanazato T (1996) Toxic cyanobacteria and the zooplankton community. In: Watanabe MF, Harada H, Carmichael WW and Fujiki H (eds) (1996) Toxic *Microcystis*. 79-102. CRC Press, London

Harada K (1996) Chemistry and detection of microcystins. In: Watanabe MF, Harada H, Carmichael WW and Fujiki H (eds) (1996) Toxic *Microcystis*. 103-148. CRC Press, London

Harada K-I, Matsuura K, Suzuki M, Watanabe MF, Oishi S, Dahlem AM, Beasley VR and Carmichael WW (1990a) Isolation and characterization of the minor components associated with microcystins-LR and -RR in the cyanobacteria (blue-green algae). Toxicon 28: 55-64

Harada K-I, Ogawa K, Matsuura K, Murata H, Suzuki M, Watanabe MF, Itezono Y and Nakayama N (1990b) Structural determination of geometrical isomers of microcystins -LR and -MR from cyanobacteria by two-dimensional NMR spectroscopic techniques. Chem Res Toxicol 3: 473-481

Harada K-I, Ohtani I, Iwamoto K, Suzuki M, Watanabe MF and Terao K (1994) Isolation of cylindrospermopsin from a cyanobacterium *Umezakia natans* and its screening method. Toxicon 32: 73-84

Hashimoto Y, Kamiya H, Yamazato K and Nozawa K (1976) Occurrence of a toxic blue-green alga inducing skin dermatitis in Okinawa. In: Ohsaka A, Hayashi K and Sawai Y (eds) Animal, Plant and Microbial Toxins. 333-338. Plenum Publishing, New York

Hawkins PR, Chandrasena NR, Jones GJ, Humpage AR and Falconer IR (1997) Isolation and toxicity of *Cylindrospermopsis raciborskii* from an ornamental lake. Toxicon 35: 341-346

Henning K, Meyer H, Kraatz-Wadsack G and Cremer J (1992) Detection of a cytotoxic substance produced by the cyanobacterium *Microcystis aeruginosa* strain PCC 7806: isolation and differentiation from the peptide toxin microcystin-LR by cytotoxicity assays. Current Microbiology 25: 129-134

Hiripi L, Nagy L, Kalmar T, Kovacs A and Voros L (1998) Insect (*Locusta migratoria migratorioides*) test monitoring the toxicity of cyanobacteria. Neurotoxicology 19(4-5): 605-608

Honkanen RE, Zwiller J, Moore RE, Daily SL, Khatra BS, Dukelow M and Boynton AL (1990) Characterization of microcystin-LR, a potent inhibitor of type 1 and type 2A protein phosphatases. J Biol Chem 265: 19401-19404

Hunter PR (1995a) Cyanobacterial toxins and their potential risk to drinking water supplies. Microbiology Europe 3: 8-10

Hunter PR (1995b) Cyanobacteria in fresh waters. Microbiology Europe 3: 10-12

Hunter PR and Roberts C (1991) Introduction. In: Codd GA and Roberts C (eds) Public Health Aspects of Cyanobacteria (blue-green algae). PHLS Microbiology Digest Supplement, 8: 80-81. London

Ito E, Kondo F, Terao K and Harada K-I (1997) Neoplastic nodular formation in mouse liver induced by repeated intraperitoneal injections of microcystin-LR. Toxicon 35: 1453-1457

Keevil CW (1991) Toxicological and detection of cyanobacterial (blue-green algal) toxins. In: Codd GA and Roberts C (eds) Public Health Aspects of Cyanobacteria (blue-green algae). PHLS Microbiology Digest Supplement 8: 91-95. London

Keleti G and Sykora JL (1982) Production and properties of cyanobacterial endotoxins. Appl Environ Microbiol 43: 104-109

Keleti G, Sykora JL, Lippy EC and Shapiro MA (1979) Composition and biological properties of lipopolysaccharides isolated from *Schizothrix calcicola* (Ag.) Gomont (cyanobacteria). Appl Environ Microbiol 38: 471-477

Kfir R, Johannsen E and Botes DP (1986) Monoclonal antibody specific for cyanoginosin-LA: preparation and characterisation. Toxicon 24: 543-552

Kiviranta J and Abdelhameed A (1994) Toxicity of the blue-green alga *Oscillatoria agardhii* to the mosquito *Aedes aegyptii* and the shrimp *Artemia salina*. World J Microbiol Biotechnol 10: 517-520

Kiviranta J, Namikoshi M, Sivonen K, Evans WR, Carmichael, WW and Rinehart KL (1992) Structure determination and toxicity of a new microcystin from *Microcystics aeruginosa*. Toxicon 30: 1093-1098

Kiviranta J, Sivonen K and Niemela SI (1991) Detection of toxicity of cyanobacteria by *Artemia salina* bioassay. J Environ Toxicol Water Qual 6: 423-436

Kotak BG, Zurawell RW, Prepas EE and Holmes CFB (1996) Microcystin LR concentration in aquatic food web compartments from lakes of varying trophic status. Can J Fish Aquat Sci 53: 1974-1985

Kuiper-Goodman T, Falconer I and Fitzgerald J (1999) Human health aspects. In: Chorus I and Bartram J (eds) (1999) Toxic Cyanobacteria in Water. A Guide to Their Public Health Consequences, Monitoring and Management. 113-153. E & FN Spon, London & New York

Kusumi K (1996) Toxicology of microcystins. In: Watanabe MF, Harada H, Carmichael WW and Fujiki H (eds) (1996) Toxic *Microcystis*. 149-174. CRC Press, London

Lanaras T, Cook CM, Eriksson JA, Meriluoto JAO and Hotokka M (1991) Computer modelling of the 3-dimensional structures of the cyanobacterial hepatotoxins microcystin-LR and nodularin. Toxicon 29: 901-906

Laurén-Määttä C, Hietala J and Walls M (1997) Responses of *Daphnia pulex* populations to toxic cyanobacteria. Freshwater Biology 37: 635-647

Lawton LA, and Codd GA (1988) Effects of cyanobacterial peptide toxins on the ciliate *Paramecium*. Br Phycol J 23: 292

Lightner DV (1978) Possible toxic effects of the marine blue-green alga, *Spirulina subsalsa*, on the blue shrimp, *Panaeus stylirostis*. J Invert Pathol 32: 139-150

Lukac M and Aegerter R (1993). Influence of trace metals on growth and toxin production of *Microcystis aeruginosa*. Toxicon 31: 293-305

Luukkainen R, Namikoshi M, Sivonen K, Rinehart KL and Niemela SI (1994) Isolation and identification of 12 microcystins from 4 strains and 2 bloom samples of *Microcystis* spp.- structure of a new hepatotoxin. Toxicon 32: 133-139

Luukkainen R, Sivonen K, Namikoshi M, Fardig M, Rinehart KL and Niemela SI (1993) Isolation and identification of 8 microcystins from 13 *Oscillatoria agardhii* strains and structure of a new microcystin. Appl Environ Microbiol 59: 2204-2209

MacKintosh C, Beattie KA, Klumpp S, Cohen P and Codd GA (1990) Cyanobacterial microcystin-LR is a potent and specific inhibitor of protein phosphatases 1 and 2A from both mammals and higher plants. Fed Eur Biochem Soc Lett 264: 187-192

Mahmood NA and Carmichael WW (1986) Paralytic shellfish poisons produced by the freshwater cyanobacterium *Anabaena flos-aquae* NH-5. Toxicon 24:175-186

Mason CP, Edwards KR, Carlson RE, Pignatello J, Gleason FK and Wood JM (1982) Isolation of chlorine-containing antibiotic from the freshwater cyanobacterium, *Scytonema hofmanni*. Science 215: 400-402

Matsunaga S, Moore RE, Niemczura WP and Carmichael WW (1989) Anatoxin-a(s), a potent anticholinesterase from *Anabaena flos-aquae*. J Amer Chem Soc 111: 8021-8023

McElhiney J, Lawton LA, Edwards C and Gallacher S (1998) Development of a bioassay employing the desert locust (*Schistocera gregaria*) for the detection of saxitoxin and related compounds in cyanobacteria and shellfish. Toxicon 36: 417-420

Meriluoto JAO, Sandstrom A, Eriksson JE, Remaud G, Grey Craig A and Chattopadhyaya J (1989) Structure and toxicity of a peptide hepatotoxin from the cyanobacterium *Oscillatoria agardhii*. Toxicon 27: 1021-1034

Mohammed AM, Jamel Al-Layl KS and Jamal Al-Lail SS (1992) Toxic effects of the freshwater cyanobacterium *Oscillatoria agardii* (blue-green algae) on the snail *Melanoides tuberculata* (Muller). J Egypt Ger Soc Zool 9: 223-235

Moiheka SN and Chu GW (1971) Dermatitis-producing alga *Lyngbya majuscula* Gomont in Hawaii. II. Biological properties of the toxic factor. J Phycol 7: 8-13

Moore RE (1981) Constituents of the blue-green algae. In: Scheuer PJ (ed.) Marine Natural Products: Chemical and Biological perspectives 4: 1-52. Academic Press: New York

Moore RE (1984) Public health and toxins from marine blue-green algae. In: Raeglis EP (ed.) Seafood Toxins. Am Chem Soc, Symposium series 262: 369-376

Moore RE, Cheuk C, Yang XG and Patterson GML (1984) Hapalindoles: new alkaloides from the blue-green alga *Hapalosiphon fontinalis*. J Am Chem Soc 106: 6456-6457

Moore RE, Patterson GML, Entzeroth M, Morimoto H, Suganuma M, Hakii H, Fujiki H and Sugimura T (1986) Binding studies of [H-3] lyngbyatoxin A and [H-3] debromoplysiatoxin to the phorbol ester receptor in a mouse epidermal particulate fraction. Carcinogenesis 7: 641-644

Mynderse JS, Moore RE, Kashiwagi M and Norton TR (1977) Antileukemia activity in the Oscillatoriaceae: isolation of debromoaplysiatoxin from *Lyngbya*. Science 196: 538-540

Nagabhushanam AK (1967) On an unusually dense phytoplankton bloom around Minicoy island (Arabian sea) and its effects on the local tuna fisheries. Current Science 22: 611-612

Nagata S, Soutome H, Tsutsumi T, Hasegawa A, Sekijima M, Sugamata M, Harada K-I, Suganuma M and Ueno Y (1995) Novel monoclonal antibodies against microcystin and their protective activity for hepatotoxicity. Natural Toxins 3: 78-86

Namikoshi M, Rinehart KL, Sakai R, Stotts RR, Dahlem AM, Beasley VR, Carmichael WW and Evans WR (1992) Identification of 12 hepatotoxins from a Homer Lake bloom of the cyanobacteria *Microcystis aeruginosa*, *Microcystis viridis* and *Microcystis wesenbergii*: nine new microcystins. J Org Chem 57: 866-872

Nishiwaki-Matsushima R, Ohta, Nishiwaki S, Sugawama M, Kohyama K, Ishikawa T, Carmichael WW and Fujiki H (1992)

Liver-tumour promotion by the cyanobacterial cyclic peptide toxin microcystin-LR. J Cancer Res Clin Oncol 118: 420-424

Oberemm A, Becker J, Codd GA and Steinberg C (1999) Effects of cyanobacterial toxins and aquaeous crude extracts of cyanobacteria on the development of fish and amphibians. Environ Toxicol 14: 77-88

Park H-D and Watanabe MF (1996) Toxic *Microcystis* in eutrophic lakes. In: Watanabe MF, Harada H, Carmichael WW and Fujiki H (eds) (1996) Toxic *Microcystis*. 57-78. CRC Press, London

Patterson GML, Larsen LK and Moore RE (1994) Bioactive natural products form blue-green algae. J Appl Phycol 6: 151-157

Pearson MJ, Ferguson AJD, Codd GA, Reynolds CS, Fawell JK, Hamilton RM, Howard SR and Attwood MR (1990) Toxic blue-green algae. NRA, Water Quality Series No. 2. Stanley L. Hunt Ltd, Rushden

Pilotto LS, Douglas RM, Burch MD, Cameron S, Beers M, Rouch GR, Robinson P, Kirk M, Cowie CT, Hardiman S, Moore C and Attewell RG (1997) Health effects of recreational exposure to cyanobacteria (blue-green algae) during recreational water-related activities. Aust N Zealand J Public Health 21: 562-566

Pomati F, Sacchi S, Rossetti C and Neiland BA (1998) ...And now saxitoxin-producing cyanobacteria in Europe. In: Carmichael WW and Paerl HW (eds) (1998) Abstracts, 4th International Conference on Toxic Cyanobacteria, Beaufort, North Carolina: 44

Pouria S, deAndrade A, Barbosa J, Caavalcanti RL, Barreto VTS, Ward CJ, Prieser W, Poon GK, Neild GH and Codd GA (1998) Fatal microcystin intoxication in haemodialysis unit in Caruaru, Brazil. Lancet 352 (9121): 21-26

Reinikainen M, Jiviranta J, Ulvi V and Nikupaavola ML (1995) Acute toxic effects of a novel cyanobacterial toxin on the crustaceans *Artemia salina* and *Daphnia pulex*. Arch Hydrobiol 133: 61-69

Repavich WM, Sonzogni WC, Standridge JH, Wedepohl RE and Meisner LF (1990) Cyanobacteria (blue-green-algae) in Wisconsin Waters acute and chronic toxicity. Wat Res 24: 225-231

Rinehart KL, Harada K-I, Namikoshi M, Chen C, Harvis CA, Munro MGH, Blunt JW, Mulligan PE, Beasley VR, Dahlem, AM and Carmichael WW (1988) Nodularin, microcystin, and the configuration of Adda. J Amer Chem Soc 110: 8557-8558

Rinehart KL, Namakopshi and Choi BW (1994) Structure and biosynthesis of toxins from blue-green algae (cyanobacteria). J Appl Phycol 6: 159-176

Robinson CP, Franz DR and Bondura ME (1991) Effects of lyngbyatoxin A from the blue-green alga *Lyngbya majuscula* on rabbit aorta contractions. Toxicon 29: 1009-1017

Rodger HD, Turnbull T, Edwards C and Codd GA (1994) Cyanobacterial bloom associated pathology in brown trout *Salmo trutta* L. in Loch Leven, Scotland. J Fish Dis 17: 177-181

Rouhiainen L, Sivonen K, Paulin L, Buikema W and Haselkorn R (1998) Peptide synthetase genes in *Anabaena* cyanobacterium. In: Carmichael WW and Paerl HW (eds) (1998) Abstracts, 4th International Conference on Toxic Cyanobacteria, Beaufort, North Carolina: 59

Runnegar M, Berndt N and Kaplowitz N (1995) Microcystin uptake and inhibition of protein phosphatase – effects of chemoprotectants and self-inhibition in relation to known hepatic transporters. Toxicol Appl Pharmacol 134: 264-272

Runnegar MTC and Falconer IR (1982) The *in vivo* and *in vitro* biological effects of the peptide hepatotoxin from the blue-green alga *Microcystis aeruginosa*. South Afr J Sci 78: 363-366

Runnegar MTC and Falconer IR (1986) Effect of toxin from the cyanobacterium *Microcystis aeruginosa* on ultrastructural morphology and actin polymerization in isolated hepatocytes. Toxicon 24: 109-115

Runnegar MTC, Falconer IR and Silver J (1981) Deformation of isolated rat hepatocytes by a peptide hepatotoxin from the blue-green alga *Microcystis aeruginosa*. Naunyn-Schmiedebergs Ach Pharmacol 317: 268-272

Runnegar MTC, Kong SM and Berndt N (1993) Protein phosphatase inhibition and *in vivo* hepatotoxicity of microcystins. Am J Physiol 265: G224-G230

Sandstrom A, Glemarec C, Meriluoto JAO, Eriksson JE and Chattopadhyaya J (1990) Structure of a hepatotoxic pentapeptide from the cyanobacterium *Nodularia spumigena*. Toxicon 28: 535-540

Sano T and Kaya K (1995) A 2-amino-2-butenoic acid (DHB) – containing microcystin isolated from *Oscillatoria agardhii*. Tetrahedron Letts 36: 8603-8606

Sawyer PJ, Gentile JH and Sasner JJ Jr (1968) Demonstration of a toxin from *Aphanizomenon flos-aquae* (L.) Ralfs. Can J Microbiol 14: 1199-1204

Seawright AA, Nolan CC, Shaw GR, Chiswell RK, Norris RL, Moore MR and Smith MJ (1999) The oral toxicity for mice of the tropical cyanobacterium *Cylindrospermopsis raciborskii* (Woloszynska). Env Toxicol Water Qual 14: (in press)

Sim ATR and Mudge LM (1994) Detection of hepatotoxins by protein phosphatase inhibition assay: advantages, pitfalls and anomalies. In: Codd GA, Jeffries TM, Keevil CW, and Potter E (eds) Detection Methods for Cyanobacterial Toxins. Proceedings of the First International Symposium on Detection Methods for Cyanobacterial Toxins, September 1983. 111-116. The Royal Society of Chemistry, Cambridge

Sivonen K (1990) Toxic Cyanobacteria in Finnish Fresh Waters and The Baltic Sea. PhD Thesis, University of Helsinki

Sivonen K and Jones G (1999) Cyanobacterial toxins. In: Chorus I and Bartram J (eds) (1999) Toxic Cyanobacteria in Water. A Guide to their Public Health Consequences, Monitoring and Management. 41-111. E & FN Spon, London & New York

Sivonen K, Kononen K, Carmichael WW, Dahlien AM, Rinehart KL, Kiviranta J and Niemala SI (1989) Occurrence of the hepatotoxic cyanobacterium *Nodularia spumigena* in the Baltic Sea and structure of the toxin. Appl Environ Microbiol 55: 1990-1995

Sivonen K, Namikoshi M, Evans WR, Carmichael WW, Sun F, Rouhiainen L, Luukkainen R and Rinehart KL (1992) Isolation and characterisation of a variety of microcystins from 7 strains of the cyanobacterial genus *Anabaena*. Appl Environ Microbiol 58: 2495-2500

Sivonen K, Niemela SI, Niemi RM, Lepisto L, Luoma TH and Rasanen LA (1990) Toxic cyanobacteria (blue-green algae) in Finnish fresh coastal waters. Hydrobiologia 190: 267-275

Skulberg OM, Carmichael WW, Andersen RA, Matsunaga, Moore RE and Skulberg R (1992) Investigation of a neurotoxic Oscillatorialean strain (Cyanophyceae) and its toxin. Isolation and characterisation of homoanatoxin-a. Env Toxicol Chem 11: 321-329

Smith MJ, Eaglesham GK, Shaw GR, Chiswell RK, Norris RL, Seawright AA, and Moore MR (1998) The cyanotoxin cylindrospermopsin, produced by *Aphanizomenon ovalisporum* in Australia. In: Carmichael WW and Paerl HW (eds) (1998) Abstracts, 4th International Conference on Toxic Cyanobacteria, Beaufort, North Carolina: 48

Smith RA and Lewis D (1987) A rapid analysis of water for anatoxin-a, the unstable toxic alkaloid from *Anabaena flos-aquae*, the stable non-toxic alkaloids left after bioreduction and related a amine which may be nature's precursor to anatoxin-a. Vet Human Toxicol 29: 153-154

Stevens DK and Krieger RI (1988) Analysis of anatoxin-a by GC/EDC. J Analyt Toxicol 12: 126-131

Swoboda UK and Dow CS (1994) Expression of cyanotoxins in environmental biomass containing species of *Oscillatoria*. In: Codd GA, Jeffries TM, Keevil CW, and Potter E (eds) Detection Methods for Cyanobacterial Toxins. Proceedings of the First International Symposium on Detection Methods for Cyanobacterial Toxins, September 1993. 164-167. The Royal Society of Chemistry, Cambridge

Swoboda UK, Dow CS, Chaivimol J, Smith N and Pound BP (1994) Alternatives to the mouse bioassay for cyanobacterial toxicity assessment. In: Codd GA, Jeffries TM, Keevil CW, and Potter E (eds) Detection Methods for Cyanobacterial Toxins. Proceedings of the First International Symposium on Detection Methods for Cyanobacterial Toxins, September 1993. 106-110. The Royal Society of Chemistry, Cambridge

Teixera MGLC, Costa MCN, Carvalho VLP, Pereira MS and Hage E (1993) In: Bulletin of the Pan American Health Organization 27: 244-253

Tencalla FG, Dietrich DR and Schlatter C (1994) Toxicity of *Microcystis aeruginosa* peptide toxin to yearling rainbow trout (*Oncorhynchus mykiss*). Aqua Toxicol 30: 215-224

Tillett D, Dittmann E, Börner T and Neilan BA (1998) Structural analysis of the *Microcystis* genome associated with microcystin biosynthesis. In: Carmichael WW and Paerl HW (eds) (1998) Abstracts, 4th International Conference on Toxic Cyanobacteria, Beaufort, North Carolina: 49

Tisdale E (1931) Epidemic of intestinal disorders in Charleston, WVa, occurring simultaneously with unprecedented water supply conditions. Am J Public Health 21: 198-200

Turell MJ and Middlebrook JL (1988) Mosquito inoculation: an alternative bioassay for toxins. Toxicon 26: 1089-1094

Turner PC, Gammie AJ, Hollinrake K and Codd GA (1990) Pneumonia associated with contact with cyanobacteria. Br Med J 300: 1440-1441

Ueno Y, Nagata S, Tsutsimi T, Hasegawa A, Watanabe MF, Park H-D, Chen GC, Chen G and Yu SZ (1996) Detection of microcystins, a blue-green algal hepatotoxin, in drinking water sampled in Haimen and Fusui, endemic areas of primary liver cancer in China, by highly sensitive immunoassay. Carcinogenesis 17: 1317-1321

Utkilen H and Gjölme N (1992) Toxin production by *Microcystis aeruginosa* as a function of light in continuous cultures and its ecological significance. Appl Environ Microbiol 58: 1321-1325

Vaitomaa J, Rapala J, Seitsonen J, Kairesalo T and Sivonen K (1998) Factors influencing growth and toxin production of cyanobacteria: field, mesocosm and laboratory experiments. In: Carmichael WW and Paerl HW (eds) (1998) Abstracts, 4th International Conference on Toxic Cyanobacteria, Beaufort, North Carolina: 51

Vasconcelos V (1998) Toxic cyanobacteria and aquatic animals: Lethal toxicity and dynamics of microcystins in mussells, crayfish and fish. In: Carmichael WW and Paerl HW (eds) (1998) Abstracts, 4th International Conference on Toxic Cyanobacteria, Beaufort, North Carolina: 52

Ward CJ, Beattie KA, Lee EYC and Codd GA (1997) Colorimetric protein phosphatase inhibition assay of laboratory strains and natural blooms of cyanobacteria: comparisons with high-performance liquid chromatographic analysis. FEMS Microbiol Lett 153: 465-473

Watanabe MF (1996) Production of microcystins. In:Watanabe MF, Harada H, Carmichael WW and Fujiki H (eds) (1996) Toxic *Microcystis*. 68-73. CRC Press, London

Watanabe MF, Harada H, Carmichael WW and Fujiki H (eds) (1996) Toxic *Microcystis*. CRC Press, London, 262 pp.

Watanabe MF and Oishi S (1985) Effects of environmental factors on toxicity of a cyanobacterium (*Microcystis aeruginosa*) under culture conditions. Appl Eviron Microbiol 24: 1342-1344

Weise G, Drews G, Jann B and Jann K (1970) Identification and analysis of lipopolysaccharides in cell walls of the blue-green algae *Anacystis nidulans*. Arch Mikrobiol 71: 89-98

WHO (1997) Report of the Working Group on Chemical Substances in Drinking Water, Geneva, 22-26 April, 1997. Section 5.2, Microcystin-LR. World Health Organization, Geneva

WHO (1998) Guidelines for Drinking-water Quality. Second Edition. Addendum to Volume 2, Health Criteria and Other Supporting Information. World Health Organization, Geneva

Wicks RJ and Thiel PG (1990) Environmental factors affecting the production of peptide toxins in floating scums of the cyanobacterium *Microcystis aeruginosa* in a hypertrophic African reservoir. Environ Sci Technol 24: 1413-1418

Wieland T and Faulstich H (1978) Anatoxins, phallotoxins, phallolysin and antamanide: the biologically active components of poisonous *Amanita* mushrooms. CRC Crit Rev Biochem 5: 185-260

Wong SH and Hindin E (1982) Detecting an algal toxin by high-pressure liquid chromatography. J Am Water Works Assoc 74: 528-529

Yamasaki H (1988) Multistage carcinogenesis: implications for risk estimation. Cancer Metastasis Rev 7: 5-18

Yu S-Z (1989) Drinking water and primary liver cancer. In: Tang ZY, Wu MC and Xia SS (eds) Primary Liver Cancer. 30-37. China Academic Publishers, New York

Yu S-Z (1995) Primary prevention of hepatocellular carcinoma. J Gasteroenterol Hepatol 10: 674-682

Zilberg B (1966) Gasteroenteritis in Salisbury European children – a five year study. Cent Afr J Med 12: 164-168

Organism Index

A

B

C

H

I

J

K

L

M

N

O

P

R

S

T

U

V

W

X

Z

Gene and Gene Product Index

H

I

K

L

M

N

O

P

R

S

U

W

Z

Subject Index

B

C

D

E

G

H

I

J

K

L

M

N

O

P

Q

R

U

V

W

X

Y